S0-EJV-278

ANNOTATED BIBLIOGRAPHIES OF MINERAL DEPOSITS IN THE WESTERN HEMISPHERE

The Geological Society of America, Inc.

Memoir 131

Annotated Bibliographies of Mineral Deposits in the Western Hemisphere

JOHN D. RIDGE

The Pennsylvania State University
College of Earth and Mineral Sciences
University Park, Pennsylvania 16802

1972

Library of Congress Catalog Card Number 72-178773

I.S.B.N. 8137-1131-2

Published by

THE GEOLOGICAL SOCIETY OF AMERICA, INC.

Colorado Building, P.O. Box 1719

Boulder, Colorado 80302

Printed in The United States of America

The publication of this volume
is made possible through the bequest of
Richard Alexander Fullerton Penrose, Jr.

CONTENTS

Page

ABSTRACT 1

INTRODUCTION 3

NORTH AMERICA 7

Canada 7

BRITISH COLUMBIA 7
Brenda Lake 7
Britannia Mines (Howe Sound) 9
Copper Mountain 11
Endako 12
Highland Valley 13
Nelson Area (General) 15
Ainsworth 17
Bluebell 17
Rossland 18
Salmo-Sheep Creek (General) 19
Salmo 19
Sheep Creek 21
Slocan 21
Nickel Plate (Hedley) 22
Pride of Emory 24
Sullivan 25
MANITOBA 27
Bird River 27
Lynn Lake 31
Sherritt Gordon-Flin Flon (General) 33
Flin Flon 34
Sherritt Gordon 35
Snow Lake 36
Thompson-Moak Lake 37
NEW BRUNSWICK 41
Bathurst-Newcastle 41
NEWFOUNDLAND 43
Buchans 43
Wabana 44
NEWFOUNDLAND-QUEBEC 46
Labrador Trough 46
NORTHWEST TERRITORIES 51
Great Bear Lake 51
Pine Point 52
Yellowknife 54
NOVA SCOTIA 58
Walton 58
ONTARIO 59
Adams Mine 59
Alexo Mine 60
Bancroft-Haliburton (General) 62
Bancroft (Nonmetallics) 63
Bancroft (Uranium) 65
Blind River (Elliot Lake) 67
Blue Mountain 69
Cobalt 70
Gowganda 73
Kerr-Addison 75
Kirkland Lake 77
Little Long Lac 79
Manitouwadge 81
Michipicoten 83

Munro 86
Pickle Crow 88
Porcupine 89
Red Lake 91
Steep Rock Lake 94
Sudbury 97
Timagami Island 108
QUEBEC 111
Allard Lake 111
Cadillac-Malartic 112
Chibougamau-Opemiska 115
Eastern Townships 118
Gaspé 120
Lacorne 122
Marbridge 125
Matagami 128
Noranda 130
Noranda Gold 135
Oka 136
St. Urbain 140
Thetford-Black Lake 142
Val d'Or 146
SASKATCHEWAN 149
Beaverlodge (Goldfields) 149
Coronation Mine 153
YUKON TERRITORY 154
Keno Hill 154

Greenland 159

Fiskenaesset 159
Ilímaussaq 161
Ivigtut 164
Mesters Vig 166

Mexico 171

Boléo, Baja California 171
Cananea, Sonora 174
Cerro de Mercado, Durango 179
Concepción del Oro-Providencia, Zacatecas 182
Guanajuato, Guanajuato 185
Pachuca-Real del Monte, Hidalgo 187
Parral, Chihuahua 191
San José, San Luis Potosí 195
Santa Eulalia, Chihuahua 196
Zacatecas, Zacatecas 200

United States 209

ALASKA 209
Juneau-Treadwell 209
Kennecott 211
Lost River 213
Ruby Creek 216
ARIZONA 218
Ajo 218
Bagdad-Massive Sulfides 220
Bagdad-Porphyry Copper 221
Bisbee 223
Castle Dome 227
Christmas 229
Globe-Miami 231
Iron King 233
Jerome 235

Johnson Camp 238
Magma (Superior) 240
Morenci 243
Pima District 245
Ray 248
San Manuel 250
Silver Bell 252
Tombstone 254
ARKANSAS 256
Arkansas Bauxite 256
Magnet Cove 258
CALIFORNIA 262
Alleghany 262
Bishop 264
Darwin 266
Eagle Mountain 268
Engels and Superior Mines 269
Foothill Copper Belt 271
Mother Lode 273
Mountain Pass 276
Nevada City-Grass Valley 279
New Almadén 281
New Idria 284
Shasta County 286
Sulphur Bank 289
COLORADO 292
Boulder County Tellurides 292
Boulder County Tungsten 294
Central City-Idaho Springs 296
Climax 299
Cripple Creek 302
Freeland-Chicago Creek 305
Gilman 307
La Plata 310
Leadville 312
Rico 315
San Juan 317
COLORADO PLATEAU (COLORADO-UTAH-NEW MEXICO-ARIZONA) 323
A. Rifle area, Colorado 323
B. Uravan District, Colorado 323
C. Slick Rock District, Colorado 324
D. Grants District, New Mexico 324
E. Shiprock District, Utah-Arizona 326
F. Monticello-Moab-Thompsons District, Utah 326
G. General 327
H. White Canyon-Elk Ridge, Utah 331
J. Monument Valley, Utah 332
K. Black Mesa-Hopi Buttes, Arizona 333
L. Cameron, Arizona 333
M. Henry Mountains-Green River, Utah 333
N. San Rafael-Cedar Mountain, Utah 334
IDAHO 339
Coeur d'Alene 339
IDAHO-MONTANA 343
Lemhi Pass 343
ILLINOIS-KENTUCKY 345
Illinois-Kentucky Fluorspar 345
MAINE 349
Newry 349
MICHIGAN 351
Keweenaw Point 351
White Pine 355

MINNESOTA 358
Minnesota Iron Ranges (General) 358
Cuyuna 363
Eastern Mesabi 366
Mesabi 368
Vermilion 371
MISSISSIPPI VALLEY TYPE (GENERAL) 373
MISSOURI 377
Iron Mountain-Pilot Knob 377
Leadbelt (Southeast Missouri) 380
MONTANA 384
Butte 384
Dillon 390
Philipsburg 392
Stillwater Complex 394
NEVADA 399
Carlin 399
Comstock Lode 400
Cordero 403
Ely 404
Eureka 407
Gabbs 410
Goldfield 412
Goodsprings 415
Mountain City 417
Oreana 419
Pioche 421
Tonopah 424
NEW HAMPSHIRE 427
Grafton-Keene 427
NEW JERSEY 431
Dover 431
Franklin-Sterling 434
NEW MEXICO 439
Magdalena 439
Questa 442
Santa Rita-Hanover 445
NEW YORK 450
Adirondacks (General) 450
Balmat-Edwards 453
Benson Mines 456
Gore Mountain 458
Lyon Mountain 460
Sanford Lake 462
Cortlandt Complex 464
NORTH CAROLINA 468
Hamme 468
Ore Knob 470
Spruce Pine 472
OKLAHOMA-KANSAS-MISSOURI 474
Tri-State 474
OREGON 480
Grants Pass 480
Riddle 483
PENNSYLVANIA 487
Cornwall 487
Lancaster Gap 490
SOUTH DAKOTA 491
Black Hills Pegmatites 491
Homestake 495
TENNESSEE 498
Ducktown 498

Mascot-Jefferson City 501
Sweetwater 505
TEXAS 507
Terlingua 507
UTAH 510
Bingham 510
Cottonwood-American Fork 513
East Tintic 517
Iron Springs 520
Marysvale 523
Park City 526
Spor Mountain 529
Tintic 531
VERMONT 535
Elizabeth 535
VIRGINIA 538
Austinville-Ivanhoe 538
Nelson and Amherst Counties 541
WASHINGTON 544
Metaline 544
Republic 546
Stevens County Magnesite 548
WISCONSIN-ILLINOIS-IOWA 550
Upper Mississippi Valley 550
WYOMING 556
Iron Mountain 556

SOUTH AMERICA 561

Argentina 561
Aguilar 561

Bolivia 563
Bolivian Tin (General) 563
Llallagua-Uncia (Catavi) 566
Oruro 568
Potosí 571
Corocoro 573
Pulacayo (Huanchaca) 577

Brazil 579
Minas Gerais 579
Morro Velho-Raposos 585
Serra de Jacobina 588

Chile 591
Braden 591
Chañarcillo 594
Chuquicamata 597

Peru 601
Central Peru (General) 601
Carahuacra 604
Casapalca 606
Cerro de Pasco 608
Colquijirca 611
Huancavelica 614
Morococha 616
Yauricocha 619

Venezuela 622
Venezuelan Guayana 622

MAPS

Western Canada 8

Central Canada 28
Eastern Canada 40
Greenland 158
Mexico 170
Alaska 203
Arizona-New Mexico 203
California-Nevada 204
Central States 205
Colorado Plateau 206
Colorado-Utah 206
Eastern States 207
Northwestern States 208
Eastern South America 560
Western South America 560

INDICES

Index of Authors 625
Alphabetical Index of Deposits 643
Index of Deposits According to Age of Mineralization 653
Index of Deposits According to Metals and Minerals Produced 659
Index of Deposits According to Categories of the Modified Lindgren Classification 667

APPENDIX I. Classification of Ore Deposits 673
APPENDIX II. Topics to be Considered in the Study of an Ore Deposit ... 679

RECENT PAPERS OF IMPORTANCE
RECEIVED TOO LATE FOR INCLUSION IN THE BIBLIOGRAPHIES
AND NOT INDEXED

Canada

Snow Lake, MANITOBA

Coats, C.J.A., and others, 1970, Geology of the copper-zinc deposits of Stall Lakes Mines Ltd., Snow Lake area, N. Manitoba: Econ. Geol., v. 65, p. 970-984

Cobalt, ONTARIO

Halls, C., and Stumpfl, E. F., 1970, Geology and ore deposition, western Kerr Lake Arch, Cobalt, Ontario: Ninth Commonwealth Min. and Met. Cong. 1969 Pr., v. 2, p. 241-284

McIlwaine, W. H., 1970, Geology of south Lorrain Township: Ont. Dept. Mines and Northern Affairs Geol. Rept. 83, 95 p.

Sudbury, ONTARIO

Thomson, James E., 1969, A discussion of Sudbury geology and sulphide deposits: Ont. Bur. Mines Misc. Paper 30, 22 p.

United States

Darwin, CALIFORNIA

Hall, W. E., and others, 1971, Fractionation of minor elements between galena and sphalerite, Darwin Lead-Silver-Zinc mine, Inyo County, California, and its significance in geothermometry: Econ. Geol., v. 66, p. 602-606

Climax, COLORADO; Bingham, UTAH; Butte, MONTANA

Roedder, E., 1971, Fluid inclusion studies on the porphyry-type ore deposits at Bingham, Utah; Butte, Montana; and Climax, Colorado: Econ. Geol., v. 66, p. 98-120

White Pine, MICHIGAN

Brown, A. C., 1971, Zoning in the White Pine copper deposit, Ontonagon County, Michigan: Econ. Geol., v. 66, p. 543-573

White, W. S., 1971, A paleohydrologic model for mineralization of the White Pine copper deposit, Northern Michigan: Econ. Geol., v. 66, p. 1-13

Mississippi Valley Type (General)

Cannon, R. S., and Pierce, A. P., 1971, Lead isotope guides for Mississippi Valley lead-zinc exploration: U. S. Geol. Surv. Bull. 1312-G, p. 20

Snyder, F. G., 1970, Structural lineaments and mineral deposits, eastern United States, *in* Rausch, D. O., and Mariacher, B. C., Editors, *World Symposium on Mining and Metallurgy of Lead and Zinc*: A.I.M.E., N. Y., p. 76-94

Butte, MONTANA; Santa Rita-Hanover, NEW MEXICO; Ely, NEVADA; Bingham, UTAH

Lange, I. M., and Cheney, E. S., 1971, Sulfur isotopic reconnaissance of Butte, Montana: Econ. Geol., v. 66, p. 63-73

Sheppard, S.M.F., and others, 1971, Hydrogen and oxygen isotope ratios in minerals from porphyry copper deposits: Econ. Geol., v. 66, p. 515-542

Bingham, UTAH

Field, C. W., and Moore, W. J., 1971, Sulfur isotope study of the "B" limestone and galena fissure ore deposits of the U. S. mine, Bingham mining district, Utah: Econ. Geol., v. 66, p. 48-62

Moore, W. J., and Lanphere, M. A., 1971, The age of porphyry-type copper mineralization in the Bingham mining district, Utah--a refined estimate: Econ. Geol., v. 66, p. 331-334

Upper Mississippi Valley, WISCONSIN-ILLINOIS-IOWA

Lavery, N. G., and Barnes, H. L., 1971, Zinc dispersion in the Wisconsin zinc-lead district: Econ. Geol., v. 66, p. 226-242

South America

Bolivian Tin (General)

Turneaure, F. S., 1971, The Bolivian tin-silver province: Econ. Geol., v. 66, p. 215-225

ABSTRACT

These bibliographies and the notes that accompany them have been prepared to aid the economic geologist in his study of ore deposits; they certainly will not do all his work for him. I have tried to include all ore districts in the Western Hemisphere for which sufficient material is available in print to permit a student of ore deposits to obtain a real understanding of any one of the deposits. I have tried to include as many references as possible in languages other than English.

The notes are designed to show (1) where the deposit is, (2) why a certain age date has been assigned to it, and (3) why it has been given the position assigned to it in the modified Lindgren classification. The notes should serve to introduce the deposit to the student and to explain my ideas about it, but he must do a great deal more than simply read these notes.

In the Introduction, the more important sources of references in the bibliographies are set down in alphabetical order with the number of references provided by each source. Several indices are provided in the back of the book; these list the deposits alphabetically, by minerals produced, by age of mineralization, and by their position in the modified Lindgren classification. An author index also is included. Short discussions on age of mineralization and the Lindgren classification are used to introduce the indices to which they apply, and I have added some remarks on the classification of ore deposits and on my modification of the Lindgren classification. Several sketch maps are included showing the approximate locations of the ore deposits discussed in this book.

Contribution No. 70-21 from the College of Earth and Mineral Sciences The Pennsylvania State University.

RÉSUMÉ

Ces bibliographies et les notes qui les accompagnent ont été préparées en vue d'aider le géologue économique dans son étude des gisements de minerai bien qu'elles ne le dispensent pas de tout travail. J'ai essayé d'inclure toutes les zones de minerai de l'hémisphère occidental pour lesquelles l'étudiant des gisements de minerai trouvera assez de publications pour lui permettre de parvenir à la compréhension exacte de n'importe quel gisement. J'ai essayé d'inclure autant de références que possible dans des langues autres que l'Anglais.

Ces notes ont pour but de montrer: (1) où se trouve le gisement; (2) pourquoi on lui a attribué tel ou tel âge; et (3) pourquoi on lui a assigné telle ou telle place dans la "classification revue de Lindgren." Chaque série de notes doit servir à presenter le gisement en question et expliquer mes idées à son sujet mais l'étudiant ne doit pas se contenter de simplement lire ces notes.

Dans l'Introduction les principales sources de références des bibliographies sont données par ordre alphabétique avec en plus de nombreuses références fournies par chacune. A la fin du livre on trouvera plusieurs index; liste des minerais par ordre alphabétique, par mineraux fournis, par âge de minéralisation et enfin par leur position dans la "classification revue de Lindgren." On trouvera également un index des auteurs avec les bibliographies dans lesquelles ils sont cités. De courtes discussions sur l'âge de minéralisation et sur la classification Lindgren servent à introduire les indices auxquelles elles s'appliquent; j'ai également ajouté quelques remarques sur la classification des gisements et sur ma propre révision de la classification Lindgren. On trouvera en fin des plans situant approximativement les gisements auxquels s'appliquent les bibliographies et les notes.

ZUSAMMENFASSUNG

Literaturverzeichnisse mit ihren Erklärungen wurden zusammengestellt, um dem Lagerstättenkundler in seinem Studium von Erzvorkommen zu helfen; sie werden ihm jedoch nicht die ganze Arbeit abnehmen. Ich habe versucht alle jene Erzlagerstätten der westlichen Hemisphäre einzuschliessen, für welche genügend

Material im Druck vorliegt, um von der Literatur aus ein wirkliches Verständnis des gegebenen Vorkommens zu erlangen. Ich habe versucht über ein gegebenes Vorkommen so viele Literaturhinweise wie möglich, auch in anderen als der Englischen Sprache, aufzunehmen.

Die Erläuterungen sind so angelegt, dass sie zeigen: (1) wo ein Vorkommen liegt, (2) warum ihm ein gegebenes Alter zugeschrieben wird, und (3) warum es eine gegebene Stellung in der modifizierten Lindgren Klassifizierung einnimmt. Es ist der Zweck dieser Erläuterungen dem Leser die einzelnen Lagerstätten vorzustellen, und meine eigenen Ansichten über die betreffenden Vorkommen darzulegen; zu ihrem wirklichen Verständnis ist jedoch mehr als ein einfaches Lesen dieser Erläuterungen Voraussetzung.

In der Einleitung werden die wichtigsten Quellen, aus denen die Literaturhinweise stammen, in alphabetischer Folge aufgeführt, nebst der Anzahl der Hinweise aus einer gegebenen Quelle. Mehrere Verzeichnisse am Ende des Buches geben die Vorkommen nach Namen in alphabetischer Folge, nach geförderten Mineralien, nach ihrem Alter und nach ihrer Stellung in der modifizierten Lindgren Klassifizierung. Ein Autorenverzeichnis mit Literaturhinweisen, in denen sie zitiert wurden, ist eingeschlossen. Kürzere Besprechungen über das Alter der Vererzung und die Lindgren Klassifizierung werden benutzt, um den Leser in die betreffenden Verzeichnisse einzuführen. Ich habe auch einige Bemerkungen über die Einteilung von Erzlagerstätten und meine Modifizierung der Lindgren Klassifizierung beigefügt. Die ungefähre Lage der Vorkommen, die in die Literaturverzeichnisse und ihre Erklärungen aufgenommen worden sind, wird an Hand von Kartenskizzen angegeben.

SUMARIO

Estas bibliografías y las notas que las acompañan han sido preparadas con objeto de auxiliar al geólogo economista en los estudios de depósitos minerales, esto ciertamente no quiere decir que ellas harán todo el trabajo del geólogo. Yo he tratado de incluír todos los distritos mineros conocido en el Hemisferio Occidental, de los cuales suficiente material impreso es disponible con objeto de permitir al estudiante de depósitos de mineral, obtener un claro entendimiento de cualquiera de ellos. Yo he tratado de incluír tantas referencias en otras lenguas como fué posible.

Las notas están diseñada para mostrar: (1) el lugar en que se encuentra el depósito; (2) la razón por la cuál cierta edad se le ha sido asignada; y (3) la razón por la cuál se le asignó una posición dada en la clasificación modificada de Lindgren. Cada grupo de notas debe servir para presentar al estudiante, el grupo al cual el depósito pertenece asi como también, para explicar mis ideas acerca de el, sin embargo, el estudiante tendrá que hacer algo mas que concretarse a leer las notas.

En la Introducción, las fuentes de información mas importantes en las bibliografiás son citadas en orden alfabético, y el número de referencias disponible es también proporcionado. Varios índices son incluídos en la parte posterior del libro; estos índices proveen la lista de depósitos en orden alfabético de acuerdo con los minerales producidos, la edad de la mineralización, y la posición de cada uno de ellos en la clasificación modificada de Lindgren, también, un índice de autores indicando las bibliografías en las cuales ellos son citados es proveído. Discusiones cortas sobre la edad de la mineralización y de la clasificación de Lindgren, son usadas para indicar los índices a los cuales ellos se aplican, y yo he agregado algunas notas a la clasificación de depósitos minerales, asi como también a mi modificación de la clasificación de Lindgren. Mapas esquemáticos son incluídos, los cuales muestran la localización aproximada de los depósitos que son objeto de estudio en las bibliografías y notas.

INTRODUCTION

This volume is a revised and expanded version of *Selected Bibliographies of Hydrothermal and Magmatic Mineral Deposits* which was published in 1958 as Memoir 75 of The Geological Society of America. In addition to bringing the bibliographies up to the end of 1969, notes have been included to explain (1) where the deposit is, (2) why a certain age date has been assigned to it, and (3) why it is given the position assigned to it in the modified Lindgren classification that was used in both Memoir 75 and this volume. In defining the locations, distances given are the lengths of straight lines between the two points in question and are *not* road distances.

I hope now, as I did in 1958, that the publication of these bibliographies will enable those who use them to spend more time on the study of the ore deposits in question and less on the somewhat mechanical process of literature search.

The first edition of this work included bibliographies of deposits on all continents except Antarctica; this one covers only the Western Hemisphere. A second volume on the Eastern Hemisphere probably will appear in 1973. The principal reason for the division into hemispheres is the difficulty of preparing so many bibliographies and their accompanying notes over a reasonable period of time. If this volume were to contain deposits from all over the world, some of the references would be incomplete and the notes out of date for some deposits. Thus, I think it best to follow this scheme of hemispheric separation. I have chosen to do the Western Hemisphere first because I have not seen deposits in southern Africa or Australia; I expect to do so during a six-month tour of these two areas from March through September of 1970.

I have included in these bibliographies all deposits that have, in my opinion, been formed in whole or in part by magmatic or hydrothermal processes, including those produced by volcanic exhalations reaching the sea floor, for which I believe a worthwhile literature exists. Each user of this volume will discover that some deposit he thinks should have been included has been omitted. Such omissions are due either to a lack of published work or to my belief that hydrothermal and/or magmatic processes had nothing to do with the formation of the deposits. Although many of the ore deposits included are or were among the hemisphere's principal ore bodies, others of small economic, but high scientific, interest also appear. The lessons to be learned from the study of the deposits in the latter class are, of course, as valid as those to be derived from the study of the deposits in the economically more important districts.

Many geologists will think that I have strained considerably to have assigned a hydrothermal or magmatic origin to some of the deposits in the volume. So far as the criticism can be met, I have done so in the notes following the bibliographies. In attempting to classify the nearly 240 deposits in this book, I have been impressed by the reasonable (in my opinion) way in which any deposit I have examined fits into the modified Lindgren classification. In placing many of the deposits, I have gone against the opinions of geologists who have studied them in detail. I have done this only after having seen the deposit in question and/or having studied the literature concerning it in detail. Until the reader has perused the notes following a given deposit's bibliography, I hope he will withhold judgment on my system of classification.

No one of the bibliographies contains all the papers written about the district with which it deals. I have, however, attempted to put into each bibliography all significant papers I have been able to find. Of necessity, I have included a larger proportion of the literature of deposits for which the number of publications is small. In the first edition, I tried to limit the citations to papers published in the last 50 years; I have been less bound by time in this edition.

The bibliographies are arranged by continent and country and, in Canada and the United States, by state or province. The state in which each Mexican deposit occurs is given, but the districts are not grouped by states. The distribution of deposits by continents and countries (including "general" bibliographies) is as follows:

North America	
Canada	69
Greenland	4
Mexico	10
United States	133
	216

South America	
Argentina	1
Bolivia	6
Brazil	3
Chile	3
Peru	8
Venezuela	1
	22

Western Hemisphere total: 238

This distribution is not to be considered as indicating the frequency or economic value of hydrothermal and magmatic mineral deposits in the various geologic or geographic subdivisions of the two continents but rather reflects the intensity with which the deposits have been sought, found, and exploited, and the degree to which the geologic results of such search, discovery, and exploitation have been published. Because most of the exploration for and exploitation of mineral deposits in the Americas has been financed by capital provided by firms and governments in English-speaking countries, most of the literature cited in this volume has been written in English. Not surprisingly, the language next most frequently met with is Spanish, but papers in Portuguese, German, French, and Danish also are encountered. The demands on the linguistic abilities of an English-speaking geologist are far less for the study of the deposits included in this volume than they will be for its companion volume on the Eastern Hemisphere.

I might say, parenthetically, that a geologist whose native tongue is English and who studies other languages is usually best advised to concentrate on German until he has mastered that tongue; then he can go on to others. I am convinced of the truth of this statement despite the huge amount of geologic literature being produced in Russian. Today, much of the work done by geologists in the U.S.S.R. is being translated into English, while essentially none of that being published in German (certainly no less valuable than that appearing in Russian) is being translated into English. French and Spanish are so easy to learn in comparison to Russian and German that the geologist needing them can easily pick them up on his own. For many years, Scandinavian geologists, fortunately, have published most of their work in English or German or have provided good summaries in one or both of these languages. The problem posed by publications in languages other than those I have mentioned is virtually insurmountable to anyone but a linguistic genius.

At the head of each bibliography is given (1) the location of the deposit, (2) the most probable age (in my opinion) of its formation, (3) the metals or mineral materials for which it has been, is being, or may be exploited, and (4) the category (or categories) in the modified Lindgren classification used in this book. (*See* Index of Deposits According to the Lindgren Classification.)

To a considerable extent, the most valuable sources for literature of the economic geology of the Americas are indicated by the frequency with which they appear in the bibliographies given here. For the citations in the 238 bibliographies, the following are sources of references at least seven times (except for the publications of the International Geological Congresses).

Source	*No. of References*
A.I.M.E. (Transactions)	124
American Journal of Science	36
American Mineralogist	79
Arizona Bureau of Mines	17
Arizona Geological Society	17
British Columbia Minister of Mines	14
California Division of Mines	21
California Journal of Mines and Geology	20
Canadian Institute of Mining and Metallurgy	266
Canadian Journal of Earth Sciences	22

Source	*No. of References*
Canadian Mineralogist	24
Canadian Mining Journal	59
Colorado School of Mines Quarterly	10
Colorado Mineral Resource Board, State of	11
Colorado Scientific Society	19
Consejo de Recursos Naturales no Renovables	9
Economic Geology	765
Engineering and Mining Journal	64
Geochimica et Cosmochimica Acta	13
Geological Association of Canada	31
Geology of the Porphyry Copper Deposits--S.W. North America	14
Geological Society of America	201
Geological Survey of Canada	218
Graton-Sales Volume	58
Illinois State Geological Survey	14
Intermountain Association of Petroleum Geologists	15
International Geological Congress (10th)	7
International Geological Congress (16th)	56
International Geological Congress (18th)	13
International Geological Congress (19th)	3
International Geological Congress (20th)	20
International Geological Congress (21st)	13
International Geological Congress (23rd)	4
Journal of Geology	45
Manitoba Department of Mines and Natural Resources	12
Meddelelser om Grønland	23
Mineralium Deposita	16
Mining and Scientific Press	13
Mining Engineering	40
Minnesota Geological Survey	14
Neues Jahrbuch für Mineralogie (all title variations)	20
New Mexico Bureau of Mines and Mineral Resources	11
New Mexico Geological Society	13
New York State Museum	13
Ontario Department (Bureau) of Mines	100
Ore Deposits as Related to Structural Features	32
Precambrian	14
Quebec Department (Bureau) of Mines	44
Quebec Department of Natural Resources	10
Royal Society of Canada	40
Science	8
Tennessee Division of Geology	11
U.N. International Conference on Peaceful Uses of Atomic Energy (2d)	9
University of California Publications (Bull. Dept. Geol. Sci.)	10
University of Nevada Bulletin	10
University of Toronto Studies	12
U.S. Bureau of Mines	36
U.S. Geological Survey	551
Utah Geological Society	16
Zeitschrift für Praktische Geologie	13

This list will serve as a guide in determining which publications are likely to be rewarding sources of information about any deposit in the Americas. Any literature survey should, of course, begin with the examination of (1) *The Annotated Bibliography of Economic Geology* that appeared from 1929 through 1965, (2) *The Bibliography of North American Geology* that covers the years since 1789 and is published by the U.S. Geological Survey (since 1965, it has been issued on a monthly basis without an annual index), and (3) *The Bibliography and Index of Geology* that covers the years since 1933 and is published by The Geological Society of America (it also has been issued on a monthly basis since 1967 and has three annual summary volumes through 1969).

For lack of space and funds, geologic maps of the various districts have been omitted; these can be found in the references listed and should be used for a better understanding of the notes appended to the bibliographies.

I am sure that no one expects me to have read all the references cited in this volume, but I have seen all but four of them. Since the publications cited are not all available in the Library of the College of Earth and Mineral Sciences at The Pennsylvania State University or in my own collection, I have received much assistance from the U. S. Geological Survey Library, the Library of Congress in Washington, the New York Public Library, and the Engineering Societies Library in New York. The Interlibrary Loan Service also has been most useful. To these institutions and their helpful staffs, I express my thanks.

To Liberata Emmerich and Emilie T. McWilliams, successively librarians of the College of Earth and Mineral Sciences, who have done so much to make this library outstanding in its fields, I owe a debt that cannot be repaid.

I must also express my gratitude for financial support in the preparation of this volume to the Central Fund for Research of The Pennsylvania State Universities and to the various deans of my college who have provided further funds from sources best known to themselves. These gentlemen are E. F. Osborn, D. R. Mitchell, R. H. Jahns, and C. L. Hosler, Jr.

To my secretaries, Barbara Poorman, Barbara Younker, and Jeanne Bruce, who have typed and retyped (as I changed my mind) bibliographies and notes, who have sorted and resorted reference cards to make the various indices, and who have caught me out in innumerable errors and inconsistencies, I am deeply grateful. I also wish to thank The Geological Society of America for its interest and help.

Finally, to that fine group of scientists and men, the mine geologists of the Americas, who have taken time and patience to show me the mines in which they work and to discuss with me their concepts of their geology, I am more deeply grateful than words can express. Where I have disagreed with them, it has been with the greatest regret, and where I have agreed with them, it has been with the greatest pleasure. To them this volume is dedicated.

Johannesburg, South Africa
29 March 1970

Note: Since my return to University Park, I have added an appreciable number of 1970 references that seem to me too important to be left out. The *Bibliography of North American Geology, 1966* and *1967*, Geological Survey Bulletins 1266 and 1267, Government Printing Office 1970, have also been published since these bibliographies were compiled.

NORTH AMERICA

CANADA

British Columbia

BRENDA LAKE

Middle Mesozoic *Copper, Molybdenum* *Hypothermal-1*

Anon., 1970, Brenda Mines Ltd.--copper-molybdenum mine now in production: Western Miner, v. 43, no. 6, p. 39-50, 52 (particularly, p. 43-44)

Carr, J. M., 1967, The geology of the Brenda Lake area: B.C. Minister Mines Ann. Rept., p. 183-212

Carr, J. M., and Smith, D., 1966, Brenda Lake: B.C. Minister Mines Ann. Rept., p. 179-187

Fountain, D. K., 1968, The application of the induced polarization method at Brenda mine, British Columbia: Canadian Inst. Min. and Met. Bull., v. 61, no. 670, p. 153-157

Little, H. W., 1961, Kettle River (west half), British Columbia: Geol. Surv. Canada Map 15-1961 (revision of Map 538A), 1:253,440

Rice, H.M.A., 1947, Geology and mineral deposits of the Princeton map-area, British Columbia: Geol. Surv. Canada Mem. 243, 136 p.

White, W. H., and others, 1968, Potassium-argon ages of some ore deposits in British Columbia: Canadian Inst. Min. and Met. Bull., v. 61, no. 679, p. 1326-1334

Notes

The Brenda Lake property has its center about 18 miles northwest of the town of Peachland that, in turn, lies on the west shore of Okanagan Lake. Peachland is about 23 miles west of north of the town of Penticton at the south end of the lake.

The deposit, though huge, is low in grade (177,000,000 tons of ore, averaging 0.183 percent Cu and 0.049 percent Mo) and is contained in the zoned and composite quartz-diorite or granodiorite Brenda stock, a phase of the Okanagan batholith. This quartz-diorite stock is made up of four north-south trending units: (1) a medium quartz diorite that has dimensions of about 500 by 1500 feet and has sharp contacts with the enclosing Nicola group (varied lavas, argillites, tuffs, limestones, and schists); (2) a speckled quartz diorite that is the host rock of the Brenda ore body and is 800 by 2000 feet; (3) a uniform quartz diorite that grades into the speckled variety within 100 to 200 feet; and (4) a porphyritic quartz diorite. A fine quartz diorite also is present but is discordant to the other units. The mineralized rock volume of speckled quartz diorite is about 1500 feet east of the contact between the stock and the Nicola group; the host rock is essentially circular and has been strongly fractured and faulted. K/Ar dating on biotite in the quartz diorite gives an age of 148 m.y. ± 6 m.y. and on hornblende of 166 m.y. ± 8 m.y. White and others (1968) suggest that the discrepancy between the two ages may be explained in one or two ways: (1) The granodiorite was crystallized about 168 m.y. ago, was jointed shortly thereafter, and the joints persisted for 20 m.y. At this time, high-temperature hydrothermal solutions invaded the area, deposited the ore, and changed the K/Ar ratios in the biotites both within and outside the area of ore deposition but essentially did not affect the hornblende. (2) The granodiorite and the ore minerals were formed about 168 m.y. ago, and subsequently a pyrogenic event (the authors postulate that it was the extrusion of plateau basalts) evenly and preferentially changed the K/Ar ratio of the biotites. The authors in question prefer explanation (2) for reasons they do not give. They do say, however, that, with depth, they would expect the mode of occurrence of the ore minerals to change from epigenetic to syn-

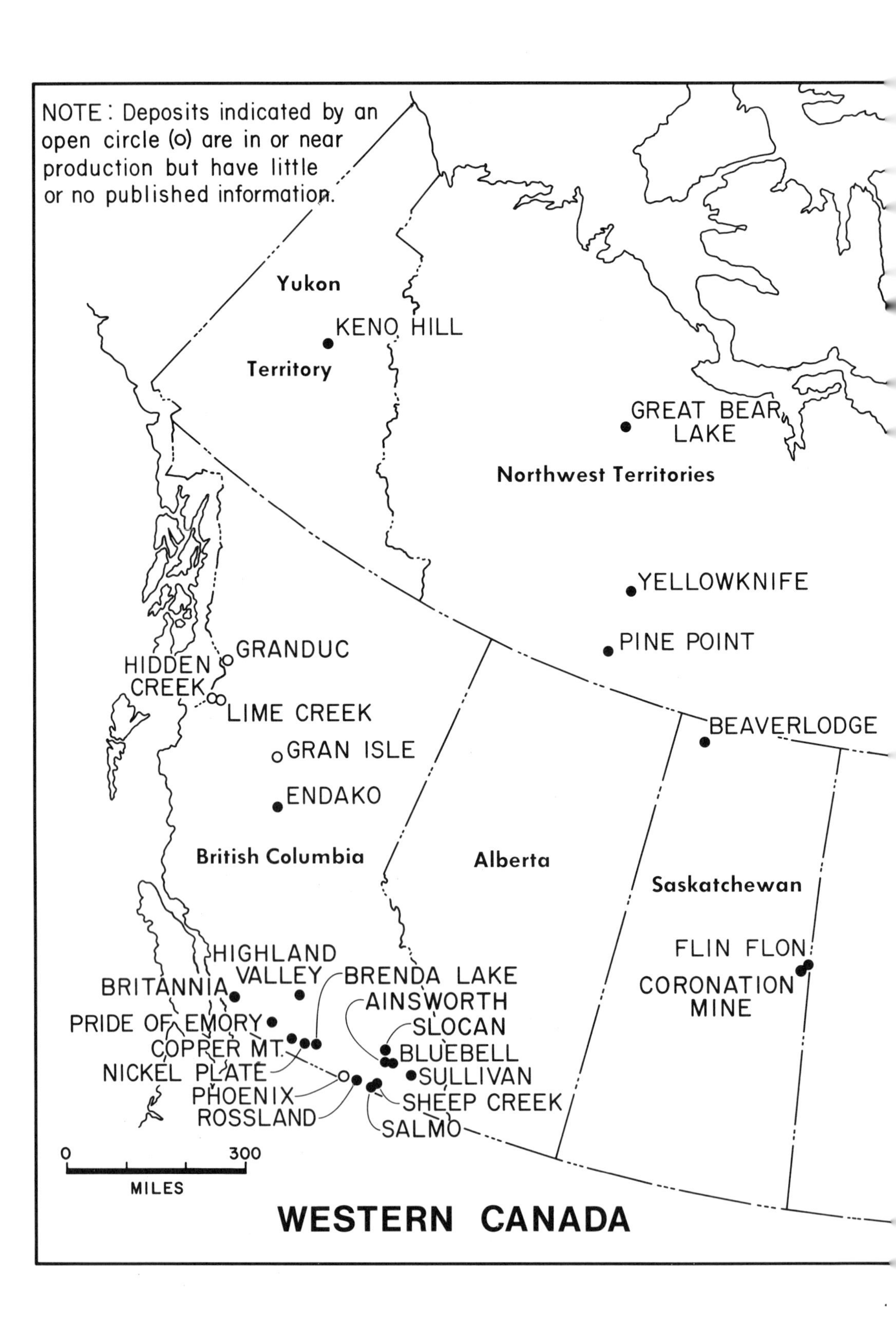
NOTE: Deposits indicated by an open circle (o) are in or near production but have little or no published information.
Yukon
Territory
KENO HILL
GREAT BEAR LAKE
Northwest Territories
YELLOWKNIFE
PINE POINT
GRANDUC
HIDDEN CREEK
LIME CREEK
BEAVERLODGE
GRAN ISLE
ENDAKO
British Columbia
Alberta
Saskatchewan
FLIN FLON
CORONATION MINE
HIGHLAND VALLEY
BRENDA LAKE
BRITANNIA
AINSWORTH
PRIDE OF EMORY
SLOCAN
COPPER MT.
BLUEBELL
NICKEL PLATE
SULLIVAN
PHOENIX
SHEEP CREEK
ROSSLAND
SALMO
0
300
MILES
WESTERN CANADA

genetic. All this seems rather fantastic; a more reasonable explanation would seem to be that the ore fluids, at the time they deposited the ores, removed, or promoted the removal of, argon from the biotites; thus, the host rock and the ore minerals, having come from the same general source, have much the same age, about 168 m.y. Whatever the explanation, either age determination suggests that the ores are Jurassic in age and are to be classified as middle Mesozoic.

The mineralized and highly broken portion of the Brenda stock shows four types of ore fractures: (1) quartz-potash feldspar-chalcopyrite-molybdenite, (2) biotite-chalcopyrite, (3) quartz-molybdenite-pyrite, and (4) epidote-magnetite-molybdenite. Most of the ore minerals are found in fractures of type (1). Practically all such fractures strike N60° to 75°E and dip from 55°S to 80°N. A later set of similar, though vuggy, fractures strikes north-south to N25°E and dips 80° to 85°W. The type (1) fractures are cut by narrow type (2) fractures that are up to 1/8 inch wide; in addition to biotite and chalcopyrite, they also contain minor potash feldspar and molybdenite. The majority of the type (2) veins strike N10° to 25°E and dip from 70°E to 35°W. The type (3) veins are found near the shear zones and are from 1/8 inch to 8 inches thick; they have N80°E strikes and dip from 75°E to vertical. The type (4) veins are not abundant but have been noted throughout the ore body; they have N55°E strikes and vertical dips. The intensity of fracturing differs from one part of the mineralized zone to another; the average spacing of fractures is about 8 to 10 inches apart, although areas are known where they are as close to each other as 2 inches. In the type (1) fractures, the vein fillings are mainly quartz and potash feldspar; chalcopyrite, pyrite, and molybdenite occur as more or less separate crystal grains, bunches, and platy networks. The chalcopyrite is much more abundant than either pyrite or molybdenite. Disseminated ore minerals are sparse outside the veins and rarely extend more than 1/4 inch from the vein walls. In the veins that are mainly quartz filled, the sulfides are finer grained and seamlike. Molybdenite is found partly in lesser, type (3), fractures that cut across the veins. Where quartz has filled faults in contrast to filling veins, the sulfides are in fine to coarse granular seams and bunches; in brecciated veins, the sulfides are in irregular networks with carbonate and a second generation of quartz. In the faults, the proportion of pyrite and molybdenite to chalcopyrite is higher than it is in the fractures. The ore body has its major dimension (2800 feet) striking northeast; it is 1300 feet wide and 900 feet deep; the grade decreases outward from the center, and presently unminable mineralization continues beyond the line of mining cutoff. The ore body does not have a pyrite-rich halo. The major alteration mineral in the ore zone is hydrothermal biotite. The shear zones, from a few inches to 30 feet wide, are characterized by argillic alteration. The remainder of the ore-bearing quartz diorite has been subjected to weak alteration that consists of chloritization of the mafic minerals. The ore body is cut by small aplitic to pegmatitic granitic dikes that strike N50° to 65°E and dip 60° to 65°NW. Some andesite dikes also are present and these strike northwest and have dips that are vertical to steeply southeast. Under the microscope, the chalcopyrite surrounds and apparently is later than the pyrite; molybdenite may be later than the chalcopyrite. The close association in time of most of the chalcopyrite with molybdenite strongly suggests that the ores were formed under hypothermal conditions, a concept confirmed by the presence of potash feldspar, magnetite, and epidote as gangue minerals. The ores are here classified as hypothermal-1.

BRITANNIA MINES (HOWE SOUND)

Late Mesozoic (pre-Laramide)	*Copper, Zinc*	*Hypothermal-1 to Leptothermal*

Alcock, F. J., 1935, Britannia mines, in *Copper resources of the world*: 16th Int. Geol. Cong., v. 1, p. 114-121

Anon., 1970, Canada update: Eng. and Min. Jour., v. 171, no. 9, p. 116

Ebbutt, F., 1935, Relationship of structure to ore deposition at the Britannia mine: Canadian Inst. Min. and Met. Tr., v. 38 (Bull. no. 276), p. 123-133

—— 1942, The Britannia mines, B.C., in Newhouse, W. H., Editor, *Ore deposits as related to structural features*: Princeton Univ. Press, p. 155-156

Irvine, W. T., 1946, Geology and development of the no. 8 orebodies, Britannia mines, B.C.: Canadian Inst. Min. and Met. Bull. no. 407, p. 191-214 (not in Tr. sec.)

—— 1948, Britannia mine, in *Structural geology of Canadian ore deposits*: Canadian Inst. Min. and Met., Montreal, p. 105-109

James, H. T., 1929, Britannia Beach map-area, British Columbia: Geol. Surv. Canada Mem. 158, 139 p.

Schofield, S. J., 1926, The Britannia mines, British Columbia: Econ. Geol., v. 21, p. 271-284

Waterman, G. C., 1970, An old mine breathes new life: Western Miner, v. 43, no. 6, p. 62, 64-65

Notes

The Britannia Mines are at the landward end of Howe Sound, 30 miles north of Vancouver.

The area underwent folding, probably in the late Jurassic, and, during the folding, tremendous volumes of granodiorite were intruded that engulfed much of the older metasediments and metavolcanics. Later erosion has left a roof pendant of these metamorphosed rocks (about 2 miles wide [N-S] and 7 miles long) in the granodiorite of the batholith. At much the same time, a 7-mile-long, 2000-foot-wide shear zone was developed along the southern margin of the pendant; this zone strikes northwest-southeast and dips steeply southeast. Because of the close temporal relationships among folding, intrusion, shear-zone formation, and ore introduction, it appears that granodiorite and ore probably came from the same general magmatic source. Radioactive age determinations suggest that the granodiorites are about 100 m.y. old; this age is much the same as that obtained for rocks taken from various locations in the Coast Range batholith. Thus, the ores probably are middle Cretaceous and are classed as pre-Laramide.

The ore bodies known at Britannia Mines prior to 1964 were confined to the 2000-foot by 7-mile shear zone mentioned above and raked steeply downward to the west. The minable ore was more continuous than the considerable number of ore body names would suggest. Work begun by Anaconda in 1962 (Waterman, 1970) suggested that the ore bodies were not concentrated in the footwall of the shear zone but were situated in positions intermediate between the hanging and footwall borders of the shear. The important structural controls appeared to be the competence of one rock unit relative to others rather than the actual composition of the units. The intensity of mineralization and localization of ore were thought to be related to (1) the percentage of a given rock volume made up of intrusive dikes, (2) the size of areas where the dikes coalesced (dacite hoods), and (3) the degree of shearing. Although sulfide seams, stringers, and veins were widespread through the shear zone, in only small volumes were they sufficiently concentrated to make ore bodies. Most of these ore bodies were found to have the same westward plunge as the dacite hoods. Further, the ore bodies were located either immediately below the westerly plunging constriction in the upper part of the shear zone or at deeper levels below the top of the constriction. These concepts provided target areas for drilling under this westerly plunging constriction and in a still deeper region extending from the footwall to the hanging wall of the shear zone. Anaconda geologists thought that previous work below the 4100 main haulage level did not extend far enough west to reach the projection of this zone of constriction nor to cut the deeper potential ore-bearing region. In addition to these two principal areas for exploration, other targets that resulted from this theory were (1) extensions of the westerly lenses of the no. 8 ore body below the 5100 level, (2) the vein areas in the Victoria ore shoot above the 4100 level, and (3) an area on the downward rake of the old Bluff ore body above the 4100 level. Implementation of an exploration program to test these areas found a sizable ore body about 2500 feet west of the no. 8 ore body that had no physical connection to the no. 8 at all. The new ore extends from a short distance above the 4100 level to below the 6300 level and is known as the 040 ore body; most of the new ore body lies below sea level. The 040 contains at least 3 million tons of ore that averages more than 1.5 percent copper and is made up of a steeply dipping concentration of mineralized stringers. During the exploration, production from the no. 8 ore body was reduced, while the Victoria and Bluff bodies were so heavily mined that they were exhausted and shut down in 1970. All of the other ore bodies described by Ebbutt (1942) and Irvine (1948) have long since been mined out. The principal ore minerals of the older ore bodies are chalcopyrite and sphalerite, and at least some of each contain oriented blebs of the other; pyrite was essentially the only iron sulfide. Galena was fairly common in the upper levels and is present in the lower parts of the ore bodies; veinlets of galena cut and replace all other sulfides. Tetrahedrite also was found in some quantity in the higher parts of the ore; it is later than the sphalerite and chalcopyrite but older than the galena. Quartz is the main gangue mineral, although locally considerable barite is found. The blebs of zinc and copper sulfide, respectively, in some of the chalcopyrite and sphalerite indicate deposition above 400°C; on the other hand, the tetrahedrite and later galena (although these two minerals are not silver-rich, they are silver-bearing) suggest that the last of the mineralization took place in the leptothermal range. Probably much of the sphalerite and chalcopyrite were formed in the mesothermal range, but enough study of these two minerals has not been done to make this concept certain. No results have been published on the new exploration at Britannia that changes these ideas on the mineralization there. Thus, the ores probably should be classed as hypothermal-1 to leptothermal.

COPPER MOUNTAIN

Middle Mesozoic *Copper* *Mesothermal*

Dolmage, V., 1929, The origin of the Copper Mountain ores: Canadian Inst. Min. and Met. Tr., v. 32 (Bull. no. 206), p. 788-802

—— 1934, Geology and ore deposits of Copper Mountain, British Columbia: Geol. Surv. Canada Mem. 171, 69 p.

Fahrni, K. C., 1951, Geology of Copper Mountain: Canadian Inst. Min. and Met. Tr., v. 54 (Bull. no. 469), p. 203-210

—— 1962, Post-production geology at Copper Mountain: Western Miner and Oil Rev., v. 35, no. 2, p. 53-54

—— 1966, Geological relations at Copper Mountain, Phoenix and Granisle mines, in *Tectonic history and mineral deposits of the western Cordillera*: Canadian Inst. Min. and Met., B.C. Sec., Spec. v. no. 8, p. 315-320

Rice, H.M.A., 1947, Geology and mineral deposits of the Princeton map-area, British Columbia: Geol. Surv. Canada Mem. 243, 136 p.

Sinclair, A. J., and White, W. H., 1968, Age of mineralization and post-ore hydrothermal alteration, Copper Mountain, British Columbia: Canadian Inst. Min. and Met. Bull., v. 61, no. 673, p. 633-636

Notes

Copper Mountain is about 115 miles slightly northeast from Vancouver and 25 miles north of the international boundary.

The most recent work on the age of the Copper Mountain ores (Sinclair and White, 1968) suggests, on the basis of K/Ar determinations, that they are 193 ± 7 m.y. old. If this is a sound age, then the ores were formed late in the Triassic. Until recently, the igneous rocks in question were thought to be, on geological grounds, Late Jurassic or early Tertiary. Certainly, the rocks of the Nicola group are Upper Triassic, and it seems unreasonable that the rocks intrusive into them, the Copper Mountain and Voight stocks and the Lost Horse intrusions, are also of that age. Rather, these intrusives should much more logically be assigned to Jurassic time. In the Highland Valley area, the wide range of ages derived from K/Ar determinations on biotites and the much narrower spread on those from hornblende strongly suggest that the latter mineral is the better one on which to make such determination. Until Rb/Sr determinations (preferably on whole rock) or at least K/Ar work on hornblende is done, it would seem reasonable to discount the K/Ar dates and still class the ores as Jurassic or middle Mesozoic. If a result from the work of Sinclair and White is to be accepted, their determination of a 150 ± 9 m.y. age on a clinopyroxene (despite the slight hydrothermal alteration of plagioclase and biotite in the rock from which the pyroxene comes) seems to fit best with the geologic facts.

The principal ore minerals in the deposit are bornite and chalcopyrite. The ore mined in the earlier days of the operation appears to have contained more bornite than chalcopyrite, but outside these bornite-rich cores, the ores were mainly chalcopyrite with marginal, apparently non-ore pyrite zones beyond the chalcopyrite shells. Some of the chalcopyrite-pyrite ores were associated with the Lost Horse intrusions and seem not to have contained bornite cores. Rice (1947) considered there to have been three types of ore: (1) bornite-chalcopyrite deposits, (2) chalcopyrite-pyrite deposits, and (3) chalcopyrite-hematite deposits. None of the deposits of type (3) ever was of economic importance (at least up to the date of Rice's work); these deposits occur only in the Voight stock. Most of the ore was found in the Triassic Wolf Creek formation of the Nicola group, but an appreciable amount also was found in the syeno-gabbroic rocks of the Copper Mountain stock. The ore minerals occurred in small veins and disseminated masses in these host rocks. Rice states that type (1) ore minerals occur in pegmatite dikes that are common in the stock; none of these pegmatites ever was large or rich enough to be mined. Fahrni (1951) reports three varieties of alteration: (1) a feldspathization of the mine rocks related to the pegmatite veins and dikes; (2) a biotitization that affected most of the rocks in the mine area to some extent and was caused by solutions (?) derived from the gabbro; and (3) chloritization and silicification associated with post-ore igneous dikes. The presence of the typical ore minerals in the dikes strongly suggests that the pegmatites and ore solutions both came from the same general source. Since no ore has been mined from pegmatite bodies, except incidentally to mining hydrothermal ore adjacent to them, they are not included in the classification of the deposits. The host sediments appear to have been converted to hornfels by the syeno-gabbro intrusion, but this effect certainly predates the ore. The chalcopyrite and bornite suggest mesothermal conditions of deposition, a concept not contradicted by the biotitization that slightly preceded it.

ENDAKO

Middle Mesozoic — *Molybdenum* — *Hypothermal-1*

Armstrong, J. E., 1949, Fort St. James map-area, Cassiar and Coast district, British Columbia: Geol. Surv. Canada Mem. 252, 210 p.

Brown, A. S., 1965, Investigation of mercury dispersion halos: B.C. Minister Mines Ann. Rept., p. 109-112

Carr, J. M., 1965, Endako: B.C. Minister Mines Ann. Rept., p. 114-135

Drummond, A. D., and Kimura, E. T., 1969, Hydrothermal alteration at Endako--a comparison to experimental studies: Canadian Inst. Min. and Met. Bull. no. 687, p. 709-714

Kimura, E. T., and Drummond, A. D., 1969, Geology of Endako molybdenum deposit: Canadian Inst. Min. and Met. Bull. no. 687, p. 699-708

Lay, D., 1927, Endako: B.C. Minister Mines Ann. Rept., p. 152-153

Tipper, H. W., 1963, Nechako River map-area: Geol. Surv. Canada Mem. 324, 59 p. (property lies just north of the northern boundary of the map area)

White, W. H., and others, 1970, Potassium-argon ages of Topley intrusions near Endako, British Columbia: Canadian Jour. Earth Sci., v. 7, p. 1172-1178

Notes

The Endako property is some 95 miles slightly north of west from Prince George and 205 miles slightly south of east from Prince Rupert.

The Endako deposit occurs entirely in the Endako quartz monzonite, which is intrusive into the Takla group of volcanic rocks of early Mesozoic age. The quartz monzonite is one of four rock types that make up the Topley intrusions; the others are (1) the Casey alaskite, (2) the Francois granite, and (3) the Glenannan granite. Dikes of the Francois and Casey cut the Endako while the Glenannan is older than the Casey, because a dike rock that correlates with the Casey intrudes the Glenannan. Thus, the Endako is older than the Francois and the Casey and probably than the Glenannan. Radio-age dating (K/Ar) suggests that the Endako is 140 m.y. old. Obviously the ore is younger than the quartz monzonite but how much younger is uncertain. The Endako is cut by (in order of decreasing age) (1) premineral aplite (probably offshoots of the Casey ?), (2) porphyritic granite (equivalent of the Francois granite ?), and (3) quartz feldspar porphyry (the most abundant dike rock in the mine area) in two distinct types. Obviously some time must have been required for the introduction of these three intrusives, so that the ore mineralization is appreciably (several million years ?) younger than the Endako. This would suggest that the ore, while probably developed as part of the same period of igneous activity that produced all the rock types discussed, is enough younger to be classed as on the border between Jurassic and Cretaceous time and is legitimately categorized as middle Mesozoic.

The minerals most abundant in the veins and fine-fracture fillings are molybdenite, pyrite, and magnetite; minor amounts of chalcopyrite and traces of bornite, bismuthinite, scheelite, and specularite also are present. Three types of alteration have affected the Endako quartz monzonite: (1) potash-feldspar selvages bordering quartz-molybdenite veins--locally, some biotite may be associated with the potash feldspar or a little quartz and biotite may be found with the feldspar; (2) quartz-sericite-pyrite selvages bordering the quartz-molybdenite veins; and (3) a pervasive kaolinization widespread through the quartz monzonite. Type (1) alteration is more commonly found around quartz-molybdenite veins, and sericitic selvages of type (2) usually are found bordering veins that are mainly quartz magnetite. Potash-feldspar selvages are cut by quartz-sericite-pyrite without any offset. The type (2) selvages are younger, however, since the potash feldspar of the type (1) selvage has been converted to sericite in the area of junction. The kaolinization lies in quartz monzonite beyond the selvages proper. The facts indicate that the ore was introduced under conditions of high intensity and should be classified as hypothermal-1.

HIGHLAND VALLEY

Middle Mesozoic *Copper* *Mesothermal*

Brown, A. S., 1969, Mineralization in British Columbia copper and molybdenum deposits: Canadian Inst. Min. and Met. Bull., v. 62, no. 681, p. 26-40

Carr, J. M., 1960, Porphyries, breccias, and copper mineralization in Highland Valley, B.C.: Canadian Min. Jour., v. 81, no. 11, p. 71-73

—— 1966, Geology of the Bethlehem and Craigmont copper deposit, in *Tectonic history and mineral deposits of the western Cordillera*: Canadian Inst.

Min. and Met., B.C. Sec., Spec. v. no. 8, p. 321-328, 353 p.

Carr, J. M., and others, 1966, Highland Valley: B.C. Minister Mines Ann. Rept., p. 149-165

—— 1967, Highland Valley: B.C. Minister Mines Ann. Rept., p. 149-165

—— 1968, Highland Valley: B.C. Minister Mines Ann. Rept., p. 179-195

Chrismas, L., and others, 1968, Rb/Sr, S, and O isotopic analyses indicating source and date of pyrometasomatic copper deposits at Craigmont, British Columbia (abs.): Econ. Geol., v. 63, p. 702-703

—— 1969, Rb/Sr, S, and O isotopic analyses indicating source and date of contact metasomatic copper deposits, Craigmont, British Columbia, Canada: Econ. Geol., v. 64, p. 479-488

Cockfield, W. E., 1948, Geology and mineral deposits of Nicola map-area, British Columbia: Geol. Surv. Canada Mem. 249, 164 p.

Coveney, C. J., 1962, The Bethlehem copper property: Western Miner and Oil Rev., v. 35, no. 2, p. 42-43

—— 1963, Bethlehem Copper Corporation--Geology: Western Miner and Oil Rev., v. 36, no. 1, p. 33-35

Drysdale, C. W., 1915, Highland Valley copper camp, Ashcroft mining division: Geol. Surv. Canada Summ. Rept., p. 85-91

Duffel, S., and McTaggart, K. C., 1952, Ashcroft map-area, British Columbia: Geol. Surv. Canada Mem. 262, 122 p.

Hansen, D. A., and Barr, D. A., 1966, Exploration case history of a disseminated copper deposit: Soc. Expl. Geophys., Mining Geophysics, v. 1, Case Histories, p. 306-312

Northcote, K. E., 1969, Geology and geochronology of the Guichon batholith, British Columbia: B.C. Dept. Mines and Petrol. Res. Bull. 56, 73 p.

Rennie, C. C., 1962, Copper deposits in the Nicola rocks, Craigmont mine: Western Miner and Oil Rev., v. 35, no. 2, p. 50-52

Rennie, C. C., and others, 1961, Geology of the Craigmont mine: Canadian Inst. Min. and Met. Tr., v. 64 (Bull. no. 588), p. 199-203

White, W. H., and others, 1957, The geology and mineral deposits of Highland Valley, B.C.: Canadian Inst. Min. and Met. Tr., v. 60 (Bull. no. 544), p. 273-289

—— 1967, Isotopic dating of the Guichon batholith, B.C.: Canadian Jour. Earth Sci., v. 4, p. 677-690

Wright, H. M., 1963-1964, The Bethlehem project of Bethlehem Copper Corporation Ltd., Highland Valley, British Columbia: Inst. Min. and Met. Tr., v. 72, p. 177-253 (particularly p. 180-187)

Notes

Highland Valley, which includes the Craigmont, Bethlehem, and Iron Mask properties, is centered about 120 miles northeast of Vancouver and nearly 40 miles southeast of Kamloops.

The Bethlehem deposits lie in the north-central part of the Guichon batholith and the Craigmont in calcareous sediments of the Nicola group along the southern margin of the batholith. White and others (1967) state that the batholith is made up of seven granitic phases plus porphyritic bodies and cataclastic dikes and that the mean age from K/Ar determinations for 24 samples from these rocks is 200 m.y. ± 5 m.y. Even using the 1964 Geological Society of London time scale rather than that of Holmes, the beginning of Jurassic time is from 190 to 195 m.y. ago. If this 200 m.y. date is a true indicator of the age of the batholith, it must have been emplaced in the latest Triassic time. The batholith intrudes the Upper Triassic Nicola rocks and is unconformably overlain by early Middle Jurassic and Upper Jurassic beds. Thus, its age must be very Late Triassic or Early Jurassic. Despite the radioactive age determinations, it is thought more reasonable to assign an Early Jurassic or middle Mesozoic age to the batholith and its attendant mineralization.

According to Carr and others (1966), most of the Bethlehem ores lie along a highly irregular contact between older and younger quartz diorite (presumably the second of the seven granitic phases of White and others [1967]). This area was intruded by a swarm of dikes of various types. Breccia bodies associated with the dikes are believed to have been formed by explosions of porphyry sheets due to gases concentrated during crystallization by the early crystallized and impervious upper surface. Still further dike intrusions after the brecciation were followed by faulting and then by mineralization. The wall-rock alteration minerals produced in the igneous rocks include quartz, sericite, chlorite, kaolin, calcite, orthoclase, biotite, tourmaline, actinolite, and epidote; zeolites were deposited after the sulfides. The main ore minerals are bornite and chalcopyrite with small amounts of molybdenite accompanying the copper sulfides. Pyrite is present in some areas but not to any great extent apparently in others. The ore bodies are emplaced in veins and disseminations, partly by fracture filling and partly by replacement. Although no detailed paragenetic information is available, it would seem that the bornite and chalcopyrite are associated with the lower-temperature alteration minerals and that the ores should be classed as mesothermal rather than placed in a higher-intensity category.

At Craigmont, the ores replace skarn rocks developed in limestone beds of the Nicola group. The skarn differs markedly in composition, the relative proportions of magnetite, specular hematite, epidote, garnet, actinolite, chlorite, calcite, orthoclase, quartz, and tourmaline changing considerably from place to place. Much of the ore is in iron-rich skarn that becomes richer in magnetite and poorer in hematite as the ore is followed from west to east. The dominant ore mineral is chalcopyrite that occurs as veins, streaks, patches, and disseminations in skarn. Where the ore is lean, some pyrite and bornite are found. The deposit appears to resemble that at Cornwall in Pennsylvania with the exceptions that (1) the iron minerals are not worth mining; (2) the chalcopyrite does not contain gold; and (3) the pyrite, what there is of it, does not contain enough cobalt to mention. But basically, the late introduction of chalcopyrite and bornite into the skarn strongly suggests that the copper minerals were introduced under appreciably less intense conditions than the skarn and that the ores should be classed as mesothermal.

NELSON AREA (GENERAL)

Middle Mesozoic and Early Tertiary	*Gold, Silver, Copper, Lead, Zinc, Tungsten*	*Telethermal, Leptothermal, Mesothermal, Hypothermal-1, Hypothermal-2*

Crosby, P., 1968, Tectonic, plutonic, and metamorphic history of the central Kootenay arc, British Columbia, Canada: Geol. Soc. Amer. Spec. Paper 99, 94 p.

Fyles, J. T., 1966, Lead-zinc deposits in British Columbia, in *Tectonic history and mineral deposits of the western Cordillera*: Canadian Inst. Min. and Met., B.C. Sec., Spec. v. no. 8, p. 231-237

—— 1967, Geology of the Ainsworth-Kaslo area, British Columbia: B.C. Dept. Mines and Petrol. Res. Bull. 53, 125 p.

Little, H. W., 1960, Nelson map-area, west half, British Columbia (82 F W1/2): Geol. Surv. Canada Mem. 308, 205 p.

Lowden, J. A., 1963, Age determinations by the Geological Survey of Canada, isotopic ages, Report 4: Geol. Surv. Canada Paper 63-17, p. 1-121 (particularly p. 22-24)

Muraro, T. W., 1966, Metamorphism of zinc-lead deposits in southeastern British Columbia, in *Tectonic history and mineral deposits of the western Cordillera*: Canadian Inst. Min. and Met., B.C. Sec., Spec. v. no. 8, p. 239-247

Nguyen, K. K., and others, 1968, Age of the northern part of the Nelson batholith: Canadian Jour. Earth Sci., v. 5, p. 955-957

Reesor, J. E., 1965, Structural evolution and plutonism in the Valhalla gneiss complex, British Columbia: Geol. Surv. Canada Bull. 129, 128 p.

Rice, H.M.A., 1941, Nelson map-area, east half, B.C.: Geol. Surv. Canada Mem. 228, 86 p.

Ross, J. V., and Kellerhals, P., 1968, Evolution of the Slocan syncline in south-central British Columbia: Canadian Jour. Earth Sci., v. 5, p. 851-872

Sinclair, A. J., 1966, Anomalous leads from the Kootenay Arc, British Columbia, in *Tectonic history and mineral deposits of the western Cordillera*: Canadian Inst. Min. and Met., B.C. Sec., Spec. v. no. 8, p. 249-262

White, W. H., and others, 1968, Potassium-argon ages of some ore deposits in British Columbia: Canadian Inst. Min. and Met. Bull., v. 61, no. 679, p. 1326-1334

Notes

The Nelson area, as used here, is the same as the Nelson map-area of the Canadian Geological Survey and extends from 49° to 50° north latitude and from 116° to 118° west longitude; this area includes the Ainsworth, Bluebell, Rossland, Salmo, Slocan, and Sheep Creek districts, and these are discussed in this volume.

The dominant igneous feature of the Nelson area is the Nelson batholith, most of the rocks of which are porphyritic granite, but some are leucocratic nonporphyritic granite and some nonporphyritic granodiorite. K/Ar age determinations on the Nelson rocks have given ages between 171 m.y. and 123 m.y., but the extremes are determinations on biotite. Hornblende has given results with a much narrower range--136 m.y. to 152 m.y., and these seem to be more dependable than those from biotite. An average age of about 143 m.y. from the hornblende determinations indicates that the Nelson rocks, or the bulk of them at least, were intruded near the end of the Jurassic and in definitely middle Mesozoic time. It has been thought that the ores of most of the deposits (certainly not Rossland) were genetically related to the Nelson magma. Fyles (1968) believes that the concordant (limestone replacement) mineralization at Salmo is older than the transgressive (vein-, vein-system-, and shear-zone-fillings) types that are dominant in the other deposits of the area. At Ainsworth, Bluebell, and Slocan, at least, the ores are younger than a series of lamprophyre dikes; these dikes are undeformed and are thought to belong to the same general suite and to be Tertiary in age. At Rossland, the ores are much the same age as a series of lamprophyre and mica-lamprophyre dikes. Thus, the Ainsworth, Bluebell, Rossland, and Slocan ores probably are even later than the Laramide deformation and should be classed as early Tertiary and not as late Mesozoic-early Tertiary. The transgressive deposits at Sheep Creek (much different from the concordant Salmo deposits of the same general area) are older than a series of lamprophyre dikes since they postdate most or all of the vein faulting and probably the vein filling as well. A few of these dikes are offset by faulting which suggests that they were affected by the vein-forming earth movements and probably are not much later than the ore; they even may be of much the same age. Thus, it would appear that the Sheep Creek deposits also should be classed as early Tertiary. The Salmo lead-zinc deposits are related to structures that are older than the Nelson batholith; this, however, well may mean nothing as to the relative ages of ore and granite. The ores probably were later than the granite-induced metamorphism of the limestones since the ore minerals have replaced silicates (one granite dike cutting ore may be a dike around which the ore replaced limestone). It would seem, however, that the mineralization and the ore formation were not far separated in time. This would require the ores to be middle Mesozoic in age. The tungsten ores of the district are at, or a short distance from, the contacts of the limestones of the Laib formation with granite stocks of the Nelson type. Their location plus their high-intensity minerals strongly suggest that they were formed by fluids generated by the Nelson magmas; the tungsten ores, therefore, are here dated as middle Mesozoic. The ore-age problem in the Nelson area has not been finally resolved.

The classification of the various Nelson-area deposits is discussed under the *Notes* for the individual deposits.

AINSWORTH

Early Tertiary — *Lead, Zinc, Silver* — *Mesothermal*

Eastwood, G.E.P., 1951, Ainsworth: B.C. Minister Mines Ann. Rept., p. 144-155

—— 1952, Ainsworth (49°116°NW): B.C. Minister Mines Ann. Rept., p. 156-162

—— 1953, Ainsworth (49°116°NW): B.C. Minister Mines Ann. Rept., p. 123-130

Rice, H.M.A., 1944, Notes on geology and mineral deposits at Ainsworth, British Columbia: Geol. Surv. Canada Paper 44-13, 6 p.

—— 1948, Ainsworth mining camp, in *Structural geology of Canadian ore deposits*: Canadian Inst. Min. and Met., Montreal, p. 216-218

Schofield, S. J., 1920, Geology and ore deposits of the Ainsworth mining camp, British Columbia: Geol. Surv. Canada Mem. 117, 73 p.

Notes

The Ainsworth district is about 280 miles east-northeast of Vancouver and 55 miles north of the international boundary.

See *Notes* under Nelson Area (General) for discussion on ore age.

The principal sulfide minerals are pyrite, galena, sphalerite, chalcopyrite, and pyrrhotite; arsenopyrite has been noted in a few places. The pyrrhotite is found in the northern part of the camp but not in the southern; where present, it is rimmed or replaced along fractures by pyrite. More than one generation of pyrrhotite may have been formed. The main gangue minerals are quartz, calcite, siderite, and fluorite; the wall-rock alteration minerals in the rocks near the veins are chlorite, sericite, and carbonates; and the iron-bearing tephroite (knebelite) is found locally. The alteration reaches out for only a foot or so from the veins. The presence of pyrrhotite in some abundance in the northern part of the camp suggests that it, at least, was deposited under hypothermal conditions; the small amounts of knebelite would seem to confirm this. The ore minerals, however, are separated in time from the pyrrhotite (or most of it) and the wall-rock alteration is typically mesothermal. The silver-rich character of the galena also agrees with deposition in the mesothermal range.

BLUEBELL

Early Tertiary — *Lead, Zinc, Silver* — *Hypothermal-2*

Irvine, W. T., 1957, The Bluebell mine, in *Structural geology of Canadian ore deposits*, v. 2: Canadian Inst. Min. and Met., Montreal, p. 95-104

Ohmoto, H., and Rye, R. O., 1970, The Bluebell mine, British Columbia. I. Mineralogy, paragenesis, fluid inclusions, and the isotopes of hydrogen, oxygen, and carbon: Econ. Geol., v. 65, p. 417-437

Walker, J. F., 1928, Kootenay Lake district, British Columbia: Geol. Surv. Canada Summ. Rept., pt. C, p. 119-135

Westervelt, R. D., 1960, An investigation of the sulphide mineralization at the Kootenay Chief ore body, Bluebell mine, British Columbia (abs.): Canadian Min. Jour., v. 81, no. 8, p. 105

Notes

The Bluebell district is about 285 miles east-northeast of Vancouver and 55 miles north of the international boundary.

See *Notes* under Nelson Area (General) for discussion on ore age.

The lead-zinc ore bodies are heavy sulfide replacements that are localized along steep cross-fractures that extend through the Bluebell limestone from foot- to hanging-wall and are, therefore, irregular tabular bodies, transverse to the bedding. According to Irvine (1957), the principal sulfide minerals in the deposits are pyrrhotite, marmatitic sphalerite, and galena in decreasing order of abundance; pyrite, chalcopyrite, and arsenopyrite are persistent but minor in amount. A little light sphalerite fills late fractures. Quartz is the most common gangue mineral, and knebelite (an iron-bearing tephroite) is found in limited amount in or near most of the ore bodies. Westervelt (1960), in a study of the Kootenay Chief ore body found two types of mineralization, the chief difference between which appear to be that (1) the gangue in one was made up of high-temperature silicates and in the other of quartz and (2) the sphalerite with quartz is low in iron and is accompanied by minor magnetite while the sphalerite in the silicated rock is marmatitic. The abundance of pyrrhotite and the presence of minor arsenopyrite, plus the close temporal association of some of the ore with knebelite, suggest that the ores were emplaced under hypothermal conditions. The light-colored sphalerite may have been formed in the mesothermal range, but this is not enough evidence to include mesothermal in the classification of the deposit. The deposit does appear to have formed under more intense conditions than the Ainsworth ores across Kootenay Lake (on the west side).

ROSSLAND

Early Tertiary — *Copper, Gold* — *Hypothermal-1*

Brock, R. W., 1906, Preliminary report on Rossland, British Columbia, mining district: Geol. Surv. Canada Rept. no. 939, 40 p.

Bruce, E. L., 1917, Geology and ore deposits of Rossland: B.C. Minister Mines, Ann. Rept. 1916, p. 214-244

Drysdale, C. W., 1915, Geology and ore deposits of Rossland, British Columbia: Geol. Surv. Canada Mem. 77, 317 p.

Gilbert, G., 1948, Rossland camp, in *Structural geology of Canadian ore deposits*: Canadian Inst. Min. and Met., Montreal, p. 189-196

Little, H. W., 1963, Rossland map-area, British Columbia (82 F/4 W1/2): Geol. Surv. Canada Paper 63-13, 11 p.

Notes

The Rossland district is 240 miles slightly south of east from Vancouver and 5 miles north of the international boundary.

See *Notes* under Nelson Area (General) for discussion on ore age.

The principal sulfides at Rossland are pyrrhotite and chalcopyrite with some pyrite, locally appreciable amounts of galena (silver-bearing), sphalerite, molybdenite, bismuthinite, and minor quantities of arsenopyrite, stibnite, and marcasite. Magnetite almost always is closely associated with chalcopyrite and pyrrhotite, though only in limited areas is it abundant. The gangue is country rock largely altered to biotite and silica and locally appreciable calcite; in places chlorite and hornblende are abundant, and muscovite, tourmaline, garnet, and wollastonite are also found. The gold occurs as impregnations in pyrrhotite, pyrite, arsenopyrite, bismuthinite, molybdenite, and, in places, chalcopyrite. Of the four types of ore present at Rossland, only one (pyrite-marcasite-silver bearing veins) seems to have formed under mesothermal conditions. Since over 90 percent of the ore shipped from Rossland came from the other three types and since these ores contain pyrrhotite in abundance, closely associated in time and space with chalcopyrite, and with high-temperature silicates and quartz as gangue and wall-rock alteration minerals, the ores probably were formed under hypothermal conditions.

SALMO-SHEEP CREEK (GENERAL)

Middle Mesozoic and Early Tertiary — *Zinc, Lead, Tungsten, Gold, Silver* — *Telethermal (Pb-Zn), Leptothermal (Au-Ag), Hypothermal-2 (W),*

Frebold, H., and Little, H. W., 1962, Paleontology, stratigraphy, and structure of the Jurassic rocks, Salmo map-area, British Columbia: Geol. Surv. Canada Bull. 81, 29 p.

Little, H. W., 1950, Salmo map-area, British Columbia (Summ. Acct.): Geol. Surv. Canada Paper 50-19, 43 p. (mimeo.)

—— 1951, The stratigraphy and structure of Salmo map-area, B.C.: Canadian Inst. Min. and Met. Tr., v. 54 (Bull. no. 467), p. 133-136

—— 1960, Nelson map-area, west half, British Columbia (82 F W1/2): Geol. Surv. Canada Mem. 308, 205 p.

—— 1965, Salmo map-area, British Columbia: Geol. Surv. Canada Map 1145A (with descriptive notes) 1:63,360

Walker, J. F., 1934, Salmo sheet, Kootenay district (B.C.): Geol. Surv. Canada Map 299A (with descriptive notes), 1:63,360

—— 1934, Geology and mineral deposits of the Salmo map-area, British Columbia: Geol. Surv. Canada Mem. 172, 102 p.

Notes

The Salmo-Sheep Creek district is 265 miles slightly south of east from Vancouver and 5 miles north of the international boundary.

SALMO

Middle Mesozoic — *Zinc, Lead, Tungsten* — *Telethermal (Pb-Zn), Hypothermal-2 (W)*

Ball, C. W., 1954, The Emerald, Feeney and Dodger tungsten ore-bodies, Salmo, British Columbia, Canada: Econ. Geol., v. 49, p. 625-638

Fyles, J. T., and Hewlett, C. G., 1957, Lead-zinc deposits of the Salmo area, British Columbia, in *Structural geology of Canadian ore deposits*, v. 2: Canadian Inst. Min. and Met., Montreal, p. 104-110

—— 1957, Reeves Macdonald mine, in *Structural geology of Canadian ore deposits*, v. 2: Canadian Inst. Min. and Met., Montreal, p. 110-116

—— 1959, Stratigraphy and structure of the Salmo lead-zinc area: B.C. Dept. Mines Bull. no. 41, 162 p.

Green, L. H., 1955, Wall-rock alteration of certain lead-zinc replacement deposits in limestone, Salmo map-area, British Columbia: Geol. Surv. Canada Bull. 29, 33 p.

Greenwood, H. J., 1967, Wollastonite - stability in H_2O-CO_2 mixtures and occurrence in a contact metamorphic aureole near Salmo, British Columbia, Canada: Amer. Mineral., v. 52, p. 1669-1680

Hedley, M. S., 1943, Nelson area, in *Tungsten deposits of British Columbia*: B.C. Dept. Mines Bull. no. 10 (revised), p. 133-146

Irvine, W. T., 1957, The H. B. mine, in *Structural geology of Canadian ore deposits*, v. 2: Canadian Inst. Min. and Met., Montreal, p. 124-132

Little, H. W., 1959, Canadian Exploration Limited, in *Tungsten deposits of Canada*: Geol. Surv. Canada Econ. Geol. Ser. no. 17, p. 105-112

Little, J. D., and others, 1953, The lead-zinc and tungsten properties of Canadian Exploration, Limited, Salmo, B.C.: Canadian Inst. Min. and Met. Tr., v. 56 (Bull. no. 496), p. 228-236

Magee, J. B., and Cummings, W. W., 1960, The Mineral King mine: Canadian Inst. Min. and Met. Tr., v. 63 (Bull. no. 578), p. 241-243

Pollock, W., and others, 1961, The Reeves Macdonald operation: Canadian Inst. Min. and Met. Tr., v. 64 (Bull. no. 586), p. 113-117

Rennie, C. C., and Smith, T. S., 1957, Lead-zinc and tungsten orebodies of Canadian Exploration Limited, Salmo, B.C., in *Structural geology of Canadian ore deposits*, v. 2: Canadian Inst. Min. and Met., Montreal, p. 116-124

Warning, G. F., 1960, Geology of the H. B. mine: Canadian Inst. Min. and Met. Tr., v. 63 (Bull. no. 582), p. 519-522

Weissenborn, A. E., Editor, 1970, Lead-zinc deposits in the Kootenay arc, northeastern Washington and adjacent British Columbia: Wash. Dept. Nat. Res. Bull. no. 61, 123 p.

Whishaw, Q. G., 1954, The Jersey lead-zinc deposit, Salmo, B.C.: Econ. Geol., v. 49, p. 521-529

Notes

See *Notes* under Salmo-Sheep Creek (General) and Nelson Area (General) for location of deposit and discussion of ore age.

The minerals in the lead-zinc ore bodies of the Mine Belt between the Waneta fault (north and west) and the Argillite fault (south and east) are sphalerite, galena, pyrite, and minor pyrrhotite; the ores contain only a fraction of an ounce of silver per ton. Sphalerite is generally the most abundant sulfide although locally galena is dominant. Pyrite is more common than sphalerite and may be present in large masses that contain only minor sphalerite and galena. Some of the pyrrhotite at least appears to have been formed by reaction between the ores and later lamprophyre dikes. The principal gangue mineral is dolomite. Although the limestones in which many of the lead-zinc ore bodies lie are altered to high-temperature silicates, these silicates appear to be contact phenomena of the Nelson batholith and were formed before the ore fluids entered the area. The low silver content of the ores and their simple mineralogy strongly suggest that they were formed in the telethermal range as were the deposits of the Metaline district immediately to the south.

The tungsten bodies of the district are contained in four varieties of host rock: (1) skarn with pyrrhotite, (2) limestone with pyrrhotite and biotite, (3) quartz with minor pyrrhotite and biotite, and (4) greisenized granite. The skarn is made up of pyroxenes, amphiboles, garnet, and minor biotite and chlorite; the greisenized granite contains mainly quartz and sericite with some apatite and a little chlorite and epidote and locally prominent quartz-tourmaline veins. The tungsten occurs almost entirely as scheelite with which a little powellite and rare wolframite are associated; there also is a little molybdenite. The ore and gangue minerals appear to have been formed at about the same time, and the assemblages are characteristic of hypothermal deposits. Since most of the mined ore has come from ore bodies in limestone, these ores are here categorized as hypothermal-2.

SHEEP CREEK

Early Tertiary *Gold, Silver* *Leptothermal*

Mathews, W. H., 1953, Geology of the Sheep Creek camp: B.C. Dept. Mines Bull. no. 31, 94 p.

McGuire, R. A., 1942, Sheep Creek gold mining camp: Canadian Inst. Min. and Met. Tr., v. 45 (Bull. no. 359), p. 169-190

Walker, J. F., 1942, Gold-quartz veins of the Sheep Creek camp, British Columbia, in Newhouse, W. H., Editor, *Ore deposits as related to structural features*: Princeton Univ. Press, 280 p.

Notes

The Sheep Creek district lies in the extreme eastern portion of the west half of the Nelson map-area, being bounded on the south by Lost Creek, on the north by Hidden Creek, on the west by the Salmo district, and on the east by the most easterly outcrops of the Laib group in the map-area.

See *Notes* under Nelson Area (General) for discussion of ore age.

In the vein fractures of the Sheep Creek district, the principal vein filling is quartz. Ore shoots occupy unevenly distributed portions of the vein space and are most common where the walls are quartzite. Although the vein structures normally widen with depth, the proportion of a given vein that can be mined is less at depth than nearer the surface. The sequence of events in the ore formation was (1) introduction of quartz and scheelite; (2) shearing and fracturing of early vein material; (3) deposition of quartz, pyrite, and arsenopyrite; (4) formation of pyrrhotite; (5) emplacement of galena, tetrahedrite, and ruby silvers; (6) precipitation of gold; and (7) introduction of late quartz and calcite. Thus, in contrast with the Nickel Plate deposits, the gold is not closely associated in time of deposition with the arsenopyrite and, instead, occurs as a late and distinct stage of mineralization after such typically leptothermal minerals as tetrahedrite and the ruby silvers. The deposit, therefore, is classified as leptothermal.

SLOCAN

Early Tertiary *Lead, Zinc, Silver* *Mesothermal to Leptothermal*

Ambrose, J. W., 1953, Ore control at Violamac mine, Slocan district, British Columbia: Geol. Assoc. Canada Pr., v. 6, pt. 1, p. 29-35

—— 1957, Violamac mine, Slocan district, B.C., in *Structural geology of Canadian ore deposits*, v. 2: Canadian Inst. Min. and Met., Montreal, p. 88-95

Bancroft, M. E., 1919, Slocan map-area, B.C.: Geol. Surv. Canada Summ. Rept., pt. B, p. 39-48

Bateman, A. M., 1925, Notes on silver-lead deposits of Slocan district, British Columbia, Canada: Econ. Geol, v. 20, p. 554-572

Cairnes, C. E., 1925, Preliminary report on Slocan mining area, B.C.: Geol. Surv. Canada Summ. Rept., pt. A, p. 182-221

—— 1926, Alps-Alturas group, Slocan mining division, B.C.: Geol. Surv. Canada Summ. Rept., pt. A, p. 45-51

—— 1934, Slocan mining camp, British Columbia: Geol. Surv. Canada Mem. 173, 137 p.

—— 1935, Descriptions of properties, Slocan mining camp, British Columbia: Geol. Surv. Canada Mem. 184, 274 p.

—— 1948, Slocan mining camp, in *Structural geology of Canadian ore deposits*: Canadian Inst. Min. and Met., Montreal, p. 200-205

Drysdale, C. W., 1916, Slocan area, Ainsworth and Slocan mining divisions (B.C.): Geol. Surv. Canada Summ. Rept., p. 56-57, and structure sections facing p. 60

Hedley, M. S., 1948, Lucky Jim mine (Slocan), in *Structural geology of Canadian ore deposits*: Canadian Inst. Min. and Met., Montreal, p. 205-215

—— 1952, Geology and ore deposits of the Sandon area, Slocan mining camp, British Columbia: B.C. Dept. Mines Bull. no. 29, 130 p.

Little, H. W., 1959, Nelson map-area, west half, British Columbia (82 F W1/2): Geol. Surv. Canada Mem. 308, 205 p.

Sinclair, A. J., 1967, Trend surface analysis of minor elements in sulfides of the Slocan mining camp, British Columbia, Canada: Econ. Geol., v. 62, p. 1095-1101

Tomlinson, W., 1911, Notes on minerals found in the Slocan district, B.C.: Canadian Min. Jour., v. 32, no. 22, p. 737-738

Uglow, W. L., 1917, Gneissic galena ore from Slocan district, British Columbia: Econ. Geol., v. 12, p. 643-662

Notes

The ore bodies of the Slocan district are almost entirely within a circular area about 15 miles in diameter that centers around the town of Sandon, some 65 miles north-northeast of Traill. The Slocan district is about 65 miles west-northwest of Kimberley and about 70 miles north of the international boundary.

See *Notes* under Nelson Area (General) for discussion of ore age.

The veins in the district are classified by Cairnes as (1) *dry ores* that consist mainly of vein quartz with sulfosalts high in silver and (2) *wet ores* in which the simple sulfides of lead and zinc and various carbonates are more important. Most of Slocan production has come from the wet ores. The upper portions of the wet ore shoots contain abundant silver-bearing galena (much of the silver is in tiny included blebs of tetrahedrite) and silver-rich tetrahedrite and lesser amounts of the ruby silvers, stephanite, argentite, and native silver, while the lower parts of the shoots have sphalerite as the principal sulfide and are low in galena and the silver minerals. Pyrite generally is present in considerable abundance, increasing with depth; the modest amounts of chalcopyrite increase in the same direction. The sphalerite contains what probably are exsolved blebs of stannite, suggesting that the zinc sulfide at depth, at least, was formed under mesothermal conditions and probably rather intense mesothermal conditions at that. This suggestion is given some confirmation by the minor amounts of pyrrhotite and arsenopyrite in the deeper ores. The abundant sulfosalts in the upper portions of the wet ores and throughout the dry ores indicate that they were formed in the leptothermal range. The short vertical extent of the deposit (probably no more than 2000 feet) makes it possible that the ore-forming solutions were undergoing rapid decreases in temperature and pressure. The rock cover at the time of ore formation, however, appears to have been at least several thousand feet, so the deposits are classified as mesothermal to leptothermal.

NICKEL PLATE (HEDLEY)

Middle Mesozoic *Gold* *Hypothermal-2*

Billingsley, P., and Hume, C. B., 1941, The ore deposits of Nickel Plate Mountain, Hedley, British Columbia: Canadian Inst. Min. and Met. Tr., v. 44 (Bull. no. 354), p. 524-590

Bostock, H. S., 1929, Geology and ore deposits of Nickel Plate Mountain, Hedley, British Columbia: Geol. Surv. Canada Summ. Rept., pt. A, p. 198-252

Camsell, C., 1910, Geology and ore deposits of the Hedley mining district, British Columbia: Geol. Surv. Canada Mem. 2, 218 p.

Dolmage, V., and Brown, C.E.G., 1945, Contact metamorphism at Nickel Plate Mountain: Canadian Inst. Min. and Met. Tr., v. 48 (Bull. no. 393), p. 27-68; disc., p. 69-85, 197-198

Knopf, A., 1942, Hedley mining district, B.C., in Newhouse, W. H., Editor, *Ore deposits as related to structural features*: Princeton Univ. Press, p. 69-70

Lamb, J., and others, 1957, Nickel Plate mine, Hedley, B.C., in *Structural geology of Canadian ore deposits*, v. 2: Canadian Inst. Min. and Met. Tr., Montreal, p. 42-46

Mayo, E. B., and Hogg, W., 1951, Orange footwall "sill," Nickel Plate mine: Canadian Inst. Min. and Met. Tr., v. 54 (Bull. no. 469), p. 211-214

Rice, H.M.A., 1947, Geology and mineral deposits of the Princeton map-area, British Columbia: Geol. Surv. Canada Mem. 243, 136 p.

Warren, H. S., and Thompson, R. M., 1945, Mineralogy of two cobalt occurrences in British Columbia: Western Miner, v. 18, no. 5, p. 34-41

Warren, H. V., and Cummings, J. M., 1937, Textural relations in gold ores of British Columbia: Mining Tech., v. 1, no. 2, p. 1-15 (particularly p. 7-8)

Notes

The Nickel Plate district is 140 miles slightly north of east from Vancouver and about 30 miles north of the international border.

The silicic igneous rocks in the area are known collectively as the Coast intrusions. Although Rice (1947) thinks the oldest of these intrusions (a slightly gneissic granodiorite) was produced largely by granitization, in places, it grades into another Coast intrusion (a coarse-grained siliceous granite) and, in others, definitely cuts the gneissic rock. The third Coast intrusive ranges from granodiorite to quartz diorite to gabbro but is not present in the mineralized area. Older than these Coast intrusives are a number of mafic bodies (mainly quartz diorite with some gabbro, quartz gabbro, and augite diorite above the quartz diorite). Dolmage and Brown (1945) think that the original magma was quartz-dioritic in composition and that the more calcic phases were produced by the assimilation of overlying limestone. The quartz diorite and its mafic congeners are so common in the ore-bearing volumes of rock and had such an important effect in localizing the ore bodies that it is tempting to assume that the source magma of these rocks had an important part in providing the ore fluids. The quartz-dioritic rocks and the later and more siliceous granite are intrusive into the Upper Triassic Nicola group composed of volcanic and sedimentary rocks, the ore-bearing portion of which has been known in the past as the Nickel Plate formation, and are overlain uncomformably (outside the immediate Hedley area) by the early to middle Tertiary Princeton group. From this it follows that these intrusives are post-Upper Triassic and pre-early to middle Tertiary. Some further evidence is provided by radioactive dating that suggests that the mafic rocks are about 185 m.y. old and that the Eagle granodiorite (a probable Coast intrusive type rock located farther west in the Princeton map-area) is about 143 m.y. old. If these dates are valid, the ores are either Late Jurassic (if genetically related to the Coast intrusives) or Early Jurassic (if related to the older mafic rocks). In either case, the ores should be classified as middle Mesozoic.

The sulfide minerals of the Nickel Plate ore bodies were introduced into the skarn masses after the silicate minerals had been emplaced and, to different degrees, fractured. Most of the gangue minerals, therefore, are earlier than the ore, but the sulfides were accompanied by sodic scapolite, dipyre (a more calcic scapolite), and chloropal (a rare montmorillonite-type mineral). The first and most abundant sulfide was arsenopyrite containing tiny inclusions of cobaltite, then came smaller but considerable amounts of pyrrhotite and generally much less chalcopyrite and sphalerite, the latter two containing oriented blebs of the other. Small amounts of loellingite and safflorite formed at about the same time as the arsenopyrite. Nearly all the gold occurs as tiny specks in the arsenopyrite, but some is present in the pyrrhotite in somewhat larger particles. Some of the gold in arsenopyrite is bounded by craters that Warren and Cummings (1937) think may have been filled with late calcite. If this is true, all of the gold may have been late or, as they think, two generations of gold may have been deposited. It seems just as reasonable to assume two generations of calcite or that the craters near the gold in arsenopyrite contained something else than calcite. Thus, the age of the gold is not certain, but so much of it is in arsenopyrite without associated craters that most of the gold probably was essentially contemporaneous with the iron-arsenic sulfide and should, therefore, be classed as hypothermal. Since the ores were formed in calcareous rocks, the designation hypothermal-2 is used here.

PRIDE OF EMORY

Early Mesozoic *Nickel, Copper* *Magmatic-2b*

Aho, A. E., 1956, Geology and genesis of ultrabasic nickel-copper pyrrhotite deposits at the Pacific nickel property, southwestern British Columbia: Econ. Geol., v. 51, p. 444-481

—— 1957, Pacific Nickel property, in *Structural geology of Canadian ore deposits*, v. 2: Canadian Inst. Min., Montreal, p. 27-36

Anon., 1962, Success finally comes to British Columbia's most famous nickel mine: Mining World, v. 24, no. 6, p. 28-31

Cairnes, C. E., 1924, Nickeliferous mineral deposit, Emory Creek, Yale mining division; British Columbia: Geol. Surv. Canada Summ. Rept., pt. A, p. 100-105

Cockfield, W. E., and Walker, J. F., 1933, The nickel-bearing rocks near Choate, British Columbia: Geol. Surv. Canada Summ. Rept., pt. A, p. 62-68

Eastwood, G.E.P., and Robinson, W. C., 1965, Pride of Emory (Giant Mascot Mines Limited): B.C. Minister Mines, Ann. Rept. for 1965, p. 213-217

Hill, H. L., and Starck, L. P., 1960, The Giant nickel mine: Western Miner and Oil Rev., v. 33, no. 11, p. 39-42

Horwood, H. C., 1936, Geology and mineral deposits at the mine of B.C. Nickel Mines, Ltd., Yale District, B.C.: Geol. Surv. Canada Mem. 190, 15 p.

—— 1937, Magmatic segregation and mineralization at the B.C. Nickel Mine, Choate, B.C.: Roy. Soc. Canada Tr., 3d Ser., v. 31, sec. 4, p. 4-15

James, A.R.C., 1961, Pride of Emory [Giant Mascot Mines, Limited]: B.C. Minister of Mines, Ann. Rept. for 1961, p. 86-88

Monger, J.W.H., 1970, Hope map-area (92H W 1/2), British Columbia: Geol. Surv. Canada Paper 69-47, 75 p.

Stephens, F. H., 1963, Giant Mascot mines: Western Miner and Oil Rev., v. 36, no. 4, p. 34-48

Notes

The deposit also has been known as Giant Mascot, B.C. Nickel Mines, and Pacific Nickel. The Pride of Emory mine is about 75 miles east-northeast of Vancouver and 35 miles north of the international border.

In the Princeton area to the east of the Pride of Emory deposits, ultramafic rocks, similar to those of the Pride of Emory area, cut the Upper Triassic Nicola group and were emplaced, therefore, no sooner than latest Triassic or earliest Jurassic time. The Princeton ultramafics are cut by the Eagle granodiorite for which an age of 143 m.y. has been obtained. An age of 186 m.y. was obtained from a biotite from a small mass of pyroxenite that probably is satellitic to the Giant Mascot stock. Thus, both lines of evidence point to the Pride of Emory ultramafics being either oldest early Mesozoic or youngest middle Mesozoic; the former is arbitrarily chosen here. Monger (1970), however, believes that the ultramafics are contemporaneous with the diorite (or quartz diorite) that encloses them. Similar quartz diorite near Hope has been dated at 76 to 102 m.y. If Monger's reasoning is correct, the ores would be Late Cretaceous, probably pre-Laramide. The age problem needs further study.

The ores at Pride of Emory are contained in two types of pipelike bodies: (1) unzoned or massive and (2) zoned. The application of the term massive to the unzoned deposits does not mean that they are composed of massive sulfides but that they lack zoning; in any ore body, zoned or unzoned, the silicates are more abundant than the sulfides. The unzoned bodies, although more irregular in outline than the zoned, exhibit sharp contacts against the various older rocks in which they are enclosed and contain inclusions of these rocks, show marked flow lines and banding, are drag-folded in places, and have minor hornblende reaction rims against the rocks containing them. If the unzoned bodies were the only ore type at Pride of Emory, the deposits would be classed as magmatic-2b without much question, on the assumption that segregated volumes of molten silicate-sulfide emulsion had been forced into the stock after it was essentially solid. The zoned ore bodies, however, are not as readily assigned to this category, being composed of concentric shells of the different ultramafic rocks, mainly peridotite and pyroxenite, that are roughly cylindrical around the ore masses. The sulfides contained in the zoned ore bodies are also zoned, there being both sulfide-rich and sulfide-poor zones and cores within any given zoned ore body. The silicates that accompany the sulfide-rich zones and cores are generally such that the rock, if it lacked sulfides, would be classed as peridotite. These zoned ore bodies grade gradually outward into the sulfide-poor ultramafic rocks that surround them, the surrounding silicate rocks containing less olivine and bronzite and more augite and hornblende than the zoned ores they enclose. These zoned ore bodies might be classed as hydrothermal replacements were it not that the differences between silicates of the zoned ore bodies and the sulfide-poor rocks surrounding them are not those which hydrothermal (deuteric) alteration has produced elsewhere in the ultramafic stock.

Instead, the relationship of the zoned ore bodies to the rock enclosing them probably is better explained by assuming that emulsions of silicates and sulfides had been segregated at depth and were on occasion introduced into considerably, but not entirely, crystallized portions of the ultramafic stock. Where these emulsified intrusions encountered solid rock, the result was the unzoned type of deposit; where volumes of still not fully crystallized ultramafics were met, there was enough mingling of the two ultramafic types to produce the gradational and zoned relationships just described. As the silicates of the later-intruded sulfide-silicate emulsion crystallized before the bulk of the molten sulfides, the existence of veins and veinlets of sulfides cutting through both the silicates of the later intrusion and of the surrounding ultramafics is not surprising. The classification of both ore types in the Pride of Emory deposit, therefore, is magmatic-2b. It is asking too much of coincidence that two ore types of essentially the same composition should have been developed in the same rock volume, one by magmatic and one by hydrothermal processes.

SULLIVAN

Late Precambrian	*Zinc, Lead, Silver, Tin*	*Hypothermal-1 to Leptothermal*

Alcock, F. J., 1930, Sullivan mine, in *Zinc and lead deposits of Canada*: Geol. Surv. Canada Econ. Geol. Ser. no. 8, p. 322-324

Freeze, A. C., 1966, On the origin of the Sullivan orebody, Kimberley, B.C., in *Tectonic history and mineral deposits of the western Cordillera*: Canadian Inst. Min. and Met., B.C. Sec., Spec. v. no. 8, p. 263-294

Leech, G. B., 1962, Metamorphism and granitic intrusions of Precambrian age in southeastern British Columbia: Geol. Surv. Canada Paper 62-13, 8 p.

—— 1963, Ages of regional metamorphism of the Aldridge formation near Kimberley, B.C. (preliminary report): Geol. Surv. Canada Paper 63-17, p. 132-135

Leech, G. B., and Wanless, R. K., 1962, Lead-isotope and potassium-argon studies in the East Kootenay district of British Columbia, in Engel, A.E.J., and others, Editors, *Petrologic studies: A volume in honor of A. F. Buddington*: Geol. Soc. Amer., p. 241-279 (particularly p. 248-256, 270-271)

Officers of the Geological Survey, 1947, The Sullivan mine, in *Geology and economic minerals of Canada*: Geol. Surv. Canada Econ. Geol. Ser. no. 1, p. 288-289

Pentland, A. G., 1943, Occurrence of tin in the Sullivan mine: Canadian Inst. Min. and Met. Tr., v. 46 (Bull. no. 369), p. 17-22

Rice, H.M.A., 1937, Cranbrook map-area, British Columbia: Geol. Surv. Canada Mem. 207, 67 p.

—— 1941, Nelson map-area, east half, British Columbia: Geol. Surv. Canada Mem. 228, 86 p.

Schofield, S. J., 1912, The origin of the silver-lead deposits of east Kootenay, British Columbia: Econ. Geol., v. 7, p. 351-355

Schwartz, G. M., 1926, Microscopic features of Sullivan ore: Eng. and Min. Jour., v. 122, no. 10, p. 375-377

Staff, Consolidated Mining and Smelting Company, Ltd., 1924, The development of the Sullivan mine: Canadian Inst. Min. and Met. Tr., v. 27 (Bull. no. 146), p. 401-465 (particularly p. 401-406)

—— 1954, Geology (of the Sullivan mine), in *"Cominco"--a Canadian enterprise*: Canadian Min. Jour., v. 75, no. 5, p. 144-153

Swanson, C. O., 1950, The Sullivan mine, Kimberley, B.C.: 18th Int. Geol. Cong. Rept., pt. 7, p. 40-46

Swanson, C. O., and Gunning, H. C., 1945, Geology of the Sullivan mine: Canadian Inst. Min. and Met. Tr., v. 48 (Bull. no. 402), p. 645-667

—— 1948, Sullivan mine, in *Structural geology of Canadian ore deposits*: Canadian Inst. Min. and Met., Montreal, p. 219-230

Notes

The Sullivan mine is at the town of Kimberley in eastern British Columbia, about 50 miles north of the international border.

Recently, the suggestion has been put forward that the Sullivan deposits (in which much of the sulfide ore is conformable to the enclosing sediments) were formed syngenetically with the enclosing sediments. Freeze (1966) argues against this by pointing out that (1) locally, fractures extend as much as 100 feet into the hanging wall of the main ore body and usually carry enough galena to be mined; (2) the footwall beds have been extensively and irregularly replaced by sulfides; (3) some strong fractures extending into the tourmalinized footwall beds contain ore and are surrounded by zones of intense post-tourmaline alteration, indicating that the ores were emplaced after the tourmalinization; (4) areas of brecciation in the footwall rocks formed after the tourmalinization are locally heavily mineralized with pyrrhotite and/or minor galena and sphalerite; and (5) areas of brecciation in the albitized hanging-wall rocks formed in two stages, one before or during albitization (early fractures are healed by albite) and the other after the albitization (late fractures are weakly sealed by chlorite that contains local pyrite, galena, and sphalerite). It is Freeze's opinion that these features are too major and drastic to have been caused by remobilization. Granted that the ores are epigenetic, there has been much controversy as to when they were introduced. Freeze favors the concept that the ore fluids came from the same general source as the Hellroaring Creek granite, which is known in a small outcrop 12 miles southwest of the mine. Only 5 miles southwest of the mine is a large area of metamorphosed sediments that locally contains sillimanite- and garnet-bearing quartz-muscovite schist; this suggests to Freeze, as it did to Leech (1962) and Swanson and Gunning (1945), that a much larger body of granite may be present at shallow depth. Several small lamprophyre dikes have been found in faults in and near the mine; one of these is known to intrude the ore zone after the main phase of pyrrhotite deposition but before the introduction of galena in that area. Both these dikes and the Hellroaring Creek granite are between 700 and 800 m.y. old, adding further indications of a Precambrian age for the ores. The Sullivan deposits are, therefore, classified as late Precambrian.

Freeze (1966) points out that the differences in concentrations of the major metals in the various parts of the deposit are roughly concentric to the central high-iron zone. Numerous smaller, but important, northerly trending linear patterns are superimposed on this major concentric arrangement, and these are, in most instances, clearly associated with certain Sullivan-type faults, fractures, or folds. The concentric pattern is not as well shown for lead and zinc as for the lesser metals because of the anticlinal warp in the north-central part of the mine, but this lead-zinc ratio map still has a definitely concentric shape. The central part of the ore body is richest in lead, relative to zinc, while the margin is higher in zinc, the zinc being several times greater than lead. Freeze does not consider these relationships to be what would be expected in a sulfide body formed in a basin of marine sedimentation but, rather, to be in agreement with an epigenetic hydrothermal origin. At Sullivan, the early minerals are pyrrhotite, pyrite (with pyrite being more abundant in the upper levels and pyrrhotite in the lower), cassiterite, and sphalerite; these certainly constitute a suite typical of high temperature conditions of deposition. With this concept the wall-rock alteration suite is in agreement even though much of the alteration seems to have been formed somewhat before the metallic minerals. The galena, even though emplaced to a greater extent toward the center of the deposit than was the sphalerite, appears to have been appreciably later than the zinc sulfide and the other minerals of the early suite, and the association of the last fraction of the galena with such typically leptothermal minerals as boulangerite, teallite, and cylindrite suggests that some of the lead sulfide was deposited in the leptothermal range. Most of the galena, however, appears to have been deposited under mesothermal conditions. Although the amount of leptothermal galena probably is only a small fraction of the total mass of metallic minerals, it is thought appropriate to include leptothermal in the classification assigned to the Sullivan deposit, and the designation hypothermal-1 to leptothermal is used here.

Manitoba

BIRD RIVER

Early Precambrian *Chromite* *Magmatic-1b*

Bateman, J. D., 1943, Bird River chromite deposits, Manitoba: Canadian Inst. Min. and Met. Tr., v. 46 (Bull. 374), p. 154-183

—— 1945, Composition of the Bird River chromite, Manitoba: Amer. Mineral., v. 30, p. 595-600

Brownell, G. M., 1943, Chromite from Manitoba: Univ. Toronto Studies, Geol. Ser. 47, p. 101-102

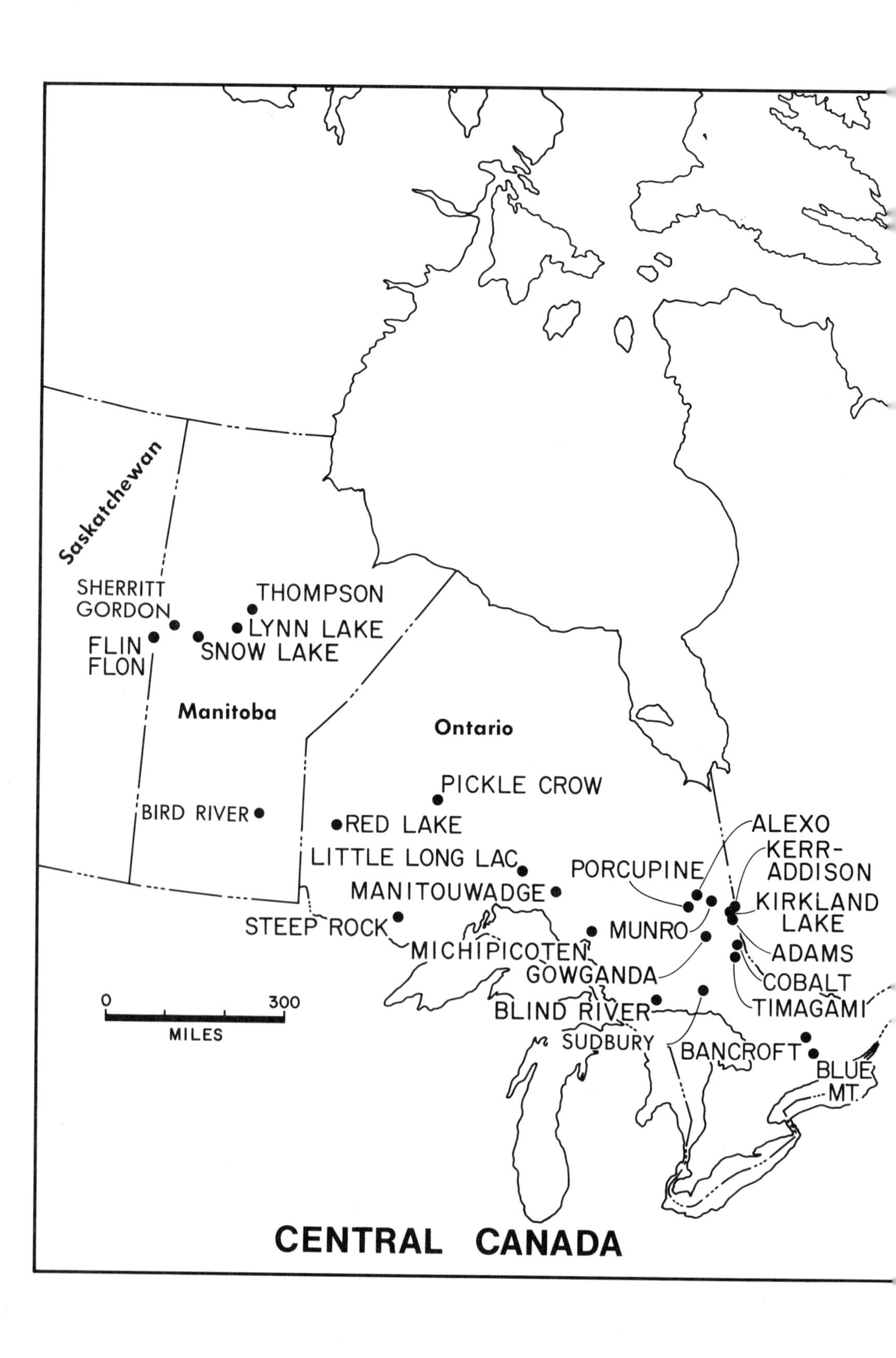

CENTRAL CANADA

Davies, J. F., 1952, Geology of the Oiseau (Bird) River area, Lac du Bonnet Mining Division, Manitoba: Manitoba Mines Br. Pub. 51-3, 24 p.

—— 1955, Geology and mineral deposits of the Bird Lake area, Lac du Bonnet Mining Division, Manitoba: Manitoba Mines Br. Pub. 54-1, 44 p.

—— 1958, Chromite deposits of southeastern Manitoba: Canadian Min. Jour., v. 79, no. 4, p. 112-114

Davies, J. F., and others, 1962, Cat Lake--Bird River--Winnipeg River area, in *Geology and Mineral Resources of Manitoba*: Manitoba Dept. Mines and Nat. Res., Winnipeg, p. 37-40

Notes

The Bird River deposits are located in the Lac du Bonnet district of Manitoba about 80 miles northeast of Winnipeg. The first discovery was made in a gabbro sill; this directed attention to the peridotites north of Bird River where most of the deposits under discussion were found.

The Bird River area is underlain by the probably early Precambrian Rice Lake group that is composed of (1) a lower segment of volcanic rocks, principally pillowed andesites and basalts, and (2) an upper series of graywackes, impure quartzites, and arkoses, with minor conglomerate, chert, and slate. To some extent the volcanic and sedimentary rocks are interbedded (Davies, 1962). These rocks were invaded, also apparently in late early Precambrian time, by an ultramafic to mafic sill (in which the rocks range from peridotite to quartz diorite) and by simple gabbro intrusives. Still later than the sill and the gabbro were batholithic bodies of granite and related pegmatite dikes. These silicic rocks may be younger than early Precambrian. Although the sediments enclosing the sill have no connection with the genesis of the chromite, they are highly useful in confirming and extending the structural deductions made from the sill.

According to Davies (1958) the Bird River sill has a length of about 20 miles along the Bird River; it lies on the south limb of a major anticline and has been appreciably cross faulted. Other chromite occurrences in the district are known along Cat Creek and at Euclid Lake, both on the north limb of the structure. This anticlinal relationship is confirmed by the sequence of rock types being repeated on each side of the crest and the presence of the chromite horizon in the same relative position on both sides of the crest. Additional evidence of the anticlinal concept has been provided from the pillow lavas in which bedding tops can be determined. The anticline plunges to the southeast, and the former axial area is now occupied by the later granitic rocks.

The sill consists of a basal peridotite sheet overlain by a sheet of hornblende gabbro; the chromite horizon is always found near the top of the peridotite, being south of the gabbro-peridotite contact on the north limb and north of it on the south. The thickness of the sill ranges up to 6000 feet and averages about 3000; the peridotite layer is from a few feet to 2500 feet in thickness, and its thickness bears no direct relationship to the thickness of the entire sill. The thickness of the chromite horizon and the total chromium content, on the other hand, are both directly proportional to the thickness of the peridotite over the entire area of the sill.

The lower section of the sill is banded and is composed mainly of peridotite and its associated chromite bands, but it also includes a few feet of olivine gabbro at several places at the base of the sill. Several feet of pyroxene-bearing rock lie between the peridotite and the hornblende gabbro; this material is either pyroxenite or olivine-pyroxene gabbro.

The original minerals of the peridotite have been altered almost completely to a fibrous aggregate of antigorite, tremolite, chlorite, talc, carbonate, and magnetite.

The thicker upper portion of the sill is made up of hornblende gabbro and related feldspar-bearing rocks; although the gabbro consists of various phases, these do not show the well-defined banding of the lower (ultramafic) portion of the sill, but fairly definite gradations can be seen. Some of the gabbroic rock actually is an anorthositic gabbro composed of 75 percent feldspar, and some is a true anorthosite 95 percent of which is labradorite and 5 percent hornblende. In many places, normal gabbro, gabbro variants, and anorthosite are intermixed. Through an increase in the amount of interstitial sodic plagioclase and quartz at and near the top of the sill, the rock becomes a quartz diorite.

The Bird River sill is displaced by several major northwest-trending cross faults, horizontal displacements being as much as several thousand feet.

What age determinations have been made in the general area (but not of the sill rocks) show ages of about 2500 m.y., indicating that the sill cannot be older than a late early Precambrian. It is possible, however, that the sill might be of middle Precambrian age; further work needs to be done on this matter, but the sill is here considered to be late early Precambrian.

The chromite horizon in the Bird River area is consistently about 175 feet below the gabbro-peridotite contact. The bands that make up the horizon are parallel to this contact and to each other. Some of the bands are massive; in others the chromite is associated with sufficient silicate gangue to be classed as disseminated. The peridotite between the chromite bands contains various amounts of chromite but not enough to be considered even as disseminated ore. In the claims north of Bird River (Chrome and Page properties), the chromite is found in three zones: (1) an upper (main) chromite band 6 to 10 feet wide that is made up of successive layers of dense ore, peridotite, and disseminated ore--all contacts are sharp; (2) a lower band, narrow but composed of dense ore; and (3) a stringer zone that lies beneath the other two. Bands (1) and (2) are separated by about 30 feet of peridotite in which there are only a few ribbons of chromite; a similar thickness of peridotite separates band (2) from band (3).

The Cr_2O_3 content of the different layers within band (1) ranges from 30 percent in the dense ore to 5 percent or less in the disseminated ore. Over the total width of the band, the average Cr_2O_3 content is between 18 to 25 percent; the foot percentages are almost the same in all the claims north of the river.

North of Bird River, a great number of closely spaced faults cut up the chromite zone into small separate blocks, the displacement on the faults being measured in a few tens of feet. The individual blocks may be from 50 to a few hundred feet long. This situation would, of course, increase mining costs, but the ore should be easy to follow.

The Bird Lake deposits (located where Bird River flows west from Bird Lake) contrasts with those north of the river in that it is made up of numerous closely spaced bands about 2.5 feet wide that alternate between dense and disseminated chromite. The dense bands may have as much as 25 percent Cr_2O_3 and the disseminated bands (actually chromitiferous peridotite) 5 to 10 percent Cr_2O_3; the Bird Lake chromite horizon is about 45 feet wide in total.

The occurrence at Euclid Lake on the north limb of the anticline (less well-known 10 years ago than the two areas just mentioned) also consists of closely spaced narrow bands that are alternatively composed of dense and disseminated ore and of chromitiferous peridotite and barren peridotite across widths of as much as 200 feet or even more.

The dense chromite ore consists of 40 to 75 percent chromite in small, irregular, and rounded octahedral grains in a matrix of serpentinized peridotite. The disseminated ore runs about 25 percent Cr_2O_3. The individual chromite grains are about 0.5 in diameter; many of them have small nuclei of pyroxene.

The worst feature of the Bird River ore from a use point of view is the low Cr:Fe ratio, which ranges from 1:1 for low-grade disseminated ore to 1.60:1 for dense, high-grade ore. Tabling should produce concentrates that contain about 40 percent Cr_2O_3, but this process does not cause a significant improvement in the Cr:Fe ratio. The formula for the Bird River chromite should be written $(Fe,Mg)(Cr,Al,Fe)_2O_4$. The substitution of Mg for Fe improves the Cr:Fe ratio, but the presence of Al^{+3} and Fe^{+3} in the 6-coordination positions reduces the chromium content and the Cr:Fe ratio. Obviously, other methods of beneficiation must be applied to the Bird River ores if they are to be used for metallurgical purposes.

It is the opinion of Davies (1958) of the Manitoba Mines Branch that the

Bird River ores constitute the largest reserves of chromite in North America. (This neglects the chromium in the lateritic soils of eastern Cuba.) Obviously, any published figures are low because they are based on shorter lengths that certainly exist and are not carried down to maximum possible depths. In the Bird Lake area, for example, the chromite-bearing horizon had been followed (in 1958) for a distance of 7000 feet. Further drilling would be needed to provide a comprehensive estimate of the chromite available in the Bird River area.

It appears probable that the deposits were concentrated in much the same manner as those of the Stillwater Complex but with not enough yet being known of the details of the Bird River occurrence to suggest what modifications, if any, must be made in theories outlined in the discussion of the Stillwater deposit. Certainly, however, the chromite horizons at Bird River were developed by concentration in a crystallizing magma and must be classified as magmatic-1b.

LYNN LAKE

Middle Precambrian *Nickel, Copper* *Magmatic-2b*

Allan, J. D., 1947, Geology of Lynn Lake area, Manitoba: Precambrian, v. 20, no. 2, p. 8-9, 17

—— 1948, Geology of Lynn Lake area, Manitoba: Precambrian, v. 21, no. 3, p. 4-8

Bateman, J. D., 1945, McVeigh Lake area, Manitoba: Geol. Surv. Canada Paper 45-14, 34 p.

Brown, E. L., 1955, Notes on discovery and financing (Lynn Lake): Canadian Inst. Min. and Met. Tr., v. 58 (Bull. no. 518), p. 335-339

Charlewood, G. H., 1954, Geology of the Lynn Lake area: Western Miner, v. 27, no. 6, p. 48-51

Davies, J. F., and others, 1962, The Granville Lake-Uhlman Lake area (the Lynn Lake district), in *Geology and Mineral Resources of Manitoba*: Manitoba Dept. Mines and Nat. Res., Winnipeg, p. 111-120

Emslie, R. F., and Moore, J. M., Jr., 1961, Geological studies of the area between Lynn Lake and Fraser Lake, Granville Lake mining division: Manitoba Dept. Mines and Nat. Res. Pub. 59-4, 76 p.

Milligan, G. C., 1957, The Lynn Lake area--the geological history: Canadian Min. Jour., v. 78, no. 7, p. 75-79

—— 1960, Geology of the Lynn Lake district: Manitoba Dept. Mines and Nat. Res. Pub. 57-1, 317 p.

Moore, J. M., Jr., and others, 1960, Potassium-argon ages in northern Manitoba: Geol. Soc. Amer. Bull., v. 71, p. 225-230

Ruttan, G. D., 1955, Geology of Lynn Lake: Canadian Inst. Min. and Met. Tr., v. 58 (Bull. no. 518), p. 191-200

—— 1957, Lynn Lake mine, in *Structural geology of Canadian ore deposits*, v. 2: Canadian Inst. Min. and Met., Montreal, p. 275-291

Turek, A., 1967, Age of sulfide mineralization at Lynn Lake, Manitoba: Canadian Jour. Earth Sci., v. 4, p. 572-574

Notes

The Lynn Lake district is about 525 miles north-northwest from Winnipeg.

Of the dozen or more variously sized nickel-copper ore bodies of the Lynn Lake area, all but one are contained in the Lynn Lake gabbro; the exception was the now mined-out EL ore body in the EL gabbro. The largest of the Lynn Lake gabbro ore bodies is the A of about 5,000,000 tons; the others, also designated by letter names, were from half as large on downward. The gabbros appear to have been intruded early in the period of metamorphism that marked the end of the middle Precambrian in the area. K/Ar determinations on biotites from an igneous body believed to have been emplaced during the metamorphism give an age of about 1700 m.y. The granite masses in the map-area show no significant post-intrusion metamorphism, and one determination on the granite gives an age of 1640 m.y. Assuming that the introduction of the gabbro and the metamorphism were at least partly contemporaneous, the mafic body would be late middle Precambrian in age. Emslie and Moore (1961) think that the ores were formed by processes internal to the gabbro; this does not now appear probable. Milligan (1960) believes that the ores were formed hydrothermally at a considerable time after the gabbro was emplaced. This seems unlikely because the ores are found only in the intrusive mass and never outside it. This spatial relationship suggests that the ores came from the same source magma chamber as one of the rocks (probably the peridotite) of the intrusive body. If this last concept is correct, the ores should be classed as late middle Precambrian in age.

The ore bodies in the Lynn Lake gabbro are irregularly scattered through the western half to two-thirds of that rock mass. The intrusives consist of gabbro and of diorite and quartz-hornblende diorite; all three rock types are irregularly distributed through the plug with gabbro (or its altered equivalent) probably occupying the largest proportion of the intrusive. Much of the gabbro has been altered to amphibolite, and the ore is consistently associated with amphibolitized gabbro; only a little ore is found in quartz-hornblende diorite or in fresh gabbro. The diorite (*sensu strictu*) is a feldspar-rich gabbro that grades into that rock type, and both rocks probably were formed from the crystallization-differentiation of a single intrusion. The quartz-hornblende diorite is mainly in the northeast part of the plug and probably was introduced soon after the gabbro-diorite mass. Although all of the amphibolite does not contain ore, it is a very good guide to ore and is the host rock of almost all of the ore. A large mass of peridotite is enclosed in the southwest part of the plug but it does not contain ore (although it may contain minor disseminated sulfides) down to the greatest depths reached in mining. The arguments favoring a magmatic origin for the Lynn Lake ores are (1) the ore bodies are confined entirely to the mafic plug; (2) the shape and attitude of the ore bodies reflect those of the intrusive complex; (3) the sulfides clearly are interstitial to the silicates where alteration, amphibolization (or uralitization), have not destroyed the original textures; (4) the ore-mineral suite (pyrrhotite, pentlandite, and chalcopyrite) is typical of magmatic sulfide deposits; and (5) the compositional and textural relations of the sulfides suggest that the first mineral to crystallize was a pyrrhotite-pentlandite solid solution from which pentlandite and perhaps a little chalcopyrite were exsolved--a later chalcopyrite appears to have formed from the remaining liquid phase. It seems most probable, though not certainly established, that emulsions of sulfides and silicates, with appreciable water content and from the same source as the peridotite (or possibly the gabbro-diorite mass), were forced into several parts of the plug and into the EL gabbro while the gabbro-diorite mass was still quite hot. The water acted on the silicates deposited from the emulsion to change them into amphiboles and also similarly affected the silicates in the surrounding original gabbro. Locally, very minor amounts of the sulfides worked their way into the margins of neighboring quartz-hornblende diorite and peridotite but produced very little ore in the former and none in the latter. The EL ore body appears to have been formed in much the same way in a similar, but much smaller, mafic mass that probably is not a faulted portion of the Lynn Lake gabbro. The mechanism here suggested for ore formation places the Lynn Lake ore bodies in the magmatic-2b category. Two deposits of the Sherritt Gordon type have been found in the Lynn Lake area, Fox Lake, and Ruttan Lake; they probably are of hydrothermal origin, being pyrrhotite-chalcopyrite-sphalerite bodies. No literature exists for them, and they are not classified or dated here.

SHERRITT GORDON-FLIN FLON (GENERAL)

Middle Precambrian *Copper, Zinc* *Hypothermal-1*

Alcock, F. J., 1923, Flin Flon map-area, Manitoba and Saskatchewan: Geol. Surv. Canada Summ. Rept., pt. C, p. 1-36

—— 1935, Northern Manitoba, Flin Flon mine, Mandy mine, Sherritt Gordon mine, in *Copper resources of the world*: 16th Int. Geol. Cong., v. 1, p. 99-110

Ambrose, J. W., 1936, Structures in the Missi series near Flin Flon, Manitoba: Roy. Soc. Canada Tr., 3d ser., v. 30, sec. 4, p. 81-98

Bruce, E. L., 1918, Amisk-Athapapuskow Lake district (Saskatchewan-Manitoba): Geol. Surv. Canada Mem. 105, 91 p.

—— 1920, Chalcopyrite deposits in northern Manitoba: Econ. Geol., v. 15, p. 386-397

Hanson, G., 1920, Some Canadian occurrences of pyritic deposits in metamorphic rocks: Econ. Geol., v. 15, p. 574-609 (particularly p. 575-590)

Harrison, J. M., 1949, Geology and mineral deposits of File-Tramping lakes area, Manitoba: Geol. Surv. Canada Mem. 250, 92 p. (area just south of Sherritt Gordon and just east of Flin Flon)

—— 1951, Possible major structural control of ore deposits, Flin Flon-Snow Lake mineral belt, Manitoba: Canadian Inst. Min. and Met. Tr., v. 54 (Bull. no. 465), p. 4-8

—— 1951, Precambrian correlation and nomenclature and problems of the Kisseynew gneisses in Manitoba: Geol. Surv. Canada Bull. 20, 53 p.

Kalliokoski, J., 1953, Weldon Bay map-area, Manitoba: Geol. Surv. Canada Mem. 270, 80 p. (areas east of Flin Flon and south-southwest of Sherridon)

—— 1953, Interpretations of the structural geology of the Sherridon-Flin Flon region, Manitoba: Geol. Surv. Canada Bull. 25, 18 p.

Robertson, D. S., 1951, The Kisseynew lineament, northern Manitoba: Precambrian, v. 24, no. 5, p. 8-11, 13, 23

—— 1953, Batty Lake map-area, Manitoba: Geol. Surv. Canada Mem. 271, 55 p. (immediately adjoins Sherridon map-area on the east)

Stockwell, C. H., Editor, 1957, Manitoba-Saskatchewan, in *Geology and economic minerals of Canada*: Geol. Surv. Canada Econ. Geol. Ser. no. 1, 4th ed., p. 85-88

Stockwell, C. H., and Harrison, J. M., 1948, Structural control of ore deposits in northern Manitoba, in *Structural geology of Canadian ore deposits*: Canadian Inst. Min. and Met., Montreal, p. 284-291

Tanton, T. L., and Harrison, J. M., 1950, The Flin Flon and Sherritt Gordon mines: 18th Int. Geol. Cong. Rept., pt. 7, p. 47-50

Wallace, R. C., 1928, Copper-zinc and gold mineralization in Manitoba: Canadian Inst. Min. and Met. Tr., v. 31 (Bull. no. 190), p. 264-273

Wright, J. F., 1929, Geology of the copper-zinc deposits of the Cold Lake area, Manitoba: Canadian Inst. Min. and Met. Tr., v. 32 (Bull. no. 203), p. 527-546

Notes

The center of the Sherritt Gordon-Flin Flon area is about 390 miles north-northwest from Winnipeg. The district can be divided into four subareas: (1) the Flin Flon, (2) the Sherritt Gordon (about 40 miles northeast of Flin Flon), (3) the Snow Lake (about 70 miles east of Flin Flon), and (4) the Coronation Mine area (the center of which is about 10 miles south-southwest of Flin Flon) just over the province boundary in Saskatchewan. The Coronation Mine area's bibliography and notes are to be found under Saskatchewan.

All of the ore bodies in these subareas, except the Sherritt Gordon, are contained in rocks of the early and/or middle Precambrian Amisk group or series (older) and the Missi series (younger). The Sherritt Gordon ores were found in the Kisseynew gneisses. It is most likely that the Kisseynew gneisses are the equivalent of the Amisk and Missi, with the Kisseynew having been derived predominantly from sedimentary rocks, while the Amisk rocks are dominantly metavolcanics; the Missi rocks are mainly metasediments (Byers and others, 1969, p. 11). Further, the Kisseynew front (the boundary between Kisseynew gneisses [north] and Amisk-Missi rocks [south]) marks a metamorphic boundary across which the grade of metamorphism increases to the north from the greenschist and lower amphibolite facies to the amphibolite facies. Byers and others (1965, p. 27) say that they believe the biotite-gneiss portion of the Kisseynew to be the equivalent of the Missi series. They think, however, that the hornblende gneiss part of the Kisseynew is of more diverse origin. Some of these gneisses, however, have a well-developed stratiform structure that can be traced into Amisk rocks. The ore deposits in the general Sherritt Gordon-Flin Flon area are apparently replacements of metamorphosed rocks of the Amisk, Missi, and Kisseynew series and, if this is correct, are younger than the rocks that contain them. The structural features that guided the emplacement of the ores are younger than the folding and metamorphism of the area and than part of the intrusive rocks of the area. The structures are, however, in part older than the latest of the intrusions and, therefore, probably were developed late in the metamorphic-orogenic cycle. Radioactive determinations indicate that the metamorphism is slightly older than 1700 m.y. and the youngest igneous rocks in the area (to which the ores would seem most reasonably to be genetically related), the Phantom Lake or Kaminis granite, are post-tectonic but not by much. The ores, therefore, seem probably to be best dated as late middle Precambrian.

See the *Notes* under the individual deposits.

FLIN FLON

Middle Precambrian | *Copper, Zinc* | *Hypothermal-1*

Alcock, F. J., 1922, Flin Flon map-area, Manitoba and Saskatchewan: Geol. Surv. Canada Summ. Rept., pt. C, p. 1-36

Badgley, P. C., 1959, Isometric prospection of the Flin Flon ore body and an analysis of the structural fabric of the Flin Flon region, in *Structural methods for the exploration geologist*: Harper and Bros., N.Y., p. 162-165

Brownell, G. M., and Kinkel, A. R., Jr., 1935, The Flin Flon mine: Geology and paragenesis of the ore deposit: Canadian Inst. Min. and Met. Tr., v. 38 (Bull. no. 279), p. 261-268

Bruce, E. L., 1918, The Amisk-Athapapuskow Lake district: Geol. Surv. Canada Mem. 105, 91 p. (particularly p. 60-65, 67-72)

Byers, A. R., and others, 1965, Geology and mineral deposits of the Flin Flon area, Saskatchewan: Sask. Dept. Mineral Res., Geol. Sci. Br., Precambrian Geol. Div., no. 62, 95 p.

Cairnes, R. B., and others, 1957, North Star and Don Jon mines, in *Structural geology of Canadian ore deposits*, v. 2: Canadian Inst. Min. and Met., Montreal, p. 247-253

—— 1957, Cuprus Mine, in *Structural geology of Canadian ore deposits*, v. 2: Canadian Inst. Min. and Met., Montreal, p. 253-258

—— 1957, Schist Lake mine, in *Structural geology of Canadian ore deposits*, v. 2: Canadian Inst. Min. and Met., Montreal, p. 258-262

Davies, J. F., and others, 1962, The Flin Flon region, in *Geology and Mineral*

Resources of Manitoba: Manitoba Dept. Mines and Nat. Res., Winnipeg, p. 64-75

Heywood, W. W., 1966, Ledge Lake area, Manitoba and Saskatchewan: Geol. Surv. Canada Mem. 337, 43 p.

Koffman, A. A., and others, 1948, Flin Flon mine, in *Structural geology of Canadian ore deposits*: Canadian Inst. Min. and Met., Montreal, p. 295-301

Smith, J. R., 1964, Distribution of nickel, copper, and zinc in bedrock of the east Amisk area, Saskatchewan: Sask. Res. Council, Geol. Div. Rept. no. 6, 36 p.

Stockwell, C. H., 1946, Flin Flon-Mandy area, Manitoba and Saskatchewan: Geol. Surv. Canada Paper 46-14, 5 p. (mimeo.)

—— 1960, Flin Flon-Mandy, Manitoba and Saskatchewan: Geol. Surv. Canada Map 1078A (with descriptive notes), 1:63,360

Tanton, T. L., 1941, Flin Flon, Saskatchewan, and Manitoba: Geol. Surv. Canada Map 632A (with descriptive notes), 1:63,360

Wallace, R. C., 1921, The Flin Flon ore body: Canadian Inst. Min. and Met. Tr., v. 24 (Bull. no. 106), p. 99-111

Notes

The center of the Flin Flon area is about 385 miles north-northwest from Winnipeg; the area includes the Flin Flon, Mandy, Schist Lake, Cuprus, North Star, and Don Jon mines.

See *Notes* under Sherritt Gordon-Flin Flon (General) for discussion on ore age.

On the basis of available data, the sulfides (and the nonmetallic minerals that bear the same textural relationships to the host rock as those exhibited by the sulfides) were metasomatically deposited by high-temperature hydrothermal solutions. Although the suggestion has been made that the deposits were formed by volcanic-exhalative process and later metamorphosed, the lack of any real evidence of metamorphism of the ore and associated gangue minerals strongly points to emplacement only after metamorphism had died away. The dark sphalerite (which contains what probably is exsolved chalcopyrite) appears to have been formed under hypothermal conditions, particularly since the chalcopyrite (at least part of which was deposited after the sphalerite) contains exsolved cubanite. The presence of modest amounts of pyrrhotite and arsenopyrite also indicates that the ores were formed in the hypothermal range. The abundant pyrite, while not diagnostic of hypothermal conditions, is not incompatible with them. About 70 percent of the ore appears to be massive sulfides, with disseminated ores occurring along the footwall side and under the keel of the solid sulfides. The minor amounts of later, and lower-temperature, metallic minerals are of such little importance as to require no mention in the classification. The ores, therefore, are assigned to the hypothermal-1 category.

SHERRITT GORDON

Middle Precambrian *Copper, Zinc* *Hypothermal-1*

Bateman, J. D., and Harrison, J. M., 1947, Sherridon, Manitoba: Geol. Surv. Canada Map 862A (with descriptive notes), 1:63,360

Bruce, E. L., 1929, The Sherritt Gordon copper-zinc deposit, northern Manitoba: Econ. Geol., v. 24, p. 457-469

—— 1929, Geology of the Sherritt Gordon mine (northern Manitoba): Eng. and Min. Jour., v. 128, no. 22, p. 853

Davies, J. F., and others, 1962, The Kississing area, in *Geology and Mineral*

Resources of Manitoba: Manitoba Dept. Mines and Nat. Res., Winnipeg, p. 92-103

Derry, D. R., 1942, The Sherritt-Gordon mine, in Newhouse, W. H., Editor, *Ore deposits as related to structural features*: Princeton Univ. Press, p. 155

Farley, W. J., 1948, Sherritt Gordon mine, in *Structural geology of Canadian ore deposits*: Canadian Inst. Min. and Met., Montreal, p. 292-295

—— 1949, Geology of the Sherritt Gordon orebody: Canadian Inst. Min. and Met. Bull. no. 441, p. 25-30 (not in Tr. sec.) (very brief summary of geology)

Wright, J. F., 1928, Kississing Lake area, Manitoba: Geol. Surv. Canada Summ. Rept., pt. B, p. 73-104

—— 1930, Sherritt-Gordon geology: Canadian Min. Jour., v. 51, no. 32, p. 762

—— 1930, The Sherritt-Gordon copper-zinc deposit, northern Manitoba: (disc.) Econ. Geol., v. 25, p. 286-289; reply (Bruce, E. L.), p. 868-870

Notes

The Sherritt Gordon mine is nearly 400 miles north-northwest from Winnipeg. This mine, now long abandoned, was the only mine in its area.

See the *Notes* under Sherritt Gordon-Flin Flon (General) for discussion on ore age.

In contrast to the ores of the Flin Flon district, which contain little pyrrhotite, in the Sherritt Gordon deposits pyrrhotite is about half as abundant as pyrite. Pyrite and pyrrhotite appear to have been in the first sulfides to have been deposited; both were replaced by chalcopyrite and sphalerite. The sphalerite is purplish-black and occurs in what Bruce (1929) described as graphic intergrowths with chalcopyrite that may indicate exsolution reactions involving these two minerals. Minor amounts of cubanite are found in the chalcopyrite. Near the surface, some marcasite was found, but it probably was produced by the action of surface waters on pyrrhotite. The nonmetallic gangue minerals of the deposit include abundant quartz, amphibole, chlorite, garnet, biotite, and scapolite. It is uncertain how much of this nonmetallic mineral assemblage was inherited from the unreplaced portions of the host pegmatite and how much was introduced by the ore-forming fluids, but probably both processes contributed. Some of the chlorite lay between amphibole and sulfides and well may have resulted from reactions between the ore fluid and the amphibole. The cubanite in the chalcopyrite and the generally high-temperature character of the gangue minerals strongly suggest that the ores were formed in the hypothermal range. On the basis of these deductions, the ores almost certainly should be classed as hypothermal-1.

SNOW LAKE

Middle Precambrian *Zinc, Copper, Lead, Silver* *Hypothermal-1*

Alcock, F. J., 1920, The Reed-Wekusko map-area, northern Manitoba: Geol. Surv. Canada Mem. 119, 47 p.

Bence, A. E., and Coleman, L. C., 1963, Temperatures of formation of the Chisel Lake sulphide deposit, Snow Lake, Manitoba: Canadian Mineral., v. 7, p. 663-666

Davies, J. F., and others, 1962, The File-Snow-Wekusko Lake areas, in *Geology and Mineral Resources of Manitoba*: Manitoba Dept. Mines and Nat. Res., Winnipeg, p. 78-92

Harrison, J. M., 1949, Geology and mineral deposits of the File-Tramping lakes area, Manitoba: Geol. Surv. Canada Mem. 250, 92 p.

Martin, P. L., 1966, Structural analysis of the Chisel Lake orebody: Canadian Inst. Min. and Met. Tr., v. 69 (Bull. no. 649), p. 208-214

Russell, G. A., 1957, Structural studies of the Snow Lake-Herb Lake area, Herb Lake mining division, Manitoba: Manitoba Br., Pub. 55-3, 33 p.

Williams, H., 1966, Geology and mineral deposits of the Chisel Lake map-area, Manitoba: Geol. Surv. Canada Mem. 342, 38 p.

Notes

The Snow Lake area lies about 70 miles east of Flin Flon and includes the Wim, Dickstone, Anderson Lake, Chisel Lake, Ghost Lake, Stall Lake, and Osborne Lake orebodies. Several of the deposits appear to be of the Sherritt Gordon type in that they contain considerable to appreciable pyrrhotite.

See *Notes* under Sherritt Gordon-Flin Flon (General) for discussion on ore age.

The gangue minerals of the Chisel Lake deposit (the only one in the area for which much information is available) are actinolite, tremolite, dolomite, and chlorite. The sulfides, in order of decreasing abundance, are sphalerite, pyrite, pyrrhotite, chalcopyrite, galena, and arsenopyrite. There appears to have been more than one generation of mineralization. Arsenopyrite was the first mineral to form; this was followed by deformation and then by the introduction of pyrite, gold, pyrrhotite, sphalerite, and chalcopyrite (with some minor reversals in that order and with some overlaps); then fracturing, followed by galena, sulfosalts, gudmundite, and pyrite. No information is available as to possible exsolution textures in sphalerite or chalcopyrite. Gangue material in places within the mineralized volumes was much broken and sheared, and pyrite crystals are commonly much broken and healed by the later precipitation of pyrrhotite, chalcopyrite, sphalerite, and/or galena. Later minerals in the sequence, particularly galena, show evidences (such as curved or broken cleavage lines) of stress after deposition. It is uncertain how much of the gangue mineral suite was introduced during the mineralization cycle, earlier in the sequence than the sulfides, and how much represents metamorphism of primary limy rocks. Nevertheless, the bulk of the ore mineralization appears to have been formed at high temperatures, so that the deposit was developed largely, if not entirely, in the hypothermal range. The association of galena with sulfosalts suggests, however, that further work may require the introduction of mesothermal and leptothermal categories into the classification of the Chisel Lake ores.

THOMPSON-MOAK LAKE

Middle Precambrian *Nickel, Copper, Cobalt, Gold(?), Platinum Metals* *Magmatic-2b*

Davies, J. F., 1960, Geology of the Thompson-Moak Lake district, Manitoba: Canadian Min. Jour., v. 81, no. 4, p. 101-104

Davies, J. F., and others, 1962, Thompson-Moak Lake area, in *Geology and Mineral Resources of Manitoba*: Manitoba Dept. Mines and Min. Res., Winnipeg, p. 103-111

Dawson, A. S., 1952, Geology of the Patridge Crop Lake area, Cross Lake mining division, Manitoba: Man. Dept. Mines and Nat. Res. Pub. 41-1, 26 p.

Goddard, J. D., 1966, Geology of the Hambone Lake area, Cross Lake mining division, Manitoba: Man. Dept. Mines and Nat. Res. Pub. 63-1, 44 p.

—— 1968, Geology of the Halfway Lake area (west half), the Pas mining district, Manitoba: Man. Dept. Mines and Nat. Res. Pub. 64-5, 20 p.

Patterson, J. M., 1963, Geology of the Thompson-Moak Lake area, Cross Lake mining division, Manitoba: Man. Dept. Mines and Nat. Res. Pub. 60-4, 50 p.

Wilson, H.D.B., and Brisbin, W. C., 1961, Regional structure of the Thompson-Moak Lake nickel belt: Canadian Inst. Min. and Met. Tr., v. 64 (Bull. no. 595), p. 470-477

Zurbrigg, H. F., 1963, Thompson mine geology: Canadian Inst. Min. and Met. Tr., v. 66 (Bull. no. 614), p. 227-236

Notes

The Thompson-Moak Lake district is about 400 miles slightly west of north from Winnipeg. The Thompson-Moak Lake deposits are located in a nickel-rich belt that extends southwest from Moak Lake (north of Thompson) to Setting Lake (near Wabowden), a distance of about 85 miles; the width of the belt is some 15 miles. Within this belt, the known deposits include those at Thompson, Birchtree, Pipe, Hambone, Soab, Moak Lake, and Mystery Lake; others may have been found since the last published material on the district (Goddard, 1968).

The ores at Thompson, about which something has been published, in contrast to the almost complete silence about the other deposits, occur in a zone of biotite schist that lies in a narrow band of altered sediments that Patterson (1963) presumes to belong to the Assean Lake group. These beds are classed as lowermost Proterozoic (in the Canadian sense). The ores, however, appear to have been introduced into the biotite schist long after these rocks had been laid down, probably during the Hudsonian orogeny toward the end of the middle Precambrian (in the sense used in this memoir). This time of ore formation receives confirmation from the presence of large bodies of serpentinized peridotite of probable Hudsonian age at both Moak Lake and Mystery Lake in which both pyrrhotite and pentlandite are disseminated in small grains and blebs and are localized in networks of tiny irregular and short stringers. Some brecciated portions of these serpentinites are cemented by massive sulfides in bands a few inches to a foot or more in width that are quite local in extent. These deposits appear to be of much lower grade than that in the schist at Thompson, but it seems reasonable to suppose that the sulfides in the Thompson body and the peridotites (now serpentinites) with their sulfides came from the same general source at the same general time. Patterson (1963) believes that the distribution of the serpentinized ultramafic rocks along the axis of a gravity low suggests that they were intruded during the Hudsonian orogeny. The serpentinite is cut by pegmatite that probably is related to the red biotite granite that cuts most of the rock units in the area and has been determined to have an age of about 1700 m.y. If, then, the serpentinite was introduced in the same orogenic epoch as the granite, their ages are not likely to be very different. Since stringers of Thompson sulfides cut pegmatites that have intruded the schists, they are younger than the pegmatites but cannot be much younger if their apparent genetic connection with the serpentinites is soundly based. Thus, the Thompson and other ores of the district probably should be classed as middle Precambrian.

The Thompson ores are quite different from those of the Sudbury or Lynn Lake districts, partly because of the low content of chalcopyrite relative to pentlandite but even more importantly because they are contained in metamorphosed sedimentary rocks almost in their entirety rather than in mafic or ultramafic rocks. At Thompson Lake, the ores lie in a more or less continuous, but irregular, sheet (Patterson, 1963) in the biotite schist of probable Assean Lake age. The ore body pinches and swells irregularly, locally being as much as 100 feet wide, but may contract, over very short distances, to a width of a few feet; the average width is 20 to 30 feet. The main ore body may split into two or more branches, thus enclosing considerable bodies of unmineralized or slightly mineralized schist. The ore is mainly pyrrhotite and pentlandite and contains many subangular to angular wall-rock fragments; these fragments are generally the biotite schist, plus pegmatized schist and, locally, pegmatite. In places, quartz is abundantly present with the sulfides. At the southwest end of the ore body, a mass of serpentinite is present that contains stringers and bands and finely disseminated grains of sulfides. Locally, pegmatite also is found in these fractures, so they must have been open well prior to the introduction of the sulfide stringers and bands; the disseminated sulfides well may be primary in the serpentinite as they appear to be in the Moak and Mystery Lake bodies. The pentlandite forms as abundant small blebs in the pyrrhotite; the chalcopyrite, while being present throughout the ore, is not abundant and occurs in fractures in the gangue minerals. Gersdorffite is locally of some importance, and the platinum metals and gold(?) are common enough to be recoverable. The ore grade is reported to be 2.9 percent combined nickel and copper, with the copper providing only about 0.2 of the 2.9.

Although the amount of information on which to base a concept of the genesis of the Thompson deposits is quite limited, the mineralogy is what would be expected if the ores had been intruded in the molten state from mafic or ultramafic sources. This concept receives confirmation from the presence of the same sulfides as disseminations in the serpentinites. The most probable explanation of these relationships is that the peridotites were intruded from some source at depth in such a state that the sulfides only rarely and locally could concentrate into massive bodies. Later, when crystallization had proceeded appreciably further in the source magma chamber, a second period of compression forced out material that was largely, though not entirely, molten sulfides. This molten sulfide material found its easiest channel of egress in the biotite schists where it solidified. Some biotite may have been taken up in the sulfide melt through reaction with the schist, or some small amounts of biotite may have been contained in the original sulfide melt. The quartz associated with the sulfides well may have been dissolved in the sulfide melt. If this explanation is correct, or approximately so, the ores must be classified as magmatic-2b.

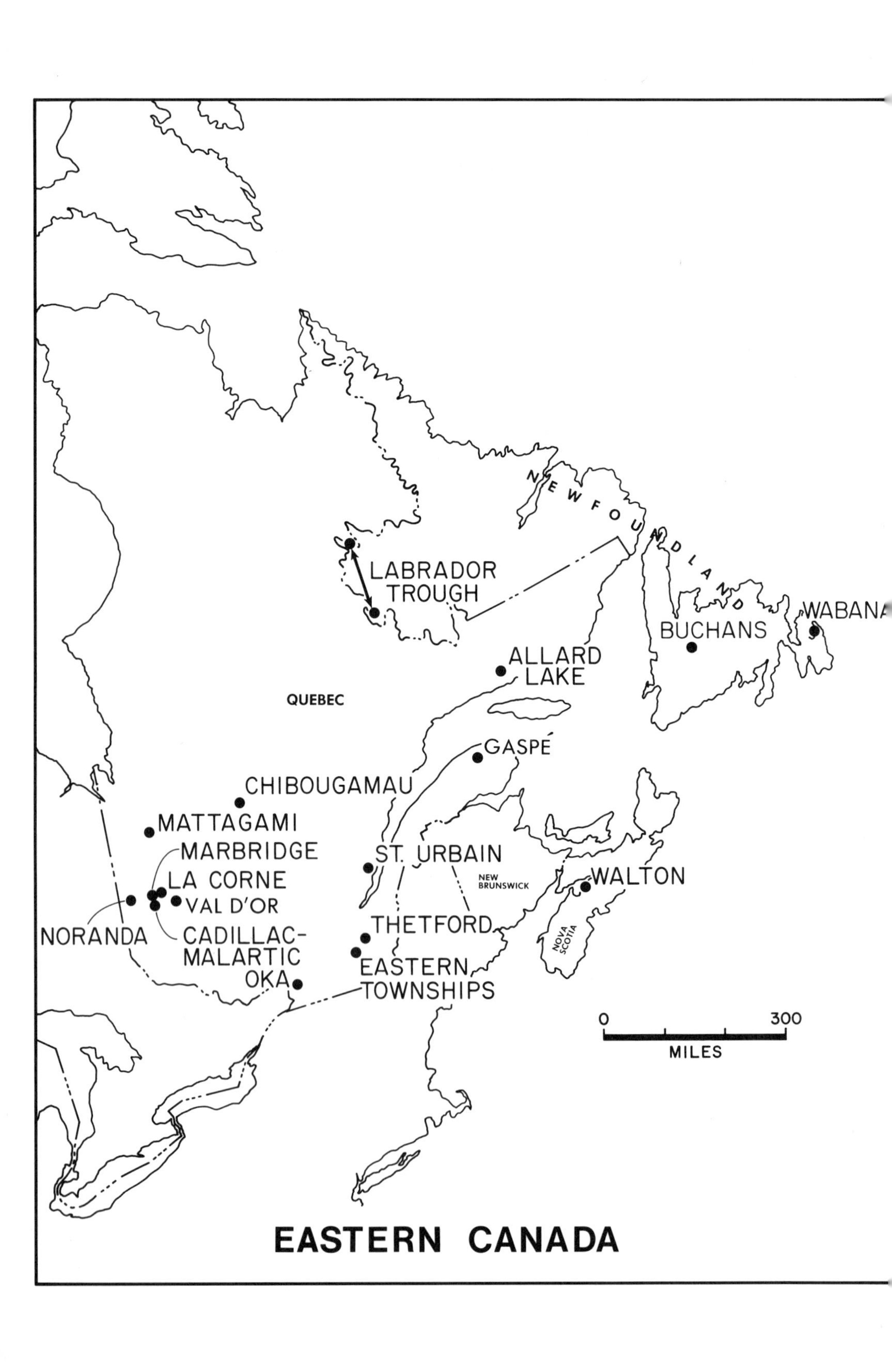

EASTERN CANADA

New Brunswick

BATHURST-NEWCASTLE

Middle Paleozoic	*Zinc, Copper, Lead*	*Hypothermal-1 to Mesothermal*

Aletan, G., 1960, The significance of microscopic investigation in the course of beneficiation of the "Brunswick ore": Canadian Inst. Min. and Met. Tr., v. 63 (Bull. no. 584), p. 653-660

Aleva, G.J.J., 1960, Geochemical and geophysical exploration of the Nigadoo base metal deposit, N.B., Canada: Geologie en Mijnbouw, n.s., 22 Jg., nr. 10, p. 492-499

Benson, D., 1960, Application of the sphalerite geothermometer to some northern New Brunswick sulfide deposits: Econ. Geol., v. 55, p. 818-826

Boyle, R. W., 1965, Origin of the Bathurst-Newcastle sulfide deposits, New Brunswick: Econ. Geol., v. 60, p. 1529-1532

—— 1969, Further remarks on the origin of massive sulfide deposits (disc.): Econ. Geol., v. 64, p. 829

Boyle, R. W., and Davies, J. L., 1963, Geology of the Austin Brook and Brunswick no. 6 sulphide deposits, Gloucester County, New Brunswick: Geol. Surv. Canada Paper 63-24, 23 p.

Cheriton, C. G., 1960, Anaconda exploration in the Bathurst district of New Brunswick, Canada: A.I.M.E. Tr., v. 217, p. 278-284

Davies, J. L., and others, 1969, Geology and mineral deposits of the Nigadoo River-Millstream River area, Gloucester County, New Brunswick (part of 21 P/12): Geol. Surv. Canada Paper 67-49, 70 p.

DeChow, E., 1960, Geology, sulfur isotopes and the origin of the Heath Steele ore deposits, Newcastle, N.B., Canada: Econ. Geol. v. 55, p. 539-556

Douglas, R. P., 1965, The Wedge mine-Newcastle-Bathurst area, N.B.: Canadian Inst. Min. and Met. Tr., v. 68 (Bull. no. 635), p. 80-86

Fleming, H. W., 1961, The Murray deposit, Restigouche County, New Brunswick, N.B., a geochemical-geophysical discovery: Canadian Inst. Min. and Met. Tr., v. 64 (Bull. no. 587), p. 163-168

Holyk, W., 1957, Structure of northern New Brunswick, in *Structural geology of Canadian ore deposits*, v. 2: Canadian Inst. Min. and Met., Montreal, p. 485-492

Hutchinson, R. W., 1962, Temperatures of formation and origin of the Nigadoo and Brunswick Mining and Smelting no. 6 deposits, New Brunswick, Canada: Econ. Geol., v. 57, p. 834-835

Jenney, C. P., 1957, Geology of New Brunswick lead-zinc: Eng. and Min. Jour., v. 158, no. 6, p. 95-96

Kalliokoski, J., 1961, Temperatures of formation and origin of the Nigadoo and Brunswick Mining and Smelting no. 6 deposits, New Brunswick, Canada: Econ. Geol., v. 56, p. 1446-1455

—— 1965, Metamorphic features in North American massive sulfide deposits: Econ. Geol., v. 60, p. 485-505 (particularly p. 492-499); disc., p. 1539-1540

Kinkel, A. R., Jr., 1966, Bathurst district, New Brunswick, in *Massive pyritic deposits related to volcanism and possible methods of emplacement*: Econ. Geol., v. 61, p. 683-684

Lea, E. R., and Rancourt, C., 1958, Geology of the Brunswick Mining and Smelting orebodies, Gloucester County, N.B.: Canadian Inst. Min. and Met. Tr.,

v. 61 (Bull. no. 551), p. 95-105

Lusk, J., 1969, Base metal zoning in the Heath Steele B-1 orebody, New Brunswick, Canada: Econ. Geol., v. 64, p. 509-518

Lusk, J., and Crocket, J. H., 1969, Sulfur isotope fractionation in coexisting sulfides from the Heath Steele B-1 orebody, New Brunswick, Canada: Econ. Geol., v. 64, p. 147-155

MacAllister, A. L., 1957, Keymet mine, in *Structural geology of Canadian ore deposits*, v. 2: Canadian Inst. Min. and Met., Montreal, p. 492-494

—— 1960, Massive sulphide deposits in New Brunswick: Canadian Inst. Min. and Met. Tr., v. 63 (Bull. no. 574), p. 50-60

MacKenzie, G. S., 1958, History of mining exploration, Bathurst-Newcastle district, New Brunswick: Canadian Inst. Min. and Met. Tr., v. 61 (Bull. no. 551), p. 84-89

Roy, S., 1961, Mineralogy and paragenesis of the lead-zinc-copper ores of the Bathurst-Newcastle district, New Brunswick: Geol. Soc. Canada Bull. 72, 19 p.

Skinner, R., 1953, Preliminary map, Bathurst--Gloucester and Restigouche Counties, New Brunswick: Geol. Surv. Canada Paper 53-29, 4 p.

Smith, C. H., 1957, Bathurst-Newcastle area, Northumberland, Restigouche, and Gloucester Counties, New Brunswick: Geol. Surv. Canada Map 1-1957, 1:126,720

Smith, C. H., and Skinner, R., 1958, Geology of the Bathurst-Newcastle mineral district, New Brunswick: Canadian Inst. Min. and Met. Tr., v. 61 (Bull. no. 551), p. 78-83

Stanton, R. L., 1959, Mineralogical features and possible mode of emplacement of the Brunswick Mining and Smelting ore bodies, Gloucester County, New Brunswick: Canadian Inst. Min. and Met. Tr., v. 62 (Bull. no. 570), p. 337-349

—— 1960, General features of conformable "pyritic" orebodies; Pt. I, field association; Pt. II, Mineralogy: Canadian Inst. Min. and Met. Tr., v. 63 (Bull. nos. 573, 574), p. 22-36

Stockwell, C. H., and Tupper, W. M., 1966, Geology of the Brunswick no. 6 and no. 12 mining area, Gloucester County, New Brunswick: Geol. Surv. Canada Paper 65-13, 8 p.

Stumpfl, E. F., and others, 1970, Tectonics, volcanism and mineralization in New Brunswick, Canada--some new aspects: IMA-IAGOD Meetings 1970, Collected Abs., Tokyo-Kyoto Meetings, Paper 3-18, p. 57

Tupper, W. M., 1960, Sulfur isotopes and the origin of the sulfide deposits of the Bathurst-Newcastle area of northern New Brunswick: Econ. Geol., v. 55, p. 1676-1707

—— 1969, The geology of Orvan Brooks sulphide deposit, Restigouche County, New Brunswick: Geol. Surv. Canada Paper 66-59, 11 p.

Tupper, W. M., and others, 1968, The geology, mineralogy, and geochemistry of the Captain sulphide deposit, Gloucester and Northumberland Counties, New Brunswick: Geol. Surv. Canada Paper 66-18, 17 p.

Notes

The ore deposits of the Bathurst area are scattered within a circular area, some 35 miles in diameter, the center of which lies a little over 25 miles southwest of the town of Bathurst.

The area is divided into two structural units by the Rocky Brook-Millstream break that strikes somewhat north of east and reaches the coast less than 10 miles north of Bathurst. The southern unit is composed of folded sedimentary and volcanic rocks that range in age from Middle Silurian to Middle Devonian; the northern unit is made up of highly folded Ordovician volcanic and sedimentary rocks; these rocks are cut by gabbro sills and granite stocks. The rocks of the southern belt are not as intensely deformed as those of the northern. Most of the silicic intrusives and all of the larger masses of these rocks are located south of the break, and these bodies are arranged peripherally to the sediment-surrounded volcanic core. These igneous rocks give K/Ar ages of 364 to 398 m.y., and the model lead ages for five galena specimens have a mean of 370 ± 20 m.y. These ages suggest that the ores were emplaced at about the same time as the granitic rocks, that is, during the Devonian. Kalliokoski (1965), however, argues that the galenas were deposited in Ordovician time (i.e., syngenetically with the rocks containing them) and were so affected by the Acadian orogeny as to receive the age imprint of that event. The close spatial relationship of the Devonian granites to most of the ore deposits, nevertheless, may be highly meaningful since the ores and the granites are the only weakly metamorphosed materials in the district. On this basis, the ores are classified as Devonian or middle Paleozoic.

Boyle (1965) points out that there is no question but that the vein-filled fractures on the north side of the Rocky Brook-Millstream break were emplaced after the rocks of the area had been folded and fractured and that these ores are essentially the same in mineral and chemical character as those south of the break. From this, it would seem reasonable to suppose that the massive deposits south of the break were formed at much the same time and from solutions from the same general source, or from similar sources, as the vein deposits north of it. The character of the mineralization, particularly the abundant pyrrhotite and common arsenopyrite, indicates that much of the deposition took place within the hydrothermal range. The late galena and chalcopyrite, however, were very possibly produced under mesothermal conditions. The differences in minerals and mineral paragenesis between deposits north and south of the break are so minor that the same classification can be applied to both vein and massive deposits, that is, hypothermal-1 to mesothermal.

Newfoundland

BUCHANS

Middle Paleozoic	*Zinc, Lead, Copper, Silver, Gold, Barite*	*Hypothermal-1 to Mesothermal*

Anger, G., 1963, Die Blei-Zink-Kupferlagerstätten von Buchans, Mittel-Neufundland: Neues Jb. f. Mineral., Mh., Jg. 1963, H. 6, S. 126-136

Baird, D. M., 1956, Base metal deposits of the Buchans-Notre Dame Bay area, Newfoundland: Geol. Assoc. Canada Pr., v. 8, p. 167-168

—— 1960, Massive sulfide deposits in Newfoundland: Canadian Inst. Min. and Met. Tr., v. 63 (Bull. no. 573), p. 39-42 (particularly p. 39-41)

George, P. W., 1937, Geology of the lead-zinc-copper deposits at Buchans, Newfoundland: A.I.M.E. Tr., v. 126, p. 488-511

Newhouse, W. H., 1931, Geology and ore deposits of Buchans, Newfoundland: Econ. Geol., v. 26, p. 399-414

Riley, G. C., 1957, Red Indian Lake (west half), Newfoundland: Geol. Surv. Canada Map 8-1957 (with descriptive notes), 1:253,440 (area lies immediately to the west of Buchans)

Snelgrove, A. K., 1928, The geology of the central mineral belt of Newfoundland--a collation and contribution: Canadian Inst. Min. and Met. Tr., v. 31 (Bull. no. 197), p. 1057-1127 (particularly p. 1106-1124)

Staff, Buchans Mining Co., Ltd., 1955, Buchans operation, Newfoundland: Canadian Inst. Min. and Met. Bull. no. 518, p. 349-353 (not in Tr. sec.) (very brief summary of geology)

Swanson, E. A., and Brown, R. L., 1962, Geology of the Buchans ore bodies: Canadian Inst. Min. and Met. Tr., v. 65 (Bull. no. 605), p. 284-292

Notes

The Buchans mine is 200 miles west-northwest from St. John's and 50 miles east of Corner Brook.

What deformation occurred in the Buchans district was mild; the pyroclastics (the least competent rocks) generally are foliated, but the "granite," conglomerates, sandstone-siltstones, arkoses, and silicic sill rocks show little effect of stress and are only slightly crushed. The regional metamorphism is quite low in grade, suggesting that the rocks were never very deeply buried nor subjected to intense earth movements. The minable ore bodies are localized by a breccia zone located in the Dacite formation made up of a series of layers of extrusive rocks and by a series of conjugate shears. This strongly indicates that the ores were not introduced into the district until after it had been affected by such earth movements as were developed. Most of the fracturing probably was, in large part, concomitant with the folding of the Buchans dome, most of the ore occurrences having been localized in its immediate vicinity; the greatest concentration of such bodies is the Lucky Strike group on the axis of the dome. Granites are found to the north and west of the Buchans district, and other granites in Newfoundland, similar to those found near Buchans, give ages based on radioactive determinations of about 360 m.y. Granted that the Buchans ore fluids came from the same general source as these granites, the Buchans ores must be Devonian or middle Paleozoic in age. The suggestion has been made that the ores are syngenetic with the formation in which they are found, but this concept does not seem to merit serious consideration when tested against the evidence for structural control of ore emplacement.

In the breccia ores, the dominant ore type at Buchans, there are three varieties of ore occurrence: (1) breccia ore proper that contains numerous rock fragments; (2) baritic ore in which the barite content is high; and (3) normal ore in which the sulfides are massive, the fragments few, and the barite moderate in amount. Most of the breccia ore falls in category (3). Swanson and Brown (1962) believe that the ores replaced volcanic breccias and their associated tuffs. Siliceous fragments within the breccias are far less readily replaced than the tuffaceous material surrounding them. Where the ore bodies consist of several sulfide lenses, the separating dacite layers normally are intensely altered to a clayey material. Sericite, however, was the most common mineral developed in the wall rocks, and the areas of alteration are far more widespread than the ore mineralization. In the mineralization cycle, the first minerals to be deposited were essentially simultaneous quartz and pyrite; these minerals were appreciably crushed and cracked, and the later sulfides (principally sphalerite, galena, and chalcopyrite) filled these cracks and replaced outward from them--the volumes replaced were much larger than those filled. The sphalerite contains what appear to be exsolved blebs of chalcopyrite; blebs of galena also are present, but these are much more probably the result of replacement than of exsolution. Barite replaces all the earlier sulfide and gangue minerals, and late fissures in the ore are filled by calcite and fluorite. A little tetrahedrite is scattered through the ores as a late mineral; it is not appreciably silver-bearing, suggesting that most of the silver at Buchans must be in the galena. The probably exsolved chalcopyrite in the sphalerite indicates that some of the sphalerite, at least, was deposited in the hypothermal range; the character of the wall-rock alteration, however, points to considerable deposition under the mesothermal conditions. The deposits are classified here as hypothermal-1 to mesothermal.

WABANA

Early Paleozoic	*Iron as Hematite, Siderite, Chamosite*	*Sedimentary-A3*

Coughlan, W. K., 1966, Geology of the Wabana deposit--with comments on exploration and development problems peculiar to a submarine deposit: Canadian Inst. Min. and Met. Tr., v. 60 (Bull. no. 646), p. 62-66

Gilliatt, J. B., 1924, Folding and faulting of Wabana ore deposits: Canadian Inst. Min. and Met. Tr., v. 27 (Bull. no. 152), p. 616-634

Gross, G. A., 1965, Clinton-type iron formation, in *Geology of iron deposits in Canada, Volume I, General geology and evaluation of iron deposits*: Geol. Surv. Canada Econ. Geol. Rept. no. 22, p. 123-125

—— 1967, Wabana iron deposits, Newfoundland, in *Geology of iron deposits in Canada, Volume II, Iron deposits in the Appalachian and Grenville regions of Canada*: Geol. Surv. Canada Econ. Geol. Rept. no. 22, p. 4-17

Hayes, A. O., 1915, Wabana iron ore of Newfoundland: Geol. Surv. Canada Mem. 78, 163 p.

—— 1920, Nova Scotian oölitic iron deposits of sedimentary origin: Canadian Min. Inst. Tr., v. 22, p. 112-122

—— 1929, Further studies of the origin of the Wabana iron ores of Newfoundland: Econ. Geol., v. 24, p. 687-690

—— 1931, Structural geology of the Conception Bay region and of the Wabana iron deposits of Newfoundland: Econ. Geol., v. 26, p. 44-64

Howell, B. F., 1926, The Cambrian-Ordovician stratigraphic column in southeastern Newfoundland: Canadian Field-Nat., v. 40, no. 3, p. 52-57

Hutchinson, R. D., 1953, Geology of the Harbour Grace map-area, Newfoundland: Geol. Surv. Canada Mem. 275, 43 p.

Rose, E. R., 1952, Torbay map-area, Newfoundland: Geol. Surv. Canada Mem. 265, 64 p.

Notes

The Wabana mine is located on Bell Island in Conception Bay about 12 miles west-northwest of St. John's. Most of the mine workings are under the sea.

Iron ore at Wabana has been mined from three beds, the Dominion or lower bed, the Scotia or middle bed, and the upper bed. All of these beds are contained in the Clinton-type iron formations of the Lower Ordovician Wabana group; Hayes (1915) identified six zones of iron formation in the Wabana group, of which zones 2, 4, and 5 correspond to the three zones just mentioned and in the order given. The Wabana group is more than 1000 feet thick and is lithologically similar to the underlying (and also Lower Ordovician) Bell Island group; it, therefore, contains sandstones, micaceous sandy shales, and ferruginous sandstones and shales plus oölitic pyrite-bearing beds as well as the iron formations. Since there appears to be no doubt but that the ores were emplaced syngenetically with the enclosing formations, the age of the deposits is Early Ordovician or early Paleozoic.

Gross (1965, p. 108) appears to think that the delicate balance of physical and chemical factors and the very special conditions needed for iron formations to have been derived from a landmass are too extraordinary and exceptional to have been repeated in a variety of areas. There is, however, (Gross, 1965, p. 125), no known association in the immediate vicinity of volcanic rocks or evidence of volcanic activity to have produced the iron now in these deposits. Gross believes, nevertheless, that direct contributions of iron in solution to ocean currents could readily account for the erratic appearance of iron formation in an otherwise normal type of sedimentation. He points out that, if currents with an abnormally high iron content came into a near-shore environment, slight changes in pH and Eh would cause precipitation of iron as the current water mingled in the neritic zone. The actual volcanism and/or thermal springs contributing to these currents might be at some considerable distance from the shallow-water zones where the iron was precipitated. Gross thinks that the shelf environment at Wabana had a pH sufficiently low to exclude the large-scale deposition of limestone and dolomite; the iron-rich oölites of the iron formation do have a carbonate matrix. A constant influx of acidic river waters with a high organic-acid content could have maintained lower pH conditions in the shallow near-shore area and thus have prevented the deposition of much limestone during the precipitation of the iron. The persistent association of the iron formations with black shales indicates, Gross believes, that there were major organic factors in the solution and deposition of the iron material in the ferruginous beds. Gross considers the hematite and chamosite to have been deposited as primary sediments with rather subtle fluctuations in the chemical environment having caused the alternate deposition of ferrous and ferric iron minerals. The primary nature of the abundant hematite-chamosite oölites is indicated by delicate algal borings that cut through the concentric spheres of the oölites. The diagenetic character of the siderite is shown by the extensive alteration of the primary oölites and their calcite matrix to siderite. Despite this not unreasonable outline of a volcanic source for the iron at Wabana, Gross (1965, p. 125) says that, in general, he favors a terrestrial source for the iron because of the association of alumina, calcium, and magnesia in quantities that might be expected from weathering in an acidic environment. Despite Gross's final conclusion, I cannot but think that submarine exhalations of hydrothermal solutions probably provided the iron for the Wabana deposits, and they are here classified as sedimentary-A3

Newfoundland-Quebec

LABRADOR TROUGH

Middle Precambrian	*Iron as Hematite,*	*Sedimentary-A1a,*
Late Precambrian	*Goethite, Magnetite*	*Metamorphic-C,*
		Hydrothermal and/or
		Ground Water-B2

Auger, P. E., 1954, The stratigraphy and structure of the northern Labrador Trough, Ungava, New Quebec: Canadian Inst. Min. and Met. Tr., v. 57 (Bull. no. 508), p. 327-330

Baragar, W.R.A., 1960, Petrology of basaltic rock in part of the Labrador Trough: Geol. Soc. Amer. Bull. 71, p. 1589-1643

—— 1963, Wakuach Lake map-area, Quebec-Newfoundland: Geol. Surv. Canada Paper 62-38, 4 p.

Bergeron, R., 1965, Geology and mineral resources of the Labrador Trough: N.Y. Acad. Sci. Tr., Ser. 2, v. 27, no. 8, p. 843-857

Blais, R. A., 1959, L'origine des minerais Crétacés du gisement de fer de Redmond, Labrador: Naturaliste Canadien, v. 86, no. 12, p. 265-299

Bonham, W. M., 1949, The Labrador Iron Range: Canadian Min. Jour., v. 70, no. 7, p. 57-61

Chakraborty, K. L., 1963, Relationship of anthophyllite, cummingtonite and manganocummingtonite in the metamorphosed Wabush iron-formation, Labrador: Canadian Mineral., v. 7, pt. 5, p. 738-750

Choubersky, A., 1957-1958, The operation of the Iron Ore Company of Canada: Canadian Inst. Min. and Met. Tr., v. 67, p. 33-88 (particularly p. 35-50)

Dimroth, E., 1970, Evolution of the Labrador geosyncline: Geol. Soc. Amer. Bull., v. 81, p. 2717-2742

Donaldson, J. A., 1966, Marion Lake map-area, Quebec-Newfoundland: Geol. Surv. Canada Mem. 338, 85 p.

Douglas, G. V., and Compton, L. P., 1957, A suggested interpretation of the Quebec-Labrador iron deposits: Econ. Geol., v. 52, p. 709-711

Fahrig, W. F., 1962, Petrology and geochemistry of the Griffis Lake ultrabasic sill of the central Labrador Trough, Quebec: Geol. Surv. Canada Bull. 77, 39 p.

—— 1967, Shabogamo Lake map-area, Newfoundland-Labrador and Quebec, 23 G E 1/2: Geol. Surv. Canada Mem. 354, 22 p.

Frarey, M. J., and Duffell, S., 1964, Revised stratigraphic nomenclature for the central part of the Labrador Trough: Geol. Surv. Canada Paper 64-25, 13 p.

Gastil, (R.) G., and Knowles, D. M., 1960, Geology of the Wabush Lake area, southwestern Labrador and eastern Quebec: Geol. Soc. Amer. Bull., v. 71, p. 1243-1254

Gastil, (R.) G., and others, 1960, The Labrador geosyncline: 21st Int. Geol. Cong. Rept., pt. 9, p. 21-38

Gross, G. A., 1961, Iron-formations and the Labrador geosyncline: Geol. Surv. Canada Paper 60-30, 7 p.

—— 1964, Mineralogy and beneficiation of Quebec iron ore: Canadian Inst. Min. and Met. Tr., v. 67 (Bull. no. 623), p. 17-24

—— 1966, Principal types of iron-formation and derived ores: Canadian Inst. Min. and Met. Tr., v. 69 (Bull. no. 646), p. 41-44

—— 1968, Iron ranges of the Labrador geosyncline, in *Geology of iron deposits in Canada, Volume III*: Geol. Surv. Canada Econ. Geol. Rept. 22, 179 p.

Gustafson, J. K., and Moss, A. E., 1953, The role of geologists in the development of the Labrador-Quebec iron ore districts: A.I.M.E. Tr., v. 196, p. 593-602; disc. (A. K. Snelgrove) in Min. Eng., v. 5, nos. 6, 11, p. 1129-1130

Harrison, J. M., 1952, Quebec-Labrador iron belt: Geol. Surv. Canada Paper 52-20, 21 p. (mimeo.)

—— 1953, Iron formations of Ungava Peninsula, Canada: 19th Int. Geol. Cong., C.R., sec. 10, p. 19-33

Hoffman, D. J., 1970, Deformational effects on iron occurrence and ore potential at Queco: Econ. Geol., v. 65, p. 583-587

—— 1970, Relation of magnetite content to geology at Queco: Econ. Geol., v. 65, p. 511-512

Klein, C., Jr., 1966, Mineralogy and petrology of the metamorphosed Wabush iron formation, southwestern Labrador: Jour. Petrol., v. 7, p. 246-305

Knowles, D. M., and Gastil (R.) G., 1959, Metamorphosed sedimentary iron formation in southwestern Labrador: Canadian Inst. Min. and Met. Tr., v. 62 (Bull. no. 568), p. 265-272

Koulomzine, T., and Jaeggin, R. P., 1961, Discovery of the iron ore deposit of Mount Wright Iron Mines Co. Limited: Canadian Inst. Min. and Met. Tr., v. 64 (Bull. no. 594), p. 456-462

MacDonald, R. D., 1960, Iron deposits of Wabush Lake, Labrador: Min. Eng., v. 12, no. 10, p. 1098-1102

Murphy, D. A., 1962, The iron-formation of Mt. Wright-Lake Carheil: A.I.M.E. Tr., v. 223, p. 285-291 (a summary appears in Min. Eng., v. 14, no. 9)

Reh, H., 1960, Erschliessung der Eisenerzvorkommen in Quebec-Labrador: Zeitsch. f. angew. Geol., Bd. 6, H. 3, S. 100-103

Retty, J. A., and Moss, A. E., 1951, Iron ore deposits of New Quebec and Labrador (abs.): Econ. Geol., v. 46, p. 799-800

Rice, H. R., 1949, Grand scale prospecting in Labrador and Quebec: Canadian Min. Jour., v. 70, no. 9, p. 65-77

Roach, R. A., and Duffell, S., 1968, The pyroxene granulites of the Mount

Wright map-area, Quebec-Newfoundland: Geol. Soc. Canada Bull. 162, 83 p.

Selleck, D. J., and Campbell, W. A., 1965, Exploration and development of the Carol ores: A.I.M.E., Minn. Sec., 38th Ann. Meeting, Minneapolis (also Univ. Minn. Center for Cont. Study, 26th Ann. Min. Symposium), p. 15-22

Sharma, T., and others, 1965, O^{18}/O^{16} ratios of minerals from the iron formations of Quebec: Jour. Geol., v. 73, p. 664-667

Stevenson, I. M., 1963, Lac Bazil, Quebec: Geol. Surv. Canada Paper 62-37, 4 p.

—— 1963, Leaf River area, New Quebec: Geol. Surv. Canada Paper 62-24, 5 p.

—— 1965, Leaf River map-area, Quebec and District of Keewatin: Geol. Surv. Canada Paper 64-28, 10 p.

Stubbins, J. B., and others, 1961, Origin of the soft iron ores of the Knob Lake Range: Canadian Inst. Min. and Met. Tr., v. 64 (Bull. no. 585), p. 37-52

Tanton, T. L., 1953, Iron ores in Canada, in *Symposium sur les gisements de fer du monde*: 19th Int. Geol. Cong., v. 1, p. 311-352 (particularly p. 325-326)

Westervelt, R. D., 1957, The Knob Lake iron ore deposits: Canadian Inst. Min. and Met. Tr., v. 60 (Bull. no. 547), p. 376-386

Notes

The Labrador Trough runs generally south-southeast almost from the northern end of the western side of Ungava Bay through eastern Quebec to cross into Labrador near the town of Schefferville and finally bends to the southwest and passes back into Quebec to end, just south of the Grenville front, in the vicinity of Mount Reed and Lac Jeannine. The southern tip is divided from the larger portion of the trough by the arc of the Grenville front (convex to the north) between Sawbill Lake on the west and Ossokmanuan Lake to the east.

The iron-bearing formations of the Labrador Trough as a whole were deposited on Keewatin rocks that had undergone their latest metamorphism prior to iron formation deposition about 2500 m.y. ago. Since the iron-bearing formations north of the Grenville front were mildly metamorphosed in the Elsonian metamorphism at about 1370 m.y. ago, it also is probable that those south of the front (involved in this and the Grenville metamorphism some 950 m.y. ago) were laid down in much the same time range as those north of the front. Thus, it is almost certain that the last of the iron-bearing formations had been deposited by the end of the middle Precambrian (1600 m.y. ago). So far as the ores in the southern part of the trough are concerned, essentially no later enrichment occurred, but the metamorphism of late Precambrian age is what made the ores minable. Therefore, the southern ores are classified as middle Precambrian (for the primary deposition) and late Precambrian (for the metamorphism). When the enrichment of the ores north of Grenville front took place is less certain. The presence of rubble ore in the Redmond formation of Cretaceous age definitely shows that all, or essentially all, of the enrichment of the primary ore had taken place prior to the end of the Mesozoic. How much before this time the enrichment was complete is less sure but, by comparison with other iron ores of this type, it would appear that the enrichment took place in Precambrian time, and probably in the late Precambrian. Thus, the ores of the Schefferville area are dated as middle Precambrian for the primary deposition and late Precambrian for the enrichment.

Because the metamorphism coarsened the grains of the iron formation in the southern portion of the trough and made other textural and mineralogical changes, much of the iron formation in that area can be mined as ore even though there was no enrichment of primary sediments after lithification (although there was some oxidation of primary ore after it was exposed at the surface). After the metamorphism, the upper unit of the Wabush formation consisted of magnetite, specular hematite, and quartz; it is the principal source of iron ore produced in the area. Magnetite is more abundant in the upper and lower portions of the upper unit and specular hematite in the central part; iron silicates also are plentiful in the upper portion. In the Wabush formation in general, the mineral assemblages indicate that the metamorphism reached the epidote-amphibolite and amphibolite facies, with an increase in grain size being concomitant with the development of higher-temperature minerals in mosaic or interlocking textures; porosity was decreased during the metamorphic process. In the oxide facies of the iron formation, quartz and iron oxides did not react to produce iron silicates. A few narrow stringers or veins of iron oxides cut across the beds and suggest that some remobilization took place; the amount of iron in these veins is negligible in comparison with the total amount of iron present. Reserves were reported in 1968 to be over 3 billion tons of 35 to 38 percent iron ore composed of magnetite and hematite and coarse crystalline quartz above the lower limits of open pit mining. The primary ores are classified as sedimentary-Ala and the metamorphism as metamorphic-C.

There is no agreement as to the initial mineral content of the iron formation either south or north of the Grenville front, but the actual elemental composition of the lithified iron formation probably was essentially the same as that of the initial sediments. In the Schefferville area, the bedded ores in the Sokoman formation almost certainly occupy positions that contained iron formation since much high-grade ore grades outward through lean ore and leached iron formation into unchanged iron formation without any change in bed thickness; therefore, it would appear certain that not all of the iron in the bedded ore bodies was present in the original iron formation. To account for the iron content of more than 50 percent in the ore, silica amounting to 30 to 50 percent of the formation must have been leached out. This leaching, if it went on independently of the introduction of iron, would have produced an initial porosity of 40 to 65 percent in the beds' now high-grade ore. Since the porosity in the ores is now about 30 percent and since the density of the ore is about that of unaltered iron formation, either iron was added or the rock was compacted. The latter explanation almost certainly is not true since the thickness of the ore beds is practically that of the beds of unenriched iron formation into which they grade. Further, there is no evidence of slumping within the ore beds as would be necessary if they had attained their present density through compaction. There is moreover considerable indication that iron and manganese oxides were introduced into the ore beds. Much of the pore space, apparently produced by the leaching of silica, has been filled by goethite and limonite. Fractures and irregular openings commonly are filled with iron and manganese oxides; the amount of iron added is, of course, much greater than that of manganese. Although much of the secondary iron in the bedded ores is goethite, that in the slaty horizons is hematite. In addition, there are considerable masses of brecciated rock cemented by iron oxides, and the slate of the Ruth formation has been extensively replaced by hematite.

Two hypotheses have been put forward to explain the development of the high-grade ores: (1) that the leaching of silica and the addition of iron and manganese were accomplished by circulating waters of meteoric origin and (2) that the leaching and adding of material were the work of hydrothermal solutions. It is possible that both processes were in some measure responsible for the formation of the high-grade ores.

Stubbins and others (1961) believe that percolating meteoric waters moved downward along bedding planes and fractures and caused not only the leaching of silica but also the enrichment in iron and manganese oxides needed to bring the leached iron formation up to its present grade in iron and manganese. Obviously, this means that sufficient iron to raise the iron content of the leached iron formation from about 35 percent to about the 57 percent of average high-grade ore must have been leached from somewhere and added to the volumes of leached iron formation. Where this somewhere may have been is difficult to say. Perhaps the iron was leached from iron formation that has since been removed by erosion. If this was the case, the silica leached from these now-eroded rocks was carried completely out of the system, but the leached iron was deposited in those porous leached volumes at lower levels and converted them to iron ore. Some such mechanism also may be assumed to have been responsible for the iron that was needed to replace such portions of ferruginous slate of the Ruth formation as have been converted to high-grade ore. The main difficulty with this concept is that there should have been some of iron formation from which iron had been removed preserved in at least some places above the ore. Such preservation does not seem to have occurred since what leached iron formation, not now ore, that can be found has been leached of silica but not of iron. The strongest argument for the downward-leaching hypothesis, nevertheless, is the close spatial relationship between the ores and the present surface. Even this relationship can be more apparent than real for had the ores been developed in the late Precambrian, as seems quite probable, they might, at their formation, have been far beneath the present surface, and their contact with the present surface would then have taken place only after considerable erosion. If the ores were deeply buried at formation and were uncovered by erosion, such uncovering would have been carried out to a considerable extent by the Late Cretaceous as is witnessed by the rubble ore of Late Cretaceous age that was formed from high-grade ore exposed at the surface. The erosion that took place between the end of the Cretaceous and the present, therefore, must have removed large tonnages of ore for every vertical foot of surface rock taken away by that process. In short, the outcropping of high-grade ore at the present surface means only that there was so much more ore available in the trough at the end of the Cretaceous that huge tonnages could be removed and huge tonnages still remain; it says nothing really as to the origin of the ore. There is no question but that high-grade ore gives way with depth to extremely hard iron formation of very low porosity. This relationship has been advanced as evidence that the leaching and enrichment proceeded from the surface downward. It is also possible, however, that the enriching solutions came from depth and only began to deposit iron when they encountered leached and porous iron formation that they had earlier produced. If the hypothesis of leaching from the surface is correct, it seems surprising that so much of the Ruth slate has been converted to hematite ore and lacks the hydrated iron oxides so common in the bedded iron ores. To have enriched the Ruth slates, it is necessary that considerable alumina and carbonaceous material have been leached from these rocks. It is also rather difficult to explain where the enriching manganese came from since the unaltered iron formation contains about one-third as much as the high-grade ore. Of course, it may have been concentrated from the hypothetical leached iron formation, now eroded away, from which the enriching iron was obtained, but larger volumes of iron formation would have to have been leached of manganese than of iron, or the process of manganese concentration would have to have been proportionally more effective than that of iron, for the present ratios of the two metals in high-grade ore to prevail.

In short, many arguments can be raised against the enriching iron and manganese having been derived from above that do not argue against the enriching elements having come from below. Many ore deposits in rocks other than iron formation contain what is accepted generally as hypothermal iron; there seems to be no good reason to assume that iron from hydrothermal solutions cannot be deposited in iron formation as well. That much of the iron was deposited as goethite suggests to most geologists that the iron must have been deposited by near-surface solutions of meteoric origin. There are, however, deposits in which goethite may have been deposited from hydrothermal solutions. Steep Rock Lake may be such a deposit, but certain recent workers even in that district believe the goethite to have been formed by near-surface processes. The question of the origin of goethite needs further study.

It is, therefore, suggested here that the high-grade ores of the Schefferville portion of the Labrador Trough received much, if not all, of their iron content above that obtained from the original iron formation from hydrothermal solutions. These solutions are suggested to have been derived from igneous activity accompanying the Elsonian (or Labrador) orogeny of some 1350 m.y. ago. Thus, the ores of this portion of the trough are here classified as sedimentary-A1a (the primary iron formation) and hydrothermal (for the enriched ores of the central part of the trough). Because of the possibility that some of the leaching and some of the enrichment, at least, were produced by circulating meteoric waters, the category ground water-B2 is added parenthetically.

Northwest Territories

GREAT BEAR LAKE

Late Precambrian *Uranium, Silver* *Mesothermal to Leptothermal*

Campbell, D. D., 1957, Port Radium Mine, in *Structural geology of Canadian ore deposits*, v. 2: Canadian Inst. Min. and Met., Montreal, p. 177-189

Collins, C. B., and others, 1954, Isotopic constitution of radiogenic leads and the measurement of geologic time: Geol. Soc. Amer. Bull., v. 65, p. 1-22 (particularly p. 5)

Donald, K. G., 1956, Pitchblende at Port Radium: Canadian Min. Jour., v. 77, no. 6, p. 77-79

Heinrich, E. W., 1958, Great Bear Lake area, Northwest Territories, in *Mineralogy and geology of radioactive raw materials*: McGraw-Hill, N.Y., p. 275-279

James, W. F., and others, 1950, Canadian deposits of uranium and thorium: A.I.M.E. Tr., v. 187, p. 239-255 (particularly p. 244-248) (in Min. Eng., v. 187, no. 2)

Kidd, D. F., 1931, Great Bear Lake--Coppermine River area, Mackenzie district, N.W.T.: Geol. Surv. Canada Summ. Rept., pt. C, p. 47-69

—— 1932, A pitchblende-silver deposit, Great Bear Lake, Canada: Econ. Geol., v. 27, p. 145-150

—— 1942, The silver-pitchblende deposit near Great Bear Lake, N.W.T., in Newhouse, W. H., Editor, *Ore deposits as related to structural features*: Princeton Univ. Press, p. 238-239

Kidd, D. F., and Haycock, M. H., 1935, Mineragraphy of the ores of Great Bear Lake: Geol. Soc. Amer. Bull., v. 46, p. 879-959

Knight, C. W., 1930, Pitchblende at Great Bear Lake: Canadian Min. Jour., v. 51, no. 41, p. 962-965, 976

Lang, A. H., 1952, Great Bear Lake region, in *Canadian deposits of uranium and thorium*: Geol. Surv. Canada Econ. Geol. Ser. no. 16, p. 46-57

Lang, A. H., and others, 1962, Eldorado Mine, in *Canadian deposits of uranium and thorium*: Geol. Surv. Canada Econ. Geol. Ser. no. 16, 2d ed., p. 186-193

Leipziger, F. D., and Croft, W. J., 1964, Geologic age determination by direct lead isotope analysis: Geochimica et Cosmochimica Acta, v. 28, p. 268-269

Murphy, R., 1946, Geology and mineralogy at Eldorado mine: Canadian Inst. Min. and Met. Tr., v. 49 (Bull. no. 413), p. 426-435

—— 1948, Eldorado mine, in *Structural geology of Canadian ore deposits*: Canadian Inst. Min. and Met., Montreal, p. 259-268

Palache, C., and Berman, H., 1933, Oxidation products of pitchblende from Great Bear Lake: Amer. Mineral., v. 18, p. 20-24

Palache, C., and others, 1944, Uraninite, in *The system of mineralogy of James Dwight Dana and Edward Salisbury Dana*, 7th ed.: Wiley, N.Y., p. 611-620

Reid, J. A., 1932, The minerals of Great Bear Lake: Canadian Min. Jour., v. 53, no. 2, p. 61-66

Ridland, G. C., 1945, Use of the Geiger-Müller counter in the search for pitchblende-bearing veins at Great Bear Lake, Canada: A.I.M.E. Tr., v. 164, p. 117-124

Spence, H. S., 1932, Character of the pitchblende ore from Great Bear Lake, N.W.T.: Canadian Min. Jour., v. 53, no. 11, p. 483-487

Thomson, J. Ellis, 1932, Mineralogy of the Eldorado mine, Great Bear Lake,

N.W.T.: Univ. Toronto Studies, Geol. Ser. no. 32, p. 43-50

—— 1934, The mineralogy of the silver-uraninite deposits of Great Bear Lake, N.W.T.: Univ. Toronto Studies, Geol. Ser. no. 36, p. 25-31

Notes

The uranium-silver ore deposits of Great Bear Lake are located at Port Radium on Labine Point on the southeastern shore of McTavish arm; the mine was known as the Eldorado, and the last ore was mined from it in 1960.

All of the several types of sedimentary and volcanic rocks in the area, now highly altered, that contain ore veins are assigned to the Echo Bay group; these rocks were intruded by several kinds of intrusives. The age of the Echo Bay rocks is uncertain, but an intrusive (hypabyssal) porphyry cutting similar rocks in an area northeast of the mine gives an age (radiometrically determined) of 1765 m.y. or late middle Precambrian. Thus, the Echo Bay rocks almost certainly are middle Precambrian or older. Since the ores are epigenetic to the host rocks, their dating simply puts an upper limit on the age of the ores. The ores, however, probably were produced by a phase of one or the other of the igneous activities from which resulted (1) the granite that is exposed along the shores of Labine Point and apophyses of which are found in the more westwardly on the mine workings and (2) the diabases. The granite and the diabases were introduced into the area at some considerable time after the emplacement of various hypabyssal intrusives. The ores are even younger than the granites, since ore veins cut the granites. Of the post-granite diabases in the area, the earlier is cut by pitchblende-bearing veins, but the later cuts through them. Although it is possible that the pitchblende-silver ores of the Eldorado mine came from the same source as these diabases, I think it more probable the ore fluids came from the same source as the granites. Radioactive dates have not been published (and may not have been determined) for the granite or the diabases, but Leipziger and Croft (1964), who made direct lead isotope analyses of lead from the uranium minerals of the mine, give an age of 1170 m.y. for the lead. Earlier work by Collins and others (1954) gave an age of about 1400 m.y. for the ores with one sample giving 1130 m.y. Although these two lead-age dates are far enough apart to cast some doubts on the validity of one or both of them, it would appear most reasonable to classify the Eldorado deposits as late Precambrian.

Campbell (1957) gives five general stages for the uranium-silver mineralization: (1) hematite and quartz; (2) pitchblende, quartz, and hematite; (3) quartz and cobalt-nickel arsenides and sulfides; (4) copper sulfides, chlorite and dolomite; and (5) carbonates, silver minerals, bismuth, and chalcopyrite. In addition to the minerals already mentioned, Kidd and Haycock (1935) recognized arsenopyrite, smalltite-chloanthite, safflorite-rammelsbergite, skutterudite and Ni-skutterudite, cobaltite, gersdorffite, glaucodot, Ni-löllingite, niccolite, polydymite, molybdenite, bornite, cubanite, tetrahedrite, chalcocite, stromeyerite, argentite, jalpaite (Cu-argentite), native silver, and hessite. It would appear reasonable to assume that the quartz-hematite-pitchblende mineralization of stage (2) took place under mesothermal conditions because pitchblende probably develops under lower intensity conditions than uraninite (the Great Bear Lake uranium was all in pitchblende). The silver minerals and native silver of stage (5), however, probably formed under less intense conditions; argentite, stromeyerite, jalpaite, tetrahedrite, and native silver are diagnostic of leptothermal conditions. The cobalt-nickel arsenides of stage (3) probably were formed in the mesothermal range. The deposits are, therefore, classified as mesothermal to leptothermal.

PINE POINT

Late Paleozoic *Lead, Zinc* *Telethermal*

Baadsgaard, H., and others, 1965, Isotopic data from the Cordillera and Liard basin in relation to the genesis of the Pine Point lead-zinc deposits (abs.): Royal Soc. Canada, Tr., 4th ser., v. 3, Appendix, p. 16

Baragar, W.R.A., 1964, Geology of the lead-zinc deposits (Pine Point): North (bimonthly publication of the northern Administration Branch, Dept. Northern Affairs and National Resources, Ottawa), v. 11, no. 3, p. 18-22

Beales, F. W., and Jackson, S. A., 1966, Precipitation of lead-zinc ores in carbonate reservoirs as illustrated by Pine Point ore field, Canada: Canadian Inst. Min. and Met. Tr., Sec. B, Applied Earth Sciences, v. 75, p. B278-B285; disc., 1966, v. 75, p. B300-B305; 1967, v. 76, p. B130-B136, p. B175-B177

—— 1968, Pine Point--a stratigraphic approach: Canadian Inst. Min. and Met. Bull., v. 61, no. 675, p. 867-878

Billings, G. K., and others, 1969, Relation of zinc-rich formation waters, northern Alberta, to the Pine Point ore deposit: Econ. Geol., v. 64, p. 385-391

Campbell, N., 1950, The Middle Devonian in the Pine Point area, N.W.T.: Geol. Assoc. Canada Pr., v. 3, p. 87-96

—— 1957, Stratigraphy and structure of Pine Point area, N.W.T., in *Structural geology of Canadian ore deposits*, v. 2: Canadian Inst. Min. and Met., Montreal, p. 161-174

—— 1966, The lead-zinc deposits of Pine Point: Canadian Inst. Min. and Met. Tr., v. 69 (Bull. no. 652), p. 288-295

—— 1967, Tectonics, reefs, and stratiform lead-zinc deposit of the Pine Point area, Canada, in Brown, J. S., Editor, *Genesis of stratiform lead-zinc-barite-fluorite deposits--a symposium*: Econ. Geol. Mono. 3, p. 59-70

Cumming, G. L., and Robertson, D. K., 1969, Isotopic composition of lead from the Pine Point deposit: Econ. Geol., v. 64, p. 731-732

Folinsbee, R. E., and others, 1966, Sulphur isotopes and the Pine Point lead-zinc deposits, N.W.T., Canada (abs.): Econ. Geol., v. 61, p. 1307-1308

Fritz, P., 1969, The oxygen and carbon isotopic composition of carbonates from the Pine Point lead-zinc ore deposits: Econ. Geol., v. 64, p. 733-742

Jackson, S. A., and Beales, F. W., 1967, An aspect of sedimentary basin evolution: the concentration of Mississippi Valley-type ores during late stages of diagenesis: Canadian Petrol. Geol. Bull., v. 15, p. 383-433

Jackson, S. A., and Folinsbee, R. E., 1969, The Pine Point lead-zinc deposits, N.W.T., Canada introduction and paleocology of the Presqu'ile Reef: Econ. Geol., v. 64, p. 711-717

Norris, A. W., 1965, Stratigraphy of Middle Devonian and older Paleozoic rocks of the Great Slave Lake region, Northwest Territories: Geol. Surv. Canada Mem. 322, 180 p.

Roedder, E., 1968, Temperature, salinity, and origin of the ore-forming fluids at Pine Point, Northwest Territories, Canada, from fluid inclusion studies: Econ. Geol., v. 63, p. 439-450

Roedder, E., and Dwornik, J., 1968, Sphalerite color banding: lack of correlation with iron content, Pine Point, Northwest Territories, Canada: Amer. Mineral., v. 53, p. 1523-1529

Sasaki, A., and Krouse, H. R., 1969, Sulfur isotopes and the Pine Point lead-zinc mineralization: Econ. Geol., v. 64, p. 718-730

Seigel, H. O., and others, 1968, Discovery case history of the Pyramid ore bodies, Pine Point, Northwest Territories, Canada: Geophys., v. 33, p. 645-656

Notes

The Pine Point property is on the south shore of Great Slave Lake, 55 miles due south of Yellowknife on the north shore.

The Pine Point ores are found in the Middle Devonian Presqu'ile formation, a porous and cavernous, coarsely recrystallized dolomite that appears to have been developed in reef-core material of a great barrier reef that may have been over 200 miles long and was from one to many miles in width. Campbell (1966) suggests that the reef was initiated by differential compaction that produced a favorable "high" above fault scarps in the underlying Precambrian basement; its continued growth, however, was related to a hingelike movement along a nearly horizontal axis that caused the floor of the basin on the northwest side to drop more rapidly than that of the platform or shelf to the southeast. This differential movement apparently did not end in the Devonian, for the reef, probably more rigid and competent than the surrounding sediments, was shattered at some time after its lithification. Probably at the same time, the reef material was intensely recrystallized with but little chemical change. The ore minerals were deposited partly as fillings of cavities and partly as replacements of dolomite. Roedder (1968) says that the lead, which is not the anomalous "J-type" of some Mississippi Valley-type deposits, could have been derived from the mantle about 250 m.y. ago. Although he points out sources from which lead of such isotopic composition could have been obtained, and although he does not consider the possibility of the source of the lead having been deep within the crust rather than from the mantle, this 250 m.y. age is the only one that can establish any definite date for the deposits. It has been suggested that the ores may have formed over a period lasting from 360 m.y. to 100 m.y. ago, but this seems extreme. Until further evidence is forthcoming, it seems best to categorize the deposits as Permian or late Paleozoic.

Campbell (1966) points out that the deposition of much of the sphalerite and galena took place in cavities formed during the brecciation of the already recrystallized dolomite and that some of it was controlled by fractures or clay seams transecting the bedding, and he concludes from this that the ore definitely was not laid down with the sediments. Where the ores came from, however, is another matter. Roedder's work (1968) indicates that the ores were deposited at temperatures between 95° and 100°C from solutions of unusually high salinity; this, he thinks, argues against the ore fluids having been low-temperature fluids of conventional magmatic-hydrothermal origin and against the deposits having been telethermal in the sense the word is used here. He says that inclusions containing definitely magmatic fluids that are strongly saline do occur but that they have been found only in high-temperature deposits. Even this objection can be obviated if the solutions were not diluted or were diluted only by strong brines. He also advances the argument of Beales and Jackson (1966) that, since deposition in the Presqu'ile reef material took place only after the solutions had traversed hundreds of meters of carbonate rocks beneath the site of deposition, a precipitation mechanism must be invoked that does not involve simple reaction with wall rock and change in temperature, pressure, Eh, and/or pH. This seems an unreasonable position to take since higher-temperature galena-sphalerite deposits throughout the world, which were almost certainly deposited from solutions of magmatic origin, lie above thick sequences of unmineralized carbonate rock. Positive evidence that the path of the ore-bearing solutions was controlled by the nearly dormant Precambrian faults is not available (Norris, 1965), but it is certainly indicated by Folinsbee's work (1966). In a situation still far from clear, the most reasonable classification of these deposits appears to be telethermal.

YELLOWKNIFE

Early Precambrian | *Gold* | *Hypothermal-1 to Leptothermal*

Ames, R. L., 1962, The origin of the gold-quartz deposits, Yellowknife, N.W.T.: Econ. Geol., v. 57, p. 1137-1140

Baragar, W.R.A., 1966, Geochemistry of the Yellowknife volcanic rocks: Canadian Jour. Earth Sci., v. 3, p. 9-30

Bateman, J. D., 1951, Application of geology to mining at Giant Yellowknife:

A.I.M.E. Tr., v. 190, p. 1051-1060 (in Min. Eng., v. 3, no. 12)

—— 1952, Some geological features at Giant Yellowknife mine: Geol. Assoc. Canada Pr., v. 5, p. 95-107

Boyle, R. W., 1954, The shear zone systems of the Yellowknife greenstone belt: Canadian Min. Jour., v. 75, nos. 6, 7, p. 59-64, 69-75

—— 1954, Structural localization of gold ore bodies of the Yellowknife greenstone belt: Canadian Min. Jour., v. 75, no. 12, p. 71-77

—— 1954, A decrepitation study of quartz from the Campbell and Negus-Rycon shear zone systems, Yellowknife, Northwest Territories: Geol. Surv. Canada Bull. 30, 20 p.

—— 1955, The geochemistry and origin of the gold-bearing quartz veins a lenses of the Yellowknife greenstone belt: Econ. Geol., v. 50, p. 51-66

—— 1959, The geochemistry, origin, and role of carbon dioxide, water, sulfur, and boron in the Yellowknife gold deposits, Northwest Territories, Canada: Econ. Geol., v. 54, p. 1506-1524

—— 1961, The geology, geochemistry, and origin of the gold deposits of the Yellowknife district: Geol. Surv. Canada Mem. 310, 193 p.

Boyle, R. W., and others, 1963, The origin of the gold-quartz deposits, Yellowknife, N.W.T.: Econ. Geol., v. 58, p. 804-807

Brown, C.E.G., and Dadson, A. S., 1953, Geology of the Giant Yellowknife mine: Canadian Inst. Min. and Met. Tr., v. 56 (Bull. no. 491) p. 59-76

Brown, C.E.G., and others, 1959, On the ore-bearing structures of the Giant Yellowknife gold mine: Canadian Inst. Min. and Met. Tr., v. 62 (Bull. no. 564), p. 107-116

Brown, I. C., 1955, Late faults in the Yellowknife area: Geol. Assoc. Canada Pr., v. 7, pt. 1, p. 123-138

Burwash, R. A., and Baadsgaard, H., 1962, Yellowknife-Nonacho age and structural relations, in Stevenson, J. S., Editor, *The tectonics of the Canadian Shield*: Roy. Soc. Canada Spec. Pubs. no. 4, p. 22-29

Campbell, N., 1947, Regional structural features of the Yellowknife area: Econ. Geol., v. 42, p. 687-698

—— 1948, West Bay fault, in *Structural geology of Canadian ore deposits*: Canadian Inst. Min. and Met., Montreal, p. 244-259

Coleman, L. C., 1953, Mineralogy of the Yellowknife Bay area, N.W.T.: Amer. Mineral., v. 38, p. 506-527

—— 1957, Mineralogy of the Giant Yellowknife gold mine, Yellowknife, N.W.T.: Econ. Geol., v. 52, p. 400-425

Dadson, A. S., 1949, The Giant Yellowknife: Western Miner, v. 22, no. 10, p. 82-90

Dadson, A. S., and Bateman, J. D., 1948, Giant Yellowknife mine, in *Structural geology of Canadian ore deposits*: Canadian Inst. Min. and Met., Montreal, p. 273-283

Henderson, J. F., and Brown, I. C., 1948, Yellowknife, Northwest Territories: Geol. Surv. Canada Paper 48-17, 6 p. (mimeo.)

—— 1952, The Yellowknife greenstone belt, Northwest Territories: Geol. Surv. Canada Paper 52-28, 41 p. (mimeo.)

—— 1966, Geology and structure of the Yellowknife greenstone belt, district of Mackenzie: Geol. Surv. Canada Bull. 141, 87 p.

Henderson, J. F., and Jolliffe, A. W., 1939, Relation of gold deposits to structure, Yellowknife and Gordon Lake areas, Northwest Territories: Canadian Inst. Min. and Met. Tr., v. 42 (Bull. no. 326), p. 314-336

Jolliffe, A. W., 1938, Yellowknife Bay-Prosperous Lake area: Geol. Surv. Canada Paper 38-21, 41 p. (mimeo.)

—— 1942, Yellowknife Bay, District of Mackenzie, Northwest Territories: Geol. Surv. Canada Map 709A (with descriptive notes), 1:63,360

—— 1946, Prosperous Lake, District of Mackenzie, Northwest Territories: Geol. Surv. Canada Map 868A (with descriptive notes), 1:63,360

Kretz, R., 1969, Study of pegmatite bodies and enclosing rocks, Yellowknife-Beaulieu Region, District of Mackenzie: Geol. Surv. Canada Bull. 159, 112 p.

Lord, C. S., 1951, Con and Rycon mines, in *Mineral industry of District of Mackenzie, Northwest Territories*: Geol. Surv. Canada Mem. 261, p. 155-171

—— 1951, Giant Yellowknife Gold Mines, Limited, in *Mineral industry of District of Mackenzie, Northwest Territories*: Geol. Surv. Canada Mem. 261, p. 155-171

McConnell, G. W., 1964, Yellowknife gold-quartz deposits: Econ. Geol., v. 59, p. 328-330; disc., p. 1176-1177

Ridland, G. C., 1941, Mineralogy of the Negus and Con mines, Yellowknife, Northwest Territories, Canada: Econ. Geol., v. 36, p. 45-70

Staff, Consolidated Mining and Smelting Company, Ltd., 1954, Geology (of the Con Mine), in *"Cominco"--a Canadian enterprise*: Canadian Min. Jour., v. 75, no. 5, p. 188-191

Wanless, R. K., 1960, Sulfur isotope investigation of the gold-quartz deposits of the Yellowknife district: Econ. Geol., v. 55, p. 1591-1621

White, C. E., and others, 1949, The Con-Rycon mine, Yellowknife, N.W.T.: Canadian Inst. Min. and Met. Tr., v. 52 (Bull. no. 446), p. 133-147

Notes

Of the two principal centers of mineralization in the Yellowknife area, the more southwestwardly is located close to the northeast shore of Great Slave Lake between 62°25' and 62°30' north latitude and centering around the town of Yellowknife, while the second is situated about 7 miles northeast of Yellowknife.

The bedded rocks of the Yellowknife group consist of (1) greenstones that were once largely mafic flows, tuffs, and agglomerates (division A) and (2) highly metamorphosed mechanical sediments in considerable variety interbedded with various igneous rock (division B). Division B probably lies conformably on division A. The bedded rocks are cut by several different igneous rock types, some of which cut only division A beds and some of which cut both divisions. The metamorphic zoning in the district has been considerably distorted from its original pattern by the two main periods of faulting and fracturing. Of the igneous rocks in the area, only the diabase dikes and sills and a composite group of peridotite-gabbro intrusive sheets are younger than the metamorphism (or most of it) and are altered only where they are cut by the late faults. There are three granitic bodies in the district that are younger than all Yellowknife group rocks but have been somewhat metamorphosed and faulted. The ores are found in or adjacent to fractures and shear zones in the Yellowknife group and in postdiabase fractures that cut all rock types and shear zones. The shear zones that transect the flows contain the major gold-bearing quartz bodies of the area. This shearing was developed, according to Boyle (1963), after all granite had been intruded; diabase dikes cut through the shear zones and through ore zones, dating the diabase as both postshearing and post-ore. Apparently, no major time lapse occurred between the various structural and igneous events in the district. Boyle believes that the gold and other materials now in, and near, the quartz lenses in the shear zones were metamorphically mobilized and migrated by diffusion over considerable distances into dilatant zones in the shear zones. Others have suggested that the ore was formed by hydrothermal solutions of magmatic origin, with the parent magma of the granites being considered as the most likely source of the ore fluids. If the granite magma chamber produced the mineralizing solutions, radioactive age determinations on the granites would help date the ores. Such work on the granite northeast of the mineralized area (the Prosperous Lake granite) gives an age of 2540 m.y., while similar work on a granodiorite that also invades rocks of the Yellowknife group gives an age of 2615 m.y. If the ores were produced by local mobilization of mineral material during the metamorphism, they probably would be slightly older than if they were genetically connected with the granites; in either case, however, they would be late early Precambrian and are so classified here.

In the greenstones, gold in economic quantities is concentrated in the shear zones cutting the epidote amphibolite facies and is present in small amounts in shear zones cutting the amphibolite facies and some of the granite. A considerable fraction of the gold occurs as microscopic blebs in the first generation of pyrite and arsenopyrite, these two sulfides being located mainly in the alteration halos surrounding the quartz lenses. Gold of the second stage is found as blebs, plates, and nuggets in fractures, crushed zones, and vugs in the quartz lenses and in cracks in the pyrite and arsenopyrite; gold of this stage also coats these two sulfides or is intermixed with them but still appears to be younger than they are. This gold is probably earlier than the next generation of sulfides (sphalerite, pyrrhotite, and chalcopyrite). These first two generations of gold in ore bodies in the greenstones appear to have been formed then under the same conditions as the typically hypothermal minerals there present, such as pyrrhotite, arsenopyrite, and sphalerite containing exsolved chalcopyrite. The third generation of gold also is associated with third generation sulfides and sulfosalts (principally jamesonite, bournonite, tetrahedrite, and berthierite), but with these minerals it is in rounded or subrounded particles that are often enclosed by one or more grains of the later minerals. No gold associated with the third generation of sulfides is definitely later than they are. There is a fourth generation of gold in calcite and ankerite that fills vugs and crushed zones in quartz-carbonate stringers that cut early quartz lenses and their mineralized halos in the Campbell fault system; this gold is accompanied by freibergite, bournonite, and chalcopyrite. A fifth generation of gold occurs in late fractures in the Giant fault system and in one known place in the Campbell system; present with it are pyrite and scalenohedral calcite. The gold of the fourth and fifth generations is of little economic importance.

In the sedimentary rocks, gold occurs in pyrite and arsenopyrite in the alteration zones adjacent to the quartz lenses and as small plates and irregular nuggests in vugs, small fractures, and crushed zones in the quartz. Pyrite, galena, sphalerite, and chalcopyrite are also present with this gold. There is a second, economically unimportant, generation of gold in carbonates in fractures that cut the mineralized quartz lenses. The recognition of three generations of gold in economically important amounts in the greenstones (and in the Western granite) and of only one of economic value in the metasediments suggests either that more detailed work has been done on the deposits in greenstone than on those in the metasediments or that the metasediment environment had a less selective effect on the ore fluids than did the greenstones. Gold is obtained also from aurostibite ($AuSb_2$); this mineral forms halos and coatings around gold in quartz lenses that also contain concentrations of antimony minerals or is found in narrow seams in gold. The aurostibite is probably later, but not much later, than the gold with which it is associated and it formed only in antimony-bearing portions of the ore bodies.

The two early generations of gold in the greenstones, those formed with or before the hypothermal minerals, probably should be classed as hypothermal-1. The gold that is closely associated with the sulfosalts (although it appears to be slightly earlier than they) probably was developed in the leptothermal range and should be classed as leptothermal. It is not clear as to whether any of the gold formed in the mesothermal range or not; possibly some of the gold that is found in otherwise barren quartz may have been formed under mesothermal conditions. This possibility, though far from a certainty, indicates that the greenstone gold ores are better classed as hypothermal-1 to leptothermal rather than hypothermal-1, leptothermal. The gold in the sedimentary rocks was in large part formed in the hypothermal range; the second (uneconomic) generation of gold in the mineralized structures in these rocks probably was formed under less intense conditions. Obviously, if Boyle's theory as to the origin of the gold ores is correct, the classification must be changed from that given here.

Nova Scotia

WALTON

Early Mesozoic	*Barite, Lead, Zinc, Copper, Silver*	*Mesothermal to Leptothermal*

Bell, W. A., 1960, Mississippian Horton group of type Windsor-Horton district, Nova Scotia: Geol. Surv. Canada Mem. 314, 59 p. (general stratigraphy)

Boyle, R. W., 1963, Geology of the barite, gypsum, manganese, and lead-zinc-silver deposits of the Walton-Cheverie area, Nova Scotia: Geol. Surv. Canada Paper 62-25, 36 p.

Boyle, R. W., and Jambor, J. L., 1966, Mineralogy, geochemistry, and origin of the Magnet Cove barite-sulphide deposit, Walton, N.S.: Canadian Inst. Min. and Met. Tr., v. 69 (Bull. no. 654), p. 394-413

Cameron, A. E., 1941, Barytes deposit at Pembroke, Hants County, Nova Scotia: Nova Scotia Inst. Sci. Pr., v. 20, pt. 3, p. 57-63

Campbell, C. O., 1942, Barytes at Pembroke, Hants County, Nova Scotia: Canadian Inst. Min. and Met. Tr., v. 45 (Bull. no. 362), p. 299-310

Crosby, D. G., Jr., 1952, Preliminary map, Wolfville (east half), Hants and Kings Counties: Geol. Surv. Canada Paper 52-18 (with descriptive notes), 1:31,680

—— 1962, Wolfville map-area: Geol. Surv. Canada Mem. 325, 67 p.

Fyson, W. K., 1964, Folds in the carboniferous rocks near Walton, Nova Scotia: Amer. Jour. Sci., v. 262, p. 513-522

Jewett, G. A., 1957, The Walton, N.S., barite deposit, in *The geology of Canadian industrial mineral deposits*: Canadian Inst. Min. and Met., Montreal, p. 54-58

Stevenson, I. M., 1959, Shubenacadie and Kenntecook map-areas, Colchester, Hants and Halifax Counties, Nova Scotia: Geol. Surv. Canada Mem. 302, 88 p. (southeast and south of the Walton area)

Tenny, R. E., 1951, The Walton barite deposit: Nova Scotia Dept. Mines Ann. Rept. 1950, pt. 2, p. 127-143

Weeks, L. J., 1948, Londonderry and Bass River map-areas, Colchester and Hants Counties, Nova Scotia: Geol. Surv. Canada Mem. 245, 86 p. (general)

Wright, W. J., 1931, Reports on Cheverie, Windsor and Shubenacadie Basins: Nova Scotia Dept. Mines Ann. Rept. 1930, p. 115-142

Notes

The Walton deposit is located about 2.5 miles southwest of the town of Walton which, in turn, is situated on the south shore of the Minas Basin, an arm of the Bay of Fundy.

Although the structures in the Mississippian rocks that contain the ores are not well exposed, these rocks appear to be highly folded and faulted; the considerably faulted Triassic rocks that lie unconformably on the Mississippian beds are tilted only slightly to the northwest. The only igneous rocks in the general area around the ore body are gabbroic sills that outcrop some 7 miles west of the ore body; Boyle (1966) believes that these gabbros were feeders of flows now long removed by erosion. These gabbros probably are equivalent to the North Mountain basalts that definitely are Triassic in age; the gabbros are known to cut Mississippian Horton Bluff rocks. There is no geologic evidence that directly connects the formation of the ore body with these Triassic rocks, but large occurrences of barite in the United States from Missouri to north-central New York are thought to be connected genetically with Triassic diabases. Lacking more definite information as to the age of the Walton deposit, it is here considered to have formed late in the Triassic and to be, therefore, early Mesozoic.

The controls over the emplacement of the Walton ore are structural. On the south, there is a footwall fault that runs about east-west and dips 70°N, and, on the north, a hanging-wall fault exists that strikes northwest-southeast and has about the same dip. The markedly wedge-shaped block of ground thus formed has its apex to the east, and the intersection of the two faults rakes 35°E, the ore following the rake. With depth, the pipelike ore body (pear-shaped in plan, small end east) becomes elongate, with its long dimension running northwest-southeast. Down to about 250 feet below the surface, the ore consisted of massive barite; at this depth, it began to grade downward into a barite-sulfide zone in which barite amounted to about 25 percent of the ore. In addition to barite, the principal nonmetallic minerals were manganiferous siderite and dolomite, mainly in the footwall of the ore body. Most of the pyrite is found in the footwall portion of the ore body in association with chalcopyrite, but the pyrite in the main sulfide mass generally is with galena and sphalerite. The large fraction of each of the sulfides, other than pyrite, is in the sulfide ore shoot. The barite was emplaced and then fractured, and the bulk of the first generation of sulfides was brought into the massive barite along these fractures. Some sulfides in the footwall zone, however, probably were deposited contemporaneously with the barite. Pyrite, galena, and sphalerite appear to have formed in the ore shoot; tennantite and chalcopyrite are intimately associated, and tennantite fills fractures in the earliest sulfides, suggesting that the chalcopyrite is also a later sulfide. Finally, proustite and argentite were formed, with proustite being appreciably present throughout the ore body and adding considerably to the silver content of the ore. The early sulfides (pyrite, galena, sphalerite), barite, siderite, and dolomite probably formed under mesothermal conditions. Tennantite normally appears to form on the mesothermal side of the boundary between mesothermal and leptothermal conditions, and its intimate association with chalcopyrite at Walton suggests that it may have formed there under low-intensity mesothermal conditions. The proustite and argentite, however, almost certainly formed under leptothermal conditions, and the proustite is of sufficient economic importance to warrant the inclusion of leptothermal in the classification of the deposit; therefore, the deposits are here classified as mesothermal to leptothermal.

Ontario

ADAMS MINE

Early Precambrian | *Iron as Magnetite* | *Sedimentary-A3, Metamorphic-C*

Abraham, E. M., 1950, Geology of McElroy and part of Boston Townships: Ont. Dept. Mines 59th Ann. Rept., v. 59, pt. 6, 66 p.

Dubuc, F., 1966, Geology of the Adams mine: Canadian Inst. Min. and Met. Tr., v. 69 (Bull. no. 646), p. 67-72

Lawton, K. D., 1957, Geology of Boston Township and part of Pecaud Township: Ont. Dept. Mines, 66th Ann. Rept., v. 66, pt. 5, 55 p.

Miller, W. G., 1905, Boston Township Iron Range: Ont. Bur. Mines, 14th Ann. Rept., v. 14, pt. 1, p. 261-268

Notes

The Adams mine is 7 miles southeast of Kirkland Lake. The basin containing the iron formations is not more than 6 miles in total length and is up to 0.75 miles wide.

Dubuc (1966) considers the various horizons of the area of the Adams mine to have been formed by sedimentary processes and believes the general sequence of sedimentary events to have been (1) deposition of lean iron formation, (2) precipitation of iron-formation proper, and (3) accumulation of cherty quartzite. These beds contain intercalated volcanic flows that appear to have invaded only portions of the basin at any one time. In the eastern portion of the range, at least three repetitions of the iron-formation sequence have been recognized, seemingly having been separated by volcanic flows almost throughout that portion of the basin--only locally does one iron horizon merge upward into another. The three iron formations are not present throughout the basin, suggesting that sedimentation differed considerably in character from one part of the basin to another, small though the basin was. Although the iron formations average only 100 to 150 feet in thickness, some enlargements were caused by folding and brecciation; these enlarged areas constitute the ore bodies, 8 of which have been recognized and 7 of which are in the eastern portion of the property. The largest of these bodies is 3000 feet long and up to 600 feet wide. The grade of the ore averages 22 percent magnetic iron. The ore bodies have been cut by numerous dikes of syenite, lamprophyre, and diabase; one ore body contains as much as 10 percent of dike material. The diabase is the youngest of the three dike rocks. Intrusions of similar syenite and diorite (some now metadiorite) also are found in the iron range. The rocks of the range have been dated as early Precambrian, and the lack of later additions of iron to the formations means that the ores are to be assigned that age exclusively.

In addition to the bedded chert and magnetite of the iron formation, hematite commonly occurs in the magnetite as small irregular lenses and as thin individual beds. Chert layers in the vicinity of this hematite always are reddish. In the higher-grade iron formation, a reddish garnet occurs as isolated masses of anhedral grains; the garnet may locally be associated with epidote. Near or within magnetite layers, tremolite and actinolite needles normally are present; in a few places, a bluish amphibole is found instead of the tremolite and actinolite. Chlorite occurs in massive beds, with or without disseminated magnetite, and is quite abundant; pyrite always is found in the chlorite. From the present character of the iron formation, it is obvious that the original rocks of the range have undergone considerable metamorphism since they were first lithified; the grain size of the iron formation, however, is much finer than that in the Wabush Lake area of the Labrador Trough with the Adams grains being less than 1 mm in diameter. Although Dubuc (1966) says that the iron formation is of sedimentary origin, he does not say from whence he believes the iron to have been derived. The determination of the source of the iron is made more difficult by the metamorphism the area has undergone, but the abundance of volcanic flows and lesser amounts of tuff strongly suggests that these local concentrations of iron derived from volcanic emanations (or hydrothermal fluids) that reached the sea floor in the area now occupied by the iron formations. On this basis, the ore bodies are classified as sedimentary-3A, and metamorphic-C is added in recognition of the considerable metamorphism they have undergone.

ALEXO MINE

Middle Precambrian — *Nickel, Copper, Gold* — *Magmatic-2b, Magmatic-2a*

Baker, M. B., 1917, The Alexo nickel mine: Royal Ont. Nickel Comm. Rept., p. 228-232

Coleman, A. P., 1910, The Alexo nickel deposit: Econ. Geol., v. 5, p. 373-376

—— 1913, The Alexo mine, in *The nickel industry*: Canadian Dept. Mines Rept. 170, p. 112-113

Naldrett, A. J., 1966, The role of sulphurization in the genesis of iron-nickel sulfide deposits of the Porcupine district, Ontario: Canadian Inst. Min.

and Met. Tr., v. 69 (Bull. no. 648), p. 147-155

—— 1967, The central portion of the Fe-Ni-S system and its bearing on pentlandite exsolution in iron-nickel sulfide ores: Econ. Geol., v. 62, p. 826-847

Uglow, W. L., 1911, The Alexo nickel deposit: Ont. Bur. Mines 20th Ann. Rept., v. 20, pt. 2, p. 34-39

Notes

The Alexo mine is 25 miles east-northeast of Timmins.

The ores of the Alexo mine and similar occurrences in the general Porcupine area are so closely associated with generally highly altered ultramafic rocks as almost certainly to be essentially the same age. With the exception of quartz-diabase dikes of probable Matachewan age and olivine-diabase dikes intruded in Keweenawan time, the ultramafic rocks are the youngest in the area, being definitely later than the Algoman intrusives and the earlier sediments through which the Algoman silicic rocks were forced. It appears probable that these ultramafics are of late middle Precambrian age, and, although the evidence is not compelling, the rocks and their associated nickel ore bodies are here classified as middle Precambrian.

Naldrett (1966) summarizes the geology of the Alexo mine by pointing out that the ore lies in a sheared zone at the base of a serpentinized peridotite lens, close to the contact of the peridotite with the underlying pillow lavas; the contact parallels the southwest strike of the lavas. At one place, the shear zone is marked by a pronounced flexure that Naldrett believes has been important in localizing the ore. The serpentinized peridotite originally contained 75 to 85 percent of olivine--olivine decreases with respect to pyroxene in the neighborhood of the contact. He points out that the underlying volcanic rocks are rich in pyrite and pyrrhotite but otherwise appear to be normal pillow lavas and lava flows that contain thin bands of intercalated breccia. The Alexo ore body is composed of a sheet of massive sulfides, ranging between a few inches and 5 feet in thickness, that was emplaced in the contact-related shear zone after that zone had been formed. The main minerals of the ore are pyrrhotite and pentlandite, accompanied by minor magnetite and a little chalcopyrite. In the immediately adjacent peridotite, a 5- to 15-foot-wide zone contains disseminated pyrrhotite and pentlandite in which the sulfides normally appear to be interstitial to the now-serpentinized olivines; where the sulfides are abundant (10 to 40 percent of the volume of the rock), the serpentinized olivines are partly to completely surrounded by sulfides. Locally, veinlets of serpentine and secondary magnetite cut across the sulfides giving the impression (to me but not to Naldrett) that the serpentinization took place after the sulfides had been emplaced. In the 5 feet nearest the massive sulfides, the larger masses of sulfides surround former grains of olivine, but the contacts between sulfides and silicates are serrated, and fine sulfide stringers and irregular blebs were developed in the silicates. It is difficult for me to believe that the relationship of disseminated ore to the massive ore body is the result of one process (as Naldrett would have it); instead, it seems to me that the disseminated sulfides, no matter how abundant they are, were produced during the final solidification of the peridotite magma. This magma well may have been intruded as a mush of olivine crystals lubricated by a residual liquid that later crystallized as pyroxene (as Naldrett thinks probable) or lubricated by such molten silicate plus molten sulfide (as I think). This molten sulfide (in my opinion) did not crystallize until after the silicates had done so, and, during that crystallization, much of the molten sulfide worked its way downward in the peridotite mass so that much more sulfide is to be found near the base of the ultramafic body than farther up. The much greater reaction between silicate and sulfide, where sulfide is most abundant, resulted entirely from the high proportion of sulfide to silicate in those areas. Later, after final solidification and fracturing to produce the shear zone, the massive molten sulfide material (derived from a source at depth) of the ore body was forced in, being essentially lacking in dissolved silicate material in contrast to the lubricating material of the peridotite intrusion. Naldrett believes that the heat of the intruded peridotite was sufficiently great to have caused the abundant pyrite in the underlying volcanic rocks to break down to pyrrhotite and sulfur gas; at the high confining pressures on the system, the sulfur would behave essentially as a liquid. This sulfur moved along a chemical gradient toward, and into, the hot peridotite. There it reacted with the iron- and nickel-bearing silicates to form iron-nickel sulfides. The marginal portions of the peridotite then solidified and then were sheared. Some though not all of these sulfides migrated toward and into the dilatant zone formed by the shearing and were solidified to form the ore body. After this, but while the peridotite still was hot, further reaction between sulfur and silicates took place to form the disseminated ore. This explanation is not entirely satisfactory since it requires a remarkable coincidence of events not likely to occur. There is a very reasonable question as to the ability of the peridotite magma to break down enough pyrite to provide the sulfur necessary for the sulfurization. Further, the ability of such sulfur, granted it was generated, to move into the peridotite is subject to question. Finally, the chances of such sulfides as formed to be moved into the sheared zone and still leave behind the perfect zoning described by Naldrett seems impossible.

A better explanation, and a modification of Naldrett's, would be to assume that a sulfide melt, from the same general source as the peridotite, was forced into the sheared zone. There the melt developed a high enough partial pressure of sulfur to produce, by reaction with the silicates of the peridotite, the aureole of disseminated sulfides.

A third explanation is that the disseminated zone was produced by gravitative settling from the original peridotite magma and that the presence of the massive ore in the shear zone in its present relationship to the disseminated zone is largely coincidence. The production of the disseminated zone by this theory requires that molten solution of iron-nickel sulfides separated from the ultramafic (peridotite) at some stage in its career and began to settle out of the magma after it had reached its present location. Much of it moved downward to produce a zonation, gradually decreasing upward, of pyrrhotite and pentlandite. The presence of pentlandite and heazelwoodite (Ni_3S_2) at the outer margins of the disseminated zone suggests that a higher proportion of the nickel was retained in the silicate melt than of iron and that, in the last stages of crystallization, essentially only pentlandite and heazelwoodite came out of solution. In the outer reaches of the disseminated zone, the nickel-rich materials cyrstallized as nickel-bearing minerals, but where they came out of solution where iron-nickel melts existed, they were absorbed into them.

Naldrett's explanation is not classifiable under the classification used in this memoir, and, if it were desired to do so, the classification would have to be modified. As things now stand, it does not seem necessary to do this. If the second explanation is correct, the ores would be classified as magmatic-2b with the disseminated zone being a deuteric by-product of the emplacement of the melt. The third explanation, the one I favor, would require a classification of magmatic-2a and magmatic-2b.

BANCROFT-HALIBURTON (GENERAL)

Late Precambrian	*Uranium, Corundum, Nepheline, Feldspar*	*Magmatic-3a, Magmatic-4, and Hypothermal-1 (U); Magmatic-1a and Magmatic-3a (Corundum); Magmatic-1a (Nepheline and Feldspar)*

Adams, F. D., and Barlow, A. E., 1910, Geology of the Haliburton and Bancroft areas, Ontario: Geol. Surv. Canada Mem. 6, 419 p.

Chayes, F., 1942, Alkaline and carbonate intrusives near Bancroft, Ontario: Geol. Soc. Amer. Bull., v. 53, p. 449-512

Hewitt, D. F., 1956, The Grenville region of Ontario, in Thomson, James E., Editor, *The Grenville problem*: Roy. Soc. Canada Spec. Pubs. no. 1, p. 22-41

Hewitt, D. F., and others, 1957, Haliburton-Bancroft area, Province of Ontario: Ont. Dept. Mines Map No. 1957b, 1:126,720

Rose, E. R., 1959, Rare earths of the Grenville sub-province, Ontario and Quebec: Geol. Surv. Canada Paper 59-10, 41 p.

Notes

The Bancroft-Haliburton area lies between 44°30' and about 45°20' north latitude and 77°30' and 78°45' west longitude. The radioactive mineralization in this area is scattered irregularly through a northeast-southwest belt about 50 miles long and nearly 20 miles wide. The most important uranium deposits are in Cardiff Township in Haliburton County and Faraday Township in adjacent Hastings County as well as in Cavendish Township in Peterborough County to the southwest of Cardiff. The nepheline syenite bodies, the source of the nepheline, feldspar, and corundum, lie in a discontinuous band, some 80 miles long, that extends from Glamorgan Township in Haliburton County northeast to Brougham Township in Renfrew County. Although the Blue Mountain nepheline syenite deposit occurs in the general Bancroft-Haliburton area, it is outside the band just mentioned and is large and geologically distinct enough to justify a separate discussion.

The uranium and nonmetallic ores of the Bancroft-Haliburton area are located in a complex of Grenville-type metasediments and metavolcanics that are cut by both mafic and silicic igneous rocks and their gneissic equivalents. During the intrusion of these rocks, both *lit-par-lit* injections and granitization were important processes. The strong gneissic structures in essentially all the rocks were produced by later folding and high-grade regional metamorphism. Radioactive age determinations from the silicic igneous rocks give formation dates of about 1000 m.y. ago. Because both the uranium and nonmetallics appear to be genetically connected with the igneous activity in the area, it is probable that the ores were formed in the same general time span as the igneous rocks and should be classed as middle late Precambrian. Although the metasediments and metavolcanics apparently were produced from the original sedimentary materials in Grenville time, they originally probably were formed long before that orogenic epoch. Even if they are much older than the intrusions, however, this has no effect on the age of the uranium ores and the nepheline syenites.

The genetic classification of the materials under discussion is given under the separate headings that follow.

BANCROFT (NONMETALLICS)

Late Precambrian	*Corundum, Nepheline, Feldspar*	*Magmatic-1a (Corundum, Nepheline, and Feldspar); Magmatic-3a (Corundum and Nepheline)*

Adams, F. D., and Barlow, A. E., 1908, The nepheline and associated alkali syenites of eastern Ontario: Roy. Soc. Canada Tr., 3d Ser., v. 2, sec. 4, p. 3-76

Appleyard, E. C., 1965, Desilication of alkali-syenite from the Wolfe nepheline belt: Canadian Mineral., v. 8, p. 159-165

Baragar, W.R.A., 1953, Nepheline gneisses of York River, Ontario: Geol. Assoc. Canada Proc., v. 6, p. 83-115

Carlson, H. D., 1957, Origin of the corundum deposits of Renfrew County, Ontario, Canada: Geol. Soc. Amer. Bull., v. 68, p. 1605-1636

Foye, W. J., 1915, Nepheline syenites of Haliburton County, Ontario: Amer. Jour. Sci., 4th Ser., v. 40, p. 413-436

Gittins, J., 1961, Nephelinization in the Haliburton-Bancroft district, Ontario, Canada: Jour. Geol., v. 69, p. 291-308

Gummer, W. K., and Burr, S. V., 1943, Nephelinized paragneisses in the Bancroft area, Ontario: Jour. Geol., v. 54, p. 137-168

Hewitt, D. F., 1953, Geology of the Brudenell-Raglan area: Ont. Dept. Mines 62d Ann. Rept., v. 62, pt. 5, 123 p.

—— 1954, Geology of Monteagle and Carlow Townships: Ont. Dept. Mines 63d Ann. Rept., v. 63, pt. 6, 78 p.

—— 1960, Nepheline syenite deposits of southern Ontario: Ont. Dept. Mines 69th Ann. Rept., v. 69, pt 8, 194 p.

Hewitt, D. F., and James, W., 1955, Geology of Dungannon and Mayo Townships: Ont. Dept. Mines 64th Ann. Rept., v. 64, pt. 8, 65 p.

Moyd, L., 1949, Petrology of the nepheline and corundum rocks of southeastern Ontario: Amer. Mineral., v. 34, p. 736-751

Moyd, L., and others, 1962, The Monteagle nepheline-corundum-mica deposit, Hastings County, Ontario: Canadian Inst. Min. and Met. Tr., v. 65 (Bull. no. 604), p. 261-268

Osborne, F. F., 1930, The nepheline-gneiss complex near Egan Chute, Dungannon Township, and its bearing on the origin of nepheline syenite: Amer. Jour. Sci., 5th Ser., v. 20, p. 33-60

Tilley, C. E., and Gittins, J., 1961, Igneous nepheline-bearing rocks of the Haliburton-Bancroft province of Ontario: Jour. Petrol., v. 2, p. 38-48

Notes

See *Notes* under Bancroft-Haliburton (General).

The most important deposits of nepheline are found in Dungannon Township in Hastings County; the individual deposits of nepheline and feldspar and, in places, of corundum are much smaller than the Blue Mountain deposit. The district can be divided into five subordinate nonmetallic areas; from east to west these are: (1) Glamorgan-Monmouth; (2) Cardiff-Faraday; (3) Dungannon-Moneagle-Carlow (York River); (4) Brudenell-Raglan; and (5) Eastern Extension in Renfrew County. The individual bands of nepheline syenite are seldom more than a few hundred feet wide or 2 to 3 miles long. Hewitt (1960) considers the nepheline syenite masses to be, in large part, truly intrusive rocks that were injected in the molten state into the rocks now containing them. Other geologists (Gittins, 1961; Tilley and Gittins, 1961; and Moyd and others, 1962) have held the nepheline-bearing rocks to have been formed by the reaction of high-temperature fluids (presumably of magmatic origin) with such metamorphic rocks as the paragneisses, marbles, and ortho- and para-amphibolites. Within the gneissic nepheline syenites of the Bancroft area, both leucocratic and melanocratic modifications exist. The most common minerals are nepheline and plagioclase, but biotite, muscovite, and magnetite generally are present. Microcline is found in minor amounts and Hewitt (1960) thinks it normally was developed by contact effects caused by the later intrusive pink syenite. Corundum commonly is found as an accessory mineral in nepheline-plagioclase-biotite-muscovite syenites. Calcite is developed to a considerable extent in all facies of the nepheline syenite and often is abundant in nepheline pegmatite. In these pegmatites, nepheline makes up 10 to 15 percent of the rock over areas of several square feet; locally, zircon may constitute 5 percent of the pegmatite. In one locality in Dungannon Township, corundum crystals make up 20 percent of a coarse-grained nepheline-albite pegmatite over a width of 10 feet. Corundum also occurs as a component of various alkali syenites and of the generally dikelike pegmatitic facies of these syenites. The main production of syenite, however, came from corundum syenites and corundum-syenite pegmatites. In those nepheline syenites in which corundum occurs, it appears, despite the extensive and intensive metamorphism that these rocks have undergone, to have been one of the primary constituents of the syenite.

The nepheline-syenite magmas, granted such existed, apparently assimilated considerable quantities of Grenville rocks, becoming even more unsaturated in the process. The calcite in the nepheline syenites probably derived in part from the parent magma in part from the assimilated Grenville limestones. In nepheline syenite pegmatites, calcite makes up the cores of these bodies in much the same manner as quartz constitutes the cores of normal zoned pegmatites.

Although the nepheline syenites show similar banding to that of the metasediments, Hewitt (1953) demonstrates that, at least in one instance, the gneissic bands gradually became reduced in number as the thickness of the syenite increased--a fact difficult to explain if the syenite originally was a sediment. The corundum syenites and syenite pegmatites, on the other hand, locally contain what appear to be vestiges of nepheline gneiss; this may demonstrate the syenites were, in minor part, the result of hydrothermal solutions adding potash to nepheline rocks. Thus, to some extent, the corundum in the syenites may have been the result of what was essentially contact metasomatism. The corundum in the syenite pegmatites, however, appears to have been a primary constituent of these rocks, so the bulk of the corundum probably should be classified as magmatic-1a or magmatic-3a; some small fraction might be categorized as hypothermal-1. Despite the considerable argument in favor of a metasomatic origin for the nepheline syenites, the local transgressive contacts of nepheline syenite with the enclosing rocks argue for their intrusive formation as do the knife-sharp contacts of nepheline syenites with their host rocks. Moreover, the banding in the nepheline syenites does not correlate with the banding in the neighboring sediments and may be, as Hewitt points out, primary. Thus, it appears that much, and probably most, of the nepheline syenite in the Bancroft area was introduced as magma and that nepheline is, in large part, best classified as magmatic-1a. Almost certainly essentially all of the feldspar in the nepheline rocks is primary, and that mineral should be classified as magmatic-1a.

BANCROFT (URANIUM)

Late Precambrian *Uranium* *Magmatic-3a, Magmatic-4, Hypothermal-1*

Armstrong, H. S., 1960, Geology of Glamorgan and Monmouth Townships: Ont. Dept. Mines 69th Ann. Rept., v. 69, pt. 8, p. 35-55

Bryce, J. D., and others, 1958, The Bicroft operation: Western Miner, v. 31, no. 4, p. 79-92

Bullis, A. R., 1965, Geology of Metal Mines Ltd. (Bancroft Division): Canadian Inst. Min. and Met. Tr., v. 68 (Bull. no. 639), p. 194-202

Chamberlain, J. A., 1964, Hydrogeochemistry of uranium in the Bancroft-Haliburton region, Ontario: Geol. Surv. Canada Bull. 118, 19 p.

Ellsworth, H. V., 1932, Rare-element minerals of Canada: Geol. Surv. Canada, Econ. Geol. Ser. no. 11, 272 p. (particularly p. 196-199)

Evans, A. M., 1966, The development of *lit par lit* gneiss at the Bicroft uranium mine: Canadian Mineral., v. 8, pt. 5, p. 593-609

Hewitt, D. F., 1957, Geology of Cardiff and Faraday Townships: Ont. Dept. Mines 66th Ann. Rept., v. 66, pt. 3, 82 p.

—— 1967, Uranium and thorium deposits of southern Ontario: Ont. Dept. Mines Mineral Res. Circ. no. 4, 76 p.

Ingham, W. N., and Keevil, N. B., 1951, Radioactivity of the Bourlamaque, Elzevir, and Cheddar batholiths, Canada: Geol. Soc. Amer. Bull., v. 62, p. 131-148

Kelly, L., 1956, The Bicroft pegmatites: Canadian Min. Jour., v. 77, no. 6, p. 87-88

Lang, A. H., 1962, Bancroft area, in *Canadian deposits of uranium and thorium*: Geol. Surv. Canada Econ. Geol. Ser. no. 16, 2d ed., p. 175-186

Robinson, S. C., 1960, Economic uranium mineralization in granitic dykes, Bancroft district, Ontario: Canadian Mineral., v. 6, pt. 4, p. 513-521

Robinson, S. C., and Hewitt, D. F., 1958, Uranium deposits of Bancroft region, Ontario: 2d U.N. International Conf. on Peaceful Uses of Atomic Energy (Geneva) Pr., v. 2, p. 498-501

Rowe, R. B., 1952, Petrology of the Richardson radioactive deposit, Wilberforce, Ontario: Geol. Surv. Canada Bull. 23, 22 p.

Satterly, J., 1943, Mineral occurrences in the Haliburton area: Ont. Dept. Mines 52d Ann. Rept., v. 52, pt. 2, 106 p.

—— 1956, Radioactive mineral occurrences in the Bancroft area: Ont. Dept. Mines 65th Ann. Rept., v. 65, pt. 6, p. 1-181

Satterly, J., and Hewitt, D. F., 1955, Some radioactive occurrences in the Bancroft area: Ont. Dept. Mines Geol. Circ. no. 2, 62 p.

Shaw, D. M., 1962, Geology of Chandos Township: Ont. Dept. Mines Geol. Rept. no. 11, p. 1-28

Thomson, James E., 1943, Mineral occurrences in the North Hastings area: Ont. Dept. Mines 52d Ann. Rept., v. 52, pt. 3, 80 p.

Notes

See *Notes* under Bancroft-Haliburton (General).

The radioactive ores of the Bancroft area are found in (1) zoned granitic pegmatites, (2) unzoned silicic rocks and pegmatites of considerable variety that have assimilated country rock and in which there has been appreciable reaction between early formed crystals and the remaining molten material, (3) hydrothermal replacement bodies in marble and in metamorphic pyroxenite, and (4) hydrothermal fissure fillings that all contain calcite and apatite but include various other minerals that differ from one vein to another. All of the economically valuable deposits of the district are of type (2) in which the host rocks generally are dikelike masses of granite or syenite. Much, though far from all, of each is pegmatitic in texture. These dikes appear to have been introduced as magmas of leucogranitic composition that have to a considerable degree assimilated rocks with which they came in contact or have converted their wall rocks to rocks of igneous appearance through metasomatic reactions. These igneous bodies, therefore, are now complex, composite granite (or syenite) masses that cut through and, in many areas, replace the country rock. Satterly (1956) has divided the deposits of type (2) into three varieties, the host rocks of which are: (1) pyroxenite or syenite pegmatites in which part of the ore of the Bicroft, Faraday, Halo, and Canadian Dyno mines occurs; (2) leucogranite and leucogranite pegmatites, with or without magnetite and/or pyroxenite, in which part of the ore of the Faraday, Greyhawk, and Canadian Dyno mines is found; and (3) cataclastic quartz-rich pegmatites in which part of the ore of the Bicroft, Greyhawk, and Halo mines is located.

The association of radioactive minerals with fluorite, apatite, biotite, and a wide variety of minerals formed under similar high-temperature conditions in all four types of Bancroft uranium deposits strongly indicates that the radioactive elements were provided by the intrusive magmas that distributed them with real impartiality among their rock products, within the surrounding wall rocks, and in their late-stage hydrothermal descendents. It follows, therefore, that the deposits of type (1) should be classified as magmatic-3a; those of type (2) as magmatic-3a; those of type (3) as hypothermal-1; and those of type (4) as magmatic-4 (deuteric) and hypothermal-1, depending on whether the site of deposition was the source pegmatite or the surrounding wall rock, respectively.

BLIND RIVER (ELLIOT LAKE)

Middle Precambrian | *Uranium* | *Hypothermal-1 to Mesothermal*

Collins, W. H., 1925, North shore of Lake Huron: Geol. Surv. Canada Mem. 143, 160 p.

Davidson, C. F., 1957, On the occurrence of uranium in ancient conglomerates: Econ. Geol., v. 52, p. 668-693; disc. by various authors in v. 53, p. 489-493, 620-622, 757-759, 887-889, 1048-1049; and in v. 54, p. 313-325, 325-329, 1316-1320, 1320-1323

—— 1960, The mineralized conglomerates of Blind River: Econ. Geol., v. 55, p. 1561-1565

Derry, D. R., 1960, Evidence of the origin of the Blind River uranium deposits: Econ. Geol., v. 55, p. 906-927

Fairbairn, H. W., and others, 1960, Mineral and rock ages at Sudbury--Blind River, Ontario: Geol. Assoc. Canada Pr., v. 12, p. 41-66

Friedman, G. M., 1958, On the uranium-thorium ratio in the Blind River, Ontario, uranium-bearing conglomerate: Econ. Geol., v. 53, p. 889-890; disc., v. 54, p. 511-512

Hart, R. C., and others, 1955, Uranium deposits of the Quirke Lake trough, Algoma district, Ontario: Canadian Inst. Min. and Met. Tr., v. 58 (Bull. no. 517), p. 126-131

Holmes, S. W., 1957, Pronto Mine, in *Structural geology of Canadian ore deposits*, v. 2: Canadian Inst. Min. and Met., Montreal, p. 324-339

Joubin, F. R., 1954, Uranium deposits of the Algoma district: Canadian Inst. Min. and Met. Tr., v. 57 (Bull. no. 510), p. 431-437

—— 1960, Comments regarding the Blind River (Algoma) uranium ores and their origin: Econ. Geol., v. 55, p. 1751-1756

Joubin, F. R., and James, D. H., 1956, Uranium deposits of the Blind River district, Ontario: Min. Eng., v. 8, no. 6, p. 611-613

—— 1957, Algoma uranium district, in *Structural geology of Canadian ore deposits*, v. 2: Canadian Inst. Min. and Met., Montreal, p. 305-316

Lang, A. H., and others, 1962, Blind River area, in *Canadian deposits of uranium and thorium*: Geol. Surv. Canada Econ. Geol. Ser. no. 16, 2d ed., p. 127-144

Mair, J. A., and others, 1960, Isotopic evidence on the origin and age of the Blind River uranium deposits: Jour. Geophysical Research, v. 65, no. 1, p. 341-348

McDowell, J. P., 1957, The sedimentary petrology of the Mississagi quartzite in the Blind River area: Ont. Dept. Mines Geol. Circ. no. 6, 31 p.

Pienaar, P. J., 1963, Stratigraphy, petrology, and genesis of the Elliot group, Blind River, Ontario, including the uraniferous conglomerate: Geol. Surv. Canada Bull. 83, 140 p.

Ramdohr, P., 1957, Die "Pronto-Reaktion": Neues Jb. f. Mineral., Mh., Jg. 1957, H. 10-11, S. 217-222

—— 1958, Die Uran- und Goldlagerstätten Witwatersrand, Blind River district, Dominion Reef, Serra de Jacobina; Erzmikroskopische Untersuchungen und ein geologischer Vergleich: Deutsche Akad. Wissen., Abh., Kl. f. Chem., Geol., Biol., Jg. 1958, nr. 3, 35 S.

Robertson, D. S., and Steenland, N. C., 1960, On the Blind River uranium ores and their origin: Econ. Geol., v. 55, p. 659-694

—— 1962, Thorium and uranium variations in the Blind River ores: Econ. Geol., v. 57, p. 1175-1184

Robertson, J. A., 1962, Geology of Townships 137 and 138: Ont. Dept. Mines Geol. Rept. no. 10, 94 p.

—— 1966, The relationship of mineralization to stratigraphy in the Blind River area, Ontario: Geol. Assoc. Canada, Spec. Paper No. 3, Precambrian Symposium, p. 121-136

—— 1967, Recent geological investigations in the Elliot Lake-Blind River uranium area, Ontario: Ont. Dept. Mines Misc. Paper MP. 9, 31 p.

—— 1967, Recent geological investigations in the Elliot Lake-Blind River area: Canadian Min. Jour., v. 88, no. 4, p. 120-126

—— 1968, Geology of Township 149 and Township 150: Ont. Dept. Mines Geol. Rept. 57, 162 p.

Roscoe, S. M., 1956, The Blind River, Ontario, uranium area, in Snelgrove, A. K., Editor, *Geological exploration*: Institute on Lake Superior Geology, Houghton, Mich., p. 40-48

—— 1957, Geology and uranium deposits, Quirke Lake-Elliot Lake, Blind River area, Ontario: Geol. Surv. Canada Paper 56-7, 21 p.

—— 1957, Stratigraphy, Quirke Lake-Elliot Lake sector, Blind River area, Ontario, in Gill, J. E., Editor, *The Proterozoic in Canada*: Roy. Soc. Canada Spec. Pubs. no. 2, p. 54-58

—— 1959, Monazite as an ore mineral in Elliot Lake uranium ores: Canadian Min. Jour., v. 80, no. 7, p. 65-66

—— 1960, Huronian uraniferous conglomerates: Econ. Geol., v. 55, p. 410-414

—— 1968, Huronian rocks and uraniferous conglomerates: Geol. Surv. Canada Paper 68-40, 205 p.

Roscoe, S. M., and Steacy, H. R., 1958, On the geology and radioactive deposits of Blind River region: 2d U.N. International Conf. on Peaceful Uses of Atomic Energy (Geneva) Pr., v. 2, p. 475-483

Traill, R. J., 1954, A preliminary account of the mineralogy of radioactive conglomerates in the Blind River region, Ontario: Canadian Min. Jour., v. 75, no. 4, p. 63-68

Van Schmus, R., 1965, The geochronology of the Blind River-Bruce Mines area, Ontario, Canada: Jour. Geol., v. 73, p. 755-780

Notes

The Blind River district consists of two principal parts, centered southwest of Quirke Lake and east of Elliot Lake, respectively, plus the Pronto deposits about 13 miles south-southwest of the town of Elliot Lake; that town is about 20 miles northeast of the town of Blind River on the north shore of the North Channel of Lake Huron.

The Blind River ores lie at the base of a folded sequence of sedimentary rocks that are grouped under the name of "Huronian"; these beds overlie the early Precambrian basement. The Huronian has been divided into the Bruce series (older) and the Cobalt series (younger), and these series are separated by an unconformity. The lowest formation of the Bruce (initially known as the Mississagi quartzite) was divided by Roscoe (1960) into four formations, the lowest of which he designated as the Matinenda formation. The Matinenda contains all of the minable ore thus far discovered in the district. Pienaar (1963) divided the Bruce somewhat differently, but his Matinenda formation (the lower of the two in the Elliot group) also is so defined as to contain all the ore. The Huronian rock sequence is cut by dikes and sills of quartz diabase and, in the southeastern part of the district, by a large body of granite.

Collins (1925) considered this granite to be connected genetically and temporally to the post-Huronian, Kilarney granite known farther east. Dikes of olivine diabase cut the granite and probably are of late Precambrian age. The ores appear to be older than the diabase since isotope work by Mair and others (1960) gives an age of about 1700 m.y. for the uraninite in the Blind River district. If the deposits were formed by hydrothermal solutions, as I believe them to have been, this uraninite age would fix the deposits as late middle Precambrian. If the deposits were of placer origin, as many believe, the age of the accumulation of the conglomerates and their contained uranium would be appreciably younger, though still well down in the Precambrian.

As is the case with the somewhat similar deposits of the Witwatersrand, the Blind River ores have been classed by some as syngenetic (placer) deposits and by others as hydrothermal ores. The arguments favoring the syngenetic origin are (1) the wide distribution of radioactive minerals in the conglomerate of the Matinenda; (2) the isolation of individual grains of the radioactive minerals; and (3) the lack of the common uranium-bearing mineral of the deposits, brannerite (an oxide of U and Ti containing various, though minor, quantities of Th, rare earths, Ca, and Fe), in known hydrothermal deposits. In support of their case, the hydrothermalists point to (1) the inability of uraninite to survive in placers; (2) the abundance of crystalline sulfides, particularly pyrite, in the ore--pyrite does not appear to have been part of the original conglomerate; (3) the high ratios of uranium to thorium and iron to titanium; and (4) the general high- to moderate-temperature mineral assemblage in the ores. The close spatial relationships of uranium and thorium minerals to the conglomeratic portions of the Matinenda may indicate (1) that the radioactive minerals were concentrated in the coarser portions of the conglomerate by fluvial processes or (2) that hydrothermal fluids followed the most porous portions of the Matinenda, the conglomeratic drainage channels. The rounded shapes of many of the particles of radioactive and other minerals have been suggested to have resulted from water-transportation rounding, but they also are thought to have been produced by the rapid deposition of these minerals in pore spaces in the conglomerate as a colloidally sized precipitate with high surface tension. The deposition of these minerals in colloidally sized particles does not, of course, provide any evidence that they were transported in that state but simply that they were precipitated rapidly. I think that the balance of probabilities indicates that the deposits were formed under hypothermal-1 to mesothermal conditions.

BLUE MOUNTAIN

Late Precambrian *Nepheline, Feldspar* *Magmatic-1a*

Derry, D. R., 1951, The Lakefield nepheline syenite evidence of a nonintrusive origin: Roy. Soc. Canada Tr., 3d Ser., v. 45, sec. 4, p. 31-40

Derry, D. R., and Phipps, C.V.G., 1957, Nepheline syenite deposit, Blue Mountain, Ontario, in *The geology of Canadian industrial mineral deposits*: Canadian Inst. Min. and Met., Montreal, p. 190-195

Edgar, A. D., 1968, Mineralogy of a zoned replacement body from the Blue Mountain litchfieldite, Peterborough County, Ontario: Amer. Mineral., v. 53, p. 1048-1053

Friedländer, C., 1952, Alkaligesteine von Blue Mountain, Ontario: Schweizer. Mineral. und Petrog. Mitt., Bd. 32, H. 2, S. 213-242

Hewitt, D. F., 1957, Blue Mountain nepheline syenite, in *The geology of Canadian industrial mineral deposits*: Canadian Inst. Min. and Met., Montreal, p. 187-189

—— 1960, Nepheline syenite deposits of southern Ontario: Ont. Dept. Mines 69th Ann. Rept., v. 69, pt. 8, 194 p. (particularly p. 105-161)

Keith, M. L., 1939, Petrology of the alkaline intrusive at Blue Mountain, Ontario: Geol. Soc. Amer. Bull., v. 50, p. 1795-1826

Payne, J. G., 1968, Geology and geochemistry of the Blue Mountain nepheline syenite: Canadian Jour. Earth Sci., v. 5, p. 259-274

Payne, J. G., and Shaw, D. M., 1967, K-Rb relations in the Blue Mountain nepheline syenite: Earth and Planetary Sci. Letters, v. 2, p. 290-292

Notes

The Blue Mountain nepheline syenite body is located mainly in Methuen Township in Peterborough County, although a narrow tongue of the deposit extends southwest into Burleigh Township in the same county; the center of Methuen Township is about 25 miles northeast of the city of Peterborough.

The nepheline syenite of Blue Mountain occurs in a series of highly altered sediments that were involved in the Grenville orogeny; the exact stratigraphy in these sediments remains to be determined. There is, as yet, no agreement on whether the 1000- to 1200-foot-thick mass of nepheline syenite was of igneous or metamorphic origin, but either way it must be later than the time of deposition of metamorphosed sediments in which it lies. The nepheline syenite was deformed by the latest (Grenville) earth movements that affected the area; this follows since the syenite body is located well within the borders of the Grenville province. K/Ar age determinations on muscovite from a granite pegmatite from Big Mountain Lake give an age of 1000 m.y. ± 70 m.y. Although the syenite is metamorphosed and the pegmatite is not, the former cannot be much the older of the two--certainly not by appreciably more than a few million years. Therefore, the nepheline rocks must have been formed about the middle of late Precambrian time and are here classified as late Precambrian.

Adams and Barlow (1910) and Keith (1939) have thought the Blue Mountain deposit to have been emplaced as an intrusive, but Friedländer (1952), Derry (1951), and Derry and Phipps (1957) believed the syenite to have been formed by the effect of hydrothermal solutions on paragneiss and, to a lesser extent, on marbles. The principal bases for Derry's opinion are that (1) the nepheline syenite does not show intrusive relationships toward any rock and (2) where contacts with paragneiss can be seen, they are gradational, and their attitudes are conformable to the strikes and dips of the gneissic layer containing them. Hewitt (1960), however, was able to find at least eight separate localities in the district where nepheline syenite bears intrusive relationships to the metasediments. He also observed that nephelitization of the paragneisses is a phenomenon of minor importance, most of the contacts between these two rock types being quite free of hybridization effects. Keith (1939) found the nepheline syenite to be, in most instances, best categorized as albite nepheline syenite in which the compositional ranges are (1) nepheline 10 to 35 percent, (2) microcline 10 to 37 percent, and (3) plagioclase (which may be perthitic or antiperthitic) 40 to 73 percent. Hewitt (1960) believes that the homogeneity shown by Keith's analyses indicates that the nepheline-bearing rock was of magmatic origin. Nephelitization of a paragneiss should have produced a rock with more marked diferences in nepheline and mafic mineral content and, at least in places, with migmatitic banding. If the nepheline syenite had been formed by the replacement of marble, the plagioclase probably would have been oligoclase or even andesine rather than albite (An_1 to An_5) that normally occurs. Although the nepheline syenite has been folded with the metasediments, this does not demonstrate anything but that it was formed before the Grenville folding. The foliation exhibited by the nepheline syenite more probably was the result of Grenville metamorphism than inherited from a replaced sediment; actually, there appear to be no relict sedimentary structures in the nepheline syenite proper (Hewitt, 1960). Finally, the banding in the syenite does not correspond with that in the neighboring metasediments. The deposits at Blue Mountain, therefore, are classified as magmatic-1a.

COBALT

Middle Precambrian *Silver, Cobalt* *Leptothermal*

Bastin, E. S., 1917, Significant mineralogical relations in silver ores of Cobalt, Ontario: Econ. Geol., v. 12, p. 219-236

—— 1925, Primary native silver ores of South Lorraine and Cobalt, Ontario: Econ. Geol., v. 20, p. 1-24

—— 1939, The nickel-cobalt-native silver ore type: Econ. Geol., v. 34, p. 1-40 (particularly p. 12-15, 29-40)

—— 1950, Significant replacement textures at Cobalt and South Lorraine, Ontario, Canada: Econ. Geol., v. 45, p. 808-817

Boydell, H. C., 1931, Geological structure disclosed in the Keeley mine (Cobalt): Canadian Inst. Min. and Met. Tr., v. 34 (Bull. no. 230), p. 726-750

Boyle, R. W., and others, 1969, Research in geochemical prospecting methods for native silver deposits, Cobalt area, Ontario, 1966: Geol. Surv. Canada Paper 67-35, 91 p.

Campbell, W., and Knight, C. W., 1906, The paragenesis of the cobalt-nickel arsenides and silver deposits of Temiskaming: Eng. and Min. Jour., v. 81, no. 23, p. 1089-1091

Ellsworth, H. V., 1916, A study of certain minerals from Cobalt, Ontario: Ont. Bur. Mines 25th Ann. Rept., v. 25, p. 200-243

Emmons, S. F., 1911, Cobalt district, Ontario: Min. and Sci. Press, v. 102, no. 11, p. 390-396

Hellens, A. D., 1952, Recent developments in the Cobalt area: Canadian Min. Jour., v. 73, no. 6, p. 73-78

Holmes, R. J., 1947, The higher mineral arsenides of cobalt, nickel, and iron: Geol. Soc. Amer. Bull., v. 58, p. 299-391

Hriskevich, M. E., 1968, Petrology of the Nipissing diabase sill of the Cobalt area, Ontario, Canada: Geol. Soc. Amer. Bull., v. 79, p. 1387-1403

Johnston, W.G.Q., 1954, Geology of the Temiskaming-Grenville contact southeast of Lake Temagami, northern Ontario: Geol. Soc. Amer. Bull., v. 65, p. 1047-1074

Keil, K., 1933, Über die Ursachen der charakteristischen Paragenesenbildung von gediegenem Silber und gediegenem Wismut mit den Kobalt-Nickel-Eisen-Arseniden auf den Gängen der Kobalt-Nickel-Wismut-Silber-Erzformation im sächsisch-böhmischen Erzgebirge und dem Cobalt-Distrikt: Neues Jb. f. Mineral., Geol. und Paläont., Abh., Beil. Bd. 66, Abt. A, S. 407-424

Knight, C. W., 1922, Cobalt and South Lorraine silver areas: Ont. Dept. Mines 31st Ann. Rept., v. 31, pt. 2, 374 p. (particularly p. 5-18, 30-40)

—— 1924, Geology of the mine workings of Cobalt and South Lorraine silver areas: Ont. Dept. Mines 31st Ann. Rept., v. 31, pt. 2, p. 1-238, 321-358

Koehler, G. F., and others, 1954, Geochemical prospecting at Cobalt, Ontario: Econ. Geol., v. 49, p. 378-388

Kulkarni, P. H., 1968, Mineralographic characteristics of nickel-cobalt-silver ores of Cobalt, Ontario, Canada: Univ. Geol. Soc. Nagpur Jour., v. 1, p. 12-15, 24-32

Mason, J., 1959, Geology of Christopher silver mine: Canadian Min. Jour., v. 80, no. 11, p. 71-77

Miller, W. G., 1906, The cobalt-nickel arsenides and silver deposits of Temiskaming: Ont. Bur. Mines (Ann.) Rept., v. 14, pt. 2, 97 p.

—— 1913, The cobalt-nickel arsenides and silver deposits of Temiskaming: Ont. Bur. Mines (Ann.) Rept., v. 19, pt. 2, 279 p.

Moore, E. S., 1934, Genetic relations of silver deposits and Keweenawan diabases in Ontario: Econ. Geol., v. 29, p. 725-756

—— 1942, Cobalt, Ontario, in Newhouse, W. H., Editor, *Ore deposits as related to structural features*: Princeton Univ. Press, p. 250-252

Petruk, W., 1967, Ore deposits of the Cobalt area: Geol. Assoc. Canada Guidebook--Geology of parts of eastern Ontario and western Quebec, p. 123-136

—— 1968, Mineralogy and origin of the Silverfields silver deposit in the Cobalt area, Ontario: Econ. Geol., v. 63, p. 512-531

Petruk, W., and others, 1969, Langisite, a new mineral, and the rare minerals cobalt pentlandite, siegenite, parkerite and bravoite from the Langis mine, Cobalt-Gowganda area, Ontario: Canadian Mineral., v. 9, p. 597-616

Phemister, T. C., 1928, A comparison of the Keweenawan sill rocks of Sudbury and Cobalt, Ontario: Roy. Soc. Canada Tr., 3d Ser., v. 22, sec. 4, p. 121-198

Reid, J. A., 1950, Silver in Ontario: Canadian Min. Jour., v. 71, no. 11, p. 129-133

Roscoe, S. M., 1968, Huronian rocks and uraniferous conglomerates: Geol. Surv. Canada Paper 68-40, 205 p.

Sampson, E., and Hriskevich, M. E., 1957, Cobalt-arsenic minerals associated with aplite, at Cobalt, Ontario: Econ. Geol., v. 52, p. 60-75

Sergiades, A. O., 1968, Silver cobalt calcite vein deposits of Ontario: Ont. Dept. Mines Mineral Res. Circ. no. 10, 498 p. (particularly p. 84-339, 418-441)

Symons, D.T.A., 1970, Paleomagnetism of the Nipissing diabase, Cobalt, Ontario: Canadian Jour. Earth Sci., v. 7, p. 86-90

Thomson, J. Ellis, 1925, Mineralographic notes on certain arsenides and sulpharsenides of cobalt, nickel, and iron: Univ. Toronto Studies, Geol. Ser. no. 20, p. 54-58

—— 1930, A qualitative and quantitative determination of the ores of Cobalt, Ont.: Econ. Geol., v. 25, p. 470-505, 627-652

Thomson, R., 1957, Cobalt camp, in *Structural geology of Canadian ore deposits*, v. 2: Canadian Inst. Min. and Met., Montreal, p. 377-388

—— 1957, The Proterozoic of the Cobalt area, in Gill, J. E., Editor, *The Proterozoic in Canada*: Roy. Soc. Canada Spec. Pubs. no. 2, p. 40-45

—— 1964, Cobalt silver area: Ont. Dept. Mines Map 2050, 2051, and 2052, 1:12,000 (no text)

—— 1965, Casey and Harris Townships: Ont. Dept. Mines Geol. Rept. 36, 77 p. (26 miles northeast of Cobalt)

Thomson, R., and others, 1967, Cobalt and district: Canadian Inst. Min. and Met., Centennial Field Excursion, Northwestern Quebec and Northern Ontario, p. 136-163

Whitehead, W. L., 1920, The veins of Cobalt, Ontario: Econ. Geol., v. 15, p. 103-135

Whitman, A. R., 1920, Diffusion in vein-genesis at Cobalt: Econ. Geol., v. 15, p. 136-142

—— 1922, Genesis of the ores of the Cobalt district, Ontario, Canada: Univ. Calif. Pubs. Bull., Dept. Geol. Sci., v. 13, no. 7, p. 253-310

Notes

The Cobalt district is about 85 miles northeast of Sudbury.

Although it was long thought that there was genetic connection between the Nipissing diabase sill that is present in much of the district and the mineralization of the Cobalt veins, it now seems probable that the ore fluids did not come from the sill mass nor even from the same magmatic hearth as the diabase. The constant presence of the ore veins in and near diabase gave considerable support to the theory that the ore fluids originated in the diabase. It now is evident, however, that the diabase must have been completely solid before any ores were deposited in the district; under such circumstances, the ore solutions cannot have come from the sill. This does not, however, eliminate the general volume from which the diabase magma came as a possible site for ore-fluid generation. Against this is placed the argument that the deposits at Cobalt resemble, among others, those silver ores in Cornwall (England), the Erzgebirge, Sardinia, Great Bear Lake, and Balmoral in South Africa where the geologic relationships strongly suggest a granitic source for the ore fluids. On the other hand, the silver ores at Gowganda (see below) and Kongsberg bear a relationship to diabase similar to that of the Cobalt ore. In the Cobalt district, no granites younger than the Nipissing diabase are known; this does not mean, of course, that post-Nipissing granites did not exist at depth nor that such magmas could not have produced the Cobalt ore fluids. Certainly, the combination of a thin, nearly horizontal covering of brittle diabase and Cobalt group sediments over the much-folded pre-Huronian terrane would have provided a locus for fracturing and faulting when post-Huronian folding and intrusion took place. The most recent work suggests a date of about 2095 m.y. for the Nipissing diabase and something less than that age for the ores in that rock. It is definitely known that the Cobalt mineralization is Precambrian; it appears, however, more probable that the ores were emplaced before the end of the middle Precambrian than that they were delayed until late Precambrian time. In favor of this concept is the dating of most of the ore mineralization in this general area of Ontario as about 1700 m.y. ago or the end of the middle Precambrian.

Cobalt ore includes a variety of arsenides and sulfarsenides as the first minerals to have been deposited. Of these, the first to form were cubic minerals: smaltite(?), chloanthite, skutterudite, cobaltite, safflorite, rammelsbergite, and loellingite. The next mineral group was arsenopyrite and pyrite, followed by niccolite and breithauptite. Then came calcite, the characteristic gangue mineral of the veins. An epoch of fracturing then affected all the earlier minerals and was followed by the deposition of the silver and bismuth minerals--native silver, dyscrasite, argentite, native bismuth, bismuthinite, and cosalite; this phase ended with the deposition of a second generation of calcite. The silver and bismuth minerals were followed by minor amounts of galena, matildite, sphalerite, and chalcopyrite, plus even less pyrrhotite, marcasite, tetrahedrite, and quartz. The next (fifth) and final primary group was made up of small quantities of sulfosalts, such as pyrargyrite, and a little native silver. The silver mined has come largely from native silver, with a minor amount being provided by argentite; the other silver-bearing minerals have been of essentially no economic value.

The presence of an abundance of cobalt-nickel arsenides and sulfarsenides in generally close temporal association with native silver and silver-bearing sulfosalts and argentite is characteristic of the leptothermal-type of ore deposits. The occurrence in these ores of minor amounts of such minerals as pyrrhotite, arsenopyrite, and bismuthinite that indicate high-temperature deposition where found in abundance does not invalidate the verdict of the arsenide minerals and native silver as indicators of moderately low intensity of deposition. The ores at Cobalt, therefore, are classified as leptothermal.

GOWGANDA

Middle Precambrian *Silver, Cobalt* *Leptothermal*

Bastin, E. S., 1935, Aplites of hydrothermal origin associated with Canadian cobalt-silver ores: Econ. Geol., v. 30, p. 715-734

—— 1949, Deposition and resolution of native silver at Gowganda, Ontario: Econ. Geol., v. 44, p. 437-444

Bowen, N. L., 1910, Diabase and granophyre of the Gowganda Lake district: Jour. Geol., v. 18, p. 658-674

Burrows, A. G., 1926, Gowganda silver area: (4th Rept. revised) Ont. Dept. Mines 35th Ann. Rept., v. 35, pt. 3, p. 1-61

Campbell, A. D., 1930, Gowganda silver area: Canadian Inst. Min. and Met. Tr., v. 33 (Bull. no. 216), p. 272-291

Campbell, W., and Knight, C. W., 1906, A microscopic examination of the cobalt-nickel arsenides and silver deposits of Temiskaming: Econ. Geol., v. 1, p. 767-776

Collins, W. H., 1909, Preliminary report on Gowganda mining division, district of Nipissing, Ontario: Geol. Surv. Canada Pub. no. 1075, 47 p.

—— 1913, The geology of the Gowganda mining division: Geol. Surv. Canada Mem. 33, 121 p.

Eakins, P. R., 1961, Colloidal jointing in diabase at Gowganda, Ontario: Geol. Assoc. Canada Pr., v. 13, p. 85-93

Hester, B. W., 1967, Geology of the silver deposits near Miller Lake, Gowganda: Canadian Inst. Min. and Met. Tr., v. 70 (Bull. no. 667), p. 277-286

Montgomery, A., 1948, Mineralogy of the silver ores of Gowganda, Ontario: Univ. Toronto Studies, Geol. Ser. no. 52, p. 23-38

Moore, E. S., 1934, Genetic relations of silver deposits and Keweenawan diabases in Ontario: Econ. Geol., v. 29, p. 725-756

—— 1955, Geology of the Miller Lake portion of the Gowganda silver area: Ont. Dept. Mines 64th Ann. Rept., v. 64, pt. 5, p. 1-141

—— 1957, Gowganda silver area, in *Structural geology of Canadian ore deposits*, v. 2: Canadian Inst. Min. and Met., Montreal, p. 388-392

Sergiades, A. O., 1968, Silver cobalt calcite vein deposits of Ontario: Ont. Dept. Mines Mineral Res. Circ. no. 10, 498 p. (particularly p. 368-416)

Thomson, J. Ellis, 1933, A mineralographic study of the minerals from the Miller Lake O'Brien Mine, Gowganda, Ontario: Univ. Toronto Studies, Geol. Ser. no. 35, p. 61-64

Todd, E. W., 1926, Gowganda vein minerals: Ont. Dept. Mines 35th Ann. Rept., v. 35, pt. 3, p. 62-78

Notes

The silver camp of Gowganda is located about 55 miles west-northwest of Cobalt and almost 85 miles north from Sudbury.

As is the case at Cobalt, the source of the ore fluids from which the Gowganda vein fillings have been deposited has been the subject of much discussion. Early work favored the sill itself, but later authors believed in a deeper source beneath the sill. This source is thought by some to have been the rest-magma chamber from which the diabase of the sill was earlier driven out. Moore (1955) is of the opinion that the Gowganda ores came from a more silicic phase of the magma from which the sill-diabase came and, therefore, from a source below the sill. He determined that the granophyre associated with the sill contained about 30 times as much silver (0.006 percent as opposed to 0.0002 percent) as the diabase. This does not prove anything as to the source of the silver in the ores, but it does indicate that silver is concentrated in late-stage differentiates as opposed to the melt from which they are derived and that, therefore, silver-bearing ore fluids are more likely to be given off in the late stages of the crystallization of a diabase magma than at an earlier time. It also, however, suggests that a source more silicic than the diabase magma or its differentiates is even more likely to have provided the ore fluids in question. Thus, although the source of the Gowganda ore fluids is far from certainly known, it seems most probable that they were given off from a silicic magma of post-Nipissing diabase age. Since the Nipissing diabase was introduced about 2095 m.y. ago, it appears reasonable that the ore fluids entered the fractures in and near the diabase sill sometime between that date and the end of the middle Precambrian for reasons outlined in the Cobalt discussion.

Considerable disagreement exists as to the order of mineral deposition at Gowganda, but Montgomery (1948) suggests that the sequence was (1) deposition of pinkish calcite in string fissures in the diabase; (2) brecciation of pink calcite and filling of the fractures so created by white to grayish calcite; (3) growth of native silver as long dendrites, composed of skeletal, by replacement of grayish calcite from centers generally located in the central portions of the veins; (4) deposition of safflorite by replacement outward from silver-calcite contacts, with more calcite being replaced than silver; (5) a change from safflorite to loellingite, cobaltite, and other late arsenides either along the silver-arsenide boundary or into calcite from the calcite-arsenide boundary; (6) filling of remaining open spaces by dark-gray calcite, especially such spaces as were left along the outer edges of the arsenide crusts; and (7) fracturing and deposition in these fractures of a second generation of silver and more dark calcite. This sequence of events is quite different from that at Cobalt with its more complex suite of minerals. Todd (1926) and Moore (1955), however, listed more Gowganda minerals than Montgomery places in his paragenetic sequence; these included loellingite, skutterudite, cobaltite, arsenopyrite, smaltite, breithauptite, niccolite, chalcopyrite, argentite, bornite, sphalerite, galena, pyrite, bismuth, dyscrasite, and amalgam. Todd believed that loellingite was the most common arsenide and smaltite the rarest. Quartz occurs in many of the veins and locally is more abundant than calcite; this is in direct contrast to Cobalt. Todd thought that the silver and some of the arsenides were deposited contemporaneously. Thus, it appears that complete agreement does not exist as to the order of the deposition of the Gowganda ores, but there is no doubt but that they are composed of minerals typical of those normally found in the leptothermal range.

KERR ADDISON

Early Precambrian *Gold* *Mesothermal*

Baker, J. W., and others, 1957, Kerr-Addison Mine, in *Structural geology of Canadian ore deposits*, v. 2: Canadian Inst. Min. and Met., Montreal, p. 392-402

Cooke, H. C., 1922, Kenogami, Round, and Larder Lake areas, Timiskaming district, Ontario: Geol. Surv. Canada Mem. 131, 64 p.

—— 1923, Recent gold discoveries at Larder Lake, Timiskaming district, Ontario: Geol. Surv. Canada Summ. Rept., pt. CI, p. 61-73

Hawley, J. E., 1952, Spectrographic studies of pyrite in some eastern Canada gold mines: Econ. Geol., v. 47, p. 260-304 (particularly p. 270-277)

Jenney, C. P., 1941, Geology of the Omega Mine, Larder Lake, Ontario: Econ. Geol., v. 36, p. 424-447

McLaren, D. C., 1945, The Kerr-Addison mine, a description of geology and milling: Canadian Min. Jour., v. 66, no. 4, p. 223-230 (particularly p. 223-226)

Murdock, J. Y., and others, 1951, Kerr-Addison Gold Mines, Limited: Canadian Min. Jour., v. 72, no. 4, p. 63-119 (particularly p. 109-117)

Thomson, James E., 1941, Geology of McGarry and McVittie Townships, Larder Lake area: Ont. Dept. Mines 50th Ann. Rept., v. 50, pt. 7, 99 p.

—— 1948, Matachewan-Kirkland Lake-Larder Lake area, in *Structural geology of Canadian ore deposits*: Canadian Inst. Min. and Met., Montreal, p. 627-643

Thomson, J. Ellis, 1941, The mineralogy of the Kerr-Addison ore, Larder Lake, Ontario: Univ. Toronto Studies, Geol. Ser. no. 46, p. 141-147

Wilson, M. E., 1912, Geology and economic resources of the Larder Lake district,

Ont.: Geol. Surv. Canada Mem. 17-E, 62 p.

Notes

The gold ore bodies of the Larder Lake area, of which Kerr Addison is the most important, are located in the extreme eastern portion of Ontario, about 2 miles from the Quebec border and less than 30 miles west-southwest of Noranda; minable deposits have been found in both McGarry and McVittie Townships, the Kerr Addison ore body being in McGarry.

The principal rocks of the district are of both Keewatin and Timmiskaming types, and these have been intruded by stocks, dikes, and irregularly shaped bodies of Algoman intrusives that range in composition from granite to diorite or lamprophyre; the larger masses are syenites or syenite porphyries with subordinate phases that represent several rock categories. Much of the Keewatin and Timiskaming sequences and of the igneous rocks have been highly sheared and carbonatized; in many places, considerable quantities of fuchsite mica are to be found in the carbonate rocks. These carbonate rocks are, in many localities, cut by quartz stringers in stock-work arrangements, and the large ore bodies of the Kerr Addison mine are found in such rocks with the greatest amount of carbonatization having taken place in Keewatin mafic volcanic rocks and their associated agglomerates and tuffs. The process of carbonatization has affected no rocks younger than the Algoman intrusives since boulders of carbonate rock are present in the basal conglomerate of the post-Algoman Cobalt series, and erosional remnants of Cobalt rocks unconformably overlie the carbonate bodies along the north shore of Larder Lake. The solutions that deposited the quartz veins almost certainly did not cause the carbonatization since large volumes of carbonated rocks lack quartz veins; these solutions, both carbonate and quartz-depositing ones, probably came from the same general source as the Algoman magmas. The main carbonate band in the mafic volcanic rocks is localized along a strong structure and should, therefore, probably persist to considerable depths; the individual carbonate bodies, however, pinch and swell vertically as well as horizontally. Because both the syenite and carbonate-quartz ore-bearing rocks probably are products of the Algoman magmatic cycle, the ores, which are so closely related in space to the carbonatized rock, probably also are Algoman in age. Radioactive age dates on biotite taken from a granite in the area that cuts Keewatin volcanic rocks and is unconformably overlain by Cobalt sediments give an age of about 2605 m.y. This indicates that the ore should be classed as late early Precambrian.

Two main ore types are found in the Kerr Addison mine. The first of these are bodies made up of carbonate material cut by irregular stockworks of quartz veins that occur within the highly broken carbonate zone; where the carbonate is bright green, ore is most likely to be found, but all green carbonate volumes do not contain ore. Gold-quartz veins also have been found in brown carbonate schists or even in talc-chlorite schists. In places, some of the igneous rock in the carbonate zone has been highly altered and may contain enough gold to be ore. The carbonate bodies contain, in addition to gold, only minor amounts of other metallic minerals, including disseminated pyrite and chalcopyrite, millerite, sphalerite, scheelite, and traces of arsenopyrite. Tetrahedrite in small quantities may be found closely associated with free gold in the quartz veins. The second type of ore body is found in lenses of mineralized and silicified tuff in somewhat carbonatized tuff and flow bands south of the carbonate zone. In the bodies, most of the gold is carried in disseminated and uniformly distributed pyrite. These bodies carry traces of chalcopyrite, galena, sphalerite, scheelite, and arsenopyrite. The close association of gold in quartz-stockwork ore type with pyrite and tetrahedrite suggests that the gold was deposited, in part at least, under leptothermal conditions; the disseminated arsenopyrite and scheelite appear to have been introduced appreciably before the gold. In the flow-type ores, the gold is largely associated with the disseminated pyrite; this relationship and the other minerals present suggest deposition in the mesothermal to leptothermal ranges. The wall-rock alteration to carbonate is compatible with either depositional range. The deposits, therefore, are classified as mesothermal to leptothermal.

KIRKLAND LAKE

Early Precambrian | *Gold* | *Mesothermal to Leptothermal*

Bain, G. W., 1933, Wall-rock mineralization along Ontario gold deposits: Econ. Geol., v. 28, p. 705-743 (particularly p. 727-740)

Burrows, A. G., and Hopkins, P. E., 1923, Kirkland Lake gold area: (rev. ed.), Ont. Dept. of Mines 32nd Ann. Rept., v. 32, pt. 4, p. 1-52

Charlewood, G. H., 1964, Geology of deep developments on the main ore zone at Kirkland Lake: Ont. Dept. Mines Geol. Circ. no. 11, 49 p.

Cooke, D. L., and Moorhouse, W. W., 1969, Timiskaming volcanism in the Kirkland Lake area, Ontario, Canada: Canadian Jour. Earth Sci., v. 6, p. 117-132

Dougherty, E. Y., 1939, Some geological features of Kolar, Porcupine and Kirkland Lake: Econ. Geol., v. 34, p. 622-653

Goodwin, A. M., 1965, Mineralized volcanic complexes in the Porcupine-Kirkland Lake-Noranda region, Canada: Econ. Geol., v. 60, p. 955-971

Hawley, J. E., 1952, Spectrographic studies of pyrite in some eastern Canada gold mines: Econ. Geol., v. 47, p. 260-304 (particularly p. 270-284)

Hewitt, D. F., 1963, The Timiskaming series of the Kirkland Lake area: Canadian Mineral., v. 7, pt. 3, p. 497-523

Hopkins, H., 1940, Faulting at the Wright-Hargreaves mine, with notes on ground movements: Canadian Inst. Min. and Met. Tr., v. 43 (Bull. no. 343), p. 685-707

—— 1949, Structure at Kirkland Lake, Ontario, Canada: Geol. Soc. Amer. Bull., v. 60, p. 909-922

Knight, C. W., 1933, Central Canada's gold belts (a comparison with Western Australia, India, and Southern Rhodesia): Canadian Min. Jour., v. 54, no. 3, p. 98-101

Lee, H. A., 1963, Glacial fans in till from the Kirkland Lake fault: a method of gold exploration: Geol. Surv. Canada Paper 63-45, 36 p.

Lovell, H. and others, 1967, Kirkland Lake and district: Canadian Inst. Min. and Met., Centennial Field Excursion, Northwestern Quebec and Northern Ontario, p. 72-100

Robson, W. T., 1936, Lake Shore geology: Canadian Inst. Min. and Met. Tr., v. 39 (Bull. no. 287), p. 99-141

Savage, W. S., 1964, Mineral resources and mining properties in the Kirkland Lake-Larder Lake area: Ont. Dept. Mines Mineral Res. Circ. no. 3, 108 p.

Thompson, R. M., 1949, The telluride minerals and their occurrence in Canada: Amer. Mineral., v. 34, p. 343-382 (particularly p. 379)

Thomson, James E., 1946, The Keewatin-Timiskaming unconformity in the Kirkland Lake district: Roy. Soc. Canada, 3d Ser., v. 40, sec. 4, p. 113-122

—— 1948, Matachewan-Kirkland Lake-Larder Lake area, in *Structural geology of Canadian ore deposits*: Canadian Inst. Min. and Met., Montreal, p. 627-643

—— 1948, Geology of Teck Township and Kenogami Lake area, Kirkland Lake gold belt: Ont. Dept. Mines 57th Ann. Rept., v. 57, pt. 5, p. 1-53

Thomson, James E., and others, 1948, Geology of the main ore zone at Kirkland Lake: Ont. Dept. Mines 57th Ann. Rept., v. 57, pt. 5, p. 54-188

Todd, E. W., 1928, Kirkland Lake gold area: Ont. Dept. Mines 37th Ann. Rept., v. 37, pt. 2, 174 p.

Tully, D. W., 1963, The geology of the Upper Canada Mine. Geol. Assoc. Canada Pr., v. 15, p. 61-86

Tyrrell, J. B., and Hore, R. E., 1926, The Kirkland Lake fault: Roy. Soc. Canada Tr., 3d Ser., v. 20, sec. 4, p. 51-63

Ward, W., and others, 1948, The gold mines of Kirkland Lake, in *Structural geology of Canadian ore deposits*: Canadian Inst. Min. and Met., Montreal, p. 644-653

Notes

The Kirkland Lake district is about 55 miles east-southeast of Noranda.

The ores of the seven mines of the Kirkland Lake district are found in a series of sedimentary and pyroclastic rocks of Timiskaming type and age (older) and a complex of Algoman intrusives (younger). The oldest rocks in the district are of Keewatin type and age but are not traversed by ore veins. A few diabase and lamprophyre dikes cut the ore and the intrusives and are the youngest rocks in the district. During the intrusion of the Algoman rocks, considerable volumes of carbonate rocks were formed by the replacement of existing rocks by carbonate and silicious materials. The carbonates usually are bright green, the color being due either to the presence of ferrous iron or nickel or of some fuchsite or mariposite micas. The carbonate rocks usually are intersected by networks of quartz veinlets and are similar to those that contain the ore at Kerr Addison; they are not ore-bearing in the Kirkland Lake area. The Algoman syenite batholith at Kirkland Lake and some of its offshoots extensively cut the carbonate rocks; in other areas, however, the syenites are carbonatized. This suggests that the carbonatization took place between stages of syenite intrusion and was caused by solutions coming from the same source as the syenite magmas. Most of the ores are in Algoman rocks cut by the N67°E Kirkland Lake fault zone; the main mass of Algoman syenite is an elongated and wedge-shaped stock that widens to the east and with depth; it dips steeply to the south and has an overall pitch to the west. The ores were introduced after the faulting and brecciation of the fault zone, but all these events probably occurred in Algoman time; there is no evidence of Kirkland Lake-type mineralization in the rocks of the Cobalt series that are present at the east and west ends of the district. As radioactive age dating gives ages of about 2600 m.y. for the Algoman intrusives, it is probable that the ore mineralization should be considered as latest early Precambrian.

Not all of the district ore zone is mineralized; the longest ore shoot in the camp had a length of about 1.25 miles, although the average ore-shoot length was much shorter. About 95 percent of the ore came from shoots in the syenite (in the general sense) or from contacts between syenite and sediments. Two-thirds of the ore was mined from syenite porphyry (in the strict sense), one-quarter came from a complex of augite syenite, syenite, and syenite porphyry, and the remainder mainly from sediments. Wall-rock alteration affected the igneous rocks much more than the sediments, but most of the gold came from the large quartz veins. Gold, however, has been profitably recovered from volumes of carbonatized, silicified, and pyritized rocks of the syenite complex that contained quartz stockworks and veins. Ore material seldom is found that contains more than 2 percent of metallic minerals; these minerals are exceptionally fine grained, and their proportions are quite constant throughout the known vertical range of mineralization. Gold at Kirkland Lake is found in the native state and in the tellurides petzite and calaverite; other common tellurides are coloradoite (HgTe) and altaite (PbTe), the latter being the most abundant telluride in the ores. The order of events in the ore-forming sequence appears to have been (1) extensive wall-rock alteration to green carbonates, plus some quartz, sericite, chlorite, and leucoxene; (2) pyritization of the wall rocks with the local formation of tourmaline and biotite; (3) main period of quartz formations, accompanied by much pyrite with which much of the native gold is associated; (4) deposition of minor and local calcite; and (5) formation of sulfides in the order--arsenopyrite and pyrrhotite, chalcopyrite with rare sphalerite and galena, molybdenite(?), tellurides, and gold. Although the textures existing among gold and the tellurides are quite complex, the two tellurides that do not contain gold (coloradoite and altaite) appear to be intergrown with gold and to replace petzite. This close association of gold with tellurides indicates that a large fraction of the gold deposition took place under leptothermal conditions. The ores were deposited too far beneath the then-existing surface to have any possibility of having been epithermal. A considerable fraction of the gold is found in the pyrite in the vein quartz with which it appears to have crystallized simultaneously. Not only is this gold much earlier than that associated with the tellurides but, after fracturing of this pyrite, gold was emplaced in probably minor amounts as fracture fillings in this pyrite and as rims around it. This second-generation gold was fractured before the tellurides were emplaced. Thus, gold deposited in at least three separate stages, the first two far enough removed in time from telluride and late gold deposition that they probably formed under mesothermal conditions. The wall-rock alteration is compatible with a mesothermal to leptothermal alteration or to both. The ores at Kirkland Lake, therefore, are classified as mesothermal to leptothermal. (The Upper Canada mine, some 10 miles east of Kirkland Lake, is appreciably different in mineralogy and ore controls from Kirkland Lake and probably was formed entirely under leptothermal conditions.)

LITTLE LONG LAC

Early Precambrian *Gold* *Mesothermal to Leptothermal*

Armstrong, H. S., 1943, Gold ores of the Little Long Lac area, Ontario: Econ. Geol., v. 38, p. 204-252

—— 1944, Mineralogy of the Little Long Lac gold area, Ontario: Amer. Mineral., v. 29, p. 305-319

Bruce, E. L., 1935, Little Long Lac gold area: Ont. Dept. Mines 44th Ann. Rept., v. 44, pt. 3, p. 1-60

—— 1936, New developments in the Little Long Lac area: Ont. Dept. Mines 45th Ann. Rept., v. 45, pt. 2, p. 118-140

—— 1939, Structural relations of some gold deposits between Lake Nipigon and Long Lake, Ontario: Econ. Geol., v. 34, p. 357-368 (particularly p. 357-359, 361-368)

Bruce, E. L., and Samuel, W., 1937, Geology of the Little Long Lac mine: Econ. Geol., v. 32, p. 318-334

Fairbairn, H. W., 1937, Geology of the northern Long Lake area: Ont. Dept. Mines 46th Ann. Rept., v. 46, pt. 3, p. 1-22

Horwood, H. C., 1948, General structural relationships of ore deposits in the Little Long Lac-Sturgeon River area, in *Structural geology of Canadian ore deposits*: Canadian Inst. Min. and Met., Montreal, p. 377-384

Horwood, H. C., and Pye, E. G., 1951, Geology of Ashmore Township: Ont. Dept. Mines 60th Ann. Rept., v. 60, pt. 5, 105 p.

Matheson, A. F., 1948, Jellicoe, Bankfield, and Tombill mines, in *Structural geology of Canadian ore deposits*: Canadian Inst. Min. and Met., Montreal, p. 399-406

Matheson, A. F., and Douglas, J. H., 1948, Hard Rock mine, in *Structural geology of Canadian ore deposits*: Canadian Inst. Min. and Met., Montreal, p. 406-413

Pye, E. G., 1951, Geology of Errington Township, Little Long Lac area: Ont. Dept. Mines 60th Ann. Rept., v. 60, pt. 6, 140 p.

Reid, J. A., 1945, The Hardrock "porphyry" of Little Long Lac: Econ. Geol., v. 40, p. 509-516; disc., v. 41, p. 282-283

Thomson, J. Ellis, 1935, Mineralization of the Little Long Lac and Sturgeon River areas: Univ. Toronto Studies, Geol. Ser. no. 38, p. 37-45

Tyson, A. E., 1945, Report on gold belts in the Little Long Lac-Sturgeon River district: Canadian Min. Jour., v. 66, no. 12, p. 839-850

Notes

The gold deposits of the Little Long Lac district center around the town of Geraldton, about 115 miles northeast of Port Arthur and slightly less than 70 miles northwest of Manitouwadge. The three townships in which gold is found are, from east to west, Ashmore, Errington, and Lindsley. The MacLeod-Cockshutt mine is on the north-south boundary between Ashmore and Errington Townships.

The oldest rocks in the Little Long Lac area are probably of Keewatin age and certainly are of Keewatin type. They are overlain unconformably by Timiskaming(?) formations, and the southern belt of these rocks and the igneous rocks intruded into them contains most of the gold mineralization of the district. This belt is divided into two parts by a narrow tuff and breccia band of Keewatin(?) rocks; the northern portion of the Timiskaming beds is known as group A and the southern as group B; all the mines that have produced worthwhile tonnages of ore lie in group A rocks. Although there is some question as to the age of the igneous rocks intrusive into the Timiskaming(?) sediments, the structural evidence suggests to Pye (1951) that the regional and local structures in the Timiskaming rocks were formed, and the intrusives then emplaced, during the same general period of orogenic activity. The Algoman rocks range from diorites and gabbros to albite porphyries and quartz-albite porphyries. The albite porphyries contain gold-bearing quartz veins so are older than the ores. Although it has been suggested that the albite porphyries were formed from sediments, Pye (1951) is convinced that they are igneous because the porphyries (1) locally contain fragments of hornblende diorite, (2) in places include pieces of slate and graywacke, and (3) occur in places within the volcanic rocks in such attitudes that bar them from ever having been sediments. The porphyries, however, (1) are generally conformable to, and interbedded with, graywacke and iron formation; (2) closely resemble the sediments with which they are associated; (3) have broken and rounded phenocrysts that might be of detrital origin; and (4) approximate in composition the graywacke matrix of an example of boulder conglomerate.

Although the district does not contain any actual Algoman granites, such rocks are known south of Little Long Lac and east of the town of Longlac. The age relationship of these granites to the albite porphyries and the ores is not known, but the granites appear to be somewhat less altered and deformed than the porphyries and are, therefore, probably younger than the prophyries. It is probable, however, that the prophyries, ores, and granites all belonged to the same stage of igneous activity. Radioactive age determinations indicate that the granites are about 2600 m.y. old; this would place the mineralization as latest early Precambrian, and that age is assigned to the Little Long Lac ores.

The gold deposits of the district have been divided by Pye (1951) and Horwood and Pye (1951) into three major categories: (1) deposits in layered volcanic and sedimentary rocks; (2) deposits in intrusive rocks; and (3) deposits along contacts between sedimentary or volcanic rocks and later intrusive rocks. Within each of these categories are included 13 different types of structural control, so it would seem to follow that almost any type of wall rock, if properly broken and connected to a source of ore fluids, permitted the deposition of gold and its associated metallic and nonmetallic minerals. Essentially all of these zones of broken rocks appear to have been formed as subsidiary breaks more or less parallel to a major fault; in most instances, this is the Tombill fault. This fault and its companion Long Lac fault appear to be related in origin to the Little Long Lac syncline.

The sequence of mineralization in the Little Long Lac ores seems to differ somewhat from one mine to another, but a general scheme that checks well with each mine shows (1) after an initial stage of fracturing, a cherty first generation of quartz that produced a general silicification of the ore zones; (2) after a second period of fracturing, a second generation of light quartz as veins and stringers with small amounts of sulfides, a little gold, and subordinate gold tellurides; minor amounts of albite and scheelite were emplaced early in this stage, and these were followed by ankeritic dolomite, after quartz, the main nonmetallic mineral of the deposits; (3) after a third period of fracturing, a third generation of quartz with pyrite, arsenopyrite, tourmaline, pyrrhotite(?), calcite, and gold; (4) after a fourth stage of transverse fractures that offset the ore zones, a fourth generation of quartz; and (5) after a fifth period of fracturing, the final deposition of a fifth generation of quartz and pink calcite. The gold is very erratic in its occurrence; it generally is found filling fractures in, and as irregular blebs within, crystals of arsenopyrite and early quartz. In time of deposition, however, it seems to be most closely related to galena. A little gold has been found along cleavage planes in the ankeritic dolomite, so some of the gold was later than the sulfides that accompanied the second generation of quartz. Some gold has been found in irregular blebs and fracture fillings in the second generation quartz and molded upon third generation quartz. The first generation of gold (associated in time with galena) was followed by very minor amounts of krennerite and coloradoite and perhaps another telluride as well. Minor quantities of sulfosalts are thought to have been deposited just before the first generation of gold. The second generation of gold appears to have been deposited just after the calcite and pyrite of stage (3). About 2 pounds of scheelite were recovered per ton of Little Long Lac ore during World War II; this scheelite appears to have been deposited with the second generation of quartz well before the precipitation of gold and overlapped in time of deposition with albite. It would appear that the first generation of gold was deposited under leptothermal conditions since its deposition was preceded by that of sulfosalts and followed by that of tellurides. The conditions of deposition of the second are not clear but probably were slightly more intense than those of the first. These ores are classified here as mesothermal to leptothermal with the understanding that appreciably more gold probably was deposited under leptothermal than mesothermal conditions.

MANITOUWADGE

Early Precambrian	*Copper, Zinc, Lead, Silver*	*Hypothermal-1 to Mesothermal*

Bray, R.C.E., 1964, Techniques used by the geology department, Geco Mines Limited: Canadian Inst. Min. and Met. Tr., v. 67 (Bull. no. 623), p. 31-40

Brown, W. L., and others, 1960, The geology of the Geco Mine: Canadian Inst. Min. and Met. Tr., v. 63 (Bull. no. 573), p. 1-9

Graham, R.A.F., 1968, Effects of diabase dike intrusion on sulfide minerals at Manitouwadge, Ontario: Canadian Jour. Earth Sci., v. 5, p. 545-547

Mookherjee, A., and Dutta, N. K., 1970, Evidence of incipient melting of sulfides along a dike contact, Geco Mine, Manitouwadge, Ontario: Econ. Geol., v. 65, p. 706-713

Pye, E. G., 1955, Preliminary report on the geology of the Manitouwadge Lake area: Ont. Dept. Mines Geol. Circ. no. 3, 9 p.

—— 1956, Geology and mineral deposits of the Manitouwadge Lake area, in Snelgrove, A. K., Editor, *Geological exploration*: Institute on Lake Superior Geology, Houghton, Mich., p. 26-39

—— 1957, Geology of the Manitouwadge area: Ont. Dept. Mines 66th Ann. Rept., v. 66, pt. 8, 114 p.

Suffel, G. G., and others, 1970, Metamorphism of massive sulphides at Manitouwadge, Ontario, Canada: IMA-IAGOD Meetings 1970, Collected Abs., Tokyo-Kyoto, Paper 4-12, p. 94

Thomson, James E., 1932, Geology of the Heron Bay-White Lake area: Ont. Dept. Mines 41st Ann. Rept., v. 41, pt. 6, p. 34-47

Tilton, G. R., and Steiger, R. H., 1969, Mineral ages and isotopic composition of primary lead at Manitouwadge, Ontario: Jour. Geophys. Res., v. 74, p. 2118-2132

Timms, P. D., and Marshall, D., 1959, The geology of the Willroy Mines base metal deposits [Ontario]: Geol. Assoc. Canada Pr., v. 11, p. 55-65

Notes

The ore deposits of the Manitouwadge area are located about 40 miles north-northeast of the town of Marathon on the north shore of Lake Superior and about 165 miles east-northeast of Port Arthur. The town is about equidistant between the main transcontinental rail lines of the Canadian National (north) and the Canadian Pacific (south).

The ore bodies in the district are grouped around the nose of a syncline (the Manitouwadge syncline) that plunges flatly northeast. the oldest rocks of the district lie south of the syncline and are metavolcanic rocks (now hornblende schists) and metasediments of Keewatin type and probably Keewatin age. These metasediments consist of a variety of gneisses and quartzite that grade into the granodiorite of the syncline core; the granodiorite is biotite rich and usually gneissic in texture. Pye (1957) did not find that the granodiorite ever had intrusive relationships to the surrounding rocks and believed that it was formed by metamorphism of various gneisses such as those that lie south of it in the area of the mines. Brown and others (1960), however, mention a projecting lobe of the granodiorite that cuts garnet-bearing hornblende-biotite-quartz feldspar gneiss and, at its southern end, quartz-muscovite schist. This relationship indicates that the biotite granodiorite may have been more of an intrusive than Pye thinks. Although no age determinations have been reported on this granodiorite gneiss nor on the metasediments containing the ore bodies of the various deposits, the deposits are in the Superior province and are in an area where the last period of metamorphic and igneous activity probably occurred about 2500 m.y. ago. The ores were formed after the metagabbro dikes, the granodiorites, and the pegmatites and aplites associated with them. The ores replace the pegmatites and, therefore, well may have been the last event to take place in the Algoman magmatic cycle. No direct evidence, at least to the present, is available to connect the hydrothermal ore fluids with an Algoman magmatic source. Nevertheless, since the granodiorite intrusion (granted it is such) and the ore formation appear to have taken place shortly after, or even in part concurrently with, late early Precambrian metamorphism, the Manitouwadge ores are here considered to be late early Precambrian. The possibility that the ores are syngenetic with the metasediments containing them seems (to me) highly unlikely and is not considered further in dating or gaging the intensity range of the ore forming process.

The ore bodies of the Manitouwadge area are from east to west: (1) the Geco, (2) the Willroy, (3) the Nama Creek, and (4) the Willecho (formerly the Lun Echo). In general, the paragenetic sequence at Manitouwadge probably can be divided into four stages: (1) the formation of pyrite; (2) the introduction of pyrrhotite and quartz; (3) the development of sphalerite, chalcopyrite, and very minor cubanite; and (4) the introduction of galena, accompanied by minor tetrahedrite, argentite, pyrargyrite, marcasite, and sericite. The pyrrhotite cements brecciated pyrite, and some pyrrhotite seals fractures in early quartz. Some of the chalcopyrite occurs as oriented blebs in sphalerite and some of the sphalerite occupies similar positions in the chalcopyrite; thus probably indicating exsolution phenomena in both minerals and considerable contemporaneity of deposition. Considerable proportions of these two minerals replaced pyrrhotite, and lesser portions replaced pyrite. Chalcopyrite continued to deposit after sphalerite and replaced the latter mineral to an appreciable degree. The tiny amounts of cubanite in the ore are located as randomly oriented streaks in chalcopyrite and may indicate exsolution effects. Galena replaced all of the older sulfides except pyrite and filled fractures in that mineral. Silver is present in close association with both galena and chalcopyrite; in galena, it probably is in solid solution but how it occurs with chalcopyrite is unknown. The probable exsolution blebs of sphalerite and chalcopyrite in each other suggest that much of these minerals was deposited under hypothermal conditions, a concept supported by the possible exsolution streaks of cubanite in the chalcopyrite. On the other hand, the galena contains considerable silver, and traces of silver sulfides and sulfosalts accompany it. This indicates that the galena formed under low intensity mesothermal or even leptothermal conditions. Chalcopyrite increases with depth, in relation to sphalerite; this may suggest that higher intensity conditions prevailed at greater depths, but this probably means only that deposition took place under somewhat higher-intensity hypothermal conditions than at higher levels. The Manitouwadge ores are, therefore, categorized as hypothermal-1 to mesothermal; possibly leptothermal should be included in the classification for at least some of the silver-rich galena.

MICHIPICOTEN

Early Precambrian — *Iron as Hematite, Siderite, Pyrite, Goethite* — *Mesothermal*

Bell, J. M., 1905, Iron ranges of Michipicoten West: Ont. Bur. Mines (Ann.) Rept., v. 14, pt. 1, p. 278-355

Brown, E. L., and Morrison, W. F., 1942, Geology of the Josephine mine--hydrothermal origin of the hematite: Canadian Min. Jour., v. 63, no. 1, p. 5-9

Bruce, E. L., 1940, Geology of the Goudreau-Lochalsh area: Ont. Dept. Mines 49th Ann. Rept., v. 49, pt. 3, p. 1-50

Coleman, A. P., 1906, The Helen iron mine, Michipicoten: Econ. Geol., v. 1, p. 521-529

—— 1906, Iron ranges of eastern Michipicoten: Ont. Bur. Mines (Ann.) Rept., v. 15, pt. 1, p. 173-206

Coleman, A. P., and Willmot, A. B., 1902, The Michipicoten iron ranges: Ont. Bur. Mines (Ann.) Rept., v. 11, pt. 1, p. 152-185

Collins, W. H., and others, 1926, Michipicoten iron ranges: Geol. Surv. Canada Mem. 147, 175 p. (particularly p. 1-141)

Douglas, G. V., 1954, Pyritic mineralization in the Goudreau area of Algoma, Ontario: Econ. Geol., v. 49, p. 310-316

Gallie, A. E., 1947, Mining methods and costs at the Josephine mine: Canadian Inst. Min. and Met. Tr., v. 50 (Bull. no. 427), p. 589-636 (particularly p. 595-600)

Goodwin, A. M., 1961, Genetic aspects of Michipicoten iron formations: Canadian Inst. Min. and Met. Tr., v. 64 (Bull. no. 585), p. 32-36

—— 1962, Structure, stratigraphy, and origin of iron formations, Michipicoten area, Algoma district, Ontario, Canada: Geol. Soc. Amer. Bull., v. 73, p. 561-586

—— 1963, Michipicoten area (42 C, 41N), district of Algoma, geology and mineral occurrences: Ont. Dept. Mines Prelim. Map P. 184, 1:126,720

—— 1964, Geochemical studies at the Helen Iron Range: Econ. Geol., v. 59, p. 684-718

—— 1966, The relationship of mineralization to stratigraphy in the Michipicoten area, Ontario: Geol. Assoc. Canada Spec. Paper No. 3, Precambrian Symposium, p. 57-73

Gross, G. A., 1965, Algoma type, in *Geology of iron deposits in Canada*: Geol. Surv. Canada Econ. Geol. Rept. No. 22, p. 116-119

Grout, F. F., 1926, Michipicoten iron ranges: (rev.) Econ. Geol., v. 21, p. 813-817

Hawley, J. E., 1942, Origin of some siderite, pyrite, chert deposits, Michipicoten district, Ontario: Roy. Soc. Canada Tr., 3d Ser., v. 36, sec. 4, p. 79-87

Kidder, S. J., and McCartney, G. C., 1948, Mining and geology at the Helen mine: A.I.M.E. Tr., v. 178, p. 240-263

Kinkel, A. R., Jr., 1966, Michipicoten-Goudreau district, Ontario, in *Massive pyritic deposits related to volcanism and possible methods of emplacement*: Econ. Geol., v. 61, p. 682-683

Matheson, A. F., 1932, Michipicoten River area: Geol. Surv. Canada Summ. Rept., pt. D, p. 1-21

Moore, E. S., 1948, Structure of the Michipicoten-Goudreau area, in *Structural geology of Canadian ore deposits*: Canadian Inst. Min. and Met., Montreal, p. 414-428

Moore, E. S., and Armstrong, H. S., 1946, Iron ore deposits of the District of Algoma: Ont. Dept. Mines 55th Ann. Rept., v. 55, pt. 4, p. 1-118 (particularly p. 1-6, 37-118)

Morrison, W. F., 1948, Josephine mine, in *Structural geology of Canadian ore deposits*: Canadian Inst. Min. and Met. Montreal, p. 429-432

Staff, Algoma Ore Properties, Limited, 1956, Helen mine: Canadian Min. Jour., v. 77, no. 11, p. 80-87

Tanton, T. L., 1948, New Helen mine, in *Structural geology of Canadian ore deposits*: Canadian Inst. Min. and Met., Montreal, p. 422-428

Notes

The Michipicoten iron ranges are located in the vicinity of the town of Wawa, lying some 6 miles northeast of the farthest indentation of Michipicoten Harbor on the north shore of Lake Superior. The area is about 105 miles slightly west of north from Sault Ste. Marie.

The mineralization almost certainly is early Precambrian whether Goodwin (1962) correctly identifies them as syngenetic or Moore and Armstrong (1946) are right that they are epigenetic. If the ores are syngenetic, they must have been formed well before the Algoman intrusions marked the end of the early Precambrian. If the ores are epigenetic, the ore solutions that produced them must have been generated in the same magma chamber from which came the Algoman granite-granodiorite intrusives. The granites are accompanied by lamprophyres, diorites (including kersantites), feldspar and quartz porphyries, granite prophyry, syenite, and aplite. The felsic dikes are cut by stringers of ore so that, to Moore and Armstrong, they are pre-ore in age. Goodwin apparently would explain these relationships by a later remobilization of the syngenetic ore. The last igneous activity in the area produced quartz and normal diabase (earlier) and olivine diabase (later); the diabases definitely are post-ore. Thus, the ores seem legitimately to be late early Precambrian in age.

Moore and Armstrong see both the siderite and hematite types of ore bodies as replacements of both banded silica and of the pyroclastic beds, the ore-forming fluids having been derived from a silicic magma of Algoman age. They believe the hematite replaced the host rocks to a lesser degree than it filled open fractures in them. On the other hand, they consider the siderite deposits to have been formed to a larger extent by replacement and by replacement of silicic lavas and pyroclastics rather than of the banded silica. In replacing the silicic pyroclastics, they believe that the ore fluids had a greater effect on the feldspars than on the quartz and that the so-called quartz eyes in the ore are remnants of the quartz from replaced tuff or rhyolite. Some of the quartz, however, they think was introduced with the siderite They consider that the pyrite was in part contemporaneous with the siderite but that generally pyrite replaces siderite. They suggest that the consistent relationship between siderite and silica in the district resulted from the ease with which ore fluids could enter the brecciated pyroclastics and, to a lesser extent, the banded silica. The brecciation centered along the contact between these two rock types and permitted the ore fluids to enter essentially the same portion of the stratigraphic column throughout the district.

On the other hand, Goodwin sees the ores as a syngenetic part of the volcanic-sedimentary sequence of the rocks in the district. He thinks that the silica in the iron formation was deposited from volcanic solutions that had leached the silica from silicic volcanic rocks through which the solutions had passed on their way from their magmatic source to the surface. He believes that the iron and sulfur the solution contained accompanied them from the volume of magma in which they had their origin. It is his opinion that where the solutions included abundant carbon dioxide, the immediate precipitation of siderite resulted. In this thinking, sulfur in considerable quantity does not appear to have become available until after precipitation of the siderite largely has been completed. Some pyrite and other sulfides locally were deposited at the same time as the siderite, but the bulk of the sulfides (mainly pyrite) deposited after siderite had ceased to deposit. In Goodwin's concept, the silica was more soluble in the volcanic solutions (or was introduced later) than siderite or pyrite and precipitated last of the three minerals in question. He considers that the brecciated pyroclastic material, later cemented by siderite, was formed by slumping after partial lithification or by the explosive power of the volcanic gases and was, therefore, produced before the siderite deposition commenced. He thinks that features (in addition to breccia cementation) that have been interpreted by others as indicating the hydrothermal origin of the deposits (e.g., siderite fillings in fractures) were produced by internal diagenetic and/or postconsolidation migration of iron and silica. Goodwin considers the alteration of the volcanic rocks above and below the ore to sericite and ottrelite (chloritoid) schists to have been accomplished by volcanic solutions that traversed the rocks during and immediately after their accumulation. This, of course, is in direct opposition to Moore and Armstrong who believe that these effects were produced by the same hydrothermal solutions that deposited the siderite and pyrite or hematite, as the case might be. Goodwin believes that the presence of hematite in the Josephine-Bartlett area resulted from differences in pH and Eh of the volcanic solutions when they reached the sea floor as compared with those that obtained in the siderite-depositing areas. Moore and Armstrong would accept the idea that pH and Eh differed from one area as opposed to the other, but they would contend that those changes took place in hydrothermal solutions in the narrow sense rather than in near-surface volcanic emanations.

Obviously, both of these hypotheses, the syngenetic-volcanic of Goodwin and the epigenetic-hydrothermal of Moore and Armstrong, cannot be correct. Douglas (1954) points out the following items of evidence that favor the hydrothermal theory for carbonate and pyrite mineralization in the adjoining Goudreau area: (1) the mineralization occupies the sheared and brecciated incompetent layer between the silicic footwall and the mafic hanging wall; (2) the silicic footwall has been sheared, and, in some of its members, ottrelite has been developed; (3) the mafic hanging wall and the silicic footwall have been cut by veins of carbonates that carry sulfides; and (4) the carbonates close to the incompetent layer are iron rich and those at a distance are less ankeritic. Douglas believes that the carbonate veins and carbonates in the foot and hanging walls are the best argument for an epigenetic hypogene origin. He does not think that remobilization after the syngenetic deposition of the siderite could have produced the veins he illustrates.

It must further be pointed out that James's definition of the carbonate and pyrite facies does not correspond to the character of the carbonate and pyrite bodies in the Michipicoten district. Thus, it would appear more probable that the iron, sulfur, and carbonate of the deposits in the area under discussion were of epigenetic hydrothermal origin rather than of syngenetic volcanic origin and that the features of the deposits, while certainly providing some grounds for Goodwin's hypothesis, are more definitely in favor of hydrothermal development of the ore bodies.

Granted that the deposits are hydrothermal in character, it appears that the siderite must have been formed under mesothermal conditions (as were the huge siderite deposits of Bilbao in Spain). The intensity range in which the hematite bodies of the Josephine-Bartlett Range were formed, however, is not as easily settled. The hematite is not the specularite of definitely hypothermal association but, instead, is a dull- to bright-red material that (as is the case with the hematite at Steep Rock Lake) probably formed under less intense conditions than would have been true had the hematite been of the specular variety. The iron deposits of the district, therefore, both the siderite and the hematite ones, are here classified as mesothermal with the understanding that the problems of their place in the classification in both the narrow and broad senses are still subject to further review.

MUNRO

Early Precambrian	*Asbestos*	*Magmatic-4, Mesothermal, Metamorphic-C*

Grubb, P.L.C., 1962, Serpentinization and chrysotile formation in the Matheson ultrabasic belt, northern Ontario: Econ. Geol., v. 57, p. 1228-1246

Hendry, N. W., 1951, Chrysotile asbestos in Munro and Beatty Townships, Ontario: Canadian Inst. Min. and Met. Tr., v. 44 (Bull. no. 465), p. 28-35

Hendry, N. W., and Conn, H.M.K., 1957, The Ontario asbestos properties of Canadian Johns-Manville Company Limited, in *The geology of Canadian industrial mineral deposits*: Canadian Inst. Min. and Met., Montreal, p. 36-45

Hopkins, P. E., 1915, The Beatty-Munro gold area: Ont. Bur. Mines Ann. Rept., v. 24, pt. 1, p. 171-184

Prest, V. K., 1951, Geology of Guibord Township: Ont. Dept. Mines 60th Ann. Rept., v. 60, pt. 9, 56 p.

Satterly, J., 1949, Geology of Garrison Township: Ont. Dept. Mines 58th Ann. Rept., v. 58, pt. 4, 33 p.

—— 1951, Geology of Harker Township: Ont. Dept. Mines 60th Ann. Rept., v. 60, pt. 7, 47 p.

—— 1951, Geology of Munro Township: Ont. Dept. Mines 60th Ann. Rept., v. 60, pt. 8, 60 p.

—— 1952, Geology of McCool Township: Ont. Dept. Mines 61st Ann. Rept., v. 61, pt. 5, 30 p.

—— 1953, Geology of the north half of Holloway Township: Ont. Dept. Mines 62d Ann. Rept., v. 62, pt. 7, 38 p.

Satterly, J., and Armstrong, H. S., 1947, Geology of Beatty Township: Ont. Dept. Mines 56th Ann. Rept., v. 56, pt. 7, 34 p.

Notes

The Munro asbestos deposits are located in the southwest corner of the township of that name and are about 10 miles east of the town of Matheson and 35 miles northwest of Kirkland Lake.

The deposits are contained in a 1000-foot-thick ultramafic sill; the sill itself, one of several large and similarly layered sills in the area, initially was composed of, from top to bottom: (1) rhythmic wavy bands of olivine and pyroxene in the area adjoining the footwall; (2) thick pyroxenite and dunite bands occupying the central portion of the sill; (3) fine, though poorly developed, bands of pyroxenite and dunite making up the top of the ultramafic portion; and (4) a gabbro layer that completes the sequence; much of this rock material is now highly altered. Contacts between the rocks of the four zones are sharp to gradational and even may be chilled; xenoliths are rare. Most geologists who have studied the area believe that the sills are Haileyburian in age, that is, late but not latest early Precambrian. If the alterations of the sill rocks, including the development of the chrysotile asbestos, are deuteric (autometamorphic), the date of the ore formation is almost certainly fixed as late (but not latest) early Precambrian. If, however, Grubb (1962) is correct in ascribing most, if not all, of the asbestos to exogenous hydrothermal fluids, the age of the ores probably is Algoman since the latest silicic intrusives in the area are of that age. In either event, the deposits can be classified as late early Precambrian.

From the central location in the sill of the huge truncated lens of asbestos-bearing serpentine, it would appear that at least part of the serpentinization was carried out by (deuteric) solutions developed in the sill mass in the late stages of its crystallization. On the other hand, this lens and the sill containing it now stand nearly vertical, indicating appreciable deformation since emplacement. The presence of the asbestos in fractures that probably were formed as a result of this deformation indicates that most, if not all, of the asbestos could not have been formed by the deuteric solutions but must have been produced by solutions of a later time and another source. Since the Keewatin rocks in the general area of the Munro mine have been widely and intensely sericitized and carbonatized, probably by solutions given off by the Algoman intrusives, it would appear most likely that the asbestos was produced from the serpentinized sill rock by these same solutions after they had acquired considerable amounts of silica from the Keewatin rocks they had so highly altered. It is probable that all of the serpentinization was not accomplished by the deuteric solutions produced in the ultramafic sill and that some of the serpentinization was carried out by these later silica-rich solutions before or during the time at which they formed the asbestos. Criteria have not been advanced that unequivocably determine when and how the serpentine was developed. It would appear, however, that the serpentinization was certainly in part automorphic and possibly in part hydrothermal (as also seems to have been the case in the Thetford-Black Lake deposits), it being very difficult to say which of the two processes produced the greater amount of serpentine. The development of sericite and carbonate in the Keewatin rocks by the later serpentinizing solutions strongly suggests that they were then within the mesothermal intensity range, and this further indicates that, when they accomplished the additional serpentinization of the ultramafics, they were also operating under mesothermal conditions. Thus, the hydrothermal serpentinization of the sill at Munro probably should be classed as mesothermal. The serpentinized bodies are bounded by irregular selvages of talc-chlorite-carbonate rock, still further indicating that the solutions responsible for the carbonatization of the Keewatin rocks also strongly affected the ultramafics. This reasoning suggests that the later serpentinization at Munro was accomplished under somewhat less intense conditions than those that obtained at Thetford-Black Lake. Grubb (1962) believes that chrysotile asbestos developed in three ways: (1) through open-space crystallization in fractures related to anticlinal crests and fault zones; (2) by replacement of picrolite (chemically and crystographically essentially the same as chrysotile but physically different according to Riordon; Deer, Howie, and Zussman hold that there are three serpentine minerals--chrysotile, lizardite, and antigorite, that these have the same general composition and structure, and that picrolite is simply a fibrous variety of antigorite); and (3) through replacement of serpentine in dunite and periodotite, more abundantly in the former. In any event, it appears probable that the actual development of the chrysotile form was the result of stresses (or relief of stresses) acting on the serpentine minerals already developed or so affecting serpentine material being deposited in open space that it formed as chrysotile. As prior serpentinization is essential for the large-scale development of chrysotile and as a considerable fraction of the serpentine at Munro is of deuteric origin, the deposits are classed as magmatic-4 in part. Because some of the serpentine probably was developed by later moderate-temperature hydrothermal solutions given off by silicic magmas, the deposit is also classified as mesothermal. Because the stress component was essential to the development of chrysotile, the deposit further is designated as metamorphic-C.

PICKLE CROW

Early Precambrian *Gold* *Mesothermal*

Bothwell, S. A., 1938, Geology of the Pickle Crow gold mine: Canadian Inst. Min. and Met. Tr., v. 41 (Bull. no. 312), p. 132-140

Chisholm, E. O., 1950, Crow River area geology: Canadian Min. Jour., v. 71, no. 11, p. 112-113

Corking, W. P., 1948, Pickle Crow mine, in *Structural geology of Canadian ore deposits*: Canadian Inst. Min. and Met., Montreal, p. 373-376

Donaldson, J. A., and Jackson, G. D., 1965, Archean sedimentary rocks of North Spirit Lake area, northwestern Ontario: Canadian Jour. Earth Sci., v. 2, no. 6, p. 622-647

Ferguson, S. A., 1966, Geology of the Pickle Crow Gold Mines Limited and Central Patricia Gold Mines Limited, no. 2 operation: Ont. Dept. Mines Misc. Paper MP-4, 97 p.

Hurst, M. E., 1930, Pickle Lake-Crow River area, district of Kenora (Patricia portion): Ont. Dept. Mines 39th Ann. Rept., v. 39, pt. 2, p. 1-35

Monette, H. H., 1949, Geological outline of Pickle Crow: Canadian Min. Jour., v. 70, no. 11, p. 99-105

Thomson, James E., 1938, Structure of gold deposits in the Crow River area: Canadian Inst. Min. and Met. Tr., v. 41 (Bull. no. 316), p. 358-374

—— 1938, The Crow River area: Ont. Dept. Mines 47th Ann. Rept., v. 47, pt. 3, p. 1-65

Notes

The Pickle Crow district is located in the Patricia portion of the district of Kenora at about 51°30'N and 90°W; it is some 215 miles north-northwest of Port Arthur.

The Pickle Crow ores occur mostly in early Precambrian layered metasediments and metavolcanics, but enough mineralization is found in the later intrusive silicic porphyries to indicate that the rock type was emplaced before the ore was deposited. Possibly some of the still later metagabbros were introduced before the ore and some after it; if this is the case, the ores are of essentially the same age as the metagabbros for which, unfortunately, no radioactive age dates have been published. Since, however, there is no evidence that metamorphism took place in this portion of the Superior province after about 2400 m.y. ago, it is probable that the age of mineralization is no younger than that. The gabbros and the porphyries do appear to have been introduced into the area during much the same time span so if the gabbros are more than 2400 m.y. old, the porphyries are also. All other gold ores in Ontario, from Red Lake to Kerr Addison, are late early Precambrian; this would seem to provide confirmation of the late Precambrian age assumed here for the Pickle Crow ores.

Because the Pickle Crow veins are widely separated in space, probably appreciable differences existed among the various ore fluids by which the veins were filled. As a result, Pye's paragenetic sequence, quoted by Ferguson (1966), must be considered to be a generalized one. Pye believes the earliest minerals were carbonate (ferruginous dolomite), arsenopyrite, and pyrite. These minerals were then fractured, and scheelite, tourmaline, quartz, and albite were introduced into the broken vein material; some of the scheelite may belong to the first stage of mineralization. A second period of fracturing was followed by the deposition of sphalerite, chalcopyrite, galena, carbonate, and gold; a third period of fracturing then occurred, and these fractures were filled by quartz and calcite. The early ore fluids almost certainly were in the hypothermal range as is witnessed by the arsenopyrite, scheelite, tourmaline, albite, and pyrrhotite of the first and second stages of mineralization. The gold and its accompanying galena and carbonate, on the other hand, probably formed under less intense conditions. Thus, the assignment of a mesothermal classification to the Pickle Crow gold deposits would seem most reasonable.

PORCUPINE

Early Precambrian — *Gold, Copper* — *Hypothermal-1 to Mesothermal (Au); Mesothermal (Cu)*

Allen, C. C., and Folinsbee, R. E., 1944, Scheelite veins related to porphyry intrusives, Hollinger mine: Econ. Geol., v. 39, p. 340-348

Bain, G. W., 1933, Wall-rock mineralization along Ontario gold deposits: Econ. Geol., v. 28, p. 705-743 (particularly p. 714-727)

Bradshaw, R. J., 1961, Geology of the Hallnor mine: Canadian Min. Jour., v. 82, no. 12, p. 35-40

Buffam, B.S.W., and others, 1948, Individual mines in the Porcupine district, Ontario, in *Structural geology of Canadian ore deposits*: Canadian Inst. Min. and Met., Montreal, p. 457-569

Burrows, A. G., 1924, The Porcupine gold area: Ont. Dept. Mines 33d Ann. Rept., v. 33, pt. 2, 112 p.

Butterfield, H. M., 1941, The Preston East Dome mine, Ontario: Canadian Min. Jour., v. 62, no. 8, p. 511-516

Dougherty, E. Y., 1934, Mining geology of the Vipond gold mine, Porcupine district, Ontario: Canadian Inst. Min. and Met. Tr., v. 37 (Bull. no. 265), p. 260-285

—— 1939, Some geological features of Kolar, Porcupine and Kirkland Lake: Econ. Geol., v. 34, p. 622-653

Dunbar, W. R., 1948, Structural relations of the Porcupine ore deposits, in *Structural geology of Canadian ore deposits*: Canadian Inst. Min. and Met., Montreal, p. 442-456

Evans, J.E.L., 1944, Porphyry of the Porcupine district, Ontario: Geol. Soc. Amer. Bull., v. 55, p. 1115-1141

Ferguson, S. A., 1966, The relationship of mineralization to stratigraphy in the Porcupine and Red Lake areas, Ontario: Geol. Assoc. Canada Spec. Paper no. 3, Precambrian Symposium, p. 99-119

—— 1968, Geology and ore deposits of Tisdale Township: Ont. Dept. Mines Geol. Rept. 58, 177 p.

Ferguson, S. A., and associated geologists, 1964, Geology of mining properties in Tisdale Township, Porcupine area: Ont. Dept. Mines P.R. 1964-65, 124 p.

George, P. T., and others, 1967, Timmins and district: Canadian Inst. Min. and Met., Centennial Field Excursion, Northwestern Quebec and Northern Ontario, p. 102-134

Goodwin, A. M., 1965, Mineralized volcanic complexes in the Porcupine-Kirkland Lake-Noranda region, Canada: Econ. Geol., v. 60, p. 955-971

Graton, L. C., and others, 1933, Outstanding features of Hollinger geology: Canadian Inst. Min. and Met. Tr., v. 36 (Bull. no. 249), p. 1-20

Griffis, A. T., 1962, A geological study of the McIntyre mine: Canadian Inst. Min. and Met. Tr., v. 65 (Bull. no. 598), p. 47-54

Gustafson, J. K., 1945, The Porcupine (Ontario) porphyries: Econ. Geol., v. 40, p. 148-152

Hawley, J. E., 1952, Spectrographic studies of pyrite in some eastern Canada gold mines: Econ. Geol., v. 47, p. 260-304 (particularly p. 284-298)

Hogg, N., 1950, The Porcupine gold area: Canadian Min. Jour., v. 71, no. 11, p. 102-106

Holmes, T. C., 1944, Some porphyry-sediment contacts at the Dome mine, Ontario: Econ. Geol., v. 39, p. 133-141

—— 1947, Structural control of ore deposits at the Dome mine: Canadian Inst. Min. and Met. Tr., v. 50 (Bull. no. 421), p. 283-297

Hurst, M. E., 1935, Vein formation at Porcupine, Ontario: Econ. Geol., v. 30, p. 103-127

—— 1936, Recent studies in the Porcupine area: Canadian Inst. Min. and Met. Tr., v. 39 (Bull. no. 291), p. 448-458

—— 1942, Gold deposits at Porcupine, Ontario, in Newhouse, W. H., Editor, *Ore deposits as related to structural features*: Princeton Univ. Press, p. 196-199

Keys, M. R., 1940, Paragenesis in the Hollinger veins: Econ. Geol., v. 35, p. 611-628

Langford, G. B., 1941, Geology of the McIntyre mine: A.I.M.E. Tr., v. 144, p. 151-171

McKinstry, H. E., and Ohle, E. L., Jr., 1949, Ribbon structure in gold-quartz veins: Econ. Geol., v. 44, p. 87-109

McLaughlin, D. B., 1956, Keewatin-Timiskaming unconformity at Porcupine, Ontario, Canada: Geol. Soc. Amer. Bull., v. 67, p. 939-940

Moore, E. S., 1953, The structural history of the Porcupine gold ores: Roy. Soc. Canada Tr., 3d Ser., v. 47, sec. 4, p. 39-53

—— 1954, Porphyries of the Porcupine area, Ontario: Roy. Soc. Canada Tr., 3d Ser., v. 48, sec. 4, p. 41-57

Ringsleben, W. C., 1935, Geology (of the Hollinger mine): Canadian Min. Jour., v. 56, no. 9, p. 364-372

Robinson, H. S., 1923, Geology of the Pearl Lake area, Porcupine district, Ontario: Econ. Geol., v. 18, p. 753-771

Smith, F. G., 1948, The ore deposition temperature and pressure at the McIntyre mine, Ontario: Econ. Geol., v. 43, p. 627-636

Stockwell, C. H., Editor, 1957, Porcupine district, in *Geology and economic minerals of Canada*: Geol. Surv. Canada Econ. Geol. Ser. no. 1, 4th ed., p. 52-59

Whitman, A. R., 1927, A syntectic porphyry at Porcupine: Jour. Geol., v. 35, p. 404-420

Notes

The Porcupine district extends east from Timmins for about 15 miles.

The ore veins in the Porcupine district are localized by fractures in the schists into which a large fraction of the Keewatin- and Timiskaming-type rocks have been converted, but most of the ore was formed by replacement of intensely sheared or foliated zones adjacent to the veins. Most of the ores were emplaced in Keewatin type rocks of the Tisdale group in and around volumes of quartz porphyry. Since the quartz porphyries also contain gold-bearing veins, the ores certainly were deposited after the quartz porphyries were introduced. Because there is no evidence of any igneous activity of any magnitude after the Algoman orogeny, it seems likely that the ore-forming fluids were derived from the same source as the porphyries themselves. Gold is known in the Algoman granite that surrounds the metavolcanic-metasedimentary host rocks of the Algoman porphyries, further indicating a genetic connection between the

Algoman intrusives and the Porcupine gold quartz veins. Muscovite from the Pearl Lake porphyry in the McIntyre mines has been dated as about 2475 m.y. Although this date is slightly on the middle Precambrian side of the age boundary (2600 m.y.) used in this volume to separate early from middle Precambrian time, it is thought better to consider that the intrusives were introduced during the Algoman orogeny, that this orogeny marked the end of the early Precambrian, and that the Porcupine ores are late early Precambrian.

Although no one mine can be said to be typical of all Porcupine mines, the Hollinger, as the largest of the group, provides a good example of the order and character of ore formation. At Hollinger the minerals appear to have been introduced in the following order: (1) after an initial period of fracturing, ankerite, tourmaline, and scheelite were deposited; locally axinite and clinozoisite also were introduced; (2) after fracturing of the type-(1) veins, albite was introduced widely in these openings, and the first pyrite was disseminated into the wall rocks; chlorite was formed in the outer part of the mineralized area; (3) following still a third period of fracturing (which did not reopen all veins) the most intense mineralization of the period took place; in this, large amounts of carbonate and pyrite were added to the wall rocks and massive pyrite and some arsenopyrite were deposited in the veins; (4) after a fourth epoch of fracturing, most of the veins (whether opened in all previous mineralizations or not) were reopened and abundant quartz was introduced; (5) this fourth epoch was followed by a fifth period of fracturing and then by the emplacement of more quartz, plus pyrrhotite, sphalerite, chalcopyrite, galena, tellurides, sericite, and gold; (6) a sixth period of fracturing was followed by more quartz and small amounts of calcite, sphalerite, chalcopyrite, sericite, and gold; (7) locally, a seventh period of fracturing allowed the entry of still more quartz and considerable calcite. Without much doubt, the scheelite (400 tons of contained WO_3 were recovered during World War II) was deposited under hypothermal conditions. By the time the gold was precipitated, however, the temporally associated minerals of (4) and (5) above indicate deposition probably in the mesothermal range. Although some of the gold is closely associated with tellurides, these latter minerals are essentially mineralogical curiosities at Hollinger, having been found in quantity at only one place in the mine, so it does not seem justifiable to include leptothermal in the classification.

At the McIntyre mine, the bulk of the gold is found in the sericitized wall-rock schists bordering the quartz veins and is almost entirely in pyrite (as is the case at Hollinger). The work done by Griffis (1962) indicates that there were two generations of gold at the McIntyre, the earlier gold being associated with slightly earlier ankerite, tourmaline, and albite, with slightly earlier and overlapping scheelite, pyrite, pyrrhotite, and quartz, with contemporaneous arsenopyrite, and with slightly later ankerite, anhydrite, and calcite. The second generation of gold occurred as the last (except for some barren gangue) in a considerable series of minerals, including early anhydrite, quartz, and chlorite; slightly later gypsum and chalcopyrite; followed by tennantite, sphalerite, and galena; and then by gold that overlaps only with the lead and zinc sulfides. The mineral association of the early gold puts it in the hypothermal range, while the later is almost certainly mesothermal. At the Dome mine, none of the minerals with which the gold is closely associated in time are too diagnostic of a particular intensity range, but mesothermal seems to be the best guess possible at this time. The ores of the Preston East Dome mine seem to be more likely to have been formed under hypothermal conditions than those of the Dome, but the probable late appearance of the gold in the paragenetic sequence may reduce the diagnostic value of the abundant tourmaline and scheelite. Enough of the gold in the district, however, appears to have been deposited in the hypothermal range to justify a classification of hypothermal-1 to mesothermal. The copper ore at the McIntyre seems to belong in the mesothermal range.

RED LAKE

Early Precambrian | *Gold* | *Hypothermal-1 to Mesothermal*

Bruce, E. L., 1924, Geology of the basin of Red Lake, district of Patricia: Ont. Dept. Mines 33d Ann. Rept., v. 33, pt. 4, p. 12-39

Bruce, E. L., and Hawley, J. E., 1927, Geology of the basin of Red Lake, district of Kenora (Patricia portion): Ont. Dept. Mines 36th Ann. Rept., v. 36, pt. 6, p. 1-52

Chisholm, E. O., 1951, Geology of Balmer Township: Ont. Dept. Mines 60th Ann. Rept., v. 60, pt. 10, 62 p.

Christopher, I. C., 1951, Geology of Cochenour Willans gold mine: Canadian Inst. Min. and Met. Tr., v. 54 (Bull. no. 470), p. 246-250

Ferguson, S. A., 1960, Balmer Township (revised): Ont. Dept. Mines Prelim. Map P. 47, 1:1000

—— 1961, Dome Township: Ont. Dept. Mines Prelim. Map P. 125, 1:1000

—— 1962, Geology of the south half of Bateman Township, district of Kenora: Ont. Dept. Mines Geol. Rept. no. 6, 31 p.

—— 1963, Heyson Township (northern half): Ont. Dept. Mines Prelim. Map P. 208, 1:1000

—— 1965, Geology of the eastern part of Baird Township: Ont. Dept. Mines Geol. Rept. no. 39, 47 p.

—— 1966, Dome Township, district of Kenora: Ont. Dept. Mines Geol. Rept. 45, 98 p.

—— 1966, The relationship of mineralization to stratigraphy in the Porcupine and Red Lake areas, Ontario: Geol. Assoc. Canada Spec. Paper no. 3, Precambrian Symposium, p. 99-119

—— 1968, Geology of the northern part of Heyson Township, district of Kenora: Ont. Dept. Mines Geol. Rept. 56, 54 p.

Gross, W. H., and Ferguson, S. A., 1965, The anatomy of an Archean greenstone belt: Canadian Inst. Min. and Met. Tr., v. 68 (Bull. no. 641), p. 254-260

Gummer, W. K., 1941, Border rocks of a granite batholith, Red Lake, Ontario: Jour. Geol., v. 49, p. 641-656

Horwood, H. C., 1940, Geology at the Cochenour Willans gold mine, Red Lake, Ontario: Canadian Inst. Min. and Met. Tr., v. 43 (Bull. no. 337), p. 217-236

—— 1940, Geology and mineral deposits of the Red Lake area: Ont. Dept. Mines 49th Ann. Rept., v. 49, pt. 2, 231 p.

—— 1940, The Keewatin-Timiskaming unconformity at Red Lake, Ontario: Roy. Soc. Canada Tr., 3d Ser., v. 34, sec. 4, p. 45-52

—— 1948, General structural relationships of ore deposits in the Red Lake area, in *Structural geology of Canadian ore deposits*: Canadian Inst. Min. and Met., Montreal, p. 322-328

Horwood, H. C., and Keevil, N. B., 1943, Age relationships of intrusive rocks and ore deposits in the Red Lake area, Ontario: Jour. Geol., v. 51, p. 17-32

Hurst, M. E., 1935, Gold deposits in the vicinity of Red Lake: Ont. Dept. Mines 44th Ann. Rept., v. 44, pt. 6, p. 1-52

Kuryliw, C. J., 1957, Cochenour Willans mine, in *Structural geology of Canadian ore deposits*, v. 2: Canadian Inst. Min. and Met., Montreal, p. 295-304

—— 1957, The structural geology of the Cochenour-Willans gold mine: Canadian Inst. Min. and Met. Tr., v. 60 (Bull. no. 540), p. 128-133

Mather, W. B., 1937, Geology and paragenesis of the gold ores of the Howey mine, Red Lake, Ontario: Econ. Geol., v. 32, p. 131-153

McIntosh, R., and others, 1948, Individual mines in the Red Lake area, Ontario, in *Structural geology of Canadian ore deposits*: Canadian Inst. Min. and Met., Montreal, p. 328-365

McLaren, D. C., 1947, Madsen Red Lake gold mines: Canadian Min. Jour., v. 68, no. 11, p. 791-808 (particularly p. 793-800)

Morrow, H. F., 1949, The geology of Hard Rock gold mines quartz stringer ore zones: Precambrian, v. 22, no. 9, p. 10-15

Nowlan, J. P., 1947, The geology of the Cochenour Willans gold mine, Red Lake, Ontario: Precambrian, v. 20, no. 5, p. 6-10

Smith, T. S., 1951, Geology of McKenzie Red Lake gold mines property: Canadian Inst. Min. and Met. Tr., v. 54 (Bull. no. 468), p. 178-183

Staff, McKenzie Red Lake Gold Mines, Limited, 1938, McKenzie Red Lake Gold Mines, Limited: Canadian Inst. Min. and Met. Tr., v. 41 (Bull. no. 316), p. 343-357 (particularly p. 346-349)

Stronach, R. S., Jr., and De Wet, J. P., 1950, The Campbell Red Lake gold mine, Balmer Township, Ontario: Precambrian, v. 23, no. 3, p. 12-17

Thomson, R., 1946, Notes on some recent work in the eastern part of Red Lake area, Ontario: Precambrian, v. 19, no. 6, p. 6-7, 9, 13

Notes

The Red Lake district is in far western Ontario, about 60 miles east of the border with Manitoba and about 160 miles slightly west of north from Fort Frances on the international border.

The oldest rocks of the Red Lake district are metamorphosed volcanic and sedimentary rocks that Horwood (1948) divided between Keewatin and Timiskaming types on the basis of their general similarities to rocks bearing these designations in other parts of Canada. Since 1951, the trend in Canada has been not to assign the terms "Keewatin" and "Timiskaming" to rocks of these types unless they actually are in the type area of these formations or can be traced uninterruptedly into them. As a result, in the latest work on the district, Ferguson (1965) dropped both of these terms, lumping all of these formations under the designation of Precambrian alone.

Gross and Ferguson (1965) have pointed out that the Red Lake altered volcanic (greenstone) belt, 35 miles long (east-west) and 18 miles wide, may be either (1) a roof pendant floating in a sea of later granite, some of which actually has intruded the greenstone belt, or (2) volcanic rocks and sediments laid down on a pre-existing crust of granitic composition. Thus, under the second concept, the greenstone belt contains the infolded remnants of younger lavas and sediments deposited on the older crust, and the belt and its surrounding granite later were intruded by younger igneous rocks. Gross and Ferguson proceed to test their two hypotheses in several ways. Absolute age determinations were made on the granodiorite of the Dome stock (one of the intrusives cutting the greenstones), and the age so found was 2400 m.y. The same age was determined for the granitic rocks outside the greenstone belt. These results would suggest that the greenstones are a roof pendant. Gross and Ferguson remark, however, that the 2400 m.y. is only the age of the last metamorphism to affect the area and that the coincidence in age between the encircling granite and the intruded granodiorite may be due to their having been metamorphosed simultaneously, with the granite actually having been much the older rock of the two. Further, they follow Wilson in believing that the greenstones are orogenic or continental basalts and that they were intruded into an earlier continental crust. They think that the silicic intrusives in the Red Lake district may have been derived from remelting of this ancient crust, the age of the original formation of which must have been at least 3200 m.y. They determined that there was a 90 percent level of confidence that there is no significant difference in composition between the granitic rocks inside and outside the greenstones. From lead obtained from the McKenzie

Red Lake mine, however, they obtained an age of about 3200 m.y. This age is about 800 m.y. older than that of the rocks containing the ores of the McKenzie mine, and they believe that this anomaly can be explained by assuming that the lead in question was obtained from remelting and redifferentiation of the 3200 m.y. crust. The results they quote from sulfur-isotope studies appear to be inconclusive. In short, it would appear that the age of ore formation is uncertain but that it is more probable that the rocks intrusive into the greenstones were emplaced some 2400 m.y. ago or at the end of the early Precambrian (end of the Archean in the Canadian sense). On this premise, the ores contained in the volcanic-sedimentary complex and in the intrusives must have been introduced there about 2400 m.y. ago, and they should be categorized as latest early Precambrian.

Horwood (1940, 1948) points out that there are three distinct types of gold mineralization in the Red Lake district: (1) quartz veins and stringers with small amounts of sulfides and different amounts of gold, (2) veins and lenses of cherty quartz and carbonate with some arsenopyrite (and minor amounts of a few other sulfides) with which gold is closely associated, and (3) silicified shear zones in tuff that contain a pyrite-dominant sulfide mineralization. Type (1) ores have been subdivided into (a) quartz veins and lenses in shear zones and fractures as are found at the McKenzie Red Lake and Gold Eagle mines and (b) quartz stringers and lenses in fracture zones as occur at the Howey and Hasaga mines. Ores of type (2) are present in the Cochenour Willans and McMarmac mines, and those of type (3) at the Madsen Red Lake mine. The New Dickenson and Campbell Red Lake ores appear to belong to type (1a), and the ores at the Starratt nickel mines are similar to the type (3) ores in the neighboring Madsen Red Lake mine.

The quartz veins and stringers of type (1) are by far the most abundant source of gold in the district. Horwood (1940a) points out that the process of ore deposition can be understood only on the basis of the following three facts: (1) there is a great diversification in the various types of mineralization; (2) the mineralization took place in at least two stages, and the stages were separated by periods of fracturing, with the later mineralizations following the fractures developed in the earlier one or ones; and (3) gold was the last mineral to have been emplaced in any of the stages of mineralization and, in most mineralized rock volumes, was deposited only at the end of the last stage.

In type (1) mineralization the gold is late enough, in relation to the sulfides, to have been formed under no more intense conditions than mesothermal; the gold might even be leptothermal. In type (2) much of the gold is so intimately associated with arsenopyrite that it almost certainly was formed under conditions of hypothermal intensity. In type (3) the gold is enough later than the sulfides to be certainly mesothermal. In each instance, the wall-rock alteration is compatible with mesothermal conditions. The ores, therefore, probably were formed mainly under mesothermal conditions and, to a slight extent, under hypothermal conditions; the deposits, then, are categorized as hypothermal-1 to mesothermal with the understanding that most of the gold probably was deposited in the mesothermal range.

STEEP ROCK LAKE

Early Precambrian	*Iron as Goethite, Hematite, Pyrite*	*Sedimentary-A1a, Residual-B1; Hypothermal-1 or -2 to Mesothermal*

Bartley, M. W., 1939, Iron deposits of the Steeprock area: Ont. Dept. Mines 48th Ann. Rept., v. 48, pt. 2, p. 35-47

—— 1948, Steep Rock iron mine, in *Structural geology of Canadian ore deposits*: Canadian Inst. Min. and Met., Montreal, p. 419-421

Brant, A., 1940, Geophysical investigation at Steep Rock Lake: Canadian Inst. Min. and Met. Tr., v. 43 (Bull. no. 338), p. 274-284

Clay, C., 1941, The Steep Rock hematite deposits: Precambrian, v. 14, no. 5, p. 2-5, 9

Gruner, J. W., 1956, Geology and iron ores of Steep Rock Lake: Econ. Geol., v. 51, p. 98-99

Hicks, H. S., 1950, Geology of the iron deposits of Steep Rock Iron Mines Limited: Precambrian, v. 23, no. 5, p. 8-10, 13

Jolliffe, A. W., 1955, Geology and iron ores of Steep Rock Lake: Econ. Geol., v. 50, p. 373-398

—— 1966, Stratigraphy of the Steep Rock Group, Steep Rock Lake, Ontario: Geol. Assoc. Canada Spec. Paper no. 3, Precambrian Symposium, p. 75-98

Lawson, A. C., 1912, Geology of Steeprock Lake, Ontario: Geol. Surv. Canada Mem. 28, 23 p.

Moore, E. S., 1938, The Steeprock series: Roy. Soc. Canada Tr., 3d Ser., v. 32, sec. 4, p. 11-23

—— 1939, Geology and ore deposits of the Atikokan area: Ont. Dept. Mines 48th Ann. Rept., v. 48, pt. 2, p. 1-34

Quirke, T. T., 1943, Hydrothermal replacement in deep seated iron ore deposits of the Lake Superior region: Econ. Geol., v. 38, p. 662-666

Roberts, H. M., and Bartley, M. W., 1943, Replacement hematite deposits, Steep Rock Lake, Ontario: Canadian Inst. Min. and Met. Tr., v. 46 (Bull. no. 378), p. 324-374

—— 1943, Hydrothermal replacement in deep seated iron ore deposits of the Lake Superior region: Econ. Geol., v. 38, p. 1-24

—— 1948, Replacement hematite deposits, Steep Rock Lake, Ontario: A.I.M.E. Tr., v. 178, p. 357-396

Royce, S., and others, 1943, The Steep Rock iron ore deposits: Canadian Min. Jour., v. 64, no. 7, p. 437-444

Smith, F. G., 1942, Notes on the iron ores of Steeprock Lake, Ontario: Univ. Toronto Studies, Geol. Ser. no. 47, p. 71-75

Smyth, H. L., 1891, Structural geology of Steep Rock Lake, Ontario: Amer. Jour. Sci., 4th Ser., v. 42, no. 250, p. 317-331

Stockwell, C. H., Editor, 1957, Steep Rock type, in *Geology and economic minerals of Canada*: Geol. Surv. Canada Econ. Geol. Ser. no. 1, 4th ed., p. 96-99

Tanton, T. L., 1925, Mineral deposits of Steeprock Lake map-area: Geol. Surv. Canada Summ. Rept., pt. C, p. 1-11

—— 1941, Origin of the hematite deposits at Steeprock Lake, Ontario: Roy. Soc. Canada Tr., 3d Ser., v. 35, sec. 4, p. 131-141

—— 1946, The iron ore at Steeprock Lake: Roy. Soc. Canada Tr., 3d Ser., v. 40, sec. 4, p. 103-110

Wright, C. M., 1965, Syngenetic pyrite associated with a Precambrian iron ore deposit: Econ. Geol., v. 60, p. 998-1019

Notes

The Steep Rock Lake iron deposits are located just north of the town of Atikokan and are slightly over 100 miles west-northwest of Port Arthur and about 85 miles east-northeast of Fort Frances on the international border.

The ore zone may be, if it actually lies on a basal Karst surface as Jolliffe (1955) thinks it does, a residual accumulation of lateritic iron oxide. It may be, however, that a carbonate bed was attacked by hydrothermal solutions and that these solutions developed a replacement contact between ore and carbonate that approximates the shape that would have obtained on a Karst surface. Finally, a considerable portion of the ore may have been derived from a sedimentary iron-bearing formation, enriched by iron provided by hydrothermal solutions. The formations of the district are cut by intrusives (the post-Steep Rock intrusives), dikes and small stocks of felsite and silicic porphyry that probably are related in time and source to the Steep Rock granite of Algoman age (it was emplaced about 2500 m.y. ago). The nearest approach of large bodies of this granite to the ore is less than 2 miles northwest of the northwest corner of the ore zone.

Granted that the post-Steep Rock intrusives are Algoman in age, all of the older formations must have been developed in early Precambrian time. If this is the case, then the ores, if residual as postulated by Jolliffe, must have been formed during the early Precambrian and perhaps before the close of that span of geologic time. On the other hand, if some or all of the ores were introduced by later hydrothermal solutions, it is difficult to assign a source to these fluids that would be younger than the Algoman magmatic activity. Thus, it follows that, whether residual or hydrothermal, the ores must have been formed in the early Precambrian, and they can safely be classified here as such. This assignment of an age to the deposits is given some confirmation by an age of 2480 m.y. determined for a paraschist from the Steep Rock district. The age so derived is probably that of the metamorphism induced by the intrusion of the Algoman granite.

Of the two widely divergent hypotheses that have been put forward to explain how the ores were emplaced between their walls of dolomite (foot) and ash rock (hanging), the first suggests that they were hydrothermal replacement deposits, but complete agreement as to what type of rock was replaced has not been reached. The original concept of Roberts and Bartley (1943) was that the ores had replaced that portion of the dolomite beds directly under the ash rock. The idea now favored by mine geologists is that the replaced formation was a chert-magnetite iron-bearing formation; whatever the replaced rock originally was, it was very thoroughly converted to ore. The presence of some hematite in the ash rock, of manganiferous paint rock in the dolomite, and of kaolinite in granite masses in the footwall is thought to indicate that wall-rock alteration occurred during the same general period of time as ore formation. The age of this hydrothermal activity is thought to have been Algoman.

The second hypothesis suggested by Jolliffe (1955) is that the ores were formed by the weathering of the dolomite prior to the deposition of any younger rocks. This weathering of the dolomite and the removal of the carbonate portion of the rock in solution allowed the residual (lateritic) accumulation of iron oxides (originally incorporated in the dolomite or in interbedded iron formation) on the Karst surface formed by these near-surface processes. The lower portion of the ore zone is composed of manganiferous paint rock that rests on the Karst-appearing surface and may have developed in the weathering cycle. The dolomite itself is well banded in large part and contains algal structures; both of these phenomena suggest a sedimentary origin for the rock as a whole and argue against the suggestion that the dolomite was hydrothermally deposited. A sedimentary origin of the dolomite does not, of course, demonstrate that the ore was of residual origin; a sedimentary dolomite could be hydrothermally replaced to give the ore zone, but the presence of goethite as the earliest mineral certainly is not what would be expected of hydrothermal processes. Nevertheless, this goethite is not the only iron mineral, and most of the goethite has been broken into angular fragments which have been largely cemented by hematite, and these fragments have, to a considerable extent, been replaced by hematite, and a minor part of the hematite is of the specular variety; the ratio of hematite to goethite is about 1:4. Finally, both the hematite and the goethite have been cut by veins of pyrite. These relationships, then, suggest that, while perhaps much of the iron oxide in the ore zone was produced by residual accumulation, additional iron probably was added by hydrothermal solutions both as hematite and as pyrite. This concept is supported by the presence in the hanging-wall volcanics of pyrite and/or hematite in siliceous replacement bodies, some of which preserve the ash-rock structure and are, therefore, most probably of hydrothermal origin.

Thus, the deposit probably is best classified as part sedimentary-Ala because a considerable portion of the iron may have been derived from original iron-bearing formation in the stratigraphic sequence, in part as residual-Bl because the concentration of a considerable fraction of the iron may have been by lateritic processes, and in part as hydrothermal (hypothermal-1 or -2 to mesothermal) because a sizable fraction of the iron appears to have been introduced hydrothermally under conditions intense enough to have caused the formation of specularite. Because it is uncertain whether the ore fluids replaced iron-bearing formation or dolomite, it is impossible to say with certainty whether the ores are hypothermal-1 or -2.

SUDBURY

Middle Precambrian — *Copper, Nickel, Cobalt, Gold, Platinum Metals* — *Magmatic-2a, Magmatic-2b*

Alcock, F. J., 1935, Sudbury area, in *Copper resources of the world*: 16th Int. Geol. Cong., v. 1, p. 89-99

Barlow, A. E., 1906, On the origin and relations of the nickel and copper deposits of Sudbury, Ontario, Canada: Econ. Geol., v. 1, p. 454-466, 545-553

Bateman, A. M., 1917, Magmatic ore deposits, Sudbury, Ont.: Econ. Geol., v. 12, p. 391-426

Burrows, A. G., and Rickaby, H. C., 1929, Sudbury basin area, Ontario: Ont. Dept. Mines 38th Ann. Rept., v. 38, pt. 3, 55 p.

—— 1934, Sudbury nickel field restudied: Ont. Dept. Mines 43d Ann. Rept., v. 43, pt. 2, 49 p.

Campbell, W., and Knight, C. W., 1907, On the microstructure of nickeliferous pyrrhotites: Econ. Geol., v. 2, p. 350-366 (particularly p. 351-359)

Card, K. D., and Meyn, H. D., 1969, Geology of the Leinster-Bowell area, district of Sudbury: Ont. Dept. Mines Geol. Rept. 65, 40 p.

Card, K. D., 1967, Geology of the Sudbury area: Geol. Assoc. Canada Guidebook (Geology of parts of eastern Ontario and western Quebec), p. 109-122

—— 1968, Geology of the Denison-Waters area, district of Sudbury: Ont. Dept. Mines Geol. Rept. 60, 63 p.

Cheney, E. S., and Lange, I. M., 1967, Evidence for sulfurization and the origin of some Sudbury-type ores: Mineral. Deposita., v. 2, p. 80-94

Clark, A. M., and Potapoff, F., 1959, Geology of McKim mine: Geol. Assoc. Canada Pr., v. 11, p. 67-80

Clark, L. A., and Kullerud, G., 1963, The sulfur-rich portion of the Fe-Ni-S system: Econ. Geol., v. 58, p. 853-885

Coleman, A. P., 1905, The Sudbury nickel region: Ont. Bur. Mines Ann. Rept., v. 14, pt. 3, 188 p. (particularly p. 11-134, 158-163)

—— 1913, The nickel industry; with special reference to the Sudbury region, Ontario: Canadian Dept. Mines, Mines Branch, no. 170, 206 p. (particularly p. 5-111)

—— 1924, Geology of the Sudbury nickel deposits: Econ. Geol., v. 19, p. 565-576

Coleman, A. P., and others, 1929, The Sudbury nickel intrusive: Univ. Toronto Studies, Geol. Ser., no. 28, p. 1-54

Collins, W. H., 1934, Life history of the Sudbury nickel irruptive--I, petrogenesis: Roy. Soc. Canada Tr., 3d Ser., v. 28, sec. 4, p. 123-177

—— 1936, Life history of the Sudbury nickel irruptive--III, environment: Roy.

Soc. Canada Tr., 3d Ser., v. 30, sec. 4, p. 29-53

—— 1937, The life history of the Sudbury nickel irruptive--IV, mineralization: Roy. Soc. Canada Tr., 3d Ser., v. 31, sec. 4, p. 15-43

Collins, W. H., and Kindle, E. D., 1935, Life history of the Sudbury nickel irruptive--II, intrusion and deformation: Roy. Soc. Canada Tr., 3d Ser., v. 29, sec. 4, p. 29-47

Cooke, H. C., 1946, Problems of Sudbury geology: Geol. Surv. Canada Bull. no. 3, 77 p.

—— 1948, Regional structure of the Lake Huron-Sudbury area, in *Structural geology of Canadian ore deposits*: Canadian Inst. Min. and Met., Montreal, p. 580-589

Cowan, J. C., 1968, Geology of the Strathcona ore deposit: Canadian Inst. Min. and Met. Bull., v. 61, p. 38-54

Craig, J. R., and Kullerud, G., 1965-1966, The Cu-Fe-Ni-S system, in *Ann. Rept. Dir. Geophys. Lab.*: Carnegie Inst. Washington Year Book 65, p. 329-336

Davidson, S., 1946, Structural aspects of the geology of the Falconbridge nickel mine, Sudbury district, Ontario: Canadian Inst. Min. and Met. Tr., v. 49 (Bull. no. 414), p. 496-504

—— 1948, Falconbridge mine, in *Structural geology of Canadian ore deposits*: Canadian Inst. Min. and Met., Montreal, p. 618-626

Desborough, G. A., and Larson, R. R., 1970, Nickel-bearing iron sulfides in the Onaping formation, Sudbury basin, Ontario: Econ. Geol., v. 65, p. 728-730

Dickson, C. W., 1904, The ore-deposits of Sudbury, Ontario: A.I.M.E. Tr., v. 34, p. 3-67

Dietz, R. S., 1963, Cryptoexplosion structures: Amer. Jour. Sci., v. 261, p. 650-664 (particularly p. 656-658)

—— 1964, Sudbury structure as an astrobleme: Jour. Geol., v. 72, p. 412-434

Dresser, M. A., 1917, Some quantitative measurements of minerals of the nickel eruptive at Sudbury: Econ. Geol., v. 12, p. 563-580

Fairbairn, H. W., and Robson, G. M., 1942, Breccia at Sudbury, Ontario: Jour. Geol., v. 50, p. 1-33

Fairbairn, H. W., and others, 1960, Mineral and rock ages at Sudbury-Blind River, Ontario: Geol. Assoc. Canada Pr., v. 12, p. 41-66

—— 1965, Re-examination of Rb-Sr whole-rock ages at Sudbury, Ontario: Geol. Assoc. Canada Pr., v. 16, p. 95-101

—— 1968, Rb-Sr whole-rock ages of the Sudbury lopolith and basin sediments: Canadian Jour. Earth Sci., v. 5, p. 707-714

Faure, G., and others, 1964, Whole-rock Rb-Sr age of norite and micropegmatite at Sudbury, Ontario: Jour. Geol., v. 72, p. 848-854

Freeman, B. C., 1933, Origin of the Frood ore deposit (Sudbury): Econ. Geol., v. 28, p. 276-288

French, B. M., 1966, Sudbury structure, Ontario--some petrographic evidence for an origin by meteorite impact, in *Shock metamorphism of natural materials*, 1st Conf. Greenbelt, Md., Pr.: Mono. Book Corp., Baltimore, Md., p. 383-412

—— 1967, Sudbury structure, Ontario--some petrographic evidence for origin by meteorite impact: Science, v. 156, no. 3778, p. 1094-1098

Goodchild, W. H., 1918, Magmatic ore deposits at Sudbury, Ontario: Econ. Geol., v. 13, p. 137-143

Hamilton, W., 1959, Form of the Sudbury lopolith: Canadian Mineral., v. 6, pt. 3, p. 437-447

Hawley, J. E., 1962, The Sudbury ores: their mineralogy and origin: Mineral. Assoc. Canada, Univ. Toronto Press, Toronto, 207 p.

—— 1965, Upside-down zoning at Frood, Sudbury, Ontario: Econ. Geol., v. 60, p. 529-575

Hawley, J. E., and Haw, V. A., 1957, Intergrowths of pentlandite and pyrrhotite: Econ. Geol., v. 52, p. 132-139

Hawley, J. E., and others, 1943, The Fe-Ni-S system: Econ. Geol., v. 38, p. 335-388

—— 1951, Spectrographic study of platinum and palladium in common sulfides and arsenides of the Sudbury district, Ontario: Econ. Geol., v. 46, p. 149-162

—— 1961, Pseudoeutectic intergrowth in arsenical ores from Sudbury: Canadian Mineral., v. 6, pt. 5, p. 555-575

Howe, E., 1914, Petrographical notes on the Sudbury nickel deposits: Econ. Geol., v. 9, p. 505-522

Keays, R. R., and Crocket, J. H., 1970, A study of precious metals in the Sudbury nickel irruptive ores: Econ. Geol., v. 65, p. 438-450

Kirwan, J. L., 1966, The Sudbury irruptive and its history: Canadian Min. Jour., v. 87, no. 7, p. 54-58

Knight, C. W., 1917, Nickel deposits of the world: Rept. of the Royal Ontario Nickel Commission, A. T. Wilgress, Printer, Ottawa, chap. 4, p. 95-286 (particularly p. 103-133, 134-209, 209-211)

Langford, F. F., 1960, Geology of Levack Township and the northern part of Dowling Township: Ont. Dept. Mines Prelim. Rept. 1960-5

Larochelle, A., 1969, Preliminary results of a study of the paleomagnetism of the Sudbury irruptive: Geol. Surv. Canada Paper 69-19, 23 p.

Lochhead, D. R., 1955, The Falconbridge ore deposit, Canada: Econ. Geol., v. 50, p. 42-50

Michener, C. E., and Peacock, M. A., 1943, Parkerite ($Ni_3Bi_2S_2$) from Sudbury, Ontario: a redefinition of the species: Amer. Mineral., v. 28, p. 343-355

Mitchell, G. P., and Mutch, A. D., 1956, Geology of the Hardy mine, Sudbury district, Ontario: Canadian Inst. Min. and Met. Tr., v. 59 (Bull. no. 526), p. 37-43

—— 1957, Hardy mine, in *Structural geology of Canadian ore deposits*, v. 2: Canadian Inst. Min. and Met., Montreal, p. 350-363

Naldrett, A. J., 1969, A portion of the system Fe-S-O between 900 and 1080°C and its application to sulfide ore magmas: Jour. Petrol., v. 10, p. 171-201

Naldrett, A. J., and Kullerud, G., 1964-1965, Investigations of the nickel-copper ores and adjacent rocks of the Sudbury district, Ontario, in *Ann. Rept. Dir. Geophys. Lab.*: Carnegie Inst. Washington Year Book 64, p. 177-188

—— 1965-1966, Investigations of the nickel-copper ore and adjacent rocks of the Strathcona mine, Sudbury district, Ontario, in *Ann. Rept. Dir. Geophys. Lab.*: Carnegie Inst. Washington Year Book 65, 1965-66, p. 302-320

—— 1965-1966, The Fe-Ni-S system, in *Ann. Rept. Dir. Geophys. Lab.*: Carnegie Inst. Washington Year Book 65, p. 320-329

—— 1967, A study of the Strathcona mine and its bearing on the origin of the nickel-copper ores of the Sudbury district, Ontario: Jour. Petrol., v. 8, p. 453-531

—— 1968, Emplacement of ore at the Strathcona mine, Sudbury, Canada, as a sulfide-oxide magma in suspension in young noritic intrusions: 23d Int. Geol. Cong. Pr., sec. 7, p. 197-213

Naldrett, A. J., and others, 1970, Cryptic variation and the petrology of the Sudbury nickel irruptive: Econ. Geol., v. 65, p. 122-155

Newhouse, W. H., 1927, The equilibrium diagram of pyrrhotite and pentlandite and their relations in natural occurrences: Econ. Geol., v. 22, p. 285-299

—— 1931, A pyrrhotite-cubanite-chalcopyrite intergrowth from the Frood mine, Sudbury, Ontario: Amer. Mineral., v. 16, p. 334-337

Phemister, T. C., 1925, Igneous rocks of Sudbury and their relation to the ore deposits: Ont. Dept. Mines 34th Ann. Rept., v. 34, pt. 8, 61 p.

—— 1937, A review of the problems of the Sudbury irruptive: Jour. Geol., v. 45, p. 1-47

—— 1956, The Copper Cliff rhyolite in McKim Township, district of Sudbury: Ont. Dept. Mines 65th Ann. Rept., v. 65, pt. 3, p. 91-116

—— 1960, The nature of the contact between the Grenville and Timiskaming sub-provinces in the Sudbury district of Ontario, Canada: 21st Inter. Geol. Cong. Rept., pt. 14, p. 108-119

Roberts, H. M., and Longyear, R. D., 1918, Genesis of the Sudbury nickel-copper ores as indicated by recent explorations: A.I.M.E. Tr., v. 59, p. 27-67

Roscoe, S. M., 1968, Huronian rocks and uraniferous conglomerates: Geol. Surv. Canada Paper 68-40, 205 p.

Schneiderhöhn, H., 1958, Sudbury, Provinz Ontario, Kanada, in *Die Erzlagerstätten der Erde*, Bd. 1: Gustav Fischer, Stuttgart, S. 151-178

Sopher, S. R., 1963, Paleomagnetic study of the Sudbury irruptive: Geol. Surv. Canada Bull. 90, 34 p.

Souch, B. E., and others, 1969, The sulfide ores of Sudbury: their particular relationship to a distinctive inclusion-bearing facies of the nickel irruptive, in Wilson, H.D.B., Editor, *Magmatic ore deposits--a symposium*: Econ. Geol. Mono. 4, p. 259-261

Speers, E. C., 1957, The age relation and origin of common Sudbury breccia: Jour. Geol., v. 65, p. 497-514

Staff, Falconbridge Nickel Mines, Ltd., 1959, The Falconbridge story; I, Geology: Canadian Min. Jour., v. 80, no. 6, p. 116-127

Staff, International Nickel Company of Canada, Limited, 1937, The operations and plants of International Nickel Company of Canada, Limited; pt. 1, chap. 2, geology: Canadian Min. Jour., v. 58, no. 11, p. 591-596

—— 1946, The operations and plants of International Nickel Company of Canada Limited; pt. 1, chap. 3, geology: Canadian Min. Jour., v. 67, no. 5, p. 322-331

Stevenson, J. S., 1961, Recognition of the quartzite breccia in the Whitewater series, Sudbury basin, Ontario: Roy. Soc. Canada Tr., 3d Ser., v. 55, sec. IV, p. 57-66

—— 1963, The upper contact phase of the Sudbury micropegmatite: Canadian Mineral., v. 7, pt. 3, p. 413-419

Stevenson, J. S., and Colgrove, G. L., 1968, The Sudbury irruptive--some petrogenetic concepts based on recent field work: 23d Int. Geol. Cong. Pr., Sec. 4, p. 27-35

Stonehouse, H. B., 1954, An association of trace elements and mineralization at Sudbury: Amer. Mineral., v. 39, p. 452-474

Thode, H. G., and others, 1962, Sulfur isotope abundances in the rocks of the Sudbury district and their geological significance: Econ. Geol., v. 57, p. 565-578

Thomson, James E., 1952, Geology of Baldwin Township, Sudbury mining district: Ont. Dept. Mines 61st Ann. Rept., v. 61, pt. 4, 33 p.

—— 1956, Geology of the Sudbury basin: Ont. Dept. Mines 65th Ann. Rept., v. 65, pt. 3, p. 1-56

—— 1957, Geology of Falconbridge Township: Ont. Dept. Mines, 66th Ann. Rept., v. 66, pt. 6, 36 p.

—— 1957, Recent geologic studies in Sudbury camps: Canadian Min. Jour., v. 78, no. 4, p. 109-112

—— 1960, Maclennan and Scadding Townships: Ont. Dept. Mines Geol. Rept. 2, 34 p.

—— 1962, Extent of the Huronian system between Lake Timagami and Blind River, Ontario, in Stevenson, J. S., Editor, *The tectonics of the Canadian Shield*: Roy. Soc. Canada Spec. Pubs. no. 4, p. 76-89

Thomson, James E., and Williams, H., 1959, The myth of the Sudbury lopolith: Canadian Min. Jour., v. 80, no. 3, p. 57-62

Tolman, C. F., Jr., and Rogers, A. F., 1916, A study of the magmatic sulfide ores: Leland Stanford, Jr., Univ. Pubs. Ser., no. 26, 76 p. (particularly p. 23-37)

Ulrych, T. J., and Russell, R. D., 1964, Gas source mass spectrometry of trace leads from Sudbury, Ontario: Geochimica et Cosmochimica Acta, v. 28, p. 455-469

Walker, T. L., 1915, Certain mineral occurrences in the Worthington mine, Sudbury, Ontario, and their significance: Econ. Geol., v. 10, p. 536-542

—— 1935, Magmatic differentiation as shown in the nickel-intrusive of Sudbury, Ontario: Univ. Toronto Studies, Geol. Ser., no. 38, p. 23-30

Wandke, A., and Hoffman, R., 1924, A study of the Sudbury ore deposits: Econ. Geol., v. 19, p. 169-204

Williams, H., 1956, Glowing avalanche deposits of the Sudbury basin: Ont. Dept. Mines 65th Ann. Rept., pt. 3, p. 57-89

Wilson, H.D.B., 1956, Structure of lopoliths: Geol. Soc. Amer. Bull., v. 67, p. 289-300 (particularly p. 296-299)

Wilson, H.D.B., and Anderson, D. T., 1959, The composition of Canadian sulphide ore deposits: Canadian Inst. Min. and Met. Tr., v. 62 (Bull. no. 570), p. 325-336

Yates, A. B., 1938, The Sudbury intrusive: Roy. Soc. Canada Tr., 3d Ser., v. 32, sec. 4, p. 151-172

—— 1948, Properties of International Nickel Company of Canada, in *Structural geology of Canadian ore deposits*: Canadian Inst. Min. and Met., Montreal, p. 596-617

Zurbrigg, H. F., and others, 1957, The Frood-Stobie mine, in *Structural geology of Canadian ore deposits*, v. 2: Canadian Inst. Min. and Met., Montreal, p. 341-350

Notes

The oval-shaped Sudbury basin is centered just north of the town of Sudbury, some 215 miles north-northwest of Toronto. The long dimension of the basin is about 37.5 miles and runs in a northeast-southwest direction; the maximum width of the structure is about 16 miles.

The distinctive feature of the Sudbury oval is that it is outlined by two irregular selvages of igneous or igneouslike rocks, the outer of which ranges from norite through quartz gabbro to quartz diorite and the inner is called micropegmatite. These two rock types are separated by a zone of transition (or hybrid) rock, and together they enclose a series of older and variously folded and faulted volcanic rocks and sediments of the Whitewater series. The name Sudbury basin would suggest that the rocks of the igneous selvages lie in a continuous basin-shaped structure under the Whitewater series, but this supposed relationship has never been confirmed by actual exploratory work. The total outcrop width of the norite-micropegmatite selvage body ranges from about 1.5 miles to some 4 miles, with the width of the micropegmatite being generally wider than that of the norite-quartz diorite; the transition zone normally is quite narrow. The genetic relationship between the norite and the micropegmatite is uncertain. It has been suggested that these rock types were derived from a single sill-like intrusion of magma that differentiated in place into the norite below and the micropegmatite above. A second explanation is that the magmas from which the two rocks were formed were intruded separately, the norite first, followed by the micropegmatite. A third concept is that there was only one intrusion, the norite, and that the micropegmatite was produced by the granitization of the basal portion of the overlying Whitewater series. The Whitewater series is middle Precambrian in age, probably being equivalent to the Animikie group of the Lake Superior area. In any event, the Whitewater rocks almost certainly belong to the upper portion of the middle Precambrian (upper Huronian) and consist, from bottom to top, of (1) the Onaping volcanics--tuffs and breccias, (2) the Onwatin slate(?), and (3) the Chelmsford arkose--a graywacke. The rhyolite breccia at the base of the Onaping is thought by Thomson to represent Pelean domes and dike feeders; the concept requires that the Whitewater series be the time equivalent of the Stobie formation, a relationship that appears unlikely. Certainly, the evidence of finer-grained mafic rock at the top of the norite body and the presence of norite-pegmatite near, but not at, the top of the norite mass indicate that the norite was introduced as a magma of essentially noritic composition. The larger amount of micropegmatite, as opposed to norite, also suggests that the former could not have been formed by differentiation from the latter; certainly the amount of material of granitic composition in the Palisades diabase and in the Skaergaard intrusive does not make up more than a small fraction of the total mass. Differences of opinion, however, exist as to the true form of the igneous mass, and, by one of these, it is possible to revise the relative volumes of the two rock types to make the volume of micropegmatite compatible with its formation as a differentiate of the norite. The early concept that the mass was intruded as a flat sill and later folded into its present form was modified to suggest that it was actually a lopolith, and this idea is supported by crystal layering at angles flatter than the attitude of the body itself. The suggestion also has been put forward that the ovals were intruded as ring dikes. The intrusive almost certainly has not been folded, eliminating the sill-concept, and the intrusion of the mass along the unconformity between the early Precambrian rocks and the middle Precambrian Whitewater series seems to rule out the ring-dike concept. Nevertheless, the presence of the igneous mass under the entire basin has never been confirmed, and no theory as to the form of the mass can be considered as correct until much more data have been obtained.

The rocks that lie to the south of the Sudbury basin appear to be appreciably older than the Whitewater series within it; some of these are of the Keewatin type and the others apparently are not much younger. Both, therefore, are probably early Precambrian in age. These rocks south of the basin are nowhere in contact with those inside the basin and consist, from bottom to top, of (1) the Keewatin type Stobie (Elsie Mountain) silicic and mafic volcanic rocks and interbedded sediments, (2) the McKim graywacke, (3) the Ramsey Lake conglomerates, (4) the Wanapitei quartzite, and (5) the Copper Cliff rhyolite that is intrusive between the Stobie on the north and the McKim graywacke on the south. Fairbairn has dated the Copper Cliff as 2200 m.y.; this would place the entire sequence as earlier than the middle of the middle Precambrian, and formations (1) through (4) probably are much older. These rocks south of the basin have no connection with the ore except that certain of the offset ore bodies are contained within them.

The norite-micropegmatite complex does not, of course, contain all the igneous rocks of the Sudbury area. There are four granites that were intruded into the early Precambrian rocks on the south side of the district: (1) the Wanapitei (to which a radioactively derived age of 2060 m.y. has been assigned), (2) the Birch Lake, (3) the Creighton (in part), and (4) the Murray (in part). On the south side of the Sudbury oval, stocks and dikes of Creighton and Murray granite extend continuously northward into norite. Yet, the Creighton granite is definitely cut by the quartz diorite (qd) Copper Cliff offset that is an intrusive tongue from the main mass of the irruptive to the north. The difference in composition between the quartz diorite and the norite is probably due to differentiation or (less reasonably) to deuteric alteration. Thus, there probably was only one norite-quartz diorite intrusion but at least two types of Creighton granite; these two types are: (1) gneissic and massive coarsely porphyritic bodies and (2) less metamorphosed or unmetamorphosed dikes. This suggests that the former are appreciably older than the latter. The concept that part of the Creighton granite is much older than the norite is confirmed by the extensive brecciation of that granite whereas the norite is essentially unbroken, and by the metamorphism imposed on the matrix of the granite breccia by the norite intruded near it. The bulk of the Murray granite probably also is older than the norite, as brecciation of the Murray granite for some distance back from norite contacts indicates, but some granite of the Murray type is found in dikes that cut the norite. Thus, Murray granite also is both older and younger than the norite. It is obvious that it is unsatisfactory to have the same name assigned to prenorite and postnorite granites, though they may be of the same general type, and Hawley's suggestion that new names should be given to the younger granites and the names Murray and Creighton retained for the older rocks should be followed. The Birch Lake granite well may be a continuation of the main mass of the Creighton granite and bears the same relationships to the norite as does the Creighton; radioactive age determinations indicate that most, if not all, of the Birch Lake is about 2150 m.y. old. On the other hand, Thomson thinks that some of the Birch Lake may be postnorite. The Wanapitei is thought to be entirely pre-Huronian and prenorite.

Considerable masses of granite are found along the north side of the basin about which far less is known than of those on the south side. The Cartier granite probably is the equivalent of the Birch Lake, while the tonalite and pink granite of the Levack complex are gradational with the Cartier, and the entire complex, except for the Cartier, may have been the result of the granitization of more mafic rocks during Cartier time.

The prenorite granites are separated from the younger rocks outside the basin by a great unconformity, and the next rocks (excluding the lower Huronian Bruce and Cobalt beds that have only a minor role in Sudbury geology and the Whitewater series) to have been developed were the Nipissing diabase dikes that have been radioactively dated as about 1800 m.y. old; they are certainly prenorite and also pre-Sudbury breccia.

The next event was the formation of the Sudbury breccias on the south side of the basin (these breccias are essentially equivalent to the Levack breccia on the north side). Outside the irruptive, these breccias are found in all prenorite rocks--volcanic and sedimentary formations and granites and gneisses; they are essentially lacking within the basin. Many of the breccias have a linear shape, following the contacts of the older formations exactly or truncating them at low angles; they show little lateral displacement. Some of the breccia bodies border the offset bodies of quartz diorite that extend south from the main norite-quartz diorite oval and were partly invaded by, and included in, the quartz diorite of the offsets. Still a third type is quite irregular in outline, branching and coalescing through their host rocks. The breccias consist of subangular to rounded fragments of all sizes, some being up to one-half mile in their longest dimension and most being composed almost entirely of the adjacent or nearby host rocks; some fragments are present, however, that must have been transported for considerable distances. The fragments are contained in fine-grained matrices that have a wide range of colors and compositions that depend on the rock types in the fragments, the intensity of metamorphism, and the degree of metasomatic replacement or even of granitization. The matrices appear to have been somewhat altered, being enriched in certain elements in respect to the host rocks. Most of the brecciation probably occurred after the intrusion of the Nipissing diabase (Sudbury gabbro). This is confirmed by the presence of fragments of that rock in abundance in the breccias and breccias intruding diabase, but some Sudbury gabbros appear to have been intruded into the breccia zones. This last relationship suggests that some, but not much pregabbro brecciation must have occurred. The lack of Sudbury breccias in the Whitewater series would seem to confirm the assignment of a younger age to the Whitewater than that given to the breccias. The breccias are of economic interest because of the ore they contain in several locations, of which the Frood-Stobie deposit is the most important. The relationship between the earth forces necessary to produce the breccias and those required to emplace the norite-micropegmatite bodies is uncertain.

The Sudbury breccias were followed by the introduction of the norite (quartz-diorite)-micropegmatite complex and the ore mineralization attendant on it. Then came the late phases of the Creighton and Murray granites, and the final igneous event in the area was the invasion by olivine diabase and trap dikes that are dated as about 1000 m.y. old. The late sphalerite-galena mineralization in the rocks of the Whitewater series was emplaced before these late diabases and probably was some 1200 m.y. old, thus placing it as far younger than the copper-nickel mineralization associated with the norite intrusion and, at most, as only remotely related to the norite hearth.

The irruptive complex at Sudbury is apparently older than the orogeny that affected the area at the end of the Huronian (middle Precambrian), and the latest radioactive age determinations place this event at about 1600 m.y. ago. This indicates that the deposits at Sudbury were developed some 1700 m.y. ago and should be categorized as late middle Precambrian.

The Sudbury ores originally were classified by Coleman (1905), on the basis of location, as (1) marginal and (2) offset types. The marginal deposits lie along the footwall of the norite and are much more important quantitatively than the offset types that occur in dikelike masses of quartz diorite that usually crosscut, but may parallel, the formations outside the basin. Hawley and Stanton (1962) place the ores into four categories, based on their physical form: (1) disseminated, (2) massive, (3) breccia, and (4) vein or stringer ore. Several or all of these ore types may be found in a single deposit. These definitions of ore types do not, in themselves, favor any particular theory of genesis. For many years, much controversy has taken place as to whether the Sudbury ores were formed by magmatic or hydrothermal processes and, if by the former, did the sulfides originate by settling in place or were they intruded into the norite after its emplacement in the original igneous intrusion? Of the minerals in the Sudbury ores, Hawley and Stanton (1962, p. 41-43) list as major metallic minerals: pyrrhotite, chalcopyrite, pentlandite, and cubanite; those in the minor category are magnetite, ilmenite, pyrite, nickeloan pyrite, gersdorffite, niccolite, maucherite, heazlewoodite, bornite, valleriite, sphalerite, and stannite. The remaining minerals, probably some 20 at least, are present to only a limited extent. Quartz and the carbonates, calcite and dolomite, are the principal nonmetallics. The probability is that all of the sulfides were originally miscible in the silicate melt; but later, as temperature fell, the sulfide material began to separate from the silicate melt (i.e., the sulfides became steadily less soluble in the molten silicate mass) in the molten state. This immiscibility of the sulfides in the silicate melt is indicated by the rounded forms of the sulfides disseminated in norite and quartz diorite. Further, the immiscibility of silicates in dominantly sulfide melts is shown by the similar, though less well-rounded and more euhedral, outlines exhibited by the minor amounts of silicate material present in the more massive sulfides. The general fact of immiscibility also is evidenced by the reaction rims of ilmenite and magnetite (or mixtures of the two) and of fine plates of biotite around whichever constituent (silicate or sulfide) is the less abundant. In disseminated ore, fine fractures and cleavages in the presumably earlier crystallized silicates, iron oxides, and quartz are filled with sulfides. In those portions of the deposit where shearing and metamorphically induced alteration have taken place, these rounded forms have been eliminated, and their outlines became so irregular as to suggest replacement, but a metamorphic origin of these textures appears to be more probable. In many instances, disseminated ores have been seen that grade gradually into more and more massive ore that contains smaller and smaller amounts of the same silicates that form the norite and quartz diorite; usually these silicates are present as isolated grains rather than as aggregates of crystals. These relationships certainly suggest ore formation by segregation in place. In other areas, however, the early introduced magma appears to have solidified with no more than a minor fraction of disseminated sulfides; this rock was then brecciated and intruded by additional molten material in which the sulfide to silicate ratio was higher and which, on solidifying, recemented the rock, the new silicate material crystallizing first and the associated sulfides filling the interstices among the silicates. Finally, on occasion, such brecciated and recemented masses were again broken and intruded by material that was almost entirely molten sulfides and which, when it solidified, provided the matrix for the fragments. Such a series of brecciations and recementations would seem to indicate that segregation also went on at depth in the parent magma chamber and that the later this segregated material was extruded from this source, the higher it would be in sulfide and the lower in silicates. Thus, massive sulfide bodies may have formed at Sudbury by segregation in place or may have been introduced late in the magmatic cycle when the material still in the molten state in the magma chamber was more largely sulfide. Further, those intrusions leaving the magma chamber in the median portion of the crystallization cycle would be more likely, on solidification, to form workable disseminations and massive bodies than a fraction which was driven out in the early stages.

From the textural relationships shown by the Sudbury ores, it is considered that the first mineral to crystallize from the immiscible sulfide melts was a nonsulfide mineral--magnetite. This early separation of the magnetite undoubtedly was due to its limited solubility in the sulfide melt; from the presence of some almost euhedral magnetite crystals in sulfide ore, it would appear that the magnetite crystallized while the sulfide complex was still in the molten state. The bulk of the magnetite particles, however, are rounded subhedral to evenly rounded, which suggests that their form derived from post-crystallization reaction between solid magnetite and molten sulfide. Minor ilmenite behaved in much the same manner as magnetite. The first of the two generations of pyrite appears to have formed while most of the sulfide material was in the molten state since it (the pyrite) is found as euhedral (though corroded) octahedral and cubic crystals (or combinations of the two forms) as much as 1 cm in size within pyrrhotite. Some pyrite is molded on magnetite and contains rounded inclusions of it that indicate that pyrite followed magnetite in the paragenetic sequence. Since the upper stability limit of pyrite is about 750°C (the exact value depending on the pressure), the pyrite must have been separated at 750°C or below. A minor part of the arsenides probably also crystallized before pyrrhotite, but iron oxides and arsenides appear to have been mutually exclusive for reasons as yet unknown; nor are pyrite and the arsenides present together.

Probably the next step in the solidification of the Sudbury ores was the precipitation of a pyrrhotite solid solution at a temperature above the upper limit of the stability of pentlandite (between 600° and 650°C). This pyrrhotite solid solution contained in it mainly the constituents of pyrrhotite and pentlandite since very few other minerals than one of these two exsolved from either one or the other, respectively. Obviously, no systematic studies have been made on a system containing all the components necessary to produce all these phases, but some not unreasonable predictions can be made from what is known of simpler systems involving three or four of the components in question. From such studies, it would appear probable that, between 650° and 600°C, pentlandite solid solution would begin to separate from the pyrrhotite solid solution. The first pentlandite to come out of solution in the solid pyrrhotite solid solution would have been close to stoichiometric pentlandite; but, as the temperature dropped, the pentlandite leaving the pyrrhotite probably departed somewhat from stoichiometry as is suggested by numerous small blebs or rods of pyrrhotite found in most Sudbury pentlandite. Some of these blebs may have been the remains of earlier pyrrhotite, but at least those blebs with sharper boundaries probably were the result of exsolution from the pentlandite solid solution that had first formed by exsolution from pyrrhotite. Although the composition of the pentlandite is essentially stoichiometric as $(Fe,Ni,Co)_9S_8$, the ratio of cobalt to nickel differs considerably from place to place, though it averages 1:34; the ratio of Ni + Co:Fe averages about 1.11:1. It appears, from the analyses of the Sudbury pentlandite and from the complete lack of chalcopyrite in exsolution particles within it, that not much copper went into the pyrrhotite solid solution and that what little was in the pyrrhotite did not transfer to the pentlandite solid solution when it exsolved from the pyrrhotite. The pentlandite ranges in size from particles 1 micron across to others 2 inches in diameter. The finer material occurs as minute exsolution lenses, lamellae, veinlets, and flame-forms within the pyrrhotite. The larger particles of pentlandite lie outside the pyrrhotite and exist as blocky grains that are much larger than the certainly exsolution pentlandite within individual pyrrhotite grains; these blocky grains never cut across pyrrhotite grain boundaries. Such blocky areas of pentlandite may include early magnetite, and the pentlandite may be partly invaded and reacted upon (replaced) by chalcopyrite. In general, the coarser the pyrrhotite the coarser the pentlandite associated with it. Experimental work by Clark and Kullerud (1963) has shown that the pentlandite will continue to exsolve from pyrrhotite down to the lowest temperatures studied so that, from a given pyrrhotite solid solution, more and more pentlandite would exsolve with falling temperatures, leaving the remaining pyrrhotite progressively enriched in sulfur. The pentlandite external to pyrrhotite probably exsolved at higher temperatures and earlier in the exsolution cycle than did the internal pentlandite that was likely to have developed under lower temperatures. A second generation of pyrite preceded the introduction of solid chalcopyrite and is generally confined to the north range deposits in the vicinity of the Levack mine. It occurs as sharp octahedral or cubic crystals 0.2 to 3.0 mm across and is found both within and around grain boundaries of coarse pyrrhotite grains and at the interfaces of pyrrhotite and magnetite. Hawley (1962) refers to this generation as "reaction" pyrite, and it may have been the result of reaction between pyrrhotite and sulfur liquid, provided some of the ferrous iron was converted to ferric iron and incorporated in magnetite [see equation (1) in these Notes].

Experimental work by Greig in 1955 (Ann. Rept. Dir. Geophys. Lab., v. 54, p. 129-134) suggests from work on the copper-iron-sulfur system that copper-bearing pyrrhotite solid solutions should form at temperatures above 912°C and that later chalcopyrite should exsolve from the pyrrhotite. The presence of nickel in the sulfide melt, however, apparently changes these reactions so that, in the Sudbury sulfide melt, a nickeliferous-copper-poor-pyrrhotite crystallized first, leaving a copper-rich residual liquid from which chalcopyrite and a wide variety of other minerals crystallized at lower temperatures. The composition of this chalcopyrite was not, of course, strictly stoichiometric, and there eventually exsolved from it such materials as (1) cubanite laths, often along three directions; (2) valleriite rods and strings, with or without cubanite and present in both chalcopyrite and cubanite (the composition of this valleriite is uncertain; it may be a Cu-Fe sulfide of Ramdohr's formula--$Cu_3Fe_4S_7$, or a Cu-Fe-Mg sulfide in which OH is probably incorporated); (3) composite spindles of valleriite and pentlandite; (4) pyrrhotite cutting both intergrown cubanite and chalcopyrite; (5) pentlandite as collars along chalcopyrite grain boundaries; and (6) very minor sphalerite asterisks. These intergrowths indicate that the chalcopyrite solid solution contained a little more iron, nickel, and/or zinc than could be held in solution in the chalcopyrite at lower temperatures or that these excess elements were acquired through replacement reactions. Both means of providing the exsolved material to the chalcopyrite probably were effective, but the former probably was more important than the latter. In places, the chalcopyrite molten solution seems to have been rich enough in iron that cubanite actually crystallized from the melt as a distinct mineral rather than later exsolving (as most of it did) from chalcopyrite solid solution.

Sphalerite, though never abundant, is commonly present in the Sudbury ores and in several generations, the first of which is as exsolution stars in chalcopyrite and, to a lesser extent, in cubanite. The second consists of quite irregularly shaped flecks, blebs, and fine aggregates in chalcopyrite that range from barely visible under the microscope to 1 mm across; quite rarely there are veinlets of sphalerite that cut both pyrite and pyrrhotite. This sphalerite also appears to have exsolved from chalcopyrite but at higher temperatures than did the sphalerite stars. Almost all the sphalerite is found in copper-rich ores, and it appears that almost all of the sphalerite was dissolved in the chalcopyrite solid solution when that material was solidified. The more sphalerite so dissolved, the more would exsolve in the chalcopyrite, or even, in instances of unusually large amounts of sphalerite in the chalcopyrite, the sphalerite might be forced entirely out of the chalcopyrite grains, as is the case with large amounts of pentlandite dissolved in pyrrhotite. With chalcopyrite, the large blebs of sphalerite have been found to occur generally between grains and along grain boundaries. In a few places, stannite is found in sphalerite, the stannite showing exsolution textures against the zinc sulfide. Thus, it would seem that the small amount of tin in the Sudbury ores entered the chalcopyrite molten solution and was precipitated with it, later to follow the zinc into sphalerite and eventually to exsolve from it. In some instances, pyrrhotite also is found exsolved from chalcopyrite, which is to be expected since there almost certainly will be tie lines between pyrrhotite and the chalcopyrite solid solution.

In addition to zinc and tin, Hawley and Stanton (1962) point out that the chalcopyrite molten solution appears to have contained essentially all of the lead, bismuth, selenium, tellurium, arsenic, and precious metals, excluding the arsenic of the arsenic minerals that crystallized before any of the sulfides. These elements, however, were essentially insoluble in solid chalcopyrite so did not precipitate with it. Instead, they remained in the molten state until the solubility products of minerals capable of containing them were reached, and they could be precipitated; lead precipitated as galena; bismuth was incorporated in michenerite, froodite, parkerite, tetradymite, matildite, and bismuthinite; tellurium in tetradymite and hessite; selenium largely as a proxy for sulfur; arsenic as arsenopyrite, gersdorffite, niccolite, and maucherite (there apparently being enough nickel in the chalcopyrite molten solution to provide the nickel needed for the last three minerals listed as well as for heazlewoodite, Ni_3S_2); gold as the native metal (electrum); silver as the native metal, hessite and matildite; platinum as sperrylite; and palladium as michenerite and froodite. Since the melt remaining by the time these unusual minerals (at Sudbury at least) began to crystallize must have lost the bulk of its metal-sulfur constituents, it is probable that water and other volatile materials, of minor importance in the original melt, might have become a noticeable fraction of the last molten material. Such a condition in the melt well might account for the association of some quartz and carbonates with the precious metal minerals in such places as the base of the deposit in the Frood mine.

From experimental work, particularly that of Clark and Kullerud (1963), it seems probable that a molten solid solution such as the pyrrhotite molten solution that contained essentially all of the Sudbury elements must have been accompanied by at least a small amount of a highly sulfur-rich liquid as the tie lines between pyrrhotite molten and pyrrhotite solid solutions demonstrate. This sulfur liquid almost certainly would react with the femic silicates present in the solid norite or quartz diorite with which the melt was in contact to produce additional pyrrhotite as is shown in a simplified form in equation

$$4FeMgSi_2O_6 + S = FeS + Fe_3O_4 + 4MgSiO_3 + 4SiO_2 \quad (1)$$

As long as the pyrrhotite molten solution (or later the chalcopyrite molten solution) remained in that state, any sulfur removed from the sulfur-rich liquid by reaction with silicates would reduce the sulfur content of the molten solution and would, thereby, change somewhat the minerals that eventually would develop in the solid state. The amount of pyrrhotite formed by this reaction between sulfur liquid and molten pyrrhotite and chalcopyrite solutions cannot be certainly estimated, but it probably was not large in relation to that produced from the pyrrhotite molten solution. At any event, the reaction certainly used up all the sulfur liquid available to it and may, in certain areas, have produced other minerals than pyrrhotite.

Dietz's (1964) suggestion that the Sudbury series of magmatic events was triggered by the impact of a meteorite and that the metallic material in the deposits was supplied from the same source, which he believes must have been a copper-rich iron-nickel meteorite, almost certainly is incorrect. No copper-rich meteorites of any kind are known, and it is unlikely that one could have been especially created for the Sudbury fall. The presence of shatter cones in the older rocks surrounding the basin, however, points up the possibility that the magmatic events that produced the ores and their associated igneous rocks followed as a result of the fall of a stoney meteorite, which, following Dietz, would have had a diameter of about 4 kilometers and would have produced some 3×10^{29} ergs of energy. Further work on this problem definitely is required.

That there was some late hydrothermal alteration of the various Sudbury ores, both from the minor amounts of water dissolved in the sulfide melts and from hydrothermal solutions from other sources, appears certain. The considerable development of marcasite and late arsenides in the Falconbridge mine, for example, probably was produced in this manner. Despite such occurrences of hydrothermal minerals, it would seem reasonable that the Sudbury ores should be classed as both magmatic-2a and magmatic-2b, the hydrothermal effects being so minor and having so little economic effect as not to be worth mentioning in the classification assigned. Nothing has been said in this discussion of such problems as the possibility of minor remobilization of the ore under the heat and stress effects of later intrusions, the details of the relationship of the quartz diorite to the norite, the effects of the later orogenies on the ores, the lead-zinc ore in the Whitewater rocks, and many other problems dear to the Sudbury geologist; these have been omitted largely because they are not directly related to the problem of classifying the ore bodies, and anyone wanting to know more about them can find a huge variety of data and opinion in the references cited.

TIMAGAMI ISLAND

Middle Precambrian	*Copper, Gold, Silver*	*Magmatic-2b (Cp lenses), Hypothermal-1 (Py ore)*

Barlow, A. E., 1903, The Timigami district: Geol. Surv. Canada Summ. Rept., pt. A, p. 120-133

Moorhouse, W. W., 1946, The northeastern portion of the Timagami Lake area: Ont. Dept. Mines 51st Ann. Rept., v. 51, pt. 6, 46 p.

Rose, E. R., 1965, Pyrite nodules of the Timagami copper-nickel deposit: Canadian Mineral., v. 8, pt. 3, p. 317-324

—— 1966, The copper-nickel deposits of Timagami Island, Ontario: Econ. Geol., v. 61, p. 27-43

Simony, P. S., 1964, Northwestern Timagami area: Ont. Dept. Mines, Geol. Rept. no. 28, 39 p.

Notes

Timagami Island lies about in the center of Timagami Lake and is some 15 miles west of the village of Timagami; the island is 55 miles northeast of the city of Sudbury and 30 miles south-southwest of the town of Cobalt.

The host rocks on Timagami Island include Keewatin-type volcanic rocks and interbedded iron formation, plus pre-Algoman and Algoman intrusives, all of which appear to be early Precambrian in age. How long after these rocks had been formed the ores were emplaced is, however, another matter. Rose (1966) suggested that the ores are related genetically to the Nipissing diabase (as he supposes the Cobalt and Gowganda ores to have been; see the discussion of these two deposits in this volume). The Nipissing diabase at Cobalt has been dated as about 2100 m.y. old, and any ore genetically related to this diabase would be middle Precambrian in age. It would seem equally possible that the ores were formed as part of the same magmatic episode that produced the Sudbury ores, in which case they would be some 1600 m.y. old or latest middle Precambrian. Since no evidence has yet been published that directly connects these ores with any period of igneous activity, any dating of them is uncertain, but the best guess that can be made at the present time is that they are not related in genesis to either the pre-Algoman or Algoman intrusives but rather to one or the other of the two middle-Precambrian magmatic episodes just mentioned. The Timagami ores, therefore, are classified here as middle Precambrian.

Rose (1966) describes two types of ores as occurring in the deposits. The first of these is pyritic ore that appears to be replacements outward from fracture zones in silicic and mafic volcanic rocks along the footwall contact of an intrusive metadiorite sill. The extensive fracturing and brecciation of these host rocks has made them highly susceptible to the replacement process. The principal mineral is nickel-bearing pyrite occurring as heavy disseminations and nearly massive aggregates. The ore also contains minor amounts of millerite, linnaeite or siegenite, and pyrrhotite, plus possibly violarite and bravoite. Locally, the ore contains considerable amounts of chalcopyrite that may, in places, be accompanied by gersdorffite. The portions of the host rocks in the ore-containing volumes that have not been converted to sulfides have been strongly dolomitized and silicified. The mafic rocks around the ore zones have been intensely saussuritized and chloritized; the rhyolite in such situations has been much less altered. The sulfide mineralization appears to have been followed by several generations of intersecting quartz veins that are practically barren but may contain small amounts of pyrite or chalcopyrite and by quartz-carbonate veinlets. This quite complex mineralization and its associated thorough and pervasive wall-rock alteration strongly suggests hydrothermal replacement rather than magmatic injection as the ore-emplacement mechanism.

The second type of ore is another matter altogether; Rose's chalcopyrite ore occurs in irregular lenticular and vein-like masses as much as 500 feet long, 40 feet thick, and of greater vertical than lateral extent. These normally are separate from, but in the same general volumes of host rocks as, the pyritic deposits. Outside these large masses, the surrounding rocks are penetrated by smaller satellite bodies and by tongues and projections from the main lenses. Some coarse chalcopyrite disseminations also are present in the country rock immediately adjacent to the massive bodies. The main metallic mineral in these bodies is chalcopyrite but locally pyrite, millerite, gersdorffite, magnetite, and sphalerite may be abundant; a little galena also is found as well as specks that may be silver and sperrylite. Some quartz and carbonate are included in the sulfide bodies as well as grains of plagioclase, clinozoisite, and chlorite. The chalcopyrite ores appear to have done far less altering of the enclosing country rock than did the pyritic ores. This suggests that the ore-forming medium was low in water; it was not lacking in this substance, however, as is shown by the presence of cavities, partly filled by chalcopyrite, in the massive chalcopyrite and by tiny vugs in the sparse quartz and carbonate gangue. It would appear that this copper-rich material probably was injected in the molten state rather than deposited from a water-rich ore fluid. The composition of this copper-rich molten phase probably would have closely resembled that developed in the late stage of the crystallization of a pyrrhotite-pentlandite-chalcopyrite melt such as that from which the chalcopyrite portions of the Sudbury deposits appear to have been developed. To have had such a melt intruded into the area of the Timagami deposits would require that, from a Sudbury type deposit at depth, after the pyrrhotite and pentlandite had been almost entirely crystallized, earth movements drove out a chalcopyrite-rich melt that contained some quartz, carbonate, and water dissolved in it and had such a proportion of sulfur to metals that, in the late stages of this melt's own crystallization in the Timagami mine area, pyrite could form. If this reasoning is correct, the pyritic ores should be classified as hypothermal-1 while the chalcopyrite ores should be assigned to the magmatic-2b category.

Quebec

ALLARD LAKE

Late Precambrian — *Iron as Hematite, Minor Magnetite, Titanium as Ilmenite* — *Magmatic-3b*

Bourret, W., 1949, Aeromagnetic survey of the Allard Lake district, Quebec: Econ. Geol., v. 44, p. 732-740

Evrard, P., 1949, The differentiation of titaniferous magmas: Econ. Geol., v. 44, p. 210-232

Gross, G. A., 1967, Allard Lake ilmenite deposits, Saguenay County, Quebec, in *Geology of iron deposits in Canada, Volume II, Iron deposits in the Appalachian and Grenville regions of Canada*: Geol. Surv. Canada Econ. Geol. Rept. no. 22, p. 55-62

Hammond, P., 1949, Geology of the Allard Lake ilmenite deposits: Canadian Inst. Min. and Met. Tr., v. 52 (Bull. no. 443), p. 64-68

—— 1952, Allard Lake ilmenite deposits: Econ. Geol., v. 47, p. 634-649

Hargraves, R. B., 1959, Magnetic anisotropy and remnant magnetism in hemoilmenite from ore deposits at Allard Lake, Quebec: Jour. Geophysical Research, v. 64, no. 10, p. 1565-1578

—— 1962, Petrology of the Allard Lake anorthosite suite, Quebec, in Engel, A.E.J., and others, Editors, *Petrologic studies: A volume in honor of A.F. Buddington*: Geol. Soc. Amer., p. 163-190

Hargraves, R. B., and Burt, D. M., 1967, Paleomagnetism of the Allard Lake anorthosite suite: Canadian Jour. Earth Sci., v. 4, p. 357-369

Osborne, F. F., 1944, Special report on the microtextures of certain Quebec iron ores: Quebec Dept. Mines Prelim. Rept. 186, 41 p.

Rechenberg, H. P., 1955, Zur Genesis der Primären Titanerzlagerstätten: Neues Jb. f. Mineral., Jg., H. 4, S. 87-96

Retty, J. A., 1944, Lower Romaine River area, Saguenay County: Quebec Dept. Mines Geol. Rept. 19, 31 p. (with map no. 582)

Rose, E. R., 1969, Lac Allard-Romaine River area, in *Geology of titanium and titaniferous deposits of Canada*: Geol. Surv. Canada, Econ. Geol. Rept. no. 25, p. 96-102

Notes

The Allard Lake deposits are located about 25 miles north of Havre St. Pierre on the north shore of the St. Lawrence, opposite the western end of Anticosti Island.

The Allard Lake ores are found entirely within a huge mass of anorthosite that is at least 70 miles long in a roughly northeast-southwest direction and has a maximum width of more than 20 miles. In the anorthosite, which very locally may contain coarse crystals or aggregates of bronzite, many of the plagioclase crystals have been bent and broken, suggesting that the material must have been moved for considerable distances as a mush of crystals. Some of the pyroxenes, however, were emplaced along well-developed joints that cut the primary foliation of the anorthosite, indicating that there was appreciable molten material in the plagioclase mush to act as a lubricant during intrusion. The relationships among the various minerals of the anorthosite, however, are so complex that the simple concept of anorthosite intrusion as a lubricated mush of feldspar crystals probably needs further refinement (Hargraves, 1962). There are three masses of oxide-rich norite within the anorthosite; in these norites, the dark minerals rarely make up more than 50 percent of the rock, and ilmenite usually is more common than magnetite, although magnetite is present in appreciable amounts--the oxides make up from 20 to locally over 50 percent of the norite. The oxides normally provide a continuous matrix for the silicates, but, if the silicates are quite abundant, the oxides may occur interstitially. Hargraves (1962) believes that the norites were immiscible oxide-rich melts formed during the late stages of the crystallization cycle of the anorthosite, and these melts were then forced into broken zones in already solid anorthosite. The ores are considered to have been developed as an extreme case of norite formation (see discussion following).

The anorthosite mass lies well within the Grenville province, and such age determinations as have been made on rocks in the general area range between 1000 and 850 m.y. The Allard Lake anorthosites probably are essentially contemporaneous with those of the Adirondacks and of other parts of Quebec as well as those of other parts of the world such as in the Egersund area of Norway. As there appears to be a direct genetic connection between the anorthosite at Allard Lake and the ores, they are here classified as late Precambrian.

The ore at Allard Lake is found mainly in massive bodies that show knife-sharp contacts with the enclosing anorthosite and have cross-cutting relations to structures within that rock. Some of the ore forms *lit-par-lit* bands in the anorthosite, and most of these have been deformed by later orogeny. Considerable masses of disseminated ore exist, however, in which the ilmenite is intimately intercrystallized with the feldspar of the anorthosite. The ore bodies are almost entirely ilmenite (hemoilmenite in Balsley and Buddington's classification) but contain normally less than 5 percent of pyroxene, feldspar, pyrite, pyrrhotite, and chalcopyrite. No hematite is megascopically visible in any of the ore. Microscopically, however, hematite appears in textures that indicate that it exsolved from the ilmenite. The host ilmenite, nevertheless, still contains from 6 to 13 percent of hematite dissolved in it, and the exsolved hematite contains up to 13 percent of ilmenite; in both the ilmenite-bearing hematite and in the hematite-bearing ilmenite there may be up to 3 percent excess TiO_2; locally some magnetite is present in the ore. The average ore contains 75 percent ilmenite and 20 percent hematite with the remaining 5 percent being the minor constituents already mentioned. The massive ore almost certainly was injected as molten material into the already solid anorthosite, having been segregated as an immiscible molten phase in the magma chamber at depth. In contrast to the massive ore, the disseminated ore appears to have crystallized at much the same time as the feldspar and probably was directly crystallized from the lubricating material of the anorthosite mush. The isolated pyroxene and feldspar crystals in the ore probably derived from minor amounts of silicate material dissolved in the molten oxide melt; the same explanation applies to the sulfides in the massive ore. Some of the disseminated ore is quite regularly banded with anorthosite and may have developed from normal disseminated ore through flowage of this variety of anorthosite under directed pressure. The local occurrence of magnetite in the ore suggests that, in such places, the oxygen pressure was sufficiently low to leave enough Fe^{+2} to produce magnetite as well as ilmenite. Although many problems exist concerning the anorthosite, such as its contact effects on the gneissic granite of the surrounding rocks, the development of the oxide-rich norite, and the relations of the sulfides in the ilmenite, enough seems to be clear to justify classifying the huge bulk of the ore as formed by the intrusion of molten oxide material into the solid anorthosite; thus the deposits are classified as magmatic-3b.

CADILLAC-MALARTIC

Early Precambrian	*Gold, Silver*	*Hypothermal-1 (Cadillac), Mesothermal to Leptothermal (Malartic)*

Bell, L. V., 1930, Central-Cadillac map-area, Abitibi County (Quebec): Quebec Bur. Mines Ann. Rept. 1930, pt. B, p. 3-17

—— 1931, Gold in Cadillac, Quebec: Econ. Geol., v. 26, p. 630-643

Bell, L. V., and MacLeod, A., 1930, Report on the geology of the Bousquet-Cadillac gold area, Abitibi district (Quebec): Quebec Bur. Mines Ann. Rept. 1929, pt. C, 71 p.

Blais, R. A., 1955, Les contrôles structuraux de la déposition d'or à la mine O'Brien, Comté d'Abitibi-est: Assoc. Canadienne-Française Av. Sci., Ann., v. 21, p. 132-137

—— 1955, L'altération hydrothermale en bordure des filons aurifères de la mine O'Brien, Comté d'Abitibi-est: Naturaliste Canadien, v. 82, nos. 4-5, p. 77-98

Brown, R. A., 1948, O'Brien mine (Quebec), in *Structural geology of Canadian ore deposits*: Canadian Inst. Min. and Met., Montreal, p. 809-816

Buchan, R., and Blowes, J. H., 1968, Geology and mineralogy of a millerite nickel ore deposit, Marbridge no. 2 mine, Malartic, Quebec: Canadian Inst. Min. and Met. Bull., v. 61, no. 672, p. 529-534

Byers, A. R., and Gill, J. E., 1948, Sladen Malartic mine (Quebec), in *Structural geology of Canadian ore deposits*: Canadian Inst. Min. and Met., Montreal, p. 858-864

Cooke, H. C., and others, 1931, Gold deposits, in *Geology and ore deposits of the Rouyn-Harricanaw region, Quebec*: Canadian Geol. Surv. Mem. 166, p. 264-282

Derry, D. R., 1939, The geology of the Canadian Malartic gold mine, N. Quebec: Econ. Geol., v. 34, p. 495-523

Dresser, J. A., and Denis, T. C., 1944, Rouyn-Bell River area, in *Geology of Quebec*, v. 2, *Descriptive geology*: Quebec Dept. Mines, p. 74-113

—— 1949, Cadillac-Malartic-Fournier belt (Quebec), in *Geology of Quebec*, v. 3, *Economic geology*: Quebec Dept. Mines, p. 195-236

Dugas, J., and Latulippe, M., 1961, Noranda-Senneterre mining belt: Quebec Dept. Nat. Res. Map no. 1388, about 1:253,440

Eakins, P. R., 1962, Geological settings of the gold deposits of Malartic district, Abitibi-East County: Quebec Dept. Nat. Res. Geol. Rept. 99, 133 p.

Gunning, H. C., 1933, Cadillac area, Quebec: Geol. Surv. Canada Mem. 206, 80 p.

—— 1942, Gold deposits of Cadillac Township, Quebec, in Newhouse, W. H., Editor, *Ore deposits as related to structural features*: Princeton Univ. Press, p. 163-165

Gunning, H. C., and Ambrose, J. W., 1937, Cadillac-Malartic area, Quebec: Canadian Inst. Min. and Met. Tr., v. 40 (Bull. no. 303), p. 341-362

—— 1939, The Timiskaming-Keewatin problem in the Rouyn-Harricanaw region, northwestern Quebec: Roy. Soc. Canada Tr., 3d Ser., v. 33, sec. 4, p. 19-49

—— 1940, Malartic area, Quebec: Geol. Surv. Canada Mem. 222, 142 p.

Halet, R. A., 1948, Malartic Goldfields mine (Quebec), in *Structural geology of Canadian ore deposits*: Canadian Inst. Min. and Met., Montreal, p. 868-875

Latulippe, M., 1963, The Val d'Or-Malartic gold area, Quebec: Canadian Min. Jour., v. 84, no. 4, p. 72-76

Mills, J. W., 1950, Structural control of ore bodies as illustrated by the use of vein contours at the O'Brien gold mine, Cadillac, Quebec: Econ. Geol., v. 45, p. 786-807

—— 1954, Vertical zoning at the O'Brien gold mine, Kewagama, Quebec: Econ. Geol., v. 49, p. 423-430

Norman, G.W.H., 1942, The Cadillac synclinal belt of northwestern Quebec: Roy. Soc. Canada Tr., Ser. 3, v. 36, sec. 4, p. 89-98

—— 1948, Major faults, Abitibi region (Quebec), in *Structural geology of Canadian ore deposits*: Canadian Inst. Min. and Met., Montreal, p. 822-839

—— 1948, The Malartic-Haig section of the southern gold belt of western Quebec, in *Structural geology of Canadian ore deposits*: Canadian Inst. Min. and Met., Montreal, p. 839-845

O'Neill, J. J., 1934, The Canadian Malartic gold mine, Abitibi County (Quebec): Quebec Bur. Mines Ann. Rept., pt. B, p. 61-84

Notes

The Cadillac-Malartic gold deposits are located in the vicinity of the two towns of those names, the center of the district being about 35 miles south of east of the cities of Noranda and Rouyn. The mineralized break is about 16 miles long and follows the Cadillac break or shear zone. The Malartic mines in the eastern part of the district are clustered around the town of Malartic where the shear zone runs in a northwest-southeast direction and the Cadillac mines are to the west along the east-west portion of the break.

The sedimentary rocks of the district consist of both Keewatin and Timiskaming types, and these have been intruded by a variety of igneous bodies. The oldest of these intrusives is a peridotite; the Marbridge nickel deposit is located immediately adjacent to a portion of this intrusive (see the *Notes* under Marbridge). The peridotite was followed by a group of less mafic intrusions that range from gabbro to granite, the last of which was made up of various types of porphyry. After these porphyries had been introduced, the rocks were sheared and fractured (the third period of such activity in the district). This shearing was, in turn, followed by fracturing and faulting during which the late-stage alteration of the rocks and the introduction of the gold deposits occurred. The porphyries, which apparently were introduced not long before the ores, probably are of the same age as the La Motte-Preissac-Lacorne granitic batholiths that border the district on the north. Radioactive age determinations on these rocks (Snelling, 1962) suggest that they were intruded about 2700 m.y. ago. The chain of reasoning, therefore, is that the ores were part of the same cycle of igneous activity that produced the porphyries; that the porphyries are about the same age as the huge batholiths to the north; that these are 2700 m.y. old; and that the ores, therefore, are of about the same age or late early Precambrian.

In the Malartic portion of the gold-bearing belt, gold has been found in essentially all rock types, with the main ore bodies in zones of shattering in graywacke, diorite, or porphyry. The gold and other hydrothermal minerals have recemented the broken rock and replaced some of the adjacent wall rock; more ore comes from altered wall rock than from veins. Shear zones, as opposed to those of shattering only, locally have been mineralized. Pyrite is the most abundant sulfide, making up as much as 20 percent of the mineralized rock; other sulfides are rare. The mineralization took place in two stages with pyrite and gold forming in both. The essential lack of arsenopyrite in these ores (except in the Malartic Gold Fields mine that may be of the Cadillac type) and the minor development of pyrrhotite suggest that the Malartic ores were formed under conditions less intense than hydrothermal. The presence of some tellurides with the third generation of gold indicates that a fraction of it (probably a small one) was precipitated in the leptothermal range. Of the four generations of gold, the first two (and major ones) probably were formed under mesothermal conditions; the wall-rock alteration is in agreement with this concept. The Malartic ores, therefore, are classified as mesothermal to leptothermal.

In the Cadillac area, the gold is found in some of the quartz veins (most of which do not contain gold) and in the wall rocks adjoining them. Next to the workable quartz veins, the wall rock has been silicified and carbonatized with the accompanying introduction of biotite, chlorite, and tourmaline. The veins and wall rock contain appreciable and locally abundant sulfides, mainly arsenopyrite, pyrrhotite, and pyrite. Most of the gold is native, and tellurides are lacking; gold was deposited after the arsenopyrite but near enough to it in time that both minerals probably were formed under essentially the same conditions. This relationship, plus a close temporal connection between gold and the high-temperature wall-rock alteration minerals--tourmaline and biotite--supports a classification of the Cadillac ore as hypothermal.

CHIBOUGAMAU-OPEMISKA

Late Precambrian	*Copper, Gold, Silver*	*Hypothermal-1 to Mesothermal*

Assad, R. J., 1957, The Chibougamau district--recent developments: Canadian Min. Jour., v. 78, no. 4, p. 96-99

Barlow, A. E., and others, 1911, Preliminary report on the geology and mineral resources of the Chibougamau mining region, Quebec: Quebec Dept. Colonization, Mines and Fisheries, Mines Br., 24 p.

Derry, D. R., and Folinsbee, J. C., 1955, Geology and structure of the Opemiska copper mine, Quebec: Canadian Inst. Min. and Met. Tr., v. 58 (Bull. no. 521), p. 333-339

—— 1957, Opemiska copper mine, in *Structural geology of Canadian ore deposits*, v. 2: Canadian Inst. Min. and Met., Montreal, p. 430-441

Dresser, J. A., and Denis, T. C., 1944, Waswanipi-Chibougamau area, in *Geology of Quebec*, v. 2, *Descriptive geology*: Quebec Dept. Mines, p. 124-154 (particularly p. 126-139)

—— 1949, Lac la Trêve-Opémisca-Chibougamau region, in *Geology of Quebec*, v. 3, *Economic geology*: Quebec Dept. Mines, p. 47-67

Graham, R. B., 1951, Geology and mineral occurrences, Chibougamau, Quebec: Canadian Min. Jour., v. 72, p. 65-71

—— 1957, Structure of the Chibougamau area, Quebec, in *Structural geology of Canadian ore deposits*, v. 2: Canadian Inst. Min. and Met., Montreal, p. 423-429

Guimond, R., 1964, The Patino mining complex in Canada--History and geology of Copper Rand: Precambrian, v. 37, no. 4, p. 22-27

Koene, J. D., 1964, Structure and mineralization of Campbell Chibougamau mines, Cedar Bay division: Canadian Inst. Min. and Met. Tr., v. 67 (Bull. no. 630), p. 213-222

MacKenzie, G. S., 1935, Mining properties of the Chibougamau-Opemisca region: Quebec Bur. Mines Ann. Rept. 1934-1935, pt. A, p. 133-145

Malouf, S. E., and Hinse, R., 1957, Campbell Chibougamau mines, in *Structural geology of Canadian ore deposits*, v. 2: Canadian Inst. Min. and Met., Montreal, p. 441-449

Malouf, S. E., and Thorpe, W., 1957, Chibougamau Explorers mine, in *Structural geology of Canadian ore deposits*, v. 2: Canadian Inst. Min. and Met., Montreal, p. 449-454

Mawdsley, J. B., and Norman, G.W.H., 1935, Chibougamau Lake map area, Quebec: Geol. Surv. Canada Mem. 185, 95 p.

—— 1938, Chibougamau map-area, east half, Abitibi Territory, Quebec: Geol. Surv. Canada Map 397A, 1:253,440

Miller, R.J.M., 1961, Wall-rock alteration at the Cedar Bay mine, Chibougamau district, Quebec: Econ. Geol., v. 56, p. 321-330

Norman, G.W.H., 1936, Geology and mineral deposits of the Chibougamau-Waswanipi district, Quebec: Canadian Inst. Min. and Met. Tr., v. 39 (Bull. no. 296), p. 767-781

—— 1936, The northeast trend of late pre-Cambrian tectonic features in the Chibougamau district, Quebec: Roy. Soc. Canada Tr., 3d Ser., v. 30, sec. 4, p. 119-128

Personnel, Producing Mines and Quebec Department of Natural Resources, 1962, A summary report on the geology and mining development of the Chibougamau, Quebec mining district: Canadian Inst. Min. and Met., Chibougamau Branch, 44 p. (particularly p. 8-40)

Retty, J. A., and Norman, G.W.H., 1938, Chibougamau map-area, west half, Abitibi Territory, Quebec: Geol. Surv. Canada Map 398A, 1:253,440

Notes

The Chibougamau-Opemiska district is located in north-central Quebec, immediately south of 50°N latitude and between 74° and 75°W longitude; it is slightly more than 300 miles north-northwest of Montreal. The Opemiska mine is some 20 miles west of Chibougamau.

Most of the ore in the Chibougamau portion of the district is found in an anorthosite complex that has largely eliminated much of the older Keewatin- and Timiskaming-type rocks that form an east-west-trending belt through the district. No anorthosite is known within several miles of the Opemiska mine. The rocks of the district apparently lie in the Superior province, but, some 15 miles east of Chibougamau Lake, the rocks of the province abut against the north-to-northeast-trending rocks of the Grenville province. North of Chibougamau Lake, a fault contact separates the Superior and Grenville provinces while, to the south of that lake, the boundary is one of metamorphic gradation that is almost lacking in faults. There are at least five large-scale northeast-trending faults that cut through the Chibougamau district, and these cut all of the formations present in the district except, perhaps, the late diabase dikes. Smaller scale northwest-to-west-trending shears and shear zones (adjacent and subordinate to the northeast faults) are of great economic importance in the Doré Lake area (immediately northwest of Chibougamau Lake) and in the Opemiska district. These northeast faults and their subordinate northwest-to-west-trending shears are almost certainly related in time to the northeast-trending boundary fault between the rocks of the Superior and Grenville provinces and were probably formed by the same earth movements. It would appear, therefore, that the Chibougamau ores, which are almost entirely localized in the subordinate shears, can only have been formed after the Grenville orogeny and must be in the neighborhood of 1000 m.y. old. The ores do not appear to show evidence of having been deformed during the Grenville orogeny so that, if the northeast faults and their associated shears had existed long before Grenville time, they were certainly given their last movement by earth forces of Grenville age. Only after these Grenville movements on the faults was the mineralization introduced, so the ores must have been post-Grenville orogeny in age, although they appear to have been introduced not long after the major earth movements of Grenville time had been completed. On this basis, the Chibougamau ores must be classified as late Precambrian.

Recently, suggestions have been made, which have not yet appeared in print, that the sulfide deposits of the area are of volcanic-exhalative origin and that their present relations to later structures are due to the metamorphic effects of the various earth movements to which they have been subjected. Although this explanation seems highly implausible, it would mean, if true, that the ores must be categorized as early Precambrian, but they are here assigned to the late Precambrian.

Although the volcanic-sedimentary rock sequence of the Opemiska portion of the district is essentially the same as that in the Chibougamau part, the location of the ore bodies in these rocks is quite different. In the Opemiska area, the oldest rocks (Keewatin-type volcanic rocks and Timiskaming(?) sediments) were intruded by a large number of now complexly folded sills that range in composition from ultramafic to granodioritic. Although there seems little doubt but that the sill rocks containing the Opemiska ore bodies were emplaced in the early Precambrian, considerable uncertainty exists as to the age of the ores themselves. If the Campbell Lake fault is assumed to be one of the northeast-trending faults of Grenville age and the fault structures controlling the ore in the Opemiska mine are believed to be subordinate to it, then the ore would be of the same general age as that in the Chibougamau portion of the district and would be late Precambrian in age. On the other hand, there is sufficient difference between the ores in the Chibougamau and Opemiska mines that they could have come from different sources at different times. Lacking firm evidence in favor of a difference in age of mineralization between the two areas, the Opemiska ores are here characterized as late Precambrian.

Despite the concentration of the ores at Opemiska in the intrusive sill material, it has been suggested that they are of a volcanic-exhalative (syngenetic) origin; this seems unlikely but, if so, would place the ores in the early Precambrian, a viewpoint not adopted here.

Although the producing mines in the district are confined to the copper-zinc (with minor gold and cobalt) deposits of Chibougamau and the copper (with minor molybdenum, tungsten, and gold) deposits of Opemiska, the area contains a wide variety of concentrations of metallic metals. These include (1) iron sulfide zones (normally barren of other metals) in sediments and volcanics; (2) asbestos within the serpentinized ultramafic rocks; (3) segregations of magnetite in the anorthosites and ultramafics, the magnetite being generally titaniferous and vanadiferous; (4) gold-quartz veins in east-west shear zones in greenstone, in granitic masses, and in faults cutting diorite; (5) a siderite-iron sulfide zone along the north side of anorthosite complex; and (6) nickel-copper mineralization in volcanic rocks. Obviously, a complete study of the Chibougamau-Opemiska district will not have been made until the geology of all the mineralization types has been studied and the results collated into an ordered concept of genesis, but, at present, there is enough information in print to justify only a consideration of the economically valuable copper-rich deposits.

As has been pointed out (Personnel, Producing Mines and Quebec Dept. of Natural Resources, 1962), the most impressive fact concerning the deposits in the district is that they are all in distinct secondary structures that lie well within the boundaries of igneous rock masses and cannot have been the results of contact effects of an intrusive on any other rock type. Further, these secondary structures lie within, or closely adjacent to, northeast-trending regional faults that probably are genetically related, and often are physically joined, to similar structures within the rocks of the Grenville province or along the boundary between the Grenville and Superior provinces.

In the Chibougamau area, it appears that both pyrrhotite and pyrite can be present, with one dominant in one deposit and the other dominant in another deposit. The presence of magnetite and locally abundant apatite point to deposition under hypothermal conditions. On the other hand, the presence in the altered wall rocks of chlorite, zoisite, sericite, talc, and carbonates and their occurrence as gangue minerals indicate some deposition in the mesothermal range. Information as to the relationships between chalcopyrite and sphalerite is lacking, but the dark color of the sphalerite in association with pyrrhotite may indicate deposition at hypothermal temperatures. The ores in which pyrite is abundant and pyrrhotite minor may have been formed under either hypothermal or mesothermal conditions. It is probable that the Chibougamau ores were formed to an appreciably greater extent in hypothermal than in the mesothermal range, but enough of both conditions may have obtained to justify the designation of the ores at hypothermal-1 to mesothermal.

The veins of the Opemiska area are confined to fractures in the rocks of an ultramafic sill that was introduced (before folding) into the upper (rhyolitic) portion of the Keewatin volcanic sequence. Within the sill, the ore-bearing sections of these vein structures are limited to the competent gabbros (Foliated and Ventures), dying out both in the less competent pyroxenites to east (stratigraphically downward) and in the rhyolites to the west (stratigraphically upward). In the Springer mine, the more westerly of the two mines in the Opemiska area, the ore zones follow generally east-west-trending faults that dip steeply north.

The only economic mineralizations in the Opemiska mines are those of the copper ore bodies. In the nos. 1, 2, and 13 veins, the principal minerals are chalcopyrite, pyrite, and magnetite; the magnetite may have been largely, or entirely, present in the gabbro host rocks in one of the magnetite segregations that are found in the gabbros. Minor amounts of molybdenite and locally bands of pyrrhotite up to several inches in width are found in the ore; quartz and carbonate are the principal gangue minerals. In the no. 3 and no. 4 ore zones, the ore minerals are mainly fine-grained mixtures of pyrite and chalcopyrite; on either side of these ore zones, the wall rocks, for as much as 60 feet, have been converted to chlorite. Although one of the pre-ore, northwest-striking faults contains an arsenopyrite mineralization that is locally quite high in gold, it is of no present economic importance.

The source of the ore fluids is uncertain but, by analogy with Chibougamau, it probably was a deep-seated magma chamber that was active in Grenville time; any rocks derived from this hypothesized magma chamber are still buried at considerable depths in the Opemiska area.

As is the case at Chibougamau, the considerable amounts of pyrrhotite suggest that the ore zones containing it must have been formed, in considerable part at least, under hypothermal conditions. The pyrite-chalcopyrite ore may have been developed in the mesothermal or hypothermal ranges. The abundant chlorite associated with this latter ore type indicates that much of the ore probably was deposited under mesothermal conditions, although the age of the chlorite relative to the ore minerals has not been reported in print. It appears most reasonable to assign the same classification--hypothermal-1 to mesothermal--to the Opemiska ores as was given to those at Chibougamau, with the understanding that the Opemiska ores probably were deposited under, on the average, somewhat less intense conditions than those at Chibougamau.

EASTERN TOWNSHIPS

Middle Paleozoic *Copper, Pyrites* *Hypothermal-1 to Mesothermal*

Adams, L. D., 1915, The Weedon or McDonald copper mine, Wolfe County, Quebec: Canadian Inst. Min. and Met. Tr., v. 18, p. 79-90

Bancroft, J. A., 1915, Report on the copper deposits of the eastern townships of the province of Quebec: Quebec Dept. Colonization, Mines and Fisheries, Mines Br., 295 p.

Béland, J., 1967, Contributions from systematic studies of minor structures in the southern Quebec Appalachians, in Clark, T. H., Editor, *Appalachian Tectonics*: Roy. Soc. Canada Spec. Pubs. no. 10, p. 48-56

Carrière, G., 1957, Huntingdon Mine, Suffield Mine, in *Structural geology of Canadian ore deposits*, v. 2: Canadian Inst. Min. and Met., Montreal, p. 462-469

Cooke, H. C., 1948, Eustis Mine, in *Structural geology of Canadian ore deposits*: Canadian Inst. Min. and Met., Montreal, p. 899-901

—— 1950, Geology of a southwestern part of the eastern townships of Quebec: Geol. Surv. Canada Mem. 257, 142 p.

—— 1957, Structure of the eastern townships of Quebec, in *Structural geology of Canadian ore deposits*, v. 2: Canadian Inst. Min. and Met., Montreal, p. 457-462

Douglas, G. V., 1941, Eustis mine area, Ascot Township: Quebec Dept. Mines, Geol. Rept. 8, 31 p.

Dresser, J. A., 1906, Copper deposits of the eastern townships of Quebec: Econ. Geol., v. 1, p. 445-453

—— 1907, Report on the copper deposits of the eastern townships of Quebec with a review of the igneous rocks of the district: Geol. Surv. Canada Rept. 974, 38 p.

Dresser, J. A., and Denis, T. C., 1949, Copper deposits in the eastern townships, in *Geology of Quebec*, v. 3, *Economic geology*: Quebec Dept. Mines, p. 383-391

Hawley, J. E., and Martison, N. W., 1948, Moulton Hill deposit, in *Structural geology of Canadian ore deposits*: Canadian Inst. Min. and Met., Montreal, p. 902-909

Hawley, J. E., and others, 1945, The Aldermac Moulton Hill deposit, eastern townships, Quebec: Canadian Inst. Min. and Met. Tr., v. 48 (Bull. no. 398), p. 367-401

Saint-Julien, P., 1967, Tectonics of part of the Appalachian region of southeastern Quebec (southwest of the Chaudière River), in Clark, T. H., Editor, *Appalachian Tectonics*: Roy. Soc. Canada Spec. Pubs. no. 10, p. 41-47

Saint-Julien, P., and Lamarche, R.-Y., 1965, Geology of Sherbrooke area, Sherbrooke County: Quebec Dept. Nat. Res. Prelim. Rept. 530, 34 p.

Stevenson, J. S., 1937, Mineralization and metamorphism at the Eustis mine, Quebec: Econ. Geol., v. 32, p. 335-363

Notes

Although copper deposits are known throughout a northeasterly striking belt that runs through the eastern townships for 130 miles from the Vermont border to the Chaudière River, three (Suffield, Eustis and Moulton Hill) of the five most important mines lie within an 8-mile radius of Sherbrooke. The fourth (Huntingdon) is some 25 miles southwest of Sherbrocke, and the fifth (Weedon) is about 40 miles to the northeast of that city.

The sequence of Paleozoic rocks in the district, ranging in age from Cambrian(?) to Lower Devonian, has been intruded by a variety of igneous rocks of which the oldest are mainly peridotites, although some of them are nearer pyroxenite in composition. Essentially all of these ultramafic rocks are intensely serpentinized. Most of the peridotite dikes and more irregular masses lie in the western portion of the district between the U.S. border and 45°45' north latitude, although a few small masses are found as far east as just west of Lake Massawippi. The largest mass of peridotite lies on both sides of Brompton Lake and covers an area of about 11 square miles; it was mined for chromite in World War II. The peridotites and pyroxenites appear to have been intruded into the area after the Taconic folding and faulting since they cut both Caldwell (Cambrian) and Beauceville (Ordovician?) rocks; they were emplaced before the Early Silurian Sherbrooke group was deposited.

The granites in the eastern townships are accompanied by numerous dikes that appear to be tributaries of the larger masses. Cooke (1950) divides the granites into two varieties. One was formed before the Acadian orogeny but at least after the lithification of the Sherbrooke group which it intrudes; in several places, this granite has been sheared as much as the enclosing rocks, and this shearing almost certainly dates from Acadian time. The second variety of granite bodies, such as the two that lie across the course of the Weedon thrust fault in the northeast of the district, appear to have been emplaced after the thrusting had occurred. These granites, therefore, would seem to be post-Acadian orogeny in age.

The youngest abundant intrusive igneous rock in the area is gabbro (both in sizable masses and as numerous dikes) that underlies much of the area between Brompton Lake and the U.S. border, with far the greater portion being in the vicinity of Brompton Lake. The gabbros intrude the Bolton lavas, and these lavas were introduced after the folding that deformed the Lower Devonian rocks. The occurrence of all the ores in early Paleozoic rocks and the presence of Devonian (in part possibly Late Silurian) igenous rocks in close association with some of the ores suggest that the mineralization probably took place in the Devonian and probably during or immediately after the Acadian revolution. This reasoning receives some confirmation from radioactive age dates of 362 and 379 m.y. obtained on rocks that appear to be indirectly related to ore formation. The five mines are quite similar in mineralization, so the ore-forming fluids all probably came from much the same type of source magma at essentially the same time. Because the hypothermal Huntingdon deposits were emplaced in Cambrian rocks and are closely associated with post-Ordovician ultramafics, the possibility exists that the ore-forming fluids involved in their formation came from a different source than those of the other four deposits and deposited the Huntingdon ore at an earlier time. Obviously, further work is needed before this point can be settled, but until more data are available, it appears more reasonable to assume that all the ore-forming fluids came from the same general magmatic source that may or may not have been the magma chamber from which the late (post-Acadian or post-Sherbrooke) granites were derived. A further argument for the Late Devonian age of the ores is that they show no evidence of having been subjected to appreciable deformation, although Hawley (Hawley and Martison, 1948) thinks that they were emplaced before the Acadian earth movements occurred and that these were dissipated in the soft schists surrounding the ore. Despite Hawley's apparent belief in an appreciably earlier time of formation of the Huntingdon ores, they and the others are here classed as middle Paleozoic.

The principal ore mineral in all five of the deposits here under discussion was chalcopyrite. In four of the deposits, the most important sulfide was pyrite (at Moulton Hill, Eustis, and Weedon pyrite was recovered for its sulfur content), but at Huntingdon pyrrhotite was most abundant and pyrite was quite minor. Sphalerite was present in all the deposits, although only at Moulton Hill was it abundant enough to warrant its recovery; galena was found in all the deposits except, perhaps, Huntingdon. In all the deposits, except Huntingdon, tetrahedrite and/or tennantite were rarely and locally developed; only at Eustis has the relationship of these two sulfosalts to the other sulfides been reported; there the sulfosalts replaced sphalerite and were replaced by chalcopyrite. In the deep levels at the Eustis mine, the ore within some 10 feet of a late camptonite dike was altered to pyrrhotite and cubanite with a gangue that was principally anthophyllite. Although Stevenson (1937) suggests that heat alone was responsible for this metamorphism, it appears that, at the very least, iron must have been added and sulfur subtracted. The gangue minerals that were introduced into the deposits were generally quartz, carbonates, chlorite, and sericite; barite, however, was the only important gangue mineral, in addition to quartz, in the Moulton Hill mine. No data are available as to the possible presence of cubanite or sphalerite in exsolution lamellae in the chalcopyrite. With the exception of Huntingdon, it appears that the metallic mineral and gangue assemblages are completely compatible with deposition under mesothermal conditions. The amounts of sulfosalts present appear to be too small to justify including a leptothermal phase in the classification. On the other hand, the ore mineralization at Huntingdon suggests, despite gangue minerals similar to those in the other deposits, that deposition occurred under conditions of hypothermal intensity. The deposits are, therefore, categorized as hypothermal-1 to mesothermal.

GASPÉ

Middle Paleozoic	*Copper, Molybdenum*	*Hypothermal-2 to Mesothermal*

Alcock, F. J., 1923, Copper prospects in Gaspé Peninsula, Quebec: Geol. Surv. Canada Summ. Rept., pt. C II, p. 1-12

Béland, J., 1969, The geology of Gaspé: Canadian Inst. Min. and Met. Bull., v. 62, no. 688, p. 811-818

Bell, A. M., 1951, Geology of the ore occurrences at the property of the Gaspé copper mines: Canadian Inst. Min. and Met. Tr., v. 54 (Bull. no. 470), p. 240-245

—— 1956, Gaspé copper mines--structural geology: Canadian Min. Jour., v. 70, no. 5, p. 61-63, 83

—— 1957, Gaspé copper mines, in *Structural geology of Canadian ore deposits*, v. 2: Canadian Inst. Min. and Met., Montreal, p. 470-477

Bell, A. M., and Scott, F. N., 1954, Alteration associated with ore at Gaspé Copper Mines: Econ. Geol., v. 49, p. 516-520

Brummer, J. J., 1966, Northwest Quarter of the Gaspé-North County: Quebec Dept. Nat. Res. Geol. Rept. 125, 95 p.

—— 1966, Northwest quarter of Holland Township: Quebec Dept. Nat. Res. Geol. Rept. 125, 95 p.

Cumming, L. M., 1959, Silurian and Lower Devonian formations in the eastern part of Gaspé Peninsula, Quebec: Geol. Surv. Canada Mem. 304, 45 p.

Dugas, J., and others, 1969, Metallogenic concepts in Gaspé: Canadian Inst. Min. and Met. Bull., v. 62, no. 688, p. 846-853

Ford, R. E., 1959, Geology of Gaspé Copper Mines: Canadian Inst. Min. and Met. Tr., v. 62 (Bull. no. 567), p. 215-221

Jones, I. W., 1932, The Bonnecamp map area, Gaspé Peninsula: Quebec Bur. Mines Ann. Rept. 1931, pt. C, p. 41-75

—— 1936, Upper York River map area, Gaspé Peninsula: Quebec Bur. Mines Ann. Rept. 1935, pt. D, p. 3-28

MacIssac, W. F., 1969, Copper Mountain geology at Gaspé Copper Mines, Limited: Canadian Inst. Min. and Met. Bull., v. 62, no. 688, p. 829-836

McGerrigle, H. W., 1946, Review of the Gaspé Devonian: Roy. Soc. Canada Tr., 3d Ser., v. 40, sec. 4, p. 4-54

Notes

The Gaspé copper mines, in the narrow sense, are located at Murdochville, some 35 miles slightly north of west from the town of Gaspé and about 55 miles from the eastern tip of the Gaspé peninsula.

The sedimentary framework of the district consists of more than 5000 feet of Lower and Middle Devonian shales, limestones, siltstones, and sandstones with minor amounts of fragmental tuffs; the sediments normally are impure representatives of the classes named. The highly altered portion of the district is confined to a rudely circular area that, on the surface, covers about 2.5 square miles. Two parts of this area are sufficiently well mineralized to be minable; the more southerly and better known of the ore volumes is the Needle Mountain area; the other is called Copper Mountain. The extensive wall-rock alteration and the ore mineralization associated with it are believed to be related genetically to the granite found at depth in the area and present in the Copper Mountain adit; rock from the adit area has been determined by radioactive methods to have an age of 395 ± 15 m.y., which would place its introduction as Late Devonian. Thus, the ores are here considered to be middle Paleozoic.

The ore bodies of the Gaspé deposit are divided into two groups: (1) the Needle Mountain ore bodies that include all those now being mined underground and (2) the Copper Mountain mineralization that is being mined as an open pit. The three Needle Mountain ore zones are: (1) ore zone C (including the small East C ore body) that occurs in the diopside-garnet skarn to which the lower limestone of the Grande Grève formation is altered; there is a minor ore zone (D) below the C; (2) ore zone B (including ore zone B-1 Central, lying above the south end of ore zone C; B-East, that is not accompanied by appreciable A or C zone ore; and certain minor ore masses) that consists of two ore types: (a) disseminated replacements in the first and second porcelanite horizons in which the ore minerals replaced the carbonates in that rock--the more carbonate originally present, the more ore, whether the carbonate was present in tiny concretions or as remnants in skarn bands in the porcelanite--and (b) replacements in the wollastonite horizon between the porcelanites, with the degree of replacement being proportional to the degree of fracturing; and (3) ore zone A, above the south end of ore zones C and B, that has formed in a white silicated quartzite that was originally a dark siltstone with lime-rich bands; most of the ore was formed by replacement of these limey bands, but some developed through the filling of closely spaced stockworks.

The Copper Mountain mineralization consists mainly of chalcopyrite and pyrite with minor molybdenite, bornite, sphalerite, galena, and chalcocite This mineralization is found in both the Grande Grève and Cap Bon Ami formations in diopside and wollastonite skarn, the skarn containing considerable garnet; also involved in the ore zone is some cherty hornfels that was developed from more siliceous beds, but it is little mineralized. Quartz-feldspar porphyry dikes are common in the Copper Mountain area but are poorly mineralized although they may contain some ore; ore in the sediments is generally richest in the vicinity of these dikes.

In all of the Needle Mountain ore bodies, chalcopyrite is the most abundant ore sulfide; pyrrhotite is the most common of all the sulfides, but in the ore zones, pyrrhotite approximately equals chalcopyrite in amount. Bornite is a minor but important source of copper in ore zone C, and primary chalcocite and tennantite are rare, while cubanite is found in minor amounts; very small quantities of cubanite occur in the chalcopyrite. These lesser copper minerals are rarely or never found in ore zones B and C. Molybdenite and scheelite are ubiquitously present, and the former is now recovered. Bismuth is produced in the smelting process and probably is derived from a bismuth mineral in solid solution in the chalcopyrite.

The wall-rock alteration in the Gaspé area probably was developed under hypothermal conditions. Because the ore appears to have been formed just after wollastonite phase of the wall-rock alteration, it also must have been deposited largely in the hypothermal range as is attested by the abundance of pyrrhotité, the cubanite in the chalcopyrite, and the appreciable amounts of scheelite and molybdenite. The important bornite in ore zone C, however, indicates that some of the mineralization occurred in the mesothermal range. The chalcopyrite-pyrite mineralization of Copper Mountain may have been formed entirely under mesothermal conditions, but this is not certain. The ore deposits, therefore, are classified as hypothermal-2 to mesothermal, with hypothermal-2 being used because the ore minerals were emplaced largely by the replacement of lime-rich minerals, although some fill fractures in the siliceous rocks.

LACORNE

Early Precambrian	*Spodumene, Beryl, Molybdenum, Bismuth*	*Magmatic-3a (Spodumene, Beryl), Hypothermal-1 (Mo, Bi)*

Brett, R., 1961, A magmatic-pegmatitic-hydrothermal sequence at Lacorne, Quebec: Econ. Geol., v. 56, p. 784-786

Clark, L. A., 1965, Geology and geothermometry of the Marbridge nickel deposit, Malartic, Quebec: Econ. Geol., v. 60, p. 792-811

Cooke, H. C., and others, 1931, Molybdenum deposits, in *Geology and ore deposits of the Rouyn-Harricanaw region, Quebec*: Canadian Geol. Surv. Mem. 166, p. 290-301

Dawson, K. R., 1953, Structural features of the Preissac-Lacorne batholith, Abitibi County, Quebec: Geol. Surv. Canada Paper 53-4, 22 p.

—— 1958, An application of multivariate variance analysis to mineralogical variation, Preissac-Lacorne batholith, Abitibi County, Quebec: Canadian Mineral., v. 6, pt. 2, p. 222-233

—— 1967, A comprehensive study of the Preissac-Lacorne batholith, Abitibi County, Quebec: Geol. Surv. Canada Bull. 142, 76 p.

Dawson, K. R., and Whitten, E.H.T., 1962, The quantitative mineralogical composition and variation of Lacorne, La Motte, and Preissac granitic complex, Quebec, Canada: Jour. Petrol., v. 3, p. 1-37

Derry, D. R., 1950, Lithium-bearing pegmatites in northern Quebec: Econ. Geol., v. 45, p. 95-104

Dresser, J. A., and Denis, T. C., 1949, Molybdenum--western Quebec, in *Geology of Quebec*, v. 3, *Economic geology*: Quebec Dept. Mines, p. 412-419

Gerry, C. N., 1927, Molybdenite in Lacorne and Malartic Townships, Quebec: Univ. Toronto Studies, Geol. Ser. no. 24, p. 37-40

Dugas, J., and Latulippe, M., 1961, Noranda-Senneterre mining belt: Quebec Dept. Nat. Res. Map no. 1388, about 1:253,440

Gussow, W. C., 1937, Petrogeny of the major acid intrusives of the Rouyn-Bell River area of northwestern Quebec: Roy. Soc. Canada Tr., 3d Ser., v. 31, sec. 4, p. 129-161

Hawley, J. E., 1931, Molybdenite deposits of Lacorne Township, Abitibi County, Quebec: Quebec Bur. Mines Ann. Rept. 1930, pt. C, p. 97-122

Ingham, W. N., and Latulippe, M., 1957, Lithium deposits of the Lacorne area, Quebec, in *The geology of Canadian industrial mineral deposits*: Canadian Inst. Min. and Met., Montreal, p. 159-163

James, W. F., and Mawdsely, J. B., 1925, LaMotte and Fournière map-areas, Abitibi County, Quebec: Geol. Surv. Canada Summ. Rept., pt. C, p. 52-77

Karpoff, B., 1960, Holmquistite occurrences in the mining property of Quebec Lithium Corporation, Barraute: 21st Int. Geol. Cong. Rept., pt. 17, p. 7-14

Leuner, W. R., 1959, Preliminary report on the west half of LaMotte Township, Abitibi-East electoral district: Quebec Dept. Mines, Mineral Deps. Branch, Prelim. Rept. no. 405, 10 p.

Mulligan, R., 1962, Origin of the lithium and beryllium-bearing pegmatites: Canadian Inst. Min. and Met. Tr., v. 65 (Bull. no. 608), p. 419-422

—— 1965, Preissac-Lacorne district, in *Geology of Canadian lithium deposits*: Geol. Surv. Canada Econ. Geol. Rept. no. 21, p. 42-51

Norman, G.W.H., 1944, La Motte map-area, Abitibi County, Quebec: Geol. Surv. Canada Paper 44-9, 13 p.

—— 1945, Molybdenite deposits and pegmatites in the Preissac-Lacorne area, Abitibi County, Quebec: Econ. Geol., v. 40, p. 1-17

—— 1948, Indian molybdenum deposit, in *Structural geology of Canadian ore deposits*: Canadian Inst. Min. and Met., Montreal, p. 845-850

—— 1948, Lacorne molybdenite deposit, in *Structural geology of Canadian ore deposits*: Canadian Inst. Min. and Met., Montreal, p. 850-852

Renault, J. R., 1963, Molybdenite paragenesis at the Lacorne Mine, Quebec: Econ. Geol., v. 58, p. 1340-1344

Rowe, R. B., 1953, Pegmatitic beryllium and lithium deposits, Preissac-Lacorne region, Abitibi County, Quebec: Geol. Surv. Canada Paper 53-3, 35 p.

Siroonian, H. A., and others, 1959, Lithium geochemistry and the source of the spodumene pegmatites of the Preissac-La Motte-Lacorne region of western Quebec: Canadian Mineral., v. 6, pt. 3, p. 320-338

Snelling, N. J., 1962, Potassium-argon dating of rocks north and south of the Grenville front in the Val d'Or region, Quebec: Geol. Surv. Canada Bull. 85, 27 p.

Tremblay, L. P., 1947, Lacorne map-area, Abitibi County, Quebec: Geol. Surv. Canada Paper 46-13, 6 p.

—— 1950, Fiedmont map-area, Abitibi County, Quebec: Geol. Surv. Canada Mem. 253, 113 p.

Vokes, F. M., 1963, Preissac-Lacorne area, in *Molybdenum deposits of Canada*:

Geol. Surv. Canada Econ. Geol. Rept. no. 20, p. 105-126

Notes

The deposits of the Lacorne area are certainly spatially, and probably genetically, related to the group of batholiths (Lacorne, La Motte, and Preissac) that make up much of an area in western Quebec about 40 miles long from east to west and nearly 16 miles wide from north to south. These batholiths are located a few miles north of the Cadillac shear zone (break) in a tier of townships immediately north of the two tiers that contain the Cadillac, Malartic, and Val d'Or deposits. The center of the area containing the batholiths is about 45 miles east-northeast of the towns of Noranda and Rouyn.

The rocks into which the batholiths and their satellite pegmatite dikes were intruded are largely those of the Keewatin Malartic group with a small proportion of metamorphosed sediments that probably are members of the Kewagama group, both groups belonging to Wilson's Abitibi group. The prebatholithic igneous rocks include dikes and lenslike sills of peridotite, now highly metamorphosed, that intruded the Keewatin rock groups in the vicinity of the present sites of the batholiths. These bodies range in size from a few hundred feet long and a few feet wide to 1 to 2 miles long and 1000 feet wide. Fresh surfaces of the peridotite are dark, and the rock appears to be fine-grained. Augite appears to have been present in the primary rock in addition to the much more abundant olivine, but this rock type now is composed of alteration products such as tremolite, serpentine, talc, chlorite, chalcite, iron oxides, and pyrite, the first four being by far the most common. The peridotite has been cut by rocks of the Lacorne batholith and by dikes of diabase. The only known ore deposit associated with the peridotite is the nickel ore body at Marbridge just north of the Preissac batholith. Moderate-sized bodies of apparently intrusive rock, now amphibolite, have been found in the Kinojevis volcanic rocks and cut across the trend of these formations. The parent rock of the amphibolite appears to have been a mafic gabbro (Tremblay, 1950). These masses of altered gabbro also occur near or in the batholiths and are not known to contain any ore deposits.

The three batholiths and their satellites (Dawson, 1953) that make up much of an area of about 650 square miles, extend from the west side of Preissac Township east just beyond Lac Fiedmont in Fiedmont Township and from the northern outlet of Lac La Motte south to a point east of the village of Vassan. The satellites range in size from a maximum diameter of 3 miles down to a few hundred feet.

The Lacorne district contains numerous granite pegmatites that are associated spatially, and probably genetically, with the three batholiths. Among the valuable minerals in these pegmatites are beryl, columbite-tantalite, molybdenite, and spodumene. The pegmatites and the quartz veins associated with them occur both in the rocks of the batholiths and in the adjacent metamorphosed Keewatin formations; they are most abundant near the contact between batholith and country rock. Most of the pegmatites known to contain beryl, spodumene, and columbite-tantalite are found in two separate belts of which one extends along the northern contacts of the La Motte and Lacorne batholiths and the second reaches from the north contact just mentioned southward along the west margin of the La Motte batholith as far as Lac Lusignan, being more or less paralleled by the road from Figuery to Val d'Or. The east-west belt contains several lithium-rich pegmatites and the north-south one at least one beryllium prospect. The north-south belt also contains most of the molybdenum mines and prospects, although some are located at the west ends of the Preissac and La Motte batholiths; most of the molybdenite deposits, however, are in hydrothermal veins rather than in pegmatites.

The batholiths appear to be multicomponent masses, the individual segments of which probably were intruded over a quite short period of geologic time but which probably were closely related in genesis. A considerable number of radioactive age determinations have been made on the rocks of the batholiths and their satellite bodies (Snelling, 1962). Uniformly, the sericite examined gives older ages than does the biotite, and Snelling thinks that the loss of argon from that mineral is negligible in comparison with that from biotite, the biotite consistently giving ages of about 10 percent lower than sericite. If this loss of argon assumption is accepted, then the probable age of the rocks of the batholiths ranges around 2700 m.y. As the ore minerals, both those in the pegmatites and those in the hydrothermal veins, are essentially contemporaneous with their igneous host rocks, the ore bodies all are categorized here as late early Precambrian.

The granitic pegmatites of the La Corne district have been divided by Rowe (1953) into two groups: (1) those that are simple aggregates of quartz, feldspar, and accessory minerals and that cannot be subdivided into contrasting units on the basis of mineralogy or texture or both and (2) those that are complex bodies that can be subdivided readily into such contrasting units. In the district, those pegmatites that have the greatest concentrations of beryl are of type (2) and generally consist of a border zone and internal pods (wall zone) surrounded by border zone material; some of these pegmatites, however, are made up of border zone, wall zone, and core. The border zone contains perthite, quartz, and plagioclase, with or without accessory garnet and biotite. The internal pods and the wall zones consist of medium-to-coarse-grained quartz and perthite; the cores are almost entirely massive quartz.

Rowe (1953) reports that all of the spodumene-bearing pegmatites that he examined, with one possible exception, had definite internal subdivisions. Several of these pegmatites had a border zone, a wall zone, and a core. The border zones are very narrow and are not continuous around the pegmatite bodies; they consist of fine-grained feldspar and quartz. The wall zones are similar in being discontinuous and very narrow and fine-grained; they contain feldspar, quartz, and spodumene. The cores (which may actually be inner zones) are composed of plagioclase, quartz, spodumene, and perthite and make up most of these pegmatite masses.

In contrast to beryl and spodumene, most of the workable deposits of molybdenite are found in quartz veins and feldspar-quartz veins (referred to in the district as pegmatitic quartz veins); minor amounts also are found in the zoned pegmatites particularly in their more siliceous portions. Molybdenite veins are known to cut spodumene-bearing pegmatite in at least one locality, the Iso-Uranium property. These molybdenite-bearing veins are of sufficient size in relation to the neighboring pegmatitic masses that it appears almost certain that they were not deposited from hydrothermal fluids generated in these adjacent bodies; instead, these fluids probably were derived from the same general source as the pegmatitic fusions themselves.

The spodumene- and beryl-bearing pegmatites of the Lacorne district probably were produced by crystallization processes within the pegmatite bodies and they should be categorized as magmatic-3a. On the other hand, although molybdenite and bismuthinite are found in the pegmatites, the commercially valuable deposits are in quartz veins that lie even farther from the major batholiths than do the spodumene-bearing pegmatites. These veins, however, cut through all rocks of the area (volcanics, sediments, pegmatites, and minor granite masses) so they appear to have been formed late in the cycle of igneous activity. Nevertheless, the presence of both molybdenite and bismuthinite in the pegmatites, as well as in the quartz veins, suggests that there must be a definite genetic connection between the two. It appears most likely that the quartz masses are true hydrothermal deposits in the sense that the ore-forming fluids probably came from the same source at depth that produced the pegmatitic fusions rather than that they came from the small amounts of hydrothermal fluid that may have been generated by the pegmatitic fusions during their crystallization. Thus, the molybdenum and bismuth deposits should best be categorized as hypothermal-1.

MARBRIDGE

Early Precambrian *Nickel* *Magmatic-2b*

Buchan, R., and Blowes, J. H., 1968, Geology and mineralogy of a millerite nickel ore deposit, Marbridge No. 2 mine, Malartic, Quebec: Canadian Inst. Min.

and Met. Bull., v. 61, no. 672, p. 529-534

Clark, L. A., 1965, Geology and geothermometry of the Marbridge nickel deposit, Malartic, Quebec: Econ. Geol., v. 60, p. 792-811

Naldrett, A. J., 1967, The central portion of the Fe-Ni-S system and its bearing on pentlandite exsolution in iron-nickel sulfide ores: Econ. Geol., v. 62, p. 826-847

Notes

The Marbridge nickel deposits are found in La Motte township about 14 miles north of the town of Malartic; the two mines are about 0.5 miles apart, with the No. 2 mine lying southeast of No. 1.

The two Marbridge nickel deposits are located between the La Motte and Preissac batholiths in the western portion of the Lacorne district in a zone of metamorphosed peridotite and gabbro, metasedimentary gneiss, greenstones, and silicic tuffs; the zone ranges from 0.5 to over 4 miles wide. The western part of this belt is narrower than that farther east, and the mines are situated about its center just west of the point where the belt narrows sharply. The No. 1 deposit is located in a minor shear zone in massive silicic tuffs at or near the contact of the tuffs with a fine-grained metagabbro; three or four metaperidotite bodies are known in the immediate mine area. In the No. 2 area, narrow lenses of ore were introduced along the steeply dipping contact (80° to 85°) between metasedimentary gneiss and metaultramafics. The peridotites are definitely older than the silicic rocks of the neighboring batholiths that have been dated as about 2700 m.y. old. This means that the ores, that are genetically related to the peridotites, or better to the source from which the peridotites came, were formed appreciably before the end of early Precambrian time but probably still in the later part of that period.

Probably no two nickel deposits situated within the same district, much less within 0.5 miles of each other, are so different. Within the immediate area of the No. 1 deposit, there are 3 or 4 metaperidotite bodies, apparently concordant with the surrounding volcanic rocks; the metaperidotites range from 30 to 200 feet in thickness. In the No. 1 mine, the shear zone appears to have been developed concomitantly with a regional drag fold, strikes about 45°NW, and dips around 55°NE. The ore body in the shear zone has an average thickness of about 50 to 10 feet and has an average strike length of some 275 feet. The ore zone plunges at 40°E, and the ore extends for 1200 feet down this plunge. The ore consists of two distinct types, massive and disseminated. The silicates of the disseminated ore have the approximate composition of a hydrated peridotite, and sulfides make up about 20 to 40 percent of this ore type. The disseminated ore has been sheared into lenses scattered through the ore zone, and the lenses are generally surrounded by 2 inches to 2 feet of talc schist. The massive ore shows no shearing and appears to have been introduced into the ore zone after the shearing had occurred; this ore forms a more or less continuous sheet through the ore zone, and a few veins extend a few feet out from it into the hanging wall at steep angles to the main massive zone. The massive ore generally is located on the hanging wall side of large lenses of disseminated ore, but some is found on the footwall. It is Clark's (1965) concept that the disseminated ore was injected as an emulsion of silicates and sulfides into the shear zone in the silicic tuffs; this shear zone was later reopened, and the area was intruded by a mass of molten material, almost entirely composed of sulfides. The forceful character of the second injection is shown by occasional inclusions in the sulfides that appear to have been torn from the adjoining walls. The massive ore is made up of intergrown pyrrhotite and pentlandite in textures that suggest that the pentlandite exsolved from the pyrrhotite or that, in Clark's opinion, it was produced largely by reactions of a nickel-rich residual liquid with the already crystallized pyrrhotite grains. Since pentlandite melts incongruently at about 610°C, this provides an upper limit on crystallization temperature of the pentlandite, if it was formed by reactions between molten and solid material; if it was formed through exsolution processes, however, it may have been incorporated in the pyrrhotite at appreciably higher temperatures. Pyrite is present in all specimens but in minor amounts, and chalcopyrite, which may be present as up to 5 percent of the ore, normally occurs only in trace amounts. Small quantities of magnetite, sphalerite, and serpentine (derived from the primary silicates) also are found in the massive ore. The paragenetic sequence in the massive ore is (1) magnetite, pyrite, pyrrhotite, pentlandite, chalcopyrite, sphalerite, and very minor carbonate gangue. In the disseminated ore, the same metallic minerals are interstitial to the now serpentinized silicates and do not replace them; it appears probable, therefore, that the original silicates crystallized before the sulfides but that the serpentinization occurred after the sulfides had solidified.

The Marbridge No. 1 ore bodies, both the massive and disseminated varieties, are intrusive into the country rock and certainly should be classified as magmatic-2b.

The No. 2 ore body contrasts remarkably with the No. 1; the ore, according to Buchan and Blowes (1968), occurs in three types: (1) massive vein sulfides, containing about 10 percent of silicate inclusions; (2) disseminated sulfides in metaultramafics; this type provides most of the ore mined; and (3) disseminated sulfides in metasediments; this type supplies only a small portion of the ore produced. In the No. 2 body, the principal nickel-bearing mineral is millerite (NiS), and the other main sulfides are pentlandite and pyrite (as compared with the pyrrhotite-pentlandite ore in No. 1). The pentlandite is far more nickel-rich than usually is the case, having an Fe:Ni ratio of 0.6:1 as compared with the 1:1 of the normal variety. The paragenetic sequence by Buchan and Blowes (1968) is (1) primary euhedral pyrite; (2) coarse grains of pentlandite; (3) millerite and primary lavender-colored violarite ($NiNiFeS_4$) that occurs with millerite as discrete grains; (4) late pyrite and gray violarite plus millerite. The gray violarite of stage (4) in most cases replaces pentlandite but may be of primary origin as well. These relationships are difficult to explain in light of the Fe-Ni-S diagrams given by Clark and Kullerud's work (1963, Econ. Geol., v. 58, p. 853-885) or by Craig and his colleagues (1966-1967, Ann. Rept., Dir. Geophys. Lab., p. 440-442), even though this latter diagram is applicable only to 400°C. It would appear probable that the composition of the No. 2 mine melt was high enough in sulfur relative to the metals that pyrite could form as the first mineral instead of the usual pyrrhotite and that the temperature of the melt was something below the incongruent melting point of pyrite (743°C at 10 bars) so that pyrite could precipitate. The removal of pyrite from the melt would shift the total composition in the direction of the nickel corner of the triangular Fe-Ni-S diagram and away from the sulfur corner, apparently reaching a point or an area in the system at a temperature at which the solid solution precipitated was of such composition that, as temperature dropped, it broke down to give the pentlandite, millerite, and violarite of Buchan and Blowes paragenetic stages (2) and (3). It is possible, of course, that the temperature of the melt at the time of the first precipitation was high enough that the solid solution formed actually was of such a composition that pyrite would exsolve from it below 743°C, developing its characteristic crystal forms even though it was developed in a solid medium. The remaining material of the solid solution eventually segregated to form pentlandite, millerite, and the lavender-colored primary violarite. Just where in the system Fe-Ni-S the composition of this melt lay is difficult to say; it had to be high enough in sulfur to permit the formation of pyrite, low enough in iron to prevent the development of pyrrhotite, and rich enough in nickel to insure the production of violarite and millerite. The work of Craig (1966-1967, Ann. Rept. Dir. Geophys. Lab., p. 434-436) on the stability relations of violarite does not answer the problem of the mineral suite at Marbridge No. 2 mine but does indicate that the violarite did not separate from the solid solution until the temperature had dropped below 500°C. Millerite, on the other hand, is stable above 500°C and should have come out of the solid solution before violarite. In addition to the lavender-colored violarite, two other types exsist. One of these is gray and is reported (Buchan and Blowes, 1968) normally to replace pentlandite, though some of it appears to be primary. The other is pink and shows cross-cutting relationships with pentlandite and gray violarite.

The delicate intergrowths of the gray violarite with millerite certainly suggest that violarite formed by unmixing from a millerite-violarite solid solution. Some, at least, of the gray violarite is said to be pseudomorphous after pentlandite and associated with secondary (i.e., exsolved) millerite; this may be simply pentlandite that has broken down into violarite and millerite of larger grain sizes than the delicate structures just mentioned. The pink violarite may be no more than the result of readjustments of earlier violarite.

The disseminated and massive ores of the No. 2 mine appear to have been formed in much the same manner as those of the No. 1, that is, by the intrusion of silicate-sulfide emulsions in which silicate was dominant over sulfide and by the later intrusion of molten sulfide material that contains only a small percentage of dissolved silicates. Thus, the No. 2 ore bodies also should be classified as magmatic-2b. The disseminated sulfides in adjacent metasediments cannot be assigned to the magmatic-2a category but can best be considered as having been transferred from melt to wall rock by the minor amounts of water and other volatiles in the molten material. Obviously, the problems of Marbridge genesis are far from solved.

MATAGAMI

Early Precambrian — *Zinc, Copper, Gold, Silver* — *Hypothermal-1*

Dresser, J. A., and Denis, T. C., 1944, Waswanipi-Chibougamau area, in *Geology of Quebec*, v. 2, *Descriptive geology*: Quebec Dept. Mines, p. 124-154 (particularly p. 140-146)

—— 1949, Kitchigama-Mattagami-Goéland lakes belt, in *Geology of Quebec*, v. 3, *Economic geology*: Quebec Dept. Mines, p. 27-30

Freeman, B. C., 1939, The Bell River complex, northwestern Quebec: Jour. Geol., v. 47, p. 27-46

Hallam, R. H., 1964, Mattagami Lake Ltd.--Some aspects of the geology and ore control: Canadian Inst. Min. and Met. Tr., v. 67 (Bull. no. 624), p. 59-66

Hallof, P. G., 1966, Geophysical results from the Orchan Mines, Ltd., property in the Mattagami area of Quebec: Soc. Expl. Geophys., Mining Geophysics, v. 1, Case Histories, p. 157-171

Jenney, C. P., 1961, Geology and ore deposits of the Mattagami area, Quebec: Econ. Geol., v. 56, p. 740-757

Joklik, G. F., 1960, The discovery of a copper-zinc deposit at Garon Lake, Quebec: Econ. Geol., v. 55, p. 338-353

Latulippe, M., 1959, The Mattagami area of northwestern Quebec: Geol. Assoc. Canada Pr., v. 11, p. 45-54

—— 1960, The Mattagami area of northwestern Quebec: Precambrian, v. 33, no. 10, p. 29-31

—— 1966, The relationship of mineralization to Precambrian stratigraphy in the Matagami Lake and Val d'Or districts of Quebec: Geol. Assoc. Canada Spec. Paper no. 3, Precambrian Symposium, p. 21-42 (particularly p. 35-41)

MacKay, D. G., and Paterson, N. R., 1960, Geophysical discoveries in the Mattagami district, Quebec: Canadian Inst. Min. and Met. Tr., v. 63 (Bull. no. 581), p. 477-483

Paterson, N. R., 1966, Mattagami Lake mines--a discovery by geophysics: Soc. Expl. Geophys., Mining Geophysics, v. 1, Case Histories, p. 185-196

Pemberton, R. H., 1961, Target Mattagami: Canadian Inst. Min. and Met. Tr., v. 64 (Bull. no. 585), p. 16-23

Roscoe, S. M., 1965, Geochemical and isotopic studies, Noranda and Matagami areas: Canadian Inst. Min. and Met. Tr., v. 68 (Bull. no. 641), p. 279-285

Sharpe, J. I., 1964, Geology of part of the east half of Daniel Township and of the west half of Isle-Dieu Township, Abitibi-East County: Quebec Dept. Nat. Res. P. R. no. 503, 13 p.

—— 1965, Field relations of Matagami sulphide masses bearing on their deposition in time and space: Canadian Inst. Min. and Met. Tr., v. 68 (Bull. no. 641), p. 265-278; disc., Miller, R.J.M., p. 305-306

—— 1968, Geology and sulfide deposits of the Matagami area, Abitibi-East County: Quebec Dept. Nat. Res. Geol. Rept. 137, 122 p.

Suffel, G. G., 1965, Remarks on some sulphide deposits in volcanic extrusives: Canadian Inst. Min. and Met. Tr., v. 68 (Bull. no. 642), p. 301-307

Tully, D. W., 1964, Notes on the geology of the No. 1 Orebody, Mattagami Lake Mines Ltd.: Canadian Inst. Min. and Met. Field Excursion Guidebook, 3 p.

Vollo, N. B., 1964, Orchan Mines Ltd.: Canadian Inst. Min. and Met. Field Excursion Guidebook, 9 p.

Notes

The Matagami district (also spelled Mattagami) is located south of Lake Matagami, with the largest known ore bodies being in the northwest quarter of Galinée Township between the Bell River (east) and the Allard River (west), both of which empty into Lake Matagami from the south. The deposits in Galinée Township are those of Matagami Lake, Orchan, Bell, Allard, and Consolidated Mining and Smelting. In the southern half of Isle-Dieu Township (directly north of Galinée) are the Radiore M "A" (west of Bell River) and the Radiore M "E", Bell Channel, and Garon Lake deposits (east of Bell River). The New Hosco mine is located in the southeast quarter of Daniel Township that adjoins Isle-Dieu Township to the west. The center of the district is at about N49°45' and W77°45' and about 85 miles north-northeast of Amos.

The oldest rocks in the district are Keewatin-type volcanic rocks that are found in an east-west belt; the northern boundary of the belt runs through Lake Matagami and the southern lies several miles south of the mineralized portion of the district. The belt also includes minor amounts of sedimentary rocks intercalated with the lavas both north and south of the area containing the ore deposits. The volcanic rocks were intruded, largely before folding, by the Bell River Complex of mafic intrusives and by later dikes that range from peridotite, through diabase, to lamprophyre. After folding, the area was again intruded, this time by appreciably more siliceous rocks, including masses of granite, granodiorite, and quartz diorite and siliceous to mafic dikes. The final igneous event was the introduction of much younger (though probably middle Precambrian) diabase and gabbro dikes.

If Sharpe's (1965) suggestion that the deposits were formed in shallow, near-surface traps in the volcanic sequence is correct, the deposits probably were not emplaced long after the volcanic rocks themselves had been accumulated. In fact, Sharpe believes that they developed during the intrusion of the rocks of the Bell River complex and its satellitic bodies. On this reasoning, the deposits must have been emplaced in late early Precambrian time, and the fluids depositing them came from the same general source as the Bell River complex of rocks and should be classified here as early Precambrian. If the suggestions of Jenney (1961) are followed, that is, that the ores are typical epigenetic replacement deposits, they may have been introduced at a somewhat later time than Sharpe would select. Jenney, however, points out that the ore bodies are cut by dikes that were the last of the Bell River rocks to have been emplaced, so the ores must be older than at least part of the Bell River complex but almost certainly came from the same source. Jenney (1961) also believes that the alteration zones that surround the ore bodies are typically hydrothermal and that the replacement follows a structural rather than a stratigraphic pattern. On the basis of Jenney's arguments, then, the ores must have been appreciably younger than the Keewatin-type volcanic rocks but must have been essentially contemporaneous with the Bell River igneous activity. Thus, although emplaced differently than under Sharpe's hypothesis, the ore deposits would have been derived, under Jenney's hypothesis, from the same source invoked by Sharpe, indicating that under either hypothesis for the origin of the ores, their age was late early Precambrian. Roscoe (1965) reports that the mean of isotopic analyses and the calculations based on them give an age of 2930 m.y. for Matagami galenas and seem to provide confirmation for the early Precambrian age assigned here to the deposits. The conclusions drawn by Roscoe as to the origin of the Matagami ores, however, differ widely from those given here.

The ore deposits of the Matagami district generally are heterogeneous masses of pyrite, pyrrhotite, sphalerite, chalcopyrite, magnetite, and various high-temperature nonmetallic minerals, although locally the metallic sulfides show considerable banding or layering. It appears that the most logical interpretation of these deposits is that they were emplaced in the structurally, stratigraphically, and lithologically most permeable portions of the completely accumulated Keewatin-type pile of lavas after the beginning of the intrusion of the Bell River complex but before all of these rocks had been introduced. Most of the folding of Bell River probably predated the deposition of the ores, but some deformation took place after most of the sulfides had been emplaced, including some post-ore movement on pre-ore faults. That the ores are concentrated beneath largely impermeable masses of lava and layers of chert is no argument for their deposition at any particular depth beneath the surface. The best argument for their deposition at great depths is that they were emplaced at much the same time as the Bell River rocks and that these rocks were intruded under a great thickness at Watson Lake and Wabassee lavas.

Although the geothermometric determinations reported by Sharpe (1965) almost certainly are meaningless, other indications point to a high-temperature origin for the ores. These indications are the high content of magnetite and pyrrhotite in the ore and the presence of exsolution blebs of chalcopyrite in the sphalerite. The conversion of the mafic minerals to chlorite, epidote, and magnetite and the considerable alteration of feldspar to saussurite are wall-rock alteration effects not inconsistent with the formation of the essentially contemporaneous ores in the hypothermal range. The late fraction of the chalcopyrite, formed after sphalerite deposition had ceased, may have been produced in the mesothermal range; this possibility, however, does not seem sufficient to warrant the inclusion of mesothermal in the classification. The Matagami deposits, therefore, are here categorized as hypothermal-1.

NORANDA

Middle Precambrian	*Copper, Zinc, Gold, Silver, Pyrites*	*Hypothermal-1 to Leptothermal*

Alcock, F. J., 1935, Western Quebec, in *Copper resources of the world*: 16th Int. Geol. Cong., v. 1, p. 76-89

Ballachey, A. G., and others, 1952, Mining at Quemont: Canadian Inst. Min. and Met. Tr., v. 55 (Bull. no. 477), p. 57-68 (particularly p. 58-60)

Baragar, W.R.A., 1968, Major-element geochemistry of the Noranda volcanic belt, Quebec-Ontario: Canadian Jour. Earth Sci., v. 5, p. 775-790

Boldy, J., 1968, Geological observations on the Delbridge massive sulphide deposit: Canadian Inst. Min. and Met. Bull., v. 61, no. 677, p. 1045-1054

Brown, W. L., 1948, Normetal mine, in *Structural geology of Canadian ore deposits*: Canadian Inst. Min. and Met., Montreal, p. 683-692

Campbell, F. A., 1962, Age of mineralization at Quemont and Horne mines: Canadian Inst. Min. and Met. Tr., v. 65 (Bull. no. 605), p. 293-296

—— 1963, Sphalerite-pyrrhotite relationships at Quemont mine: Canadian Mineral., v. 7, pt. 3, p. 367-374

Campbell, F. A., and Imrie, A. S., 1965, Composition of diabase dikes at Quemont mine: Canadian Jour. Earth Sci., v. 2, p. 324-328

Campbell, F. A., and Williams, K. L., 1968, Composition of sphalerite from Quemont mine, Quebec: Econ. Geol., v. 63, p. 824-831

Cooke, H. C., 1928, Ore relations at the Horne and Aldermac mines, Quebec: Canadian Inst. Min. and Met. Tr., v. 31 (Bull. no. 198), p. 57-82

Cooke, H. C., and others, 1931, Geology and ore deposits of Rouyn-Harricanaw region, Quebec: Geol. Surv. Canada Mem. 166, 314 p. (particularly p. 175-233)

Dresser, J. A., and Denis, T. C., 1949, Aldermac Copper Corporation, Limited, in *Geology of Quebec*, v. 3, *Economic Geology*: Quebec Dept. Mines, p. 135-139

Dugas, J., 1966, The relationship of mineralization to Precambrian stratigraphy in the Rouyn-Noranda area, Quebec: Geol. Assoc. Canada, Spec. Paper No. 3, p. 43-55

Gill, J. E., and Schindler, N. R., 1932, Geology of the Waite-Ackerman-Montgomery property, Duprat and Dufresnoy Townships, Quebec: Canadian Inst. Min. and Met. Tr., v. 35 (Bull. no. 246), p. 398-416

Gilmour, P., 1965, The origin of massive sulphide mineralization in the Noranda district, northwestern Quebec: Geol. Assoc. Canada Pr., v. 16, p. 63-81

Hawley, J. E., 1948, The Aldermac copper deposit, in *Structural geology of Canadian ore deposits*: Canadian Inst. Min. and Met., Montreal, p. 719-730

Machairas, G., 1966, La métallogénie du cuivre dans la région de Noranda (Québec): Bur. Recherches Géologique et Minieres Bull. no. 4, p. 113-133

Machairas, G., and Blais, R. A., 1966, La transformation de l'hédenbergite manganésifère en ilvaite dans les sulfures de cuivre et de zinc de la région de Noranada: Soc. Francaise Mineral. Bull., no. 5. 79, p. 372-376

Mawdsley, J. B., 1928, Desmeloizes map-area, Abitibi district, Quebec: Geol. Surv. Canada Summ. Rept., pt. C, p. 28-82

Mookherjee, A., and Suffel, G. C., 1968, Massive sulfide-late diabase relationships, Horne mine, Quebec--Genetic and chronological implications: Canadian Jour. Earth Sci., v. 5, p. 421-432

Peale, R., 1931, The geology of the Waite-Ackerman-Montgomery ore deposit: Canadian Inst. Min. and Met. Tr., v. 34 (Bull. no. 233), p. 1069-1086

Price, P., 1934, The geology and ore deposits of the Horne mine, Noranda, Quebec: Canadian Inst. Min. and Met. Tr., v. 37 (Bull. no. 263), p. 108-140

—— 1948, Horne mine, in *Structural geology of Canadian ore deposits*: Canadian Inst. Min. and Met., Montreal, p. 763-772

—— 1949, Noranda Mines, Limited, in *Geology of Quebec*, v. 3, *Economic Geology*: Quebec Dept. of Mines, p. 338-361

—— 1949, Waite Amulet Mines, Limited, in *Geology of Quebec*, v. 3, *Economic Geology*: Quebec Dept. of Mines, p. 360-383

Price, P., and Bancroft, W. L., 1948, Waite Amulet mine; Waite section, in *Structural geology of Canadian ore deposits*: Canadian Inst. Min. and Met., Montreal, p. 748-756

Robinson, W. G., 1951, Structural geology and ore deposits of the Rouyn-Noranda district: Geol. Assoc. Canada Pr., v. 4, p. 61-67

Roscoe, S. M., 1965, Geochemical and isotopic studies, Noranda and Matagami areas: Canadian Inst. Min. and Met. Tr., v. 68 (Bull. no. 641), p. 279-285

Rosen-Spence, Andrée de, 1969, Genèse des roches à cordierite-anthophyllite des gisements cupro-zinofères de la région de Rouyn-Noranda, Québec, Canada: Canadian Jour. Earth Sci., v. 6, p. 1339-1345

Ryznar, G., and others, 1967, Sulfur isotopes and the origin of the Quemont ore body: Econ. Geol., v. 62, p. 664-678

Spence, C. D., and others, 1967, Noranda and district: Canadian Inst. Min. and Met., Centennial Field Excursion, Northwestern Quebec and Northern Ontario, p. 36-68

Suffel, G. G., 1935, Relations of later gabbro to sulphides at the Horne mine, Noranda, Quebec: Econ. Geol., v. 30, p. 905-915

—— 1948, Waite Amulet mine; Amulet section, in *Structural geology of Canadian ore deposits*: Canadian Inst. Min. and Met., Montreal, p. 757-763

—— 1965, Remarks on some sulphide deposits in volcanic intrusives: Canadian Inst. Min. and Met. Tr., v. 68 (Bull. no. 642), p. 301-307

Taylor, B., 1957, Quemont mine, in *Structural geology of Canadian ore deposits*, v. 2: Canadian Inst. Min. and Met., Montreal, p. 405-413

Tolman, C., 1951, Normetal mine area Abitibi-West County: Quebec Dept. Mines Geol. Rept. 34, 34 p.

Webber, G. R., 1962, Variation in composition of the Lake Dufault granodiorite: Canadian Inst. Min. and Met. Tr., v. 65 (Bull. no. 598), p. 55-62

Wilson, M. E., 1934, The multiple and complementary sills and dykes at the Waite-Ackerman-Montgomery mine, Noranda district, Quebec: Roy. Soc. Canada Tr., 3d Ser., v. 28, sec. 4, p. 65-74

—— 1935, Rock alteration at the Amulet mine, Noranda district, Quebec: Econ. Geol., v. 30, p. 478-492

—— 1941, Noranda district, Quebec: Geol. Surv. Canada Mem. 229, 162 p. (particularly p. 5-6, 49-58, 59-81)

—— 1948, Structural features of the Noranda-Rouyn area, in *Structural geology of Canadian ore deposits*: Canadian Inst. Min. and Met., Montreal , p. 672-683

—— 1956, Early Precambrian rocks of the Timiskaming region, Quebec and Ontario: Geol. Soc. Amer. Bull., v. 67, p. 1397-1430; disc., 1959, v. 70, p. 935-940

—— 1962, Rouyn-Beauchastel map-areas, Quebec: Geol. Surv. Canada Mem. 315, 140 p.

Notes

The district here designated as Noranda includes Beauchastel, Rouyn, and Dufresnoy Townships; Dufresnoy Township lies directly north of Rouyn and Beauchastel directly west. The Normetal mine in Desmeloizes Township, some 60 miles west of north from the city of Noranda, also is considered a part of the district. The twin cities of Noranda and Rouyn are about 22 miles from the Ontario border and some 260 miles northwest of Ottawa.

Beauchastel and Rouyn Townships are divided roughly in half by the more or less east-west Cadillac-Bouzan fault or break; the bedded rocks north of this break are largely Keewatin-type volcanic rocks that range from massive and brecciated rhyolite to andesite. To the south of the break, the first rocks encountered on the surface are either Timiskaming-type sediments that are principally graywackes and conglomerates or late Precambrian Cobalt-group rocks that also are mainly conglomerates and graywackes. South of these sediments, the bedded rocks are thought to be Keewatin in the sense that they are probably of the same age as the Keewatin-type rocks north of the break, but they are largely mica schists, derived from sediments, with some minor amounts of included volcanic rocks.

The district has been subjected to both pre- and post-Timiskaming intrusions that range from rhyolites and granites to andesites to originally mafic rocks that are now amphibolites, to lamprophyres. All of these rocks have been appreciably altered. Into this framework of folded, faulted, intruded and altered rocks, two types of mineralization were introduced: (1) copper-zinc-gold, principally in the Normetal (60 miles west of north from Rouyn); Aldermac (about 11 miles west of Rouyn); and Horne, Delbridge, Lake Dufault, Quemont, Waite, and Amulet mines (the last six in a belt about 10 miles long running more or less north from Rouyn); and (2) straight-gold mines (with locally appreciable silver), such as Arntfield, Francoeur, Powell Rouyn, Pontiac Rouyn, Stadacona, and many others, in which base metal sulfides are normally minor in amount and always economically unimportant. (References that discuss only the straight gold mines are given in the following bibliography which is headed *Noranda Gold.*) Not only are the absolute ages of these two types of mineralization uncertain but also their ages relative to each other. The late diabase dikes definitely cut the gold-quartz (straight-gold) veins in a number of mines (such as Stadacona and Powell Rouyn); this mineralization is, therefore, older than the diabase dikes. On the other hand, the relationship between these dikes and the copper-zinc-gold mineralization is far less certainly understood. In the Horne mine, Price (1934, 1948) believes that the ore mineralization (certainly the chalcopyrite-magnetite-gold phase and probably the earlier, but closely associated, pyrite-sphalerite phase) is later than the dikes. His reasons are (1) numerous fractures in the diabase contain chalcopyrite veinlets; in at least three locations much larger tongues of chalcopyrite and pyrrhotite ore, from 18 inches to 3 feet in width, extend from 4 to 8 feet into the diabase; (2) in one location the ore cuts completely across a diabase dike offshoot; (3) ore bodies on the two sides of the dikes do not match and, therefore, were not parted by the intrusive; (4) the abrupt termination of the ore against the dikes in many places probably means that the fine-grained chilled margins of the diabase were a barrier to the ore-forming fluid; (5) in the C ore body of the Horne mine the chalcopyrite-pyrrhotite ore has been found intimately to cut and replace the east-west diabase dike. Price considers any apophyses of diabase in the ore to be antecedent veins or veinlets, that is, offshoots of the diabase in the rhyolite were so much less susceptible to replacement than the rhyolite that they were left essentially unaffected by the ore-forming process. Against this, Campbell (1962) argues that Hawley's work has shown that chalcopyrite becomes quite mobile at low temperatures and will penetrate other minerals on reheating; he assumes that the diabase was introduced at a temperature of some 1150°C, far higher than the 400°C which Hawley found sufficient to remobilize chalcopyrite. Further, he presents K/Ar age determinations to show that sericite in the altered rhyolite has an average age of 2360 m.y., while hornblende in the diabase shows an age of 1740 m.y. Campbell, therefore, holds that (1) the penetration of diabase by chalcopyrite indicates remelting of the chalcopyrite by the heat of the diabase intrusion and, therefore, that the chalcopyrite remained molten until after the diabase had solidified; and (2) the age relations show that the chalcopyrite ore cannot be genetically related to both the alteration and to the diabase and that it is more reasonable to suppose that the ore is genetically connected to the alteration than to the diabase intrusion. Finally, it should be pointed out that the age of the alteration agrees fairly well with the general age of ore mineralization from Porcupine on the west to Val d'Or on the east, while the age of the diabase agrees with that of the Nipissing diabase at Cobalt and of the Sudbury intrusives. Obviously, then, the case is not perfect for either theory, and it is difficult to decide between the two proposed ages for ore mineralization. At the Normetal mine, Brown believes that the sulfides were deposited before the introduction of the major diabase dike but after the massive porphyritic rhyolite that cuts across the sheared rhyolite agglomerate that has been penetrated and altered by the sulfides. Nevertheless, Brown thinks that the sulfide mineralization was emplaced only a short time, geologically speaking, before the diabase.

Assuming that the radioactive age dates are correct, Campbell's case depends largely on the close time relationship of the alteration to the ores. Price believes that all three types of hydrothermal alteration (silicification, sericitization, and chloritization) both in the rhyolite and in the andesites (where alteration is generally less intense) are interrelated in age and in location. The relationships of these alterations to ore, however, are less clear. As the radioactive age determinations given by Campbell were made on sericite, the time relationship of this mineral to ore is of obvious importance in giving an age to the ore-forming period. Price recognizes four types of sericite in the Horne mine: (1) that produced by stresses attendant on folding (formed long prior to ore emplacement); (2) that formed by shearing stresses in the early stages of the development of the highly faulted and sheared zones (most faults were first produced during the early folding); (3) that originating as veinlets much later than (1) and (2) and at the same time as pyrite, chlorite, and the ore minerals were formed; and (4) that developed as massive sericite during the replacement of brecciated and sheared rhyolite, also at the same time as the ore emplacement. Although the sericite used by Campbell was taken from contacts between ore and sericite in both the Quemont and Horne mines, there is no guarantee that the sericite so sampled was not formed in Price's categories (1) or (2); such sericite might well have been formed long before the ore with which it is in close spatial juxtaposition. It would seem, therefore, that the age evidence supplied by the sericite is questionable. Further, the ore that cuts the diabase in the Horne mine is usually chalcopyrite-pyrrhotite ore, while (in Hawley's experiments) pyrrhotite and diopside were fractured and small veinlets of chalcopyrite entered into them. This suggests that both chalcopyrite and pyrrhotite in veins in diabase are less likely to have been introduced as remelted ore than to have been emplaced by normal hydrothermal processes after the diabase has been intruded. In short, although the problem is far from solved, it appears the more probable that copper-zinc-gold mineralization in the Horne mine was introduced after the diabase rather than before it. In the Quemont mine, in contrast to the Horne, the copper and zinc mineralizations appear to have been emplaced as closely connected events in the same sequence of ore deposition, and the contacts between ore and diabase provide no incontrovertible evidence as to their relative ages. On the balance of probabilities, then, the Noranda ore mineralization is dated as late middle Precambrian. On the other hand, the straight-gold mineralization, which is almost certainly older than the diabase, very probably was introduced during the early part of the igneous history of the area at much the same time as the gold ores of the Cadillac-Malartic and Val d'Or districts. The exact age and genetic relationship of the gold mineralization of the copper-zinc-gold type to the sulfides and magnetite, on the one hand, and to the straight-gold mineralization, on the other, will be considered below.

The copper-zinc-gold mineralization in the Noranda district differs in detail from one deposit to another, but all the deposits contain magnetite, pyrite, pyrrhotite, chalcopyrite, and sphalerite in major amounts; the appreciable quantities of gold usually are in the native state, but in the Horne mine tellurides are quite common. The mineralization in the Horne mine differs from that in the other deposits in that it has been considered (Price, 1948) to have been introduced in three stages: (1) pyrite, followed by sphalerite--a considerable amount of the chlorite and an appreciable part of the sericite were deposited in close space and time association with the pyrite and sphalerite; (2) magnetite, pyrrhotite, and chalcopyrite (which last contains some probably exsolved laths of cubanite) and a minor amount of pyrite; and (3) the bulk of the gold (with which some silver is alloyed) plus some tellurides as hessite, petzite, sylvanite, krennerite, calaverite, altaite, tetradymite, and rickardite. Definite time breaks appear to take place between stages (1) and (2) and (2) and (3). An appreciable fraction of the stage (2) minerals were deposited in different rock volumes from those containing the pyrite-sphalerite ore. The gold-bearing solutions followed different paths from those that brought in the earlier minerals as is shown by the rarity of high-grade gold ore in high-grade copper ore. Nevertheless, the gold is present in the copper and zinc ores and has replaced all of the early minerals except perhaps the late pyrite. In the other mines similarly mineralized, the definite time break between stages is lacking but the order of mineral introduction seems to have been much the same. From the Horne mine sequence it appears reasonable to classify the early pyrite-sphalerite as mesothermal, the later magnetite-pyrrhotite-chalcopyrite as hypothermal-1, and the gold-telluride as leptothermal. How much of the gold in the other deposits is leptothermal is unknown, but enough is present in the Horne mine to justify including that term in the classification.

I am aware of recent suggestions (Campbell, 1963; Gilmour, 1965; Suffel, 1965) that the deposits in the Noranda district were not formed by hydrothermal solutions acting on rocks deeply buried beneath the earth's surface but were developed (1) by deposition from a melt (Webber, 1962) introduced into the rock volume in question or (2) by deposition in sea water at intervals between the emplacement of the lava flows on the sea floor (Gilmour, 1965; Suffel, 1965). It is argued in support of suggestion (2) that the consistent occurrence of these deposits (and of others like them) in volcanic rocks indicates that the ores must also have been formed by volcanic processes and that any departures from conformity with volcanic stratigraphic relationships is due to late remobilization under orogenic stress. It seems to me, however, that the ores are in volcanic rocks because the district contains such large volumes of them and because they were more readily fractured than the intrusive igneous masses. The departure of such a large proportion of the ores in these deposits from interflow contacts is convincing evidence to me of their emplacement by hydrothermal fluids and not by movement under earth stresses. Obviously the problem needs further study through which definite criteria might be formulated as to the manner of emplacement.

NORANDA GOLD

Early Precambrian *Gold, Silver* *Mesothermal*

Davidson, S., and Banfield, A. F., 1944, Geology of the Beattie gold mine, Duparquet, Quebec: Econ. Geol., v. 39, p. 535-556

Davidson, S., and others, 1948, Structural relations of the Noranda straight gold deposits, in *Structural geology of Canadian ore deposits*: Canadian Inst. Min. and Met., Montreal, p. 692-701, 701, 710, 711-719, 730-734, 735-739, 739-747, 776-782, 783-789, 789-796

Dresser, J. A., and Denis, T. C., 1949, Beauchastel and Rouyn Townships, in *Geology of Quebec*, v. 3, *Economic geology*: Quebec Dept. Mines, p. 131-135, 138-147, 148-173

Halet, R. A., 1957, Quesabe mine, in *Structural geology of Canadian ore deposits*, v. 2: Canadian Inst. Min. and Met., Montreal, p. 413-415

Hawley, J. E., 1932, The Granada gold mine and vicinity, Rouyn Township: Quebec Bur. Mines Ann. Rept. 1931, pt. B., p. 3-57

—— 1934, McWatters mine gold belt, East Rouyn and Joannes Townships: Quebec Bur. Mines Ann. Rept. 1933, pt. C., p. 3-74

—— 1942, The Granada gold mine, in Newhouse, W. H., Editor, *Ore deposits as related to structural features*: Princeton Univ. Press, p. 100-191

—— 1942, The McWatters gold deposit, in Newhouse, W. H., Editor, *Ore deposits as related to structural features*: Princeton Univ. Press, p. 99-100

O'Neill, J. J., 1933, The Beattie gold mine, Duparquet Township, western Quebec: Quebec Bur. Mines Ann. Rept. 1932, pt. C., p. 3-28

—— 1934, Beattie-Galatea map-area, parts of Duparquet and Destor Townships: Quebec Bur. Mines Ann. Rept. 1933, pt. C., p. 75-109

—— 1934, Geology of the Beattie gold mine, Duparquet Township, Quebec: Canadian Inst. Min. and Met. Tr., v. 37 (Bull. no. 266), p. 299-315

Robinson, W. G., 1943, Flavrian Lake area, Beauchastel and Duprat Townships, Temiscamingue and Abitibi Counties: Quebec Dept. Mines, Geol. Rept. 13, 21 p.

Notes

The straight gold mines of the Noranda district are located mainly in Beauchastel and Rouyn Townships with most, but far from all, of them being south of the copper-zinc-gold mines.

The absolute age of the straight gold deposits and the relationship of that age to that of the copper-zinc-gold deposits have been discussed under the preceding heading of *Noranda*.

The gold normally is closely associated in space and probably in time with pyrite. The gangue minerals are a typical mesothermal assemblage of quartz, chlorite, sericite, and carbonates; only locally are minor tellurides known. Although the suggestion has been made that the gold is much later than the other vein minerals (in an effort to bring it in line in time of emplacement with the gold in the copper-zinc-lead deposits), no sound evidence seems to have been offered to support this concept. Very small quantities of high-temperature minerals such as arsenopyrite have been found in the veins but not enough to affect the classification of the deposits. These ores are here assigned to the mesothermal range.

OKA

Late Mesozoic (pre-Laramide)	*Columbium, Rare Earths*	*Hypothermal-2 (Cb), Magmatic-4 (RE)*

Deines, P., 1970, The carbon and oxygen isotopic composition of carbonates from the Oka carbonatite complex, Quebec, Canada: Geochim. et Cosmochim. Acta, v. 34, p. 1199-1225

Dresser, J. A., and Denis, T. C., 1944, The Monteregian Hills, in *Geology of Quebec*, v. 2, *Descriptive geology*: Quebec Dept. Mines, p. 455-482 (the Oka area may be the ninth Monteregian Hill; see p. 463)

Fairbairn, H. W., and others, 1963, Initial ratio of strontium 87 to strontium 86, whole rock age, and discordant biotite in the Monteregian igneous province: Jour. Geophys. Res., v. 68, p. 6515-6522

Girault, J., 1966, Sur la genèse des cristaux d'apatite des carbonatites d'Oka (Canada): Acad. Sci. C.R., Ser. D, v. 263, no. 2, p. 97-100

Girault, J., and Chaigneau, M., 1967, Sur les inclusions fluides présentes dans les cristaux d'apatite des roches de la région d'Oka (Canada): Acad. Sci. C.R., Ser. D, v. 264, no. 4, p. 529-532

Gold, D. P., 1966, The minerals of the Oka carbonatite and alkaline complex, Oka, Quebec: Int. Mineral. Assoc., 4th Gen. Mtg., Papers and Pr., p. 109-125

—— 1967, Alkaline ultrabasic rocks in the Montreal area, Quebec, in Wyllie, P. J., Editor, *Ultramafic and related rocks*: John Wiley & Sons, Inc., N.Y., p. 288-301

—— 1969, The geology of the diatreme breccia pipes and dykes, and related alnoite, kimberlite, and carbonatite intrusions occurring in the Montreal and Oka area, P.Q.: Part II: The Oka carbonatite and alkaline complex, in *The Monteregian Hills*: Geol. Assoc. Canada Guidebook, p. 43-62

Gold, D. P., and Marchand, M., 1969, The diatreme breccia pipes and dykes, and the related alnoite, kimberlite, and carbonatite intrusions occurring in the Montreal and Oka areas, P.Q.: Part I: The alnoite, kimberlite, and diatreme breccia pipes and dykes, in *Geology of the Monteregian Hills*: Geol. Assoc. Canada Guidebook, p. 5-42

Gold, D. P., and others, 1967, Economic geology and geophysics of the Oka alkaline complex, Quebec: Canadian Inst. Min. and Met. Tr., v. 70 (Bull. no. 666), p. 245-258

—— 1967, Field guide to the mineralogy and petrology of the Oka area, Quebec, in *Geology of parts of eastern Ontario and western Quebec*: Geol. Assoc. Canada Guidebook, p. 147-166

Harvie, R., Jr., 1909, On the origin and relationships of the Paleozoic breccia in the vicinity of Montreal: Roy. Soc. Canada, Tr., 3d Ser., v. 3, sec. 4, p. 249-299

Heinrich, E. W., 1966, Oka Quebec, in *The Geology of Carbonatites*: Rand McNally and Co., Chicago, p. 371-376

Hogarth, D. D., 1961, A study of pyrochlore and betafite: Canadian Mineral., v. 6, p. 610-633

Howard, W. V., 1922, Some outliers of the Monteregian Hills: Roy. Soc. Canada Tr., 3d Ser., v. 16, sec. 4, p. 47-95

Hurley, P. M., 1960, Summary of potassium-argon ages from the Monteregian Hills, Quebec: N.Y.O. 3941, 8th Ann. Prog. Rept. for 1960, Dept. Geol. and Geophys., Mass. Inst. Tech., p. 283

Maurice, O. D., 1957, Preliminary report on Oka area, electoral district of Deux-Montagnes: Quebec Dept. Mines, Mineral Deps. Br. Prelim. Rept. no. 351, 12 p.

Nickel, E. H., 1956, Niocalite, a new calcium-niobium silicate mineral: Amer. Mineral., v. 41, p. 785-786

Nickel, E. H., and McAdam, R. C., 1963, Niobian perovskite from Oka, Quebec; a new classification of minerals from the perovskite group: Canadian Mineral., v. 7, pt. 5, p. 683-697

Osborne, F. F., 1938, Lachute map-area; Pt. I, general and economic geology: Quebec Bur. Mines Ann. Rept. 1936, pt. C, p. 3-40

Perrault, G., 1968, La composition chimique et la structure cristalline du pyrochlore d'Oka, P.Q.: Canadian Mineral., v. 9, p. 383-402

Pouliot, G., 1970, Study of carbonatitic calcites from Oka, Quebec: Canadian Mineral., v. 10, p. 511-540

Rowe, R. B., 1955, Notes on columbium mineralization, Oka district, Two Mountains County, Quebec: Geol. Surv. Canada Paper 54-22, 18 p.

—— 1958, Deposits of the Oka region, Quebec, in *Niobium (Columbium) deposits of Canada*: Geol. Surv. Canada Econ. Geol. Ser. no. 18, p. 65-88

Sclar, C. B., and Smerchanski, M. G., 1958, Columbium-rare earth-titanium mineralization at St. Joseph du Lac, Oka region, Quebec (abs.): Econ. Geol., v. 53, p. 926-927

Shafiqullah, M., and others, 1970, K-Ar age of the carbonatite complex, Oka, Quebec: Canadian Mineral., v. 10, p. 541-552

Stansfield, J., 1923, Extensions of the Monteregian petrographical province to the west and north-west: Geol. Mag., v. 60, p. 433-453

Watkinson, D., 1970, Experimental studies bearing on the origin of the alkalic rock--carbonatite complex and niobium mineralization at Oka, Quebec: Canadian Mineral., v. 10, pt. 3, p. 350-362

Notes

The Oka district is situated on the north shore of the widened portion of the Ottawa River known as the Lake of Two Mountains (Lac des Deux Montagnes); Oka is about 22 miles due west of the center of Montreal.

With the exception of the unusual rocks of the Oka complex proper, the rocks of the district in the northwestern half consist of anorthosite, gabbroic anorthosite, and mangerites of the Morin series, and in the southeastern half consist of Grenville quartz-feldspar gneisses and minor quartzites and marble. The area is part of the Beauharnois arch that connects the late Precambrian rocks of the Adirondacks with those of the southern part of the Canadian Shield. The Oka complex is located at the intersection of two structural trends, faults associated with the Ottawa-Bonnechère graben and the linear trend of the Monteregian Hills. The rocks of the complex were introduced as ring dikes, cone sheets, plugs, breccia pipes, and dikes, and the pattern of these rocks and the flow layering they exhibit point to emplacement through twin conduits in the upper zone of metamorphism. Paleozoic sedimentary rocks surround the complex, and their presence in the breccia pipes shows that they once overlay the area and provided a cover probably as much as 8000 to 10,000 feet thick.

It has been suggested that Oka constitutes the ninth (tenth if Mount Megantic is included) and most westerly of the Monteregian Hills. These hills lie on a somewhat curved, southeasterly trending line and are spaced from 8 to 10 miles apart with Brome and Shefford Hills being the most easterly and Mount Royal (Mons Regius) the most westwardly of the eight. In a general way, the hill masses increase in size to the east with Brome being largest of all.

The Oka area, on the other hand, has had a quite different history from that of the other Monteregian Hills and contains rather different rocks. The Oka complex is a northwest-trending oval about 4 miles long and 1.5 miles wide at its greatest width; the rocks of the complex definitely are discordant to the Precambrian metamorphic rocks by which they are enclosed. The major carbonate-rock portion of the complex lies in the center of the northwestern two-thirds of the entire mass; this carbonate body is about 1.5 by 0.5 miles in size. A smaller carbonate body is located near the southeastern end and is about 0.5 miles across. The complex is both magnetic and radioactive, the magnetism being caused by disseminated magnetite and the radioactivity deriving principally from betafite $[(U,Th,Ca)(Cb,Ta,Ti)_3O_9 \cdot nH_2O]$; the betafite is unusual in that the thorium content is much higher than that of uranium. The betafite is concentrated in biotized rocks of the ijolite (essentially nepheline and a sodic pyroxene, aegerine-augite) and microijolite series and, therefore, is not as widely distributed as the magnetite. Bodies of explosion-vent breccia (one of which is abnormally radioactive) are found outside the carbonate mass, and the country rocks have been fenitized along the contact between carbonate and Precambrian metamorphosed rocks.

Granted that the Oka complex is the ninth Monteregian Hill, its age, on stratigraphic grounds, is at least post-Middle Devonian since the rocks of some of the hills intrude formations of Middle Devonian age. It has been suggested that the intrusive rocks in the hills were a phase of the igneous activity that accompanied the Acadian orogeny. An early radioactive age determination (made in 1936), however, indicated that the Monteregian rocks were early Tertiary. This age has not been confirmed by recent analyses; instead, the work of Fairbairn and others (1963) gives the age of the Oka deposits as about 114 m.y. (by the Rb/Sr method) and a probably less reliable age of 95 m.y. is given (by the K/Ar method) on biotite. The ages of the rocks of the Monteregian Hills (by the Rb/Sr method) range between 114 m.y. (Oka) to 84 m.y. (Brome). Certainly, the ages, if correct, indicate that the rocks of the different hills were not introduced at quite the same time, but all of them would appear to have been emplaced during the Cretaceous. Oka itself probably was developed early enough in the Cretaceous to be categorized as late Mesozoic (pre-Laramide) rather than Laramide and is so recorded here.

The valuable materials of the Oka district occur in two types of deposits: (1) lenticular ore shoots of columbium minerals that range from 50 feet in length and 10 feet in width to 1700 feet and 200 feet in the same dimensions and (2) sharp-walled veins (1 mm to 25 cm thick) and a body of irregular shape, both containing columbium- and rare earth-bearing perovskite and both occurring in alnöite; the rare-earth elements are mainly cerium and lanthanum.

Although Rowe (1958) believes that the Oka complex is fundamentally a part of the Monteregian petrographical province, he points out certain important differences: (1) post-Precambrian alkalic gabbro (essexite) is absent from the Oka area; (2) nepheline syenite is scarce in the complex; and (3) carbonate rock is abundant at Oka and is, except for invaded limestones, essentially lacking in the other Monteregian complexes. The Oka complex also differs from the carbonatites of northern Europe in that the proportion of calcite rock to alkalic rocks is much larger at Oka than in Alnö or the Fen. It is probable, however, that the alkaline rock-carbonate complex is of magmatic origin and has been worn down to a greater depth, relative to its original upper surface, than is usual in carbonatites or than is true of the other Monteregian Hills. No alnöite or kimberlite are directly associated with the carbonatite, but there are three bodies of explosion-vent breccia, circular to oval in shape, that lie within 4000 feet of the northeast margin of the Oka complex proper. The fragments in these bodies are a collection of Precambrian rocks a foot or more in maximum diameter that are cemented by fine-grained carbonate material or by alnöite. At least one pipe is filled with massive alnöite.

Opinions have been expressed by competent observers (Rowe, 1958) that the calcite rock at Oka resembles (1) Grenville marble and (2) African carbonatites. It is possible, however, that the calcite rock of Oka could have formed in one of four ways: (1) from post-Precambrian sedimentary carbonate rock that had been forced into its present relationship to the Precambrian rocks by post-Precambrian earth movements and had been metamorphosed by the alkalic intrusives; (2) from Precambrian Grenville limestone that was forced into its present position relative to the other Precambrian rocks by Precambrian or post-Precambrian earth movements and was later metamorphosed by the alkalic intrusives; (3) from magmatic carbonate that was intruded from depth prior to the introduction of the alkalic intrusives that, when they entered the area, metamorphosed the carbonate material; and (4) from carbonate that hydrothermally replaced either Precambrian or post-Precambrian rocks; this hydrothermal carbonate was later metamorphosed by the alkalic intrusives.

The fourth suggestion above seems impossible since the carbonate material shows no relict structures, and it is difficult to see how the so completely replaced host rock could have been so markedly discordant with the surrounding Precambrian rocks unless it had already been a plastically introduced carbonate rock that did not require hydrothermal replacement to convert it to carbonate. The main arguments against the third suggestion above are (1) the overabundance of carbonate in relation to alkalic intrusives and (2) the probable emplacement of the carbonate material before most of the alkalic intrusives--the reverse of the usual sequence in carbonatite complexes. Heinrich (1966), however, reports that some exposures of carbonate rock contain angular xenoliths of some of the alkalic rocks, indicating that at least part of the alkalic igneous activity preceded the emplacement of the carbonate rock body. The main bases for preferring concept (2) to concept (1) are that the rock closely resembles Grenville marble and that there is no such carbonate material in any volume in the other Monteregian Hill complexes.

Thus, although possibly valid arguments can be brought forward against the carbonate rock at Oka being carbonatite, it shows typically carbonatite relationships with its associated alkalic rocks and with the fenetized Precambrian rocks surrounding it. Further, the accessory minerals in the carbonatite, such as monticellite, perovskite, magnetite, apatite, and pyrochlore, are normal accessories of carbonatites all over the world. The alkalic rocks in the Oka complex consist of alnöite breccias and other lamprophyres and rocks of the melteigite-ijolite-urtite series (including microijolite) and of the okaite-jacupirangite series, typical members of many carbonatite complexes. It appears that the ijolite and jacupirangite series were late differentiates from a Monteregian-type essexite-nepheline syenite magma since tendencies toward such differentiation are shown in other Monteregian complexes and because of the enrichment of the Oka complex rocks in columbium, thorium, uranium, and the rare earths.

The most economically valuable concentrations of columbium minerals are found in calcite rock that contains soda pyroxene and biotite in addition to pyrochlore. Since pyrochlore was always a later mineral to form and since the pyrochlore ore shoots normally are in calcite near to, or between, bodies of microijolite and biotitized microijolite, it would seem probable that the pryochlore was developed in the calcite rock by reactions between the minerals already present and hydrothermal solutions given off by the crystallizing microijolite. Where other columbium-bearing minerals are present in the ore shoots, they undoubtedly were formed through similar reactions, the differences in mineral species being attributable to slight differences in host rock and ore-forming fluid content. The columbium ores, therefore, are here classified as hypothermal-2.

The consistent confinement of the rare-earth-bearing perovskite veins and of the irregular mass of similar material to alnöite bodies suggests that they were produced by reactions between the solid alnöite and the late-stage water-rich fluids generated in the alnöite magma. Such reactions should be considered as deuteric rather than placed in the hydrothermal category. The rare-earth deposits at Oka, therefore, are here classified as magmatic-4.

Although a decision has been reached in this discussion that the calcite-rich rocks at Oka are carbonatites of a somewhat unusual type, the problem certainly has not been sufficiently studied, and further work may bring to light facts that will require a reversal of the opinion expressed here. At the moment, however, the evidence is definitely in favor of the concept that carbonate and alkaline-rock magmas intruded a host mass of Precambrian silicate rocks.

ST. URBAIN

Late Precambrian — *Iron as Hematite, Titanium as Ilmenite, Rutile* — *Magmatic-3b*

Dresser, J. A., and Denis, T. C., 1949, St. Urbain ilmenite deposits (Quebec), in *Geology of Quebec*, v. 3, *Economic geology*: Quebec Dept. Mines, p. 428-432

Evrard, P., 1949, The differentiation of titaniferous magmas: Econ. Geol., v. 44, p. 210-232

Gillson, J. L., 1932, Genesis of the ilmenite deposits of St. Urbain, County Charlevoix, Quebec: Econ. Geol., v. 27, p. 554-577

Karpoff, D., 1953, Contribution to the study of the St. Urbain ilmenite deposit: Canadian Inst. Min. and Met. Tr., v. 56 (Bull. no. 496), p. 240-246

Mawdsley, J. B., 1927, St. Urbain area, Charlevoix district, Quebec: Geol. Surv. Canada Mem. 152, 58 p.

Osborne, F. F., 1928, Certain magmatic titaniferous iron ores and their origin: Econ. Geol., v. 23, p. 724-761, 895-922 (particularly p. 743-745)

Rechenberg, H. P., 1955, Zur Genesis der primären Titanerzlagerstätten: Neues Jb. f. Mineral., Mh., H. 4, S. 87-96

Robinson, A.H.A., 1922, Titanium: Canadian Dept. Mines, Mines Branch, no. 579, p. 41-44, 46-53

Rose, E. R., 1961, Iron and titanium in the anorthosite, St. Urbain, Quebec: Geol. Surv. Canada Paper 61-7, 20 p.

—— 1969, St. Urbain area, in *Geology of titanium and titaniferous deposits of Canada*: Geol. Surv. Canada, Econ. Geol. Rept. no. 25, p. 77-85

Ross, C. S., 1941, St. Urbain, Quebec, in *Occurrence and origin of the titanium deposits of Nelson and Amherst Counties, Virginia*: U.S. Geol. Surv. Prof. Paper 198, p. 23

Tuttle, O. F., 1943, Orientation of ilmenite and andesine from St. Urbain, Quebec, titaniferous iron-ore deposit (abs.): Amer. Geophys. Union Tr., 24th Ann. Mtg., pt. 1, p. 280-281

Warren, C. H., 1912, The ilmenite bearing rocks near St. Urbain, Quebec; a new occurrence of rutile and sapphirine: Amer. Jour. Sci., 4th Ser., v. 33, no. 195, p. 263-277

—— 1918, On the microstructure of various titanic iron ores: Econ. Geol., v. 13, p. 419-446 (particularly p. 433-434)

Notes

The center of the St. Urbain district is located 8 miles slightly west of north from the town of Baie St. Paul that, in turn, lies directly on the north shore of the St. Lawrence some 51 miles northeast of the city of Quebec.

The oldest rocks of the St. Urbain area are highly metamorphosed silicic to mafic gneisses that were intruded by granite masses and mafic dikes. In this rock mass the St. Urbain complex was emplaced; it consists of an anorthosite core with a complete marginal rim of gabbroic anorthosite and diorite that includes some syenite and granite. No Grenville quartzite or limestone are found in the area. The anorthosite core is some 18 miles long in a north-south direction and is about 10 miles wide; it consists of two phases, the older containing labradorite feldspar and the younger andesine; blocks of the labradorite phase are contained in the andesine anorthosite. Rounded bodies of anorthosite and fragments of andesine crystals (of the type found in the anorthosite) are to be found locally in the diorite-granite series.

The presence of the anorthosites among the intruded rocks of the district suggests that the age of these rocks is about 1000 m.y. (Grenville) as is true of almost all anorthositic complexes the world over, and this is confirmed by a radioactive age determination on biotite found in the ore which gives an age of 890 m.y. Although the ores are almost entirely intrusive into the anorthosite of the complex, the presence of minor disseminated ilmenite in the anorthosite and of abundant titaniferous-magnetite-bearing gabbro or norite in the outer rim of the complex suggests that the ore was generated within the complex during its crystallization cycle and was emplaced somewhat before the end of the late Precambrian. Some minor amounts of ilmenite ore in what appear to be early Paleozoic rocks are probably lithified placer deposits and not primary introductions of ilmenite into these formations.

Although either ilmenite-hematite bodies or float are known in all parts of the anorthosite mass, the minable ore is in a group of deposits west of the town of St. Urbain. These deposits are dikelike or sill-like or occur as massive veins a few inches to 50 feet in width and up to 300 feet long. These bodies branch and split around horses of essentially barren anorthosite. Some of the deposits have a nearly vertical attitude and appear to extend to considerable depths, and others are close to horizontal; most are lenticular in plan.

There are a few small areas of disseminated ilmenite-hematite grains in anorthosite at, or near, contacts of ore and the feldspathic rock. Bands and seams of the ilmenite-hematite, normally connected to massive ore, in many places penetrate the host rock and lie generally parallel to the long dimensions of the larger ore bodies. Along the ends and sides of some of the deposits, biotite selvages, pyrite veins, and slickensides were developed after postmineralization movement had fractured the ore and further fractured the host rocks. In a narrow zone near the contact with the enclosing anorthosite, the Bignell ore is quite vuggy, and the vugs are partly filled with pyrite, hematite, and limonite.

The principal ore mineral is ilmenite that contains microintergrowths of hematite; there is probably also hematite in unexsolved solid solution in the ilmenite and similarly of ilmenite in the hematite. Titaniferous magnetite is sparingly present in the ore, and rutile is sporadically distributed through it, mainly in fractures between ore and anorthosite, but there is a little rutile disseminated locally in the anorthosite and in the ilmenite-hematite ore aggregate. Microscopically, the ore is a mosaic of inter-locking grains of ilmenite in which exsolved blades, lenticles, and blebs of hematite occur. Some hematite is quite coarse-grained, but a considerable portion of it is much finer, suggesting to Rose (1961) two generations of exsolution.

In the ilmenite-hematite grains, hematite particles usually form less than 25 percent of the total; near the hanging wall of the Bignell deposit, however, hematite makes up about 50 percent of the ore and is in coarse-grained platy crystals. Rounded grains of silicate minerals, mainly plagioclase, occur in the ore and are generally cut or penetrated by veinlets or tongues of ilmenite-hematite. Between ilmenite-hematite and plagioclase, there are narrow rims of biotite, chamosite, chlorite, and a little sapphirine. Both the rounding and the minerals in the rims were probably caused by a reaction between the ilmenite-hematite melt and the plagioclase crystals that solidified from the melt early in its crystallization cycle. Pyrite is present as a late mineral in the ore, some pyrite being late enough in solidifying to cut previously deposited pyrite; minor chalcopyrite and pyrrhotite are associated with the pyrite, and all of these sulfides invade and cement ilmenite grains.

The presence of ilmenite-hematite in the ore, with minor rutile and very little titaniferous-magnetite, suggests that the PO_2 of the iron-titanium-oxygen melt was even higher than that of the gabbroic anorthosite in which titaniferous magnetite is the dominant mineral and than that at Allard Lake where essentially no rutile is present. This high availability of oxygen was undoubtedly responsible for so complete an oxidation of the ferrous iron originally present in the melt that there was not enough Fe^{+2} present to neutralize all of the TiO_3^{-2}, forcing some of it to precipitate late in the cycle as TiO_2 by the loss of one screening oxygen to the melt. Rose (1961) suggests that the exsolution textures exhibited by the ilmenite show that the ores probably crystallized at about 600°C, a temperature compatible with those achieved experimentally by Tuttle for iron-phosphorus-oxygen melts. Thus, it appears probable that the ores were introduced as water-poor (but water-bearing) melts, rich in iron, titanium, and oxygen, that were somewhat hotter than 600°C. It is probable that these melts were immiscible fractions separated from the parent magma of the anorthosite complex at a late stage in the crystallization cycle and that the melts were introduced into the anorthosite portion of the igneous complex after the silicate rocks had already solidified. The quite small amounts of disseminated ilmenite-hematite in the anorthosite probably were formed as direct crystallization segregations from the anorthosite magma, but some of these disseminated grains may have been originally thin intrusive bands of ilmenite broken up by later movements within the anorthosite mass. The ores are, therefore, dominantly the products of crystallization from the molten state, and should be classified as magmatic-3b.

THETFORD-BLACK LAKE

Middle Paleozoic | *Asbestos* | *Magmatic-4, Hypothermal-1, Metamorphic-C*

Bain, G. W., 1932, Chrysotile asbestos: II, Chrysotile solutions: Econ. Geol., v. 27, p. 281-296

—— 1939, Thetford, Disraeli, and eastern half of Warwick map-areas, Quebec: (rev.) Econ. Geol., v. 34, p. 235-237

Cooke, H. C., 1935, The composition of asbestos and other fibers of Thetford district: Roy. Soc. Canada Tr., 3d Ser., v. 29, sec. 4, p. 7-19

—— 1936, Asbestos deposits of Thetford district, Quebec: Econ. Geol., v. 31, p. 355-376

—— 1937, Thetford, Disraeli, and eastern half of Warwick map-areas, Quebec: Geol. Surv. Canada Mem. 211, 160 p.

Denis, B. T., 1930, Asbestos occurrences in southern Quebec: Quebec Bur. Mines Ann. Rept., pt. D, p. 147-193

Dresser, J. A., 1913, Preliminary report on the serpentine and associated rocks of southern Quebec: Geol. Surv. Canada Mem. 22, 103 p. (particularly p. 54-73)

Dresser, J. A., and Denis, T. C., 1944, The Eastern Townships, part II--Intrusive rocks--The serpentine belt, in *Geology of Quebec*, v. 2, *Descriptive geology*: Quebec Dept. Mines, p. 413-447

Faessler, C., and Badollet, M. S., 1947, Epigenesis of the minerals of the serpentine belt, Eastern Townships of Quebec: Canadian Min. Jour., v. 68, no. 3, p. 157-167

Graham, R.P.D., 1917, Origin of massive serpentine and chrysotile asbestos, Black Lake-Thetford area, Quebec: Econ. Geol., v. 12, p. 154-202

Olsen, E. J., 1961, High temperature acid rocks associated with serpentinite in eastern Quebec: Amer. Jour. Sci., v. 259, p. 329-347

Paige, S., 1937, Asbestos deposits of Thetford district, Quebec: Econ. Geol., v. 32, p. 108-109

Poitevine, E., and Graham, R.P.D., 1918, Contributions to the mineralogy of Black Lake area, Quebec: Geol. Surv. Canada Mus. Bull. no. 27 (Geol. Ser. no. 35), 103 p. (particularly p. 73-80)

Riordon, P. H., 1954, Preliminary report on Thetford mines-Black Lake area, Frontenac, Megantic, and Wolfe Counties: Quebec Dept. Mines Prelim. Rept. no. 295, 23 p.

—— 1955, The genesis of asbestos in ultrabasic rocks: Econ. Geol., v. 50, p. 67-81

—— 1957, The structural environment of the Thetford-Black Lake asbestos deposits: Geol. Assoc. Canada Pr., v. 9, p. 83-93

—— 1962, Geology of the asbestos belt in southeastern Quebec: Canadian Inst. Min. and Met. Tr., v. 65 (Bull. no. 601), p. 182-184

Riordon, P. H., and others, 1957, The asbestos belt of southeastern Quebec, in *The geology of Canadian industrial mineral deposits*: Canadian Inst. Min. and Met., Montreal, p. 3-36 (including descriptions of several individual mines, mainly by Riordon)

Notes

The asbestos deposits of southern Quebec are found in a belt that lies generally west of the Eastern Townships copper belt and reaches some 55 miles through the Eastern Townships of Quebec from the city of Asbestos on the southwest to East Broughton on the northeast. Most of the deposits are clustered in the Thetford-Black Lake area that is about one-third of the way from East Broughton to Asbestos.

The Cambrian and Ordovician sedimentary rocks of this portion of the Eastern Townships consist of the Cambrian(?) Caldwell group and the Ordovician Beauceville group. During either the Taconic or the Acadian disturbance, probably during the former, the ultramafic and mafic rocks of the asbestos belt were intruded, essentially concordantly, along the folded erosional contact between the Caldwell and Beauceville groups. In the middle portion of the intrusive belt, where the greatest amount of igneous activity occurred, three separate intrusive complexes have been identified. These are (from oldest to youngest): (1) peridotite and dunite (the early complex); (2) dunite, pyroxenite, and gabbro (the late complex); and (3) syenite and granite (which may not have been introduced until Acadian time). The asbestos deposits always are found in the earliest (peridotite-dunite) complex and, within that complex, are confined to its peridotite portions. This peridotite makes up a large proportion of the ultramafic belt, and in the Thetford area the main mass of peridotite is 15 miles by 3 miles in size. Northeast of the town of Thetford Mines, the early ultramafics are confined to a narrow sill and consist almost entirely of peridotite. Peridotite is also the most abundant ultramafic in the town of Asbestos end of the belt, although that area does contain rocks of the late complex. The peridotite is classified by Riordon (1957) as harzburgite containing from 10 to 15 percent of enstatite. Segregations of dunite are scattered through the peridotite; these olivine-rich bodies range in length from one foot up to several hundred feet. A belt of dunite also extends for 4 miles along the southeast side of the large peridotite body in the Thetford area. All the dunite masses in the peridotite appear to have been formed from the early ultramafic magma by gravitative differentiation. More than limited differentiation of this earlier ultramafic magma probably was prevented by a quite rapid rate of crystallization.

The later granitic and syenitic rocks were emplaced almost entirely in the ultramafic and mafic bodies; the granitic rock masses are quite small in comparison with the ultramafic and mafic bodies, and the syenites are even smaller. The silicic rocks are arranged in no recognizable pattern; generally, they are tabular in form and often are quite flat lying. The syenitic bodies normally are those that bear close spatial relations to the asbestos deposits, and the syenites are made up mainly of albite with various quantities of orthoclase and biotite. They often are highly altered, and the principal alteration mineral is secondary quartz. It appears that the contacts between silicic and ultramafic rocks are usually sharp, although faulting may conceal the true relationships. The silicic rocks are so irregularly distributed within the femic rocks that they almost certainly could not have resulted from the ultramafic magma by any process of differentiation. The silicic rocks do tend to have produced shearing in the femic rocks adjacent to them, and some of the small silicic bodies also have been highly sheared or even brecciated. No large masses of silicic rocks are known in the district from which these small bodies might have been derived so Riordon (1957) believes that they travelled a long way from their parent magma chamber. The granite and syenite masses have not been observed to intrude one another, but the granite bodies (always larger than those of syenite) grade into marginal facies or apophyses of syenite, so Riordon thinks that the syenite bodies, mainly dikes, were late-stage (pegmatitic) differentiates of the larger granite masses driven out from them late in their crystallization cycle. The syenite bodies in the ultramafics are surrounded by aureoles of completely serpentinized ultramafic rocks, the serpentinization probably having been accomplished by hydrothermal fluids given off by the crystallizing syenite pegmatites whether they were in the form of dikes or more massive bodies. Riordon thinks that the lack of quartz in the syenites probably was due to much of the silica needed for this mineral having been transferred in water-rich fluids from the syenitic pegmatitic fusion to the surrounding ultramafics. In these aureoles of complete serpentinization around the syenites, brucite (a common mineral in the deuterically altered peridotites) is lacking and probably was converted to serpentine through the addition of silica from such a silica-rich fluid.

Initially, about 30 to 40 percent of the peridotite was converted to serpentine, the serpentine being distributed quite uniformly throughout the ultramafic rock. Many of the small dunite segregations also have been altered to serpentine, but in contrast to the peridotites these dunites have been completely changed to serpentine. In addition, most of the belt of dunite on the southeast side of the Thetford area peridotite has been converted to serpentine. Further, the dunite of the late complex, associated with pyroxenite, has been completely serpentinized, but in the pyroxenites only the minor olivines and rhombic pyroxenes have been changed to serpentine. All of this serpentine is believed to have been formed as the last stage in the crystallization cycles of the rocks in which it is found; the earliest differentiates to crystallize were the most altered. The peridotite of the early complex is completely serpentinized only (1) in aureoles that surround the late silicic intrusives and (2) around shatter zones, faults, and joints through which solutions from the same source (or sources) as the silicic rocks (or from the silicic rocks themselves) could have had access to the partially autometamorphosed (serpentinized) peridotite. The broken rock volumes of type (2) were developed at the same time as, or closely after, the entry of the silicic intrusives into the ultramafics. Most of the earth movements were concentrated in the margins of the intrusives, and the broken zones there developed provided channelways through which serpentinizing solutions might rise and from which they might work their way into the surrounding ultramafics. In addition to these major controlling structures, the aureoles of serpentinized rock that had been developed around the silicic intrusives behaved as incompetent masses; within them considerable shearing took place, and outside them, in the more competent peridotite, fracturing occurred. Bodies of serpentinized dunite had much the same effect on the peridotites surrounding them. Because of the ease of solution movement through these more broken rock volumes, the serpentinized contact zones around the silicic bodies and the serpentinized dunites commonly are barren of asbestos; the only dunites to contain asbestos are those that were not completely serpentinized during the stage of autometamorphic serpentinization.

It appears probable that the silicic intrusives were introduced into the area during the general time of the Acadian disturbance, since there are considerable masses of silicic intrusives in the Eastern Townships copper belt that, while introduced before the disturbance, intruded the Silurian Sherbrooke group of formations. In the Weedon area two granite bodies are known that were intruded after the Weedon thrust fault had been formed and are, therefore, at least a short time later than this expression of the Acadian disturbance. No granite rocks are found that are older than the Sherbrooke group. Thus, since the silicic rocks appear to have been necessary concomitants to asbestos formation, the asbestos deposits must have been formed just before, during, or just after the Acadian disturbance, and perhaps their time of formation extended from before to after the disturbance. The asbestos deposits, therefore, probably were developed in Middle or Late Devonian time and should be categorized as middle Paleozoic. The ultramafic rocks, the hosts and at least one of the parents of the asbestos deposits, almost certainly were introduced in Late Ordovician time, but granted that this is so, the ultramafics themselves probably are more logically classified as middle rather than early Paleozoic.

The workable deposits of asbestos in the district are found in those volumes of peridotite adjacent to highly fractured zones. The first step in asbestos formation, serpentinization of the peridotite, requires most importantly, assuming no volume change, the addition of water (Riordon, 1955, 1957); such additions of water would release magnesia and silica from the peridotite. This displaced magnesia and silica can be deposited as serpentine veins in the open fractures in peridotite only if a further increment of silica is made available. Riordon thinks that the serpentine along fractures in the massive serpentine and the serpentine filling the fractures can be converted to asbestos only if still additional amounts of silica are brought into the volumes in question. Both the silica needed for completing serpentinization and for asbestos production are believed to come from the silicic intrusives or from the magma chamber from which these intrusives came.

Magnetite was a by-product of complete serpentinization and usually is concentrated along the larger fractures; it is thus a common constituent of the asbestos veins and of the wall rocks immediately adjacent to them. In addition to the magnetite that is associated directly with serpentine veins and veinlets, the completely serpentinized dunites and highly shattered peridotites also contain considerable magnetite but seldom are sites of appreciable asbestos formation. Thus, magnetometer surveys may find anomalies in the district that have no relationship to asbestos deposits at all. Magnetite also usually is abundant in the wall rocks immediately adjacent to the asbestos veins and in inclusions in the veins.

In addition to chrysotile asbestos, a common variety of fibrous material in the Thetford-Black Lake area is slip-fiber picrolite (Riordon, 1955); it is found along fault planes, particularly those in broken ground. Asbestos veins may grade into picrolite veins. Picrolite is a fibrous antigorite, chemically the same as chrysotile but crystographically and physically different from it; in picrolite the y axis is consistently parallel to the fiber axis, while in chrysotile the fiber axis almost always is parallel to the x axis. One variety of picrolite is composed of bundles and sheaves of coarse fibers that roughly parallel the plane of the fracture containing them and generally is soft and has a splintery fracture; the fibers are difficult to separate and generally break easily. This type of picrolite grades into a harsh, compact, brittle form that, under the microscope, also is fibrous. There also is a cross-fiber variety of picrolite that is found in compound veins that are banded parallel to their walls and columnar normal to them. These veins range from microscopic up to several inches in width. In part, the cross-fiber picrolite generally is hard and compact, but some of it is quite soft and may be mistaken for talc. These banded veins of picrolite are found both in fault planes and in the same types of fractures as those in which the asbestos occurs.

The hypotheses offered to account for the generation of asbestos are of three general types: (1) the fibers were formed in open space between the walls of the original fracture in the host rock, (2) the fibers formed by recrystallization or reorganization of the material of the host rock immediately adjacent to the initial fracture, and (3) the fibers formed by a combination of fissure filling and wall-rock replacement.

It would appear that the sequence of events in the production of chrysotile asbestos was (1) partial serpentinization of the peridotite by solutions autometamorphically generated; this serpentine should be classed as magmatic-4; (2) locally complete serpentinization of the peridotite by solutions generated in the silicic igneous magmas; the production by these solutions of zoisite, clinozoisite, albite, and quartz within the silicic rocks they encountered suggests that they lay in the hypothermal intensity range and that the serpentine they developed should be classed as hypothermal-1; (3) development of picrolite through open-space filling and replacement of serpentine by hydrothermal solutions, probably generated in the same silicic magma chambers that produced the earlier serpentinizing solutions; this picrolite should be categorized as hypothermal-1, although some of it may have been developed in the upper mesothermal range; and (4) development of asbestos due to the relief of stresses attendant on cooling; this asbestos should be classed as metamorphic-C. Thus, the Thetford-Black Lake deposits should be considered as the result of three distinct processes, acting successively, and should be classified as magmatic-4, hypothermal-1, and metamorphic-C.

VAL D'OR

Early Precambrian | *Gold* | *Mesothermal to Leptothermal*

Ames, H. G., 1948, Perron mine, in *Structural geology of Canadian ore deposits*: Canadian Inst. Min. and Met., Montreal, p. 893-898

Auger, P. E., 1947, Siscoe map-area, Dubuisson and Vassan Townships, Abitibi-East County: Quebec Dept. Mines, Mineral Deps. Branch, Geol. Rept. 17, 40 p.

Backman, O. L., 1936, Geology of the Siscoe gold mine: Canadian Min. Jour., v. 57, no. 10, p. 467-475

Bell, L. V., 1933, Mining properties of Pascalis-Louvicourt area: Quebec Bur. Mines Ann. Rept. 1932, pt. B, p. 3-59

—— 1935, Lamaque-Sigma mines and vicinity, western Bourlamaque Township: Quebec Bur. Mines Ann. Rept. 1934, pt. B, p. 3-60

Bell, L. V., and Bell, A. M., 1935, Structural features of gold deposits in certain intrusives of Western Quebec: Econ. Geol., v. 30, p. 347-369 (particularly p. 353-359)

Cooke, H. C., and others, 1931, Gold deposits, in *Geology and ore deposits of the Rouyn-Harricanaw region, Quebec*: Geol. Surv. Canada Mem. 166, p. 242-262

Dresser, J. A., and Denis, T. C., 1947, Dubuisson-Bourlamaque-Louvicourt district, in *Geology of Quebec*, v. 3, *Economic geology*: Quebec Dept. Mines, p. 237-279

Dugas, J., and Latulippe, M., 1961, Noranda-Senneterre mining belt: Quebec Dept. Nat. Res. Map no. 1388, about 1:253,440

Gussow, W. C., 1937, Petrogeny of the major acid intrusives of the Rouyn-Bell River area of northwestern Quebec: Roy. Soc. Canada Pr., 3d Ser., v. 31, sec. 4, p. 129-161 (particularly p. 134-144)

Hawley, J. E., 1931, Gold and copper deposits of Dubuisson and Bourlamaque Townships, Abitibi County: Quebec Bur. Mines Ann. Rept. 1930, pt. C, p. 3-95

—— 1932, The Siscoe gold deposit: Canadian Inst. Min. and Met. Tr., v. 35

(Bull. no. 245), p. 368-386

—— 1942, The Siscoe gold deposit, in Newhouse, W. H., Editor, *Ore deposits as related to structural features*: Princeton Univ. Press, p. 96-99

Hoyles, N.J.S., and others, 1967, Val d'Or and district: Canadian Inst. Min. and Met., Centennial Field Excursion, Northwestern Quebec and Northern Ontario, p. 12-33

Ingham, W. N., 1953, Ore deposits of the Val d'Or district: Geol. Soc. Amer., 66th Ann. Mtg. (1953), Guidebook of Excursions, p. 21-27

Ingham, W. N., and Keevil, N. B., 1951, Radioactivity of the Bourlamaque, Elzevir, and Cheddar batholiths, Canada: Geol. Soc. Amer. Bull., v. 52, p. 131-148

James, W. F., 1948, Siscoe mine, in *Structural geology of Canadian ore deposits*: Canadian Inst. Min. and Met., Montreal, p. 876-882

Kempthorne, H. R., 1957, Bevcon mine, in *Structural geology of Canadian ore deposits*, v. 2: Canadian Inst. Min. and Met., Montreal, p. 416-419

Latulippe, M., 1963, The Val d'Or-Malartic gold area, Quebec: Canadian Min. Jour., v. 84, no. 4, p. 72-76

—— 1966, The relationship of mineralization to Precambrian stratigraphy in the Matagami Lake and Val d'Or districts of Quebec: Geol. Assoc. Canada, Spec. Paper no. 3, Precambrian Symposium, p. 21-42 (particularly p. 26-35)

McDougall, D. J., 1954, The marginal luminescence of certain intrusive rocks and hydrothermal ore deposits: Econ. Geol., v. 49, p. 717-726

Norman, G.W.H., 1943, Bourlamaque Township, Abitibi County, Quebec: Geol. Surv. Canada Paper 43-2, 14 p.

—— 1948, The Malartic-Haig section of the southern gold belt of western Quebec, in *Structural geology of Canadian ore deposits*: Canadian Inst. Min. and Met., Montreal, p. 839-845

Sharpe, J. I., 1968, Louvicourt Township, Abitibi-East County: Quebec Dept. Res. Geol. Rept. 135, 53 p.

Smith, F. G., 1954, Direction of flow of late stage solutions in the Lamaque no. 6 vein: Econ. Geol., v. 49, p. 530-536

Snelling, N. J., 1962, Potassium-argon dating of rocks north and south of the Grenville front in the Val d'Or region, Quebec: Geol. Surv. Canada Bull. 85, 27 p.

Wilson, H. S., 1936, The geology of the Lamaque mine: Canadian Min. Jour., v. 57, no. 10, p. 511-516

—— 1948, Lamaque mine, in *Structural geology of Canadian ore deposits*: Canadian Inst. Min. and Met., Montreal, p. 882-891

Notes

The ore deposits of the Val d'Or district center around the town of that name, 16 miles east-southeast of Malartic and about 265 miles northwest of Montreal. Immediately west of the district are the gold deposits of the Cadillac-Malartic district and a short distance farther west are the copper-zinc-gold and straight gold deposits of the Noranda district.

The ore bodies of the Val d'Or district are, in a general way, related to the Cadillac-Larder Lake fault zone, the same major structure that can be followed from west of Kirkland Lake to several miles east of Val d'Or. Most of the Val d'Or mines are located in the townships of Dubuisson (west), Bourlamaque (center), and Louvicourt (east). The older nonintrusive rocks in the district are typical of those of the Keewatin-type in much of eastern Canada. The elucidation of the structure north of the Cadillac break still is far from complete, the complex structure of the Keewatin rocks having been complicated even beyond that produced by the earth movements at the end of Keewatin and Timiskaming time by the intrusion of post-Timiskaming igneous rocks that range from peridotite through gabbro, diorite, and granodiorite to syenite and granite.

The first intrusives to enter the district appear to have been ultramafic magmas from which peridotites were crystallized; these rocks form several belts essentially concordant with the strike of the host rocks; they are most abundant in Dubuisson Township but also occur to a limited extent in Bourlamaque and Louvicourt Townships. The ultramafics were followed by gabbros, diorites, and diabases that are only slightly represented in the Val d'Or district proper but are quite abundant in the area of the Lacorne batholith to the north. Following these mafic rocks, the volcanic rocks north of the Cadillac break were intruded by the Bourlamaque batholith, a pear-shaped mass with the stem at its west end; the batholith has a small lobe at its eastern end that extends almost to the Cadillac break; the lobe is bisected by the Bourlamaque River. An outlier, west of the stem, forms the north part of Siscoe Island in Lac Montigny. The dimensions of the main igneous mass are 14 miles east-west and a maximum of 7 miles north-south. The batholith is referred to as granodiorite, but it originally may have been as mafic as quartz gabbro. If this is the case, it is quite possible that the Bourlamaque batholith more properly belongs with the mafic rocks just mentioned than with the latter silicic rocks of the Lacorne, Preissac, and La Motte batholiths to the north and northwest. The minerals of the Bourlamaque batholith are now highly altered, and much of the quartz is thought to be secondary (Gussow, 1937). Dikes of various types, including aplite, "porphyry," syenite, and diorite also are present in the district. The youngest igneous rocks in the district are diabase dikes of the Keweenawan type.

In the Val d'Or district the Cadillac break is a more complex structure than is true farther to the west. Here it consists of a series of faults and shear zones that are found over a belt 4 miles wide between the north side of Siscoe Island and the Piche River, which is near but still slightly north of the southern edge of the break. The faulting along the Piche River is the direct continuation of the main Cadillac break in the Malartic district with which the Malartic ore bodies are associated.

The ore deposits of the area lie in or near zones of faulting and shearing; and in turn, these zones are localized by the contacts between the altered volcanics and the igneous rocks. Almost no minable deposits are found well within the Bourlamaque batholith; instead, the workable ores are confined to its margins and (more importantly) to the volcanic rocks beyond them. This close spatial relationship of ore and igneous rock suggests that the source of the ore-forming fluid was the same general volume of magma that provided the batholithic magma.

Snelling (1962) made numerous age determinations on the granitic rocks of the Preissac-Lacorne batholiths, and the southern boundary of the Lacorne batholith lies only some 5 or 6 miles northwest of the northern edge of the Bourlamaque batholith. Nevertheless, the larger granitic masses are so much more silicic in composition that it seems unlikely that they came from the same source as the more mafic rocks to the south. The only geologic aid in dating the mafic in relation to the silicic rocks is the cutting of the Bourlamaque by silicic dikes. Assuming that these silicic rocks come from the same general source as those of the Preissac-Lacorne batholith, then the Bourlamaque is definitely the older of the two masses. The ages for the granites (as determined by Snelling) range from 2735 m.y. to 2205 m.y., and it seems reasonable to assume that the age of the Bourlamaque mass is at least as old as the oldest granites. This certainly would place the intrusion of the Bourlamaque in the late early Precambrian and would give the gold ores about the same age.

Among the most important gold mines of the Val d'Or district are Siscoe, Sullivan, Lamaque, Sigma, and Perron. As a general rule in the Val d'Or district, most of the native gold, which has a wide range of silver content, is found in fractures in quartz; some native gold occurs in fractures in pyrite, the iron sulfide being later than the quartz. The pyrite normally is accompanied by small quantities of chalcopyrite and less commonly by even smaller amounts of pyrrhotite, magnetite, and sphalerite. A small part of the gold in the ore is contained in gold tellurides that also appear to have formed late in the ore sequence. The intense wall-rock alteration extends for various distances out from the vein and consists of quartz, tourmaline, carbonates (also a late gangue mineral), albite, sericite, and less abundantly epidote, chlorite, and talc. The early quartz-tourmaline-scheelite mineralization probably was formed under hypothermal conditions and so probably was the pyrite (plus the minor pyrrhotite-magnetite-chalcopyrite suite). That the native gold was formed under such high intensity conditions, however, is doubtful because of the considerable time lag between the formation of quartz, tourmaline, scheelite, and pyrite, and the deposition of gold. Because some of the wall-rock alteration minerals are compatible with mesothermal conditions and because these minerals (sericite and carbonate--carbonate particularly) are late, it would seem reasonable to categorize the bulk of the native gold, at least as mesothermal in origin. The minor tellurides, however, probably were deposited under leptothermal conditions, suggesting that some of the gold may have been formed in the leptothermal range. The Val d'Or deposits, therefore, are classified as mesothermal to leptothermal, with the understanding that a small fraction of the native gold may have been developed in the hypothermal range and perhaps a good deal of it under the leptothermal conditions.

So little has been published about the copper-zinc and copper-zinc-lead ores in the Val d'Or area that they are not discussed here. The deposits of East Sullivan and Manitou-Barvue are of these base-metal types.

Saskatchewan

BEAVERLODGE (GOLDFIELDS)

Middle Precambrian *Uranium* *Mesothermal*

Alcock, F. J., 1936, Geology of the Lake Athabasca region: Geol. Surv. Canada Memo. 196, 41 p.

—— 1936, The gold deposits of Lake Athabasca: Canadian Inst. Min. and Met. Tr., v. 39 (Bull. no. 292), p. 531-546

Allen, R. B., 1950, Fracture systems in the pitchblende deposits of the Beaverlodge area, Saskatchewan: Canadian Inst. Min. and Met. Tr., v. 53 (Bull. no. 460), p. 299-300

Beck, L. S., 1964, Structural environment of uranium mineralization in the Athabasca region: Canadian Min. Jour., v. 85, no. 4, p. 98-102

—— 1970, Genesis of uranium in the Athabasca region and its significance in exploration: Canadian Inst. Min. and Met. Bull., v. 63, no. 695, p. 367-377

Beecham, A. W., 1970, The ABC fault, Beaverlodge, Saskatchewan: Canadian Jour. Earth Sci., v. 7, p. 1264

Brooker, E. J., and Nuffield, E. W., 1952, Studies of radioactive compounds: IV--Pitchblende from Lake Athabaska: Amer. Mineral., v. 37, p. 363-385

Buffam, B.S.W., and others, 1957, Beaverlodge mines of Eldorado Mining and Refining Ltd., in *Structural geology of Canadian ore deposits*, v. 2: Canadian Inst. Min. and Met., Montreal, p. 220-235

Campbell, D. D., 1957, Geology and ore control at the Verna mine, Beaverlodge, Saskatchewan: Canadian Inst. Min. and Met. Tr., v. 60 (Bull. no. 545), p. 310-317

Chamberlain, J. A., 1959, Structural history of the Beaverlodge area: Econ. Geol., v. 54, p. 478-494; disc., v. 54, p. 1577; v. 55, p. 617-618; v. 56, p. 614-618

Christie, A. M., 1953, Goldfields-Martin Lake map-area: Geol. Surv. Canada Mem. 269, 126 p.

Christie, A. M., and Kesten, S. N., 1949, Pitchblende occurrences of the Goldfields area, Saskatchewan: Canadian Inst. Min. and Met. Tr., v. 52 (Bull. no. 452), p. 285-293

—— 1949, Goldfields and Martin Lake map-areas, Saskatchewan: Geol. Surv. Canada Paper 49-17, 22 p. (mimeo.)

Conybeare, C.E.B., and Campbell, C. D., 1951, Petrology of the red radioactive zones north of Goldfields, Saskatchewan: Amer. Mineral., v. 36, p. 70-79

Conybeare, C.E.B., and Ferguson, R. B., 1950, Metamict pitchblende from Goldfields, Saskatchewan, and observations on some ignited pitchblendes: Amer. Mineral., v. 35, p. 401-406

Dawson, K. R., 1951, A petrographic description of the wall-rocks and alteration products associated with pitchblende-bearing veins in the Goldfields region, Saskatchewan: Geol. Surv. Canada Paper 51-24, 58 p. (mimeo.)

—— 1956, Petrology and red coloration of wall-rocks, radioactive deposits, Goldfields region, Saskatchewan: Geol. Surv. Canada Bull. 33, 46 p.

Eckelmann, W. R., and Kulp, J. L., 1956, Uranium-lead method of age determination--part 1: Lake Athabasca problem: Geol. Soc. Amer. Bull., v. 67, p. 35-53

Edie, R. W., 1952, Studies in petrology, Goldfields area, Saskatchewan: Canadian Inst. Min. and Met. Tr., v. 55 (Bull. no. 487), p. 406-415

—— 1953, Hydrothermal alteration at Goldfields, Saskatchewan: Canadian Inst. Min. and Met. Tr., v. 56 (Bull. no. 493), p. 118-123

Fahrig, W. F., 1961, Geology of the Athabasca formation: Geol. Surv. Canada Bull. 68, 41 p.

Heinrich, E. W., 1958, Goldfields region, Saskatchewan: in *Mineralogy and geology of radioactive raw materials*: McGraw-Hill, N.Y., p. 304-315

Hill, P. A., 1959, Geology of the Pitch group, Beaverlodge Lake, Sask.: Precambrian, v. 29, no. 1, p. 6-10

James, W. F., and others, 1950, Canadian deposits of uranium and thorium: A.I.M.E. Tr., v. 187, p. 239-255 (particularly p. 248-252) (in Min. Eng., v. 187, no. 2)

Jolliffe, A. W., 1956, The Gunnar "A" orebody: Canadian Inst. Min. and Met. Tr., v. 59 (Bull. no. 528), p. 181-185

Jolliffe, A. W., and Evoy, E. P., 1957, Gunnar mine, in *Structural geology of Canadian ore deposits*, v. 2: Canadian Inst. Min. and Met., Montreal, p. 240-246

Joubin, F. R., and James, D. H., 1957, Rix Athabasca mine, in *Structural geology of Canadian ore deposits*, v. 2: Canadian Inst. Min. and Met., Montreal, p. 235-240

Koeppel, V., 1968, Age and history of the uranium mineralization of the Beaverlodge area, Saskatchewan: Geol. Surv. Canada Paper 67-31, 111 p.

Koster, F., and Baadsgaard, H., 1970, On the geology and geochronology of northwestern Saskatchewan. I. Tazin Lake region: Canadian Jour. Earth Sci., v. 7, p. 919-930

LaBine, J. S., 1963, Mine geology [Gunnar Mine]: Canadian Min. Jour., v. 84, no. 7, p. 56-59

Lang, A. H., 1952, Goldfields region, in *Canadian deposits of uranium and thorium*: Geol. Surv. Canada, Econ. Geol. Ser. no. 16, p. 68-106

Lang, A. H., and others, 1962, Beaverlodge area, in *Canadian deposits of uranium and thorium*: Geol. Surv. Canada Econ. Geol. Ser. no. 16, 2d ed., p. 145-175

Macdonald, B. C., and Kermeen, J. S., 1956, The geology of Beaverlodge: Canadian Min. Jour., v. 77, no. 6, p. 80-83, 156

Macdonald, J. A., 1969, An orientation study of the uranium distribution in Lake Waters, Beaverlodge district, Saskatchewan: Colo. Sch. Mines Quart., v. 64, no. 1, p. 357-376

Palache, C., and others, 1944, Uraninite, in *The System of mineralogy of James Dwight Dana and Edward Salisbury Dana*: 7th ed., Wiley & Sons, Inc., N.Y., p. 611-620

Robinson, S. C., 1950, Mineralogy of the Goldfields district, Saskatchewan: Geol. Surv. Canada Paper 50-16, 38 p. (mimeo.)

—— 1952, The occurrence of uranium in the Lake Athabasca region: Canadian Inst. Min. and Met. Tr., v. 55 (Bull. no. 480), p. 150-153

—— 1955, Mineralogy of uranium deposits, Goldfields, Saskatchewan: Geol. Surv. Canada Bull. 31, 128 p.

Robinson, S. C., and Brooker, E. J., 1952, A cobalt-nickel-copper selenide from the Goldfields district, Saskatchewan: Amer. Mineral., v. 37, p. 542-544

Staff, Eldorado Mining and Refining Limited, 1960, Eldorado Beaverlodge operation--geology: Canadian Min. Jour., v. 81, no. 6, p. 84-98

Tremblay, L. P., 1957, Ore deposits around Uranium City, in *Structural geology of Canadian ore deposits*, v. 2: Canadian Inst. Min. and Met., Montreal, p. 211-220

—— 1958, Geology and uranium deposits of Beaverlodge region, Saskatchewan: 2d U.N. International Conf. on Peaceful Uses of Atomic Energy (Geneva) Pr., v. 2, p. 491-497

—— 1967, Geology of the Beaverlodge mining area, Saskatchewan: Geol. Surv. Canada Mem. 367 (adv. ed.), 468 p.

—— 1970, The significance of uranium in quartzite in the Beaverlodge area, Saskatchewan: Canadian Jour. Earth Sci., v. 7, p. 280-305

Turek, A., 1965, Geology of the Lake Cinch mine, Uranium City, Saskatchewan: Canadian Inst. Min. and Met. Tr., v. 68 (Bull. no. 635), p. 80-86

Notes

The Beaverlodge (Goldfields) district is located in the northwestern corner of Saskatchewan and extends from the Gunnar mine on the northeast shore of Lake Athabasca to the Ace and Verna mines about 20 miles to the northeast, just beyond Beaverlodge Lake; Edmonton lies about 450 miles south-southwest of the district.

The bedrock in the Beaverlodge area consists of three principal rock series: (1) the Tazin group that is composed chiefly of quartzite, granitized quartzite, quartz-feldspar gneiss, argillite, amphibolite, chlorite schists, femic schist, and garnet-bearing schist with lesser amounts of ferruginous and dolomitic quartzites, dolomite, and conglomerate; most of the Tazin beds have been feldspathized, silicified, chloritized, epidotized, or hematized; (2) granitic and related rocks that may be in considerable part granitized portions of the Tazin group; in addition to typical granite and pegmatite, the group contains much banded granite gneiss and other rocks of what may be a hybrid character since the contacts between the granites and altered Tazin group rocks are normally gradational; and (3) the Martin (Athabasca) group that is composed of arkose, arkosic sandstone, conglomerate, siltstone, basalt and andesite flows, and a basal conglomerate of angular fragments of Tazin and granitic rocks; this group rests unconformably on the older rocks. Finally, there were introduced dikes and sills of gabbro and basalt. On structural and metamorphic grounds, the first two groups have been considered to be Archean and the third to be Proterozoic. The rocks of the group have been intricately folded, and the axes of these folds trend generally northeast.

Although Robinson (1955) has presented evidence for two periods of uranium mineralization, the finding of mineralogically similar uranium-bearing veins (some 3000 have been discovered in the area) in the rocks of all three groups suggests that all the veins were formed at essentially the same time. Radioactive age dates indicate this time of ore deposition to have been about 1800 m.y. ago, that is, from muscovite from a pegmatite in the Gunnar mine that is post-ore. The time lapse between ore deposition and pegmatite formation probably was not more than 100 m.y., so the ore should be classified as late middle Precambrian.

The Beaverlodge district contains deposits that appear to have been formed from hydrothermal solutions and others (far less important economically) that were developed in pegmatites. The hydrothermal deposits, in turn, can be divided into two types: (1) replacement bodies and (2) pitchblende-calcite veins in clean fractures. The replacement deposits of type (1) are the major sources of uranium in the area and are thought by Tremblay (1957) to have been localized along early folds and faults. These structures were formed in the Tazin group and its associated granites before the Martin group of rocks was laid down, so ore bodies of this type cannot be expected in the Martin rocks. This does not mean, of course, that the uranium deposits in the Martin group are of a much younger age than those in the Tazin rocks since these older folds and faults may not have been mineralized until post-Martin time. In replacement deposits, the pitchblende may be massive or disseminated and normally replaced brecciated material or worked its way outward into the wall rocks of zones of intense fracturing in, or related to, faults; some ore was localized in flexures or drag folds on the limbs of larger fold structures.

The localization of the hydrothermal replacement deposits appears to have been strongly influenced by the character of the wall rock, the most favorable type being that which contains mafic minerals, such as chlorite, amphibole, and pyroxene of their alteration products. Although brecciated rock is a common host for ore, dense, chertlike breccias appear to have been unfavorable to replacement by pitchblende.

The vein deposits in clean fractures, type (2), are thought to have been localized in late faults and joints and even along the unconformity between Tazin and Martin rocks (Tremblay, 1957). Although a few of these deposits are large enough to be mined and although there are hundreds of them, they are usually small, narrow, lenticular, very erratic in their distribution, and not constant in grade.

None of the pegmatites, which contain small amounts of uraninite, monazite, thorite, pyrochlore, and zircon, has been of a high enough grade to be mined.

Although the pitchblende deposits in the Beaverlodge area take many forms (microscopic disseminations and veinlets, coarse disseminations, stringers, pods, irregular masses, lenses, and true veins), the two largest producers are: (1) the large disseminated Gunnar deposit and (2) the Ace-Verna complex of disseminations and stringers, so closely grouped as to be minable as ore. These, and other workable concentrations of ore, are associated with zones of fracturing, mylonization, and brecciation. Pitchblende (in the sense of being uraninite with appreciable change in the unit cell parameters due to the inclusion of UO_3) is the main primary uranium material and comprises four main forms: (1) colloform, (2) massive, (3) sooty, and (4) euhedral. Some of the pitchblende has been brecciated and cemented by later pitchblende, indicating deposition over a considerable period of time. Thucolite also is present as a primary mineral in several deposits, and there are at least 14 secondary uranium-bearing minerals, including liebigite [$Ca_2(UO_2)(CO_3)_3 \cdot 10H_2O$] and uranophane [$Ca(UO_2)_2(SiO_3)_2(OH)_2 \cdot 5H_2O$] as the most common. Although most of the supergene material is found within a few feet of the surface, some supergene minerals have been deposited, along major fault zones, as much as 400 feet beneath the surface. Some of the pitchblende contains inclusions of such minerals as calcite, chlorite, and hematite. Hematite (and minor ilmenite) began to deposit before the pitchblende and colored the early feldspar to produce the prepitchblende red alteration. Hematite deposition, however, continued after that of feldspar had ceased, and the later hematite deposition considerably overlapped that of pitchblende. Nolanite--$3FeO \cdot V_2O_3 \cdot 3V_2O_4$--preceded and overlapped with the pitchblende. With the later (hematite-free portion of the pitchblende) a variety of arsenides, pyrite, chalcopyrite, bornite, galena, and gold were deposited; none is normally present in more than accessory amounts. Several selenides, including clausthalite, klockmannite, and tiemannite, were essentially the last metallic minerals to form and are definitely later than the pitchblende. Calcite continued to deposit throughout the mineralizing process, and a little native copper was associated with the latest calcite. Calcite and chlorite are the main gangue minerals, and quartz is present in moderate amounts.

The early feldspar-hematite (red) wall-rock alteration may have formed under hypothermal conditions, but the pitchblende (and the late hematite precipitated with the early pitchblende), the bulk of the metallic minerals, and the later gangue minerals (calcite, chlorite, and quartz) probably deposited in the mesothermal range. The selenides may well have formed under leptothermal conditions, but they are not of economic importance and are not included in the mesothermal classification assigned to the deposit.

CORONATION MINE

Middle Precambrian *Copper* *Hypothermal-1*

Byers, A. R., Editor, 1969, Symposium on the geology of the Coronation mine, Saskatchewan: Geol. Surv. Canada Paper 68-5, 329 p. (contains 17 papers on various aspects of the geology of the Coronation mine)

Byers, A. R., and Dahlstrom, C.D.A., 1954, Geology and mineral deposits of the Amisk-Wildnest Lakes area, Saskatchewan: Sask. Dept. Mineral Res., Geol. Br., Rept. no. 14, 169 p.

Gilliland, J. A., 1965, A proposed ore control at the Coronation mine, Saskatchewan: Canadian Inst. Min. and Met. Tr., v. 68 (Bull. no. 637), p. 157-164

Hajnal, Z., 1965, Paleomagnetic study of the Coronation mine area: Canadian Inst. Min. and Met. Tr., v. 68 (Bull. no. 637), p. 165-168

Smith, J. R., 1964, Distribution of nickel, copper, and zinc in bedrock of the East Amisk area, Saskatchewan: Sask. Res. Coun., Geol. Div. Rept. 6, 36 p.

Stockwell, C. H., 1960, Flin Flon-Mandy, Manitoba and Saskatchewan: Geol. Surv. Canada Map 1078A (with descriptive notes), 1:63,360

Notes

The center of the Coronation mine area is about 10 miles south-southwest from Flin Flon and includes the Coronation, Birch Lake, and Flexar ore bodies.

See *Notes* under Sherritt Gordon-Flin Flon (General).

The ores of the Coronation mine lie in a zone of shearing that cuts across the bedding of the enclosing metavolcanic rocks of the Amisk group, the host rocks being mafic volcanics that, in places, have been converted to cordierite-anthophyllite rocks. The deposit is made up of three ore shoots in a zone of lower-grade mineralization some 900 feet long and 120 feet wide; their vertical range is about 1000 feet. The ore shoots and the mineralized zone plunge steeply south, following the plunge of the linear structures found in the surrounding volcanic rocks. Most of the ore is disseminated in the gangue and wall rock, but some is quite massive. Two ages of mineralization (Byers, 1969, p. 3) have been recognized; the older consists principally of pyrite, pyrrhotite, and chalcopyrite with which are associated hornblende, tremolite-actinolite, cordierite, anthophyllite, chlorite, and quartz as well as altered plagioclase from the volcanic rocks. Also present are minor amounts of magnetite, sphalerite, garnet, zoisite, and gahnite. Some of the pyrite has been broken and cemented and replaced by pyrrhotite and chalcopyrite. It has been suggested that, in the last stages of the orogeny, the sulfides were recrystallized and partially remobilized to produce stringers into local areas of dilation in the zone of shearing. The preferential orientation of the pyrrhotite (the *c*-axis of a considerable percentage of that mineral being oriented perpendicularly to the foliation) has been considered as being caused by the regional metamorphism but may be a replacement-induced orientation. Late sulfide and quartz veins have been attributed to postmetamorphic movements of the Ross Lake fault system, aided by low-temperature solutions. It seems equally possible that the ores are all later than the major orogeny and that what breaking and other structural effects occurred were due to the Ross Lake faulting; the late mineralization well may have been due to the late, low-temperature phases of the main mineralization. A high-temperature of deposition for all but the minor late sulfide-calcite veins is indicated by the presence locally of lamellae of cubanite (probably produced by exsolution) in the chalcopyrite, the blebs of chalcopyrite in the sphalerite, the abundant pyrrhotite, the moderate amounts of magnetite, the not uncommon ilmenite, and the rare arsenopyrite. The late sulfide-calcite mineralization consisted of calcite, chalcopyrite, tetrahedrite, and pyrite; this mineralization is not insufficient to justify consideration in the classification. The deposits are here considered to be hydrothermal-1.

Yukon Territory

KENO HILL

Late Mesozoic (pre-Laramide) — *Silver, Lead, Zinc* — *Mesothermal to Leptothermal*

Aho, A. E., 1963, Silver in the Yukon: Canadian Inst. Min. and Met. Bull., v. 56, no. 611, p. 232-239 (particularly p. 236-239) (not in Tr.)

Bostock, H. S., 1947, Mayo, Yukon Territory: Geol. Surv. Canada Map 890A (with descriptive notes), 1:253,440

—— 1948, Mayo district, Yukon, in *Structural geology of Canadian ore deposits*: Canadian Inst. Min. and Met., Montreal, p. 110-112

Boyle, R. W., 1956, Geology and geochemistry of silver-lead-zinc deposits of Keno Hill and Sourdough Hill, Yukon Territory: Geol. Surv. Canada Paper 55-30, 78 p.

—— 1957, Lead-zinc-silver lodes of the Keno Hill-Galena Hill area, Yukon, in *Structural geology of Canadian ore deposits*, v. 2: Canadian Inst. Min. and Met., Montreal, p. 51-65

—— 1957, The geology and geochemistry of the silver-lead-zinc deposits of Galena Hill, Yukon Territory: Geol. Surv. Canada Paper 57-1, 41 p.

—— 1960, Occurrence and geochemistry of native silver in the lead-zinc-silver lodes of the Keno Hill-Galena Hill area, Yukon, Canada: Neues Jb. f. Mineral., Abh., Bd. 94 (Festband Ramdohr), Erste Hälfte, S. 280-297

—— 1965, Geology, geochemistry, and origin of the lead-zinc-silver deposits of the Keno Hill-Galena Hill area, Yukon Territory: Geol. Surv. Canada Bull. 111, 302 p.

Boyle, R. W., and Cragg, C. B., 1957, Soil analysis as a method of geochemical prospecting in Keno Hill-Galena Hill area, Yukon Territory: Geol. Surv. Canada Bull. 39, 27 p.

Boyle, R. W., and Jambor, J. L., 1963, The geochemistry and geothermometry of sphalerite in the lead-zinc-silver ores of the Keno Hill-Galena Hill area, Yukon: Canadian Mineral., v. 7, pt. 3, p. 479-496

Boyle, R. W., and others, 1955, Geochemical investigation of the heavy metal content of stream and spring waters in the Keno Hill-Galena Hill area, Yukon Territory: Geol. Surv. Canada Bull. 32, 34 p.

—— 1970, Sulphur isotopic investigations of the lead-zinc-silver-cadmium deposits of the Keno Hill-Galena Hill area, Yukon, Canada: Econ. Geol., v. 65, p. 1-10; disc. p. 731

Cairnes, D. D., 1915, Mayo area, Yukon Territory: Geol. Surv. Canada Summ. Rept., pt. A, p. 10-34

Carmichael, A. D., Jr., 1957, United Keno Hill mines, in *Structural geology of Canadian ore deposits*, v. 2: Canadian Inst. Min. and Met., Montreal, p. 66-77

Cockfield, W. E., 1923, Geology and ore deposits of Keno Hill, Mayo district, Yukon: Geol. Surv. Canada Summ. Rept., pt. A, p. 1-21

Gleeson, C. F., 1967, The distribution and behavior of metals in stream sediments and waters of the Keno Hill area, Yukon Territory: Geol. Surv. Canada Paper 66-54, p. 134-144

Green, L. H., and McTaggart, K. C., 1960, Structural studies in the Mayo district, Yukon Territory: Geol. Assoc. Canada Pr., v. 12, p. 119-134

Kindle, E. D., 1955, Keno Hill, Yukon Territory: Geol. Surv. Canada Paper 55-12 (map with marginal notes), 1:63,360

McTaggart, K. C., 1950, Keno and Galena hills, Yukon Territory: Geol. Surv. Canada Paper 50-20 (two maps with marginal notes), 1:12,000

—— 1960, Geology of Keno and Galena hills, Yukon Territory: Geol. Surv. Canada Bull. 58, 37 p.

Stockwell, C. H., 1925, Galena Hill, Mayo district, Yukon: Geol. Surv. Canada Summ. Rept., pt. A, p. 1-14

Van Tassel, R. E., 1969, Exploration by overburdening drilling at Keno Hill Mines Limited: Quart. Colo. Sch. Mines, vol. 64, no. 1, p. 457-478

Notes

The Keno Hill district is located in the central Yukon, some 220 miles slightly west of north from Whitehorse and about 160 miles west of the boundary with Alaska.

The oldest rocks in the area are those of the Yukon group, the age of which was originally placed by Boyle (1957) as between Precambrian and Paleozoic; on the basis of work done more recently, Boyle (1965) recognizes that the sequence may have been greatly interrupted by thrust faults and that some of the Yukon group rocks may be as young as late Paleozoic or early Mesozoic. In the immediate vicinity of Keno and Galena hills, the Yukon group can be divided lithologically into three units: (1) a lower schist, (2) a central quartzite, and (3) an upper schist. The central quartzite contains three ore-bearing members, the Galkeno (lowest), the Hector-Calumet, and the Silver King (uppermost), each bounded both below and above by unnamed less-competent members.

The central quartzite is the most important ore host in the district (the ore being mainly in the thick-bedded quartzites); where the veins pass into the thin-bedded quartzites, they are much narrower and are far less well mineralized. In the immediate neighborhood of the ore bodies, the igneous rocks are limited to diorites and gabbros that are now greenstones, to biotite lamprophyres, and to quartz-feldspar porphyry or rhyolite. These greenstones occur as conformable elongated lenses and sills, principally in the schists but to some extent in the quartzites. The quartz-feldspar porphyries were introduced as sills and are now most commonly exposed on Keno and Galena hills and on Mount Haldane. They cut the greenstones and must, therefore, be younger than the more mafic rocks; they are, in turn, cut and offset by veins of the siderite type of mineralization and are, therefore, pre-ore.

A few discontinuous dikes and sills of lamprophyre are found in the schists and the quartzites. They are cut and altered by the siderite type of ore veins and are, therefore, pre-ore; their age relationship to the greenstones and porphyries is unknown, but they must be of much the same general age. Although no evidence has yet been discovered as to the absolute age of these igneous rocks, radioactive age determinations by the Geological Survey of Canada have established outside the Keno Hill area itself that the Yukon formation was subjected to Jurassic metamorphism and intrusion. For lack of more positive evidence, it is here assumed that the Keno Hill igneous rocks were emplaced in part, at least, in the Jurassic.

Some 12 miles east and 8 miles northwest of the Keno Hill town, respectively, are the nearest outcrops of two groups of granitic-textured rocks; these outcrops are of appreciable size and range in composition from granite through granodiorite to quartz diorite. The principal minerals of these igneous rocks are quartz, plagioclase (oligoclase to andesine), potash feldspar, biotite, muscovite, and hornblende; pyroxene is common in some of these bodies. Plagioclase has been considerably saussuritized, and biotite and hornblende have been notably altered to chlorite. The age of these rocks in relation to the igneous rocks of the Keno Hill district is unknown, but it is probable that they were all introduced as part of the late Mesozoic igneous activity that affected the Coast Ranges throughout their length. On this basis, the granites are probably Early Cretaceous and the mafic rocks Jurassic.

The vein faults are later than all rock formations in the district, since all formations are cut by the faults. It follows, therefore, that the faulting episode occurred after the last igneous rocks were introduced into the area and that the vein filling was even later. Thus, assuming that the ore fluids were derived from the major granitic mass of which the granitic-textured rock outcrops are the uppermost expression, the ore fluids must have entered the area at some appreciable time after the actual cycle of intrusion had been completed. On this basis, it is thought that the ore bodies were probably formed in the Cretaceous but early enough in that period to be categorized best as late Mesozoic (pre-Laramide). It should be pointed out that Boyle (1965) believes that the hypogene deposits were formed by the diffusion of ore and gangue elements from the country rocks into the vein faults. Since the ore minerals are found in the vein faults but not in the later faults, it would seem reasonable to assume that the diffusion process if it occurred, took place before the later faults were available. If this represents Boyle's thinking correctly, the ores, under his hypothesis, should have been formed at about the same time as under the hydrothermal concept.

The vein faults of the Keno Hill district contain ore bodies where (1) two or more vein faults join; (2) a vein fault joins a subsidiary fracture; and (3) a vein fault passes from quartzites or greenstones into overlying schists, phyllites, or thin-bedded quartzites, these type (3) ore bodies being in the first two rocks named. Since the vein faults cannot be correlated across the valleys between Galena Hill and Keno Hill (due to the complex cross faulting and drift filling of the valleys), the two areas have to be studied as essentially different structural entities. Boyle (1965) thinks that at least three stages of hypogene mineralization occurred at Keno Hill, plus a supergene stage. The first stage consisted of two substages: (1) in sedimentary rocks, early quartz stringers and lenses containing some pyrite and carbonates; in greenstones, similar lenses of quartz with chlorite, epidote, and carbonates and (2) quartz lenses and stringers with wolframite and scheelite. The age relationships between these two stages are not clear--perhaps they were contemporaneous. The second stage resulted in the deposition of quartz in lenses in dilatant zones in northeast-trending faults, accompanied by pyrite, arsenopyrite, and minor sulfosalts and tourmaline; silver is not abundant. The third and economically major stage consists of siderite as the principal gangue mineral, plus pyrite, chalcopyrite, galena, sphalerite, Ag-bearing tetrahedrite, boulangerite, and a few other sulfosalts. The siderite lodes show rude vertical and lateral zoning. At the outer edges, pyrite and siderite are the main minerals; inward, pyrite decreases and the veins contain mainly siderite, galena, sphalerite, tetrahedrite, and other sulfosalts. Sphalerite increases and galena decreases with depth; tetrahedrite decreases with depth but less so than galena; pyrite and siderite increase with depth and are the main minerals at the bottoms of most lodes. Most of these third stage ores have been brecciated and locally recemented by dolomite, calcite, quartz, and small amounts of sphalerite, pyrargyrite, native silver, acanthite, and galena; Boyle thinks some or all of these minerals may be supergene. Boyle thinks that all the sulfosalts were deposited at the same time. This suggests that where the sulfosalts are associated with the quartz-pyrite-arsenopyrite of the second stage, these second-stage minerals were appreciably earlier than the sulfosalts.

The deposits seem to me to be typical hydrothermal ores formed under conditions of slow loss of heat and pressure, but Boyle is convinced that they were developed through diffusion of ore- and gangue-forming materials from the country rock. Boyle (1965) can be consulted for a full presentation of his point of view, one that is not shared by me.

It would seem that the quartz-pyrite-arsenopyrite mineralization, with its minor tourmaline, was formed under hypothermal conditions, but this phase of the mineralization is of no economic value and is not included in the classification here given to the Keno Hill deposits. It would appear to be almost certain that the siderite-galena-sphalerite-sulfosalt mineralization was formed under more moderate conditions, the only question being as to whether or not the generally somewhat earlier sphalerite and galena were formed under mesothermal or leptothermal conditions; certainly, the sulfosalts were products of the leptothermal range. The presence of siderite as the principal gangue mineral is definitely not diagnostic as it has been found in other ore districts in association with ore minerals formed under both mesothermal and leptothermal conditions. It appears that more silver may have been recovered from tetrahedrite than from galena; the fact that an ounce of silver is recovered for each 4 or 5 pounds of lead does not demonstrate, however, that there is a direct relationship between abundance of galena and abundance of silver, but simply states a fact of smelter experience. Another appreciable fraction of the silver appears to have been produced from the pyrargyrite, which is certainly a later mineral and which may be in part secondary. The massive pyrargyrite found in siderite clearly is primary and essentially all of it may be. In any event, the recovery of much silver from the silver sulfosalts is another argument for much of the ore having been formed in the leptothermal range. As sphalerite consistently increases with depth, however, it is thought best to include the term mesothermal in the classification, and the deposits of the Keno Hill district are here categorized as mesothermal to leptothermal.

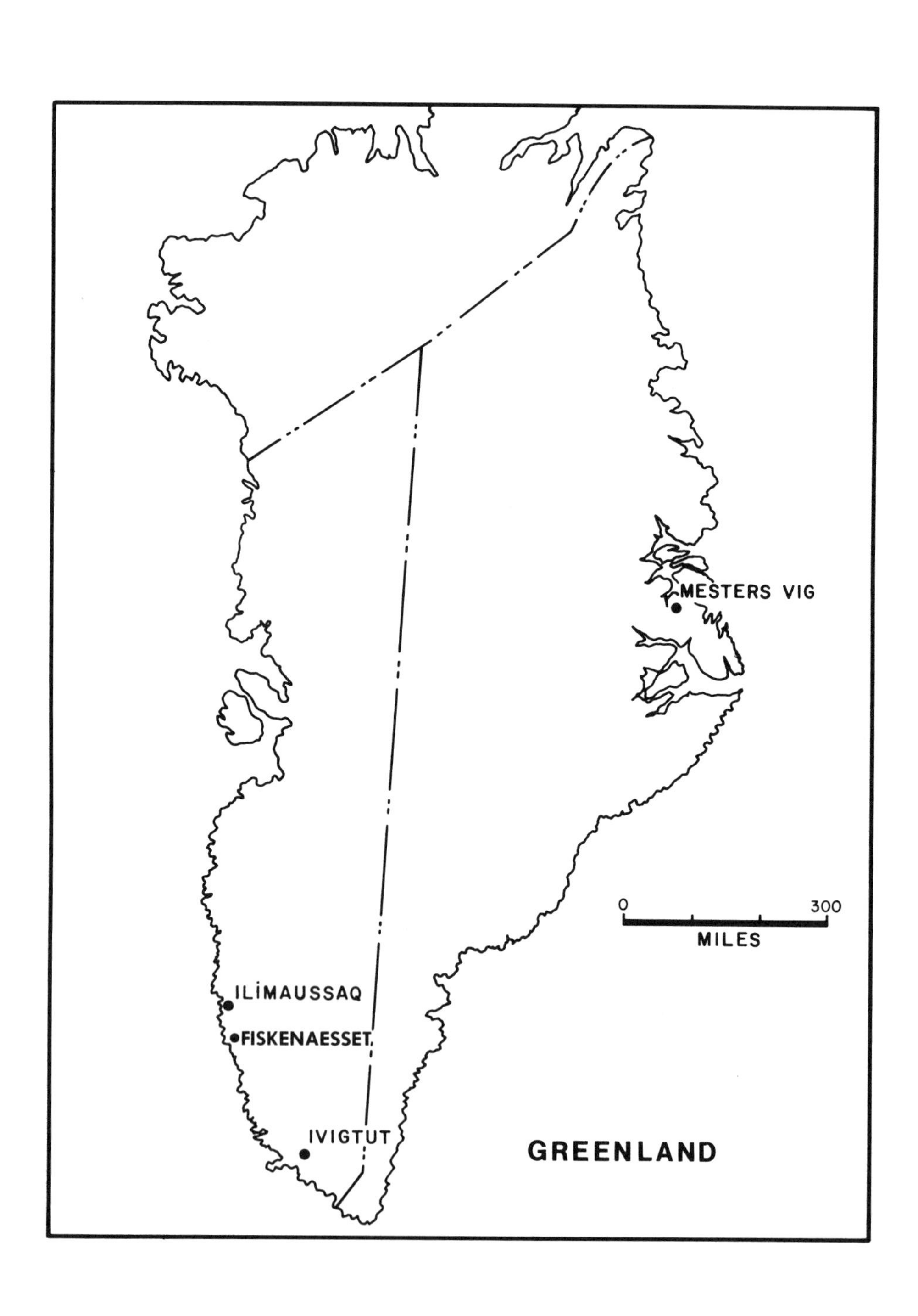

MESTERS VIG
0
300
MILES
ILÍMAUSSAQ
FISKENAESSET
IVIGTUT
GREENLAND

FISKENAESSET

Early Precambrian *Chromite* *Magmatic-1b*

Berthelsen, A., 1965, The country between Frederikshaab Glacier and Buksefjorden, in *The Precambrian of Greenland*, in Rankama, K., Editor, *The geologic systems--The Precambrian*, v. 2: Interscience Pubs., N.Y., p. 164-165

Ghisler, M., 1966, The chromite deposits at Fiskenaesset: Grönlands Geol. Undersögelse Rapp. 11, p. 31-32

—— 1970, Pre-metamorphic folded chromite deposits of stratiform type in the early Precambrian of west Greenland: Mineralium Deposita, v. 5, p. 223-236

Ghisler, M., and Sharma, P. V., 1969, On the applicability of magnetic prospecting for chromite in the Fiskenaesset region, west Greenland: Grönlands Geol. Undersögelse Rapp. 20, 25 p.

Ghisler, M., and Windley, B. F., 1967, The chromite deposits of the Fiskenaesset region: Grönlands Geol. Undersögelse Rapp. 12, 39 p.

Herd, R. K., and others, 1969, The mode of occurrence and petrogenesis of sapphirine-bearing and associated rocks of west Greenland: Grönlands Geol. Undersögelse Rapp. 24, 44 p.

Windley, B. F., 1966, Anorthosites and polymetamorphism between Ravens Storö and Sukkertoppen, west Greenland: Grönlands Geol. Undersögelse Rapp. 11, p. 27-29

—— 1968, New field relations of the early Precambrian of west Greenland: Grönlands Geol. Undersögelse Rapp. 15, p. 27-31

—— 1969, The anorthosites of southern west Greenland: Amer. Assoc. Petrol. Geols. Mem. 32, p. 899-915

Notes

The Fiskenaesset chromite deposits are located on the southwestern coast of Greenland, about 150 miles south-southeast of the town of Godthaab. Some of the chromite horizons occur along the shores of the fjords, and others are found only a few kilometers from the coast in easily accessible terrain that rarely exceeds 500 m in altitude. The harbor at Fiskenaesset is ice-free the year-round.

The more than 2600 m.y. old central gneiss unit of west Greenland characteristically contains relict layers and inclusions of meta-anorthosite and mafic and ultramafic rocks (Ghisler, 1970). Ghisler considers that these included rocks are the remains of pre-orogenic layered igneous complexes in which chromite layers are associated with the anorthosites. Minor amounts of chromite are found in layers in ultramafic rocks some 400 km north-northwest from Fiskenaesset (at Fiskefjord). The original chromite-bearing layered complex now forms a metamorphosed anorthosite-pyroxene amphibolite-ultramafic sequence known as the Fiskenaesset complex and occurring as conformable stratigraphic horizons within the gneisses. The width of these steeply dipping horizons normally ranges between 100 and 1000 m, with the average width being about 400 m and the maximum some 2000 m. These variations in thickness are due mainly to orogenic effects but to some extent were the result of original differences in strata thickness.

Within the complex, anorthosites are the dominant rock type, but pyroxene amphibolite and amphibolite are intercalated in layers that usually are between 25 and 100 m thick but may be much thinner. Ultramafic rocks are even less common, being present as discontinuous layers and lenses and layers that are between 30 cm and 20 m wide, with a maximum thickness of 200 m; the maximum length of these ultramafic bands is 1500 m.

Since the Fiskenaesset complex was formed, it has undergone two episodes of regional metamorphism; a granulite facies metamorphism that was followed by an amphibolite facies, the latter downgrading the minerals formed in the earlier and higher intensity facies. Ghisler (1970) reports that the remarkable state of preservation of the complex, despite the orogenic stresses applied, is due to essentially isochemical recrystallization in both metamorphic events. The original form of the complex has been changed by the tectonic thickening and thinning; the complex as a whole now presents a complicated double-folded interference pattern on a large scale. The dips of the strata normally are steep to vertical but may be as low as 20°. Although some 750 km^2 have been investigated, the complex is known to extend out of this area into the highlands to the northeast where it passes under the ice cap. The anorthosites themselves normally contain 75 to 90 percent plagioclase; of the dark minerals, hornblende is the most abundant. Within homogeneous anorthosites with less than 10 percent dark minerals there may be layers of varied thicknesses that are much higher in mafic constituents. The plagioclase ranges between An_{80} and An_{95} and has been recrystallized, with the glomeroblasts lying oblique to the layering. Locally, the anorthosites (actually, of course, meta-anorthosites) contain considerable amounts of quartz. The pyroxene amphibolites are composed mainly of hypersthene and hornblende in a stable assemblage that indicates to Ghisler (1970) hornblende-granulite sub-facies conditions. The hornblende-dominant mineral assemblages in the ultramafic rocks originally were formed under igneous conditions as dunites, peridotites, and pyroxenites. Typical, but quantitatively unimportant, chromite-layered bronzites and magnetite-layered pyroxenites occur in the sequence; several sapphirine localities are known (Herd and others, 1969) that were first reported in 1809.

The rocks of the area are probably the metamorphosed equivalents of the Sortis group, a subdivision of the Ketilidian supracrustals known farther south in the Ivigtut region, and are considered (Ghisler, 1970) to have been formed more than 2600 m.y. ago, although Berthelsen (1965) says that they are older than 2200 m.y., which may not actually be in disagreement with Ghisler. Although the metamorphic events may have been later than the end of the early Precambrian, the chromite horizons certainly were formed concomitantly with the rocks enclosing them, so they must be dated as (probably late) early Precambrian.

Except for what probably are uneconomic chromite layers in the ultramafic rocks, all the chromite horizons of the Fiskenaesset are enclosed in anorthosite. Each chromite horizon is made up on several tens of chromitite layers (bands composed of chromite and hornblende) that range between a few mm and 10 cm wide; seams up to 1 m have been observed (Ghisler, 1970); each chromitite layer is separated from the next by a band of anorthosite of about the same width. Throughout most of the anorthosite complex two chromite horizons can be found, and in several places, four are to be seen. The two major chromite horizons lie close to each other in relation to the total thickness of the mass of anorthosite; these horizons never are at contacts between anorthosite and other rocks nor in contact with amphibolite or ultramafic layers in the anorthosite. In places the lower contact of the chromite horizon may be basinlike and the upper one straight. Ghisler (1970) thinks that these basin structures may have been caused by erosion by magmatic currents as has been suggested for similar structures in the Bushveld. Ghisler supports the idea of a magmatic sedimentary origin for the chromite horizons because of the mutual sharp contacts between chromite and anorthosite layers and because of rhythmic layering within chromitite. The recrystallized textures of the chromitites have appreciably concealed their original structures, but this does not cast doubt on their ultimately magmatic origin. Although the chromite horizons have been followed for as much as 4 km, they have been disrupted, boudinaged, and faulted; folds are common on both a major and a minor scale, and all have been tectonically thickened and thinned.

The chromitite bands that make up the chromite horizons contain between 50 and 75 percent of chromite, with a silicate matrix that is mainly hornblende but also includes biotite and plagioclase; rutile, ilmenite, magnetite, and sulfides are accessory minerals. The chromite grains are generally euhedral to subhedral, usually with rounded corners, but in close packing they may be definitely anhedral. The grains normally are about 0.3 mm in diameter but may be as much as 0.7 mm wide or as small as 0.05 mm. Typically the chromite grains are arranged in elongate chainlike aggregates 2 to 5 mm long. Tiny inclusions of silicates are found in the chromite grains, mainly in the interiors of the grains. After hornblende and plagioclase, rutile is the most common mineral associated with chromite and forms 0.1 to 2 percent of the chromitite. Lesser amounts of ilmenite are associated with the rutile, suggesting that the oxygen fugacity of the magma was such that ferrous ion was not available in sufficient quantities to convert more than a fraction of the titanium-oxygen complexes to ilmenite. Magnetite occurs as what are probably tiny exsolution grains in a minority of the chromite. The chromites contain about 34 percent Cr_2O_3, 23 percent Al_2O_3, 26 percent FeO, 8 percent Fe_2O_3, and 5 percent MgO. The Cr/Fe ratio is about 0.93:1. Since metallurgical chromite must have not less than 45 percent Cr_2O_3, the Fiskenaesset ores cannot be placed in this category. Because refractory grade chromite should have about 60 percent Cr_2O_3 and Al_2O_3 combined, and not more than 12 percent Fe, these ores are at most barely passable for this grade. Ghisler and Windley (1967) estimate that deposits at about 2,500,000 tons of chrome ore containing 1,000,000 tons of chromite or 350,000 tons of Cr_2O_3 and 300 tons of V_2O_5. This estimate covers only a small section of the deposit, which should be far larger in total.

Ghisler (1970) points out the striking similarity between the premetamorphosed Fiskenaesset ores and those of the Critical Zone of the Bushveld complex, especially its anorthosite-rich upper part. He considers chromite horizons of the Fiskenaesset complex closely analogous to the alternating layers of chromitite and anorthosite in the Dwars River section of the Bushveld. In the latter sequence, however, the chromitite bands are less closely spaced than at Fiskenaesset and are less regularly arranged, especially since connecting veins between bands are not uncommon in the Dwars River area. Although the only original magmatic cumulate features preserved in the Fiskenaesset chromites are the poikilitic chromite crystals with euhedral to subhedral outlines and regular silicate inclusions, the primary character clearly was magmatic. The deposits are, therefore, here classified confidently as magmatic-1b.

ILÍMAUSSAQ

Late Precambrian | *Columbium* | *Hypothermal-1*

Andersen, S., 1967, Beryllite and bertrandite from the Ilímaussaq alkaline intrusion, south Greenland: Medd. Grønland, v. 181, no. 4, p. 11-27

Ball, S. H., 1923, The mineral resources of Greenland: Medd. Grønland, v. 63, p. 1-60

Berthelsen, A., and Noe-Nygaard, A., 1965, The Ilímaussaq intrusion, in *The Precambrian of Greenland*, in Rankama, K., Editor, *The geologic systems--The Precambrian*, v. 2: Interscience Pubs., N.Y., p. 155-160

Bondam, J., and Sørensen, H., 1959, Uraniferous nepheline syenites and related rocks in the Ilímaussaq area, Julianehaab district, Southwest Greenland: 2d International U.N. Conf. on Peaceful Uses of Atomic Energy (Geneva) Pr., v. 2, p. 555-559

Bridgwater, D., 1965, Isotopic age determinations from South Greenland and their geological setting: Medd. Grønland, v. 179, no. 4, p. 1-56

Ferguson, J., 1964, Geology of the Ilímaussaq alkaline intrusion, South Greenland: Part I, Description of map and structure: Medd. Grønland, v. 172, no. 4, p. 1-81

—— 1970, The differentiation of agpaitic magmas: the Ilímaussaq intrusion, South Greenland: Canadian Mineral., v. 10, pt. 3, p. 335-350

Hamilton, E. I., 1964, The geochemistry of the northern part of the Ilímaussaq intrusion S.W. Greenland: Medd. Grønland, v. 162, no. 10, p. 1-104

Hansen, J., 1968, A study of radioactive veins containing rare-earth minerals in the area surrounding the Ilímaussaq alkaline intrusion in South Greenland [with Russian abs.]: Medd. Grønland, v. 181, no. 8, 47 p., reprinted as Grönlands Geol. Undersögelse Bull. 76, 1968

—— 1968, Niobium mineralization in the Ilímaussaq alkaline complex, southwest Greenland: 23d Int. Geol. Cong. Rept., Sec. 7, p. 263-273

Oen, I. S., and Sørensen, H., 1964, The occurrence of nickel-arsenides and nickel-antimonide at Igdlunguaq in the Ilímaussaq alkaline massif, South Greenland: Medd. Grønland, v. 172, no. 1, p. 1-50

Semenov, E. I., and others, 1968, Ilímaussite, a new rare-earth-niobium-barium silicate from Ilímaussaq, South Greenland (Contribution to the Mineralogy of Ilímaussaq no. 10) [with Russian abs.]: Medd. Grønland, v. 181, no. 7, p. 1-7

—— 1968, On the mineralogy of pyrochlore from the Ilímaussaq alkaline intrusion, South Greenland (Contribution to the Mineralogy of Ilímaussaq no. 11) [with Russian abs.]: Medd. Grønland, v. 181, no. 7, p. 9-25

Sørensen, H., 1958, The Ilímaussaq batholith, a review and discussion: Medd. Grønland, v. 162, no. 3, p. 1-48

—— 1968, På spor af själdne mettaler in Sydgrönland--[pt.] 7, Zirconium, et af fremtidens metaller: Grønland, 1968, no. 12, p. 353-365 (Danish, Eng. summ.)

—— 1970, Internal structures and geological setting of the three agpaitic intrusions: Khibina, and Lovozero of the Kola Peninsula and Ilímaussaq, South Greenland: Canadian Mineral., v. 10, p. 299

Ussing, N. V., 1912, Geology of the country around Julianehaab, Greenland: Medd. Grønland, v. 38, p. 1-376

Wegemann, E., 1937, Sur la genèse des roches alcalines de Julianehaab (Groenland): Acad. Sci. Paris C. R., v. 204, p. 1125-1127

Notes

The Ilímaussaq alkaline intrusive complex is located on the coast of southwest Greenland. The center of the complex appears to be beneath the waters of the fiord (Tunugdliarfik) at about 60°56'N and 45°52'W. The complex is about 130 km east-southeast of Ivigtut. The long dimension of the body is about 16 km in a northwest-southeast direction.

The country rocks into which the complex was intruded were the Julianehaab granite and the Gardar continental series of sandstones and basalts. These Gardar rocks form the roof of the complex, and they are injected by dikes and sills from the complex. The intrusion is made up of an early augite syenite and a later agpaitic suite. Sørensen (21st Int. Geol. Cong., Pt. 13, p. 323) defines agpaitic rocks as (1) being per-alkaline nepheline syenites; (2) containing aegirine, soda-amphibole, and/or aenigmatite instead of biotite, diopsidic pyroxene, and hornblende; that is, these rocks are low in calcium and magnesium; (3) containing complex zirconium and titanium silicates instead of zircon and sphene; and (4) being rich in fluorine, chlorine, and water; such complex silicates as eudialyte and rinkolite are present; carbonates are rare. At Ilímaussaq, the agpaitic rocks are mainly of four types: (1) naujaite (poikilitic sodalite syenite); (2) foyaite; (3) kakortokite (layered agpaitic rocks in which the layers are black [arfvedsonite-aegirine], red [eudialyte], and white [feldspar]; their nepheline content is about 18 percent in all types; perthitic feldspar ranges from 18 to 50 percent; aegirine 13 to 9 percent; arfvedsonite 40 to 13 percent; and eudialyte 11 to 29 percent); and (4) lujavrite (in green lujavrite aegirine is dominant and in black, arfvedsonite). The black lujavrite contains the following percentages: microcline 18, albite 13, aegirine 5, arfvedsonite 30, eudialyte 9, nepheline 24, and sodalite 1; the green lujavrite contains the following percentages: microcline 25, albite 15, aegirine 28, arfvedsonite 8, eudialyte 10, nepheline 13, and sodalite 1.

The lujavrites are accompanied by hydrothermal veins and were emplaced after strong shearing of the rocks of the roof and, in places, of the augite syenite.

The agpaitic rocks have high contents of columbium, beryllium, uranium, and thorium and a low content of titanium. From the pre-agpaitic rocks to the agpaitic rocks to the hydrothermal veins, a progressive enrichment of columbium, beryllium, uranium, and thorium takes place.

At Ilímaussaq, the lavas and dikes are very low in columbium and tantalum (maximum 25 ppm of Cb_2O_5). The augite syenite contains up to 300 ppm Cb_2O_5 and the agpaitic rocks 550 to 1500 ppm Cb_2O_5. Each rock type in the agpaitic series shows wide differences in columbium content, but the ranges are about the same in each rock type. The Cb/Ta ratios are between 14:1 and 64:1. The columbium minerals are found principally in the late hydrothermal veins and in sheared rocks in the Kvanefjeld area in the northwestern part of the intrusive complex. In such sheared rocks the columbium content may be as much as 6000 ppm. Pyrochlore is the most abundant columbium mineral with members of the epistolite-murmanite group being common.

Rubidium-strontium age dates for the Ilímaussaq intrusion are 1086 m.y. (Moorbath and others, 1960), indicating that the complex was introduced in about the middle of the late Precambrian, during the Gardar cycle in south Greenland. From the abundance of columbium in the rocks of the Ilímaussaq intrusion and the close spatial association of the columbium veins with the intrusives, the introduction of the rocks and veins almost certainly took place at much the same time. The columbium mineralization, therefore, is dated as late Precambrian.

The columbium hydrothermal veins are most abundant in naujaites and coarse- to medium-grained lujavrites in the Kvanefjeld area; locally this mineralization also is found in more or less irregular bodies, a few millimeters to as much as 2 m wide. Sørensen (1968) divides the veins into six types, based on the principal mineral they contain: (1) analcime, (2) albite, (3) tetragonal natrolite, (4) chkalovite, (5) ussingite, and (6) quartz. Type (1) veins also contain pyrochlore, igdloite, and epistolite [$(Na,Ca)(Cb,Ti,Mg,Fe,Mn)SiO_4(OH)$] and occur in naujaites and lujavrites; the total mineral assemblage is large and various, and pyrochlore may make up 10 percent of the vein volume. Type (2) veins contain epistolite, gerassimovskite, pyrochlore, and igdloite, plus a wide variety of other minerals; the pyrochlore occurs in platy aggregates up to 2 cm across interstitially to the other minerals and fills fractures in them. Type (3) veins also have epistolite, gerassimovskite, and pyrochlore; the grains of pyrochlore are small and not abundant. Type (4) veins have, as main constituents, chkalovite and epistolite; epistolite is found in plates up to 10 cm across, usually filling spaces between chkalovite crystals. Type (5) veins contain epistolite, gerassimovskite, pyrochlore, niobophyllite, and nenadkevichite; epistolite occurs as plates up to 10 cm across or as "veins" surrounded by borders of albite; only small grains of pyrochlore are present. Type (6) veins contain pyrochlore along with microcline, arfvedsonite, lithium mica, eudidymite, monazite, and neptunite. The country rocks containing the veins show hydrothermal alteration; in the naujaite and lujavrite, the aegirine and arfvedsonite largely have disappeared but have been replaced in part by acmite; eudialyte has been changed to katapleite; and the main minerals of the veins have been added to the country rocks.

The principal mineral in the sheared pre-agpaitic rocks is murmanite [$Na_2Ti_2(OH)_4Si_2O_7$]; the mineralized shear zones are close to contacts with lujavrites. Murmanite is not found in the lujavrites.

It is apparent that the columbium in the differentiating alkaline magma at Ilímaussaq was concentrated in the residual fractions, probably entering hydrothermal solutions produced in the late stages of its solidification. The columbium-bearing veins, however, are located mainly in the naujavites (though near to their contacts with lujavrites). Only in the Kvanefjeld area are such veins found in appreciable numbers in lujavrites. The titanium-bearing minerals, principally murmanite, are found only in the pre-agpaitic rocks; thus, the titanium minerals required outside help, not being able to form directly from the lujavrite-produced ore fluids. Accompanying the columbium minerals in the veins are beryllium, rare-earth, lithium, and thorium minerals. Although Sørensen thinks the epistolite, niobophyllite, and some of the pyrochlore are primary, he believes that the gerassimovskite and some of the pyrochlore and nenadkevichite were replacements of epistolite. Although Sørensen thought that these minerals suggest a depositing temperature of 400°C or less, it seems to me that the entire assemblage is typically hypothermal, and the veins are classed here as hypothermal-1. The veins in coarse- and medium-grained lujavrites probably should be classed as deuteric (magmatic-4); they are not common enough, I think, to justify inclusion in the classification of the deposit.

IVIGTUT

Late Precambrian	*Cryolite, Lead, Silver*	*Magmatic-3a and Magmatic-4*

Ayrton, S. N., 1963, A contribution to the geological investigations in the region of Ivigtut, SW Greenland: Medd. om Grønland, Bd. 167, nr. 3, 139 p. (general)

Baldauf, R., 1910, Über das Kryolith-Vorkommen in Grönland: Zeitsch. f. Prakt. Geol., Jg. 18, H. 11/12, S. 432-440

Ball, S. H., 1922, The mineral resources of Greenland: Medd. om Grønland, Bd. 63, 60 p. (particularly p. 17-31)

Beck, R., 1910, Ergebnisse einer mikroskopischen Untersuchung von Ivigtut-Gesteinen: Zeitsch. f. Prakt. Geol., Jg. 18, H. 11/12, S. 440-443

Berthelsen, A., 1960, An example of a structural approach in the migmatite problem: 21st Int. Geol. Cong. Rept., pt. 14, p. 149-157

—— 1962, On the geology of the country around Ivigtut, S.W.-Greenland: Geol. Rundschau, Bd. 52, S. 269-280

Berthelsen, A., and Noe-Nygaard, A., 1965, The Ivigtut region, in *The Precambrian of Greenland*, in Rankama, K., Editor, *The geologic systems--The Precambrian*, v. 2: Interscience Pubs., N.Y., p. 123-128, 130-131, 137-141, 143-154, 160-161

Böggild, O. B., 1953, The mineralogy of Greenland: Medd. om Grønland, Bd. 149, nr. 3, 442 p. (particularly p. 74-98)

Bøgvad, R., 1942, Magnetkis fra Ivigtut: Dansk Geol. Fören., Medd. Bd. 10, H. 2, p. 115-118 (Eng. Summ.)

Bondesen, E., and Henriksen, N., 1965, On some pre-Cambrian metadolerites from the central Ivigtut region, SW Greenland: Medd. om Grønland, v. 179, no. 2, 42 p. (reprinted as Grønlands Geol. Undersøgelse Bull. 52)

Callisen, K., 1943, Igneous rocks of the Ivigtut region, Greenland, Part I: Medd. om Grønland, Bd. 131, nr. 8, 74 p.

Gordon, S. G., 1926, Mining cryolite in Greenland: Eng. and Min. Jour.-Press, v. 121, no. 6, p. 236-240

Harry, W. T., and Emeleus, C. H., 1960, Mineral layering in some granite intrusions of S.W. Greenland: 21st Int. Geol. Cong. Rept., pt. 14, p. 172-181

Jõrgensen, O., 1968, K/Ar determinations from western Greenland--[pt.] 2, The Ivigtut region, in *Report of Activities, 1967:* Grönlands Geol. Undersögelse Rap. 15, p. 87-91

Karup-Møller, S., 1966, Berryite from Greenland: Canadian Mineral., v. 8, pt. 4, p. 414-423

Legraye, M., 1938, L'association galena-chalcopyrite-blende dans la cryolite du Greenland: Soc. Geol. Belgique, t. 61, nos. 4, 5, p. B109-B113

Moorbath, S., and Pauly, H., 1962, Rubidium-strontium and lead isotope studies on intrusive rocks from Ivigtut, south Greenland, in *Variations in isotopic abundances of strontium, calcium, and argon and related topics*: N.Y. 3943, 10th Ann. Prog. Rept. for 1962, p. 99

Moorbath, S., and others, 1960, Absolute age determinations in South-West Greenland. I. The Julianehaab granite, the Ilímaussaq batholith and the Kûngnât syenite complex: Medd. om Grønland, Bd. 162, nr. 9, 40 p.

Pauly, H., 1960, Paragenetic relations in the main cryolite ore of Ivigtut, South Greenland: Neues Jb. f. Mineral., Abh., Bd. 94, I. Hälfte (Festband Ramdohr), S. 121-139

Schneiderhöhn, H., 1961, Der Kryolithpegmatit von Ivigtut, in *Die Erzlagerstätten der Erde*, Bd. 2: Gustav Fischer, Stuttgart, S. 102-103

Soen, O. I., and Pauly, H., 1967, A sulphide paragenesis with pyrrhotite and marcasite in the siderite-cryolite ore of Ivigtut, South Greenland: Medd. om Grønland, v. 175, no. 5, 55 p.

Ulrych, T. J., 1964, The anomalous nature of Ivigtut lead: Geochimica et Cosmochimica Acta, v. 28, p. 1389-1396

Wegmann, C. E., 1938, Geological investigations in southern Greenland, pt. 1, On the structural divisions of southern Greenland: Medd. om Grønland, Bd. 113, nr. 2, 148 p. (general)

—— 1947, Note sur la chronologie des formations précambriennes du Groenland meridional: Eclogae Geol. Helvetiae, v. 40, nr. 1, p. 7-14

Notes

The Ivigtut granite in which the Ivigtut cryolite deposit is contained is located on the south shore of Arsuk Fjord at about latitude 61°21'N and longitude 48°12'W on the coast of southwest Greenland.

In the Ivigtut region (Berthelsen and Noe-Nygaard, 1965), the basement rocks are remnants of the old Ketilidian mountain chain and consist of migmatites or gneisses. In places, the much less metamorphosed superstructure is separated from the basement by a structural break; locally, however, a tectonically undisturbed contact can be seen between the migmatites and the overlying metavolcanics. The pink Ivigtut granite forms a stocklike intrusion with a diameter of about 300 m. The granite contains porphyritic quartz and microcline microperthite, plus appreciable amounts of albite. Mafic minerals are scarce, but the most prominent is biotite with some aegirine-augite and soda-amphibole. The granite in turn is surrounded by an eruptive breccia that holds blocks of gneiss as angular fragments in a granite porphyry matrix. At shallow depth, the granite cuts through a gabbro-anorthosite-bearing gneiss, but no fragments of it are found in the breccia. From this, it is thought that the xenoliths in the breccia were derived from overlying gneisses through overhead stoping. The stock shows nearly vertical borders at depth, and the original roof of the intrusion is thought to have been not far above the present-day erosion surface. Just to the east is a smaller explosion breccia, separated from the granite by a narrow barrier of gneiss; this breccia, the Bunke breccia, is made up entirely of wall-rock material that includes angular gneiss blocks. It is Berthelsen's (1962) opinion that the granite stock and the associated breccias were emplaced in late Gardar time. Whole-rock rubidium-strontium dating gives the granite an age of about 1250 m.y. according to work by Moorbath and Pauly, 1962. Later both the granite and the breccias were cut by tiny dikes of tinguaite. The last event in the granite cycle was the development of the cryolite body in the upper part of the Ivigtut granite. Rubidium-strontium dating on biotite from the greisen rocks surrounding the cryolite gives a date of about 1150 m.y. that should be essentially that of the formation of the cryolite. These dates and the geologic character of the granite and cryolite strongly indicate that a late Precambrian age can be assigned to the Ivigtut cryolite.

The cryolite body, now entirely mined out though a considerable stockpile still remains, was an irregular mass some 50 m by 115 m that had a depth of about 70 m. The cryolite body consisted, in the main, of a siderite-cryolite rock in which the cryolite made up over 75 percent of the mass, siderite about 20 percent, and sulfides and quartz 1 to 2 percent. The sulfides included sphalerite, galena, chalcopyrite, and pyrite; some of the sphalerite was marmatitic and the galena was silver rich. Locally, cryolithionite ($Na_3Al_2Li_3F_{12}$) is present; several secondary minerals, such as thomsenolite ($NaCaAlF_6 \cdot H_2O$) and jarlite ($NaSr_3Al_3F_{16}$), have been developed from the cryolite. The cryolite mass is, in turn, almost entirely enclosed by pegmatite in which cryolite and feldspar occur in the graphic intergrowths similar to those typically found between quartz and feldspar in granitic pegmatites. In the western part of the cryolite body, fluorite and a fine-grained potash mica, topaz, and other minerals are present. At depth, sulfide-bearing, quartz-rich rocks lie between the cryolite and the granite. At higher levels, the granite surrounding the cryolite was altered to greisen.

This strange mineral assemblage is thought to have been a product of the enclosing granite (Callisen, 1943); on the other hand, the suggestion has been made that the cryolite came from the same peralkaline magma chamber that produced the tinguaitic dikes (Berthelsen, 1962). Ratios of Sr^{87}/Sr^{86} in barytocelestite from the cryolite body indicate to Moorbath and Pauly, 1962, that the source of the cryolite was a mafic magma rather than the granite.

From the examination of a considerable number of Ivigtut specimens, I believe that the graphic intergrowths of feldspar and cryolite were formed by the simultaneous precipitation of these two minerals from a magma high in fluorine and lower than normal (for a granite) in silica. By the time the precipitation of these intergrowths had been largely or entirely completed, I conceive that cryolite, siderite, and their associated sulfides had begun to deposit in the upper portions of the magma chamber from the water-rich fluid generated from the magma during the pegmatitic stage of its crystallization, much, if not all of this fluid had been prevented from leaving that chamber (as it normally would have done) by the resistance afforded by the essentially unbroken and impervious formations enclosing the magma body. The sulfides (particularly the argentiferous galena) and the siderite of the cryolite mass form a mineral suite typical of those developed from hydrothermal fluids; in this instance, however, the fluid depositing these minerals did not leave the magma chamber and is, therefore, to be classed as deuteric rather than hydrothermal. Thus, the graphically intergrown feldspar-cryolite rocks are classified as magmatic-3a, and the deuterically formed cryolite mass is designated as magmatic-4.

Many specimens of the surrounding granite contained enough disseminated chalcopyrite to suggest that the district may have contained a minable porphyry-copper deposit.

MESTERS VIG

Middle Tertiary — *Lead, Zinc, Barite, Silver* — *Mesothermal*

Bierther, W., 1941, Vorläufige Mitteilung über die Geologie des östlichen Scoresbylandes in Nordostgrönland: Medd. om Grønland, Bd. 114, no. 6, 19 p.

Bondam, J., and Brown, H. (C.T.), 1955, The geology and mineralization of the Mesters Vig area, east Greenland: Medd. om Grønland, Bd. 135, no. 7, 40 p.

Brown, H. (C.T.), 1955, The lead-zinc occurrence in east Greenland: Royal School of Mines Jour., no. 4, p. 33-45

Fischer, B., and others, 1958, Der Blei-Zink-Bergbau Mesters Vig in Ostgrönland: Berg- und Hüttenmännisches Mh., Jg. 103, H. 8, S. 145-153

Gross, W. H., 1956, The direction of flow of mineralizing solutions, Blyklippen mine, Greenland: Econ. Geol., v. 51, p. 415-426

Kock, L., 1929, The geology of east Greenland: Medd. om Grønland, Bd. 73, no. 1, 204 p. (general)

Lehnert-Thiel, K., and Vohryzka, K., 1968, Untersuchungen an Hg-Aureolen um polymetallische Sulfidgänge im Permafrostboden Ostgrönlands 72° nl. Breite: Montan-Rundschau, Jg. 16, H. 5, S. 104-108

Witzig, E., 1954, Stratigraphische und tektonische Beobachtungen in der Mesters

Vig-Region (Scoresby Land, Nordostgrönland): Medd. om Grønland, B. 72, Anden Afdeling, nr. 5, 26 p.

Notes

The Mesters Vig district is located in the northeast corner of Scoresby Land, in eastern Greenland, at latitude 72°15'N and longitude 24°W. Immediately southeast of the deposits is the Mesters Vig arm of King Oscar's Fjord.

The oldest rocks in the general area may be sandstones and shales of Devonian age; however, their presence in the area is not certain. The oldest certain occurring rocks are of Westphalian (Lower Pennsylvanian) age. The oldest of these is the Blyklippen series that includes sandstones, arkoses, and conglomerates with small shale and limestone zones. These beds contain the ores of the Mesters Vig district and of the Sorte Hjørne area also; their age is determined from plant remains that are not sharply diagnostic. Next in order is the Lebachia series of gray sandstones with limestone and shale bands; it is partly marine; the boundary between these and the underlying beds has not been accurately determined, and the Lebachia might be lower Permian. Above the Lebachia is the Domkirken series of red arkosic sandstones and conglomerates of continental origin; these beds are quite similar to those of the upper Blyklippen, but they contain shale beds that are not present in the Blyklippen. These beds might be the equivalent of the Permian lower New Red sandstones. These probably upper Carboniferous beds are unconformably overlain by marine upper Permian rocks that are mainly shales, the lower ones limy and the upper ones sandy to marly. The discordance between the Permian rocks and the Triassic ones overlying them is slight, not more than 5°. This Triassic series contains a slight basal conglomerate and then shales, sandy marls, and thin sandstone beds.

Volcanic activity of Tertiary age has produced a wide variety of rocks in the Mesters Vig area. Several dikes intersect the Paleozoic and Triassic sediments, and thick sills appear in essentially all the beds. Much of these rocks are basalt and probably are part of the huge Arctic basalt province of Tertiary age. Associated in age with these volcanic rocks are syenitic intrusives, the largest of which is a syenite complex (Werner Bjerge) just south of Mesters Vig. The age of this rock is certainly Tertiary and probably middle Tertiary.

It has been suggested that the age of the Mesters Vig ores is that of the Werner Bjerge complex since a small lead-zinc contact-metamorphic body has been found in the carbonate rocks adjacent to the syenite. On the other hand, all of the ores are found in the Paleozoic sediments. In the vicinity of Lumensö, a zone of ore development has been where the veins are filled with hematite and quartz, and hematite pebbles are found in the Permian basal conglomerate, suggesting that the material in these veins is Carboniferous. No lead or zinc minerals, however, have been found in the hematite-quartz mineralization, hematite is not found at Mesters Vig. It appears more reasonable to assume that no genetic connection exists between the two mineralizations and that the lead-zinc ores are Tertiary and probably middle Tertiary; they are so categorized here.

The first event in the mineralization was the development of the quartz-sphalerite-galena vein filling, followed by the intrusion of mafic dike rocks, and then the barite mineralization. The veins at Mesters Vig are filled open fissures in which the gangue is normally quartz. The ore mineralization is confined to the veins, usually in shoots or pockets close to the footwall. The rocks beyond the veins can be extensively silicified. Barite is the only other common gangue mineral. The principal sulfides are galena and sphalerite; locally small amounts of chalcopyrite are present; other sulfides are rare. The sphalerite always shows twinning lamellae and often is crumpled, and the galena is often tectonically curved or recrystallized. Pyrite and tetrahedrite, though present, are uncommon. The grade of the ore is about 10 percent lead and 8 percent zinc. The ore-element content of the ore is not constant, being higher in zinc at depth and near the north cross fault and higher in lead up and away from the fault. Although the silver in the ore ordinarily runs less than 1 oz. per ton, the silver content decreases with depth and toward the fault, being apparently directly related to the lead content. Temperatures obtained by Gross (1956) in quartz crystals indicate that the quartz (and presumably the sulfides) were deposited between 95° and 120°C. It is, however, difficult to understand why the temperatures shown in his Figure 8 are higher, the farther the ore lies from the north cross fault from which he thinks the ore fluids moved outward and upward. Everything known about lead-zinc-silver ore bodies argues against this temperature relationship being correct. In any event, however, the abundance of quartz as a gangue mineral argues for a higher temperature of deposition than the decrepitation work gives. On the other hand, the silver content is low enough to cast some doubt on the deposits having been formed in the mesothermal range. In fact, Gross refers to them as epithermal, which hardly seems to be possible considering the thick cover of rock over the ore-containing beds at the time of ore formation. The deposits, therefore, are here classified as mesothermal, but obviously more study of them is required.

CANANEA
BOLEO
SANTA EULALIA
PARRAL
CONCEPCION DEL ORO
CERRO DE MERCADO
ZACATEGAS
GUANANUATO
SAN JOSÉ
PACHUCA
0
400
MILES
MEXICO

BOLÉO, BAJA CALIFORNIA

Late Tertiary *Copper* *Telethermal*

Beyschlag, F., and others, 1916, (Truscott, S. J., Translator), The copper deposits at Boléo in Lower California, in *The deposits of the useful minerals and rocks--their origin, form, and content*: Macmillan, London, v. 2, p. 1181-1184

De Launay, L., 1913, Boléo (Basse California), in *Traité de metallogénie--gîtes minéraux et métallifères*: Librairie Polytechnique, Paris and Liége, v. 2, p. 790-799

Fuchs, E., 1885-1886, Note sur le gîte de cuivre du Boléo: Soc. Géol. France, 3e Ser., v. 14, p. 79-92

Krusch, P., 1899, Über eine Kupfererzlagerstätte in Nieder-Californien: Zeitschr. f. Prakt. Geol., Jg. 7, H. 3, S. 83-86

Locke, A., 1935, The Boléo copper area, Baja California, Mexico, in *Copper resources of the world*: 16th Int. Geol. Cong., v. 1, p. 407-412

Nishihara, H., 1957, Origin of the "manto" copper deposits in Lower California, Mexico: Econ. Geol., v. 52, p. 944-951

Ordoñez, G., and Ulloa, S., 1956, Notas sobre la geología del estado de Sonora y la península de Baja California: 20th Int. Geol. Cong., Guidebook for Excursions A-1 and C-4, p. 11-31 (particularly p. 21-31)

Pošepný, F., 1893, The genesis of ore deposits: A.I.M.E. Tr., v. 23, p. 197-369 (particularly p. 317-318)

Touwaide, M. E., 1930, Origin of the Boléo copper deposit, Lower California, Mexico: Econ. Geol., v. 25, p. 113-144

Wilson, I. F., 1948, Buried topography, initial structures, and sedimentation in Santa Rosalía area, Baja California, Mexico: Amer. Assoc. Petroleum Geologists Bull., v. 32, p. 1762-1807

---- 1956, Geología del distrito cuprífero del Boléo, Baja California: 20th Int. Geol. Cong., Guidebook for Excursions A-1 and C-4, p. 53-68

Wilson, I. F., and Rocha, V. S., 1955, Geology and mineral deposits of the Boléo copper district, Baja California, Mexico: U.S. Geol. Surv. Prof. Paper 273, 134 p.

---- 1957, Geología y yacimientos minerales del distrito cuprífero del Boléo, Baja California: Consejo de Recursos Naturales no Renovables Bol. 152, 424 p.

Wisser, E., 1954, Geology and ore deposits of Baja California: Econ. Geol., v. 49, p. 44-76 (Boléo is discussed on p. 72-74 only, but the entire article is worth reading for its presentation of the general geology of the region)

Notes

The Boléo copper deposits are located in the area immediately surrounding the town of Santa Rosalía that is situated on the east (Gulf of California) coast of Baja California about halfway between the mouth of the Colorado River and the southern tip (Cabo de San Lucas) of the peninsula.

All ore bodies of commercial importance in the district are found above the upper surface of the middle and late Miocene Comondú beds no matter how great the relief on that surface may be. The copper content of the volcanic rocks averages only 0.004 percent of copper with the range being between 0.002 and 0. 007 percent. Although no fossils have been reported from the Comondú beds, they are known, outside the Boléo area, to lie unconformably above the Isidro formation that probably is middle Miocene. Thus, the Comondú cannot be older than late middle Miocene. The Comondú cannot be as young as Pliocene since it is unconformably overlain by the lower Pliocene Boléo formation.

The outcrops of the Boléo formation normally occur on the sides of the arroyos; on the mesas a thick blanket of younger rocks overlies the Boléo beds. The unconformity between the Boléo and the Comondú is remarkably conspicuous; in places the Comondú may dip as steeply as 45°, while the overlying Boléo beds are nearly horizontal and, on the other hand, horizontal Comondú rocks may be overlain by Boléo beds dipping as steeply as 30°, this dip being an initial one and not due to deformation.

In two areas the Boléo contains thick lenses of gypsum (20 to 80 m thick) above the limestone member, but locally these lenses rest directly on the Comondú formation or on the basal conglomerate; in a few places the gypsum is intercalated with tuff beds near the base of the formation. The rock volumes containing ore generally are located to the southwest of the gypsum beds. Although the gypsum seems more probably to have been formed by evaporation in minor basins isolated from the sea, it may be, as Touwaide (1930) suggests, precipitated from hydrothermal submarine springs. The gypsum almost certainly was not formed by the surface alteration of anhydrite.

The basal portions of the Boléo formation thus far described make up only a small fraction of its total bulk. Most of the Boléo is made up of alternating layers of tuff and tuffaceous conglomerate within which the beds of greatest economic importance are layers of clayey tuff that contain the ore bodies. There are five of these layers, and they have been numbered 0, 1, 2, 3, and 4; each layer is underlain by a conglomerate to which the same number is given. Ore bed 0 is stratigraphically the highest of these ore-bearing layers and 4 is the lowest. Most of the fragments in the conglomerates were derived from the Comondú volcanic beds, and the matrix is similar to that of the tuffs; crossbedding and channeling are common features both at the bases of the conglomerates and within the beds.

The five beds of the Boléo formation to which the ore is confined do not have the same areal distribution. Thus, the irregular topography of the rocks immediately beneath the Boléo formation blocked the formation of early Boléo beds over much of the general area; therefore, of the ore beds, number 4 and its associated conglomerate and tuffs are the most restricted in extent. Furthermore, pre-Gloria erosion has removed much of ore beds 0 and 1 and their associated rocks and has even penetrated down into ore bed 2. These factors have prevented a full representation of the Boléo formation in any one area, and the middle ore bed 2 is areally most widespread even though it too is missing in some places.

Unconformably above the Boléo lies the marine middle Pliocene Gloria formation that is composed of (1) a locally present basal conglomerate, (2) a fossiliferous marine sandstone that thins toward the land and thickens toward the gulf where it grades into siltstones and clays, and (3) an overlying conglomerate that thickens inland and passes gradually into a nonmarine facies. Above a slight unconformity, the late Pliocene Infierno formation lies on the Gloria and is a sequence of fossiliferous marine sandstone (below) and conglomerate (above). The Infierno is overlain unconformably by the nonmarine Santa Rosalía formation of Pleistocene age; it is composed of calcareous fossiliferous conglomerate with some sandy layers.

The mineralization in the Boléo beds must have taken place either during the time these beds were being laid down or at least before the deposition of the Gloria formation began for there are no traces of copper mineralization in the Gloria or any of the younger formations. It would follow, therefore, no matter what the origin of the ores, that they must have been formed in early Pliocene time, and they are here classified as late Tertiary.

The copper deposits of the Boléo district are thin, gently dipping tabular bodies that are confined to certain clayey tuff beds within the Boléo formation that, while they are quite impervious to vertically moving solutions, well may have been definitely permeable to ore fluids moving laterally through them as appears to have been the case in the shaly beds in the lead deposits of southeast Missouri. The mined area (excluding minor marginal deposits) is 11 m long from northwest to southeast and from 0.5 to 3 m wide. The principal primary ore minerals in the clayey tuff are finely disseminated chalcocite and subordinate chalcopyrite, bornite, covellite, and native copper. The main gangue mineral is montmorillonite with which are associated manganese and iron oxides and smaller amounts of gypsum, calcite, chalcedony, and jasper; pyrite is generally uncommon, galena is not abundant, and primary zinc minerals have not been identified although the zinc content of the ore probably averages about 0.5 percent.

Touwaide (1930) believed that the copper in the ore beds was derived from the tuff which he thought had originally contained 0.2 percent of copper which he believed to have been the same as the copper content of the Comondú andesites. Wilson and Rocha (1957), however, determined the copper content of the Comondú to average 0.004 percent. Thus, the reasoning that the tuff must have contained 0.2 percent copper because the andesite did is invalid. Touwaide also thought that mineralization in the ore beds was very uniform and that this argued to him that the copper must have come from the overlying rocks. This supposed uniformity, however, Wilson and Rocha have found not to exist so this argument too must be dropped. Their principal objections to the theory of ore deposition by descending ground water are that (1) it is improbable that the tuff originally contained enough copper to have been the source of the copper in the ore beds; (2) there is no correlation between the thickness of the supposed source beds and the distribution of the ore; (3) it is difficult to imagine the processes by which copper would be both dissolved and deposited by the same solutions, both processes taking place beneath the water table; and (4) the ground water theory does not explain the vein deposits in the underlying Comondú beds.

When Wilson and Rocha did their work, the ideas of Nishihara (1957) had not yet been published. Nishihara believes that the deposits were formed by the precipitation of copper and iron sulfides in portions of the shallow Pliocene seas. The source of these materials he thinks to have been the pre-Pliocene volcanic rocks, presumably the more mafic varieties, in the general neighborhood. These materials he would have transported as fine detritus or in solution as the sulfate, either being accompanied by tuffaceous detritus. Reduction, precipitation, concentration, and deposition of copper sulfides he believes probably occurred on the carbonaceous, calcareous, gypsiferous, and muddy floor of the subsiding coastal plain. Why such materials on the floor of the sea would cause the formation of the ore beds, he does not explain; certainly the gypsum, which is almost entirely outside the ore-bearing rock volumes, could hardly have been responsible for the precipitation of copper sulfides, mainly chalcocite. Neither is the reduction of copper from the cupric to the cuprous state compatible with the reduction of sulfur from the S^{+6} state to S^{-2}, nor does the presence of carbonaceous matter aid in the simultaneous accomplishment of these two reactions. The presence of the Boléo copper in the same general region as the Lucifer manganese deposits demonstrates that appreciable surface oxidation has occurred in the area, which is hardly an argument for the reduction of copper and sulfur to give copper sulfides.

Wilson and Rocha (1957) list the following items of evidence as supporting the concept of ore deposition from ascending solutions: (1) the copper-bearing veins and stockworks in the Comondú volcanic rocks can hardly have been filled from above, particularly since some of them at least were never covered by Pliocene sediments at all; (2) the presence of high-grade ore not only in clayey layers but also along the unconformity between Comondú and Boléo beds indicates that clayey material was not necessary for the ore minirals to deposit; (3) the greater concentration of high-grade ore on the gulfward side of hills of Comondú volcanic rocks and their gradual change to low-grade ore and then to waste as they are followed toward the gulf--this is not a pattern which would be expected from solutions that had moved downward through a volume of tuff of essentially uniform copper content; (4) the best ore is found in structural and compositional traps that would have impeded the upward flow of ore-fluids--the ore is on the wrong side of these traps to have come from above; (5) the localization of the ores in such a small proportion of the tuffs--not what would be expected from the downward movement of ground water through an essentially homogeneous source of copper; (6) the repetition of ore from one bed to another appears to be a more reasonable result of hydrothermal as opposed to ground water solutions; and (7) the concentration of copper in the lower portions of each clayey ore bed--a relationship more likely to occur from upward than downward moving solutions.

It would appear the hydrothermal origin of the Boléo ores is a more reasonable explanation of the observed facts than any other that has been put forward. The presence of small amounts of native copper is easily explained by the oxidation of enough of the screening sulfur ions of the copper-sulfur complexes in which the copper was transported so that some of the copper was reduced beyond the cuprous state of the dominant chalcocite to native copper.

Although the mineral assemblage of the Boléo deposits strongly suggests that the ore and gangue minerals were deposited under telethermal rather than epithermal conditions, the thin cover over the ore beds at the time of ore deposition makes it necessary to consider the possibility that they were, nevertheless, formed from low-intensity solutions under conditions of rapid loss of heat and pressure. Against this concept it can be argued that the lack of silver- and gold-bearing minerals points to the solutions, before their arrival in the Boléo formation at least, having been governed by conditions of slow loss of heat and pressure with almost all of the gold and silver having been removed during mesothermal-intensity reactions at much greater depths. This does not mean, of course, that the last portions of such solutions cannot have been brought under rapid loss of heat and pressure conditions after having spent all of their previous career under slow heat and pressure loss. Nevertheless, the clayey beds of the Boléo formation, permeable though they may have been in their long dimensions, may have been so impermeable in a vertical direction that the loss of heat and pressure from the ore fluids was slow and the Boléo ores were deposited under telethermal conditions despite the near-surface environment in which they were formed. On the basis of this argument, the Boléo ores are here classified as telethermal, but it is quite possible that further work may show them actually to be an unusual variant of the epithermal type.

CANANEA, SONORA

Early Tertiary | *Copper, Molybdenum, Silver* | *Hypothermal-1 to Telethermal*

Elsing, M. J., 1913, Relation of outcrops to ore at Cananea: Eng. and Min. Jour., v. 95, no. 7, p. 357-362

—— 1930, Secondary enrichment at Cananea: Eng. and Min. Jour., v. 130, no. 6, p. 285-288

Emmons, S. F., 1910, The Cananea mining district of Sonora, Mexico: Econ. Geol., v. 5, p. 312-356

Kelley, V. C., 1935, Paragenesis of the Colorada copper sulphides, Cananea, Mexico: Econ. Geol., v. 30, p. 663-688

Lee, M. L., 1912, A geological study of the Elisa mine, Sonora, Mexico: Econ. Geol., v. 7, p. 324-339

Mitchell, G. J., 1924, Primary chalcocite at Cananea, Mexico: Eng. and Min. Jour.-Press, v. 117, no. 22, p. 880-882

—— 1925, Ore injection at the Cananea-Duluth mine: Eng. and Min. Jour.-Press, v. 119, no. 2, p. 45-48

Mulchay, R. B., and Velasco, J. R., 1954, Sedimentary rocks at Cananea, Sonora, Mexico, and tentative correlation with the sections at Bisbee and the Swisshelm Mountains, Arizona: A.I.M.E. Tr., v. 199, p. 628-632 (in Min. Eng., v. 6, no. 6)

Perry, V. D., 1933, Applied geology at Cananea, Sonora, in *Ore deposits of the western states* (Lindgren Volume): A.I.M.E., p. 701-709

—— 1935, Copper deposits of the Cananea district, Sonora, Mexico, in *Copper resources of the world*: 16th Int. Geol. Cong., v. 1, p. 413-418

—— 1961, The significance of mineralized breccia pipes: Min. Eng., v. 13, no. 4, p. 367-376 (particularly p. 367-372)

Ramsey, R. H., 1944, How Cananea develops newest porphyry copper: Eng. and Min. Jour., v. 145, no. 12, p. 74-87 (essentially no geology)

Schwartz, G. M., 1947, Hydrothermal alteration in the "porphyry copper" deposits: Econ. Geol., v. 42, p. 319-352 (particularly p. 345-352)

Valentine, W. G., 1936, Geology of the Cananea Mountains, Sonora, Mexico: Geol. Soc. Amer. Bull., v. 47, p. 53-86

Velasco, J. R., 1956, Geología del mineral de Cananea, Sonora, Mexico: 20th Int. Geol. Cong., Guidebook for Excursions A-1 and C-4, p. 43-51

—— 1966, Geology of the Cananea district, in Titley, S. R., and Hicks, C. L., Editors, *Geology of the porphyry copper deposits--southwestern North America*: Univ. Ariz. Press, Tucson, p. 245-249

Warren, H. V., 1932, Relation between silver content and tetrahedrite in the ores of the North Cananea Mining Co., Cananea, Sonora, Mexico: Econ. Geol., v. 27, p. 737-743

White, C. H., 1924, Supergene enrichment of copper below a lean pyritic zone (Cananea): Econ. Geol., v. 19, p. 724-729

Notes

The mineralized belt at Cananea is located in the north-central part of the state of Sonora, just over 25 miles south of the international boundary and 38 miles southwest of the twin towns of Naco, Sonora, and Naco, Arizona. This belt has a length of about 10 km and width of 3.5 km.

The oldest known rocks in the Cananea district are Paleozoic and were deposited on an unknown Precambrian basement complex. The oldest of these Paleozoic formations is the Capote quartzite and it probably is Cambrian. Conformably above the Capote quartzite lies the Esperanza limestone that is now a series of very thin-bedded, highly altered and mineralized limestones. Mulchay and Velasco (1954) consider the Capote to be the equivalent of the Cambrian Bolsa quartzite and the Esperanza to be essentially equal to the Upper Cambrian Abrigo. Above the Esperanza is the Devonian and Mississippian Crystalline limestone that despite the lack of Ordovician and Silurian rocks in the Cananea district, appears to lie conformably on the Upper Cambrian beds and even to grade into them. Conformably above the Crystalline limestone is the Chivatera zone that was originally composed of limestone; it is also of Mississippian age. Conformably above the Chivatera zone is the Mississippian-Pennsylvania Puertecitos limestone; it was originally a generally thick-bedded limestone that has now been largely garnetized. Mulchay and Velasco believe that these beds are equivalent to the Escabrosa and Naco limestone at Bisbee.

The volcanic rocks that overlie the Puertecitos formation are divided into three formations that are, from oldest to youngest, the Elenita, the Henrietta, and the Mesa. The oldest of the Cananea abyssal intrusives is the Cuitaca granodiorite that outcrops over a considerable area in the northwest portion of the area mapped by Valentine (1936). The El Torro syenite is the next of the intrusive rocks and outcrops in a small area near the western edge of Valentine's map. The Tinaja diorite follows the El Torro in time extending to the north as far as a point just west of Capote Pass. The Cananea granite is next in the line of intrusives. Numerous dikes of Campana diabase are even younger than the granite. With the possible exception of a few mafic dikes, the Coloradá quartz porphyry is the youngest intrusive rock in the district. The rock occurs in a double belt of stocks of moderate to small size that extends from the Teocali (itself a moderate-sized Colorada stock) southeastward to the vicinity of Sonora Hill, passing through the main area of ore mineralization and including Cerro de Cobre (a stock of about the outcrop size of the Teocali); there is also a group of small porphyry plus in the Puertecitos-Henrietta area west-northwest of the Teocali.

The Cananea ore deposits certainly were introduced into the district after the quartz porphyry stocks and plugs and after the development of the breccia pipes since all of these rocks are mineralized to some extent. The ore is younger than all of the extrusive volcanic rocks, and it has already been pointed out that these rocks probably were emplaced after the deposition of Mesozoic sediments in the general region had ceased. This would place the earliest time at which the first of the volcanic formations could have been extruded as Cretaceous, and each of these three formations is appreciably more than 5000 feet thick and is separated from the one immediately above it by a considerable period during which earth movements and erosion took place. It would not be unreasonable, therefore, to believe that at least all of Late Cretaceous time was occupied by the accumulation, deformation, and erosion of these volcanic formations and that the intrusion of the igneous rocks into them did not begin until Tertiary time had begun. The principal questions to be asked are how much of Tertiary time was taken up by the intrusion of the wide variety of Cananea igneous rocks and how quickly after their intrusion was completed did the mineralization occur? Almost certainly, the Cananea ores are younger than those at Bisbee that are dated as late Mesozoic (pre-Laramide), but were they formed within the time of Laramide orogeny as were those in the Pima district or even later as were those of central Utah? For lack of better evidence, it is assumed that the abundant intrusive rocks were introduced late in the Laramide orogeny and that the ores should be dated as early Tertiary.

Three general types of ore bodies have been mined at Cananea; these are: (1) replacement deposits in favorable (partly or complete garnetized) limestone beds or along contacts between limestone and impervious rocks; (2) disseminated porphyry copper deposits, and (3) deposits in brecciated rock masses. The replacement deposits in limestone are localized in two general ways: (1) along favorable beds within the limestone sequence into which the ore solutions penetrated along a variety of minor fractures and folds and (2) along steeply dipping contacts between limestone and impervious rocks (different varieties of the latter may be involved from one ore body to another) and along such contacts as are capped unconformably by impervious volcanic rocks so as to trap the upward-moving ore fluids and force them out into the limestone.

The porphyry copper deposits are localized (1) in highly fractured portions of quartz porphyry plugs (in the contact breccias between these plugs and the surrounding volcanic rocks) and in highly fractured portions of the volcanic rocks outside the contact breccias and (2) in highly fractured volcanic rocks overlying breccia pipes. The contact breccias between plugs and volcanic rocks should not be confused with breccia pipes proper that are not included in the disseminated porphyry-copper category; the contact breccias were formed by stresses developed during the intrusion of the quartz porphyry pipes; the breccia pipes were formed by subsidence in a manner discussed below.

The most productive Cananea deposits, at least in terms of copper recovered per ton of ore mined, were the more-or-less-vertical breccia pipes. These structures probably acted as collecting channelways for ore solutions that eventually deposited their loads of ore minerals in structurally favorable portions of the Cananea limestones or, more importantly, served as both channelways and sites of major ore deposition. Because of the limited depth to which mining has been carried in the breccia pipes, their structures are not completely known, but through the 1600 feet of known vertical range, the breccia pipes are isolated, sharply defined, range between oval and nearly circular in plan, and are not connected with any major lateral fissures; they appear to grade into the less violently fractured rocks in which disseminated porphyry copper deposits may be found.

In the East-Breccia type of pipe deposits, much of the ore is chalcocite that secondarily replaced the primary pyrite and sparse chalcopyrite and is located (1) in the breccia at the top of the pipe and (2) in a stockwork of highly fractured but unbrecciated rock immediately above the top of the pipe. The principal difference between breccia pipes and the true disseminated porphyry copper type of deposits is that the fracturing containing localized porphyry copper deposits is largely due to the intrusion of small stocks or plugs of porphyry, while the breccia pipes were produced in rock volumes into which abyssal igneous rocks were not introduced or were emplaced there after the broken pipe-rock had formed.

In the vicinity of the Cananea-Duluth pipe, there are no igneous rocks at all, and the forces that brecciated the pipe-volume were not those generated by an igneous magma in its immediate vicinity. The Cananea-Duluth pipe is an elliptically shaped structure about 1200 by 300 feet in plan that cuts steeply through bedded tuffs that dip at low angles; it is known to continue to at least 2000 feet beneath the surface, and the ore is confined to the periphery of the pipe where the rocks are highly brecciated. The breccia is cemented by quartz, chalcopyrite, carbonates, adularia, and minor galena, sphalerite, and tetrahedrite; the sulfides are zoned with galena decreasing and chalcopyrite increasing with depth; the sphalerite content remains essentially constant. Inward from the highly brecciated rim of the pipe, the rocks become progressively less broken, and mineralization becomes markedly less, consisting mainly of quartz and calcite crystals lining large vugs. The alteration in the pipe was mild, consisting only of small amounts of chlorite and sericite.

The Colorada pipe was discovered in 1926 during exploration of a disseminated porphyry copper deposit and has been the most productive of the breccia pipes in the district. The pipe contains a ringlike ore body 600 feet long and 500 feet wide at the 6th level; below this level, the ring ore body contracts downward in a funnel-shaped form and becomes solidly mineralized at the 10th level. The pipe axis is vertical from the 5th to the 10th levels but changes direction sharply at the 10th level and plunges down to the northeast. In plan, the ore body is outlined by a broad ring of massive glassy quartz, one facies of which consists of quartz and phlogopite, the texture of which is pegmatitic. The sulfides are normally later than the quartz and occur in (1) networks of stringers in the wall rock outside the quartz ring, (2) networks of veinlets cutting the quartz into small fragments, (3) a ring of massive sulfides (partly brecciated by post-ore movements) that lies immediately inside the quartz ring, and (4) (below the 11th level) disseminations or irregular masses and lenses in a pipe of coarsely crystalline quartz and phlogopite.

The sequence of mineralization in the Colorada pipe appears to have been (1) pyrite; (2) quartz and phlogopite, and some feldspar; (3) molybdenite, chalcopyrite, and bornite; (4) blue chalcocite (much or all of which is probably digenite); (5) covellite; (6) core alteration and brecciation; (7) veins of intimately intergrown enargite and tennantite cutting across aggregates of early sulfides that include some sphalerite, galena, pyrite, covellite, and chalcopyrite--the total amount of sulfide material deposited at this stage appears to have been large enough to have made the copper sulfosalts the second most important source of copper in the Colorada pipe; (8) some reaction rims between the different copper sulfides--of minor economic importance; (9) alunite; (10) further brecciation of the core; and (11) secondary chalcocite.

According to Perry (1961), the Cananea pipes were primarily the result of subsidence. He points out that there was a considerable downward displacement of rock fragments in the Cananea pipes and that the great increase in volume of these fragments after their brecciation required the removal of large volumes of rock material. Further, the essentially unbroken rock over the tops of some of the pipes shows that the missing material could not have been removed upward; it must have been removed from the lower end of the pipes. He thinks that this was accomplished by the subsidence of deep-seated columns of rock, the subsidence having been caused by the withdrawal of magma support from underlying magmatic cupolas of restricted extent and the outline largely having been controlled by pre-existing lines of crustal weakness. The tendency to subside and cave would become less, the less the load of overlying rock, so that no pipe body should extend higher than a probably appreciable distance beneath the then-existing surface. The characteristics of the pipe are such that it well may have been formed by a gigantic underground explosion, the effects of which did not reach the surface. The formative mechanism, however, is certainly not yet completely understood; it is apparent, nevertheless, that these pipes provided excellent channelways for the upward movement of ore-forming fluids and were perfectly designed sites for ore deposition from such fluids.

The mineralization of the Cananea deposits appears to have been emplaced under a considerable range of intensity conditions. The quartz-phlogopite-minor feldspar masses formed in the Colorada pipe, and the widespread garnetization of the Paleozoic limestones almost certainly were formed within the hypothermal range. On the other hand, the chalcopyrite, (generally minor) bornite, pyrite, quartz, sericite mineralization (which was definitely later than the hypothermal gangue mineral development) may have been emplaced in the less intense portions of the hypothermal range or in the mesothermal; the molybdenite closely associated in space and in time of deposition with the chalcopyrite appears to have been sufficiently abundant also to suggest deposition on the borderline between hypothermal and mesothermal conditions. The bornite to some extent may have been initially deposited in solid solution in the chalcopyrite, and some may have formed as a separate and later mineral; more quantitative studies are needed on the relationships between bornite and the chalcopyrite. The abundant sphalerite of the Chivatera zone may have been deposited under hypothermal conditions, but not enough has been published concerning it to justify a definite opinion.

With the exception of the Colorada pipe, the chalcopyrite-minor bornite mineralization seems to have provided almost all of the primary copper minerals found in the Cananea ore bodies; thus, the great bulk of the copper in the entire district can reasonably be properly categorized as hypothermal-1 to mesothermal with most of it probably mesothermal. Within the Colorada pipe, however much of the copper was derived from covellite and digenite (chalcocite?), and these appear to have been deposited at a somewhat later time than the chalcopyrite and bornite. Essentially all of the digenite (chalcocite?) was deposited as a solid solution of some covellite in digenite, the covellite having later exsolved to give the typical exsolution textures reported by Kelley (1935). On the other hand, the bulk of the covellite shows what must be replacement relations with digenite (chalcocite?) and other and earlier sulfides and contains no exsolved digenite (chalcocite?). This covellite cannot have exsolved from digenite (chalcocite?) and must have been the last mineral to form in the pipes. Minor primary simple copper sulfide also occurs in some of the other ore bodies in the Cananea district, but in these instances it adds little to the value of the ores that are ores only through secondary chalcocite enrichment. Because of the great economic importance of the Colorada pipe, however, it is necessary to include that ore body in the classification of the district as a whole and, therefore, to add the designations leptothermal (for the digenite-covellite solid solution) and telethermal (for the late covellite) to the classification already assigned. A second generation of mesothermal mineralization appears to have taken place only in the Colorada pipe with the introduction of the intimately intergrown enargite (luzonite) and tennantite.

It would seem most reasonable, therefore, to classify the entire district as hypothermal-1 to telethermal, considering that probably the great bulk of the ore was formed in the more intense portion of the mesothermal range. The lower-temperature mineralization is almost entirely confined to the Colorada pipe.

CERRO DE MERCADO, DURANGO

Late Tertiary *Iron as Martite, minor Magnetite* *Magmatic-3b*

Farrington, O. C., 1904, Observations on the geology and geography of western Mexico, including an account of the Cerro Mercado: Field Columbian Museum, Geol. Ser., v. 2, no. 5, p. 197-228 (particularly p. 210-228)

Flores, T., 1950, Geologic and structural environment of the iron ore deposits of Mexico: Econ. Geol., v. 45, p. 105-126 (particularly p. 112-115)

Foshag, W. F., 1929, Mineralogy and geology of Cerro Mercado, Durango, Mexico: U.S. National Mus. Pr., v. 74, art. 23, 27 p.

Geijer, P., 1930, The iron ores of the Kiruna type: Sveriges Geol. Undersök., ser. C, no. 367, Årsbok 24, no. 4, p. 1-39 (particularly p. 18)

Ortiz-Asiáin, R., 1956, Notes on Cerro de Mercado: 20th Int. Geol. Cong., Guidebook for Excursion A-2 (Eng. translation), p. 104-109 (Spanish ed., p. 199-224)

Rangel, M. F., 1902, Criadero de fierro del Cerro de Mercado de Durango: Inst. Geol. México Bull. no. 16, p. 3-14

Salazar Salinas, L., and others, 1923, El Cerro de Mercado,-Durango: Inst. Geol. México Bol. no. 44, 90 p. (particularly p. 14-46)

Santillan, M., 1936, Durango, in *Carta geológico-minera del estado de Durango*: Inst. Geol. México, Cartas Geol. y Geol.-Mineras de la Republica Mexicana, no. 2, p. 81-89

Terrones Benitez, A., 1944, Cerro de Mercado, Mexico's iron mountain: Eng. and Min. Jour., v. 145, no. 9, p. 88-89

Torón Villegas, L., and Esteve Torres, A., 1946, Yacimientos del grupo del norte: Estudio de los yacimientos ferríferos de México, f. 2, Banco México, S. A., Mongrafías Industriales, 147 p. (particularly p. 33-52 and plans 24-28)

Notes

The iron deposits of Cerro de Mercado are located 3 km north of the city of Durango, capital of the state of the same name, and situated roughly halfway between El Paso and Mexico City.

The hill-mass of the Cerro de Mercado is rudely circular in outline, has a diameter of about 1.75 km, and is largely composed of volcanic rocks that have been uparched to a maximum elevation of about 175 m above the surface of the surrounding plain. Although the oldest rock to outcrop in the Cerro proper appears to be a latite; it, in turn, overlies the ore layer and beneath the ore is another, probably earlier, rhyolite (the underlying rhyolite). This rhyolite has been encountered only in drill holes that have penetrated completely through the ore, the ore having been emplaced essentially along the contact between latite (above) and underlying rhyolite (below). The cause of the uparching of the beds of the Cerro proper is not certainly known.

Any study of the district will be confused by the lack of agreement among published papers as to the designations appropriate to the various formations. What Foshag (1929) calls latite is referred to by Flores (1960) as rhyolite, and Torón Villegas and Esteve Torres (1946) call it rhyolite on one map and latite on another.

The rocks of the Cerro de Mercado appear to have been domed up so as to give the beds and the ore included within them quaquaversal dips. The cause of this doming may well have been the intrusion of an igneous plug beneath the underlying rhyolite. It is just possible that the underlying rhyolite is a pluglike intrusive body that was introduced into the area after the other volcanic rocks had been laid down. Faulting must have attended the doming of the Cerro, but it is uncertain whether the doming occurred before or after the emplacement of the ore. On the balance of probabilities, it seems more reasonable to assume that the ore was introduced at much the same time as the rocks of the Cerro were domed but that the doming did not occur until the ore was completely emplaced.

Geijer quotes Ordoñez (Instituto Geológico de México, Bol. no. 14) as dating the volcanic rocks of the region as early Pliocene. The ores were certainly introduced after the volcanic rocks, but the ores also almost certainly were emplaced within Pliocene time. The ore deposits of the Cerro de Mercado are, therefore, here classified as late Tertiary.

The hematite (martite) ore of the Cerro de Mercado appears to consist of a massive body from 5 to 65 m in thickness that underlies essentially the entire Cerro and may extend outward from its margins in all directions. Since the contact of the ore mass with the underlying rhyolite is much more regular than with the overlying latite and since the ore is much more resistent to erosion than the silicic volcanic rocks in which it is contained, those portions of the ore of greatest thickness tend to stand out as impressive outcrops above the level of the surrounding volcanic beds and to give the impression that the ore mass is made up of several ore bodies separated from each other by volcanic rocks. It would appear, however, from the cross sections given in Villegas and Torres (1946) that the ore actually forms an almost completely continuous layer within the rocks of the Cerro, the layer being somewhat displaced by faults; in places the displacements were sufficiently great to break the continuity of the ore layer over short distances. In at least two places on the plain beyond the foot of the Cerro slopes minable ore bodies have been found which confirm the idea that the ore layer is wider in extent than the Cerro proper.

The character of the ore within the layer, however, is not constant, some of the ore being far less compact and massive than the remainder. Locally veins of hematite branch out from the ore layer to cut up into the overlying latite, and considerable volumes of brecciated latite are cemented with hematite and its accessory minerals. The principal mineral of the ore bodies is reported to be martite. Martite, of course, is defined as hematite occurring in dodecahedral or octahedral crystals pseudomorphous after magnetite (or pyrite). In the Cerro de Mercado ores, crystals of iron oxide projecting into vugs are found in well-formed or flattened octahedra. On this evidence, it has been assumed that the massive iron oxide is also martite and must have been deposited as a replacement of original magnetite. The frequently flattened octahedra resemble stout rhombohedral crystals of hematite, but they probably are truly pseudomorphous after magnetite. The iron oxide throughout the deposit is partly magnetic even where no visible magnetite can be seen in the ore. This last fact suggests that some of the ore may be maghemite (=alpha hematite), the spinel structure of which accounts for its magnetic properties even though it lacks ferrous iron, or that some very fine-grained magnetite is present. Goethite (FeO·OH) is widespread in the deposit but is never abundant; it forms thin coatings on martite crystals in vugs and locally small crystals of goethite have been found; the goethite is probably a primary mineral. Limonite is sparingly present as thin films developed by surface weathering. Although magnetite is essentially lacking in most of the ore of the Cerro proper, the ore body reported to underlie the flat country to the north of the Cerro is said to be composed mainly of magnetite. Although it appears probable that most of the hematite of the Cerro was originally deposited as magnetite, it cannot be said that this has been definitely proved; some of the massive hematite may have been formed as a primary constituent of the ore.

Apatite crystals are found throughout the ore in which they are not uncommon; they occur in a variety of ways: (1) most usually in cavities in compact hematite (martite) ore and in association with martite crystals and/or with fine-grained silica and carbonate minerals that essentially fill the cavities,

(2) less commonly imbedded in massive ore, (3) in small veinlets in the ore, (4) in association with quartz and chalcedony the masses of which usually are broken and imbedded in masses of sepiolite, (5) with augite in cracks in brecciated rock where these minerals coat breccia fragments--the augite and apatite crystals are often broken and cemented by silica. In some places there appear to have been two generations of apatite crystals; early yellow apatite crystals are coated with goethite and the goethite by colorless apatites. The two varieties differ in form as well as in color.

In addition to apatite, the carbonate apatite, dahllite [$Ca_{10}(PO_4)_6(CO_3)\cdot H_2O$] also is sparsely present in vugs in martite as coatings on martite crystals. An unknown phosphate mineral also is found with apatite in vugs in martite, but it exists in such small amounts as not to have been identifiable by Foshag. Recent X-ray analyses of what is probably the same mineral have failed to provide an identity for it.

Diopsidic pyroxene is a common constituent of altered wall rocks in the vicinity of ore bodies where it may be associated with clayey altered latite or may form massive pyroxene bodies. Some quantities of apatite and magnetite are intergrown with the diopside, and sepiolite [$Mg_3Si_4O_{11}\cdot nH_2O$] is found with the diopside, probably as an alteration product of it.

Quartz is not abundant and is one of the late minerals to form; it occurs in cavities with sepiolite or as coatings of brecciated fragments; in a few places, it occurs with sparse amounts of barite and calcite. A little quartz is found as crystals in cavities in massive hematite ore. Silica is also present as chalcedony in diopside-rich rock and as an important constitutent of impregnated tuff. Cherty silica considerably replaces rhyolite porphyry, and some opal is found on apatite crystals in vugs in martite ore.

Although there is some alteration of the wall rock surrounding the ore, it is small in comparison with that which would be expected if the ores had been introduced by iron-bearing hydrothermal solutions as were, for example, those of Cornwall, Pennsylvania, or even Iron Springs, Utah. This relationship of alteration volume to ore volume was recognized by Foshag, and he accepted it as an indication of a magmatic origin for the ores. He rejected the magmatic concept, however, on the grounds that a magma of the composition of the Cerro de Mercado ores could not logically be derived from the possibly available silicate magmas. Since that time much work has been done both in the field and in the laboratory on this problem, and it is now nearly certain that such magmas can be derived as immiscible fractions from much larger masses of highly silicic magmas, provided these magmas initially contained iron-oxide material in considerable abundance. It also appears probable that iron-oxygen-rich melts containing minor quantities of phosphates, carbonates, sulfates, and silicates can remain molten down to temperatures of 500° or 600°C. Further, at least moderate amounts of water probably can be dissolved in such melts, to be released as the crystallization of the iron oxides sufficiently reduces the volume of remaining molten material.

Probably the major problem in reconciling the characteristics of the Cerro de Mercado ore with a magmatic origin is explaining the formation of martite from magnetite. Most of the martite crystals in vugs probably were originally deposited as magnetite and were later converted pseudomorphically to hematite. From this it has been reasoned that the entire mass of ore was initially magnetite which was later converted to hematite. Since the massive hematite does not show the crystal outlines of magnetite, it is not, in the strict sense, martite but is simply hematite which may or may not have replaced magnetite. Since the martite crystals in the vugs in the hematite ore, however, were the last iron-oxide minerals to form, it seems reasonable to assume that, since they were originally magnetite, the rest of the ore must have been also; thus, the principal problem remaining is to explain how the primary magnetite was converted to secondary hematite. One not unreasonable explanation is shown in equation (1):

$$2FeFe_2O_4 + 2Fe^{+3} + H_2O = 3Fe_2O_3 + 2Fe^{+2} + 2H^{+1}. \quad (1)$$

This would mean that the late hydrothermal fluids added ferric iron to the already solid magnetite and carried the ferrous ion so released out into the wall rocks to form diopside.

Another explanation for the oxidation is that, in the near-surface environment in which the oxidation took place, the lowered confining pressure (after faulting had opened somewhat the cover over the ore mass) allowed the dissociation of nitrogen-oxygen gases (and possibly of water gas) dissolved in the melt and/or the ore-magma-generated hydrothermal fluid to produce nascent oxygen plus nitrogen (or plus hydrogen which might diffuse from such a system before it could react to limit the oxidation of magnetite). The oxygen would react with the magnetite as is shown in equation (2):

$$4FeFe_2O_4 + O_2 = 6Fe_2O_3. \qquad (2)$$

Perhaps both reactions were of some importance since the amount of ferrous iron transferred to the wall rock certainly was much less than the amount of that ion that would have been produced had all the magnetite been oxidized by the reaction of equation (1).

Assuming that these two equations explain both the oxidation and the production of sufficient ferrous iron to form the known quantities of diopside as a wall-rock alteration mineral around the ore, it seems safe to classify the Cerro de Mercado deposits as magmatic-3b.

CONCEPCIÓN DEL ORO-PROVIDENCIA, ZACATECAS

Middle Tertiary	*Copper, Gold, Lead, Zinc, Silver, Iron*	*Hypothermal-2 to Mesothermal*

Bergeat, A., 1909, Granodiorit von Concepción del Oro im Staate Zacatecas (Mexiko) und seine Kontaktbildungen: Neues Jb. f. Mineral., Geol. und Paläont., Abh., Beil. Bd. 67, Abt. A. S. 421-573 (Spanish translation in Inst. Geol. Mexico Bol. no. 27)

Buddington, A. F., 1959, Granite emplacement with special reference to North America: Geol. Soc. Amer. Bull., v. 70, p. 671-747 (particularly p. 689-690)

Burckhardt, C., 1906, Géologie de la Sierra Mazapil et Santa Rosa: 10th Int. Geol. Cong., Guide des Excursions no. 26, 40 p.

—— 1906, Géologie de la Sierra del Oro (Mexico): 10th Int. Geol. Cong., Guide des Excursions no. 24, 24 p.

Buseck, P. R., 1966, Contact metasomatism and ore deposition: Concepción del Oro, Mexico: Econ. Geol., v. 66, p. 97-136

Casteñedo, J., 1927, Los distritos cupríferous de Mazapil y Concepción del Oro del Estado de Zacatecas: Boletín Minero, v. 24, p. 443-453

Ibarra, J., 1931, Informe general de la región minera comprendida en las municipalidades de Mazapil, Concepción del Oro y San Pedro de Ocampo del Estado del Zacatecas: Boletín Minero, v. 29, p. 255-278

Imlay, R. W., 1938, Studies of the Mexican geosyncline: Geol. Soc. Amer. Bull., v. 49, p. 1651-1693

Mapes-Vásques, E., 1964, Geología y yacimientos minerales de los distritos de Concepción del Oro y Avalos, Zacatecas: Consejo de Recursos Naturales no Renovables, Pub. 10 E, 133 p.

Maugher, R. L., and Damon, P. E., 1965, Lead-zinc deposition in light of fluid inclusion studies, Providencia mine, Zacatecas, Mexico: Econ. Geol., v. 60, p. 1542

Ohmoto, H., and others, 1966, Studies in the Providencia area, Mexico II, K-Ar and Rb-Sr ages of intrusive rocks and hydrothermal minerals: Econ. Geol., v. 61, p. 1205-1213; disc., 1967, v. 62, p. 862-863

Puchner, H. F., and Holland, H. D., 1966, Studies in the Providencia area, Mexico, III, Neutron activation and analysis of fluid inclusions from Noche Buena: Econ. Geol., v. 61, p. 1390-1398

Rogers, C. L., and others, 1956, General geology and phosphate deposits of Concepción del Oro district, Zacatecas, Mexico: U. S. Geol. Surv. Bull. 1037-A, p. 1-102

—— 1962, Tectonic framework of an area within the Sierra Madre Oriental and adjacent Mesa Central, north central Mexico: U.S. Geol. Surv. Prof. Paper 450-C, p. C21-C24

Rye, R. O., and Haffty, J., 1969, Chemical composition of the hydrothermal fluids responsible for the lead-zinc deposits at Providencia, Zacatecas, Mexico: Econ. Geol., v. 64, no. 6, p. 629-643

Rye, R. O., and O'Niel, J. R., 1968, The O^{18} content of water in primary fluid inclusions from Providencia, north-central Mexico: Econ. Geol., v. 63, p. 232-238

Sawkins, F. J., 1964, Lead-zinc ore deposition in the light of fluid inclusion studies, Providencia mine, Zacatecas, Mexico: Econ. Geol., v. 59, p. 883-919

Sawkins, F. J., and Huebner, J. S., 1965, Postmagmatic ore deposition in light of fluid inclusion studies, Providencia mine, Zacatecas, Mexico, in Kutina, J., Editor, *Symposium--Problems of postmagmatic ore deposition*, v. 2: Geol. Surv. Czech., p. 424-428

Triplett, W. H., 1952, Geology of the silver-lead-zinc deposits of the Avalos-Providencia district of Mexico: A.I.M.E. Tr., v. 193, p. 583-593 (also in Min. Eng., v. 4, no. 6, and disc., no. 11)

Notes

The abundantly and variously mineralized district of Concepción del Oro-Providencia is located near the northern border of the State of Zacatecas about 165 km southwest of the city of Monterrey and about 100 km south-southwest of Saltillo, the capital of Coahuila. The lead-zinc-silver mining is centered around the town of Providencia and the gold and copper mining around Concepción del Oro itself.

The district lies in the Sierra Madre Oriental on the southwestern border of a belt of eastward-trending folds. This departure from the normal northwest-southeast trend of the Sierra Madre was caused by the former Coahuila highland areas to the north-northwest acting as buttresses against the northeast-southwest compression that affected this general portion of Mexico in middle Mesozoic time. Of the numerous east-west anticlinal structures in the region, the one of greatest economic geologic interest is that composed of the Sierra de la Caja which runs essentially east-west, and at its eastern end, turns southeast and becomes the Sierra de Concepción del Oro. The core of the Sierra de Concepción del Oro is made up of two granodiorite bodies, actually connected at depth into one, the more northwesterly of which is known as the Providencia stock and the more southeasterly as the Concepción del Oro stock.

The only igneous bodies that show direct spatial, and perhaps genetic, connection with the ore deposits are the Concepción del Oro and Providencia granodiorite stocks and, to a lesser extent, the Noche Buena diorite stock. The carbonate sediments adjacent to the granodiorite and the granodiorite (in much smaller volume) of the Concepción del Oro stock have been strongly altered so that, in places, the original boundary between them cannot be certainly determined.

The ore mineralization, although occurring after the granodiorites had been solidified (or at least after their outer margins were solid), must have taken place at much the same time as the actual intrusions themselves. For Concepción del Oro, Buseck (1966) gives age determinations made from adularia, the last mineral in the ore paragenetic sequence, and from biotite, an integral component of the granodiorite of the stock. The age obtained for the biotite was 40 m.y. and for the adularia 38 m.y. This would strongly suggest that the ores were formed during the early Oligocene and should be classified here as middle Tertiary.

Further work on radioactive dating (Ohmoto and others, 1966) gives ages by the K/Ar method of (1) 40.0 m.y. from biotite in the granodiorite stock, (2) 38 m.y. from adularia from the late hydrothermal veins, and (3) 34.5 m.y. from muscovite associated with the sulfide ore. The Rb/Sr method gives ages, with one exception of 44.0 m.y., of 41.0 m.y. These data appear to confirm those of Buseck (1966) for granodiorite and hydrothermal adularia and considerably strengthen the assignment of early Oligocene (or latest Eocene) as the age of the granodiorite.

In the Concepción del Oro-Providencia district, there are three principal types of ore deposits: (1) the lead-zinc sulfide pipes associated with the Providencia stock, (2) the lead-zinc sulfide pipes bearing similar relations to the Noche Buena stock, and (3) the silicate-oxide-sulfide-copper-bearing ore bodies in the skarn zones associated with the Concepción del Oro stock and its satellites. Several areas in the Concepción del Oro portion of the district have been mined for iron oxides, although only one, the Sol y Luna mine near the southwest tip of the stock, was mined as late as 1960. Essentially all the ore bodies in the district are found in silicated or marmorized limestone host rocks near to the granodiorite masses. In the Providencia area there are at least 10 ore pipes in marmorized limestone of sufficient size to have been mined, and a total of more than 30 such pipes are known, all but one outcrop at the surface. The pipes mined in the Noche Buena area are in general similar to those around the Providencia stock, while the ore bodies in the Concepción del Oro area are principally massive replacement bodies in which the sulfides of iron and copper are in the same rock volumes as the lime silicate and iron oxide minerals of the skarn belts.

The paragenesis of the Concepción minerals, according to Buseck (1966), began with the formation of garnet (and the other silicates associated with it); the garnet was overlapped by magnetite and the magnetite by pyrite. Specularite overlapped with pyrite but not with magnetite, and chalcopyrite overlapped with specularite; specularite appears to have continued to deposit after the precipitation of chalcopyrite had stopped. Late in the specularite cycle, a minor second generation of magnetite and a little powdery hematite were formed. The small quantities of pyrrhotite were deposited during the pyrite cycle. The minor amounts of enargite, tetrahedrite-tennantite, sphalerite, and two or three bismuth minerals followed the chalcopyrite and the specular hematite. Quartz and adularia were late gangue minerals, with the adularia having been the last mineral of all to form.

In the Providencia area, studies by Sawkins (1964) strongly suggest that the massive sphalerite was emplaced at temperatures in excess of 350°C but below 425°C and that the zoned sphalerite crystals in vugs were formed in a range between 350° and 200°C. Calcite and quartz, largely later to form than the sphalerite, were precipitated below 350°C. Since galena was deposited at much the same time as sphalerite, it almost certainly was formed over much the same temperature range. The late sulfosalts probably were deposited at temperatures between 300° and 200°C. It would appear to follow from these determinations by Sawkins that the Providencia ores were formed in the lower intensity segment of the hypothermal range and in the upper limits of the mesothermal range. The deposits of this area, therefore, are classified as hypothermal-2 to mesothermal.

In the Concepción del Oro area, Buseck (1966) believes that the ores were deposited in the range between 500° and 350°C and considers that this places them in Lindgren's contact metasomatic classification. He goes on to say that this class should, by definition, have an origin distinct from hydrothermal deposits. Yet he finds the genesis of the Concepción ores so similar to that of the ores at Providencia that he thinks there must be some discrepancy in the genetic distinction commonly drawn between deposits of these two types.

It is for exactly Buseck's reasons that the term "contact metasomatic" has been dropped from the classification used in this volume and the term "hypothermal in calcareous rocks" used instead to emphasize that high-temperature deposits in calcareous rocks are simply a variant of other types of hypothermal ores. The close association in time of formation of the magnetite and the high-temperature silicates requires that the magnetite was deposited under hypothermal conditions. The overlap of pyrrhotite with pyrite and of pyrite with magnetite indicates strongly that the pyrite was also formed under hypothermal conditions. The overlap of specularite and chalcopyrite with the pyrite, in turn, suggests that they were, in large part, formed in the hypothermal range. The last of the chalcopyrite may have been developed in the upper intensity portions of mesothermal deposition. The late enargite also was almost certainly mesothermal, but the tetrahedrite-tennantite and the bismuth minerals probably were produced mainly under leptothermal conditions; these sulfosalts are of so little importance, however, as to negate the use of the term leptothermal in the classification of the deposits; thus, the classification assigned to Concepción ores is hypothermal-2 to mesothermal, the same given to the Providencia ores. It should be emphasized, however, that an appreciable fraction of the Providencia ores probably was developed in the mesothermal range while almost all of those at Concepción formed in the hypothermal range; thus, the Concepción ores were introduced from ore-forming fluids at a definitely higher intensity level than those at Providencia.

GUANAJUATO, GUANAJUATO

Late Tertiary | *Silver, Gold* | *Epithermal*

Antúnez Echagaray, F., 1964, Monografía historica y minera sobre el distrito de Guanajuato: Consejo Recursos Naturales no Renovables Pub. 17E, 588 p. (particularly p. 113-155)

Bengoechea, A., and others, 1964, Distrito minero de Guanajuato, Mexico--Plano fotogeológico: Consejo Recursos Naturales no Renovables Pub. 8-E, 1:25,000

Blake, W. P., 1902, Notes on the mines and minerals of Guanajuato, Mexico: A.I.M.E. Tr., v. 32, p. 216-223

Botsford, C. W., 1909, Geology of the Guanajuato district, Mexico: Eng. and Min. Jour., v. 87, no. 14, p. 691-694

Edwards, J. D., 1955, Studies of some early Tertiary red conglomerates of central Mexico: U.S. Geol. Surv. Prof. Paper 264-H, p. 153-183 (particularly p. 155-172)

González Reyna, J., 1959, El intrusivo granítico de Arperos y su influencia en la mineralización de Guanajuato, Gto.: Soc. Geol. Mexicana Bol., t. 22, no. 2, p. 9-18

Guiza, R., Jr., 1949, Estudio geológico del distrito minero de Guanajuato, Gto. (zona de la Veta Madre): Consejo Recursos Naturales no Renovables (formerly Mex. Inst. Nac. Ins. Rec. Mineral.) Bol. 22, 75 p.

——— 1956, El distrito minero de Guanajuato: 20th Int. Geol. Cong., Guidebook to Excursions A-2 and A-5, p. 141-148

Milton, C., and others, 1951, Estudio mineralógico de los minerales auro-argentiferos del distrito de Guanajuato, México (abs.): Conv. Interamet. Recoursos Mineral., Primero, México, p. 72-73

Monroy, P. L., 1888, Las minas de Guanajuato: Ministerio de Fomento (Mexicana), Anales, t. 10, p. 69-738 (particularly p. 106-342)

Mullerried, F. K. G., 1947, Historia de la geología Guanajuatense (México): Acad. Nac. Cienc. Mem. y Rev., t. 56, no. 1, p. 87-109

Villarello, J. D., and others, 1906, Étude de la Sierra de Guanajuato: 10th

Int. Geol. Cong., Guide des Excursions no. 15, 33 p.

Wandke, A., and Martínez, J., 1928, The Guanajuato mining district, Guanajuato, Mexico: Econ. Geol., v. 23, p. 1-44

Wisser, E., 1960, Guanajuato, Guanajuato, Mexico: in *Relation of ore deposition to doming in the North American Cordillera*: Geol. Soc. Amer. Mem. 77, p. 77-82

Notes

Guanajuato is located in the central Mexican plateau, some 280 km northwest of Mexico City.

The district lies in the long anticlinally folded and block-faulted Sierra de Guanajuato. The oldest rocks in the district are clay shales (lutites) of marine origin (the Esperanza formation) that have been, in places, converted to phyllites and schists. Through correlation with similar rocks in the state of Zacatecas, these unfossiliferous beds are thought to be Triassic in age. In some parts of the district, the lutites are intercalated with and covered by bodies of calcareous clay shale, limestone, and sandstone. These calcareous rocks are thought (Bengoechea and others, 1964) to be at least partly of Jurassic age. The last event in the Triassic-Jurassic sequence appears to have been the emplacement of the La Luz basalts conformably above the older sediments and volcanic rocks. No sedimentation seems to have occurred during the remainder of the Mesozoic (or its results were removed by erosion), but small masses of intrusive diorites cut the Triassic rocks after they had been folded into gentle folds in the Late Cretaceous orogeny but before the first Tertiary sediments were laid down. These Tertiary rocks were, in contrast to the marine conditions of the Mesozoic, laid down on land. The first of these terrestrial formations was the Red Conglomerate of Guanajuato that contains a wide variety of pre-Tertiary rock fragments. On the basis of fossils described by Edwards (1955), the conglomerate is considered to be late Eocene or early Oligocene. Conformably overlying the Red Conglomerate is a tuffaceous sandstone that is known locally as "losero."

The conglomerate and its associated rocks are overlain by a series of extrusive volcanic rocks of late Tertiary age. The oldest of these is the thin Bufa rhyolite, mainly a banded, silicic volcanic ash with some flow rocks. The next formation to be emplaced was the Cedro andesites, and these cover certain hills with a thickness on the order of 100 m. The next younger flow rock is the Chichíndaro rhyolites; these rocks show flow lines and locally have a spherulitic texture.

Although there are some dikes of intermediate composition in the area, the major Tertiary intrusive is the Arperos granite that not only contains a considerable length of the Veta Madre vein but also underlies much of the district and sends off apophyses into many of the older rocks. The granite is, however, probably older than the late Tertiary volcanic rocks.

The assignment of a late Miocene-Pliocene age to the igneous activity that culminated in the formation of the ore deposits in the Guanajuato veins is based largely on the similarity of these events to those that can be so dated in other parts of central Mexico; no direct evidence of such age has been reported from the Guanajuato rocks. Granted that the igneous activity in the Guanajuato area began in late Miocene time, the total volume of materials introduced into the area and the extent of the earth movements strongly suggest that the actual emplacement of the ores did not occur until Pliocene time, and the ore deposition is dated here as late Tertiary.

Volcanism in the Guanajuato area continued after the formation of the ores, flows of various types having been interspersed between periods of volcanic quiescence until quite recent time; most of the post-ore flow rocks in the mineralized area have been removed by erosion.

Guiza (1956) divides the district into areas in which deposits of major economic importance were formed. These are (1) the central part of the Veta Madre, (2) La Luz mine in the northwestern part of the district, and (3) the

Sierra veins in the Villalpando-El Cubo area in the southeast part of the district. The ores of the Guanajuato district are found in two types of veins: (1) well-defined veins and (2) stockworks that are masses of stringers in the hanging walls of the strongest veins. All of the veins of type (1), including the Veta Madre, can be classed as fissure veins, but the considerable development of stockwork ore in its hanging wall puts the Veta Madre into type (2) as well. In these fissure veins probably not more than 10 percent of the total ore and gangue mix was emplaced by the filling of open space.

Alteration around the various veins depends largely on the character of the rocks making up the walls. Where the footwalls of the veins are composed of schists, and they abut against a rich ore shoot or are themselves considerably fractured, this grayish-black-to-green rock may have been whitened or bleached. The mineral change has been largely the conversion of chlorite to sericite and the addition of a little quartz and carbonate. The Red Conglomerate is often exposed in the hanging walls of the veins and is extensively altered, the red color being changed to green (Fe^{+3} reduced to Fe^{+2}) or even locally to a buff color. The green rock is colored by chlorite and a little quartz and carbonate have been added, while in the buff-colored rock, the chlorite has been converted to sericite and tiny veinlets of adularia have been introduced.

In the veins, the main gangue minerals are quartz, carbonates, and adularia with minor fluorite, barite, and certain zeolites. The most common ore minerals are argentite and polybasite ($Ag_{16}Sb_2S_{11}$), but several other silver sulfosalts also are to be found, for example, myrargyrite ($AgSbS_2$), pyrargyrite (Ag_3SbS_3), and stephanite (Ag_5SbS_4). Base-metal sulfides are insignificant in amount, and pyrite, chalcopyrite, galena, sphalerite, and tetrahedrite are the most important of these. Some native gold occurs as thin incrustations, but the proportion of silver to gold normally is about 100 to 1. Native silver also occurs as wires, threads, and incrustations and probably is primary since it is found far below the usual depth of oxidation, which reaches between 90 to 150 m below the vein outcrops.

The close space and time association of the ore-filled veins with the Arperos granite strongly suggests that the granite and ore-forming fluid came from the same source. The geologic setting, vein character, and minerals present all are typical of epithermal deposition, and the Guanajuato deposit here is categorized as epithermal.

PACHUCA-REAL DEL MONTE, HIDALGO

Late Tertiary *Silver, Gold* *Epithermal*

Bastin, E. S., 1948, Mineral relationships in the ores of Pachuca and Real del Monte, Hidalgo, Mexico: Econ. Geol., v. 43, p. 53-65

—— 1948, Mineral relationships in the ores of Pachuca and Real del Monte, Hidalgo, Mexico (reply): Econ. Geol., v. 43, p. 525

Friedrich, G. H., and Hawkes, H. E., 1966, Mercury as an ore guide in the Pachuca-Real del Monte district, Hidalgo, Mexico: Econ. Geol., v. 61, p. 744-753

Geyne, A. R., 1949, Mineral relationships in the ores of Pachuca and Real del Monte, Hidalgo, Mexico (disc.): Econ. Geol., v. 44, p. 233-234

—— 1956, Las rocas volcánicas y los yacimientos argentíferos del distrito minero de Pachuca-Real del Monte, estado de Hidalgo: 20th Int. Geol. Cong., Guidebook for Excursions A-3 and C-1, p. 47-57

Geyne, A. R., and others, 1961, Pachuca mining district, Hidalgo, Mexico: U.S. Geol. Surv. Prof. Paper 424-D, p. D221-D222

—— 1963, Geology and mineral deposits of the Pachuca-Real del Monte district, Hidalgo, Mexico: Consejo Recursos Naturales no Renovables Pub. 5E, 203 p.

Hulin, C. D., 1929, Geology and mineralization at Pachuca, Mexico (abs.):

Geol. Soc. Amer. Bull., v. 40, p. 171-173

Ordoñez, E., 1902, The mining district of Pachuca, Mexico: A.I.M.E. Tr., v. 32, p. 224-241

Rove, O. N., 1964, Geology and mineral deposits of the Pachuca-Real del Monte district, Hidalgo, Mexico (rev.): Econ. Geol., v. 59, p. 177-180

Segerstrom, K., 1956, Estratigrafía y tectónica del Cenozoico entre Mexico, D. F. y Zimápan, Hgo.: 20th Int. Geol. Cong., Guidebook for Excursions A-3 and C-1, p. 11-22

Thornburg, C. L., 1945, Some applications of structural geology to mining in the Pachuca-Real del Monte area, Pachuca silver district, Mexico: Econ. Geol., v. 40, p. 283-297

—— 1952, The surface expression of veins in the Pachuca silver district of Mexico: A.I.M.E. Tr., v. 193, p. 594-600 (in Min. Eng., v. 4, no. 6)

Winchell, H. V., 1922, Geology of Pachuca and El Oro, Mexico: A.I.M.E. Tr., v. 66, p. 27-41

Wisser, E., 1937, Formation of the north-south fractures of Real del Monte-Pachuca silver district, Mexico: A.I.M.E. Tr., v. 126, p. 442-487

—— 1942, The Pachuca silver district, Mexico, in Newhouse, W. H., Editor, *Ore deposits as related to structural features*: Princeton Univ. Press, p. 229-236

—— 1946, Some applications of structural geology to mining in the Pachuca-Real del Monte area, Pachuca silver district, Mexico (disc.): Econ. Geol., v. 41, p. 77-86

—— 1948, Mineral relationships in the ores of Pachuca and Real del Monte, Hidalgo, Mexico (disc.): Econ. Geol., v. 43, p. 280-292

—— 1951, Tectonic analysis of a mining district: Pachuca, Mexico: Econ. Geol., v. 46, p. 459-477

—— 1964, Geology and mineral deposits of the Pachuca-Real del Monte district, Hidalgo, Mexico (rev.): Econ. Geol., v. 59, p. 725-732

Notes

The Pachuca-Real del Monte district is located in the south-central portion of the state of Hidalgo, some 100 km north-northeast of Mexico City and covers an area of about 130 sq km; Real del Monte is 6 km east-northeast of Pachuca.

Although the district is almost certainly underlain unconformably by a pre-Tertiary basement of strongly folded and highly eroded Mesozoic marine formations, the structural trend of which is north-northwest, the basement is not exposed in the district. On this probable Mesozoic erosion surface lies a thick sequence of continental rocks; above this lies the thick Pachuca group that is composed of volcanic lava flows interbedded with tuff and breccia layers and including occasional layers of water-deposited sediments. Above the Pachuca group are younger Tertiary volcanic formations of dacite, andesite, rhyolite, and a series of clastic deposits made up of debris from older Tertiary rocks. The maximum thicknesses of all these Tertiary volcanic formations added together would be more than 3700 m; because the maximum thicknesses never coincide, the greatest thickness at any one place is appreciably less than the total just given. The source of these volcanic formations appears to have been in the Sierra de Pachuca itself.

The stratified and faulted Tertiary volcanic rocks of the Pachuca district have been intruded by great swarms of dikes and by numerous irregular bodies; at the surface these range in width from a few meters to more than 100 m and are between a few hundred meters to some 4 km long. At depth, some of the dikes may be as much as 10 to 14 km in length. Several funnel-shaped bodies fan out upward from the dikes and, where truncated at the surface, are from 1000 m in diameter. All of the Tertiary volcanic rocks, except the San Cristobal formation, are cut by dikes of one type or another.

The great majority of the dikes strike between N45°W and N75°W, with the principal direction being about N70°W; another fairly frequent direction is about east-west. The dikes coincide in trend with the "east-west" vein system at Pachuca and Real del Monte; the north-south fracture system at Real del Monte contains essentially no dikes and probably was, therefore, developed largely after the dikes were emplaced. The high-angle normal faults in which the dikes are now found appear to have been developed when the Tertiary rocks were deformed by moderate to strong tilting. These earth movements probably began in late Miocene time, with the intrusions having taken place in latest Miocene and earliest Pliocene time. Dike emplacement was followed by strong faulting and tilting in which the maximum throw was nearly 500 m and the dips were as much as 40°. The continued development of these tectonic features was accompanied by widespread normal faulting and then by the ore mineralization. After the ore had been emplaced, further, but minor, normal faulting occurred, followed by significant wrench faulting. After this strong deformation, fractures and faults of small displacement that strike essentially north-south and dip steeply either west or east were formed as a result of moderate uparching along a north-south axis; these minor structures might instead have been caused by compression from the north against a convex buttress in the south of the district.

Alteration of considerable magnitude affected the wall rocks of both the earlier and the later faults, the former to an appreciably greater degree, and the quartz-ore mineral veins were formed shortly after the development of the north-south fracture system. This would appear to confirm the age of the deposits as concluded from the relationship of the ores to the volcanic stratigraphy, thus placing the period of ore formation in the middle or early late Pliocene, and they are here categorized as late Tertiary.

In late Pliocene or early Pleistocene time, after the mineralization had been completed, strike-slip movement took place either along, or parallel to, east-west veins. This resulted in some crushing of vein-filling material and the displacement in the Real del Monte area of the north-south veins where they intersect the east-west structures.

The ore bodies in the district are found in two distinct areas: Pachuca (west) and Real del Monte (east). The vertical range of mineralization in both areas is about 600 m, and the original cover appears to have been not more than 300 m; thus, the deepest mineralization occurred not more than 1000 m below the then-existing surface. The localization of the ores within the fracture zones seems to have been due to the minor structural features. Fractures that were regular in strike and dip were accompanied by little shattering of the wall rocks and, therefore, little space for ore deposition was developed. On the other hand, irregularities in strike and dip normally were usually joined with broken or shattered zones of different widths and with splits off the main faults; ore was largely deposited in the open spaces so produced.

The veins in the Pachuca-Real del Monte district are grouped into an east system and a north system. The east system contains most of the veins of the district, but those of the north system have been very productive despite their smaller number. The east system, individual veins of which strike between N45°E and S45°E, occurs in both the Pachuca (west) and the Real del Monte (east) areas and includes 16 major veins at Pachuca and 10 at Real del Monte; there are numerous smaller veins and fractures in both areas. The east-system veins normally change in strike along their lengths, some even changing from northeast to southeast, but in most the strike is southeasterly. The north system, occurring only in the Real del Monte area, has many minor veins in addition to 6 major ones; most of these minor veins are splits or branches off the main ones. No extension of the north-system veins south of the Santa Gertrudis-Pinta-Regia vein has been recognized with one minor and isolated exception. The maximum length of the entire district is about

10 km from east to west and about 6 km from north to south; the Real del Monte district occupies a zone 4.5 km from north to south and 3 km from east to west within the northeastern corner of the district as a whole.

The mineralogy of the two systems is much the same, although base-metal sulfides are more abundant in the north veins, and more sulfides are disseminated in the walls of the north system than in those of the east veins. The width of the north veins decreases with depth, in contrast to the east veins. The ore bodies of the north system, however, are more continuous along strike and dip than those in the east system.

In the Pachuca-Real del Monte district, some 75 minerals have been identified, of which the most common metallic ones are pyrite, sphalerite, galena, chalcopyrite, argentite, and acanthite. Polybasite [$(Ag,Cu)_{16}Sb_2S_{11}$] and stephanite (Ag_5SbS_4) are found in less than half the veins of the district (mainly in the Pachuca area), and miargyrite, pyrargyrite, proustite, and sternbergite are rare. Native gold is rare and what is present probably occurs as extremely fine dust. Secondary chalcocite and covellite are common but occur in small quantities. The most abundant gangue mineral is quartz, including amethyst, opal, and chalcedony; it occurs in all veins. Calcite also is present in all veins but generally in small amounts; it may, however, be locally absent or present in large quantities. Albite is found in most of the veins in small amounts but in a few places may be quite abundant. Rhodonite and bustamite ($CaMnSi_2O_6$) are present in many veins but generally in small amounts; they are more common in the north-south veins of Real del Monte and where abundant are associated with rich silver ore. Chlorite, prehnite, kaolin, barite, sericite, adularia, and epidote are present in various amounts in veins in different parts of the district.

The wall-rock alteration in the district is strong and consists of propylitization, albitization, silication, silicification, sericitization, and carbonitization. Generally, wall-rock alteration is more pronounced in the vicinity of the larger veins, but the walls of some veins are less altered than rock volumes at some distance from any known vein. The types of wall-rock alteration in the district are largely controlled by the rocks themselves but are typical of those formed under epithermal conditions.

The process of vein formation in the district took place over a short period of time and was quite uniform and regular throughout the area. Sulfides and gangue minerals overlap extensively in time of deposition. Pyrite was the first sulfide and quartz the first gangue mineral, but calcite and rhodonite-bustamite were abundant in certain locations, and smaller amounts of albite, epidote, and chlorite also were deposited in the early stages. Although most of the deposition was by open-space filling, some replacement of wall rock and wall-rock fragments occurred. The deposition of these minerals was partly followed and partly accompanied by that of small amounts of sphalerite, galena, and chalcopyrite; the silver minerals to some extent accompanied, but also followed, the base metal sulfides. Of the principal silver minerals, argentite and acanthite, acanthite began to deposit first, but the deposition of argentite, which overlapped and followed it, continued for a longer period of time; the deposition of the minor silver minerals, including native silver, accompanied that of argentite and acanthite, and none of them continued after argentite had stopped. Most of the silver minerals filled open space in successive layers; replacement by silver minerals was minor. Native gold was deposited in about the middle of the silver mineral period. All of the gangue minerals except quartz and calcite had ceased to form by the middle of the silver mineral phase; calcite actually continued to deposit after all sulfide mineralization had stopped.

The type of mineralization and wall-rock alteration exhibited at Pachuca-Real del Monte is typically that of the epithermal range, and this is confirmed by the near-surface loci of deposition and by the recent age of the mineralization. The deposits are, therefore, here classified as epithermal. The time and character of mineralization is so much the same in both areas of the district that, though Pachuca and Real del Monte are several kilometers apart, they can be considered as one deposit.

PARRAL, CHIHUAHUA

Late Tertiary	*Zinc, Lead, Copper, Silver, Gold*	*Xenothermal to Kryptothermal*

Allen, V. T., and Fahey, J. J., 1957, Some pyroxenes associated with the pyrometasomatic zinc deposits in Mexico and New Mexico: Geol. Soc. Amer. Bull., v. 68, p. 881-895

Barnes, H. L., 1959, The effect of metamorphism on metal distribution near base metal deposits: Econ. Geol., v. 54, p. 919-943 (particularly p. 934-941

Dauth, H. L., and others, 1936, Mining and geology of San Francisco mines of Mexico: Eng. and Min. Jour., v. 137, no. 8, p. 377-422 (particularly p. 381-382

González Reyna, J., 1946, La industria minera en el estado de Chihuahua: Comité Directivo para la Investigación de los Recursos Minerales de México Bol. no. 7, 152 p. (particularly p. 24-29, 75-77, 86-90)

—— 1956, Santa Barbara-San Francisco del Oro-Hidalgo del Parral: 20th Int. Geol. Cong., Memoria geológico-mineral del Estado de Chihuahua, p. 234-254

Kierans, M. D., 1956, The mines of Cia. Minera Asarco, S. A., at Santa Barbara, Chihuahua: 20th Int. Geol. Cong., Guidebook for Excursion A-2 (Eng. translation), p. 89-95 (Spanish ed., p. 101-108)

Koch, G. S., Jr., 1956, The Frisco mine, Chihuahua, Mexico: Econ. Geol., v. 51, p. 1-40

—— 1956, Geology of the area near San Francisco del Oro and Santa Barbara: 20th Int. Geol. Cong., Guidebook for Excursion A-2 (Eng. translation), p. 84-95 (Spanish ed., p. 97-100)

—— 1956, The Frisco and Clarines mines: 20th Int. Geol. Cong., Guidebook for Excursion A-2 (Eng. translation), p. 96-103 (Spanish ed., p. 109-118)

Koch, G. S., Jr., and Link, R. F., 1960, Zoning of metals in two veins of the Frisco mine, Chihuahua, Mexico: 21st Int. Geol. Cong. Rept., pt. 16, p. 192-199

—— 1963, Distribution of metals in the Don Tomás vein, Frisco mine, Chihuahua, Mexico: Econ. Geol., v. 58, p. 1061-1070

Link, R. F., and Koch, G. S., Jr., 1967, Linear discriminant analysis of multivariate assay and other mineral data: U.S. Bur. Mines R.I. 6898, 25 p.

Lowther, G. K., and Bell, E. B., 1956, Geology of the Esmeralda mine, Parral, Chih.: 20th Int. Geol. Cong., Guidebook for Excursion A-2 (Eng. translation), p. 80-83 (Spanish ed., p. 93-96)

Lowther, G. K., and Marlow, G. C., 1956, Geology of the Parral area: 20th Int. Geol. Cong., Guidebook for Excursion A-2 (Eng. translation), p. 64-79 (Spanish ed., p. 79-88)

Robles, R., 1906, Etude minière de la "Veta Colorada" de Minas Nuevas a Hidalgo del Parral (Etat de Chihuahua): 10th Int. Geol. Cong., Guide des Excursions, no. 22, 15 p.

Schmitt, H. (A.), 1928, Geologic notes on the Santa Barbara area in the Parral district of Chihuahua, Mexico: Eng. and Min. Jour., v. 126, no. 11, p. 407-411

—— 1931, Geology of the Parral area of the Parral district, Chihuahua, Mexico: A.I.M.E. Tr., v. 96 (1931 General Volume), p. 268-290

—— 1935, Structural associations of certain metalliferous deposits in the southwestern United States and northern Mexico: A.I.M.E. Tr., v. 115, p. 36-58 (particularly p. 45-52)

Scott, J. B., 1958, Structure of the ore deposits at Santa Barbara, Chihuahua, Mexico: Econ. Geol., v. 53, p. 1004-1037

Smith, F. W., 1910, Conditions at the Palmilla mine, Parral, Mexico: Eng. and Min. Jour., v. 90, no. 6, p. 259-262

Spurr, J. E., 1923, The ore magmas: McGraw-Hill, N. Y., 2 vols., 915 p. (particularly v. 2, 669-674)

Waitz, P., 1906, Esquisse géologique et pétrographique des environs de Hidalgo de Parral: 10th Int. Geol. Cong., Guide des Excursions, no. 21, 21 p.

Notes

The Parral district is located about 545 km slightly east of south from El Paso and about 900 km northwest of Mexico City. The district actually consists of two subdistricts, one of which lies within an area of 10 by 10 km immediately west and north of the city of (Hidalgo del) Parral and the other of which surrounds the towns of Santa Bárbara and San Francisco del Oro about 20 km southwest of Parral. There is very little known of the geology of the area that lies between Parral proper and Santa Bárbara-San Francisco; therefore, the two subdistricts must be largely discussed as separate geological units.

In the Parral area proper, the oldest rocks are Mesozoic sediments that outcrop in the southern part of the subdistrict and are covered by volcanic rocks in the northern part. At some time after their formation, the Mesozoic sediments in the Parral area were strongly folded, uplifted, and eroded into mountains of considerable relief; this deformation probably was of Laramide age. On the irregular surface of the Mesozoic sediments, volcanic materials of the Escobedo series were accumulated to a maximum depth of as much as 1.6 km with the more mafic rocks being found in the lower part of the sequence and the more silicic in the upper part.

The major intrusive body in the Parral area proper is a stock of quartz monzonite porphyry located in the central portion of the southern half of the subdistrict; it is an irregularly oval-shaped body about 7 km long from north to south and about 4 km wide in its widest portion. Although the monzonite is designated as a stock, it may be much wider at and near the present surface than it is at depth. There are a number of rhyolite dikes in the area that strike between north-south and N22°E and stand almost vertically; they range in width from less than 1 m to more than 30 m and one of them has a length of over 6 km. This longest dike cuts silver-lead-zinc veins in two places but is itself cut by later calcite-fluorite veins.

In the San Francisco-Santa Bárbara area, the principal rock is, as in the Parral area itself, a calcareous shale that contains minor amounts of noncalcareous shale and almost pure limestone; there are a few sand lenses. No intrusive rocks are known in this sainted portion of the district. Rhyolite is present in the San Francisco-Santa Bárbara area as well as around Parral, but none of these dikes (so far as is known) cuts the ore veins. Unconformably overlying the shale (and probably younger than the rhyolite) are small patches of caliche-cemented gravel up to 10 m thick that contain pebbles and other fragments of shale, limestone, rhyolite, and andesite (a rock essentially absent from the immediate limits of the sainted portion of the district). The gravel is unusual in that it is covered by the very recent basalt flows and their diabase-dike feeder-fissures. As is true in the Parral portion of the district, the shale formation was considerably deformed prior to the extrusion of the pre-ore volcanic rocks.

It has been suggested by Schmitt (1928) that the district was covered by 1000 m or less of volcanic rocks (mainly andesites) at the time of ore deposition. This rock almost certainly was broken, at least to some extent, by the same forces that fractured the shales, but the removal of the volcanic rocks by later erosion prevents any determination of the extent of this fracturing or the degree to which such fractures, if they existed, were mineralized. The mineralization, of course, occurred after the fractures had been opened and probably after no more than a moderate interval. The rhyolite and basalt intrusive and extrusive bodies did not develop until after the ore mineralization had taken place; the gravels also were post-ore.

Thus, in the entire district, the ore appears to have formed before most of the rhyolite in the Parral area proper and apparently before all of it in the sainted area. If, therefore, the age of the rhyolites were known, this would almost certainly provide the age of the ores; unfortunately, this information is not available. What is known is that, after the ores had been emplaced, essentially all of the andesite cover in the southwest part of the district and much of it in the Parral area was removed by erosion. Following this erosion the cemented gravels and the late basaltic flows were formed. Since the gravels and basalts probably are Pleistocene or Recent, the end of the pre-gravel erosive period probably was late Pliocene or even early Pleistocene. If the period of erosion was not overly long, not more than a million years, the ores could have been emplaced in early Pliocene time. The alternative is that the erosive period was considerably longer and that the ores were emplaced in the Miocene; they hardly could have been introduced into the area at an earlier time. Because of the considerable relief on the surface on which the post-ore and post-rhyolite gravels were deposited, it would seem that the period of erosion was shorter rather than longer and that the mineralization was probably early Pliocene in age. In central Mexico wherever it has been possible to determine the age of igneous activity generally similar to that which affected the Parral area in post-Cretaceous time, that age has been Miocene to Pliocene. Such an age for the igneous rocks at Parral would be compatible with an early Pliocene age for the Parral ore mineralization; therefore, these ores are here classified as late Tertiary.

The principal mineralized veins of the Parral district proper have been, in order of discovery (1) the Prieta-Tajo-Europa, (2) the Veta Colorada, (3) the Palmilla, and (4) the Esmeralda. There were many other veins of less importance found throughout the district. La Prieta vein is the most important vein in the Prieta subarea and in the entire district for that matter. The minable ore in the Prieta vein is continuous for over a mile, but there are three high-grade ore-shoots within the vein, known (from southwest to northeast) as Prieta, Tajo, and Europa (= T-P-E), respectively. The Esmeralda mine lies on a small westward-trending nose of the quartz monzonite intrusive in the southwest part of the subdistrict.

In the Parral area proper, the principal gangue mineral is silica that normally makes up more than 50 percent of the primary vein material. Although much of the silica is crystalline quartz, chalcedony, jaspery material, silicified wall rock, and a little opal are present. The chalcedony and opal may be supergene. Fluorite is the next most important gangue mineral, having provided 12 to 15 percent of the Veta Colorada vein filling. Barite is quite abundant in some places; it was generally earlier than the quartz and is not closely associated with the sulfides as is fluorite. Galena may be the most abundant sulfide although in the deeper workings reached since Schmitt's work (1931) sphalerite has become more important and well may be more abundant overall than galena. In the primary ore much silver is present; in the Veta Colorada deposits (where sphalerite averaged about 3 percent of the ore as mined) sphalerite was an important source of silver--essentially microscopically pure sphalerite concentrates would carry about 20 ounces of silver per ton. In the southern part of the Parral area (Prieta and Esmeralda mines), the sphalerite does not appear to have been an important source of silver. In these mines the silver recovered from the upper levels came largely from galena and proustite (Ag_3AsS); Schmitt (1931) thinks that the proustite may have been of secondary origin. As the amount of galena, and therefore of silver from it, dropped off with greater depths, enough silver was present in the increasing amounts of copper minerals (chalcopyrite, bornite, and tetrahedrite) to maintain an essentially constant silver content in the ore. The bulk of the copper-associated silver probably is derived from the tetrahedrite, but some may be in the chalcopyrite itself. No primary silver mineral, however, has been identified in either sphalerite or chalcopyrite; galena contains microscopically small blebs of what is probably a silver mineral; this mineral may be either argentite or matildite.

The sphalerite contains microscopic-sized blebs of chalcopyrite, suggesting its deposition under conditions of very high intensity. Some specularite was found in the Veta Colorada ore; it was associated with the quartz in time of deposition and was not important in amount. Pyrite was important in the Prieta, Palmilla, and San Juanico mines but was not common in the others. The age of the massive chalcopyrite (as opposed to that in blebs in the sphalerite) may be younger than the sphalerite and galena, but this relationship has not been certainly determined.

In the San Francisco area the veins are in a system of short fractures, while at Santa Bárbara the fractures are much longer and more persistent. The San Francisco-Santa Bárbara veins are almost entirely confined to fault zones within the Jurassic-Cretaceous shale, the only exception being a Santa Bárbara vein that is in andesite at its outcrop but shortly passes downward into shale. This contrasts sharply with the Parral area where the veins are essentially confined to monzonite and die out on entering shale.

The most common sulfides in the Santa Bárbara area are sphalerite, galena, chalcopyrite, and pyrite, and the most abundant gangue minerals are quartz, calcite, fluorite, and several high-temperature silicates. All the sulfides become coarser with depth (although the chalcopyrite is the finest-grained of the sulfides), and arsenopyrite is much more abundant on the lower levels. The bulk of the silver occurs in the galena and may be argentite or matildite; the locus of gold concentration is uncertain. Schmitt (1928) considered that there were three ore types in the Santa Bárbara area: (1) the lead-silver-zinc type in which the sphalerite and galena are massive and chalcopyrite and pyrite minor in amount; (2) the siliceous gold-silver type in which massive sulfides are essentially lacking and the gangue minerals include abundant silicates; and (3) the massive sulfide-silicate type in which sulfides of type (1) ore are accompanied by abundant high-temperature silicates. The silicates in the ore include garnet, clinoenstatite, manganiferous hedenbergite, epidote, zoisite, and idocrase; orthoclase has been found in several locations within the area. In the San Francisco area, the Frisco mine, comprising some 32 fissure veins, is typical of the subdistrict; the veins are grouped into four sets.

The common metallic minerals of the Frisco mine are sphalerite, galena, chalcopyrite, pyrite, arsenopyrite, gold, hematite, and scheelite; there may be small amounts of bornite, tetrahedrite, and covellite. The silver is probably largely in the galena, but the form it takes there is unknown. The sphalerite averages about 6 percent iron and contains appreciable copper (0.6 percent) in typical exsolution textures (blebs and rods). Much of the chalcopyrite that is not contained in sphalerite veins or rims pyrite or arsenopyrite; it also replaces silicates. Pyrite ranges in amount from 1 to 10 percent. Arsenopyrite is more abundant with depth and, on the lower levels, may compose 50 percent or more of the material in a vein. Hematite is negligible in amount, and scheelite is sparsely disseminated in portions of the veins; it seems to be more common on the lower levels. Quartz probably was deposited through most of the mineralization cycle, but calcite and fluorite appear largely to have been formed late. The silicates were apparently developed early in the mineralization sequence; diopside is probably the most common of the silicates, particularly on the lower levels, and is close to salite in composition. Actinolite is widespread in the veins but is probably less abundant than diopside; it may locally be near tremolite in composition. Epidote is the most widespread silicate but it is not abundant; it may be found in the shales as far into these wall rocks as several meters. Ilvaite [$CaFe_2^{2}Fe^{3}(OH)Si_2O_8$] is common though it is nowhere found in abundance; it normally has replaced silicated fragments of shale within the veins.

The abundance of arsenopyrite in the Santa Bárbara-San Francisco ores, the considerable development of silicate minerals at no great time before the deposition of the ores, the presence of exsolution particles of chalcopyrite in sphalerite, and the presence of scheelite in the San Francisco area ores all point toward at least part of the ore bodies having been formed under high intensity conditions. On the other hand, the presence of considerable tetrahedrite and bornite in the lower levels of the Prieta mine suggests that the final stages of deposition were carried out under much less intense conditions. If the proustite in the upper levels of that mine was primary, it would confirm the existence of a low-intensity, late-stage mineralization. The galena probably was deposited under conditions intermediate between those represented by the high-temperature sulfides and silicates and the lower intensity tetrahedrite and proustite (?). This would suggest that some of the galena at least (and probably some of the sphalerite) was deposited under moderately intense conditions. Because of the rather thin cover, less than 1000 m, over the area at the time of ore deposition and its probably highly fractured character, it is thought that the ores were emplaced by solutions undergoing rapid loss of heat and confining pressure. The Parral deposits, therefore, are here categorized as xenothermal (San Francisco-Santa Bárbara certainly and Parral probably) to kryptothermal (both San Francisco-Santa Bárbara and Parral).

SAN JOSÉ, SAN LUIS POTOSÍ

Late Tertiary *Antimony* *Epithermal*

Archibald, J. C., Jr., 1952, Estudio de los depositos de antimonio y su extracción en las minas de San José, Wadley, San Luis Potosí, México: Conven. Interamer. Recursos Minerales, Mem. 1, p. 64-269

White, D. E., and González Reyna, J., 1946, San José antimony mines near Wadley, State of San Luis Potosí, Mexico: U.S. Geol. Surv. Bull. 946-E, p. 131-153; also published as González Reyna, J., and White, D. E., 1947, Los yacimientos de antimonio de San José, Sierra de Catorce, Estado de San Luis Potosí: Consejo Recursos Naturales no Renovables Bol. 14, 39 p.

Notes

The San José deposits are located in the state of San Luis Potosí on the west flank of the Sierra de Catorce at an altitude of 2500 m; the area is about 660 km nearly due south of Laredo in Texas.

The oldest formation in the area is the Caliza del Fondo that is made up of dark limestone and limey shale. Above the Caliza del Fondo is the Santa Emilia formation that is essentially all limestone and contains the principal antimony ore bodies that are confined almost entirely to the upper 40 m of the formation. Next in order is the San José formation that consists of limey shale in the lower part and of interbedded shale and limestone in the upper portion and conformably overlies the Santa Emilia. The San José grades upward into the Corona formation that is lithologically quite similar to the Santa Emilia but is only slightly mineralized.

Although the age of the beds in the San José area is almost certainly Jurassic and probably Upper Jurassic, this tells nothing as to the age of the mineralization. The fact that the ores appear not to have been involved in the folding probably dates them as post-Jurassic but how long after the Jurassic cannot be certainly determined. The folding probably took place during the Laramide orogeny that affected northern Mexico at the end of the Mesozoic. The Laramide orogeny in northern Mexico, however, does not appear to have been a time of much ore mineralization, whereas the late Tertiary does. The deposits at Zacatecas not far to the northwest were formed in late Tertiary time, and deposits such as those in Santa Eulalia in Chihuahua were emplaced in Cretaceous rocks by late Tertiary mineralizing solutions. This strongly suggests that the San José antimony ores were formed in late Tertiary time, and their near-surface loci of deposition indicate that had they been formed much before the late Tertiary, they would have been removed by erosion. They are, therefore, here classed as late Tertiary.

The most important antimony deposits in the district are found within an area 2 km long by 1 km wide, with the longer dimension running north-south in essentially the direction of the trend of the major folds. The ores are largely, though not entirely, confined to four mineralized limestone beds known as mantos in the upper 40 m of the Santa Emilia formation; these mantos are located at or near intersections with north-west-trending faults and fractures. In the San José district the term *manto* is given to any recrystallized limestone bed containing ore, no matter what the dip may be; in certain restricted rock volumes, the recrystallized volume of limestone may shift from one limestone bed to one adjacent to it. The recrystallization of the limestone of the manto beds probably occurred appreciably before the antimony mineralization for recrystallization is found over a much wider area than that occupied by antimony ore bodies.

The actual development of mantos (ore bodies) within the manto beds appears to have been structurally controlled. Ore solutions probably moved up major faults and fractures until their upward movement was barred by impermeable shale in one wall of the fault, usually the hanging wall, and by similar shale overlying limestone in the other wall. When the upward flow of the ore fluids in the fault channelway was so stopped, it was normally diverted to the limbs of anticlinal structures in the limestones cut by the fault in question, favoring such beds as were broken or contained abundant solution cavities. Once the solutions had moved up the limb to the crest of an anticline, they then changed direction to follow the anticlinal axes upward to the southwest.

The principal primary mineral is stibnite, and it was deposited both by open-space filling and by replacement of limestone; the relative importance of the two processes is uncertain. The stibnite was originally very coarse-grained and was associated with chalcedonic silica and calcite as gangue minerals; the quartz of the gangue probably came from the siliceous zone above the manto bed. After the formation of the primary stibnite, the ore bodies were brought near the surface, well above the water table, and by the action of ground water were converted to antimony oxides, principally cervantite ($Sb_2O_3 \cdot Sb_2O_5$), valentinite (Sb_2O_3), and stibiconite [$Sb_3O_6(OH)$(?)]. A small fraction of the oxides is pseudomorphic after stibnite, but the bulk of the oxide material is hard, earthy, and light colored. After the oxides had been formed in a near-surface environment, the formations were brought appreciably below the water table, and a considerable fraction of the oxide ore was covered by a thin film of cinnabar. This is the only deposit of which I have knowledge in which a hypogene sulfide was deposited after early primary minerals had been brought near enough to the surface for oxidized ore to develop above the water table.

The stibnite and the simple suite of gangue minerals associated with it in time of deposition certainly were deposited close to the then existing surface under what must have been rapid loss of heat from and pressure on the ore fluid. The simple mineralogy also argues for low intensity conditions so that there can be little doubt but that the ore minerals were deposited in the epithermal range, and the deposits are here categorized as epithermal--certainly the late cinnabar was epithermal.

SANTA EULALIA, CHIHUAHUA

Middle Tertiary	*Lead, Silver, Zinc, Tin, Vanadium*	*Xenothermal to Kryptothermal*

Argall, P. (B.), 1903, Notes on the Santa Eulalia mining district, Chihuahua, Mexico: Colo. Sci. Soc. Pr., v. 7, p. 117-126 (abs. in Eng. and Min. Jour., v. 76, no. 10, p. 350-351)

Brinker, A. C., 1913, Geology at Santa Eulalia, Chihuahua: Min. and Sci. Press, v. 106, no. 24, p. 895-896

Fletcher, A. R., 1929, Mexico's lead-silver manto deposits and their origin: Eng. and Min. Jour., v. 127, no. 13, p. 509-513

González Reyna, J., 1946, La industria minera en el estado de Chihuahua:

Comité Directivo para la Investigación de los Recursos Minerales de México Bull. no. 7, 152 p. (particularly p. 24-29, 55-58)

—— 1956, Santa Eulalia--San Antonio: 20th Int. Geol. Cong., Memoria geológico--minera del Estado de Chihuahua, p. 198-211

Hewitt, W. P., 1943, Geology and mineralization of the San Antonio mine, Santa Eulalia district, Chihuahua, Mexico: Geol. Soc. Amer. Bull., v. 54, p. 173-204

—— 1968, Geology and mineralization of the main mineral zone of the Santa Eulalia district, Chihuahua, Mexico: A.I.M.E. Tr., v. 241, p. 228-260

Horcasitas, A. S., and Snow, W. E., 1956, Geological resume of the Santa Eulalia district, Chihuahua, Mexico: 20th Int. Geol. Cong., Guidebook for Excursion A-2 (Eng. translation) p. 37-45

Knapp, M. A., 1906, The fault system of eastern Santa Eulalia: Eng. and Min. Jour., v. 81, no. 21, p. 993-994

Merrill, F.J.H., 1909, Santa Eulalia mines, Chihuahua: Min. and Sci. Press, v. 98, no. 1, p. 37-39

Prescott, B., 1916, The main mineral zone of the Santa Eulalia district, Chihuahua: A.I.M.E. Tr., v. 51, p. 57-99

—— 1925, Sampling and estimating Cordilleran lead-silver limestone replacement deposits: A.I.M.E. Tr., v. 72, p. 666-676

—— 1926, The underlying principles of the limestone replacement deposits of the Mexican province: Eng. and Min. Jour., v. 122, nos. 7, 8, p. 246-253, 289-296 (particularly p. 248-253)

Rice, C. T., 1908, The ore deposits of Santa Eulalia, Mexico: Eng. and Min. Jour., v. 85, no. 25, p. 1229-1233

Signer, C. M., and Hewitt, W. P., 1952, San Antonio mine--landmark on the path of the Conquistadores: Min. Eng., v. 4, no. 5, p. 459-463

Walker, H. A., 1934, Mining methods and costs at El Potosí mine, Chihuahua, Mexico: U. S. Bur. Mines I. C. 6804, 38 p. (particularly p. 1-6)

Weed, W. H., 1902, Notes on certain mines in the states of Chihuahua, Sinaloa, and Sonora, Mexico: A.I.M.E. Tr., v. 32, p. 396-443 (particularly p. 396-399)

Notes

The Santa Eulalia mining district is located about 25 km slightly south of east of the city of Chihuahua, the capital of the state of the same name, and about 345 km slightly east of south of El Paso, Texas. The town of Santa Eulalia (Aquilas Serdan) is at the base of a mountain range, near the 2000 m crest of which are scattered the mine openings of the West Camp (Campo Oeste). The center of the San Antonio mine (Campo Este) lies about 7 km east-northeast of the town of Santa Eulalia; the center of the West Camp is a little over 2 km northeast of Santa Eulalia.

The range in which both camps are located lies near the western edge of the limestone province of the Mexican plateau, trends in a generally north-south direction, and is bounded on the east and west by wide valleys. Of the ore mined from the two camps, 90 percent has come from the West Camp, and 50 percent of the total from both has been mined since 1908; the two camps are separated by 2 km of unproductive ground. The base of the geologic column in the Santa Eulalia district is a quartz monzonite that was intruded into the oldest of the known sedimentary rocks in the area.

Essentially all of the igneous rocks in the district appear to have been introduced before the ores were emplaced. Although there is some controversy as to the channels followed by the ore fluids and although Hewitt (1967) would have the ore fluids derived from magma contaminated by the assimilation of brines trapped in the Evaporite series, it appears probable that the ore fluids were a late development in the source magma chamber at depth, but they may have been derived similarly from the silicic felsite magmas of the sills and the sill-like bodies. Thus, the age of ore emplacement undoubtedly is that of the intrusives, whether or not the intrusive in question was the quartz monzonite magma that underlies the district or the felsites included in the stratigraphic column. Unfortunately, however, no direct evidence is known as to the age of the igneous rocks of the district, nor is it possible to be certain that the quartz monzonite was not intruded into the sedimentary sequence at an earlier time than the sills and sill-like masses of moderately mafic rocks and of silicic felsites. It is not possible, from their relations to each other, to say what the relative ages of the intrusives and the extrusives may be, but there is some mineralization in the Capping series, which makes it almost certain that the Capping series was in place before the sills were intruded. Because the Capping series is Paleocene or Eocene and this series is slightly mineralized, the ore must be later than at least part of the Eocene. If it is considered that the quartz monzonite, or the magma chamber from which the quartz monzonite came, is the source of the ore-forming fluids, the monzonite must have been introduced after the silicic felsite of the sills and sill-like masses in the Limestone series. If, on the other hand, as Hewitt (1967) would prefer, the ores came from the silicic felsites, contaminated by the assimilation of residual brines from the Evaporite series, the age of the quartz monzonite relative to that of the silicic felsites is of no importance. Since Hewitt appeals to the incorporation-of-brine theory to explain how such thin bodies of the no. 1 and no. 2 rhyolites (silicic felsite) sills could furnish all the ore material found in the Santa Eulalia ore bodies, the elimination of the sills as the sources of the ore fluids would obviate the need to believe that the ores were provided by residual brines. It seems much more reasonable to me to assume that the ores came from depth, probably from the quartz monzonite magma chamber, and found their way upward along the paths provided by the no. 1 and no. 2 rhyolites. The no. 3 rhyolite (the main sill) did not provide as ready a path for the ore fluids and, as a result, there are no ores spatially related to the no. 3 rhyolite.

If the quartz monzonite magma chamber was the source of the ore fluids, the deposits bear some similarity in origin to those at Concepción del Oro where the granodiorite magma chamber almost certainly provided the ore fluids. Thus, it would appear best to assume, until the appropriate radioactive dating conforms or refutes it, that the ores at Santa Eulalia came from the quartz monzonite, that the quartz monzonite was middle Tertiary (or the oldest early Tertiary), and that the ores are middle Tertiary rather than late Tertiary as are those of the nearer surface deposits of such districts as Guanajuato, Zacatecas, Cerro de Mercado, and Parral.

The ore deposits in both camps at Santa Eulalia were emplaced by the replacement of limestone and take the forms of mantos, chimneys, replacement veins, and replacements associated with these veins. The district is a classic locality for the study of mantos and chimneys. The replacement veins in the district are less abundant and contain far less ore than the chimneys and mantos. In many instances mantos cross the courses of fissures, and mineralization is best developed in the fissure immediately adjacent to the manto. In places mantos may depart drastically from that form to follow a fissure, with the fissures being mineralized almost exclusively only where they are directly joined to a manto. In some areas ore has extended outward from the fissures into limestone beds that, for some unknown reason, are more susceptible to replacement reactions than are those of the remainder of the stratigraphic sequence. In the massive sulfide replacement bodies of the district, the contacts between ore and limestone are megascopically knife-sharp, but the bedding planes can be seen to pass uninterruptedly from limestone to ore and back to limestone again.

The normal ore of the deep zone and of the southern part of the upper zone is a mixture of pyrite, pyrrhotite, marmatite, and galena; silver sulfide minerals are not present so the silver probably is in the galena (although it will be remembered that at Parral much of it is inexplicably in the sphalerite). Fluorite is present in minor amounts, and chalcopyrite and arsenopyrite are sparsely developed. Gold is present only rarely and then only in traces. Although pyrrhotite is an abundant sulfide throughout the district, in the upper levels, pyrite was predominant. At somewhat greater depths, the two iron sulfides were more or less equal in amount, and still farther down, nearly all the iron sulfide was pyrrhotite, but all pyrrhotites at depth are not equally magnetic. The sphalerite is marmatite with 48 to 49 percent zinc; densely distributed dust in the sphalerite appears, under the microscope, to be pyrrhotite (Hewitt, 1967).

The paragenetic sequence (Hewitt, 1967) appears to have arsenopyrite as the earliest mineral, followed by the iron sulfides and, still later, by marmatite and then galena. Pyrite has been found in microscopic veinlets cutting pyrrhotite, but marmatite, galena, and chalcopyrite show so-called mutual boundaries; there is, in many places, an unidentified mineral along the boundaries between sphalerite and galena that appears to be within the galena; it may be silver-rich tetrahedrite. Calcite is a late mineral, and much of it is not spatially related to ore.

In addition to the primary ores that are almost massive sulfides, there are sulfide-bearing masses of hard black silicate material that contain about 0.8 to 1.6 percent lead and 1.4 to 1.6 percent zinc with 50 to 150 gm of silver per ton. The silicate minerals included ilvaite, knebelite [$(Fe,Mn)_2SiO_4$], and fayalite; some magnetite was found with the fayalite and knebelite. The ilvaite was reported to have been altered to iddingsite (?), chlorite, and chalcedony. This type of mineralization was found in mantos, chimneys, and replacement veins, principally in the upper parts of the district but also, to some extent, in the deeper levels. So far as can be told from Hewitt's (1967) discussion, these silicate-rich portions of the ore probably were formed in the same general period of time as the massive sulfides but may have been formed somewhat earlier.

In the East (San Antonio) Camp, most of the ore has been mined from chimneys (Hewitt, 1943) although moderate tonnages have been recovered from replacement veins. In contrast to the West Camp, the mantos are small and scarce and have been derived from the larger chimneys. Much of the mineralization is concentrated along the contacts of dikes and sills with the limestone. The rhyolite dikes were the most important of such sites of ore deposition, but the andesite and diorite sills also probably exerted control over the movement of the ore-forming fluids.

In contrast to the West Camp, the San Antonio ore bodies, now highly oxidized in the upper levels, consisted originally of skarn minerals and sulfides; in some places the ore minerals were disseminated in a silicate gangue, while in others the sulfides occurred as a massive core, with only minor amounts of silicates, surrounded by a shell of silicates. In still other locations, the ore sulfides were much more massive, and the silicates were present only in small patches.

Another important difference between the East and West Camp ores was the presence of tin in those of the East Camp; some of the East Camp ores averaged 1.5 percent tin, and a few averaged 4.0 or even 5.0 percent tin without any tin mineral being visible megascopically. Tin was found in all types of ore bodies, but in some of each type of ore body, tin was lacking; in those ore bodies that contain it, the tin may have been quite uniformly distributed or it may have been irregularly present. Some areas that include much tin are notably high in fluorite, but many tin-bearing areas lack fluorite, and many fluorite-rich ores have no tin. In areas where calcite-magnetite veinlets are present, tin usually is higher than in areas that lack such veinlets. The tin is largely, if not entirely, in cassiterite. Cassiterite is associated with hematite, topaz, quartz, and, to a lesser extent, magnetite, ilmenite, columbite, muscovite, tourmaline, and calcite; hematite in particular appeared to be closely related in space and time to cassiterite. Hewitt believes that the cassiterite was deposited after the sulfides but probably was part of the same sequence of mineral deposition.

The vanadium in the San Antonio ores was concentrated in the process of ore oxidation. Hewitt believes that the vanadium was introduced as a minor component of some of the silicates and some of the sulfides and was leached from them. The most common secondary vanadium mineral appears to have been vanadinite [$Pb_5(VO_4)_3Cl$], but some descloizite and mimetite also were present.

The ores in the two camps seem to be those of normal high-temperature deposits that were introduced in the East Camp along much the same channels as the earlier hydrothermal fluids that produced the contact silicates. In the West Camp, the ore fluids entered ground that had been reached only in a few places (the masses of hard black silicate material) by solutions capable of producing high-temperature silicates; thus, the ore-forming solutions deposited their loads in limestones in which only local and minor contact silicates had been developed. Even in the West Camp the presence of abundant pyrrhotite and high-iron sphalerite strongly suggests that these ores were deposited in the hypothermal range. The association of the ores in the East Camp with high-temperature silicates not much older than the sulfides and with cassiterite and its accompanying minerals not much younger seems to confirm the diagnosis of high-temperature deposition of the ores of both camps. Hewitt's (1967) suggestions that the ore elements in the ore fluids for the West Camp had been derived from brines trapped in the Evaporite series and had been differentiated out of the highly silicic intrusives that absorbed them seem improbable when the similarity of the sulfide ores of the two camps is taken into account. The East Camp gangue and ore minerals probably were deposited entirely under high-temperature conditions; in the West Camp there is a possibility that some of the ore was formed under intermediate temperatures, as is suggested by the dominance of pyrite over pyrrhotite in the upper levels, the increase of sphalerite in relation to galena with depth, and the lack of exsolved chalcopyrite in the sphalerite or exsolved sphalerite in the chalcopyrite. Hewitt thinks that the ore minerals were deposited in a single period, with the differences between ores at higher and lower elevations being due to progressive reactions and concomitant cooling that took place as the ore fluids moved upward and north through the district. The total vertical distance over which the ore mineralization was deposited appears to have been about 600 m; much of the ore in the upper levels, of course, was highly oxidized before mining began. The Capping series over the mineralized beds probably averaged about 200 m in thickness. It is probable, therefore, that the ores at both Santa Eulalia camps were formed under such near-surface conditions that the ore fluids rapidly lost heat and pressure; therefore, the high- and intermediate-temperature deposits of the district should be categorized as xenothermal to kryptothermal.

ZACATECAS, ZACATECAS

Late Tertiary *Silver* *Epithermal*

Bastin, E. S., 1941, Paragenetic relations in the silver ores of Zacatecas, Mexico: Econ. Geol., v. 36, p. 371-400

Botsford, C. W., 1909, The Zacatecas district and its relations to Guanajuato and other camps: Eng. and Min. Jour., v. 87, no. 25, p. 1227-1228

Burckhardt, C., and Scalia, S., 1906, Géologie des environs de Zacatecas: 10th Int. Geol. Cong., Guide des Excursions, no. 16, 26 p.

Edwards, J. D., 1955, Studies of some early Tertiary red conglomerates of central Mexico: U.S. Geol. Surv. Prof. Paper 264-H, p. 153-183 (particularly p. 179-181)

Flores, T., 1906, Étude minière du district de Zacatecas: 10th Int. Geol. Cong., Guide des Excursions, no. 17, 25 p.

González Reyna, J., 1946, La industria minera en el estado de Zacatecas: Comité Directivo para la Investigación de los Recursos Minerales de México Bol. no. 4, 127 p. (particularly p. 12-20, 105-111)

Mapes-Vázques, E., 1949, Los criaderos minerales de "El Bote," Zacatecas, Zac.: Mex. Inst. Nac. Inves. Rec. Min. Bol. no. 24, 39 p.

Mapes-Vázquez, E., and Stone, J. B., 1956, Notes on the geology of the Zacatecas mining district: 20th Int. Geol. Cong., Guidebook for Excursion A-2 (Eng. translation), p. 114-122

Pérez Martínez, J. J., and others, 1961, Bosquejo geológico de distrito mimero de Zacatecas: Consejo Recursos Naturales no Renovables Bol. 52, 38 p.

Rice, C. T., 1908, Zacatecas, a famous silver camp of Mexico: Eng. and Min. Jour., v. 89, no. 9, p. 401-407

Stone, J. G., 1956, Geology and ore deposits of the Cantera mine, Zacatecas, Mexico: Econ. Geol., v. 51, p. 80-95

Notes

The Zacatecas district, which centers around the city of that name, is located about 320 km northwest of Mexico City and about 375 km south-southwest of Laredo, Texas. The mines of the district essentially surround the city, although minable veins are most abundant south, southwest, west, and north to northeast of Zacatecas.

The rocks exposed in the mountainous area around the city include (1) metamorphosed sediments, believed to be Paleozoic (?) and Upper Triassic (?) of the Carnian stage; (2) marine sediments of Jurassic age (?); (3) continental sediments that are certainly Tertiary and probably are Oligocene to Miocene in age; and (4) similar continental materials of Quaternary age.

La Cantera fault has been mineralized along most of its length, and in La Cantera mine contains workable lead-zinc-silver ore. Because much of the fault has been silicified, it stands above the surrounding country in a series of prominent ridges. The fault strikes about west-northwest; its various segments range in strike from west-southwest to northwest; this segmentation is due to the series of north- to northeast-striking faults that Stone (1956) thinks to be much younger than La Cantera fault and to be post-mineral.

The bulk of the ore in the district, however, has been obtained from a widespread development of N50°W fissure veins and from a conjugate series of east-west veins. These veins, Pérez Martínez' (1961) third structural event, include those of the Loma de Bolsas, El Bote, Mala Noche, and Veta Grande areas and appear to have been formed before the rhyolites were poured out over the red conglomerate.

The last fractures to form in the district, Pérez Martínez' (1961) fifth tectonic event, are the north- to northeast-striking faults that cut La Cantera fault into segments and displaced and rotated the segments of La Cantera fault into the positions they now occupy. These cross faults contain veins of white, glassy, nearly barren quartz that locally carry some gold; included in the veins are silicified fragments of breccia or silicified gouge. As the last event in the tectonic sequence, these faults must be no older than late Miocene and probably are Pliocene. Stone (1956) considers them to be post-ore because (1) the quartz veins cut cleanly across the massive sulfides of the ore-bearing portions of La Cantera fault and (2) the quartz veins, where they cross La Cantera ore segments, contain essentially no lead, silver, or zinc minerals.

The latest movement on La Cantera fault displaces the late (Miocene or Pliocene) rhyolites at La Buia. Since movement along the fault has not crushed or shattered the ore mineralization in it, this movement must have been pre-ore. The ores, therefore, must have been younger than the late rhyolites. On this basis the ores appear to have been introduced in the Pliocene and are, therefore, late Tertiary in age.

Pérez Martínez (1961) divides the mineralization of the district into five zones based on origin, location, and type of mineralization: (1) the southern auriferous zone that includes the vein systems of El Orito, Monte de Oro, and Santa Rita, among others; these veins lie to the southwest of Zacatecas and are confined to the roca verde and the Pliocene rhyolite; they strike generally in a northerly direction; (2) the San Rafael zone that encompasses the northwest-striking veins in the Loma de Bolsas area immediately southeast of the city; these veins are located almost entirely in the red conglomerate; (3) the El Bote zone that is northwest of Zacatecas and consists of a portion of La Veta de la Cantera and less impressive veins of more or less similar strike; on the surface these veins are found in the roca verde, the Paleozoic rocks, and even in the porphyritic rhyolites, but with depth they are mainly in the Paleozoic schists; (4) the Mala Noche zone that is made up of more or less east-west veins located some 2.5 km north of the center of the city; these veins are contained mainly in the roca verde but do cut the porphyritic rhyolite in a few places; and (5) the Veta Grande zone that is about 7.5 km north-northeast of the center of Zacatecas; the veins strike generally northwest and are contained entirely in the roca verde. In addition to the five zones defined above, the deposits of La Cantera mine, near the eastern end of La Cantera fault, should be counted as a sixth zone; this fault-vein strikes about west-northwest and lies along the contact of the red conglomerate (south) and the roca verde (north).

Bastin (1941) studied the paragenesis of mines in the various mineralized zones and produced a generalized paragenetic diagram for the district as a whole. This diagram shows pyrite as the earliest mineral, followed without overlap by sphalerite; about halfway through the period of sphalerite deposition, quartz began to precipitate. The quartz was followed without overlap by a group of minerals including tetrahedrite, galena, chalcopyrite, and argentite; about halfway through the deposition of these minerals, sulfosalts of silver and antimony began to form, and these were followed by calcite. The calcite first deposited before the last of the sulfide minerals and continued to form after all other precipitation had ceased.

The early sphalerite contains tiny inclusions of chalcopyrite; these suggest that the early mineralization took place in the xenothermal rather than the epithermal range. The next group of minerals, the galena, tetrahedrite (locally bournonite is found with tetrahedrite), chalcopyrite, and argentite group may have been formed under kryptothermal or epithermal conditions, although the presence of tetrahedrite and argentite certainly indicates that the intensity of the solutions was no higher than that of the lowest portion of the kryptothermal range. The antimony-bearing sulfosalts of silver, however, overlap appreciably with the galena, chalcopyrite, tetrahedrite, and argentite, which strongly suggests that these four minerals were deposited more probably in the epithermal than in the kryptothermal range. Because of the marked difference in intensity between the ore fluids depositing the early pyrite-sphalerite-galena sequence and those responsible for the later sulfides and sulfosalts, it would seem that the hiatus between the early and later mineralizations was greater than Bastin's (1941, p. 308) diagram would indicate. The silver-antimony sulfosalts in the district include miargyrite, stephanite, pyrargyrite, and polybasite and the lead-silver-antimony sulfosalt freieslebenite; fluorite is a gangue mineral with the sulfosalts. Locally, a little chalcopyrite accompanies the sulfosalts. A study of Bastin's paragenetic diagrams for individual mines and specimens shows numerous differences in sequence, but none of these does appreciable violence to his general diagram.

In the Cantera mine, Stone (1956) considers the paragenesis of the vein minerals to have been quartz and sphalerite with exsolution blebs of chalcopyrite; these last are somewhat surprising since the sphalerite is light yellow-brown and is reported in places to have been formed after galena. The galena is followed by pyrite and quartz and then by second generations of pyrite and quartz. Stone is uncertain as to the position of the silver minerals in the paragenetic sequence.

The primary deposits are here classified as epithermal because the early (xenothermal) sphalerite is not of economic value and because the tetrahedrite and argenite appear to be assigned more properly to the epithermal than to the kryptothermal range.

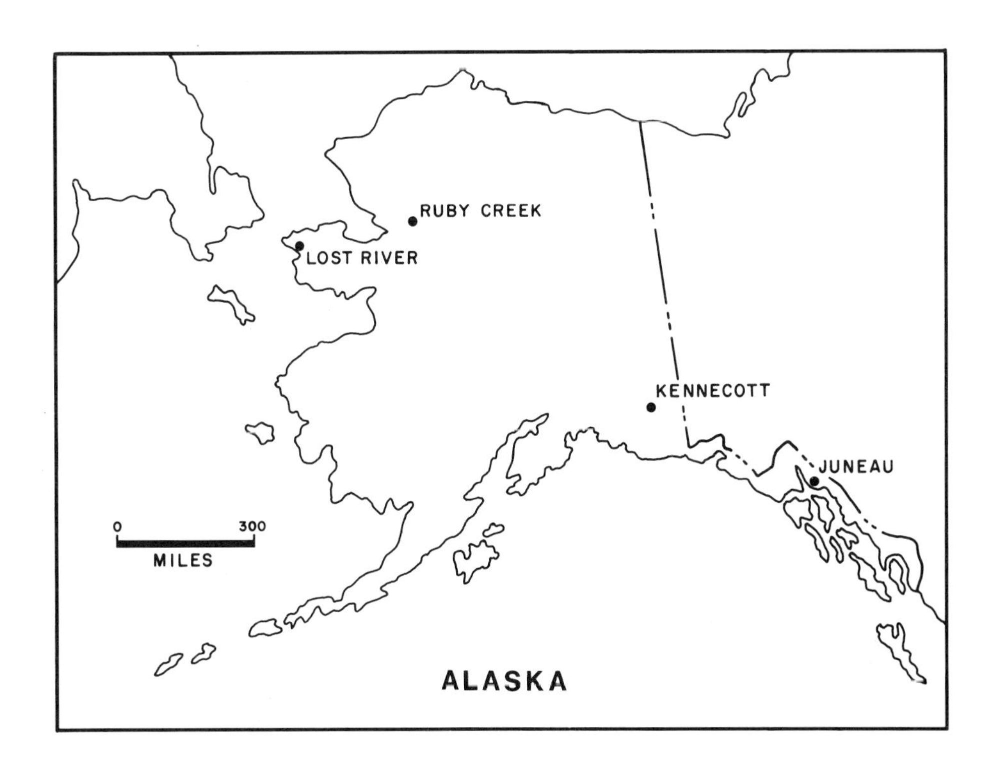
RUBY CREEK
LOST RIVER
KENNECOTT
JUNEAU
0
300
MILES
ALASKA

QUESTA
ARIZONA
NEW MEXICO
BAGDAD
JEROME
IRON KING
MAGDALENA
PHOENIX
CASTLE DOME
MAGMA
GLOBE-MIAMI
RAY
MORENCI
CHRISTMAS
SANTA RITA
AJO
SILVER BELL
SAN MANUEL
TUCSON
PIMA
JOHNSON CAMP
TOMBSTONE
BISBEE
0
100
MILES
TERLINGUA
ARIZONA – NEW MEXICO

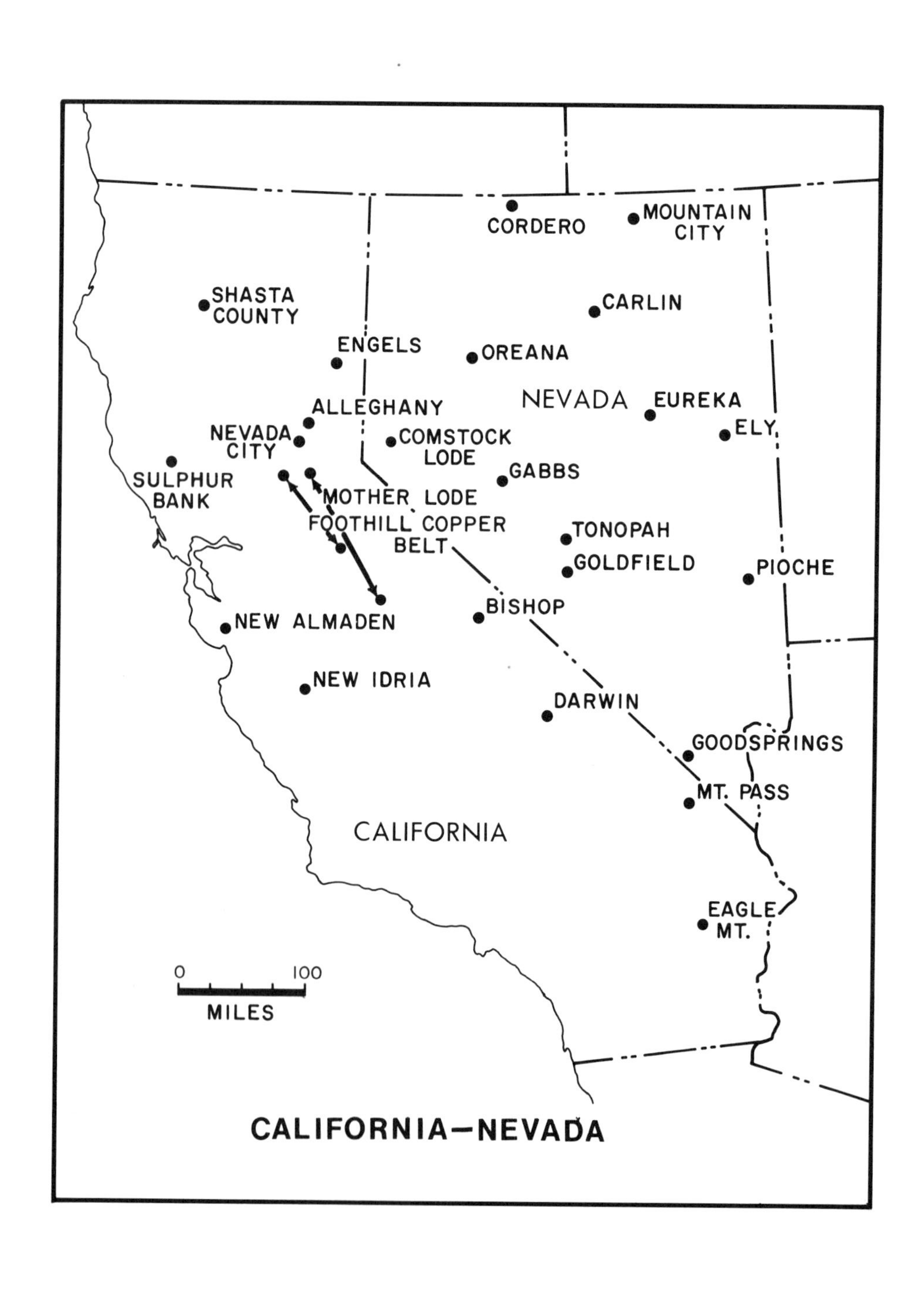

CALIFORNIA—NEVADA

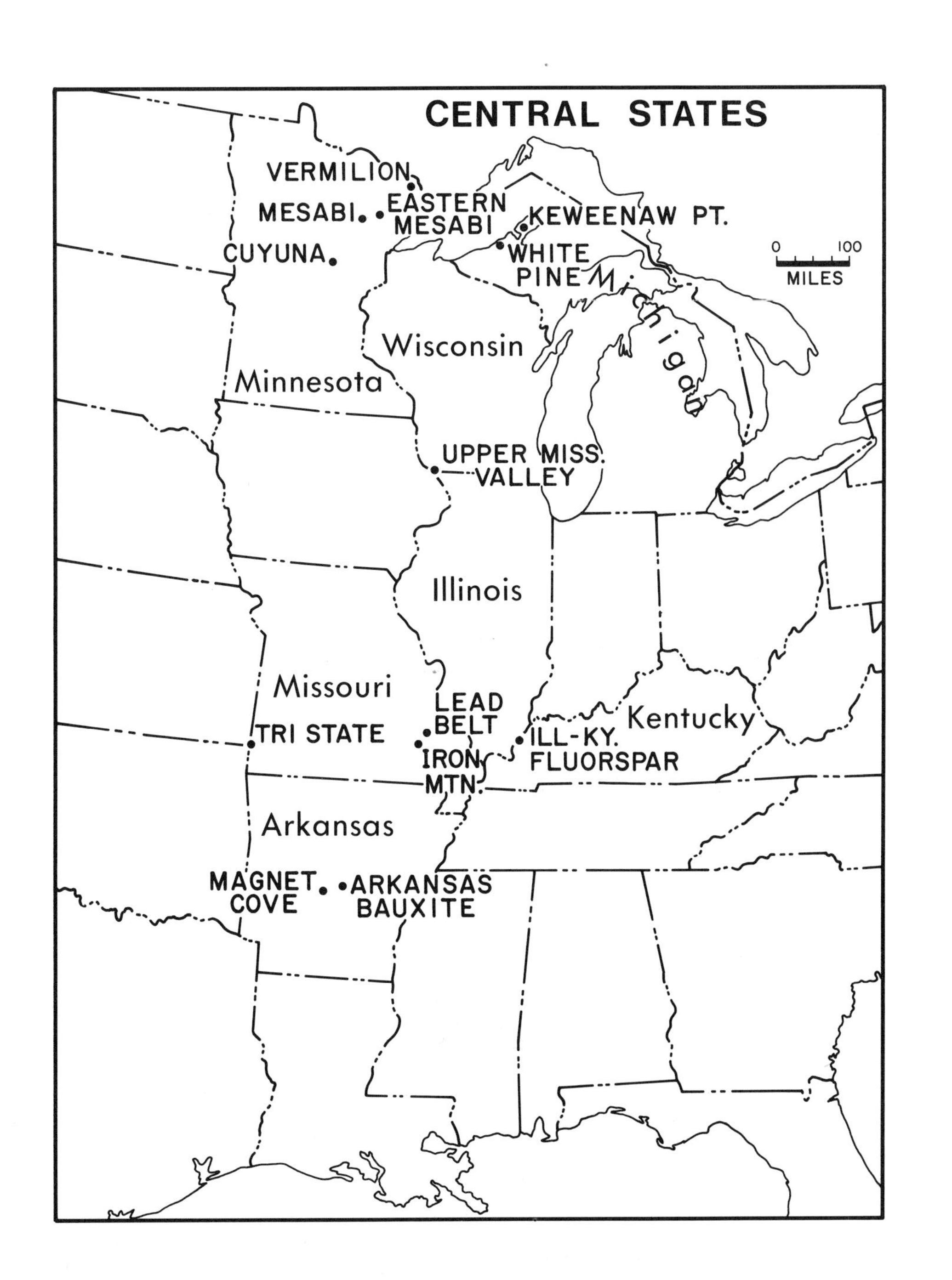
CENTRAL STATES
VERMILION
MESABI
EASTERN MESABI
KEWEENAW PT.
CUYUNA
WHITE PINE
0
100
MILES
Michigan
Wisconsin
Minnesota
UPPER MISS. VALLEY
Illinois
Missouri
LEAD BELT
Kentucky
TRI STATE
ILL-KY. FLUORSPAR
IRON MTN.
Arkansas
MAGNET COVE
ARKANSAS BAUXITE

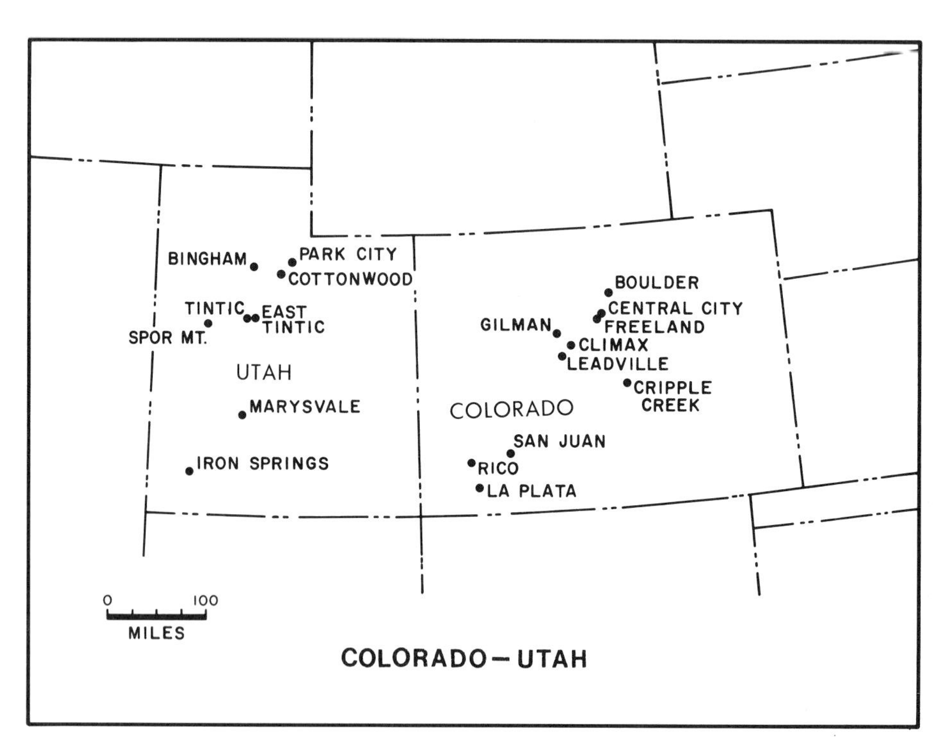
BINGHAM
PARK CITY
COTTONWOOD
TINTIC
EAST TINTIC
SPOR MT.
UTAH
MARYSVALE
IRON SPRINGS
BOULDER
CENTRAL CITY
FREELAND
GILMAN
CLIMAX
LEADVILLE
CRIPPLE CREEK
COLORADO
SAN JUAN
RICO
LA PLATA
0
100
MILES
COLORADO – UTAH

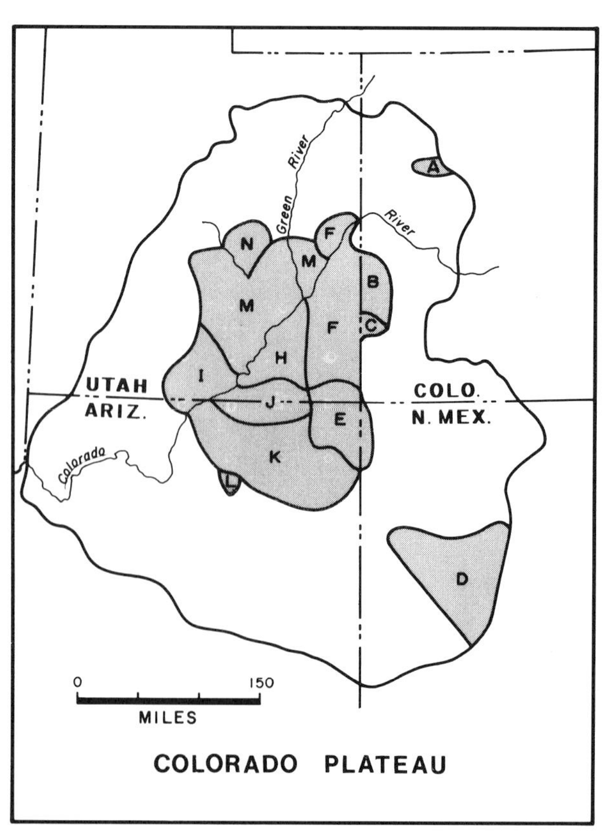
A
Green River
River
N
F
M
B
M
C
F
H
I
J
UTAH
ARIZ.
COLO.
N. MEX.
E
K
L
Colorado
D
0
150
MILES
COLORADO PLATEAU

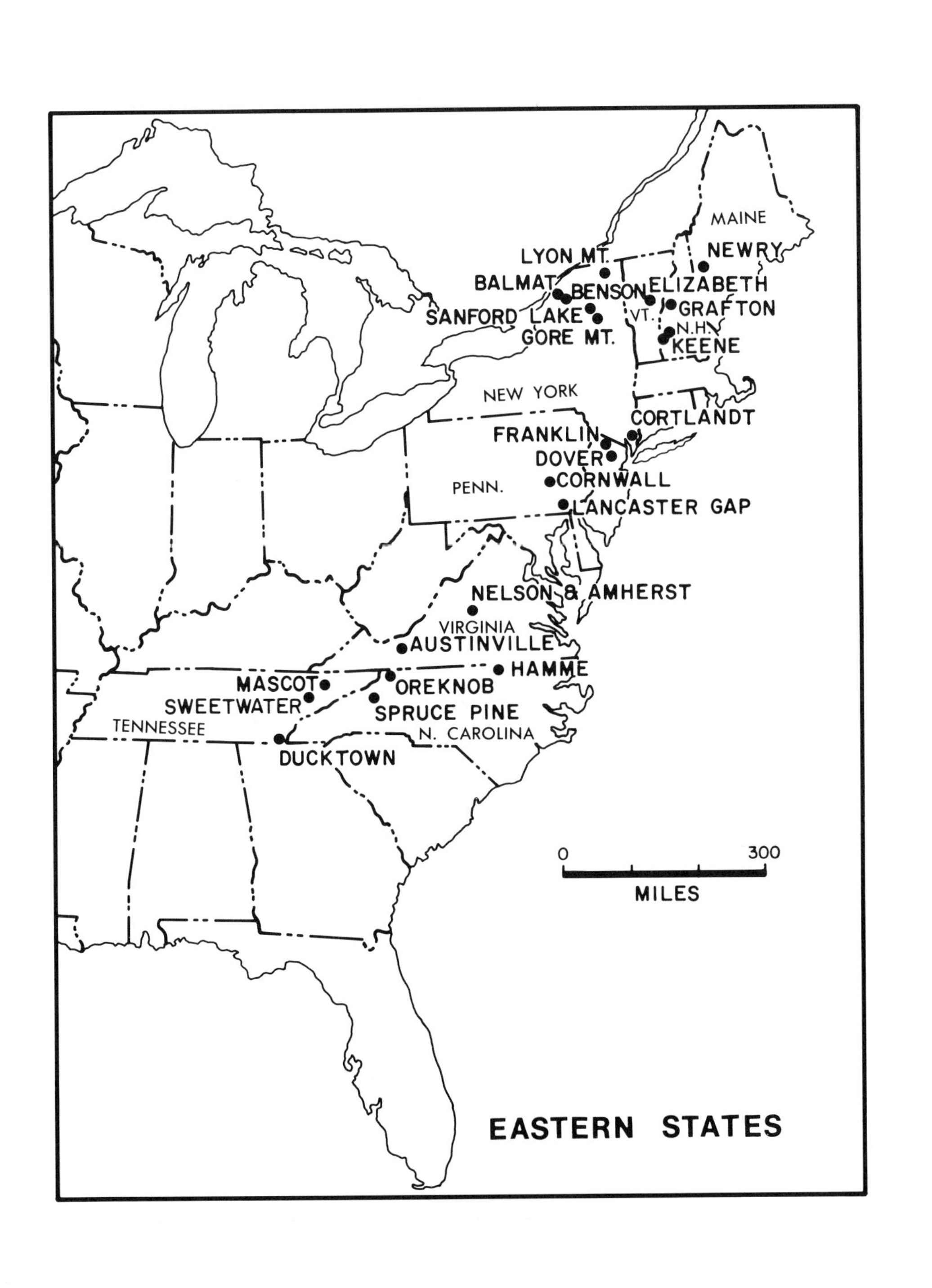
MAINE
NEWRY
LYON MT.
BALMAT
BENSON
ELIZABETH
GRAFTON
SANFORD LAKE
VT.
GORE MT.
N.H.
KEENE
NEW YORK
CORTLANDT
FRANKLIN
DOVER
PENN.
CORNWALL
LANCASTER GAP
NELSON & AMHERST
VIRGINIA
AUSTINVILLE
HAMME
MASCOT
OREKNOB
SWEETWATER
SPRUCE PINE
TENNESSEE
N. CAROLINA
DUCKTOWN
0
300
MILES
EASTERN STATES

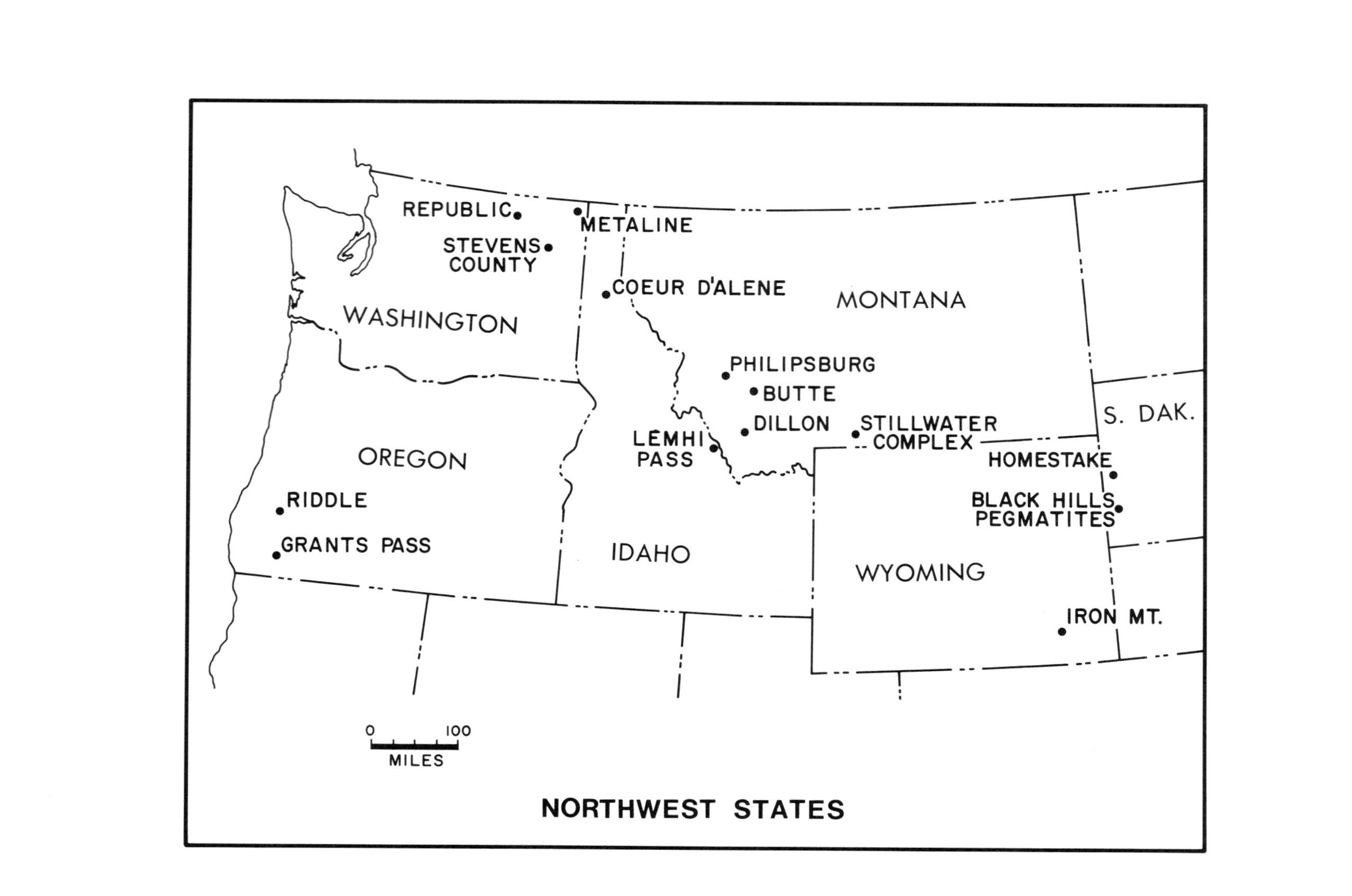

NORTHWEST STATES

Alaska

JUNEAU-TREADWELL

Early Tertiary | *Gold* | *Hypothermal-1 to Mesothermal*

Barker, F., 1957, Geology of the Juneau (B-3) quadrangle, Alaska: U.S. Geol. Surv. G. Q. 100, 1:5280 (immediately west of the Juneau-Treadwell area)

Buddington, A. F., 1927, Coincident variations of types of mineralization of Coast Range intrusives: Econ. Geol., v. 22, p. 158-179 (general)

Buddington, A. F., and Chapin, T., 1929, Geology and ore deposits of southeastern Alaska: U.S. Geol. Surv. Bull. 800, 398 p. (particularly p. 70-72, 157-158, 298-299, 364-366)

Forbes, R. B., and Engles, J. C., 1969, K^{40}/Ar^{40} age relations of the Coast Range batholith and related rocks of the Juneau ice field area, Alaska: Geol. Soc. Amer. Bull., v. 81, p. 579-584

Lathram, E. H., 1959, Progress map of the geology of the Juneau quadrangle, Alaska: U.S. Geol. Surv. Misc. Geol. Invest. Map I-303, 1:250,000

Mertie, J. B., Jr., 1921, Lode mining in the Juneau and Ketchikan districts: U.S. Geol. Surv. Bull. 714, p. 105-128

Simmersbach, B., 1919, Der Quarzgoldbergbau in Küstenstreifen von Südost-Alaska: Zeitsch. f. prakt. Geol., Jg. 27, H. 4, S. 62-69 (largely mining and historical)

Spencer, A. C., 1906, The Juneau gold belt, Alaska: U.S. Geol. Surv. Bull. 287, p. 1-137 (particularly p. 69-73 [Juneau], 86-116 [Treadwell])

Wayland, R. G., 1960, The Alaska-Juneau gold ore body: Neues Jb. f. Mineral., Abh. Bd. 94 (Festband Ramdohr), 1. Hälfte, S. 267-279

Wernecke, L., 1932, Geology of the ore zones, Alaska Juneau mine: Eng. and Min. Jour., v. 133, no. 9, p. 493-499

Notes

The Juneau and Treadwell gold properties lie on opposite sides of Gastineau Channel, immediately south of the city of Juneau in the Alaskan Panhandle; the Juneau mine is on the mainland and the Treadwell mine on Douglas Island.

The Juneau ore body lies in the Triassic or older Perseverance slate, mainly a carbonaceous, graphite phyllite. Several irregular gabbroic dikes and sills (now schistose and altered to hornblende and andesine) cut through the phyllite. The Perseverance formation appears to be younger than the Clark Peak schist to the northeast and older than the Gastineau volcanic rocks to the southwest. In the Juneau area the beds strike northwest and dip steeply northeast, indicating that folding has overturned the beds; faults are generally nearly parallel to the bedding or to the schistosity and are post-ore. One transverse fault strikes east-west and displaces the north end of the ore about 1800 feet to the west and probably has dropped it downward by nearly the same amount. In the mine area, the slate has been subjected to a low-grade regional metamorphism and the basic flows and tuffs have been converted to greenstone. The Coast Range batholith that intruded the Clark Peak schist to the east of the mine area may have been responsible for this metamorphism.

The Treadwell mine is on the west side of Gastineau Channel also in the Perseverance formation. The ore has been largely localized in irregular and highly shattered dikes of albite diorite that are up to several feet wide and that largely parallel the northwest-southeast strike of the beds. The albite diorite (and unmineralized diorite porphyry) dikes probably are of the same general age and origin as the Coast Range batholith.

The mineralization in the area probably is genetically related to the granodiorites and quartz monzonites of the Coast Range batholiths. In areas far removed from southeastern Alaska, similar batholiths are known to cut fossiliferous rocks of the latest middle Cretaceous. Radioactive age determinations on the Coast Range igneous rocks in the Juneau area (Miller, 1946) suggest an age of 52.8 ± 2.6 m.y. to 46.9 ± 1.4 m.y. which, although taken from rocks at least 30 miles north-northeast from Juneau in the East Marginal pluton of the Coast Range batholith, probably indicate the general time of emplacement of the entire composite intrusive mass. Crystalline schists in the Juneau area give ages of 57 to 60 m.y., and the formation of the ores probably occurred not long after the metamorphism had been completed. Granted these K/Ar determinations are correct, this would place the mineralization in the early Eocene, late enough to be categorized as early Tertiary rather than late Mesozoic to early Tertiary, so the ores are here designated early Tertiary.

The ore in the Juneau deposit is found in four zones of abundant quartz veins, a few inches to a few feet wide and a few feet to several tens of feet long; from northwest to southeast these ore zones are the Ebner, the North, the South, and the Perseverance. The movement of ore fluids and the deposition of the ores have been concentrated where there are (1) small ore medium-sized altered mafic igneous rocks in the phyllites, (2) zones of extensive hydrothermal alteration, (3) shear and drag folds, and (4) shears or secondary cleavage opened by fracturing. Almost all the gold is in or along quartz veins, with a little being found in hydrothermally altered igneous rocks away from the veins; essentially no gold is in the phyllite. Sulfides, especially pyrrhotite, are more widely distributed than the gold and exhibit a definite zoning from north to south. Pyrrhotite is the dominant sulfide throughout the length of the ore zones, while pyrite and sphalerite appear in the northwest end of the North ore body; farther south in that ore body galena is found. In the South ore body, the proportion of galena and sphalerite to pyrrhotite rises and arsenopyrite is locally important. In the Perseverance ore body, galena, sphalerite, and arsenopyrite reach their greatest abundance in relation to pyrrhotite.

The paragenetic sequence was biotite, quartz, ankerite (with minor chlorite, garnet, sericite, albite, carbonates, tourmaline, and rutile), pyrite, arsenopyrite, pyrrhotite, sphalerite, galena, and gold (with minor early specularite and marcasite and later tetrahedrite, chalcopyrite, and possibly aikenite). Exsolution blebs of pyrrhotite and chalcopyrite are found in sphalerite indicating that it was formed under hypothermal conditions. Gold is the latest ore mineral, replacing all sulfide and gangue minerals but most commonly attacking sphalerite. Gold is often found in the same areas as galena but does not replace that mineral as commonly as it does sphalerite; pyrrhotite often contains considerable gold. The late deposition of the gold which was even later than the galena, suggests that it was formed under less intense conditions than the sulfides, probably on the borderline between the mesothermal and hypothermal ranges, and it is so classified here.

The Treadwell deposits are located largely in fracture-veinlets in the shattered albite-diorite dikes; these are filled mainly with calcite and quartz, but some masses of the diorite are replaced by sulfides. The quartz-calcite veinlets provided the bulk of the ore, and the veinlets usually contain about 2 percent of metallic minerals. The principal metallic mineral is pyrite, although considerable amounts of pyrrhotite and magnetite are present. Chalcopyrite, galena, and sphalerite occur in small quantities. The gold is present, largely if not entirely, in the native state and apparently was deposited late in the mineralization cycle. The principal hydrothermal alteration of the host rock dikes has been the conversion of their original albite-oligoclase to albite, although minor zoisite, calcite, and pyrite probably were introduced at the same general time. The early metallic and nonmetallic minerals (albite, pyrite, pyrrhotite, and magnetite) suggest hypothermal conditions, but the gold probably was deposited in the same intensity range as that in the Juneau mines--hypothermal-1 to mesothermal.

KENNECOTT

Early Tertiary *Copper* *Telethermal*

Bateman, A. M., 1942, The ore deposits of Kennecott, Alaska, in Newhouse, W. H., Editor, *Ore deposits as related to structural features*: Princeton Univ. Press, p. 188-193

Bateman, A. M., and Lasky, S. G., 1932, Covellite-chalcocite solid solution and ex-solution: Econ. Geol., v. 27, p. 52-86 (particularly p. 52-61, 81-86)

Bateman, A. M., and McLaughlin, D. H., 1920, Geology of the ore deposits of Kennecott, Alaska: Econ. Geol., v. 15, p. 1-80

Forbes, R. B., and Barsdate, R. J., 1969, Trace metal zonation in a native copper nugget from the McCarthy district, Alaska: Econ. Geol., v. 64, p. 455-458

Lasky, S. G., 1929, Transverse faults at Kennecott and their relation to the main fault system: A.I.M.E. Tr., v. 85 (1929 Yearbook), p. 303-317

—— 1930, A colloidal origin of some of the Kennecott ore minerals: Econ. Geol., v. 25, p. 737-757

MacKevett, E. M., Jr., 1965, Factors of probable significance in the genesis of copper deposits in the Kennecott district, Alaska (abs.): Geol. Soc. Amer. Program 1965 Ann. Meetings, p. 98-99

MacKevett, E. M., Jr., and Radtke, A. S., 1966, Hydrothermal alteration near the Kennecott copper mines, Wrangell Mountains area, Alaska--a preliminary report: U.S. Geol. Surv. Prof. Paper 550-B, p. B165-B168

MacKevett, E. M., Jr., and Smith, J. G., 1968, Distribution of gold, copper, and some other metals in the McCarthy B-4 and B-5 quadrangles, Alaska: U.S. Geol. Surv. Circ. 604, 25 p.

Miller, D. J., 1946, Copper deposits of the Nizina district, Alaska: U.S. Geol. Surv. Bull. 947F, p. 95-120

Moffit, F. H., 1911, Geology and mineral resources of the Nizina district, Alaska: U.S. Geol. Surv. Bull. 448, 111 p. (particularly p. 75-97)

—— 1924, The metalliferous deposits of Chitina Valley, in *Mineral resources of Alaska, 1922*: U.S. Geol. Surv. Bull. 755, p. 57-72

—— 1935, Copper River basin, in *Copper resources of the world*: 16th Int. Geol. Cong., v. 1, p. 139-142

—— 1938, Geology of the Chitina Valley, Alaska: U.S. Geol. Surv. Bull. 894, 137 p. (particularly p. 107-121 [descriptive geology, p. 19-107])

Notes

The Kennecott deposits are located on the south slopes of the Wrangell Mountains, about 100 miles northeast of Cordova.

The oldest rocks in the area are those of the early Carboniferous Streina formation consisting of argillite, slate, limestone, basic lava flows and tuffs, and granular intrusives, all more or less metamorphosed. The Streina is unconformably overlain by 3000 to 5000 feet of Permian or Triassic Nikolai greenstone, originally made up of diabasic lava flows. It was followed conformably by the Upper Triassic Chitistone limestone, some 3000 feet thick. It, in turn, was followed by about 3000 feet of Upper Triassic McCarthy shale and by some 7000 feet of Upper Jurassic Kennecott formation, made up of shales, sandstones, and conglomerates. At some time after the end of sedimentation in the Upper Jurassic, the area was intruded by stocks, dikes, and sills of quartz-diorite porphyry. The ores are found in the basal portion of the Chitistone limestone but appear to have been hydrothermally introduced at a later time. Although no definite evidence relates these ores to a specific igneous source, it appears most probable that they were formed in connection with the intrusion of the post-Lower Cretaceous quartz diorites that occur in the general area of the mines. Dikes of this igneous rock actually cut the ore and are not displaced by post-ore faulting, so the dikes are not only later than the ore but are also later than the post-ore faulting. A quartz-diorite porphyry stock lies to the south of the mines; its age in relation to the ores is unknown. The quartz-diorite dikes are entirely post-ore, but this dike material may represent a last stage of a period of igneous activity of which the stock may be an earlier manifestation. Granted this reasoning may be correct, the ores would be of the same general age as the Coast Range batholith. The latest work on that composite intrusion (Forbes and Engles, 1969) suggests that it actually is early Eocene rather than belonging to the older epoch to which it was assigned by such workers as Buddington and Chapin (1929); therefore the Kennecott ores are here categorized as early Tertiary. If the ore was formed from copper remobilized from the Nikolai greenstone, it probably is of essentially the same age.

The ore bodies in the gently northeastwardly dipping Chitistone limestone are of three types: (1) veins formed in and around steeply inclined fissures that strike at right angles to the bedding, partly by open-space filling but mainly by replacement of the wall rock; the veins are usually widest at their bases which are normally bedding-plane faults; (2) massive replacements of favorable beds of dolomitized limestone; their bases are formed by flat bedding-plane faults and their sides and tops irregularly bounded by unreplaced carbonate rocks; and (3) stockwork veinlets mineralized by open-space filling, with only a minor amount of replacement associated with them; most of these stockworks resulted from the feathering out of the upper segments of the replacement veins. The massive replacements extend out from fissures and, in favorable beds, may be surrounded by disseminated ore; some of the replacement deposits may consist of disseminated ore only. Some talus ore was recovered. All of the ore bodies were restricted to the basal portion of the Chitistone limestone, 50 to 400 feet above its contact with the underlying Permo-Triassic Nikolai greenstone. None of the overlying formations contains ore. The first minerals developed in the ore bodies were widespread, though small, amounts of chalcopyrite and bornite; these minerals were followed by minor luzonite and tennantite, and then by additional small quantities of chalcopyrite that often rings the tennantite, and still later by a little enargite and calcite. The rocks and their contained minerals were then fractured, and large amounts of covellite were introduced as replacements both of limestone and of sulfides. Finally chalcocite (orthorhombic) and digenite were emplaced in large amounts, replacing all earlier minerals but favoring the limestone. There is essentially no pyrite, and there are, for all practical purposes, no gangue minerals. Although the minor quantities of the copper sulfosalts probably were deposited under mesothermal to leptothermal conditions, the only economically valuable materials, the low-temperature simple copper sulfides, were formed under telethermal conditions. It has been suggested that the copper minerals were derived from the underlying greenstone (which does contain up to 0.8 percent of copper), but the mechanisms put forward to achieve extraction and removal of the Chitistone formation seem inadequate; the ores are, therefore, here classed as telethermal.

The massive ore bodies are composed almost entirely of sulfides, but the sulfides in the less thoroughly mineralized rock volumes have been considerably altered to azurite and malachite; perhaps 25 percent of the copper occurs as these copper carbonates. The oxidation has been found to affect the ores as far as 6800 feet down the dip of the beds or down the pitch of the lodes; oxidation certainly was preglacial in age, and there is no more than negligible supergene enrichment.

LOST RIVER

Early Tertiary | *Tin, Beryllium* | *Hypothermal-1 (Sn), Hypothermal-2 (Be)*

Fearing, F. C., 1920, Alaska tin deposits: Eng. and Min. Jour., v. 110, no. 4., p. 154-158

Harrington, G. L., 1919, Tin mining in Seward Peninsula, Alaska: U.S. Geol. Surv. Bull. 692, p. 353-363

Heide, H. E., 1946, Investigation of the Lost River tin deposit, Seward Peninsula, Alaska: U. S. Bur. Mines R. I. 3902, 57 p. (mimeo.)

Hess, F. L., 1906, The York tin region: U.S. Geol. Surv. Bull. 284, p. 145-147

Knopf, A., 1908, Geology of the Seward Peninsula tin deposits, Alaska: U. S. Geol. Surv. Bull. 358, 71 p. (particularly p. 44-58)

Mertie, J. B., Jr., 1917, Lode mining and prospecting on Seward Peninsula, Alaska: U.S. Geol. Surv. Bull. 662, p. 425-449

Sainsbury, C. L., 1960, Metallization and post-mineral hypogene argillization, Lost River tin mine, Alaska: Econ. Geol., v. 55, p. 1478-1506

—— 1964, Geology of Lost River mine area, Alaska: U.S. Geol. Surv. Bull. 1129, 80 p.

—— 1965, Beryllium deposits of the western Seward Peninsula, Alaska: U.S. Geol. Surv. Circ. 479, 18 p.

—— 1968, Tin and beryllium deposits of the central York Mountains, western Seward Peninsula, Alaska, in Ridge, J. D., Editor, *Ore deposits of the United States, 1933-1967* (Graton-Sales Volumes): Chap. 74, v. 2

—— 1969, Geology and ore deposits of the central York Mountains, western Seward Peninsula, Alaska: U.S. Geol. Surv. Bull. 1287, 101 p.

Steidtmann, E., and Cathcart, S. H., 1922, Geology of the York tin deposits, Alaska: U.S. Geol. Surv. Bull. 733, 130 p. (particularly p. 44-81)

Notes

The tin-tungsten-beryllium deposits of the Lost River area are located on the south slope of the York Mountains about 28 miles east-southeast of the western tip on the Seward Peninsula and about 90 miles northwest of Nome. The amount of lode tin that has been mined from the area is not much over 350 tons, and tungsten has never been recovered in amounts worth recording, but appreciable tonnages of this ore might be mined in the future. The potential for beryllium production, however, appears to be large.

Although it was thought in 1922 (Steidtman and Cathcart) that the Ordovician limestones of the Lost River area were centrally located in a huge, north-trending syncline surrounded on three sides by older slates, recent work (Sainsbury, 1968) has shown that the slate-limestone contacts are thrust faults. These pre-Ordovician slates, as they were designated (Knopf, 1908), actually consist of two formations: (1) slates, phyllites, siltstones, slaty limestones, graywackes, and minor argillaceous limestones that are moderately to intensely deformed and (2) younger thin-bedded argillaceous and dolomitic, silty, and shaly limestones that contain quartz-carbonate veins and veinlets. In pre-Ordovician times, these rocks were intruded by plugs, sills, and dikes of mafic rocks, originally mainly gabbro (but now largely converted to greenstones). Later, Early to Middle Ordovician rocks that are equivalent to part of the Ordovician and Silurian Port Clarence limestone were deposited in the area. These formations include a thick basal section of thin-bedded argillaceous and siliceous limestone, dolomitic limestone, and carbonaceous limestone. This thin-bedded rock was succeeded by a massive to thick-bedded micritic limestone with local chert nodules. The next and last of the Early Ordovician rocks is, at its base, a black shale-siltstone that grades upward into a black limestone and then into a largely dolomitized gray limestone. The final (probably Middle) Ordovician formation is a gray medium-bedded limestone. Prior to the introduction of much later igneous rocks, the area was the scene of extensive thrust faulting. Thus, not only are the contacts of the pre-Ordovician rocks with those of Ordovician age along thrust planes, but the Ordovician rocks also are cut by thrusts as is particularly evident in the Lost River area (Sainsbury, 1968). In places, the thrusts break down into zones of imbricate thrusting, and the thrust plates are cut by several sets of normal faults, on some of which the vertical displacement is more than 1000 feet. Folding on any appreciable scale is confined to the pre-Ordovician rocks and occurs in the younger beds only near the low-angle thrusts.

In the Lost River area of the central York Mountains, two main thrusts have been identified (Sainsbury, 1968): (1) the Mint River that marks the northern margin of the York Mountains; it has brought Ordovician rocks over pre-Ordovician formations that are exposed to the north of the range and (2) the Rapid River that can be followed through the central part of these mountains; the intersections of this fault with later lamprophyre dikes localize beryllium-bearing fluorite deposits. Where younger, thick-bedded Ordovician limestones are thrust over older thin-bedded (also Ordovician) limestones, the latter generally are much drag folded below the thrust. Where a thrust passes through thick-bedded limestones, the fault plane is marked by macrobreccias of jumbled, angular blocks of limestone that were very porous and, where intruded by dikes, were appreciably mineralized. Of the four sets of normal faults that cut the thrust-faulted limestones, the one that has localized the most ore strikes N65°E to N85°E through the central part of the York range. Some of the dikes intruded into these faults were broken by later earth movements, the last of which was post-mineralization.

After (and during) thrusting, but before the ore was emplaced, much of the limestone in the vicinity of the major thrusts was dolomitized; since this dolomite exhibits no regular relationship to ore, Sainsbury (1968) believes it to be genetically unrelated to the ore-forming process; the dolomitization may, however, be connected with some phase of the igneous activity.

After the thrust faulting stopped, the region was intruded by stocks of biotite granite (earlier) and dikes and plugs of rhyolite, rhyolite porphyry, diabase, and lamprophyre (later). One of the granite stocks underlies the Lost River mine; locally, it is mineralized. Thermal metamorphism around the stocks produced biotite-andalusite hornfels in the shaly rocks and calc-silicate rocks or marble from the carbonates. In places where material was brought into the limestone through igneous activity, tactites were formed that range from garnet-hedenbergite-diopside-fluorite skarn to a mainly vesuvianite-fluorite type. Here and there, the tactite is largely a magnetite-fluorite-amphibole rock that shows pronounced banding (ribbon rock).

During and after the granite intrusions, the rhyolites were introduced, generally as dikes in fractures near the granites. Partly contemporaneous with the rhyolites were lamprophyres and diabase dikes; these generally are in fractures of regional extent but may be grouped around the stocks. The lamprophyres probably are younger than the granites and contain unusually large amounts of tin, beryllium, boron, and lithium (as do the granites). Sainsbury (1968) believes that the lamprophyres acquired these elements through contact with granite at depth, having been diabase prior to this. These lamprophyre dikes and the fractures they occupy were the principal channelways along which the ore fluids migrated to the loci of ore deposition.

The granite stock at Brooks Mountain (about 7 miles northwest of the Lost River mine) has been given a biotite-derived K/Ar date by the U.S.G.S. that places it in the Late Cretaceous. Sainsbury (1968) does not say what that date is. After the granite was intruded, much of the normal faulting and the emplacement of the two types of dikes occurred before the ores were deposited. How much time was required by these events is uncertain. Sainsbury (1968) assigns the ore to the last of the Laramide stage (Paleocene), but here, in agreement with the dates given to deposits far to the south in the Alaskan Panhandle, the Lost River ores are placed in the early Tertiary (Eocene).

The tin-tungsten-beryllium mineralization of the Lost River area is far more complex (Sainsbury, 1968) than was earlier thought (Steidtman and Cathcart, 1922). The recognition (Sainsbury, 1964) of important quantities of beryllium minerals in the limestone surrounding the tin-tungsten deposits in, and adjacent to, the greisenized granite has added greatly to the potential worth of the district.

In the Lost River mine, the locus of ore deposition appears to have been the intersection of a minor thrust fault with the Cassiterite (rhyolite) dike. Four distinct types of tin-tungsten ore were formed: (1) high-grade tin ore in the quartz-mica-topaz greisen to which the Cassiterite dike was altered; the ore is contained principally in replacement veins of greisen in earlier massive greisen; (2) tungsten-rich ore in the greisenized dike below the main tin ore shoot; the wolframite seems to occur in veinlets similar to those that contain the cassiterite ore; some overlap between the two types exists--none of the tungsten ore in this category appears to have been exploited; (3) low-grade tin ore in irregular masses of greisen in the buried granite under the dike or in numerous small greisen veinlets in joints in the outer shell of the granite; and (4) low-grade tin-tungsten ore in limestone overlying the granite in a complex zone of small veins and veinlets that have filled fractures in altered limestone and have replaced that rock outward from them.

The Lost River greisen shows less than the usual amounts of mica and is made up largely of quartz and topaz, with quartz generally being more abundant; appreciable amounts of fluorite are associated with the white mica, and some tourmaline is present. The higher the sulfide content in the greisen, the lower is that of white mica and vice versa. The later greisen veinlets that cut the massive greisen often are rich in cassiterite and include a wide variety of sulfide minerals. The paragenetic sequence of these latter minerals is pyrite (replacing mica); then cassiterite as euhedral crystals and as anhedral grains; the sulfides follow the cassiterite and are essentially contemporaneous; they include arsenopyrite, iron-rich sphalerite, galena, chalcopyrite, stannite, and molybdenite and are accompanied by rutile and magnetite. The deposition of wolframite probably began slightly later than that of cassiterite and may have continued for an appreciable time after. Late pyrite formed as euhedral crystals independently of the other sulfides.

The greisenization of the granite, the silication of the limestone, and the ore mineralization were followed by a stage of extreme argillaceous alteration in which both granite and limestone were affected. In the Cassiterite dike, the argillization extended far beyond the limits of commercial ore, and in places, the limestone, whether silicated or not, was extensively converted to clay minerals. The granite mass in the mine area also was much argillized but in a most irregular pattern. The clay minerals were accompanied by widely differing amounts of lithium mica and muscovite and of fluorite, the fluorite probably having been derived from topaz. Most of the iron-bearing sulfides seem to have been removed in the zones of intense argillization, but cassiterite, sphalerite, and arsenopyrite were not affected by the process.

In addition to the tin-tungsten mineralization, a strong concentration of beryllium minerals was developed in replacement veins, pipes, and stockworks in the carbonate rocks above the buried granite of the Lost River mine and around other granite bodies. In the mine the beryllium mineralization probably lies farther away from the contact with granite than does the low-grade tin-tungsten ore. Sainsbury (1968) shows that there are three main zones of beryllium minerals at different distances from the granite bodies: (1) at or near the limestone-granite contacts, the beryllium mineral is helvite [$(Mn,Fe,Zn)_4(BeSiO_4)_3S$] in earlier banded magnetite-fluorite skarn; (2) farther out, in premineralization silicated and marmorized limestone, the beryllium mineral is chrysoberyl ($BeAl_2O_4$) in fluorite veins in which it is associated with tin and sulfide minerals; and (3) still farther into the limestone, the main beryllium mineral again is chrysoberyl, but euclase [$BeAlSiO_4(OH)$] and bertrandite [$Be_4Si_2O_7(OH)_2$] also are included in fluorite veins that lack tin and sulfide minerals. The materials of zones (1) and (2) are quantitatively unimportant; those of zone (3) are common and occur in large bodies, not only at Lost River but also at Camp Creek and Rapid River. Zone (3) ores are made up of 45 to 65 percent of fluorite; 5 to 15 percent diaspore; 0 to 10 percent tourmaline; 3 to 10 percent chrysoberyl; and 0 to 5 percent white mica. Small but varied amounts of hematite, euclase, and sulfides are found in the zone (3) ores as are traces of cassiterite, phenakite (Be_2SiO_4) bertrandite, todokorite, and probably others. Where the fluorite-beryllium veinlets have resulted partly or largely from replacement, the original limestone textures are preserved even where the faint diffusion (?) banding in the fluorite veinlets might tend to conceal it.

The tin-tungsten mineralization in the greisen and the tin-tungsten-beryllium mineralization in the skarn probably developed at essentially the same time and were all formed under high-intensity conditions. Possibly the clay alteration also took place within the lower limits of the hypothermal range, but this alteration definitely was later than the ore; conditions that obtained during the formation of the clay minerals, therefore, do not directly affect the classification of the ores. The minable ore in the greisen of the Cassiterite dike must be considered, therefore, as hypothermal in noncalcareous rocks and is here categorized as hypothermal-1. On the other hand, the cassiterite and wolframite in the lime-silicate skarns have never been mined and are not included in the classification despite their hypothermal-2 character. The beryllium deposits, however, are of such high economic potential that they are made part of the classification of the Lost River deposit. Because the principal beryllium mineral (chrysoberyl) and the tourmaline and white mica with which it is associated are diagnostic of high-temperature conditions and because the fluorite and diaspore do not contradict this intensity range, the beryllium-rich portion of the deposits is categorized as hypothermal-2.

Placer cassiterite, derived from the primary deposits, has been of much greater economic importance than the tin from the Lost River mine.

RUBY CREEK

Late Mesozoic (pre-Laramide) — *Copper* — *Mesothermal to Telethermal*

Brosgé, W. P., 1960, Metasedimentary rocks in the south-central Brooks Range, Alaska: U.S. Geol. Surv. Prof. Paper 400-B, p. B351-B352

Herreid, G., 1961, Geology and ore deposits of Alaska: Min. Eng., v. 13, no. 12, p. 1316-1325

Patton, W. W., Jr., and others, 1968, Regional geologic map of the Shungnak and southern part of the Ambler River quadrangles: U.S. Geol. Surv. Misc. Geol. Invest. Map I-554, 1:250,000

Runnells, D. D., 1969, The mineralogy and sulfur isotopes of the Ruby Creek Copper Prospect, Bornite, Alaska: Econ. Geol., v. 64, p. 75-90

Smith, P. S., and Mertie, J. B., Jr., 1930, Geology and mineral resources of northwestern Alaska: U.S. Geol. Surv. Bull. 815, 351 p.

Notes

The Ruby Creek deposit (the post office is Bornite) is located on the north flank of the Cosmos Hills, near the southwestern margin of the Brooks Range. The deposit is about 300 miles northwest of Fairbanks.

The Cosmos Hills (Runnells, 1969) were produced on a northwest-southeast trending symmetrical anticline that plunges gently northwest. In the core of the anticline are Paleozoic schists, phyllites, and Middle Devonian carbonates; the pelitic rocks have reached the greenschist facies of regional metamorphism. The carbonate rocks probably are equivalent of the Alaskan Devonian Skajit limestone. Farther down the flanks of the anticline are Jurassic or Cretaceous graywacke, siltstone, conglomerate, basalt, and mafic tuff. The rocks of the area have been intruded by a few mafic intrusives, now altered, that range from andesite to gabbro. The intrusive nearest to the ore body is a chloritized gabbro plug or sill nearly 2 miles to the southeast. About 7 miles east of Ruby Creek is gneissic muscovite-biotite-albite granite; based on K/Ar determinations made on an associated soda aplite (Patton and others, 1968), an age of 121 ± 3.8 m.y. was assigned to the granite.

The mineralization at Ruby Creek is enclosed in a complex series of interbedded dolomites, limestones, and phyllites, with the dolomite containing the most ore, followed in order by impure limestone and albitic lime phyllite. These strata appear to be part of a reef complex that the abundant and varied fossils date as Middle Devonian.

No source of the ore fluids has been established for the Ruby Creek ores; the ores, however, appear definitely to be epigenetic, and the isotopic composition (Runnells, 1969, p. 79) indicates the mineralizing solutions to have been of magmatic-hydrothermal origin. Further, the only important silicic igneous activity in the area was in the Early Cretaceous. On this basis the Ruby Creek ores are here assigned to late Mesozoic (pre-Laramide) time. Runnells thinks that the possibility exists that submarine volcanoes or hot springs supplied the ore fluids, yet envisions the ore fluids as moving through solid rock. If this should be proved correct, the ores probably would be middle Paleozoic in age.

The most abundant sulfide by far at Ruby Creek is pyrite; this is true both in carbonates and phyllites. The normal range of pyrite content in phyllites is 1 to 20 percent, but as much as 50 percent has been found in some of them. Runnells (1969) suggests, but does not say, that pyrite in dolomites and limestones is less abundant than in the phyllites. In both phyllites and carbonate rocks, the pyrite exhibits framboidal textures, textures that Runnells quotes Love as saying are indicative of sedimentary or early diagenetic origin. These textures are, however, less common in carbonate pyrite than that in the phyllites; carbonate pyrite also may be colloform. The most important primary copper sulfides are chalcopyrite, bornite, tennantite-tetrahedrite, and chalcocite. Normally, only two or three of these are found together, and tennantite-tetrahedrite is rare or absent in most specimens containing bornite or chalcocite. The mutual intergranular textures make it difficult to determine relative ages; Runnells seems certain that all, or most, of the sulfides are younger than the pyrite. Chalcopyrite is widely disseminated through the carbonate host rocks but in places massively replaces them; locally thin veinlets cut the carbonates, and copper sulfide crystals are found in vugs. Bornite is of importance only in the areas most heavily mineralized in copper. The primary steely chalcocite is confined to centers of mineralization, while the sooty secondary chalcocite is much more widely distributed, as are the other secondary sulfides. Tennantite-tetrahedrite is sparsely present and generally is intimately intergrown with chalcopyrite. Runnells is convinced that the copper sulfides are epigenetic and thinks that the pyrite may be epigenetic in the carbonates if not in the phyllites. The ore contains small amounts of sphalerite and even lesser amounts of galena; these two minerals generally are closely associated. Some pyrrhotite is present in certain barren phyllites and is intergrown with pyrite in some deeper, iron-rich dolomites; it is not found in the main copper-bearing bodies. A little carrollite is present, and some little germanite is found in bornite. Runnells attempts to work out the paragenesis of the deposit on the basis of its vertical zoning. Assuming that the central core of the ore body was the main conduit for the ore fluids and if the ore fluids spread out from this conduit, the oldest assemblage would be by Runnells' reasoning, the one farthest from the conduit and the youngest would be the one in the central conduit. It seems equally reasonable to me that the central one would be the oldest and that those farther out would be progressively younger since they could not be deposited where the oldest sulfides had preempted the space, except by replacing them. Runnells reaches the conclusion from this, however, that the paragenetic sequence would be pyrite, chalcopyrite, tennantite (minor), bornite, and primary chalcocite. This certainly is the sequence I would expect in a deposit such as this. The wall-rock alteration in the deposit is minor, the most remarkable fact about it being the presence of cymrite ($BaAl_2Si_2O_8 \cdot H_2O$), but this mineral is replaced by copper sulfides so must have been earlier than they. Chlorite, dickite, sericite, and bementite occur at Ruby Creek. The sulfides and wall-rock alteration minerals (except for the early cymrite) certainly seem compatible with deposition in the ranges from mesothermal through telethermal, and the deposits are so classified here.

Arizona

AJO

Late Mesozoic to Early Tertiary — *Copper* — *Mesothermal*

Bütler, B. S., 1947, The Ajo mining district, Arizona (rev.): Econ. Geol., v. 42, p. 583-587

Creasey, S. C., 1959, Some phase relations in the hydrothermally altered rocks of porphyry copper deposits: Econ. Geol., v. 54, p. 351-373 (general)

DeKalb, C., 1918, Ajo copper mine: Min. and Sci. Press, v. 116, 26 Jan., p. 115-118; 2 Feb., p. 153-156

Dixon, D. W., 1966, Geology of the New Cornelia mine, Ajo, Arizona, in Titley, S. R., and Hicks, C. L., Editors, *Geology of the porphyry copper deposits--southwestern North America*: Univ. Ariz. Press, Tucson, p. 123-132

Gilluly, J., 1935, Ajo district (Arizona), in *Copper resources of the world*: 16th Int. Geol. Cong., v. 1, p. 228-233

—— 1937, Geology and ore deposits of the Ajo quadrangle, Arizona: Ariz. Bur. Mines, Geol. Ser. no. 9, Bull. 141, 83 p.

—— 1942, The mineralization of the Ajo copper district, Arizona: Econ. Geol., v. 37, p. 247-309

—— 1946, The Ajo mining district, Arizona: U.S. Geol. Surv. Prof. Paper 209, 112 p., with revised sections

Joralemon, I. B., 1914, The Ajo copper-mining district: A.I.M.E. Tr., v. 49, p. 593-609

Schwartz, G. M., 1947, Hydrothermal alteration in the "porphyry copper" deposits: Econ. Geol., v. 42, p. 319-352 (particularly p. 323, 346-352)

Wadsworth, W. B., 1968, The Cornelia pluton, Ajo, Arizona: Econ. Geol., v. 63, p. 101-115

Notes

The Ajo mining district is situated about 110 miles slightly north of west from Tucson and about 85 miles south-southwest of Phoenix.

The oldest rocks in the Ajo area are probably Precambrian and consist chiefly of the largely igneous Cardigan gneiss; the gneiss is composed of highly contorted rocks. The post-Precambrian Chico Shunie quartz monzonite contains both large and small inclusions of the Cardigan gneiss that have been appreciably altered by the later intrusive. There are a few randomly distributed bodies of hornblendite in the gneiss, and most form pipes or bosses in the Cardigan rocks. The Precambrian rocks in the general area are unconformably overlain by probably Paleozoic hornfels that originally was made up of sandstone, shale, andesite, and rhyolite; these rocks were converted to hornfels by the Chico Shunie quartz monzonite. Much of the hornfels in the immediate mine area occurs as inclusions in the Chico Shunie, but farther south a series of bedded rocks includes hornfelsed dolomites, shales, and sandstones (presumably altered by an intrusive mass not exposed at the surface). This Precambrian-Paleozoic sequence was invaded by the Chico Shunie quartz monzonite that Gilluly (1946) thought to be Mesozoic (?) but which the Arizona Survey now considers to be certainly Mesozoic; Butler (1947) pointed out that igneous activity was not widespread in southern Arizona in the Paleozoic, indirect evidence of the Mesozoic age of the Chico Shunie. The best guess that can be made probably is that the Chico Shunie is of the same age (pre-Laramide) as the Sacramento stock at Bisbee. Andesite and rhyolite flows, tuffs, and breccias of the Concentrator volcanic rocks (now much altered) were deposited unconformably on the Chico Shunie and probably can be correlated with the Cretaceous "earlier volcanics" of other portions of southern Arizona. The Cornelia quartz-monzonite pluton (in which the ore is contained) was intruded into these rocks; it cuts across the structures of the host rocks and consists (Wadsworth, 1968) of five lithologic varieties, all of which were introduced in a closely spaced sequence. Soon after the pluton had consolidated, fissures developed in its apex into which quartz-orthoclase pegmatitelike rocks were emplaced; this process included a potash-feldspar and quartz impregnation of the wall rocks and an appreciable hydrothermal alteration. Further shattering of the apex cupola provided channelways for the ore fluids that followed shortly after the pegmatites. The Cornelia is appreciably less metamorphosed than the Chico Shunie and certainly is intrusive into the Concentrator volcanic rocks.

The Cornelia pluton was considerably eroded before it was unconformably covered by the Locomotive fanglomerate; the fanglomerate probably is of middle Tertiary age. The post-ore fanglomerate grades upward into tuffs, breccias, and flows that are termed the Ajo volcanic rocks; these are overlain by the Sneed andesite and a variety of later volcanic materials that were emplaced between the middle Tertiary and the end of the Pliocene. The younger volcanic rocks are separated from the Sneed andesite by the Daniels conglomerate.

It appears that the mineralization of the New Cornelia ore body took place before the Gibson normal fault was formed so that the mineralized quartz monzonite mass was, at the time of mineralization, a cupola on the mass of the Cornelia quartz monzonite to the northwest. The Gibson fault cuts post-Cornelia dikes and is parallel to a regional set of faults that does not appear to have been controlled by the monzonite intrusion. The lack of mineralization in the Locomotive fanglomerate shows that the ore was older than that middle Tertiary formation. Thus, the mineralization probably occurred between the early Late Cretaceous (probably Chico Shunie time) and the early Tertiary, with the greater probability that the ore was Paleocene; it is classed here as late Mesozoic to early Tertiary. This checks with an age of 63 m.y. on the quartz-orthoclase pegmatites late in the Cornelia cycle.

Although a huge development of faulting took place in the area in the middle and late Tertiary, the Cornelia quartz monzonite (which has notable dioritic border phases) had been highly fractured during Laramide time and was invaded by hydrothermal solutions that first produced a pegmatitically textured orthoclase-biotite-quartz rock in the vicinity of main fracture zones but also caused much of the ground mass to be replaced by fine-grained orthoclase or by quartz; some magnetite was developed in this portion of the alteration cycle. After the completion of this phase, the ore, mainly pyrite, chalcopyrite, and bornite, with some minor tetrahedrite, molybdenite, and sphalerite, was introduced in quartz veinlets and as disseminations in the monzonite. The ore was accompanied by the alteration of the plagioclase to aggregates of albite and sericite and of the dark silicates to chlorite, epidote, and small quantities of other minerals; the quartz and orthoclase were essentially unaffected.

The boundaries of the disseminated ore are generally indefinite, and there are no readily available visual criteria to be used in differentiating ore from waste rock.

Although the orthoclase-quartz-magnetite alteration almost certainly was emplaced under hypothermal (if not deuteric) conditions, the ore minerals and the accompanying alteration of the wall rock appears definitely to have been developed within the mesothermal range, and the primary ores are categorized here as mesothermal.

The ore bodies were oxidized almost down to the water table; although there are local variations in the base of the oxidized zone, it is remarkably level, the depth of oxidation ranging between 20 to 90 feet (55 feet average). With one rather minor exception, only insignificant amounts of chalcocite were developed below the water table, and the average grade of the oxidized ore is essentially that of the underlying primary material, indicating that practically no copper was removed in the process of oxidation. The copper minerals in the oxidized zone are malachite, a little azurite, cuprite, tenorite, chrysocolla, plus hematite and limonite (the last two derived largely from the magnetite). Locally, the oxidized zone and the very minor zone of supergene enrichment have been appreciably displaced by the late Tertiary faults. The oxidized ore should, of course, be classified as ground water-B2.

BAGDAD-MASSIVE SULFIDES

Middle Precambrian	*Lead, Zinc, Copper, Gold, Silver*	*Hypothermal-1 to Mesothermal*

Anderson, C. A., 1950, Lead-zinc deposits, Bagdad area, Yavapai County, Arizona, in *Arizona zinc and lead deposits*, pt. 1: Ariz. Bur. Mines, Geol. Ser. no. 18, Bull. no. 156, p. 122-138

—— 1968, Arizona and adjacent New Mexico, in Ridge, J. D., Editor, *Ore deposits of the United States, 1933-1967* (Graton-Sales Volumes): Chap. 56, v. 2

Anderson, C. A., and others, 1955, Geology and ore deposits of the Bagdad area, Yavapai County, Arizona: U.S. Geol. Surv. Prof. Paper 278, 103 p. (particularly p. 85-89)

Baker, A., III, and Clayton, R. L., 1968, Massive sulfide deposits of the Bagdad district, Yavapai County, Arizona, in Ridge, J. D., Editor, *Ore deposits of the United States, 1933-1967* (Graton-Sales Volumes): Chap. 62, v. 2

Notes

The massive sulfide deposits of the district include the Hillside mine (over 3 miles north of Bagdad town), the Copper King (1.5 miles south of Bagdad), the Old Dick (2.75 miles south-southwest of Bagdad), and the Copper Queen (about 0.25 miles south of the Old Dick).

The area of the King, Dick, and Queen mines contains three major rock groups: (1) the Bridle formation of the older Precambrian Yavapai series; (2) the Dick rhyolite; and (3) diabase, lamprophyre, and gabbro. The Dick rhyolite trends parallel to the strike of the Bridle formation that encloses it; the much younger mafic rocks occur as dikes or sills in both the Dick and the Bridle. The Bridle is made up of andesite flows, tuffs, and sediments; the sediments are now schists that are neither mineralized nor hydrothermally altered and make up over half of the Bridle formation proper. Three recognizable members compose the andesitic portion of the Bridle, the middle one of which (immediately below the intruded Dick rhyolite) contains the Dick mine ores in a quartz-sericite schist band. Above the Dick rhyolite is a quartz-sericite schist member that includes the Copper Queen ore body. The Copper King ore is confined at the surface to a tuffaceous sedimentary unit of the Bridle, which is within 100 feet of the edge of a rhyolite probably equivalent to the Dick rhyolite; the ore location was controlled by a zone of small interlacing faults parallel to the bedding of the tuff. Some ore was emplaced by replacement of a small diorite porphyry dike.

The Hillside mine actually is a typical fissure vein in the Hillside mica schist, the uppermost of the three formations that make up the Yavapai series; the Hillside and the Bridle are separated by the Butte Falls tuff.

Anderson and others (1955) considered the mineralization of these deposits to have been Laramide in age, but Anderson (1968) now believes that lead-isotope studies on galena from the Old Dick mine show the ores to be middle Precambrian (± 1700 m.y.). Anderson thinks that this age is supported by the resemblance of these deposits to those at Jerome. In the Copper King mine, copper sulfides vein a Laramide diorite porphyry dike, but Anderson now considers this relationship to have been caused by remobilization during dike intrusion. Baker and Clayton (1968), however, still consider the ores as more probably Laramide than Precambrian; they are here categorized as middle Precambrian, but lead-isotope ratios are a weak basis for age dating.

At the Copper King mine, the ore occurs in three major lenses, the locations of which are fault controlled; on lower levels, however, some of the massive ore, at least, grades into disseminated mineralization in which the grade is 3 to 8 percent zinc. The ore in both the Old Dick and Copper Queen mines is in tabular bodies lying parallel, or nearly so, to the steeply dipping bedding of the host rocks; locally, however, cross-cutting relationships are known. The Dick ore consists of both massive (80 percent) and disseminated (20 percent) material, but that of the Queen is essentially all massive.

The sulfides in the Dick, Queen, and King ores are pyrite, sphalerite, and chalcopyrite with rare galena and arsenopyrite. Closely joined to the sulfides are small amounts of calcite, chlorite, sericite, and quartz, but the gangue is mainly unreplaced fragments of the host rocks. The sulfides are more uniformly distributed in the Queen ore than in that of the Dick. Baker and Clayton (1968) point out that they have no accurate information on the distribution of galena and arsenopyrite but they have the impression that galena decreases with depth and arsenopyrite increases.

Pyrite and arsenopyrite were the first minerals formed, followed by sphalerite containing oriented blebs of chalcopyrite and by chalcopyrite in stringers and bands; minor silver-rich galena occurs in pods irregularly distributed through the sphalerite. Small quantities of chlorite, sericite, and calcite were developed at the same time as the sulfides. The pyrite-arsenopyrite-sphalerite phase of the mineralization probably occurred under hypothermal conditions; that of the later chalcopyrite and silver-rich galena probably was mesothermal.

In the Hillside mine, Anderson (1950) recognized four periods of quartz-sulfide deposition; some faulting between ore stages is shown by microbrecciation of the sulfides and the cementation of the fragments by later quartz. The common sulfide minerals he reports to be pyrite, arsenopyrite, galena, and sphalerite that occur as blebs or bunches in quartz; chalcopyrite is minor in amount. Argentiferous tetrahedrite is found as coatings on other sulfides in vugs in quartz. Some oxidized ore was mined in which the main minerals were cerussite, andesite, smithsonite, and hemimorphite; this ore was low grade. Evidence is lacking as to whether or not the tetrahedrite was sufficiently abundant to justify adding lepothermal to the classification of the Hillside ores; it is not done here, and they are considered to be hypothermal-1 to mesothermal as are the other ores of the area outside the porphyry copper body.

BAGDAD-PORPHYRY COPPER

Late Mesozoic to Early Tertiary	*Copper, Molybdenum*	*Hypothermal-1 to Mesothermal*

Anderson, C. A., 1948, Structural control of copper mineralization, Bagdad, Arizona: A.I.M.E. Tr., v. 178, p. 170-180

—— 1950, Alteration and metallization in the Bagdad porphyry copper deposit, Arizona: Econ. Geol., v. 45, p. 609-628

—— 1969, Arizona and adjacent New Mexico, in Ridge, J. D. Editor, *Ore deposits of the United States, 1933-1967* (Graton-Sales Volumes): Chap. 56, v. 2

Anderson, C. A., and others, 1955, Geology and ore deposits of the Bagdad area, Yavapai County, Arizona: U.S. Geol. Surv. Prof. Paper 278, 103 p.

Butler, B. S., and Wilson, E. D., 1938, Bagdad mine, Eureka district, in *Some Arizona ore deposits*: Ariz. Bur. Mines, Geol. Ser. no. 12, Bull. no. 145, p. 98-103

Creasey, S. C., 1959, Some phase relations in the hydrothermally altered rocks of the porphyry copper deposits: Econ. Geol., v. 54, p. 351-373 (particularly p. 355, but the entire paper should be read)

Hardwick, W. R., and Jones, E. L., III, 1959, Open-pit copper mining methods and costs at the Bagdad mine, Bagdad Copper Corp., Yavapai County, Ariz.: U. S. Bur. Mines I.C. 7929, 30 p. (mimeo.) (particularly p. 2-10)

Huttl, J. B., 1943, Bagdad--Arizona's latest porphyry copper: Eng. and Min. Jour., v. 144, no. 6, p. 62-66 (essentially no geology)

Moxham, R. M., and others, 1965, Gamma-ray spectrometer studies of hydrothermally altered rocks: Econ. Geol., v. 60, p. 653-671

Schwartz, G. M., 1947, Hydrothermal alteration in the "porphyry copper" deposits: Econ. Geol., v. 42, p. 319-352 (particularly p. 324-325, 346-352)

Notes

The ore deposits of the Bagdad area are located in west central Arizona about 45 miles slightly north of west from Prescott. Because of the probably considerable difference in age between the disseminated ores of the Bagdad porphyry copper and massive sulfides in the general vicinity, these two ore types are discussed separately.

The oldest rocks in the Bagdad area are the three formations of the Precambrian Yavapai series; they are highly metamorphosed volcanic and clastic rocks of terrestrial origin. These rocks were intruded, still in Precambrian time, by a wide variety of igneous materials, including rhyolite, gabbro, quartz diorite and diabase, alaskite porphyry, granodiorite gneiss, and ending with two facies of granite, the porphyritic Lawler Peak granite and the fine-grained Cheney Gulch granite; both granites are intruded by aplite-pegmatite. No Paleozoic rocks are known in the area, the next formation in the sequence being, by analogy with similar rocks elsewhere in Arizona, a Late Cretaceous to early Tertiary rhyolite tuff, the Grayback Mountain; the tuff is intruded by rhyolite dikes, and the complex then was intruded by a series of stocks and plugs of quartz monzonite (perhaps mainly granodiorite) that probably also is of Late Cretaceous to early Tertiary age. The largest of these stocks is that at Bagdad that contains the ore body now being mined from the Bagdad pit. The quartz monzonite is cut by dikes of quartz-monzonite porphyry and diorite porphyry. By Pliocene time an erosional surface of considerable relief had been developed in the area and on this was deposited a conglomerate, the Gila (?), in which are intercalated basalt flows, the flows being divided into two formations. Although there is no stratigraphic evidence as to the absolute age of the Yavapai formation and of the granites intrusive into it, radioactive age determinations (Creasey, 1959) indicate an age of 1600 m.y. for the Lawler Peak granite, making it late middle Precambrian and the metamorphosed rocks even older. One line of evidence available for the age of the quartz monzonite is that in other regions of Arizona Late Cretaceous igneous activity produced lava flows and pyroclastic accumulations and, at the same time or somewhat later, intrusions of monzonitic to granitic composition that took the form of batholiths, stocks, plugs, and dikes. This less than certain evidence has now been supplemented by radioactive age dating by Damon and Mauger on the quartz monzonite that gives this rock an age of 71 m.y. The Bagdad ores, which are almost certainly genetically related to the quartz monzonite, are considered, therefore, to be late Mesozoic to early Tertiary.

In the Bagdad mine proper, the primary ore is contained in minute veinlets, the older veinlets containing quartz, pyrite, and chalcopyrite and the younger ones containing quartz, orthoclase, molybdenite, and pyrite; a little late sphalerite and galena is found in the wider quartz veins. Most of the ore is in the veinlets rather than truly disseminated in the quartz monzonite. Apparently contemporaneously with the mineralization, the quartz monzonite was considerably altered, mainly by the development of additional orthoclase, the conversion of the plagioclase to more sodic varieties (albite and albite-oligoclase), and the conversion of much of the hornblende and perhaps some of the primary biotite to a pale, leafy biotite. Some sericite was developed in the plagioclase between quartz and orthoclase crystals and interleaved with secondary biotite; a little chlorite is found in this biotite. The deposition of the molybdenite-bearing veinlets and the changes in the feldspar and biotite probably took place under hypothermal conditions, perhaps near the boundary line with mesothermal, while the quartz-chalcopyrite-pyrite veinlets might have been formed in the mesothermal or hypothermal ranges. Here the primary copper-molybdenum mineralization is categorized as hypothermal-1 to mesothermal.

BISBEE

Late Mesozoic (pre-Laramide)	*Copper, Zinc, Lead, Silver, Gold*	*Mesothermal to Telethermal*

Bain, G. W., 1952, The age of the "Lower Cretaceous" from Bisbee, Arizona, uraninite: Econ. Geol., v. 47, p. 305-315

Bonillas, Y. S., and others, 1916, Geology of the Warren mining district: A.I.M.E. Tr., v. 55, p. 284-355

Bryant, D. G., 1968, Intrusive breccias associated with ore, Warren (Bisbee) mining district, Arizona: Econ. Geol., v. 63, p. 1-12

Bryant, D. G., and Metz, H. E., 1966, Geology and ore deposits of the Warren mining district, in Titley, S. R., and Hicks, C. L., Editors, *Geology of the porphyry copper deposits--southwestern North America*: Univ. Ariz. Press, Tucson, p. 189-203

Douglas, J., 1899, The Copper Queen mine: A.I.M.E. Tr., v. 29, p. 511-546

Hogue, W. G., and Wilson, E. D., 1951, Bisbee or Warren district, in *Arizona zinc and lead deposits*, pt. 1: Ariz. Bur. Mines, Geol. Ser. no. 18, Bull. no. 156, p. 17-29

Livingston, D. E., and others, 1968, Geochronology of the emplacement, enrichment, and preservation of the Arizona porphyry copper deposits: Econ. Geol., v. 63, p. 30-36

Ransome, F. L., 1904, The geology and ore deposits of the Bisbee quadrangle, Arizona: U.S. Geol. Surv. Prof. Paper 21, 168 p.

—— 1914, Bisbee folio, Arizona: U.S. Geol. Surv. Geol. Atlas, Folio 112, 17 p.

Rove, O. N., 1942, Bisbee district, Arizona, in Newhouse, W. H., Editor, *Ore deposits as related to structural features*: Princeton Univ. Press, p. 211-215

Schwartz, G. M., 1947, Hydrothermal alteration in the "porphyry copper" deposits: Econ. Geol., v. 42, p. 319-352 (particularly p. 344-345, 346-352)

Schwartz, G. M., and Park, C. F., Jr., 1932, A microscopic study of ores from the Campbell mine, Bisbee, Arizona: Econ. Geol., v. 27, p. 39-51

Tenney, J. B., 1927, The Bisbee mining district: Eng. and Min. Jour., v. 123, no. 21, p. 837-841

—— 1933, The Bisbee mining district, in *Ore deposits of the southwest*: 16th Int. Geol. Cong., Guidebook 14, p. 40-67

—— 1935, Bisbee district, in *Copper resources of the world*: 16th Int. Geol. Cong., v. 1, p. 221-228

Trischka, C., 1938, Bisbee district, in *Some Arizona ore deposits*: Ariz. Bur. Mines, Geol. Ser. no. 12, Bull. no. 145, p. 32-41

Wisser, E., 1927, Oxidation subsidence at Bisbee, Arizona: Econ. Geol., v. 22, p. 761-790

Notes

The Bisbee district is located in the extreme southeastern corner of Arizona, less than 10 miles from the Mexican border and about 80 miles southeast of Tucson.

At Bisbee, the oldest rocks are those of the older Precambrian Pinal schist, unconformably over which lie more than 4000 feet of Paleozoic sediments. The Paleozoic formations include the Cambrian Bolsa quartzite, the Late Cambrian thin-bedded Abrigo limestone, the Devonian Martin limestone, the Mississippian Escabrosa limestone, and the Pennsylvanian Naco limestone. During the early and middle Mesozoic, a highly irregular erosion surface was developed on the Precambrian schist and the various Paleozoic rocks; on this surface Early Cretaceous sediments were deposited of which the oldest is the Glance conglomerate, followed by the Morita shales and sandstones, the fossil-bearing Mural limestone, and the Cintura shales and sandstones. Between the end of Paleozoic sedimentation and the beginning of that in the Cretaceous, a large stock of granite, with satellite sills and dikes of granite and rhyolite porphyry, was intruded into the northwestern part of the district; these silicic rocks show little alteration. Later than the granite, a stocklike mass of quartz monzonite porphyry (now highly altered) intruded the central part of the Bisbee area; its contact metamorphic effects on the surrounding limestone were neither intense nor extensive. This quartz monzonite (Sacramento) stock cut through the area of the post-Pennsylvanian pre-Cretaceous Dividend fault, which drops the rocks' southwest side some 5000 feet below their counterparts of the northeast. Thus, on the present erosion surface, the monzonite is in contact with Precambrian schists on the northeast and with Paleozoics on the southwest. In the Paleozoics numerous sills and dikes extend out from the stock, while such apophyses are uncommon in the schist. A highly silicified, pyrite-rich zone was formed adjacent to the stock, probably in the course of the primary mineralization and not at the time of intrusion. Certain intrusive breccias were developed in the area about this time, presumably after the pyrite mineralization, since the breccias contain pyrite fragments. These breccias were followed by the base and precious metal mineralization that used the same channels of entrance as did the igneous rocks and the pyrite-depositing solutions, as well as those avenues provided by the intrusive breccias. Next followed a pronounced period of erosion that developed, among other things, a secondarily enriched deposit of chalcocite in the quartz monzonite (stock). Further movement of the Dividend fault then dropped the southwestern side of the fault sufficiently for it to be covered by a shallow sea, while the northern side was subjected to intense erosion, removing the Paleozoics almost entirely and the chalcocite blanket completely from that area, filling the arm of the sea on the southwest with detritus, and eventually depositing a thin layer of basal Cretaceous conglomerate on the northeast side. After the Cretaceous sediments had been deposited on both sides of the fault, the area was elevated with additional movement on the Dividend fault, the district was regionally tilted to the east, and the Cretaceous cover over the area was largely removed by erosion.

This scheme of events indicates that the ore was formed at an appreciable time before the first of the Cretaceous sediments was deposited in the area. This concept is confirmed by (1) pebbles in the Glance conglomerate of altered porphyry of iron-stained silica (probably derived from an oxidized portion of the silicified, pyrite-rich zone), and of silicified limestone, (2) complete lack of alteration of Glance conglomerate in contact with quartz monzonite, (3) conformity of the chalcocite blanket in the porphyry with the pre-Glance surface rather than with any later surface, and (4) lack of any ore in the Glance or any of the younger rocks. Although there is a little post-Cretaceous mineralization, it is quite different from that in the porphyry or in the limestone.

The Mural limestone contains fossils that date it as being in the lower half of the upper third of the Lower Cretaceous. The Mural lies above about 2000 feet of Lower Cretaceous shales, sandstones, and basal (Glance) conglomerate, which would indicate that Cretaceous sedimentation, while it may have begun as late as the first third of Trinity time, probably commenced still earlier in the Cretaceous. It is, therefore, probable that the mineralization must have been introduced, at the latest, in the lowermost Cretaceous and well may have been brought in toward the end of the Jurassic. For the purposes of classification in this study, the age of late Mesozoic (pre-Laramide) is assigned, but it may be that this will have to be changed later to middle Mesozoic.

The ores at Bisbee consist of two main types: (1) massive replacement bodies in limestone and (2) disseminated replacements and veinlet fillings in quartz monzonite porphyry. In the limestones, the copper ore bodies occur in close association with the larger and earlier emplaced masses of pyrite that coincide in time with the silicification of the wall rock. Most commonly, the masses of copper ore are found around the outer margins of the pyrite bodies, as replacements of both pyrite and limestone. Copper ores are also present as open space fillings and as replacements in brecciated early pyrite. The lead-zinc ore bodies normally occur in zones bordering pyrite and copper mineralization, although some lead-zinc masses lie completely out of contact with the copper ores. Locally, the ore solutions appear to have used the same channelways traveled by the monzonite dikes and sills, and ore bodies occupy areas in limestone immediately adjacent to irregularities or embayments in the limestone-porphyry contacts; essentially no ore occurs in the porphyry apophyses. The limestone bedding generally has a strong influence on the shape of replacement ores, the ore masses largely reflecting the character of the bedding in the formation in question.

Particularly favorable to the deposition of the ore bodies were the intersections of the various faults, and the breccia pipes appear largely to have been developed at fault intersections. The most productive portions of the Paleozoic sequence (the upper Abrigo, all the Martin, and the lower Escabrosa) are found at progressively greater depths as they are followed to the east because of rotation of the fault blocks.

The highly fractured Sacramento stock also was a locus of mineralization that converted 15 to 18 percent of the rock to sulfides, most of which was pyrite with minor chalcopyrite and less bornite. This pyrite too was fractured, and when the monzonite was exposed at the surface, the upper portions of the stock were leached of the copper they contained, and this copper was transferred beneath the water table where it, by replacement, produced thin films of chalcocite on the pyrite fragments. The chalcocite blanket so formed is from 50 to 400 feet thick, being thickest in areas of greatest fracturing, and dips east in conformity with the pre-Glance erosion surface above it. The leaching of the oxidized capping above the chalcocite is very thorough, there being no mixed oxidized-enriched ore belt between the oxidized zone and the blanket. No evidence has been reported to indicate that the two essentially simultaneous events: (1) sericite, hydromuscovite, kaolinite, allophane, and alunite alteration of the monzonite and (2) the pyrite-chalcopyrite primary mineralization occurred under anything but the mesothermal conditions they so strongly suggest.

In the ore in limestone, the interpretation of the paragenesis is complicated by the presence of the large portion of the lead-zinc in ore masses distinct from the copper ore bodies. In the copper ores, quartz and normally slightly later pyrite were the first minerals to form; the pyrite was then considerably replaced by chalcopyrite with which bornite is ordinarily so closely associated that most of it probably formed at much the same time; some bornite, however, appears to have replaced the chalcopyrite and was, therefore, later than the chalcopyrite in part. In some instances, sphalerite forms thin borders around chalcopyrite grains; this relationship may be the result of replacement or of unmixing of a previously formed solid solution of the two minerals; not enough evidence is available to determine this point. Rounded, probably replaced, grains of sphalerite occur in bornite. Although galena is essentially lacking in the copper ores, it is of much the same age as the sphalerite; some veinlets of galena are known to cut sphalerite in the lead-zinc ores. Considerable specular hematite and some magnetite are found in the copper ore bodies, and they appear to have formed by the replacement of pyrite and, locally, of chalcopyrite. Although the galena is silver-bearing, the silver in the copper ores apparently is derived mainly from stromeyerite (CuAgS), at least in the Campbell mine. The stromeyerite occurs almost entirely in chalcocite, although some stromeyerite has been found in bornite; in chalcocite, the stromeyerite is present as irregularly shaped blades that may have been introduced as replacements of chalcocite or as replacements of bornite, the bornite not replaced by stromeyerite having been later largely replaced by chalcocite. Within the chalcocite, the stromeyerite blades normally cut off against irregularly rounded grains of arsenic and antimony sulfosalts that probably developed before, and were replaced by, the chalcocite. Primary chalcocite and digenite seem to have been the last of the main copper-bearing minerals to have been formed. Some chalcocite may have been produced from unmixing of a bornite-chalcocite solid solution, but most of the chalcocite appears to have replaced other sulfides, principally bornite, as irregular veinlets and larger masses. There are, moreover, rounded spots of tetrahedrite, tennantite, enargite, and lesser famatinite present in the chalcocite; similar occurrences are more rarely found in bornite, and all these spots are probably replacement remnants. Digenite has not been identified in published work on Bisbee, but it occurs in what appears to be much the same manner as chalcocite and is, therefore, of essentially the same age. Some of the chalcocite is of the sooty variety and probably is of secondary origin, but the greater part of this mineral is massive and almost certainly is primary.

From the mineral relationships found in the Bisbee district, it would appear that the chalcopyrite formed under mesothermal conditions; this interpretation is based on the lack of exsolved blebs of sphalerite in the chalcopyrite (and on considering the sphalerite surrounding some chalcopyrite grains as replacement and not exsolution textures), on the absence of high-temperature silicates in the limestone (attributable to the hydrothermal solutions), and on the lack of certainly high-temperature sulfides such as pyrrhotite and arsenopyrite. The hematite replacing pyrite and chalcopyrite suggests hypothermal conditions prevailed at that stage, but this material is not, of course, of economic importance. The chalcopyrite-bornite-stromeyerite-chalcocite-digenite sequence appears to span the intensity ranges from mesothermal to telethermal. The sphalerite-galena mineralization in the lead-zinc deposits appears to have formed under mesothermal to leptothermal conditions, but not enough work has been done on the silver content of galenas from various parts of the area and from various geologic situations to make this certain. The primary deposits at Bisbee are, therefore, here classified as mesothermal to telethermal.

The wall-rock alteration at Bisbee--silification and pyrite dissemination in limestone and sericitization in quartz monzonite porphyry--is compatible with mesothermal conditions of ore deposition and would not have been changed appreciably during the later deposition of lower intensity ore minerals. The Sacramento porphyry was altered to a dominantly sericitized rock in which most of the sericite, some hydromuscovite (muscovite in which K^{+1} is largely replaced by H^{+1}), kaolinite, allophane ($Al_2SiO_5 \cdot nH_2O$), and alunite [$KAl_3(OH)_6(SO_4)_2$] were developed from the feldspars. Much sericite and some leucoxene and a little chlorite and kaolinite were derived from the biotite; the quartz was essentially unchanged. The quartz monzonite porphyry in the dikes, sills, and other apophyses was also highly altered; in addition to the minerals formed in the stock itself, locally abundant calcite and considerable epidote (originating from biotite) were developed. Since the altering solutions had become highly enriched in calcium (and carbonate) ions in their passage through the Bisbee limestones, it is not unreasonable that they would have produced these calcium-rich minerals on encountering the quartz monzonite bodies. The loss of potash during the solutions' travels through limestone probably accounts for locally heavy concentrations of hydromuscovite and kaolinite in portions of the smaller quartz-monzonite masses.

The original ore mined in the district was the oxidized ore from the limestones in which malachite and azurite were the principal minerals but which also contained some copper oxides and even more metallic copper. This ore owed its origin to the downward movement of copper, derived from the primary ore, through the limestone in which it was reprecipitated mainly as basic carbonates before it could pass beneath the water table. As the oxidized ore was followed downward and eastward, the primary ore was encountered, although local patches of oxidized ore have been mined as far down as the 2700 level of the Junction mine. Both the oxidized ore in the limestone and the secondary enrichment ore in the stock are classed as ground water-B2.

CASTLE DOME

Late Mesozoic to Early Tertiary — *Copper* — *Mesothermal*

Creasey, S. C., 1959, Some phase relations in the hydrothermally altered rocks of the porphyry copper deposits: Econ. Geol., v. 54, p. 351-373 (general)

Peterson, N. P., 1947, Phosphate minerals in the Castle Dome copper deposits, Arizona: Amer. Mineral., v. 32, p. 547-582

—— 1948, Geology of the Castle Dome copper deposit, Arizona: A.I.M.E. Tr., v. 178, p. 195-205

—— 1962, Castle Dome, in *Geology and ore deposits of the Globe-Miami district, Arizona*: U.S. Geol. Surv. Prof. Paper 342, p. 88

Peterson, N. P., and others, 1946, Hydrothermal alteration in the Castle Dome copper deposit, Arizona: Econ. Geol., v. 41, p. 820-840

—— 1951, Geology and ore deposits of the Castle Dome area, Gila County, Arizona: U.S. Geol. Surv. Bull. 971, 134 p.

Schwartz, G. M., 1947, Hydrothermal alteration in the "porphyry copper" deposits: Econ. Geol., v. 42, p. 319-352 (particularly p. 328-329, 346-352)

Notes

The Castle Dome area is a subdivision of the Globe-Miami mining district with its center some 5 miles west of the town of Miami; mining has been stopped in the area but leaching operations continue.

The oldest rock in the vicinity of Castle Dome is the older Precambrian Pinal schist; a much younger Precambrian formation, the Apache group, lies above a pronounced unconformity. A somewhat metamorphosed granite is intrusive in the Pinal schist but is not found in the Apache group. The first Cambrian formation was the Troy quartzite that was deposited disconformably on an erosion surface from which all of the uppermost formation of the Apache group, the Mescal limestone, had been removed. Whatever Paleozoic formations were deposited before the Devonian Martin limestone, except for small remnants of the Troy quartzite, were removed by pre-Martin erosion. Unconformably on the Martin lies the Mississippian Escabrosa limestone; and probably unconformably on the Escabrosa, the Pennsylvania Naco limestone was laid down.

The oldest post-Precambrian igneous rock in the area is a small mass of Mesozoic granodiorite lying to the south of the Castle Dome pit; its intrusion was followed by two bodies of mainly porphyritic quartz monzonite. The larger of these two intrusions lies to the north of the granodiorite and contains all the minable ore of the Castle Dome mine, while the smaller is about a mile southeast of the open pit on the south side of a belt of Pinal schist that lies between the two monzonite bodies. The quartz monzonite in the Castle Dome area probably is equivalent to the Lost Gulch quartz monzonite of the Globe quadrangle. The first major post-Precambrian faulting occurred after the intrusion of the quartz monzonite. This faulting was followed by the intrusion of huge volumes of diabase that entered the sediments as sills, filled many of the older faults, and created new ones west of the mine. The last intrusion of this period of igneous activity was of dikes and small bodies of granite porphyry now exposed in, and for some distance southeast of, the open pit. This granite probably is the equivalent of the Schultze granite in the main Globe-Miami area and differs in composition from it only slightly. The next formation in time was the Whitetail conglomerate that is absent from the immediate Castle Dome area and is composed mainly of material derived from the diabase bodies; it is probably early Tertiary in age, although it contains no fossils. It appears to be conformably overlain by probably middle Tertiary basal dacite tuff and massive dacite. After this eruption, the area was considerably faulted and eroded, and much of the dacite was removed before the deposition of the Gila conglomerate that may have formed in Pleistocene as well as Pliocene time. A final set of faults was produced after the deposition of the Gila beds.

The ore mineralization is confined to the southern half of the quartz monzonite body; the ore-bearing fractures also are mineralized where they cut the diabase sills in this portion of the quartz-monzonite bodies. From this it follows that the mineralization is post-diabase; however, the lack of fragments of altered and mineralized rock in the Whitetail conglomerate fails to put an upper bracket around the period of mineralization in the Castle Dome area. In the Miami and Inspiration areas, the zones of secondary enrichment are related to pre-Whitetail topography (and not to that of the present-day surface as is true at Castle Dome). The Miami-Inspiration primary ore certainly was appreciably pre-Whitetail, and the quite similar Castle Dome primary mineralization probably occurred at the same time. As the granite porphyry (the Schultze granite) is uneconomically mineralized southeast of the ore area and is nearest in age to the mineralization of all the intrusions; its source magma probably was also the source of the Castle Dome ore fluids. The granite porphyry contains altered diabase fragments and is pre-mineralization. Even if the granite porphyry is a late product of the Schultze granite magma chamber, the ore fluids apparently came from that source at a still later time. No exact age is firmly established for the Schultze granite, but it was almost certainly intruded at some time in the late Mesozoic to early Tertiary, and the ores are here given that age.

The hypogene mineralization at Castle Dome is quite simple, with chalcopyrite (later) and pyrite (earlier) being the most abundant sulfides and the ratio between them differing from place to place. The ore is either in, or in the immediate vicinity of, narrow, closely spaced generally parallel veins that contain quartz as the principal mineral. Some consist solely of quartz; others are composed of quartz and pyrite, and some mainly of pyrite; these last normally have a selvage of coarse sericite. The amount of chalcopyrite in these veins is small, the unenriched ore containing 0.3 percent copper as a maximum, much of which occurs as threadlike stringers or disseminated in the wall rock; very little is in the quartz-pyrite veins. Molybdenite in the deposit was formed before chalcopyrite but long after the first pyrite was deposited, and the ore runs between 0.01 and 0.02 percent MoS_2; no attempt has ever been made to recover it. A little sphalerite and galena are present in a few widely scattered veins related to the more or less north-south Dome fault and were developed well after the pyrite-chalcopyrite mineralization. Gold and silver are almost negligible constituents of the ore.

The mineralized areas would be unminable if they had not been secondarily enriched. The upper surface of the zone of supergene enrichment is largely parallel to the present surface of the area; on the other hand, the base of the zone bears no relation to topography whatsoever. Almost all of the zone is well above the present water table, and the zone increases generally from south to north both in thickness and amount of enrichment. Peterson (1951) suggests that the monzonite in the southern area was longer exposed to the leaching action of surface water and, therefore, greater amounts of secondary copper minerals were developed there than in the longer protected northern section.

Chalcocite is by far the most abundant secondary mineral, and it was almost entirely emplaced by rarely complete pseudomorphic replacements of chalcopyrite; only in areas unusually rich in primary chalcopyrite did chalcocite replace pyrite. The rare sphalerite was even more readily and completely replaced by chalcocite than was chalcopyrite. The minor covellite in the enriched zone appears to have been derived largely by oxidation of chalcocite; only a little was formed by the replacement of chalcopyrite.

The primary mineralization appears to fulfill all the requirements of deposition under mesothermal conditions, and the deposit is so classified here.

CHRISTMAS

Late Mesozoic to Early Tertiary	*Copper, Lead, Silver, Gold, Zinc*	*Hypothermal-2 to Mesothermal*

Creasey, S. C., and Kistler, R. W., 1962, Ages of some copper-bearing porphyries and other igneous rocks in southeastern Arizona: U.S. Geol. Surv. Prof. Paper 450-D, p. D1-D5

Eastlick, J. T., 1958, New developments at the Christmas mine, Arizona: Ariz. Geol. Soc. Digest, v. 1, p. 1-6

—— 1968, Geology of the Christmas mine and vicinity, Banner mining district, Arizona, in Ridge, J. D., Editor, *Ore deposits of the United States, 1933-1967* (Graton-Sales Volumes): Chap. 57, v. 2

Perry, D. V., 1969, Skarn genesis at the Christmas mine, Gila County, Arizona: Econ. Geol., v. 64, p. 255-270

Peterson, N. P., and Swanson, R. W., 1956, Geology of the Christmas copper mine, Gila County, Arizona: U.S. Geol. Surv. Bull. 1027-H, p. 351-373

Ross, C. P., 1925, Ore deposits of the Saddle Mountain and Banner mining districts, Arizona: U.S. Geol. Surv. Bull. 771, 72 p.

Silver, L. T., 1960, Age determinations on Precambrian diabase differentiates in the Sierra Ancha, Gila County, Arizona (abs.): Geol. Soc. Amer. Bull., v. 71, p. 1973-1974

Wiliden, R., 1964, Geology of the Christmas quadrangle, Gila and Pinal Counties, Arizona: U.S. Geol. Surv. Bull. 1161-E, p. E1-E64

Notes

The Christmas mine is located in the southwest corner of Gila County, about 60 miles north-northwest of Tucson.

The oldest rocks that outcrop in the immediate vicinity of the mine lie to the northwest and consist of the various subdivisions of the younger Precambrian Apache group. These rocks have not yet been encountered in the mine, although the lowest explored levels have reached the basal Cambrian beds.

The rocks of the Apache group are cut by tabular masses of diabase and diorite that generally parallel the bedding of the sediments. The diorites contain pegmatites, but the diabases do not. Age determination work by Silver (1960) on uranium and thorium minerals from these intrusives probably dates them as 1200 m.y. old.

The Cambrian rocks in the area are undifferentiated and consist of a wide variety of rock types. A considerable erosional unconformity separates the Cambrian beds from the Precambrian beneath. Unconformably overlying the Cambrian rocks in the mine area is the Devonian Martin limestone that is divided into three members. The Martin is conformably overlain by the massive Mississippian Escabrosa limestone. The next formation in the mine area is the Pennsylvania Naco limestone, the uppermost portion of which might have been deposited in Permian time although Wiliden (1964) does not think so.

Overlying the Paleozoic sediments is a thick series of undifferentiated volcanic rocks that includes andesite flows, tuffs, flow breccias, and conglomerates; in the mine area, these volcanic rocks are Cretaceous in age and almost completely surround the mine area on the surface; only to the west and in an isolated patch to the south do Paleozoic rocks appear at the surface.

Both the sedimentary sequence and the volcanic rocks are cut by three varieties of igneous rocks. The first of these is a microdiorite which, although it occurs in the country around the mine, does not appear in the mine area proper. The second type has been called quartz-mica diorite until the recent USGS bulletin (Wiliden, 1964) in which it is referred to as a feldspar-mica porphyry. The quartz-mica diorite occurs in a number of different shapes including numerous plugs, dikes, and sills, many of which are in the mine area itself. The third igneous rock type is a hornblende andesite that is found in dikes no nearer than about 2 miles from the mine; it also occurs in sills and plugs.

The main mass of quartz-mica diorite in the mine area is a plug, poorly exposed at the surface. The Christmas fault cuts irregularly north-northwest across the central portion of the stock; the block to the northeast (at the surface composed of Cretaceous volcanic rocks and quartz-mica diorite) has been dropped down relative to that on the southwest (made of Paleozoic beds surrounding the diorite core).

The structural relationships demonstrate that the igneous intrusion took place not only after the Cretaceous volcanic rocks had been deposited in the area but also after much of faulting had occurred. This would seem to indicate that the diorite probably was intruded in Tertiary time; since the intrusion apparently was a part of the same orogenic epoch as the faulting, this would date the intrusion as early Tertiary. This date has recently been confirmed (Creasey and Kistler, 1962) by K/Ar dating of biotite in one of the diorite dikes in the Christmas mine that gives an age for that intrusion of 62 m.y. Thus, it appears reasonable to categorize the deposits as late Mesozoic to early Tertiary.

The ore bodies of the Christmas mine occur as replacement deposits in the Naco, Escabrosa, and Martin formations, with the type and intensity of mineralization depending on the distance of the ore from the contact with the diorite, on the degree of fracturing (particularly that fracturing that occurred after the limestones had been altered to skarn), and on the nature of the various beds. Before (but apparently not long before) the ore minerals were introduced, the Naco and Escabrosa limestones were altered to marble and skarn.

The middle member of the Naco limestone was the first ore zone to be extensively mined, there being nine distinct ore horizons on the northwest contact and eleven on the south. Some ore is known in the Naco where it has been encountered in drill holes on the northeast (downdropped) side of the Christmas fault.

In the Escabrosa the ores occur as massive, irregular replacements near the diorite contacts; these ores are generally higher in grade than those in the Naco. The lower part of the Martin limestone has provided the most extensive of the replacement deposits. In this horizon the ore is found in a flat-dipping, massive, tabular bed. The most favorable rock for replacement was the lower 30 feet of the Martin.

Magnetite is the most abundant metallic mineral in the mine, making up 15 to 25 percent of the mineralized rock and increasing in amount with depth. The sulfides occur with the magnetite but show a lateral and a less well-developed vertical zoning. Near the intrusive borders, the sulfides are mainly pyrite and chalcopyrite; farther away from the contact they change to chalcopyrite and bornite; and still farther out to pyrrhotite, pyrite, sphalerite, and chalcopyrite. In the vertical direction, in the thicker ore volumes, the core of the mineralized mass is chalcopyrite and bornite and is surrounded by pyrrhotite, pyrite, chalcopyrite, and sphalerite, and sometimes galena; some molybdenite is present but its relations with the other sulfides are uncertain.

It appears certain that most of the magnetite-sulfide mineralization occurred under hypothermal conditions. Although the ore minerals are definitely later than the silicates, fractures in the silicated rock apparently being necessary to permit entry of the ore-forming fluids, the suite of magnetite, pyrrhotite, chalcopyrite, and sphalerite is typical of the hypothermal intensity range. The time relationship of the bornite to the other sulfides has not been discussed in the literature but its appearance in hand specimens suggests that it was introduced after the chalcopyrite. It may be, therefore, that the bornite and the late and minor galena were the last minerals of the sulfide assemblage to form and that they were deposited in the mesothermal range. Because bornite is so consistently diagnostic of ore formation in the mesothermal range, the Christmas ores are here classified as hypothermal-2 to mesothermal.

In the upper levels of the mine, the ores were highly oxidized, with both oxide minerals and supergene sulfides having been developed, the former having resulted from reactions above the water table between the downward moving solutions and the abundant carbonate rocks.

GLOBE-MIAMI

Late Mesozoic to Early Tertiary	*Copper, Molybdenum*	*Mesothermal (por-Cu), Hypothermal-1 and Hypothermal-2 (veins)*

Jespersen, A(nna), 1964, Aeromagnetic interpretation of the Globe-Miami copper district, Gila and Pima Counties, Arizona: U.S. Geol. Surv. Prof. Paper 501-D, p. D70-D75

Kuellmer, F. J., 1960, Compositional variation in alkali feldspars in some intrusive rocks near Globe-Miami, Arizona: Econ. Geol., v. 55, p. 557-562

Olmstead, H. W., and Johnson, D. W., 1966, Inspiration geology, in Titley, S. R., and Hicks, C. L., Editors, *Geology of the porphyry copper deposits--southwestern North America*: Univ. Ariz. Press, Tucson, p. 143-150

Peterson, N. P., 1950, Lead and zinc deposits in the Globe-Miami district, Arizona, in *Arizona zinc and lead deposits*, pt. 1: Ariz. Bur. Mines, Geol. Ser. no. 18, Bull. no. 156, p. 98-112

—— 1952, Structural history of the Globe-Miami district: Ariz. Geol. Soc., Guidebook I, Southern Arizona, p. 123-127

—— 1954, Copper Cities copper deposit, Globe-Miami district, Arizona: Econ. Geol., v. 49, p. 362-377

—— 1954, Geology of the Globe quadrangle, Arizona: U.S. Geol. Surv. Geol. Quad. Map GQ 41, 1" = 2000' (with text)

—— 1962, Geology and ore deposits of the Globe-Miami district, Arizona: U.S. Geol. Surv. Prof. Paper 342, 151 p.

Ransome, F. L., 1903, Geology of the Globe copper district, Arizona: U.S. Geol. Surv. Prof. Paper 12, 168 p. (particularly p. 114-163)

—— 1904, Globe folio, Arizona: U.S. Geol. Surv. Geol. Atlas, Folio 111, 17 p.

—— 1919, The copper deposits of Ray and Miami, Arizona: U.S. Geol. Surv. Prof. Paper 115, 192 p. (particularly p. 107-121 and 144-176)

Schwartz, G. M., 1921, Notes on textures and relationships in the Globe copper ores: Econ. Geol., v. 16, p. 322-329

—— 1947, Hydrothermal alteration in the "porphyry copper" deposits: Econ. Geol., v. 42, p. 319-352 (particularly p. 329-331, 346-352)

Simmons, W. W., and Fowells, J. E., 1966, Geology of the Copper Cities mine, in Titley, S. R., and Hicks, C. L., Editors, *Geology of the porphyry copper deposits--southwestern North America*: Univ. Ariz. Press, Tucson, p. 151-156

Tenney, J. B., 1935, Globe-Miami district (Arizona), in *Copper resources of the world*: 16th Int. Geol. Cong., v. 1, p. 189-201

Notes

The Globe-Miami district is some 75 miles east of Phoenix. The oldest formation is the Pinal schist, and the Madera diorite and the Ruin granite are intrusive into it. These rocks were followed by the late Precambrian Apache group containing conglomerate, arkosic quartzite, quartzite shale, and limestone. The Apache rocks are appreciably less metamorphosed than the Pinal schists. The probably Precambrian Troy quartzite was deposited disconformably on a moderately eroded surface of older Precambrian rocks; there followed a period of little or no sedimentary deposition that lasted until the Late Devonian when the Martin limestone was first laid down in areas of low relief and later was extended to cover the entire district. The Mississippian Escabrosa limestone and the Pennsylvanian Naco limestone are conformable on the Martin, although the sequence lacks upper Mississippian rocks; there appear to have been no angular disconformities developed during Paleozoic time. There are no Permian or Mesozoic sediments in the area. The end of the Mesozoic was marked by a series of igneous intrusions of which the first was the Lost Gulch monzonite (the host rock of the Copper Cities ore body). It was followed by huge volumes of diabase, introduced primarily as sills and mainly into the rocks of the Apache group and the Troy quartzite. The area had been considerably faulted, prior to the injection of the diabase, and much additional faulting appears to have been caused by that intrusion. The next igneous event was the introduction of a great many small dikes and irregular bodies of diorite porphyry that entered the area generally along diabase contacts. The final premineralization igneous rocks were the Schultze granite and related dikes and small masses of granite porphyry; the Schultze granite is, in part, the host rock of the great Miami-Inspiration ore body, the remainder having been formed in Pinal schist. The igneous rocks appear to have been, in large part, brought into the area along a northeast-southwest zone of weakness that had some earlier expression at least in the Precambrian. The primary ore mineralization must have been introduced shortly after the intrusion of the Schultze granite but before the formation of the early Tertiary Whitetail conglomerate, since the huge blanket of secondarily enriched sulfide ore in the Miami-Inspiration deposit is definitely related to the pre-Whitetail erosion surface. The period of erosion and Whitetail deposition was interrupted by the extrusion of a dacite lava. Large-scale faulting followed shortly after the dacite eruption, and the elevated blocks were rapidly worn down with the formation of the Gila conglomerate in the lower lying areas. This conglomerate, in turn, has now been largely removed from the higher elevations, but thousands of feet of Gila beds remain in the huge graben valley between Miami and Globe. The Inspiration block (west) and the Globe Hills block (east), the former including such prominent structural features as the Castle Dome and Copper Cities horsts, lie on either side of the graben valley.

It appears that the mineralization was, at the latest, developed early in the early Tertiary and probably during the late Mesozoic to early Tertiary Rocky Mountain (Laramide) orogeny. Because the Schultze granite was the last igneous rock to have been intruded before the mineralization, its parent magma chamber is considered to have been the most likely source of the ore forming fluids for both the porphyry copper type deposits (Miami-Inspiration and Copper Cities) and for the vein type ore bodies of the Globe Hills area.

There are two types of ore deposits in the Globe-Miami area: (1) the huge porphyry copper type deposits of the Miami-Inspiration area and of the Copper Cities area about two miles north of the eastern end of the Miami-Inspiration workings and (2) the vein deposits in the Globe Hills area. The mineralization in the two porphyry copper deposits is simple, consisting mainly of veinlets of quartz, pyrite, minor chalcopyrite, and molybdenite, plus the products of host-rock alteration (quartz, sericite, and clay minerals) generally developed somewhat earlier than the sulfides. Although the Copper Cities mineralization is in the Lost Gulch quartz monzonite, that at Miami (east) is largely in Schultze granite and that at Inspiration (west) mainly in Pinal schist; the mineralization and wall-rock alteration is essentially the same in both areas. The ore bodies are only a small fraction of the total mineralized area that may average 0.4 and 0.5 percent copper, while the ore bodies themselves are enriched to less than 1 percent copper. The enrichment consists of the conversion of the primary chalcopyrite and much of the pyrite to chalcocite and minor covellite. After the present erosion cycle had removed most of the cover (dacite, Gila conglomerate, and leached capping) from the enriched zone, the exposed copper sulfides were extensively altered to carbonates and silicates.

The vein deposits in the Globe Hills area, now long abandoned, lie in northeast-striking fault fractures in both diabase and late Precambrian and Paleozoic sediments. The only economically important vein fillings were the copper deposits of the Old Dominion system and other similar lodes. The primary minerals in these veins were principally quartz, pyrite, chalcopyrite, and specular hematite. Where the veins cut the Paleozoic limestones, massive replacement deposits, that carry the same sulfides as the veins, were formed in certain favorable beds. Secondary enrichment that produced chalcocite and covellite and subsequent oxidation were effective down to 1000 feet or even more below the surface; most of the mining in the Globe Hills area was done in the secondarily enriched material.

The mineralization in the primary ores of the porphyry coppers is typically that of the mesothermal range both as to sulfides and host rock alteration minerals, and the primary deposits can safely be classified as mesothermal. The presence of specular hematite in considerable quantity in close association with the pyrite-chalcopyrite primary mineralization and the development of high-temperature wall-rock alteration minerals in the limestones of the bedded replacements at about the same time as the sulfide mineralization strongly suggest that the Old Dominion type vein and replacement deposits were formed in the hypothermal range. These deposits, therefore, are classified as hypothermal-1 and hypothermal-2, but it must be remembered that the production from these veins has had only a small fraction of the value of that from the porphyry-copper deposits in the western part of the Globe-Miami district.

IRON KING

Middle Precambrian	*Zinc, Lead, Gold, Silver, Copper*	*Hypothermal-1 to Leptothermal*

Creasey, S. C., 1950, Iron King mine, Yavapai County, Arizona, in *Arizona zinc and lead deposits*, pt. 1: Ariz. Bur. Mines, Geol. Ser. no. 18, Bull. no. 156, p. 112-122

—— 1952, Geology of the Iron King mine, Yavapai County, Arizona: Econ. Geol., v. 47, p. 24-56

—— 1958, Iron King mine, in *Geology and ore deposits of the Jerome area, Yavapai County, Arizona*: U.S. Geol. Surv. Prof. Paper 308, p. 155-169

Gilmour, P., and Still, A. R., 1968, The geology of the Iron King mine, in Ridge, J. D., Editor, *Ore deposits of the United States, 1933-1967* (Graton-Sales Volumes): Chap. 59, v. 2

Krieger, M. H., 1965, Geology of the Prescott and Paulden quadrangles, Arizona: U.S. Geol. Surv. Prof. Paper 467, 127 p. (particularly p. 109-110)

Lindgren, W., 1926, Iron King mine, in *Ore deposits of the Jerome and Bradshaw Mountains quadrangles, Arizona*: U.S. Geol. Surv. Bull. 782, p. 127-128

Mills, H. F., 1941, Ore occurrence at the Iron King mine: Eng. and Min. Jour., v. 142, no. 10, p. 56-57

—— 1947, Occurrence of lead-zinc ore at Iron King mine, Prescott, Arizona: A.I.M.E. Tr., v. 178, p. 218-222

Notes

The Iron King deposit is located some 75 miles north of Phoenix and about 12 miles south of east of Prescott.

The ore veins lie in a group of older Precambrian volcanic rocks that contain one dike of foliated quartz diorite, also of older Precambrian age. A dike of rhyolite (some 700 feet east of the Iron King veins) has not been dynamo-metamorphosed. Similar rhyolite dikes in the Bradshaw Mountains are mineralized and Lindgren considered them to be Late Cretaceous to early Tertiary. The older of the volcanic rocks in the area is the Spud Mountain, made up of medium- to fine-grained tuffaceous andesite, rhyolitic tuff and conglomerate, and interbedded coarse- to medium-grained andesitic tuff and breccia. The ore veins are contained in the coarse- to medium-grained tuff of the upper unit of the Spud Mountain formation. The adjacent and younger Iron King volcanic rocks dip northwest at steep angles, indicating that the beds in the mine area are slightly overturned. The rocks of the area have been intensely deformed, the beds being isoclinally folded, with uniformly steep dips, the beds in much of the area being overturned. Foliation appears to have formed at the same time as the folding, and the axial planes of the folds and the foliation are approximately parallel. Faults in the mine area are probably more abundant than the maps (Creasey, 1958) show because of coincidence of the fault planes with the foliation and the lack of recognizable marker beds in the volcanic rocks.

It is, as yet, impossible to assign an absolute age to the older Precambrian formations in the Iron King (and general Jerome) area. In the Bagdad area, the older Precambrian rocks have been tentatively correlated with the Yavapai series; into these rocks near Bagdad the Lawler Peak granite was intruded, and radioactive age determinations assign an age of 1600 m.y. to that silicic rock. The Lawler Peak granite probably was equivalent in time to the intrusives that constituted the last stage in the older Precambrian portion of geologic time in the Jerome area, the stage to which the mineralization at Iron King (and at Jerome) belongs. Thus, it appears that the best age that can be given the Iron King mineralization is middle Precambrian.

The ore-bearing veins of the Iron King mine occupy a considerable portion of a zone about 2500 feet long and 100 feet wide that has the same strike as the foliation (N20°E), dips west at 70° to 80°, and plunges northward at 50° to 60°. In places the parallel veins can be readily distinguished one from the others; in other places they are so poorly defined as to require classification as zones of veins. On the lower levels (2100 was the lowest level in 1960), the ore is less massive than on levels above, and is found in several narrow veins (about six inches in width), separated by chloritized and sericitized schistose tuff. The vein structures in the hanging wall trend toward the main vein or vein zone at low angles, the hanging wall structures being farther from the vein as they are followed south.

The first stage of mineralization (and wall-rock alteration) was the deposition in the sheared tuff of pyrite, quartz, ankerite, chlorite (at least on the lower levels), and sericite, with some apatite. Along the footwall (east) side of the vein-bearing structure, these minerals formed well-defined veins, while to the west they were developed as disseminations and narrow veinlets. This stage of mineralization was followed by a second period of shearing that mylonized the northern part of the veins. Into the sheared first generation of minerals was introduced a series of minerals that apparently began with pyrite, overlapped and followed by arsenopyrite that is quite widespread but minor in amount. The next mineral was a brown sphalerite that well may overlap with the arsenopyrite but also replaces it; the brown sphalerite makes up 90 percent of the zinc sulfide. The other 10 percent of the sphalerite is a rosin jack, veinlets of which cut the brown sphalerite. At least a major fraction of the brown sphalerite contains microscopic blebs of chalcopyrite, though the rosin sphalerite apparently does not; veinlets of this rosin sphalerite, in fact, cut galena. Galena and tetrahedrite-tennantite are closely associated, being generally in contact with each other; galena is interstitial to pyrite and contains replacement remnants of that mineral; veinlets of galena cut arsenopyrite and brown sphalerite. Tennantite shows the same relationships to other minerals that are exhibited by the lead sulfide. Chalcopyrite, in addition to its occurrence as blebs in brown sphalerite, appears as small grains in association with various minerals and in small masses with tetrahedrite-tennantite. Both the galena and the tetrahedrite contain silver, but the greater amount of silver is recovered with copper and therefore must come from the tetrahedrite.

In addition to the wall-rock alteration (first generation) minerals, a second generation of sericite was developed later than the ore minerals. Some ankerite appears to have been deposited with the ore minerals as well as in the first or wall-rock alteration generation.

The presence of small amounts of arsenopyrite throughout the ore and of blebs of chalcopyrite in much of the brown sphalerite indicates that part of the mineralization at least took place under low-intensity hypothermal conditions; on the other hand, the wall-rock alteration (admittedly earlier than the ore) suggests mesothermal deposition. Finally the presence of considerable silver-rich tetrahedrite-tennantite associated with silver-bearing galena argues for some deposition in the leptothermal range. The deposit, therefore, is here classified as hypothermal-1 to leptothermal.

The outline of ore processes just given is completely different from that of Gilmour and Still (1968). Their concept of a volcanic-exhalative source for the ores and their remobilization under metamorphic conditions seems unreasonable to me but must be considered in detail by any student.

JEROME

Middle Precambrian	*Copper, Silver, Gold, Zinc*	*Hypothermal-1 to Mesothermal*

Anderson, C. A., and Creasey, S. C., 1958, Geology and ore deposits of the Jerome area, Yavapai County, Arizona: U.S. Geol. Surv. Prof. Paper 308, 185 p.

Fearing, J. L., Jr., 1926, Some notes on the geology of the Jerome district, Arizona: Econ. Geol., v. 21, p. 757-773

Fearing, J. L., Jr., and Benedict, P. C., 1925, Geology of Verde Central mine: Eng. and Min. Jour.-Press, v. 119, no. 15, p. 609-611

Finlay, J. R., 1918, The Jerome district of Arizona: Eng. and Min. Jour., v. 106, no. 13, 14, p. 557-562, 605-610

Hansen, M. G., 1930, Geology and ore deposits of the United Verde mine: Min. Cong. Jour., v. 16, no. 4, p. 306-311

Lindgren, W., 1926, Ore deposits of the Jerome and Bradshaw Mountains quadrangles, Arizona: U.S. Geol. Surv. Bull. 782, 192 p. (particularly p. 7-97)

Moxham, R. M., and others., 1965, Gamma-ray spectrometer studies of hydrothermally altered rocks: Econ. Geol., v. 60, p. 653-671

Ransome, F. L., 1933, Jerome (Arizona), in *Ore deposits of the southwest*: 16th Int. Geol. Cong., Guidebook 14, p. 20-22

Reber, L. E., Jr., 1922, Geology and ore deposits of the Jerome district: A.I.M.E. Tr., v. 66, p. 3-26

—— 1938, Jerome district, in *Some Arizona ore deposits*: Ariz. Bur. Mines, Geol. Ser. no. 12, Bull. no. 145, p. 41-65

Rice, M., 1920, Petrographic notes on the ore deposits of Jerome, Arizona:

A.I.M.E. Tr., v. 61, p. 60-65

Schwartz, G. M., 1938, Oxidized copper ores of the United Verde Extension mine: Econ. Geol., v. 33, p. 21-33

Tenney, J. B., 1935, Jerome district (Arizona), in *Copper resources of the world*: 16th Int. Geol. Cong., v. 1, p. 179-189

Notes

The deposits of the United Verde and the United Verde Extension mines are found in the immediate vicinity of Jerome, some 90 miles due north of Phoenix and some 15 miles northeast of Prescott.

The oldest rocks in the area are the older Precambrian Yavapai series which is made up of volcanic and sedimentary rocks, now all considerably metamorphosed. In the Jerome area, the Yavapai series (which well may be equivalent to the older Precambrian rocks in the Bagdad area and of the Pinal schist in southern Arizona) is divided into two groups of the north-south trending Shylock fault. The east of this fault is the Ash Creek group that consists of basaltic, dacitic, and andesitic flows and pyroclastic rocks; associated with these is a sequence of sedimentary tuff rocks with which are interbedded jasper-magnetite and chert; the ore bodies at Jerome are on this side of the fault. The Ash Creek has been divided into seven formations and neither base nor top is known. The early Precambrian rocks to the west of the Shylock fault are known as the Alder group that contains, among others, the Spud Mountain volcanic rocks.

An impressive unconformity divides the older Precambrian rocks from the Paleozoic sediments that lie above them in the Jerome area. The oldest of these is the Cambrian Tapeats(?) sandstone on which the Upper Devonian Martin limestone appears to have been conformably laid down; above the Martin is the Mississippian Redwall limestone and the Pennsylvanian-Permian Supai formation. During most of the Mesozoic and the early Tertiary, the Jerome area was being markedly eroded, a cycle interrupted by the extrusion of some 1400 feet of late Tertiary Hickey basalt interbedded with gravels in the Jerome area. The last deposition in the area was that of the Verde Lake beds which were deposited in the Verde valley after that structure was formed by late Tertiary faulting.

As the massive sulfides are largely emplaced in a remnant of the youngest of the Precambrian volcanic materials, the structure of the Ash Creek is of concern in a study of the ores. In the United Verde area the once overlying cover of essentially flat-lying Paleozoics and Tertiary sediments and volcanic rocks has been removed by erosion; therefore, their structure is of only academic interest so far as the major deposit of the area is concerned. On the other hand, on the east side of the Verde normal fault, in the area of the United Verde Extension (U.V.X.) mine, the Precambrian was buried under hundreds of feet of gently dipping Tertiary (Hickey formation) gravels, volcanic rocks, and Paleozoic limestone and sandstone, but no evidence of any extension of the U.V.X. mineralization has been found in these rocks. The United Verde ore deposits appear to have been localized in the nose of a north-northwestward plunging anticline, with the plunge of the pipes being parallel to the plunge of the minor folds and lineation as marked by cleavage-bedding intersections. In the U.V.X. mine the evidence is uncertain as to the location of the axial plane of the Mingus anticline for the ore may be, depending on the interpretation put on the data, on either the eastern or western limb of that structure.

The two deposits are located one on either side of the west-of-north-trending Verde normal fault zone that dips east at about 60°. The latest displacement on the fault was in Tertiary time (after the deposition of the Hickey formation) and caused a vertical separation of about 1500 feet; there appears, however, to have been an earlier vertical separation of about 1000 feet before the deposition of the first Paleozoic beds but after the mineralization. The possibility that the U.V.X. ore body is the faulted upper extension of the United Verde mass has been thoroughly discussed by Norman and others (Anderson and Creasey, 1958, p. 145-149), and these authors have reached the conclusion that this relationship is unlikely to have obtained.

Within the United Verde and U.V.X. areas, it is evident that the ore was introduced after the Precambrian gabbro but before the somewhat later andesite dikes. This places the mineralization appreciably before the last of the older Precambrian igneous rocks of the area had been emplaced and, therefore, probably dates the mineralization as occurring toward the end of the early Precambrian. The dating of the older Precambrian rocks of Arizona as early Precambrian is not, however, firmly established, but the age of the Jerome deposits is most probably late middle Precambrian.

Most of the metal produced from the immediate Jerome area has been from the massive sulfide deposits of the United Verde and the United Verde Extension mines that occur in form of pipes, lenses, or tabular bodies. The massive pyrite is located almost entirely in a highly contorted band of sedimentary tuff of the Grapevine Gulch formation (probably the youngest of the Ash Creek group) that lies on the southeast contact of a semi-concordant body of gabbro, this latter rock forming the hanging wall of the sulfide bodies. The first step in the mineralization process was the introduction of large and small lenticular to irregular masses of nearly pure quartz in the Grapevine Gulch formation. The next stage in the mineralization was the replacement of a huge pipelike volume of the Grapevine Gulch by massive pyrite containing some quartz and carbonate; some of the quartz may be residual. The pipe bears about N20°W and plunges about 65°N. The area of the pipe differs markedly from level to level. On the 4500 level the sulfide mass consists of only six rootlike masses, the largest of which has an area of less than 2000 square feet. Although the contact of the sulfide pipe with the surrounding rocks is usually sharp, tongues of massive sulfide often extend out into the bordering rocks.

The third phase of the mineralization was the formation of nearly pure chlorite rock (locally called black schist) mainly on the footwall of the pyrite body probably by replacement of quartz porphyry. That the chlorite is later than the pyrite is indicated by the lack of chlorite in the remnants of quartz porphyry and volcanic rocks protected within the pyrite of the pipe.

The fourth stage was the introduction (1) into the massive pyrite of chalcopyrite and some sphalerite as fillings of fractures and as replacements of pyrite cut from the fractures; intersecting veinlets of chalcopyrite indicate that additional fracturing of pyrite occurred after the early chalcopyrite had been emplaced; (2) into the chlorite (black schist) of chalcopyrite and pyrite as fillings of intersecting and branching veinlets; and (3) into the quartz porphyry of chalcopyrite, with some sphalerite and pyrite, as fillings of intersecting and branching veins; chlorite is common along the margins of these veins. The final period of mineralization was the formation of abundant quartz-carbonate veins and nodules that included a variety of sulfide minerals; locally some of these veins carried tennantite in sufficient quantities to double or treble the silver content of the copper ore with which they were associated.

The ore veinlets were not emplaced uniformly throughout the pyrite pipe but formed ore shoots of various shapes and sizes that plunge northward more steeply than does the pipe. Two subsidiary bodies, the Haynes and the North, are generally similar in character to the main pipe, although the Haynes ore body contains pyrrhotite (and some magnetite is present in the black schist and the quartz porphyry), and the North ore body appears to be richer in sphalerite and galena than the main ore and also contains minor arsenopyrite.

The chalcopyrite of the United Verde contains tiny grains or blebs of sphalerite 0.05 to 0.1 mm in diameter; if these blebs are the result of unmixing of sphalerite that was originally in solid solution in the chalcopyrite, they indicate deposition of the original solid solution at temperatures well above the lower limit of the hypothermal range. The chloritic alteration that precedes and probably to some extent accompanies the emplacement of the chalcopyrite ore is not incompatible with hypothermal conditions of ore formation. The United Verde ores, therefore, are here classified as hypothermal-1.

The United Verde sulfide mass was overlain by an oxidized zone averaging

100 feet in thickness, almost all of which was of economic value. Some sulfide enrichment also appears to have taken place below the water table.

The rocks associated with the U.V.X. ore body are much the same as those in the vicinity of the United Verde pipe with the exception of the addition of Deception rhyolite and, of course, the post-ore covering of Paleozoic and Tertiary rocks. A quartz-sericite alteration in the quartz porphyry and the Deception rhyolite is present only between the ore body and the Verde fault to the west; the remaining volcanic rocks are much fresher. On the lower levels of the mine, the massive sulfide body was similar to that of the United Verde mine. In the zone of supergene enrichment from which the bulk of U.V.X. production came, chalcopyrite and sphalerite probably were almost completely replaced by chalcocite, and pyrite largely so. The enrichment in the U.V.X. mine was completed before the deposition of the Cambrian Tapeats (?) sandstone.

There appears to be no published information on the presence or absence of sphalerite blebs in the primary ore of the U.V.X.; it appears doubtful, however, that two such similar primary ore bodies as those of the U.V.X. and the United Verde would have been formed under different conditions. The primary U.V.X. ore is therefore also categorized as hypothermal-1.

JOHNSON CAMP

Late Mesozoic to Early Tertiary	*Copper, Zinc*	*Hypothermal-2 to Mesothermal*

Baker, A., III, 1953, Localization of pyrometasomatic ore deposits at Johnson Camp, Arizona: A.I.M.E. Tr., v. 196, p. 1272-1277 (in Min. Eng., v. 5, no. 12)

—— 1960, Chalcopyrite blebs in sphalerite at Johnson Camp, Arizona: Econ. Geol., v. 55, p. 387-398

Cooper, J. R., 1950, Johnson Camp area, Cochise County, Arizona, in *Arizona lead and zinc deposits*, pt. 1: Ariz. Bur. Mines, Geol. Ser. no. 18, Bull. no. 156, p. 30-39

—— 1957, Metamorphism and volume losses in carbonate rocks near Johnson Camp, Cochise County, Arizona: Geol. Soc. Amer. Bull., v. 68, p. 577-610

—— 1959, Some geologic features of the Dragoon quadrangle: Ariz. Geol. Soc. Guidebook II, Southern Arizona, p. 139-158

Cooper, J. R., and Huff, L. C., 1951, Geological investigations and geochemical prospecting experiment at Johnson, Arizona: Econ. Geol., v. 46, p. 731-756

Cooper, J. R., and Silver, L. T., 1964, Geology and ore deposits of the Dragoon quadrangle, Cochise County, Arizona: U.S. Geol. Surv. Prof. Paper 416, 196 p.

Gilluly, J., and others, 1954, Late Paleozoic stratigraphy of central Cochise County, Arizona: U.S. Geol. Surv. Prof. Paper 266, 49 p.

Kellogg, L. O., 1906, Sketch of the geology and ore deposits of the Cochise mining district, Cochise County, Arizona: Econ. Geol., v. 1, p. 651-659

Romsio, T. M., 1949, Investigation of Keystone and St. George copper-zinc deposits, Cochise County, Arizona: U. S. Bur. Mines R.I. 4504, 21 p. (mimeo.)

Notes

The Johnson Camp ore deposits lie in the eastern slopes of the Little Dragoon Mountains some 50 miles south of east from Tucson and have produced ores valued at about twenty million dollars. Recent exploration in the area appears to have made sufficient discoveries to justify large-scale mining.

The oldest rocks in the area are the ubiquitous Pinal schists of older Precambrian age that here consist of highly metamorphosed arkosic sandstones, shales, and volcanic rocks that were intruded by a variety of igneous rocks also of older Precambrian age. The Pinal schist is overlain, above a strongly angular unconformity, by the ever-present Apache group of younger Precambrian age. A thick sequence of Paleozoic sedimentary rocks lies unconformably above the Precambrian beds, the lowest of which is the Bolsa quartzite, a formation of highly varied thickness due to the irregular erosion surface on which it was deposited. Conformably over the Bolsa quartzite is the Abrigo formation composed of dolomitic sandstone, sandy dolomite, quartzite, limestone, calcareous sandstone, and shale, containing the bulk of the ore. The time span between the Cambrian Abrigo and the overlying upper Devonian Martin formation apparently was one in which essentially no deposition of sediments occurred, or what sediments did form have since been eroded. Above the Devonian beds are the Mississippian Escabrosa limestone, the Pennsylvanian or Mississippian Black Prince limestone, and the Pennsylvanian and Permian Naco group. No later sediments are known in the general area until the Lower Cretaceous Bisbee group that has a fault contact with the older rocks; some andesitic and rhyolitic tuff breccia and conglomerate were, however, developed locally during the lower Mesozoic. The Bisbee group consists of the basal Glance conglomerate and the Morita-Cintura formation. The Cretaceous and older rocks were thrust faulted before the intrusion of the Texas Canyon quartz monzonite stock, the northeastern contact of which lies within about one-half mile of the southern part of the mineralized zone. The monzonite has exerted a considerable contact effect on the surrounding Paleozoics, particularly in the area to the northeast in which the ore deposits were later formed. The Texas Canyon quartz monzonite is considered to be 50 m.y. old or middle Eocene. Its intrusion into already deformed lower Cretaceous rocks indicates that this dating is reasonable. The monzonite magma chamber is the most reasonable source both for the solutions that produced the early, high-temperature silicates and the ore-forming fluids. The ores are, therefore, here dated as early Tertiary.

The Johnson Camp ore bodies occur at or near the intersections of certain of the fractures with favorable beds in the Abrigo formation. There are three main types of these ore bodies: (1) chimneys, generally large, with more or less oval cross sections and with their long and intermediate axes in the plane of the beds and their short axes perpendicular to the bedding plane; (2) mantos, again with an oval cross section, but with their axes at high angles to the dips of the beds, their intermediate axes normally lying parallel to the strike of the beds, and their short axes perpendicular to the long axes; these mantos may have tabular extensions that follow the bedding of favorable beds; and (3) tabular replacements of beds in which the long axes are in the plane of the bedding and the short axes perpendicular to the bedding plane.

The ores occur in highly altered portions of the Abrigo formation in which high-temperature silicates were developed largely before the ore fluids reached the area.

The quartz-monzonite source magma appears to have been responsible for both the contact-metamorphic type of change in the Abrigo rocks and for the later mineralization. Both of these effects appear to have been produced after the monzonite had cooled and probably had solidified, at least in its upper portions. The most intense metamorphism took place away from the actual monzonite-sedimentary contact, with a bed usually being far less changed at the actual contact than was the same bed thousands of feet from it. It would seem probable, therefore, that emanations from the same general source as the monzonite magma, rather than that material transferred directly across the contact, were responsible for the metamorphic effects.

The metamorphosed carbonate rocks can be divided into four zones. If these are examined outward from the contact, the characteristic minerals are (1) garnet, with local wollastonite and idocrase; (2) forsterite and diopside; (3) tremolite; and (4) chlorite and local talc. The outer three zones developed in abundance only in impure dolomites; the inner zone formed only in impure limestones (impure dolomites in the inner zone contain the minerals of the zone 2, the zone second from the contact).

No matter what the processes involved in forming the early metamorphic minerals, the bulk of the ore mineralization is later than the higher-temperature silicates. The ore minerals are chalcopyrite (with local bornite), sphalerite, pyrite, scheelite, and a little magnetite and molybdenite. Tremolite, chlorite, calcite, and in places, quartz and traces of fluorite and biotite were deposited at the same time as the ore minerals. Not only are the ore stage minerals interstitial to the earlier garnet, diopside, and wollastonite, but they also cut and replace them, thus separating the nonmetallics into two groups: one pre-ore and the other contemporaneous with it.

The high-temperature character of the ore stage minerals is indicated not only by the presence of tremolite in particular but also by the scheelite, magnetite, and molybdenite. Even more diagnositc of hypothermal conditions, however, is the occurrence in the sphalerite of microscopic exsolution blebs of chalcopyrite, indicating deposition of the original solid solution at temperatures at or above the lower limit of the hypothermal range. Where bornite is present in the ore, however, the associated sphalerite contains essentially no chalcopyrite blebs even though chalcopyrite is present in quantity in the rock volume involved. It would seem, therefore, that a minor portion of the ore may have been emplaced under mesothermal conditions. The deposits, therefore, are here classified as hypothermal-2 to mesothermal.

MAGMA (SUPERIOR)

Late Mesozoic to Early Tertiary	*Copper, Silver, Gold, Zinc, Lead*	*Hypothermal-2 to Telethermal*

Ettlinger, I. A., and Short, M. N., 1935, The Magma mine Superior (Arizona), in *Copper resources of the world*: 16th Int. Geol. Cong., v. 1, p. 207-213

Hammer, D. F., and Peterson, D. W., 1968, Geology of the Magma mine area, Arizona, in Ridge, J. D., Editor, *Ore deposits of the United States, 1933-1967* (Graton-Sales Volumes): Chap. 61, v. 2

Hammer, D. F., and Webster, R. N., 1962, Some geologic features of the Superior area, Pinal County, Arizona (with a section by D. C. Lamb, Jr.): New Mex. Geol. Soc. 13th Ann. Field Conf. Guidebook, p. 148-152

Michell, W. D., 1948, Applied geology at the Magma mine, Superior, Arizona: A.I.M.E. Tr., v. 178, p. 158-169

Park, C. F., Jr., and MacDiarmid, R. A., 1964, Magma mine, Arizona: in *Ore deposits*: Freeman, San Francisco, p. 297-302

Ransome, F. L., 1913, Copper deposits near Superior, Arizona: U.S. Geol. Surv. Bull. 540, p. 139-158

Sell, J. D., 1960, Diabase at the Magma mine, Superior, Arizona: Ariz. Geol. Soc. Digest, v. 3, p. 93-98

Short, M. N., and Ettlinger, I. A., 1927, Ore deposition and enrichment at the Magma mine, Superior, Arizona: A.I.M.E. Tr., v. 74, p. 174-222

Short, M. N., and Wilson, E. D., 1938, Magma mine area, Superior, in *Some Arizona ore deposits*: Ariz. Bur. Mines, Geol. Ser. no. 12, Bull. no. 145, p. 90-98

Short, M. N., and others, 1943, Geology and ore deposits of the Superior mining area, Arizona: Ariz. Bur. Mines, Geol. Ser. no. 16, Bull. no. 151, 159 p.

Webster, R., 1958, Exploration extends Magma's future: Min. Eng., v. 10, no. 10, p. 1062-1065

Wilson, E. D., 1950, Superior area, in *Arizona zinc and lead deposits*, pt. 1: Ariz. Bur. Mines, Geol. Ser. no. 18, Bull. no. 156, p. 84-98

Notes

The Magma mine in the Superior mining district is located about 55 miles south of east from Phoenix.

The oldest formation in the area is the ever-present Pinal schist which is classed as older Precambrian in Arizona. The Pinal schist probably is middle Precambrian, but there is a real possibility that further age determinations will demonstrate that it is early Precambrian. Near Magma, the Pinal schist is mainly sedimentary material with some greenstone derived from mafic igneous rocks. To the east, toward Miami, the schist has been invaded by large older Precambrian granitic intrusions, but none of these is exposed in the Superior district. Unconformably above the Pinal formation lies the younger Precambrian Apache group that here is composed of the basal Scanlan conglomerate, the Pioneer shale, the Barnes conglomerate, the Dripping Spring quartzite, the Mescal limestone upon which, locally, is found a basalt flow, and the Troy quartzite (work of the USGS has removed this from the Cambrian). Disconformably above the Apache group is the Paleozoic sequence of Devonian Martin limestone, Mississippian Escabrosa limestone, and Pennsylvanian Naco limestone. The dip and strike of the beds from Scanlan conglomerate to Naco limestone is the same despite the presence of four disconformities. At some time between the end of Troy deposition and the beginning of that of the Martin limestone, the area was intruded by huge sills of diabase that occasionally break across the beds; the total thickness of diabase in the Superior district is over 3000 feet, and much of the ore in the Magma mine proper is found within the diabase sills. No sedimentary rocks of Mesozoic age are present, but the area was intruded by stocks and dikes of quartz diorite and quartz-monzonite porphyry at some time after the close of Paleozoic sedimentation. These igneous masses appear to be related to the Late Cretaceous-early Tertiary intrusions of the Globe-Miami area that extend to within 8 miles of Superior. In the early Tertiary the area was strongly eroded and local basins were filled with Whitetail conglomerate. This erosion was ended by the pouring out of later Tertiary dacite flows. From this dacite a dacite conglomerate (probably equals the Gila) was formed; locally it contains basalt flows and may be of the same general age as certain minor sialic rocks. The last igneous activity was the intrusion of dikes and plugs of basalt that are certainly late Tertiary and may actually be Quaternary in age. The ore bodies of the Magma mine lie east of the Concentrator fault, a more or less northwest-southeast striking normal fault; it is roughly paralleled by the nearer north-south normal Main fault, to the east of which lie almost all of the Magma ore bodies. Still farther east is another north-south fault that cuts the east replacement ore body. The principal mineralized faults are the east-west Magma fault and its counterpart, the Koerner fault, some 1100 to 1300 feet to the south. Both are in general normal faults with a steep dip to the east, although the dip of the Magma fault on the upper levels is steeply west.

These faults are pre-dacite in age so that they must be younger than the late Tertiary and older than the Pennsylvanian. As the only igneous activity between the Pennsylvanian and the late Tertiary to which the ore-forming fluids could be related is that of the quartz diorite and the quartz monzonite correlated with the same general cycle of igneous activity as the Schultze granite of the Miami area, the mineralization is thought also to have formed at that time (60 m.y. ago) and to be best categorized as late Mesozoic to early Tertiary.

There are two principal physical expressions of mineralization in the Magma mine: (1) vein fillings in, and replacements out from, the Magma and the Koerner fault zones of the east-west fracture system; ore is present in these veins from the lower portions of the Martin limestone down to the upper levels of the Pinal schist, with a large fraction of the ore being in the diabase sills intercalated through the formations; and (2) a replacement zone down dip to the east in the lower Martin limestone, a zone that extends north and south from the south split of the Main (Magma fault) vein.

In the vein-type bodies, of which that along the Magma fault is by far the more important, the mineralization in the Main ore shoot (east of the Main fault) above the 1200 level on the west and continuing down to the 3200 level to the east consisted mainly of sphalerite, silver minerals (such as stromeyerite), and a little galena. Between the 1300 and 1400 levels to the west and continuing down to the 4000 level to the west, the ore became bornite-rich, with little to no sphalerite and galena. In addition to the bornite, this segment of the Main ore shoot contained silver-rich tennantite, locally abundant chalcopyrite, chalcocite, and digenite. Between the 3000 and the 3200 levels on the west and the lowest mineralized levels reached on the east, enargite became an important ore mineral, although chalcopyrite and bornite also contributed appreciably to the copper content, and tennantite gradually died out. Pyrite is common in all vein ores but is less abundant with bornite than with chalcopyrite.

The West ore body lies between the Main and Concentrator faults in the Magma fault and contains a minor shoot of bornite ore with subordinate sphalerite. The East ore bodies, also on the Magma fault, lie at a considerable distance east of the Main ore shoot and are sphalerite-rich above the 2550 level and chalcopyrite-rich below. The Koerner ore body is quite similar to the Main ore shoot on the Magma vein and contains the same ore minerals at essentially the same levels. The Main vein also has a north branch and a south split that are locally of economic importance, and it is to this south split that the replacement ore appears to be directly related.

The replacement ore associated with the lower portion of the Martin limestone below the 2550 level appears to constitute a downward extension of the East ore bodies or at least a downward extension of the south split of the Main vein on which the East ore bodies are located. The mineralization of the replacement zone is much the same as in the middle and lower portions of the Main vein, with the exception that the most abundant mineral is specular hematite and that there are considerable amounts of hematite and minor ones of magnetite. Pyrite is the next most abundant mineral above the 3200 level, followed by chalcopyrite, bornite, tennantite, chalcocite, and digenite. Below the 3200 level, tennantite disappears and enargite comes in.

In the western part of the mine in the vein type ores, there appears to have been a zoning of the mineralization both in space and in time. The first mineralization appears to have consisted of sphalerite and silver sulfosalt minerals in the upper part of the vein, with bornite and chalcopyrite below the zinc area; enargite with bornite and chalcopyrite appear below the upper copper area. The tennantite probably is later than the bornite and chalcopyrite in the middle zone. The chalcocite and digenite, in turn, appear to have been formed later than the tennantite and the earlier copper minerals, with primary simple copper sulfides not having been deposited in the zinc ore. This sequence of mineralization indicates that ore deposition occurred between intense mesothermal, as is shown by the enargite; through less intense mesothermal, as is suggested by the bornite; through leptothermal to telethermal as suggested by the tennantite; to primary chalcocite and digenite sequence. Some of the chalcocite is intimately intergrown with the bornite, and this chalcocite may have been formed originally as part of a bornite-chalcocite solid solution, but the bulk of the chalcocite appears to be later than the bornite as does the digenite. The main vein ore, therefore, appears to have been formed from the upper mesothermal to the telethermal range and is so classified here. The silicification and sericitization of diabase and of siliceous and aluminous sediments agree with the mesothermal part of the classification.

Were it not for the abundant specular hematite (and the associated hematite and magnetite) in the replacement ore, it would be classified similarly to that of the vein type. Because bornite, chalcopyrite, and pyrite have been noted as finely divided particles in the hematite, it may well be that the iron oxides were replacements of earlier sulfides; however, the iron oxides indicate deposition under higher intensity conditions than the copper sulfides so that it is quite probable that a portion of the replacement ore was formed under hypothermal conditions. On the other hand, the enargite and bornite are characteristic of the mesothermal range, tennantite of the leptothermal, and chalcocite and digenite of the telethermal. The replacement ores are, therefore, here categorized as hypothermal-2 to telethermal.

MORENCI

Late Mesozoic to Early Tertiary — *Copper, Molybdenum* — *Mesothermal*

Butler, B. S., and Wilson, E. D., 1938, Clifton-Morenci district, in *Some Arizona ore deposits*: Ariz. Bur. Mines, Geol. Ser. no. 12, Bull. no. 145, p. 72-80

Creasey, S. C., 1959, Some phase relations in the hydrothermally altered rocks of porphyry copper deposits: Econ. Geol., v. 54, p. 351-373 (general)

Lindgren, W., 1905, Clifton folio, Arizona: U.S. Geol. Surv. Geol. Atlas, Folio 120, 13 p.

—— 1905, The copper deposits of the Clifton-Morenci district, Arizona: U.S. Geol. Surv. Prof. Paper 43, 375 p. (particularly p. 51-223)

Lindgren, W., and Hillebrand, W. F., 1904, Minerals from the Clifton-Morenci district, Arizona: Amer. Jour. Sci., 4th Ser., v. 18, no. 108, p. 448-460

Moolick, R. T., and Durek, J. J., 1966, The Morenci district, in Titley, S. R., and Hicks, C. L., Editors, *Geology of the porphyry copper deposits--southwestern North America*: Univ. Ariz. Press, Tucson, p. 221-231

Reber, L. E., Jr., 1916, Mineralization at Clifton-Morenci: Econ. Geol., v. 11, p. 528-573

Schwartz, G. M., 1947, Hydrothermal alteration in the "porphyry copper" deposits: Econ. Geol., v. 42, p. 319-352 (particularly p. 339-342, 346-352)

Tenney, J. B., 1935, Morenci district, in *Copper resources of the world*: 16th Int. Geol. Cong., v. 1, p. 213-221

Tovote, W. L., 1910, The Clifton-Morenci district of Arizona--I: Min. and Sci. Press, v. 101, no. 24, p. 770-773

Notes

The Morenci district is located in eastern Arizona, about 15 miles from the New Mexico border and about 110 miles northeast of Tucson. Much of the mining in the district has come from rocks west of Chase Creek, but there are some indications that enriched ore of minable grade is present east of that stream.

The principal Precambrian rock in the Morenci area is a granite that has been considerably metamorphosed in the various orogenic cycles through which it has passed. Associated with the granite is a perhaps even older granodiorite that lies to the south of the more silicic rock. The oldest Paleozoic formation is the Coronado quartzite that is followed by the Ordovician Longfellow limestone, then by the Devonian Morenci shale, which includes shaly limestone, and finally by the Carboniferous Modoc limestone. Cretaceous shale and sandstone lie unconformably on the Paleozoic rocks. At or near the close of the Cretaceous a generally northeast-southwest fracture zone was developed into which a stocklike mass of porphyritic material, ranging from granite through quartz monzonite to diorite, was intruded. Dikes, sills, and laccoliths or plugs were forced from the stock into the surrounding rocks. The quartz monzonite is cut by later granite porphyry dikes, but the principal masses of granite porphyry are plugs similar to those of the quartz diorite; locally the granite porphyry grades into the quartz monzonite, there being, for example, a 50 to 100 foot transition zone between the two rocks on the north side of the present pit. There is an appreciable amount of diabase (some of which may be lamprophyre) in the area; most of it appears to have been introduced in Laramide time. The diabase usually is mineralized. Small to large xenoliths of the older rocks were incorporated in the marginal portions of the silicic intrusives. After these igneous rocks had solidified, they were subjected to further stresses that produced prominent northeast-southwest fractures and caused further movement on those already in existence. The Clifton-Morenci area is bordered on the west side by major northeast-southeast faults and by northeast-southwest on the east side. Both of these types of fractures are normal faults and dip away from each other; their strikes cause them to coalesce to the south, and the principal area of mineralization is in the apex between the faults. Some of the northwest-southeast fractures developed after the mineralization but before the oxidation-secondary enrichment; they do not displace the enrichment blanket nor do they cut the Tertiary volcanic rocks. These faults were followed by a period of erosion that exposed the mineralized rock volumes to surface oxidation that developed the oxidized ores in limestone and the secondarily enriched material in the porphyry. The area was then covered by a series of Tertiary volcanic flows, tuffs, and breccias that range in composition from basalt to rhyolite. After the volcanic episode additional northwest-southeast faults were formed and certain of these (e.g., the Copper Mountain fault) displace the enrichment blanket; these faults are responsible for the Basin-Range structures of the district. In the basins thus formed, poorly sorted gravel deposits (the Gila conglomerate) were developed. Thus, the porphyritic rocks probably were formed after Cretaceous sedimentation had ended in the area and before the probably middle Tertiary volcanism. No absolute age determinations are available, but it appears probable that the mineralization (which was somewhat later than the igneous intrusion, but not greatly so) was late Mesozoic to early Tertiary, and the deposits are so classified here.

The two types of primary mineralization at Morenci consisted of: (1) replacement deposits in favorable limestone beds, previously largely converted to high-temperature silicates of which garnet and epidote were the most common, and (2) fissure fillings and disseminations in the igneous rocks (all of which are mineralized, although none is completely so) and in the quartzite. The sulfide minerals in limestone were principally pyrite, chalcopyrite, and sphalerite, although the copper content was so low that the mineralized beds probably could not have been profitably mined. The oxidation of these minerals produced higher grade deposits of basic copper carbonates (azurite and malachite) and oxides. The most important mine of this type was the Longfellow mine at Morenci, but there were many others.

Because the mineralized veins in the porphyry were leached to depths of some 200 feet, early miners confined their operations to deposits in the sedimentary rocks. As the veins in porphyry were explored in depth, however, it was found that they contained considerable secondary chalcocite, mainly as replacements of pyrite. The sparse primary chalcopyrite and sphalerite of the veins and veinlets appear also to have been largely replaced by secondary copper sulfides, with covellite, rather than chalcocite, replacing chalcopyrite. The minor molybdenite does not seem to have been affected by secondary processes. The primary sulfides, other than pyrite, are generally confined to the veins and veinlets in which they are thinly distributed among the dominant quartz, pyrite, and sericite. Pyrite, however, is widely disseminated in the altered igneous rocks (and to some extent in the quartzite) through which the veinlets run, and this pyrite also has been considerably replaced by chalcocite. After vein mining in the monzonite had begun, it did not take long for mining and milling techniques to improve to the point at which much of the mass of altered and mineralized porphyry veins and wall rock alike could be mined at a profit, and the result was the opening of the Morenci pit. The primary copper content must have been low (0.1 to 0.2 percent copper), and apparently none of it was sufficiently high in grade to have been minable without secondary enrichment. The water table during the enrichment process must have been much higher than it is now because its present position is well below the enriched ore; it is probable that most of the enrichment was accomplished before the eruption of the middle Tertiary volcanic rocks. There has been some enrichment, however, since the erosional removal of the lavas.

The pattern exhibited by the high-temperature alteration silicates in the limestones indicates that these minerals were developed by hydrothermal fluids that entered the rock volumes involved along intrusive-sedimentary contacts. Since the sulfide emplacement was controlled by the fracture systems and not by these contacts, it would appear that the silicate alteration was not formed at an early stage of the primary sulfide mineralization but preceded it by an appreciable period of time. The fractures, however, affected limestone as well as igneous rocks, so sulfides were deposited in limestone in considerable amount. Where sulfides were emplaced in altered limestone, they replaced limestone far more readily than the alteration silicates. This lapse of time between silicate and sulfide development plus the lack of exsolution blebs in both sphalerite and chalcopyrite and the deposition of quartz-sericite-clay alteration minerals before and during sulfide emplacement strongly suggest that the sulfides were deposited in the mesothermal range. The Morenci deposits, therefore, are here classified as mesothermal.

PIMA DISTRICT

Late Mesozoic to Early Tertiary	*Copper, Lead, Zinc, Silver, Gold, Molybdenum*	*Hypothermal-2 to Mesothermal*

Argall, G. O., Jr., 1962, ASARCO's Mission copper--ore body: Mining World, v. 24, no. 1, p. 22-23

Bowman, A. B., 1955, Banner Mining Co. opens the Mineral Hill copper property in Arizona: Min. Eng., v. 7, no. 11, p. 1022-1025

Cooper, J. R., 1960, Some geologic features of the Pima mining district, Pima County, Arizona: U.S. Geol. Surv. Bull. 1112-C, p. 63-103

Cummings, J. B., and Romslo, T. M., 1950, Investigation of Twin Buttes copper mines, Pima County, Arizona: U. S. Bur. Mines R. I. 4732, 12 p. (mimeo.)

Huff, L. C., 1970, A geochemical study of alluvium-covered copper deposits in Pima County, Arizona: U.S. Geol. Surv. Prof. Paper 1312-C, p. C1-C31

Irvin, G. W., 1959, Pyrometasomatic deposits at San Xavier mine: Ariz. Geol. Soc. Guidebook II, Southern Arizona, p. 195-197

Journeay, J. A., 1959, Pyrometasomatic deposits at Pima mine: Ariz. Geol. Soc. Guidebook II, Southern Arizona, p. 198-199

Kinnison, J. E., 1966, The Mission copper deposit, Arizona, in Titley, S. R., and Hicks, C. L., Editors, *Geology of the porphyry copper deposits--southwestern North America*: Univ. Ariz. Press, Tucson, p. 281-287

Lacy, W. C., 1959, Structure and ore deposits of the east Sierrita area: Arizona Geol. Soc. Guidebook II, Southern Arizona, p. 185-192, 206

Lacy, W. C., and Titley, S. R., 1962, Geological developments in the Twin Buttes district: Min. Cong. Jour., v. 48, no. 4, p. 62-64

Lootens, D. J., 1966, Geology and structural environment of the Sierrita Mountains, Pima County, Arizona: Ariz. Geol. Soc. Digest, v. 8, p. 33-56

Lynch, D. W., 1966, The economic geology of the Esperanza mine and vicinity, in Titley, C. R., and Hicks, C. L., Editors, *Geology of the porphyry copper deposits--southwestern North America*: Univ. Ariz. Press, Tucson, p. 267-279

—— 1968, The geology of the Esperanza mine: Ariz. Geol. Soc. Guidebook III, Southern Arizona, p. 125-136

MacKenzie, F. D., 1959, Pyrometasomatic deposits at the Mineral Hill and Daisy mines: Arizona Geol. Soc. Guidebook II, Southern Arizona, p. 193-194

—— 1963, Geological interpretation of the Palo Verde mine based upon diamond drill core: Geol. Soc. Ariz. Digest, v. 6, p. 41-48

Mitcham. T. W., and Arnold, L. C., 1964, The West San Xavier mine, keys to a district: Ariz. Geol. Soc. Digest, v. 7, p. 147-152

Ransome, F. L., 1922, Ore deposits of the Sierrita Mountains, Pima County, Arizona: U.S. Geol. Surv. Bull. 725, p. 407-440

Richard, K., and Courtright, J. H., 1959, Some geologic features of the Mission copper deposit: Ariz. Geol. Soc. Guidebook II, Southern Arizona, p. 201-204

Schmitt, H. A., and others, 1959, Disseminated deposits at the Esperanza copper mine: Ariz. Geol. Soc. Guidebook II, Southern Arizona, p. 205

Thurmond, R. E., and others, 1954, Geophysical discovery and development of the Pima mine, Pima County, Arizona: A.I.M.E. Tr., v. 199, p. 197-202 (in Min. Eng., v. 6, no. 2)

—— 1958, Pima: A three-part story--geology, open pit, milling: Min. Eng., v. 10, no. 4, p. 453-462 (particularly p. 457)

Wilson, E. D., 1950, Pima district, in *Arizona zinc and lead deposits*, pt. 1: Ariz. Bur. Mines, Geol. Ser. no. 18, Bull. no. 156, p. 39-51

Notes

The Pima mining district (15 to 30 miles south-southwest of Tucson) was an active mining area long before the low-grade copper deposits, now its most important resource, were discovered. These recent finds include the Pima and Mission pits, essentially one deposit just east of the old Mineral Hill mine, the Esperanza mine about nine miles south-southwest of the Pima pit, and the Twin Buttes area between Mission and Esperanza.

The rocks in the area are essentially those of the Tucson mountains north of the district although local thickening and thinning have been effected by bedding plane faults, and interruptions in continuity have been caused by higher-angle faulting. The oldest rock in the area is the Precambrian Sierrita granite that is present in several large masses scattered along the north-south center line of the district. This granite apparently has been encountered in drill holes put down as far to the east as the Mission pit, although ASARCO geologists consider the granite in the Mission pit to be Laramide. The rock itself is a coarse-grained oligoclase granite that contains xenoliths of diorite, schist, and gneiss that locally show gradational contacts against the granite. Unconformably above the granite is the Cambrian Abrigo formation; disconformably above the Abrigo is the Devonian Martin limestone that is followed by the Mississippian Escabrosa limestone and then by the Pennsylvanian-Permian Naco group. Other formations than those named may later be identified. Unconformably on the Paleozoics lies a complex of post-Paleozoic arkosic sandstones and volcanic rocks, red shale, siltstone, sandstone, and almost any other type of non-marine sedimentary or volcanic rock that can be imagined with the exception of basaltic materials. Paleontologic evidence indicates that these beds are younger than mid-Jurassic, but they might be as young as Recent.

The Cretaceous(?) rocks were considerably deformed before they were intruded by a variety of igneous magmas that range in composition (Cooper, 1960) from diorite to andesite to granodiorite to quartz-monzonite porphyry. Cooper considers the igneous activity to have taken place in the Late Cretaceous or early Tertiary. Lacy, on the other hand, draws a different picture of the rock types developed in this cycle, the two different concepts probably reflecting the need for further work on the entire problem.

The last of this group of igneous rocks is the quartz-monzonite porphyry that outcrops as plugs and stocks in the Esperanza mine and also is known in the Pima mine, the Mission area, and near Mineral Hill. West of the Esperanza mine, a dike of quartz monzonite extends a short distance into what is probably a variety of the granodiorite, and other, smaller dikes that probably are monzonite are found cutting typical granodiorite. This evidence well may establish the quartz monzonite as the youngest of these intrusives. Mineralization in the quartz monzonite and the andesite indicates that it occurred after the completion of this igneous cycle.

Except for some small amounts of early Tertiary(?) rhyolitic tuff in the Helmet Peak area, the next formation in the district is the probably middle Tertiary Helmet fanglomerate that is later than the mineralization but has been in part appreciably deformed.

Radioactive age dating on the quartz monzonite phase of the granodiorite from the Esperanza area suggests that it was emplaced about 62 m.y. to 56 m.y. ago. This would indicate that it was emplaced in the Paleocene as part of the Laramide sequence of igneous events; the ore mineralization followed shortly after the introduction of the quartz monzonite.

The evidence so far presented, though obviously tentative, indicates that the mineralization is, at the oldest, Late Cretaceous, but is more probably very early Tertiary. It appears most reasonable, however, to classify the deposits as late Mesozoic to early Tertiary.

The ore deposits of the Pima area consist of two main types: (1) massive and disseminated deposits in limestone converted to tactite, hornfels, or marble and (2) disseminated and veinlet deposits in broken and altered sediments, volcanic rocks, and igneous rocks. The first type includes not only the high-grade deposits of such mines as Mineral Hill and San Xavier but also the lower-grade deposits in altered limestone found in the Pima and Mission pits. The principal differences between the higher- and lower-grade deposits of type (1) are the spottier character of the low-grade mineralization and its intimate association, largely through the movement of fault blocks relative to each other, with much lower-grade mineralization in noncarbonate rocks. The low-grade deposits of type (1) have been categorized as porphyry coppers, although they justify this designation mainly because their grade is about the same as that of the typical porphyry copper deposit in such rocks as quartz monzonite. The ore minerals in these type (1) low-grade deposits were generally not emplaced as fissure fillings but as replacements of highly broken volumes of reactive rocks. Type (2) deposits are much more closely allied to the typical porphyry copper ore body in that they consist both of fissure fillings and disseminated replacements in noncarbonate, igneous, volcanic, or clastic sedimentary rocks.

In the Mission-Pima area the mineralization is essentially unbroken from the Pima ore body into that of the Mission property to the northeast; the mineralized area continues onto the Banner (Anaconda) property to the north and then onto the Indian Reservations still farther north where it apparently ends. The entire mineralized block lies above a prominent post-mineralization thrust fault.

The carbonate rocks in this block were altered to tactite (garnet, wollastonite, and/or tremolite), hornfels (diopside, wollastonite, and tremolite), and recrystallized limestone or marble. The alteration of the argillaceous rocks and of the quartz monzonites consisted of the introduction of quartz and orthoclase into the monzonites as irregular blebs and veinlets that replace the ground mass and the plagioclase phenocrysts; the plagioclases were also converted to sericite and, in a lesser extent, to clay minerals. The argillites were altered to fine-grained aggregates of sericite and/or quartz and feldspar, and the feldspathic material of the arkoses was converted to fine-grained sericite that provides a matrix for the clastic quartz grains.

The ore mineralization in the tactites, hornfelses, and limestones consisted of magnetite, chalcopyrite, minor bornite, sphalerite, and molybdenite; locally gypsum is present as a gangue mineral slightly later than the sulfides (brecciated chalcopyrite is found in the gypsum) and may contain noticeable, but tiny, amounts of native copper. The gypsum and native copper both probably are primary minerals.

The grade of the mineralization in the altered clastic rocks is considerably lower than in the tactites and hornfelses, and the minerals in the noncarbonate sediments appear not to include magnetite or more than minor sphalerite. Pyrite, chalcopyrite, and rarely molybdenite occur in the monzonite as disseminated grains and in veinlets bordered by quartz and/or orthoclase; the grade is even lower than that of the argillite. The ores are quite different from those in the Esperanza area. Not only does the Pima-Mission ore appear to have formed at higher temperatures as is witnessed by the abundant magnetite and the higher temperature alteration minerals, but it also is appreciably higher in grade; primary ore is mined almost exclusively in the Mission-Pima area, while in the Esperanza area proper (not including the higher-grade primary Sierrita ore), the mineralized rock is minable only because it has been enriched.

The ores in the Mission-Pima portion of the district apparently were formed under both hypothermal and mesothermal conditions, with probably the greater proportion having been formed in the higher intensity range. The ores in the generally underlying limestones are largely hypothermal, while those in the argillaceous rocks and the monzonites were formed nearer the boundary between the hypothermal and mesothermal ranges, with more of the mineralization probably having been mesothermal than hypothermal. Because most, though probably not all, of the hypothermal mineralization was formed in carbonate rocks, the Mission-Pima deposits are here classified as hypothermal-2 to mesothermal.

The first mines worked in the district were such higher-grade deposits as those of Mineral Hill, but these ore bodies appear to have been formed under much the same conditions as those in the larger and lower-grade deposits of the Mission-Pima area proper.

These higher-grade deposits must have been formed under much the same conditions as the deposits in tactite and hornfels in the Pima and Mission pits. The presence of iron oxides in intimate association with the chalcopyrite strongly suggests that they were formed under hypothermal conditions, although the bornite and silver-rich galena as late minerals indicate an appreciable period of mesothermal deposition. The limited occurrence of silver-rich tetrahedrite probably was not extensive enough to justify including leptothermal in the classification.

Among the more typical disseminated deposits in the area are the Esperanza, Sierrita, and Twin Buttes ore bodies with undoubtedly others to be discovered. The Esperanza and the associated Western Extension and Sierrita deposits much more closely resemble porphyry copper deposits than do the Pima or Mission ore bodies.

The primary mineralization at Esperanza consists mainly of pyrite and chalcopyrite (usually in a ratio of about 1:1) as disseminated grains in the host rocks and in quartz-sericite veinlets cutting through them. On the lower benches of the present pits, chalcopyrite is the dominant copper sulfide. In the topographically higher rocks, chalcopyrite, while present, has been extensively replaced by chalcocite (as has the pyrite); all chalcocite appears to have been of supergene origin, and pseudomorphs of chalcocite after pyrite are locally common. Although chalcocite is the more abundant supergene copper sulfide, covellite is common as a coating on pyrite, chalcopyrite, and secondary chalcocite. Molybdenite is widespread, being particularly abundant in the Sierrita area, and is a significant primary constituent of the quartz-sericite-chalcopyrite-pyrite veinlets; it is only rarely disseminated in the host rocks. Galena and sphalerite, although present, are never of economic importance. Gangue minerals include purple anhydrite and selenite gypsum. The selenite replaces the anhydrite and is intimately intergrown with chalcopyrite and is almost certainly primary. The oxidized capping over the secondary blanket is essentially devoid of copper.

The Esperanza ores probably were deposited under mesothermal conditions. Further work on this problem is necessary. It would appear that the ores in the Pima district have been formed under hypothermal and mesothermal conditions, with essentially all of the hypothermal minerals having been emplaced in carbonate rocks. Mesothermal conditions prevailed in part in the deposition in carbonate rocks and largely, though probably not entirely, in the deposition in noncarbonate (clastic, volcanic, and igneous) rocks. The Pima deposits, then, are here classified as hypothermal-2 to mesothermal.

RAY

Late Mesozoic to Early Tertiary — *Copper* — *Mesothermal*

Butler, B. S., and Wilson, E. D., 1938, Ray district, in *Some Arizona ore deposits*: Ariz. Bur. Mines, Geol. Ser. no. 12, Bull. no. 145, p. 80-86

Clarke, O. M., Jr., 1952, Structural control of ore deposition at Ray, Arizona: Ariz. Geol. Soc. Guidebook I, Southern Arizona, p. 91-95

—— 1953, Geochemical prospecting for copper at Ray, Arizona: Econ. Geol., v. 53, p. 39-45

Metz, R. A., and Rose, A. W., 1966, Geology of the Ray copper deposit, Ray, Arizona, in Titley, S. R., and Hicks, C. L., Editors, *Geology of the porphyry copper deposits--southwestern North America*: Univ. Ariz. Press, p. 177-188

Metz, R. A., and others, 1968, Recent developments in the geology of the Ray area: Ariz. Geol. Soc. Guidebook III, Southern Arizona, p. 137-146

Ransome, F. L., 1919, The copper deposits of Ray and Miami, Arizona: U.S. Geol. Surv. Prof. Paper 115, 192 p. (particularly p. 122-131, 144-176)

---- 1923, Ray folio, Arizona: U.S. Geol. Surv. Geol. Atlas Folio 217, 24 p.

Schwartz, G. M., 1947, Hydrothermal alteration in the "porphyry copper" deposits: Econ. Geol., v. 42, p. 319-352 (particularly p. 331-332, 346-352)

---- 1952, Chlorite-calcite pseudomorphs after orthoclase phenocrysts, Ray, Arizona: Econ. Geol., v. 47, p. 665-672

Tenney, J. B., 1935, Ray-Christmas district, in *Copper resources of the world*: 16th Int. Geol. Cong., v. 1, p. 201-206

Tolman, C. F., Jr., 1909, Disseminated chalcocite deposits at Ray, Arizona: Min. and Sci. Press, v. 99, no. 19, p. 622-624

Notes

The ore deposit at Ray is located about 12 miles southeast of Superior and slightly over 60 miles south of east of Phoenix.

The oldest rock in the area is the older Precambrian Pinal schist that has been intruded by various older Precambrian granites. Unconformably above the Pinal lies a complete representation of the beds of the younger Precambrian Apache group from the basal Scanlan conglomerate through the Pioneer shale, the Barnes conglomerate, and the Dripping Strings quartzite to the Mescal limestone that locally is unconformably overlain by a vesicular basalt. The Precambrian probably was terminated by the deposition of the Troy sandstone or quartzite that apparently is unconformable, though not markedly so, on the Apache group. This Precambrian series has been intruded by huge diabase sills that cut across the formations in such a highly irregular manner that, though the beds retain essentially a common dip and strike, they are now largely surrounded and enclosed in diabase. Although relations of the diabase to the Paleozoic formations are not certain, it is probable that the Ray diabase, as is true of the diabase at Superior, is post-Devonian. There is no evidence of Ordovician or Silurian sedimentation but, nevertheless, the Martin(?) limestone probably is Devonian and is conformable on the Troy. Conformably above the Martin is the Carboniferous Tornado limestone that apparently represented essentially uninterrupted sedimentation during the entire period. Deposition did not recommence until the Cretaceous when sediments probably were deposited, although there is now no remnant of them left in the Ray area. Eruptions of andesite followed the Cretaceous sedimentation in the adjacent Deer Creek area so the minor amount of andesite south of Ray is probably of the same age. Younger than the andesite, but of uncertain age, were intrusions of fairly large dikes and small irregular masses of quartz diorite, followed by bodies of granite, quartz-monzonite porphyry, and granodiorite that may be as much as several miles in diameter. The Granite Mountain quartz-monzonite porphyry is the only one of these igneous rocks to be mineralized in the Ray area; the unmineralized Teapot Mountain granite porphyry lies north of the present Ray pit and the Schultze granite of the Globe-Miami area does not extend as far south as Ray.

The period of igneous activity was followed by active erosion in which the early(?) Tertiary Whitetail conglomerate was formed in local basins. Then the area was covered by dacite flows, also known in the Superior and Globe-Miami districts, after which the area was highly faulted and then again heavily eroded to produce the middle or late Tertiary Gila conglomerate. Further faulting then affected the Gila conglomerate and the older rocks of the area.

The greater part of the secondary enrichment of the Ray primary mineralization appears to have taken place before the deposition of the Whitetail conglomerate, but the present topography is such as to suggest that the enrichment process was also operative, but to a much lesser extent, in Recent time after the removal of the Whitetail and the dacite.

The actual development of the primary mineralization certainly did not occur until after the intrusion of the Granite Mountain porphyry in which part of the Ray mineralization is found (primary sulfides, later secondarily enriched to minable ore, are also found in the Pinal schist and the diabase); thus, the primary mineralization was post-Granite Mountain and pre-Whitetail. Since the Whitetail is probably early Tertiary and the Granite Mountain, with an age of 63 m.y., is probably Late Cretaceous or very early Tertiary, the mineralization almost certainly is correctly classed as late Mesozoic to early Tertiary.

The primary, copper-bearing mineralization (protore) at Ray consists of disseminated pyrite and of innumerable quartz-pyrite veinlets with minor and erratically distributed chalcopyrite and sparse molybdenite in fractures developed in the Pinal schist, the diabase, and the Granite Mountain quartz monzonite porphyry shortly after the last material was intruded and consolidated. In the diabase sill that makes up much of the eastern portion of the Ray pit, the primary mineralization was, on the average, rich enough for the mineralized material to be classed as hypogene ore. There is also a minor tonnage of hypogene ore east of the Ray fault in Granite Mountain porphyry. With the exception of these two masses of hypogene ore, only the secondarily enriched portion of the protore is minable. To the south and west of the secondary ore body, there is a considerable mass of weakly mineralized protore (0.2 percent copper, on the average, probably as chalcopyrite), but such protore is not found beneath the enriched zone nor on the north side of the enriched material. Essentially copper-free, pyrite-bearing rock is present beneath the secondary ore and the outlying protore and outward from the protore; one type of mineralization grades gradually into that or those adjacent to it. The wall-rock alteration attendent on the sulfide mineralization is typical of that associated with porphyry copper deposits in the southwestern United States. Quartz is present in considerable amounts in the veinlets and some silicification extends for short distances out from the veinlets. The bulk of the alteration, however, is sericite. Outward from the sericite, probably earlier argillic minerals are strongly developed. The enriched zone at Ray has a maximum thickness of 400 feet and a minimum of 50, with the average being about 150 feet.

At Ray, the less fractured diabase acted to prevent the downward movement of ground water carrying copper in much of the eastern part of the area so the greatest enrichment is generally just above the diabase sill. The enrichment in the schist and to a lesser extent in the quartz monzonite has been by the almost complete replacement of chalcopyrite and the marked, but not as complete, replacement of pyrite by chalcocite; the molybdenite in the protore appears to have been unaffected by the enrichment process. Enrichment in the less permeable diabase was less impressive than in the schist and porphyry, but what enrichment there was, plus the higher-grade primary mineralization in the diabase, allow that material also to be mined.

The primary mineralization bears all the characteristics of formation under mesothermal conditions, and the deposit is so classified here.

SAN MANUEL

Late Mesozoic to Early Tertiary | *Copper, Molybdenum* | *Mesothermal*

Chapman, T. L., 1947, San Manuel copper deposit, Pinal County, Ariz.: U.S. Bur. Mines, R.I. 4108, 93 p. (mainly drill-hole logs)

Creasey, S. C., 1959, Some phase relations in the hydrothermally altered rocks of porphyry copper deposits: Econ. Geol., v. 54, p. 351-373 (general)

—— 1967, General geology of the Mammoth quadrangle, Pinal County, Arizona:

U.S. Geol. Surv. Bull. 1218, 94 p.

Creasey, S. C., and Pelletier, J. D., 1965, Geology of the San Manuel area, Pinal County, Arizona: U.S. Geol. Surv. Prof. Paper 471, 64 p.

Heindl, L. A., 1963, Cenozoic geology in the Mammoth area, Pinal County, Arizona: U.S. Geol. Surv. Bull. 1141-E, p. E1-E41

Lovering, T. S., 1948, Geothermal gradients, recent climatic changes, and rate of sulfide oxidation in the San Manuel district, Arizona: Econ. Geol., v. 43, p. 1-20

Lovering, T. S., and others, 1950, Dispersion of copper from the San Manuel copper deposit, Pinal County, Arizona: Econ. Geol., v. 45, p. 493-514

Lowell, J. D., 1968, Geology of the Kalamazoo ore body, San Manuel district, Arizona: Econ. Geol., v. 63, p. 645-654

Pelletier, J. D., 1957, Geology of the San Manuel mine: Min. Eng., v. 9, no. 7, p. 760-762

Peterson, N. P., 1938, Geology and ore deposits of the Mammoth mining camp area, Pinal County, Arizona: Ariz. Bur. Mines, Geol. Ser. no. 11, Bull. no. 144, 63 p. (the Mammoth area lies immediately north of the San Manuel property)

Schwartz, G. M., 1947, Hydrothermal alteration in the "porphyry copper" deposits: Econ. Geol., v. 42, p. 319-352 (particularly p. 342-344, 346-352)

—— 1949, Oxidation and enrichment in the San Manuel copper deposit, Arizona: Econ. Geol., v. 44, p. 253-277

—— 1953, Geology of the San Manuel copper deposit, Arizona: U.S. Geol. Surv. Prof. Paper 256, 65 p.

Steele, H. J., and Rubly, G. R., 1948, San Manuel prospect: A.I.M.E. Tr., v. 178, p. 181-194

Thomas, L. A., 1966, Geology of the San Manuel ore body, in Titley, S. R., and Hicks, C. L., Editors, *Geology of the porphyry copper deposits--southwestern North America*: Univ. Ariz. Press, Tucson, p. 133-142

Wilson, E. D., 1957, Geologic factors related to block caving at San Manuel copper mine, Pinal County, Arizona: U.S. Bur. Mines R.I. 5336, 78 p.

Notes

The San Manuel mine is located some 35 miles northeast of Tucson and about a mile south of the small town of Tiger, once the site of St. Anthony polymetallic mine.

There is no great variety of rocks exposed either on the surface or underground; the oldest rock is a Precambrian quartz-monzonite, known as the Oracle granite, which is part of a much larger granite mass on the north slope of the Santa Catalina Mountains. The quartz monzonite is intruded by a quartz-monzonite porphyry (with quartz only in the ground mass), but it is locally called monzonite to avoid confusion with the Precambrian rock of similar composition. The age of the monzonite porphyry is uncertain; it intrudes Precambrian rocks and is overlain by those of the late Tertiary. From analogy with igneous activity at other mining camps in central Arizona, it is reasoned that the monzonite is of essentially the same age as the Schultze granite at Miami and the quartz-monzonite and granite porphyries at Ray; this places the monzonite somewhere in the Laramide period of orogeny--Late Cretaceous to early Tertiary. Damon and Mauger report an age of 67 m.y. for altered granodiorite porphyry at San Manuel (undoubtedly the monzonite mentioned here) that confirms the age just given. The monzonite porphyry is cut by numerous diabase dikes and irregular masses; and, although the diabase is altered and mineralized, it appears to have acted somewhat as a barrier to the movement of ore solutions. The diabase is overlain by Gila conglomerate and must be Tertiary in age; it is here thought to have been introduced toward the end of the same Late Cretaceous to early Tertiary time span as the monzonite porphyry. A few andesite porphyry dikes that cut the rocks already mentioned have been exposed in the mine; although some of them are slightly mineralized, the mineralization and the alteration are different from that of the main ore masses. These mafic volcanic rocks are cut by numerous dikes and other masses of rhyolites. This rhyolite is also older than the Gila conglomerate. With the exception of recent alluvium, the Gila conglomerate is the youngest rock in the area, it normally being unconformable on all the older rocks; locally it exhibits fault contacts against the other rocks. The Gila is probably late Tertiary and Quaternary in age.

The primary ore mineralization was developed after the intrusion of the diabase and before that of the andesite porphyry. The mineralization probably occurred shortly after the intrusion of the diabase. The deposits are, therefore, here categorized as late Mesozoic to early Tertiary.

The Kalamazoo ore body is a faulted segment of the San Manuel. It was discovered through the assumption (Lowell, 1968) that an original cylindrical ore body with concentric alteration zoning had first been titled approximately 70°, then bisected by the flat San Manuel normal fault into two halves, the San Manuel ore body in the footwall plate and the Kalamazoo in the upper. The mineralization in this body is about equally distributed, according to Lowell (1968), between the Precambrian quartz monzonite and the Laramide monzonite porphyry; the ore in the Precambrian rock is slightly higher in grade.

The primary metallic minerals in the combined San Manuel-Kalamazoo ore body are pyrite, chalcopyrite, minor molybdenite, and rare bornite--a typical mesothermal assemblage. The wall-rock alteration in the low-grade core is mainly biotite-potash feldspar; in the surrounding ore zone, the alteration is an overlapping of the biotite-potash feldspar type by a quartz-sericite type; in the surrounding marginal ore zone, the alteration is almost entirely quartz-sericite. With depth, the biotite-potash feldspar of the low-grade-ore core grades downward into quartz-sericite, and the alteration in the ore zone shell contains more quartz-sericite with depth and less biotite-potash feldspar. This would seem to indicate that the biotite-potash feldspar alteration was formed in a lower intensity range than the quartz-sericite. If this is the case, the ores almost certainly formed in the mesothermal range.

SILVER BELL

Late Mesozoic to Early Tertiary	*Copper, Molybdenum*	*Hypothermal-2 to Mesothermal*

Kerr, P. F., 1951, Alteration features at Silver Bell, Arizona: Geol. Soc. Amer. Bull., v. 62, p. 451-480

McClymonds, N. E., and others, 1959, Stratigraphy of the Waterman and Silver Bell Mountains, Trip II, Road Log: Ariz. Geol. Soc. Guidebook II, Southern Arizona, p. 212-217

Richard, K., and Courtright, J. H., 1954, Structure and mineralization at Silver Bell, Ariz.: A.I.M.E. Tr., v. 199, p. 1095-1099 (in Min. Eng., v. 6, no. 11)

—— 1966, Structure and mineralization at Silver Bell, Arizona, in Titley, S. R., and Hicks, C. L., Editors, *Geology of the porphyry copper deposits--southwestern United States*: Univ. Ariz. Press, Tucson, p. 157-163

Stewart, C. A., 1912, The geology and ore-deposits of the Silver Bell mining district, Arizona: A.I.M.E. Tr., v. 43, p. 240-290

—— 1913, Magmatic differentiation at Silver Bell, Ariz.: Science, new series, v. 37, no. 948, p. 338-340

Tolman, C. F., Jr., 1909, Copper deposits of Silverbell, Arizona: Min. and Sci. Press, v. 99, no. 22, p. 710-712

Notes

The Silver Bell mining district is situated some 35 miles northwest of Tucson, and mining in the area dates from 1865.

The oldest rocks in the vicinity of the mine are those of the Cambrian Bolsa quartzite, conformably above which is the Cambrian Abrigo limestone; unconformably above the Abrigo is the Devonian Martin limestone, followed by the Mississippian Escabrosa limestone, the Pennsylvanian Naco limestone, and the Permian blue and white limestone, with some interbedded shale. The probably Permian sedimentary series consists of blue limestone, sandstone, and another thickness of blue limestone with shale and hornstone. Younger than the Paleozoics is a series of probably Cretaceous rocks that includes conglomerates, red shales, and arkosic sandstones that may be as much as 5000 feet thick. These Paleozoic and Mesozoic sediments are intruded by a series of igneous rocks that probably were introduced in Laramide time, both before and after the end of the Mesozoic. The first of these rocks was an elongate stock of alaskite that extends west-northwest for some 4 miles; it was at first thought to be Precambrian, but this has been disproved by the presence in it of xenoliths of Paleozoic limestone and the finding of places where it intrudes post-Cambrian rocks. The alaskite was followed by the intrusion of a 3- by 6-mile, northwest-trending group of sill-like bodies and feeders of dacite porphyry that penetrated the Paleozoic sediments. The intrusion of the alaskite probably followed a fault contact between Paleozoic sediments (northeast) and Cretaceous sediments (southwest) and that of dacite entered through the same zone of weakness which then became the contact between dacite (northeast) and alaskite (southwest); the dacite contains huge pendant masses of Paleozoic sediments. A mass of ignimbrite or welded tuff present in the area south of the Oxide pit is probably somewhat later than the dacite. On the northeast, the volcanic rocks are partly in fault contact with younger syenodiorite porphyry and monzonite.

This phase of intrusive activity was followed by a period of extreme erosion during which considerable thicknesses of coarse conglomerates, sandstones, and arkoses (Clafin Ranch sediments) were produced. This period of erosion was followed by several thousand feet of volcanic flows and pyroclastics, which appear to correspond to the Silver Bell volcanic rocks in the Esperanza district. Following this erosion and volcanic activity, several major faults were formed along a west-northwest line. The faulted area was broken into horsts and grabens. After the faulting came an intrusion of monzonite stocks and of syenodiorite porphyry dikes associated with them into the area northwest of that now containing the faults just mentioned. The monzonite may have followed such faults in the area they intruded and so removed all traces of them. Most of the syenodiorite porphyry dikes associated with the monzonite were emplaced in east-northeast fractures but a few dikes trend northwest; further fracturing followed that both paralleled the dikes and the major west-northwest structures (locally, fracturing can be found in almost any direction). After this fracturing, the mineralizing and rock-altering solutions entered the area. The last igneous event in the immediate mine area was the intrusion of andesite dikes along the major structures rather than through the east-northeast cross structures that had directed the flow of the ore fluids. Late Tertiary and Quaternary erosion brought the lean primary mineralization to the surface, and ground waters developed oxidized ores in the limestones and enriched zones in siliceous rocks.

It appears certain from the sequence of events just recorded that the mineralization occurred after the end of the Cretaceous period but probably took place in the Paleocene. The ores had definitely been emplaced before the period of profound erosion that probably occurred in the middle Tertiary. Thus, although the ores are certainly early Tertiary, they still were formed during the Laramide orogeny and should be categorized as late Mesozoic to early Tertiary. Although the results of radioactive age dating are confusing in that rocks definitely older on geologic grounds are dated as being younger than ones that intrude, all ages range between 67 and 63 m.y., confirming a late Mesozoic-early Tertiary age for the ores that came in as part of this intrusive cycle.

The first ores mined at Silver Bell came from mines such as the Union that produced copper ores (mainly nonsulfide) from the belt of metamorphosed limestone in the pendants of carbonate rock in the dacite; locally the full width of a pendant may be metamorphosed. The ore bodies are mainly of the disseminated type with blebs and small veinlets as well as individual disseminated grains. The old mines were in the higher-grade areas, mainly in the oxidized zone, but some primary ore was also taken. The primary ores, unaffected by ground water attack, are mainly composed of pyrite and chalcopyrite, with small and irregularly distributed amounts of sphalerite and galena, small stringers of molybdenite, and very minor bornite. The minerals of the tacite and hornfels, developed before the ore minerals, are mainly garnet and diopside, plus calcite and including minor wollastonite and quartz, which last continued to deposit during the emplacement of the ore. The ore minerals normally occur in veinlets in the garnet-diopside-quartz-recrystallized calcite mass or as disseminations in that same material. It is reported that pyrite and chalcopyrite are often intergrown with sphalerite, and at least some of these intergrowths are of the type to indicate unmixing of chalcopyrite from solid solution in sphalerite.

Much of the ore mined from the high-grade areas in limestone came from those portions that had been oxidized to such minerals as malchite, chrysocolla, copper pitch ore, azurite, and native copper (minor). Overall, oxidation normally concentrated the copper only slightly, although locally primary ore averaging 2 to 3 percent copper was enriched to 10 to 12 percent by the oxidation process. Very little copper passed below the water table to be deposited as secondary enrichment of the primary sulfides.

About half of the primary sulfide mineralization in both igneous and altered (silicated) limestones occurs in narrow, parallel veins and veinlets that stand essentially vertically and strike in a generally northeast direction; the other half of the primary minerals is present as disseminated grains. The average vein is a thin seam containing pyrite and chalcopyrite and usually quartz surrounded by sericite that grades outward into clay minerals; enough molybdenite is present to justify its recovery. Enriched ore appears to have been about equally developed in altered zones of the various igneous rock types; some ore is now mined from the sediments that underlie the dacite on the east side of the El Tiro pit.

The two ore deposits (Oxide and El Tiro) of the disseminated type are roughly tabular masses of secondarily enriched ore in which chalcocite is the principal new mineral; it replaces chalcopyrite preferentially over pyrite. The primary disseminated copper ores in the altered igneous rocks at Silver Bell have all the characteristics of formation in the mesothermal range, but there is some question as to whether the ores in altered limestone were formed under mesothermal or under hypothermal conditions or partly in both ranges. Although the evidence is not yet conclusive, the presence of exsolution chalcopyrite in at least some of the sphalerite indicates that the deposits in altered carbonate rocks should be designated as hypothermal-2 to mesothermal and that the district as a whole should be so classified.

TOMBSTONE

Late Mesozoic to Early Tertiary	*Silver, Gold, Lead, Copper*	*Xenothermal to Epithermal*

Blake, W. P., 1882, The geology and veins of Tombstone, Arizona: A.I.M.E. Tr., v. 10, p. 334-345

Butler, B. S., and others, 1938, Geology and ore deposits of the Tombstone district, Arizona: Ariz. Bur. Mines, Geol. Ser. no. 10, Bull. no. 143, 114 p.

Church, J. A., 1903, The Tombstone, Arizona, mining district: A.I.M.E. Tr., v. 33, p. 3-37

Ransome, F. L., 1920, Bisbee and Tombstone district, Cochise County: U.S.

Geol. Surv. Bull. 710, p. 101-103, 113-116, 118-119

Tenney, J. B., 1938, Geology and ore deposits of the Tombstone district, Arizona (rev.): Econ. Geol., v. 33, p. 675-678

Notes

The Tombstone deposits (primarily worked for silver that furnished over 80 percent of the dollar value of the ores mined) are situated some 20 miles northwest of Bisbee.

The oldest rock in the area is the older Precambrian Pinal schist that appears to have been intruded in Precambrian times by a granodiorite at some stage during the various deformations of the middle Precambrian. The area, as is evidenced by the absence of rocks of the younger Precambrian Apache series, probably was considerably eroded during late Precambrian and Early Cambrian time; on the quite even surface thus produced, the Cambrian Bolsa quartzite was deposited. Although there were appreciable periods of time, particularly from the Upper Cambrian to the Middle Devonian, during which no sedimentation took place in the Tombstone area, there is essentially no evidence of angular unconformity between the various Paleozoic limestones that overlie the Bolsa. The first of the carbonate rocks was the Upper Cambrian Abrigo limestone, followed by the Mississippian Escabrosa limestone, and the Pennsylvanian Naco limestone. After another lengthy hiatus in sedimentation, the Lower Cretaceous Bisbee group was laid down with only a slight angular unconformity to the older rocks.

After the close of lower Mesozoic sedimentation, the rocks of the area were deformed and intruded by two igneous rocks of essentially the same age, the Uncle Sam quartz-latite porphyry to the west and the Schieffelin granodiorite to the east. The granodiorite produced strong contact-metamorphic effects on the rocks it intruded, particularly on the Naco and Blue limestones. The Uncle Sam quartz latite had essentially no effect on the rocks it encountered. Outside the Tombstone area, the Schieffelin granodiorite cuts off dikes that intrude the Uncle Sam; thus, though the two intrusions have not been found in contact in the district, the former probably is the younger. All the intrusives contain more or less ore, establishing the pre-ore age of these intermediate rocks. In 1938, Butler and Wilson thought that the porphyries were introduced during the Laramide orogeny. This is confirmed by Creasy and Kistler's determination of the age of the Schieffelin granodiorite to be about 71 m.y. old; this places it in the Late Cretaceous and categorizes the ores as late Mesozoic to early Tertiary.

If my reasoning is correct, these ore bodies are unique in Arizona in having been deposited near enough to the surface to be categorized as having formed under conditions of rapid loss of heat and pressure. This concept is supported by the near-surface intrusion of the Uncle Sam quartz-latite porphyry shortly before ore emplacement, the large amount of faulting, the breaking of minor folds into faults, and the narrow vertical range of the mineralization. The extensive masses of high-temperature lime silicates in the wall rock almost certainly were developed before, or almost entirely before, the ore minerals began to deposit; however, the abundant sphalerite in part contains oriented blebs of chalcopyrite, suggesting high-intensity conditions of deposition for that sphalerite. The sphalerite was followed by chalcopyrite (which does not contain exsolved sphalerite); sphalerite was then replaced by silver-bearing galena and tetrahedrite. Native gold was deposited in the same general period as the last two sulfides; an appreciable portion of the gold and silver were contained in probably still later hessite. Primary stromeyerite is present in some quantity, and bournonite is uncommon. Alabandite also appears to be abundant, much more than is usual, but its position in the paragenesis is uncertain. Thus, if the deposit was formed near the surface, the sphalerite containing chalcopyrite blebs suggests the xenothermal range; the tetrahedrite and silver-bearing galena can be categorized as kryptothermal, and the hessite and stromeyerite as epithermal.

Arkansas

ARKANSAS BAUXITE

Late Mesozoic (pre-Laramide) (syenite), Early Tertiary (bauxite)	*Bauxite*	*Magmatic-1a, Residual-B1, Placers-IB and IC*

Behre, C. H., Jr., 1932, Origin of bauxite deposits: Econ. Geol., v. 27, p. 678-680 (general)

Bramlette, M. N., 1936, Geology of the Arkansas bauxite region: Ark. Geol. Surv. I.C. 8, 68 p.

Branner, J. C., 1897, The bauxite deposits of Arkansas: Jour. Geol., v. 5, p. 263-289

Branner, G. C., 1932, The Arkansas bauxite deposits, in *Mining districts of the eastern states*: 16th Int. Geol. Cong. Guidebook II, p. 92-103

Goldman, M. I., 1955, Petrography of bauxite surrounding a core of kaolinized nepheline syenite in Arkansas: Econ. Geol., v. 50, p. 586-609

Goldman, M. I., and Tracey, J. I., Jr., 1946, Relations of bauxite and kaolin in the Arkansas bauxite deposits: Econ. Geol., v. 41, p. 567-575

Gordon, M., Jr., and Murata, K. J., 1952, Minor elements in Arkansas bauxite: Econ. Geol., v. 47, p. 169-179

Gordon, M., Jr., and Tracey, J. I., Jr., 1952, Origin of the Arkansas bauxite deposits, in *Problems of clay and laterite genesis*: A.I.M.E. Symposium in St. Louis, 1951, New York, p. 12-34

Gordon, M., Jr., and others, 1958, Geology of the Arkansas bauxite region: U.S. Geol. Surv. Prof. Paper 299, 268 p.

Harder, E. C., 1949, Stratigraphy and origin of bauxite deposits: Geol. Soc. Amer. Bull., v. 60, p. 887-907 (general)

Hartman, J. A., 1959, The titanium mineralogy of certain bauxites and their parent minerals: Econ. Geol., v. 54, p. 1380-1405 (particularly p. 1383-1388) (disc., v. 55, p. 1313)

Hayes, C. W., 1901, The Arkansas bauxite deposits: U.S. Geol. Surv. 21st Ann. Rept., pt. 3, p. 435-472

Malamphy, M. C., and Vallely, J. L., 1944, Geophysical survey of the Arkansas bauxite region: Geophysics, v. 9, p. 324-366

Mead, W. J., 1915, Occurrence and origin of the bauxite deposits of Arkansas: Econ. Geol., v. 10, p. 28-54

Thoenen, J. R., and others, 1945, Geophysical survey of the Arkansas bauxite region: U. S. Bur. Mines R.I. 3791, 49 p.

Williams, J. F., 1891, The igneous rocks of Arkansas: Ark. Geol. Surv. Ann. Rept. 1890, v. 2, p. 1-391, 429-457

Notes

The bauxite deposits of Arkansas are concentrated in two areas: (1) in Pulaski County, immediately south of Little Rock, and (2) in Saline County about 15 miles southwest of that city; both areas are in the region of Tertiary coastal plain sediments, just southwest of the outcrop-boundary with Paleozoic rocks.

The deposits in both bauxite areas are closely associated with hills of nepheline syenite that rise as bosses above a much larger batholith; the syenite was intruded into the Paleozoics and was later buried by the Tertiary sediments. Although the Paleozoic formations in the district were indispensible as the host rocks for the nepheline syenites and other related alkalic bodies, they had nothing to do with the development of the bauxite, and all that needs to be said here about this thick series of formations is that they probably range in age from Ordovician to Pennsylvanian, are mainly sandstones and shales, have a definite strike to the northwest, are steeply dipping, and become older and more tightly folded to southwest. The boundary between the Interior Highlands to the northwest and the Gulf Coastal Plain to the southeast is the contact between these hard Paleozoic beds and the soft Tertiary beds that overlap them from the southeast.

The Coastal Plain sediments are marine and nonmarine rocks of Paleocene and Eocene age, plus gravels of the late Tertiary and Quaternary and terrace deposits and alluvium of the Quaternary. The Tertiary beds lie unconformably on both the Paleozoics and the alkalic intrusives, with some of the nepheline syenite masses never having been covered by Tertiary sediments at all. The Tertiary strata generally strike to the northeast and dip to the southeast at low angles.

Although the nepheline syenite from which the bauxite is derived cannot be closely dated through its own relations to sedimentary rocks, the peridotite dikes of Pike County farther southwest are similar to some of the monchiquite dikes rocks found in the bauxite region. These peridotite dikes intrude the Lower Cretaceous Trinity formation and are overlain unconformably by the younger, early Upper Cretaceous Tokio beds. Granted that the similarity in composition of the peridotite and monchiquite indicate a similarity in age, the nepheline syenites, genetically related to the monchiquites, also probably were introduced in the early Upper Cretaceous. The bauxites, derived from the nepheline syenite (and to a lesser extent from its subtype pulaskite), appear to have been formed during a long and essentially uninterrupted period of weathering that followed the deposition of the marine sediments of the Paleocene Wills Point formation and ended before the beginning of the deposition of the continental early Eocene Saline formation. Thus, the bauxitization of the nepheline syenite was certainly early Eocene, with the possibility that the process actually began in the Paleocene. The movement of the clay, from which the secondary (as opposed to the residual bauxite formed on the syenite hills) bauxite was formed, probably commenced at the very beginning of Wilcox time as almost all of these deposits lie on the upper surface of the Wills Point formation and, therefore, in the basal portion of the Berger. The alluvial bauxite placers are found interstratified with the beds of the Berger and the lower Saline and are, therefore, of early Eocene age. The syenite source rocks, therefore, are classified as late Mesozoic (pre-Laramide) and the actual bauxite deposits as early Tertiary.

The nepheline syenite and its subtype called pulaskite, from which almost all of the bauxite was formed, were introduced as magmas intrusive into Paleozoic sediments. These igneous source rocks are classed as magmatic-1a. The bauxite, of which the principal mineral is gibbsite--$Al(OH)_3$, appears to have been formed through tropical or subtropical weathering during which bauxite was developed directly from nepheline syenite above the water table. At the same time that this type one bauxite was forming above the water table, kaolin appears to have been forming below that surface. Thus, both bauxitization and kaolinization took place at the same time in the same general areas.

Although kaolin lies between most of the bauxite and its source rock, the nepheline syenite, this underlying kaolin does not appear to grade into the bauxite. Instead, the zone of kaolin formation probably was separated from that of bauxite by the water table, and the contact between them, though irregular, was quite sharp. After large amounts of bauxite (above) and kaolin (below) had formed, it appears that the water table rose appreciably, probably as the drainage system was largely filled with sediment, and kaolinization became the dominant process. This later stage of kaolinization in the rock volumes that already had been converted largely to bauxite is evidenced by the presence of kaolin in pores in the bauxite and as veinlets cutting through it. Further, what were unaltered nepheline syenite boulders in the bauxite were partly to entirely converted to kaolin during this later kaolinization. Thus, the sequence of events in the production of this residual (type one) bauxite and its associated kaolin appears to have been (1) bauxitization of the syenite to produce the so-called granitic-textured bauxite, with microcrystalline gibbsite as the main mineral; (2) formation within granitic-textured bauxite of pisolites and other concretionary structures and the production of cryptocrystalline gibbsite from the microcrystalline variety; (3) local dehydration, resulting in the conversion of gibbsite in some pisolites to boehmite--AlO(OH), accompanied by the conversion of some iron oxide to magnetite; (4) downward migration of the constituents of gibbsite and its deposition in fissures of various sizes within both bauxite matrix and enclosed pisolites; (5) following a rise of the water table that accompanied Eocene sedimentation, derivation of further kaolin from the still-fresh nepheline syenite beneath the bauxite and from residual nepheline syenite in the bauxite; some of the silica set free in this process, on encountering bauxite, converted some of it to kaolin; some siderite also was formed and filled fractures and pores in the bauxite. Thus, in its original formation from the syenite, the type one bauxite is best categorized as formed by rock decay and weathering and should be classified as residual-B1 while the minor amounts formed as under (4) above would, if of economic value, be classified as ground water-B2. Much of this residual (type one) bauxite deposit (formed through rock decay and weathering) remains essentially where it was developed and has been mined in such locations. While bauxite was forming on the upper slopes, kaolin was being produced in the valleys, and both were available for transport to the growing deposits of detritus further toward the sea. Bauxite probably was not moved downslope in large amounts until the Berger sediments had filled a considerable portion of the drainage basin, and the loci of the collection of these alluvial masses of bauxite gradually moved back up the slope as the depth of the early Eocene Berger sediments increased.

Much of the kaolinitic debris probably was not moved into position by running water but, instead, was carried downslope by soil creep, mass wasting, slump, mudflow, and landslide. Thus, the kaolinitic debris can be considered at least as much eluvial as alluvial and probably can better be classified as eluvial rather than alluvial. The type two (secondary) bauxite that developed from the clayey debris, however, appears to have formed almost entirely in situ so that it is as much a residual bauxite as that still overlying the nepheline syenite. The only difference between these two types of bauxite was the material from which they were formed--nepheline syenite for the bauxite above that rock type and eluvial kaolinitic debris for that lying on the clays.

Siderite also is present in this type two bauxite as veinlets and as fillings of voids and pores. The amounts of bauxite mined from types one and two of residual deposits appear to have been about equal, and type two also should be classified as residual-B1. The third type of bauxite deposits is much less important in tonnage developed than either of the first two and exists as stratified beds of bauxite pisolites, pebbles, and grains in the Berger formation. These cut, overlie, or occupy channels in the second type of deposit and are reasonably classified as alluvial placers (placers-IC). The fourth (and also economically unimportant) variety of bauxite deposits occupies much the same position in relation to the Saline formation (middle Wilcox) that the second type does in relation to the Berger formation. These basal Saline deposits, however, are made up of rubble of variously sized fragments of bauxite and clay and appear to have moved so short a distance, mainly under the influence of gravity, as to be best classified as eluvial placers (placers-IB). To summarize, the residual type (type one and type two) is classed as residual-B1; type four is classed as placers-IB; and type three as placers-IC. The minor amounts of gibbsite carried downward by ground water were of no economic importance.

MAGNET COVE

Late Mesozoic (pre-Laramide)	*Barite, Titanium, Vanadium*	*Rutile, Magmatic-3b; Brookite, Hypothermal-1; Barite, Mesothermal*

Erickson, R. L., and Blade, L. V., 1956, Map of bedrock geology of Magnet Cove igneous area, Hot Spring County, Arkansas: U.S. Geol. Surv. Mineral Investigations, Field Studies Map MF 53, 1:6000

—— 1963, Geochemistry and petrology of the alkalic igneous complex at Magnet Cove, Arkansas: U.S. Geol. Surv. Prof. Paper 425, 95 p.

Fryklund, V. C., Jr., and Holbrook, D. F., 1950, Titanium deposits of Hot Spring County, Arkansas: Ark. Res. and Dev. Comm., Div. Geol. Bull. 16, 173 p.

Fryklund, V. C., Jr., and others, 1954, Niobium (columbium) and titanium at Magnet Cove and Potash Sulphur Springs, Arkansas: U.S. Geol. Surv. Bull. no. 1015-B, p. 23-57

Holbrook, D. F., 1947, A brookite deposit in Hot Spring County, Arkansas: Ark. Res. and Dev. Comm., Div. Geol. Bull. 11, 21 p.

—— 1948, Molybdenum in Magnet Cove, Arkansas: Ark. Res. and Dev. Comm., Div. Geol. Bull. 12, 16 p.

Hollingsworth, J. S., 1967, Geology of the Wilson Springs vanadium deposits, in *Central Arkansas, economic geology and petrology*: Geol. Soc. Amer. Field Conf. 1967 Guidebook, Ark. Geol. Comm., p. 22-24 (This deposit is associated with the Potash Sulfur Springs intrusive bodies 6 miles west of the Magnet Cove complex where similar ores are to be found.)

Landes, K. K., 1931, A paragenetic classification of the Magnet Cove minerals: Amer. Mineral., v. 16, p. 313-326

Landes, K. K., and others, 1933, Magnet Cove, Arkansas, in *Mining districts of the eastern states*: 16th Int. Geol. Cong., Guidebook 2, p. 104-112

McElwaine, R. B., 1946, Exploration for barite in Hot Spring County, Arkansas: U.S. Bur. Mines, R.I. 3963, 21 p. (mimeo.)

Miser, H. D., and Stevens, R. N., 1938, Taeniolite from Magnet Cove, Arkansas: Amer. Mineral., v. 23, p. 104-110

Parks, B., and Branner, G. C., 1932, A barite deposit in Hot Spring County, Arkansas: Ark. Geol. Surv. Inf. Circ. 1, 52 p. (mimeo.)

Purdue, A. H., and Miser, H. D., 1923, Hot Springs, Arkansas folio: U.S. Geol. Surv. Geol. Atlas, Folio 215, 12 p. (Magnet Cove lies just east of eastern edge of quadrangle.)

Rechenberg, H. P., 1955, Zur Genesis der primären Titanerzlagerstätten: Neues Jb. f. Mineral., Mh., H. 4, S. 87-96

Reed, D. F., 1949, Investigation of Magnet Cove rutile deposit, Hot Spring County, Ark.: U.S. Bur. Mines, R.I. 4593, 9 p. (mimeo.)

Ross, C. S., 1941, Magnet Cove, Arkansas, in *Occurrence and origin of the titanium deposits of Nelson and Amherst Counties, Virginia*: U.S. Geol. Surv. Prof. Paper 198, p. 23-26

Scull, B. J., 1958, Origin and occurrence of barite in Arkansas: Ark. Geol. and Cons. Comm. I.C. 18, 101 p. (particularly p. 29-51, 54, 56-70)

—— 1967, The Chamberlin Creek barite deposit, in *Central Arkansas, economic geology and petrology*: Geol. Soc. Amer. Field Conf. 1967 Guidebook, Ark. Geol. Comm., p. 19-21

Taylor, I. R., 1969, Union Carbide's twin-pit vanadium venture at Wilson Springs: Min. Eng., v. 21, no. 4, p. 82-85

Washington, H. S., 1900, Igneous complex of Magnet Cove, Arkansas: Geol. Soc. Amer. Bull., v. 11, p. 389-416

—— 1901, The foyaite-ijolite series of Magnet Cove; a chemical study in differentiation: Jour. Geol., v. 9, p. 607-622, 645-670

Williams, J. F., 1891, The igneous rocks of Arkansas: Ark. Geol. Surv. Ann. Rept. 1890, v. 2, 457 p. (particularly p. 163-343)

Zartman, R. E., and Marvin, R. F., 1965, Age of alkalic intrusive rocks from eastern and mid-continent states: U.S. Geol. Surv. Prof. Paper 525A, p. A-162

Zimmermann, R. A., 1970, Sedimentary features in the Meggen barite-pyrite-sphalerite deposit and a comparison with the Arkansas barite deposits: Neues Jb. f. Mineral., Abh., Bd. 113, H. 2, S. 179-214

Zimmermann, R. A., and Amstutz, G. C., 1964, Small scale sedimentary features in the Arkansas barite district, in Amstutz, G. C., Editor, *Sedimentology and ore genesis*, v. 2 of *Developments in sedimentology*: Elsevier, Amsterdam, 184 p. (particularly p. 157-163)

Notes

The Magnet Cove district lies about 15 miles south of east of Hot Springs in central Arkansas.

The oldest known rock in the area is the Ordovician (Trenton) Bigfork chert, with which minor amounts of limestone and shale are interbedded. The Bigfork is overlain by the Polk Creek graptolitic shale of Richmond age. The next formation is the Blaylock sandstone, but in some places shale predominates, and the beds have been mapped as Missouri Mountain shale; the formation is probably lower Middle Silurian. A considerable period of deformation followed the deposition of the Blaylock. The Arkansas novaculite, an extremely fine-grained siliceous rock originally probably deposited as a chemical precipitate of silica, unconformably overlies the Blaylock; its age is uncertain as it is essentially lacking in fossils, but it has been tentatively correlated with rocks ranging from Lower Devonian to Lower Mississippian in age and may cover that entire span of time. The novaculite is overlain by the Stanley shale, a dark shale with thick siltstone and sandstone members, with some local tuffs near its base. It is Mississippian in age. The Stanley, in turn, was followed by the Jackfork sandstone (lacking in the Magnet Cove area); its time of deposition probably overlapped the Mississippian-Pennsylvanian boundary. The Magnet Cove alkalic rocks and the barite and rutile deposits are located in the eastern end of the Mazarn basin, a synclinorium between the Trap Mountains (south) and the Zig Zag Mountains (northeast), in a minor structure known as the Chamberlain Creek syncline. No rocks younger than the Paleozoic are known in the Magnet Cove area proper except the alkalic intrusives and some Quaternary valley alluvium.

The alkalic rocks are younger than the Stanley shale (the youngest rock intruded by them) and older than the Quaternary valley alluvium. To the northeast of Magnet Cove, however, eroded surfaces of nepheline syenite are overlain by Eocene sediments. Pebbles of varieties of the igneous rocks known in Magnet Cove are found in the lowest Upper Cretaceous beds elsewhere in central Arkansas, and the Murfreesboro peridotite, almost certainly from the same general source as the Magnet Cove igneous rocks, intrudes the Trinity formation (Lower Cretaceous). This evidence strongly suggests that the alkalic intrusives were emplaced during the Cretaceous, probably near the lower-Upper Cretaceous boundary. These alkalic rocks are comparatively rich in barium and titanium, among other elements; these are the principal metals in the two varieties of mineral deposits of economic importance in the Magnet Cove area, the bedded barium deposits and the rutile-rich veins and irregular masses. The age of the economic mineral deposits, therefore, is probably very nearly that of the alkalic rocks, which indicates that they should be classified as late Mesozoic (pre-Laramide).

The alkalic igneous complex of Magnet Cove covers an area of 4.6 square miles and is a ring-dike complex that consists of (1) an inner core of ijolite (a member of a series of nepheline-pyroxene rocks, essentially feldspar-free), carbonatite, and lime-silicate rocks; (2) an intermediate ring of fine-grained trachytes and phonolites and breccias of the same materials, rutile veins, and irregular masses of metamorphosed sedimentary rocks; and (3) an outer ring of nepheline syenite. In addition to the rocks of the rings, most of which are arcuate bodies and only one of which forms a complete ring, there are two masses of jacupirangite intrusive into the outer ring and a large number and variety of dikes within the complex (such as tinguaite, analcime olivine melagabbro, nepheline syenite pegmatite, and trachyte porphyry) and other dikes outside it (such as pegmatite, aplite, syenite, trachyte, porphyry trachyte, tinguaite, andesite, and diorite). An area of metamorphosed sediments almost exactly the size of the complex itself surrounds the alkalic rocks. The alkalic rocks of the Cove are high in sodium, as is true of the rocks of similar complexes throughout the world, but no zone of fenite was developed in the Paleozoic sediments where they have been exposed to contact metamorphic effects from the intrusives as is true of the sediments surrounding such alkalic rocks as those of the Fen area and of Alnö Island. Nor, and this well may be related to the lack of fenite, are there any representatives of the alnoite-kimberlite series in Magnet Cove as there are in both Scandinavian localities.

The rocks now known in the complex were intruded in the following order: (1) the phonolite and trachyte of the intermediate ring, (2) the jacupirangite now surrounded by the rocks of the outer ring and by altered sediments, (3) the alkalic syenites of the outer ring, (4) the ijolite series of the inner core, and (5) the carbonatite of the core and the minor dikes and veins throughout and beyond the complex.

Among the veins and irregular masses introduced during the last stage of activity in Magnet Cove were ones containing deposits of rutile, brookite, and molybdenite. A small production has come from the rutile bodies, but the brookite and molybdenite veins have not been worked commercially. The four principal titanium mineralization types are: (1) sugary-textured albite-dolomite masses, (2) microcline-calcite bodies, (3) coarse-grained albite-perthite-carbonate veins, and (4) albite-ankerite veinlets. Types (1), (2), and (4) are the only ones that contain important amounts of rutile; small, but possibly economically recoverable, amounts of columbium and vanadium are associated with the titanium minerals. The dominantly high-temperature mineral content of these rutile bodies, their generally irregular outlines, the lack of large-scale alteration of the surrounding phonolite, and the abundance of carbonate suggest that they are carbonatites rather than hydrothermal veins and probably are related in ultimate origin to the carbonate masses in the inner core. They should, therefore, be classified as magmatic-3c rather than hypothermal-1.

On the other hand, the brookite deposits are in recrystallized Arkansas novaculite in veins that are composed mainly of quartz with brookite, rutile, taeniolite ($KLiMg_2Si_4O_{10}F_2$), pyrite, and probably hydrothermal clay minerals. These veins, in contrast to the rutile-bearing masses, are probably of high-temperature hydrothermal origin and are here classed as hypothermal-1.

The most important mineral raw material recovered from the Magnet Cove area is barite. This barite is in deposits in the lower part of the Stanley formation (shale) just above its contact with the underlying Arkansas novaculite in the Chamberlain Creek syncline some 2 miles northeast of the town of Magnet. The maximum length of the barite-bearing volume of the Stanley shale is about 3200 feet and the width is about 1800 feet in an area of the syncline where the trend is nearly east-west. The average thickness of the zone containing at least some barite is about 300 feet, while the minable thickness averages some 60 feet and may reach 80 to 100 feet; locally the ore body may be divided into upper and lower members by 5 to 15 feet of barren shale. The barite in this deposit is present as (1) finely crystalline material in massive beds, (2) dense cryptocrystalline materials in massive beds, (3) nodular zones, and (4) coarsely crystalline material in lentils. The types are listed in order of decreasing abundance. What appear to be replacement remnants of the host rock occur in the finely crystalline barite. The suggestion has been made (Scull, 1958) that differential slippage of the Stanley units over the more massive novaculite during deformation created openings that allowed ore-forming fluids to enter and move through the basic portions of the Stanley beds along fractures and joints in those rocks. The concentration of the ore in synclines is less readily explained but may simply be due to synclinal rock volumes being those first encountered by ore solutions rising through the Arkansas novaculite. The unusually high concentration of barium in the rocks of Magnet Cove suggests that the source magma for these rocks was also the hearth from which the barium-rich ore fluids were generated. The presence of barite veins in the Alnö area is an interesting confirmation of the genetic connection postulated between the Magnet Cove rocks and the barite deposits. The lack of sulfides and other hydrothermal minerals associated with the barite and the absence of wall-rock alteration other than the introduction of barite make it difficult to classify the barite deposits, but other massive barite deposits, normally accompanied by considerable amounts of sulfide minerals, were formed in the mesothermal range, so the Magnet Cove barite bodies are here categorized as mesothermal. The strata-bound deposits of the Magnet Cove area are so closely related to a single formation that the possibility of their having been formed by syngenetic processes must be considered. Their close spatial relationship to the alkaline rocks of the Magnet Cove complex and the common occurrence of abundant barite with carbonatites in many parts of the world strongly suggest that these are hydrothermal replacements of the Stanley shale.

California

ALLEGHANY

Late Mesozoic (pre-Laramide) *Gold* *Leptothermal*

Carlson, D. W., and Clark, W. B., 1956, Lode gold mines of the Alleghany-Downieville area, Sierra County, California: Calif. Jour. Mines and Geol., v. 52, no. 3, p. 237-272 (particularly p. 240-245, 247-268)

Cooke, H. R., Jr., 1947, The original Sixteen to One gold-quartz vein, Alleghany, California: Econ. Geol., v. 42, p. 211-250

Ferguson, H. G., 1915, Lode deposits of the Alleghany district, California: U.S. Geol. Surv. Bull. 570, p. 153-182

Ferguson, H. G., and Gannett, R. W., 1929, Gold-quartz veins of the Alleghany district, California: A.I.M.E. Tech. Pub. 211-1-24, 40 p. (not in Tr.)

—— 1932, Gold-quartz veins of the Alleghany district, California: U.S. Geol. Surv. Prof. Paper 172, 139 p.

Knopf, A., 1933, Gold-quartz veins of the Alleghany district, California (rev.): Econ. Geol., v. 28, p. 399-401

Lindgren, W., 1900, Colfax folio, California: U.S. Geol. Surv. Geol. Atlas, Folio 66, 10 p.

McKinstry, H. E., and Ohle, E. L., Jr., 1949, Ribbon structure in gold-quartz veins: Econ. Geol., v. 44, p. 87-109

Notes

The Alleghany district lies almost 70 miles north-northeast of Sacramento on the west flank of the Sierra Nevada about 20 miles west of the crest, being bounded on the south by the Middle Fork of Yuba River.

The oldest rocks of the area are those known through much of the California gold belt as the Calaveras formation (usually assigned to the Carboniferous); in the Alleghany district, however, these highly tilted and metamorphosed sediments and volcanic rocks have been divided into five formations. The oldest of these is the Blue Canyon formation (clay slate and quartzitic sandstone). The next is the Tightner (mainly amphibolite schist derived from andesite lava), followed by the Kanaka (mappable units of conglomerate, chert, and interbedded slate and tuff), the Relief (chiefly schistose quartzite), and the Cape

Horn slate. All of these lie between the base and the top of the Calaveras formation as mapped elsewhere; therefore, all are probably Carboniferous in age; they strike generally north-northwest and dip at high angles, normally to the east. These formations were intruded by a variety of rocks, all probably of Mesozoic age, the oldest of which is the serpentine that originally probably was composed of peridotite, pyroxenite, and rocks intermediate between the two. The serpentine is closely associated in space with the gabbro, the marginal position of the former suggesting that it was a differentiate of the gabbro magma and that the two rocks, therefore, were intruded at much the same time. The youngest Mesozoic intrusives are granite and minor aplite that provide only a small portion of the rock outcrops in the district. These silicic rocks were intruded after the alteration of the gabbro to amphibolite and after the development of its schistosity; where the granite intruded amphibole-bearing rocks, the amphibole has been converted to biotite. The granite and aplite show essentially no schistosity, nor have they undergone changes in mineralogy due to regional metamorphism as have the gabbro and diorite; this strongly suggests that the granite is appreciably younger than the mafic rocks. Although Ferguson and Gannett (1932) conclude that the igneous activity was probably Late Jurassic to earliest Cretaceous, later work has cast doubt on this age assignment. It appears, according to Curtis and his colleagues, that the various silicic intrusions (at least eight that range from granodiorite to granite) of the Sierra Nevada batholith in the Yosemite area were intruded over a span of nearly 20 m.y. (76.9 to 95.3 m.y.) which places them in the Upper Cretaceous (Santa Lucian orogeny) but definitely before the Laramide orogeny. The granite and aplite of the Alleghany area probably were intruded during this time span, while the mafic rocks, on the other hand, probably were introduced during the orogeny at the end of the Jurassic (Nevadan). As the ore mineralization is later than the granite, and as there appears to have been no igneous activity later than the granite-aplite intrusion to which the ore fluids possibly were related, the age of the deposits is here designated as Late Cretaceous (pre-Laramide).

The veins in the Alleghany district occur in two main groups: (1) those that strike northwest to west and dip 25° to 40° north and (2) those that strike northwest to north and dip west at high angles (never less than 60°). In any given portion of the district, one vein type will predominate over the other. The veins cut all of the pre-Tertiary rocks except for the larger bodies of serpentine, with the majority of the productive veins being in the Tightner formation (amphibolite schist), but some are found in the gabbro, the Kanaka formation, and the Cape Horn formation. Although vein systems cross the entire area, individual veins persist over much shorter distances; one of the longest is the Sixteen to One vein that extends for nearly a mile. The veins, on the average, contain very little gold (normally less than 0.05 ounces of gold per ton), but the gold content of the scattered high-grade shoots is as much as thousands of ounces per ton, making mining possible; sulfides usually form less than 2 percent by volume of the vein material.

The first stage of the mineralization may have been the formation of chlorite on a regional scale, although it developed more intensely in the neighborhood of the veins; at least part of the serpentinization of the peridotite may have occurred at this stage. The next stage was the combined quartz-pyrite-arsenopyrite mineralization during which quartz both preceded and followed the sulfides. This stage was followed by a large-scale formation in the wall rock of carbonate (mainly ankerite) with lesser sericite, mariposite, and locally developed talc. During the later stages of alteration, a series of sulfides was formed by fracture filling and by replacement of carbonates, followed by the introduction of gold as the latest mineral of the sequence. The first of these sulfides was sphalerite which contains exsolved blebs of chalcopyrite; this was followed by tetrahedrite and then by chalcopyrite (definitely later than that which exsolved from the sphalerite); next came galena and finally the free gold which overlapped to some extent with the galena. There are no gold or other tellurides in the ore.

The sphalerite, containing exsolved chalcopyrite, probably was deposited under hypothermal conditions, although the wall-rock alteration developed at much the same time would, taken alone, be considered more characteristic of mesothermal than hypothermal conditions. The sphalerite, tetrahedrite, and most chalcopyrite (that occurs as large grains) appear to have replaced carbonate, while the galena and gold and some chalcopyrite occur mainly as open space fillings in veinlets and tiny wisps of mineral material. This later part of the sequence of mineralization is not too readily classified. There is no evidence that the chalcopyrite and tetrahedrite ever were in solid solution in each other, and almost all of these two sulfides appears to have formed after the precipitation of sphalerite had ceased; this suggests that there may have been a considerable change in ore fluid intensity at this stage and that the two copper-bearing minerals were deposited under far less intense conditions than the sphalerite. Thus, the later chalcopyrite and the galena and gold probably were deposited in the leptothermal stage.

BISHOP

Late Mesozoic (pre-Laramide)	*Tungsten, Molybdenum, Copper*	*Hypothermal-2 to Mesothermal*

Bateman, P. C., 1945, Pine Creek and Adamson tungsten mines, Inyo County, California: Calif. Jour. Mines and Geol., v. 41, no. 4, p. 231-249

—— 1956, Economic geology of the Bishop tungsten district, California: Calif. Div. Mines Spec. Rept. 47, 87 p.

—— 1961, Granitic formations in the east-central Sierra Nevada, near Bishop, California: Geol. Soc. Amer. Bull., v. 72, p. 1521-1537

Bateman, P. C., and others, 1950, Geology and tungsten deposits of the Tungsten Hills, Inyo County, California: Calif. Jour. Mines and Geol., v. 46, no. 1, p. 23-42

—— 1963, The Sierra Nevada batholith--a synthesis of recent work across the central part: U.S. Geol. Surv. Prof. Paper 414-D, p. D1-D46

—— 1965, Geology and tungsten mineralization of the Bishop district, California: U.S. Geol. Surv. Prof. Paper 470, 208 p.

Chapman, R. W., 1937, The contact-metamorphic deposit of Round Valley, California: Jour. Geol., v. 48, no. 8, p. 859-871

Gilbert, C. M., 1941, Late Tertiary geology southeast of Mono Lake, California: Geol. Soc. Amer. Bull., v. 52, p. 781-815

Gray, R. F., and others, 1968, Bishop tungsten district, California, in Ridge, J. D., Editor, *Ore deposits of the United States, 1933-1967* (Graton-Sales Volumes): Chap. 73, v. 2

Hess, F. L., and Larsen, E. S., Jr., 1921, Contact-metamorphic tungsten deposits of the United States: U.S. Geol. Surv. Bull. 725-D, p. 245-309 (particularly p. 268-274)

Kerr, P. F., 1946, Tungsten mineralization in the United States: Geol. Soc. Amer. Mem. 15, 241 p. (particularly p. 17-19, 142-147)

Knopf, A., 1917, Tungsten deposits of northwest Inyo County, California: U.S. Geol. Surv. Bull. 640, p. 229-249

Lemmon, D. M., 1941, Tungsten deposits in the Tungsten Hills, Inyo County, California: U.S. Geol. Surv. Bull. 922-Q, p. 497-514

—— 1941, Tungsten deposits of the Benton Range, Mono County, California: U.S. Geol. Surv. Bull. 922-S, p. 581-593

—— 1941, Tungsten deposits in the Sierra Nevada near Bishop, California: U.S. Geol. Surv. Bull. 931-E, p. 79-104

Mayo, E. B., 1941, Deformation in the interval Mt. Lyell-Mt. Whitney, California: Geol. Soc. Amer. Bull., v. 52, p. 1001-1084

Norman, L. A., Jr., and Stewart, R. M., 1951, Mines and mineral resources of Inyo County: Calif. Jour. Mines and Geol., v. 47, no. 1, p. 17-223 (particularly p. 85-98)

Rinehart, C. D., and Ross, D. C., 1956, Economic geology of the Casa Diablo Mountain quadrangle, California: Calif. Div. Mines Spec. Rept. 48, 17 p.

Notes

The Bishop area lies in east-central California about halfway between Los Angeles and Reno, Nevada.

Over 50 tungsten prospects have been located in the district, of which the Pine Creek mine is the only one now in production, and it is operating only because it produces considerable amounts of molybdenum and copper in addition to tungsten. Essentially all the deposits of the district are found in tactite or skarn derived from marble near the contact of such rocks with intrusive igneous masses in which the altered sedimentary rocks are present as inclusions or pendants. The carbonate rocks from which the tactite of the Pine Creek mine was derived probably are of Paleozoic age, but the metamorphism they have undergone has made it difficult, if not impossible, to date these rocks more exactly. During the Mesozoic, these Paleozoic limestones and other sediments were extensively intruded by silicic rocks, the immediate effect of which was the conversion of pure limestone to marble, impure limestone to calc-hornfels (which locally may contain marble lenses), and noncalcareous rocks to hornfels and schist. Following the development of the marble and other contact-metamorphosed rocks, the area was invaded by hydrothermal solutions that converted the marble to tactite and light-colored calc-silicate rock, and deposited quartz veins and associated silicified zones; the solutions then deposited the ore minerals.

The oldest of the igneous rocks of the Bishop area are metavolcanic rocks, mainly andesites and rhyolites. Some metasediments are intercalated with the volcanic rocks, and one such marble lens contains the Hanging Valley mine. These volcanic rocks are probably Mesozoic in age, but there appears to be no definite evidence of this. Later than the metavolcanic rocks, several varieties of dark-colored Mesozoic igneous rocks were introduced into the area; these are lumped under the designation of quartz diorite and hornblende gabbro; they may include some masses of amphibolite of metamorphic origin. A few tungsten deposits in the Deep Canyon area are found in inclusions of calcareous rocks in the gabbro. Next, still younger Mesozoic igneous rocks that range from granitic through quartz monzonitic to granodioritic (with the granites being slightly the youngest and the granodiorites the oldest) intruded the area.

In 1958, Larsen and his colleagues made lead-alpha age determinations on granites from the Bishop area. These determinations ranged from 88 to 116 m.y.; the ages of the individual intrusions, however, did not correlate well with the order of their intrusion. More recent K/Ar studies of rocks from the neighboring Yosemite area give ages between 77 and 95 m.y., ages which agree well with the observed order of intrusion. One of these rocks, the Cathedral Peak granite, has been correlated with similar masses in the Bishop area, and the age of this rock in Yosemite is about 84 m.y. This places the time of the intrusion of the Bishop granites some 15 to 25 m.y. before the end of the Cretaceous, and as the ore formation appears to be directly connected with these intrusions, the ores have about the same age. This date is sufficiently earlier than the end of the Cretaceous for the ores to be classified as Late Mesozoic (pre-Laramide).

The tungsten deposits of the Bishop area are all in tactite in close juxtaposition to igneous rocks. Almost 75 percent of the deposits are in tactite in direct contact with granite while 10 percent are in tactite in contact with granodiorite, and the remainder are in inclusions enclosed in quartz diorite or granodiorite; these deposits are always close, though not directly adjacent, to granite. The Pine Creek ore bodies all contain tungsten ore shoots, but only upper parts of two of them contain molybdenum ore shoots; the boundaries of both types of ore shoots are determined by assays. Some of the molybdenum ore shoots may contain as much tungsten as the tungsten ore shoots, and all the molybdenum shoots contain substantial amounts of copper. The tungsten ore shoots, on the other hand, contain about 0.2 percent of MoS_2. Although the molybdenum associated with the scheelite is calculated as MoS_2, much of it actually is present as powellite in solid solution in the scheelite or as powellite itself, the latter derived from the alteration of scheelite. In the upper part of the main ore body, about 25 percent of the ore shoots contained tungsten and molybdenum mineralizations, both of ore grade.

The scheelite probably was deposited contemporaneously with the garnet and pyroxene (usually hedenbergite) of the tactite; the molybdenite in part appears to present the same relationships to the tactite minerals that the scheelite does, but some of the molybdenite is in quartz veins and veinlets that are younger than the bulk of the tactite minerals. Of the two main primary copper minerals (chalcopyrite and bornite), the chalcopyrite apparently is closely associated in time with the molybdenite, in large part at least; some chalcopyrite may be later than the molybdenite and all of the bornite probably is, but not enough work has been done on the paragenesis of the Pine Creek minerals for positive statements about age relations of the copper ore minerals to be made. The primary tungsten and molybdenum minerals probably were formed under high-temperature conditions, as was much of the chalcopyrite; the bornite, however, probably indicates that the solutions were in the mesothermal range when this last ore mineral was formed. Thus, it seems most reasonable to classify the deposits as hypothermal-2 to mesothermal, with the understanding that the mesothermal portion of the mineralization was minor in comparison with the hypothermal part.

DARWIN

Late Mesozoic (pre-Laramide)	*Lead, Zinc, Silver*	*Hypothermal-2 to Mesothermal*

Davis, D. L., and Peterson, E. C., 1949, Anaconda operation at Darwin mines, Inyo County, California: A.I.M.E. Tr., v. 181, p. 137-147

Hall, W. E., 1959, Geochemical study of Pb-Ag-Zn ore from the Darwin mine, Inyo County, California: A.I.M.E. Tr., v. 214, p. 940 (in Min. Eng., v. 11, no. 9)

Hall, W. E., and MacKevett, E. M., 1958, Economic geology of the Darwin quadrangle, Inyo County, California: Calif. Div. Mines Spec. Rept. 51, 73 p.

—— 1962, Geology and ore deposits of the Darwin quadrangle, Inyo County, California: U.S. Geol. Surv. Prof. Paper 368, 87 p.

Kelley, V. C., 1937, Origin of the Darwin silver-lead deposits: Econ. Geol. v. 32, p. 987-1008

—— 1938, Geology and ore deposits of the Darwin silver-lead mining district, Inyo County, California: Calif. Jour. Mines and Geol., v. 34, p. 503-562

Knopf, A., 1915, The Darwin silver-lead mining district, California: U.S. Geol. Surv. Bull. 580, p. 1-18

Norman, L. A., Jr., and Stewart, R. M., 1951, Mines and mineral resources of Inyo County: Calif. Jour. Mines and Geol., v. 47, no. 1, p. 17-223 (particularly p. 59-68)

Notes

The Darwin mines are located in east-central California about 35 miles southeast of Lone Pine.

Although the oldest rocks in the general area of the Darwin quadrangle are Ordovician in age, the oldest exposed in the Darwin Hills proper (as a small outcrop of marble) is probably the Devonian Lost Burro formation. Conformably over the Lost Burro is the Mississippian Tin Mountain limestone and conformably over that is the Mississippian Perdido formation (limestone and bedded chert). The upper part of the Perdido formation is equivalent to the lower part of the Mississippian-Pennsylvanian Lee Flat limestone and interfingers with it. The upper part of the Lee Flat probably interfingers, in turn, with the Rest Spring shale, which latter is entirely Pennsylvanian in age. Most of the sedimentary rocks outcropping in the Darwin Hills area are the lower and upper units of the Pennsylvanian-Permian Keeler Canyon formation that rests conformably on the Lee Flat. Conformably overlying the Keeler Canyon is the Permian Owens Valley formation (three units containing several rock types, in the main wholly or partly calcareous). No evidences of Mesozoic sedimentation have been found, but the Paleozoic rocks were strongly deformed before the intrusion of the Mesozoic igneous rocks further warped the area. These intrusives were mainly biotite-hornblende-quartz monzonite and leucocratic quartz monzonite, with leucogranite, aplite, and pegmatite being present as minor masses in the border regions of the quartz monzonite and as thin dikes cutting through it. The quartz monzonite forms the core of the Darwin Hills immediately adjacent to, and west of, the mines. In the Darwin Hills, the quartz monzonite intrudes rocks of Permian age (specifically the upper unit of the Owens Valley formation) and is overlain by late Cenozoic materials. In the Inyo Mountains to the northwest, similar rocks intrude Upper Triassic shales and volcanic rocks. The quartz monzonites probably were introduced at much the same time as similar rocks in the Yosemite area for which Bateman (1965) quotes Evernden and others as reporting ages ranging from 77 to 95 m.y., indicating their emplacement during the late Mesozoic but in pre Laramide time. A determination for a specimen from the Hunter Mountain quartz monzonite in the Ubehebe quadrangle, immediately north of the Darwin quadrangle, gives (by the suspect lead-alpha method) an age of 99 m.y., which, at least, is in the range given for the Yosemite rocks. The balance of probabilities, therefore, indicates that the Darwin deposits, which are probably genetically related to the quartz monzonite, are late Mesozoic but pre-Laramide.

As the most productive mines are in the thin-bedded Pennsylvanian and Permian rocks surrounding the quartz monzonite, the ore-forming fluids are presumed to have been channeled along the contacts between the igneous bodies and the sediments and to have entered the area within a short time after the intrusives themselves. This dates the ore formation, tentatively at least, as late Mesozoic (pre-Laramide).

The silty, sandy, or arenaceous limestones of the Keeler Canyon and Owens Valley formations, the main hosts of the ore bodies, were altered extensively during the intrusion of the quartz monzonites to calc-hornfels and tactite. The major minerals of the calc-hornfels are diopside, wollastonite, idocrase (vesuvianite), garnet, plagioclase, orthoclase, quartz, tremolite, and epidote, with relics of unreplaced calcite, while the tactite contains mainly wollastonite, idocrase, and garnet; the two rocks grade into each other. The ore bodies are concentrated in these altered sediments near the Darwin Hills stock. The ores occur as (1) bedded deposits, (2) irregular replacement bodies close to major faults, (3) fissure or vein deposits, and (4) small ore bodies in flat-lying fractures. Type (1) is the economically most important.

The ore deposits favor altered limestone as opposed to the altered facies of the other sediments. The first of the hypogene, as opposed to the contact-metasomatic minerals, were scheelite, fluorite, pyrite, arsenopyrite, and pyrrhotite--a typical hypothermal suite, but one of no economic value in this instance. The next stage in the sulfide mineralization was the formation of bornite (as thin selvages on blebs of pyrrhotite oriented in sphalerite), sphalerite in important amounts, and chalcopyrite (mainly as oriented inclusions in sphalerite); this mineral group also appears to have formed under hypothermal conditions, judging by the oriented blebs of pyrrhotite and chalcopyrite in sphalerite. Both sphalerite and chalcopyrite were replaced by silver-rich galena, the galena being the most abundant sulfide in the Darwin mines. Tetrahedrite-tennantite, also silver-rich, is present in small amounts and appears to be of the same age as the galena. Small quantities of clausthalite and matildite are present in such relations to galena as to suggest that they exsolved from it. The amounts of the sulfosalts associated with the galena are so small as to suggest that the lead sulfide formed in the mesothermal range, an intensity level compatible with the gangue minerals--calcite, fluorite, and sericite--associated with the sulfide mineralization. The only real suggestion of the existence of leptothermal conditions in the formation of the Darwin ores is the andorite-rich stope on the 400-level of the Darwin mine; one development of a sulfosalt in some abundance is not considered sufficient justification for including leptothermal in the classification assigned to the deposit. The bleb-filled sphalerite indicates hypothermal conditions, while the silver-rich galena, with minor exsolved sulfosalts, suggests mesothermal deposition; the deposits, therefore, are here categorized as hypothermal-2 to mesothermal.

EAGLE MOUNTAIN

Late Mesozoic (pre-Laramide)	*Iron as Primary Magnetite, Pyrite; Secondary Hematite, Limonite, Goethite*	*Hypothermal-2*

Davis, C. E., 1967, Electric computation of the Eagle Mountain ore reserves: Univ. Ariz. Coll. Mines, Computer short course and symposium on mathematical techniques and computer applications in mining and exploration, p. E2-1-E2-13

DuBois, R. L., and Brummett, R. W., 1968, Geology of the Eagle Mountain mine area, in Ridge, J. D., Editor, *Ore deposits of the United States, 1933-1967* (Graton-Sales Volumes): Chap. 76, v. 2

Hadley, J. B., 1945, Iron-ore deposits in the eastern part of the Eagle Mountains, Riverside County, California: Calif. Div. Mines Bull. 129, pt. A, p. 1-24

Harder, E. C., 1912, Iron-ore deposits of the Eagle Mountains, California: U.S. Geol. Surv. Bull. 503, 81 p.

Notes

The Eagle Mountain iron deposits are located in Riverside County about 180 miles slightly south of east from Los Angeles.

The oldest rocks in the area consist of Precambrian gneisses, schists, and quartzites that lie south of another group of metasediments of probably Paleozoic age (the Eagle Mountain group). These two groups are nowhere in contact, being separated by various thicknesses of a Mesozoic porphyritic quartz monzonite. These Paleozoic metasediments in which the ores are developed are divided into six units. Three of these are quartzites, the first and second quartzites being separated by a schistose meta-arkose zone and an ore zone, and the second and third by an ore zone. All units are conformable with those in which they are in contact. The rocks of the ore zones are limestones and dolomites. The thicker (50 to 400 feet) of these two zones, the north ore zone, lies between the second and third (middle and upper) quartzite units; the thinner zone (20 to 120 feet), the south ore zone, lies between the meta-arkose and the second quartzite unit. Some ore occurs in the middle quartzite zone and some in intruded masses of quartz monzonite in the ore zones. The main outcrops of the quartz monzonite are in two areas, respectively north and south of the Eagle Mountain metasediments; they probably are sills because they tend to parallel the sedimentary structures. Numerous minor masses of the monzonite have been intruded into the section, mainly along sedimentary boundaries. The close spatial relationships of ore and quartz monzonite strongly suggest close time and genetic relationships as well; therefore, the age of the intrusive probably is the age of the ore. The quartz monzonite is appreciably younger than even the Paleozoic metasediments and probably was introduced at much the same time as the intrusive rocks of the Sierra Nevada batholith. If this is the case, the monzonite is late Mesozoic but pre-Laramide in age, and the same age almost certainly should be assigned to the ores.

The principal iron minerals in the deposits (magnetite and pyrite) were introduced into the highly altered carbonate rocks of the Paleozoic metasedimentary series. The principal alteration minerals are actinolite and tremolite with local epidote and serpentine and some phlogopite and orthoclase and minor tourmaline. Where the texture of the original rock has been obliterated, Dubois and Brummett refer to the product as a granofels, but appreciable volumes of the altered carbonate rocks retain their primary textures; the one type grades into the other. Some actinolite and tremolite also were developed in the quartzite adjacent to the middle unit and in the quartz monzonite. Whether this alteration took place at the same time as the alteration of the carbonates or was produced from the debris picked up by the ore fluids as they were depositing the iron minerals is uncertain; Dubois and Brummett (1968) favor the latter explanation largely because of the finer grain of the alteration minerals and the lack of magnetite in these noncarbonate rocks. Magnetite appears to have replaced the silicate minerals as well as residual carbonate. The close time relationship of the silicate alteration minerals and the abundance of magnetite as a primary mineral seem definitely to force the assignment of the deposits to the hypothermal range, and since the ores were emplaced almost entirely in altered carbonate rocks, they are here classed as hypothermal-2.

Thus far, the mining at Eagle Mountain has been confined to the secondary ore of the oxidized zone where the magnetite has been changed to martitic hematite and goethite and the pyrite to a limonite-hematite-goethite mixture. Some magnetite has survived the oxidation process. Eventually, the primary ore will be mined; present-day technology guarantees that it will be convertible to a highly usable product.

ENGELS AND SUPERIOR MINES

Late Mesozoic (pre-Laramide) — *Copper* — *Hypothermal-1* (*Walker*), *Hypothermal-1 to Mesothermal* (*Engels*), *Mesothermal* (*Superior*)

Anderson, C. A., 1931, The geology of the Engels and Superior mines, Plumas County, California, with a note on the ore deposits of the Superior mine: Univ. Calif. Pubs., Bull. Dept. Geol. Sci., v. 20, no. 8, p. 293-330

Donnay, J. D. H., 1930, Genesis of the Engels copper deposit; a field study and microscopic investigation of a late magmatic deposit: Cong. Int. Mines, Mét. et Géol. Appl., 6th Sess., Sec. Géol., Liége, p. 99-111

Graton, L. C., 1918, Further remarks on the ores of Engels, California: Econ. Geol., v. 13, p. 81-99

Graton, L. C., and McLaughlin, D. H., 1917, Ore deposition and enrichment at Engels, California: Econ. Geol., v. 12, p. 1-38

Greig, J. W., 1932, Temperature of formation of the ilmenite of the Engels copper deposits--a discussion: Econ. Geol., v. 27, p. 25-38

Knopf, A., 1935, The Plumas County copper belt, California, in *Copper resources of the world*: 16th Int. Geol. Cong., v. 1, p. 241-245

Knopf, A., and Anderson, C. A., 1930, The Engels copper deposits, California: Econ. Geol., v. 25, p. 14-35

Tolman, C. F., Jr., 1917, Ore deposition and enrichment at Engels, California: Econ. Geol., v. 12, p. 379-386

Tolman, C. F., Jr., and Rogers, A. F., 1916, A study of the magmatic sulfide ores: Leland Stanford, Jr., Univ. Pubs., Univ. Ser., no. 26, 76 p. (particularly p. 61-64)

Turner, H. W., and Rogers, A. F., 1914, A geologic and microscopic study of a magmatic copper sulphide deposit in Plumas County, California, and its modification by ascending secondary enrichment: Econ. Geol., v. 9, p. 359-391

Notes

The deposits of the Engels, Superior, and Walker mines are located some 120 miles northeast of Sacramento the north-central part of Plumas County.

The oldest (pre-Mesozoic ?) rocks exposed in the area of the Engels and Superior mines are keratophyre flows and tuffs with minor admixtures of sediments; bedding and flow structure are not normally observed. Two meta-andesites constitute the next and predominant pre-Mesozoic(?) rocks in the geologic sequence. There is no direct evidence in the Engels area as to the age of either the keratophyre or the andesites; the andesites are quite similar to formations in the Taylorsville region to the south that were designated by Diller as Carboniferous. After appreciable metamorphism of the pre-Mesozoic (Paleozoic ?) rocks, there followed a series of intrusions that apparently range from gabbro (oldest) through quartz diorite to quartz monzonite to granite (youngest). Dikes of all these rocks cut the Paleozoic(?) volcanic rocks. To the northwest of the area of the mines, several dikes of quartz monzonite have been found that cut the quartz diorite, while dikes of the granite intrude both the quartz diorite and the quartz monzonite. Several varieties of younger dikes--aplite, pegmatite, soda-granite porphyry, and granite porphyry--are known; these are related to the same general period of igneous activity as the somewhat older quartz-bearing intrusives.

The most recent age determinations in the Sierra Nevadas suggest that the batholith is the result of several successive intrusions that occurred over much of pre-Laramide upper Cretaceous time, and this age is here assigned to the gabbro and the quartz-bearing intrusives of the Engels and Superior area.

In the Engels mine, the ore mineralization is found in gabbro, quartz diorite, and the altered Paleozoic(?) volcanic rocks; in the Superior mine, the ore bodies were formed in the quartz monzonite; and in the Walker mine (at the south end of the Plumas County copper belt), the ore was developed in highly altered, fossiliferous Carboniferous(?) sediments near their contact with quartz diorite. These relationships strongly indicate that the age of ore formation was at least post-quartz monzonite but do not guarantee that it also was later than the granite. It appears probable, moreover, that the ore-forming fluids were derived from one of the magmatic hearths from which the igneous materials came. This would strongly suggest that the ore mineralization was, equally with the igneous activity, of late Mesozoic (pre-Laramide) age.

The ore deposits at the three principal Plumas County mines were formed in quite different environments. The mineralizations at the three mines show considerable similarities and appreciable differences. In all three the principal copper mineral is chalcopyrite; the Engels mine contains considerable bornite, the Superior mine some, and the Walker mine essentially none. Magnetite (irregularly distributed at Walker) is abundant in all three mines; ilmenite (not normally a hydrothermal mineral) is common in the Engels mine but does not occur in the other two. Pyrite is extremely rare at Engels, rare at Walker, and minor at Superior. A little late primary chalcocite is found in the Engels mine, minor amounts of secondary chalcocite are present at Superior, while none has been reported from Walker. Tourmaline is a prominent gangue mineral at the Superior and Walker mines but is lacking at Engels.

In the Engels mine the ore occurs not only in gabbro but also in quartz diorite and in the metamorphosed volcanic rocks; the early wall-rock alteration differs from one rock type to the next, but that in the gabbro consists of a conversion of the hornblende of that rock to actinolite and biotite and a recrystallization of the coarse-grained calcic feldspars to fine-grained andesine. This alteration was followed by the introduction of the metallic minerals, magnetite, ilmenite, chalcopyrite, and bornite, mainly as replacements of biotite and of feldspar; the oxide minerals locally may be somewhat older than the sulfides (locally sulfides fill fractures in the oxides) but generally the sulfide minerals appear to be only slightly younger than the oxides. After the main ore deposition, chlorite, stilbite, analcite, and siderite were formed, locally accompanied by minor amounts of primary chalcocite. The chalcopyrite shows no exsolved lamellae of bornite, so the two minerals probably were formed between somewhat above and slightly below the border between the hypothermal and mesothermal intensity ranges. The ores probably should be classified as hypothermal-1 to mesothermal with the chalcocite being too minor in amount to enter the classification.

In the Superior mine the gangue minerals of the veins are mainly green mica and tourmaline, with tourmaline extensively developed in the wall rock; also present in the gangue are actinolite, titanite, apatite, and quartz. Magnetite also appears to have been emplaced concomitantly with the gangue. Following the emplacement of the early gangue minerals and magnetite, chalcopyrite and subordinate bornite and pyrite were introduced in fractures in magnetite and as replacements of it along with siderite, chlorite, sericite, and epidote gangue; locally brecciated fragments of the earlier gangue and of the quartz monzonite are crusted over with ore minerals and siderite. A few small fractures in the ore sulfides are filled with hematite; the minor amounts of chalcocite present almost certainly are secondary. In the vicinity of one of the main veins, veins of quartz and dolomite, with minor amounts of pyrite, chalcopyrite, and tetrahedrite, cut the earliest minerals; small amounts of galena and sphalerite are locally found in these veins. Under no circumstances is this late mineralization of economic importance. From the time-separation between magnetite and copper sulfide deposition and from the gangue minerals associated with those of copper, it would appear that the Superior ores were formed under mesothermal conditions.

In the Walker mine the main gangue mineral is quartz, accompanied by erratically distributed barite; this quartz was followed by the deposition of tourmaline, actinolite, and magnetite in fractures in the quartz. On further fracturing, the sulfides were emplaced, mainly chalcopyrite with minor pyrrhotite and cubanite. Here, although the sulfides and the magnetite are of different ages, the presence of cubanite lamellae in the chalcopyrite and of pyrrhotite strongly suggests that the Walker ore minerals were formed under hypothermal conditions.

Greig (1932) has conclusively shown that the ilmenite-hematite relationships observed by the early workers do not require the Engels ores to have been formed at temperatures higher than those normally assigned to the hypothermal range.

FOOTHILL COPPER BELT

Late Mesozoic (pre-Laramide)	*Copper, Zinc, Gold, Silver*	*Hypothermal-1 to Mesothermal*

Aubury, L. E., 1908, The copper resources of California: Calif. Min. Bur. Bull. 50, 366 p. (particularly p. 189-289)

Clark, W. B., and Lydon, P. A., 1962, Mines and mineral resources of Calaveras County, California: Calif. Div. Mines and Geol. County Rept. no. 2, 217 p. (particularly p. 11-17, 23-32)

Forstner, W., 1908, Copper deposits in the western foothills of the Sierra Nevada: Min. and Sci. Press, v. 96, no. 22, p. 743-748

Hershey, O. H., 1908, Foothill copper belt of the Sierra Nevada: Min. and Sci. Press, v. 96, no. 12, p. 388-393; v. 97, no. 2, p. 48-49

Heyl, G. R., and others, 1948, Copper in California: Calif. Div. Mines Bull. 144, 429 p. (particularly pt. 1, p. 11-157)

Knopf, A., 1906, Notes on the Foothill copper belt of the Sierra Nevada: Univ. Calif. Pubs., Bull. Dept. Geol., v. 4, no. 17, p. 411-423

Lang, H., 1907, The copper belt of California: Eng. and Min. Jour., v. 84, nos. 21, 22, 23, p. 909-913, 963-966, 1006-1010 (disc. 1908, v. 85, no. 8, p. 420-421)

Lindgren, W., and Turner, H. W., 1895, Smartsville folio, California: U.S. Geol. Surv. Geol. Atlas, Folio 18, 6 p.

McDonald, G. A., 1941, Geology of the western Sierra Nevada between Kings and San Joaquin Rivers, California: Univ. Calif. Pubs., Bull. Dept. Geol. Sci., v. 26, no. 2, p. 215-273 (particularly p. 267-269)

Reid, J. A., 1907, The ore-deposits of Copperopolis, Calaveras Co., California: Econ. Geol., v. 2, p. 380-417 (disc. 1908, v. 3, p. 340-342)

—— 1908, Foothill copper belt of the Sierra Nevada: Min. and Sci. Press, v. 96, no. 12, p. 388-393; v. 97, no. 2, p. 48-49

Taliaferro, N. L., 1942, Geologic history and correlation of the Jurassic of southwestern Oregon and California: Geol. Soc. Amer. Bull., v. 53, p. 71-112

Tolman, C. F., Jr., 1935, The Foothill copper belt of California, in *Copper resources of the world*: 16th Int. Geol. Cong., v. 1, p. 247-250

Turner, H. W., 1907, The ore deposits of Copperopolis, California: Econ. Geol., v. 2, p. 797-799

Notes

The Foothill copper belt forms a 250-mile-long zone of sulfide mineralization in the western foothills of the Sierra Nevadas that runs southeastwardly from the Big Bend mine north of the Feather River on the north to the Copper King mine in Fresno County on the south. The major production has come from that portion of the belt between the Spencerville mine, east of Marysville, and the Quail Hill mine in Tuolumne County.

The rocks of the Foothill copper belt are divided into two major groups: the complexly folded and faulted Paleozoic and Jurassic metasediments and metavolcanic rocks and a series of flat-lying (or nearly so) Tertiary and Quaternary volcanic rocks and sediments. The oldest of the units of the older group is the Calaveras formation, which has usually been designated as Carboniferous but which may contain rocks appreciably older and younger than that period; a large number of metamorphosed rock types are included in the Calaveras in the Foothill belt. The Jurassic rocks have been divided into two formations, the older of which is the Monte de Oro formation consisting of clay slate, with some conglomerate and sandstone; it is probably Middle Jurassic in age, and most of it has been removed by erosion. The younger is the Mariposa, predominantly a slate, with some siltstone, graywacke, sandstone, conglomerate, and locally volcanic rocks; it is probably Upper Jurassic. Taliaferro has separated a section of sediments and volcanic rocks, ranging from less than 5000 to more than 15,000 feet in thickness, from the basal portion of the Mariposa; to these rocks he has assigned the name Amador, leaving the term Mariposa to apply to the upper part of that group as originally defined. The Amador generally grades upward into the Mariposa, but occasionally the basal Mariposa contains pebbles and cobbles of the Amador. The relationship of the Amador to the Monte de Oro is uncertain, but Taliaferro considers the Amador to be Upper Jurassic and the younger of the two.

After the formation of the Mariposa sediments, the area was strongly folded, faulted, and dynamo-metamorphosed in the building of the ancestral Sierra Nevadas. Some of the igneous rocks in the Foothill belt were intruded prior to the Nevadan revolution and were deformed and metamorphosed during it; these included rhyolite and andesite porphyries, peridotites, and dunites (now almost entirely serpentine), all of which are pre-ore. After the Nevadan deformation, the granodioritic core of the Sierra Nevadas was intruded, accompanied by smaller masses of genetically related rocks. These intrusive rocks interrupt and truncate folds in the metamorphic rocks and occur along zones of structural weakness parallel to regional trends of folding and cleavage and, therefore, definitely are later than the deformation.

The ores certainly are later than the Jurassic rocks in which they are found and also were introduced after the granitic-type intrusions, since normal copper veins occur in granodiorite (Copperopolis district) and since certain atypical ore bodies are found at contacts with granodiorite masses. Moreover, no ore is present in the Tertiary rocks, so the mineralization probably lies in the same general time span as that during which the post-Nevadan revolution intrusives of Yosemite were emplaced, that is, between 95 and 75 m.y. ago. This would date the ore deposits as late Mesozoic (pre-Laramide).

Most of the deposits of the Foothill belt are massive to disseminated replacement bodies developed in zones of shearing, faulting, and crushing where crushing and brecciation were intense. Some irregular lenticular quartz veins, however, do occur. In both types of deposits, replacement was the dominant process but some open-space filling also occurred. On the basis of the primary minerals present, the ore deposits have been divided into four types: (1) sphalerite, chalcopyrite, and pyrite, with some galena and considerable gold and silver; (2) pyrrhotite and chalcopyrite, with or without pyrite and sphalerite and generally with some gold and silver; (3) pyrite and chalcopyrite with practically no gold, silver, or zinc; and (4) pyrite, chalcopyrite, and quartz, with or without pyrrhotite and sphalerite and generally with some gold and silver. Bornite and tetrahedrite have been seen in minor amounts in ores of types (1) and (3); occasionally other sulfides have been reported, but none is of economic importance.

The wall-rock alteration associated with most of the ore bodies is of sericite-silica-pyrite type, although at one of the main type (3) ore bodies (Copperopolis), chlorite is the principal alteration mineral, sericite being present only locally. The character of the alteration has remained constant to the greatest depths reached in mining. Sericite appears to have begun to form before silica and both before pyrite that continued on into the sulfide stage of deposition; clay minerals are uncommon except at the Quail Hill mine. Gangue minerals are not abundant, quartz and barite being the most abundant, calcite being present, and chlorite, epidote, and a few other minerals being minor in amount. Some chalcocite and a little covellite were developed by secondary enrichment, but they were not usually, at least, of economic importance.

It is probable that the type (2) deposits and perhaps the type (4) deposits, in part at least, were formed under hypothermal conditions because of the abundance of pyrrhotite associated with all of the type (2) and some of the type (4) ores. So far as the type (1) and type (3) deposits are concerned, however, the absence of microscopic data prevents certain classification of their conditions of deposition, but the wall-rock alteration and gangue minerals associated with them strongly suggest that they were formed in the mesothermal range. The deposits of the belt are here classified as hypothermal-1 to mesothermal, with the probability that the greater amounts of ore sulfides were developed in the hypothermal range.

There are at least two small deposits in the belt that were emplaced in tactite. The tactite was formed by contact metamorphic action of the granodiorite on limestone interbedded with the predominant hornfels. These deposits are of such slight economic importance that their designation is not included in the classification of the district as a whole.

MOTHER LODE

Late Mesozoic (pre-Laramide)	*Gold*	*Hypothermal-1 to Mesothermal*

Bowen, O. E., Jr., and Gray, C. H., Jr., 1957, Mines and mineral deposits of Mariposa County, California: Calif. Jour. Mines and Geol., v. 53, p. 35-343 (particularly p. 69-189)

Carlson, D. W., and Clark, W. B., 1954, Mines and mineral resources of Amador County, California: Calif. Jour. Mines and Geol., v. 50, p. 149-285 (particularly p. 164-200)

Clark, L. D., 1960, Foothills fault system, western Sierra Nevada, California: Geol. Soc. Amer. Bull., v. 71, p. 483-496

Clark, W. B., and Lydon, P. A., 1962, Mines and mineral resources of Calaveras County, California: Calif. Div. Mines and Geol. County Rept. no. 2, 217 p. (particularly p. 11-17, 32-76)

Cloos, E., 1935, Mother Lode and Sierra Nevada batholith: Jour. Geol., v. 43, p. 225-249

Eric, J. H., and others, 1955, Geology and mineral deposits of the Angels Camp and Sonora quadrangles: Calif. Div. Mines Spec. Rept. 41, 55 p.

Hulin, C. D., 1930, A Mother Lode gold ore: Econ. Geol., v. 25, p. 348-355

Jenkins, O. P., 1948, Sierra Nevada province and geologic history of the Sierran gold belt, in *The Mother Lode country*: Calif. Div. Mines Bull. 141, p. 20-30

Knopf, A., 1929, The Mother Lode system of California: U.S. Geol. Surv. Prof. Paper 157, 88 p.

—— 1933, The Mother Lode system, in *Middle California and western Nevada*: 16th Int. Geol. Cong., Guidebook 16, p. 45-60

Logan, C. A., 1934, Mother Lode gold belt of California: Calif. Div. Mines Bull. 108, 240 p.

—— 1938, Mineral resources of El Dorado County, Calif.: Calif. Jour. Mines and Geol., v. 24, no. 3, p. 206-280

McKinstry, H. E., and Ohle, E. L., Jr., 1949, Ribbon structure in gold-quartz veins: Econ. Geol., v. 44, p. 87-109

McLaughlin, D. H., 1930, The Mother Lode system of California (rev.): Econ. Geol., v. 25, p. 225-227

O'Brien, J. C., 1949, Geology and mineral resources of Butte County, California: Calif. Jour. Mines and Geol., v. 45, p. 417-454 (particularly p. 426-433)

Ransome, F. L., 1900, Mother Lode district folio, California: U.S. Geol. Surv. Geol. Atlas, Folio 63, 11 p.

Whitehead, W. L., 1942, The Mother Lode system in southern El Dorado and Amador Counties, California, in Newhouse, W. H., Editor, *Ore deposits as related to structural features*: Princeton Univ. Press, p. 178-182

Notes

The Mother Lode comprises a belt 120 miles long and from 1000 yards to 2 to 3 miles wide that extends from north of Georgetown in Eldorado County south to Mormon Bar in Mariposa County. More than half of the gold produced from the Mother Lode has come from the 10-mile portion of the belt between Plymouth (north) and Jackson (south) in Amador County.

The oldest rocks in the area are considered to be Paleozoic in age and have been grouped under the designation of Calaveras formation; these include black phyllite with fine-grained quartzite or quartzitic schist, various types of quartz mica schist, quartz-amphibole-albite schist, chert, and marble; much of the Calaveras in the Mother Lode belt has been converted to tectonic breccia or crush conglomerate. The exact age of these beds is uncertain; they are probably upper Paleozoic, either Mississippian or Pennsylvanian. The Calaveras formation in the gold belt is overlain unconformably by Mesozoic sediments that probably are all of Middle and/or Upper Jurassic age. The earlier Jurassic beds (based on work done in Calaveras and Tuolumne Counties) are placed in the Amador group that is subdivided into the older Cosumnes formation and the younger Logtown Ridge formation. The later Jurassic beds are assigned to the Mariposa formation; its age is certainly Late Jurassic, although there is no firm agreement as to whether it is Oxfordian or Kimmeridgian. In addition to the formations to which names are given, there are two groups of rocks: (1) the greenschists and related rocks and (2) the phyllites, stretched conglomerate, and related rocks; the age and stratigraphic relationships of the two groups are uncertain.

Between the end of Upper Jurassic sedimentation (not necessarily the end of Jurassic time) and the beginning of Upper Cretaceous sedimentation on eroded Jurassic rocks to the west, a wide variety of igneous rocks was introduced into the Mother Lode country. The earliest of these igneous materials were stocks, long lenses, and belts of peridotite and related rocks, now converted to chlorite-rich serpentinites. Small masses and dikes of gabbro and hornblendite occur throughout the gold belt, with far the greater number of them being in, or closely associated with, serpentine. Associated in time with these mafic intrusives are dikes and sills of porphyritic diabase and dikes of diabase and hornblende lamprophyre. The Mother Lode ends, on the south, against a mass of granodiorite that protrudes to the west from the main mass of the Sierra Nevada batholith. Bodies of similar rock (mapped as metadiorite on Ransome's Mother Lode map) are scattered through the length of the gold belt from the mass southeast of Coulterville (south) to those east and southwest of Georgetown (north). These granodiorites actually range in composition between granite and diorite and are now all thought to be post-Mariposa and later than the serpentinites and the gabbros.

Although all of the intrusives are thought to be post-Mariposa, only some of them are known actually to cut rocks of that age; however, the intrusives appear to have been intruded over a geologically short period of time, not more than 10 to 20 m.y., between the end of Mariposa sedimentation and the beginning of that in the Upper Cretaceous. It would follow from this reasoning that the ages of the granodiorites and related rocks may well lie within the same limits as those of the similar rocks in the Yosemite area (east of the southern part of the gold belt). This would make the silicic Mother Lode intrusives between 95 and 75 m.y. old or late Mesozoic (pre-Laramide) in age. The ore, though slightly younger than the intrusives, probably should be given the same age designation.

The Mother Lode fault system, rather than the host rocks, probably exerted the greater control on the deposition of the ore bodies. The large majority of the ore bodies are found in, or adjacent to, these faults; a few, generally of short to moderate length, are associated with branching or discontinuous faults. The gold deposits of the Mother Lode are found in both quartz veins and in zones of mineralized country rock; the mineralized zones may adjoin the quartz veins or may be centered around fissure zones containing little or no quartz. The quartz veins appear to have formed predominantly by cavity filling because the fragments caught up in the veins show no evidence of replacement by that mineral. As is indicated by the presence of some of the gold as fillings of crushed quartz, the cavity fillings probably took place in stages, new openings being developed by earth movements after those previously formed had been filled by vein material. On the other hand, the ore bodies formed by mineralization of the country rock, mineralized greenstone (gray ore) and mineralized schist, were emplaced by replacement of wall-rock minerals by the ore and gangue suite. The gray ore is composed largely of fine-grained ankerite, with more or less sericite, albite, and quartz, and 3 to 4 percent of pyrite and arsenopyrite, minor sulfides, and native gold; these ores are commonly cut by veinlets of quartz, ankerite, and albite. The gray ore bodies are found in shoots that normally lie adjacent to and in the foot or hanging wall of quartz veins, though they are not necessarily in contact with the veins; the boundaries between the outer margins of the gray ore bodies and the country rock are usually gradational. Ore in slaty varieties of the greenstone generally contains unreplaced black slaty material, but those ores that were developed in slate-free greenstone are often almost solid ankerite rock.

The ore bodies in mineralized chlorite and amphibolite schists generally are of appreciably lower grade than the quartz veins or the mineralized greenstones (gray ore) and contain various proportions of new minerals, the exact ratios depending on the original composition of the schist. These normally consist of dominant ankerite, with lesser pyrite, sericite, quartz, albite, and accessory rutile; apparently the more ankerite and the less sericite in the altered schist, the more likely it is that the schist will contain minable amounts of gold.

Metallic minerals ordinarily make up 2 to 3 percent of the ore; pyrite is universally present and arsenopyrite is common; the next most abundant metallics are chalcopyrite, sphalerite, and gold, and tetrahedrite frequently is found. Although present in minor amounts relative to native gold, several tellurides are known in the ore bodies. The quartz veins in certain areas, such as the Angels Camp and Sonora quadrangles, are quite poor in gold in comparison with the adjacent mineralized schist and greenstone.

In contrast to the Nevada City-Grass Valley and Alleghany mines, the gold in the Mother Lode appears to have been formed at the same time as the sulfides rather than only with the lower temperature and later sulfides and sulfosalts. This close time relationship of gold and sulfide deposition would suggest that these gold ore bodies were formed under mesothermal conditions. The suite of hydrothermal alteration minerals developed in connection with the formation of the ores--ankerite, sericite, quartz, chlorite, epidote, pyrite, and mariposite--also indicates that mesothermal conditions prevailed over at least a considerable portion of the time involved in the ore-forming process. The presence of arsenopyrite in gray ore and in quartz veins and of albite in the gray ore suggests some deposition under the low intensity hypothermal conditions. The deposits, therefore, are here classified as hypothermal-1 to mesothermal with most of the gold probably having been mesothermal. The minor amounts of tellurides probably do not require the addition of leptothermal to the classification.

MOUNTAIN PASS

Late Precambrian *Rare Earths, Barite* *Magmatic-3b*

Anon., 1952, Rare-earth bonanza: Eng. and Min. Jour., v. 152, no. 1, p. 100-102

Evans, J. R., 1966, California's Mountain Pass mine now producing europium oxide: Calif. Div. Mines and Geol., v. 18, no. 2, p. 23-32

Glass, J. J., and others, 1958, Cerite from Mountain Pass, San Bernardino County, California: Amer. Mineral., v. 43, p. 460-475

Hewett, D. F., 1956, Geology and mineral resources of the Ivanpah quadrangle, California and Nevada: U.S. Geol. Surv. Prof. Paper 275, 172 p. (This is the quadrangle in which Mountain Pass lies, and the report covers the general geology of the area although Mountain Pass is mentioned only on p. 164.)

Jaffe, H. W., 1955, Precambrian monazite and zircon from the Mountain Pass rare-earth district, San Bernardino County, California: Geol. Soc. Amer. Bull., v. 66, p. 1247-1256

Olson, J. C., and Pray, L. C., 1954, The Mountain Pass rare-earth deposits: Calif. Div. Mines Bull. 170, Chap. 8, p. 23-29

Olson, J. C., and others, 1954, Rare-earth mineral deposits of the Mountain Pass district, San Bernardino County, California: U.S. Geol. Surv. Prof. Paper 261, 75 p.

Pray, L. C., 1957, Rare-earth elements: Calif. Div. Mines Mineral Information Serv., v. 10, no. 6, p. 1-8

Sharp, W. N., and Pray, L. C., 1952, Geologic map of bastnaesite deposits of the Birthday claims, San Bernardino County, California: U.S. Geol. Surv. Mineral Investigations, Field Studies Map MF 4, 1:6000

Walker, G. W., and others, 1956, Radioactive deposits in California: Calif. Div. Mines Spec. Rept. 49, 38 p. (particularly p. 22-23)

Notes

The Mountain Pass district comprises an area of unusual mineralization some 6 miles long by 1.5 miles wide, bisected by U.S. Highway 91 and lying about 60 miles southwest of Las Vegas, Nevada.

The rare earth-barite deposits of Mountain Pass occur within a block of metamorphosed rocks of Precambrian age from 4 to 5 miles wide and 18 miles long. These metamorphic rocks include gneisses and schists, amphibolite, migmatites, granitic pegmatites, and minor bodies of foliated mafic rocks; of these a granitic augen gneiss is the most abundant. Their exact age is unknown, but it is certainly middle or early Precambrian. This mass of older rocks has been intruded by a group of unfoliated igneous rocks that can be divided into two general groups on the basis of relative age: (1) an older group of potash-rich rocks (stocks and dikes) among which are included biotite, shonkinite, hornblende and biotite syenites, and granite (and related dike rocks) and (2) a younger group of dike rocks that range from basalt to rhyolite, although most of them are andesitic, plus a group of carbonate dikes and masses. Through most of the area, the dikes of the older group trend from north to N30°W, while the younger dikes strike generally east-west, with the latter strike changing toward N60°E in the southern part of the district. Inasmuch the shonkinite dikes cut the major shonkinite body, they are appreciably younger than the large mass of that material. In at least one locality a shonkinite dike is cut by a granite dike; therefore, these latter bodies are younger than the shonkinite; in turn the carbonate dikes cut through the granite dikes, establishing the carbonate bodies as the youngest of the three.

In addition to the carbonate dikes or veins that cut both the unfoliated igneous rocks and the earlier foliated metamorphic rocks, a vaguely pear-shaped mass of carbonate rock (the Sulphide Queen), some 2400 feet long (north-south) by up to 700 feet wide (east-west), was intruded into the area just south of the main concentration of carbonate dikes. There are several small carbonate bodies satellitic to the main mass, and the massive body, its satellites, and the dikes contain essentially the same mineral suites in which carbonate, barite, quartz, and bastnaesite are the principal minerals. Although there are some 200 of the dikes or veins, their total area is less than one-tenth of that of the main carbonate mass.

Seven shonkinite-syenite-granite stocks are known in the area, and in the vicinity of these, some of the gneisses have been converted to quartz-poor syenites in which albite and quartz replace some of the microcline, and biotite, soda amphibole, and soda pyroxene have been produced by the metamorphic process. This soda enrichment suggests that the effect of the intrusions on the gneisses was, to a minor degree, similar to the fenitization of the rocks of Alnö Island and the Fen area. The presence of the carbonate mass increases this resemblance, but the complex assemblage of rock types known in the Scandinavian occurrences is lacking at Mountain Pass.

So far as the field relationships in the area go, the older potash-rich rocks and the carbonate rocks could be of any age from Precambrian to late Mesozoic (the assumed age of the younger dikes). Neither the carbonate rocks nor the older potash-rich rocks have been found in any of the Paleozoic or younger rocks of the general area, suggesting that the Paleozoic rocks were not present when the carbonatites were introduced. On the other hand, it is possible that the andesitic dikes served as feeders for probably Cretaceous dacites in the Mescal Range south of Mescal Spring, 1.5 miles south of the carbonate mass. This suggests, but does not prove, a late Mesozoic age for the andesite dikes. Radioactive age determinations on zircons from the shonkinite give an age for that rock of 800 to 900 m.y., while similar determinations on monazite from the carbonate mass give ages of from 900 to 1000 m.y. As it is certain that the carbonate mass is younger than the shonkinite, despite the indication to the contrary given by the age dating, these ages cannot be accepted without hesitation; however, the evidence strongly suggests that the potash-rich rocks, the carbonate mass, the carbonate veins, and the carbonate in shear zones are all of late Precambrian age, and the carbonate mass is so classified here.

Although the carbonate mass at Mountain Pass was first recognized as something unusual by the radioactivity of the thorium contained in its minor content of monazite, the value of the deposit lies in the barite and bastnaesite found in it. Carbonates, including rare-earth carbonates, are more abundant than any other mineral in the various carbonate bodies, making up about 60 percent of these masses; calcite is the most common of these but dolomite and ankerite also are found in considerable amounts; even a little lead carbonate (cerussite) is present at the north end of the main mass. Other carbonates present include strontianite, aragonite, siderite, and sahamalite (a cerium-rare earth, magnesium, iron carbonate); these are found only sparsely and locally. Barite is the most abundant sulfate and averages 20 to 25 percent of the Sulphide Queen mass; the strontium content of the barite differs widely, with some of the sulfate being nearer celestite than barite in composition. Dark-red barite is associated in general with crocidolite (fibrous, blue soda-amphibole) and chlorite, while that in silicified carbonate is both pink and white and that near silicified zones is white. The quartz content is irregular, generally being from 5 to 40 percent, while silicate minerals, though present in some variety, occur only in small quantities. The principal silicates are crocidolite, several micas, chlorite, aegirite, epidote, titanite, thorite, and rare-earth-bearing allanite and cerite. Monazite occurs mainly in the dolomitic parts of the main carbonate mass; apatite is found in the shonkinite but is essentially lacking in the carbonate bodies. The principal rare earth minerals, however, are bastnaesite ($RFCO_3$, where R is mainly cerium, lanthanum, neodymium, and praeseodymium) and much lesser amounts of parisite ($2RFCO_3 \cdot CaCO_3$); the other rare earth minerals have already been mentioned. Bastnaesite normally composes 5 to 15 percent of the Sulphide Queen body, but it is much more varied in the amount in the carbonate veins.

The Sulphide Queen carbonate body contains numerous inclusions, ranging from fragments less than an inch in length to massive bodies, mainly gneiss and syenite, up to 150 feet long. Olson and his colleagues (1954) list 21 facts for which any hypothesis to explain the origin of carbonate bodies must account. These range from the lack of Precambrian sedimentary carbonate in the area to the discordant relationship of the Sulphide Queen mass to the Precambrian foliation. They dismiss as possible explanations for the origin of the carbonate bodies both a Paleozoic sedimentary origin and a Precambrian sedimentary origin. This leaves them with no alternative but to suggest that the carbonate bodies are of magmatic origin, but they do not finally decide between deposition from hydrothermal solutions or from a rather viscous magmatic differentiate; they do favor most of the carbonate rock in the area being classed as carbonatite. The most important resemblances between the Mountain Pass deposits and other carbonatite areas of the world are (1) the presence of several varieties of carbonate rock, (2) the lack of bedding and the presence of steeply dipping foliation or banding that conforms to the enclosing walls, (3) inclusions of moderately altered wall rock that deflect flow structures, (4) the occurrence of veinlike bodies of carbonate material in addition to large masses of similar rock, (5) the markedly large concentrations of rare elements and of barium, (6) (some) fenitic alteration, including the widespread development of soda-amphibole both in the carbonate masses and the wall rocks, and (7) the lack of definite evidence that any of the veins were formed by hydrothermal solutions. There are, however, certain differences between the Mountain Pass deposit and certain other carbonatites; these include (1) essential lack of feldspathoids, (2) absence of ring dike-structure, (3) absence of lime-silicate minerals, (4) the concentration of the barite in the carbonate mass instead of in replacement masses or veins adjacent to it, (5) the scarcity of titanium- and columbium-bearing minerals, and (6) the lack of magnetite.

Weighing these items, it would appear that the evidence in favor of the Mountain Pass carbonate material being classified as a carbonatite outweighs the differences between Mountain Pass and other carbonatites, differences that can readily be explained by variations in the process of carbonatite development and by the amount of erosion the deposit has undergone. The Mountain Pass deposit, therefore, is here assumed to have developed by immiscible separation of a dense, water-poor carbonate-rich magma from a mafic magma at considerable depth beneath the present location of the carbonatite. How the magma acquired its carbonate content is unknown. The deposits, therefore, are here categorized as magmatic-3c.

NEVADA CITY-GRASS VALLEY

Late Mesozoic (pre-Laramide) — *Gold* — *Leptothermal (Nevada City), Mesothermal (Grass Valley)*

Farmin, R., 1938, Dislocated inclusions in gold-quartz veins at Grass Valley, California: Econ. Geol., v. 33, p. 579-599

—— 1941, Host rock inflation by veins and dikes at Grass Valley, California: Econ. Geol., v. 36, p. 143-174

Howe, E., 1924, The gold ores of Grass Valley, California: Econ. Geol., v. 19, p. 595-619; disc. (A. M. Bateman) p. 619-622

Johnston, W. D., Jr., 1938, Vein-filling at Nevada City, California: Geol. Soc. Amer. Bull., v. 49, p. 23-33

—— 1940, The gold-quartz veins of Grass Valley, California: U.S. Geol. Surv. Prof. Paper 194, 101 p. (particularly p. 5-62)

Johnston, W. D., Jr., and Cloos, E., 1934, Structural history of the fracture system at Grass Valley, California: Econ. Geol., v. 29, p. 39-54

Knaebel, J. B., 1931, The veins and crossings of the Grass Valley district, California: Econ. Geol., v. 26, p. 375-398

Lindgren, W., 1895, Characteristic features of California gold-quartz veins: Geol. Soc. Amer. Bull., v. 6, p. 221-240

—— 1895-1896, The gold-quartz veins of Nevada City and Grass Valley districts, California: U.S. Geol Surv. 17th Ann. Rept., pt. 2, p. 1-262

—— 1896, Nevada City special folio, California: U.S. Geol. Surv. Geol. Atlas, Folio 29, 7 p.

—— 1900, Colfax folio, California: U.S. Geol. Surv. Geol. Atlas, Folio 66, 10 p. (east edge of Nevada City area)

Lindgren, W., and Turner, H. W., 1895, Marysville folio, California: U.S. Geol. Surv. Geol. Atlas, Folio 17, 2 p.

Raucq, P., 1951-1952, Considérations sur la minéralisation aurifère de Grass Valley (Californie): Soc. Géol. Belgique, Ann. t. 75, Bull. no. 1-3, p. B35-B47

Notes

The Nevada City-Grass Valley deposits are located on the western slopes of the Sierra Nevadas about 50 miles northeast of Sacramento.

The oldest rocks in the area are those of the probably largely Carboniferous Calaveras formation that here consists mainly of schists, phyllites, slates, and quartzites in an eastern belt (striking generally northwest) and of argillites and cherts in a western belt (having the same strike). Some volcanic rocks appear to have been interbedded with the Calaveras sediments. Following the deposition of the Calaveras formation, flows and near-surface intrusions of andesitic composition (locally known as diabase and porphyrite and differing essentially only in texture) took place; they were probably of Late Triassic or Early Jurassic age. The volcanic rocks were followed by the deposition of the original shales and sandy shales of the Mariposa formation (now slates and clay slates with quartz). The amount of the probably Upper Jurassic Mariposa formation mapped in the area is small, and it may actually be part of the Calaveras formation. After the deposition of the Mariposa formation, igneous magmas in considerable variety were intruded into the older rocks of the district. The first intrusives were mafic materials--peridotites (now serpentinites), gabbros, and diorites. The relative times of intrusion are uncertain, but the rocks probably were introduced in the order named.

The ores at both Nevada City (northeast) and at Grass Valley (southwest) are closely associated in space and in time of origin with two bodies of probably early Upper Cretaceous granodiorite. The two granodiorite bodies lie about two and a half miles apart on the surface, but they well may be connected at depth. The granodiorite is cut in several places by narrow, intermediate to mafic dikes and by lamprophyre dikes of various types. These dikes are parallel to the northeast-trending "crossings" but were intruded before the veins were formed and filled.

The intrusion of the granodiorite probably should be dated as between 95 and 75 m.y. ago, as seems to be established for other portions of the complexly intruded Sierra Nevada batholith. The close time relationship of the mineralization of the veins to the introduction of the batholith is shown by the presence of rich fossil gold-bearing placers in early Tertiary river channels, reworked in Oligocene to Miocene time and then covered by Miocene andesite tuff and breccia. Much of this andesite was eroded in the Pliocene, and new placers were developed in Pleistocene and Recent time.

The contemporaneity of the granodiorite intrusion and the Late Cretaceous uplift of the Sierra Nevada Mountains, plus radioactive dating of the intrusions in other but similar areas strongly indicate that the ore-forming process should be dated as late Mesozoic (pre-Laramide).

Although the ore-forming process and the rocks in which it took place were much the same in the Nevada City and Grass Valley areas, enough difference exists between the structures into which the ore-forming fluids were introduced and the minerals they deposited to justify each being discussed separately.

In the Grass Valley area, veins are found in both the granodiorite and the rocks that surround it, principally in serpentinite and diabase and porphyrite. The veins of the North Star and New York Hill mines are in diabase and porphyrite; those of the Idaho-Maryland mine are in serpentinite. The veins of the Pennsylvania, Empire, and Osborne Hill mines are in granodiorite and diabase and porphyrite. The veins do not fill simple fractures but instead are found in fracture zones within which the veins pass from one fracture in the zone to another with numerous changes in dip and strike; thus the general geologic maps of the area show only the average trends of the veins.

In addition to the types of veins just mentioned, the area is cut by northeast-striking fractures, known as crossings; these may be simple fractures or sheeted zones, and they may be open or tightly closed and are usually very short. Few crossings contain quartz, and they are of economic importance only because changes in width, character, and gold content of the vein quartz take place where crossings intersect veins. The importance of these structures appears to lie in their breaking the veins into segments, each of which could be opened or closed essentially independently of those adjacent to it, as the earth stresses dictated.

Except for the occurrence, in a single vein, of magnetite and pyrrhotite with pyrite, pyrite is the oldest of the vein minerals; this is evidenced by broken crystals of pyrite that are cemented by early quartz. Pyrite is the most abundant of the metallic minerals and was deposited over a very considerable length of time, about that required for quartz deposition, but pyrite began to deposit first and stopped slightly before the quartz did. Chlorite, sericite, and epidote were formed during the period of the deposition of the early quartz and pyrite; all three minerals were strongly developed in the wall rock, and the first two were common vein minerals as well. Arsenopyrite was the first sulfide to have begun to precipitate after pyrite; its deposition ended before that of quartz as is witnessed by the presence of late quartz in fractures in arsenopyrite. The next sulfide was sphalerite which contains numerous blebs of chalcopyrite (almost the only occurrence of that copper sulfide). The precipitation of gold and galena began at some time after that of sphalerite; in fact there may have been no overlap at all as some gold replaces sphalerite (as well as pyrite and arsenopyrite) along narrow fractures.

Galena behaves in much the same way as gold and probably is of the same age. Before gold, and probably galena, had ceased to deposit, the precipitation of carbonates (ankerite, then calcite) began; ankerite is found in both the walls and the veins, but calcite is mainly a vein mineral that often replaces ankerite. The deposition of quartz did not end entirely when that of the carbonates began, for crystals of quartz surrounded by ankerite are further rimmed by another generation of quartz. Locally a little hematite was deposited with the calcite, and the final mineral to form was chalcedony (or occasionally opal). Tetrahedrite is rare at Grass Valley. From the blebs of chalcopyrite in the sphalerite and the presence of arsenopyrite, it is probable that the early sulfides (pyrite, arsenopyrite, and sphalerite) were deposited under hypothermal conditions. The gold, on the other hand, is not directly associated with these minerals, being contemporaneous with the galena. The galena is normally silver-poor, the bulk of the silver being alloyed with the gold. The essential lack of tetrahedrite removes the opportunity to use that mineral to judge the intensity range of gold deposition (as is possible at Nevada City). Because the sulfides that preceded gold probably were deposited under hypothermal conditions and because the time gap between them and gold and galena does not appear to have been long, the Grass Valley ores are here categorized as mesothermal.

As is true at Grass Valley, the Nevada City veins do not fill simple fractures but change from one fracture to another within fracture zones up to 40 feet wide. The Nevada City veins are mainly in granodiorite or the beds of the Calaveras formation, although there are a few in the porphyrite.

At Nevada City, as at Grass Valley, the mineralization can be divided into an early stage in which quartz was the dominant mineral and a late stage in which carbonates were the important vein material. Essentially all of the sulfides, which normally make up from 2 to 4 percent of the vein fillings, belong to the quartz stage. The early sulfides were pyrite, arsenopyrite, sphalerite (containing tiny blebs of chalcopyrite), and massive chalcopyrite (lacking at Grass Valley); chalcopyrite and sphalerite were later than the pyrite. Normally the early sulfides were cracked and broken, and the fractures were healed by the deposition of quartz, tetrahedrite (quite common), galena, and gold. Alternate concentric shells of quartz and carbonate indicate the same gradual transition from quartz to carbonate deposition that was the case at Grass Valley. As is true at Grass Valley, the last of the gold overlapped in time with ankerite, the earlier carbonate; calcite was the last mineral to form in any abundance. Some molybdenite is present in the ores, but its relationships to other sulfides are not known. The wall-rock alteration is similar to that at Grass Valley.

The main difference between the Grass Valley and Nevada City ores is the common occurrence of tetrahedrite in close association with gold and galena in the latter area; most of the silver not alloyed with the gold is found in the tetrahedrite, while the galena is usually, though not always, silver-poor. The early sulfides at Nevada City probably were deposited under hypothermal conditions as is evidenced by the presence of arsenopyrite and the blebs of chalcopyrite in the sphalerite. The break between the early sulfide mineralization at Nevada City seems to have been more definite and possibly of longer duration than at Grass Valley; this plus the close association of gold and tetrahedrite in time and space suggest strongly that the gold at Nevada City was deposited under leptothermal conditions, and the deposit is so categorized here.

NEW ALMADÉN

Late Tertiary *Mercury* *Epithermal*

Bailey, E. H., 1951, The New Almadén quicksilver mines: Calif. Div. Mines Bull. 164, p. 263-270

—— 1952, Suggestions for exploration at New Almadén quicksilver mine, California: Calif. Div. Mines Spec. Rept. 17, 2 p.

Bailey, E. H., and Everhart, D. L., 1964, Geology and quicksilver deposits of the New Almadén district, Santa Clara County, California: U.S. Geol. Surv. Prof. Paper 360, 206 p.

Becker, G. F., 1888, Geology of the quicksilver deposits of the Pacific slope: U.S. Geol. Surv. Mono. 13, 486 p. (particularly p. 310-330)

Bradley, W. W., 1918, Quicksilver resources of California: Calif. Min. Bur. Bull. 78, p. 160-164

Forstner, W., 1903, The quicksilver resources of California: Calif. Min. Bur. Bull. 27, p. 174-186

Ransome, A. L., and Kellogg, J. L., 1939, The quicksilver resources of California: Calif. Jour. Mines and Geol., v. 35, no. 4, p. 353-486 (particularly p. 452-455)

Schuette, C. N., 1931, Occurrence of quicksilver orebodies: A.I.M.E. Tr., v. 96 (1931 General Volume), p. 403-488 (particularly p. 411-417)

Notes

The New Almadén area is located a few miles south of San Jose with the majority of the mines being found in a northwest-southeast trending belt on the eastern slopes of Los Capitanillos Ridge; farther east there is a minor development of mercury deposits in the Santa Teresa Hills that form the west wall of the Santa Clara Valley.

The oldest rocks in the area are those of the Upper Jurassic to Upper Cretaceous Franciscan group that consists mainly of medium- to fine-grained graywacke and dark shale with a smaller quantity of altered mafic volcanic rocks normally referred to as greenstones; there also are small amounts of conglomerate, limestone, and chert.

Serpentinite is present in the Franciscan in considerable amounts and consists of masses that are only slightly younger than the rocks containing them. This is evidenced by the presence of abundant serpentine detritus in the Upper Jurassic Knoxville formation outside the New Almadén area. Some serpentine was emplaced later than the Knoxville as is shown by intrusions of serpentinized peridotite that cut the Knoxville. Still younger Coast Range rocks also contain intruded serpentine. The serpentine was developed by the alteration of ultramafic rock (in the New Almadén area harzburgite and, to a lesser extent, lherzolite) but how, where, and when the alteration was achieved is another matter. At New Almadén the development of the serpentinites now located there probably took place at depth, and the serpentinite bodies in the area appear to have been introduced as plastic masses that have all been more or less sheared, probably by earth movements taking place after their introduction into the rocks of the area.

Minor developments of Upper Cretaceous rocks younger than the Franciscan group are known in the area, but they are not mineralized and are of no economic importance. They consist, in the Santa Teresa Hills, of arkose and fissile shale and near Sierra Azul of interbedded conglomerate, graywacke, and shale.

The Cretaceous rocks are unconformably overlain by Tertiary (Eocene to Miocene) sediments that range from fissile shale, sandstone, and limestone, through conglomerate and sandstone and some silicic to intermediate volcanic rocks, to thin-bedded diatomaceous shale.

The formation of these Tertiary rocks was followed by the hydrothermal alteration of the folded sill-like serpentine to silica-carbonate rock (often the host to mercury deposits in the California Coast Ranges) that is composed mainly of quartz, chalcedony or opal, and carbonate (usually ferroan-magnesite). The silica-carbonate rock is far less widespread than the serpentinite, being largely confined to the highly sheared serpentinite. The portions of the Franciscan formation immediately adjacent to the silica-carbonate rock areas in the serpentinite are usually highly sheared, and the resulting material, originally siltstone or shale, is known as alta. More competent bands in the rock volumes converted to alta often are drawn out into lenticular pods to produce boudinage structures. The alta is found on the hanging wall of the silica-carbonate rock when that material lies along the upper margin of a serpentine body and in the footwall when it was developed on the lower margin of a serpentine sill.

The silica-carbonate rock is probably no older than late Miocene because (1) pebbles and boulders of serpentinite, without any silica-carbonate alteration, are found in middle Miocene conglomerate; (2) dolomite veins, similar to those probably formed in the late stages of the mineralization in the Guadalupe mine area, cut middle to late Miocene rocks; and (3) quartz veins that contain cinnabar are known to cut middle to upper Miocene rocks.

On the other hand, detrital cinnabar is found in perched gravel in the Blossom Hill area that has been tentatively correlated with the Pliocene to early Pleistocene Santa Clara formation; this age is confirmed by the amount of erosion and attendant chances in physiography that have affected the gravel since it was formed. Thus, the cinnabar mineralization cannot be older than middle to late Miocene and must be younger than the earliest Pleistocene. The most probable age for the mineralization, therefore, is Pliocene or late Tertiary.

The first stage in the formation of the ore bodies at New Almadén was the development, in late Miocene or Pliocene time, of the silica-carbonate rock in the margins of the serpentine masses. This hydrothermal activity was followed by the formation of steep, northeast-trending fractures in the silica carbonate shells; these fractures do not appear to have continued into the overlying or underlying alta. The next step was the Pliocene introduction of the cinnabar as a replacement of the silica-carbonate rock, the replacement being demonstrated by the nearly perfect preservation of the original rock textures by the fine-grained cinnabar and by the essential absence of distinct veins of cinnabar. The usual ore mass contains so much cinnabar that it seems impossible that it could have formed simply by filling fractures and small openings. Generally the sheared textures in the silica-carbonate rock indicate that it was formed by the replacement of sheared serpentine, and where small quantities of ore were formed in unsheared silica-carbonate rock, relict pseudomorphs of bastite can be recognized. In the silica-carbonate rock the magnesite was most easily replaced by cinnabar, although quartz also was considerably replaced; any remaining serpentine minerals appear to have been quite resistent to replacement. Still later in the Pliocene, the open fractures in the silica-carbonate rock (through which the solutions depositing the cinnabar moved) were filled by dolomite-quartz veins that locally contain a little cinnabar. During Pleistocene and Recent erosion, some of the ore bodies were exposed at the surface, and some placers containing considerable quantities of cinnabar nuggets were formed. Although 99 percent of the ore was developed in the silica-carbonate rock, a little has been found as replacements of Franciscan graywacke and shale. Although most of the ore bodies in silica-carbonate rock are near contacts with the alta, a few are found in the middle of altered serpentine sill material, unrelated directly to any contact.

The principal mineral of the deposits is cinnabar; metacinnabar has been reported from the area, but these occurrences cannot now be confirmed. Native mercury occurs only locally, but where it is found it is abundant. Pyrite is uncommon, stibnite is sparingly present, late sphalerite is found in a few places, and galena and chalcopyrite associated with the zinc sulfide are even more rare. Dolomite is the main gangue mineral, and much of it is appreciably younger than the mercury sulfide. Magnesite and quartz appear to be present only as original constituents of the silica-carbonate rock. Calcite, though present in some amount, is probably not connected in time with the cinnabar mineralization. Hydrocarbons are common in and near the ores in thin quartz-dolomite veins.

Although the New Almadén ores extend over the greatest vertical range known among cinnabar deposits (at least 2000 feet), their highest upward extension, based on the recent history of the San Francisco Bay area, is thought to have been within a few hundred feet of the surface. Despite the impermeable character of the overlying Franciscan rocks, it is probable that the New

Almadén ores were deposited under conditions of rapid loss of heat and pressure. As is true of all cinnabar deposits, the simple character of the mineralization, the dearth of other sulfides, and the essential lack of wall-rock alteration suggest strongly that deposition occurred below 200°C. It is, therefore, considered probable that the ores were deposited in the epithermal range.

NEW IDRIA

Late Tertiary *Mercury* *Epithermal*

Becker, G. F., 1888, Geology of the quicksilver deposits of the Pacific slope: U.S. Geol. Surv. Mono. 13, 486 p. (particularly p. 301-309, 379-381)

Bradley, W. W., 1918, Quicksilver resources of California: Calif. Min. Bur. Bull. 78, p. 93-95

Coleman, R. G., 1961, Jadeite deposits of the Clear Creek area, New Idria district, San Benito County, California: Jour. Petrol., v. 2, p. 209-246

Eckel, E. B., and Meyers, W. B., 1946, Quicksilver deposits of the New Idria district, San Benito and Fresno Counties, California: Calif. Jour. Mines and Geol., v. 42, no. 2, p. 81-124

Forstner, W., 1903, The quicksilver resources of California: Calif. Min. Bur. Bull. 27, p. 127-147

Linn, R. K., 1968, New Idria mining district, in Ridge, J. D., Editor, *Ore deposits of the United States, 1933-1967* (Graton-Sales Volumes): Chap. 78, v. 2

Linn, R. K., and Dietrick, W. F., 1961, Mining and furnacing mercury ore at the New Idria mine, San Benito County, Calif.: U.S. Bur. Mines I.C. 8033, 35 p. (particularly p. 1-14)

Ransome, A. L., and Kellogg, J. L., 1939, Quicksilver resources of California: Calif. Jour. Mines and Geol., v. 35, no. 4, p. 353-486 (particularly p. 421-428)

Schuette, C. N., 1931, Occurrence of quicksilver orebodies: A.I.M.E. Tr., v. 96 (1931 General Volume), p. 403-488 (particularly p. 417-420)

Notes

The New Idria mining district is located in the Diablo Range about 140 miles southeast of San Francisco near the southern end of San Benito County.

The oldest exposed rock in the area is the Jurassic Franciscan group, mainly massive, arkosic sandstone, with minor shale, chert, and greenstone. The Franciscan rocks largely surround an elliptical mass of serpentinite some 14 miles long by about 4 miles wide and probably not less than 2000 feet thick. The serpentine rock apparently was originally an ultramafic intrusive into the Franciscan group; the presence of detrital serpentine in the Panoche formation of lower Upper Cretaceous age shows that the initial intrusion occurred either late in the Jurassic or early in the Cretaceous. The Upper Cretaceous Panoche formation, shale and massive concretionary sandstone, entirely surrounds the serpentinite and its interrupted selvage of Franciscan sandstone. Except for surficial rocks, including landslide debris, the youngest rocks in the area are three small stocks of soda-syenite that intrude the serpentinite and the Panoche formation; the largest is under a half mile in length. No indication exists that they had any genetic connection with the ore, but they may, of course, be the surface expression of much larger bodies at depth from which ore fluids may have come.

The principal structure of the New Idria area is the elongate dome (striking generally northwest-southeast) of the intruded serpentinite mass with its selvage of Franciscan rocks perched on its outer margins and separated from it by fault contacts. The serpentinite certainly was intruded into the Franciscan before the Upper Cretaceous Panoche shale was laid down, but the original igneous contacts with the Franciscan appear to have been destroyed in the later movements of the two rocks relative to each other. At some time after the Upper Cretaceous had been laid down, the serpentinite was forced upward through the Cretaceous beds toward, or perhaps even to, the surface. The force that moved the serpentinite upwards appears to have been directed from the south at a steep angle as is suggested by the strongest fault development being in the northwest corner of the intrusion and the tear faulting occurring along the north margin of the dome. The movement also had a large horizontal component, the west side having moved relatively northward. The presence of serpentine debris in both late middle Miocene and Pliocene rocks indicates that the intrusion probably took place in three stages, the first being the pre-Upper Cretaceous initial introduction of the ultramafic material (at this time probably largely molten or at least well lubricated) that was altered, probably mainly automorphically, to the serpentinite. The second intrusion of the probably plastic serpentinite occurred before the late middle Miocene, and the third and last in the Pliocene.

The New Idria thrust fault brings the serpentine and its partial capping of Franciscan sandstone over the overturned or upturned and crumpled edges of the Panoche beds (shale and sandstone) that form the footwall of the fault. In places the fault is a complex zone, interrupted by a variety of cross faults, but in other areas it is a simple thrust fault.

Although some movement occurred along the faults during the period of ore deposition, as is evidenced by the brecciated and recemented vein fillings, this movement was minor in nature; certainly if the ore had been deposited after the late middle Miocene earth movements, renewed movement during the Pliocene would have had a far more drastic effect on the ores than the minor amount of brecciation and recementing of the vein fillings would suggest. It would seem quite certain, therefore, that the ores were formed after the Pliocene movement of the serpentine and Franciscan sandstone plug and must therefore be late Tertiary or perhaps even Quaternary. They are here classed as late Tertiary, but the subject is not finally settled.

The mercury deposits of the district are found mainly in the shaly parts of the lower Panoche formation, but some of the ore is found both in Panoche sandstone and in serpentinite. Only three deposits of any economic importance are known in Franciscan sandstone. In the Panoche shale the cinnabar coats rock fragments and cavity walls and fills fractures but makes almost no penetration of solid rock. In the Panoche sandstone the cinnabar enters solid rock along fractures and between grains more deeply than in shale. In the Franciscan sandstone the cinnabar of the West Idria ore body is found intergrown with the calcium and magnesium carbonates that fill the fissures.

In the sediments, ore deposition was preceded by wall-rock alteration which, in its weaker phases, produced an argillized and slightly indurated material. Where alteration was more strongly developed, an intense induration was accompanied by sparse to moderate amounts of pyrite in veins or veinlets. The deposits in the sediments occupy veins and stockworks in the altered rocks; these are filled mainly with cinnabar, although they also contain a little metacinnabar and native mercury. Much of the cinnabar is found in indurated and pyritized rock and is free of nonmetallic gangue minerals. Some of the deposits along the edge of the plug consist of veins in which cinnabar is associated with pyrite and marcasite and quartz or calcite. The ore shoots show considerable diversity in size.

Cinnabar was normally the last mineral to form, though in places, it was followed by the iron sulfides. If the early iron sulfides largely filled the fracture openings, cinnabar was weakly developed or may not have formed at all. A light, pulverulent cinnabar was later than the more massive variety; in a few places postmineral movement is evidenced by slickensided and polished cinnabar. In the uncommon occurrences of metacinnabar, it appears to have been deposited before the cinnabar.

The mineralization in the ore bodies both in sediments and in serpentine is typical of low-temperature mercury mineralizations throughout the world, and the geologic setting at New Idria definitely indicates deposition near the surface. The ores are, therefore, here classified as epithermal.

SHASTA COUNTY

Late Mesozoic (pre-Laramide)	*Copper, Zinc, Pyrites (Silver, Lead, Gold)*	*Mesothermal to Leptothermal (East Shasta), Mesothermal (West Shasta)*

Albers, J. P., 1953, Geology and ore deposits of the Afterthought mine, Shasta County, California: Calif. Div. Mines Spec. Rept. 29, 18 p.

—— 1959, Soda metasomatism in the East-Shasta copper-zinc district, northern California: Geol. Soc. India Jour., v. 1, p. 31-43

Albers, J. P., and Robertson, J. F., 1961, Geology and ore deposits of East Shasta copper-zinc district, Shasta County, California: U.S. Geol. Surv. Prof. Paper 338, 107 p.

Averill, C. V., 1935, The Shasta County copper belt, California, in *Copper resources of the world*: 16th Int. Geol. Cong., v. 1, p. 237-240

Boyle, A. C., 1914, The geology and ore deposits of the Bully Hill mining district: A.I.M.E. Tr., v. 48, p. 67-117

Diller, J. S., 1906, Redding folio, California: U.S. Geol. Surv. Geol. Atlas, Folio 138, 14 p.

Friedrich, G. H., and Hawkes, H. E., 1966, Mercury dispersion halos as ore guides for massive sulfide deposits, West Shasta district, California: Mineralium Dep., v. 1, p. 77-88

Graton, L. C., 1909, The occurrence of copper in Shasta County, California: U.S. Geol. Surv. Bull. 430, p. 71-111

Hinds, N.E.A., 1933, Geologic formations of the Redding-Weaversville districts, northern California: Calif. Jour. Mines and Geol., v. 29, nos. 1-2, p. 76-122

Kett, W. F., 1947, Fifty years of operation by the Mountain Copper Company, Ltd., in Shasta County, California: Calif. Jour. Mines and Geol., v. 43, no. 2, p. 105-162

Kinkel, A. R., Jr., 1955, Structural and stratigraphic control of ore deposition in the West Shasta copper-zinc district, California: A.I.M.E. Tr., v. 202, p. 167-174 (in Min. Eng., v. 7, no. 2)

Kinkel, A. R., Jr., and Albers, J. P., 1951, Geology of the massive sulfide deposits of Iron Mountain, Shasta County, California: Calif. Div. Mines Spec. Rept. 14, 19 p.

Kinkel, A. R., Jr., and Hall, W. E., 1951, Geology of the Shasta King mine, Shasta County, California: Calif. Div. Mines Spec. Rept. 16, 11 p.

—— 1952, Geology of the Mammoth mine, Shasta County, California: Calif. Div. Mines Spec. Rept. 28, 15 p.

Kinkel, A. R., Jr., and others, 1956, Geology and base-metal deposits of West Shasta copper-zinc district, Shasta County, California: U.S. Geol. Surv. Prof. Paper 285, 156 p.

Walker, R. T., and Walker, W. J., 1956, Structural and stratigraphic control of ore deposition in the West Shasta copper-zinc district, California (disc.): A.I.M.E. Tr., v. 205, p. 322-324 (with author's reply in Min. Eng., v. 8, no. 3)

Notes

The Shasta County copper district is located in the foothills of the Klamath Mountains and lies on an arc some 45 miles in length that half encircles the town of Redding from N65°E counterclockwise to S50°W at distances from 10 to 20 miles. The district is divided into east and west portions that differ appreciably in geologic column, geologic history, and manner of ore occurrence. There is a gap of some 12 miles of essentially unmineralized ground (except for some gold-bearing quartz veins not directly related to the mineralization discussed here) in the center of the arc due north of Redding. The eastern portion of the district is mineralized in two main centers, one west of Squaw Creek and the other south of Pit River; in the western area the most impressive mineralization is in the 10-mile section southwest of Backbone Creek.

In both areas the oldest exposed formation is the probably Middle Devonian Copley greenstone composed mainly of volcanic flows, breccias, and tuffs of intermediate and mafic composition. It is conformably overlain by the Balaklala rhyolite, consisting of quartz keratophyre flows and pyroclastics; it also is Middle Devonian and is the host rock of the West Shasta ore bodies. The Middle Devonian Kennett formation conformably overlies the Balaklala. Probably the Mississippian Bragdon formation unconformably overlies the Kennett; it consists mainly of shale and mudstone with numerous lenses of coarser materials. In the West Shasta district the Bragdon is the last member of the sedimentary portion of the stratigraphic column until the depositon of the Upper Cretaceous Chico formation. In the East Shasta district the Bragdon is unconformably overlain by the Mississippian Baird formation of which the upper and lower parts are mainly volcanic rocks and the middle part largely siliceous mudstone. Unconformably over the Baird is the Permian McCloud limestone, with local layers of chert and chert nodules. The next formation in the eastern area is the Nosoni, consisting principally of tuffaceous mudstone and tuff; it is unconformable on the McCloud. The Permian Dekkas andesite lies unconformably on the Nosoni; it is mainly keratophyre and spillite lavas and pyroclastics. Overlying the Dekkas is the Triassic Bully Hill rhyolite that is principally quartz keratophyre flows and pyroclastics; some of the Bully Hill rock occurs as dikes in the Dekkas and Dekkas-type material. Bully Hill flows are interbedded with Dekkas andesite so a contact is hard to draw. The major portion of the East Shasta ores is to be found in the Bully Hill. The Bully Hill is partly overlain by, and is partly contemporaneous with, the Pit formation that mainly consists of shales, mudstones, and pyroclastics of middle and late Triassic age; some of the East Shasta ores are found in the Pit. Overlying the Pit is the late Triassic Hosselkus limestone. In the East Shasta area the Paleozoic and early Mesozoic rocks were intruded by three major rock types during Late Jurassic or, more probably, Early Cretaceous time: (1) a small stock chiefly made up of leucocratic granodiorite and related rocks--the Pit River stock; (2) an irregular mass and numerous tabular bodies of quartz diorite; and (3) a few tabular bodies of granodiorite and aplite. Almost all of these rocks occur at least 3 miles west of the westernmost mineralization in the East Shasta district (the Bully Hill mine). The quartz diorite is probably younger than the Pit River stock because dikes of fine-grained mafic quartz diorite cut the stock. That part of the area west of the zone of ore mineralization is cut by a great many dikes of quartz diorite, diorite, metadiabase, and dacite porphyry and by a few dikes of granodiorite; their dominant strike is northwest and they generally dip steeply. These dikes appear to have been the last rocks developed in the East Shasta area except for rather minor amounts of volcanic materials in the Pliocene and Pleistocene that are much younger than the ores.

In the West Shasta district the apparent stratigraphic hiatus after the deposition of the Mississippian Bragdon formation was ended by the intrusion probably Late Jurassic or Early Cretaceous Mule Mountain stock of trondjemite and albite granite that includes some quartz diorite. The next major event was the intrusion, southwest of the Mule Mountain stock, of the probably Early Cretaceous Shasta Bally batholith of which a few satellitic bodies intrude both the stock and the Copley greenstone between the batholith and the stock. A number of minor intrusive bodies are related to the Shasta Bally batholith; these intrusive bodies include diorite porphyry, dacite porphyry, quartz-latite porphyry, andesite porphyry, and lamprophyre dikes. Several of the rock types in the West Shasta district--keratophyre, soda-rich rhyolite, albite diabase, albite granite, and spilite proper--constitute a spilitic suite in which both primary and secondary albite were formed by several processes. Practically all of the igneous basement rocks in the East Shasta district contain metasomatic albite and also form a spilitic suite. All of these intrusives and the ore bodies are older than the Upper Cretaceous Chico formation (shale, sandstone, and conglomerate) that is known to rest unconformably on all the older rocks.

All of the folding and faulting that can be identified in the two areas, plus the igneous intrusives, rock alteration, and mineralization, seem certainly to have taken place in the Late Jurassic and Early Cretaceous with the balance of evidence now seeming to favor an Early Cretaceous age for most, if not all, of these events.

The evidence from the Chico formation in the West Shasta district is that this folding all took place before the Upper Cretaceous for the Chico is uninvolved in it. The Pit River stock in the East Shasta area was intruded after the north-northeast folding but before the cross folding, while the later dike rocks occupy faults and fractures formed during the cross folding. This strongly suggests that the entire period of igneous activity was Lower Cretaceous rather than Late Jurassic. In any event, the similar Pit River and Mule Mountain stocks could well be cupolas rising above a larger mass of igneous rock underlying the area between the cupolas. These two masses and their hypothesized larger downward extension seem to be the most probable source from which the ore-forming solutions that produced the ores of the two districts could have come. The deposits of Shasta County are, therefore, here classified as late Mesozoic (pre-Laramide).

Although the form of the ore bodies (massive sulfide lenses) and the principal metallic minerals themselves (pyrite, chalcopyrite, sphalerite, tetrahedrite-tennantite, and galena) are much the same in the two Shasta County districts, the details of structural control, gangue mineral content and character, and proportions of sulfides to each other are appreciably different in the one district from the other.

In the East Shasta district the sulfide deposits consist of massive lenses or groups of lenses of sulfides that replaced the Bully Hill rhyolite along shear zones and faults and, to a much lesser extent, the Pit formation along fault contacts between the rhyolite and that formation; locally metadiabase in the rhyolite also was replaced. These lenses are generally tabular and, for the most part, are steeply inclined; a few are cigar-shaped; they range between a few inches and 400 feet in their longest dimension. The shear zones and faults may have acted as channelways for the ore-forming fluids. The crushed siliceous rock within these broken rock volumes probably was favorable to replacement, and the shale of the Pit formation slowed down solution movement and gave time for replacement reactions to occur.

The various wall-rock alteration and gangue minerals developed before the sulfides and formed over much wider areas than the metallics. There appears to have been no fixed order of deposition among the gangue minerals, although quartz was generally the first to form and calcite and anhydrite were last. Near the present surface a considerable fraction of the anhydrite has been altered to gypsum, apparently by surface processes. The first step in the production of the massive pyrite bodies of the district was the large-scale introduction of pyrite itself, mainly as a replacement of quartz. The next step was the irregular metasomatic emplacement of sphalerite, chalcopyrite, and minor bornite to convert some of the massive pyrite to copper-zinc ore. The sphalerite generally contains irregular blebs and tiny masses of chalcopyrite, only a small fraction of which is arranged in a geometric pattern. These irregular blebs are essentially identical to similar occurrences of sulfides in the gangue, and such blebs certainly were formed by replacement; thus the mottled texture of chalcopyrite in sphalerite probably was the result of metasomatism, and only the few examples of geometrically arranged blebs were due to exsolution. The wide range in the amount of chalcopyrite in sphalerite (1 to 15 percent) also argues for a replacement origin for the iron-copper sulfide, making the chalcopyrite probably in large measure younger than the sphalerite. Bornite forms small veinlets and irregular masses in sphalerite but shows little evidence of replacement of chalcopyrite. The two copper sulfides were probably largely contemporaneous. Galena and tetrahedrite normally show mutual boundary relations and probably were formed at essentially the same time, but some tetrahedrite occurs along irregular boundaries between sphalerite and galena and may, therefore, in part be later than the galena. Both galena, which is found in an appreciable amount through most of the deposits, and tetrahedrite, which is common but not abundant, are silver-rich and account for practically all the silver recovered. Assuming that the major fraction of the chalcopyrite was formed by replacement of sphalerite, the sphalerite and galena probably were formed under mesothermal conditions. The galena and tetrahedrite appear more likely to have formed under leptothermal than mesothermal conditions; therefore the East Shasta deposits are here classified as mesothermal to leptothermal.

In the West Shasta district the sulfide deposits consist mainly of pods, lenses, and pancake-, kidney-, or cigar-shaped masses in which the length and width are usually from 2 to 10 times the thickness; the deposits are normally flat-lying, so the vertical dimension is the shortest one. The ore fluids apparently rose along steep feeder channels and left these to follow fractured zones in gently plunging folds, although some ore has been formed where no feeder channels have been recognized. The solutions also appear to have been slowed down by the thick shale cover that overlay the Balaklala at the time of ore formation.

The first minerals of the hydrothermal sequence appear to have been quartz and pyrite, with the pyrite being disseminated through the Balaklala rhyolite in amounts up to 10 percent by weight of the total rock. This stage was followed first by an intense sericite-hydromica development and then by that of chlorite. The next stage was the deposition of the bulk of the massive pyrite of the ore bodies, with small amounts of magnetite and pyrrhotite having been formed locally at much the same time. The pyritic bodies thus formed were then fractured, and quartz filled the fractures and entered along the boundaries between gangue minerals. The quartz was followed by chalcopyrite and minor amounts of sphalerite and even of pyrite; these minerals mainly replaced the residual gangue associated with the pyrite. The next stage introduced much larger amounts of sphalerite, additional chalcopyrite, and small amounts of galena and tetrahedrite. The tetrahedrite appears to have formed near the end of the deposition of sphalerite and chalcopyrite and before the deposition of the galena; tetrahedrite is found as small inclusions in sphalerite, and galena is seen along boundaries between sphalerite and gangue minerals. Sulfides make up from 65 to 98 percent of the mineralized zones. The final hypogene stage consisted of small amounts of quartz and calcite in veinlets and small masses cutting the massive sulfides. Although there are large amounts of chalcopyrite in much of the sphalerite, the chalcopyrite inclusions seem to bear no relationship to sphalerite crystal geometry; the erratic distribution and the varied proportions of chalcopyrite to sphalerite strongly suggest that this chalcopyrite was emplaced by replacement and did not result from exsolution from an original copper-iron-rich sphalerite. There seems to be no bornite in the ore, and the amount of galena and tetrahedrite is far smaller than in the East Shasta district. On the basis of these facts, it is here thought that the West Shasta ores should be classified as having been formed entirely in the mesothermal range. Although Kinkel (1955, 1956) once argued for a hydrothermal origin of the Shasta County ores, he now believes them to have been folded with the containing rocks (Econ. Geol., v. 61, p. 681) so that the textures that he originally considered to have been caused by hydrothermal replacement must be inferred to have resulted from remobilization due to directed earth pressures. This opinion must be given serious consideration, although I believe his original interpretation to have been the correct one.

SULPHUR BANK

Pleistocene to Recent	*Mercury, Sulfur*	*Epithermal*

Anderson, C. A., 1936, Volcanic history of the Clear Lake area, California:

Geol. Soc. Amer. Bull., v. 47, p. 629-663

Becker, G. F., 1888, Geology of the quicksilver deposits of the Pacific slope: U.S. Geol. Surv. Mono. 13, 486 p. (particularly p. 251-264, 269-270)

Bradley, W. W., 1918, Quicksilver resources of California: Calif. Min. Bur. Bull. 78, p. 63-68

Everhart, D. L., 1946, Quicksilver deposits at the Sulphur Bank mine, Lake County, California: Calif. Jour. Mines and Geol., v. 42, no. 2, p. 125-153

Forstner, W., 1903, The quicksilver resources of California: Calif. Min. Bur. Bull. 27, p. 61-72

LeConte, J., 1883, On mineral vein formation now in progress at Steamboat Springs compared with the same at Sulphur Bank: Amer. Jour. Sci., 3rd Ser., v. 25, no. 150, p. 424-428

LeConte, J., and Rising, W. B., 1882, The phenomena of metalliferous vein-formation now in progress at Sulphur Bank, California: Amer. Jour. Sci., 3rd Ser., v. 24, no. 139, p. 23-33

Pošepný, F., 1893, The genesis of ore-deposits: A.I.M.E. Tr., v. 23, p. 197-369 (particularly p. 225-229)

Ransome, A. L., and Kellogg, J. L., 1939, The quicksilver resources of California: Calif. Jour. Mines and Geol., v. 35, no. 4, p. 353-486 (particularly p. 395-399)

Ross, C. P., 1940, Quicksilver deposits of the Mayacmas and Sulphur Bank districts, California--a preliminary report: U.S. Geol. Surv. Bull. 922-L, p. 327-353

Schuette, C. N., 1931, Occurrence of quicksilver orebodies: A.I.M.E. Tr., v. 96 (1931 General Volume), p. 403-488 (particularly p. 425-426)

White, D. E., 1955, Sulphur Bank, California, in *Economic Geology--Fiftieth Anniversary Volume*: pt. 1, p. 117-120

White, D. E., and Roberson, C. E., 1962, Sulphur Bank, California--a major hot-spring quicksilver deposit, in Engel, A. E. J., and others, Editors, *Petrologic studies: A volume in honor of A. F. Buddington*: Geol. Soc. Amer., p. 397-428

Notes

The Sulphur Bank mercury deposits lie a few hundred feet east of the east shore of Clear Lake in Lake County and are located some 95 miles slightly west of north from San Francisco.

The oldest rocks in the area are those of the Upper Jurassic to Upper Cretaceous Franciscan formation that is composed of shale and graywacke, with lenses and veinlets of quartz; these rocks have been highly folded. This formation is unconformably overlain by lake sediments that are made up of poorly stratified conglomerate and breccia that include lenses of cross-bedded sandstone and beds of silt and sand. Rather uncertain carbon-14 dating places the age of the lake beds at some 50,000 years or late Pleistocene. Wood fragments of *Sequoia sempervirens*, a species of Pleistocene or post-Pleistocene age, found in the lake beds do not conflict with a late Pleistocene age for the sediments. Overlying the lake sediments is a flow of augite andesite lava that covers an area of approximately 1 square mile; the fragmental upper surface of much of the flow suggests that it was quite viscous when extruded. Associated with the flow are small hillocks of pyroclastic agglutinate that probably are the remains of spatter cones that developed when the flow poured out over the water-saturated lake sediments; the base of the flow appears to have been about at the then lake level. White (1962) reports that the andesite has been much more extensively eroded than the latest Pleistocene moraines in the Sierras, and the margins of the flow seem to have been beveled by erosion.

White, therefore, considers that the flow was emplaced shortly after the lake sediments and that it is also late but not latest Pleistocene in age. In places the andesite is overlain by patches of sand and gravel that may not be appreciably younger than the andesite.

Evidence that mercury was deposited in the area within historic time is shown by the development of cinnabar and matacinnabar on pebbles that had been agitated in spring pools and by the presence of sooty metacinnabar found as coatings on fragments in a spring vent in a newly opened underground working at a depth greater than any other known occurrence of quicksilver minerals. It would appear, however, that most of the ore and gangue minerals were emplaced more nearly to the late Pleistocene time of the andesite extrusion than to the present. Although White assumes that deposition took place over a span of 10,000 years, he does not place that time period in the 50,000 years between lake sediment deposition and the present. On the basis of these facts and assumptions, the mercury ores are here classified as late Pleistocene to Recent; it should, however, be emphasized that it is probable that much more ore was formed in the Pleistocene than in the Recent.

Ore minerals are found in all of the rocks of the sequence at Sulphur Bank, with the mineralization being confined to a highly fractured and irregularly shaped area, the maximum dimension of which is about a half mile. Present-day hot springs and fumaroles are scattered along the east-northeast-trending fault zone, and the most intense solfataric activity is associated with the intersection of that fault zone with the most easterly of a series of northwest-striking faults; this line probably was also the center of hydrothermal ore-forming activity. The andesite is hydrothermally altered in and near, but not appreciably beyond, the mine area, and the alteration consists of (1) a near-surface zone of bleached rock; (2) an intermediate or boulder zone in which joint blocks were altered progressively inward so that the boulders now consist, from the inside outward, of fresh andesite grading to opalized white or gray rocks surrounded by shells of exfoliated material; and (3) a basal zone in which andesite ranges from fresh through volumes altered along joints to complete alteration to montmorillonite or zeolite; the basal zone andesite usually exhibits columnar jointing and makes up the lowest portion of the 100-foot-thickness of the andesite. In the lake sediments under the andesite, the alteration is also essentially limited to the mineralized area and consists mainly of the considerable conversion of the beds to kaolinite, while the Franciscan rocks are altered essentially only along fault zones where the principal effect was formation, by replacement and open-space filling, of carbonate veins.

Although sulfur is now being precipitated at the orifices of gas vents and hot springs, at most only trace amounts of cinnabar are being formed near the surface at the present time. The bulk of the cinnabar appears to have been emplaced as veins and disseminated masses in the lower portion of the andesite lava and in the lake beds immediately below the andesite contact. In the Franciscan rocks, cinnabar forms finely crystalline coatings in fractures and on blocks of sandstone, but the most important mineralization in these Mesozoic rocks is that of disseminated cinnabar in fault gouge. The major ore bodies are localized mainly by the intersections of the major fault zones, but some ore bodies occur in fault zones at some distance from such intersections. The ores in the Franciscan formation are generally in nearly vertical chimneys, while the ores in the sediments are normally mushroom-shaped bodies, the tops of which abut against the base of the lavas. The deposits in the andesite are more widely dispersed in the altered rock but are most abundant in joints and sheeting planes in the clay that encloses essentially fresh andesite; the richest concentrations are at fracture intersections.

Originally, the mines were operated to recover sulfur that appears to have formed, and to be forming at present, through the reaction of hydrogen sulfide with oxygen in the atmosphere or in the vadose water above the water table; the water table apparently has fluctuated through only a narrow vertical range in the last 50,000 years. In the upper part of the bleached zone, cinnabar was essentially lacking, but the mining of sulfur was stopped in the basal part of the bleached zone largely because minor amounts of cinnabar were found in the sulfur. Minable cinnabar occurred in veins and open spaces between original joint blocks in the upper parts of the boulder zone in association with opal, cristobalite, kaolinite, alunite, iron oxides, hydrocarbons, native sulfur, and sulfates. Rich ore bodies were found as veins and disseminated masses in the basal andesite and in the lake sediments lying below it. In the Franciscan formation cinnabar was found down to a few feet below the 262-foot level, with marcasite, pyrite, metacinnabar (locally scarce to abundant), stibnite, zeolite, montmorillonite, dolomite, calcite, and a little opal. In the lake sediments pyrite is the principal iron sulfide, but marcasite is more abundant in the Franciscan rocks and in the andesite below the boulder zone; only a little pyrite is found in the rocks of the boulder zone and none in the bleached zone. Stibnite has been found in many places in the mine, but it appears that nowhere is it abundant. In the strictly hydrothermal portions of the deposit, the paragenetic sequence was opal as the first mineral, followed by hydrothermal clay, then by quartz, pyrite, cinnabar, stibnite, and a second generation of opal. Above the water table sulfur was the principal primary mineral, although a little cinnabar is present with sulfur in the lower parts of the vadose zone; the numerous sulfate species (except alunite and very minor barite) appear to have been developed secondarily from sulfur and iron sulfides. The depths to which sulfur and secondary sulfates deposited at any given time were controlled by the water table; neither sulfur nor sulfates formed appreciably below it.

There appears to be no doubt but that the Sulphur Bank ores were formed in a near-surface environment, and the ore, gangue, and wall-rock alteration minerals are all indicative of moderately low to low temperatures and chemical intensities. Since there are no important sulfides other than cinnabar and the two iron sulfides and no sulfosalts, the ores are here classified as epithermal.

Colorado

BOULDER COUNTY TELLURIDES

Late Mesozoic to Early Tertiary | *Gold, Silver, Lead, Zinc, Fluorspar* | *Leptothermal*

Bray, J. M., 1942, Spectroscopic distribution of minor elements in igneous rocks from Jamestown, Colorado: Geol. Soc. Amer. Bull., v. 53, p. 765-814

Goddard, E. N., 1935, The influence of Tertiary intrusive structural features on mineral deposits at Jamestown, Colorado: Econ. Geol., v. 30, p. 370-386

—— 1940, Preliminary report on the Gold Hill mining district, Boulder County, Colorado: Colo. Sci. Soc. Proc., v. 14, p. 103-139

—— 1946, Fluorspar deposits of the Jamestown district, Boulder County, Colorado: Colo. Sci. Soc. Proc., v. 15, p. 1-47

Goddard, E. N., and Lovering, T. S., 1942, Nickel deposit near Gold Hill, Boulder County, Colorado: U.S. Geol. Surv. Bull. 931-O, p. 349-362

Kelly, W. C., and Goddard, E. N., 1969, Telluride ores of Boulder County, Colorado: Geol. Soc. Amer. Mem. 109, 237 p.

Wahlstrom, E. E., 1950, Melonite in Boulder County, Colorado: Amer. Mineral., v. 35, p. 948-953

Wilkerson, A. S., 1939, Telluride-tungsten mineralization of the Magnolia mining district, Colorado: Econ. Geol., v. 34, p. 437-450

Notes

The Boulder County telluride ores lie in a generally north-south-trending belt some 13 miles in length and less than 5 miles wide at its maximum width; the center of the district is nearly 7 miles northwest of the city of Boulder. The principal structures (breccia reefs and fissures) in the district, however, trend northwest and northeast, respectively. The most important mining districts in the telluride belt are the Jamestown, Gold Hill, Magnolia, and Sugar Loaf in that order from north to south.

The oldest rocks in the district are the schists and gneisses of the Precambrian Idaho Springs formation that lie in an irregular belt between the Silver Plume granite (younger) to the north and the Boulder Creek granite (older) on the south. The Boulder Creek granite contains most of the ore veins in the Magnolia, Sugar Loaf, and Gold Hill districts and a few of those in the Jamestown district. Most of the Jamestown veins are in the Silver Plume granite. The Boulder Creek ranges in composition from quartz monzonite through granodiorite to granite, while the Silver Plume is a normal granite which Kelly and Goddard (1969) believe is a very favorable host rock for ore. They also think that the vein fissures parallel to or crosscutting the flow structure are equally favorable sites for ore deposition. In addition to the large granite bodies, the district contains dikes and irregular bodies of pegmatite and aplite in both granites and in the schists and gneisses. In the Gold Hill district telluride veins in places follow the walls of pegmatite dikes; most of the vein structures are barren in pegmatite. The complex of rocks so far described was cut by a series of Late Cretaceous-early Tertiary stocks and dikes; these rocks are generally porphyritic and range in composition from diorite to sodic granite. Among these rocks are inconspicuous dikes of biotite latite that Kelly and Goddard (1969) think have a close genetic relationship to the telluride ores. Locally thin seams of biotite-latite intrusion breccia are known; these contain small fragments of granite and pegmatite. Since these intrusives are definitely stated to be Late Cretaceous-early Tertiary by Kelly and Goddard and since they emphasize the close genetic connection between intrusives and ores, even listing the associated intrusives for each ore type, the age of the ore-forming process almost certainly must have been late Mesozoic-early Tertiary, and the ores are so classified here.

Kelly and Goddard describe five types of ore deposits of which one (the youngest) is the tungsten mineralization discussed under the heading "Boulder County Tungsten" in this volume; only the four oldest types are considered here. Of these, the oldest is the lead-silver type, which is made up of Ag-bearing galena, tetrahedrite, and lead-silver sulfantimonides with sphalerite, chalcopyrite, and minor free gold; pyrite was the main gangue sulfide. The lead-silver ore type was important in the main fluorite zone at Jamestown and at several locations at Gold Hill. Next in order is the fluorite type in which the principal minerals are fluorite, quartz, sericite, and clay minerals, plus included fragments of lead-silver ore. This fluorite ore occurs in rather short breccia zones and veins; the breccias are large lenticular bodies of broken granite and earlier fluorite with a matrix of fine-grained later fluorite and clay minerals. The fragments of lead-silver ore in the veins indicate that the brecciated area (a small portion of the Jamestown district) had been affected by the earlier lead-silver mineralization. The first and second types are spatially, and appear to be genetically, connected with stocks and dikes of granite and quartz monzonite. The third type is pyritic gold ore in which the main minerals are pyrite, quartz, free gold, chalcopyrite, and roscoelite. This type was widespread with important veins in the Jamestown and Gold Hill districts but with nonminable veins in the Magnolia and Tungsten districts. They are associated spatially and probably genetically with dikes of bostonite. The fourth type (and youngest considered here) was composed of interlaced quartz seams that surround breccia fragments and lenticular-shaped masses of wall rock (highly sheared and altered). The telluride minerals are found as blades or small masses in the quartz, one being complexly intergrown in general with the others. The chief ore minerals are petzite, hessite, sylvanite, and native gold, though calaverite and krennerite were important in a few mines. In addition to the tellurides named, 10 others were identified by Kelly and Goddard; these were essentially of no value. Native tellurium also was present. Pyrite and marcasite were normally disseminated in the ore and pyrite in the wall rock. Sphalerite and galena were present but in small amounts; other sulfides and sulfosalts were found. Almost all the dollar-value of the ore produced from all ore types was derived from gold and silver, and about two-thirds of this came from the telluride ores of the fourth type. In places the age relationships between types were more complex than outlined here.

Kelly and Goddard conclude that the ores were all formed below 350°C and most of them below 250°C. They estimate that the pressure during deposition ranged between 360 and 78 bars. The temperature range of the tellurides was between 200°C and 100°C with the maximum amount of deposition being at about 130°C; even the minor sulfides appear to have formed below 200°C. These data would indicate that ore deposition occurred in the leptothermal range, particularly since Kelly and Goddard estimate that the deposition took place from 2600 to 4600 feet beneath the reconstructed Flattop surface that existed at the time of ore deposition. The deposits are, therefore, here categorized as leptothermal.

BOULDER COUNTY TUNGSTEN

Late Mesozoic to Early Tertiary — *Tungsten* — *Mesothermal*

George, R. D., 1909, The main tungsten area of Boulder County, Colorado: Colo. Geol. Surv., 1st Rept., 1908, p. 7-103 (particularly p. 16-36, 61-76)

Hess, F. L., and Schaller, W. T., 1914, Colorado ferberite and the wolframite series: U.S. Geol. Surv. Bull. 583, 75 p. (particularly p. 8-39)

Kirk, C. T., 1916, Tungsten district of Boulder County, Colorado: Min. and Sci. Press, v. 112, no. 22, 27 May, p. 791-795

Lindgren, W., 1907, Some gold and tungsten deposits of Boulder County, Colorado: Econ. Geol., v. 2, p. 453-463

Loomis, F. B., Jr., 1937, Boulder County tungsten ores: Econ. Geol., v. 32, p. 952-963

Lovering, T. S., 1940, Tungsten deposits of Boulder County, Colorado: U.S. Geol. Surv. Bull. 922-F, p. 135-156

—— 1941, The origin of the tungsten ores of Boulder County, Colorado: Econ. Geol., v. 36, p. 229-279

Lovering, T. S., and Goddard, E. N., 1938, Laramide igneous sequence and differentiation in the Front Range, Colorado: Geol. Soc. Amer. Bull., v. 49, p. 35-68 (General discussion of Laramide igneous rocks of the Front Range)

—— 1950, Boulder County tungsten district, in *Geology and ore deposits of the Front Range, Colorado*: U.S. Geol. Surv. Prof. Paper 223, p. 214-227

Lovering, T. S., and Tweto, O., 1953, Geology and ore deposits of the Boulder County tungsten district, Colorado: U.S. Geol. Surv. Prof. Paper 245, 199 p. (particularly p. 6-90)

Tweto, O., 1947, Scheelite in the Boulder district, Colorado: Econ. Geol., v. 42, p. 47-56

—— 1947, The Boulder tungsten district, Boulder County, in Vanderwilt, J. W., Editor, *Mineral resources of Colorado*: State of Colo. Mineral Res. Bd., p. 328-336

Notes

The Boulder County tungsten district lies in a narrow belt nearly 10 miles long that extends from Arkansas Mountain (4 miles west of the town of Boulder) west-southwest to Sherwood Flats (a mile northwest of Nederland); the mineralized area is some 30 miles northwest of Denver.

The oldest rocks in the area are those of the Precambrian Idaho Springs formation that occupies the portion of the belt between Sherwood Flats and Hurricane Hill and consists mainly of biotite schist and schistose quartzite.

From Hurricane Hill to Arkansas Mountain (slightly over 7 miles), the country rock is the Boulder Creek granite (which actually is mainly quartz monzonite with minor granodiorite) with related aplites and pegmatites. All of these rocks are part of the Boulder Creek batholith that extends irregularly some 18 miles from north to south and some 10 miles from east to west. The aplite constitutes a considerable fraction of the Boulder Creek batholith in the mineralized area and also is largely developed in the ore-containing portion of the Idaho Springs formation. Pegmatite is somewhat less abundant than aplite and is generally the younger of the two; many aplite bodies, however, grade into a mixture of aplite and pegmatite and then into pegmatite. Aplites were intruded over a considerable time span, and the older aplites are the more gneissic.

The younger, but still Precambrian, Silver Plume granite is found in the tungsten district only as dikes and small, irregular masses not more than a few hundred feet in diameter. These masses strike north to northeast. Some of them parallel adjacent bodies of Boulder Creek aplite and others do not.

No Paleozoic or Mesozoic rocks have been found in the mineralized area, and the next rocks to be introduced after the Silver Plume were a series of very early Tertiary porphyry dikes that range in composition from silicic felsite to limburgite, with monzonite and hornblende-monzonite porphyries being the most abundant; these last are located mainly in the western part of the district. The large, continuous, and early Iron Dike in the eastern part of the area is composed of gabbro; dikes of other species are irregularly distributed in the area. Lovering and Goddard (1938) concluded that the ultimate source of the various porphyry magmas was a deep-seated gabbro on the east side of the range and that the tungsten and associated gold-telluride deposits were to be correlated with the biotite-latite group of porphyries.

In one locality a probably slightly older biotite-latite intrusion breccia and the granite enclosing it are altered on the footwall side of the dike in which they both occur, but both are fresh on the hanging wall side, strongly suggesting that the intrusive breccia was introduced before the argillic alteration had begun.

The rather late occurrence of the igneous activity within the Laramide orogenic cycle is indicated by at least five periods of Laramide faulting that preceded the introduction of the intrusion breccia and the two periods that followed it. As the mineralization was essentially contemporaneous with the intrusion breccia, it also developed late in the Laramide orogeny; as a phase of the Laramide cycle, therefore, the mineralization must be included within the late Mesozoic to early Tertiary time span, and this age is here assigned to the Boulder County tungsten ores.

Throughout the district the vertical range over which ore has been found has been about 700 feet. This apparently does not mean that tungsten mineralization was confined to a 700-foot belt of equal elevation because, in the extreme eastern edge of the district, tungsten has been mined between 6400 and 7100 feet above sea level while, in the western part, ore has been mined between 8200 and 8800 feet. Further, the majority of the individual ore shoots have been less than 100 feet in vertical extent, but the shape of the ore shoots depends almost entirely on the shape of the open space present in the veins at the time when ferberite was deposited; vein volumes lacking such open space are essentially barren of tungsten. Many of the ore shoots mined, particularly during World War II, were blind ore bodies found by underground development, and it has been pointed out by Lovering and Goddard that the future of the district depends on the finding of more of such blind bodies, something that they consider would be readily possible, though uneconomic at present tungsten prices, to do. Apparently two interpretations can be placed on these spatial relationships, the simpler of which is that the mineralization was confined to a vertical range of 700 feet, with the base of the mineralized belt rising from a depth of 6400 feet in the east of the district to 8200 feet in the west. Acceptance of this hypothesis would suggest that the ores must have been formed under conditions of rapid loss of heat and pressure and, therefore, near the surface. On the other hand, it well may be that mineralization extended over the entire vertical range from what is now 6400 feet above sea level to 8700 feet, with the possibility that further blind ore bodies may be found (particularly in the west) at depths greater than those now reached by mining. From this reasoning the mineralization would be considered to have a minimum vertical range of 2300 feet, with the possibility of still greater ranges being found by further prospecting. Although rapid loss of heat and pressure deposits are known in which the vertical range of mineralization was over 3000 feet (Comstock and Cripple Creek for example), the vertical range of the average deposit in this category rarely exceeds 2000 feet. It is, therefore, considered to be more appropriate, based on the data now at hand, to classify the tungsten deposits of Boulder County as formed under conditions of slow loss of heat and pressure rather than rapid.

The mineralization of the average tungsten vein is simple; it normally consists of generally vuggy, brecciated, and porous quartz (known locally as horn) and ferberite that is usually present as the matrix of horn fragments, although some solid seams of ferberite are found free from fragments and a few instances of brecciated ferberite cemented by later quartz are known. The ferberite accomplished essentially no replacement of quartz or country rock despite the highly sheared character of the host rock and the occurrence of local shearing during ferberite deposition.

In the tungsten veins, the first mineral, in addition to the ubiquitous quartz, to be formed was hematite; this was followed by the deposition of dickite, marcasite, and then goyazite (a strontium aluminum phosphate that may contain some sulfate and is a member of the alunite group); some dickite overlapped the ferberite deposition as locally did minor amounts of pyrite, marcasite, siderite, and barite, and small amounts of other sulfides and several carbonates followed the ferberite.

The wall rock near the main tungsten veins was strongly altered to the clay minerals beidellite and dickite for distances of from 10 to 50 feet at about the same time as the first quartz was deposited in the veins. The clay minerals nearest the veins were later partly converted to quartz, sericite, and hydromica. The argillic alteration probably occurred before the tungsten mineralization, and in many places bears no direct spatial relationship to it. Some ferberite veins occur in almost unaltered country rock, and some veins with strongly argillized walls contain no ferberite.

Although it has been suggested that dickite has a higher stability limit than kaolinite and that its presence indicates formation under somewhat higher temperatures than those required by the formation of kaolinite, recent work by Roy and Osborn implies that the upper stability limit is essentially the same for kaolinite, dickite, and nacrite. Thus, the presence of abundant alteration clay minerals in the wall rocks, not far removed in time from the ferberite mineralization, and the deposition of scattered and small amounts of simple sulfides with the ferberite strongly suggest the formation of the ferberite under conditions of moderate (or mesothermal) intensity. Normally, the presence of abundant wolframite or huebnerite in ore deposits automatically categorizes them as hypothermal; in such deposits the wall-rock alteration minerals also agree with this classification. In the Boulder County veins, however, the tungstate is the iron-rich end member of the series, and it was deposited with appreciably lower-temperature gangue minerals than normally are associated with huebnerite and wolframite. Therefore, it appears that these deposits must be categorized as mesothermal (or possibly kryptothermal if Lovering and Goddard's suggestion of their rapid loss of heat and pressure origin is accepted) despite the abundant development of a tungstate as the principal ore mineral.

CENTRAL CITY-IDAHO SPRINGS

Late Mesozoic to Early Tertiary	*Gold, Silver, Lead, Zinc, Uranium*	*Mesothermal to Leptothermal*

Alsdorf, P. R., 1916, Occurrence, geology and economic value of the pitchblende deposits of Gilpin County, Colorado: Econ. Geol., v. 11, p. 266-275; disc., p. 681-685

Bastin, E. S., and Hill, J. M., 1917, Economic geology of Gilpin County and adjacent parts of Clear Creek and Boulder Counties, Colorado: U.S. Geol. Surv. Prof. Paper 94, 379 p.

Drake, A. A., Jr., 1957, Geology of the Wood and East Calhoun mines, Central City district, Gilpin County, Colorado: U.S. Geol. Surv. Bull. 1032-C, p. 129-170

Lovering, T. S., and Goddard, E. N., 1950, Central City-Idaho Springs district, in *Geology and ore deposits of the Front Range, Colorado*: U.S. Geol. Surv. Prof. Paper 223, p. 167-191

Moench, R. H., 1964, Geology of Precambrian rocks, Idaho Springs district, Colorado: U.S. Geol. Surv. Bull. 1182-A, p. A1-A70

Moench, R. H., and Drake, A. A., Jr., 1966, Economic geology of the Idaho Springs district, Clear Creek and Gilpin Counties, Colorado: U.S. Geol. Surv. Bull. 1208, 91 p.

Sims, P. K., 1955, Uranium deposits in the Eureka Gulch area, Central City district, Gilpin County, Colorado: U.S. Geol. Surv. Bull. 1032-A, p. 1-32

—— 1956, Paragenesis and structure of pitchblende-bearing veins, Central City district, Colorado: Econ. Geol., v. 51, p. 739-756

—— 1960, Geology of the Central City-Idaho Springs area, Front Range, Colorado: Geol. Soc. Amer. Guidebook for Field Trips (Guide to the geology of Colorado), Field Trip C-3, p. 279-285

—— 1963, Geology of uranium and associated ore deposits, central part of the Front Range mineral belt, Colorado: U.S. Geol. Surv. Prof. Paper 371, 119 p.

Sims, P. K., and Barton, P. B., Jr., 1961, Some aspects of the geochemistry of sphalerite, Central City district, Colorado: Econ. Geol., v. 56, p. 1211-1237

—— 1963, Hypogene zoning and ore genesis, Central City district, Colorado, in Engel, A.E.J., and others, Editors, *Petrologic studies: A volume in honor of A. F. Buddington*: Geol. Soc. Amer., p. 373-396

Sims, P. K., and Gable, D. J., 1965, Geology of Precambrian rocks, Central City district, Colorado: U.S. Geol. Surv. Prof. Paper 474-C, p. C1-C52

—— 1967, Petrology and structure of Precambrian rocks, Central City quadrangle, Colorado: U.S. Geol. Surv. Prof. Paper 554-E, p. E1-E56

Sims, P. K., and Tooker, E. W., 1956, Pitchblende deposits in the Central City district and adjoining areas, Gilpin and Clear Creek Counties, Colorado, in *Contributions to the geology of uranium and thorium*: U.S. Geol. Surv. Prof. Paper 300, p. 105-111

Sims, P. K., and others, 1963, Economic geology of the Central City district, Gilpin County, Colorado: U.S. Geol. Surv. Prof. Paper 359, 231 p.

Tooker, E. W., 1956, Altered wall rocks along vein deposits in the Central City-Idaho Springs region, Colorado, in Swineford, A., Editor, *Clays and clay minerals*: 4th Natl. Conf. on Clays and Clay Minerals, Natl. Acad. Sci.-Natl. Research Council Pub. 456, p. 348-361

—— 1963, Altered wall rocks in the central part of the Front Range mineral belt, Gilpin and Clear Creek Counties, Colorado: U.S. Geol. Surv. Prof. Paper 439, 102 p.

Wells, J. D., 1960, Petrography of radioactive Tertiary rocks, central part of Front Range mineral belt, Gilpin and Clear Creek Counties, Colorado: U.S. Geol. Surv. Bull. 1032-E, p. 223-272

Notes

The Central City-Idaho Springs district is located less than 30 miles west of Denver in the center of the Front Range mineral belt. The district is part of a larger, well-zoned area of mineralization that also includes the Freeland-Lamartine and Chicago Creek districts (discussed together in this volume as Freeland-Chicago Creek) and the Lawson-Dumont-Fall River district. The zoned area is about 8 miles long in a northeast-southwest direction and some 4.5 miles wide in a northwest-southeast direction. This larger area is divided into three zones: (1) a central zone that contains quartz-minor pyrite veins lacking in base metals, and quartz-pyrite-copper sulfide-gold veins; (2) a narrow intermediate zone containing quartz-pyrite-copper sulfide-lead and zinc sulfide-gold veins, in which the lead and zinc minerals are far more abundant than the copper and iron sulfides; and (3) a broad peripheral zone in which the veins contain minor quartz, pyrite, and copper sulfides and abundant galena and sphalerite; in these veins silver sulfosalts may be common. All three zone types are found in the Central City-Idaho Springs district.

The oldest rocks in the district are metasediments, most of which were grouped in the older literature as the Idaho Springs formation. The Idaho Springs gneisses are migmatized throughout the region with various quantities of granite gneiss and related pegmatite; the large masses of pegmatite shown by Lovering and Goddard on their plate 2 (1950) are of a later age. South of Idaho Springs the migmatization is more intense, and the granite gneiss and its related pegmatites occur in large enough bodies to be mapped. The migmatized metasedimentary complex was intruded, in later Precambrian times, by a number of igneous rock types, of which the oldest is a granodiorite (probably equivalent to the Boulder Creek granite of Lovering and Goddard), a quartz diorite and associated hornblendite, a biotite-muscovite granite (probably equivalent to the Silver Plume granite) and varieties of pegmatites, most of which are unzoned and of simple mineral composition. Most of these Precambrian rocks have a low natural radioactivity, and what activity there is is mainly caused by thorium in pegmatitic xenotime and monazite.

The area contains no rocks of Paleozoic or Mesozoic age, and the only younger rocks introduced into the district are a series of earliest Tertiary porphyries that are divided into four groups that contain a total of 14 rock types. The Tertiary igneous rocks have a general high radioactivity (though there may be a considerable variation from one rock type to another), and the quartz bostonite is among the most radioactive rocks known in the world, its radioactivity being 15 to 25 times that of the average granite. This radioactivity is due to both thorium and uranium (Th being appreciably more abundant than U), with the radioactive elements being associated in minerals with zirconium. The age of this complex appears to be between 70 and 57 m.y., although the available age determinations were made from lead-uranium isotope ratios obtained from uraninite in the ore bodies rather than from the radioactive minerals in the rocks themselves. These ages indicate that the rocks in question were formed in the early Tertiary; although the magmatic material was largely introduced along already existing planes of weakness in Precambrian rocks, their opening at this time appears to have been related to the Laramide orogeny, and the intrusives appear to have entered them while that orogeny was still in progress. This would place the intrusives in the late Mesozoic to early Tertiary. Insomuch as the ages of the intrusives have been inferred from those of uraninite in the vein deposits, the age of the uraninite mineralization is the same as that of the igneous rocks; although the base metal sulfide-quartz, gold mineralization was later than the uraninite, the time interval between the two appears to have been so short that both uraninite and sulfide mineralizations are classified under the same age group--late Mesozoic to early Tertiary.

Within the concentric zones of mineralization of the district, the central zone contains two types of veins: (1) quartz and pyrite with traces of copper sulfides and gold in some of the pyrite and (2) quartz-pyrite veins with substantial amounts of copper, mainly as chalcopyrite or tennantite but with local enargite; sphalerite is not common, galena is rare, and moderate amounts of gold are irregularly distributed in the pyrite. The deposition of the base metal sulfides was separated from that of quartz and pyrite by a period of fracturing and brecciation. What sphalerite is present appears to have begun to form in general before the copper sulfides, but some exsolved chalcopyrite is found in sphalerite; tennantite was contemporary with and overlapped the chalcopyrite in time of deposition, and enargite, where present, shows much the same relations to tennantite as does the chalcopyrite.

The intermediate zone contains one vein type (3) in which quartz and pyrite are still abundant but in which the amounts of sphalerite, particularly, and galena are much larger and those of the copper minerals appreciably less. The paragenetic relationships in this type of vein are much the same as in types (1) and (2). Silver sulfosalts provide much of the silver; the remainder probably is in the galena. The silver to gold ratio in vein type (3) is much higher than in the central zone. The veins in the peripheral zone [vein type (4)] are predominantly composed of sphalerite and galena while pyrite and the copper minerals are sparse; silver sulfosalts are common and the silver to gold ratio has values up to 200:1. In addition to quartz, the gangue minerals in the peripheral zone veins include a variety of carbonates and rare barite and fluorite. Outside the peripheral zone is the barren zone, but it has been known to contain veins in which small pockets of silver ore have been found. Although most of the tonnage mined has come from veins, some has been produced from stockworks.

In addition to the zoned area of pyrite-base metal sulfide veins, the district contains both telluride and uranium mineralizations, the patterns of which are much different from those of the sulfide veins. The telluride deposits occur in a narrow north-northeast zone along the Dory Hill fault. The two main areas of uranium ores, as opposed to those containing tellurides, are grouped near to, or within a few hundred feet of, exposed dikes of (late) quartz-bostonite porphyry; the larger deposits are not more than 500 feet from a quartz-bostonite dike, this rock being itself highly radioactive. In addition to this spatial relationship, the quartz bostonite is high in zirconium, and most of the uranium deposits contain more than 1 percent zirconium, some of them running more than 7.0 percent. On the other hand, the minerals of the base-metal sulfide deposits are uniformly low in zirconium. This strongly suggests that the uranium deposits were derived from a different source than the sulfides.

The wall-rock alteration associated with the sulfide veins is typically that developed in connection with ore deposition under mesothermal conditions, having produced a sericitized zone next to the veins that grade outward into argillic material that, in turn, grades into fresh rock. The argillite is divided into three zones: (1) farthest from the vein is a montmorillonite-rich rock, (2) next is a kaolinite-rich material which contains some illite and sericite, and (3) nearest the vein is illite-rich, and sericite-bearing rock. The width of the alteration envelope is greater in the central and intermediate zones than in the peripheral one.

The copper sulfide-gold mineralization of the central zone contains typical mesothermal minerals, while the intermediate zone minerals range from mesothermal copper sulfides, sulfosalts, and sphalerite to leptothermal galena and silver-sulfosalts. The pitchblende mineralization is probably mesothermal, while the tellurides are definitely characteristic of the leptothermal range. The depth at which mineralization took place probably was sufficiently great to insure deposition under conditions of slow loss of heat and pressure. The deposits of the entire district are, therefore, here categorized as mesothermal to leptothermal.

CLIMAX

Late Mesozoic to Early Tertiary — *Molybdenum, Tungsten* — *Hypothermal-1*

Blanchard, R., and Boswell, P. F., 1935, "Limonite" of molybdenite derivation: Econ. Geol., v. 30, p. 313-319

Butler, B. S., and Vanderwilt, J. W., 1931, The Climax molybdenum deposits of Colorado: Colo. Sci. Soc. Pr., v. 12, no. 10, p. 309-353

—— 1933, The Climax molybdenum deposit, Colorado: U.S. Geol. Surv. Bull. 846-C, p. 195-237

Carpenter, R. H., 1960, Molybdenum, in Rio, S. M. del, Editor, *Mineral resources of Colorado--first sequel*: State of Colo., Mineral Res. Bd., p. 317-321

Hess, F. L., 1923, Colorado, in *Molybdenum deposits--A short review*: U.S. Geol. Surv. Bull. 761, p. 9-13 (see also plates 1-9)

Staples, L. W., and Cook, C. W., 1931, A microscopic investigation of molybdenite ore from Climax, Colorado: Amer. Mineral., v. 16, p. 1-17

Vanderwilt, J. W., 1933, The molybdenum deposit at Climax, in *Colorado*: 16th Int. Geol. Cong., Guidebook 19, p. 92-102

—— 1947, Climax, in Vanderwilt, J. W., Editor, *Mineral resources of Colorado*: State of Colo., Mineral Res. Bd., p. 223-225

Vanderwilt, J. W., and King, R. U., 1946, Geology of the Climax ore body: Min. and Met., v. 27, no. 474, p. 299-302

—— 1955, Hydrothermal alteration at the Climax molybdenum deposit: A.I.M.E. Tr., v. 202, p. 41-53 (in Min. Eng., v. 7, no. 1)

Walker, M. S., and Taylor, R. H., 1955, Engineering and geology (Climax): Min. Eng., v. 7, no. 8, p. 730-731 (not in Tr.) (valuable only for sections on p. 731)

Wallace, S. R., and others, 1960, Geology of the Climax molybdenite deposit; a progress report: Geol. Soc. Amer. Guidebook for Field Trips (Guide to the geology of Colorado), Field Trip B-3, p. 238-252

—— 1968, Multiple intrusion and mineralization at Climax, Colorado, in Ridge, J. D., Editor, *Ore deposits of the United States, 1933-1967* (Graton-Sales Volumes): Chap. 29, v. 1

Notes

The Climax deposit lies just above Fremont Pass on the west slope of the Ten Mile Range, some 12 miles northeast of Leadville and about 65 miles west-southwest of Denver.

The oldest rocks in the area probably correlate with the Idaho Springs formation. The Idaho Springs is the most abundant Precambrian rock in the mine and outlines the domal structure, the center of which is the Climax stock. Some of the Idaho Springs contains small masses (up to 750,000 square feet in outcrop) of granite gneiss and pegmatite that were developed during the first stage of Precambrian folding. The next Precambrian igneous event was the intrusion of diorite (now metadiorite) that cuts granite gneiss and pegmatite but none of the three phases of the somewhat younger Silver Plume granite.

Although in general the area in which the Climax mine is situated was the site of extensive sedimentation in the Paleozoic, the Climax footwall block on the east side of the normal Mosquito fault (north-northeast strike, 70°W dip) has been eroded essentially clear of Paleozoic sediments.

The next recorded geologic event was the intrusion of a variety of igneous rocks that have been classed as early Tertiary, largely because of their similarity in composition and occurrence to the certainly early Tertiary rocks of the Front Range mineral belt. The first of the Tertiary Climax igneous rocks consisted of steeply dipping dikes and flat to gently dipping sills of quartz-monzonite porphyry that is generally well foliated as a result of the later intrusion of the various rocks of the Climax stock. The central and older core of the Climax stock is composed of porphyritic granite, and the surrounding and younger shell is mineralogically similar to the core and is known as the Climax porphyry. It has been suggested recently (Wallace and others, 1960) that the core was intruded first and formed a chilled margin against the enclosing Precambrian rocks. The shell was then intruded along the core-Precambrian contact and developed chilled contacts in turn on its upper and lower surfaces. Thus, the inner surface of the shell and the outer surface of the core show chilled contacts, the one against the other. The next igneous material entering the area was the intra-mineral rhyolite that generally occupies an imperfect pattern of radial fractures probably produced by the intrusion of the Climax stock. These dikes are confined to the upper portion of the stock, a situation for which there is no completely satisfactory explanation. In the lower (and younger) ore body, these dikes are mineralized in MoS_2 and in the upper (and older) ore body they are essentially barren; this argues convincingly that the dikes came in during the mineralization or between two stages of mineralization. As the dikes are mineralogically similar to the shell porphyry, it would appear that the stock and dike intrusion and the MoS_2 mineralization occurred over a geologically short period, and the probability that they were all genetically related appears to be good. The last intrusive rock was the late rhyolite porphyry that occurs in a system of dikes.

Accepting the reasoning that the post-Precambrian rocks are, by analogy, of the same age (earliest Tertiary) as those in the mineral belt of the Front Range, it follows that the mineralization that occurred concurrently with the introduction of the igneous rocks was early Tertiary as well. The Climax ores are, therefore, here classified as late Mesozoic to early Tertiary.

The first event in the hydrothermal mineralization at Climax was the beginning of the deposition of quartz; quartz is the most abundant hydrothermal mineral at Climax and continued to form throughout the mineralization process, and five stages have been recognized. The next most abundant of the hydrothermal minerals was potash feldspar that was developed in three principal forms. Most of the potash feldspar was introduced before the molybdenite mineralization, although orthoclase is common in the quartz-molybdenite veinlets where it may have been contemporaneous with the other minerals or may have been relict from an earlier formation.

Topaz is found in all vein types, but it is present in the quartz-molybdenite veinlets only in minor amounts and is much more abundant in the quartz-pyrite-huebnerite veinlets. Some veinlets containing little but topaz are known to cut the earlier molybdenite- and huebnerite-containing veinlets. Sericite is common and is locally abundant; it is most frequently associated with quartz-topaz-pyrite veinlets. Kaolinite and montmorillonite are confined essentially to the late, post-molybdenite rhyolite dikes. Fluorite is abundant as disseminated grains in the ore zone and is found in all veinlet types.

Molybdenite was the first of the sulfide minerals to form; it is essentially confined to veinlets with quartz that range from 1/16 to 1/4 inches in width. For all practical purposes, the deposition of molybdenite ceased before that of huebnerite began, even though traces of molybdenite are found in the quartz-pyrite-huebnerite veinlets. Molybdenite deposition appears to have continued through the period of the intrusion of the intra-mineral dikes that cut ore in the upper ore body but are well mineralized in the lower, suggesting that mineralization began in the upper levels of the present ore bodies and gradually worked its way downward as conditions changed.

The fractures that contain the quartz-pyrite-topaz-huebnerite mineralization cut those containing the molybdenite but apparently were formed by similar earth forces. Essentially all the pyrite in the ore bodies (2 to 3 percent of the mill feed) was introduced with the huebnerite. The minor amounts of cassiterite in the Climax ore, amounts too small to be detected by assay, are thought to be of the same general age as the huebnerite, but this is far from certain. A few veins of rhodochrosite and fluorite, with appreciable chalcopyrite, sphalerite, and galena, are found both in the ore zone and in fresh Precambrian rocks; their relationship to the molybdenite ore is not definitely known, but they may be the last phase of the early Tertiary mineralization.

The molybdenite ore occurs in two stockwork ore bodies that lie one above the other; they are circular to ring-shaped in plan and arcuate in section. These circular structures are eccentric to each other and coalesce in the eastern part of the mine. Beneath each ore body is a large zone composed almost entirely of quartz that appears to have formed by the replacement of portions of what probably was originally a much larger volume of ore-bearing rock than now remains. In detail the patterns of the two ore zones are quite complex, and they occur in essentially all the rock types present at the time of molybdenite mineralization. The huebnerite mineralization is also found in two distinct zones; the inner tungsten zone lies largely between the upper and lower molybdenite bodies, but where the two ore bodies merge, the inner tungsten body is found in the same rock volume as the earlier molybdenite. The outer tungsten zone coincides with the northwest portion of the upper ore body, but to the south it accompanies the hanging wall of this ore zone and on the southwest side has moved well outside this molybdenite ore body; some additional spots of tungsten mineralization are known just outside the hanging wall of the upper ore body.

The Climax ore bodies, both molybdenite and huebnerite, are mainly associated with ubiquitous minerals, quartz with the molybdenite and quartz and pyrite with the huebnerite. The molybdenite was preceded, accompanied, and followed by quartz deposition, and this mineral gave no indication of the conditions under which the molybdenite actually was deposited. The huebnerite, on the other hand, was accompanied by topaz, in addition to quartz and pyrite, and the combination of huebnerite and topaz suggest strongly that the tungsten mineralization took place under hypothermal conditions despite the close association of sericite with huebnerite veinlets. Certainly the formation of quartz and feldspar, which preceded the quartz-molybdenite deposition, indicates that that portion of the mineralizing process at Climax also occurred under hypothermal conditions. Because the molybdenite was deposited between two almost certainly hypothermal stages, it would appear probable that the molybdenite was also formed in this intensity range. Thus, although the deposition may well have taken place in the less intense portion of that category of mineralization, it appears necessary to classify the Climax ores as hypothermal in noncalcareous rocks (hypothermal-1).

CRIPPLE CREEK

Middle Tertiary *Gold, Silver* *Epithermal*

Gott, G. B., Jr., and others, 1967, Distribution of gold, tellurium, silver, and mercury in part of the Cripple Creek district, Colorado: U.S. Geol. Surv. Circ. 543, 9 p.

—— 1969, Distribution of gold and other metals in the Cripple Creek district, Colorado: U.S. Geol. Surv. Prof. Paper 625-A, p. A1-A17

Koschmann, A. H., 1941, New light on the geology of the Cripple Creek district, Colorado, and its practical significance: Colo. Min. Assoc., Denver, Colo., (Jan. 25) 28 p.

—— 1947, The Cripple Creek district, Teller County, in Vanderwilt, J. W., Editor, *Mineral resources of Colorado*: State of Colo., Mineral Res. Bd., p. 387-395

—— 1949, Structural control of the gold deposits of the Cripple Creek district, Teller County, Colorado: U.S. Geol. Surv. Bull. 955-B, p. 19-60

—— 1960, Cripple Creek district: Geol. Soc. Amer. Guidebook for Field Trips (Guide to the geology of Colorado), Field Trip A-3, p. 185-187

Lindgren, W., and Ransome, F. L., 1906, Geology and gold deposits of the Cripple Creek district, Colorado: U.S. Geol. Surv. Prof. Paper 54, 516 p. (particularly p. 18-113, 153-232)

Loughlin, G. F., 1927, Ore at deep levels in Cripple Creek district, Colorado:

A.I.M.E. Tr., v. 75, p. 42-73

—— 1933, Cripple Creek mining district, in *Colorado*: 16th Int. Geol. Cong., Guidebook 19, p. 113-122

Loughlin, G. F., and Koschmann, A. H., 1935, Geology and ore deposits of the Cripple Creek district, Colorado: Colo. Sci. Soc. Pr., v. 13, p. 217-435

Loughlin, G. F., and others, 1940, Paragenetic study of hypogene gold and silver telluride ores of Cripple Creek, Colo. (abs.): Pan. Amer. Geol., v. 74, no. 1, p. 36-37

Lovering, T. S., and Goddard, E. N., 1950, Cripple Creek district, in *Geology and ore deposits of the Front Range, Colorado*: U.S. Geol. Surv. Prof. Paper 223, p. 289-312

Rickard, T. A., 1900, The telluride ores of Cripple Creek and Kalgoorlie: A.I.M.E. Tr., v. 30, p. 708-718

Notes

The Cripple Creek area lies some 20 miles southwest of Colorado Springs between the towns of Cripple Creek and Victor.

The oldest rocks in the district are variously sized masses of schists and gneisses that lie largely west of the mine area and probably are to be correlated with the Idaho Springs formation. Small bodies of these ancient rocks are also enclosed in the Pikes Peak granite that surrounds the mineralized area on the north, east, and south. This granite generally is gneissic in character near its contacts with the schists; normally, however, the granite is a typical coarse-grained rock containing only a sparse development of dark silicates. The Pikes Peak granite is, in turn, intruded by the Cripple Creek granite that probably can be correlated with the Silver Plume granite of the northern Front Range. The Cripple Creek granite is the dominant rock southwest, west, and northwest of Cripple Creek, and an island of this granite is found in the mineralized area. There is a stocklike mass of olivine-syenite composed of several facies and lying between the Pikes Peak and Cripple Creek granites; it is younger than the Pikes Peak granite. These rocks (all of which are Precambrian in age) were once covered by Paleozoic and Mesozoic sediments, but these have been removed by erosion, leaving a gently undulating surface of Precambrian rocks around the mineralized area.

The oldest Tertiary rocks in the district are rhyolites that are younger than the late Oligocene or early Miocene Florissant Lake beds that lie to the west of the town of Cripple Creek. The rhyolite was followed by an eruption of andesite, of which only two small remnants remain in the area. In valleys and local basins on the extrusive surface, bodies of sand and gravel were accumulated that consolidated into the High Park beds; some minor extrusions of the rhyolite took place during the development of these beds. Only minor traces of these conglomerate beds are to be found today.

All the rocks thus far mentioned had no direct connection with the formation and emplacement of the Cripple Creek ores. Essentially all the gold-silver bearing veins of the area lie within a highly eroded, composite, volcanic crater that occupies a generally northwest-trending elliptical area about 4 miles long and 2 miles wide between Victor (south) and Cripple Creek (northwest). A few minable veins are found in the granite immediately adjacent to the crater. The crater is cut into three fractions of unequal size (the largest one to the south) by a continuous rib of Precambrian rocks running generally northwest and from which rise two islands (one granite and one schist) that emphasize the division of the crater into three parts. The walls of the crater in general dip steeply but irregularly inward. There are also three minor and essentially unmineralized subcraters northwest of the volcano and one to the southwest.

These craters are filled with both volcanic and nonvolcanic materials, a considerable fraction of which is well-bedded and sorted rocks that almost certainly were deposited as shallow-water sediments. Accumulation of sediments in this manner indicates strongly that the crater sank gradually, space for sediments being developed at about the same rate as the rate at which they were being supplied to the basin. The first sediments to form were essentially nonvolcanic in origin, but the later mechanically transported material was largely consolidated as latite-phonolite and phonolite breccia and tuff; stratification was equally well developed in nonvolcanic and volcanic rocks. The lack of volcanic necks or vents from which the igneous fragments could have been explosively expelled suggests that the volcanic material was ejected quietly through fissures that cut through the earlier sediments. The only true explosive vent in the crater appears to be the Cresson "blowout," a small crater of volcanic breccia in the central part of the southern half of the crater which was the last event in the igneous cycle that followed the fissure eruptions. After the great bulk of the surface-transported material had been deposited in the crater, some 12 stages of igneous activity introduced a variety of igneous materials into the area as dikes and irregular masses, apparently in the following order: (1) two varieties of latite phonolite, (2) two types of syenite, (3) two of phonolite, (4) one of trachydolerite, (5) four of lamprophyres and related rocks, and (6) one of basalt. During these intrusive cycles, portions of the area were broken by minor explosions that extensively fractured the igneous rocks and obscured their boundaries and structural relationships.

The outer form of the crater appears to have been controlled by structures developed in the area during the Laramide orogeny. At this time a broad bulging dome formed over Cripple Creek and much of the surrounding area, and an extensive fracture pattern was developed. This pattern consisted of (1) compression fracture zones of northwest and northeast trends that were, respectively, parallel and oblique to the major northwest axis of the dome; (2) fracture zones of north-northeast trend (parallel to the minor fold axes); and (3) east-southeast zones (perpendicular to the minor fold axes). These fracture zones, in turn, probably were initially controlled by similar zones that had existed in the basement rocks since Precambrian time; movement along these fractures was most impressive during the accumulation of the fragmental materials, but appreciable movement took place on them coincident with the introduction of the later mineralizing solutions.

The oldest Tertiary rock in the Cripple Creek district is the rhyolite which originally extended as far west as the Florissant Lake area and there overlies the late Oligocene to early Miocene beds. This same rhyolite is interbedded with the High Park conglomerate that is, therefore, also probably of early Miocene age. The formation of the crater began shortly after the extrusion of the rhyolite and the deposition of the conglomerate and, on these grounds, can be dated as Miocene. Miocene plant remains found in the bedded material in the crater were formed as part of the same series of geologic events that produced the crater; it appears reasonable, therefore, to date the ore deposition as Miocene or middle Tertiary.

The ore deposits at Cripple Creek occur in (1) veins formed by fissure filling, (2) irregular masses in shattered rock masses, and (3) pipelike bodies in collapse breccias. The actual ore minerals appear to have been derived from the same general source as the igneous rocks and to have been developed in three stages: (1) The first stage was characterized initially by intense corrosion of the country rock over a considerable portion of the total crater volume and then by the deposition (largely by open-space filling) of quartz, adularia, massive dark-purple fluorite, and rather coarse-grained pyrite; many channelways were blocked by the deposition of the first-stage minerals, so not all corroded rock volumes were later reached by ore solutions depositing the second and third mineralizations. (2) The second stage minerals include most of those of the first, although adularia is lacking and the quartz is smokey to milky, the fluorite is lighter purple, and the pyrite is fine-grained and inconspicuous; additional minerals are dolomite, ankerite, or celestite, and the gold and silver tellurides that provide the valuable metals of the deposits; the most common telluride is calaverite ($AuTe_2$), but sylvanite [$(Ag,Au)Te_2$] and krennerite ($AuTe_2$) also are important; tellurides present in minor amounts include hessite, petzite, and a silver-copper telluride; some roscoelite (the vanadium mica), a little barite, and small amounts of base-metal sulfides also are found in this stage. Again, deposition was almost entirely by open-space filling. (3) The third stage consists of insignificant amounts of smokey to colorless quartz, chalcedony, fine-grained pyrite, calcite, and even less cinnabar and fluorite.

The highly broken character of the rocks of the crater and the near coincidence of mineralization and volcanism strongly suggest that the ores were formed near the surface under conditions of rapid loss of heat and pressure. This is further confirmed by the narrowing of fissures and their decrease in number with depth by the lack of persistence of individual fractures. Nevertheless, the character of the mineralization shows little or no change over the 3300 vertical feet over which mining workings have been opened.

The combination of typically low-temperature minerals and of near-surface conditions of ore deposition confirms the long-standing assignment of this deposit to the epithermal category.

FREELAND-CHICAGO CREEK

Late Mesozoic to Early Tertiary	*Gold, Silver, Copper, Lead, Zinc*	*Mesothermal to Leptothermal*

Ball, S. H., 1906, Pre-Cambrian rocks of the Georgetown quadrangle, Colorado: Amer. Jour. Sci., 4th Ser., v. 21, p. 371-389

Harrison, J. E., 1955, Relation between fracture pattern and hypogene zoning in the Freeland-Lamartine district, Colorado: Econ. Geol., v. 50, p. 311-320

Harrison, J. E., and Wells, J. D., 1956, Geology and ore deposits of the Freeland-Lamartine district, Clear Creek County, Colorado: U.S. Geol. Surv. Bull. 1032-B, p. 33-127

—— 1959, Geology and ore deposits of the Chicago Creek area, Clear Creek County, Colorado: U.S. Geol. Surv. Prof. Paper 319, 42 p.

Lovering, T. S., and Goddard, E. N., 1950, Freeland-Lamartine area, in *Geology and ore deposits of the Front Range, Colorado*: U.S. Geol. Surv. Prof. Paper 223, p. 191-193

Sims, P. K., and others, 1963, Geology of the uranium and associated ore deposits, central part of the Front Range mineral belt, Colorado: U.S. Geol. Surv. Prof. Paper 371, 119 p. (particularly p. 110-114)

Spurr, J. E., and others, 1908, Lamartine-Trail Creek group, in *Economic geology of the Georgetown quadrangle, Colorado*: U.S. Geol. Surv. Prof. Paper 63, p. 314-341

Wells, J. D., 1960, Petrography of radioactive Tertiary rocks, central part of Front Range mineral belt, Gilpin and Clear Creek Counties, Colorado: U.S. Geol. Surv. Bull. 1032-E, p. 223-272

Notes

The area discussed here as the Freeland-Chicago Creek district is composed of the contiguous Freeland-Lamartine and the Chicago Creek districts that lie west and southwest, respectively, of Idaho Springs and make up the southern part of the larger zoned area of which the Central City-Idaho Springs district constitutes the northeastern half. Both the Freeland and the Chicago Creek subdivisions of the combined district contain small areas in which the same mineralization types are found as make up the much larger central zone of the Central City-Idaho Springs district. Similarly in this district both the intermediate and peripheral mineralization types also are present.

The oldest rocks in the area, as is true in the Central City-Idaho Springs district, are a series of metasediments that probably correlate with the Precambrian Idaho Springs formation in the sense that is used by Lovering and

Goddard. Truly igneous rocks are known that are younger than all of the metasediments and include granodiorite (probably the Boulder Creek granite), quartz diorite and associated hornblendite, biotite-muscovite granite (probably the Silver Plume granite), and granite pegmatite.

Of the 14 members of the early Tertiary group of porphyritic intrusives, dikes and small, irregular plutons, recognized in the entire Central City zoned area, only eight are actually mapped in the district here under discussion. The intrusion of all of these rocks appears to have occurred during the Laramide orogeny, since the youngest of the porphyries has been cut by post-mineral faults.

The mineralized veins in the Freeland-Chicago Creek district bear so close a spatial relationship to the porphyry dikes and plutons that they are considered to have been developed at about the same time as the dikes and from the same general source magmas. The dikes show a considerable amount of variation in composition between the oldest and the youngest that probably is due to differentiation in the source magma chambers; the variation in the vein fillings well may be due to the same process in the same locations, differentiation in the magma appreciably affecting the material transferred to the water-rich fluid generated during the process. Although no dates have been obtained by radioactive determinations in this district, such ages obtained in the Central City-Idaho Springs district immediately to the north (70 to 57 m.y.) strongly suggest that the deposits were formed during the early Tertiary portion of the Laramide orogeny and they are, therefore, here classified as late Mesozoic to early Tertiary.

The veins in the two portions of the district can be fitted into the classification given in the discussion of the Central City-Idaho Springs (CC-IS) area although there are differences both in the Freeland area and that of Chicago Creek from the standard division given by Sims (1963, p. 23). In the Chicago Creek area, Harrison and Wells (1959, p. 39) divided the veins into five types (plus a telluride-bearing subtype) instead of the four given by Sims and his colleagues. The principal difference between the two classifications, however, results from the unusual character of one vein in the Chicago Creek area (the Kitty Clyde) that could be classed as a variant of type (3) in the CC-IS classification. In the Freeland veins, on the other hand, three types of mineralization were recognized by Harrison and Wells (1956, p. 74-75), the first type includes both types (1) and (2) of the CC-IS classification, and the composite type, intermediate between the two is essentially type (3) of the CC-IS classification plus a fraction of the CC-IS galena-sphalerite category.

In the Chicago Creek area the earliest [type (1)] mineralization was the introduction of quartz and pyrite (some of it gold-bearing); in type (2), quartz and pyrite continued to deposit in large amounts and moderate quantities of chalcopyrite and minor amounts of free gold, copper sulfosalts, galena, sphalerite, and silver sulfosalts also were brought in. In one vein of type (1) (West Gold), the quartz-pyrite, auriferous-pyrite mineralization was followed by minor amounts of gold and silver tellurides and more free gold; this one vein has been denoted as type 1A by Harrison and Wells. The type (3) mineralization in the Chicago Creek area is much like that in the CC-IS district except that the amounts of probably primary silver sulfosalts in the Chicago Creek mines appear to have been greater than those in the CC-IS district. This type (3) mineralization consists of similar amounts of pyrite (first to deposit), sphalerite (next to form and containing blebs of chalcopyrite), and galena (last to form). Between the beginning of sphalerite deposition and that of galena, copper sulfides (mainly chalcopyrite), copper sulfosalts, and silver sulfosalts were deposited in amounts that were appreciable but less than in type (2); gold was formed to a much lesser extent than in types (1) and (2). The last galena-sphalerite type (4) mineralization is composed almost entirely of those two minerals; the galena is quite silver-rich. In the Chicago Creek area veins, the sequence of mineralization can readily be determined from the differences between the minerals developed along the vein walls and those deposited more centrally and from the differences between the minerals in the breccia fragments (many veins have been fractured more than once during the depositional sequence) and the matrix minerals cementing them. These changes in mineral character with time would indicate a steady variation in the composition of the ore-forming fluid.

In the Freeland area, the earliest mineralization corresponds to types (1) and (2) in the CC-IS district and in the Chicago Creek area; the differentiation later made between quartz-pyrite-gold ore and quartz-pyrite-gold-copper ore in the other two areas was developed after publication of the Freeland study. The zoning in the Freeland area is different from that in the other two areas in that type (3) ore is never found centrally placed in, or cutting, types (1) and (2) ore; instead, type (3) is so zoned in the individual veins that it occurs transitionally between the central vein area that contains the early ore types and the far extremities of the same vein composed of the galena-sphalerite type. In short, type (3) in the Freeland area is developed by the deposition of the sphalerite-galena ore in portions of the veins already containing types (1) and (2) ore. To some extent, this is also the explanation of the type (3) ore in the other parts of the zoned area, but in those areas type (3) ore could be developed in veins containing no ore of types (1) and (2).

Some emphasis is placed, in the discussions of the Freeland and Chicago Creek ores, on the presence of two types of sphalerite, an earlier dark brown to black zinc sulfide that contains blebs of chalcopyrite and a later red-brown mineral that appears to be free of the copper-iron sulfide. Because the blebs are not illustrated, it is difficult to determine whether or not they reasonably can be considered the result of exsolution. The lack of any other evidence, such as high-temperature metallic, gangue, or wall-rock alteration minerals, suggests that the chalcopyrite in the sphalerite may have been emplaced metasomatically and not by exsolution. Probably the early copper-gold-zinc mineralization seems to have been formed within the mesothermal range, while the later silver sulfosalts and argentiferous galena suggest deposition under leptothermal conditions. It seems, therefore, that these deposits, as is true of those of the CC-IS district, should be classified as mesothermal to leptothermal.

GILMAN

Late Mesozoic to Early Tertiary	*Copper, Zinc, Lead, Gold*	*Hypothermal-2 to Leptothermal*

Borcherdt, W. O., 1931, The Empire Zinc Company's operation at Gilman, Colorado: Eng. and Min. Jour., v. 132, nos. 3, 6, p. 99-105, 251-261 (particularly p. 99-100)

Crawford, R. D., and Gibson, R., 1925, Geology and ore deposits of the Red Cliff district, Colorado: Colo. Geol. Surv. Bull. 30, 89 p.

Lovering, T. G., 1958, Temperatures and depth of formation of the sulfide ore deposits at Gilman, Colorado: Econ. Geol., v. 53, p. 689-707

Lovering, T. S., and Behre, C. H., Jr., 1933, Battle Mountain (Red Cliff, Gilman) mining district, in *Colorado*: 16th Int. Geol. Cong., Guidebook 19, p. 69-76

McDougall, D. J., 1959, Temperatures and depths of formation of sulphide ore deposits at Gilman, Colorado: Econ. Geol., v. 54, p. 140-141

Means, A. H., 1915, Geology and ore deposits of Red Cliff, Colorado: Econ. Geol., v. 10, p. 1-27

Pearson, R. C., and others, 1962, Age of Laramide porphyries near Leadville, Colorado: U.S. Geol. Surv. Prof. Paper 450-C, p. C78-C80

Radabaugh, R. E., 1953, Geology and ore occurrence (Gilman, Colo.): Min. Eng., v. 5, no. 12, p. 1223-1224

Radabaugh, R. E., and others, 1968, Geology and ore deposits of the Gilman (Red Cliff, Battle Mountain) district, Eagle County, Colorado, in Ridge, J. D., Editor, *Ore deposits of the United States, 1933-1967* (Graton-Sales

Volumes): Chap. 30, v. 1

Roach, C. H., 1960, Thermoluminescence and porosity of host rocks at the Eagle Mine, Gilman, Colorado: U.S. Geol. Surv. Prof. Paper 400-B, p. B107-B111

Silverman, A. (J.), 1965, Studies of base metal diffusion in experimental and natural systems; Part II, Diffusion sytems at Gilman, Colorado: Econ. Geol., v. 60, p. 325-350

Tweto, O., and Lovering, T. S., 1947, The Gilman district, Eagle County (Colo.), in Vanderwilt, J. W., Editor, *Mineral resources of Colorado*: State of Colo., Mineral Res. Bd., p. 378-387

Wehrenberg, J. P., and Silverman, A. (J.), 1965, Studies of base metal diffusion in experimental and natural systems: Econ. Geol., v. 60, p. 317-350

Notes

The Gilman district lies about 20 miles north-northwest of the town of Leadville and about 80 miles west of Denver.

The ores of the Gilman mines are largely confined to the Cambrian to Mississippian portion of the Paleozoic sequence, although there are some mineralized veins in the underlying Precambrian granite, schist, and minor gneissic diorite. The mineralized Paleozoic section is composed of some 600 feet of a variety of sedimentary rocks that, from bottom to top, are the Cambrian Sawatch quartzite, made up of the Quartzite member and of the Peerless shale member, the Ordovician Harding sandstone; the Devonian Chaffee formation, made up of the Parting quartzite member and the Dyer dolomite member; and the Mississippian Leadville limestone that includes the basal Gilman sandstone member. Above the Leadville lies at least 4500 feet of the Pennsylvanian to Permian Maroon formation that contains essentially no ore. The only known igneous rock of post-Precambrian age in the area is a continuous sill of quartz-latite porphyry that was intruded a few feet above the top of the Leadville limestone.

No direct evidence connects the Gilman ores with any known body of intrusive rocks. The emplacement of the ore, however, was controlled by Laramide fractures, and the only igneous rock is the quartz-latite sill of Laramide age. It is, therefore, thought reasonable to class the ores as late Mesozoic to early Tertiary, particularly as this appears to be the age of the generally similar ore mineralization in the neighboring Leadville district.

There are three types of minable veins at Gilman: (1) narrow fissure veins that occur mainly in Precambrian rocks and only rarely persist upward into the Sawatch quartzite; (2) replacement deposits in the Sawatch quartzite; and (3) replacement deposits in the carbonate rocks, mainly the Leadville limestone and the Dyer dolomite.

Although many of the fissure veins contained base-metal sulfides, normally they were of economic value only for the precious metals they contained, and these fissure veins have accounted for only a small fraction of the total dollar value of the production of the district.

The replacement deposits in quartzite were essentially manto or bedding-vein deposits in the Rocky Point breccia zone (which lies in the quartzite member of the Sawatch about 180 feet above its base). The first stage in the formation of these deposits was extensive solution of the quartzite through which all types of openings in the broken quartzite were enlarged. In these, pyrite was deposited as the first hydrothermal mineral, and it also slightly replaced the quartzite as disseminated grains; this pyrite contained very little of the base or precious metals. The pyrite and its associated quartzite were again brecciated and fractured, and then, in this newly broken rock and in the remaining older channelways, manganosiderite, pyrite, chalcopyrite, iron-rich sphalerite, and barite were deposited both as open-space fillings and as replacements of pyritized quartzite within a much more limited and well-defined area than that occupied by the early pyrite. Very little production has been taken from the unoxidized portions of these mantos; most of it came from the oxidized zone.

The replacement deposits in the carbonate rocks occur either as chimneys or mantos, the locations of which were largely determined by channelways and caves dissolved in the carbonate rocks by early, but barren, hydrothermal solutions that were introduced along rather weak joints. Even before the advent of these barren solutions, the Leadville limestone was widely dolomitized by hydrothermal solutions; following dolomitization, the altered Leadville beds and, to a lesser extent, the Dyer dolomite were converted to alternating bands of dark, fine-grained dolomite and white, coarsely crystalline and vuggy dolomite to produce what is locally known as zebra rock. This rock, when attacked by the still-later barren solutions, was extensively dissolved, the first stage of which was a loosening of the individual dolomite grains to produce a gamut of materials from friable dolomite to free-running sand. With a continuation of this attack, enough dolomite was removed and sufficiently large channelways and caves were developed to provide sites for stratified deposition of much of the sand to occur. Blocks of dolomite in the walls and backs of the open spaces were loosened and fell into the channelways as loosely packed rubble; such caves as reached the top of the Leadville formation also were provided with rubble from the weak cap rocks. Some channels made contact with Pre-Pennsylvanian ground-water courses thereby increasing the length of connected channelways appreciably. Many of the channelways ended down-dip in chimney-like holes that on occasion descended into the Parting quartzite.

When hydrothermal solutions capable of depositing ores entered the carbonate rocks of the Gilman area, they were largely directed into these chimneys and the channelways into which they connected and there deposited their loads as replacements of sand, rubble, and adjacent bedrock and as open-space fillings.

No break in continuity occurs between the ore in chimneys and that in mantos, but the mineral characteristics are quite different. A chimney usually consists of a core of pyrite surrounded by a discontinuous inner shell of sphalerite and a nearly continuous outer shell of manganosiderite. Within the pyrite, later solutions eratically distributed chalcopyrite, minor galena, with inclusions of hessite and petzite, and still smaller amounts of tetrahedrite, stromeyerite, and a few lead and silver sulfosalts and a little native gold. The chimneys are raggedly round in outline, being as much as 300 feet in diameter at their upper ends. They reach from the top of the Leadville down to or into the Parting quartzite where they degenerate into pyrite-coated joint surfaces.

The manto ores are composed of iron-rich sphalerite, pyrite, manganosiderite, minor galena, accessory chalcopyrite, barite, dolomite, and quartz. Some of the mantos, at their lower ends, contain barren cores of pyrite that are the upward continuations of the pyrite cores of the chimney ore bodies. The manto ore, in contrast to that of the chimneys, is low in silver and gold. The mantos are confined to the Leadville limestone.

At Gilman, much of the more iron-rich sphalerite contains randomly oriented blebs of chalcopyrite while that lower in iron shows no exsolved chalcopyrite. This would indicate that part of the sphalerite was deposited in the hypothermal range, but that some of it probably was formed under mesothermal conditions. That none of the sphalerite formed at higher intensities than those of the lower portion of the hypothermal category is suggested by the apparent lack of exsolved blebs of sphalerite in the chalcopyrite. Although tetrahedrite is present in moderate amounts in the ore, none of it is present as exsolved blebs in chalcopyrite. On the other hand, the inclusions of anisotropic hessite in the late galena indicate that both the telluride and the galena were probably deposited below 150°C. It would appear, then, that part of the Gilman ore formed in the lower intensity portion of the hypothermal range but that a second portion of it was formed under mesothermal conditions. The tetrahedrite, silver-rich galena, and the tellurides probably are of sufficient economic importance to justify adding a leptothermal phase to the Gilman classification; therefore, the ores are here categorized as hypothermal-2 to leptothermal.

LA PLATA

Late Tertiary *Gold, Silver* *Leptothermal*

Atwood, W. W., and Mather, K. F., 1923, Physiography and Quaternary geology of the San Juan Mountains, Colo.: U.S. Geol. Surv. Prof. Paper 166, 176 p.

Cross, W., and others, 1894, La Plata folio, Colorado: U.S. Geol. Surv. Geol. Atlas, Folio 60, 14 p.

Eckel, E. B., 1936, Resurvey of the geology and ore deposits of the La Plata mining district, Colorado: Colo. Sci. Soc. Pr., v. 13, no. 9, p. 507-547

—— 1938, Copper ores of the La Plata district and their platinum content: Colo. Sci. Soc. Pr., v. 13, no. 12, p. 647-664

—— 1947, La Plata district, La Plata and Montezuma Counties, in Vanderwilt, J. W., Editor, *Mineral resources of Colorado*: State of Colo., Mineral Res. Bd., p. 416-419

Eckel, E. B., and others, 1949, Geology and ore deposits of the La Plata district, Colorado: U.S. Geol. Surv. Prof. Paper 219, 179 p. (particularly p. 6-84)

Galbraith, F. W., 1941, Ore minerals of the La Plata Mountains, Colorado, compared with other telluride districts: Econ. Geol., v. 36, p. 324-334

Wisser, E., 1960, La Plata, Colorado, in *Relation of ore deposition to doming in the North American Cordillera*: Geol. Soc. Amer. Mem. 77, p. 36-40

Notes

The La Plata mining district is situated in southwestern Colorado between the San Juan Mountains and the Colorado Plateau, its center being some 12 miles northwest of the town of Durango.

Although several hundred feet of Cambrian, Devonian, and Mississippian and about 3000 feet of Pennsylvanian rocks are known in the area, essentially no ore is found in them. Most of the ore in sedimentary rocks is confined to the Upper Jurassic Entrada sandstone, the Upper Jurassic Pony Express member (limestone) of the Wanakah formation, and the Junction Creek sandstone; somewhat lesser amounts occur in two Permian formations, the Cutler (arkosic sandstones, conglomerates, limy shales, and mudstones) and the Rico (shale, sandstone, and thin beds of sandy limestone). Considerable thicknesses of younger Upper Cretaceous and Tertiary sediments overlay the Dakota sandstone in the district, but they have all been removed by erosion. The igneous rocks in the La Plata district are all intrusive and of considerably different compositions (though mainly diorite to monzonite), are both porphyritic and nonporphyritic, and occur as stocks, sills, and dikes. Much of the telluride ore is in diorite-monzonite, and no ore deposits are found in syenite, syenite porphyry, and monzonite. The nonporphyritic rocks are normally younger than the porphyries; the latter were intruded between the sedimentary beds and were thereby largely responsible for the doming of the area, the dome later having been eroded into the present La Plata mountains. The nonporphyritic rocks cut the sediments and the porphyritic sills with negligible disturbance of the attitude of the invaded beds. The nonporphyritic magmas attacked their wall rocks during intrusion, and the resulting stocks are generally surrounded by contact-metamorphic aureoles; the porphyritic magmas did not attack their hosts, and stocks derived from them have no contact aureoles.

The similarity of the La Plata igneous rocks in form and composition to those intruded into the San Juan Mountain area during late Mesozoic to early Tertiary time suggests a similar age for the La Plata rocks. On the other hand, it is at least equally possible that the La Plata intrusions occurred at the same time as the main San Juan volcanism (which was post-Potosi volcanic series; the fossil content of the Potosi demonstrates its Miocene age). Pebbles of La Plata Mountains porphyry have been found in gravels on the San

Juan peneplain surface, so the porphyries were emplaced before the peneplanation that occurred in the Pliocene. The ore mineralization probably took place shortly after the end of igneous activity in the strict sense and was, therefore, either late Miocene or early Pliocene. It appears that the arguments in favor of a late Tertiary age for the ore mineralization are better than those for an earlier one and that the ores should be classified as late Tertiary with the qualification that age determinations on the igneous rocks of the area may demonstrate that the mineralization took place in the late Mesozoic to early Tertiary.

Although the La Plata area contains a wide variety of ore-deposit types, the only one of any real economic importance is that of telluride veins and replacement masses; of all the mines in the district that yielded at least $100,000 worth of ore, only one was certainly not of this category. Telluride ore bodies are known throughout the La Plata Mountains, but they are largely concentrated in three rather small areas: (1) in the northeast part of the central mountain mass near Lewis Mountain, (2) south of Madden Peak and Deadwood Mountain, and (3) in the north-central part of the district, centering around Diorite Peak. The only deposits within the central metamorphic area of the La Plata dome are essentially all in one or the other of the two diorite stocks that have not been silicified as have most of the rocks of the metamorphic area. Two of the three main groups of deposits lie along the hinge fold that surrounds the central metamorphic core, while the third is found associated with the belt of east-west faults that take the place of the hinge fold across the south margin of the La Plata dome.

The telluride veins in general consist of discontinuous veins and veinlets that run through the breccia and sheared rock of the fault zones; some of the veins are well-defined but lenticular, whereas most of them are composed of interlaced small veins and veinlets. Much of the ore was emplaced by filling open space in broken rock, but other ore fills actual fractures and, to some extent, replaces the bordering country rock. Only rarely do veins completely fill the zone of broken rock in which they are found.

Quartz is the principal vein mineral and was the first to begin to deposit; it was followed by barite, ankerite, and calcite, all of which ceased to deposit before quartz precipitation had stopped. Long before the carbonates had ceased to deposit, sulfide precipitation began with pyrite (two generations), arsenopyrite (widespread but not abundant), and locally gold disseminated in some of the second-generation pyrite. Next in the sequence were sphalerite, chalcopyrite, tetrahedrite, and galena. Except for some galena, these four minerals had ceased to deposit by the time telluride deposition began. Native gold and a little cinnabar followed the tellurides and were the last primary minerals to deposit. Tetrahedrite (which contains some arsenic) is abundant in most of the telluride ores; it does not appear to contain any appreciable amount of silver. Galena followed tetrahedrite and is also low in silver. Argentite and several silver sulfosalts have been identified in La Plata ores, but these minerals are so rare as to be of no economic importance. The late gold was emplaced mainly by the replacement of tetrahedrite and the various tellurides. Hessite (Ag_2Te) is the most abundant telluride, which accounts for the dominance of silver over gold in the La Plata ores; coloradoite (HgTe) is the telluride next in abundance after hessite. Of the minor tellurides, petzite is the most common, followed by sylvanite, krennerite, and calavarite.

The replacement bodies are confined to the Pony Express limestone member of the Wanakah formation and extend out into that rock from producing veins that cut through it. In these replacements the ore minerals normally decrease uniformly away from the veins and are rarely economically valuable more than 50 feet from the veins. These bodies consist mainly of calcite, ankerite, some barite, and a little quartz, with tellurides and native gold generally scattered through the carbonate gangue although occasionally large masses of nearly solid tellurides are found. Minor amounts of base-metal sulfides are associated with the tellurides and native gold.

The tellurides and the late native gold almost certainly were deposited under leptothermal conditions. At the time of the deposition of these minerals, the gangue minerals being precipitated were minor quartz and ankerite, minerals not incompatible with this intensity range. Although the deposits contain characteristic minerals of epithermal deposits, the depth of burial of the ore zone before the development of San Juan peneplain was great enough almost certainly to insure that the ores were formed under conditions of slow loss of heat and pressure. The only mineralogical question in the classification of the deposits is the economic importance of the gold disseminated in second-generation pyrite. If it was a major factor in the worth of the ore produced in the La Plata area, it appears that the district should probably be classified as mesothermal to leptothermal. Although published works are not clear on this point, it appears that such gold in pyrite is of value only after it has been concentrated in the zone of oxidation. For this reason the term mesothermal is omitted from the classification of the La Plata ores, and they are designated as leptothermal.

LEADVILLE

Late Mesozoic to Early Tertiary	*Silver, Lead, Zinc, Copper, Gold*	*Mesothermal to Telethermal*

Banks, N. G., 1970, Nature and origin of early and late cherts in the Leadville limestone, Colorado: Geol. Soc. Amer. Bull., v. 81, no. 10, p. 3033-3048

Behre, C. H., Jr., 1929, Revision of structure and stratigraphy in the Mosquito Range and the Leadville district, Colorado: Colo. Sci. Soc. Pr., v. 12, no. 3, p. 37-57

—— 1932, The Weston Pass mining district, Lake and Park Counties, Colorado: Colo. Sci. Soc. Pr., v. 13, no. 3, p. 53-75

—— 1933, Physiographic history of the upper Arkansas and Eagle Rivers, Colorado: Jour. Geol., v. 41, p. 785-814

—— 1939, Preliminary geological report on the west slope of the Mosquito Range in the vicinity of Leadville, Colorado: Colo. Sci. Soc. Pr., v. 14, no. 2, p. 49-79 (with geological map)

—— 1953, Geology and ore deposits of the west slope of the Mosquito Range: U.S. Geol. Surv. Prof. Paper 235, 176 p. (particularly p. 18-119)

Chapman, E. P., 1941, Newly recognized features of mineral paragenesis at Leadville, Colorado: A.I.M.E. Tr., v. 144, p. 264-275

Chapman, E. P., and Stevens, R. E., 1933, Silver and bismuth-bearing galena, Leadville: Econ. Geol., v. 28, p. 678-685

Davidson, R. N., 1950, Hydrothermal alteration effects in the Leadville limestone and their relation to metalization (abs.): Geol. Soc. Amer. Bull., v. 61, p. 1551

Ebbley, N. E., Jr., and Schumacher, J. I., 1949, Examination, mapping and sampling of mine shafts and underground workings, Leadville, Lake County, Colo.: U.S. Bur. Mines R.I. 4518, 115 p. (mimeo.) (details of condition of mine workings and shafts as of 1946)

Emmons, S. F., and others, 1927, Geology and ore deposits of the Leadville mining district, Colorado: U.S. Geol. Surv. Prof. Paper 148, 368 p. (particularly p. 21-108, 145-150, 177-272)

Engel, A. E. J., and others, 1958, Variations in the isotopic composition of oxygen in the Leadville limestone (Mississippian) of Colorado as a guide to the location and origin of its mineral deposits: 20th Int. Geol. Cong., Symposium de Exploracion Geoquimica, t. 1, p. 3-20

Loughlin, G. F., 1918, The oxidized zinc ores of Leadville, Colorado: U.S. Geol. Surv. Bull. 681, 91 p.

—— 1927, Guides to ore in the Leadville district, Colorado: U.S. Geol. Surv.

Bull. 779, 37 p.

Loughlin, G. F., and Behre, C. H., Jr., 1933, Leadville mining district, in *Colorado*: 16th Int. Geol. Cong., Guidebook 19, p. 77-91

—— 1934, Zoning of ore deposits in and adjoining the Leadville district, Colorado: Econ. Geol., v. 29, p. 215-254

—— 1942, Leadville district, Colorado, in Newhouse, W. H., Editor, *Ore deposits as related to structural features*: Princeton Univ. Press, p. 203-206

—— 1947, Leadville mining district, Lake County, in Vanderwilt, J. W., Editor, *Mineral resources of Colorado*: State of Colo., Mineral Res. Bd., p. 350-370

Pearson, R. C., and others, 1962, Age of Laramide porphyries near Leadville, Colorado: U.S. Geol. Surv. Prof. Paper 450-C, p. C78-C80

Tweto, O., 1960, Pre-ore age faults at Leadville, Colorado: U.S. Geol. Surv. Prof. Paper 400-B, p. B10-B11

—— 1968, Leadville district, Colorado, in Ridge, J. D., Editor, *Ore deposits of the United States, 1933-1967* (Graton-Sales Volumes): Chap. 32, v. 1

Notes

The town of Leadville is situated on the west edge of the Leadville mining district on the west slope of the Mosquito Range, some 80 miles southwest of Denver; altitudes range between 10,000 and 13,000 feet.

The Leadville area is underlain by a Precambrian basement composed of granite, gneiss, and schist containing essentially no ore; the granite is probably to be correlated with the Silver Plume granite of the Front Range. The Paleozoic sedimentary rocks in the area include the (1) Upper Cambrian Sawatch quartzite, divided into the Sawatch quartzite (lower) and the Peerless shale member (upper); (2) Lower Ordovician Manitou dolomite (originally called the White limestone); (3) Upper Devonian Chaffee formation, divided into the Parting quartzite member (lower) and the Dyer dolomite member (upper); (4) Mississippian Leadville dolomite with the Gilman sandstone member at its base; and (5) Pennsylvanian Molas, Belden, and Minturn formations (shale and grit with minor limestone, quartzite, and coal). The Dyer and the Leadville were once grouped together as the Blue or Leadville limestone. Ore was found in all of the Paleozoic formations, although that in the Leadville was of greatest economic importance.

The igneous rocks of the area (Tweto, 1968) are now divided into nine units that are, from bottom to top: (1) the Pando porphyry; (2) the Sacramento porphyry; (3) the Evans Gulch porphyry; (4) the Johnson Gulch porphyry; (5) the Lincoln porphyry (these five are early Tertiary in age); (6) younger porphyries of at least six varieties, including the Iowa Gulch porphyry (this group appears to be mainly late Tertiary in age); (7) the Little Union quartz latite; (8) the Rhyolite porphyry; and (9) rhyolite and rhyolite explosion breccia. Number (1) was known as the Early White porphyry. Numbers (2) through (6) were formerly known as the Gray porphyry and number (8) as the Late White porphyry.

The igneous rocks of the area are intrusive into all the formations in the sedimentary column and are subdivided into the nine units on the basis both of relative time of intrusion and of petrographic character. The first igneous unit (Pando porphyry) is confined to a series of remarkably extensive sills, the most conspicuous and continuous of which is up to 1000 feet thick and is found to exist over most of the Leadville district. This sill lies in the basal Pennsylvanian rocks with only about 30 feet of grit and shales between the base of the sill and the top of the Leadville dolomite. The Early White porphyry was fractured at the same time as the ore-bearing fractures were developed in the sedimentary rocks, and some of the fractures in the igneous rock contain ore. It is, therefore, certainly of pre-ore age; the only mineralization in the Rhyolite porphyry (Late White [no. 8] porphyry) is irregularly distributed pyrite. It may, therefore, be pre-ore but its structural relations suggest that it is post-ore. The Gray porphyries also form conspicuous sills but were introduced also in large numbers as dikes and occasionally as plugs. The Gray porphyries were intruded in the same orogenic cycle as the White, so the Gray porphyries are pre-ore.

On the basis of the relations of the Early White and Gray porphyries to the fault pattern, these rocks are considered to have been intruded during the Laramide orogeny. This is confirmed by radioactive age determinations on igneous rocks from the general Leadville area that have been correlated with various members of the Early White and of the Gray porphyries. The Early White porphyry studied gives an age of 70 m.y., and each of the two Gray porphyries gives an age of 64 m.y. The Late White porphyry and the other younger igneous rocks (all of which are probably post-ore) may have been introduced appreciably later in the Tertiary. Thus, it seems almost certain that the age of the Leadville ores, the deposition of which followed closely after the emplacement of the Gray porphyries, must have been late Mesozoic to early Tertiary.

Although the bulk of the ore from the Leadville district has been mined from the area bounded by Leadville on the west, the Ball Mountain fault on the east, Evans Gulch on the north, and Iowa Gulch on the south, appreciable quantities have come from the outlying areas such as the Mount Sheridan area, as well as from the Weston Pass area about 10 miles south-southeast of Leadville.

The hypogene ores at Leadville occur in three principal ways: (1) as silicate-oxide deposits; (2) as veins of mixed sulfides, mainly in siliceous rocks; and (3) as sulfide replacement deposits in dolomite. The widespread introduction of ore-forming fluids into the Leadville area was made possible by structures developed in four periods of faulting.

That silicate-oxide deposits consist of a mixture of magnetite and hematite in a gangue now composed almost entirely of serpentine (that formerly was pyroxene or olivine or both) and manganosiderite; a little wollastonite, epidote, and sericite also are present. The ore of this type is largely restricted to dolomite in the vicinity of the Breece Hill stock, although some was also found in the neighborhood of a small stock in Iowa Gulch. The iron oxide, dark silicate mineralized rock was later fractured to some degree, and the fractures were filled by gold-bearing pyrite veins. These mineralized masses were, to a minor extent, mined for gold ore and for smelter flux.

The veins of mixed sulfides occur mainly in siliceous rocks but intersect all bedrock formations in the Leadville area. These veins are far more common in the eastern part of the district than farther west because of the much greater abundance in the eastern area of siliceous rocks (both the porphyry sills and the Pennsylvanian formations). In siliceous rocks the veins consist principally of pyrite with some interstitial chalcopyrite and appreciable gold in a gangue of quartz. In some places these veins expand into replacement deposits. The pyrite and quartz of the veins are present for short distances into the surrounding replacement mass, but the mineralization changes laterally into a mixture of sphalerite and galena in dense quartz or jasperoid gangue. When unenriched primary ore of this category has been mined, gold has been the most important metal recovered; silver and copper have also been recovered from these veins. The pyrite ore, however, where enriched by secondary chalcocite, silver, and gold, has been extensively exploited. The lead-zinc portions of the replacement bodies are high in sphalerite and galena and carry some silver but generally were too small to justify extensive mining.

The sulfide replacement bodies in dolomite are dominant in the western part of the district where carbonate rocks make up a much larger portion of the rock volume than they do in the eastern part. These replacement bodies (almost entirely in carbonate rocks) were developed outward from fractures and sheeted zones that underlay impervious covers such as the thick porphyry sill above the Leadville limestone.

The ore from the replacement masses proper consists of sphalerite and galena primarily, with appreciable amounts of silver. The earliest and generally main gangue mineral in these deposits is manganosiderite (the same mineral associated with magnetite in the silicate-oxide deposits), and it was replaced to a considerable degree by the sulfides and quartz; a little barite formed in the outer portions of those bodies. In many instances the ore bodies are surrounded by selvages of dense quartz or jasperoid. Although sphalerite and galena were the principal ore sulfides, pyrite was the most abundant sulfide, its deposition beginning before and ending after that of the other sulfides. The mixed sulfide ore contains a few ounces of silver per ton and a few hundredths of an ounce of gold. Irregularly throughout the replacement ore, considerable amounts of later bismuth-bearing minerals are found, accompanied by chalcopyrite (actually the most abundant mineral of this phase), hessite (Ag_2Te), altaite (PbTe), silver-bismuth-rich galena that consistently approximates $Pb_{11}Ag_2Bi_2S_{15}$, argentite, and native gold. Although the amounts of those minerals are small in comparison with those of sphalerite and galena, their precious metal (though not their copper) content made them of appreciable economic importance.

Superficially, the ores deposited in the ore bodies more distant from the center of the Leadville district are similar to those of the main replacement deposits. The tabular ore bodies in those distant areas, however, are appreciably smaller than those of the main zone, and the mineralogy is much simpler. These deposits (such as the Continental Chief and Hill Top bodies--the Sherman type) contain a little early barite, later dark- to light-colored sphalerite and galena, and a little pyrite and quartz; there are a few ounces of silver to the ton, but the silver mineral, if any, has not been identified. Still farther from the main ore areas, as in the Ruby mine in the Weston Pass area (10 miles south-southeast of Leadville), the sphalerite is lighter, barite is lacking, and the gangue consists of dolomite, calcite, quartz, and light-colored jasperoid; silver is essentially absent.

When the mineralization over the entire Leadville area is considered, it seems to have been deposited under a wide range of temperatures and pressures. The magnetite-hematite ores in the Breece Hill area, with their silicate gangue, almost certainly were formed under hypothermal conditions. Because of the minor economic worth of these high-temperature Breece Hill ores in comparison with those of lead and zinc, their correct designation, hypothermal-2, is not included in the Leadville classification. The high-grade replacement deposits in the main ore area and the veins of mixed sulfides in siliceous rocks probably were formed under mesothermal conditions, the huge envelope of hydrothermal dolomite alteration surrounding them appears certainly (on the basis of oxygen-isotope ratios) to have been formed in the mesothermal range, the mineral assemblage (though not diagnostic) is certainly characteristic of mesothermal deposits, and the size of the deposits indicates their introduction under conditions of appreciable chemical intensity. The minerals of the so-called bismuth stage form a typical leptothermal suite, and the tellurides and the argentite of this stage are diagnostic of leptothermal conditions. The deposits appear to have been formed under too great a cover of essentially impervious rocks for the bismuth stage minerals to have been formed under epithermal conditions. The deposits of the more distant areas (the Sherman and Weston Pass types) are simple enough in mineralogy and poor enough in silver to have been formed under telethermal conditions; the Weston Pass type probably was formed under somewhat less intense conditions than those of the Sherman types, but both types fit within the limits of the telethermal division of the classification.

On the basis of the discussion just presented, it appears reasonable to categorize the Leadville ores as mesothermal to telethermal.

RICO

Late Tertiary	*Silver, Zinc, Lead, Gold, Copper*	*Mesothermal to Leptothermal*

Collins, G. E., 1931, Localization of ore bodies at Rico and Red Mountain, Colorado, as conditioned by geologic structure and history: Colo. Sci. Soc. Pr., v. 12, no. 12, p. 407-424

Cross, W., and Ransome, F. L., 1905, Rico folio, Colorado: U.S. Geol. Surv. Geol. Atlas, Folio 130, 20 p.

Cross, W., and Spencer, A. C., 1900, Geology of the Rico Mountains, Colorado: U.S. Geol. Surv. 21st Ann. Rept., pt. 2, p. 7-165

Hubbell, A. H., 1927, Rico revived: Eng. and Min. Jour., v. 123, no. 8, p. 317-321

McKnight, E. T., 1933, Rico district, in *Colorado*: 16th Int. Geol. Cong., Guidebook 19, p. 63-66

Ransome, F. L., 1901, The ore deposits of the Rico Mountains, Colorado: U.S. Geol. Surv. 22nd Ann. Rept., pt. 2, p. 229-397

Rickard, T. A., 1896, The Enterprise mine, Rico, Colorado: Amer. Inst. Min. and Met. Tr., v. 26, p. 906-980

Varnes, D. J., 1947, Rico mining district, Dolores County, in Vanderwilt, J. W., Editor, *Mineral resources of Colorado*: State of Colo., Mineral Res. Bd., p. 414-416

Wisser, E., 1960, Rico, Colorado, in *Relation of ore deposition to doming in the North American Cordillera*: Geol. Soc. Amer. Mem. 77, p. 40-46

Notes

The ore deposits of the Rico Mountain area center around the town of Rico that lies about 30 miles north of Durango on a N15°W bearing. Almost all the production in the Rico district came from a circular area some 3 miles in diameter, the center of which coincides approximately with the apex of the Rico Mountain dome; the center of the Rico dome is nearly 20 miles due north of the La Plata dome.

The rock sequence in the Rico area begins with Precambrian rocks, schist, quartzite, and diorite, and these are overlain by Cambrian Ignacio quartzite, Devonian Ouray limestone, Pennsylvanian Hermosa formation (shale and limestone), Permian Rico formation (shale and sandstone), Permian Cutler formation (red beds), Upper Triassic and Jurassic(?) Dolores formation (red beds), Upper Jurassic Entrada sandstone, Upper Jurassic Morrison formation (sandstones and shales), Upper Cretaceous Dakota sandstone, and Upper Cretaceous Mancos shale. All of these formations are exposed on the flanks of the dome; the younger beds are at the greatest distances from the center of the dome, except for an isolated remnant of Upper Jurassic rocks still present in the middle of the mineralized area.

All of the sedimentary rock units have been intruded by numerous sills of hornblende-monzonite porphyry that were responsible for 700 feet of the 4500-foot rise of the dome. The next igneous activity was the intrusion of a quartz-monzonite stock that probably provided most of the force needed to complete the development of the dome. As Wisser pointed out, the ratio of vertical rise to diameter is 1:16, while that of the basically similar La Plata dome was much the same--1:14. The dome is elongated eastwardly and is intensely fractured and faulted, particularly in the apex area. Doming produced three varieties of faults: (1) those roughly concentric to the dome circumference, (2) radial faults in the northeast and southeast quadrants of the dome, and (3) faults parallel to the east-west trending axis of elongation of the dome. One major fault, the Blackhawk (strike roughly N40°W), falls into none of these categories. Long blocks and wedge-shaped slices have dropped in steplike blocks away from the center of the dome.

The igneous intrusions and structures of the Rico area were definitely developed after the end of the Mesozoic, but whether these events should be categorized as late Mesozoic to early Tertiary--the date of the earlier mineralization in the main San Juan area--or late Tertiary--when the principal ore development in the San Juan occurred--is uncertain. The problem probably could be resolved by radioactive age determinations on the igneous rocks of the Rico dome, but such have not, to my knowledge, been made. At the present time it appears more reasonable to consider the Rico ores to have been emplaced in the late Tertiary, which appears to have been the time of greatest igneous and ore-forming activity in this part of Colorado. On the balance of probabilities, then, the Rico ores are here categorized as late Tertiary.

Although some of the ore at Rico came from veins in faults or in fissures parallel, or at various angles, to the main faults, most of these veins are too small to be mined at a profit, and most of the mined ore has come from large, irregular replacement bodies formed where the veins cut through limestone or gypsum beds or brecciated zones in the shales of the Hermosa formation (manto or blanket ores).

Mineralization in the area commenced shortly after the dome began to form; in the blanket area, cross jointing, plus bedding-plane faulting, enlarged and extended the openings, making this rock volume even more acceptable as a site for ore emplacement. In many of the fissured areas, where there was only one period of fracturing, openings in the rocks were not sufficiently developed to make possible the formation of manto-type replacement bodies.

In both the replacement bodies and the fissure veins, pyrite and quartz were the first minerals to be deposited, followed by sphalerite, galena, and chalcopyrite--normally in that order; there were, however, at least two generations of sulfides, and the order of deposition differed not only from generation to generation but from place to place within a given generation. Thus, galena generally was the last of the four sulfides to begin to deposit, but vein and replacement volumes can be found in which galena appears to have been the first sulfide to have formed; the same order variations were true of the other sulfides as well. Argentite and the sulfosalts begin to deposit at an appreciably later time than the base-metal sulfides, usually having been precipitated in open spaces among the older ore minerals. Tetrahedrite, although nowhere abundant, was the most common of the sulfosalts and was of considerable economic importance because of its high silver content. The other silver-bearing sulfosalts at Rico include proustite (Ag_3AsS_3), polybasite [$(Ag,Cu)_{16}Sb_2S_{11}$], stephanite(?) (Ag_5SbS_4), and probably pyrargyrite (Ag_3SbS_3); the sulfosalts were accompanied by argentite, and all contributed to the generally high silver content of the ores. Free gold appears to have been rare, and there may have been some gold tellurides; the quantity of gold recovered from the ores, however, indicates that more gold was present than the reported amounts of native gold and tellurium would suggest.

The abundance of sulfosalts and argentite and the considerable thickness of rocks then existing over the rock volume in which the ores were formed strongly suggest that much of the economically valuable portions of the Rico ores were deposited under leptothermal conditions. The rather narrow vertical range over which the ores were developed mught be thought to point to epithermal conditions of deposition, but the short vertical distance involved appears to have resulted from structural conditions that limited the ores to blanket (or manto) forms and, therefore, to a short vertical range. The base-metal sulfides deposited before the sulfosalts probably were formed under more intense conditions than the sulfosalts and probably should be categorized as mesothermal. The gangue of quartz and rhodochrosite, with locally important calcite and fluorite, are characteristic of either the mesothermal or leptothermal ranges or both. The Rico ores are, therefore, here classified as mesothermal to leptothermal.

SAN JUAN

Late Mesozoic to Early Tertiary (minor part), Late Tertiary (major part)	*Gold, Silver, Lead, Zinc, Copper*	*Mesothermal to Leptothermal*

Baars, D. L., and See, P. D., 1968, Pre-Pennsylvanian stratigraphy and paleotectonics of the San Juan Mountains, southwestern Colorado: Geol. Soc. Amer. Bull., v. 79, p. 333-349

Bastin, E. S., 1923, Silver enrichment in the San Juan Mountains, Colorado:

U.S. Geol. Surv. Bull. 735-D, p. 65-129

Bejnar, W., 1949, Lithologic control of ore deposits in the San Juan Mountains, Colorado: Compass, v. 26, no. 2, p. 117-130

Burbank, W. S., 1930, Revision of geologic structure and stratigraphy in the Ouray district of Colorado, and its bearing on ore deposition: Colo. Sci. Soc. Pr., v. 12, no. 6, p. 151-232

—— 1933, Vein systems of the Arrastre Basin and regional geologic structure in the Silverton and Telluride quadrangles, Colorado: Colo. Sci. Soc. Pr., v. 13, no. 5, p. 135-214

—— 1933, The western San Juan Mountains, in *Colorado*: 16th Int. Geol. Cong., Guidebook 19, p. 34-63

—— 1940, Structural control of ore deposition in the Uncompahgre district, Ouray County, Colorado: U.S. Geol. Surv. Bull. 906-E, p. 189-265

—— 1941, Structural control of ore deposition in the Red Mountain, Sneffels, and Telluride districts of the San Juan Mountains, Colorado: Colo. Sci. Soc. Pr., v. 14, no. 5, p. 141-261

—— 1950, Problems of wall-rock alteration in shallow volcanic environments, in Van Tuyl and Kuhn, Editors, *Applied Geology*: Colo. Sch. Mines Quart., v. 45, no. 1B, p. 287-319

—— 1951, The Sunnyside, Ross Basin, and Bonita fault systems and their associated ore deposits, San Juan County, Colorado: Colo. Sci. Soc. Pr., v. 15, no. 7, p. 285-304

—— 1960, Pre-ore propylitization, Silverton caldera, Colorado: U.S. Geol. Surv. Prof. Paper 400-B, p. B12-B13

—— 1961, Pre-ore propylitization, Silverton caldera, Colorado: U.S. Geol. Surv. Prof. Paper 400-B, p. B12-B13

Burbank, W. S., and Luedke, R. G., 1961, Origin and evolution of ore and gangue-forming solutions, Silverton caldera, San Juan Mountains, Colorado: U.S. Geol. Surv. Prof. Paper 424-C, p. C7-C11

—— 1966, Geologic map of the Telluride quadrangle, southwestern Colorado: U.S. Geol. Surv. Geol. Quad. Map, GQ 504, 1:24,000

—— 1968, Geology and ore deposits of the western San Juan Mountains, Colorado, in Ridge, J. D., Editor, *Ore deposits of the United States, 1933-1967* (Graton-Sales Volumes): Chap. 34, v. 1, 1880 p.

—— 1970, Geology and ore deposits of Eureka and adjoining districts, San Juan Mountains, Colorado: U.S. Geol. Surv. Prof. Paper 535, 73 p.

Burbank, W. S., and others, 1947, The San Juan region, in Vanderwilt, J. W., Editor, *Mineral resources of Colorado*: State of Colo., Mineral Res. Bd., p. 396-451

Collins, G. E., 1931, Localization of ore bodies at Rico and Red Mountain, Colorado, as conditioned by geologic structure and history: Colo. Sci. Soc. Pr., v. 12, no. 12, p. 407-424

Cross, W., and Larsen, E. S., 1935, A brief review of the geology of the San Juan region of southwestern Colorado: U.S. Geol. Surv. Bull. 843, 138 p.

Cross, W., and Purrington, C. W., 1899, Telluride folio, Colorado: U.S. Geol. Surv. Geol. Atlas, Folio 57, 19 p.

Cross, W., and others, 1905, Silverton folio, Colorado: U.S. Geol. Surv. Geol. Atlas, Folio 120, 34 p.

—— 1907, Ouray folio, Colorado: U.S. Geol. Surv. Geol. Atlas, Folio 153, 20 p.

Emmons, W. H., and Larsen, E. S., Jr., 1923, Geology and ore deposits of the

Creede district, Colorado: U.S. Geol. Surv. Bull. 718, 198 p.

Irving, J. D., 1905, Ore deposits of the Ouray district, Colorado: U.S. Geol. Surv. Bull. 260, p. 50-77

Kelley, V. C., 1946, Geology, ore deposits, and mines of the Mineral Point, Poughkeepsie, and Upper Uncompahgre districts, Ouray, San Juan, and Hinsdale Counties, Colorado: Colo. Sci. Soc. Pr., v. 14, no. 7, p. 287-466

Kelley, V. C., and Silver, C., 1946, Stages and epochs of mineralization in the San Juan Mountains, Colorado, as shown at the Dunmore mine, Ouray County, Colorado: Econ. Geol., v. 41, p. 139-159

Kottlowski, F. E., Editor, 1957, Guidebook to the geology of the southwestern San Juan Mountains, Colorado: New Mex. Geol. Soc., 8th Field Conf., Guidebook, p. 102-207, 217-221

Larsen, E. S., Jr., 1929, Recent mining developments in the Creede district, Colorado: U.S. Geol. Surv. Bull. 811, p. 89-112

Larsen, E. S., Jr., and Cross, W., 1956, Geology and petrology of the San Juan region, southwestern Colorado: U.S. Geol. Surv. Prof. Paper 258, 303 p.

Lipman, P. W., and others, 1970, Volcanic history of the San Juan Mountains, Colorado, as indicated by potassium-argon dating: Geol. Soc. Amer. Bull., v. 81, p. 2329-2352

Luedke, R. G., and Burbank, W. S., 1962, Geology of the Ouray quadrangle, Colorado: U.S. Geol. Surv. Geol. Quad. 152, 1" = 5280'

—— 1963, Tertiary volcanic stratigraphy in the western San Juan Mountains, Colorado: U.S. Geol. Surv. Prof. Paper 475-C, p. C39-C44

—— 1968, Volcanism and cauldron development in the western San Juan Mountains, Colorado: Colo. Sch. Mines Quart., v. 63, no. 3, p. 175-208

Meeves, H. C., and Darnell, R. P., 1968, Study of the silver potential, Creede district, Mineral County, Colorado: U.S. Bur. Mines I.C. 8370, 58 p.

Moehlman, R. S., 1936, Ore deposition south of Ouray, Colorado: Econ. Geol., v. 31, p. 377-397, 488-504

Patton, H. B., 1917, Geology and ore deposits of the Platoro-Summitville mining district, Colorado: Colo. Geol. Surv. Bull. 13, 122 p.

Ransome, F. L., 1901, A report on the economic geology of the Silverton quadrangle, Colorado: U.S. Geol. Surv. Bull. 182, 265 p.

Ratté, J. C., and Steven, T. A., 1964, Magmatic differentiation in a volcanic sequence related to the Creede caldera, Colorado: U.S. Geol. Surv. Prof. Paper 475-D, p. D49-D53

—— 1967, Ash flows and related volcanic rocks associated with the Creede caldera, San Juan Mountains, Colorado: U.S. Geol. Surv. Prof. Paper 524-H, p. H1-H58

Siems, P. L., 1968, Volcanic geology of the Rosita Hills and Silver Cliff district, Custer County, Colorado: Colo. Sch. Mines Quart., v. 63, no. 3, p. 89-124

Spurr, J. E., 1925, The Camp Bird compound veindike: Econ. Geol., v. 20, p. 115-152

Steven, T. A., 1963, Geologic setting of the Spar City district, San Juan Mountains, Colorado: U.S. Geol. Surv. Prof. Paper 475-D, p. D123-D127

—— 1968, Ore deposits in the central San Juan Mountains, Colorado, in Ridge, J. D., Editor, *Ore deposits of the United States, 1933-1967* (Graton-Sales Volumes): Chap. 33, v. 1, 1880 p.

—— 1969, Possible relation of mineralization to thermal springs in the Creede district, San Juan Mountains, Colorado (disc.): Econ. Geol., v. 64, no. 6, p. 696-698

Steven, T. A., and Ratté, J. C., 1960, Geology and ore deposits of the Summitville district, San Juan Mountains, Colorado: U.S. Geol. Surv. Prof. Paper 343, 70 p.

—— 1960, Relation of mineralization to caldera subsidence in the Creede district, San Juan Mountains, Colorado: U.S. Geol. Surv. Prof. Paper 400-B, p. B14-B17

—— 1964, Revised Tertiary volcanic sequence in the central San Juan Mountains, Colorado: U.S. Geol. Surv. Prof. Paper 475-D, p. D54-D63

—— 1965, Geology and structural control of ore deposition in the Creede district, San Juan Mountains, Colorado: U.S. Geol. Surv. Prof. Paper 487, 90 p.

Steven, T. A., and others, 1967, Age of volcanic activity in the San Juan Mountains, Colorado: U.S. Geol. Surv. Prof. Paper 575-D, p. D47-D55

Varnes, D. J., 1947, Recent developments on the Black Bear vein, San Miguel County, Colorado: Colo. Sci. Soc. Pr., v. 15, p. 135-146

—— 1962, Analysis of plastic deformation according to Von Mises' theory with application to the south Silverton area, San Juan County, Colorado: U.S. Geol. Surv. Prof. Paper 378-B, 49 p.

—— 1963, Geology and ore deposits of the south Silverton mining area, San Juan County, Colorado: U.S. Geol. Surv. Prof. Paper 378-A, p. A1-A56

Vhay, J. S., 1962, Geology and mineral deposits of the area south of Telluride, Colorado: U.S. Geol. Surv. Bull. 1112-G, p. 209-310

Wisser, E., 1960, Creede, Colorado, in *Relation of ore deposition to doming in the North American Cordillera*: Geol. Soc. Amer. Mem. 77, p. 63-68

—— 1960, Silverton, Colorado, in *Relation of ore deposition to doming in the North American Cordillera*: Geol. Soc. Amer. Mem. 77, p. 26-36

Notes

The San Juan mineralized area lies in southwestern Colorado some 200 miles southwest of Denver and includes nearly 30 mining districts of generally similar character that extend from Ouray on the northwest to Summitville on the southeast. Other famous districts in the San Juan area include Telluride, Sneffels, Ophir, Silverton, Red Mountain, Needle Mountain, Creede, Lake City, and Bonanza. Most of the San Juan deposits are located around the margin of an elliptical area within the San Juan Mountains some 75 miles long (northwest-southeast) and some 30 miles wide (northeast-southwest). The Rico and La Plata districts, often discussed with the San Juan proper, are here considered separately because they are related to distinct centers of doming, igneous activity, and mineralization at some distance from the main San Juan deposits.

The San Juan area contains a wide variety of rocks ranging in age from Precambrian to Recent. The principal Precambrian rocks are schists, gneisses, granites, granodiorites, gabbros, quartzites, conglomerates, and greenstones; the famous alkaline rock complex of Iron Hill in the far northeastern San Juans, thought by many to be of the same origin as the alkaline igneous complexes of Alnö Island and the Søve area, is also Precambrian in age. Cambrian rocks are mainly quartzites; Ordovician principally limestones; Devonian generally limestone, with some sandstones; Mississippian mostly limestones; Pennsylvanian dominantly shale, limestone, and sandstone, some of which is red; and Permian mainly reddish sandstones and grits, with some shale and limestone. The Mesozoic formations are Jurassic and Upper Cretaceous; the Jurassic rocks are mainly sandstones, with a little limestone and shale; the Upper Cretaceous Mesa Verde group consists of sandstones and shales, while younger than these are more sandstones and shales, followed by the first volcanic activity since the Precambrian, the principal present trace of which is volcanic debris in the Upper Cretaceous and Paleocene Animas formation. Tertiary beds range from Paleocene in age to Pliocene(?) and include a basal till in the Paleocene, overlain by sandstone, with some shale; the Eocene consists mainly of sandstone with some shale; the Oligocene of sandstone and conglomerate; the Miocene was a time of volcanic activity in which the rocks developed were quartz latites and latites and rhyolites and some basalts, while tuffs and breccias were not uncommon. Intrusive igneous rocks (discussed below) were developed at the same time and from the same general sources as the volcanic rocks. The Miocene volcanic rocks were followed by Miocene and Pliocene(?) gravels that appear to have been cut by the Pliocene San Juan peneplain. Further volcanic material was then spread over the peneplain and consists of Pliocene(?) rhyolite, latite, and basalt. In the early Quaternary, the area was domed by some 3000 feet, and erosion accompanied and followed the doming to produce the San Juan Mountains. Some latite and basalt flows are younger than the doming and much of the erosion; the last materials to be laid down were recent alluvial deposits.

Late Mesozoic to early Tertiary igneous rocks were intruded in the Ouray area concomitantly with the initial rise of the ancestral San Juan Mountains and consisted of generally porphyritic diorite, granodiorite, and granite laccoliths and sills. These intrusives are unconformably overlain by the Oligocene(?) Telluride conglomerate, almost certainly dating them as Laramide. A minor stage of ore formation closely followed these intrusions; from these Laramide deposits has come some 10 percent of the value of San Juan metal production. The next intrusive activity occurred in conjunction with the recommencement of volcanism in the middle Miocene. The rocks produced at this time fall into two major groups: (1) the laccoliths and stocks that intrude in the main sedimentary rocks in the western part of the San Juans and (2) the less abundant stocks (and the dikes associated with them) that intrude the volcanic rocks of the Miocene Conejos quartz latite in the eastern part. In addition to these principal occurrences, a few small intrusive masses can be found to have been a minor part of almost every period of eruption. Most of the intrusive masses can be correlated with one of the volcanic formations, but it has been impossible to determine the ages of some intrusives except to place them within the broad boundaries of Tertiary age. Those that can be dated range in age from Miocene(?) through Pliocene and in composition from rhyolite and granite porphyry through granodiorite, quartz latite, and diorite to gabbro and intrusive basalts, plus diabases and lamprophyres.

Near the close of the Miocene-Pliocene volcanism, the San Juan region was highly deformed; in most of the area this deformation consisted of an accentuation of the collapse of the crust that had begun during the earlier volcanic epochs, to produce extensive faulting and fissuring. About 80 percent of the metals extracted from the San Juan area have come from Miocene and Pliocene volcanic rocks in the immediate neighborhood of the larger late Tertiary intrusive centers, the veins cutting both volcanic and intrusive rocks.

The bulk of the San Juan ores was deposited after the end of late Miocene to early Pliocene volcanism and before the development of the San Juan peneplain and the extrusion of the late Pliocene Hinsdale basalt; this dates this large fraction of San Juan ores as late Tertiary. The minor amount of ore developed during Laramide time requires also the mention of a late Mesozoic to early Tertiary age in categorizing the deposits.

The numerous systems of strong fissure veins that cut the huge uplifted and eroded dome of the San Juan Mountains often continue uninterruptedly for several miles and, in many instances, have been mined over vertical distances of several thousand feet. Locally, the fissuring appears to be aligned between intrusive bodies, although elsewhere the fracturing probably is related to the broader regional tilting.

Most of the earlier (Laramide) ores are found in the Ouray (Uncompahgre) district and exhibit a wide diversity of forms that range from typical fissure veins to flat-lying bedding-plane replacement masses. The amount of ore of this type still undiscovered in the area may be large because along the Laramide intrusive zone, the potentially ore-bearing lower Paleozoic limestones lie 2000 feet or more below the valley floor. These early ores appear to have been zoned outward from the intrusive center. The economically most valuable ores of Laramide age in the Ouray area appear to have been formed under leptothermal conditions (they were emplaced under too thick and too impervious a rock cover to have been epithermal). Although the pyritic copper-gold ores and the magnetite-pyrite ores appear to have been formed under mesothermal and hypothermal conditions, respectively, they were of too little economic worth to be mentioned in the classification.

The late Tertiary ores of the San Juan were introduced into the area after the emplacement of large bodies of intrusive rock that, in turn, had followed the end of the extrusion of the late Miocene to early Pliocene volcanic rocks. Although fissure veins are probably the predominant structure in which the ores were deposited, important ore bodies have been found in chimneys, mantos, stockworks, and volcanic pipes.

In the Silverton-Ouray area, the mineralogy of the ores is comparatively simple; the bulk of the sulfides in the average vein is composed of pyrite, sphalerite, silver-rich galena, and chalcopyrite in a gangue of quartz, with some calcite, rhodochrosite, and rhodonite(?). The veins were normally formed in several stages, with the base-metal sulfides decreasing from each earlier to the next later stage so that the final stages are usually nearly all barren quartz or calcite. Tetrahedrite-tennantite often rivals galena in abundance and usually is silver-rich; polybasite [$(Ag,Cu)_{16}Sb_2S_{11}$], stephanite (Ag_5SbS_4), and argentite often add appreciably to the economic value of the ores, although the ruby silvers (pyrargyrite and proustite) ordinarily are of little economic importance.

In the Lake City area the ores were emplaced in several stages, not all of which occurred in all veins: (1) quartz and pyrite; (2) sphalerite, silver-rich galena, chalcopyrite, and rhodochrosite; (3) silver-rich tetrahedrite, (4) quartz and locally tellurides and free gold.

At Creede the primary ores, though considerably banded, are quite simple in mineralogy, with quartz and iron-rich chlorite being early gangue minerals, followed by silver-rich galena and lesser amounts of argentite and chalcopyrite; some native gold is present, apparently in close association with pyrite. Barite is moderately abundant, adularia and fluorite are minor, and rhodochrosite is rare. The primary ores appear to have been formed under mesothermal to leptothermal conditions with argentite being the only important mineral diagnostic of the leptothermal range.

At Bonanza the veins in the northern part of the district are high in quartz and, near the surface, contain less zinc than at depth. The principal sulfides, in order of formation, are pyrite (probably two generations), sphalerite, bornite, minor enargite, chalcopyrite, silver-rich galena, and silver-rich tennantite; stromeyerite, covellite, and chalcocite are minor and late. In addition to quartz, sericite is common, and rhodonite and rhodochrosite much less so. These northern ores appear almost certainly to have formed under mesothermal to leptothermal conditions, although some telescoping suggests kryptothermal to epithermal conditions may have obtained in places. In the southern sector, the sulfide content is much lower, and the economically valuable minerals in the primary ores appear to have been silver-rich galena, silver-rich tennantite, and the ruby silvers. No evidence of deposition under mesothermal conditions has been found, and these southern ores appear to have formed entirely within the leptothermal range.

At Summitville, gold was by far the principal mineral recovered, and as a primary mineral, it appears to have been associated with enargite (the principal copper sulfide) rather than with pyrite. Tennantite and covellite also were present in the primary ores in some quantity, and chalcopyrite and chalcocite were found in probably minor amounts. So little is known of the mineralogy of the primary Summitville ores that their classification is uncertain, but they probably fit into the mesothermal to leptothermal categories.

This brief review of the major areas of mineralization in the San Juan Mountains strongly suggests that the ores were formed under mesothermal to leptothermal conditions, with the importance of the leptothermal range being established by the common presence of silver-rich tetrahedrite-tennantite and the less widespread development of silver sulfosalts; much of the silver-rich galena formed at the same time as the tetrahedrite-tennantite and is, therefore, also leptothermal. The late Tertiary ores in the San Juan, therefore, are here classified as mesothermal to leptothermal.

Colorado-Utah-New Mexico-Arizona

COLORADO PLATEAU

Late Mesozoic to Early Tertiary (major part), Late Tertiary (minor part)	*Uranium, Vanadium*	*Mesothermal to Telethermal and/or Ground Water-B2*

A. *RIFLE AREA, COLORADO*

Botinelly, T., and Fischer, R. P., 1959, Mineralogy and geology of the Rifle and Garfield mines, Garfield County, Colorado: U.S. Geol. Surv. Prof. Paper 320, p. 213-218

Coffin, R. C., 1921, Radium, uranium, and vanadium deposits of southwestern Colorado: Colo. Geol. Surv. Bull. 16, 231 p. (particularly p. 29-186)

Fischer, R. P., 1960, Vanadium-uranium deposits of the Rifle Creek area, Garfield County, Colorado: U.S. Geol. Surv. Bull. 1101, 52 p.

Fischer, R. P., and Hilpert, L. S., 1952, Geology of the Uravan mineral belt: U.S. Geol. Surv. Bull. 988-A, p. 1-13

B. *URAVAN DISTRICT, COLORADO*

Boardman, R. L., and others, 1956, Sedimentary features of upper sandstone lenses of the Salt Wash member and their relation to uranium-vanadium deposits in the Uravan district, Montrose County, Colorado, in *Contributions to the geology of uranium and thorium*: U.S. Geol. Surv. Prof. Paper 300, p. 221-226

Coffin, R. C., 1921, Radium, uranium and vanadium deposits of southwestern Colorado: Colo. Geol. Surv. Bull. 16, 231 p. (particularly p. 29-186)

Elston, D. P., and Botinelly, T., 1959, Geology and mineralogy of the J. J. Mine, Montrose County, Colorado: U.S. Geol. Surv. Prof. Paper 320, p. 203-211

Fischer, R. P., and Hilpert, L. S., 1952, Geology of the Uravan mineral belt: U.S. Geol. Surv. Bull. 988-A, p. 1-13

Heyl, A. V., Jr., 1957, Zoning of the Bitter Creek vanadium-uranium deposits near Uravan, Colorado: U.S. Geol. Surv. Bull. 1042-F, p. 187-201

Motica, J. E., 1968, Geology and uranium-vanadium deposits in the Uravan mineral belt, southwestern Colorado, in Ridge, J. D., Editor, *Ore deposits of the United States, 1933-1967* (Graton-Sales Volumes): Chap. 39, v. 1, 1880 p.

Newman, W. L., and Elston, D. P., 1959, Distribution of chemical elements in the Salt Wash member of the Morrison formation, Jo Dandy area, Montrose County, Colorado: U.S. Geol. Surv. Bull. 1084-E, p. 117-150

Phoenix, D. A., 1956, Relation of carnotite deposits to permeable rocks in the Morrison formation, Mesa County, Colorado, in *Contributions to the geology of uranium and thorium*: U.S. Geol. Surv. Prof. Paper 300, p. 213-219

—— 1958, Uranium deposits under conglomeratic sandstone of the Morrison formation, Colorado and Utah: Geol. Soc. Amer. Bull., v. 69, p. 403-417

Roach, C. H., and Thompson, M. E., 1959, Sedimentary structures and localization and oxidation of ore at the Peanut mine, Montrose County, Colorado:

U.S. Geol. Surv. Prof. Paper 320, p. 197-202

Shawe, D. R., 1962, Localization of the Uravan mineral belt by sedimentation: U.S. Geol. Surv. Prof. Paper 450-C, p. C6-C8

—— 1969, Possible exploration targets for uranium deposits, south end of the Uravan mineral belt, Colorado--Utah: U.S. Geol. Surv. Prof. Paper 650-B, p. B73-B76

Wood, H. B., and Lekas, M. A., 1958, Uranium deposits of the Uravan mineral belt: Intermount Assoc. Petrol. Geols. 9th Ann. Field Conf. Guidebook, p. 208-215

Wright, R. J., and Everhart, D. L., 1960, Uravan district, in Rio, S. M. del, Editor, *Mineral resources of Colorado--first sequel*: State of Colo., Mineral Res. Bd., p. 333-343

C. SLICK ROCK DISTRICT, COLORADO

Bowers, H. E., and Shawe, D. R., 1961, Heavy minerals as guides to uranium-vanadium ore deposits in the Slick Rock district, Colorado: U.S. Geol. Surv. Bull. 1107-B, 49 p.

Rogers, W. B., and Shawe, D. R., 1962, Exploration for uranium-vanadium deposits by U.S. Geological Survey 1948-56 in western Disappointment Valley, Slick Rock district: U.S. Geol. Surv. Mineral Invest., Field Studies Map MF-241 (3 sheets), 1:12,000

Shawe, D. R., 1968, Petrography of sedimentary rocks in the Slick Rock district, San Miguel and Dolores Counties, Colorado: U.S. Geol. Surv. Prof. Paper 576-B, p. B1-B34

—— 1970, Structure of the Slick Rock district and vicinity, San Miguel and Dolores Counties, Colorado: U.S. Geol. Surv. Prof. Paper 567-C, p. C1-C18

Shawe, D. R., and others, 1958, Geology and uranium-vanadium deposits of the Slick Rock district, San Miguel and Dolores Counties, Colorado: 2d U.N. International Conf. on Peaceful Uses of Atomic Energy (Geneva) Pr., v. 2, p. 515-522

—— 1959, Geology and uranium-vanadium deposits of the Slick Rock district, San Miguel and Dolores Counties, Colorado: Econ. Geol., v. 54, p. 395-415

—— 1961, Preliminary geologic map of the Slick Rock district, San Miguel and Dolores Counties, Colorado: U.S. Geol. Surv. Mineral Invest., Field Studies Map MF-203, 1:4,000

—— 1969, Stratigraphy of the Slick Rock district and vicinity, San Miguel and Dolores Counties, Colorado: U.S. Geol. Surv. Prof. Paper 576-A, p. A1-A108

D. GRANTS DISTRICT, NEW MEXICO

Bassett, W. A., and others, 1963, Potassium-argon ages of volcanic rocks near Grants, New Mexico: Geol. Soc. Amer. Bull., v. 74, p. 221-225

Bell, K. G., 1963, Uranium in carbonate rocks: U.S. Geol. Surv. Prof. Paper 474-A, p. A1-A29 (particularly p. A12-A17)

Birdseye, H. S., 1957, The relation of the Ambrosia Lake uranium deposits to a pre-existing oil pool: Four Corners Geol. Soc., 2d. Field Conf. Guidebook, p. 26-29

Dooley, J. R., and others, 1966, Uranium-234 fractionation in the sandstone-type uranium deposits of the Ambrosia Lake district, New Mexico: Econ. Geol., v. 61, p. 1362-1382

Gabelman, J. W., 1956, Uranium deposits in limestone, in *Contributions to the geology of uranium and thorium*: U.S. Geol. Surv. Prof. Paper 300, p. 387-404

Granger, H. C., 1962, Clays in the Morrison formation and their spatial relation to the uranium deposits at Ambrosia Lake, New Mexico: U.S. Geol. Surv. Prof. Paper 450-D, Art. 124, p. D15-D20

—— 1968, Localization and control of uranium deposits in the southern San Juan mineral belt, New Mexico--an hypothesis: U.S. Geol. Surv. Prof. Paper 600-B, p. B60-B70

Granger, H. C., and Santos, E. S., 1963, An ore-bearing cylindrical collapse structure in the Ambrosia Lake uranium district, Nex Mexico: U.S. Geol. Surv. Prof. Paper 475-C, p. C156-C161

Granger, H. C., and others, 1961, Sandstone-type uranium deposits at Ambrosia Lake, New Mexico--an interim report: Econ. Geol., v. 56, p. 1179-1210

Hilpert, L. S., 1961, Structural control of epigenetic uranium deposits in carbonate rocks of northwestern New Mexico: U.S. Geol. Surv. Prof. Paper 424-B, p. B5-B8

—— 1969, Uranium resources of northwestern New Mexico: U.S. Geol. Surv. Prof. Paper 603, 166 p.

Hilpert, L. S., and Freeman, V. L., 1956, Guides to uranium deposits in the Gallup-Laguna area, New Mexico, in *Contributions to the geology of uranium and thorium*: U.S. Geol. Surv. Prof. Paper 300, p. 299-302

Hilpert, L. A., and Moench, R. H., 1958, Uranium deposits of the southern part of the San Juan Basin, New Mexico: 2d U.N. International Conf. on Peaceful Uses of Atomic Energy (Geneva) Pr., v. 2, p. 527-538

—— 1960, Uranium deposits of the southern part of the San Juan Basin, New Mexico: Econ. Geol., v. 55, p. 429-464

Kelley, V. C., General Chairman, 1963, Geology and technology of the Grants uranium region: New Mex. Bur. Mines and Mineral Res. Mem. 15, 277 p. (contains 34 papers on various aspects of Grants geology plus a few on technical operations)

Kelley, V. C., and others, 1968, Uranium deposits of the Grants region, in Ridge, J. D., Editor, *Ore deposits of the United States, 1933-1967* (Graton-Sales Volumes): Chap. 36, v. 1, 1880 p.

Kittel, D. F., and others, 1967, Uranium deposits of the Grants region: New Mex. Geol. Soc., 18th Field Conf., New Mex. Bur. Mines and Mineral Res., p. 173-183

Laverty, R. A., 1967, Geomorphology and structure in the Grants mineral belt: New Mex. Geol. Soc. Guidebook, 18th Field Conf., p. 188-194

Lovering, T. G., 1956, Radioactive deposits of New Mexico: U.S. Geol. Surv. Bull. 1009-G, p. 315-390 (covers entire state, not Grants district alone)

Mergrue, G. H., and Kerr, P. F., 1965, Alteration of sandstone pipes, Laguna, New Mexico: Geol. Soc. Amer. Bull., v. 76, p. 1347-1360; disc., 1968, v. 79, p. 787-790; reply, 1968, v. 79, p. 791-794

Moench, R. H., 1962, Properties and paragenesis of coffinite from the Woodrow mine, New Mexico: Amer. Mineral., v. 47, p. 26-33

Moench, R. H., and Schlee, J. S., 1967, Geology and uranium deposits of the Laguna district, New Mexico: U.S. Geol. Surv. Prof. Paper 519, 117 p.

Nash, J. T., 1968, Uranium deposits in the Jackpile sandstone, New Mexico: Econ. Geol., v. 63, p. 737-750

Nash, J. T., and Kerr, P. F., 1966, Geologic limitations on the age of the uranium deposits in the Jackpile sandstone, New Mexico: Econ. Geol., v. 61, p. 1283-1287

Santos, E. S., 1970, Stratigraphy of the Morrison formation and structure of the Ambrosia Lake district, New Mexico: U.S. Geol. Surv. Prof. Paper 1272-E

Tanner, W. F., 1965, Upper Jurassic paleogeography of the Four Corners region: Jour. Sed. Petrol., v. 35, p. 564-574

Truesdell, A. H., and Weeks, A. D., 1960, Paragenesis of uranium ores in Todilto limestone near Grants, New Mexico: U.S. Geol. Surv. Prof. Paper 400-B, p. B52-B54

Zitting, R. T., and others, 1957, Geology of the Ambrosia Lake area uranium deposits, McKinley County, New Mexico: Mines Mag., v. 47, no. 3, p. 53-58

E. SHIPROCK DISTRICT, UTAH-ARIZONA

Blagbrough, J. W., 1967, Cenozoic geology of the Chuska Mountains, in *Guidebook of Defiance-Zuni-Mt. Taylor region, Arizona and New Mexico*: New Mex. Geol. Soc., 18th Field Conf., New Mex. Bur. Mines and Mineral Res., p. 70-77

Chenoweth, W. L., 1955, The geology and uranium deposits of the northwest Carrizo area, Apache County, Arizona: Four Corners Geol. Soc. 1st Field Conf. Guidebook, p. 177-185

—— 1967, The uranium deposits of the Lukachukai Mountains, Arizona, in *Guidebook of Defiance-Zuni-Mt. Taylor region, Arizona and New Mexico*: New Mex. Geol. Soc., 18th Field Conf., New Mex. Bur. Mines and Mineral Res., p. 78-85

Gavasci, A. T., and Kerr, P. F., 1968, Uranium emplacement at Garnet Ridge, Arizona: Econ. Geol., v. 63, p. 859-876

Lowell, J. D., 1955, Applications of cross-stratification studies to problems of uranium exploration, Chuska Mountains, Arizona: Econ. Geol., v. 50, p. 177-185

Masters, J. A., 1955, Geology of the uranium deposits of the Lukachukai Mountains area, northeastern Arizona: Econ. Geol., v. 50, p. 111-126

O'Sullivan, R. B., and Beikman, H. M., 1963, Geology, structure and uranium deposits of the Shiprock quadrangle, New Mexico and Arizona: U.S. Geol. Surv. Misc. Geol. Invest. Map I-345, 1:250,000

F. MONTICELLO-MOAB-THOMPSON DISTRICT, UTAH

Carter, W. D., and Gualtieri, J. L., 1965, Geology and uranium-vanadium deposits of the La Sal quadrangle, San Juan County, Utah, and Montrose County, Colorado: U.S. Geol. Surv. Prof. Paper 508, 82 p.

Davidson, D. M., Jr., and Kerr, P. F., 1966, Uranium deposits at Kane Creek, Utah: Soc. Min. Eng. Tr., v. 235, p. 127-132

—— 1968, Uranium-bearing veins in Plateau strata, Kane Creek, Utah: Geol. Soc. Amer. Bull., v. 79, p. 1503-1525

Gross, E. B., 1956, Mineralogy and paragenesis of the uranium ore, Mi Vida mine, San Juan County, Utah: Econ. Geol., v. 51, p. 632-648

Holland, H. D., and others, 1957, The use of leachable uranium in geochemical prospecting on the Colorado Plateau. I. The distribution of leachable uranium in core samples adjacent to the Homestake ore body, Big Indian Wash, Utah; II. The distribution of leachable uranium in surface samples in the vicinity of ore bodies: Econ. Geol., v. 52, p. 546-569, v. 53, p. 190-209

Huff, L. C., and Lesure, F. G., 1962, Diffusion features of uranium-vanadium deposits in Montezuma Canyon, Utah: Econ. Geol., v. 57, p. 226-237

—— 1965, Geology and uranium deposits of Montezuma Canyon area, San Juan County, Utah: U.S. Geol. Surv. Bull. 1190, 102 p.

Huff, L. C., and others, 1958-1959, Preliminary geologic maps of the Verdure quadrangle, San Juan County, Utah: U.S. Geol. Surv. Mineral Invest., Field

Studies Maps MF-163, 164, 166 (3 sheets), 1:24,000

Hunt, C. B., 1958, Structural and igneous geology of the La Sal Mountains, Utah: U.S. Geol. Surv. Prof. Paper 294-I, p. 305-364

Jacobs, M. B., and Kerr, P. F., 1965, Hydrothermal alteration along the Lisbon Valley fault zone, San Juan County, Utah: Geol. Soc. Amer. Bull., v. 76, p. 423-440

Johnson, H. S., Jr., and Thorardson, W., 1966, Uranium deposits of the Moab, Monticello, White Canyon, and Monument Valley districts, Utah and Arizona: U.S. Geol. Surv. Bull. 1222-H, p. H1-H53

Lekas, M. A., and Dahl, H. M., 1956, The geology and uranium deposits of the Lisbon Valley anticline, San Juan County, Utah: Intermountain Assoc. Petrol. Geols. 7th Ann. Field Conf. Guidebook, p. 161-168

Rasor, C. A., 1952, Uraninite from the Grey Dawn mine, San Juan County, Utah: Science, v. 116, no. 3004, p. 89-90

Steen, C. A., and others, 1953, Uranium mining operations of the Utex Exploration Company in the Big Indian district, San Juan County, Utah: U.S. Bur. Mines I.C. 7669, 11 p. (mimeo.)

Stern, T. W., and others, 1965, Zircon uranium-lead and thorium-lead ages and mineral potassium-argon ages of La Sal Mountains rocks, Utah: Jour. Geophys. Res., v. 70, p. 1503-1507

Stokes, W. L., 1952, Uranium-vanadium deposits of the Thompson area, Grand County, Utah, with emphasis on the origin of the carnotite ores: Utah Geol. and Mineral. Surv. Bull. 46, 48 p.

Stokes, W. L., Editor, 1954, Uranium deposits and general geology of southeastern Utah: Utah Geol. Soc. Guidebook to the Geology of Utah, no. 9, 115 p. (particularly p. 12-105)

Weir, G. W., and Puffett, W. P., 1960, Similarities of uranium-vanadium and copper deposits in the Lisbon Valley area, Utah-Colorado, U.S.A.: 21st Int. Geol. Cong. Rept., pt. 15, p. 133-148

Williams, P. L., 1964, Geology, structure, and uraniferous deposits of the Moab quadrangle, Colorado and Utah: U.S. Geol. Surv. Misc. Geol. Invest. Map I-360 (2 sheets), 1:250,000

Witkind, I. J., 1958, The Abajo Mountains, San Juan County, Utah: Intermountain Assoc. Petrol. Geols. 9th Ann. Field Conf. Guidebook, p. 60-65

—— 1964, Geology of the Abajo Mountains area, San Juan County, Utah: U.S. Geol. Surv. Prof. Paper 453, 110 p.

Wood, H. B., 1968, Geology and exploitation of uranium deposits in the Lisbon Valley area, Utah, in Ridge, J. D., Editor, *Ore deposits of the United States, 1933-1967* (Graton-Sales Volumes): Chap. 37, v. 1, 1880 p.

G. GENERAL

Abdel-Gawad, A. M., and Kerr, P. F., 1961, Urano-organic mineral association: Amer. Mineral., v. 46, p. 402-419

—— 1963, Alteration of Chinle siltstone and uranium emplacement, Arizona and Utah: Geol. Soc. Amer. Bull., v. 74, p. 23-46; disc., 1964, v. 75, p. 775-776, 777-780

Adler, H. H., 1963, Concepts of genesis of sandstone-type uranium ore deposits: Econ. Geol., v. 58, p. 839-852

Adler, H. H., and Sharp, B. J., 1967, Uranium ore rolls--occurrence, genesis and physical characteristics: Utah Geol. Soc., Guidebook to the Geology of Utah, no. 21, p. 53-77

Bain, G. W., 1960, Patterns to ores in layered rocks: Econ. Geol., v. 55, p. 695-731 (particularly p. 720-728)

Bates, R. C., 1959, An application of statistical analysis to exploration for uranium on the Colorado Plateau: Econ. Geol., v. 54, p. 449-466

Botinelly, T., and Weeks, A. D., 1957, Mineralogical classification of uranium-vanadium deposits of the Colorado Plateau: U.S. Geol. Surv. Bull. 1074-A, p. 1-15

Breger, I. A., and Duel, M., 1956, The organic geochemistry of uranium, in *Contributions to the geology of uranium and thorium*: U.S. Geol. Surv. Prof. Paper 300, p. 505-510

Cadigan, R. A., 1959, Characteristics of the host rocks (Colorado Plateau): U.S. Geol. Surv. Prof. Paper 320, p. 13-24

—— 1967, Petrology of the Morrison formation in the Colorado Plateau region: U.S. Geol. Surv. Prof. Paper 556, 113 p.

Chew, R. T., III, 1956, Uranium and vanadium deposits of the Colorado Plateau that produced more than 1,000 tons of ore through June 30, 1955: U.S. Geol. Surv. Mineral Invest., Field Studies Map MF-54, 1:750,000

Coleman, R. G., 1957, Mineralogical evidence on the temperature of formation of the Colorado Plateau uranium deposits: Econ. Geol., v. 52, p. 1-4

Craig, L. C., and others, 1955, Stratigraphy of the Morrison and related formations, Colorado Plateau, a preliminary report: U.S. Geol. Surv. Bull. 1009-E, p. 125-168

Davidson, C. F., 1955, Concentration of uranium by carbon compounds: Econ. Geol., v. 50, p. 879-880

Dodd, P. H., 1956, Some examples of uranium deposits in the Upper Jurassic Morrison formation on the Colorado Plateau, in *Contributions to the geology of uranium and thorium*: U.S. Geol. Surv. Prof. Paper 300, p. 243-262

Evans, H. T., Jr., and Garrels, R. M., 1958, Thermodynamic equilibria of vanadium in aqueous systems as applied to the interpretation of Colorado Plateau ore deposits: Geochimica et Cosmochimica Acta, v. 15, p. 131-149

Finch, W. I., 1955, Preliminary geologic map showing the distribution of uranium deposits and principal ore-bearing formations of the Colorado Plateau region: U.S. Geol. Surv. Mineral Invest., Field Studies Map MF-16, 1:500,000

—— 1959, Geology of uranium deposits in Triassic rock of the Colorado Plateau region: U.S. Geol. Surv. Bull. 1074-D, p. 125-164

—— 1964, Epigenetic uranium deposits in sandstone: U.S. Geol. Surv. Prof. Paper 501-D, p. D76-D79

—— 1967, Geology of epigenetic uranium deposits in sandstone: U.S. Geol. Surv. Prof. Paper 538, 121 p.

Fischer, R. P., 1937, Sedimentary deposits of copper, vanadium-uranium and silver in the southwestern United States: Econ. Geol., v. 32, p. 906-951

—— 1942, Vanadium deposits of Colorado and Utah: U.S. Geol. Surv. Bull. 936-P, p. 363-394

—— 1947, Deposits of vanadium-bearing sandstone, in Vanderwilt, J. W., Editor, *Mineral resources of Colorado*: State of Colo., Mineral Res. Bd., p. 451-456

—— 1950, Uranium-bearing sandstone deposits of the Colorado Plateau: Econ. Geol., v. 45, p. 1-11

—— 1956, Uranium-vanadium-copper deposits on the Colorado Plateau, in *Contributions to the geology of uranium and thorium*: U.S. Geol. Surv. Prof. Paper 300, p. 143-154

—— 1968, The uranium and vanadium deposits of the Colorado Plateau region, in

Ridge, J. D., Editor, *Ore deposits of the United States, 1933-1967* (Graton-Sales Volumes): Chap. 35, v. 1, 1880 p.

—— 1970, Similarities, differences, and some genetic problems of the Wyoming and Colorado Plateau types of uranium deposits in sandstone: Econ. Geol., v. 65, p. 778-784

Fischer, R. P., and Stewart, J. H., 1961, Copper, vanadium, and uranium deposits in sandstone--their distribution and geochemical cycles: Econ. Geol., v. 56, p. 509-520

Frondel, C., 1958, Systematic mineralogy of uranium and thorium: U.S. Geol. Surv. Bull. 1064, 400 p.

Garrels, R. M., 1953, Some thermodynamic relations among the vanadium oxides and their relation to the oxidation states of the uranium ores of the Colorado Plateaus: Amer. Mineral., v. 38, p. 1251-1265

—— 1955, Some thermodynamic relations among the uranium oxides and their relations to the oxidation states of the uranium ores of the Colorado Plateau: Amer. Mineral., v. 40, p. 1004-1021

Granger, H. C., and Warren, C. G., 1969, Unstable sulfur compounds and the origin of roll-type uranium deposits: Econ. Geol., v. 64, p. 160-171

Gruner, J. W., 1954, The origin of the uranium deposits of the Colorado Plateau and adjacent regions: Mines Mag., v. 44, no. 3, p. 53-56

—— 1956, Concentration of uranium in sediments by multiple migration-accretion: Econ. Geol., v. 51, p. 495-520

—— 1956, Mineral associations in the continental-type uranium deposits of the Colorado Plateau and adjacent areas: Intermountain Assoc. Petrol. Geols. 7th Ann. Field Conf. Guidebook, p. 151-160

Hail, W. J., Jr., and others, 1956, Uranium in asphalt-bearing rocks of the western United States, in *Contributions to the geology of uranium and thorium*: U.S. Geol. Surv. Prof. Paper 300, p. 521-526

Heinrich, E. W., 1958, Epigenetic stratiform deposits in sedimentary rocks, in *Mineralogy and geology of radioactive raw materials*: McGraw-Hill, N.Y., p. 357-420

Hess, F. L., 1933, Uranium, vanadium, radium, gold, silver, and molybdenum sedimentary deposits, in *Ore deposits of the western states* (Lindgren Volume): A.I.M.E., p. 450-481

Hostetler, P. B., and Garrels, R. M., 1962, Transportation and precipitation of uranium and vanadium at low temperatures, with special reference to the sandstone-type uranium deposits: Econ. Geol., v. 57, p. 137-167

Hunt, C. B., 1956, Cenozoic geology of the Colorado Plateau: U.S. Geol. Surv. Prof. Paper 279, 99 p.

Isachsen, Y. W., and Evensen, C. G., 1956, Geology of uranium deposits of the Shinarump and Chinle formations on the Colorado Plateau, in *Contributions to the geology of uranium and thorium*: U.S. Geol. Surv. Prof. Paper 300, p. 263-280

Isachsen, Y. W., and others, 1955, Age and sedimentary environments of uranium host rocks, Colorado Plateau: Econ. Geol., v. 50, p. 127-134

Jensen, M. L., 1967, Stable isotopes and the origin of uranium deposits of Utah: Utah Geol. Soc., Guidebook to the Geology of Utah, no. 21, p. 78-90

Jobin, D. A., 1956, Regional transmissivity of the exposed sediments of the Colorado Plateau as related to distribution of uranium deposits, in *Contributions to the geology of uranium and thorium*: U.S. Geol. Surv. Prof. Paper 300, p. 207-211

—— 1962, Relation of the transmissive character of the sedimentary rocks of the Colorado Plateau to the distribution of uranium deposits: U.S. Geol. Surv. Bull 1124, 151 p.

Kelley, V. C., 1955, Monoclines of the Colorado Plateau: Geol. Soc. Amer. Bull., v. 66, p. 789-804

—— 1955, Regional tectonics of the Colorado Plateau and relationship to the origin and distribution of uranium: Univ. New Mex. Pubs. in Geol., no. 5, 120 p.

—— 1956, Influence of regional structure and tectonic history upon the origin and distribution of uranium on the Colorado Plateau, in *Contributions to the geology of uranium and thorium*: U.S. Geol. Surv. Prof. Paper 300, p. 171-178

Kelley, V. C., and Clinton, N. J., 1960, Fracture systems and tectonic elements of the Colorado Plateau: Univ. New Mex. Pubs. in Geol., no. 6, 104 p.

Kerr, P. F., 1958, Criteria of hydrothermal emplacement in Plateau uranium strata: 2d U.N. International Conf. on Peaceful Uses of Atomic Energy (Geneva) Pr., v. 2, p. 330-334

—— 1958, Uranium emplacement in the Colorado Plateau: Geol. Soc. Amer. Bull., v. 69, p. 1075-1111

Keys, W. S., and Dodd, P. H., 1958, Lithofacies of continental sedimentary rocks related to significant uranium deposits in the western United States: 2d U.N. International Conf. on Peaceful Uses of Atomic Energy (Geneva) Pr., v. 2, p. 367-378

Laverty, R. A., and Gross, E. B., 1956, Paragenetic studies of uranium deposits of the Colorado Plateau, in *Contributions to the geology of uranium and thorium*: U.S. Geol. Surv. Prof. Paper 300, p. 195-201

Maugher, R. L., 1967, A summary of isotopic ages of Colorado Plateau, Utah, mineral deposits: Utah Geol. Soc., Guidebook to the Geology of Utah, no. 21, p. 91-98

McKay, E. J., 1955, Criteria for outlining areas favorable for uranium deposits in Colorado and Utah: U.S. Geol. Surv. Bull. 1009-J, p. 265-282

Miesch, A. T., 1963, Distribution of elements in Colorado Plateau uranium deposits--a preliminary report: U.S. Geol. Surv. Bull. 1147-E, p. E1-E57

Miesch, A. T., and others, 1960, Chemical composition as a guide to the size of sandstone-type uranium deposits in the Morrison formation on the Colorado Plateau: U.S. Geol. Surv. Bull. 1112-B, p. 17-61

Miller, D. S., and Kulp, J. L., 1958, Isotopic study of some Colorado Plateau ores: Econ. Geol., v. 53, p. 937-948

—— 1963, Isotopic evidence on the origin of the Colorado Plateau uranium ores: Geol. Soc. Amer. Bull., v. 74, p. 609-630

Miller, L. J., 1958, The chemical environment of pitchblende: Econ. Geol., v. 53, p. 521-545

Noble, E. A., 1960, Genesis of uranium belts of the Colorado Plateau: 21st Int. Geol. Cong. Rept., pt. 15, p. 26-39

Page, L. R., 1960, The source of uranium in ore deposits: 21st Int. Geol. Cong. Rept., pt. 15, p. 149-164

Reinhardt, E. V., 1954, Structural controls of uranium ore deposits: Min. Cong. Jour., v. 40, no. 10, p. 49-52, 56

Rosenzweig, A., and others, 1954, Widespread occurrence and character of uraninite in the Triassic and Jurassic sediments of the Colorado Plateau: Econ. Geol., v. 49, p. 351-361

Routhier, P., 1963, Les gisements d'uranium de type "red beds"--Plateau du Colorado, in *Les gisements métallifères--Géologie et principes de recherches:*

t. 1, Masson et Cie, Paris, p. 398-400, 443

Shawe, D. R., 1966, Zonal distribution of elements in some uranium-vanadium roll and tabular ore bodies on the Colorado Plateau: U.S. Geol. Surv. Prof. Paper 550-B, p. B169-B171

Shoemaker, E. M., 1956, Structural features of the central Colorado Plateau and their relation to uranium deposits, in *Contributions to the geology of uranium and thorium*: U.S. Geol. Surv. Prof. Paper 300, p. 155-170

Shoemaker, E. M., and others, 1959, Elemental composition of the sandstone-type deposits (Colorado Plateau): U.S. Geol. Surv. Prof. Paper 320, p. 25-54

Stieff, L. R., and Stern, T. W., 1952, Identification and lead-uranium ages of massive uraninites from the Shinarump conglomerate, Utah: Science, v. 115, no. 3000, p. 706-708

—— 1956, Interpretation of the discordant age sequence of uranium ores, in *Contributions to the geology of uranium and thorium*: U.S. Geol. Surv. Prof. Paper 300, p. 549-555

—— 1959, Isotopic study of some Colorado Plateau ores: Econ. Geol., v. 54, p. 752

Stieff, L. R., and others, 1953, A preliminary determination of the age of some uranium ores of the Colorado Plateau by the lead-uranium method: U.S. Geol. Surv. Circ. 271, 19 p.

—— 1956, Coffinite, a uranous silicate with hydroxyl substitution; a new mineral: Amer. Mineral., v. 41, p. 675-688

Stokes, W. L., 1967, Stratigraphy and primary sedimentary features of uranium occurrences of southeastern Utah: Utah Geol. Soc., Guidebook to the Geology of Utah, no. 21, p. 32-52

Thomson, K. C., 1967, Structural features of southeastern Utah and their relations to uranium deposits: Utah Geol. Soc., Guidebook to the Geology of Utah, no. 21, p. 23-31

Waters, A. C., and Granger, H. C., 1953, Volcanic debris in uraniferous sandstones and its possible bearing on the origin and precipitation of uranium: U.S. Geol. Surv. Circ. 224, 26 p.

Weeks, A. D., 1956, Mineralogy and oxidation of the Colorado Plateau uranium ores, in *Contributions to the geology of uranium and thorium*: U.S. Geol. Surv. Prof. Paper 300, p. 187-193

Weeks, A. D., and Garrels, R. M., 1959, Geologic setting of the Colorado Plateau ores: U.S. Geol. Surv. Prof. Paper 320, p. 3-11

Weeks, A. D., and others, 1959, Geochemistry and mineralogy of the Colorado Plateau uranium ores; summary of the ore mineralogy: U.S. Geol. Surv. Prof. Paper 320, p. 65-79

Witkind, I. J., 1964, Age of the grabens in southeastern Utah: Geol. Soc. Amer. Bull., v. 75, p. 99-106

Wright, R. J., 1955, Colorado Plateau uranium deposits: Econ. Geol., v. 50, p. 884-885

—— 1955, Ore controls in sandstone uranium deposits of the Colorado Plateau: Econ. Geol., v. 50, p. 135-155

H. WHITE CANYON-ELK RIDGE, UTAH

Benson, W. E., and others, 1952, Preliminary report on the White Canyon area, San Juan County, Utah: U.S. Geol. Surv. Circ. 217, 10 p.

Campbell, R. H., and Lewis, R. Q., 1961, Distribution of uranium ore deposits in the Elk Ridge area, San Juan County, Utah: Econ. Geol., v. 56, p.

111-131

Finnell, T. L., and Gazdik, W. B., 1958, Structural relations at the Hideout No. 1 uranium mine, Deer Flat area, San Juan County, Utah: Econ. Geol., v. 53, p. 949-957

Finnell, T. L., and others, 1963, Geology, ore deposits, and exploratory drilling in the Deer Flat area, White Canyon district, San Juan County, Utah: U.S. Geol. Surv. Bull. 1132, 114 p.

Grundy, W. D., and Oertell, E. W., 1958, Uranium deposits in the White Canyon and Monument Valley mining districts, San Juan County, Utah, and Navajo and Apache Counties, Arizona: Intermountain Assoc. Petrol. Geols. 9th Ann. Field Conf. Guidebook, p. 197-207

Johnson, H. S., Jr., and Thorardson, W., 1959, The Elk Ridge-White Canyon Channel system, San Juan County, Utah; its effects on uranium distributions: Econ. Geol., v. 54, p. 119-129

—— 1966, Uranium deposits of the Moab, Monticello, White Canyon, and Monument Valley districts, Utah and Arizona: U.S. Geol. Surv. Bull. 1222-H, p. H1-H53

Lewis, R. Q., Sr., and Campbell, R. H., 1958-1959, Preliminary geologic maps of the Elk Ridge quadrangle, San Juan County, Utah: U.S. Geol. Surv. Mineral Invest., Field Studies Maps MF-190, -191, -192, -194, -199, -200, -201 (7 sheets), 1:24,000

—— 1965, Geology and uranium deposits of Elk Ridge and vicinity, San Juan County, Utah: U.S. Geol. Surv. Prof. Paper 474-B, p. B1-B65

Miller, L. J., 1955, Uranium ore controls of the Happy Jack deposit, White Canyon, San Juan County, Utah: Econ. Geol., v. 50, p. 111-126

Mitcham, T. W., and Evensen, C. G., 1955, Uranium ore guides, Monument Valley district, Arizona: Econ. Geol., v. 50, p. 170-176

Thaden, R. E., and others, 1964, Geology and ore deposits of the White Canyon area, San Juan County, Utah: U.S. Geol. Surv. Bull. 1125, 166 p.

Trites, A. F., Jr., and Chew, R. T., III, 1955, Geology of the Happy Jack mine, White Canyon area, San Juan County, Utah: U.S. Geol. Surv. Bull. 1009-H, p. 235-248

Trites, A. F., Jr., and others, 1956, Uranium deposits in the White Canyon area, San Juan County, Utah, in *Contributions to the geology of uranium and thorium*: U.S. Geol. Surv. Prof. Paper 300, p. 281-284

—— 1959, Mineralogy of the uranium deposit at the Happy Jack mine, San Juan County, Utah: U.S. Geol. Surv. Prof. Paper 320, p. 185-195

Young, R. G., 1964, Distribution of uranium deposits in the White Canyon-Monument Valley district, Utah-Arizona: Econ. Geol., v. 59, p. 850-873

J. MONUMENT VALLEY, UTAH-ARIZONA

Beaumont, E. C., and Dixon, G. H., 1965, Geology of the Kayenta and Chilchinbito quadrangles, Navajo County, Arizona: U.S. Geol. Surv. Bull. 1202-A, p. A1-A28

Evensen, C. G., and Gray, I. B., 1958, Evaluation of uranium ore guides, Monument Valley, Arizona and Utah: Econ. Geol., v. 53, p. 639-662

Finnell, T. L., 1957, Structural control of uranium ore at the Monument No. 2 mine, Apache County, Arizona: Econ. Geol., v. 52, p. 25-35

Grundy, W. D., and Oertell, E. W., 1958, Uranium deposits in the White Canyon and Monument Valley mining districts, San Juan County, Utah, and Navajo and Apache Counties, Arizona: Intermountain Assoc. Petrol. Geols. 9th Ann. Field Conf. Guidebook, p. 197-207

Johnson, H. S., Jr., and Thorardson, W., 1966, Uranium deposits of the Moab, Monticello, White Canyon, and Monument Valley districts, Utah and Arizona: U.S. Geol. Surv. Bull. 1222-H, p. H1-H53

Lewis, R. Q., Sr., and Trimble, D. E., 1959, Geology and uranium deposits of Monument Valley, San Juan County, Utah: U.S. Geol. Surv. Bull. 1087-D, p. 105-131

Malan, R. C., 1968, The uranium mining industry and geology of the Monument Valley and White Canyon districts, Arizona and Utah, in Ridge, J. D., Editor, *Ore deposits of the United States, 1933-1967* (Graton-Sales Volumes): Chap. 38, v. 1, 1880 p.

Mitcham, T. W., 1957, Uranium ore guides, Monument Valley, Arizona: Econ. Geol., v. 52, p. 586-588

Mitcham, T. W., and Evensen, C. G., 1955, Uranium ore guides, Monument Valley district, Arizona: Econ. Geol., v. 50, p. 170-176

Witkind, I. J., 1956, Channels and related swales at the base of the Shinarump conglomerate, Monument Valley, Arizona, in *Contributions to the geology of uranium and thorium*: U.S. Geol. Surv. Prof. Paper 300, p. 233-237

—— 1956, Uranium deposits at the base of the Shinarump conglomerate, Monument Valley, Arizona: U.S. Geol. Surv. Bull. 1030-C, p. 99-130

Witkind, I. J., and Thaden, R. E., 1963, Geology and uranium-vanadium deposits of the Monument Valley area, Apache and Navajo Counties, Arizona: U.S. Geol. Surv. Bull. 1103, 171 p.

K. BLACK MESA-HOPI BUTTES, ARIZONA

Anderson, R. Y., and Harshbarger, J. W., Editors, 1958, Guidebook of the Black Mesa Basin, northeastern Arizona: New Mex. Geol. Soc. 9th Ann. Field Conf. Guidebook, p. 65, 88-149, 161-163

Shoemaker, E. M., 1956, Occurrence of uranium in diatremes on the Navajo and Hopi Reservations, Arizona, New Mexico, and Utah, in *Contributions to the geology of uranium and thorium*: U.S. Geol. Surv. Prof. Paper 300, p. 179-185

Shoemaker, E. M., and others, 1963, Diatremes and uranium deposits in the Hopi Buttes, Arizona, in Engel, A.E.J., and others, Editors, *Petrologic studies: A volume in honor of A. F. Buddington*: Geol. Soc. Amer., p. 327-356

L. CAMERON, ARIZONA

Akers, J. P., and others, 1962, Geology of the Cameron quadrangle, Arizona: U.S. Geol. Surv. Geol. Quad. Map GQ-162, 1":5208' (with text)

Barrington, J., and Kerr, P. F., 1961, Breccia pipe near Cameron, Arizona: Geol. Soc. Amer. Bull., v. 72, p. 1661-1674; disc., v. 74, p. 227-232, 233-237

—— 1963, Collapse features and silica plugs near Cameron, Arizona: Geol. Soc. Amer. Bull., v. 74, p. 1237-1258

Bollin, E. M., and Kerr, P. F., 1958, Uranium mineralization near Cameron, Arizona: New Mex. Geol. Soc. 9th Ann. Field Conf. Guidebook, p. 164-168

M. HENRY MOUNTAINS-GREEN RIVER, UTAH

Davidson, E. S., 1959, Geology of the Rainy Day uranium mine, Garfield County, Utah: Econ. Geol., v. 54, p. 436-448

Doelling, H. H., 1967, Uranium deposits of Garfield County, Utah: Utah Geol. and Mineral. Surv. Spec. Studies 22, 113 p.

Hunt, C. B., and others, 1953, Geology and geography of the Henry Mountains, Utah: U.S. Geol. Surv. Prof. Paper 228, 234 p.

Johnson, H. S., Jr., 1959, Uranium resources of the Green River and Henry Mountains districts, Utah--a regional synthesis: U.S. Geol. Surv. Bull. 1087-C, p. 59-104

Steed, R. H., 1954, Geology of Circle Cliffs anticline: Intermountain Assoc. Petrol. Geols. 5th Ann. Field Conf. Guidebook, p. 99-102

N. SAN RAFAEL-CEDAR MOUNTAIN, UTAH

Breger, I. A., and Duel, M., 1959, Association of uranium with carbonaceous materials, with special reference to Temple Mountain region: U.S. Geol. Surv. Prof. Paper 320, p. 139-149

Clark, E. L., and Million, I., 1956, Uranium deposits in the Morrison formation of the San Rafael district: Intermountain Assoc. Petrol. Geols. 7th Ann. Field Conf. Guidebook, p. 155-160

Hawley, C. C., and others, 1965, Geology and uranium deposits of the Temple Mountain district, Emery County, Utah: U.S. Geol. Surv. Bull. 1192, 154 p.

—— 1968, Geology, altered rocks, and ore deposits of the San Rafael Swell, Emery County, Utah: U.S. Geol. Surv. Bull. 1239, 115 p.

Johnson, H. S., Jr., 1957, Uranium resources of the San Rafael district, Emery County, Utah: U.S. Geol. Surv. Bull. 1046-D, p. 37-54

—— 1959, Uranium resources of the Cedar Mountain area, Emery County, Utah--a regional synthesis: U.S. Geol. Surv. Bull. 1087-B, p. 23-58

Kelley, D. R., and Kerr, P. F., 1957, Clay alteration and ore, Temple Mountain, Utah: Geol. Soc. Amer. Bull., v. 68, p. 1101-1116

—— 1958, Urano-organic ore at Temple Mountain, Utah: Geol. Soc. Amer. Bull., v. 69, p. 701-756

Kerr, P. F., and Kelley, D. R., 1956, Urano-organic ores of the San Rafael Swell, Utah: Econ. Geol., v. 51, p. 386-391

Kerr, P. F., and others, 1957, Collapse features, Temple Mountain uranium area, Utah: Geol. Soc. Amer. Bull., v. 68, p. 933-981

Keys, W. S., 1956, Investigation of the Temple Mountain collapse, San Rafael Swell, Emery County, Utah, in *Contributions to the geology of uranium and thorium*: U.S. Geol. Surv. Prof. Paper 300, p. 285-298

Wood, H. B., 1959, Uranium ore controls and guides in the San Rafael Swell, Utah: 20th Int. Geol. Cong., Sec. 13, p. 415-434

Notes

The Colorado Plateau is a rudely circular region some 150,000 square miles in area centering around Monument Valley in Utah just north of the Arizona line; it is composed of many smaller plateaus that are separated from each other by mountains, valleys, plains, and mesas. Although as a tectonic unit the Plateau does not exactly coincide with the Plateau as a physiographic unit, it is a region of appreciably greater structural stability than the more active tectonic units of the Rocky Mountains that border it. Although uranium-vanadium ore bodies are present over much of the Plateau belts, along its western and southern margins there is little mineralization.

The uranium-bearing Plateau has been divided by various authors into a considerable number of districts that attempt to group deposits geographically and/or genetically related together. The grouping adopted here is based largely on that shown in *Guidebook to the Geology of Utah*, No. 21 (p. 17, Fig. 2). It has been expanded to include districts in Arizona, New Mexico, and Colorado that are not shown in Figure 2. The division made in Colorado does not follow the terminology of Figure 2, and the districts in Colorado and Utah are arbitrarily divided by the state boundary. The boundaries of all the districts are shown on the map of the Colorado Plateau in this volume. The letters A through F and H through N apply to the districts. The Kaiparowits district is designated by the letter I on the map but no literature is cited for it. The letter G in the foregoing references cited lists general papers on the Plateau as a unit.

The rocks of the Plateau range in age from Precambrian (strongly deformed and metamorphosed) to Tertiary, but the principal uranium and vanadium deposits are located in Mesozoic rocks, mainly, but not entirely, those of Triassic and Jurassic age; a few minable deposits have been found in Pennsylvanian and Permian rocks. Tertiary formations are abundantly present only on the Plateau margins and are of limited extent in the interior; they are largely unmineralized. The oldest Triassic formation in the Plateau region is the Lower and Middle Triassic Moenkopi that is separated by a pronounced unconformity from the Permian formations beneath it; it is not an important ore-bearing formation. Unconformably above the Moenkopi is the Upper Triassic Shinarump conglomerate that contains important ore bodies. The Shinarump may be the basal conglomerate of the overlying Upper Triassic Chinle formation that outcrops widely over the Plateau and contains a considerable number of valuable deposits. The Chinle contains lenticular and inter-tonguing conglomerates, sandstones, shales, and limestones and has a basal sandstone-conglomerate member (the Moss Back) that can be confused with the Shinarump. The formations lying between the Chinle and the Upper Jurassic Todilto limestone of the San Rafael group are unimportant as hosts to uranium deposits (although the Upper Jurassic Entrada sandstone contains some uranium deposits). The most important loci of ore deposition on the Plateau are in the uppermost Jurassic Morrison formation (lying immediately above the San Rafael group) which, over most of the Plateau, is divisible into four members (two upper and two lower) of quite different lithologies. The lowest member, the Salt Wash, is composed of clastic material which becomes increasingly finer grained to the northeast; it intergrades with the Recapture member (centering around the Four Corners area) that is essentially equivalent to the Salt Wash in age. The upper Morrison is composed of the Brushy Basin and the Westwater Canyon members; the Westwater also centers in the Four Corners area and grades northward into the Brushy Basin. Cretaceous formations are quite varied in character and are widely exposed on the Plateau but are unimportant as hosts for ore deposits.

Although the large proportion of the rocks of the Plateau are sedimentary in origin, there has been widespread igneous activity throughout the region, most of which occurred from Laramide time into the Pliocene or even Quaternary. Within the borders of the Plateau, igneous activity took place within centers of two main types: (1) the porphyritic laccolith-bysmalith-sill type and (2) the plug-volcano type. The main laccolithic centers on the Plateau are in the La Sal, Ute, Abajo, and Henry Mountains, with Navajo Mountain probably being a similar center, although no igneous rocks are exposed there; the laccoliths range in age from Late Cretaceous to Oligocene or Miocene. The principal centers of plug-volcanos are in the Taylor, Zuni, and San Francisco Mountains (where the rocks are chiefly basalt, with some andesite and rhyolite) and Hopi Buttes and Navajo centers (where the rocks are alkaline mafic and ultramafic types such as minette, monchiquite, and kimberlite). (The Navajo plug-volcano center should not be confused with Navajo Mountain which lies farther west.) The rocks of the plug-volcanos are mainly Pliocene in age. Cryptovolcanic structures that resemble diatremes, but lack igneous rocks, are widespread through the Plateau; in places both diatremes and cryptovolcanic structures carry uranium ores.

The present structural features of the rocks of the Plateau were developed in Laramide time, although minor faulting occurred during the Pliocene. During early Miocene time, the Plateau was raised above the level of the valleys in the adjacent basin and range country, and a pronounced erosional period began. Intrusive igneous activity of the laccolith-bysmalith-sill type in the central Plateau appears to have ended in the Miocene, and volcanic activity in the southern Plateau began at about the same time. Uplift of the

Plateau along border faults continued into the middle Pliocene, and the Plateau was generally tilted to the northeast; lavas were extruded on the southern and western edges of the Plateau into the upper Pliocene and even into the Recent.

From the geologic evidence, the age of the uranium mineralization of the Colorado Plateau is difficult to determine. Primary (unoxidized) uranium deposits occur in the diatremes of the Hopi Buttes area or closely adjacent to them; thus, these deposits probably were introduced after the diatremes were formed, an event that occurred in Pliocene times. From this it follows that some of the uranium of the Plateau was emplaced as late as the end of the Tertiary. On the other hand, intrusives of the laccolithic type, such as those of the Henry and La Sal Mountains, probably were introduced at an appreciably earlier time than those of the plug-volcano type associated with the diatremes. Further, in the Henry Mountains, bedded uranium deposits occur in the highly distorted rocks on the south flank of Mount Hillers where the beds dip off to the south at 85° and are highly fractured but lack any mineralization in these fractures. It would seem almost certain that, if the uranium in the bedded deposits had been supplied from the same source as the magmas of the laccoliths, some mineralization would have been deposited in the fractured rocks and that not all of it would have been emplaced in the bedded rocks. From this it follows that there must have been at least two periods of uranium mineralization on the Colorado Plateau, one in Pliocene time associated with plug volcanos, and the other of Miocene or earlier age, the second having occurred prior to the latest laccolithic intrusions.

Throughout the Plateau area, however, the lead-uranium isotopic ages that have been determined range from 22 to 255 m.y. For most of the samples studied, the U^{238}-Pb^{206} ages agree with the U^{235}-Pb^{207} ages within 5 percent, but the latter ages are consistently greater than the former and sometimes are much greater. If these results are to be taken at face value, they indicate that the deposition of uranium in the Plateau rocks took place between the middle Permian and the early Miocene.

Still unexplained, moreover, is the wide spread of age determination values between 255 m.y. at Haystack Butte in the Grants area and 22 m.y. at the Happy Jack mine in Utah. One possible explanation, of course, is that the ores were actually deposited at the times indicated by the determinations. The 255 m.y. ores, however, occur in the Westwater Canyon member of the Morrison which, as Upper Jurassic, cannot have an age of more than 150 m.y. It seems obvious that the pitchblende involved cannot have been deposited in the Grants area some 100 m.y. before the formations, and some other explanation must be put forward.

Perhaps this explanation is that there is something wrong with the major assumptions. As Miller and Kulp (1958) say, "...discordance among the three isotopic ages (and the discordance between Pb^{207}/Pb^{206} and U^{238}/Pb^{206} ages usually is a few hundred m.y.) is a very clear index that one of the major assumptions had been violated." It seems reasonable to leave it at that and to attempt to find criteria on which to base the absolute age of the deposits other than those supplied by isotopic ratios.

On geologic grounds (as has already been demonstrated), it appears certain that the uranium and vanadium ore bodies of the Plateau were formed in at least two stages, one of which occurred appreciably before the intrusion of the Henry Mountains laccoliths and the other shortly after the formation of the Hopi Buttes diatremes. In the adjacent San Juan Mountains, where the laccoliths bear a considerable resemblance to those on the Plateau, there were also two periods of mineralization, one of which occurred during the Laramide orogeny and the other during the early Pliocene. Reasoning by analogy, it appears possible that the two periods of primary Plateau mineralization (whether the ore fluids were of magmatic origin or were uranium-bearing ground water), despite the wide differences between the San Juan and Plateau ores, may have occurred at the same times as those in the San Juan Mountains. Until some more firm basis is provided than at present exists for accepting the ages established by the lead-uranium isotope work, it will be considered that the great majority of the primary (pitchblende) ore bodies of the Plateau were introduced in, or immediately following, Laramide time and that a few of them, including some addition of ore minerals to deposits already in existence, occurred in the Pliocene. The ages assigned to the Plateau deposits, therefore, are late Mesozoic to early Tertiary (major part) and late Tertiary (minor part).

Much of the ore mined on the Colorado Plateau has come from deposits in which the primary uranium and vanadium minerals have been oxidized to a wide variety of mineral species. These highly colored secondary minerals probably were formed above the water table by the action of oxygen-rich ground water on the highly susceptible minerals of the primary ores. Not only is the chemistry of the conversion of the lower oxidation states of uranium and vanadium to higher ones by near-surface phenomena well known but so are the relations of the physical characteristics of the rocks to the oxidation and solution of these two elements and their precipitation as secondary U^{+6}-bearing minerals. Although it has been suggested that near-surface conditions can also result in the reduction of U^{+6} to U^{+4}, the chemistry of such reactions under natural conditions is far less well understood. It is, therefore, here proposed to assume the formation of the secondary uranium minerals by ground water activity and to concentrate this discussion on the development of the primary pitchblende from which these secondary minerals were derived.

Most of the primary uranium deposits are tabular to lenticular masses of greater or lesser degrees of irregularity in outline. Although there is some connection between ore deposits and major tectonic features, such as salt anticlines, laccolithic domes, uplifts, monoclines, synclinal axes, and diatremes, the most important controls were provided by structures of a lower order of magnitude such as (1) lenticular masses of sandstone, pure to half mudstone, deposited in a stream-flood plain environment and (2) paleostream scour channels cut into underlying beds and filled by sandstone to sandstone-mudstone materials. Within such permeable rocks, deposition often appears to have been controlled by (1) uniform grain size and sorting, (2) carbonaceous (fossil plant) or asphaltic content, (3) lithofacies boundaries, (4) low to moderate concentrations of calcite, and (5) high kaolinite concentrations. Such controls either aid ready movement of solutions or promote precipitation of ore minerals but say nothing as to the character or source of the ore fluids involved.

Several authors (Fischer, 1956; Weeks, 1956; Botinelly and Weeks, 1957) have indicated that the primary Plateau ores may be divided into four main types: (1) vanadiferous with V:U = 15:1, (2) vanadiferous with V:U = 15:1 to 1:1, (3) nonvanadiferous, sulfide-poor, and (4) nonvanadiferous, sulfide-rich including the U-Cu type. Type (1) deposits are generally black, and their minerals include pitchblende [If a mineral can be designated as uraninite only if its composition, disregarding impurities, is UO_2, there is no uraninite in the world, because the least oxidized uranium oxide known (Frondel, U.S.G.S. Bull. 1064) has an average composition of $UO_{2.15}$. On the Plateau the ratio of uranium to oxygen is appreciably less than 1:2.15, and there is some question as to whether or not it should be referred to as uraninite or pitchblende. The primary uranium oxide of the Plateau has a much smaller content of rare earths and thorium than does uraninite from pegmatites, is much more finely crystalline, and is often botryoidal in outline; it is, therefore, here designated as pitchblende, with the reminder that it is called uraninite in much of the literature on the Plateau deposits. The Colorado Plateau pitchblende appears to average about $2UO_2 \cdot UO_3$ or U_3O_7.] and coffinite [$U(SiO_4)_{1-x}(OH)_{4x}$], and low-valence vanadium minerals, especially micaceous ones and montroseite [(Fe,V)O(OH)], a diaspore-type mineral; minor amounts of iron, copper, lead and zinc sulfides, selenides, and arsenides also may be present. Type (2) ores differ from type (1) in their lower vanadium-uranium ratios and in including carbonized wood and other plant debris. Type (3) ores consist of pitchblende, with coffinite scarce or lacking, and little else. Type (4) ores usually contain two generations of pyrite, one before and one after the pitchblende, nickel-cobalt minerals, chalcopyrite, bornite, chalcocite, covellite, and some sphalerite and galena.

In addition to the original detrital minerals (quartz, feldspars, clays, and carbonate) and authogenic quartz, certain minerals (pyrite and marcasite, illite, calcite, dolomite, barite, and fluorite) are probably pre-ore and may have been deposited early in the mineralization cycle or may have been introduced diagenetically. The primary ore minerals are pitchblende, coffinite, and (in the vanadiferous deposits) roscoelite (vanadium muscovite), vanadium clays, doloresite ($H_8V_6O_{16}$), montroseite, sulfides, and calcite; galena may have deposited before pitchblende, but much of it is younger than the uranium mineral and the chalcopyrite. Pitchblende may have continued to form after the iron and copper sulfides had ceased to deposit, though this does not seem to have been the usual situation.

Four theories have been advanced to explain the origin of the Plateau uranium-vanadium deposits: (1) that they were emplaced syngenetically with the enclosing sediments in a nonoxidizing environment as low-valence uranium and vanadium minerals--the syngenetic theory; (2) that the uranium and vanadium were obtained by the supergene leaching of volcanic ash, transported by ground water, and deposited where they are now found, presumably beneath the water table as low-valence uranium and vanadium minerals that were later oxidized when brought above the water table--the ash-leach theory; (3) that the uranium and vanadium were derived by the solvent action of ground water on the appropriate minerals in igneous rocks and were transported and redeposited in their present sites over a considerable space of time--the circulatory ground water theory; and (4) that the uranium was brought into the area by magma-generated solutions--the hydrothermal theory.

The Plateau uranium (and vanadium) minerals either fill open spaces or replace minerals already present. This demonstrates that they were not detrital and must have been deposited from solutions that moved through the previous rocks in which they are found. This fact, of course, tells nothing about the origin and character of these solutions, but the association of pitchblende in many deposits with sulfides (particularly with chalcopyrite, bornite, and enargite) that are characteristic and/or diagnostic of mesothermal conditions suggests that they must have been at temperatures above 200 degrees and well may have been appreciably above that limit during the deposition of the bulk of the pitchblende. The presence of the vanadium micas and clays also indicates mesothermal conditions, and at least this degree of intensity is confirmed by the hydroxyl-bearing uranium silicate, coffinite. On the other hand, the rather considerable amounts of chalcocite and covellite with some of the pitchblende probably were deposited within the telethermal range of solution intensity.

Much of the pitchblende of the Plateau is found in deposits surrounded by halos of bleached rock, suggesting that ferric ion was involved in the precipitation of uranium oxide. If this is the case, then the following half reactions must have governed the formation of the pitchblende:

$$U^{+4} + 2H_2O = UO_2^{+2} + 4H^{+1} + 2e^{-1} \qquad -0.62\ v. \qquad (1)$$

$$Fe^{+2} = Fe^{+3} + e^{-1} \qquad -0.77\ v. \qquad (2)$$

These two half reactions can go forward only in the direction of Fe^{+3} and U^{+4} (which may have been present as UO_3^{-2}) to $U^6O_2^{+2}$ and Fe^{+2}. It would appear, therefore, that the pitchblende in such bleached-halo deposits probably was developed by some such reaction as that which follows:

$$3UO_3^{-2} + 2Fe^{+3} + 4H^{+1} = U_3O_7 + 2Fe^{+2} + 2H_2O. \qquad (3)$$

On the other hand, L. J. Miller (1958) provides a basis for saying that, if the uranyl ion (UO_2^{+2}) is brought into contact with hydrogen sulfide, the following reaction will occur:

$$6UO_2^{+2} + S^{-2}\ (\text{from } H_2S) + 6H_2O = 2U_3O_7 + SO_4^{-2} + 12H^{+1}. \qquad (4)$$

Thus, if it is considered likely that uranyl-bearing solutions would consistently have encountered rock volumes in which hydrogen sulfide was being produced in reasonably adequate amounts and at a fairly steady rate, much of the Plateau pitchblende may have been formed in the manner suggested by equation (4).

If the uranium ions in the ore-forming fluid were in the plus-four state, these ions and the solutions containing them probably were ultimately of magmatic origin. If the uranium ions were in the plus-six state, they probably were derived by ground-water leaching of volcanic ash beds or uranium-bearing igneous rocks. The difficulty lies in saying which state prevailed in the Plateau ore fluids.

The common association of bleached zones around the Plateau ore bodies suggests that the uranium was precipitated with the aid of ferric ion [equation (3)]. The presence of fossilized organic debris (coalified wood) indicates that there may have been a source of hydrogen sulfide in certain of the Plateau rocks and that some of the uranium may have been precipitated through such a reaction as that given in equation (4).

It appears impossible, at this stage, to say for certain if the primary (pitchblende) deposits of the Plateau were formed from solutions that derived their uranium content from magmatic sources or from ground waters that leached their uranium from rocks of igneous origin. I favor the concept that the uranium was obtained from magmatic sources because (1) of the apparent importance of ferric ion in the precipitation of the pitchblende, (2) the association of moderate-temperature minerals, such as sulfides, selenides, and arsenides of iron, copper, lead, and zinc, with some of the pitchblende ores, and (3) the difficulty of explaining how the necessary hydrogen sulfide required in such reactions as that of equation (4) could have been consistently available for pitchblende formation. The possibility, however, certainly exists that the Plateau pitchblende deposits were formed in both the manner of equation (3) and in that of equation (4). The uranium deposits of the Colorado Plateau, therefore, here are categorized as mesothermal to telethermal and/or ground water-B2 for the primary (pitchblende) ores and ground water-B2 for the secondary deposits.

Idaho

COEUR D'ALENE

Late Mesozoic to Early Tertiary	*Silver, Lead, Zinc, Copper*	*Hypothermal-1 to Leptothermal*

Anderson, A. L., 1949, Monzonite intrusion and mineralization in the Coeur d'Alene district, Idaho: Econ. Geol., v. 44, p. 169-185

Anderson, R. A., 1967, Graben structure in the Coeur d'Alene district: Econ. Geol., v. 62, p. 1092-1094

Anderson, R. J., 1940, Microscopic features of ore from the Sunshine mine: Econ. Geol., v. 35, p. 659-667

Arnold, R. R., and others, 1962, Temperature of crystallization of pyrrhotite and sphalerite from the Highland-Surprise mine, Coeur d'Alene district, Idaho: Econ. Geol., v. 57, p. 1163-1174

Chan, S.S.M., 1969, Suggested guides for exploration from geochemical investigation of ore veins at the Galena mine deposits, Shoshone County, Idaho: Colo. Sch. Mines Quart., v. 64, no. 1, p. 139-168

Clark, B. R., 1970, Origin of slaty cleavage in the Coeur d'Alene district, Idaho: Geol. Soc. Amer. Bull., v. 81, no. 10, p. 3061-3072

Crosby, G. M., 1956, Geology of the Hercules mine, Burke, Idaho: Min. Cong. Jour., v. 42, no. 6, p. 43-45, 82

—— 1959, The Gem stocks and adjacent ore bodies, Coeur d'Alene district, Idaho: A.I.M.E. Tr., v. 214, p. 697-700 (in Min. Eng., v. 11, no. 7)

—— 1969, A preliminary examination of trace mercury in rocks, Coeur d'Alene district, Wallace, Idaho: Colo. Sch. Mines Quart., v. 64, no. 1, p. 169-194

Folwell, W. T., 1958, Lucky Friday mine; history, geology, and development: Min. Eng., v. 10, no. 4, p. 1266-1268

Fryklund, V. C., Jr., 1960, Origin of the main period veins, Coeur d'Alene district, Idaho: U.S. Geol. Surv. Prof. Paper 400-B, p. B29-B30

—— 1964, Ore deposits of the Coeur d'Alene district, Shoshone County, Idaho: U.S. Geol. Surv. Prof. Paper 445, 103 p.

Fryklund, V. C., Jr., and Fletcher, J. D., 1956, Geochemistry of sphalerite from the Star mine, Coeur d'Alene district, Idaho: Econ. Geol., v. 51, p. 223-247

Fryklund, V. C., Jr., and Hutchinson, M. W., 1954, The occurrence of cobalt and nickel in the Silver Summit mine, Coeur d'Alene district, Idaho: Econ. Geol., v. 49, p. 753-758

Hobbs, S. W., and Fryklund, V. C., Jr., 1968, The Coeur d'Alene district, Idaho, in Ridge, J. D., Editor, *Ore deposits of the United States, 1933-1967* (Graton-Sales Volumes): Chap. 66, v. 2

Hosterman, J. W., 1956, Geology of the Murray area, Shoshone County, Idaho: U.S. Geol. Surv. Bull. 1027-P, p. 725-748

Kennedy, V. C., 1961, Geochemical studies in the Coeur d'Alene district, Shoshone County, Idaho: U.S. Geol. Surv. Bull. 1098-A, p. 1-55

Kerr, P. F., and Kulp, J. L., 1952, Pre-Cambrian uraninite, Sunshine mine, Idaho: Science, v. 115, no. 2978, p. 86-88

Kerr, P. F., and Robinson, R. F., 1953, Uranium mineralization in the Sunshine mine, Idaho: A.I.M.E. Tr., v. 196, p. 495-511 (in Min. Eng., v. 5, no. 5)

Kullerud, G., 1956, Geochemistry of sphalerite from the Star mine, Coeur d'Alene district, Idaho: Econ. Geol., v. 51, p. 828-830

Long, A., and others, 1960, Isotopic composition of lead and Precambrian mineralization of the Coeur d'Alene district, Idaho: Econ. Geol., v. 55, p. 645-658

McConnel, R. H., 1939, Bunker Hill ore deposits in complex fractures: Eng. and Min. Jour., v. 140, no. 8, p. 40-42

McKinstry, H. E., and Svendsen, R. H., 1942, Control of ore by rock structure in a Coeur d'Alene mine: Econ. Geol., v. 37, p. 215-230

Mitcham, T. W., 1952, Significant spatial distribution patterns of minerals in the Coeur d'Alene district, Idaho: Science, v. 115, no. 2975, p. 11

—— 1952, Indicator minerals, Coeur d'Alene silver belt: Econ. Geol., v. 47, p. 414-450

Ransome, F. L., and Calkins, F. C., 1908, Geology and ore deposits of the Coeur d'Alene district, Idaho: U.S. Geol. Surv. Prof. Paper 62, 203 p. (particularly p. 21-77, 104-143)

Reid, R. R., Editor, 1961, Guidebook to the geology of the Coeur d'Alene mining district: Idaho Bur. Mines and Geol., Bulletin 16, 37 p.

Searls, F., Jr., 1960, Isotopic composition of lead and Precambrian mineralization of the Coeur d'Alene district, Idaho: Econ. Geol., v. 55, p. 1565

Shenon, P. J., 1938, Geology and ore deposits near Murray, Idaho: Idaho Bur. Mines and Geol., Pamphlet no. 47, 44 p. (mimeo.) (particularly p. 3-26)

Shenon, P. J., and McConnel, R. H., 1939, The Silver Belt of the Coeur d'Alene district, Idaho: Idaho Bur. Mines and Geol., Pamphlet no. 50, 9 p. (mimeo.)

—— 1940, Use of sedimentation features and cleavage in the recognition of overturned strata: Econ. Geol., v. 35, p. 430-444

Silverman, A. J., and others, 1960, Age of the Coeur d'Alene mineralization; an isotopic study: Min. Eng., v. 12, no. 5, p. 470-471 (a discussion of Fryklund, 1960)

Sorenson, R. E., 1947, Deep discoveries intensify Coeur d'Alene activities: Eng. and Min. Jour., v. 148, no. 10, p. 70-78

—— 1948, Silver Summit opens rich ore: Eng. and Min. Jour., v. 149, no. 7, p. 70-73, 151

—— 1951, Shallow expression of Silver Belt ore shoots, Coeur d'Alene district, Idaho: A.I.M.E. Tr., v. 190, p. 605-611 (in Min. Eng., v. 3, no. 7)

Stringham, B., and others, 1953, Mineralization and hydrothermal alteration at the Hercules mine, Burke, Idaho: A.I.M.E. Tr., v. 196, p. 1278-1282 (in Min. Eng., v. 5, no. 12)

Thurlow, E. E., and Wright, R. J., 1950, Uraninite in the Coeur d'Alene district, Idaho: Econ. Geol., v. 45, p. 395-404; disc. (R. F. Robinson) p. 818-819

Umpleby, J. B., and Jones, E. L., Jr., 1923, Geology and ore deposits of Shoshone County, Idaho: U.S. Geol. Surv. Bull. 732, 156 p.

Waldschmidt, W. A., 1925, Deformation in ores, Coeur d'Alene district, Idaho: Econ. Geol., v. 20, p. 577-586

Wallace, R. E., and others, 1960, Tectonic setting of the Coeur d'Alene district, Idaho: U.S. Geol. Surv. Prof. Paper 400-B, p. B25-B27

Warren, H. V., 1934, Silver-tetrahedrite relationships in the Coeur d'Alene district, Idaho: Econ. Geol., v. 29, p. 691-696

Weis, P. L., 1960, Bleaching in the Coeur d'Alene district, Idaho: U.S. Geol. Surv. Prof. Paper 400-B, p. B27-B28

Willard, M. E., 1941, Mineralization at the Polaris mine, Idaho: Econ. Geol., v. 36, p. 539-550

Notes

The Coeur d'Alene district is located in the panhandle of northern Idaho with its center some 60 miles west of Spokane and nearly 100 miles south of the Canadian border. It is part of the highly dissected plateau region of Idaho and is characterized by deep, narrow valleys and high, well-wooded divides.

Most of the rocks encountered in the Coeur d'Alene are Precambrian sediments that have been intruded by a modest volume of igneous rocks. Except for unconsolidated Tertiary sand, gravel, and silt, there are no post-Precambrian sedimentary rocks now present in the area.

The Precambrian Belt series is divided into six formations, consisting of over 20,000 feet of sediments, along which quartzites, argillites, and limestones predominate. The oldest of these Precambrian units is the Prichard formation that is almost entirely composed of dark argillite. The Prichard grades into the Burke formation above it; the Burke is mainly a gray argillaceous quartzite. The Burke passes gradually into the Revett formation overlying it, the change taking place over several hundred feet. The Revett is uniformly composed of white quartzite but also contains, particularly near the base, argillaceous beds similar to the Burke. The white quartzite of the Revett gradually changes to the purplish gray quartzite and argillite that make up the St. Regis formation. The last of the purplish beds in the upper part of the St. Regis marks its transition into the Wallace formation that is quite diverse in character; most of the clastic rocks of which it is composed are calcareous. The uppermost formation of the Belt series is the Striped Peak that is of minor importance in the Coeur d'Alene; it is mainly variously colored quartzites.

The monzonite intrusives in the Coeur d'Alene area are considered to be outliers of the Idaho batholith that covers some 25,000 square miles in central and southern Idaho. Although the monzonite ranges in composition from granite to syenite to monzonite to quartz monzonite, the amount of plagioclase feldspar, on the average, is about equal to that of potash feldspar, and the quartz content is less than 5 percent; thus, monzonite is the best term to apply to the rock mass as a whole. It is possible that much of the potash was introduced deuterically or hydrothermally and that the original igneous rock was more mafic than monzonite. Diabase and lamprophyre are also found in the area.

The monzonite bodies (stocks and dikes) all lie north of the Osburn fault; this is the principal fault of the area that strikes about N80°W across the region and dips to the south at 55° to 60°. The fault has displaced the area north of it from 12 to 16 miles to the east, and the ground to the south of it has been dropped as much as 15,000 feet below the north side. Because the Gem group of stocks (the other group of stocks clusters around Dago Peak) is cut by the Dobson fault, which probably moved contemporaneously with the Osburn fault, it is thought that the monzonite intrusions occurred before the last, and principal, movement on the Osburn fault. The Osburn fault is certainly older than middle Miocene because it is covered by flows of Columbia basalt of that age; it is probable, therefore, that the fault was developed fairly late in Laramide time. As the Coeur d'Alene monzonite is probably an outlier of the Idaho batholith, it is almost certainly also of Laramide age, but it must have been intruded quite early in that orogeny to account for its involvement as a solid mass in Osburn-age faulting.

Hobbs and Fryklund (1968) consider that the main-period veins in the Coeur d'Alene are later than the monzonitic rocks and that the monzonites are Late Cretaceous in age. Lead-lead ages on the galena in the deposits give an age of 1250 m.y. Here, of course, the wall rocks of the veins in the metasediments are older than 1250 m.y.; therefore, it is vaguely possible that the ores are as old as the lead-lead ages suggest. As Hobbs and Fryklund point out, however, work by Murthy and Patterson (1961) at Butte shows that the Butte lead is older than its wall rocks. Thus, they suggest that the Coeur d'Alene lead came from a deeper reservoir of B-type lead through some process of remobilization that took place in late Mesozoic-early Tertiary time. They give geologic evidence that compels belief in the concept that the ores are post-monzonite. Only one period of monzonite intrusion is known in the district, and dikes occur in the border zones of the stocks that are mineralogically similar to the main igneous bodies. In several mines and prospects, sulfide-bearing veins (many of which are major producers) cut through these dikes. These authors do not believe that such relationships could have been caused by remobilization of Precambrian veins by late Mesozoic intrusions nor were the silicate minerals in the veins near the stocks formed by thermal metamorphism of pre-existing vein minerals. Ore bodies are lacking in the monzonite stocks probably because the stocks did not break as readily or in the same manner as the Belt series rocks since the central portions of the stocks appear to still have been plastic when the fracturing took place, even though their margins were hard enough to break to permit minor mineralization in them as is shown in several mines. The ore-forming fluids for all ore and gangue minerals probably came from the same general source as the monzonites of the stocks. The ores, therefore, should be late Mesozoic to early Tertiary.

As would be expected in a district that covers some 375 square miles, there is a considerable variety in the ores found in the Coeur d'Alene. Nor is the structural pattern less complex than the mineralization. The rocks of the area are intimately folded into a highly contorted and complex system of anticlines and synclines, the axial planes of some of which are overturned. Usually the axes of folds north of the Osburn fault strike northwest to nearly north, while those south of the fault trend in general slightly north of west. The district is much faulted, and the faults are complex and are of marked displacement, both vertically and horizontally; mineralization is most intense in the more faulted portions of the area, although the major faults are themselves essentially barren.

The principal mine areas of the Coeur d'Alene south of the Osburn fault are (1) the Pine Creek, southwest of the town of Kellogg and including the

Sidney and Highland Surprise mines; (2) the Kellogg, between the Pine Creek and the town of Kellogg and including the Bunker Hill and Page mines; (3) the Silver Belt, west of Wallace, between Placer Creek (east) and Big Creek (west) and including the Sunshine, Silver Summit, and Galena mines; and (4) the East Silver Belt, between Wallace and Mullan and including the Atlas and Gem State mines. North of the Osburn fault, the main mine areas are (1) the Lucky Friday, north of Mullan and including the Lucky Friday and Gold Hunter mines; (2) the Mullan Canyon Creek, northwest of the Lucky Friday area and including the Hecla and Morning mines; (3) the Sunset Peak-Nine Mile (Gem stock), northwest of the Mullan Canyon Creek area and including the Interstate and Hercules mines. The Dayrock mine is just west of the Dobson Pass fault, outside any of the areas just listed.

The characteristics of the Coeur d'Alene mineralization point to two high-temperature areas: (1) in the Pine Creek area and (2) around the Gem stocks. South of the Osburn fault, the mineralization changes from hypothermal to mesothermal in the Pine Creek area to mesothermal in the Kellogg area to mesothermal to leptothermal in the Silver Belt. In the vicinity of the Gem stocks, north of the Osburn fault, there is a considerable development of hypothermal mineralization even though it, in itself, is not of economic importance; the economic mineralization in the mines around the stocks is probably mesothermal. Southeast of the Gem stock (Sunset Park-Nine Mile) area, the mineralization of the Mullan Canyon Creek is probably mesothermal, while that of the Lucky Friday area still farther to the southeast is leptothermal. If the Gem stocks were displaced some 15 miles to the west (the reverse in distances and direction to the displacement given them--relative to the other side of the fault--by the Osburn fault), the Pine Creek and Gem stock areas would be essentially opposite each other, and the two high-temperature centers would be in much the same position relative to the lower temperature mineralizations to the east. This suggests that the mineralizing solutions entered the district near its western margins, where they first produced ore deposits, and then moved eastward and upward, depositing ores under less and less intense conditions as they moved away from their source. The lateral extent of the ore bodies was greater south of the area where the Osburn fault was to develop than it was north of that potential fault belt. It would follow, from this reasoning, that the Coeur d'Alene ore fluids came from the same general magma volume as the Gem and Dago stock rocks--this despite any arguments put forward on the basis of radioactive ages that the ores are far older than the monzonite stocks.

It seems most reasonable to categorize the Coeur d'Alene deposits as having formed under conditions ranging from hypothermal-1 to leptothermal with the proviso that the amount of economically valuable hypothermal ore was far less than that formed in either the mesothermal or leptothermal range.

Idaho-Montana

LEMHI PASS

Late Mesozoic to Early Tertiary | *Thorium, Rare Earths* | *Mesothermal*

Anderson, A. L., 1958, Uranium, thorium, columbium, and rare earth deposits in the Salmon region, Lemhi County, Idaho: Idaho Bur. Mines and Geol. Pamph. 115, p. 45-74

—— 1959, Geology and mineral resources of the North Fork quadrangle, Lemhi County, Idaho: Idaho Bur. Mines and Geol. Pamph. 118, 92 p.

—— 1961, Thorium mineralization in the Lemhi Pass area, Lemhi County, Idaho: Econ. Geol., v. 56, p. 177-197

Geach, R. D., 1966, Thorium deposits of the Lemhi Pass district, Beaverhead County, Montana: Mont. Bur. Mines and Geol. Spec. Pub. 41, 22 p.

Sharp, W. N., and Cavender, W. S., 1962, Geology and thorium-bearing deposit of the Lemhi Pass area, Lemhi County, Idaho, and Beaverhead County, Montana: U.S. Geol. Surv. Bull. 1126, 76 p.

Trites, A. F., Jr., and Tooker, E. W., 1953, Uranium and thorium deposits in east-central Idaho and southwestern Montana: U.S. Geol. Surv. Bull. 988-H, p. 157-208

Umpleby, J. B., 1913, Geology and ore deposits of Lemhi County, Idaho: U.S. Geol. Surv. Bull. 528, 182 p.

Vhay, J. S., 1950, Reconnaissance examination for uranium at six mines and properties in Idaho and Montana: U.S. Geol. Surv. Trace Element Memo Rept. 30, 31 p.

Notes

The Lemhi Pass district lies on the Idaho-Montana border in the Beaverhead Mountains; the center of the district is about 25 miles southeast of Salmon, Idaho, and nearly 50 miles west-southwest of Dillon, Montana.

The oldest rocks in the district belong to the Precambrian Belt series, but they have not been sufficiently studied for these clastic and non-clastic sediments to be assigned to any specific subdivision of that series although Sharp and Cavender (1962) suggest that they may tie closely to the Coeur d'Alene or Blackfoot Canyon facies. They indicate, however, that these rocks are closely similar to the Ravalli group of Calkins in the Philipsburg quadrangle. The Belt rocks in the Lemhi area are mainly micaceous quartzite with thin interbeds of argillite. No evidence has been put forward that these beds have been intruded by any igneous material until that introduced in what probably was Late Cretaceous to early Tertiary time. These intrusives are small masses and dikes of diorite that Umpleby (1913) considered to be slightly younger than the granite of the Idaho batholith but related to the magmatic cycle of which that batholith was the main expression. The next rocks to form in the area are lavas, both basalt and rhyolite, and rhyolitic welded tuffs and tuffaceous materials. Umpleby dated these volcanic rocks as later than Eocene because the lavas fill valleys that were cut by Eocene erosion; some of the younger lavas are interbedded with Miocene lake beds. In any event, the volcanic rocks are younger than the ores. On this basis, the ores can be related, granted they are of hydrothermal origin, only to diorites or the magmatic hearth of which the diorites were one expression. Thus, rather tentatively, the thorium-rare-earth ores are dated as late Mesozoic to early Tertiary.

The quartz veins and replacement masses of the Lemhi Pass district are divided by Sharp and Cavender (1962) into four types: (1) quartz-hematite-thorite, (2) quartz- and copper-bearing sulfide-thorite, (3) quartz- and copper-bearing sulfide veins, and (4) quartz-hematite. Type (1) veins are most abundant in the district and contain most of the thorite reserves. These veins are made up of white quartz with zones rich in pink and white barite. The thorite generally is evenly distributed through the veins as small streaks and masses and as small disseminated flakes. Where the area was more complexly cross-fractured than for the normal veins, the quartz masses are quite irregularly shaped through replacement of the surrounding Belt series rocks by quartz. All four vein types in the area appear to have been developed at about the same time, but in those veins containing thorite, it is a late mineral. Thorite occurs as fracture fillings in specularite (the next most abundant mineral to quartz) in the type (1) veins. In type (2) veins the thorite not only fills fractures in hematite but also in the copper minerals. Much of the thorite appears to be metamict, and the hydrated form--thorogummite--is found in the near-surface, oxidized portions of the veins. Other minerals in the veins include siderite, chlorite, generally sparse chalcopyrite and bornite, and rare chalcocite. The last three minerals occur only in the type (2) and type (3) veins. Geach (1966) reports that rare-earth elements in the veins have been detected by X-ray analysis but that the mineral form, distribution, and quantities are not established. The rare-earth minerals probably are associated spatially with the thorite, and it seems that yttrium is the dominant rare earth present, followed by elements of the cerium and terbium groups in that order of abundance. Monazite has not been recognized, at least as yet, as a mineral in the veins.

The presence of specular hematite in the thorium-bearing veins suggests that it was formed under hypothermal conditions; that the thorite fills fractures in the specularite indicates that the thorium mineral was formed later and possibly under less intense conditions than the hematite. In the copper-sulfide-containing veins, the thorite is also later than the copper minerals; this also points to lower intensity conditions than hypothermal since one of the sulfides is bornite, a diagnostic mesothermal mineral. The wall-rock alteration consists of the addition of quartz and the development of sericite in the argillites of the Belt series. Such wall-rock alteration seems to conform to a mesothermal range of deposition. Thus, it seems safest to classify the ores as having been deposited under mesothermal conditions.

Illinois-Kentucky

ILLINOIS-KENTUCKY FLUORSPAR

Late Mesozoic (pre-Laramide) — *Fluorite, Zinc, Lead* — *Telethermal*

Amstutz, G. C., and Park, W. C., 1967, Stylolites of diagenetic age and their role in the interpretation of the southern Illinois fluorspar deposits: Mineralium Deposita, v. 2, p. 44-53

Bain, H. F., 1905, The fluorspar deposits of southern Illinois: U.S. Geol. Surv. Bull. 255, 75 p.

Bastin, E. S., 1931, The fluorspar deposits of Hardin and Pope Counties, Illinois: Ill. Geol. Surv. Bull. 58, 116 p.

—— 1933, The fluorspar deposits of southern Illinois, in *Mining districts of the eastern states*: 16th Int. Geol. Cong., Guidebook 2, p. 32-44

Baxter, J. W., and Desborough, G. A., 1965, Areal geology of the Illinois fluorspar district. Part 2. Karbers Ridge and Rosiclare quadrangles: Ill. Geol. Surv. Circ. 385, 40 p.

Baxter, J. W., and others, 1963, Areal geology of the Illinois fluorspar district. Part 1. Saline mines, Cave-in-Rock, Dekoven, and Repton quadrangles: Ill. Geol. Surv. Circ. 342, 44 p.

—— 1967, Areal geology of the Illinois fluorspar district. Part 3. Herod and Shetlersville quadrangles: Ill. Geol. Surv. Circ. 413, 41 p.

Bradbury, J. C., 1959, Barite in the southern Illinois fluorspar districts: Ill. Geol. Surv. Circ. 265, 14 p.

Brecke, E. A., 1962, Ore genesis of the Cave-in-Rock fluorspar district, Hardin County, Illinois: Econ. Geol., v. 57, p. 499-535

—— 1964, Barite zoning in the Illinois-Kentucky fluorspar district: Econ. Geol., v. 59, p. 299-302

—— 1964, A possible source of solutions of the Illinois-Kentucky fluorspar district: Econ. Geol., v. 59, p. 1293-1297

—— 1967, Sulfide and sulfur occurrences of the Illinois-Kentucky fluorspar district: Econ. Geol., v. 62, p. 376-389

Brown, J. S., and others, 1954, Explosion pipe in test well on Hicks dome, Hardin County, Illinois: Econ. Geol., v. 49, p. 891-902

Clegg, S. K., and Bradbury, J. C., 1956, Igneous intrusive rocks in Illinois and their economic significance: Ill. Geol. Surv. R.I. 197, 19 p.

Currier, L. W., 1923, Fluorspar deposits of Kentucky; a description and interpretation of the geologic occurrence and industrial importance of Kentucky·

fluorspar: Ky. Geol. Surv., Ser. 6, v. 13, 189 p. (particularly p. 35-132)

—— 1937, Origin of the bedding replacement deposits of fluorspar in the Illinois field: Econ. Geol., v. 32, p. 364-386

Currier, L. W., and Wagner, O. E., Jr., 1944, Geology of the Cave-in-Rock district: U.S. Geol. Surv. Bull. 942, pt. 1, p. 1-72

Erickson, A. J., Jr., 1965, Origin of the Illinois-Kentucky fluorspar deposits: Econ. Geol., v. 60, p. 384-385; disc., p. 1070-1073

Frease, D. H., 1961, Temperature of mineralization by liquid inclusions, Cave-in-Rock fluorspar district, Illinois: Econ. Geol., v. 56, p. 542-556

Grawe, O. R., and Nackowski, M. P., 1949, Strontianite and witherite associated with southern Illinois fluorite: Science, v. 110, no. 2857, p. 331

Grogan, R. M., 1949, Structures due to volume shrinkage in the bedding-replacement fluorspar deposits of southern Illinois: Econ. Geol., v. 44, p. 606-616

Grogan, R. M., and Bradbury, J. C., 1967, Origin of the stratiform fluorite deposits of southern Illinois, in Brown, J. S., Editor, *Genesis of stratiform lead-zinc-barite-fluorite deposits--a symposium*: Econ. Geol. Mono. 3, p. 40-51

—— 1968, Fluorite-zinc-lead deposits of the Illinois-Kentucky mining district, in Ridge, J. D., Editor, *Ore deposits of the United States, 1933-1967* (Graton-Sales Volumes): Chap. 19, v. 1

Grogan, R. M., and Shrode, R. S., 1952, Formation temperatures of southern Illinois bedded fluorite as determined from fluid inclusions: Amer. Mineral., v. 37, p. 555-566

Hall, W. E., and Friedman, I., 1963, Composition of fluid inclusions, Cave-in-Rock fluorite district, and Upper Mississippi Valley zinc-lead district: Econ. Geol., v. 58, p. 886-911

Hall, W. E., and Heyl, A. V., 1968, Distribution of minor elements in ore and host rock, Illinois-Kentucky fluorite district and Upper Mississippi Valley zinc-lead district: Econ. Geol., v. 63, p. 655-670

Hardin, G. C., Jr., and Trace, R. D., 1959, Geology and fluorspar deposits, Big Four fault system, Crittenden County, Kentucky: U.S. Geol. Surv. Bull. 1042-S, p. 699-724

Heyl, A. V., and Brock, M. R., 1961, Structural framework of the Illinois-Kentucky mining district and its relation to mineral deposits: U.S. Geol. Surv. Prof. Paper 424-D, p. D3-D6

Heyl, A. V., and others, 1966, Isotopic study of galenas from the Upper Mississippi Valley, the Illinois-Kentucky, and some Appalachian Valley mineral deposits: Econ. Geol., v. 61, p. 933-961 (particularly p. 947-951)

—— 1966, Regional structure of the southeast Missouri and Illinois-Kentucky mineral districts: U.S. Geol. Surv. Bull. 1202-B, p. B1-B20

Nackowski, M. P., 1958, Physical and chemical environment of Illinois-Kentucky fluorspar deposits (abs.): Econ. Geol., v. 53, p. 925-926; disc., 1959, v. 54, p. 751

Oesterling, W. A., 1952, Geologic and economic significance of the Hutson zinc mine, Salem, Kentucky--its relation to the Illinois-Kentucky fluorspar district: Econ. Geol., v. 47, p. 316-338

—— 1959, Illinois-Kentucky fluorspar deposits: Econ. Geol., v. 54, p. 751

Park, W. C., 1967, Early diagenetic framboidal pyrite, bravoite, and vaesite from the Cave-in-Rock fluorspar district, southern Illinois: Mineralium Deposita, v. 2, p. 372-375

Park, W. C., and Amstutz, G. C., 1968, Primary "cut-and-fill" channels and gravitational diagenetic features. Their role in the interpretation of the southern Illinois fluorspar deposits: Mineralium Deposita, v. 3, p. 66-80

Pinckney, D. M., and Haffty, J., 1970, Content of zinc and copper in some fluid inclusions from the Cave-in-Rock district, southern Illinois: Econ. Geol., v. 65, p. 451-458

Schwerin, M., 1928, An unusual fluorspar deposit: Eng. and Min. Jour., v. 126, no. 9, p. 335-339

Stonehouse, H. B., and Wilson, G. M., 1955, Faults and other structures in southern Illinois: Ill. Geol. Surv. Circ. 195, 4 p.

Trace, R. D., 1960, Significance of unusual mineral occurrence at Hicks dome, Hardin County, Illinois: U.S. Geol. Surv. Prof. Paper 400-B, p. B63-B64

—— 1962, Geology and fluorspar deposits of the Levias-Keystone and Dike-Eaton areas, Crittenden County, Kentucky: U.S. Geol. Surv. Bull. 1122-E, p. E1-E26

—— 1962, Geology of the Salem quadrangle, Kentucky: U.S. Geol. Surv. Geol. Quad. Map GQ 206, 1:2,000 (with text)

Ulrich, E. O., and Smith, W.S.T., 1905, The lead, zinc, and fluorspar deposits of western Kentucky: U.S. Geol. Surv. Prof. Paper 36, 218 p.

Weller, J. M., and Grogan, R. M., 1945, An occurrence of granite in Pope County, Illinois: Jour. Geol., v. 43, p. 398-402

Weller, J. M., and others, 1952, Geology of the fluorspar deposits of Illinois: Ill. State Geol. Surv. Bull. no. 76, 147 p.

Weller, S., and Sutton, A. H., 1951, Geologic map of the western Kentucky fluorspar district: U.S. Geol. Surv. Mineral Invest. Field Studies, Map MF-2, 1:62,500

Weller, S., and others, 1920, The geology of Hardin County: Ill. Geol. Surv. Bull. 41, 416 p.

Williams, J. S., and others, 1954, Fluorspar deposits in western Kentucky: U.S. Geol. Surv. Bull. 1012-A-E, 127 p.

Notes

The Southern Illinois Fluorspar district covers some 700 square miles both in southern Illinois and western Kentucky; over 60 percent of the production has come from the Illinois side of the Ohio River that bisects the district.

The geologic formations on the two sides of the Ohio are essentially the same, although minor differences in lithology and thickness exist. Deep-hole drilling, mainly carried out to look for petroleum, has demonstrated that the depth of the Precambrian basement from the rock surface in the center of the area is at least 7500 feet. Although this deep drilling has disclosed the presence of rocks of all periods in the Paleozoic sequence, the oldest exposed rocks on the Illinois side are Upper Devonian limestones and shales that grade upward into the predominantly limestone rocks of the Mississippian Iowa series that includes the Meremac, Osage, and Kinderhook groups (from youngest to oldest) that are mainly limestones with some sandstone and shale. The St. Louis limestone, the oldest formation of the Meremac group, is the oldest formation exposed on the Kentucky side of the river. Above the Iowa lies the Chester series that is composed of the Elvira, Homber, and New Design groups (from youngest to oldest) that are mainly sandstones with limestone and shale. Above the Chester is the lowest Pennsylvanian group, the Caseyville, that is mainly sandstone with some sandy shale and clay shale; it contains one thin coal bed. The Caseyville is the youngest formation exposed on the Kentucky side of the river. Above the Caseyville in Illinois is the Pennsylvanian Tidewater group that is a variable succession of shales and sandstones with several thin coal seams and two thin limestone beds; it is the youngest group exposed in the

Illinois portion of the district.

The igneous rocks of the district are peridotites and lamprophyres and are found mainly in northwest-trending dikes that, on the Illinois side of the river, are concentrated in the Hicks dome area; in Kentucky, the dikes are more widely scattered but have the same northwest-trend. Contrary to the general trend, one Hicks dome dike strikes northeast-southwest and is highly enriched in elements such as barium, beryllium, columbium, gallium, scandium, and thorium not normally found in highly mafic rocks; the thorium content is directly related to that of the other unusual elements. The domed area also contains a considerable number of explosion breccias that are grouped in two main areas, the center of the dome and its outer flanks. These explosion breccias contain fragments of widely different sizes and have a matrix of finely ground rock material, none of which (fragments or matrix) has been demonstrated to be of igneous or metamorphic origin; igneous fragments, however, are known in similar breccia masses outside the dome itself. If these breccias within the dome were produced by igneous activity, any igneous material involved in the process has remained at depths below any of those reached by drilling in the dome area. Although not too much is known of what minerals may have been added to the breccias after they were formed, one drill hole in the dome found the breccia to be continuously, but erratically, mineralized with fluorspar in amounts ranging from 5 percent in the upper portions of the breccia to 2 percent at the bottom. Other minerals present in the breccia include calcite, pyrite, quartz, and very minor traces of sphalerite and galena. The fluorite and the minerals associated with it provide some of the matrix, are in veinlets in rock fragments, and have to some extent replaced fragments. This fluorite mineralization strongly suggests hydrothermal activity, and radioactive age determinations on the thorium minerals of the breccia should date both the igneous and the essentially contemporaneous hydrothermal activity.

Brown (1954) considers the Hicks dome to be an incipient or imcompleted structure of the type that Bucher designated as crytovolcanic and as examples of which he cited the Wells Creek Basin in Tennessee, the Jeptha Knob in Kentucky, and the Serpent Mound structure in Ohio. No volcanic rocks are associated with these structures, at least at the surface. Their circular shape, their upheaved, broken, and in places brecciated character, however, plus the presence of igneous dikes cutting the essentially horizontal sedimentary rocks of the disturbed areas suggested to Bucher that they were of volcanic origin. The geologic character of these areas bears some resemblance to that of certain European structures of Tertiary age, for example, the Ries Kessel and the Steinheim Basin (that are thought by some to have been caused by meteoritic impact); these structures have been appreciably less eroded than those in the United States. Bucher believed that the greater length of time required for the structures in the central United States to reach their present state of erosion meant that they must have been formed in the late Paleozoic or in the Mesozoic. Heyl and Brock (1961), however, have demonstrated a middle Cretaceous age for monazite from the Hicks dome; this age coincides with the younger limit of the ages suggested by Bucher.

Another evidence of the age of mineralization in the area is provided by the ore-bearing and ore-controlling faults that cut both peridotite dikes and the Pennsylvanian and older rocks but extend beneath unfaulted Upper Cretaceous rocks to the southwest. Since the probably middle Cretaceous Hicks dome was formed slightly before the fault pattern was developed, the faults must have been essentially the last geologic event of the middle Mesozoic in the district. It would seem, therefore, that the ore mineralization, which is present in both the breccias of the Hicks dome and in the later faults, must have formed within a short time after the structures containing it were produced. Thus, the evidence provided by all the features just discussed points to the fluorspar mineralization having occurred at much the same time as that of the igneous activity in the Magnet Cove area, and this would place the ore formation in pre-Laramide time.

The fluorspar area in Illinois and Kentucky is situated in the northern part of a collapsed, block-faulted, sliced, and partly rotated, north to northwest-trending anticline.

The doming of the beds into the major anticlinal structure has been thought by many to have been due to igneous intrusion, with the faulting having been triggered by a later collapse of the dome due to withdrawal or solidification of the magma.

The deposits of fluorspar and associated minerals in the district occur in three forms, the first two of which are primary and the third residual: (1) steeply inclined veins formed by the filling of the fractures in fault zones, aided by a minor amount of replacement of wall-rock limestone; (2) flat-lying or blanket-bedded replacement deposits in certain limestone beds that are generally immediately overlain by thin shale beds; and (3) fluorite rubble created by the solution by ground water of limestone and vein calcite and the mixing of the insoluble fluorite with undissolved clay and wall rock fragments. In the Kentucky portion of the district, the deposits are mainly of the fissure type, although there are bedded replacement bodies associated with a few of the veins. In the Illinois area the fissure deposits are also important and have their greatest concentration in the Rosiclare portion of the district.

The paragenesis of the fluorspar ores in the vein deposits is different from that in the bedded replacements. The first mineral to form in the veins was calcite, and it is the dominant vein mineral. Before the deposition of calcite had ceased, all the other primary minerals of the deposits, except galena, and the second generations of pyrite and calcite had begun to form. Of these pre-galena minerals, abundant fluorite was the first to precipitate, followed by locally important amounts of sphalerite, minor chalcopyrite, and some quartz. Only quartz had ceased to deposit before the formation of galena began; the second generation of quartz and pyrite did not begin to form until after galena precipitation had stopped.

In the bedded replacements the first mineral to replace limestone was fluorite, followed by locally important sphalerite and minor amounts of chalcopyrite, pyrite (or marcasite), galena, and quartz. Apparently only the deposition of chalcopyrite overlapped that of fluorite. The early minerals were followed by second generations of fluorite and chalcopyrite and by a first generation of bitumen. Finally, there was a minor development of calcite and second generations of bitumen, pyrite (or marcasite), and quartz; the last mineral of all was barite. Grogan (1949) has suggested that the replacement of limestone by fluorite resulted in a reduction of volume because of the smaller volume occupied by one unit cell of fluorite as opposed to that of one unit cell of calcite. It appears to be equally possible, at least, that the areas in which solutions had thinned the limestone beds were most favorable for replacement activity; the presence in the district of thinned but unmineralized beds supports this argument.

The mineral assemblage of the fluorspar deposits is typically one formed under low intensity conditions, and the rock cover beneath which the ores were deposited probably was sufficiently thick and unbroken to insure deposition with slow loss of heat and pressure. The ores, therefore, are here classified as telethermal.

Maine

NEWRY

Late Paleozoic | *Lithium, Feldspar, Rare Alkalis, Beryl* | *Magmatic-3a*

Bastin, E. S., 1911, Geology of the pegmatites and associated rocks of Maine: U.S. Geol. Surv. Bull. 445, 152 p. (particularly p. 76-78)

Berman, H., and Gonyer, F. A., 1930, Pegmatite minerals of Poland, Maine: Amer. Mineral., v. 15, p. 375-387 (conclusions apply to Newry)

Cameron, E. N., 1954, Pegmatite investigations, 1942-45, New England: U.S. Geol. Surv. Prof. Paper 255, 352 p. (particularly p. 13-15, 80-83)

Fraser, H. J., 1930, Paragenesis of the Newry pegmatite, Maine: Amer. Mineral., v. 15, p. 349-364

Gregory, C., 1967, The famous pegmatites of Newry, Maine: Gems and Minerals no. 359, p. 22-24

Hess, F. L., 1943, The rare alkalis in New England: U.S. Bur. Mines I.C. 7232, 51 p. (particularly p. 12-20)

Palache, C., and Shannon, E. V., 1928, Beryllonite and other phosphates from Newry, Maine: Amer. Mineral., v. 13, p. 392-396

Shainin, V. E., and Dellwig, L. F., 1955, Pegmatites and associated rocks in the Newry Hill area, Oxford County, Maine: Maine Geol. Surv. Bull. 6, 58 p.

Shaub, B. M., 1940, On the origin of some pegmatites in the two of Newry, Maine: Amer. Mineral., v. 25, p. 673-688

Notes

The Newry area lies near the western border of Maine, almost 100 miles slightly west of north from Portland.

The country rocks in the Newry area consist of nine types: (1) quartz-muscovite schist, (2) quartz-biotite schist, (3) actinolite schist, (4) gabbro, (5) diorite-pegmatite, (6) hornblendite, (7) granodiorite, (8) granite-pegmatite, and (9) greisen. Types (1) and (2) were formed by the metamorphism of sediments as is indicated by the occasional limey and quartzite layers they contain. Type (3) has resulted from metamorphism of type (4), the gabbro. Of the igneous rocks, types (4) and (7) are the most widespread, while the other igneous rocks form much smaller bodies both in the igneous rocks and in the sediments. The granite-pegmatite, type (8), is present in the area as 37 irregularly shaped bodies up to 4000 feet long mainly on the slopes and crests of Newry Hill (east) and Plumbago Mountain (west). The most important pegmatites are (1) the Dunton (Gem) on the crest of Newry Hill, (2) the Main on the east slopes of Newry Hill, (3) the Red Crossbill immediately southeast of the Main pegmatite, (4) the Crooker almost due north of the Dunton, and (5) the Kinglet southwest of the Dunton.

The sediments from which the metamorphosed rocks were derived probably were formed in Cambrian through Silurian times; the metamorphism was probably accomplished in the Devonian. The conversion of portions of the gabbro to actinolite schist appears to have been due mainly to the pegmatite intrusions and is, therefore, probably post-Devonian. The granodiorite does not appear to have been involved in the Devonian metamorphism to any appreciable extent and may have been emplaced in the late Paleozoic. The Newry pegmatites have not been metamorphosed and probably are, therefore, the youngest rocks in the Newry area, and of the same age as the major granite masses of southwestern Maine that Fisher (G.S.A. Bull., v. 52, p. 153) thinks to be Carboniferous. He cites a radioactive age determination by Gonyer (1937) made on samarkskite from the Topshaw pegmatite (near Bath) as confirming this age. At least, it appears almost certain that the granite-pegmatites are post-Devonian and are here classified as late Paleozoic.

In general, the Newry pegmatites are disconformable to the schists into which they have been intruded. As an example, the Dunton pegmatite contains six zones. The border zone is an albite pegmatite (1 to 3 inches thick) with minor actinolite and black tourmaline. The wall zone is a quartz-albite pegmatite (8 inches to 4 feet thick) with muscovite, black tourmaline, and minor garnet. The first intermediate zone is a quartz-cleavelandite-muscovite pegmatite (1 to 2 feet thick) with minor amounts of green and black tourmaline. The second intermediate zone is an albite-quartz-perthite pegmatite with some amblygonite [$LiAl(F,OH)PO_4$] and minor triphylite, beryl, tantalite, lepidolite, and pollucite ($CsSi_2AlO_6$) that forms a discontinuous series with analcite; this pollucite was not mined. The core-margin zone is a cleavelandite-quartz-perthite pegmatite (5 to 10 feet thick) with muscovite and minor lepidolite, beryl, pollucite, amblygonite, tantalite, cassiterite, and green, blue and pink tourmaline; the bulk of the mined pollucite is reported to have come from this zone.

The core is a quartz-cleavelandite-perthite pegmatite (8 by 12 feet) with about 2 percent of spodumene and white beryl, with minor dark blue and green gem tourmaline, cassiterite, tantalite, and amblygonite. The other Newry pegmatites are of generally similar, but less complex, character than those described.

Suggestions as to the origin of the Newry pegmatites range from (1) the concept that they were introduced as a single injection of material that through fractional crystallization and reaction between solid and still fluid materials, produced the mineral assemblages now present and (2) the theory that they were once simple pegmatites that were modified into their present species by the action of later hydrothermal fluids. There appears to be no evidence of such widespread hydrothermal activity that, had it occurred, certainly would not have limited its effects to the pegmatites and their immediate wall rocks but would have also changed the adjacent rocks as well. Actually the only effect of the pegmatites on their schist-rock host environment appears to have been the development of black (deep indigo in thin section) tourmaline as "suns," irregular clusters, and individual crystals within the schist. Similar tourmaline formed an irregular, discontinuous crust or bank of black tourmaline immediately outside the fine-grained quartz-albite pegmatite of the wall zone. The wide variety of Newry pegmatite minerals probably was caused by reactions among the solidified minerals, the remaining molten silicate material, and such water-rich hydrothermal fluids as separated from the silicate melt as the early anhydrous minerals crystallized.

It would appear that the pegmatites in the Newry area fill all the requirements of pegmatites of the class magmatic-3a, and they are so categorized here.

Michigan

KEWEENAW POINT

Late Precambrian | *Copper* | *Mesothermal to Leptothermal*

Broderick, T. M., 1929, Zoning in Michigan copper deposits and its significance: Econ. Geol., v. 24, p. 149-162, 311-326

—— 1931, Fissure vein and lode relationships in Michigan copper deposits: Econ. Geol., v. 26, p. 840-856

—— 1935, Differentiation in lavas of the Michigan Keweenawan: Geol. Soc. Amer. Bull., v. 46, p. 503-558

—— 1952, The origin of Michigan copper deposits: Econ. Geol., v. 47, p. 215-220

—— 1956, Copper deposits of the Lake Superior region: Econ. Geol., v. 51, p. 285-287

Broderick, T. M., and Hohl, C. D., 1935, Differentiation in traps and ore deposition: Econ. Geol., v. 30, p. 301-312

—— 1935, The Michigan copper district, in *Copper resources of the world*: 16th Int. Geol. Cong., v. 1, p. 271-284

Broderick, T. M., and others, 1946, Recent contributions to the geology of the Michigan copper district: Econ. Geol., v. 41, p. 675-725

Butler, B. S., and others, 1929, The copper deposits of Michigan: U.S. Geol. Surv. Prof. Paper 144, 238 p. (particularly p. xi-xii, 21-62, 101-156)

Cornwall, H. R., 1951, Ilmenite, magnetite, hematite, and copper in lavas of Keweenawan series: Econ. Geol., v. 46, p. 51-67

—— 1951, Differentiation in the lavas of the Keweenawan series and the origin of the copper deposits of Michigan: Geol. Soc. Amer. Bull., v. 62, p. 159-201

—— 1951, Differentiation in magmas of the Keweenawan series: Jour. Geol., v. 59, p. 151-172

Cornwall, H. R., and Rose, H. J., Jr., 1957, Minor elements in the Keweenawan lavas, Michigan: Geochimica et Cosmochimica Acta, v. 12, p. 209-224

Drier, R. W., 1954, Arsenic and native copper: Econ. Geol., v. 49, p. 908-911

Hamblin, W. K., and Horner, W. J., 1961, Sources of the Keweenawan conglomerates of northern Michigan: Jour. Geol., v. 69, p. 204-211

Lane, A. C., 1935, Differentiation in traps and ore deposition (disc.): Econ. Geol., v. 30, p. 924-927

Lindgren, W., 1933, The Lake Superior copper deposits, in *Mineral deposits*: 4th ed., McGraw-Hill, N.Y., p. 517-526

McKinstry, H. E., 1951, Differentiation in . . . Keweenawan series (rev.): Econ. Geol., v. 46, p. 658-659

Routhier, P., 1963, Les gites de cuivre natif du Lac Superieur (Michigan), in *Les gisements métallifères--Géologie et principes de recherches*, pt. 1: Masson et Cie, Paris, p. 640-643, 646

Ruotsala, A. P., and others, 1969, Trace elements in accessory calcite--a potential exploration tool in the Michigan copper district: Colo. Sch. Mines Quart., v. 64, no. 1, p. 451-456

Singewald, J. T., Jr., 1928, A genetic comparison of the Michigan and Bolivian copper deposits: Econ. Geol., v. 23, p. 55-61

Stoiber, R. E., and Davidson, E. S., 1959, Amygdule mineral zoning in the Portage Lake lava series, Michigan copper district: Econ. Geol., v. 54, p. 1250-1277, 1444-1460

Weege, R. J., and Schillinger, A. W., 1962, Footwall mineralization in Osceola amygdaloid, Michigan, native copper district: A.I.M.E. Tr., v. 223, p. 344-350

White, W. S., 1956, Regional structural setting of the Michigan native copper district, in Snelgrove, A. K., Editor, *Geological exploration*: Institute on Lake Superior Geology, Houghton, Mich., p. 3-16; disc. p. 18-19

—— 1960, The Keweenawan lavas of Lake Superior, an example of flood basalts: Amer. Jour. Sci., v. 258A (Bradley Volume), p. 367-374

—— 1966, Tectonics of the Keweenawan basin, western Lake Superior region: U.S. Geol. Surv. Prof. Paper 524-E, p. E1-E23

—— 1968, The native-copper deposits of northern Michigan, in Ridge, J. D., Editor, *Ore deposits of the United States, 1933-1967* (Graton-Sales Volumes): Chap. 16, v. 1

Notes

The copper deposits of Upper Michigan are located on Keweenaw Point about 175 miles slightly north of east of Duluth. The deposits extend along the Point in a northeast-trending belt that is from 2 to 4 miles wide and is more than 100 miles long.

The ores are contained in rocks of the Keweenawan series that are late Precambrian in age and are composed of silicic and mafic intrusive and extrusive rocks, ash, tuff, shale, sandstone, and conglomerate. Sediments are quite unimportant in the lower portions of the Keweenawan rocks but make up almost all of the sequence above the Eagle River group at the top of the basaltic sequence. Most of the lower Keweenawan rocks are basaltic flows of the plateau type, although there are many examples of lavas grading from typical olivine basalts, basalts, and basaltic andesites through andesites to rhyolites. The total thickness of the dominantly basaltic rocks below the Great Conglomerate is over 15,000 feet and may be as much as 18,500 feet. The sediments above and including the Great Conglomerate are over 10,000 feet thick and, although they are essentially unmineralized in the Keweenaw Point area, do contain the White Pine chalcocite-native copper deposits nearly 60 miles to the southwest.

Intercalated in the mafic flows are over 20 beds of felsitic conglomerate the thickness of which ranges from inches to several feet; above the mafic flows the dominant rock is conglomerate, with the Great Conglomerate (nearly 1800 feet thick) being the thickest. Normally a felsite conglomerate is underlain by a fragmental layer composed of basic sand and pebbles and boulders of amygdaloid, and such an amygdaloidal conglomerate usually directly overlies the amygdaloidal upper portion of a basalt flow. Many locations are known, however, where no felsitic material is associated with the amygdaloidal conglomerate.

Intrusive igneous rocks occur in the district throughout the lower part of the series and comprise gabbro and gabbro aplite (that are similar to the gabbro and red rock of Duluth laccolith or lopolith), quartz porphyry and felsite (that are probably fine-grained equivalents of the Duluth red rock), and a few chloritized mafic dike rocks. The presence of these rocks in the area suggests that the Duluth laccolith probably extends under the Lake Superior basin, and that the magmatic source (or sources) of the gabbro and its associated red rock well may have been the source of the ore fluid that deposited the copper in the Keweenawan ore bodies. If this is the case, the 1000 to 1050 m.y. age determined for the Duluth gabbro would place the ores in the late Precambrian. If, as Cornwall (1951) has suggested, the ores are the result of a combination of syngenetic and epigenetic processes or were developed largely syngenetically, these activities must have taken place at much the same time as the deposition of the lavas and of the Duluth gabbro so that the age of the ores would still be late Precambrian, and they are so categorized here.

The copper deposits of the Keweenaw Point district can be divided into two general types: (1) bedded (or lode or tabular) deposits and (2) fissure deposits. The bedded deposits can be further subdivided into lode deposits in (1) felsite conglomerate beds and (2) the amygdaloidal portions of lava flows. The amygdaloidal types can be broken down still further into lodes in (1) brecciated volumes of amygdaloidal rock, (2) scoriaceous volumes of amygdaloidal rock, and (3) volumes of coalescing amygdaloidal rock. The principal horizons containing productive lode deposits have been (from the base of the section upward) (1) the Baltic amygdaloid, (2) the Isle Royale amygdaloid, (3) the Kearsarge amygdaloid, (4) the Osceola amygdaloid, (5) the Calumet and Hecla conglomerate, (6) the Allouez conglomerate, (7) the Pewabic amygdaloid, and (8) the Ash bed amygdaloid. The fissure deposits were developed in the same lower portion of the Keweenawan series as the lodes although, in general, the fissure deposits were found in different areas along the strike than were the lodes. As of 1925, about 45 percent of the total production had come from the Calumet and Hecla conglomerate; the remaining 55 percent of production was divided among the various loci of copper deposition in the following order: Kearsarge, Baltic, Pewabic, Osceola, Isle Royale, and all fissure deposits; all the remaining loci accounted for less than 8 percent of the total.

The mineralization in the conglomerates is quite simple in comparison with that in the flow rocks. In order of deposition, the minerals introduced were a red alkali feldspar, then abundant epidote and a little pumpellyite [$Ca_2Al_3(OH)Si_3O_{12}\cdot H_2O$?], then copper after the bulk of the epidote and pumpellyite had formed; quartz and calcite deposited over essentially the same period of time as the copper, as did a little chlorite and silver; the last minerals were small amounts of sulfides and barite. Zeolites and related minerals such as prehnite are almost entirely absent from the conglomerates, suggesting strongly that the formation of zeolites is a function of the character of host rock as well as that of the ore-forming fluid.

The mineralization in the flow rocks is much more complex than in the conglomerates. The earliest minerals were the red feldspar and epidote in much the same amounts as in the conglomerates, plus much more abundant chlorite and pumpellyite. The deposition of copper began toward the end of the deposition of those earlier minerals, and the principal species accompanying the copper were quartz and calcite. During this time, datolite ($CaOHSiBO_4$), ankerite, and sericite were locally common. Analcite was widespread but not common, while laumontite, a complex calcium-sodium alumino-silicate with 25 molecules of water, was the most abundant zeolite, but it usually was found in copper-poor areas. Silver was somewhat more abundant than in the conglomerates, but normally not enough so to justify separate recovery. Arsenides and sulfides were present in veins that cut through the lodes at a variety of angles--some native copper is associated with the sulfides and arsenides. Part of the minerals deposited in the porous portions of the flow rocks filled vesicles to form amgydules and the remainder replaced and, to some extent cemented, broken flow fragments while the wall rocks were being altered to epidote, calcite, pumpellyite, chlorite, and quartz; the degree of bleaching in the wall rocks depends largely on the amount of chlorite formed. With little chlorite, the rocks were given a gray-green, bleached color; with much chlorite the rocks changed from the original deep red to dark green.

The fissure veins are divided mineralogically into three types: (1) those containing only native copper (all the commercial important fissures), (2) those composed of copper arsenides, and (3) those consisting of copper sulfides. Some overlap exists among the three types. The gangue minerals were the same as those associated with the porous flow rocks, but far more variation occurred in the mineralogy of the fissure veins than in the other types of deposits. Wall-rock alteration was appreciably less intensive and extensive.

Assuming that the copper in the deposits was introduced by hydrothermal solutions of magmatic origin and not by redistribution of copper already present in the flows as suggested by Cornwall (1951) and refuted by Broderick (1952), it is necessary to explain why the copper is present in the native state instead of in the far more usual sulfide forms. This problem appears to have been satisfactorily solved by Wells (1925, U.S. Geol. Surv. Bull. 778) through his suggestion that the reaction of sulfur in the S^{-2} state with the Fe^{+3} of the abundant hematite in the wall rocks and with the cupric or cuprous copper of the ore fluids was responsible for almost exlusive precipitation of copper in the native state. Equation (1) shows this reaction may have occurred; the reactants are assumed to have included CuS_2^{-2} as the source of copper:

$$CuS_2^{-2} + 14Fe^{+3} + 8H_2O = Cu^\circ + 14Fe^{+2} + 2SO_4^{-2} + 16H^{+1}. \quad (1)$$

Had part of the sulfur been oxidized only as far as S_2^{-2}, the appreciable amounts of Fe^{+2} still produced by the reaction probably would have made possible the precipitation of some pyrite, a mineral lacking in the deposits. This complex oxidation of sulfur, probably accounts for the deposition of minor amounts of anhydrite late in the mineralization cycle. The occurrence of arsenic in various alloys with copper, probably present in the ore fluid as some such complex as AsS_4^{-3}, indicates that the reduction of this element to As° probably aided in the oxidation of sulfur as is shown in equation (2):

$$AsS_4^{-3} + 27Fe^{+3} + 16H_2O = As^\circ + 4SO_4^{-2} + 27Fe^{+2} + 32H^{+1}. \quad (2)$$

The reduction of silver must have been accomplished in much the same manner as that of arsenic and copper.

It is, of course, impossible to say what sulfides of copper, arsenic, and silver might have been developed in the district had huge amounts of ferric iron not been present in the conglomerates and lavas; even the formation of chalcocite in the fissure veins, does not guarantee that chalcocite would have been precipitated had the ore fluid traversed appreciably less iron-rich rocks than those containing the fissure veins. Instead, in such a low-iron environment, the degree of reduction of copper, arsenic, and silver might have been sufficiently low to have forced the deposition of a sulfosalt of copper and arsenic or of chalcopyrite or bornite rather than of chalcocite as is suggested in equations (3a-3d):

$$CuS_2^{-2} + Fe^{+2} = FeCuS_2. \quad (3a)$$

$$10CuS_2^{-2} + 88Fe^{+3} + 48H_2O = 2FeCu_4{}^1Cu^2S_4 + 88Fe^{+2} + 12SO_4^{-2} + 96H^{+1}. \quad (3b)$$

$$10CuS_2^{-2} + 110Fe^{+3} + 60H_2O = 10Cu_2S + 110Fe^{+2} + 15SO_4^{-2} + 96H^{+1}. \quad (3c)$$

$$3CuS_2^{-2} + AsS_4^{-3} + 45Fe^{+3} + 24H_2O = Cu_3AsS_4 + 45Fe^{+2} + 6SO_4^{-2} + 48H^{+1}. \quad (3d)$$

Thus, the chalcocite in the fissure veins may indicate deposition from a high-iron environment (though one somewhat lower than obtained in the main lodes) under mesothermal or leptothermal conditions rather than telethermal conditions of which chalcocite is normally diagnostic.

The gangue minerals that were deposited in the lodes before the native copper (red potash feldspar, epidote, chlorite, and pumpellyite) probably are indicative of the more intense portion of the mesothermal range. On the other hand, the principal gangue minerals introduced at the same time as the copper (quartz and calcite) could have been formed either under mesothermal or leptothermal conditions, while the zeolites and related minerals that accompanied copper probably show deposition within the leptothermal range. The extreme vertical distance, over which minable copper was formed (upwards of 10,000 feet in certain lodes), however, suggests a rate of change of temperature and pressure more appropriate to mesothermal than leptothermal conditions. The mesothermal to leptothermal classification here assigned to the Keweenaw Point deposits, therefore, is a compromise between the somewhat contradictory conclusions that can be drawn from the available data.

WHITE PINE

Late Precambrian *Copper* *Telethermal*

Ayer, F. A., 1950, White Pine mine, potential major copper producer: Min. Cong. Jour., v. 36, no. 12, p. 26-30, 55 (history and exploration)

Brecke, E. A., 1968, Copper mineralization in the upper part of the Copper Harbor conglomerate at White Pine, Michigan: Econ. Geol., v. 263, p. 294

Brooks, E. R., and Garbutt, P. L., 1969, Age and genesis of quartz-porphyry near White Pine, Michigan (disc.): Econ. Geol., v. 64, no. 3, p. 342-347

Butler, B. S., and others, 1929, The copper deposits of Michigan: U.S. Geol. Surv. Prof. Paper 144, 238 p. (particularly p. 169-174)

Carpenter, R. H., 1963, Some vein-wall rock relationships in the White Pine mine, Ontonagon County, Michigan: Econ. Geol., v. 58, p. 643-666; reply, 1964, v. 59, p. 1179-1180

Chaudhuri, S., and Faure, G., 1967, Geochronology of the Keweenawan rocks, White Pine, Michigan: Econ. Geol., v. 62, p. 1011-1033

Cornwall, H. R., and Wright, J. C., 1956, Geologic map of the Hancock quadrangle, Michigan: U.S. Geol. Surv. Mineral Invest. Field Studies, Map MF-46

Ensign, C. O., Jr., and others, 1968, Copper deposits in the Nonesuch shale, White Pine, Michigan, in Ridge, J. D., Editor, *Ore deposits of the United States, 1933-1967* (Graton-Sales Volumes): Chap. 22, v. 1

Hamilton, S. K., 1967, Copper mineralization in the upper part of the Copper Harbor conglomerate at White Pine, Michigan: Econ. Geol., v. 62, p. 885-904

—— 1969, Copper mineralization in the upper part of the Copper Harbor conglomerate at White Pine, Michigan--a reply (disc.): Econ. Geol., v. 64, p. 462-468

Joralemon, I. B., 1959, The White Pine copper deposit: Econ. Geol., v. 54, p. 1127

—— 1963, Vein-wall rock relationships, White Pine mine: Econ. Geol., v. 58, p. 1345-1346

Nishio, K., 1919, Native copper and silver in the Nonesuch formation, Michigan: Econ. Geol., v. 14, p. 324-334

Ohle, E. L., 1962, Thoughts on epigenetic vs. syngenetic origin for certain copper deposits: Econ. Geol., v. 57, p. 831-834

—— 1968, Copper mineralization in the upper part of the Copper Harbor conglomerate at White Pine, Michigan: Econ. Geol., v. 63, p. 190-191

Ramsay, R. H., 1953, White Pine copper: Eng. and Min. Jour., v. 154, no. 1, p. 72-87 (no geology)

Rand, J. R., 1964, Vein-wall rock relationships, White Pine mine: Econ. Geol., v. 59, p. 160-161

Sales, R. H., 1959, The White Pine copper deposit: Econ. Geol., v. 54, p. 947-951

White, W. S., 1970, A paleohydrologic model for mineralization of the White Pine copper deposit, northern Michigan: Geol. Soc. Amer. Abs. with Programs, v. 2, no. 7, p. 721

White, W. S., and Wright, J. C., 1954, The White Pine copper deposit, Ontonagon County, Michigan: Econ. Geol., v. 49, p. 675-716

—— 1960, The White Pine copper deposit: Econ. Geol., v. 55, p. 402-410

—— 1960, Lithofacies of the Copper Harbor conglomerate, northern Michigan: U.S. Geol. Surv. Prof. Paper 400-B, p. B5-B8

—— 1966, Sulfide-mineral zoning in the basal Nonesuch shale, northern Michigan: Econ. Geol., v. 61, p. 1171-1190

Notes

The White Pine copper deposits are located at the eastern end of the Porcupine Mountains, some 45 to 75 miles southwest of the copper-bearing portion of Keweenaw Point and about 120 miles east of Duluth.

The Porcupine Mountains are a conspicuous dome-shaped uplift on the south limb of the Lake Superior syncline. The ores are found in the upper portion of the late Precambrian Keweenawan series, specifically in a 25-foot section that spans the uppermost part of the Copper Harbor conglomerate and the lowest portion of the Nonesuch shale. The Copper Harbor has, as its youngest member, a fine- to coarse-grained sandstone composed of angular to subrounded fragments of mafic lava, quartz, and rhyolite, the interstices of which (20 to 25 percent of the rock) are filled with calcite or, to a lesser extent, calcite and chlorite. The uppermost 5 to 20 feet of the Copper Harbor is gray, while the formation as a whole is red to brown, the color difference being due to the higher ratio of chlorite to calcite in the gray rock; the hematite content of the two phases appears to be much the same. The Nonesuch shale generally consists of well-bedded to laminated gray siltstones in which current ripples and mud cracks indicate that it was deposited in shallow water. The 25 feet of ore-bearing rock is divided lithologically into four sections: (1) a 5-foot-thick lower sandstone (the uppermost Copper Harbor) that is fine- to coarse-grained and contains local pebbles and minor interbedded shale; (2) a 6.5-foot-thick parting shale (the lowest Nonesuch) that consists of 1.5 feet of siltstone and shale at the bottom, followed by 4 feet of massive gray siltstone and 2 feet of laminated gray siltstone with shale partings; (3) a 4.5-foot-thick upper sandstone that is gray and fine- to medium-grained and is locally interbedded with gray siltstone and red shale; and (4) a 10-foot, evenly laminated gray siltstone and green-gray shale (the lowest portion of the upper shale).

In the White Pine area, the structure pattern is largely controlled by the White Pine fault that strikes northwest and dips steeply in either direction; the fault shows a right-handed horizontal displacement of over a mile, and the southwest side has been dropped about 1500 feet. Immediately adjacent to the fault on the northeast, the beds have been somewhat folded, probably largely as a result of drag along the major fault; similar, though somewhat offset, structures were developed southwest of the White Pine break. Numerous minor faults are related to the White Pine, which strike at angles between

N20°E and N50°W and dip steeply in either direction; their strikes become more and more westwardly as they approach the White Pine fault, and the number of faults increases as the main fault structure is approached. On most of these faults, the strike slip is appreciably more impressive than that in the vertical direction.

No igneous rocks are known in the immediate vicinity of the mine, although the Porcupine Mountain uplift to the west of the mines may have been caused by the intrusion of igneous material; gabbro and gabbro aplite (similar to the rocks of the Duluth lopolith) are known to have intruded upper and middle Keweenawan series rocks in the Porcupine Mountain area. As the age of these intrusives is about 1000 m.y., the ores, if syngenetic, are somewhat older, although still probably of late Precambrian age; if the ores are epigenetic, they probably are genetically related to the gabbro-aplite series and are, therefore, also late Precambrian in age and are so classified here.

The principal ore minerals in the White Pine deposits are chalcocite and native copper, with the sulfide being many times as abundant as the native metal. In the old White Pine mine, operated in the World War I period, native copper was the principal ore mineral, and the ore was found almost exclusively in the upper and lower sandstones in rock volumes that contained interstitial carbonaceous material and chlorite. In the present mine, chalcocite is much the more abundant ore mineral, and both copper minerals occur nearly entirely in the Parting and Upper shales as fine disseminations in the rocks and as somewhat larger nodules on the bedding planes. To a lesser extent, the ore minerals are present as thin seams in joints and along bedding planes and in veins and veinlets that fill the minor fractures formed during the development of the White Pine fault. In these fractures the mineralization consists of early quartz, followed by calcite and then by chalcocite and a little native copper; some chlorite is present in the veins. In a few instances native copper in the minor fractures is accompanied by cuprite and malachite, and all three may have been formed from primary chalcocite by supergene processes. In the primary ore small grains of native silver are usually associated with the native copper, although silver amounts to less than 0.01 percent of the copper. The bulk of the primary native copper appears to be distributed through the host rocks in much the same manner as the chalcocite disseminations, but Carpenter holds that narrow halos (from a fraction of an inch to 20 inches in width) around the chalcocite-bearing veins and veinlets are barren of native copper. These halos, as Joralemon points out, cannot involve more than a minor fraction of the copper and can be explained by more than one hypothesis. The amount of copper sulfides, other than chalcocite, in the White Pine ores is essentially negligible, White and Wright reporting that only about 0.0025 percent of the copper occurs as chalcopyrite, bornite, and what Carpenter calls blue chalcocite. Carpenter suggests that these minor copper sulfides were formed late in the mineralization cycle.

The quartz-calcite-chalcocite veins and veinlets give every appearance of having been formed by hydrothermal solutions. So far as native copper is concerned, White and Wright believe that it was introduced into the upper and lower sandstones by hydrothermal fluids after the faulting. If these two vein types of mineralization were produced by hydrothermal solutions, it seems reasonable to assume that the remainder of the copper mineralization, mainly chalcocite in the apparently weakly permeable siltstones, was also introduced by such ore fluids. The presence of the same minerals--chalcocite, native copper, calcite, and chlorite--in the veins and in the siltstones (though in different proportions in the two types) also would indicate a similarity in origin.

Carpenter considers that the halos locally barren of native copper were the result of the migration of syngenetic copper, originally deposited as native copper, toward the veins and its precipitation in and near these structures as chalcocite by reaction with sulfur (probably from an igneous source) diffusing outward from the veins. Syngenetically deposited chalcocite apparently did not move. This hypothesis requires that (1) chalcocite and copper originally were deposited either contemporaneously with the sediments or diagenetically before they were lithified; (2) after fracturing, the copper of the native copper, though apparently not that of chalcocite, was dissolved and moved down concentration gradients toward the veins where it encountered S^{-2} ion coincidentally provided at exactly the time it was needed to precipitate chalcocite (instead of arriving sooner or later than required); and (3) that none of the chalcocite was emplaced by replacement of native copper because Carpenter considers the native copper grains to have been consistently of larger size than those of chalcocite. It would appear that the complexity of Carpenter's hypothesis and the degree of coincidence required for it to have been carried out at all make it far less attractive as an explanation for the development of the White Pine ores than that of simple hydrothermal origin for the ores under which the controlling factor in emplacement was ferric ion content rather than permeability.

Carpenter points out that all crystals of chalcocite which he examined showed orthorhombic outlines and were, therefore, deposited below 100°C. This fact, added to the rather vuggy character of the ore bodies, strongly suggests that the ore minerals were formed under telethermal conditions, and the deposits are so classified here.

Minnesota

MINNESOTA IRON RANGES (GENERAL)

Early, Middle and Late Precambrian — *Iron, Manganese* — *Sedimentary, Metamorphic, Hydrothermal*

Bailey, S. W., and Tyler, S. A., 1960, Clay minerals associated with the Lake Superior iron ores: Econ. Geol., v. 55, p. 150-175

Bruce, E. L., 1945, Pre-Cambrian iron formations: Geol. Soc. Amer. Bull., v. 56, p. 589-602

Cloud, P. E., Jr., and Licari, G. R., 1968, Microbiotas of the banded iron formations: Nat. Acad. Sci. Pr., v. 61, p. 779-786

Curtis, C. D., and Spears, D. A., 1968, The formation of sedimentary iron minerals: Econ. Geol., v. 63, p. 257-270

Dunn, J. A., 1941, The origin of banded hematite ores in India: Econ. Geol., v. 36, p. 355-370

—— 1954, The Kolhan series, and banded hematite quartzite: Econ. Geol., v. 49, p. 332-334

Flaschen, S. S., and Osborn, E. F., 1957, Studies of the system iron oxide-silica-water at low oxygen partial pressures: Econ. Geol., v. 52, p. 923-943

Goldich, S. S., and others, 1961, The pre-Cambrian geology and geochronology of Minnesota: Minn. Geol. Surv. Bull. 41, 193 p. (particularly p. 36-100, 150-168)

Govett, G.J.S., 1966, Origin of banded iron formations: Geol. Soc. Amer. Bull., v. 77, p. 1191-1211

Gross, G. A., 1965, Iron formations and bedded iron deposits, in *Geology of iron deposits in Canada*, v. 1, General geology and evaluation of iron deposits: Geol. Surv. Canada, Econ. Geol. Rept. no. 22, p. 82-132

Grout, F. F., and others, 1951, Precambrian stratigraphy of Minnesota: Geol. Soc. Amer. Bull., v. 62, p. 1017-1078

Huber, N. K., 1958, The environmental control of sedimentary iron minerals: Econ. Geol., v. 53, p. 123-140

James, H. L., 1953, Origin of the soft iron ores of Michigan: Econ. Geol., v. 48, p. 726-728

—— 1954, Sedimentary facies of iron formation: Econ. Geol., v. 49, p. 235-293

James, H. L., and Clayton, R. N., 1963, Oxygen isotope fractionization in metamorphosed iron formations of the Lake Superior region and other iron-rich rocks, in Engel, A.E.J., and others, Editors, *Petrologic studies, a volume in honor of A. F. Buddington*: Geol. Soc. Amer., p. 217-240

Krumbein, W. C., and Garrels, R. M., 1952, Origin and classification of chemical sediments in terms of pH and oxidation-reduction potentials: Jour. Geol., v. 60, p. 1-33

LaBerge, G. L., 1964, Development of magnetite in iron-formations of the Lake Superior region: Econ. Geol., v. 59, p. 1313-1342

Leith, C. K., and others, 1935, Pre-Cambrian rocks of the Lake Superior region: U.S. Geol. Surv. Prof. Paper 184, 34 p.

Lepp, H., 1968, The distribution of manganese in the Animikian iron formations of Minnesota: Econ. Geol., v. 63, p. 61-75

Lepp, H., and Goldich, S. S., 1964, Origin of Precambrian iron formation: Econ. Geol., v. 59, p. 1025-1060; disc., 1965, v. 60, p. 1063-1065, 1065-1070, 1731-1734

Mann, V. I., 1953, The relation of oxidation to the origin of the soft iron ores of Michigan: Econ. Geol., v. 48, p. 251-281

Marsden, R. W., 1968, Geology of the iron ores of the Lake Superior region in the United States, in Ridge, J. D., Editor, *Ore deposits of the United States, 1933-1967* (Graton-Sales Volumes): Chap. 23, v. 1

Owens, J. S., 1965, Origin of the Precambrian iron formations by Lepp and Goldich ... and development of magnetite in iron formations of the Lake Superior region by LaBerge ....: Econ. Geol., v. 60, p. 1731-1734

Pettijohn, F. J., 1943, Archean sedimentation: Geol. Soc. Amer. Bull., v. 54, p. 925-972

Royce, S., 1942, Iron ranges of the Lake Superior district, in Newhouse, W. H., Editor, *Ore deposits as related to structural features*: Princeton Univ. Press, p. 54-63 (particularly p. 61-63)

Symons, D.T.A., 1966, A paleomagnetic study of the Gunflint, Mesabi, and Cuyuna iron ranges in the Lake Superior region: Econ. Geol., v. 61, p. 1336-1361

Tanton, T. L., 1950, The origin of iron range rocks: Roy. Soc. Canada Tr., 3d Ser., v. 44, sec. 4, p. 1-19

Tyler, S. A., 1949, Development of Lake Superior soft iron ore from metamorphosed iron formation: Geol. Soc. Amer. Bull., v. 60, p. 1101-1124

—— 1952, Sedimentary iron deposits, in Trask, P. D., Editor, *Applied sedimentation*: John Wiley & Sons, Inc., N.Y., p. 506-523 (General discussion, not specifically directed to any range.)

Tyler, S. A., and others, 1940, Studies of the Lake Superior pre-Cambrian by accessory-mineral methods: Geol. Soc. Amer. Bull., v. 51, p. 1429-1537

Van Hise, C. R., and Leith, C. K., 1911, The geology of the Lake Superior region: U.S. Geol. Surv. Mono. 52, 641 p.

Woolnough, W. G., 1941, Origin of banded iron deposits--a suggestion: Econ. Geol., v. 36, p. 465-489

Yoder, H. S., 1957, Isograd problems in metamorphosed iron-rich sediments: Carnegie Inst. Washington Ann. Rept., 1956-1957, p. 232-237

Notes

The Iron Ranges of northeastern Minnesota include (from west to east) the Cuyuna, the Mesabi, the Eastern Mesabi, and the Vermilion, and these extend to the northeast in an en echelon pattern from the vicinity of the town of Crosby, in Crow Wing County, 15 miles northeast of Brainerd, into the western edge of St. Louis County, a few miles northeast of Ely. The Mesabi and the Eastern Mesabi constitute one continuous district that grades from the more metamorphosed Eastern Mesabi into the less metamorphically affected Mesabi proper.

The Ely district is contained in early Precambrian rocks, and the others are found in those of the middle Precambrian, but there is considerable divergence of opinion as to the ages of the various post-depositional processes that affected the iron formations and converted them into minable ore.

The earliest Precambrian rocks of northeastern Minnesota are Keewatin volcanic rocks (in the Canadian sense) known as the Ely greenstone; they are of early Precambrian age. The Ely formation consists of mafic extrusive rocks (mainly) and intrusive rocks (in minor part) that are now schistose greenstones. Near contacts with later granite, the greenstones have been converted to hornblende schist and near gabbro contacts to hornfels. The Ely formation is present in discontinuous belts as far west as the Mesabi district proper, but the great bulk of it is found in the Vermilion area.

The Ely contains, interbedded with the greenstones, iron-bearing beds known as the Soudan member; these beds consist of alternating bands of chert (recrystallized to fine-grained quartz) and iron oxides. In the Vermilion range, the Ely greenstone has been intruded by the Saganaga granite. Age determinations on this rock indicate that it has an age of some 2700 m.y. No granites of similar age are known in the Mesabi or Cuyuna districts. The intrusion of the Laurentian granites was followed by Algoman granites, some 2500 m.y. in age, that are known only in the Vermilion range.

The youngest of the early Precambrian formations in northeastern Minnesota is the Knife Lake group in which Gruner recognized 19 members made up of slates, graywackes, and conglomerates with some associated mafic flows, tuffs, and agglomerates; the group contains a few thin and locally developed lenses of iron formation. The Knife Lake is present in the Mesabi and Vermilion ranges, but it appears not to have been found in the Cuyuna range.

The Knife Lake group is intruded by such silicic rocks as the Vermilion granite (in the Vermilion range) and the Giants range granite (in the Mesabi range); no such granites have been reported from the Cuyuna range. It is considered by Lepp and Goldich (1964) that these granites are about 2500 m.y. old and are referred to as Penokean; their age dates them as late early Precambrian.

During the middle Huronian (middle Precambrian), the lower portion of the Animikie group was deposited; in the Mesabi range, the rocks formed in this time were the Pokegama quartzite (below) and the Biwabik iron formation (above). In the Vermilion district, only the Gunflint iron formation was developed, and in the Cuyuna range, the Mahnomen and Trommald formations appear to be the equivalents of the Pokegama and the Biwabik. In upper Huronian (upper Animikie) time, the Virginia, Rove, and Rabbit Lake formations were deposited in the Mesabi, Vermilion, and Cuyuna ranges, respectively. It appears that the middle Huronian formations are conformable on each other and that the upper Huronian has the same general dip as the middle Huronian rocks, but some doubt as to whether or not the rocks of the upper portion of the formation are strictly conformable on those beneath. The characteristics of these Animikie formations are discussed in some detail under the districts in which they occur.

An appreciable time interval probably occurred in northeastern Minnesota between the end of the upper Huronian and the introduction of the late Precambrian Keweenawan rocks, and a wide diversity of contact relationships exists between the Huronian and Keweenawan formations.

The only rocks of the Keweenawan group in the Minnesota Iron Ranges are those of, and related to, the Duluth gabbro, and even these are present only in the Mesabi and Vermilion districts. The age of these rocks is about 1050 m.y., which places them in about middle late Precambrian time. No other rocks of late Precambrian age are found in the iron ranges discussed here. The only consolidated rocks in the Minnesota ranges of post-Precambrian age are Cretaceous conglomerates (containing ore fragments) shales, and sands in the Mesabi district and consolidated Pleistocene gravels in the Cuyuna area.

The iron-bearing formations were deposited in the northeastern Minnesota area during both early and middle Precambrian time, with the iron formation in the Vermilion range having been formed more than 2500 m.y. ago and those in the other ranges between 2500 and 1600 m.y. ago, probably about 2000 m.y. ago.

In the Vermilion district the huge amounts of iron added to the broken iron formation probably were introduced in connection with the intrusion of the Duluth gabbro and thus the deposits in their final form were late Precambrian. The processes that affected the Eastern Mesabi to produce the magnetite taconite of that area also seem definitely to have been related to the intrusion of the Duluth gabbro, so that range also was brought into essentially its present form in the late Precambrian.

In the Mesabi range proper, however, the alteration of the primary iron formation to the higher-grade secondary ores (if the solutions that removed silica were of surface origin) may have taken place at any time between the end of the middle Huronian (lower Animikie) and the Cretaceous when conglomerates were developed that carry fragments of ore of the present Mesabi type. If the leaching of silica from the Mesabi ores was accomplished by heated water (entirely or partially of magmatic origin), the most likely source of the hydrothermal solutions or of the heat necessary to produce hot ground water was the Duluth gabbro; on the basis of this concept, the conversion of the Mesabi iron formation to higher-grade ore is thought to have occurred in the late Precambrian. In the Cuyuna range, the time required for the conversion of the primary ore is even less certainly known than in the other ranges, but it also appears probable that here the secondary ores had been completely developed by the end of the late Precambrian. Thus, the formation of the iron ores of the Minnesota Iron Ranges must have required much of Precambrian time--from somewhere in the early portion of that epoch through at least Keweenawan time in the late Precambrian, but the secondary processes may have extended into the Paleozoic, if they were accomplished by surface waters.

Although the classification of the various Minnesota Iron Ranges treated in this volume (Cuyuna, Eastern Mesabi, Mesabi, and Vermilion) are examined in some detail in the individual discussions of each range, it should be pointed out that certain fundamental problems of Precambrian iron ore genesis are not considered in them. Iron formation had been defined by James (1954) as a chemical sediment, typically bedded and commonly laminated, containing 15 percent or more iron of sedimentary origin and commonly, but not necessarily, containing layers of chert. Little doubt exists but that such primary iron formations are sedimentary rocks, but little agreement has been reached as to the processes and locations in which the sedimentation took place. Two basic hypotheses have been put forward to explain the character and composition of the iron formations. These are (1) that the constituents of the iron formations were removed by erosion from adjacent land surfaces, transported to the sea, and there precipitated, or (2) that the constituents were added as volcanic exhalations or hydrothermal solutions of magmatic origin that rose through fissure channelways to pour out on the sea floor.

Under the sedimentation hypothesis, one of the major problems is to explain how so much iron could have been transported by surface processes. Ferric iron compounds are quite insoluble under natural conditions, and iron oxides and hydroxides tend to accumulate in lateritic soils. On the other hand, although ferrous ion is readily soluble, it is easily oxidized to the ferric state and, in that state, is generally quickly precipitated in ferric compounds. Despite this apparent difficulty in moving large quantities of iron from one place to another under oxidizing conditions, it has been demonstrated that impressive amounts of iron are being brought into the sea at the present time; Gruner (1922) showed that the Amazon River alone, in 176,000 years, would have brought 1.94×10^{12} tons of iron into the sea. Further, James (1954) has pointed out that the accumulation of an iron formation does not depend on the quantity or concentration of iron (above a modest minimum amount) in the sea water but on pH and oxidation-reduction potential obtaining in the waters in question. The principal problem of iron deposition according to Pettijohn (*Sedimentary Rocks*, 2d ed., Harper and Bros., p. 461) is to define those geologic conditions that allow iron minerals to deposit but largely prevent the deposition of calcium carbonate and of clastic materials in that portion of the sea in which the iron formation is developed. James (1954) believes that the clastic material can be denied to the basin of sedimentation by low relief of the land mass from which flow the iron-bearing streams and calcium carbonate inhibited by the proper combination of pH and of calcium and carbonate and bicarbonate ion concentration in the water. Lepp and Goldich (1964), however, favor a source region of moderate relief since they consider that low relief favors kaolinization rather than laterization and that moderate relief is necessary to provide drainage and to permit the rapid removal of silica and soluble elements from the weathered source rocks.

Lepp and Goldich, moreover, believe that, had the Precambrian iron formations been accumulated under present-day conditions, they should be much richer in aluminum, titanium, and phosphorus than they are. They explain this apparently anomalous condition by assuming that the atmosphere of that earlier portion of Precambrian time in which iron formations were developed as having had a reducing atmosphere (one lacking, or essentially lacking, free oxygen). This suggestion that the Precambrian atmosphere was much different from that of later time is, of course, not original with Lepp and Goldich, but they have considered the implications of it in great detail. They also emphasize that in Precambrian iron formations silica and iron minerals are intimately associated, while later geologic time finds iron and silica (generally, but not always) separated in sediments. Further work is necessary to determine whether or not Lepp and Goldich's suggestion is sound.

James (1954) has put forward the concept that Precambrian iron formation can be separated into four facies: (1) sulfide, (2) carbonate, (3) oxide, and (4) silicate, with the major control for the first three, at least, being the oxidation-reduction potential obtaining in the sedimentary basin at the time of deposition. Huge rock volumes, however, are known in the northeastern Minnesota iron formations that are mixtures of silicates, oxides, and iron carbonate, suggesting that certain limited Eh-pH regions exist within which all three of these mineral species can form. Gruner (Schwartz, 1956, p. 201) points out that it is not certain if the original precipitate was made up of an amorphous mud or of distinct minerals. He believes it more probable that the crystallization closely followed precipitation, that chert was originally the last mineral to form, and that, in many of the rock volumes, chert segregated into chert-rich bands, separated by bands rich in iron-bearing minerals, usually oxides, silicates, or carbonates depending on the chemical character and composition of the sea water from which the precipitate came. Chert may in some places, however, be unimportant enough in amount that chert-rich bands were not developed and what chert was formed became intimately intergrown with the iron-bearing minerals. Although Gruner may be correct in claiming that the segregation of the iron-bearing formation into chert and iron mineral-rich bands was caused by diagenetic processes, it is also possible that this banding was caused by cyclical differences in primary sedimentation. These differences may have been the result of the climatic changes from winter to summer or from dry season to wet or may, instead, have been caused by cyclical differences in the character of the volcanic exhalations reaching the bottom of the iron-formation-depositing sea. In either event, the direct controls over such cyclical deposition would have been the pH and oxidation-reduction potentials of the sea water. The different ion concentrations on which these controls acted, however, would have been due to changes in the character of the solutions brought into the depositing environment from whatever was the supplying source.

The pyrite facies (in which pyrite may make up as much as 40 percent of the rock) are not normally considered as iron formation since they are not minable as iron ore in their primary state nor are they converted to iron ore by later geologic processes.

Van Hise and Leith (1911) were so impressed by the difficulties of removing sufficient iron from adjoining land surfaces for the development of the Lake Superior iron formations that they had recourse to volcanic sources for the iron (and silica) required. This theory has the great advantage of assuming that the volcanic exhalations contained only those elements that were deposited in the iron formations and that they lacked those which are not present. Obviously this hypothesis does not explain why only the required elements were contained in the volcanic exhalations or hydrothermal solutions involved nor does it explain why the exhalations would reach the only portions of the sea bottom where just the right conditions of sea depth, salinity, pH, Eh, and relationship to adjacent land surfaces obtained to produce typical iron formation. The hypothesis is not strongly defended at the present time.

Even these brief remarks on the conditions under which the primary iron formations were developed should, if nothing else, serve to illustrate the lack of agreement on the fundamental problems of iron-formation genesis and the complexity of the problems themselves. The primary iron formations almost certainly were developed by sedimentary processes in sea water (although the suggestion has been made that they were formed in fresh water environments); some diagenetic alteration of the original precipitates is generally accepted. No real consensus has been reached as to the mechanisms by which, and the conditions under which, the raw materials of the iron formations were made available in, and were precipitated from, sea water. It seems most reasonable, however, to classify the primary deposits as sedimentary-A1a, although the possibility exists that they might, if the precipitated materials were added from igneous sources by exhalations or solutions, be categorized as sedimentary-A3. Further, the diagenetic changes may have been sufficiently impressive to warrant the inclusion of sedimentary-4 in the classification.

The changes produced in the iron formation of the Minnesota ranges by metamorphic, hydrothermal, and ground water activity are considered in some detail in the discussions of the individual districts.

CUYUNA

Middle Precambrian, Late Precambrian	*Iron as Goethite, Hematite; Manganese as Manganite, Pyrolusite*	*Sedimentary-A1a, Hydrothermal-1 to Mesothermal, and Ground Water-B2*

Adams, F. S., 1910-1911, The iron formation of the Cuyuna range: Econ. Geol., v. 5, p. 729-740; v. 6, p. 6-70, 156-180

Emmons, W. H., and Grout, F. F., Editors, 1943, The Cuyuna range, in *Mineral resources of Minnesota*: Minn. Geol. Surv. Bull. 30, p. 23-26

Grout, F. F., 1946, Acmite occurrences on the Cuyuna range, Minnesota: Amer. Mineral., v. 31, p. 125-130

Grout, F. F., and Wolff, J. F., 1955, The geology of the Cuyuna district, Minnesota--a progress report: Minn. Geol. Surv. Bull. 36, 144 p.

Gruner, J. W., 1947, Groutite, $HMnO_2$--a new mineral of the diaspore-goethite group: Amer. Mineral., v. 32, p. 654-659

Han, T.-M., 1968, Ore relations in the Cuyuna sulfide deposit, Minnesota: Mineralium Deposita, v. 3, p. 109-134

Harder, E. C., 1918, Manganiferous iron ores of the Cuyuna district, Minnesota: A.I.M.E. Tr., v. 58, p. 453-486

Harder, E. C., and Johnston, A. W., 1917, Notes on the geology of east central Minnesota, including the Cuyuna iron ore district: Minn. Geol. Surv. Bull. no. 15, 178 p. (particularly p. 94-135)

Leith, C. K., 1907, The geology of the Cuyuna iron range: Econ. Geol., v. 2, p. 145-152

Schmidt, R. G., 1958, Map of bedrock geology of the southwestern part of the North range, Cuyuna district, Minnesota: U.S. Geol. Surv. Mineral Invest. Field Studies, Map MF-181 (sheets 4-6), 1:7200

—— 1958, Titaniferous sedimentary rocks in the Cuyuna district, central Minnesota: Econ. Geol., v. 53, p. 708-721

—— 1959, Map of bedrock geology of the northern and eastern parts of the North range, Cuyuna district, Minnesota: U.S. Geol. Surv. Mineral Invest. Field Studies, Map MF-182 (sheets 7-11), 1:7200

—— 1963, Geology and ore deposits of the Cuyuna North range, Minnesota: U.S. Geol. Surv. Prof. Paper 407, 96 p.

Schmidt, R. G., and Dutton, C. E., 1957, Map of bedrock geology of the south-central part of the North range, Cuyuna district, Minnesota: U.S. Geol. Surv. Mineral Invest. Field Studies, Map MF-99 (sheets 1-3), 1:7200

Thiel, G. A., 1924, High temperature manganese veins of the Cuyuna range: Econ. Geol., v. 19, p. 377-381

—— 1924, Iron sulphides in magnetic belts near the Cuyuna range: Econ. Geol., v. 19, p. 466-472

—— 1924, The manganese minerals; their identification and paragenesis: Econ. Geol., v. 19, p. 107-145 (particularly p. 132-145)

—— 1926, Phosphorus iron ores on the Cuyuna range: Eng. and Min. Jour., v. 121, no. 17, p. 687-690

—— 1927, Geology of the Cuyuna range: Geol. Soc. Amer. Bull., v. 38, p. 785-793

Winchell, N. H., 1907, The Cuyuna iron range: Econ. Geol., v. 2, p. 565-571

Zapffe, C., 1925, Manganiferous iron ores of Cuyuna district, Minnesota: A.I.M.E. Tr., v. 71, p. 372-385

—— 1928, Geologic structure of the Cuyuna iron district, Minnesota: Econ. Geol., v. 23, p. 612-646

—— 1933, The Cuyuna iron ore district, in *Lake Superior region*: 16th Int. Geol. Cong., Guidebook 27, p. 72-88

Notes

The Cuyuna district lies essentially in the geographic center of Minnesota and covers a distance, from northeast to southwest, of nearly 70 miles. The district is divided into north and south ranges, with the North range having an area of about 50 square miles and centering around the town of Crosby and the South range extending for the full length of the district. The Emily district, by some included with the Cuyuna ranges, is almost directly north of the North range and occupies a somewhat greater area. Almost all of the production of the Cuyuna district has come from the North range; no mines have ever been developed in the Emily district.

The latest work (Schmidt, 1963) in the Cuyuna district divides the metamorphosed Animikie sedimentary rocks into three formations: (1) the Mahnomen formation that consists of fine clastic, low-iron material with some quartzitic lenses near the top; (2) the Trommald formation that is made up of iron formation and shows considerable differences in textures and mineral content between thin- and thick-bedded facies; and (3) the Rabbit Lake formation that is largely gray to black ferruginous argillite with lenses of low-iron formation--the lowest 300 feet of the Rabbit Lake in the North range are unique in that they contain about 2.0 percent titania. Neither top nor bottom of this Animikie sequence is known. On part of the North range, the base of the Rabbit Lake includes up to three sills or flows of basalt and also probably contains beds of water-laid tuff. Thin Keweenawan flows overlie a small area of the folded and metamorphosed Animikie; the flows are essentially horizontal. If these flows are actually Keweenawan, they indicate that the erosion surface on the Animikie beds was developed before or during Keweenawan time and strongly suggests that the various stages of ore development had occurred before the flows were extruded. The relationship of the flows to the deep weathering of the Animikie surface, however, has not been studied, and the possibility still exists that the final stages of ore formation were later than the Precambrian.

In addition to the flow rocks in the Rabbit Lake area, igneous rocks that range from intermediate to mafic composition intrude as dikes (locally as sills) in an extensive belt along the southeast side of the North range. These rocks are now highly chloritized but appear initially to have been gabbro or diorite; the plagioclase in these igneous rocks has been albitized while the other primary minerals have been converted to chlorite, hornblende, epidote, clinozoisite, calcite, and leucoxene with some residual sphene.

The rocks designated as Animikie in the Cuyuna district almost certainly are middle Precambrian in age; the only question that remains unsettled is the length of time that was required to convert the primary iron formation rocks into the ores mined today. What little evidence there is, the occurrence of horizontal Keweenawan flow rocks over the eroded Animikie, suggests that all of the alteration processes had been completed before the end of the Precambrian; the ores are, therefore, here categorized as middle Precambrian and late Precambrian.

The primary iron-bearing Animikie sediments of the Trommald formation were deposited in two marine or lacustrine environments: (1) deep, quiet waters in which the oxidation-reduction potential was such that the thin-bedded facies that developed were composed of chert and silicates, oxides, and carbonates of iron, and (2) shallower, probably somewhat agitated waters in which the oxidation-reduction potential was such that the thick-bedded facies formed contained chert and iron oxides. The first of these two facies corresponds with the carbonate and silicate facies of James (1954) and the second with his oxide facies. At the beginning of Trommald deposition, the thick-bedded portion covered about one-third of the North range basin and two-thirds of that area by the end of Trommald time. Thus, the thick-bedded facies steadily overlapped the thin bedded with the passage of Trommald time. The primary manganese minerals were mostly finely disseminated in the iron formation in and near the transition zones between the thick- and thin-bedded facies; the fine-grained and disseminated character of the manganese minerals now in the ore makes their identification generally impossible, and it is, therefore, impossible to say what primary manganese minerals are dominant in the ore now being mined, much less what manganese minerals were deposited in the primary sediments and in what proportions. The manganese minerals that probably have been formed by weathering include psilomelane [$BaMn^{2}Mn^{4}_{6}O_{16}(OH)_4$], pyrolusite ($MnO_2$), rhodocrosite, and stilpnomelane (a chlorite-type Mn-bearing silicate). The deposition of the iron formation was ended when large amounts of silt and clay were brought into the basin by stream action and tuffs and flows were added through volcanic activity.

The Trommald and the older and younger Animikie beds were still later intruded by mafic to intermediate rocks that are generally discordant to the bedding; the dikes, however, roughly parallel the fold axes and probably were introduced after much or all of the folding had been completed.

The conversion to ore of the metamorphosed iron formation, in the thin-bedded facies, and probably in the thick-bedded rocks as well, appears to have occurred in two stages: (1) when hot (or warm) waters, probably to an appreciable extent of hydrothermal character and magmatic origin, circulated through the fractured iron formation oxidizing it over a wide area and converting some of it to ore through the leaching of silica (both chert and silica from silicates); ore bodies formed in this manner are largely composed of red-brown hematite and are tabular or lenticular in the plane of the bedding, and (2) when ores developed in the first stage were subjected to further alteration, probably by surface waters that carried oxygen deep beneath the then-existing surface but possibly by ground water (energized by hydrothermal solutions) that produced brown goethitic ores as irregular blankets from such thin-bedded ores as were exposed and probably from most of the thick-bedded ore as well; brown iron oxides, and locally brown ores, were formed on almost all the remaining masses of unoxidized iron formation. That there has been redistribution of iron and manganese and probably have been additions of these two elements from sources outside the iron formation are indicated by the common occurrence of replacement textures in the ores and the presence of both iron and manganese minerals as fracture fillings. How much of the iron and manganese now in the ores and altered iron formations was added and how much derived from the iron formation itself is uncertain. Schmidt (1963) appears to believe that the additions were essentially negligible, but the evidence of replacement and of open-space filling, though far less clear than in the Vermilion district, suggests that appreciable additions may have come from sources outside the iron formation.

In any event, the process of ore formation in the Cuyuna district probably required that the primary iron formation be metamorphosed, hydrothermally altered, and further altered by ground water (in which process non-ore material was removed to a much greater extent than ore material was added) before the ores now being mined could be produced. The deposits, therefore, are here classified as sedimentary-Ala, metamorphic-C, hydrothermal (probably hypothermal-1 to mesothermal), and ground water-B2.

EASTERN MESABI

Middle Precambrian *Late Precambrian* — *Iron as Magnetite* — *Sedimentary-A1a, Metamorphic-C, and Hypothermal-1*

Broderick, T. M., 1919, Detailed stratigraphy of the Biwabik iron-bearing formation, east Mesabi district, Minnesota: Econ. Geol., v. 14, p. 441-451

Grout, F. F., and Broderick, T. M., 1919, The magnetite deposits of the Eastern Mesabi range, Minnesota: Minn. Geol. Surv. Bull. no. 17, 58 p.

Gruner, J. W., 1946, Mineralogy and geology of the Mesabi range: Iron Range Resources and Rehabilitation, St. Paul, 127 p. (particularly p. 60-62, 71-84)

Gunderson, J. N., 1960, Lithologic classification of taconite from the type locality: Econ. Geol., v. 55, p. 563-573

—— 1960, Stratigraphy of the Eastern Mesabi district, Minnesota: Econ. Geol., v. 55, p. 1004-1029

Gunderson, J. N., and Schwartz, G. M., 1961, Magnetic taconites of the Eastern Mesabi district, Minnesota: A.I.M.E. Tr., v. 222, p. 227-233

—— 1962, The geology of the metamorphosed Biwabik iron-formation, Eastern Mesabi district, Minnesota: Minn. Geol. Surv. Bull. 43, 139 p.

Richarz, S., 1927, Grunerite rocks of the Lake Superior region and their origin: Jour. Geol., v. 35, p. 690-707

Notes

The eastern portion of the Mesabi range, the area of magnetic taconites, extends from near the village of Mesaba northeastward to just beyond the town of Old Babbitt, a distance of almost 20 miles; the northeast tip of the district is overlain and transgressed by the Duluth gabbro, and nowhere is the Duluth gabbro more than a mile or two south of the taconite belt. The city of Duluth is 55 miles south of the Eastern Mesabi.

The taconites lie within the Biwabik iron formation, the middle of the three subdivisions of the Animikie in the general Mesabi area. In the Eastern Mesabi the Animikie rocks lie unconformably on the Giants range granite and consist of the Pokegama quartzite, overlain by the Biwabik iron formation which, in turn, lies beneath the Virginia slate or argillite; locally the Biwabik is in direct contact with the Giants range granite. Gunderson and Schwartz (1962) were able to break the Biwabik down into 22 members, 5 in the lower cherty, 2 in the lower slaty, 8 in the upper cherty, and 7 in the upper slaty; in most instances these 22 members can be recognized only in drill core. The contacts between one member and those above and below it are normally gradational, and the lateral variations in mineralogy and mineral assemblages appear mainly to be the result of thermal metamorphism and hydrothermal replacement and do not generally reflect differences in mineral content of the primary sediments. The original composition of the Biwabik probably was much the same as that of the Biwabik in the main Mesabi area, but these eastern rocks have been more drastically modified than farther to the west. The Biwabik strata here are of two quite different kinds: (1) bedded taconite and (2) massive taconite. These strata alternate over short vertical distances, and the rocks of the formation are divided into five types, depending on the proportions of massive and bedded taconite contained in them: (1) massive taconite--less than 10 percent bedded taconite strata, (2) layered taconite--10 to 40 percent bedded taconite strata, (3) laminated taconite--40 to 60 percent bedded taconite strata, (4) shaly bedded taconite--60 to 90 percent bedded taconite strata, and (5) shaly taconite--less than 10 percent massive taconite strata. Most of the iron oxides are found in the bedded taconite strata, although important amounts may occur in the massive taconite strata.

The latest interpretation of the origin of taconites from the primary sediments (Gunderson and Schwartz, 1962) is that they were first thermally metamorphosed due to the intrusion of the late Precambrian Duluth gabbro, the age of which is about 1100 m.y., and then retrogressively altered by hydrothermal solutions given off by crystallizing pegmatites derived from the Duluth gabbro complex in the late stages of its crystallization cycle.

The primary sedimentation of the Biwabik formation took place in the middle Precambrian, while the thermal metamorphic and hydrothermal effects were produced in the late Precambrian. As these rocks were essentially unaffected by any later geologic processes, the deposits should be dated as middle Precambrian and late Precambrian.

The origin of the primary Biwabik beds is discussed in the classification section of the main Mesabi, and the same scheme applies to the Eastern Mesabi except that the original Biwabik sediments in their eastern extension appear to have contained more silicates and less carbonates than they did farther to the west. Gruner (Schwartz, 1956, p. 186) considers the oldest silicates--minnesotaite, stilpnomelane, and greenalite, layer silicates similar to talc, chlorite, and serpentine, respectively--to have been formed diagenetically. James (1954, p. 266) appears to believe that, although greenalite is primary, stilpnomelane and minnesotaite were produced by low-grade metamorphism. White (1954, p. 36-38) insists that, because the three silicates are so intimately intergrown and occur together over a horizontal distance of at least 80 miles, they must be of the same age and probably were formed diagenetically. Thus, before the Biwabik in the Eastern Mesabi was affected by the intense thermal metamorphism induced by the intrusion of the Duluth gabbro, it consisted of the three silicates just mentioned, and chert, magnetite, siderite, and a little hematite and other carbonates. The thermal metamorphism appears to have converted the hydrous layer silicates, the carbonates, and some of the chert and magnetite into fayalite with local traces to subordinate amounts of ferrohypersthene and poikilitic cummingtonite and to have caused the recrystallization of chert and magnetite into coarser-grained intergrowths of quartz and magnetite. Fayalite seems to have been formed only in rock volumes containing sufficient carbonaceous material to permit the reduction of iron from the ferric to the ferrous state, while, in the rock volumes lacking primary silicates and carbonates, the only effect of the thermal metamorphism was the recrystallization to a coarser texture of the primary magnetite and chert.

Not long after the thermal metamorphism, if not during the last stages of it, hydrothermal solutions caused a widespread change of the mineral composition of the Biwabik formations through the replacement of fayalite, quartz, and magnetite by such minerals as hedenbergite, less ferrohypersthene, various calcium-rich amphiboles, and cummingtonite, plus feldspars, biotite, apatite, sulfides, and loellingite. Detailed field studies by Gunderson and Schwartz (1962) suggest that the solutions that produced these metasomatic effects came from pegmatite veins, derived from the acid phase of the Duluth gabbro and intruded into the Biwabik. Cummintonite is the most abundant of the metasomatic silicates and mainly replaces quartz and magnetite, although it is known to replace all the anhydrous silicates in the rock. As the only valuable iron-bearing mineral in the Eastern Mesabi taconites is magnetite, it is apparent the greater the conversion of magnetite and quartz to silicates, the lower is the grade of the Biwabik beds in recoverable iron. Tyler (1949) argued that magnetite had been added by these hydrothermal solutions, and the demonstrated additions of iron to iron-bearing formations in districts both in Minnesota and elsewhere suggests that some of the magnetite in these taconites may be of hydrothermal origin, probably emplaced as later replacement of the early hydrothermal silicates.

In any event, it is certain that much, if not all, of the iron in the Eastern Mesabi taconites was placed there through primary sedimentation; therefore, the classification sedimentary-A1a must be assigned to them. Since a pronounced change in grain size was affected by thermal metamorphism, the category metamorphic-C must be added to the classification and because of the metasomatic emplacement of anhydrous silicates and possibly of additional magnetite, the term hydrothermal also must be included. From the character of the minerals produced by the hydrothermal reactions, it appears probable that these solutions were in the hypothermal range, so the hypothermal category that applies is hypothermal-1.

MESABI

Middle Precambrian *Late Precambrian*	*Iron as Hematite (same as Martite), Goethite*	*Sedimentary-A1a, Hypothermal-1 to Mesothermal, and/or Ground Water-B2*

French, B. M., 1968, Progressive contact metamorphism of the Biwabik iron formation, Mesabi range, Minnesota: Minn. Geol. Surv. Bull. 45, 103 p.

Gruner, J. W., 1922, Paragenesis of the martite ore bodies and magnetites of the Mesabi range, Minnesota: Econ. Geol., v. 17, p. 1-14

—— 1922, The origin of sedimentary iron formations: the Biwabik formation of the Mesabi range: Econ. Geol., v. 17, p. 407-460

—— 1924, Contributions to the geology of the Mesabi range, with special reference to the magnetites of the iron-bearing formation west of Mesabi: Minn. Geol. Surv. Bull. no. 19, 71 p.

—— 1930, Hydrothermal oxidation and leaching experiments; their bearing on the origin of Lake Superior hematite-limonite ores: Econ. Geol., v. 25, p. 697-719, 837-867 (particularly p. 850-858)

—— 1932, Additional notes on secondary concentration of Lake Superior iron ores: Econ. Geol., v. 27, p. 189-205 (particularly p. 189-200, 203-204)

—— 1933, The Mesabi range, in *Lake Superior region*: 16th Int. Geol. Cong., Guidebook 27, p. 88-97

—— 1937, Hydrothermal leaching of iron ores of the Lake Superior type--a modified theory: Econ. Geol., v. 32, p. 121-130

—— 1946, Mineralogy and geology of the Mesabi range: Iron Range Resources and Rehabilitation, St. Paul, 127 p. (particularly p. 26-59, 64-70, 89-116)

Gunderson, J. N., 1959, Lithologic classification of taconite from the type locality: Econ. Geol., v. 54, p. 563-573

Leith, C. K., 1903, The Mesabi iron-bearing district of Minnesota: U.S. Geol. Surv. Mono. 43, 316 p.

—— 1931, Secondary concentration of Lake Superior iron ores: Econ. Geol., v. 26, p. 274-288

Schwartz, G. M., Editor, 1956, Precambrian of northeastern Minnesota: Geol. Soc. Amer. Guidebook for Field Trips, Field Trip no. 1, 235 p. (particularly p. 1-9, 20-28, 159-235)

White, D. A., 1954, The stratigraphy and structure of the Mesabi range: Minn. Geol. Surv. Bull. 38, 92 p.

Wolff, J. F., 1916, Recent geologic developments on the Mesabi iron range, Minnesota: A.I.M.E. Tr., v. 46, p. 142-169

Notes

The center of the Mesabi range lies some 50 miles northwest of Duluth, the Mesabi range being the designation of the preglacial outcrop area of the Biwabik formation that extends from southwest to northeast in a flat reverse "S" for 120 miles from easternmost Cass County to Birch Lake. The eastern Mesabi includes the 20 miles between Mesaba and Birch Lake. The main Mesabi is situated between Mesaba and Nashwauk and includes over 60 miles along the formation. The remaining 40 miles of Biwabik formation is divided between the west Mesabi and the westernmost Mesabi. The bend in the reverse "S" is known as the Virginia Horn and is located in the vicinity of Virginia and Eveleth; the bend is the result of folding. The Mesabi range is located on the middle slopes of the south flanks of the Giants range that reaches from Birch Lake in the northeast to Grand Rapids to the southwest; west of Grand Rapids the entire ridge is covered by glacial debris.

The older rocks of the Mesabi range consist of the Giants range granite that is intrusive into the Ely greenstone and into the Knife Lake group. The Animikie group of rocks lies unconformably on an eroded surface composed of these older rocks and consists of the Pokegama, Biwabik, and Virginia formations. These beds form part of the north limb of the Lake Superior syncline and dip generally to the south-southeast at angles between 5° and 15°; thus, the Pokegama outcrops lie north of those of the Biwabik, while the Virginia argillite lies to the south of the Biwabik under various thicknesses of glacial drift. The Animikie rocks are essentially unmetamorphosed (in the opinion of most, though not of all, geologists who have studied the area) except for the Eastern Mesabi where the Duluth gabbro was intruded diagonally across the Animikie beds in late Precambrian time and essentially eliminated the rocks of this group for the 50 miles between Birch Lake and about 10 miles west of Gunflint Lake. The Animikie group (under different formational names) then continues on to the northeast as the Gunflint range.

The central rocks of the Animikie group are the Biwabik iron formation which outcrops quite conspicuously to the east of Mesaba but which was exposed, before mining began, in no more than five or six places west of the town. The Biwabik formation, according to Gruner (Schwartz, 1956) was deposited as a chemical precipitate containing a large percentage of silica that probably consisted, prior to diagenesis, of a mud of colloidally sized particles that was of essentially the same chemical composition throughout its vertical and horizontal extent. This material is thought to have crystallized into an intimately intergrown mixture of minerals in a wide variety of shapes, grain sizes, textures, and bedding and banding, these minerals being, in order of decreasing abundance: (1) chert; (2) iron silicates--minnesotaite (talc), stilpnomelane (chlorite), and greenalite (serpentine); (3) iron oxides, mainly magnetite but with a little hematite; and (4) carbonates, chiefly siderite but including calcite and dolomite; pyrite and graphite are accessory minerals. The crystallized rock is known as taconite, and much of it has a granule texture of about pinhead size; this texture is particularly common in cherty layers from 0.5 to 1.0 foot in thickness. The granules may be formed of any mineral or mixture of minerals, have no regular internal structure, and are rarely greater than 1 mm in diameter (except in the metamorphosed rocks of the Eastern Mesabi where they may be much larger). In the main, where taconite consists principally of chert, with iron silicates or magnetite, the rock is normally granular, but taconite composed of siderite and/or iron sulfides is usually slaty.

The slaty and cherty varieties of the taconite were used to form the initial stratigraphic divisions of the Biwabik (from bottom to top): (1) lower cherty, (2) lower slaty, (3) upper cherty, and (4) upper slaty, with the contact between (2) and (3) being gradational; further subdivision among the rocks of these divisions has been made. It is White's (1954) opinion that most of the features of Animikie sediments can be explained by normal sedimentary processes, apparently including diagenesis. Following James, he believes that little or no iron has been introduced into the rocks since their consolidation. If this is the case, then the age of the primary iron formation definitely is middle Precambrian.

For years the taconites of the main Mesabi were not minable, the only minable deposits in that area being those in which the silica content of the rocks has been reduced sufficiently by a leaching process that essentially has not affected (though it has mineralogically changed) the iron content. Iron, however, has been increased, relative to the other constituents of the altered iron formation, to an extent that converted the leached rock to minable ore. The time during which this leaching occurred has not been definitely determined. If the leaching is thought to have been accomplished by hydrothermal solutions or by solutions composed of a mixture of magmatic emanations and ground water, the most likely source of the heat of such solutions was the Duluth gabbro, and the age of the alteration must be late Precambrian. If, however, the leaching is believed to have been accomplished by ground water alone, the process may have extended considerably into post-Precambrian time. Were the leached Mesabi ore bodies overlain by Keweenawan flows, the problem of dating the alteration would be solved, but no such stratigraphic relationships are known on the Mesabi, so the time span of the alteration is uncertain. As will be discussed under the classification note, however, it appears probable that the leaching was accomplished largely by solutions of hydrothermal origin that, it appears most probable, were derived from the same cycle of magmatic activity that produced the Duluth gabbro. On this basis, it seems most logical to assign the leaching process to the late Precambrian, and the Mesabi ores are, therefore, dated as middle Precambrian and late Precambrian.

Most, if not all, of the iron in the Mesabi ores was introduced during the deposition of the primary sediments. It is equally certain, however, that the high-iron ores required the interposition of other geologic processes to develop them from the original taconites of the iron formation. Some authors, such as James, have suggested that some of the iron silicates, specifically minnesotaite and stilpnomelane, are the result of low-grade metamorphism, but the present consensus probably is that all the iron silicates and the other minerals of the taconite date from no later in the rock-forming process than the diagenetic stage. Therefore, the unaltered iron-formation (taconite) is classified here as sedimentary-A1a.

The first author to suggest that the leaching of the taconites to produce the high-iron ores was the result of action other than that of ground water was Gruner in 1930, and he has summarized his reasons (Schwartz, 1956) for believing that meteoric waters, heated by magmatic emanations, were responsible for the leaching of silica (both from chert and from the iron silicates) and the oxidation of iron as follows: (1) large ore bodies extend to a maximum depth of almost 800 feet below the glacial drift; (2) no change in quality or physical condition of the ore is noted with increasing depth; (3) the leaching of fine-grained quartz and iron silicates has taken place on so large a scale that the solvent action of normal ground water can hardly have been responsible; (4) ferrous iron has been completely converted to the ferric state in rock volumes largely leached of silica, whereas the chert in primary taconite containing hematite has not been so affected; (5) there is essentially no difference today between the silica content of ground water in the drift and that in the mines; (6) if the alkali necessary for the leaching of the silica has entered the ground water through the decomposition of feldspars, such ground water would have been saturated in silica on encountering the iron formation and incapable of its further solutions; (7) the probability that the leaching was completed before the end of Precambrian time; (8) the axes of elongation of the trough ore bodies in the main Mesabi are normally not in positions favorable to their production by ground water flowing down the dip of the formation; (9) many ore bodies do not extend *up-dip* to the surface but are capped in that direction by unleached taconite; (10) the minor cross folding and attendant fracturing, which was more intense at the base of the Biwabik and within this portion of the formation most intense in the rock volumes later converted to high-iron ore, probably provided channelways for the altering solutions; (11) there are no portions of the Biwabik that could have promoted artesian circulation through the formation; (12) the slump structures of the ore bodies give the impression of being bounded by faults caused by the sudden collapse of material undermined from beneath; had the solutions done their work from above downward, it is unlikely that such structures would have developed other than gradually; and (13) the presence of highly decomposed Virginia formation rocks on the tops of some of the high-iron ore bodies indicates that the surface at the time of silica leaching was considerably above what it is now, making it even less likely that unaided ground water was responsible for the removal of silica. The lack of leached-silica ore bodies in the Eastern Mesabi may be due to the failure of ground water to pass through the probably overlying Duluth gabbro, but such lack may instead have been the result of the more intense alteration (see Eastern Mesabi discussion) that hydrothermal solutions nearer their source produced in the taconite, alteration that resulted in new silicates rather than the removal of silica as took place farther west.

The arguments against hot-water leaching are principally (1) the lack of an immediately apparent igneous source for the magmatic emanations--although solutions from magmatic sources almost certainly affected both the Eastern Mesabi and the Cuyuna districts that lie to either side of the main Mesabi; (2) the abundance of goethite (about one-half of the iron oxides)--which may limit the temperatures of the solutions to something less than 200°C; and (3) the ore bodies appear to be generally at the present erosion surface and none has been found completely covered by Virginia formation. This is not proof, however, that no rocks overlay the ores at the time they were developed.

The balance of probabilities, then, would seem to indicate that the Mesabi high-iron ores were produced by the leaching of silica through the activity of hot water solutions that were, initially at least, low in silica. These solutions well may have been in part derived from ground water, but it appears equally probable that the leaching of silica took place at depths sufficiently beneath the surface that ground water would have played no more than a minor part in their composition. The most logical source from which the hydrothermal leaching solutions (= magmatic emanations in Gruner's sense of the term) appears to have been the Duluth gabbro from which also came the hydrothermal solutions responsible for the metasomatic alteration of the taconites of the Eastern Mesabi. If these solutions were able to make their way westward after having completed their work in the Eastern Mesabi, they probably would have been little, if any, higher in silica than they were initially and well might have been able to carry out the leaching of silica in much the same manner as did such hydrothermal solutions as those that attacked the rocks in the Cripple Creek crater in the early stages of its alteration. The high-iron ores of the Mesabi are here categorized as sedimentary-A1a, hydrothermal-1 to mesothermal, and/or ground water-B2, with the hydrothermal classification being more likely than the ground water one. It may develop that the hydrothermal alteration of at least part of the Mesabi ores may have taken place under conditions as low in intensity as leptothermal.

VERMILION

Early Precambrian	*Iron as Hematite, Minor Magnetite, Siderite*	*Sedimentary-A1a and Hypothermal-1 to Mesothermal*

Abbott, C. E., 1907, Geology of the Ely trough iron-ore deposits: Eng. and Min. Jour., v. 83, no. 13, p. 601-605

Clements, J. M., 1903, The Vermilion iron-bearing district of Minnesota: U.S. Geol. Surv. Mono. 45, 463 p.

Emmons, W. H., and Grout, F. F., Editors, 1943, The Vermilion range, in *Mineral resources of Minnesota*: Minn. Geol. Surv. Bull. 30, p. 26-29

Grout, F. F., 1926, The geology and magnetite deposits of northern St. Louis County, Minnesota: Minn. Geol. Surv. Bull. no. 20, 220 p.

Gruner, J. W., 1926, The Soudan formation and a new suggestion as to the origin of the Vermilion iron ores: Econ. Geol., v. 21, p. 629-644

—— 1941, Structural geology of the Knife Lake area of northeastern Minnesota: Geol. Soc. Amer. Bull., v. 52, p. 1577-1642

Kemp, J. F., 1906, Vermilion range, Minnesota, in *The ore deposits of the United States and Canada*: McGraw-Hill, N. Y., p. 144-150

Machamer, J. F., 1968, Geology and origin of the iron ore deposits of the Zenith mine, Vermilion district, Minnesota: Minn. Geol. Surv. Spec. Pub. Ser. SP-2, 56 p.

Schwartz, G. M., 1924, The contrast in the effect of granite and gabbro intrusions on the Ely greenstone: Jour. Geol., v. 32, p. 89-138

—— Editor, 1956, Precambrian of northeastern Minnesota: Geol. Soc. Amer. Guidebook for Field Trips, Field Trip no. 1, 235 p. (particularly p. 109-150)

Smyth, H. L., and Finlay, J. R., 1895, The geological structure of the western part of the Vermilion range, Minnesota: A.I.M.E. Tr., v. 25, p. 595-645

Van Hise, C. R., 1889, The chemical origin of the Vermilion Lake iron ores: Amer. Geol., v. 4, p. 382-383

Notes

The Vermilion range, in the broad sense, extends from the town of Tower, south of Vermilion Lake, some 90 miles northeast across the Canadian border as far as Saganaga and Gunflint Lakes, but the economically valuable portion consists of the 25-mile segment from Tower to some 7 or 8 miles northeast of Ely. Northeast of Snowbank Lake (25 miles northeast of Ely), the iron formation is so deeply buried that outcrops of it are rare.

The ores in the Vermilion range lie in the Soudan formation, a sedimentary inclusion in the Ely greenstone. The structure of the range is thought to be synclinal, with the greenstone and the included Soudan formation having been protected from erosion by downfolding into still older rocks. The location of the iron formation within the greenstone indicates that the rocks in the synclinal structure were closely folded along an axis that generally paralleled the strike of the region. The rocks appear to have been strongly cross folded as is demonstrated by the definite pitch of the primary folds (to the east or west) in extensive volumes of the iron formation. The rocks of the range have long been considered on geological grounds to be early Precambrian (Archean) in age, and this is confirmed by radioactive dating which places the age of the last major orogeny in the area at 2500 m.y. The ores in the Vermilion range probably were not affected by this metamorphic epoch because they show much less deformation than is true of the greenstones and the iron-bearing formation. Sericite from the wall-rock alteration zone associated with the mineralization, however, gives an age of 2500 m.y. which, if correct and the wall-rock alteration is directly connected with the hydrothermal introduction of much of the iron into broken volumes of the Soudan formation (as it certainly appears to be), means that the ores are also late early Precambrian and probably genetically related to the Laurentian or Algoman intrusive cycles, both of which are represented in the area by the Saganaga or the Giants range and Vermilion granites, respectively. The process of ore enrichment, nevertheless, may have been connected with the magmatic cycle of the Duluth gabbro, and the age might be late Precambrian; it almost certainly cannot be younger than the Precambrian.

The ore bodies of the Vermilion range are largely steeply dipping, tabular masses that replaced Soudan iron formation; a little ore was formed by the replacement of Ely greenstone immediately adjacent to the iron formation. Although the iron-bearing formation and the enclosing greenstones have been subjected to low-grade metamorphism, the ores themselves appear to be post-metamorphic in age.

In some of the mines the minable depth of ore was determined by a decrease in hematite and an increase in carbonate content and by a decrease in ore width. In the Zenith mine, however, magnetite was found at depth instead of hematite, but carbonate and pyrite became so abundant that the magnetite-rich mineralization was below ore grade; what further changes there may have been in the Zenith deposit with still greater depth have not been determined. The bulk of the ore in the Vermilion ore bodies is composed of hematite, the major fraction of which was added to that originally present in the iron-bearing formation; most of this first generation of epigenetic hematite was brecciated and then recemented by a second generation of hematite. The second generation of ferric oxide often contained tiny vugs lined with minute crystals of hematite. Almost all the ore was either hard, dense hematite or brecciated and recemented hematite.

Despite the dominance of hematite over magnetite, the first deposition of iron oxide anywhere in the deposits was as magnetite; the magnetite was later oxidized to, or (to say the same thing differently) was replaced by hematite. Thus, almost all the hematite, when examined under the microscope, shows remnant cores of magnetite. Some pyrite also formed during most of the epoch of mineralization; pyrite is enclosed by and encloses hematite, and pyrite is veined by fractures that are filled with hematite. In turn, much of the pyrite contains remnant cores of pyrrhotite. Thus, both iron and sulfur were more strongly oxidized with the passage of time. There are rare copper sulfides that were introduced after the pyrite. Hausmannite ($MnMn_2O_4$) occurs in both early hematite fragments and in the later carbonate veins and cementing material, but the abundance of hausmannite, relative to hematite, is much greater in the carbonate veins.

The ore bodies are normally surrounded by alteration halos that can each be divided into two main zones. In the outer zone, the rocks differ from normal greenstone in that they do not contain any amphibole and epidote, the main minerals of the greenstone; the principal minerals in this zone (which may be over 100 feet wide) are chlorite and lesser amounts of fine-grained muscovite, quartz, albite, and even less leucoxene. The rocks of this outer zone are cut by veins of quartz, chlorite, and carbonates, the carbonates being the latest minerals to form and often carrying small amounts of hematite. In the inner zone the principal alteration mineral is chlorite in several varieties. The red to brown color is due to hematite impregnation that gradually dies out toward the boundary of the inner with the outer zone. In many localities kaolinite and fine-grained muscovite are present; kaolinite may constitute the bulk of small rock volumes, and some illite probably is present. Quartz and carbonate are largely confined to veins but may be present as grains in the chlorite mass; apatite, zircon, zeolites, and talc are found in tiny to rare amounts. Leucoxene and pyrite accompany the hematite. The hematite of the ore appears to have been formed after the chlorite; in fact the chlorite appears to have been fractured or brecciated before the earlier, impregnating hematite was introduced; the later hematite is associated with the quartz and carbonate veins. Kaolinite occurs only immediately adjacent to the ore bodies, while muscovite is always separated from the ore by thin layers of kaolinite or chlorite.

The early magnetite-pyrrhotite mineralization must have been formed under hypothermal conditions, and at least the first and major portion of the hematite probably was formed under such conditions as well. The late hematite and the quartz and carbonate with which it is associated may have been developed under mesothermal conditions. The wall-rock alteration is compatible with both the less intense portions of the hypothermal range and the more intense of mesothermal conditions. Although the iron of the sedimentary iron-bearing formation must be classified as sedimentary-Ala, the bulk of the iron was introduced hydrothermally and is categorized here as hypothermal-1 to mesothermal.

Mississippi Valley Type (General)

Late Paleozoic (4), *Middle Mesozoic (1),* *Late Mesozoic (5)* *(pre-Laramide)*	*Lead, Zinc, Fluorite,* *Barite*	*Telethermal*

Barton, P. B., Jr., 1967, Possible role of organic matter in the precipitation of the Mississippi Valley ores, in Brown, J. S., Editor, *Genesis of stratiform lead-zinc-barite-fluorite deposits--a symposium*: Econ. Geol. Mono. 3, p. 371-378

Behre, C. H., Jr., and Heyl, A. V., Jr., 1958, Erzvorkommen vom Typus "Mississippi-Tal" in den Vereinigten Staaten: Zeitsch. der deutschen Geologischen Gesellschaft, Bd. 110, Teil 3, S. 514-558

Bernard, A., and Folierini, F., 1967, Etude méthodologique sur la génèse des gisements stratiformes du plomb-zinc en environment carbonate, in Brown, J. S., Editor, *Genesis of stratiform lead-zinc-barite-fluorite deposits--a symposium*: Econ. Geol. Mono. 3, p. 267-277

Boyle, R. W., and Lynch, J. J., 1968, Speculations on the source of zinc, cadmium, lead, copper, and sulfur in Mississippi Valley and similar types of lead-zinc deposits: Econ. Geol., v. 63, p. 421-422

Brown, J. S., 1962, Ore leads and isotopes: Econ. Geol., v. 57, p. 673-720

—— 1970, Mississippi Valley type lead-zinc ores: Mineralium Deposita, v. 5, p. 103-119

Callahan, W. H., 1964, Paleophysiographic premises for prospecting for strata bound metal mineral deposits in carbonate rocks: CENTO Symposium on Mining Geology and Base Metals, Ankara, p. 191-248

—— 1967, Some spatial and temporal aspects of the localization of Mississippi Valley-Appalachian type ore deposits, in Brown, J. S., Editor, *Genesis of stratiform lead-zinc-barite-fluorite deposits--a symposium*: Econ. Geol. Mono. 3, p. 14-19

Duhovnik, J., 1967, Facts speaking for and against a syngenetic origin of the stratiform deposits of lead and zinc, in Brown, J. S., Editor, *Genesis of stratiform lead-zinc-barite-fluorite deposits--a symposium*: Econ. Geol. Mono. 3, p. 108-125

Emmons, W. H., 1929, The origin of the deposits of sulphide ores of the Mississippi Valley: Econ. Geol., v. 25, p. 221-271

Heyl, A. V., 1967, Some aspects of genesis of stratiform lead-zinc-barite-fluorite deposits in the United States, in Brown, J. S., Editor, *Genesis of stratiform lead-zinc-barite-fluorite deposits--a symposium*: Econ. Geol. Mono. 3, p. 20-32

Heyl, A. V., and others, 1966, Isotopic study of galenas from the upper Mississippi Valley, the Illinois-Kentucky, and some Appalachian Valley mineral deposits: Econ. Geol., v. 61, p. 933-961 (particularly p. 942-947)

Jensen, M. L., and Dessau, G., 1967, The bearing of sulfur isotopes on the origin of Mississippi Valley type deposits, in Brown, J. S., Editor, *Genesis of stratiform lead-zinc-barite-fluorite deposits--a symposium*: Econ. Geol. Mono. 3, p. 400-409

Kautzsch, E., 1967, Genesis of the stratiform lead-zinc deposits in Central Europe, in Brown, J. S., Editor, *Genesis of stratiform lead-zinc-barite-fluorite deposits--a symposium*: Econ. Geol. Mono. 3, p. 133-137

Kulp, J. L., and others, 1956, Lead and sulfur isotopic abundances in Mississippi Valley galenas: Geol. Soc. Amer. Bull., v. 67, p. 123

Maucher, A., and Schneider, H. J., 1967, The Alpine lead-zinc ores, in Brown, J. S., Editor, *Genesis of stratiform lead-zinc-barite-fluorite deposits--a symposium*: Econ. Geol. Mono. 3, p. 71-89

McKnight, E. T., 1967, Bearing of isotopic composition of lead on the genesis of Mississippi Valley ore deposits, in Brown, J. S., Editor, *Genesis of stratiform lead-zinc-barite-fluorite deposits--a symposium*: Econ. Geol. Mono. 3, p. 392-399

Ohle, E. L., Jr., 1959, Some considerations in determining the origin of ore deposits of the Mississippi Valley type: Econ. Geol., v. 54, p. 769-789

—— 1967, The origin of ore deposits of the Mississippi Valley type, in Brown, J. S., Editor, *Genesis of stratiform lead-zinc-barite-fluorite deposits--a symposium*: Econ. Geol. Mono. 3, p. 33-39

Pélissonnier, H., 1967, Analyse paléohydrogéologique des gisements stratiformes de plomb, zinc, baryte, fluorite du type "Mississippi Valley," in Brown, J. S., Editor, *Genesis of stratiform lead-zinc-barite-fluorite deposits--a symposium*: Econ. Geol. Mono. 3, p. 234-252

Roedder, E., 1967, Environment of deposition of stratiform (Mississippi Valley type) ore deposits, from studies of fluid inclusions, in Brown, J. S., Editor, *Genesis of stratiform lead-zinc-barite-fluorite deposits--a symposium*: Econ. Geol. Mono. 3, p. 349-362

Sangster, D. F., 1968, Some chemical features of lead-zinc deposits in carbonate rocks: Geol. Surv. Canada, Paper 68-39, 17 p.

Sawkins, F. J., 1968, The significance of Na/K and Cl/SO_4 ratios in fluid inclusions and subsurface waters, with respect to the genesis of Mississippi-Valley-type ore deposits: Econ. Geol., v. 63, p. 935-942

Skinner, B. J., 1967, Precipitation of Mississippi Valley type ores; a possible mechanism, in Brown, J. S., Editor, *Genesis of stratiform lead-zinc-barite-fluorite deposits--a symposium*: Econ. Geol. Mono. 3, p. 363-370

Smirnov, V. I., 1967, Problem of the origin of stratiform lead and zinc deposits lying in the territory of the USSR, in Brown, J. S., Editor, *Genesis of stratiform lead-zinc-barite-fluorite deposits--a symposium*: Econ. Geol. Mono. 3, p. 219-220

Snyder, F. G., 1967, Criteria for the recognition of stratiform ore bodies with application to southeast Missouri, in Brown, J. S., Editor, *Genesis of stratiform lead-zinc-barite-fluorite deposits--a symposium*: Econ. Geol. Mono. 3, p. 1-13

Taupitz, K. C., 1967, Textures in some stratiform lead-zinc deposits, in Brown, J. S., Editor, *Genesis of stratiform lead-zinc-barite-fluorite deposits--a symposium*: Econ. Geol. Mono. 3, p. 90-107

White, D. E., 1967, Outline of thermal and mineral waters as related to origin of Mississippi Valley ore deposits, in Brown, J. S., Editor, *Genesis of stratiform lead-zinc-barite-fluorite deposits--a symposium*: Econ. Geol. Mono. 3, p. 379-382

Notes

The ores of the Mississippi Valley type for which there are bibliographies in this volume include: (1) Austinville, (2) Mascot-Jefferson City, (3) Sweetwater, (4) Illinois-Kentucky Fluorspar, (5) Leadbelt, (6) Tri-State, (7) Upper Mississippi Valley, (8) Metaline, (9) Salmo, and (10) Pine Point.

The ages assigned to these deposits depend on the hypothesis of genesis favored by the authors in question. If the deposits are thought to have been formed by volcanic exhalations reaching the sea floor at the same time the sediments of the host formation were being accumulated, the age of the deposit is the age of the rocks. If the deposits are believed to have resulted from the chemical or mechanical precipitation of ore minerals with the enclosing sediments or to have been added during diagenesis, the age of ore formation is the age of the rocks involved. If the ore minerals were deposited from heated brine solutions (hydrothermal in the broad sense) that derived their salt content from the dissolution of evaporite deposits and their ore metal content from a general scavenging of the rocks through which they pass, the age of the deposits can be any date after the consolidation of the host rocks or the evaporites, depending on which stands higher than the stratigraphic sequence. Finally, if the deposits were formed by hydrothermal activity (in the narrow sense of magmatic origin for some of the ore fluid and for the ore elements at least), the age of the ores probably is that of the most closely related igneous activity. For each of the 10 deposits listed above, ages have been assigned that can be determined by consulting the appropriate bibliographies in this volume. Nearly as much controversy could be generated about these ages as about the genetic aspects of these deposits. Detailed notes are given as to why each age assigned was selected with each bibliography.

The principal theories as to the origin of Mississippi-Valley type deposits have been listed in the preceding paragraph. Snyder (1967) discusses the evidence as to origin under three headings. The first of his topics deals with the evidence for syngenetic origin, and he points to two criteria most approved by proponents of this type of origin: (1) uniform mineralization at a given stratigraphic position or within a restricted stratigraphic interval and (2) close, but widespread, correlation between mineralization and particular sedimentary lithologies, facies, and features. He believes that these criteria can be considered diagnostic only if there is no conflicting evidence. He mentions that remobilization generally is used to explain those features that seem to have resulted from epigenetic processes, such as ore crosscutting the bedding. He believes that remobilization has occurred but that the amount of metal remobilized is minor in comparison with the total in the deposit. The second topic is concerned with diagenetic changes that may add ore minerals to unlithified rocks. Snyder considers diagenetic iron sulfide, at least so far as framboidal spheres and individual grains are concerned, as well documented and beyond question for argillites, mudstones, and gray shaly carbonates (whether or not they contain ores) in eugeosynclinal and miogeosynclinal environments. Such sulfides also occur in the gray shales and shaly limestones of the midcontinent craton. He emphasizes, however, that other sulfides are not necessarily diagenetic because they are found in rocks with diagenetic iron sulfide. As he points out, large lead-zinc deposits are found in structural traps, with the host rocks beyond the trap containing as much iron sulfide as in the trap; thus, the trap seems far more important as the cause of ore localization and the trap could have developed only long after the diagenetic period was over. His third topic involves the localization of ores by post-lithification structures, for example, marked changes in ore height, width, and tenor that cannot be related to sedimentary features; extensive filling of open space; and lack of correlation on a district-wide basis of sedimentary features and mineralization. Certainly, the direct relationship of all or much of the ore to post-lithification features establishes the epigenetic character of the deposit in question. Snyder also includes a category of permissive evidence, evidence that is of use in establishing the history of ore development in the district but that does not affect the choice among the three possible origins (syngenetic, diagenetic, or epigenetic). These evidences are: (1) most paragenetic and textural relationships; (2) conformity with ancient shore lines; (3) local fitting of ore minerals with sedimentary structures or features; (4) the presence or absence of nearby igneous intrusives or volcanic rocks; (5) and the independence of mineralization from later geologic events.

Snyder also discusses lead-isotope data as a tool in deciphering the origin and history of such lead-zinc deposits. At the present time it is considered that the ratios can follow a fairly straightforward model and indicate directly the time of formation of the deposit or they can contain an excess (over model requirements) of one or more of the radiogenic leads. These so-called anomalous leads, known as J-type leads in many instances, reflect, judging by the model, unusual conditions of source or of behavior during movement to the deposit in which they are found, often giving a wide range of isotopic ratios within a district. It is Snyder's opinion that an essential requirement of a syngenetic or diagenetic deposit containing lead minerals is uniformity in isotopic ratios, and these isotopic ratios must conform to those obtaining in the sea water from which the lead minerals were precipitated. Because the isotopic ratios of lead in sea water differ with time, the ratios in a syngenetic or diagenetic deposit formed from such water must be such that an age can be assigned from the ratios. Snyder does say that young deposits cannot be accurately dated by this method. On the other hand, if the lead in the deposits shows a wide range in radiogenic character or departs appreciably from ordinary lead-isotope ratios valid for the time of its formation, the deposit must be epigenetic. An epigenetic deposit can, however, have lead of uniform isotopic ratios, depending on the behavior of the lead in the transporting medium and the character of its source.

In discussing deposits (2), (4), (5), (6), and (7) of the list given at the beginning of these notes, Snyder says that ore bodies or parts of ore bodies show a high degree of conformability to sedimentary features but that on a district-wide basis they show disconformable characteristics; all districts also show much open space filling and marked differences, over short distances, in ore height, width, and extent. Fluid inclusions in these districts suggest that they were formed from concentrated brines in the lower portions of the range of hydrothermal temperatures. It should be emphasized, however, that Snyder does not equate epigenetic with hydrothermal in the narrow sense. He believes that the ore fluid (specifically in the Leadbelt) was a concentrated, low-temperature brine that apparently came from nearby sedimentary basins. The metals in the brines were extracted from basin sediments in which they, the metals, had remained long enough to acquire their radiogenic character. He says that the sulfur possibly was not brought with the metals but does not say where it came from; two possible sources, in addition to the brines, can be visualized for this sulfur: (1) it was trapped in the host rocks and (2) it came from magmatic sources. Alternative (1) seems unlikely because of the difficulty of retaining such huge quantities of sulfur in any host rock. Water saturated with hydrogen sulfide at the temperatures and pressures obtaining in deposits of the Mississippi Valley type could contain, at a maximum, only about 3.6 percent of the sulfur needed to fill the open space with zinc sulfide; the percentage available for galena would be even smaller because of its greater density. If alternative (2) is correct, any fluid of magmatic origin that could bring in the sulfur also could bring in the metals, probably complexed with sulfur, so why appeal to one source for sulfur and another source for metals? I think that the possibilities of isotope fractionation during ore-fluid travel never has been sufficiently tested and that the anomalous lead-isotope ratios may be the result of nothing more drastic than the amount of fractionation possible during the movement of a high-temperature ore fluid from a magma at depth to a low-temperature state far from its source. The high-brine content of the fluids in inclusions in these low-temperature ores may well reflect only the high-brine content of the fluids when they left the magma chamber, high-brine fluids being better able to retain metal-sulfur complexes for low-temperature deposition than those lower in brine. In any event, far too little is known about the behavior of isotopes in ore fluids for us to be certain about anything connected with them. The deposits of this type are, therefore, in this volume, classified as telethermal in the magmatic sense of that term.

Missouri

IRON MOUNTAIN-PILOT KNOB

Late Precambrian — *Iron as Hematite, Minor Magnetite* — *Magmatic-3b (Iron Mountain), Hypothermal-1 (Pilot Knob)*

Allen, V. T., and Fahey, J. J., 1952, New occurrences of minerals at Iron Mountain, Missouri: Amer. Mineral., v. 37, p. 736-743

Anderson, J. E., Jr., and others, 1969, Some age relations and structural features of the Precambrian volcanic terrane, St. Francois Mountains, southeastern Missouri: Geol. Soc. Amer. Bull., v. 80, p. 1815-1818

Crane, G. W., 1912, Specular hematite in porphyry, in *The iron ores of Missouri:* Mo. Bur. Mines and Geol., 2d Ser., v. 10, p. 107-145

Desborough, G. A., 1963, Mobilization of iron by alteration of magnetite-ülvospinel in basic rocks in Missouri: Econ. Geol., v. 58, p. 332-346 (particularly p. 341)

Geijer, P., 1930, The iron ores of the Kiruna type: Sveriges Geol. Undersök., Ser. C, no. 367, Årsbok 24, no. 4, p. 1-39 (particularly p. 15-16)

Hayes, W. C., 1961, Guidebook to geology of St. Francois Mt. area: Mo. Geol. Surv. R.I. 26, 137 p. (particularly p. 127-137)

Johnson, C. H., 1961, A brief description of Pilot Knob: Assoc. Mo. Geols. 8th Ann. Field Trip, Mo. Div. Geol. Surv. and Water Res. R.I. 26, p. 127-128

Kisvarsanyi, G., and Proctor, P. D., 1967, Trace elements content of magnetites and hematites, southeast Missouri iron metallogenic province, U.S.A.: Econ. Geol., v. 62, p. 449-471

Lake, M. C., 1933, The iron-ore deposits of Iron Mountain, Missouri, in *Mining districts of the eastern states*: 16th Int. Geol. Cong., Guidebook 2, p. 56-67

Meyer, C., and Tolman, C., 1939, Structural geology of the felsites of Iron Mountain, Missouri (abs.): Geol. Soc. Amer. Bull., v. 50, p. 1922-1923

Murphy, J. E., and Mejia, V. M., 1961, Underground geology at Iron Mountain: Assoc. Mo. Geols. 8th Ann. Field Trip, Mo. Div. Geol. Surv. and Water Res. R.I. 26, p. 129-137

Murphy, J. E., and Ohle, E. L., Jr., 1968, The Iron Mountain mine, Iron Mountain, Missouri, in Ridge, J. D., Editor, *Ore deposits of the United States 1933-1967* (Graton-Sales Volumes): Chap. 15, v. 1

Ridge, J. D., 1957, The iron ores of Iron Mountain, Missouri: Mineral Industries, v. 26, no. 9, p. 1-6

Singewald, J. T., Jr., and Milton, C., 1929, Origin of iron ores of Iron Mountain and Pilot Knob, Missouri: A.I.M.E. Tr., v. 85 (1929 Yearbook), p. 330-340

Spurr, J. E., 1927, Iron ores of Iron Mountain and Pilot Knob: Eng. and Min. Jour., v. 123, no. 9, p. 363-366

Steidtmann, E., 1933, The iron ore deposits of Pilot Knob, Missouri, in *Mining districts of the eastern states*: 16th Int. Geol. Cong., Guidebook 2, p. 68-73

Tolman, C., and Meyer, C., 1939, Pre-Cambrian iron mineralization in southeastern Missouri (abs.): Geol. Soc. Amer. Bull., v. 50, p. 1939-1940

Notes

The Iron Mountain and Pilot Knob mines are located in the southwest corner of St. Francois County, about 85 miles southwest of St. Louis; Pilot Knob is about 5 miles south of Iron Mountain.

At Iron Mountain the ore-containing rocks are Precambrian in age, probably having formed late in that period; they consist of intrusive granite, andesite porphyry, and rhyolite and of lava flows of rhyolite, dacite, and diorite, while diorite, dacite, and andesite dikes cut the older rocks. The Precambrian rocks were eroded to an irregular surface on the lower portions of which Cambrian sediments were laid down. Although these sediments are essentially flat-lying, they are often tilted adjacent to the erosional knobs of granite or porphyry, partly due to initial dip and partly to consolidation during lithification. All of the primary ore at Iron Mountain occurs within the andesite porphyry, but conglomerate ore, derived from erosion of the primary iron deposits, occurs at the base of the Cambrian sediments and contains fragments that range from sand-grain size to masses several tons in weight. This relationship demonstrates the Precambrian age of the primary ores which, having originally been formed at considerable depths beneath the surface, must have been brought to the surface by erosion, probably by the end of Precambrian time. Thus, the primary Iron Mountain ores must have been emplaced appreciably before the end of the Precambrian era; they are here assumed to be late Precambrian in age, although this dating is certainly not definitely established.

Pilot Knob is composed of a series of Precambrian lava flows and mechanically deposited sediments, the latter forming the upper portions of the knob; the Precambrian rocks are surrounded by flat-lying Cambrian limestones. The porphyries dip gently in a southwesterly direction, and the ore has been emplaced in the sediments that dip essentially parallel to the contacts between the flows. The porphyries underlying the lower ore bed, the foot-wall porphyries, are fine-grained, reddish-brown felsites, some of which are classified as rhyolites; phenocrysts are scarce. The smooth upper surface of these felsite flows was shortly covered by a shallow lake in which fine-grained clastic sediments were deposited; probably the material was provided by a series of ash falls, an hypothesis supported by the angular character of the fragments. That the water was shallow is indicated by the presence of raindrop prints and sun cracks and of very minor ripple marks. After an appreciable amount of this fine-grained and finely banded sediment had formed, much coarser angular fragments from the surrounding slopes were washed out over the sediments in poorly sorted masses. In places on these masses of debris, new and smaller pools were formed in which additional finely laminated sediments were laid down. This area of diversified sediments was covered by later flows of silicic lavas. After the sediments had been consolidated and buried to considerable depths, the ores were deposited principally as replacements of the fine-grained sediments. A talus deposit of hematite ore and felsite porphyry lies between the eroded Precambrian surface and the Cambrian limestone, demonstrating that the primary ore was formed in Precambrian time at some considerable time before the end of that era. Although there is no definite information to date the ores and the rocks that contain them within the Precambrian time, they are here assumed to have been formed in the late Precambrian, but this dating is subject to possible future correction.

Although the principal ore mineral in both Iron Mountain and Pilot Knob is hematite and although the two deposits are situated within 5 miles of each other, they appear to have been formed by two quite different processes that affected the area at about the same time. The Iron Mountain ores, in addition to hematite, contain important amounts of magnetite, which mineral increases with depth and probably amounts to 15 to 20 percent of the total mass of iron oxides. The most important nonmetallic minerals appear to be an amphibole (actinolite or tremolite) and andradite garnet; apatite is locally present in appreciable quantities. Quartz and calcite are quite common, and red jasper (presumably a fine-grained mixture of hematite and quartz) is frequently found between massive hematite and quartz. Salite (a pyroxene intermediate between diopside and hedenbergite) has been reported (Allen and Fahey, 1952) as growing outward from the andesite porphyry walls of the ore body into the ore. Some of the actinolite appears to have altered to chlorite, but chlorite and epidote are much more abundant in the altered porphyry adjacent to the ore than in the ore itself. Microscopic work (Singewald and Milton, 1929) has shown that some plagioclase phenocrysts in the andesite immediately adjacent to the ore have been largely converted to fine-grained aggregates of quartz and chlorite. In masses of andesite porphyry in, and adjacent to, the ore, the groundmass has been converted to hematite, and at least enough iron has been introduced into the feldspar phenocrysts to turn them bright pink. Geijer (1930) suggested that the ores were formed from the crystallization of a water-poor, iron-rich melt and based his hypothesis on (1) the coarse texture of the hematite, (2) the presence of considerable martite, (3) the development of apatite (and salite) crystals growing out from enclosed porphyry fragments or porphyry walls of the ores, and (4) minor reaction rims of alteration products between ore and porphyry in the walls or in inclusions. If the apatite and salite crystals were deposited from hydrothermal solutions as fissure fillings, growing outward from the original fissure walls, all the later hematite (if hydrothermally formed) must have been deposited either (1) outward into the open space of a wide fissure or (2) by replacement of porphyry in only one direction into the walls of narrow fissures, away from the walls on which the apatite or salite crystals had grown.

From the relationships just described, it would appear more probable that the ores were introduced as an iron-oxide-silicate-phosphate-carbonate melt rather than in hydrothermal fluids. The melts presumably entered fissures that were, or developed into, essentially the size of the known Iron Mountain ore bodies.

The principal points that have been advanced as favoring a hydrothermal replacement origin for the Iron Mountain deposits are (1) the contact between ore and wall rock is an irregular one along which some alteration of the porphyry to chlorite and epidote obviously has taken place and (2) garnet (and in some places quartz) has replaced amphibole. These facts can, however, be reconciled with the hypothesis that the ore was introduced as a melt (developed as an immiscible fraction in the late stages of igneous activity) which, while dominantly composed of iron oxides, also contained appreciable silicate-carbonate-phosphate material, plus water and other constituents, such as fluorine ions, dissolved in it. These two reactions (chlorite and epidote from the wall rock and garnet from the amphibole) would be at least equally possible in a molten (as opposed to a hydrothermal) environment, and the minor extent to which these reactions took place would be more reasonably explained if they were carried out between solid rock and a melt than between rock and a hydrothermal fluid. The presence of most of the iron oxides as hematite indicates that a high oxygen pressure existed in the melt, particularly in its upper portions. On the basis of this reasoning, the Iron Mountain ores are classified here as magmatic-3b.

The ore mineral at Pilot Knob is largely hematite but with some magnetite; the proportion of magnetite to hematite increases with depth. The ore, near the surface, is found in a 6- to 30-foot thick ore bed, separated from a 10- to 20-foot thick upper ore bed by a 1- to 3-foot sheared sericite layer. The ore beds as they were first known were finely banded alterations of dark-gray crystalline hematite and a little quartz and sericite with brownish-red ferruginous, fine-grained quartz, and a little sericite. The ore and gangue minerals of the ore beds seem to have been introduced by the replacement of the original ash fragments and not to have been deposited syngenetically with them. In the near-surface ores, tourmaline, apatite, and barite were found in minor amounts; the phosphorus and sulfur in these ores were extremely low. At greater depths, appreciable amounts of chalcopyrite and pyrite are present at least locally; whether there is enough chalcopyrite in the ore to warrant its recovery is uncertain. Apatite appears to be present in a considerable fraction of the ore, and dark-purple fluorite occurs frequently with the hematite. The upper ore bed is overlain by a ferruginous silicified breccia from which small, local lenses, similar in character and origin to the ore beds, have been mined. Locally, the brecciated rocks are embayed by hematite and cut by tiny hematite veinlets, and the fine breccia fragments are much more thoroughly replaced by hematite than are the larger ones; the total amount of hematite in the breccias is minor. The introduction of the hematite by replacement of the ore beds and the breccias and the presence of the moderate amounts of high-temperature replacement minerals (apatite, tourmaline, dark fluorite) in the ore suggests that they were emplaced after lithification and probably at considerable depth beneath the then-existing surface. The widespread sericitization of the higher-level volcanic rocks suggests that, in contrast to Iron Mountain, the Pilot Knob ores were introduced by hydrothermal solutions, and the general character of the mineralization indicates that the ores were probably formed in the hypothermal range, although the pryite, chalcopyrite, and sericite may point to an overlap into the upper portion of the mesothermal range. The Pilot Knob ores are here classified as hypothermal-1 because the mesothermal(?) minerals are not, as yet, of demonstrated economic importance.

LEADBELT (SOUTHEAST MISSOURI)

Late Mesozoic (pre-Laramide)	*Lead, Copper, Cobalt, Nickel, Silver*	*Leptothermal (minor) to Telethermal*

Behre, C. H., Jr., and others, 1950, Zinc and lead deposits of the Mississippi Valley: 18th Int. Geol. Cong., Rept., pt. 7, p. 51-69 (particularly p. 59-60)

Brown, J. S., 1958, Southeast Missouri lead belt: Geol. Soc. Amer. Guidebook for Field Trips, Field Trip no. 1, p. 1-7

—— 1967, Isotopic zoning of lead and sulfur in southeast Missouri, in Brown, J. S., Editor, *Genesis of stratiform lead-zinc-barite-fluorite deposits--a symposium*: Econ. Geol. Mono. 3, p. 410-426

Buehler, H. A., 1933, The disseminated-lead district of southeastern Missouri, in *Mining districts of the eastern states*: 16th Int. Geol. Cong., Guidebook 2, p. 45-55

Dake, C. L., 1930, The geology of the Potosi and Edgehill quandrangles: Mo. Bur. Geol. and Mines, v. 23, p. 198-219

Davis, J. H., 1958, Distribution of copper, zinc, and minor metals in the southeast Missouri lead district (abs.): Geol. Soc. Amer. Bull., v. 69, p. 1551

Emmons, W. H., 1929, The origin of the deposits of sulphide ores of the Mississippi Valley: Econ. Geol., v. 24, p. 221-271 (particularly p. 234-239)

Fenoll, P., and Ottemann, J., 1970, Linneit und Bravoit von Fredericktown, Missouri (eine Mikrosonde-Untersuchung): Aufschluss, H. 1, S. 29-33

Heyl, A. V., and others, 1966, Regional structure of the southeast Missouri and Illinois-Kentucky mineral districts: U.S. Geol. Surv. Bull. 1202-B, p. B1-B20

James, J. A., 1949, Geologic relationships of the ore deposits in the Fredericktown area, Missouri: Mo. Geol. Surv. and Water Res., R.I. no. 8, 25 p.

—— 1952, Structural environments of the lead-deposits in the southeastern Missouri mining district: Econ. Geol., v. 47, p. 650-660

Kulp, J. L., and others, 1956, Lead and sulfur isotopic abundances in Mississippi Valley galenas: Geol. Soc. Amer. Bull., v. 67, p. 123

Ohle, E. L., Jr., 1952, Geology of the Hayden Creek lead mine, southeast Missouri: A.I.M.E. Tr., v. 193, p. 477-483 (in Min. Eng., v. 4, no. 5)

Ohle, E. L., Jr., and Brown, J. S., 1954, Geologic problems in the southeast Missouri lead district: Geol. Soc. Amer. Bull., v. 65, p. 201-221

Snyder, F. G., 1967, Criteria for origin of stratiform ore bodies, with application to southeast Missouri, in Brown, J. S., Editor, *Genesis of stratiform lead-zinc-barite-fluorite deposits--a symposium*: Econ. Geol. Mono. 3, p. 1-13

Snyder, F. G., and Emery, J. A., 1956, Geology in development and mining, southeast Missouri lead belt: A.I.M.E. Tr., v. 208, p. 1216-1224 (in Min. Eng., v. 8, no. 12)

Snyder, F. G., and Gerdemann, P. E., 1968, Geology of the southeast Missouri lead district, in Ridge, J. D., Editor, *Ore deposits of the United States 1933-1967* (Graton-Sales Volumes): Chap. 17, v. 1

Snyder, F. G., and Odell, J. W., 1958, Sedimentary breccias in the southeast Missouri lead district: Geol. Soc. Amer. Bull., v. 69, p. 899-925

Spurr, J. E., 1926, The southeast Missouri ore-magmatic district: Eng. and Min. Jour., v. 122, no. 25, p. 968-975

Tarr, W. A., 1936, Origin of the southeastern Missouri lead deposits: Econ. Geol., v. 31, p. 712-743, 832-866

—— 1939, Southeastern Missouri (lead) district, in *Lead and zinc deposits of the Mississippi Valley region* (Bastin): Geol. Soc. Amer., Spec. Paper no. 24, p. 18-25, 64-65

Taylor, C. M., and Radtke, A. S., 1969, Micromineralogy of silver-bearing sphalerite from Flat River, Missouri: Econ. Geol., v. 64, p. 306-318

Notes

The Leadbelt of southeast Missouri, now that it has been extended northwest into the Viburnum-Salem area by recent exploratory work, covers a wide area on both sides of the St. Francois Mountains, some 75 miles south to southwest of St. Louis. In the past, the term "Leadbelt" was confined to the deposits in St. Francois County, but it has now been broadened to include all of the lead deposits of southeastern Missouri.

The Leadbelt rocks consist of (1) a basement complex (devoid of nonferrous metal ore bodies) composed of porphyritic felsite flows (mainly rhyolites) that were intruded first by granite masses and later by mafic dikes (the dikes are known to cut Paleozoic rocks in several places) and (2) a series of lower Paleozoic sediments, deposited on the irregularly eroded Precambrian surface, that are largely dolomites and limestones. The oldest rock is the upper Cambrian La Motte sandstone that fails to cover the higher knobs on the Precambrian surface. Conformably above the La Motte is the Bonneterre dolomite; the lower units of the formation may be absent where it lies above the higher Precambrian erosional elevations. The Bonneterre is conformably overlain by the Davis formation that contains subordinate amounts of shale interbedded with thin layers of dolomite and limestone. The Derby-Doe Run formation of finely crystalline to earthy dolomite is found conformably above the Davis. Unconformably above the Derby-Doe Run, although still of Cambrian age, are the Potosi and Eminence dolomites. The probably Ordovician rocks that lie above the Eminence are dolomites, followed by a sandstone, but are not known to contain ore. Most of the ore in the area has come from the lowest 250 feet of the Bonneterre, but minor amounts have been mined from rock volumes of restricted size higher in the formations.

There is no direct evidence as to the time at which the ores were introduced into the Bonneterre. For many years, almost no geologist familiar with the Leadbelt has suggested that the ores were introduced syngenetically into the sediments, and the large majority of geologists who have studied the ores believe that they were emplaced through reactions between the rocks and hydrothermal (in the broad sense) solutions. The various alternatives to the hydrohypothesis, and that hypothesis itself, are discussed in detail by Ohle (1959) who reaches the conclusion that the low silver content of the galena and the anomalous lead-isotope ratios indicate that the ores, while probably hydrothermal, diverge somewhat from the normal hydrothermal pattern--a divergence, in my opinion, that can probably be explained by the long distance that the solutions must have travelled and the wide variety of reactions that they must have undergone. I think that the metal-sulfur content of these solutions, at least, was derived from an igneous source at depth, so the determination of the age of the ores requires the dating of the period of igneous activity that produced the ore solutions. The most common post-Precambrian igneous rocks in the district are peridotites that have been found in dikes and plug-like masses (probably in cryptovolcanic structures) throughout the central Mississippi valley. This igneous activity appears to have occurred throughout the area, with dating being possible from (1) the peridotites of the Hicks Dome in southern Illinois (probably related to the igneous cycle that produced the fluorspar deposits of that area) radioactively dated as Cretaceous in age and (2) the peridotites of the coastal plain of Arkansas and elsewhere in the valley that intrude Early Cretaceous sedimentary rocks and are unconformably overlain by Upper Cretaceous sediments; some of the Upper Cretaceous sediments contain abundant eroded mafic volcanic debris; in addition to the peridotites, alkaline rocks of the type made famous in the Magnet Cove area are present in the region, and pebbles of similar alkaline rock types have been found in Upper Cretaceous sediments to the northeast of Magnet Cove. These relationships provide compelling, but not incontrovertible, evidence that Leadbelt ore-forming fluids were developed during the igneous cycle that produced the Illinois-Kentucky fluorspar deposits. As the Leadbelt ore mineralization is in the same general area as the fluorite deposits and the middle Cretaceous peridotites and alkaline rocks of Arkansas, the Leadbelt ores are here categorized as late Mesozoic (pre-Laramide).

The rocks in which the Leadbelt ores are found are normally essentially horizontal but may show local departures of a few degrees from that datum; where they rest on buried hills and knobs, the rocks may have dips of as much as 20 to 25 degrees. Jointing and minor faulting are prevalent throughout the district and undoubtedly directed the movement of the ore fluids, but other physical features have exerted important influences on the localization of ore. The principal structure in the area is a broad, gentle anticline that trends northwest, the long dimension of the district. At right angles to the anticlinal axis, the carbonate rocks of the lower portion of the Bonneterre formation are arranged in a series of northeast-trending ridges (some as much as 12,000 feet long) composed mainly of calcite sands (calcarenite), with occasional shale bands intercalated among them. These ridges are separated from each other by basins filled with calcite silts (calcilutites). These two classes of sediments interfinger back and forth along the flanks of the ridges, and the calcite silts, having been the more compactible, form thinner beds than the comparable volumes of calcite sands into which they interfinger. The ridge and basin structures, then, are perhaps due as much to compaction as to the presence of initial highs and lows in the original sediments. After appreciable thicknesses of Bonneterre sediments had developed in the Bonneterre sea, algae began to grow on the upper surfaces of most of the ridges. Reefs also developed on buried Precambrian ridges, as at Indian Creek, where the depth of the rock beneath the surface was conducive to algal growth. Although the gelatinous material secreted by the algae has long been replaced by calcite and then (largely, but not entirely) by dolomite, the vertically elongate patterns and growth lines of the algal fingers can still be recognized in the rock. Over the tops of the algal reefs, the upper beds of the Bonneterre lie essentially horizontally.

The character of the Bonneterre sediments is further complicated by a change in the character of the beds of which the reefs and basins are composed. The Bonneterre has been divided into a number of zones, not all of which need be present throughout the area. These zones are (1) the 19, (2) the 15, (3) the 12, (4) the 10, (5) the 7, and (6) the 5. The reefs make up the 7 zone, and the sediments of the 5 zone horizontally blanket the entire Bonneterre formation.

An unusual characteristic of the rocks of the zones is that the contacts between them cut across the bedding planes (themselves essentially horizontal). Thus, in the one area the contact between two zones may be along a given bedding plane, but a few tens of feet away, the contact will be appreciably higher in the stratigraphic column. These changes of the contact between zones are directly related to, and mark the boundaries of, the sedimentary ridges.

Within the reef structures of the 7 zone, there are many sub-reefs, each of which rests on the essentially horizontal upper surface of the 10 zone and rises convexly upward so that the sub-reefs, at right angles to the northeast trend of the main ridges, give a washboard pattern to the upper surface of the reef itself. Between each two steep-sided sub-reefs, then, is an appreciable wedge-shaped depression which, though also filled with carbonate material, contains no algal structures. Another but less important structural control of ore deposition (Snyder and Odell, 1958) appears to have been sedimentary breccias developed in the Bonneterre formation by submarine slides that moved down the flanks of calcarenite and reef-deposit ridges into nearby depositional basins containing fine-grained calcilutite carbonate and argillaceous muds. The slopes in the border zones between ridges and basins were over-steepened by differential compaction, resulting in the downward movement of unconsolidated and partly lithified sediments. Large thicknesses of breccias in the lower parts of the Bonneterre were produced by repeated slides that occurred at close stratigraphic intervals. The loci of mineralization in the district are directly related to these several types of sedimentary structures.

In the reef structures, mineralization was developed (1) along the basal contact of the shales, that is between the 10 and 7 (reef) zones, and (2) along the convex outer boundaries (roll-edges) of reef material. Not all roll-edges are mineralized, and those that are loci of ore deposition may be strongly mineralized for hundreds or even thousands of feet while others are weakly mineralized over a few tens of feet.

The paragenesis of the Leadbelt ores has been studied by Tarr (1936) in detail and by Davis (1958). Both agree that marcasite was the earliest metallic mineral to form (Tarr demonstrates that appreciable dolomitization preceded the marcasite), but they give different sequences after that. Tarr says that, in the usual sequence, siegenite followed and probably overlapped with marcasite; he holds that chalcopyrite was essentially contemporaneous with siegenite (although he thinks it less likely that siegenite overlapped marcasite), that sphalerite followed chalcopyrite without overlap, and that galena was last of all and did not overlap with sphalerite. Davis, on the other hand, believes that sphalerite followed marcasite, siegenite followed sphalerite, galena followed siegenite, and chalcopyrite was last of all. Tarr indicates, however, that there was enough variation in conditions of deposition that Davis's sequence could have been developed in certain areas without invalidating Tarr's concept of the general paragenesis for the district. The amounts of chalcopyrite, siegenite, and sphalerite are generally minimal in comparison with galena, and there is no real agreement among present-day geologists in the area as to whether or not all of the nickel and cobalt are in the siegenite alone or are in part present as inclusions of siegenite in chalcopyrite or as substitutions of these ions for copper in chalcopyrite; some nickel is found in bravoite (nickeliferous pyrite). Surprisingly, what silver is present is some 20 times as abundant in sphalerite (probably as tiny inclusions) as in galena, but since sphalerite is not now recovered, silver is obtained only from the galena in amounts so small as barely to pay the costs of extracting it. Taylor and Radtke (1969) suggest that the silver in sphalerite may be present as silver chloride. If this is a correct interpretation, then some support is added to the concept that the metal ions were transported as chloride complexes, although far more data are needed before this theory can be fully evaluated. What cadmium is present is in the sphalerite. With the essential exhaustion of the minable cobalt-nickel ores in the Fredericktown area in the early 1960s, production of these two metals has ceased; minor quantities of copper continue to be produced, mainly in the southeast part of the district from rock volumes low in the Bonneterre formation.

The abundance of galena in relation to the other ore sulfides and the low silver content of the lead sulfide strongly suggests that the ores were formed under telethermal conditions. Only in the Fredericktown area in the southeastern part of the Leadbelt are siegenite, bravoite, and chalcopyrite present in sufficient abundance to indicate that deposition under more intense conditions occurred to an appreciable extent. This concentration of higher-temperature minerals probably was achieved in the leptothermal range, and their economic importance probably was great enough to justify the inclusion of the term leptothermal in the classification of the ores of the district. On this reasoning, therefore, the Leadbelt is here categorized as leptothermal (minor) to telethermal.

Montana

BUTTE

Late Mesozoic to Early Tertiary	*Copper, Zinc, Lead, Manganese, Silver*	*Hypothermal-1 to Telethermal*

Agar, W. M., 1926, The minerals of the intermediate zone, Butte, Montana: Econ. Geol., v. 21, p. 695-707

Allsman, P. L., 1956, Oxidation and enrichment of the manganese deposits of Butte, Mont.: A.I.M.E. Tr., v. 205, p. 1110-1112 (in Min. Eng., v. 8, no. 11)

Atwood, W. W., 1916, The physiographic conditions at Butte, Montana, and Bingham Canyon, Utah, when the copper ores in these districts were enriched: Econ. Geol., v. 11, p. 697-740 (particularly p. 698-727)

Doe, B. R., and others, 1968, Lead and strontium isotope studies of the Boulder batholith, southwestern Montana: Econ. Geol., v. 63, p. 884-906

Fritzche, H., 1938, Die Kupfererzlagerstätten und der Bergbau von Butte in Montana: Metall und Erz, Bd. 35, H. 14, S. 367-375

Garlick, G. D., and Epstein, S., 1966, The isotopic composition of oxygen and carbon in hydrothermal minerals at Butte, Montana: Econ. Geol., v. 61, p. 1325-1335

Grunig, J. K., and others, 1961, Lead isotopes in the ores and rocks of Butte, Montana: Econ. Geol., v. 56, p. 215-217

Hart, L. H., 1935, The Butte district, Montana, in *Copper resources of the world*: 16th Int. Geol. Cong., v. 1, p. 287-305

Kirk, C. T., 1912, Conditions of mineralization in the copper veins at Butte, Montana: Econ. Geol., v. 7, p. 35-82

Knopf, A., 1957, The Boulder batholith of Montana: Amer. Jour. Sci., v. 255, no. 2, p. 81-103

Lindgren, W., 1927, Paragenesis of minerals in the Butte veins: Econ. Geol., v. 22, p. 304-307

Locke, A., and others, 1924, Role of secondary enrichment in genesis of Butte chalcocite: A.I.M.E. Tr., v. 70, p. 933-963

Meyer, C., 1965, An early potassic type of wall-rock alteration at Butte, Montana: Amer. Mineral., v. 50, p. 1717-1722

Meyer, C., and others, 1968, Ore deposits at Butte, Montana, in Ridge, J. D., Editor, *Ore deposits of the United States, 1933-1967* (Graton-Sales Volumes): Chap. 65, v. 2

Murthy, V. R., and Patterson, C., 1961, Lead isotopes in ores and rocks of Butte, Montana: Econ. Geol., v. 56, p. 59-67, 217-218

Perry, E. S., 1933, The Butte mining district, Montana: 16th Int. Geol. Cong., Guidebook 23, 25 p.

Ray, J. C., 1914, Paragenesis of the ore minerals in the Butte district, Montana: Econ. Geol., v. 9, p. 436-482

—— 1931, Age and structure of the vein systems at Butte, Montana: A.I.M.E. Tr., v. 96 (1931 General Volume), p. 252-267

Rogers, A. F., 1913, Upward secondary sulphide enrichment and chalcocite formation at Butte, Montana: Econ. Geol., v. 8, p. 781-794

Sales, R. H., 1910, Superficial alteration of the Butte veins: Econ. Geol., v. 5, p. 15-21

—— 1913, Ore deposits at Butte, Montana: A.I.M.E. Tr., v. 46, p. 3-109

Sales, R. H., and Meyer, C., 1948, Wall-rock alteration at Butte, Montana: A.I.M.E. Tr., v. 178, p. 9-35

—— 1949, Results from preliminary studies of vein formation at Butte, Montana: Econ. Geol., v. 44, p. 465-484

—— 1950, Interpretation of wall-rock alteration at Butte, Montana: Colo. Sch. Mines Quart., v. 45, no. 1B, p. 261-273

—— 1951, Effect of post-ore dike intrusion on Butte ore minerals: Econ. Geol., v. 46, p. 813-820

Weed, W. H., 1912, Geology and ore deposits of the Butte district, Montana: U.S. Geol. Surv. Prof. Paper 74, 262 p. (particularly p. 26-72, 86-112)

Notes

The Butte ore deposits are situated in and near the town of that name which lies on the 42nd parallel of latitude, some 50 miles south-southwest of Helena, the capital city of Montana.

The ore deposits are contained in the Boulder batholith, an irregular mass of silicic rock that covers nearly 2000 square miles and is intrusive into Precambrian, Paleozoic, and Mesozoic sediments and late Mesozoic andesite lavas. The granite was intruded, probably not long after it had solidified, by many dike- and sill-like bodies of aplite that, in some locations, grades into pegmatite. The introduction of the aplites was followed by further fracturing of the rocks and the intrusion of dikes of fine- to coarse-grained quartz porphyry; the quartz porphyry dikes are largely confined to the area now containing the copper mineralization.

The first earth movements to affect the batholith occurred before the intrusion of the aplites and quartz porphyries, but the widespread and thoroughgoing fracturing through which the Butte ore fluids entered the area were not fully developed until after the quartz porphyry had been emplaced. The fact that the older vein systems are offset by younger ones indicates that the fracturing occurred over a measurable period of time. This is confirmed by the moderate differences in character between the mineralizations in the different vein systems and degree to which each is mineralized. All of the fractures, however, may have developed over a geologically short period of time and may have been formed by the same general torsional stresses.

The three main fissure systems in the Butte deposits were formed in the following order: (1) the Anaconda or east-west, (2) the Blue or northwest, and (3) the Steward-Rarus or northeast that Sales divided into the Steward, Rarus, and Middle groups; of these three, only the Steward veins are mineralized. The strikes of the various veins on occasion depart rather radically from the directions implied by their alternate names. The Anaconda veins were formed by the filling of open fissures and are generally uniform and continuous in their mineralization, while the Blue and Steward veins are filled largely with gouge and crushed monzonite and are mineralized only in shoots. The Rarus and Middle faults, forming after the three mineralized systems, increase the structural complexity of the deposit without adding anything to its value. The strange Mountain View breccia faults developed after the Blue veins, and possibly after the Steward, but before the Rarus faults; they are unmineralized. The last fractures to have formed in the district are those of the Continental fault system; they are unmineralized and have essentially no effect on the ore-bearing portion of the district. The concentration of geological study at Butte on the veins and their mineralogy should not obscure the fact that there is appreciable ore mineralization throughout the entire block of monzonite included within the vein systems and that the open-pit mining in the Berkeley pit is taking the entire mass of altered monzonite.

From the fact that the monzonite cuts Mesozoic sediments and probably Late Cretaceous lavas, the deposits are thought to have been formed in the Laramide orogeny, probably during the early Tertiary rather than in the late Mesozoic. Radioactive age determinations on the various Butte rocks tend to confirm this concept. The quartz monzonite appears to have been emplaced between 78 and 82 m.y. ago, while the sericite of the wall-rock alteration is dated as 61 to 63 m.y. old, and the quartz porphyry shows an age of 63 to 65 m.y. The late (post mineralization) rhyolite gives an age of about 53 m.y. These dates place the introduction of the batholith as Upper Cretaceous and the ore mineralization in the Paleocene. On this basis the mineralization is here categorized as late Mesozoic to early Tertiary.

Probably no paragenesis of a sulfide ore is more complicated than that of Butte where the ore minerals appear to have been zoned in both space and time. The first ore mineral deposition in the rock volumes that now contain the Butte mines probably was as veinlets of quartz, pyrite, and molybdenite. These veinlets are confined to a cone-shaped volume of rock that has its highest elevation at about the 2800 level; the cone widens downward as far as exploration has reached into it, that is, about the 4500 level. The fracture system containing these molybdenite-bearing veinlets was formed before any of the main-stage Butte veins (Anaconda, Blue, and Steward) since the veinlets consistently are cut by the veins. Some of the molybdenite veinlets are bordered by alteration selvages made up of biotite, sericite, and potash feldspar that may be as much as 8 inches wide but normally are much thinner. The wider alteration borders of this type may include disseminated chalcopyrite that may give the altered material a copper content of as much as 2 percent.

In the main-stage Butte veins, there were several stages of mineralization, the first of which was composed of quartz, pyrite, and sphalerite, with the sphalerite being in considerable part later than the pyrite. The quartz and pyrite of this early main-stage mineralization appear to have been deposited throughout the central, intermediate, and peripheral zones into which the mineralized portion of the district is divided (Perry, 1933, p. 296); important amounts of the zinc sulfide now are found only in the two outer zones. Sphalerite may have been deposited in abundance in the central zone during this early stage of mineralization, but if this is so, this sphalerite has been replaced almost entirely by later copper sulfides. Some of the primary chalcocite in the central zone is pseudomorphous after sphalerite, and essentially all chalcocite in this zone contains relics of sphalerite, suggesting that sphalerite well may have been far more abundant initially in that zone than it is now. In the intermediate zone the zinc sulfide increases steadily away from the central zone; whether this relationship is due to a decreasing replacement of sphalerite the farther the rock volume in question may be from the central zone or whether it was caused by a steady increase in sphalerite deposition from the inner to the outer margin of the intermediate zone is uncertain. The sphalerite of the intermediate zone (or at least of the inner part of that zone) contains abundant inclusions of chalcopyrite as rods and dots that are more probably the result of exsolution than of replacement. This may indicate that the sphalerite of the intermediate zone (or at least of the inner portion of it) was formed at or above 400°C and, therefore, in the hypothermal range. Sphalerite was deposited in important amounts in the peripheral zone, but nothing has been published as to its chalcopyrite content. It may be sphalerite that was removed from the central zone and redeposited in the peripheral area.

The relationships of the copper sulfides to each other are most complex; they appear, however, to be later than the sphalerite in the rock volumes (outer central zone and inner intermediate zone) where they and the zinc sulfide are found together. In order of decreasing abundance, the copper minerals appear to be chalcocite (even though only primary chalcocite is considered), bornite, enargite, chalcopryite, covellite, tennantite, and tetrahedrite. At the time that Perry (1933) made the list just given, the presence of digenite and djurleite ($Cu_{1.96}S$) in the ore had not been recognized, but it appears probable that they would fit into the list just given between covellite and tetrahedrite-tennantite or possibly between chalcopyrite and covellite. Chalcopyrite was of small importance as a source of copper in the upper levels, but recent development work in the lower reaches in the intermediate zone has found it to be of considerable and rising abundance with depth though it probably remains, at least until now, less abundant than enargite. In addition to the deep-level chalcopyrite, this iron-copper sulfide has been found in considerable abundance in the outer portions of the intermediate zone. This chalcopyrite, marginal to the intermediate zone, probably can be followed downward to make connection with the deep-level occurrences of that mineral.

Enargite was consistently the first copper sulfide to have been deposited. Much of it is found in fractures in older quartz-pyrite veins or as replacements of the altered wall rock or of pyrite; some pyrite and quartz and barite appear to have formed contemporaneously with the enargite. Although there are essentially no rock volumes in which sphalerite and enargite are both present in quantity, Lindgren (1927) believes the sphalerite to have been deposited first; with this Agar (1926) agrees. The primary chalcocite and bornite probably are later than the enargite, but their time relationships with each other are not certainly known. Bornite and chalcocite are found filling fractures in quartz and pyrite and as replacements of altered wall rock and pyrite; the degree to which the minerals have replaced enargite is uncertain but probably was not as great as that by which they replaced pyrite. Brecciated enargite, cemented by chalcocite, however, is known.

In the Butte mines it appears that the proportions of enargite and chalcopyrite in the central and intermediate zones, respectively, increase with depth, although even in the deepest levels now reached in the intermediate zone, bornite may still be more abundant than chalcopyrite.

Some bornite appears to have formed sufficiently above 200°C that small amounts of chalcopyrite were often contained in it in solid solution--chalcopyrite that exsolved on cooling to form Widmannstätten patterns in the bornite. Some of the early bornite probably was deposited in the cubic form, some of which may have failed to invert to the isometric variety. Other bornite probably developed when the bulk composition of the system was on the copper-rich rather than the iron-rich side of bornite composition in the Fe-Cu-S system since it contains exsolved chalcocite and digenite. There appears to have been a gradual transition during the course of deposition from iron-rich bornite to copper-rich bornite to copper-poor chalcocite-digenite solid solutions since in some exsolution textures bornite predominates, while in others chalcocite and/or digenite is the more abundant material.

Thus, it would appear the copper-iron and copper sulfides were deposited over an appreciable range in ore fluid temperature and composition. It is probable that the chalcopyrite was developed first and that it rather closely approximated stoichiometric composition throughout its depositional cycle. In contrast to bornite, it was confined to the intermediate zone and was not developed to any notable extent in the central zone. Bornite, on the other hand, was introduced in both the central and intermediate zones as a filling of open space and as a replacement of pyrite and enargite in the central zone and of chalcopyrite in the intermediate zone. As ore fluid temperature dropped and iron became less abundant, the iron-copper-sulfur mineral being formed changed from bornite to chalcocite or digenite (locally covellite) from which, on cooling, the excess material above the appropriate stoichiometric composition exsolved. This exsolution process produced chalcocite with exsolved digenite, bornite, and/or covellite; digenite with exsolved chalcocite and bornite and locally covellite; and covellite with exsolved digenite. Experimental studies indicate that covellite should not exsolve from chalcocite and that chalcocite, digenite, and covellite should not occur in the same specimen if the sulfide volume in question was deposited as a single solid solution. In the Butte ores, however, exsolution covellite is found in chalcocite and all three simple sulfides are found in the same specimen, with the two in minor quantities probably having exsolved from solid solution in the dominant third; the explanation for such occurrences is, as yet, unknown. Where covellite is a primary mineral (and not the result of exsolution), the ore fluid in that vicinity must have had an oxidation-reduction potential sufficiently high to retain essentially all the copper in the cupric state.

Were the relationships just described the only ones developed, they would be complex enough, but early bornite, with exsolved chalcocite and/or digenite, probably was replaced by later chalcocite or digenite from which still later exsolution occurred. Thus, a single specimen may contain bornite that replaced enargite, and this bornite, from which digenite or chalcocite was later exsolved, may still later have been replaced by chalcocite from which minor bornite later exsolved. Sales and Meyer (1948) have discussed this problem and have arrived at somewhat different conclusions.

The late chalcocite, digenite, and covellite did not extend as far out from the central zone of mineralization as did the bornite, so that bornite is found in association with both enargite and chalcopyrite while the simple copper sulfides themselves are found only in the central zone.

Possibly a second generation of chalcopyrite was developed since it has been reported as occurring in cracks, vugs, and cavities within older vein fillings and as thin coatings on, and replacements of, older copper and iron sulfides.

In addition to the tennantite in exsolution intergrowths with the main generation of chalcopyrite, considerable tennantite was developed in the lower portion of the intermediate zone as vein fillings and as replacements of chalcopyrite and bornite. The arsenic needed for this mineral may have been that liberated when enargite was replaced by bornite and chalcocite solid solutions in the central zone. Tennantite, however, also is found as a replacement of enargite in the central zone, but the amount of tennantite in the enargite does not provide a home for anywhere near all the arsenic liberated by the replacement of enargite by the sum-total of later copper-bearing sulfides. Whether the tennantite that replaces enargite is of the same age as the tennantite in exsolution textures in the chalcopyrite or of that of the tennantite that is later than bornite is uncertain.

Tetrahedrite is an unimportant copper mineral at Butte but Sales (1913) reports that it is found in small quantities in nearly every mine at Butte; its position in the paragenetic scheme is uncertain, but it is probably late.

As the inner boundary of the peripheral zone is defined as the outer limit of economically valuable copper deposits, copper sulfides are essentially lacking in the peripheral zone and are of minimal economic value in the outer portions of the intermediate zone. As copper sulfides decrease in abundance outward in the intermediate zone, the importance of zinc sulfide rises. In part, this increase is due to a greater abundance of sphalerite in this portion of the intermediate zone than farther inward, but it also is caused by the reduced amount of sphalerite replaced by earlier minerals. The consensus seems to be that there was only one generation of sphalerite, all of which was deposited before the copper sulfides. The presence of exsolved blebs and rods of chalcopyrite in the sphalerite in part of the intermediate zone, at least, indicates that it was deposited under hypothermal conditions. By inference, the central pyrite of the central and intermediate zones also was deposited under these intense conditions. There is considerable doubt, however, that the sphalerite of the outer portions of the intermediate zone and of the peripheral zone was deposited in the hypothermal range; it was much more probably formed under mesothermal and less intense conditions.

In the intermediate zone, it is thought by Perry (1933) and Agar (1926) that galena, although later than most of the copper sulfides, was never later than chalcocite or covellite. Lindgren (1927) thought that galena certainly was later than bornite and tennantite but believed it to have been deposited before chalcocite. It does appear certain that the galena is appreciably later than sphalerite even though it is normally associated spatially with that mineral rather than with the copper sulfides. Thus, the relationship of galena to sphalerite provides no clues as to the conditions under which either of the two minerals was deposited. After the deposition of galena, the rocks were again broken, and rhodochrosite and quartz were emplaced as breccia cements, mainly in the zinc-rich areas. The carbonate was accompanied by minor amounts of pyrite and chalcopyrite; some carbonates formed in druses in the copper mineralization, including a late dolomite and still later calcite. Rare barite was the last mineral to have formed.

The complicated interrelations among the Butte minerals need much more study if the sequence is ever to be fully understood. (The publication of the studies carried out by the Anaconda Company would be helpful.) Probably, however, the chalcopyrite-bearing sphalerite was deposited under hypothermal conditions, the enargite and chalcopyrite were formed in the mesothermal range, and the bornite-chalcocite complex probably deposited over from the less intense mesothermal through the leptothermal and telethermal stages. The conditions under which galena was formed are not definitely understood; if it was deposited before chalcocite, it probably was formed in the leptothermal to telethermal ranges; its appreciable silver content suggests leptothermal conditions. The considerable amounts of tennantite deposited after the bornite indicate that the ore fluids at that time were in the low-intensity portion of the mesothermal range. It is quite possible that much of the sphalerite, particularly that in the outer intermediate and peripheral zones, was formed under conditions less intense than hypothermal, but no published information exists as to the presence or lack of exsolved chalcopyrite in the sphalerite in the outer portions of the Butte deposit. The classification of the Butte ores appears, from this line of reasoning, to have been formed under hypothermal-1 through telethermal conditions.

The wall-rock alteration associated with the copper mineralization at Butte, superlatively well described by Sales and Meyer (1948), is that typically developed under mesothermal conditions in porphyry copper deposits and does not conflict with the classification given here. Since Sales's and Meyer's work was published, much deeper levels have been developed in which the considerable abundance of alteration biotite suggests that wall-rock alteration grows more intense with depth.

DILLON

Late Mesozoic to Early Tertiary | *Graphite* | *Xenothermal*

Bastin, E. S., 1912, The graphite deposits of Ceylon; a review of present knowledge with a description of a similar graphite deposit near Dillon, Montana: Econ. Geol., v. 7, p. 419-443 (particularly p. 435-443)

Cameron, E. N., and Weis, P. L., 1960, Vein deposits (and) graphite districts--Montana, in *Strategic graphite--a survey*: U.S. Geol. Surv. Bull. 1082-E, p. 247-249, 252-255

Ford, R. B., 1954, Occurrence and origin of the graphite deposits near Dillon, Montana: Econ. Geol., v. 49, p. 31-43

Heinrich, E. W., 1949, Pegmatites in Montana: Econ. Geol., v. 44, p. 307-334 (particularly p. 326-330)

Perry, E. S., 1948, Talc, graphite, and vermiculite in Montana: Mont. Bur. Mines and Geol. Mem. no. 27, 44 p. (mimeo.) (particularly p. 13-18)

Winchell, A. N., 1911, A theory for the origin of graphite as exemplified by the graphite deposit near Dillon, Montana: Econ. Geol., v. 6, p. 218-230

—— 1911, Graphite near Dillon, Montana: U.S. Geol. Surv. Bull. 470, p. 528-532

Notes

The Dillon graphite deposits are located at the southern end of the Ruby Range about 60 miles almost due south of Butte.

The principal rocks of the area consist of the Cherry Creek and Pony series of pre-late Precambrian age. In the vicinity of the Crystal Graphite mine, the only property that has been operated on any appreciable scale in the Dillon area, the rocks are of the Cherry Creek series. The lower part of the Cherry Creek outcrops in the mine area and includes biotite-garnet gneiss, hornblende gneiss, quartzite, and marble. Pegmatitic quartz-feldspar masses are conformable to the regional foliation; these masses, however, are not large in size, being from a few inches to several feet thick and rarely more than 100 feet long; they are normally surrounded by narrow selvages of mafic minerals and are generally coarse and varied enough to be classed as pegmatites. These pegmatites contain a rather standard mineral assemblage of quartz, orthoclase, microcline, plagioclase, biotite, garnet, and graphite, plus small amounts of sericite, chlorite, and epidote.

Heinrich (1949) thought the pegmatites to be of two ages, one pre-late Precambrian and one Laramide, because he believed some of them to have been metamorphosed and some not. Ford (1954), on the other hand, pointed out that many of the pegmatites, each within itself, showed gradation from highly granulitic (therefore, well foliated) through less granulitic to massive, convincing him that all of the pegmatites were of the same age. Since the foliation in the pegmatites is parallel to that in the rocks surrounding them, both probably were developed at the same time; the granulation may have taken place during magmatic flow before the pegmatites had solidified but more probably during the later regional metamorphism.

Because the rocks of the Cherry Creek series are more metamorphosed than the younger Beltian rocks and because the pegmatites probably have been affected by the same metamorphism as the country rocks containing them, Ford believes that the pegmatites, as well as their host rocks, were emplaced in pre-Beltian time and are, therefore, either early or middle Precambrian.

In the Crystal mine, however, the graphite occupies fissures that certainly were formed at a later time than any of the associated rocks, including the pegmatites. The length of the time that elapsed between the foliation of the gneisses and their contained pegmatites and the introduction of the graphite is unknown. Because of the striking contrast between the textural and structural features of the country rocks and of the graphite in the veins, the lack of wall-rock alteration, and the short horizontal and vertical extent of the veins, it would seem probable that the ores were deposited only after a long period of time and probably of erosion had brought the originally deeply buried metamorphic rocks near to the surface. Because the first extensive post-Precambrian earth movement in the area appears to have resulted from the Laramide orogeny and because the graphite probably was deposited shortly after the fissures were formed, it is thought that the age of the mineralization was late Mesozoic to early Tertiary, and the deposits are so dated here.

The graphite veins of the Crystal mine have widths between 1/8th inch and 2 feet, and their lengths are short; only occasionally is a vein as much as 50 feet long. The veins often are arranged in a stockwork pattern and are filled with irregular patches and networks of flake graphite that form comb structures composed of parallel platy needles and of rosettes of such needles irregularly oriented. Although the walls of the veins usually are sharp, some flake graphite is disseminated in the wall rocks, the flakes being arranged parallel to the rock foliation; the stronger and better mineralized the fissure veins, the more abundant the disseminated graphite in the adjoining wall rocks. The disseminated graphite does not appear to favor any particular type of wall-rock material. Some quartz-albite masses in the graphite veins might be thought, at first glance, to be gangue minerals, but they appear (Ford, 1954) to be wall-rock fragments mechanically incorporated in the vein graphite and to have been partly replaced by it. The graphite is never enclosed by any other mineral and is, therefore, probably later than all igneous and metamorphic minerals.

Many of these veins are discordant to the regional foliation, but others were located in planes of weakness that border the pegmatite bodies; except for the introduction of disseminated graphite into the rocks, the mineralizing process does not seem to have caused any alteration of any rock type. The veins contain numerous open cavities, crustification, and comb structures; these features added to the short horizontal and vertical extent of the veins and to the stockworks in which many of them are arranged, point to their near-surface origin.

Because no minerals in the veins were deposited syngenetically with the graphite, no direct clue is known as to the temperature at which mineral formation took place. It is, therefore, necessary to consider how the graphite may have been formed to see if a reasonable theory of origin might give a clue as to the temperatures at which deposition occurred. The most popular genetic suggestion is that of Winchell (1911, 1911a) who thought that the ore fluid contained abundant carbon monoxide that reacted with water to produce carbon dioxide and nascent hydrogen--equation (1).

$$CO + H_2O = CO_2 + H_2. \quad (1)$$

He further supposed that the hydrogen reacted with the carbon dioxide, not to reverse the reaction, but to produce carbon and water--equation (2).

$$CO_2 + 2H_2 = C + 2H_2O. \quad (2)$$

This is in the nature of lifting yourself by your chemical bootstraps, particularly because equation (2) requires twice as much hydrogen as is produced by equation (1). Winchell's rationale for both of these reactions going forward was that the first took place above 500°C and the second below that temperature. As these reactions must have occurred near the surface, as the geologic conditions demonstrate, and as the reaction of equation (2) is one favored by increased pressure, it seems unlikely that these near-surface conditions would have driven equation (2) to the right even granted that the additional hydrogen needed could have been supplied from the volcanic emanations that Winchell assumes provided the original carbon monoxide and the water of equation (1).

It appears more reasonable to assume that the carbon was introduced as methane, the principal component of natural gas. It has been suggested that methane can be converted to carbon by reaction, in a water-rich medium, with ferric ion--equation (3), the water serving to remove the debris of the reaction.

$$CH_4 + 4Fe^{+3} = C + 4Fe^{+2} + 4H^{+1}. \quad (3)$$

Because of the lack of minerals other than graphite in the Dillon deposits, it appears unlikely that enough iron would have been present in the ore-forming medium to have permitted graphite to have been precipitated from it by reaction with ferric iron, although this mechanism may have been that by which the graphite disseminated in the wall rocks was introduced.

Again, because the veins contain nothing but graphite, it appears reasonable to hypothesize that the ore fluids contained essentially nothing but methane, water (perhaps of meteoric origin), and (as will be shown helpful) a source of nascent oxygen. Although methane is the most stable of the hydrocarbons, it will, if heated by itself, slowly decompose between 650° and 700°C and much more rapidly at higher temperatures, according to the reaction given in equation (4).

$$CH_4 = C + 2H_2. \quad (4)$$

Only at 1000°C, however, is the reaction rapid, except in the presence of a catalyst when the effective temperature is appreciably lowered. What such a catalyst might be under natural conditions is uncertain, but nascent oxygen to combine with the hydrogen produced in equation (3) to form water would act to drive the reaction of equation (4) to the right without the aid of a catalyst. Such nascent oxygen might be derived from the dissociation of nitrogen-oxygen compounds, a dissociation that would have been favored by the near-surface conditions under which the Dillon graphite was deposited.

If the ore fluids consisted only of methane, nitrogen-oxygen compounds, and water (which may have been heated meteoric water), they can hardly be considered as hydrothermal fluids in the strict sense since they almost certainly would not have been of magmatic origin. The graphite deposited from the solutions, however, appears to have formed at high temperatures (at least 500°C) and in a rock volume where the confining pressure on the ore-forming gas (all the components of which almost certainly were above their critical temperatures and pressures) was being rapidly lowered. This environment not only fits the requirements for driving equation (4) to the right but also those of xenothermal deposition in the broad sense; the deposits, therefore, are here classified as xenothermal even though the basic definition of that category is somewhat strained in the process.

PHILIPSBURG

Late Mesozoic to Early Tertiary	*Silver, Lead, Zinc, Copper, Manganese*	*Mesothermal to Leptothermal (primary), Ground Water-B2 (secondary)*

Calkins, F. C., and Emmons, W. H., 1915, Description of the Philipsburg quadrangle, Montana: U.S. Geol. Surv. Atlas, Folio 196, 26 p.

Cole, J. W., 1949, Core-drill testing for base-metal mineralization below the Hope silver mine, Granite County, Montana: U.S. Bur. Mines R.I. 4399, 9 p.

Emmons, W. H., 1907, The Granite-Bimetallic and Cable mines, Philipsburg quadrangle, Montana: U.S. Geol. Surv. Bull. 315-A, p. 31-55

Emmons, W. H., and Calkins, F. C., 1913, Geology and ore deposits of the Philipsburg quadrangle, Montana: U.S. Geol. Surv. Prof. Paper 78, 271 p.

Fleischer, M., and Richmond, W. E., 1943, The manganese oxide minerals; a preliminary report: Econ. Geol., v. 38, p. 269-286 (General)

Goddard, E. N., 1940, Manganese deposits at Philipsburg, Granite County, Montana: U.S. Geol. Surv. Bull. 922-G, p. 157-204

Hewett, D. F., and Fleischer, M., 1960, Deposits of the manganese oxides: Econ. Geol., v. 55, p. 1-55 (General)

Holser, W. T., 1950, Metamorphism and associated mineralization in the Philipsburg region, Montana: Geol. Soc. Amer. Bull., v. 61, p. 1053-1090

Larson, L. T., 1964, Geology and mineralogy of certain manganese oxide deposits: Econ. Geol., v. 59, p. 54-78

Lorain, S. H., 1950, Investigation of manganese deposits in the Philipsburg mining district, Granite County, Montana: U.S. Bur. Mines R.I. 4723, 57 p.

McNabb, J. S., 1955, Manganese exploration in the Philipsburg district, Granite County, Montana: U.S. Bur. Mines R.I. 5173, 25 p.

Pardee, J. T., 1921, Deposits of manganese ore in Montana, Utah, Oregon, and Washington: U.S. Geol. Surv. Bull. 725-C, p. 141-243 (particularly p. 146-174)

Prinz, W. C., 1961, Manganese oxide minerals at Philipsburg, Montana: U.S. Geol. Surv. Prof. Paper 424-B, p. B296-B297

—— 1967, Geology and ore deposits of the Philipsburg district, Granite County, Montana: U.S. Geol. Surv. Bull. 1237, 66 p.

Zwicker, W. K., and others, 1962, Nsutite--a widespread manganese oxide mineral: Amer. Mineral., v. 47, p. 246-266

Notes

The Philipsburg district is located in southwestern Montana, about 45 miles N55°W of Butte and nearly 50 miles S60°E of Missoula. The mines extend eastward from Philipsburg for about 3 miles.

The oldest rocks in the district, exposed in the core of the north-trending anticline, are quartzites of the Missoula group (probably equivalent to the Belt series). Above a slight unconformity is the middle Cambrian Flathead quartzite, followed by the also middle Cambrian Silver Hill formation. The upper Cambrian consists of the Hasmark and Red Lion formations that are mainly dolomite and limestone, respectively. Unconformably above the Red Lion are the Devonian Maywood formation (mainly dolomitic limestone and sandy shale) and the Jefferson limestone; the Jefferson is the youngest formation in the district that includes significant amounts of ore. The various formations higher in the Paleozoic contain none of the Philipsburg-type ore although the Mississippian Madison is a limestone.

In early Tertiary time, apparently as part of the events of Laramide time, the sediments were intruded by the Philipsburg batholith, mainly a medium-grained granodiorite. Although most of this rock is quite fresh, it is intensely altered near veins. The intrusion followed Laramide folding and faulting and was, after solidification, itself fractured. Ore deposits were formed in the fissures and as replacements out from them; the manganese-rich replacement deposits were the last event in the sequence of ore formation, but all stages appear to have been essentially continuous. Post-ore normal faulting along bedding planes was the last structural event and appears to have marked the end of the Laramide orogeny. On this basis it seems reasonable to date the ores as late Mesozoic-early Tertiary; this checks with the age of the Boulder batholith to the east and south.

Prinz (1967) recognizes four groups of metalliferous ore deposits in the district: (1) steeply dipping quartz veins, (2) quartz veins along bedding,

(3) manganese-rich replacement deposits, and (4) contact-metasomatic magnetite deposits (these last are of no economic importance and are not considered further). Veins of type (1) are found in both granodiorite and the sediments; only rarely, however, can a type-(1) vein be followed from granodiorite into sediments. The veins of type (2) are in limestones and marbles only. The type-(1) veins have produced silver, zinc, and lead and minor copper, while the bedding-plane deposits have been worked essentially only for silver. Those of type (3) are found in favorable beds adjacent to, or near, the veins of type (1). Veins of this type (3) contain little quartz and are quite irregular in shape. Although principally composed of rhodochrosite and manganese-bearing dolomite, enough sulfides are present in those bodies in the southeastern part of the district that primary replacement ores can be mined for silver and zinc. Manganese, however, can be profitably recovered only where the deposits have been secondarily altered and enriched. The first of the primary minerals to be deposited during the single period of ore mineralization was magnetite; it was followed by pyrite and barite (some of the barite was later dissolved and redeposited). Then came sphalerite, followed by galena that did not overlap with the zinc sulfide; galena is not abundant in either type (1) or type (2) veins. The galena was accompanied during the latter part of its deposition by lead sulfosalts (?). Enargite was the first of the copper sulfides and overlapped with the questionable lead sulfosalts. Tennantite followed the enargite and may have begun to deposit as soon as the enargite; it overlapped with chalcopyrite and bornite; the chalcopyrite continued to deposit after bornite and enargite had stopped. Bornite may have deposited again later, but this is not certain. Proustite and pyrargyrite, which are the principal sources of silver in the high-grade silver-sulfide ores, followed the chalcopyrite and the early bornite. Chalcocite and covellite, which are not present in important amounts, were the latest sulfides to form except for a possible second generation of pyrite. The rhodochrosite did not begin to deposit until all of the sulfides (except the second generation of pyrite) had done so. Quartz probably deposited almost throughout the entire mineralization epoch; fluorite came in after rhodochrosite. Except for the magnetite, most of the minerals appear to have been formed under mesothermal conditions, the enargite, tennantite, and bornite being diagnostic of that range. The proustite, however, is characteristic of the leptothermal range and is important enough to deserve mention in the classification. The primary Philipsburg ores, therefore, are classified as mesothermal or leptothermal.

The oxidized manganese ores all overlie unoxidized manganese carbonate ores that essentially are too low grade to mine; the base of the oxidized zone is very irregular. The primary manganese minerals were emplaced almost entirely by replacement, deposition in open space having been minor and largely confined to cavities produced by dissolution during ore deposition. Since the veins were largely blocked by sulfides deposited earlier in the mineralization cycle, the fluids that deposited the manganese carbonates had to travel through the host rocks rather than through fractures, although some manganese carbonates did deposit in the few vein centers that remained open. The secondary minerals developed in the manganese-rich rock volumes are, in probable order of abundance: cryptomelane, todorokite, nsutite, pyrolusite, and manganite; near veins where sphalerite probably was located, chalcophanite and hetaerolite are present. It seems certain that these manganese ores were produced by circulating ground water, and they are here classified as ground water-B2.

STILLWATER COMPLEX

Early Precambrian *Chromite* *Magmatic-1b*

Allsman, P. T., and Newman, E. W., 1948, Exploration on the Stillwater chromite deposits, Stillwater and Sweetgrass Counties, Montana: A.I.M.E. Tr., v. 178, p. 327-338

Fenton, M. D., and Faure, G., 1969, The age of the igneous rocks of the Stillwater complex of Montana: Geol. Soc. Amer. Bull., v. 80, p. 1599-1604

Hess, H. H., 1939, Extreme fractional crystallization of a basaltic magma; the Stillwater igneous complex: Amer. Geophys. Union Tr., 20th Ann. Meeting, pt. 3, p. 430-432

—— 1941, Pyroxenes of common mafic magmas. Part 1: Amer. Mineral., v. 26, p. 515-535 (particularly p. 528-530)

—— 1960, Stillwater igneous complex, Montana: Geol. Soc. Amer. Mem. 80, 230 p.

Howland, A. L., 1955, Chromite deposits in the central part of the Stillwater complex, Montana: U.S. Geol. Surv. Bull. 1015-D, p. 99-121

Howland, A. L., and others, 1936, The Stillwater igneous complex and associated occurrences of nickel and platinum group metals: Mont. Bur. Mines and Geol. Misc. Contrib. 7, 15 p.

—— 1949, Chromite deposits of the Boulder River area, Sweet Grass County, Montana: U.S. Geol. Surv. Bull. 948-C, p. 63-82

Jackson, E. D., 1961, Primary textures and mineral associations in the ultramafic zone of the Stillwater complex: U.S. Geol. Surv. Prof. Paper 358, 106 p.

—— 1963, Stratigraphic and lateral variation of chromite composition in the Stillwater complex, in *Mineral. Soc. Amer. Spec. Paper 1*: Int. Mineral. Assoc., 3rd Gen. Meet., p. 46-54

—— 1967, Ultramafic cumulates in the Stillwater, Great Dyke, and Bushveld intrusions, in Wyllie, P. J., Editor, *Ultramafic and related rocks*: John Wiley & Sons, Inc., N. Y., p. 20-38

—— 1968, The chromite deposits of the Stillwater complex, Montana, in Ridge, J. D., Editor, *Ore deposits of the United States, 1933-1967* (Graton-Sales Volumes): Chap. 70, v. 2.

—— 1969, Chemical variation in coexisting chromite and olivine in chromitite zones of the Stillwater complex, in Wilson, H. D. B., Editor, *Magmatic ore deposits--a symposium*: Econ. Geol. Mono. 4, p. 47-71

Jones, W. R., and others, 1960, Igneous and tectonic structures of the Stillwater complex, Montana: U.S. Geol. Surv. Bull. 1071-H, p. 281-337

Page, N. J., and Jackson, E. D., 1967, Preliminary report on sulfide and platinum-group minerals in the chromites of the Stillwater complex, Montana: U.S. Geol. Surv. Prof. Paper 575-D, D123-D126

Page, N. J., and others, 1969, Platinum, palladium, and rhodium analyses of ultramafic and mafic rocks from the Stillwater complex, Montana: U.S. Geol. Surv. Circ. 624, 12 p.

Peoples, J. W., 1936, Gravity stratification as a criterion in the interpretation of the structure of the Stillwater complex, Montana: 16th Int. Geol. Cong. Rept., v. 1, p. 353-360

Peoples, J. W., and Howland, A. L., 1941, Chromite deposits of the eastern part of the Stillwater complex, Stillwater County, Montana: U.S. Geol. Surv. Bull. 922-N, p. 371-413

Powell, J. L., and others, 1969, Whole-rock Rb-Sr age of metasedimentary rocks below the Stillwater complex, Montana: Geol. Soc. Amer. Bull., v. 80, p. 1605-1612

Price, P. M., 1963, Mining methods and costs, Mouat mine, American Chrome Co., Stillwater County, Montana: U.S. Bur. Mines I.C. 8204, 58 p.

—— 1964, Primary features of stratiform chromite deposits, in Woodtli, R., Editor, *Methods of prospection for chromite*: O.E.C.D., Paris, p. 111-134

—— 1964, Mining methods and costs, Mouat mine, American Chrome Company, Stillwater County, Montana: U.S. Bur. Mines I.C. 8204, 58 p.

Schafer, P. A., 1937, Chromite deposits of Montana: Montana Bur. Mines and Geol. Mem. 18, 35 p.

Thayer, T. P., 1946, Preliminary chemical correlation of chromite with containing rocks: Econ. Geol., v. 41, p. 202-217 (general)

—— 1960, Some critical differences between Alpine-type and stratiform peridotite-gabbro complexes: 21st Int. Geol. Cong. Rept., pt. 13, p. 247-259

Turner, F. J., and Verhoogen, J., 1960, The Stillwater complex of Montana, in *Igneous and metamorphic petrology*: McGraw-Hill, 2nd ed., N. Y., p. 300-302

Westgate, L. G., 1922, Deposits of chromite in Stillwater and Sweet Grass Counties, Montana: U.S. Geol. Surv. Bull. 725-A, p. 67-84

Wimmler, N. L., 1948, Investigation of chromite deposits of the Stillwater complex, Stillwater and Sweet Grass Counties, Montana: U.S. Bur. Mines R.I. 4368, 41 p. (mimeo.)

Notes

The highly differentiated and largely gravity-stratified basic rocks of the Stillwater complex make up the northern portion of the Precambrian crystalline core of the Beartooth Mountains in the extreme southern part of central Montana, about equidistant between the towns of Livingston and Billings.

The complex appears originally to have been intruded as a sill into a group of metamorphosed sediments that consists mainly of hornfels but also contains quartzite and iron-bearing formation. The hornfels is normally a fine-grained rock, quite similar macroscopically to the basal rock (norite) of the Stillwater complex. The hornfels, however, can be distinguished from the norite by its granular texture and its lack of feldspar laths. The pale-blue quartzite is of such coarse grain that it has been confused with vein quartz. The iron formation is made up of alternating layers of quartz, magnetite, and iron silicates that range in thickness from a fraction of an inch to several inches.

The Stillwater complex is known over an exposed distance of 30 miles, but because it abuts at both ends against faults, it was initially probably appreciably longer. The complex has a measured thickness of 16,000 to 18,000 feet, but Hess estimates on compositional grounds that its original thickness was 25 to 40 percent greater. As a result of both Precambrian and Laramide folding and Laramide rotational faulting, the complex now strikes in a N60°W direction and dips north at angles of from 40 to 80 degrees; dips of 55 to 60 degrees are most common. Before the first folding, the complex was intruded by granite; it was then tilted, beveled by erosion, and covered by sediments of Middle Cambrian through Mesozoic age. The present large-scale structure of the complex was developed during the Laramide folding. In detail there are departures from the regional trend that are often too extensive to be accounted for by the cross faults with which they are associated; perhaps some of them may be explained by folding in the already-solid lower part of the sill before the remainder of the magma had solidified.

According to Hess (1960) the complex can be grossly divided by mineral content into six zones: (1) the 400-foot thick basal border zone, (2) the 2500-foot thick ultramafic zone, (3) the 2750-foot norite zone, (4) the 2200-foot lower gabbro zone, (5) the 6200-foot anorthosite zone which contains three anorthosite subzones, separated by two gabbro subzones, and (6) the 2100-foot upper gabbro zone. Because the top of the upper gabbro is covered by Paleozoic rocks, the sequence above zone (6) is unknown.

The ultramafic zone overlies the basal rocks and is of greatest interest to economic geologists because it contains the chromite horizons. This zone consists (Jackson, 1961) of a lower (peridotite) member and an upper (bronzitite) member. The peridotite member is composed of some 15 complete or partial repetitions of the series; (1) poikilitic harzburgite, (2) chromitite, (3) olivine chromitite, (4) poikilitic harzburgite, (5) granular harzburgite, and (6) bronzitite. The bronzitite (upper) member is made up almost exclusively of bronzite, although a few percent of chromite is first noted about 100 feet beneath the top of this member; this chromite gradually decreases in amount upward and dies out completely before the top of the bronzitite member is reached.

At the top of the bronzitite member, plagioclase, which previously had been formed only from trapped interstitial material, is present in large quantities as settled crystals in association with bronzite to produce the norite zone.

At the top of the norite member, clinopyroxene, which previously had been formed only from trapped interstitial material, is present in large quantities as settled crystals in association with orthopyroxene and plagioclase to form gabbro. Of the five principal minerals of the complex: (1) olivine, (2) orthopyroxene, (3) plagioclase, (4) clinopyroxene, and (5) chromite, clinopyroxene was the last to be deposited as settled crystals.

From the top of the lower gabbro to the uppermost known rocks of the complex, the major rock divisions (zones and members) are alternations of gabbro and anorthosite, with small amounts of olivine locally turning the gabbro into olivine gabbro. Within any one of these major divisions, the rocks are distinctly layered, though the layering is on a larger scale than was the case in the peridotite member of the ultramafic zone. These major layers are achieved by abrupt changes in mineralogical composition, and within these major layers, subordinate layers can be recognized due to the presence of a particular mineral and the absence of another. Within these subordinate layers, still further variations are produced by changes in proportions of minerals, mineral grain size, and mineral habit.

As the materials that make up the complex almost certainly were introduced in a single magmatic invasion, the remarkable layering on several levels of magnitude must have been due to processes that went on within the sill-like body of magma after its intrusion. Much of the layering of the complex probably is due to fractional crystallization, that is, the precipitation of crystals of the various minerals at different times and in different proportions and their settling to the floor of the magma mass at different rates. The different types and scales of layering, however, are too complicated and too often repeated to be explicable by fractional crystallization alone. The wide degree of the departure from the pattern that would have been produced by simple fractional crystallization is clearly shown by the cyclical nature of the layering, particularly in the peridotite member of the ultramafic zone.

To account for the marked differences between the peridotite zone and the less strongly layered zones above it, Jackson believes that conditions of convective overturn and bottom crystallization, repeated cyclically, gave way to continuous convection and deposition in the upper part of the magma volume; perhaps the principal cause of this change in the modus operandi was the decrease in magma thickness effected by crystal accumulation.

Although the complex was introduced in the Precambrian, no direct evidence has been found as to whether or not the intrusion took place in the middle or late Precambrian. From the considerable deformation and erosion of the sill before the end of Precambrian time, it is probable that it was intruded long before the end of that era. Jackson (1968) has reported preliminary results from K/Ar determinations on the chilled marginal rocks of the complex that indicate an absolute age of at least 3200 m.y. On the other hand, K/Ar ratios from biotites in the ultramafic zone (Jackson, 1968) give ages of 2600 m.y. This later age may be due to these biotites having been affected by a metamorphism of that age, a metamorphism that did not appreciably change the rocks of the basal chilled zone. Even if the age of 2600 m.y. is accepted as the age of the sill, the chromite deposits would have been deposited in the early Precambrian; if the age of 3200 m.y. is found to be the true one, then the Stillwater chromite deposits are among the oldest known ore deposits in the world.

Although a maximum of 13 chromitite zones have been recognized in the ultramafic zone of the Stillwater complex, only one--the G zone-- has been mined in all three areas of worthwhile chromitite development: the Benbow area at the eastern end of the complex, the Mountain View area (Mouat mine) just west of the Stillwater River, and the Gish area near the western end of the complex. The H zone was mined in the Mountain View area, but all other chromitite zones, also given letter designations beginning with A, were not of minable character. In the Mountain View area, the G and H zones were both mined, both lying in a wedge-shaped block of peridotite and bronzitite members that is included between two more-or-less northwest-southeast striking thrust faults that dip to the south (the Lake fault north and the Bluebird fault south). Although the surface geology is quite complex, due to landslides and surficial gravity faults, at depth the chromitite zones and their enclosing peridotite are quite continuous. The geology is, however, somewhat complicated by the thickening of the peridotite member and its component layers downdip toward the north and east; at the southwestern edge of the area, the peridotite member is about 2500 feet thick, while at the northeast, along the Lake fault, it is 4000 feet thick. The G zone has been followed along strike for some 4000 feet. Its footwall is quite sharp, but the chromite content dies out quite gradually upward. The basal portion of the G zone is a 2- to 6-foot massive layer of chromitite that is overlain by a section in which chromitite, olivine chromitite, and chromite harzburgite layers alternate. If chromitite is any portion of the G zone averaging more than 10 percent chromite, the zone rises from 4 feet thick (SW) to 18 feet thick (NE). If the zone must be 50 percent chromite to justify the designation of chromitite, the thickness ranges from 3 feet (SE) to 12 feet (NW).

From 45 to 60 feet stratigraphically above the G zone lies the hanging wall G chromitite, which also consists of alternating layers of chromitite, olivine chromitite, and chromite harzburgite. It ranges in thickness from 1 to 2 feet, but it is too thin for present-day mining.

The next zone, from 200 to 450 feet (SW-NE) stratigraphically above the G, is the H in the Mountain View portion of the district. The contact with the underlying ultramafic rocks is quite sharp but that with the ultramafics above it is gradational. The massive chromitite layer at the base is from 9 inches to 2.5 feet thick, and it is overlain by the usual olivine chromitite and chromite harzburgite in thicknesses of 2 to 6 feet. If the chromitite is considered to be all of the rock thickness that averages 10 percent chromite or more, the zone is 3 feet thick in the southwest portion of the district and 9 feet in the northeast. If the zone must be 50 percent chromite to justify the designation of chromitite, then the thicknesses are 1.5 feet (SW) and 2.5 feet (NE). The H zone has been followed along strike for about 3000 feet. As is true of the G zone, the H zone has a hanging wall chromite concentration about 15 feet stratigraphically above the base of the zone. In the hanging wall zone, the basal zone is only 1 or 2 inches thick while the chromite harzburgite above it is about 1.5 feet thick; this zone cannot be mined.

Other zones in the Mountain View area that are worthy of mention are (1) the B zone that is some 1200 feet stratigraphically below the G zone in the southwestern part of the area and about 2500 feet below it in the northwest. This zone probably has three narrow and locally divided chromitite layers that are separated by two barren harzburgite layers 40 to 45 feet thick; it is unminable; (2) the A zone some 200 feet stratigraphically below the B zone (SW) and 450 feet below (NE); the layering is irregular; (3) the K zone is about 400 feet stratigraphically above the H zone (SW) and about 675 feet above it (NE). It is made up of two thin (2.5 inches) but persistent chromitites separated by an olivine chromitite layer 7 inches to 1 foot thick. The other chromitite zones are too restricted in thickness and lateral extent to be worth description here.

Accepting the essential correctness of the basic premises of Hess and Jackson, it would seem nearly certain that the various rock types of the Stillwater complex, in general, and the chromite layers, in particular, were formed by segregation processes internal to the magma. On this basis, the chromite ores of the complex are here classified as magmatic-1b.

Nevada

CARLIN

Middle Tertiary *Gold* *Epithermal*

Akright, R. L., and others, 1969, Minor elements as guides to gold in the Roberts Mountains formation, Carlin gold mine, Eureka County, Nevada: Colo. Sch. Mines Quart., v. 64, no. 1, p. 49-66

Hardie, B. S., 1966, Carlin gold mine, Lynn district, Nevada: Nevada Bur. Mines Rept. 13, Part A, p. 73-83

Hausen, D. M., and Kerr, P. F., 1968, Fine gold occurrence at Carlin, Nevada, in Ridge, J. D., Editor, *Ore deposits of the United States, 1933-1967* (Graton-Sales Volumes): Chap. 47, v. 1

Radtke, A. S., and Scheiner, B. J., 1970, Studies of hydrothermal gold deposition (I). Carlin gold deposit, Nevada: The role of carbonaceous materials in gold deposition: Econ. Geol., v. 65, p. 87-102

—— 1970, Influence of organic carbon on gold deposition at the Carlin and Cortez deposits, Nevada: Geol. Soc. Amer. Abs. with Programs, v. 2, no. 7, p. 660

Roberts, R. J., and others, 1967, Geology and mineral resources of Eureka County, Nevada: Nevada Bur. Mines Bull. 64, 152 p.

Notes

The Carlin mine is located about 28 miles slightly north of west of Elko and some 18 miles northwest of Carlin in the Lynn Mining District in northwestern Eureka County.

The host rock of the gold ores in the Lower Silurian Roberts Mountains formation consists of thin-bedded argillaceous to dolomitic sandstones that have been variously bleached and silicified in the ore-forming process. The Roberts Mountains formation is disconformably overlain by the Devonian "Popovich formation" that is made up of medium-bedded, blocky dolomitic limestones and occasional beds of dolomite. A moderate-sized intrusive mass of quartz diorite is exposed some 3 miles north of the Carlin mine; it has an age of 121 ± 5 m.y. (Early Cretaceous). Some mineralization in the district, composed dominantly of barite, appears to be related to this intrusion, but the gold was introduced appreciably later. In the actual pit area, dikes of dacite or quartz latite are present in considerable number. They are pre-ore and probably are of much the same age as the quartz monzonite; locally gold ore has been emplaced in them. No firm data on which to base an age for the gold mineralization are available. Reasoning by analogy with other gold deposits of the general area, Hausen and Kerr (1968) think that the ore is Tertiary; here it is dated as middle Tertiary largely because of the middle Tertiary dates established for Goldfield and Tonopah. Further work obviously is needed on the age problem.

The main ore body at Carlin is an irregular, inclined deposit located near the top of the Roberts Mountains formation; its thickness ranges from a few feet to nearly 100 feet; the boundaries of the ore are determined by assay. The upper boundary is found a few feet to a few tens of feet below the "Roberts Mountains-Popovich" boundary. Horizontally, the ore body is elongate, striking northeast; on the incline, it continues downward to a fault contact. The favorable horizons for ore deposition were silty, dolomitic limestones; the gold in these rocks ranges from a few ounces to practically nothing per ton. The more silicified the formation, the less the gold content normally is.

Radtke and Scheiner (1970) point out that the host rocks that contained 0.25 to 0.3 weight percent organic carbon were not oxidized before ore deposition. They believe that the gold complexes (type not specified) were removed from the hydrothermal ore solutions by the organic carbon with the formation of various gold-organic compounds. Only after subsequent oxidation, which destroyed the organic compounds, was native gold produced. This gold is of micron to submicron size; none of it can be seen by the naked eye, and some of it is so fine as to be visible only under the electron microscope. Gold often borders detrital quartz grains or is found in fractures in them or it may be disseminated in the matric clays or contained in late quartz veinlets in silicified siltstones. Hausen and Kerr (1968) are convinced that the gold was not transported mechanically. The gold and the minerals associated with it were deposited over much the same time span, although pyrite began to deposit first, followed by gold and then, essentially simultaneously, by realgar, stibnite, native arsenic, jordanite, and tennantite; cinnabar and jarosite were slightly later but overlapped with the gold. The close spatial and temporal association of the gold with the other hydrothermal minerals is difficult to explain by Radtke's and Scheiner's organic gold compound-plus oxidation hypothesis; further work is needed on the problem. Hausen and Kerr (1968) consider the deposit has many similarities to the Getchell deposit that is located some 65 miles west-northwest of the Carlin mine. They consider Carlin to be an epithermal-type deposit, feeders in the deposit having some resemblance to siliceous sinters. The deposits appear to have formed so near to the then-obtaining surface as almost certainly to have been formed under conditions of rapid loss of heat and pressure. The deposits here are classified as epithermal.

COMSTOCK LODE

Middle Tertiary | *Silver, Gold* | *Epithermal*

Bastin, E. S., 1923, Bonanza ores of the Comstock Lode, Virginia City, Nevada: U.S. Geol. Surv. Bull. 735-C, p. 41-63

Becker, G. F., 1882, Geology of the Comstock Lode and the Washoe district: U.S. Geol. Surv. Mono. 3, 422 p. (now mainly of historical interest)

Calkins, F. C., 1944, Outline of the geology of the Comstock Lake district, Nevada: U.S. Geol. Surv. Prelim. Rept. 105154, 35 p. (mimeo.)

Coats, R., 1940, Propylitization and related types of alteration of the Comstock Lode: Econ. Geol., v. 35, p. 1-16

Cornwall, H. R., and others, 1967, Silver and mercury geochemical anomalies in the Comstock, Tonopah, and Silver Reef districts, Nevada-Utah: U.S. Geol. Surv. Prof. Paper 575-B, p. B10-B20

Gianella, V. P., 1934, New features of the geology of the Comstock Lode: Min. and Met., v. 15, no. 331, p. 298-300

—— 1936, Geology of the Silver City district and the southern portion of the Comstock Lode, Nevada: Univ. Nev. Bull., v. 30, no. 9, 105 p.

—— 1959, Period of mineralization of the Comstock Lode, Nevada (abs.): Geol. Soc. Amer. Bull., v. 70, p. 1721-1722

Knochenhaver, B., 1939, Das Gutachten Ferdinand von Richthofens über den Comstockgang und seine Bedeutung für die Gegenwart: Zeitsch. prakt. Geol., Jg. 47, H.3, S. 42-53

Milton, C., and Johnston, W. D., Jr., 1938, Sulfate minerals of the Comstock Lode, Nevada: Econ. Geol., v. 33, p. 749-771

Reid, J. A., 1905, The structure and genesis of the Comstock Lode: Univ. Calif. Pubs., Bull. Dept. Geol., v. 4, no. 10, p. 177-199

Smith, G. H., 1943, The history of the Comstock Lode: Univ. Nev. Bull., v. 37, no. 3, 297 p. (historical rather than geological) (Geol. and Min. Ser. no. 37)

Stoddard, C., and Carpenter, J. A., 1950, Mineral resources of Storey and

Lyon Counties, Nevada: Univ. Nev. Bull., v. 44, no. 1, 115 p. (particularly p. 54-73)(Geol. and Min. Ser. no. 49)

Thompson, G. A., 1956, Geology of the Virginia City quadrangle, Nevada: U.S. Geol. Surv. Bull. 1043-C, p. 45-77

Thompson, G. A., and White, D. E., 1964, Regional geology of the Steamboat Springs area, Washoe County, Nevada: U.S. Geol. Surv. Prof. Paper 458-A, p. A1-A52

White, D. E., and others, 1964, Rocks, structure, and geologic history of Steamboat Springs thermal area, Washoe County, Nevada: U.S. Geol. Surv. Prof. Paper 458-B, p. B1-B63

Notes

The Comstock Lode is located nearly 20 miles south-southeast of Reno and slightly over 20 miles from the California line; the district centers around the town of Virginia City.

The oldest rocks in the general area are folded and regionally metamorphosed pre-Tertiary sedimentary and volcanic rocks that were intruded by intermediate to silicic magmas prior to Tertiary time but constitute a small fraction of the total. On the basis of rather poor fossil evidence, these sediments are no younger than Jurassic and include a wide variety of rock types. Near the contacts of these rocks with the granites, hornfels, schist spotted with andalusite and cordierite, marble, and tactite have been developed. The pre-Tertiary metavolcanic rocks were largely basalt and andesite lavas and pyroclastics that have been altered to albite-epidote-amphibole rocks; near the granites, these have been converted to andesine-hornblende-biotite rocks. The intrusives in the general area were mainly granodiorite, but some granitic rocks were introduced; the granodiorite contains about twice as much plagioclase as orthoclase. Small pre-Tertiary intrusions near the Comstock Lode itself are of quartz-monzonite porphyry. These granodiorites and related types show little foliation, apparently having entered the area after the regional metamorphism and probably in Early Cretaceous time. The major portion of the Comstock rock sequence is composed of Tertiary volcanic agglomerates, tuffs, and lavas, plus a small amount of sedimentary rocks derived from the volcanic material; the volcanic rocks are largely andesites but range in composition from basalt to rhyolite.

Quartz veins of the Comstock Lode cut dikes and other intrusions of biotite-andesite porphyry that Thompson (1956) considers to be the probably Miocene Kate Peak, and the Kate Peak rocks are altered over wide areas, but the next younger formation, the Pliocene Truckee sediments--deposited in lakes and streams and containing intercalated tuff beds--is essentially unaltered. This indicates that the Comstock mineralization was, at the youngest, very earliest Pliocene and was more probably late Miocene; the process may, however, have begun much earlier in the Miocene. The alteration and siliceous sinter associated with hot-spring activity at the neighboring Steamboat Springs and the hot water, hydrogen sulfide, and carbon dioxide encountered in the Comstock mines suggest that the mineralizing process may, though to a minor degree, have continued into the present time. The bulk of the mineralization, however, must have occurred during the Miocene and is here classified as middle Tertiary.

The area to which the Comstock Lode is confined is bounded on the east by the Flowery range and on the west by the Virginia range. The Flowery range is bordered on the east mainly by normal faults; on the west side of the Virginia range, however, the strata dip off to the west as the limb of a broad syncline, the axis of which is in the Truckee meadows and the other limb of which lies along the east front of the Carson range, still farther west. The rocks on the west side of the Virginia range are also cut by normal faults (developed at the same time as the syncline) the east side of which is downdropped. Since the vertical movement on the faults is, therefore, generally in the opposite direction from the changes in strata elevation caused by the folding, the structural relief is much less than would have resulted if the folding had been the only process affecting the rocks.

The ore deposits of the Comstock Lode are associated with the Comstock normal fault and its southeasterly branch, the Silver City fault, which continues the fault structure to the south, the total mineralized distance being about 22,000 feet; both faults dip at some 45° to the east. The throw on the Comstock fault was about 2400 feet before mineralization and 1600 feet afterward, bringing Tertiary rocks in the hanging wall into contact with Mesozoic rocks in the foot. The veins, particularly in the Silver City fault, have been broken by cross faults, some of which are of considerable magnitude. Some production has come from many of the branches and cross faults of the Silver City fault. A minor amount of silver and gold has been taken from the Occidental vein that parallels the Comstock fault 1.5 miles farther to the east. Other veins in the area have provided a little silver and gold. Most of the ore was mined above the 2000 foot level; although considerable erosion has taken place since the ores were formed, it is doubtful if more than 2000 feet of overlying rock have been removed. How far mineralization extended into this eroded rock is not known, but it probably was for a considerable fraction of the 2000 feet.

In the Comstock fault portion of the Lode, centering around Virginia City, the base metal sulfide content of the ores was much larger than that in the Silver City area to the south where such sulfides make up only 1 or 2 percent of the vein material. The ores of the area generally are fine-grained, metallic minerals commonly having diameters of less than 1.0 mm, and the texture is granular with essentially no banding or crustification. The most abundant minerals are quartz, sphalerite, galena, chalcopyrite, and pyrite; calcite is sparse to abundant. The principal ore minerals are argentite, gold, and polybasite. These minerals appear to have formed at much the same time largely, if not entirely by the filling of open space. No veinlets of one mineral cutting another nor any strong evidence that one replaces another are known; it appears most probable that the primary ore and gangue minerals were deposited at much the same time. Perhaps those whose boundaries are convex outward may have formed somewhat before those concave toward them, but even this is not certain. In any event, the chalcopyrite, galena, argentite, gold, and polybasite occupy matrix-like or interstitial positions in relation to quartz and sphalerite.

Throughout the mineralized portion of the Comstock Lode, from the highest levels to nearly 3000 feet beneath the surface, the mineral association is essentially constant, although gold and argentite grow less abundant below the 2500 foot level. The gold is a very pale yellow and probably is largely alloyed with silver. The precious-metal minerals locally are joined by polybasite that is probably primary and bears exactly the same relationships to the other ore and gangue minerals as do gold and argentite. Tiny metallic-mineral inclusions in galena may be composed of a silver-bearing mineral. Native silver is probably a secondary mineral in the Comstock, being formed by the replacement of argentite; it is the most important secondary mineral, but some argentite, in the uppermost levels, is found as irregular borders around galena and probably is also of secondary origin. Most of the argentite, however, is primary. Most of the ore minerals were crushed by later movements on the faults.

Much of the Comstock alteration is propylitization in which the original rock was converted to epidote and albite (replacing plagioclase) and chlorite, calcite, and epidote (replacing ferromagnesian minerals). Pyrite is not an essential constituent of propylite, and it usually is not present in Comstock propylites. The widespread nature of the propylite and its lack of spatial relationship to vein fractures suggest that it was not caused by the ore fluid. Coats (1940) thinks that the propylitization was due to hydrothermal alteration independent of, and probably occurring before, the deposition of the Comstock ore; this alteration may have been carried out at higher temperatures than the ore deposition. In the immediate vicinity of the veins, the chief alteration has been the introduction of silica into the rocks, obliterating what previous alteration, such as propylitization, may have been present.

The mineralization in the Comstock Lode, though simpler than that of many epithermal ore deposits, is definitely diagnostic of epithermal conditions; the alteration is also typical of epithermal deposits. There appears to be no doubt but that the Comstock Lode should be classified as epithermal.

CORDERO

Late Tertiary *Mercury* *Epithermal*

Bailey, E. H., and Phoenix, D. A., 1944, Quicksilver deposits in Nevada: Univ. Nev. Bull., v. 38, no. 5, 206 p. (particularly p. 95-100) (Geol. and Min. Ser. no. 41)

Fisk, E. L., 1961, Cinnabar at Cordero: Min. Eng., v. 13, no. 11, p. 1228-1230

—— 1968, Cordero mine, Opalite mining district, in Ridge, J. D., Editor, *Ore deposits of the United States, 1933-1967* (Graton-Sales Volumes), Chap. 75, v. 2

Gilbert, J. E., and Haas, V. P., 1959, Cordero--Nevada's largest Hg mine: Eng. and Min. Jour., v. 160, no. 3, p. 88-90

Ross, C. P., 1942, Quicksilver deposits in the Steens and Pueblo Mountains, southern Oregon: U.S. Geol. Surv. Bull. 931-J, p. 227-258

Schuette, C. N., 1938, Quicksilver in Oregon: Ore Dept. Geol. and Mineral Inds. Bull. 4, 172 p. (particularly p. 147-169)

Yates, R. G., 1942, Quicksilver deposits of the Opalite district, Malheur County, Oregon, and Humboldt County, Nevada: U.S. Geol. Surv. Bull. 931-N, p. 319-348

Notes

The Cordero mine is located in north-central Nevada about 10 miles south-southwest of the town of McDermitt. The smaller Opalite and Bretz mines are across the boundary in Oregon, the former being about 15 miles northwest of the Cordero mine and the latter just over 10 miles north-northwest.

The geologic column in the area (Fisk, 1968) consists, except for young alluvium, of middle and late Tertiary extrusive rocks, tuffs, and tuffaceous lake beds. The volcanic material ranges from scoriaceous basalt, to andesitic basalt, to porphyritic andesite, to basalt, to intrusive (Footwall) rhyolite, to (the Cordero) rhyolite porphyry, to pitchstone and other glassy rhyolites, to tuffs, and finally to tuffaceous lake beds. The major ore body at Cordero occurs entirely in the Cordero rhyolite, and all ore bodies there are in rhyolite of some kind. The ores at the Opalite mine occur in silicified lake beds and those of the Bretz mine in tuffaceous shales and sandstones that lie outside the main belt of silicified rocks and apparently have no equivalent in the Cordero area.

All of the rocks listed are of Miocene age. Of Pliocene age is a riebeckite rhyolite that makes up a large dike or volcanic neck just south of the Cordero mine. If any flow rocks were extruded from this vent, they have since been removed by erosion. Bodies of soda-rich rhyolites are exposed southwest and northeast of the Cordero mine and Fisk considers them related to the riebeckite rhyolite.

Since the ores at Cordero were deposited in Miocene rocks (the Cordero rhyolites) and in Miocene lake beds at the Opalite mine, the ores are no older than late Miocene. Fisk (1968) believes that the hydrothermal fluids that deposited the ores of the Cordero mine used the same channelway, in whole or in part, that had been followed by the intrusive rhyolite. He points out, however, that an appreciable time interval may have elapsed between the rhyolite emplacement and the introduction of the ore. On the other hand, the ore has been in place long enough for oxidation along the major faults to have reached 500 feet below the surface. Nevertheless, hot water still is found in wells in the Cordero area. Balancing these various factors, Fisk considers the ores to have been deposited in the Pliocene, so they are categorized here as late Tertiary.

The main ore body at Cordero has the form of an irregular pipe with associated small pockets and lenses that rake toward it and probably are offshoots from it. All of the bodies die out about 250 feet below the present surface, but no evidence exists that any impermeable barrier existed to account for this relationship. The nearest opalite over this pipe is 150 feet above the top of the ore. On the other hand, the ore in the Opalite and Bretz mines either is in, or in contact with, silicified rocks. Obviously the problem of ore controls is far from solved.

Since almost all the cinnabar at Cordero is in rhyolite and very little in andesite, the former rock certainly is far more susceptible to replacement by mercury sulfide. The highly silicified rocks of the other deposits are quite comparable in composition to rhyolite. The only mercury minerals are cinnabar and a little native mercury; the gangue minerals are quartz, chalcedony, barite, and marcasite; wall rocks have been converted to alunite and kaolinite and other clay minerals.

Some of the quartz and chalcedony probably were leached from the high-silica rocks at depth, but some well may have been brought in by the ore fluids from their original source. The cinnabar appears to have formed both by replacement and open-space deposition.

The first event in the formation of the ores, according to Fisk (1968) was the silicification and argillic alteration of the host rocks. Before this stage had ended, but after the process began to decline in intensity, cinnabar commenced to deposit. Then came the first of the barite that deposited during and after the late quartz that began after barite.

Since the ores are now, and probably were when formed, quite close to the surface and were deposited under low-temperature conditions in highly broken rocks, they almost certainly were formed in the epithermal range and are so classified here.

ELY

Late Mesozoic to Early Tertiary	*Copper, Zinc, Molybdenum*	*Mesothermal*

Bateman, A. M., 1935, The copper deposits of Ely, Nevada, in *Copper resources of the world*: 16th Int. Geol. Cong., v. 1, p. 307-321

Bauer, H. L., Jr., and others, 1960, Porphyry copper deposits in the Robinson mining district, White Pine County, Nevada: Intermountain Assoc. Petrol. Geols. 11th Ann. Field Conf. Guidebook, p. 220-228, plus fig. 4A, revision of fig. 4, which places the contact between Pennsylvanian and Permian as the base of the Riepe Springs limestone, the lowest member of the Rib Hill formation

—— 1964, Origin of the disseminated ore in metamorphosed sedimentary rocks, Robinson mining district, Nevada: A.I.M.E. Tr., v. 229, p. 131-140

—— 1966, Porphyry copper deposits in the Robinson mining district, Nevada, in Titley, S. R., and Hicks, C. L., Editors, *Geology of the porphyry copper deposits--southwestern North America*: Univ. Ariz. Press, Tucson, p. 233-244

Billingsley, P., and Locke, A., 1941, Ely, in *Structure of ore districts in the Continental framework*: A.I.M.E. Tr., v. 144, p. 33-36

Brokaw, A. L., 1967, Geologic map and sections of the Ely quadrangle, White Pine County, Nevada: U.S. Geol. Surv., Quad Map GQ-697, 1:24,000

Brokaw, A. L., and Barosh, P. J., 1968, Geologic map of the Riepetown quadrangle,

White Pine County, Nevada: U.S. Geol. Surv. Quad. 578, 1:24,000

Brokaw, A. L., and Heidrick, T., 1966, Geologic map and sections of the Giroux Wash quadrangle, White Pine County, Nevada: U.S. Geol. Surv. Geol. Quad. 476, 1:24,000

Creasey, S. C., 1959, Some phase relations in the hydrothermally altered rocks of porphyry copper deposits: Econ. Geol., v. 54, p. 351-373 (general)

Fournier, R. O., 1967, The porphyry copper deposit exposed in the Liberty open-pit mine near Ely, Nevada. Part I. Syngenetic formation: Econ. Geol., v. 62, p. 57-81; Part II. The formation of hydrothermal alteration zones: Econ. Geol., v. 62, p. 207-227

Gott, G. B., Jr., and others, 1965, Preliminary report comparing geochemical halos around an exposed and a concealed copper deposit: Seminar on field techniques for mineral investigation, Central Treaty Organization (CENTO), Isfahan, Iran, Oct. 1965, p. 133-144

Lawson, A. C., 1906, The copper deposits of the Robinson mining district, Nevada: Univ. Calif. Pubs., Bull. Dept. Geol., v. 4, no. 14, p. 287-357

Lindgren, W., 1907, The copper deposits of the Robinson mining district, Nevada (rev.): Econ. Geol., v. 2, p. 195-204

McDowell, F. W., and Kulp, J. L., 1967, Age of intrusion and ore deposition in the Robinson mining district of Nevada: Econ. Geol., v. 62, p. 905-909

Parsons, A. B., 1933, Nevada Consolidated, in *The porphyry coppers*, A.I.M.E., p. 114-133

Pennebaker, E. N., 1932, Geology of the Robinson (Ely) mining district, Nevada: Min. and Met., v. 13, no. 304, p. 163-168

—— 1942, The Robinson (Ely) mining district, Nevada, in Newhouse, W. H., Editor, *Ore deposits as related to structural features*: Princeton Univ. Press, p. 128-132

Schwartz, G. M., 1947, Hydrothermal alteration in the "porphyry copper" deposits: Econ. Geol., v. 42, p. 319-352 (particularly p. 332-336, 346-352)

Spencer, A. C., 1913, Chalcocite enrichment (Ely district): Econ. Geol., v. 8, p. 621-652

—— 1917, The geology and ore deposits of Ely, Nevada: U.S. Geol. Surv. Prof. Paper 96, 180 p. (particularly p. 23-91, 98-130)

Whitman, A. R., 1914, Notes on the copper ores at Ely, Nevada: Univ. Calif. Pubs., Bull. Dept. Geol. Sci., v. 8, no. 17, p. 309-318

Notes

The deposits of the Ely (Robinson mining) district are located in east-central Nevada, some 185 miles southwest of Salt Lake City.

The pre-Pennsylvanian sequence in the Ely district is not well understood. On the other hand, the Pennsylvanian and Permian formations have been thoroughly studied because of their close relationship to the bulk of the ore and because they also are somewhat mineralized. The Pennsylvanian rocks are those of the Ely limestone that, except for the uppermost massive limestone member, includes abundant chert nodules up to 12 inches thick or chert bands of considerable continuity that are less than 1 inch thick. The oldest Permian rocks are those of the Rib Hill formation, the basal member of which is the Riepe Springs limestone that rests unconformably on the Ely. Conformably above the Riepe Springs is the Rib Hill sandstone that contains intercalated limestone beds. The sandstone is conformably overlain by the Arcturus limestone member and the Arcturus is capped by the Permian Kaibab formation. No younger sedimentary rocks, other than very young gravels, are known in the area.

The older igneous rocks in the district are a series of monzonite stocks (or chonoliths) of widely different shapes and sizes that outcrop in an east-west trending belt nearly 6 miles long; these stocks probably all connect to one large intrusive mass at depth. The monzonites are all porphyritic, though they differ widely in the degree to which this texture is developed; the ground masses of the various monzonites range from microcrystalline to granitic. Although the porphyry masses appear to be of essentially the same age, they probably are the result of multiple intrusions as is indicated by intrusive relationships exhibited among the various porphyry types. The main mass of the intrusive at depth probably was localized by a major east-west trending fault in the highly folded Paleozoic formations, but the individual stocks probably were guided into place by north-trending tension faults developed after the termination of the compressional forces. Thus, the apparent north-south fault control of the monzonites at the surface probably is of far less structural importance than the older east-west faulting through which igneous magma almost certainly entered the area.

Because it is almost certain that the mineralization and the alteration that preceded it occurred close in time to the Laramide igneous intrusions, no choice remains but to date the mineralization as of that time. The Ely ore deposits are, therefore, categorized as late Mesozoic to early Tertiary in age.

The primary mineralization in the Ely district followed shortly after the hydrothermal alteration; this alteration affected both sedimentary and igneous rocks. In the sedimentary rocks, the alteration caused bleaching and recrystallization of the limestone and the introduction of notable amounts of silicate minerals. Of these silicates, tremolite needles were the first to develop, followed by garnet, diopside, idocrase, chlorite, and epidote in the more intensely metamorphosed zones. In the limestone immediately adjacent to, or included in, the monzonite, a dense skarn was formed, in which garnet and chlorite were the principal silicates and to which specularite, magnetite, pyrite, and chalcopyrite were later added.

In the shales the intrusions converted the rocks to hornfels, bleaching them, destroying their original fissile character, and changing them to rocks resembling blocky mudstones. In the more strongly altered shales, sericite is common and silicification is often so intense that the rocks are difficult to distinguish from altered monzonite. In the sandstones the alteration consists of silicification and some sericitization of the minor shaly material between the sand grains.

The general order of mineral development in the altered monzonites was the conversion of hornblende to biotite and of plagioclase to clay minerals, the alteration of some of these clay minerals to sericite, the development of sericite directly from some plagioclase and from potash feldspar, and the introduction of secondary quartz in appreciable amounts as the youngest alteration mineral.

As is true of many other porphyry copper deposits where both igneous and sedimentary rocks are involved in the alteration zone, the minerals developed in the limestones appear to indicate a higher intensity of alteration than do the minerals formed in the shales or the porphyries. How these anomalous relationships are to be explained is, as yet, uncertain. It seems unlikely that higher intensity solutions could have reached only the limestones and not shales or sandstones or the monzonite itself. At the moment the most reasonable answer appears to be that sericite and the clay minerals, or sericite at least, can form not only in the mesothermal but also in the hypothermal range. Probably, therefore, the sericite (at least in part) in the noncalcareous rocks and the silicates in the limestones formed at temperatures within the hypothermal ranges, as did the magnetite and specularite in these limestone skarns; the clay minerals, however, may have formed under mesothermal conditions. The late quartz probably formed under mesothermal conditions.

Although no definite correlation exists between ore mineralization and type of alteration, chalcopyrite appears to have been formed in greater abundance in the probably more porous clay-bearing monzonite than in that converted to quartz and sericite. On the other hand, the quartz-sericite-bearing monzonite is more thoroughly enriched with chalcocite than the clay-rich portions of that rock.

The argillic and quartz-sericite portions of the altered porphyries are cut by quartz stringers and locally are intensely silicified; hydrothermal biotite was strongly developed in portions of the altered monzonite but is not widely distributed. After the clay and mica alterations, pyrite was developed throughout the altered porphyry in considerable and quite uniform amounts, while chalcopyrite was introduced most irregularly at a slightly later time. Pyrite is present as veinlets, as biebs in quartz veinlets, and as disseminated grains; chalcopyrite occurs as blebs in quartz veinlets and as disseminated grains but seldom is found in veinlets composed of that mineral alone. The richest primary copper ore, in argillic-type monzonite, may contain as much as 1.5 percent copper. The altering and ore-depositing solutions that moved through the thoroughly broken and shattered altered porphyries appear to have deposited molybdenite on the fracture surfaces well before the deposition of chalcopyrite; the mineral character and distribution of the minor amounts of gold and silver is, as yet, uncertain.

Above the water table the sulfide minerals in the ore bodies in monzonite have been oxidized and leached of most of their copper, generally to depths of 100 to 300 feet and locally to appreciably greater depths. Beneath the water table much of the copper has been redeposited, mainly as chalcocite that irregularly replaced grains of pyrite and chalcopyrite.

Ore in sediments adjacent to mineralized monzonite is usually of lower grade than that in the altered porphyry, and the total amount of ore recovered from sedimentary rocks is small in comparison with that obtained from the monzonite. Those portions of the Ely limestone converted to skarn provide the best sedimentary loci of ore deposition, but some ore has come from shales and a little from sandstones. Generally supergene enrichment is lacking in sedimentary rocks, although there were some important exceptions to the generalization.

The quartz-pyrite-chalcopyrite mineralization in the Ely district was definitely later than both the sericitic and argillic alteration in the monzonite and the skarn development in the limestone. Thus, although a possibility exists that the pre-ore alteration may have taken place in part under hypothermal conditions, it is probable that the chalcopyrite was deposited in the mesothermal range. The minor amounts of molybdenite may have been formed under hypothermal conditions, but this is not certain. It appears, therefore, that the copper ores of the district should be classified as mesothermal.

Fournier (1967) believes that the chalcopyrite is distributed as grains in the groundmass with no trace of alignment along earlier cracks. This groundmass is made up of aplitic-textured K-feldspar and quartz that was formed, in his opinion, at the time that the rock originally crystallized and was not produced by a later hydrothermal alteration process. He considers that the formation of hydrous alteration minerals after feldspar and hornblende may effectively have kept the fugacity of water low so that only an attenuated gas phase remained in the pores of the rock. If this concept is correct, (1) the ores should be classed as magmatic-1a or magmatic-4 and (2) all other porphyry copper deposits should be examined to see if they too were formed in this manner. I still prefer the hydrothermal origin despite the apparent lack of fracturing, since Fournier's description of the textural relations of the chalcopyrite does not agree with earlier work.

EUREKA

Late Mesozoic to Early Tertiary	*Silver, Gold, Lead*	*Hypothermal-2 to Mesothermal*

Binyon, E. O., 1946, Exploration of the gold, silver, lead, and zinc properties, Eureka Corporation, Eureka County, Nevada: U.S. Bur. Mines R.I. 3949, 18 p. (mimeo.)

Curtis, J. S., 1884, Silver-lead deposits of Eureka, Nevada: U.S. Geol. Surv. Mono. 7, 200 p.

Hague, A., 1892, Geology of the Eureka district, Nevada: U.S. Geol. Surv. Mono. 20, 419 p.

Ingalls, W. R., 1907, The silver-lead mines of Eureka, Nevada: Eng. and Min. Jour., v. 84, no. 23, p. 1051-1058

Miesch, A. T., and Nolan, T. B., 1958, Geochemical prospecting studies in the Bullwhacker mine area, Eureka district, Nevada: U.S. Geol. Surv. Bull. 1000-H, p. 397-408

Nolan, T. B., 1962, The Eureka mining district, Nevada: U.S. Geol. Surv. Prof. Paper 406, 78 p.

Nolan, T. B., and Hunt, R. N., 1968, The Eureka mining district, Nevada, in Ridge, J. D., Editor, *Ore deposits of the United States, 1933-1967* (Graton-Sales Volumes): Chap. 48, v. 1

Nolan, T. B., and others, 1956, The stratigraphic section in the vicinity of Eureka, Nevada: U.S. Geol. Surv. Prof. Paper 276, 77 p.

Roberts, R. J., and others, 1958, Paleozoic rocks of north-central Nevada: Amer. Assoc. Petrol. Geol. Bull., v. 42, p. 2813-2857

Sharp, W., 1948, The story of Eureka: A.I.M.E. Tr., v. 178, p. 206-217

Vanderburg, W. O., 1938, Reconnaissance of mining districts in Eureka County, Nevada: U.S. Bur. Mines I.C. 7022, 66 p. (mimeo.)

Wheeler, H. E., and Lemmon, D. M., 1939, Cambrian formations of the Eureka and Pioche districts, Nevada: Univ. Nev. Bull., v. 33, no. 3, 60 p. (Geol. and Min. Ser. no. 31)

Notes

The Eureka district is located in east-central Nevada, 60 miles west-northwest of Ely and 205 miles east of Reno.

The oldest rocks in the district are those of the early Cambrian Prospect Mountain quartzite composed of fractured quartzite, the base of which is not exposed. It is conformably overlain by the early Cambrian Pioche shale, a micaceous shale with some interbedded sandstone and limestone. The Pioche is conformably followed by the middle Cambrian Eldorado dolomite that is the economically more important of the two formations that are the host rocks of the Eureka ores. The Eldorado is a massive, thick-bedded dolomite; it contains some limestone beds near its base. The Eldorado is, in turn, conformably overlain by the middle Cambrian Geddes limestone of dark, carbonaceous, fine-grained limestone with, in places, small amounts of nodular black chert.

Conformably over the Geddes is the middle Cambrian Secret Canyon shale that is divided into the Lower Shale member and the Clarks Spring member, the former a fissile shale and the latter a thin-bedded, platy, and silty limestone with shale partings. The Secret Canyon lies conformably beneath the Middle and Late Cambrian Hamburg dolomite, a massively bedded dolomite with some limestone at its base; the Hamburg is the other important ore-bearing formation in the district, and its distribution is somewhat similar to that of the other ore-bearing formation, the Eldorado. The Hamburg rests unconformably beneath the Late Cambrian Dunderberg shale with interbedded nodular limestone. The Dunderberg is conformably overlain by the Late Cambrian Windfall formation that is composed of the Catlin member (massive limestone interbedded with thin sandy limestone--the massive beds are somewhat cherty) and the Bullwhacker member (thin-bedded, sandy limestone). The Cambrian rocks are followed by a considerable sequence of Ordovician rocks, followed by a considerable hiatus and then Middle and Late Devonian limestones; a second hiatus is followed by late Mississippian shale, conglomerate, limestone, and sandstone; a third hiatus is followed by Permian beds; then comes another long break and the strictly sedimentary sequence ends with an Early Cretaceous formation. Some Miocene rhyolite is known in the district, as is some Quaternary alluvium.

The first intrusive rock to be introduced into the area was a small plug of quartz diorite, the center of which is located about 1500 feet south of the crest of Ruby Hill. The quartz diorite has considerably altered the surrounding rocks. The alteration was much more intense within a few hundred feet of the quartz diorite than farther away; some contact metamorphic minerals such as manganese epidote, garnet, diopside, and serpentine were formed in the Hamburg dolomite near the igneous mass as well as appreciable quantities of magnetite, pyrrhotite, and some pyrite and chalcopyrite. No minable ore deposits, however, are known in the quartz diorite or in the altered zone immediately around it. A radioactive age determination by the lead-alpha method (now highly suspect) points to an age of 62 m.y. ± 12 m.y., which would place the quartz diorite in the Eocene.

The next intrusive is a quartz porphyry that is found in patches east and north of Adams Hill (north of Ruby Hill). The main mass of quartz porphyry is a sill-like body that extends southward from Mineral Point towards Adams Hill. The quartz porphyry is highly altered, but the surrounding rocks are far less affected than those in contact with the quartz diorite. The quartz porphyry is essentially a granite porphyry. The quartz porphyry is probably of about the same age as the quartz diorite, but no radioactive age determinations on it are available.

The Tertiary rocks (a small fraction of which may be Quaternary) are of volcanic origin and are post-ore; they include a hornblende andesite that has both intrusive and extrusive aspects; a series of rhyolite flows, dikes, and plugs that lead-alpha radioactive age determinations suggest to be Oligocene; a rhyolite tuff (closely related to rhyolite extrusives and intrusives just mentioned) that indirect correlations indicate is Miocene; and andesite and basalt, largely flow rocks but also occurring as intrusive dikes and irregular masses that have been considerably eroded and probably are late Pliocene or early Pleistocene in age.

It appears that the Cretaceous Newark Canyon formation, although it shows gentle to close folding and often is cut by normal faults, was not involved in the thrust movements affecting the area. Thus, the bulk of the thrusting occurred before the lithification of the Newark Canyon. On the other hand, the quartz diorite has cut through the structural pattern and probably was introduced after the Newark Canyon formation had been lithified and after the structures had been almost, if not entirely, developed. As the ore-forming fluids appear to have come from the same general source as the quartz diorite (and probably as the quartz porphyry), these solutions are likely to have been introduced in Late Cretaceous time. The question to which there is no definite answer is: were the ore solutions introduced in pre-Laramide or Laramide time? From the general location of the deposits in Nevada, it appears more reasonable that the ores should be categorized as late Mesozoic to early Tertiary rather than late Mesozoic (pre-Laramide), but further work is necessary to settle this point.

The workable ore deposits in the Eureka district can be grouped in five clusters or belts; from north to south to northeast, these are located (1) on the north slope of Adams Hill, (2) around Ruby Hill (the economically most important area), (3) on Prospect Ridge south of Ruby Hill, (4) east and southeast of the third cluster in a linear belt, and (5) northeast of group (4) near the mouth of New York Canyon (a subordinate group). The concentrations of ore bodies in these five areas resulted mainly from the presence in them of one or more major structures and the occurrence there of favorable beds.

The primary ores were formed as replacement bodies in carbonate rocks and show the wide variety of shapes and sizes characteristic of such deposits. The five general types of deposits are: (1) irregular replacement deposits, (2) bedded replacement deposits, (3) fault-zone replacement deposits, (4) disseminated deposits, and (5) contact-metasomatic deposits. Most of the Eureka production has come from the irregular replacement type, the distinctive features of which are (1) nearly complete replacement of the host rock by ore minerals, (2) restriction to massive carbonate rocks, mainly dolomites, and (3) lack of relationship to bedding in the host rocks. Deposits of this type are found in each of the five geographic areas just mentioned and appear as (1) small pods, (2) gently to steeply dipping ore chimneys and pipes, and (3) flat, mantolike bodies; many are associated with open caves.

Because of the intensive and extensive oxidation of the primary ores at Eureka, the mineralogy of the primary ores is not well known nor has it been thoroughly studied. The primary ores appear to have been compact masses of pyrite, arsenopyrite, galena, and sphalerite; the appreciable to important amounts of silver and gold probably were associated mainly with galena and pyrite, respectively, although some of the gold may have occurred with the arsenopyrite. Small amounts of molybdenite must have been present. Pyrite was the most abundant sulfide and contained numerous diamond-shaped crystals of arsenopyrite.

The presence of arsenopyrite in considerable quantity in the primary ore strongly suggests that it was deposited, at least in part, under hypothermal conditions, but a portion of the zinc and lead sulfides may also have formed in the mesothermal range. The ores, where they are associated with high-temperature alteration silicates in the limestone, were emplaced later than the silicates and probably were formed under less intense conditions. It appears probable, on the available evidence, that the best classification that can be applied to the primary ores is hypothermal-2 to mesothermal. The primary ores may have been emplaced near enough to the then-existing surface that the possibility of their having formed in the xenothermal to kryptothermal ranges should be considered, and the rather narrow vertical range over which they are known might be thought to add weight to this suggestion. The primary ores, however, appear to have been developed with slow loss of heat and pressure, so the classification hypothermal-2 to mesothermal is here preferred.

In the oxidized ores, galena seems to have largely remained unchanged, although much lead was recovered from fine-grained oxidation products and occurred principally in such minerals as plumbojarosite [$PbFe_6(OH)_{12}(SO_4)_4$], mimetite [$Pb_5(AsO_4)_3Cl$], and cerussite; anglesite also probably was abundant. Less common lead minerals included wulfenite ($PbMoO_4$) and bindheimite [$Pb_2Sb_2O_6(O,OH)$]. Most of the oxidized ore contained little zinc, the zinc from the oxidized sphalerite apparently having been carried out of the general ore area; hemimorphite (calamine) [$Zn_4(OH)_2Si_2O_7 \cdot H_2O$] was the principal oxidized zinc mineral; some smithsonite was recovered as was cerargyrite (AgCl), while the gold probably was entirely in the native state, having been somewhat concentrated in the secondary ore by the removal of most of the pyrite. The occurrence of azurite and malachite in the oxidized ore indicates that there must have been some primary copper mineral, but it cannot have been present in any abundance. There seems to be no doubt but that the oxidized ores were formed by the action of circulating ground water on the primary ore and that these ores should be classified as ground water-B2.

GABBS

Late Mesozoic to Early Tertiary	*Magnesite, Brucite*	*Mesothermal (Magnesite), Hypothermal-2 (Brucite)*

Callaghan, E., 1933, Brucite deposit, Paradise range, Nevada--a preliminary report: Nev. Bur. Mines Bull. 14, 34 p.

Ferguson, H. G., and Muller, S. W., 1949, Structural geology of the Hawthorne and Tonopah quadrangles, Nevada: U.S. Geol. Surv. Prof. Paper 216, 53 p.

Kral, V. E., 1951, Mineral resources of Nye County, Nevada: Nev. Bur. Mines Bull. 50, p. 103-108

Martin, C., 1956, Structure and dolomitization in crystalline magnesite deposits, Paradise range, Nye County, Nevada (abs.): Geol. Soc. Amer. Bull., v. 67, p. 1774

—— 1960, Origin of crystalline magnesite deposits (abs.): Geol. Soc. Amer. Bull., v. 71, p. 1921-1922

Muller, S. W., and Ferguson, H. G., 1939, Mesozoic stratigraphy of the Hawthorne and Tonopah quadrangles, Nevada: Geol. Soc. Amer. Bull., v. 50, p. 1573-1624

Rubey, W. W., and Callaghan, E., 1936, Paradise range, in Hewett, D. F., and others, *Mineral resources of the region around Boulder Dam*: U.S. Geol. Surv. Bull. 871, p. 142-143

Schilling, J. H., 1968, The Gabbs magnesite-brucite deposit, Nye County, Nevada, in Ridge, J. D., Editor, *Ore deposits of the United States 1933-1967* (Graton-Sales Volumes): Chap. 77, v. 2

Vitaliano, C. J., and Callaghan, E., 1956, Geologic map of the Gabbs magnesite and brucite deposits, Nye County, Nevada: U.S. Geol. Surv. Mineral Inv. Field Study Map MF-35, 1:24,000

—— 1963, Geology of the Paradise Peak quadrangle, Nevada: U.S. Geol. Surv. Geol. Quad. Map GQ-250, 1:62,500

Vitaliano, C. J., and Taylor, L. A., 1970, Brucite-granodiorite metasomatism at Gabbs, Nevada: IMA-IAGOD Meetings 1970, Collected Abs., Tokyo-Kyoto, Paper 6-5, p. 154

Vitaliano, C. J., and others, 1957, Geology of Gabbs and vicinity, Nye County, Nevada: U.S. Geol. Surv. Mineral Inv. Field Study Map MF-52, 1:24,000

Notes

The Gabbs deposit is located about 120 miles east-southeast of Reno and about 70 miles north-northwest of Tonopah.

The sedimentary rocks in which the magnesite-brucite deposits are contained are regionally metamorphosed Triassic and Jurassic limestones, dolomites, shales, and volcanic materials that have been subdivided into the Excelsior(?), Luning, and Gabbs formations of Triassic age and the Sunrise and Dunlap of Jurassic; the Late Triassic Luning formation contains the ore bodies. A large variety of igneous rocks is known in the area; in order of decreasing age these are (1) andesite hornfels (dikes); (2) biotite granite (laccolith and small stocks); (3) aplitic granophyre (dikes); (4) aphanitic andesite (dikes); (5) biotite-hornblende granodiorite (large stock and apophyses); (6) aplite (dikes); (7) porphyritic andesite (dikes); (8) lamprophyre, dacite, and rhyolite (dikes). The granodiorite (5) in the stock is probably equivalent to the granodiorite in a remarkably similar sequence of igneous rocks in the Sand Springs range (some 25 miles northwest of Gabbs); this Sand Springs rock has a K/Ar age of 76 m.y. A granite similar to the granite (2) at Gabbs has an age of 80 m.y. The development of the magnesite (but not the brucite) Schilling (1968) thinks was caused by hydrothermal solutions that got their heat, at least, from the granite (2). Thus, the magnesite must be Late Cretaceous in age and is classified here as late Mesozoic to early Tertiary. The brucite was formed by contact-metasomatic processes where the magnesite was partly enclosed by the intrusion of granodiorite (5). Although slightly younger than the magnesite, the brucite was developed in the same general time span as the magnesium carbonate.

The commercial magnesite and brucite deposits are confined to an area of about 2 square miles, although magnesite occurrences are known over a much larger area. The magnesite appears to be randomly mixed with the dolomite; much of the magnesite is massive, and jointing is minor and irregular. The brucite is found mainly in two large bodies and several smaller ones which are more regular in shape and have sharper contacts than the magnesite ores. No obvious structural control of the magnesite ores has been noted; they are, however, definitely confined to the Luning formation, which has suggested to Martin (1956, 1960) that the magnesite was formed syngenetically with the carbonate rocks that contain it.

The deposits contain four types of dolomite: (1) sedimentary (diagenetic?) now metamorphosed, (2) recrystallized dolomite (formed by hydrothermal solutions), (3) post-brucite (hypogene), and (4) supergene. The first two are premagnesite; the latter two are postmagnesite. The bulk of the magnesite is much like the recrystallized (type 2) dolomite in appearence and shows intricate replacement textures with the dolomite; neither mineral has exsolved from the other. Some minor amounts of ("bone") magnesite were produced as dense nodules at the contact between granodiorite and dolomite or magnesite and between brucite and magnesite or dolomite; this magnesite is of the same age as the brucite. The brucite is definitely later than the "recrystallized" magnesite and (from its spatial relationship to granodiorite contacts) probably was produced at higher temperatures. This suggestion gains weight from the development of magnesium-bearing silicates in the deposit that are most abundant adjacent to the granodiorite stock and the granodiorite and granophyre dikes and are most common in brucite. Some of these silicates may have been produced by the regional metamorphism or by solutions from the granite, but most of them probably were formed at the same time as the brucite and by the same solutions.

A consensus appears to have been reached among the workers in the district that the magnesium for the production of magnesite came from the dedolomitization of dolomite at greater depths rather than having been directly introduced into hydrothermal solutions during their formation in a magma chamber. No dedolomitized beds, however, appear to have been seen in the district, either in outcrops or in drill core. Since the dolomite seems to have been formed with essentially no other minerals, it is difficult to assess the temperature-pressure range in which it may have formed. If the ore fluids were at least heated by the granite, which was some distance away, the ore fluids probably were in the mesothermal rather than the hypothermal range, and the magnesites are here classified as mesothermal. The brucite, on the other hand, probably formed at higher temperatures and is thought to be hypothermal in carbonate rocks (hypothermal-2).

GOLDFIELD

Middle Tertiary *Gold, Silver, Copper* *Epithermal*

Ferguson, H. G., 1929, The mining districts of Nevada: Econ. Geol., v. 24, p. 115-148 (general discussion of ore occurrences in the state)

—— 1949, A contribution to the published information on the geology and ore deposits of Goldfield, Nevada (rev.): Econ. Geol., v. 44, p. 455-457

Harvey, R. D., and Vitaliano, C. J., 1964, Wall-rock alteration in the Goldfield district, Nevada: Jour. Geol., v. 72, p. 564-579

Locke, A., 1912, The ore deposits of Goldfield (Nevada): Eng. and Min. Jour., v. 94, nos. 17, 18, p. 797-802, 843-849

Ransome, F. L., 1907, The association of alunite with gold in the Goldfield district, Nevada: Econ. Geol., v. 2, p. 667-692

—— 1910, Geology and ore-deposits of the Goldfield district, Nevada: Econ. Geol., v. 5, p. 301-311, 438-470

Ransome, F. L., and others, 1909, The geology and ore deposits of Goldfield, Nevada: U.S. Geol. Surv. Prof. Paper 66, 258 p. (particularly p. 27-31, 150-201)

Searls, F., Jr., 1948, A contribution to the published information on the geology and ore deposits of Goldfield, Nevada: Univ. Nev. Bull., v. 42, no. 5, 21 p. (Geol. and Min. Ser. no. 48)

Silberman, M. L., and Ashley, R. P., 1970, Age of ore deposition at Goldfield, Nevada, from potassium-argon dating of alunite: Econ. Geol., v. 65, p. 352-354

Tolman, C. F., and Ambrose, J. W., 1934, The rich ores of Goldfield, Nevada: Econ. Geol., v. 29, p. 255-279

Wilson, H.D.B., 1944, Geochemical studies of the epithermal deposits at Goldfield, Nevada: Econ. Geol., v. 39, p. 37-55

Wisser, E., 1960, Goldfield, Nevada, in *Relation of ore deposition to doming in the North American Cordillera*: Geol. Soc. Amer. Mem. 77, p. 46-52

Notes

The deposits of the Goldfield district are situated in south-central Nevada, 25 miles south of Tonopah and about 190 miles southeast of Reno.

The oldest rocks in the immediate mine area are shales that probably are of Cambrian age; this dating depends on uncertain correlations with Cambrian rocks that encircle the district. The oldest igneous rocks of the district intrude the shale and are alaskite and granite, the transition between alaskite and granite being a gradational one. These rocks are younger than the Cambrian shales and older than the Tertiary volcanic rocks; most probably they are of late Mesozoic age, but it is uncertain whether they formed in pre-Laramide or Laramide time.

The oldest of the Tertiary volcanic rocks is the highly altered Vindicator rhyolite that rests on the eroded surface of Cambrian shale and Cretaceous granite and alaskite; the basal portion of the rhyolite includes a large number of inclusions of the older rocks. The rhyolite was followed by an unnamed latite, also highly altered; the alteration makes it difficult to distinguish the latite from both later andesite and dacite. It underlies a large part of the district but is exposed only where extensive deformation and erosion have removed the overlying rocks. The latite was followed by the Kendall tuff that contains some fragments of alaskite, rhyolite, and latite. The tuff probably derived from the same source as the almost contemporaneous Sandstorm rhyolite, the latite fragments having been acquired when the rhyolite broke through the latite with explosive violence. The tuff appears to have been collected in isolated basins at more than one period during the eruptions of the Sandstorm rhyolite. The Sandstorm rhyolite overlies the latite and most of the Kendall tuff; it shows conspicuous flow laminations. The Morena rhyolite is intrusive into the Sandstorm rhyolite and older rocks; it ordinarily shows no banding such as characterizes the Sandstorm. Apparently a period of erosion intervened between the introduction of the Morena rhyolite and that of the Milltown andesite; the latter probably occupies more of the surface than any other rock of the geologic column, only the older unnamed latite and the younger unnamed dacite approaching it in outcrop area. The Milltown overlies most of the older formations, particularly the Sandstorm and Morena rhyolites, mainly as gently inclined flows. The next volcanic formation in Ransome's sequence was the unnamed dacite that contains most of the minable lodes in the district; it is, at least in part, intrusive into the Milltown andesite. Searls (1948) believes that most of the dacite is of the same age as the Milltown andesite. A vitrophyric dacite is of the same general age as the dacite, and probably it is a phase of that eruptive; interbedded with the dacite vitrophyre is the Chispa andesite; both rocks are well removed from the mineralized area. Minor quantities of other volcanic rocks of no importance in the ore area were developed after the dacite vitrophyre and before the deposition of the Siebert lake beds. These lake beds (or tuffs) contain intercalated coarse gravel and boulder horizons that include water-worn fragments of vein material that replaced earlier Miocene lavas; thus the lake beds are definitely post-ore. The lake beds have been established as upper Miocene, while the ores were emplaced in early Miocene lavas; this would date the formation of the ores as about middle Miocene or middle Tertiary, and they are so classified here.

The dominant structure of the ore-bearing portion of the district is a dome about 6 miles in diameter, a dome that Wisser (1960) points out is somewhat flatter than those at Rico and La Plata; there are satellitic domes on the southwest flank of the main dome. The Columbia Mountain (CM) fault follows the structural contours of the dome in its northwest quadrant, the fault dipping inward toward the apex of the dome. This fault completes a belt of intense minor fracturing (and alteration) that almost encircles the dome; such minor fracturing, however, is almost lacking in the portion of the arc occupied by the CM fault. In the productive southwest side of the dome what are minor fractures at or near the surface coalesce at depth into a few strong, concentric fractures that dip, as does the Columbia Mountain fault, toward the apex of the dome. The Goldfield Consolidated Main (GCM) vein is an example of such a coalescing concentric fracture. Wisser (1960) believes that the Columbia fault encircles the dome completely but reaches the surface in its ring fault-type of structure only in the northwest quadrant. Wisser thinks that the uplifting force was not centered under the apex of the dome but was so distributed throughout the domed area that the hanging wall block on the CM fault dropped in a normal fault manner rather than rose as a thrust.

Structurally most of the mineralized bodies in the Goldfield district are irregular masses of altered and mineralized rock traversed by large numbers of small, irregularly intersecting fractures, the fracturing often being so severe as to be better classed as brecciation. The ore bodies (ledges in Goldfield terminology) can hardly be called veins for they do not possess the tabular character normally expected of veins nor do they generally have any definite trend or strike; dips at all angles are known, though most ledges appear to have moderate dips. These ledges are mainly composed of silicified or otherwise altered rock, and since they are more resistant to erosion than the rocks that enclose them, they usually stand out above the surrounding terrain. The main belt of mineralized (but not necessarily ore-bearing) ledges extends from south of Columbia Mountain (the area where the CM fault first changes direction) southeasterly (passing to the east of Goldfield) and then turning east through Preble Mountain as far as, or even beyond, Blackcap Mountain. Ninety-five percent of the gold has come from a few ledges in the rock volumes between Milltown and the south end of Columbia Mountain, of which the GCM vein is by far the most important. The ledges are almost entirely confined to two rocks--dacite and Milltown andesite--and where the ledges encounter the pre-andesite or pre-dacite basement complex, they either die out or follow the volcanic rock-basement contact.

In some of the ledges, the whole mass of altered rock contains ore, but usually only a small part of the entire ledge is minable ore. The shape of the ore bodies within the ledges is seldom predictable for the ore shoots show no consistent relationship to the shape of the ledge, and the ledge itself normally follows no distinct and definite pattern within the enclosing rocks; the boundaries between ore and barren ledge are visible only when the assayer's results are available.

The typical primary ore as it occurs in the dacite or Milltown andesite, particularly in the mines south of the Mohawk, consists of pyrite, bismuthinite, and famatinite(?) containing some arsenic, in a dark, flinty quartz gangue. Native gold is generally associated with bismuthinite and famatinite but may be emplaced in the quartz; the gold probably is late in the sequence. The only published work based on the study of polished specimens is that of Tolman and Ambrose (1934) in which they establish a paragenetic sequence in which the mineralization begins with the alteration of the country rock (generally dacite, but sometimes andesite) to a mixture of silica, kaolinite, and alunite. Before this alteration was complete, the first metallic minerals, pyrite and marcasite, began to deposit. Shortly after the first iron sulfides were emplaced, the deposition of famatinite and then tetrahedrite, sphalerite, and wurtzite followed. Toward the end of this stage of deposition, bismuthinite began to deposit; it was followed by minor amounts of the suspect species, goldfieldite (possibly $Cu_{12}Sb_4Te_3S_{16}$), then by two minor minerals, then by gold and silver tellurides in minute amounts, and finally by gold with some silver. Although the specimens studied by Tolman and Ambrose did not include samples from all ore ledges in the district, their results probably are generally applicable to most, if not all, of the ore mineralization. Famatinite was the most abundant mineral in Tolman's and Ambrose's specimens; the amount of tennantite associated with it is small. Bismuthinite is typically emplaced by replacement of famatinite and is usually present in rich ore masses. Famatinite and bismuthinite are cut by veinlets of gold and, less commonly, of goldfieldite; the economically unimportant gold tellurides, mainly hessite, petzite, and sylvanite, have selectively replaced bismuthinite. Beyond the bismuthinite boundaries, veinlets that contained goldfieldite in the bismuth mineral are filled with gold. This certainly suggests a definite control of bismuthinite over telluride precipitation. Gold is present in the district in two varieties, distinguished from each other by their colors. A yellow type of gold is present only in the Mohawk mine while the other, a reddish gold, is characteristic of all other ore bodies. The yellow gold contains some silver, the red gold apparently has little or none; the red color may in part be due to inclusions of famatinite, but some gold that lacks famatinite also is red.

The Goldfield ores probably were deposited in a near-surface environment, and the mineral suite, typical of low intensity conditions, almost certainly indicates deposition in the epithermal range. The Goldfield ores, therefore, are here classified as epithermal.

At depth (immediately above the Cambrian shale basement) the ore in the Grizzly Bear mine was of exceedingly high grade, not only in gold but in silver and copper as well; about 22,500 tons of this ore averaged 1.07 ounces of gold, 5.58 ounces of silver, and 4.21 percent copper. This is one of the few known examples of epithermal ore being of sufficient base metal content that not only were the base metals recovered but also that the ore was smelted as a base-metal ore. Perhaps this remarkable abundance of famatinite should be taken as an indication that the conditions under which it was deposited approached, if they had not reached, the kryptothermal.

GOODSPRINGS

Early Tertiary | *Zinc, Lead, Silver* | *Mesothermal*

Albritton, C. C., Jr., and others, 1954, Geologic controls of lead and zinc deposits in Goodsprings (Yellow Pine) district, Nevada: U.S. Geol. Surv. Bull. 1010, 111 p.

Barton, P. B., Jr., 1956, Fixation of uranium in the oxidized ores of the Goodsprings district, Clark County, Nevada: Econ. Geol., v. 51, p. 178-191

Geehan, R. W., and Benson, W. T., 1949, Investigation of the Yellow Pine zinc-lead mine, Clark County, Nevada: U.S. Bur. Mines R.I. 4613, 15 p. (mimeo.)

Hale, F. A., Jr., 1918, Ore deposits of the Yellow Pine mining district, Clark County, Nevada: A.I.M.E. Tr., v. 59, p. 93-111

Hazard, J. C., and Mason, J. F., 1953, The Goodsprings dolomite at Goodsprings, Nevada: Amer. Jour. Sci., v. 251, no. 9, p. 643-655

Hewett, D. F., 1931, Geology and ore deposits of the Goodsprings quadrangle, Nevada: U.S. Geol. Surv. Prof. Paper 162, 172 p.

—— 1956, Geology and mineral resources of the Ivanpah quadrangle, California and Nevada: U.S. Geol. Surv. Prof. Paper 275, 172 p.

Hill, J. M., 1914, The Yellow Pine mining district, Clark County, Nevada: U.S. Geol. Surv. Bull. 540-f, p. 223-274

Knopf, A., 1915, A gold-platinum-palladium lode in southern Nevada: U.S. Geol. Surv. Bull. 620-a, p. 1-18

Longwell, C. R., 1926, Structural studies in southern Nevada and western Arizona: Geol. Soc. Amer. Bull., v. 37, p. 551-584

Notes

The bulk of the mineralization of the Goodsprings district lies along a belt, some 16 miles long, that stretches along the east slopes of Spring Mountains, from the Potosi mine about 10 miles north-northwest of the town of Goodsprings to the Valentine mine about 5.5 miles south of the town. Considerable concentration of mineralization is known west and southwest of the town, but only one prominent mine (the Root or Bonanza mine) lies in that area. Goodsprings itself is about 27.5 miles southwest of Las Vegas.

The oldest rocks that outcrop in the area are thin-bedded dolomites from the lower portion of the Devonian Goodsprings dolomite. Conformably on the Goodsprings is the Devonian Sultan limestone that consists of three members; the Sultan outcrops occupy a much smaller area than do those of the Goodsprings. The Sultan is overlain conformably by the Mississippian Monte Cristo limestone, and the areal distribution of the Monte Cristo coincides closely with that of the Sultan. The Monte Cristo is divided into five members. The Yellow Pine limestone member (the youngest) is completely altered to dolomite; it has provided about 85 percent of the ore mined in the district. The Monte Cristo is unconformably overlain by the Pennsylvanian Bird Spring formation that consists of limestone and sandstone; in the northern part of the area, the formation has a basal conglomerate. Two Permian formations, the Supai (shaly sandstone below, then reddish sandstone, then gypsum-bearing, shaly sandstone) and the Kaibab limestone are known in the eastern, unmineralized portion of the district. The Lower Triassic is represented by the Moenkopi formation that rests unconformably on the Kaibab; these rocks are exposed along the eastern margin of the mineralized area but are not themselves mineralized. The Shinarump conglomerate and the Chinle formation make up the Upper Triassic; they also are exposed to the east of the mineralized area. The Jurassic consists of the Aztec sandstone; it is unmineralized and of small areal extent; except for Pleistocene gravels and Recent alluvium, the Aztec is the youngest sedimentary material in the district.

The first igneous rocks to be introduced are divided into two types: (1) a coarse-grained granite porphyry that forms a number of dikes and sills in the central part of the district; the total area of these rocks is probably not more than several hundred acres, most of which is concentrated in the vicinity of the Yellow Pine mine, and (2) a dark rock of finer grain than the porphyry that is found as dikes in only three mines; it probably should be classed as a lamprophyre. The next igneous rocks are of middle Tertiary, probably Miocene, age and include several varieties that occur as volcanic necks, dikes, and surface flows and bedded tuffs; they range in composition from latite through andesite to basalt and probably occupy no more than 3 square miles in an area that is adjacent to, but does not overlap, that of the coarse-grained igneous rocks. These fine-grained rocks are not only much younger than the coarse-grained but were emplaced long after the ore deposits had formed.

The faulting, early igneous activity, and ore mineralization certainly took place in that order after the Jurassic Aztec formation was lithified and before the Miocene (?) volcanic rocks were introduced into the area. It is possible, however, to place the mineralization and the igneous events that preceded it as Late Cretaceous to early Tertiary by analogy with other districts throughout the southwest in which similar igneous rocks have been intruded and similar ore deposits definitely were formed within that time range. Further, the overthrusting in Nevada appears to be a southern extension of a belt of overthrust faults that extends from northern Montana southeastward into western Wyoming and eastern Idaho and Utah; these more northerly thrusts have been dated as early to late Eocene. Thus, it would appear that the most probable age for the Goodsprings ores is early Tertiary.

In the Goodsprings district, the composition of the primary ores and the relationships of the ore and gangue minerals has been difficult to interpret because of the widespread oxidation, particularly of the sphalerite. Although some of the ore was emplaced as the fillings of cavities that range from a few inches to several feet across, most of the ore was deposited by the replacement of country rock in or near zones of fracturing and shearing. In most instances the host rock was dolomite, but limestone around the margins of some of the larger ore bodies has been somewhat replaced. Where the host rocks are flat-lying, the ore bodies generally are tabular and parallel the bedding; where the bedding dips at moderate to steep angles, the ore bodies are normally flattish pipes that range from a few tens of feet to a thousand feet long and have cross-sections that are between 100 and 3500 square feet in area. Many of these pipes largely follow the plane of the bedding, others cut across it at low angles, and a few of the ore bodies are localized by fractures that cut across the bedding at high angles. No regularity exists in the orientation of the axes of the pipes; these may be linear, curved, or sinuous and bear little relation to the directions of dip and strike of the beds in which they are enclosed.

The concentration of mines in the central part of the district can probably be explained by the peculiarities of the fault pattern. In the southern part of the Goodsprings quadrangle, the major feature of the structural pattern is the northwest-trending high-angle faults. This arrangement also obtains in the northwestern part of the district. Between these two areas, the district is broken by a network of faults, fractures trending both northeast (tears) and northwest (rifts) being prominent. This area includes Prairie Flower and Yellow Pine mines, and from it has come over 60 percent of the lead and zinc mined in the district as well as almost all the gold. Thus, the rock volumes that are most thoroughly broken are those that were most thoroughly mineralized.

The lead-zinc ores in the Goodsprings district are located within a narrow vertical range, being confined, with a few exceptions, to the Monte Cristo limestone.

Although four categories of ore deposits have been recognized in the Goodsprings district: (1) gold, (2) silver in excess over gold, (3) copper with some gold or platinum metals but little silver, and (4) lead and zinc with appreciable silver and some vanadium. Only type (4) was of real economic importance. The primary mineralogy appears to have been quite simple. Although little pyrite (or marcasite) as such has been found in the ore, considerable quantities of it were present in the lead-zinc ore bodies and even more in those containing chalcopyrite. The ore sulfides were undoubtedly silver-bearing galena and sphalerite, accompanied by minor quantities of stibnite and cinnabar. Chalcopyrite was not important in the lead-zinc ore bodies, but it was present in some abundance in the copper properties; thin films of bornite were found on some of the chalcopyrite, but it is uncertain whether the bornite was primary or secondary. In the gold ores, a little tennantite and prousite were found, but their relationships to the other primary minerals are uncertain. The primary source of the secondary vanadium minerals associated with the oxidized lead-zinc ores is unknown; the most common locus of deposition of vanadium in sulfides appears to be in enargite where VS_4^{-3} appears to proxy for AsS_4^{-3}.

The principal gangue minerals are dolomite and less abundant calcite with some chert, barite, and a little hydrocarbon; these minerals are less common than is usual for this type of lead-zinc deposit, particularly as much of the carbonate appears to be of secondary origin.

From the simple primary mineralogy, it is difficult to classify the deposits with certainty. The actually known primary minerals probably could have been formed from the mesothermal to the telethermal range. The presence of considerable silver in the galena appears to rule out the telethermal range, and the notable amounts of vanadium in the ores points toward mesothermal rather than to leptothermal conditions of formation. On these bases, the primary deposits are here categorized as mesothermal.

In the secondary ores, the galena has remained largely unchanged by surface processes although a little anglesite and cerussite have been found. On the other hand, the sphalerite has been widely affected, little of it being found in the ores exposed in the deeper mines. The principal oxidized zinc minerals, which do not, from their quite close spatial relations to galena, appear to have been formed far from the source of the zinc, are hydrozincite $[(ZnCo_3)\cdot 2Zn(OH)_2]$ in abundance, hemimorphite (calamine) $[Zn_4(OH)_2Si_2O_7\cdot H_2O]$ in considerable amount, and smithsonite $(ZnCo_3)$ in small quantities. The most common vanadates are cuprodescloizite, descloizite, and vanadinite, but not enough work has been done to make certain which of these is the most important vanadium mineral. These minerals, and the numerous other secondary minerals, probably were formed through the action of ground water on the primary ores; therefore, the secondary ores should be classified as ground water-B2.

MOUNTAIN CITY

Late Paleozoic *Copper* *Mesothermal*

Axelrod, D. I., 1966, The Eocene copper basin flora of northwestern Nevada: Calif. Univ. Pubs. Geol. Sci., v. 59, 125 p.

Coats, R. R., 1968, Upper Paleozoic formations of the Mountain City area, Elko County, Nevada: U.S. Geol. Surv. Bull. 1274-A, p. A22-A27

Coats, R. R., and Stephens, E. C., 1968, Mountain City copper mine, Elko County, Nevada, in Ridge, J. D., Editor, *Ore deposits of the United States 1933-1967* (Graton-Sales Volumes): Chap. 52, v. 2

Coats, R. R., and others, 1965, Reconnaissance of mineral ages of plutons in Elko County, Nevada, and vicinity: U.S. Geol. Surv. Prof. Paper 525-D, p. D11-D15

Crawford, A. L., and Frobes, D. C., 1932, Microscopic characteristics of the Rio Tinto, Nevada, copper deposit: Mines Mag., v. 22, no. 8, p. 7-9

Emmons, W. H., 1910, A reconnaissance of some mining camps in Elko, Lander, and Eureka Counties, Nevada: U.S. Geol. Surv. Bull 408, 130 p.

Granger, A. E., and others, 1957, Geology and mineral resources of Elko County, Nevada: Nev. Bur. Mines Bull. 54, 190 p.

Roberts, R. J., and others, 1958, Paleozoic rocks of north-central Nevada: Amer. Assoc. Petrol. Geols. Bull., v. 42, p. 2813-2857

Notes

The Mountain City copper deposit is located some 75 miles slightly west of north from Elko and only about 10 miles south of the Idaho line.

The rocks of the district consist of the Ordovician Valmy (or Rio Tinto) formation and several upper Paleozoic units that lie unconformably on the Valmy. The Valmy belongs to Roberts' (1958) detrital volcanic (western) assemblage of eugeosynclinal rocks; the other two assemblages are: (1) the carbonate (eastern) miogeosynclinal and (2) transitional intermediate in character between the eastern and western. The upper Paleozoic rocks consist of clastics and largely intermediate to mafic volcanic rocks that range in age from Devonian or Mississippian to Pennsylvanian or Permian. None of the formations in the sequence resembles either the Precambrian or Triassic rocks of the general area. Within this series of formations, the probably Pennsylvanian, certainly Carboniferous, Nelson formation (originally designated as an amphibolite) consists largely of flows and tuff breccias of andesitic composition, with minor sills of diabase and one lens of rhyolitic tuff. It is a greenschist in the mine area. This formation is of particular importance since Coats and Stephens (1968) believe that the ore fluids came from the same magmatic source as the mafic volcanic materials of which this formation is composed. If their reasoning is correct, then the ores are late Paleozoic in age. They base their decision on comparison of the deposit with a number of others from various parts of the world such as those of the Huelva province; Shasta County and the Foothill belt in California; several in the Urals, of which Pychmisko-Klutchevsky is most like Mountain City; Cyprus; and most of the massive copper deposits of Scandinavia and Canada; Ergani Maden in Turkey; and Rammelsburg and others of that type. Just what the reasoning is that justifies this conclusion is not clear. If the deposits are related in time and source to this mafic volcanism, then this source magma underlies and presumably is younger than the regionally metamorphosed Mountain City and Reservation Hill formations. Further, Coats and Stephens argue that the regional metamorphism of the area is entirely pre-ore. Thus, the ore can hardly have been introduced until after the Reservation Hill was lithified and metamorphosed, presumably appreciably after the mafic volcanism that produced the Nelson formation. Despite the stratabound character of the ores, Coats and Stephens believe that the texture of the ores, the structure of the ore body, and the relationship of the ore to structures in the country rock all indicate that the ores are epigenetic; this would seem to be confirmed by the presence of an envelope of wall-rock alteration around the ore that is most intense near the sulfide lenses and gradually dies out away from the ore. Granted the nonmetamorphosed character of the ore and its epigenetic characteristics, it must have been introduced after the metamorphism within the Paleozoic and before the extrusion of the late Eocene to early Pliocene volcanic rocks. Although a Late Cretaceous quartz monzonite pluton is known on the surface less than a mile from the mine, and the ore fluids may have come from the same source as this body, it seems more reasonable to assume that the ore fluids came from the magma chamber from which the andesite of the Pennsylvanian Nelson formation came. The ores, therefore, here are considered to be late Paleozoic in age.

The primary ore bodies of the district occur as disc-shaped lenses in a portion of the Valmy formation that consists of dark shales with a little interbedded quartzite. This horizon has a maximum thickness of 200 feet. Although the ore lenses commonly are parallel to the stratification, they cut across it in places. Many of the lenses are massive bodies of quartz, pyrite, and chalcopyrite; no great difference exists in sulfide ratios from one lens to another, and no significant change in ratios is found within any single lens. In addition to the massive ore type, Coats and Stephens also recognize (1) a dark-gray quartzite type in which silicification is erratic, chalcopyrite is more abundant than pyrite, and the sulfides cut the quartzite in random directions; (2) a dark-quartz type in which the lens has been completely silicified and the sulfides occur as streaks and blobs and only rarely are accompanied by later white quartz; (3) banded white-quartz type in which the ore is in distinct layers, with the attitude of the layers seldom parallel to the attitude of the lens; later this quartz was crackled and glassy quartz, pyrite, and chalcopyrite were emplaced in that order.

The primary mineralization is of so simple a character that it might have been produced in any intensity range. The wall-rock alteration, however, consists of chlorite and clay minerals and suggests that the ores probably were formed in the mesothermal range.

The deposit has been tremendously enriched by secondary processes. The massive secondary ore has largely removed the primary ore textures, replacing both chalcopyrite (first) and pyrite (later) with sooty and massive secondary chalcocite. The oxidized zone above the then-existing water table seems to have been almost completely leached of copper, so the gossan gave essentially no clue to the high-grade ore below. The enrichment would, of course, be classified as ground water-B2.

OREANA

Late Mesozoic to Early Tertiary | *Tungsten* | *Magmatic-3a*

Cameron, E. N., and others, 1949, Internal structure of granitic pegmatites: Econ. Geol. Mono. 2, 115 p.

Hess, F. L., 1933, Pegmatites: Econ. Geol., v. 28, p. 447-462

Jenney, C. P., 1935, Geology of the central Humboldt range, Nevada: Univ. Nev. Bull., v. 29, no. 6, 73 p.

Kerr, P. F., 1938, Tungsten mineralization at Oreana, Nevada: Econ. Geol., v. 23, p. 390-427

—— 1946, Tungsten mineralization in the United States: Geol. Soc. Amer. Mem. 15, 241 p. (particularly p. 38-41, 189-192)

Landes, K. K., 1933, Origin and classification of pegmatites: Amer. Mineral., v. 18, p. 33-56, 95-103

Schaller, W. T., 1933, Pegmatites, in *Ore deposits of the western states* (Lindgren Volume): A.I.M.E., p. 144-151

Shand, S. J., 1947, The genesis of pegmatite, in *Eruptive rocks*, 3d. ed.: Wiley & Sons, Inc., N.Y., p. 178-189

Notes

The Oreana district is situated on the western slopes of the Humboldt range in northwestern Nevada, 95 miles northeast of Reno and about 55 miles southwest of Winnemucca.

The oldest rocks in the Oreana area are keratophyres and rhyolite tuffs, flows, and breccias that make up the major part of the formation originally known as the Koipato; the Koipato rocks range in age from probably Pennsylvanian to Middle Triassic. Although the formation has been subdivided by Knopf (U.S. Geol. Surv. Bull. 762), Koipato rocks do not occur in the immediate mine area, and the exact designation of its members is not of direct concern to a study of the Oreana deposit. The next rocks in order of decreasing age are those of the Star Peak formation, of Middle and Upper Triassic age. These late Paleozoic and Triassic volcanic and sedimentary rocks have been intruded by at least six different igneous materials between the Late Triassic or Early Jurassic and the late Mesozoic to early Tertiary. The first of these intrusives was a metadiorite (or metagabbro) that cuts through limestones in the Star Peak formation and is, therefore, at least late Upper Triassic or Early Jurassic in age. In the general Oreana area, the metadiorite follows the long, narrow belt of Star Peak limestone. The metadiorite may occur as a single layer several hundred feet thick or as numerous thin sheets more or less parallel to each other. It may owe its metamorphic characteristics to the post-Triassic deformation. Near the Oreana mine, the next intrusive to be introduced was the early aplite, dikes of which cut the Star Peak limestones. These dikes do not cut the metadiorite but are cut by a granite porphyry. Some early aplite dikes near the metadiorite have been broken into isolated and rounded segments, presumably during the metamorphism that converted the original diorite or gabbro into metadiorite. On the other hand, many of the aplite dikes are unbroken and unmetamorphosed. Two periods of aplite intrusion, therefore, may have taken place, the first occurring before the post-Triassic deformation and the other after it.

Some 5 miles southeast of the mine at Oreana, several granite porphyry dikes cut aplite; nearer the mine, similar granite porphyry has been cut by Rocky Canyon granite (or quartz monzonite). These relationships indicate that the age sequence was early aplite, granite porphyry, and Rocky Canyon granite. The Rocky Canyon granite occurs in a large body about 3.5 miles long by 2.5 miles wide located on the west slope of the Humboldt range that, on the surface, approaches within less than 1000 feet to the east of the Oreana mine. The granite caused considerable contact metamorphism of the rocks into which it was intruded. No geologic relationship exhibited by the granite dates it absolutely. Although it probably was emplaced in Laramide time, this is not certain, and the Rocky Canyon granite may have been introduced as late as middle Miocene time.

In the vicinity of the Oreana mine, dikes of pegmatite and of a second generation of aplite are found either in the metadiorite or along the contacts between metadiorite and limestone. The pegmatite dikes that occur along the limestone-metadiorite contacts are lenticular and follow the strike and dip of the contact diagonally along the contact plane.

The principal (east) pegmatite dike in the Oreana area, along which the best ores are found, has a length of about 2000 feet, although it has been displaced to a minor extent by cross faults; in its southern extension it divides into two essentially parallel dikes. Some of the less prominent pegmatite dikes cross into the limestone, but if they do, they terminate within a short distance.

Although the pegmatite dikes normally cut the aplites, the two dike-types are so closely connected in space as to suggest that they must have come from a common igneous source; this concept is supported by the presence of scheelite in both dike-types. The granite contains no scheelite, indicating that it came from a different source than the pegmatites and aplites or, if it came from the same source, that it left that source at an appreciably earlier time. No available information suggests how long the interval between granite and pegmatite-aplite intrusions may have been. It is probable that the dike materials were introduced as a late phase of the Laramide igneous activity to which it is thought that the granite must have belonged, but they may be as young as Miocene. On a balance of probabilities, it is thought best here to date the pegmatites and aplites as late Mesozoic to early Tertiary, but their age is not yet certainly determined; further work is necessary on this problem.

The scheelite ores in the Oreana area appear to have been formed as primary minerals of the pegmatites, although the abundance of calcium ion in the pegmatites may be related to their being emplaced close to, and possibly having passed through, the limestone of the Star Peak formation. The pegmatites are extremely irregular, not only in their content of scheelite but also in that of the other minerals of which they are composed. In the ore-bearing pegmatites in metadiorite, the most abundant minerals are beryl, oligoclase and albite feldspar (albite is appreciably less abundant than oligoclase), quartz, scheelite, and phlogopite; these minerals appear to have begun to crystallize in approximately this order. Of the minor minerals present, orthoclase and microcline seem to have formed during the time between the beginning of beryl deposition and that of plagioclase. Rutile, sphene, and apatite began to form after albite and before scheelite, while fluorite (locally quite abundant), muscovite, sericite, calcite, and chlorite followed phlogopite. In the limestone, the minor pegmatite bodies that penetrate that rock consist of orthoclase, microcline, albite, quartz, muscovite, and traces of scheelite. The pegmatites were responsible for some contact metamorphism of the limestones where they (the pegmatites) were emplaced adjacent to the carbonate rock. Such contact zones contain garnet, epidote, diopside, zoisite, clinozoisite, idocrase, and tremolite.

The principal (east) pegmatite changes mineral composition quite drastically over very short distances. In places it is no more than a solid quartz body about 5 feet thick that contains no scheelite but has some tourmaline as small black needles; the tourmaline may also be developed in the neighboring wall rock. In another section, the pegmatite is made up of almost solid fluorite about 4 feet wide. Many of the pegmatite minerals show that they have reacted with the still molten portions of the pegmatite mass and have been replaced, largely or in part, by minerals later in the paragenetic sequence. Examples of such replacements include tongues and veinlets of albite that invade oligoclase; phlogopite containing isolated grains of scheelite that fills fractures in oligoclase; veinlets of fluorite that cut through oligoclase; quartz that appears to surround and replace such minerals as beryl and feldspar; and broken crystals of tourmaline that are recemented by quartz. Probably the latest minerals developed in the Oreana pegmatites are the sulfides, pyrrhotite and pyrite, that fill fractures in the older minerals such as quartz, beryl, oligoclase, and phlogopite; a few grains of scheelite have been found enclosed in the sulfides.

The most common type of scheelite ore consisted of coarse, irregular crystals of scheelite in oligoclase feldspar with which streaks of phlogopite were associated. In places, beryl was present in the scheelite-oligoclase type of ore. In the pegmatite lenses along the metadiorite-limestone contacts, plagioclase feldspar was much less abundant and phlogopite much more so than in pegmatite in the metadiorite; some specimens found along the contacts contained intimate mixtures of fine-grained scheelite and phlogopite.

The Oreana scheelite ore bodies almost certainly are true magmatic pegmatites that are somewhat (though not uniquely) unusual in their mineral content and should be categorized here as magmatic-3a.

PIOCHE

Late Mesozoic to Early Tertiary	*Copper, Lead, Zinc, Silver, Gold*	*Mesothermal*

Anderson, J. C., 1922, Ore deposits of the Pioche district, Nevada: Eng. and Min. Jour., v. 113, no. 7, p. 279-285

Gemmill, P., 1968, The geology of the ore deposits of the Pioche district, Nevada, in Ridge, J. D., Editor, *Ore deposits of the United States 1933-1967*

(Graton-Sales Volumes): Chap. 54, v. 2

Gillson, J. L., 1929, Petrography of the Pioche district, Lincoln County, Nevada: U.S. Geol. Surv. Prof. Paper 158-D, p. 77-86

Hill, J. M., 1919, Bristol district, in *Notes on some mining districts in eastern Nevada*: U.S. Geol. Surv. Bull. 648, p. 124-137

Merriam, C. W., 1964, Cambrian rocks of the Pioche mining district, Nevada: U.S. Geol. Surv. Prof. Paper 469, 59 p.

Park, C. F., Jr., and others, 1958, Geologic map and sections of the Pioche Hills, Lincoln County, Nevada: U.S. Geol. Surv. Mineral Invest., Field Studies Map MF-136, 1:12,000 (no text)

Tschanz, C. M., 1960, Geology of northern Lincoln County, Nevada: Intermountain Assoc. Petrol. Geols., 11th Ann. Field Conf. Guidebook, p. 198-208

Westgate, L. G., and Knopf, A., 1927, Geology of Pioche, Nevada, and vicinity: A.I.M.E. Tr., v. 75, p. 816-836

—— 1932, Geology and ore deposits of the Pioche district, Nevada: U.S. Geol. Surv. Prof. Paper 171, 79 p. (particularly p. 6-53)

Wheeler, H. E., 1940, Revisions in the Cambrian stratigraphy of the Pioche district, Nevada: Univ. Nev. Bull., v. 34, no. 8, 40 p. (Geol. and Min. Ser. no. 34)

Young, E. B., 1950, The Pioche district: 18th Int. Geol. Cong. Rept., pt. 7, p. 111-120

Notes

The Pioche district is situated about 125 miles north-northeast of Las Vegas and some 155 miles slightly south of east of Tonopah; it is only about 20 miles west of the Utah border. The Bristol district is some 15 miles northwest of the Pioche district proper and is geologically and mineralogically similar to it.

The oldest rocks in the area are probably the Early Cambrian Prospect Mountain quartzite, the base of which is unexposed in the area; the Prospect Mountain grades upward into the Early to Middle Cambrian Pioche shale; the Pioche is economically the most important rock in the district, most of the ore being emplaced in the lowest of its limestone members. The Pioche is divided into six conformable members; at least 90 percent of the ore produced in the Pioche district has come from the Combined Metals (CM) member, the second oldest member. The Pioche is overlain, apparently unconformably by the Middle Cambrian Lyndon limestone that is a massive, cliff-making, fine-grained or porcelaneous limestone; it is the host rock of some replacement ore bodies in the Ely Valley and Prince mines, some 3 miles northeast and southwest, respectively, of the town of Pioche. The Lyndon passes, probably abruptly, but conformably, into the Chisholm shale. In the vicinity of Mount Ely, some small, now-oxidized replacement deposits have been found in the limestone members of the Chisholm shale. Probably conformably above the Chisholm shale is the Highland Peak formation; the Highland Peak is divided into two parts, the lower of which contains six named members, all of which are found in the Pioche and Bristol districts. The upper part of the Highland Peak formation is broken down into seven units, numbered sequentially with those of the lower part of the formation, not all of which have as yet been found to be mappable. Above the Highland Peak lies the probably conformable 2000 feet of Mendha formation (both Late Cambrian and Early Ordovician).

In the Pioche district, Paleozoic rocks younger than the Mendha are known, but these, although they all were formed before the ores were emplaced, have essentially no spatial, and certainly no genetic, connection with the ore-forming process.

The igneous rocks of the area include intrusive stocks of quartz monzonite that outcrop in an area about 2.5 miles northwest to southeast and 1.0 miles wide just southwest of Blind Mountain and about 10 miles northwest of Pioche town. Although contact metamorphism or high-temperature hydrothermal alteration was most impressively developed around the Blind Mountain quartz-monzonite body, other areas were also affected, suggesting that the Blind Mountain quartz monzonite is a cupola on a much larger batholithic mass deeper beneath the surface. The solutions that cause this alteration do not seem to have been those that introduced the ores since the spatial patterns of the two phenomena are quite different. Aplite dikes are associated with the quartz monzonite, and narrow pegmatite stringers occur in some of the quartz monzonite. North and east of the quartz-monzonite masses are dikes of darker, finer-grained porphyritic rocks that differ appreciably from one occurrence to the next in mineral composition but that generally resemble the quartz monzonites. Somewhat later than the quartz monzonites are a number of granite-porphyry dikes (such as the Yuba dike), and a few lamprophyre dikes have cut through the limestones, particularly in Lyndon Gulch. Two diabase dikes are known that probably are younger than the granite porphyry dikes.

Many square miles of the surface of the Pioche district are covered by lavas and tuffs that are certainly post-Paleozoic and pre-Pliocene. Many of the lavas in the district were involved in thrust faulting; thus, much of the extrusive rock was emplaced before the thrusting. This thrust faulting, by analogy with other districts in Nevada, probably took place in the late Cretaceous and early Tertiary. It appears most probable that some of the lavas were emplaced in the late Mesozoic (before the thrust faulting); that the quartz monzonite was intruded after the thrusting but probably was no younger than early Eocene; and that further lava flows, perhaps of Miocene age, were erupted after the quartz monzonites were introduced. The ore-forming solutions probably were derived from the same source as the quartz monzonite, which probably, in general, was that of the lavas as well; the ore fluids, therefore, probably deposited the ore bodies during the early Tertiary. The ore deposits, on this reasoning, are here categorized as late Mesozoic to early Tertiary.

The district is one in which the ore deposits appear to have been localized largely in favorable beds (the Combined Metals member of the Pioche shale being the most important host rock) that the ore solutions reached through the huge number and variety of faults developed in the area. The first faulting to affect the rocks of the district was post-Pennsylvanian but was earlier than the late Mesozoic thrust faulting since some of these faults are cut off by the thrust faults above them. The thrusting was followed by a long period of normal faulting that is more emphatically developed in the Ely Range (Pioche Hills) than in the Bristol range to the northeast, although economically valuable mineralization occurred in both areas. In the Bristol range, at least three periods of normal faulting occurred. In the Ely range, the fault pattern is even more complex.

The result of this intricate fault pattern is that the district is divided into many fault blocks, tilted at different angles that seldom exceed 30° from the horizontal. Probably, in the major fraction of the blocks, the beds have been tilted to the northeast or east, but tilting in any given direction can be found if the proper block is selected for study.

The veins in the quartzite were no more than 9 feet wide and dipped steeply; they were filled with broken quartzite, and the ore minerals were deposited between the fragments. The primary minerals appear to have been black sphalerite, silver-rich galena, and pyrite. The veins were oxidized to depths of 1200 feet beneath the surface, and the principal secondary silver mineral was cerargyrite (AgCl), but both galena and lead carbonate were recovered, as was some gold; oxidized zinc minerals probably were present but do not appear to have been recovered.

The replacement fissures in limestone cut across the bedding and dip steeply through the formations. The ores were localized in shoots, the position of which appears to have been determined by the presence of primary fault breccias; the primary mineralization consisted of a quartz gangue with sulfides of copper, lead, zinc, and silver [the sulfide minerals containing silver were not specified (Young, 1950) but probably were tennantite and pyrargyrite]. The primary ores were oxidized to depths of 1700 feet. The ore minerals probably were introduced in more than one stage since iron-copper minerals were characteristic of east-west fractures, lead-zinc minerals were found in the northwest-southeast fissures, and lead-silver in the north-south fractures. Copper in the oxidized ores occurred as chrysocolla and copper pitch ore, the galena was largely altered to cerussite, and the silver minerals probably to cerargyrite. Considerable manganese was present in both the primary and secondary ores.

The bedded replacements in limestone and dolomite frequently merged into the fissure replacements, and both types often were mined from the same stope. The primary minerals of the bedded deposits were intimately intergrown pyrite, sphalerite, and galena, the last of which almost certainly was silver-rich. Near the edges of these replacement masses, there was usually a considerable development of manganosiderite; some quartz was also formed as a gangue mineral. Where the replacement-bed ore has been ozidized, manganese-iron deposits have been formed in which the manganese is present in such minerals as pyrolusite, braunite, and wad and the iron as goethite, limonite, or hematite. Small, but different, amounts of lead, zinc, and silver were present in the manganese ores that were mined largely in World War I for use as a fluxing ore; the 1 percent of zinc in the ore debarred its use as a manganiferous iron ore. The fissure veins in the Yuba granite porphyry dike are of little economic importance.

The simple mineralogy of the deposit and the moderate temperature alteration of the rocks in and near the ore bodies suggest that the ores probably were emplaced under mesothermal conditions. The high-temperature silicates associated with the Blind Mountain quartz monzonite are not directly connected with ore mineralization. The ore fluids probably were not derived from the Blind Mountain stock of the Yuba dike, but rather from the deep source magma of these two igneous masses. Some suggestion exists that some of the silver in the primary ores may have come from a silver sulfide or from one or more silver sulfosalts, but only very minor amounts of tennantite and pyrargyrite have been identified in the ores. Unless further work on the primary ores should show a greater content of silver-bearing minerals, it seems unlikely that these were present in sufficient amount to justify the inclusion of the term leptothermal in the classification. The primary Pioche ores, therefore, are here classified as mesothermal.

The oxidized ores were formed by circulating, oxygen-charged ground water and should be classified as ground water-B2.

TONOPAH

Middle Tertiary *Silver, Gold* *Epithermal*

Bastin, E. S., and Laney, F. B., 1918, The genesis of the ores at Tonopah, Nevada: U.S. Geol. Surv. Prof. Paper 104, 50 p.

Burgess, J. A., 1909, The geology of the producing part of the Tonopah mining district: Econ. Geol., v. 4, p. 681-712

—— 1911, The halogen salts of silver and associated minerals of Tonopah, Nevada: Econ. Geol., v. 16, p. 13-21

Campbell, I., 1933, Some interrelations between phases of alteration and types of rocks in the Tonopah mining district, Nevada (abs.): Geol. Soc. Amer. Bull., v. 44, p. 164

Cornwall, H. R., 1967, Silver and mercury geochemical anomalies in the Comstock, Tonopah, and Silver Reef districts, Nevada-Utah: U.S. Geol. Surv. Prof. Paper 575-B, p. B10-B20

Cornwall, H. R., and others, 1967, Silver and mercury geochemical anomalies in the Comstock, Tonopah, and Silver Reef districts, Nevada-Utah: U.S. Geol. Surv. Prof. Paper 575-B, p. B10-B20

Eakle, A. S., 1912, The minerals of Tonopah, Nevada: Univ. Calif. Pubs., Bull. Dept. Geol., v. 7, no. 1, p. 1-20

Ferguson, H. G., and Muller, S. W., 1949, Structural geology of the Hawthorne and Tonopah quadrangles, Nevada: U.S. Geol. Surv. Prof. Paper 216, 55 p. (particularly p. 4-14)

Locke, A., 1912, The geology of the Tonopah mining district: A.I.M.E. Tr., v. 43, p. 157-166

Muller, S. W., and Ferguson, H. G., 1939, Mesozoic stratigraphy of the Hawthorne and Tonopah quadrangles, Nevada: Geol. Soc. Amer. Bull., v. 50, p. 1573-1624

Nolan, T. B., 1930, The underground geology of the western part of the Tonopah mining district, Nevada: Univ. Nev. Bull., v. 24, no. 4, 35 p.

—— 1935, The underground geology of the Tonopah mining district, Nevada: Univ. Nev. Bull., v. 29, no. 5, 49 p.

Spurr, J. E., 1905, Geology of the Tonopah mining district, Nevada: U.S. Geol. Surv. Prof. Paper 42, 295 p. (particularly p. 30-82, 83-104, 207-266)

—— 1915, Geology and ore-deposition at Tonopah, Nevada: Econ. Geol., v. 10, p. 713-769

Notes

The Tonopah district is situated in south-central Nevada, 25 miles north of Goldfield and 170 miles southeast of Reno.

The oldest known rocks in the actual mine area are of Tertiary age, having been found to depths of nearly 1900 feet in the western part of the district. This oldest formation is the Tonopah that is composed of an interbedded sequence of volcanic breccias, massive tuffs of widely different grain sizes, porphyritic flows, banded flows, and water-laid deposits of several kinds, including well-laminated silts that resemble those in the Esmeralda formation (a much younger formation, equivalent of the Siebert Lake beds in the Goldfield district). The nonintrusive character of these rocks, disputed by earlier workers, is demonstrated by the bedding planes that are present in many places and by the numerous examples of progressive reduction in particle size as a given bed is followed from bottom to top; one example of a ripple-marked surface has been observed. The next formation was the Sandgrass andesite that was emplaced as intercalated flows that are entirely confined to the Tonopah formation. The next formation in order of decreasing age was the Mizpah trachyte, so named although it is actually an albitized andesite or keratophyre rather than a trachyte. The Mizpah trachyte is widely distributed throughout the district and makes up the greater part of the rocks exposed in mining.

The next rock type to be developed was the Extension breccia that occurs only in the northwestern part of the district; it is an intrusion that was localized along an earlier fault within the Tonopah formation; the fault trended east-west and dipped north.

The Extension breccia is intruded by the West End rhyolite, but the rhyolite occurs mainly as a westward-dipping mass between the Tonopah breccia (below) and the Mizpah trachyte (above); it thins considerably toward the northwest and locally may be absent. In addition to intruding the Extension breccia, the rhyolite sends small apophyses into the Mizpah trachyte and locally contains inclusions of the Mizpah.

The Fraction breccia is a generally south-dipping mass that unconformably overlies the older formations and the veins as well. Its irregular lower contact shows that it was deposited on an erosion surface of moderate relief. The rock, although post-ore, has been subjected to some hydrothermal alteration. Ferguson (in his work on the Manhatten district north of Tonopah) considered the Fraction breccia to be the equivalent of the basal member of the Esmeralda formation that he also thought to be the equivalent of the Siebert Lake beds at Goldfield. The Esmeralda has been assigned an age of late Miocene.

The youngest rocks found in the mine workings at Tonopah are dikelike masses of rhyolite, but only three of them are of appreciable size; one of these dikes cuts across ore veins and contains no ore, so it must be post-ore in age, and another cuts the Fraction breccia.

From the surface studies of the Tonopah quadrangle, it appears almost certain that the earliest of the volcanic formations is of Tertiary age. The Tonopah formation, the Sandgrass andesite, the Mizpah trachyte, the Extension breccia, and the West End rhyolite are all pre-ore; therefore, a considerable portion of Tertiary time had elapsed before the ores were introduced into the area. As the ores were formed before the Fraction breccia, which is probably upper Miocene, the mineralization took place before the end of the Miocene. As the Fraction breccia was deposited on an irregular erosion surface, it would appear that some time had passed between the formation of the ore bodies and the introduction of the Fraction breccia. This strongly suggests that the ores, although probably Miocene, were deposited early in that period. On this basis, the ores are here categorized as middle Tertiary.

Generally, the formations in the Tonopah area are considerably tilted; only in the central part of the district where they were extensively faulted before mineralization are they essentially flat-lying. Essentially all the movement along the faults had occurred before the mineralization as is indicated by the generally unbroken character of the quartz in the faults. This quartz is continuous with the quartz of ore-bearing veins, and some of the fault quartz was well enough mineralized to have been mined.

The ore bodies at Tonopah are replacement veins that are located in minor faults or fractures and grade gradually into the host rock of their walls; filling of open space was rare and never occurred on more than a small scale. Such banding as is found in the ores almost certainly took place through unequal rates of the diffusion of the added ions into the wall rocks and was not due to crustification. The size of the minable ore shoots within the veins themselves ranges within wide limits, being determined not only by the mineral content but also by the current price of silver. The longest continuous ore shoot covered 1500 feet, and the widest reached 40 feet. The ore-forming solutions probably entered the area through the fault system and the localization of the ore bodies within that system depended mainly on temperature of the ore fluids and the chemical character of the wall rock.

Bastin and Laney (1918) concluded that the ores were emplaced in essentially one period, but they do recognize that there were two substages within the mineralizing epoch. Bastin and Laney note that the first (alpha) substage ore minerals formed by direct replacement of wall-rock minerals or of the common hydrothermal alteration minerals developed in the wall rocks before the ore minerals entered them. The second substage (beta) minerals are those that clearly show that they formed by replacement of the minerals of the first substage; first substage minerals possibly were still being deposited in certain areas while second substage minerals were being formed in others.

No criteria have been advanced to show certainly the order of placement of the minerals of the first substage, but quartz and sphalerite may have been the first hypogene minerals to form. Sphalerite, galena, and chalcopyrite were the most abundant of the first substage minerals. Pyrite is generally less abundant than the base-metal sulfides but forms larger grains, while arsenopyrite was noted in one specimen. Arsenic-bearing pyrargyrite [$Ag_3(Sb,As)S_3$], though rare in comparison with the base-metal sulfides, was present in all veins studied by Bastin and Laney and locally was very abundant; a small fraction of it may conceivably be supergene. In the first substage, some polybasite [$(Ag,Cu)_{16}(Sb,As)_2S_{11}$] and argyrodite (?) [$Ag_8GeS_6$] were intergrown with quartz and pyrargyrite, but the amounts of these two minerals were minor; electrum was present in small amounts, and some argentite may have belonged to the first substage. Quartz was the most abundant gangue mineral of the first substage, and there was considerable carbonate that was mainly intimately intergrown with quartz. The second substage of hypogene minerals included polybasite in considerable amounts; it normally was emplaced by the replacement of galena and occasionally of sphalerite. Argentite also was formed mainly by the replacement of galena, almost certainly largely under hypogene conditions; some chalcopyrite often accompanied the argentite, but the copper sulfide also replaced galena by itself. Some of the electrum in the ores also formed by the replacement of galena, and small amounts of an unknown mineral, probably a lead-bearing silver sulfosalt, intervened between argentite or polybasite and galena, probably as an intermediate product of the replacement reactions.

The primary ores of Tonopah and the wall-rock alteration minerals that preceded them were formed under epithermal conditions. The mineral suites, both of wall-rock alteration and ore, are typical of the epithermal category. The suite of silver sulfide and sulfosalt minerals in association with minor amounts of the ubiquitous base-metal sulfides is diagnostic of the epithermal range. The hypogene deposits are, therefore, here classified as epithermal.

Bastin and Laney believed that an appreciable fraction of the silver sulfosalts and most, if not all, of the native silver at Tonopah were formed under supergene conditions. Native silver, argentite, polybasite, pyrargyrite, native gold, chalcopyrite, and pyrite, plus quartz and perhaps some carbonates occur in fractures that cut the minerals of the first and second substages of the certainly hypogene ores. Whether or not these minerals were the products of a third substage of hypogene deposition or were the result of deposition below the water table of materials dissolved from higher reaches of the veins probably will never be known for certain. It appears likely, however, that the fact that these late minerals were not formed by replacement but by open space deposition would argue that they were deposited from hypogene rather than supergene solutions; they are considered here to be largely, if not entirely, hypogene and to represent a third hypogene substage.

On the other hand, there is no question but that the silver halides such as cerargyrite, embolite [$Ag_2(Br,Cl)_2$], and iodyrite [AgI], and the hydrous oxides of iron, the oxides of manganese, calamine, malachite, epsomite, and some kaolin, which are all found in veins that outcrop at the surface, were formed by oxygen-charged descending ground water. The position in the scheme of things of barite and selenite gypsum is uncertain.

Further, however, some minor replacement of chalcopyrite by bornite, covellite, and argentite, of sphalerite and galena by argentite and/or covellite, and of argentite by native silver well may have been produced below the water table by solutions of surface origin. In any event, the definitely oxidized minerals, such as cerargyrite, and the probably supergene replacement minerals, such as native silver that belong in the replacement category, indicate that supergene waters played a considerable part in the development of minable ores. These supergene reactions should be classified as ground water-B2.

New Hampshire

GRAFTON-KEENE

Middle Paleozoic *Feldspar, Mica, Beryl* *Magmatic-3a*

Bannerman, H. M., 1943, Structural and economic features of some New Hampshire pegmatites: Natl. Plan. and Dev. Comm., Mineral Res. Surv., pt. 7, 22 p.

Berman, H., 1927, Graftonite from a new locality in New Hampshire: Amer. Mineral., v. 12, p. 170-172

Cameron, E. N., 1954, Pegmatite investigations, 1942-1945, New England: U.S. Geol. Surv. Prof. Paper 255, 352 p. (particularly p. 15-19, 22-56, 107-139, 139-265)

Chapman, C. A., 1939, Geology of the Mascoma quadrangle, New Hampshire: Geol. Soc. Amer. Bull., v. 50, p. 127-180

—— 1941, The tectonic significance of some pegmatites in New Hampshire: Jour. Geol., v. 49, p. 370-381

Fowler-Billings, K., 1942, Geological map of the Cardigan quadrangle, New Hampshire: Geol. Soc. Amer. Bull., v. 53, p. 177-178

Fowler-Lunn, K., and Kingsley, L., 1937, Geology of the Cardigan quadrangle, New Hampshire: Geol. Soc. Amer. Bull., v. 48, p. 1363-1386

Frondel, C., 1936, Oriented inclusions of tourmaline in muscovite: Amer. Mineral., v. 21, p. 777-799

—— 1941, Whitlockite; a new calcium phosphate, $Ca_3(PO_4)_2$: Amer. Mineral., v. 26, p. 145-152

Frondel, C., and Lindberg, M. L., 1948, Second occurrence of brazilianite: Amer. Mineral., v. 33, p. 135-141

Holden, E. F., 1918, Famous mineral localities; Beryl Mountain, Acworth, New Hampshire: Amer. Mineral., v. 3, p. 199-200

Kruger, F. C., 1946, Structure and metamorphism of the Bellows Falls quadrangle of New Hampshire and Vermont: Geol. Soc. Amer. Bull., v. 57, p. 161-206

Kruger, F. C., and Linehan, D., 1941, Seismic studies of floored intrusives in western New Hampshire: Geol. Soc. Amer. Bull., v. 52, p. 633-648

Megathlin, G. R., 1929, The pegmatite dikes of the Gilsum area, New Hampshire: Econ. Geol., v. 24, p. 163-181

Olson, J. C., 1950, Feldspar and associated pegmatite minerals in New Hampshire: N. H. Plan. and Dev. Comm., Mineral Res. Surv., pt. 14, 50 p.

Page, J. J., and Larrabee, D. M., 1962, Beryl resources of New Hampshire: U.S. Geol. Surv. Prof. Paper 353, 49 p.

Shaub, B. M., 1937, Contemporaneous crystallization of beryl and albite versus replacement: Amer. Mineral., v. 22, p. 1045-1051

—— 1938, The occurrence, crystal habit, and composition of the uraninite from the Ruggles mines, near Grafton Center, New Hampshire: Amer. Mineral., v. 23, p. 334-341

Sterrett, D. B., 1915, Grafton County and Cheshire County: U.S. Geol. Surv. Bull. 580, p. 70-94

Stoll, W. C., 1945, The presence of beryllium and associated chemical elements in the wall rocks of some New England pegmatites: Econ. Geol., v. 40, p. 136-141

Notes

Although the state of New Hampshire contains many pegmatite deposits that have been worked sporadically in the past, the economically most important and geologically most interesting are those of the Palermo mine near North Groton in Grafton County and of the Beryl Mountain and Chickering mines near South Acworth in Sullivan County and Walpole in Cheshire County, respectively. The Palermo mine is about 40 miles north-northwest of Concord, the Beryl Mountain mine some 15 miles north of Keene, and the Chickering mine about 10 miles north-northwest of Keene.

Intrusive igneous rocks make up about half of the Cardigan and Rumney quadrangles and are now of gneissic texture and range from granite to quartz diorite in composition. These gneisses are divided into two groups, the older of which is the Oliverian magma series; the only representative of this series in the general area of the Palermo pegmatite is the Smarts Mountain group that is composed mainly of quartz diorite, with some granodiorite, and makes up the Smarts Mountain dome. The Oliverian series intrudes (or displaces) the Littleton formation and must be younger than the Lower Devonian and was introduced before the Late Devonian folding. The younger of the two magma series is known as the New Hampshire and is represented in the general Palermo area by sill-like masses of the Bethlehem gneiss and its attendant pegmatites, by the Insman quartz monzonite, and by the Concord granite. These rocks, although they were intruded at the end of, or shortly after, the Late Devonian folding, are usually foliated and contain considerable muscovite and biotite. The Bethlehem gneiss (granite to granodiorite) outcrops less than 2 miles from the Palermo pegmatite, and there is a broad zone of genetically related pegmatites several thousand feet in width in the Littleton formation to the east of the Bethlehem gneiss; the pegmatite belt extends through all of the Cardigan and much of the Rumney quadrangles and contains (in addition to the Palermo mine) such well-known deposits as the Ruggles mine near Grafton Center. The Palermo mine pegmatites, therefore, probably are middle Paleozoic in age.

In the Keene district, the Oliverian magma series is represented by granite and quartz-monzonite gneisses that occupy the cores of a series of domes. The Oliverian gneisses cut the Partridge formation and probably the Littleton, which would make the gneisses post-Lower Devonian in age. The New Hampshire magma series is represented in part by the Bethlehem gneiss (granite to granodiorite) in which the foliation is due to alignment of the large content of micas. Biotite-muscovite granite also is included in the New Hampshire series, but it occupies an appreciably smaller surface area in the district than does the Bethlehem gneiss. Although the Chickering and Beryl Mountain pegmatites are at some distance from the nearest outcrops of Bethlehem gneiss, they are far closer to masses of that rock than to the New Hampshire granite and probably are genetically related to the Bethlehem gneiss. As these pegmatites are known to cut the Littleton formation, they are certainly younger than that group of rocks, but they also appear to be of much the same age as the gneiss that probably is no younger than Late Devonian. Some confirmation of the age is given by radioactive age determinations made by Schaub (1938). Thus, the pegmatites in both areas are categorized as middle Paleozoic.

Six mineralogically different units have been identified (Page and Larrabee, 1962) in the Palermo pegmatite; the outermost of these is a quartz-muscovite-plagioclase border zone on the north and east margins of the pegmatite that is ordinarily less than 4 inches thick. Along the footwall contact, the pegmatite grades so gradually into the wall rock that no contact zone can be recognized. The border-zone unit contains 40 to 50 percent quartz, 30 to 40 percent muscovite, and 10 to 20 percent An_{10}-plagioclase, biotite is a minor accessory mineral and probably derived from the wall rock. The second zone from the outer margin of the body is a medium-grained An_{4-7}-plagioclase-quartz-muscovite wall zone from which most of the commercial muscovite of the deposit was produced, although it probably does not exceed 5 feet in thickness. The composition of the zone averages 50 percent plagioclase, 35 percent quartz, and 15 percent muscovite, with biotite and black tourmaline as accessory minerals. The next inner zone is the first intermediate that is a medium- to coarse-grained An_{4-7}-plagioclase-quartz-perthite pegmatite containing sheet muscovite; it has black tourmaline and biotite as the most common accessories, with the biotite being intergrown in strips with the muscovite. The zone is as much as 25 feet thick. The second intermediate zone, quite similar in composition to the first, is also a medium- to coarse-grained quartz-An_{4-7}-plagioclase-perthite pegmatite. On the east side, this zone is as much as 40 feet thick, but it thins to nothing at depth on the west side. Its average composition is 35 percent plagioclase, 35 percent quartz, and 30 percent perthite; the unit is varied in composition, and any one of the three major constituents may make up 75 percent of the total in certain localities. Perthite was commercially mined from this unit. The third intermediate (core-margin) zone is composed of An_{2-4}-plagioclase, quartz, muscovite, and perthite, with accessory beryl and phosphate minerals. An outcrop of beryl-bearing pegmatite on the top of the core suggests that the core-margin zone once surrounded the entire core. The composition of this zone is markedly different from one place to another but averages 30 percent plagioclase (cleavelandite), 25 percent quartz, 25 percent muscovite, and 20 percent perthite. This zone is 30 feet thick on the footwall side but is only discontinuously present on the hanging wall. Large crystals of beryl are most abundant near the core margin of this zone, but smaller beryl crystals are found in the muscovite aggregates as are the secondary uranium minerals. Lenses of the phosphate minerals, of which triphylite $[Li(Fe^{2},Mn^{2})(PO_4)]$ is the most important, are also present in the rock near the core. Although there appears to have been considerable replacement of earlier minerals by later ones, the primary deposition and the replacement reactions were not appreciably separated in time and were carried out between the still molten portion of the pegmatite fusion and the solid minerals. The core consists of massive quartz (60 percent) surrounding large euhedral crystals of perthite, with plagioclase (An_4) as thin rims between quartz and perthite. Potash-feldspar, sheet mica, and beryl were the main minerals recovered in the mining of the Palermo pegmatite.

The Beryl Mountain mine consists of two pegmatite masses that are discordant to the enclosing biotite gneiss and amphibolite of the Ammonoosuc formation; for 3 feet outward from the contact, the amphibolite has been altered to a granulitic rock. The west pegmatite consists of a thin border zone, a wall zone of plagioclase-quartz-perthite pegmatite and what appears to be the core made up of perthite, quartz, and plagioclase. It has been prospected only in a minor way. The main (or easterly) pegmatite body plunges at 10° to 15° in a N30°E direction; the maximum depth to the bottom of the pegmatite is probably about 100 feet. The border zone is very narrow and, in places, is lacking, but it surrounds a thick wall zone that consists of 60 percent plagioclase, 30 percent quartz, and 10 percent perthite; there are minor quantities of muscovite books less than 2 inches in diameter and black tourmaline crystals less than 4 inches long. The intermediate zone is made up of 35 percent plagioclase, 30 percent muscovite, 25 percent quartz, 5 percent perthite, and 5 percent beryl (about 14 percent BeO) in crystals that may be as large as 1.5 by 6 feet. This zone has an average thickness of only 2 feet but may be as much as 7 feet thick; it is about 300 feet long and has an average depth of 25 feet. Wedge-shaped crystals of muscovite are characteristic of the zone. Both composition and grain sizes are widely different from one part of the zone to another. The core is composed of white, granular quartz with irregular streaks of pale rose quartz in grains less than 1/8 inch in diameter, but the outer portion of the core contains perthite crystals up to 20 feet long. Veinlets of quartz run out from the core into the intermediate zone, and where the perthite margin of the core is absent, beryl crystals may extend as far as 7 feet into the core. The core has a length of about 500 feet and averages 25 feet in thickness, though this dimension may be as much as 60 feet. A unit of perthite-quartz plagioclase pegmatite, lithologically similar to the core of the western pegmatite, appears to have invaded the east side of the beryl-bearing pegmatite and probably was introduced after the beryl-bearing pegmatite. The pegmatite does not appear to have been largely affected by replacement reactions. Potash feldspar, muscovite, and beryl were the principal products of mining.

Five units have been mapped in the Chickering pegmatite. The border zone is made up of very fine-grained quartz-muscovite-An_{2-4}-plagioclase; it is no more than a few inches thick. The wall zone is a fine- to medium-grained An_{2-3}-plagioclase-quartz-perthite pegmatite; the plagioclase includes cleavelandite, and the total feldspar is about as abundant as quartz, with each of these minerals being about twice as common as perthite. Muscovite is the principal accessory mineral, and tourmaline (blue, black, and green) and apatite are minor accessories. The cleavelandite occurs as veinlets cutting the perthite and forms rims around it, suggesting both deposition in fractures and replacement of the earlier perthite. The quartz and muscovite make up a finer-grained matrix between the masses of perthite and cleavelandite. The second intermediate zone is composed of the same essential minerals as the first but a number of new accessory minerals--spodumene, lepidolite, beryl, columbite, and several phosphate minerals--are present. This zone is discontinuous; on the footwall side of the pegmatite it averages about 2 feet in thickness, but on the hanging wall it is as much as 10 feet thick, although the average thickness is much less. Of the accessory minerals, spodumene was originally the most abundant, but it has nearly all been altered to mixtures of quartz and muscovite in fine-grained aggregates. At one time the second intermediate zone locally contained 15 to 20 percent of spodumene in crystals as much as 2 feet long, 6 inches wide, and 4 inches thick. Lepidolite is much less abundant than altered spodumene, and the white beryl is difficult to distinguish megascopically from quartz, feldspar, and amblygonite. Spodumene and lepidolite are more abundant in the inner part of the zone and beryl in the outer portion; only in a few places do these three minerals all occur together. Triphylite is the most common of the phosphate minerals, although both amblygonite and apatite are present; the phosphate minerals are usually found in isolated patches. Some sulfides and carbonates (mostly siderite) also are present and are usually associated with the phosphates. The core is made up of medium- to coarse-grained perthite, quartz, and plagioclase; perthite makes up more than 50 percent of the core, and quartz is about twice as abundant as plagioclase. The mine has been worked for potash feldspar and might be workable for beryl, although the average beryl content in the second intermediate zone is only 0.11 percent and the white color of the beryl makes it difficult to distinguish.

Although the three New Hampshire pegmatites described here are only a small fraction of the total present in the two districts and in the entire state, they appear to be reasonably typical examples of the form and mineral character of such pegmatites. There seems to be little doubt as to the igneous origin of these pegmatite bodies; although these three pegmatites occur outside the intrusive masses of the district, they are so different in composition from their host rocks that they hardly can have been formed from them by any process other than the intrusion of an aqueo-igneous fusion of magmatic origin. From the location of these pegmatites in relation to the igneous rocks of the area, it would seem most probable that their source was the magma chamber of the Bethlehem gneiss, but this certainly has not been established beyond further argument.

Although the development of replacement minerals in the pegmatites in question is less impressive than in other pegmatites in other districts, there appears to have been considerable reaction between the already solid portions of the pegmatites and the still molten mass of igneous material. The cleavelandite in the Palermo and Chickering mines is almost certainly the product of such reactions; the indications of replacement reactions involving internally generated hydrothermal fluids, however, are not spectacular.

It appears highly probable, therefore, that these pegmatites are the products of magmatic activity and should be classified as magmatic-3a.

New Jersey

DOVER

Late Precambrian | *Iron as Magnetite* | *Hypothermal-1*

Bayley, W. S., 1910, Iron mines and mining in New Jersey: Geol. Surv., N. J., Final Rept., v. 7, p. 89-193

—— 1941, Pre-Cambrian geology and mineral resources of the Delaware Water Gap and Easton quadrangles, New Jersey and Pennsylvania: U.S. Geol. Surv. Bull. 930, 98 p. (particularly p. 54-77)

Buddington, A. F., 1966, The Precambrian magnetite deposits of New York and New Jersey: Econ. Geol., v. 61, p. 484-510

Collins, L. G., 1968, Trace ferrides in the magnetite ores of the Mount Hope mine and the New Jersey Highlands: Econ. Geol., v. 63 (disc.), p. 193-195

—— 1969, Regional recrystallization and the formation of magnetite concentrations, Dover magnetite district, New Jersey: Econ. Geol., v. 64, no. 1, p. 17-33

Hotz, P. E., 1953, Magnetite deposits of the Sterling Lake, N. Y. - Ringwood, N. J. area: U.S. Geol. Surv. Bull. 982-F, p. 153-244

—— 1954, Some magnetite deposits in New Jersey: U.S. Geol. Surv. Bull. 995-F, p. 201-253

James, A. H., and Dennen, W. H., 1962, Trace ferrides in the magnetite ores of the Mount Hope mine and the New Jersey highlands: Econ. Geol., v. 57, p. 439-449

Klemic, H., and others, 1959, Radioactive rare-earth deposit at Scrub Oaks mine, Morris County, New Jersey: U.S. Geol. Surv. Bull. no. 1082-B, p. 29-59

Sims, P. K., 1953, Geology of the Dover magnetite district, Morris County, New Jersey: U.S. Geol. Surv. Bull. 982-G, p. 245-305

Sims, P. K., and Leonard, B. F., 1952, Geology of the Andover mining district, Sussex County, New Jersey: N. J. Div. Plan. and Dev. Bull. 62, Geol. Ser., 46 p.

Smith, L. L., 1933, Magnetite ores of northern New Jersey: Econ. Geol., v. 28, p. 658-677

Williams, R. L., 1967, Reconnaissance of yttrium and rare-earth resources in northern New Jersey: U.S. Bur. Mines R.I. 6885, 34 p.

Notes

The Dover district in central and northern Morris County covers an area of about 80 square miles in a belt about 18 miles long, from southwest to northeast, that runs from the village of Ironia (6 miles southwest of Dover) to the Pequannock River; the belt is from 4 to 5 miles wide. Deposits similar to those in the Dover area are found in Warren County, some 20 miles to the west of the town of Dover, in northern Hunterdon County, and in the Mount Olive district in southwestern Morris County.

The oldest rocks in the Dover district proper are metasediments of probably early Precambrian age, and they amount to about 25 percent of the bed rock of the district. The minor areal extent of these classes of metamorphosed rocks suggests that calcareous beds made up only a small part of the sediments of the Dover district. Hornblende skarn, which contains important ore bodies at Mount Hope and Richard of the Wharton ore belt, appears to have been produced from pyroxene skarn and probably represents a more extreme case of hydrothermal alteration than does the pyroxene variety. The lathlike deposits of ore in hornblende skarn probably are not parallel to the original bedding, and it appears likely that hornblende (or pyroxene) skarn has, therefore, not been developed exclusively from carbonate-bearing beds but has been derived from metamorphosed sediments of all types. The close association of amphibolite with the certainly sedimentary rocks suggests that this rock type was derived at least largely, if not entirely, from sedimentary rocks and was not originally mafic igneous material.

Six large bodies of the biotite-quartz-feldspar gneiss exist and make excellent marker beds. This type of gneiss contains little ore. The igneous rocks consist of two general types: (1) older hypersthene quartz diorites in considerable volume and lesser quantities of diorite and (2) much later hornblende granite of which five facies are present in the district. The quartz diorites are not important as hosts to ore.

The hornblende granite is the youngest of the Precambrian rocks of the area and makes up about 40 percent of the bedrock in the Dover district. Five principal facies of the granite have been recognized: (1) hornblende granite, (2) alaskite, (3) biotite granite, (4) microantiperthite, and (5) granite pegmatite. The hornblende granite outcrops over large areas but does not appear to have direct connections with the ore bodies; it may correlate with the Storm King granite of southern New York.

The igneous rocks of the area (except the diabase) are Precambrian in age; it also appears highly probable that the ores were introduced into the area as late-stage effects of that igneous activity. The question that remains to be answered is that of how late in the Precambrian did the igneous activity and the ore emplacement occur. The only data that give some idea of the absolute ages of these events were obtained from radioactive age determinations made on zircons found in association with certain radioactive rare-earth minerals present in the Scrub Oaks mine. These minerals, such as doverite, xenotime, chevkinite, and bastnaesite, are intimately intergrown with both magnetite and hematite and must be of the same general age of formation. The ages obtained from the zircons range between 545 and 600 m.y., but two of the three determinations were made by the suspect lead-alpha method. The third was the result of the uranium-lead and thorium-lead isotope methods and agrees well with the other two. Placing reliance on these ages would indicate that the ores were formed in the very latest Precambrian--there appears to be no chance of their being Cambrian. The nearby ores at Franklin-Sterling, on the other hand, appear to be appreciably older, perhaps with an age of 1000 m.y. Thus, further work is necessary to make certain just how old the Dover ores may be, but the information available definitely seems to make them late Precambrian, and they are so classified here. If Collins' (1969) suggestion is correct, and the ores are the result of regional recrystallization of amphibolite and gneiss during the deformation of the region that produced the Hibernia anticline, the ores still probably are late Precambrian in age.

The ores in the Dover district are found in three varieties of host rocks: (1) oligoclase-quartz-biotite gneiss, (2) skarn (both hornblende and pyroxene), and (3) albite-oligoclase gneiss (granite); from each of these has come a considerable percentage of the ore mined in the district. Magnetite concentrations often occur in both amphibolite and granite pegmatite, but these are never of ore grade. The deposits in oligoclase-quartz-biotite gneiss are massive magnetite replacements with which are associated only minor amounts of gangue minerals of which apatite is the most abundant, followed by pyroxene, amphibole, and a little quartz and feldspar; the gangue minerals are arranged in rude laminae parallel to the ore contacts. The contacts of this ore-type with the walls are sharp, there being only small amounts of disseminated magnetite in the wall rocks or perhaps a few magnetite veinlets may enter the wall rocks for short distances. Biotite selvages, a few inches to 2 feet thick, may occur along one or both walls of an ore body.

Hornblende-skarn ores are important, though not dominant, in the Mount Hope and Richard mines; the ore in skarn may be massive as in the gneiss, but most of it contains abundant granules of pyroxene, hornblende, and biotite. This ore usually is laminated in alternating layers of different thicknesses made up, respectively, of massive magnetite and of skarn that is partially replaced by magnetite. Ore in the pyroxene skarn normally is massive and forms irregular bunches and veinlets. Most contacts between skarn ore and the wall rocks are sharp, though some magnetite may replace the wall rocks for a few inches beyond the skarn-wall rock contact.

In the ore bodies in albite-oligoclase gneiss, the magnetite (and locally abundant hematite) are present in finely disseminated grains or irregular veinlets and lenses; these ores have much less apatite than the more massive deposits. The walls of the albite-oligoclase gneiss ore bodies are assay walls.

The ore bodies in the Dover area are grouped into seven ore belts: (1) Wharton (by far the most important economically), (2) Hibernia Pond-Hibernia, (3) White Meadow-Cobb, (4) Dalrymple, (5) Beach Glen, (6) Green Pond, and (7) Splitrock Pond-Charlottesburg. The Wharton belt extends from Ironia northeast for about 10 miles and includes the Mount Hope, Scrub Oaks, Mount Pleasant, and Richard mines, of which none is now in production.

The mineralogy of the ore is simple; magnetite is the principal ore mineral while hematite is present in abundance (to about 15 percent of the ore) only in the Scrub Oaks mine; in the other ore bodies, hematite makes up not more than 1 percent of the ore. Ilmenite never exceeds 1 percent of the ore, and sulfides are essentially absent except in the Green Pond ore belt where there are a few percent of pyrrhotite, chalcopyrite, and pyrite. The gangue minerals are mainly unreplaced grains of the host rocks (including skarn); the unreplaced minerals are mainly pyroxene, hornblende, feldspar, quartz, and biotite. The minerals introduced at the same general time as the ores include apatite, calcite, chlorite, serpentine, biotite, quartz, pumpellyite, tourmaline, sphene, and spinel. In the massive deposits, apatite content ranges from 1 to 40 percent of the ore; in the disseminated deposits (Scrub Oaks), it constitutes less than 1 percent of the ore.

In 1955 a group of unusual rare-earth minerals was discovered in the Scrub Oaks mine; these minerals are appreciably radioactive, radioactivity deriving largely from thorium (the ores contain 0.07 percent ThO_2) rather than from uranium (U = 0.009 percent). Both cerium- and yttrium-group elements are abundant in the rare-earth minerals; the average RE_2O_3 content of the rare-earth portion of the Scrub Oaks ore is about 1.5 percent where the Fe_2O_3 content is about 35.5 percent. The principal rare-earth minerals are doverite (an yttrium fluocarbonate) and xenotime [$Y(PO_4)$] that occur in brick-red aggregates that resemble red jasper; hematite and quartz are usually intimately associated with the rare-earth minerals; the color of the aggregates is due to hematite, as doverite is pale yellow and xenotime deep yellow to brown. These aggregates usually make up about 3 percent of the ore in the areas where they occur. Bastnaesite (a cerium fluocarbonate) and chevkinite (a titano-silicate of the cerium earths) also are present. The rare-earth minerals usually occur in irregularly shaped bodies near the upper margins of the ore shoots, but these minerals may occur far below the upper reaches of the ore. The first rare-earth minerals to form (doverite and xenotime) were introduced in the middle of the period when magnetite and hematite were being emplaced; the bastnaesite and chevkinite were somewhat later, chevkinite overlapping with both magnetite and hematite but bastnaesite only with hematite.

From the replacement character of the ore minerals, it appears almost certain (to me) that they were emplaced by hydrothermal solutions, and the mineral suite, ore and gangue, is typically that of high-temperature deposition. Only the late pumpellyite, chlorite, calcite, sericite, and the sulfides suggest any lower intensity conditions; and when these minerals had formed, deposition of the ore minerals had been over for a considerable period of time. Reason exists, therefore, to categorize these deposits as hypothermal-1. Although several important ore bodies are found in hornblende skarn, which probably was developed in some part from impure carbonate-bearing sediments, their carbonate content is here considered to have been small enough that the ores are classified as hypothermal-1 and not hypothermal-2. Collins (1969) thinks that the ores were formed through recrystallization of Fe-rich amphibolite by earth forces. These earth forces established pressure gradients between the Fe-rich amphibolite and adjacent shear zones. The elements released by recrystallization of the rocks just mentioned moved down the pressure gradients to deposit in the shear zones either as magnetite ores plus Mg-rich silicates. This brief summary does not do justice to Collins' detailed presentation, which should be consulted by anyone interested, but his failure to provide a detailed mechanism to mineral element movement and concentration seems to me to force suspension of judgment of his concept until he does so.

FRANKLIN-STERLING

Late Precambrian — *Zinc, Manganese; Iron in Franklinite* — *Hypothermal-2*

Baker, D. R., and Buddington, A. F., 1970, Geology and magnetite deposits of the Franklin quadrangle and part of the Hamburg quadrangle, New Jersey: U.S. Geol. Surv. Prof. Paper 638, 73 p.

Baum, J. L., 1957, Precambrian geology and structure of the Franklin-Sterling area, New Jersey: Geol. Soc. Amer. Guidebook for Field Trips, Field Trip no. 3, p. 100-111

Callahan, W. H., 1966, Genesis of the Franklin-Sterling, New Jersey ore bodies: Econ. Geol., v. 61, p. 1140-1141

Frondel, C., 1970, Scandium content of ore and skarn minerals at Franklin, New Jersey: Amer. Mineral., v. 55, p. 1051-1054

Frondel, C., and Klein, C., Jr., 1965, Exsolution in franklinite: Amer. Mineral., v. 50, p. 1670-1680

Hague, J. M., and others, 1956, Geology and structure of the Franklin-Sterling area, New Jersey: Geol. Soc. Amer. Bull., v. 67, p. 435-473

Kerr, P. F., 1932, Zinc deposits near Franklin, New Jersey, in *Mineral deposits of New Jersey and eastern Pennsylvania*: 16th Int. Geol. Cong., Guidebook 8, p. 2-14

King, H. F., 1958, Notes on ore occurrences in highly metamorphosed Precambrian rocks: Aust. Inst. Min. and Met., Stillwell Anniv. Vol., p. 143-167; disc., Pr. no. 193, p. 125-130

Metsger, R. W., 1965, Notes on the Sterling Hill ore body, Ogdensburg, New Jersey: Inter. Mineral. Assoc., Knoxville Meeting, Northern Field Excursion Handbook, p. 12-18

Metsger, R. W., and others, 1958, Geochemistry of the Sterling Hill zinc deposit, Sussex County, New Jersey: Geol. Soc. Amer. Bull., v. 69, p. 775-788

Milton, C., 1947, Diabase dikes of the Franklin Furnace, New Jersey, quadrangle: Jour. Geol., v. 55, p. 522-526

Palache, C., 1929, Paragenetic classification of the minerals of Franklin, New Jersey: Amer. Mineral., v. 14, p. 1-18

—— 1929, A comparison of the ore deposits of Långban, Sweden, with those of Franklin, New Jersey: Amer. Mineral., v. 14, p. 43-47

—— 1935, The minerals of Franklin and Sterling Hill, Sussex County, New Jersey: U.S. Geol. Surv. Prof. Paper 180, 135 p. (particularly p. 14-24 and bibliography p. 2-14)

Pinger, A. W., 1948, Geology of the Franklin-Sterling area, Sussex County, New Jersey: Geol. Soc. Amer., 61st Ann. Meeting (1948), Guidebook of Excursions, p. 1-14 (also in 18th Int. Geol. Cong. Rept., 1950, pt. 7, p. 77-87)

Ridge, J. D., 1952, The geochemistry of the ores of Franklin, New Jersey: Econ. Geol., v. 47, p. 180-192

Ries, H., and Bowen, W. C., 1922, Origin of the zinc ores of Sussex County, New Jersey: Econ. Geol., v. 17, p. 517-571

Spencer, A. C., and others, 1908, Franklin Furnace folio, New Jersey: U.S. Geol. Surv. Geol. Atlas, Folio 161, 27 p.

Spurr, J. E., and Lewis, J. V., 1925, Ore deposition at Franklin Furnace, New Jersey: Eng. and Min. Jour.-Press, v. 119, no. 8, p. 317-328

Takahashi, T., and Myers, C. E., 1963, Nature of ore-forming fluid for the Franklin and Sterling Hill deposits in New Jersey, U.S.A., in Kutina, J., Editor, *Symposium--Problems of postmagmatic ore deposition*: vol. 1, p. 459-465

Tarr, W. A., 1929, The origin of the zinc deposits at Franklin and Sterling Hill, New Jersey: Amer. Mineral., v. 14, p. 207-221

Notes

The zinc-manganese deposits of Franklin and Sterling are located in northwestern New Jersey, some 40 miles northwest of the center of New York City and within the area of the New Jersey highlands.

The rocks of the Franklin district consist largely of a series of Precambrian metasediments and metavolcanics intruded by a number of silicic igneous rocks. The oldest rock unit is the interlayered Hamburg Mountain gneiss series that was followed by the two members of the Franklin marble; the lower of these two portions of the marble is, because of the duplication due to folding, of unknown thickness. The lower portion is separated from the upper by the 50- to 300-foot thick Median biotite and quartz gneiss; it is overlain by the upper portion of the Franklin marble, a coarsely crystalline limestone with local banding provided by concentrations of such minerals as mica, tremolite, chondrodite, norbergite, and other silicates, and pyrite, pyrrhotite, and graphite; none of these minerals is found in the ore bodies. The Franklin ore horizon is located just beneath the top of the upper Franklin marble, while the Sterling ore zone is some 600 feet below the upper contact. The Franklin marble is followed by the Cork Hill gneiss zone composed of a variety of gneissic types such as graphitic gneiss, garnet gneiss and pyroxene gneiss. The Wildcat marble, indistinguishable from the Franklin marble, except where its stratigraphic position can be established, overlies the Cork Hill gneiss. The uppermost unit of the sequence is the Pochuck Mountain gneiss series, quite similar to the Hamburg Mountain gneiss in texture and composition.

The Precambrian metasediments and metavolcanics were intruded by some half-dozen igneous rocks, four of which have been so highly metamorphosed as to be designated as gneisses; the two youngest were essentially unaffected by the earth movements that produced the gneisses. The oldest of these igneous rocks appears to be the generally concordant, faintly foliated Losee gneiss. It was followed by the Oligoclase gneiss which, although similar to the Losee in mineralogy, was intruded in sill-like masses instead of in the phacolithic form of the Losee. The next intrusion was that of the Granodiorite gneiss, an appreciably more mafic gneiss than the first two; it was followed by the Byram gneiss, all three types of which are normally concordant to the rocks they intrude. The next intrusion was of the two largely unmetamorphosed types of granite--the Mt. Eve and Type II--of which the Mt. Eve forms a stock with sills and dikes as apophyses. Finally, pegmatites are common in all Precambrian rocks in the area and range in thickness from a fraction of an inch to 100 feet or more. Both potassic and sodic types are present, and many of the unusual Franklin minerals were formed in the zones of reaction between ore and pegmatite. Few pegmatites occur in the Sterling area; this paucity of pegmatites at Sterling accounts for much, though far from all, of the differences in the mineralogies of the two deposits.

After the Precambrian rock series had been formed, folded, and metamorphosed, the area was eroded, and the Cambrian Hardyston quartzite (conglomerates, sandstones, and sandy or shaly dolomites) were laid down on the irregular Precambrian eroded surface. The Hardyston was followed by the Ordovician Kittatiny formation (mainly dolomite, in part shaly or sandy and with a few limestone beds near the top). Post-Cambrian camptonite dikes transect the banding in the gneisses and the beds in the Paleozoic sediments at high angles; they are the youngest rocks in the area.

The main structural feature of the area is a major, northeast-trending syncline with superimposed smaller anticlines and synclines locally present. The major structures are complicated by drag and cross folds. Lineations in the area conform to the northeast trend, and they plunge to the northeast except in areas where they have been rotated by the cross folds. The major, post-Ordovician faults probably represent renewed movement in fractures first developed in the Precambrian. A major result of this faulting was the down dropping of a large graben that trends northeast through the Wallkill valley and brings a block of basal Paleozoic sediments into elevational coincidence with Franklin marble to the northwest. In the lower levels of the Sterling mine, the east limb (vein) of the ore body has been sheared and perhaps truncated by the northwest-bounding (Zero) fault zone of the graben. Extensive exploration, however, has not discovered any downfaulted portion of the east vein.

Although the ores do not appear to have been folded (Hague and others, 1956), the age of the mineralization is fixed as Precambrian by the presence of detrital Franklin ore in the Hardyston quartzite. Such radioactive age determinations as have been made on minerals derived from the rocks of the area give ages between about 1400 m.y. and 1000 m.y. Although the various ages are not in good agreement, they more or less are compatible with the final metamorphism in the area having occurred in Grenville time. As the ores were formed later in the Precambrian than the metamorphism and before the beginning of the Paleozoic, it appears almost certain that the deposits should be categorized as late Precambrian. Even if the deposits should be shown to be metamorphosed syngenetic deposits, they still would be late Precambrian.

The ore bodies at Franklin and at Sterling have a rough similarity in structure; the ore at Franklin outcropped in an open synclinal fold in the

Franklin marble that pitches northeast at about 25° to the horizontal. On the surface, the west limb of the ore body was cut off against the post-Precambrian sediments about one-half mile north of the south end of the keel but continued downward under the overlying Paleozoics. The maximum depth of mineralization at Franklin was reached at 1150 feet beneath the surface. The surface expression of the east ore band outcropped for about 200 yards to the northeast of the south end of the syncline but died out at depth because it had a steeper plunge than that of the trough of the fold. The suggestion has been made that the east limb was cut off by a fault, but no downfaulted continuation of the east ore has ever been found.

At Sterling, in the other hand, the ore outcropped in another open synclinal fold in the Franklin marble that pitches northeast at about 50° to the horizontal; the maximum vertical depth of the ore at Sterling is about 2500 feet. The west ore band was exposed for about 700 feet and the east for about 1600. At a point about 300 feet north from the south end of the ore body, a cross vein leaves the west vein but runs parallel to it for some 400 feet; then it doubles back in a complexly cross-folded fashion to meet the east vein somewhat farther north. The structure well may have been a simple isoclinal fold, later folded into its present synclinal form with the separation of the original limbs to form the outer keel and the cross vein. The east branch of the west vein, however, does not merge smoothly with the east vein as would be expected if it had once been part of the same bed of an isoclinal fold; instead the east end of the cross vein appears to have been forced against the east vein in the area where they join.

At first glance, it would appear that the Franklin and Sterling ore bodies are both truly synclinal and are overturned to the west, thus accounting for the eastward dips of both limbs of both deposits. Actually, however, the rocks lying beneath the west limbs of the ore bodies are younger than those containing the ore. This means that the entire section in the mine area has been overturned and that the structures of the ore bodies are not synclinal but merely synclinal in form and are the result of movements within the more plastic marble that are not reflected in the more competent gneiss that originally overlay it.

The ore in the two mines differs somewhat in composition; that at Franklin contained 40 percent franklinite, 23 percent willemite, and less than one percent zincite while that at Sterling is composed of 33 percent franklinite, 16 percent willemite, and 1 percent zincite. The remaining percentages at both deposits are supplied by gangue which is principally calcite. Nevertheless, this gangue contains over 150 other minerals, more than 100 of which have never been found in any other district in the world.

It seems reasonable that deposits with such unusual mineralogy must have been formed in an unusual manner or by unusual ore-forming fluids in any event. Many, though far from all, of the minerals unique to Franklin proper were formed in the reactions that took place between the pegmatites and their skarn minerals (earlier) and the ore fluids (later), but many cannot be explained by such a conjunction of materials since they were formed, as were most Sterling minerals, at a considerable distance from any known masses of pegmatite.

Typical mineral classes found in high-temperature hydrothermal deposits--spinels, oxides, garnets, olivines, pyroxenes, melilites, epidotes, micas, and vesuvianites--are all found at Franklin, although their representatives are unusual in composition. Thus, granted that the Franklin deposit was formed by an ore fluid, this fluid appears to have been hypothermal in temperature, pressure, and probably pH, but it departed markedly from the average of such ore fluids in composition. It has been suggested (Ridge, 1952) that this unique composition can be explained almost entirely as a result of a high manganic ion content in the parent magma of the ore fluid. When this abundant manganic ion content was brought into a position where it could react with ferrous iron (probably in the magma chamber itself), the result would have been the oxidation of ferrous ion to ferric and the reduction of manganic ion to manganous. That this reaction was not carried to completion is shown by the presence of minor Mn^{+3} in the franklinite and some Fe^{+2} in the same and other minerals; but most of the iron is in the ferric state and most of the manganese in the manganous. The initial result of the scarcity of ferrous ion in the magma chamber was that minerals normally formed in the magma by the neutralization of an appropriate anion by ferrous ion could not be precipitated and that other cations, principally Mg^{+2} but also Cu^{+2} to a considerable extent, took the place of Fe^{+2} in the mafic minerals. The essential lack of Mg^{+2} in the Franklin ores and the small amount (in late serpentine minerals and dolomite) at Sterling, therefore, is interpreted to mean that Mg^{+2} was so largely used up in neutralizing dark-silicate anions in the magma chamber that very little was left over for incorporation in the primary ore fluid. The late serpentinization and dolomitization at Sterling probably were caused by hydrothermal fluids from a different source than those that produced the primary minerals, one in which magnesium was available for transfer from the silicate melt to the hydrothermal fluid. The lack of copper in the ores also probably was due to a similar introduction of that element into the magmatic dark silicates as a proxy for Fe^{+2}. The paucity of lead in the deposits is thought to have been caused by the difficulty of inserting the large lead ion in the structure of the silicate minerals present; a little lead did enter the structures of larsenite and calcium-larsenite, but the instability of these structures is attested by the identification of minor amounts of native lead in association with these minerals. The deficiency of sulfur in the Franklin deposits probably is also due to the abundance of Mn^{+3} which would have reacted with S^{-2} so strongly to produce S^{+6} in SO_4^{-2} that essentially all screening sulfur ions would have been removed from their complexes with the metals thus freeing such ions as zinc and lead to act as free cations in the hydrothermal fluid.

It has also been suggested that the Franklin deposit was formed by metamorphism from a mass of oxidized iron, manganese, and zinc minerals that had been developed from a normal hypogene deposit of iron and zinc sulfides and manganese carbonate. Pinger (1948, 1950) has pointed out that, in limestone (as at Gilman, Colorado) the oxidation of such a sulfide deposit almost certainly would have resulted in the separation of the iron from the manganese and probably of the zinc from the iron. This argument, plus the failure to find any appreciable evidence of metamorphism of the ores themselves, would seem to cast doubt on the possibility of the Franklin deposit having been formed from an oxidized deposit. Further, it appears probable that, since the deposits are confined to two definite but different horizons in the Franklin marble, a two-stage process (hypogene and supergene) is less probable than for one requiring only one stage (hypogene).

It would appear, from the reasoning here outlined, that the Franklin portion of the deposits was formed by hypothermal solutions reacting with carbonate rocks (except to the minor extent that they acted on pegmatites or pegmatite skarns) and should be classified as hypothermal-2.

The mineralization of the Sterling ore body is sufficiently different from that of the Franklin deposit to justify a brief separate treatment. The earliest ore mineral to form at Sterling was willemite, the interstices of rounded willemite grains being filled with franklinite and/or tephroite. Zincite also appears to have replaced calcite interstitial to willemite. Metsger (1958) believes that the willemite (Zn_2SiO_4) was replaced by tephroite (Mn_2SiO_4) and that the zinc so removed reacted with calcite to form zincite (ZnO). The main gangue minerals are augite, diopside, and biotite. At Sterling, the willemite, instead of being green as at Franklin, is either black, brown, or red, the exact color depending on the color of inclusions of franklinite. Black franklinite inclusions produced black willemite and red franklinite red willemite; brown willemite is due to the presence of both red and black franklinite inclusions. Tephroite shows color variations also caused by franklinite inclusions. The differences in structural character of the franklinite and of the willemite and tephroite make it almost certain that the inclusions were not the result of exsolution. It is possible that these inclusions were developed as the first stages in the serpentinization of tephroite (an olivine) and of willemite (a phenacite-type mineral), although they may be remnants of partly replaced franklinite or partial replacements of willemite and tephroite by a second generation of franklinite. The concept of incipient serpentinization being responsible for the franklinite inclusions is given support by the presence of considerable amounts of chrysotile, antigorite, and friedelite [$Mn_8(OH,Cl)_{10}Si_6O_{15}$] in which inclusions of franklinite, similar to those in the willemite and tephroite, are found. That hydrothermal solutions containing the magnesium necessary for the production of these serpentine minerals were actually available in the Sterling area (in contrast to Franklin where no serpentinization occurred) is indicated by the considerable amounts of hydrothermal dolomite developed at Sterling at much the same period of time as that in which the serpentinization occurred. The principal difference between normal serpentinization and that at Sterling is that the inclusions in normal serpentine are magnetite and that those in the serpentine-type minerals at Sterling are franklinite.

Thus, the primary ores at Sterling probably were formed under much the same conditions as those at Franklin, except that the primary mineralogy at Sterling was much less complex than that at Franklin because there were essentially no pegmatites or pegmatite skarns with which the Sterling ore fluids could have reacted.

As the primary mineralization at Sterling appears to have brought in all the economically valuable elements and the later serpentinization and dolomitization merely added magnesium, the deposits at Sterling also are categorized as hypothermal-2.

New Mexico

MAGDALENA

Middle Tertiary — *Zinc, Silver, Lead, Copper* — *Hypothermal-2 to Mesothermal*

Anderson, E. C., 1957, Magdalena district, in *The metal resources of New Mexico and their economic features through 1954*: New Mex. Bur. Mines and Mineral Res. Bull. 39, p. 135-139

Argall, P. (B.), 1908, The ore deposits of Magdalena, New Mexico: Eng. and Min. Jour., v. 86, no. 8, p. 366-370

Belt, C. B., Jr., 1960, Intrusion and ore deposition in New Mexico: Econ. Geol., v. 55, p. 1244-1271 (particularly p. 1258-1263)

Gordon, C. H., 1910, Magdalena district, in Lindgren, W., and others, Editors, *The ore deposits of New Mexico*: U.S. Geol. Surv. Prof. Paper 68, p. 241-258

Loughlin, G. F., and Koschmann, A. H., 1942, Geology and ore deposits of the Magdalena mining district, New Mexico: U.S. Geol. Surv. Prof. Paper 200, 168 p.

Titley, S. R., 1959, Geological summary of the Magdalena mining district, Socorro County, New Mexico: New Mex. Geol. Soc., 10th Ann. Field Conf. Guidebook, p. 144-148

—— 1961, Genesis and control of the Linchburg orebody, Socorro County, New Mexico: Econ. Geol., v. 56, p. 695-722

—— 1963, Lateral zoning as a result of a monoascendent hydrothermal process in the Linchburg mines, New Mexico, in Kutina, J., Editor, *Symposium--Problems of postmagmatic ore deposition*: I.A.G.O.D., Prague, vol. 1, p. 312-316

Notes

The Magdalena district is located in west-central New Mexico, just over 70 miles south-southwest of Albuquerque and centers around the town of Kelly, some 18 miles west of Socorro. The mines are concentrated on the northern end of the Magdalena range, a basin-range type of mountain.

The oldest rocks in the district are of Precambrian age and consist, in order of decreasing age, of argillite and schist, early gabbro or diabase, felsite, granite, and late diabase; the granite is exposed over by far the greatest area. Tentative correlations, based on the rock character and degree of metamorphism, have been made between these metasediments and the Pinal schists of Arizona and the Idaho Springs formation of the Colorado Front Range; the late diabases probably were formed at the same time as those in southern Arizona and probably are about 1200 m.y. old.

The oldest Paleozoic rocks in the area belong to the Carboniferous system. The oldest of the Carboniferous rocks is the Kelly limestone; this Kelly limestone contains essentially all the ore bodies in the district. Although the Pennsylvanian Magdalena group appears to lie conformably on the Kelly, these rocks probably are separated by a considerable interval of nondeposition during late Mississippian time. The Magdalena group is subdivided into the Sandia formation and the Madera limestone. Unconformably above the Magdalena group is the Abo sandstone, the lowest member (and the only one present in the district) of the Manzano group.

The district contains a wide variety of extrusive and intrusive igneous rocks that have been dated as Tertiary (?); the intrusive rocks are probably all younger than the extrusives with the possible exception of the youngest extrusive of all--the white felsite tuff. The extrusives were restricted in their distribution, and the lower formations interfinger and overlap to a considerable extent. The intrusives are divided into three groups: (1) the earliest that consists of sills and dikes of latite porphyry (older) and rhyolite porphyry (younger), (2) the middle that is made up of stocks of monzonite and granite (with some associated aplite and pegmatite) and possibly small stocks of andesite and a sill on Stendal Ridge, and (3) dikes of lamprophyre and white rhyolite. The first group of intrusives was introduced into the area before the main period of faulting; as the extrusives also predate these earth movements, the early intrusives probably are not much older than the extrusives. The second group of intrusives entered the area after the main period of faulting, and it is suggested that these masses may have been localized by the intersections of these faults. The third group cuts the stocks and older rocks and appears to be closely related in age and structural control to the ore bodies in the district.

Since no Mesozoic rocks are known in the area and since the series of extrusives contains essentially no intercalated sediments, little sound information is available on which to base an estimate of the age of the igneous activity or of the ore deposition. Loughlin and Koschmann (1942) are of the opinion that the principal structural elements in the area were formed in Laramide time, basing this on the general agreement that the significant stages of tilting and faulting in New Mexico took place during and immediately after the Laramide revolution. A firm correlation between the Magdalena district and the remainder of New Mexico is difficult because the Magdalena area lacks well-defined overthrusts and folds.

Granting that the deformations in the district began with the first of Laramide time, more time than that provided by the Laramide revolution may have been required to have accomplished all of the stages of deformation and the igneous and ore-forming activities spaced among them. The large number and variety of extrusive and intrusive rocks that were accumulated in the area prior to ore formation provides a resemblance to those of other ore districts in the basin-range country. From this, it well may follow that the Magdalena ores were deposited later in the Tertiary than would be covered by Laramide time and perhaps should be categorized as early Tertiary or middle Tertiary. Arbitrarily, they are here assigned to the middle Tertiary.

Three types of ore bodies are found in the Magdalena district: (1) high-temperature deposits of specularite and magnetite that are too small to be of economic importance; these deposits are limited to the immediate vicinity of the quartz monzonite and granite centers of igneous intrusion (except for the specularite-magnetite bodies in the Linchburg mine), (2) high-temperature deposits (possibly not of as high temperature as the specularite-magnetite bodies) of sphalerite and galena in a gangue generally consisting of andradite garnet and hedenbergite; these are economically the most important deposits, and (3) lower-temperature deposits of light-colored sphalerite and galena in a gangue of fluorite, barite, and calcite.

The high-temperature character of the magnetite-specularite type of deposits is supported by their close spatial and temporal relationships to such high-temperature silicates as garnet, hedenbergite, actinolite, tremolite, diopside, wollastonite, and epidote; andradite garnet and hedenbergite are the silicates most closely associated with the iron oxides. Much of the specularite-magnetite mineralization is also closely related to the sulfide mineralization of category (2) south of the Nitt stock.

Titley's work (1961) on the Linchburg ore body provides much detailed information on the specularite-magnetite, plus sulfide, type of mineralization. Titley has recognized five characteristic mineral assemblages which indicate a definite zoning of the ore body; each zone is gradational with those adjacent to it. Pyrite is essentially absent from the Linchburg ore body; minor amounts are found only in zone (1). The spatial relationships of the ore zones to each other do not establish the intensity relationships of one zone to the others. It appears highly probable that the specularite-magnetite of zone (5), for example, was formed at higher temperature than the sphalerite and galena of zone (3). If one or more of the zones is missing from a given location, the missing zones are those at either end of the zonal sequence rather than zones (2), (3), or (4).

It would appear from Titley's work that the first phase of mineralization was the production, from the feeding fractures outward, of andradite, hedenbergite, and hematite and the marbleization of the limestone; this phase may have been modified before ore formation by a magnesium alteration of the silicates, mainly to cummingtonite and chlorite. The initial interface between the andradite and hedenbergite phases seems to have moved gradually outward with the passage of time (and the heating of the rock volumes involved). Thus, in the deposit as a whole, garnet and pyroxene formed at the same time. In a given rock volume containing both andradite and hedenbergite, however, most of the pyroxene formed before the garnet, although the first of the andradite probably developed contemporaneously with the last of the hedenbergite. Sulfides appear to have begun to develop before silicate deposition had ceased, with galena forming with hedenbergite and sphalerite with andradite. In the same manner that the hedenbergite-andradite interface moved outward, so did that between sphalerite and galena; thus, although some galena was forming at the same time as sphalerite and in the same rock volume, most of the galena was forming farther from the ore-solution entry points at any given time than was sphalerite. Thus, the expanding sphalerite-galena interface made possible the replacement of galena by sphalerite in large amounts; the replaced lead from the galena probably was, to some extent, redeposited as galena farther out.

Those rock volumes to which the ore-forming fluids still had access after silicate-iron oxide deposition (of which the Linchburg area was one) were next affected by magnesium alteration and then by high-temperature sulfide deposition. Rock volumes that remained open to the ore fluids as the temperature of sulfide deposition changed from high to moderate then saw the further deposition of light-colored sphalerite, galena, quartz, barite, and fluorite. It would seem, from the less detailed description of Loughlin and Koschmann (1942), that much the same sequence of events as that of the Linchburg ores must have occurred in the development of the other primary zinc-lead ore deposits of the Magdalena district.

From Titley's detailed description of the Linchburg mineralogy, it appears probable that much of the sphalerite and perhaps a good deal of the galena were formed under hypothermal conditions; the exsolution blebs of chalcopyrite in much, though perhaps not all, of the sphalerite argues strongly for deposition under hypothermal conditions. On the other hand, probably a sizeable fraction of the galena of zone (4) may have been formed under mesothermal conditions. The light-colored, late sphalerite, the galena associated with it, and the quartz, fluorite, barite gangue indicate that that portion of the ore was deposited under mesothermal conditions. Thus, the classification of the ores as hypothermal-2 to mesothermal seems more nearly correct than that of hotter mesothermal and cooler mesothermal (leptothermal) applied to the dark and light-sphalerite ores, respectively, by Loughlin and Koschmann.

Until 1906 the main output of the district came from oxidized ores; some oxidized and some related copper sulfide ore in which the copper was largely of secondary origin were of considerable value. The main production, however, came from oxidized zinc and lead ores in which smithsonite was the principal zinc mineral and unoxidized galena, cerussite, and anglesite were the main lead minerals; cerargyrite was the principal source of silver. The oxidized ores were undoubtedly developed by oxygen-charged ground water, and these same solutions, below the water table, must have deposited the usually minor supergene sulfides. These oxidized deposits should be classed as ground water-B2.

QUESTA

Late Tertiary *Molybdenum* *Hypothermal-1 to Mesothermal*

Anderson, E. C., 1957, Red River district, in *The metal resources of New Mexico and their economic features through 1954*: New Mex. Bur. Mines and Mineral Res. Bull. 39, p. 147-148

Carpenter, R. H., 1960, A résumé of hydrothermal alteration and ore deposition at Questa, New Mexico, U.S.A.: 21st Int. Geol. Cong. Rept., pt. 16, p. 79-86

—— 1968, Geology and ore deposits of the Questa molybdenum mine area, Taos County, New Mexico, in Ridge, J. D., Editor, *Ore deposits of the United States 1933-1967* (Graton-Sales Volumes): Chap. 63, v. 2

Clark, K. F., 1968, Structural controls in the Red River district, New Mexico: Econ. Geol., v. 63, p. 553-566

Gustafson, W. G., and others, 1966, Geology of the Questa molybdenite deposit, Taos County, New Mexico: New Mex. Geol. Soc. Guidebook, 17th Ann. Field Conf., p. 51-55

Ishihara, S., 1967, Molybdenum mineralization at Questa mine, New Mexico, U.S.A.: Japan Geol. Surv. Rept. 218, 64 p.

Kuellmer, F. J., 1961, Alkali feldspars from some intrusive porphyries of southwestern United States: New Mex. Bur. Mines and Mineral Res. Circ. 62, 15 p.

Larsen, E. S., Jr., and Ross, C. S., 1920, The R and S molybdenum mine, Taos County, New Mexico: Econ. Geol., v. 15, p. 567-573

Laughlin, A. W., and others, 1969, K-Ar chronology and sulfur and strontium isotope ratios at the Questa mine, New Mexico: Econ. Geol., v. 64, p. 903-909

McKinley, P. F., 1957, Geology of Questa quadrangle, Taos County, New Mexico: New Mex. Bur. Mines and Mineral Res. Bull. 53, 23 p.

Schilling, J. H., 1956, Geology of the Questa (Moly) mine area, Taos County, New Mexico: New Mex. Bur. Mines and Mineral Res. Bull. 51, 87 p.

—— 1960, Mineral resources of Taos County, New Mexico: New Mex. Bur. Mines and Mineral Res. Bull. 71, 124 p. (particularly p. 36-42)

—— 1965, Questa molybdenum mine, in *Molybdenum resources of New Mexico*: New Mex. Bur. Mines and Mineral Res. Bull. 76, p. 28-34

Vanderwilt, J. W., 1938, Geology of the "Questa" molybdenite deposit, Taos County, New Mexico: Colo. Sci. Soc. Pr., v. 13, no. 11, p. 599-643

Notes

The molybdenum deposits of the Questa district are located near the northern boundary of New Mexico, about 80 miles west-southwest of Trinadad, Colorado, and slightly less than 90 miles north-northeast of Santa Fe, New Mexico. The deposits are situated on the western slope of the Taos range of the Sangre de Cristo Mountains in Taos County.

The principal formations in the Taos range are highly metamorphosed Precambrian rocks, and these are overlain by Tertiary volcanic rocks and intruded by middle Tertiary soda granites and other rocks. Probably the oldest of the Precambrian rocks is an amphibolite complex that covers considerable areas both east and west of the mine location but is present in only minor amounts near the actual mine itself.

Probably the next rock in the stratigraphic sequence is the Cabresto metaquartzite; it is largely absent from the vicinity of the mine but is present at the head of Blind Gulch. Because of its highly metamorphosed character, it is considered to be Precambrian in age.

The district was intruded in Precambrian time by large volumes of granite, two massive bodies of which outcrop in the mine neighborhood. One is exposed in the canyon of the Red River between the mouth of Sulphur Gulch and the main haulage adit to the southwest; the other is north of the highway and east of Sulphur Gulch, being opposite to the mouth of Columbine Canyon. The granite is dated as Precambrian, mainly on the basis of the extensive and intensive metamorphism it has undergone. There are several diabase dikes in the neighborhood of the mine; since they have been observed to cut the Precambrian rocks but not those of younger age, they are thought to be Precambrian.

The Precambrian rocks are unconformably overlain by the Sangre de Cristo(?) formation that consists of massive conglomerate, sandstone (=arkosic graywacke), and siltstone; the various rock types grade into each other, and the series includes some thin limestone lenses. A narrow and discontinuous belt of these beds extends north from the vicinity of the main haulage adit to just south of Sulphur Gulch. These sediments have been altered where they have been intruded by the soda granite, chlorite and epidote being developed in them up to a hundred feet out from the contact.

The thick Tertiary volcanic complex covers the higher slopes north of the Red River Canyon; these rocks, therefore, essentially surround the soda granite of the Sulphur Gulch stock that contains most of the Questa ore deposits.

Many dikes (at least some of which were feeders for the flows) and sills of andesite and latite(?) cut through the older rocks (including the volcanic rocks just mentioned) in the vicinity of the mine. None of these intrusions is known to cut the soda granite, and the andesite intrusives do not cut the rhyolite flows, suggesting that the andesite intrusives were essentially contemporaneous with the andesite flows and predated the rhyolites. The rhyolites are cut by soda-granite dikes, which dates the soda granite as later than any of the flows just discussed.

The next rock of igneous origin (Schilling, 1965) to have been intruded into the district was a granite porphyry plug that underlies the ridge between Sulphur Gulch and Iron Gulch in the area of the west edge of the present pit. At depth, this plug has been cut off by the Sulphur Gulch stock.

The Sulphur Gulch stock is one of several Tertiary granitic stocks in the Taos range; it has an elliptical form, extends 1.25 miles north from the mouth of Sulphur Gulch, and has a maximum width of 0.75 miles. There is a smaller body, the Goat Hill stock, 2 miles to the west of Sulphur Gulch; these two stocks apparently are connected beneath the overlying andesite by a sill-like mass of the Tertiary granite.

The fissure-vein molybdenum mineralizations of the Questa area were centered above the west margin of the Sulphur Gulch stock, mainly in the soda granite (aplite) and to a minor extent outside it.

The district also contains a plug and numerous dikes of quartz porphyry that are younger than the granite (aplite) of the Sulphur Gulch stock. The district also contains several breccia dikes, 5 to 20 feet wide, that intrude aplite on the ridge east of the pit and aplite and andesite in the pit due east of the granite porphyry plug. These breccias are made up of both rounded and angular fragments of aplite, quartz porphyry, and andesite in a matrix of fragments of quartz and granitic rocks; one of these dikes cuts through the open pit.

Since most of the vein-type ore, and of the lower-grade fracture-filling molybdenite ore as well, is found in the Sulphur Gulch stock and its aplite, sill-like extensions to the west, it is obvious that the ore was emplaced after the stock had been formed. It is far more probable that the ore fluids that deposited the ores were genetically related to the magmatic hearth from which the Sulphur Gulch soda granite came than that they were derived from the later, and much farther removed, Hinsdale basalts. Laughlin and his colleagues assign K/Ar dates for the soda granite and the molybdenum mineralization of between 23.5 and 22.3 m.y. From this it would seem to follow that the ores were introduced at much the same time as the soda granite and that they should, therefore, be classified as middle Tertiary.

Until recently, the molybdenite ore mined in the Questa district came from fissure veins along an east-west-trending south-dipping portion of the generally north-south-striking, west margin of the Sulphur Gulch soda-granite stock. It has been found, however, that enough molybdenite occurs in thin veinlets in stockworks in rock broken during the downfaulting of the east-west fault block, that appreciable volumes of the more fractured portions of the aplite and the overlying andesite, particularly in the flatter-lying portions adjacent to the Sulphur Gulch and Goat Hill stocks, can be mined as ore.

The Questa mine proper (also known as the Moly or R and S mine) was unusual among molybdenum mines in that the ore came from major fissure veins of considerable continuity and was of much higher grade in MoS_2 than obtained at Climax or in the porphyry copper deposits. Most of the fissure veins, now mined out, were on the contact between granite and propylitized volcanic rock or were in the granite within 50 feet of the contact; although several of the veins passed from granite into the altered volcanic rocks, they died out shortly after having left the granite.

Although it was known that brecciated zones containing molybdenite-bearing stockworks were present in the general area of the fissure veins, it was not until 1954 that systematic work was put in train to see if economically profitable bodies of this type could be found in the area of the Questa mine. By the end of 1961, two Climax-type bodies containing, in total, appreciably more than a quarter of a billion tons had been found; these are (1) the northeast zone that lies north-northwest of the lower workings of the old mine and (2) the southwest zone that lies west-southwest of the lower workings of the old mine. Since the northeast zone extends to, and can be mined by open-pitting from, the surface and since the southwest zone lies under more than 1000 feet of barren rock, it was decided to open-pit the northeast zone first. Later underground mining may recover the lower portions of the northeast zone and all of the southwest zone.

The stockwork-filling, Climax-type deposits in the Questa area actually are part of the even larger mineralized volume that includes not only the Questa mine and the two disseminated zones but also a huge volume of still lower-grade molybdenum mineralization that is not minable at present; it even may be that the lower-grade mineralized aplite and andesite that lie between them can be recovered. Although all the mineralized ground is considerably broken, the most broken volumes are so thoroughly fragmented by short, tight, irregular fissures that, when mineralized, they make up the two huge molybdenite-bearing stockworks. The fissure veins, on the other hand, were much fewer in number, much wider, more open, and longer and provided a home for so much molybdenite mineralization that they could be mined individually or in pairs.

The first mineral to have been deposited in the fissure veins was quartz; it is probable that the next mineral was the first generation of fluorite. The precipitation of biotite probably began before the end of that of the first generation of fluorite and was followed shortly by that of molybdenite; molybdenite may have overlapped with fluorite. Before molybdenite had ceased to deposit, pyrite, chalcopyrite, sphalerite, and galena had begun to form; they all probably stopped depositing at about the same time as the molybdenite and the quartz. The sulfides were followed (and perhaps overlapped slightly with) the second generation of fluorite. The fluorite, in turn, was followed, but not overlapped, by calcite and rhodochrosite; a little late molybdenite completed the mineral sequence. The molybdenite is normally found in irregular aggregates in the veins and less commonly forms banded arrangements with quartz; molybdenite also fills fractures in the quartz.

In the stockwork-type veinlets and disseminations in the surrounding wall rock, the mineral content is essentially the same as that of the larger fissure veins, although these lower-grade occurrences have not been studied to the extent of those in the larger fissure veins.

Much of the rock in the vicinity of the fissure and disseminated ore bodies was intensely altered during the sequence of hydrothermal events that included the deposition of the ore minerals. The altered areas are spatially associated with stocks, plugs, and dikes of soda granite and intrusive rhyolite.

The association in time and space of the molybdenite with quartz and biotite suggests that these minerals were deposited in the less intense portion of the hypothermal range. The alteration, however, appears to be typical of that developed under mesothermal conditions. Although the Questa deposits are appreciably both different in structure and mineralogy from those at Climax, the two deposits were probably formed under much the same conditions. Questa, nevertheless, lacks such high intensity minerals as topaz, orthoclase, and huebnerite that are present at Climax and may have formed under somewhat less intense conditions. It is, however, considered best to classify Questa as hypothermal-1 to mesothermal.

SANTA RITA-HANOVER

Late Mesozoic to Early Tertiary	*Copper, Zinc, Lead, Molybdenum*	*Hypothermal-2 to Mesothermal*

Allen, V. T., and Fahey, J. J., 1957, Some pyroxenes associated with pyrometasomatic zinc deposits in Mexico and New Mexico: Geol. Soc. Amer. Bull., v. 68, p. 881-898

Anderson, E. C., 1957, Central area, in *The metal resources of New Mexico and their economic features through 1954*: New Mex. Bur. Mines and Mineral Res. Bull. 39, p. 48-67

Barnes, H. L., 1959, The effect of metamorphism on metal distribution near base metal deposits: Econ. Geol., v. 54, p. 919-943 (particularly p. 922-932)

Belt, C. B., Jr., 1960, Intrusion and ore deposition in New Mexico: Econ. Geol., v. 55, p. 1244-1271 (particularly p. 1247-1257)

Creasey, S. C., 1959, Some phase relations in the hydrothermally altered rocks of porphyry copper deposits: Econ. Geol., v. 54, p. 351-373 (general)

Graf, D. F., and Kerr, P. F., 1950, Trace-element studies, Santa Rita, New Mexico: Geol. Soc. Amer. Bull., v. 61, p. 1023-1052

Hernon, R. M., and Jones, W. R., 1968, Ore deposits of the Central mining district, Grant County, New Mexico, in Ridge, J. D., Editor, *Ore deposits of the United States 1933-1967* (Graton-Sales Volumes): Chap. 58, v. 2

Hernon, R. M., and others, 1953, Some geologic features of the Santa Rita quadrangle, New Mexico: New Mex. Geol. Soc. 4th Ann. Field Conf. Guidebook, p. 117-130

—— 1964, Geology of the Santa Rita quadrangle, New Mexico: U.S. Geol. Surv. Geol. Quad. Map GQ-306, 1:24,000

—— 1965, Some geologic features of the Santa Rita quadrangle, New Mexico: New Mex. Geol. Soc. Guidebook, 16th Ann. Field Conf., p. 175-183 (plus map in pocket)

Horton, J. S., 1953, The Hanover mine--geology: Min. Eng., v. 5, no. 12, p. 1228-1229

Jones, W. R., Editor, 1955, Geologic map of the Central mining district, Grant County, New Mexico (with summary of strata and intrusive rocks): U.S. Geol. Surv. Open File Rept., 1:10,000

Jones, W. R., and others, 1961, Geologic events culminating in primary metalization in the Central mining district, Grant County, New Mexico: U.S. Geol. Surv. Prof. Paper 424-C, p. C11-C16

—— 1967, General geology of Santa Rita quadrangle, Grant County, New Mexico: U.S. Geol. Surv. Prof. Paper 555, 144 p.

Kerr, P. F., and others, 1950, Hydrothermal alteration at Santa Rita, New Mexico: Geol. Soc. Amer. Bull., v. 61, p. 275-347

Knopf, A., 1942, Central mining district, New Mexico, in Newhouse, W. H., Editor, *Ore deposits as related to structural features*: Princeton Univ. Press, p. 67

Lasky, S. G., 1930, Geology and ore deposits of the Ground Hog mine, Central district, Grant County, New Mexico: New Mex. Bur. Mines and Mineral Res., Circ. no. 2, 14 p.

—— 1936, Geology and ore deposits of the Bayard area, Central mining district, New Mexico: U.S. Geol. Surv. Bull. 870, 144 p.

Lasky, S. G., and Hoagland, A. D., 1950, Central mining district, New Mexico: 18th Int. Geol. Cong. Rept., pt. 7, p. 97-110

Leroy, P. G., 1954, Correlation of copper mineralization with hydrothermal alteration in the Santa Rita porphyry copper deposit, New Mexico: Geol. Soc. Amer. Bull., v. 65, p. 739-767

Lindgren, W., and others, 1910, The ore deposits of New Mexico: U.S. Geol. Surv. Prof. Paper 68, 361 p. (particularly p. 61-62, 305-318)

Nielsen, R. L., 1968, Hypogene texture and mineral zoning in a copper-bearing granodiorite porphyry stock, Santa Rita, New Mexico: Econ. Geol., v. 63, p. 37-50

—— 1970, Mineralization and alteration in calcareous rocks near the Santa Rita stock, New Mexico: New Mex. Geol. Soc., 21st Field Conf., New Mex. Bur. Mines and Mineral Res., p. 133-139

Ordoñez, G., and others, 1955, Geologic structures surrounding the Santa Rita intrusive, New Mexico: Econ. Geol., v. 50, p. 9-21

Paige, S., 1916, Silver City folio, New Mexico: U.S. Geol. Surv. Geol. Atlas, Folio 199, 19 p.

—— 1922, Copper deposits of the Tyrone district, New Mexico: U.S. Geol. Surv. Prof. Paper 122, 53 p. (particularly p. 4-40)

—— 1933, The region around Santa Rita and Hanover, New Mexico, in *Ore deposits of the Southwest*: 16th Int. Geol. Cong., Guidebook 14, p. 23-40

—— 1935, Santa Rita and Tyrone, New Mexico, in *Copper resources of the world*: 16th Int. Geol. Cong., vol. 1, p. 327-335

Rose, A. W., 1961, The iron content of sphalerite from the Central district, New Mexico, and the Bingham district, Utah: Econ. Geol., v. 56, p. 1363-1364

—— 1967, Trace elements in the sulfide minerals from Central district, New Mexico, and Bingham district, Utah: Geochim. et Cosmochim. Acta, v. 31, p. 547-585

—— 1970, Origin of trace element distribution patterns in sulfides of the Central and Bingham districts, Western U.S.A.: Mineralium Deposita, v. 5, p. 157-163

Rose, A. W., and Baltosser, W. W., 1966, The porphyry copper deposit at Santa Rita, New Mexico, in Titley, S. R., and Hicks, C. L., Editors, *Geology of the porphyry copper deposits--southwestern North America*: Univ. Ariz. Press, Tucson, p. 205-220

Schmitt, H., 1935, The Central mining district, New Mexico: A.I.M.E. Tr., v. 115, p. 187-208

—— 1939, The Pewabic mine: Geol. Soc. Amer. Bull., v. 50, p. 777-818

—— 1942, Certain ore deposits in the southwest, in Newhouse, W. H., Editor, *Ore deposits as related to structural features*: Princeton Univ. Press, p. 73-79 (particularly p. 73-77)

Schwartz, G. M., 1947, Hydrothermal alteration in the "porphyry copper" deposits: Econ. Geol., v. 42, p. 319-352 (particularly p. 332-336, 346-352)

Spencer, A. C., and Paige, S., 1935, Geology of the Santa Rita mining area, New Mexico: U.S. Geol. Surv. Bull. 859, 78 p.

Notes

The various deposits of the Santa Rita-Hanover area (the Central mining district) are located nearly 120 miles northwest of El Paso in Texas and less than 60 miles east of the Arizona line.

The oldest rocks in the district are Precambrian granite, gneiss, and schist, overlain by Paleozoic rocks that include representatives of all geologic periods within that era; the thicknesses of the formations lie with quite wide limits because of (1) erosion of parts of certain beds before the deposition of the next in the sequence, (2) stretching and thinning of beds near intrusive masses, and (3) appreciable relief on erosion surfaces at the time of deposition of the next succeeding bed. The oldest Paleozoic rock is the Upper Cambrian Bliss sandstone; it was followed by the Lower Ordovician El Paso limestone (and dolomite). Then follows the Upper Ordovician Montoya limestone (and dolomite) and the overlying Silurian Fusselman limestone (actually a cherty dolomite). Above an appreciable unconformity is the Upper Devonian Percha shale, followed by the Mississippian Lake Valley limestone that is the important sedimentary host rock to zinc ore in the district. It is unconformably overlain by the Pennsylvanian Magdalena group that includes the Syrena formation (limestone with interbedded shale) and the Oswaldo formation; the group contains iron and/or zinc ores at many places in the district. The Permian Abo formation (red shale with some thin-bedded limestone) ends the Paleozoic sequence. The area contains no rocks of Triassic, Jurassic, and Early Cretaceous age, but the Upper Cretaceous is represented by the Beartooth quartzite and the Colorado formation. Except for Miocene(?), Miocene (?) to Pliocene, and Pleistocene to Recent sediments that are post-ore, the Upper Cretaceous rocks mark the end of sedimentary processes in the district.

The igneous activity that began in the Late Cretaceous continued in the Central mining district at least into the Miocene and encompassed a wide variety of igneous magmas introduced into the area as a broad spectrum of types in a wide variety of physical relationships. The first event was the intrusion of a multitude of sills and laccoliths composed of porphyritic rocks ranging from rhyolite to andesite and quartz diorite. The propylitization that occurred as a result of these intrusions was essentially confined to the porphyries themselves. The introduction of these rocks was followed by major normal faulting and differential subsidence to produce the Santa Rita horst and several other small horsts and grabens and then by a swarm of dikes of mafic porphyry, plus andesite breccia, agglomerate, and mafic flows and plugs or irregular masses of gabbro to syenite. Andesite breccia in the early flows was extensively propylitized around the mafic plutons. The next event was the intrusion of discordant, composite stocks and plutons of granodiorite (Hanover-Fierro), quartz monzonite (Santa Rita), and quartz latite (Copper Flat) and the large amount of deformation caused thereby; any alteration that accompanied these intrusions was later covered by that caused by hypogene solutions. After the stocks and plutons had been emplaced, a swarm of granodiorite dikes was intruded with generally north and northeast trends; alteration appears to have been unimportant. These dikes were followed by the main epoch of primary mineralization during which existing faults were reactivated, breccia pipes developed, and the plutons fractured. The mineralization probably was deposited from hypogene ore fluids that rose along stock margins and fractures connected to those margins. High-temperature alteration minerals were developed for considerable distances into the wall rocks, and magnetite-chalcopyrite ores were formed in carbonate rocks near the plutons; chalcopyrite-molybdenite-adularia-quartz-epidote veins and veinlets formed in the propylitized part of the Santa Rita stock before the argillic alteration of that mass occurred. Sphalerite-minor galena-minor chalcopyrite replacement bodies formed in carbonate rocks both between Santa Rita and Hanover and around the south end of the Hanover-Fierro pluton. To the west and southwest of the plutons, sphalerite-galena and sphalerite-galena-chalcopyrite veins and replacement bodies formed along the major faults, along most granodiorite dikes, and around the pluton in Copper Flat. In the late stages of mineralization, much of the sill and stock volume at Santa Rita was altered to montmorillonite and biotite and to kaolinite-alunite-illite-quartz rock; in these altered zones, pyrite and chalcopyrite were introduced as disseminations and as veinlets. The alteration was later intensified by supergene processes and the ores enriched. The final event in the Late Cretaceous-early Tertiary sequence was the intrusion of a swarm of quartz monzonite porphyry dikes that were mildly to moderately altered and mineralized.

The igneous rocks (dikes and plugs) introduced in the middle and late Tertiary were essentially unaccompanied by mineralization and were separated from each other by the development of the Wimsattville formation in a steep-walled, crater-like hole; the rocks of older ages once in that area were either blown out or sank more than 1000 feet.

The events outlined above probably began in the Late Cretaceous and continued into the Tertiary. The principal age question to be answered concerns the length of Tertiary time over which the latter stages of intrusion, alteration, and mineralization took place. It is probable that the mineralization phase had been long concluded when the first of the Miocene (?) formations were laid down, and these followed the development of the Wimsattville formation that was post-ore and was even formed after the post-ore quartz monzonite dikes. It would seem reasonable, then, to assume that the mineralization had been emplaced quite early in the Tertiary and that the Santa Rita-Hanover ores should be classified as late Mesozoic to early Tertiary.

The ore mineralization in the Central mining district includes (1) the huge porphyry copper deposit of the Chino mine in the Santa Rita stock; (2) the nearly lead-free zinc deposits around the south lobe of the Hanover stock (Hanover and Pewabic mines), along the Barringer fault close to the Hanover stock (Continental mine), and north and northwest of the Santa Rita stock (Oswaldo No. 2 and Kearney mines); (3) the zinc replacement deposits, with lesser amounts of lead and copper, extending southwest from Hanover in a broad belt (the Groundhog, Bullfrog, and Blackhawk mines) and along the northeast end of the Barringer fault; (4) minor vein deposits of zinc, lead, and copper in the southeast quarter of the Santa Rita quadrangle; and (5) the magnetite-chalcopyrite deposits near the north end of the Fierro-Hanover stock.

The complexity of events attendant on the mineralization in the district and the widespread distribution of the ore bodies in themselves indicate a complicated geologic history that detailed study of the area has confirmed. The major emplacement of ore deposits occurred between the intrusion of the swarm of granodiorite porphyry dikes (earlier) and that of the quartz monzonite dike swarm (later). The first stage of the mineralization cycle was a high-temperature alteration that, in part, converted low-magnesium limestones of the Hanover area to silicates (garnet, salite, tremolite, actinolite, and epidote) and in part recrystallized them; the pyroxenes formed nearer the igneous bodies than the amphiboles. Some magnetite also was introduced into the rock volumes at this time. In this same stage, magnesium-rich limestones around the northern part of the Hanover-Fierro stock were converted to silicates (wollastonite, serpentine, and tremolite), with abundant magnetite.

Shales were converted to hornfels or epidotized; granodiorite dikes (many of which served as channelways for the altering solutions) and older porphyries were chloritized and epidotized; porphyries in the Bayard area were changed to calcite-sericite-clay-quartz close to the veins and to chlorite-sericite-epidote farther away. Magnetite ores with some hematite and minor (and at least in part later) chalcopyrite formed around the Hanover and Santa Rita stocks, and a quartz-chalcopyrite-molybdenite-pyrite-chlorite-adularia mineralization filled fractures in the propylitized portions of the Santa Rita igneous mass and around the south lobe of the Hanover stock.

The next stage was the development of the various types of zinc mineralization in essentially all parts of the district except in the Santa Rita stock and a zone extending outward from it for 500 feet. Sphalerite, with only minor galena and chalcopyrite, formed replacement masses in the altered carbonate rocks around the south lobe of the Hanover stock and between that stock and the Santa Rita igneous mass. Sphalerite-galena and sphalerite-galena-chalcopyrite veins and replacement bodies were introduced along the major faults (such as the Groundhog, Hobo, and Barringer), along most granodiorite dikes, and around the Copper Flat pluton.

An apparently later stage of chalcopyrite mineralization than that which is present in the ore bodies just mentioned was deposited in the Santa Rita stock after that mass had first been altered to montmorillonite and biotite and then to kaolinite-alunite-illite-quartz rock; small volumes of the rock were completely silicified. The chalcopyrite and much more abundant pyrite were emplaced as disseminations and as veinlets; some of these two minerals are found in the later quartz monzonite as well as in the granodiorite. These relationships can be interpreted as the result of either (1) two periods of pyrite-chalcopyrite mineralization, one before and one after the intrusion of the quartz monzonite dikes or (2) one long-continued period of pyrite-chalcopyrite mineralization only slightly affected by the intrusion of the later dike swarm. The chalcopyrite in general is spatially associated with those alterations that bleached the igneous mass at Santa Rita by converting epidote and chlorite to sericite, argillite, and quartz; however, in detail chalcopyrite is actually most abundant in slightly altered rock, particularly where hornblende has been biotized.

The relationships almost certainly demonstrate that the zinc mineralization, so closely related in time to the high-temperature silicates and the magnetite, must have been formed under hypothermal conditions. Most of the zinc ore is emplaced in carbonate rocks, as are the galena and minor chalcopyrite associated locally with it, so it is here classified as hypothermal-2. The molybdenum of the Santa Rita stock occurred in veinlets with quartz, chlorite, adularia, pyrite, and minor chalcopyrite; this mineralization probably was formed near the boundary between hypothermal and mesothermal conditions. Finally, the main pyrite-chalcopyrite mineralization seems to have taken place just after the argillic alteration of much of the stock, which strongly suggests that it was developed under mesothermal conditions. The iron ores at Fierro, not actually included in the classification here, probably formed under essentially the same conditions as the sphalerite and should be classified as hypothermal-2; the chalcopyrite found in the magnetite ore is probably of the same general age as that associated with the hypothermal sphalerite; the high-temperature deposition of this chalcopyrite is confirmed by the presence of some minor cubanite with the iron-copper sulfide. Inclusions of sedimentary rocks in the Santa Rita stock are as highly altered as those sediments outside the stock, and this alteration probably is connected in time to that of early chalcopyrite-molybdenite-pyrite-quartz-chlorite-adularia mineralization rather than to the later chalcopyrite-pyrite-quartz mineralization.

New York

ADIRONDACKS (GENERAL)

Late Precambrian	*Iron, Titanium, Zinc, Lead, Garnet, Pyrites*	*Magmatic, Hydrothermal, Metamorphic*

Alling, H. L., 1925, Genesis of the Adirondack magnetites: Econ. Geol., v. 20, p. 335-363

—— 1929, The ages of the Adirondack gabbros: Amer. Jour. Sci., 5th Ser., v. 18, p. 472-476

—— 1939, Metasomatic origin of the Adirondack magnetite deposits: Econ. Geol., v. 34, p. 141-172

—— 1942, The Adirondack magnetite deposits, in Newhouse, W. H., Editor, *Ore deposits as related to structural features*: Princeton Univ. Press, p. 143-146

Alling, H. L., and Kemp, J. F., 1925, The Adirondack anorthosite and its problems: Jour. Geol., v. 40, p. 193-237

Balk, R., 1931, Structural geology of the Adirondack anorthosite: Mineral. u. Petrog. Mitt., Bd. 41, H. 3-6, S. 308-434

—— 1944, Comments on some eastern Adirondack problems: Jour. Geol., v. 52, p. 289-318

Bowen, N. L., 1917, Adirondack intrusives: Jour. Geol., v. 25, p. 509-512; disc., p. 512-514

—— 1917, The problem of the anorthosites: Jour. Geol., v. 25, p. 209-243

Buddington, A. F., 1939, Adirondack igneous rocks and their metamorphism: Geol. Soc. Amer. Mem. 7, 354 p.

—— 1966, The Precambrian magnetite deposits of New York and New Jersey: Econ. Geol., v. 61, p. 484-510; disc., p. 1290-1291

Buddington, A. F., and others, 1955, Thermometric and petrogenic significance of titaniferous magnetite: Amer. Jour. Sci., v. 253, no. 9, p. 497-532

—— 1963, Degree of oxidation of Adirondack iron oxide and iron-titanium oxide minerals in relation to petrogeny: Jour. Petrol., v. 4, p. 138-169

Cushing, H. P., and others, 1915, Age of the igneous rocks of the Adirondacks region: Amer. Jour. Sci., 4th Ser., v. 39, p. 288-294

Engel, A.E.J., 1956, Apropos the Grenville, in Thomson, James E., Editor, *The Grenville problem*: Roy. Soc. Canada Spec. Pubs. no. 1, p. 74-96

Engel, A.E.J., and Engel C. G., 1953, Grenville series in the northwest Adirondack Mountains, New York; Part I: General features of the Grenville series: Geol. Soc. Amer. Bull., v. 64, p. 1013-1047

—— 1963, Metasomatic origin of large parts of the Adirondack phacoliths: Geol. Soc. Amer. Bull., v. 74, p. 349-352; disc., p. 353

Evrard, P., 1949, The differentiation of titaniferous magmas: Econ. Geol., v. 44, p. 210-232

Hills, A., and Gast, P. W., 1964, Age of pyroxene-hornblende granitic gneiss of the eastern Adirondacks by the rubidium-strontium whole-rock method: Geol. Soc. Amer. Bull., v. 75, p. 759-766

Kemp, J. F., 1897-1898, The titaniferous iron ores of the Adirondacks: U.S. Geol. Surv. 19th Ann. Rept., pt. 3, p. 377-422

Miller, W. J., 1918, Adirondack anorthosite: Geol. Soc. Amer. Bull., v. 29, p. 399-462

Newland, D. H., 1908, Geology of the Adirondack magnetic iron ores: N. Y. State Mus. Bull. no. 119, 182 p.

Osborne, F. F., 1928, Certain magmatic titaniferous iron ores and their origin: Econ. Geol., v. 23, p. 724-761, 895-922

Simmons, G., 1964, Gravity survey and geological interpretation, northern New York: Geol. Soc. Amer. Bull., v. 75, p. 81-98

Turner, F. J., and Verhoogen, J., 1960, The anorthosite massif of the Adirondacks, in *Igneous and metamorphic petrology*, 2d ed.: McGraw-Hill, New York, p. 323-328

Waard, D. de, 1969, Facies series and P-T conditions of metamorphism in the Adirondack Mountains: Ned. Akad. Wetensch. Pr., Ser. B, v. 50, no. 2, p. 124-131

Walton, M., and Waard, D. de, 1963, Geologic evolution of the Precambrian in the Adirondack Highlands, a new synthesis: Kon. Akademie van Wetenschappen, Sec. Sci. (Afdeeling Naturkunde), Pr., v. 66, p. 98-106

Notes

The Adirondacks occupy a rudely circular area of some 3000 square miles in the northeast corner of New York State.

The rocks of the region are essentially all Precambrian in age and are connected to those of the Canadian Shield by the Frontenac axis that crosses the St. Lawrence River near Kingston, Ontario. The northwestern portion of the Adirondacks (the Grenville lowland) is a gently rolling terrain (with an average elevation of about 1000 feet) and the southeastern is a ruggedly mountainous area (with many ridges standing above 3000 feet and the highest summit, Mt. Marcy, reaching nearly 5350 feet above sea level). The Adirondacks are bordered (except for the Frontenac axis) entirely by Cambrian and Ordovician sediments that dip gently away from the massif; small, uneroded patches of these sediments are found within the uplifted Precambrian area. The already complexly folded rocks of the Adirondacks were moderately uplifted during the late Precambrian and by late Cincinnatian time were arched into a broad domed structure; during the Taconic orogeny, the dome was broken along its eastern side by block faults. Since that time, the area has been largely one of erosion.

The oldest rocks in the Adirondack complex are those of the Grenville series, consisting of marbles, quartzites, amphibolites, and paragneisses, with volcanic rocks, pillow lavas, conglomerates, and graywackes being local variations from the normal character of the series. Nevertheless, these latter rock types have been found interfingering with the diagnostic marble of the Grenville series and may well be acceptable members of the series.

Although the Adirondacks as a structural unit have been uplifted in a broad dome, the Grenville rocks of the region have been moderately to highly deformed and injected by a series of igneous rocks; in order of decreasing age, these are anorthosites, gabbros, diorites, syenites, and a complex of granites. Some of the amphibolites in the Grenville may have been derived from the metamorphism of gabbros earlier than those so closely associated in time with the anorthosites. Other amphibolites may have been formed by metasomatism of limestone or marble during the intrusion of the granites and still others may have come from lava flows and tuffs or even from argillaceous dolomite layers; at least half the amphibolites cannot be assigned a definite mode of origin.

The controversy over the origin of the Grenville, great though it is, is less intense than the dispute that has arisen over the origin of the igneous rocks that have intruded it. Although there appears, until recently, to have been general agreement that the order of introduction into the area was anorthosite, gabbro, syenites, and granites, even this has not been fully accepted for there are those who have thought that the gabbros preceded the anorthosite. Further, there is much argument as to whether the igneous rocks were derived entirely from one parent magma, of a composition intermediate between gabbroic anorthosite on the one hand and syenite-granite on the other, or from three magmas, one gabbroic anorthosite, one essentially quartz syenite, and a third granite. Questions also have arisen as to whether required differentiation took place at depth or in the general rock volume into which the magma or magmas were intruded. At present it appears more probable that there were three magma sources, one for the gabbroic-anorthosite magma, another for the quartz-syenite magma, and a third for the late granites.

Buddington (1939) points out that the introduction of the gabbroic anorthosite magma resulted in the development of composite gabbroic-anorthosite border facies as chilled zones around the anorthosite massifs and that the settling of mafic minerals during the consolidation of the magma resulted in true anorthosite lying above minor amounts of more gabbroic material beneath. He also believes that, as crystallization of the gabbroic anorthosite progressed, gabbroic and mafic-gabbroic magmas (and even iron-titanium-rich immiscible phases) developed as late-stage differentiates and that these gabbros were olivine-free rocks. He also considers that the olivine-bearing gabbros, generally found in the area as sheets and dikes, were not differentiates from the gabbroic anorthosite but were intruded after the gabbroic anorthosite and from a somewhat different source. That these relationships of saturated and unsaturated gabbros to the gabbroic anorthosite did not always hold is suggested by Levin's (1950) work in the Gore Mountain area where gabbro that grades into anorthosite is definitely olivine-bearing. The possibility exists that some of the unsaturated gabbros were products of the differentiation of the gabbroic anorthosite magma. Buddington believes that the intrusion of the syenite magma, although not much later than that of the gabbroic anorthosite, was from a separate source and that this magma entered the area in sheets and local stock-like masses.

The intrusion of the syenite magma was followed, and probably late in its history accompanied by, orogenic deformation and dynamic metamorphism. In the later stages of metamorphism and deformation, granitic intrusions of batholithic size, as well as smaller phacoliths and sheets, were introduced into the area. At contacts between syenite and granite and between anorthosite and granite, transitional rocks were formed that suggest that the region was still at an elevated temperature when the granite magma entered it.

This scheme for the development of the Adirondack igneous rocks differs somewhat in detail from those of Bowen, Miller, and Balk, but all of these authors agree in the essentially magmatic character of the events involved in the production of the igneous masses. Recently, Engel and Engel (1963) and Walton and de Waard (1963) have suggested that much of the igneous rock of the area is either of metasomatic origin or was present in the area prior to Grenville sedimentation. Buddington (1963), however, questions whether the "mystique" of the potentiality of emanations to produce any results one wants for any time, place, or physical conditions is really applicable without limitation.

Recent gravity surveys are interpreted by Simmons (1964) with considerable elegance to show that the anorthosite massif cannot be a batholith or a large tilted lens but must be essentially a slab or giant sill with two roots that extend downward for about 10 km. The anorthosite sill is between 3.0 and 4.5 km thick and is a layered body as Buddington has suggested it would be; the amount of gabbro does not appear to be significant as Buddington's concept also requires.

Whole-rock Rb/Sr analysis on 8 samples of granitic gneiss from the eastern Adirondacks are in close agreement in suggesting an age for the gneiss of 1035 m.y. (± 20 m.y.). The age of the Grenville sedimentation probably, but not definitely, is older than that of the granites, and the age of the anorthosites, gabbros, and syenites apparently lies between that of the primary sedimentation and of the development of the granite. Throughout the Grenville province, age determinations on lead and uranium in zircon, apatite, and sphene in granites agree with the results of Hills and Gast (1964). These results almost certainly indicate that the last important metamorphism in the area occurred about 1035 m.y. ago. This may mean that (1) the granite is actually 1035 m.y. old; (2) the original rock from which the granite was derived was so thoroughly homogenized by melting and mixing or by diffusion 1035 m.y. ago as to give the same Rb/Sr ratios today as would a granite magmatically emplaced at that time; or (3) most or all of the Rb present was introduced during the pervasive granitization, 1035 m.y. ago, of a homogeneous host rock. Obviously the ore deposits of the Adirondacks which were developed during the last orogeny to affect the area must, therefore, be of Grenville age (1035 m.y. [± 20 m.y.]) and should be classified as late Precambrian no matter what the original age of the metasedimentary and igneous (or pseudo-igneous) rocks may have been or what their geologic history had been prior to the Grenville orogeny.

Although the classifications of the various Adirondack ore deposits treated in this volume (Balmat-Edwards, Benson Mines, Gore Mountain, Lyon Mountain, and Sanford Lake) are considered in the individual discussions of these districts, it should be said here that these discussions are based on the assumption that Buddington's concept for the development of the Adirondack igneous rocks and their time relations to the Grenville metasediments is correct. It will be shown that the intimate associations of these ore bodies to Grenville igneous activity strongly suggest that each is a product of processes that had their beginnings, at least, in Grenville magmas. The Balmat-Edwards deposits were formed by hydrothermal solutions that probably were generated in Grenville granite magmas; the Gore Mountain deposits by deuteric alterations in Grenville gabbro (derived from gabbroic anorthosite magmas), aided by dynamic metamorphic effects of intrusive syenite; the Lyon Mountain deposits by hydrothermal solutions probably generated in Grenville granite magmas, and the Sanford Lake deposits from late-stage iron-titanium-rich masses developed as immiscible phases in the late stages of the crystallization of the Grenville gabbroic anorthosite. The reasons for each of these classifications are given in the appropriate places in the discussions of the deposits.

BALMAT-EDWARDS

Late Precambrian *Zinc, Lead, Pyrites* *Hypothermal-2*

Brown, J. S., 1936, Structure and primary mineralization of the zinc mine at Balmat, New York: Econ. Geol., v. 31, p. 233-258

—— 1936, Supergene sphalerite, galena, and willemite at Balmat, New York: Econ. Geol., v. 31, p. 331-354

—— 1941, Factors of composition and porosity in lead-zinc replacements of metamorphosed limestone: A.I.M.E. Tr., v. 144, p. 250-263 (particularly p. 253-256)

—— 1942, Edwards-Balmat zinc district, New York, in Newhouse, W. H., Editor, *Ore deposits as related to structural features*: Princeton Univ. Press, p. 171-174

—— 1947, Porosity and ore deposition at Edwards and Balmat, New York: Geol. Soc. Amer. Bull., v. 58, p. 505-546

—— 1959, Occurrence of jordanite at Balmat, N.Y.: Econ. Geol., v. 54, p. 136-137

Brown, J. S., and Engel, A.E.J., 1956, Revision of Grenville stratigraphy and structure in the Balmat-Edwards district, northwest Adirondacks, New York: Geol. Soc. Amer. Bull., v. 67, p. 1599-1622

Brown, J. S., and Kulp, J. L., 1959, Lead isotopes from Balmat area, New York: Econ. Geol., v. 54, p. 137-139

Buddington, A. F., 1929, Granite phacoliths and their contact zones in the northwest Adirondacks: N.Y. State Mus. Bull. no. 281, 7, p. 51-107

Cushing, H. P., and Newland, D. H., 1925, Geology of the Gouverneur quadrangle: N.Y. State Mus. Bull. no. 259, 122 p. (particularly p. 14-105)

Doe, B. R., 1962, Distribution and composition of sulfide minerals at Balmat, N.Y.: Geol. Soc. Amer. Bull., v. 73, p. 833-854

—— 1962, Relationships of lead isotopes among granites, pegmatites, and sulfide ores near Balmat, New York: Jour. Geophys. Res., v. 67, p. 2895-2906

Engel, A. E. J., and Engel, C. G., 1958, Progressive metamorphism and granitization of the major paragneiss, northwest Adirondack Mountains, New York, Part I, Total rock: Geol. Soc. Amer. Bull., v. 69, p. 1369-1414; 1960, Part II, Mineralogy: Geol. Soc. Amer. Bull., v. 71, p. 1-58

—— 1962, Hornblendes formed during progressive metamorphism of amphibolites, northwest Adirondack Mountains, New York: Geol. Soc. Amer. Bull., v. 73, p. 1499-1514

Lea, E. R., and Dill, D. B., 1968, Zinc deposits of the Balmat-Edwards district, New York, in Ridge, J. D., Editor, *Ore deposits of the United States, 1933-1967* (Graton-Sales Volumes): Chap. 2, v. 1

Lessing, P., 1969, Jordanite at Balmat, New York: Econ. Geol., v. 64, p. 932

Newland, D. H., 1916, The new zinc mining district near Edwards, N. Y.: Econ. Geol., v. 11, p. 623-644

Reynolds, P. H., and Russell, R. D., 1968, Isotopic composition of lead from Balmat, New York: Canadian Jour. Earth Sci., v. 5, p. 1239-1245

Smythe, C. H., Jr., 1918, Genesis of the zinc ores of the Edwards district, St. Lawrence County, N. Y.: N. Y. State Mus. Bull., no. 201, 41 p.

Notes

The ore deposits of the Balmat-Edwards district lie in a northeast-southwest trending zone of the zinc-lead and talc mineralization that begins just west of Balmat Corners, runs between Fowler and Fullerville and through Talcville to the vicinity of Edwards. Although the Balmat mine has been the largest producer in the area, considerable tonnages of zinc-lead ore have been mined at Edwards and smaller amounts from the Hyatt mine, less than a mile southwest of Talcville. A number of talc mines also operate in the area; these include the American, Gouverneur, and Wright mines, all located within about 4000 feet of the Balmat No. 2 mine. The general area of mineralization is some 100 miles north-northeast of Syracuse.

The district is a part of the Grenville lowlands of the northwest Adirondacks not far from the line of transition between the metasedimentary lowlands and the silicic igneous and metamorphic rocks that form the perimeter of the Adirondack massif. In the general region, the rocks are those of the late Precambrian Grenville series; the series is divided into five major divisions, the lower two of which are not present in the immediate Balmat-Edwards district. These divisions are, from top to bottom: (1) feldspathic gneiss, (2) upper marble belt which contains the zinc and talc deposits of the Balmat-Edwards district, (3) quartz-biotite-oligoclase gneiss, (4) siliceous dolomite marble, the Gouverneur marble belt, and (5) siliceous and gneissic marble, the Black Lake sedimentary belt.

The oldest rock in the Balmat-Edwards area is a quartz-biotite oligoclase gneiss, referred to by Brown and Engel (1956) as a migmatitic paragneiss; the gneiss largely surrounds the district and encloses the younger rocks of the upper marble belt; the gneiss is about 3000 feet thick. The upper marble belt in the Balmat area consists of 15 units.

The sedimentary sequence, in addition to the complex folding to which it was subjected, was intruded by alaskitic and biotitic gneissic granites (which Doe considers to be synkinematic) and associated pegmatites and by gabbro. Probably none of the gabbro is unaltered; that which most resembles the original rock appears to have been somewhat granitized, while the remainder has been converted to amphibolite. Not all of the amphibolite, however, seems to have been derived from gabbro; some of it undoubtedly has come from the alteration of volcanic rocks and some from the alteration of highly silicated carbonate rocks.

Much work has been done on the dating of the pegmatites and granite gneisses that appear to be intrusive into the Grenville rocks. Several isotopic techniques have been used, and these indicate that the igneous masses were either introduced into the area between 950 and 1200 m.y. ago or were highly changed during that period. A recent date by the K/Ar method on phlogopite in marble from the general Balmat belt indicates that it was formed at least 1000 to 1200 m.y. ago. The ores probably were introduced after the silicate minerals had been developed in the marbles, but it would also seem that the introduction of the ores occurred not long after the metamorphic alteration of the carbonate rocks of the Grenville metasediments to their present composition. It is Doe's opinion (1962) that 1050 m.y. ± 50 m.y. represents the most recent episode of severe metamorphism in the Balmat area, and it is most probable that the process of ore mineralization took place in the declining stages of that period of igneous and metamorphic activity. It, therefore, seems reasonable to date the emplacement of the Balmat-Edwards ores as having happened about 1000 m.y. ago and to classify them as late Precambrian. If the ores are metamorphosed syngenetic deposits (Lea and Dill, 1968), they were part of the original Grenville sediments and may be as old as early Precambrian.

Pyrite is the oldest mineral of the sulfide suite; anthophyllite began to deposit at about the same time as pyrite and continued almost until the end of the deposition of the certainly primary sphalerite. Pyrite was followed by brecciation and then by a first generation of galena which, in turn, was followed with slight overlap by sphalerite--this sphalerite contains innumerable tiny blebs of chalcopyrite that almost certainly were exsolved from solid solution after the deposition of the sphalerite. Pyrrhotite fills cracks in pyrite and probably also was formed after most of the sphalerite was emplaced; it was largely confined to the certral ore body of the No. 2 mine and overlapped with the sphalerite during the first part of its period of deposition. The second generation of galena overlapped with the last of the pyrrhotite. The pyrrhotite does not appear to contain exsolved pyrite or sphalerite, nor does sphalerite contain any exsolved pyrrhotite. The deposition of galena was followed by a period of brecciation and then by the deposition of some chalcopyrite, mainly in association with pyrrhotite in the No. 2 mine. Further brecciation was followed or accompanied by the deposition of minor marcasite, barite, and anhydrite and then by a second generation of sphalerite; this latter sphalerite overlapped serpentine and talc produced in the late stages from early hydrothermal silicates. A still later brecciation was followed by a new generation of sphalerite, galena, and chalcopyrite, accompanied by chlorite, iron oxides (mainly earthy hematite), and some ilvaite. Brown (1936) thought these minerals were formed by surface processes because of their spatial relation to the present surface, but Dill (1962) was not so sure they were secondary; they certainly are not typical supergene minerals, and probably are primary.

As evidence of high-temperature deposition of the ore sulfides in the district, it should be pointed out that there is certainly exsolution chalcopyrite in the sphalerite, a definite indication of hypothermal conditions. It is uncertain as to whether or not the primary chalcopyrite contains exsolved sphalerite; it probably does. Thus, the bulk of the Balmat mineralization, at least, appears to have taken place in the hypothermal range. Almost certainly, however, the marcasite, barite, and anhydrite did not form in that high-intensity range nor, probably, did the sphalerite deposit with and after these minerals. It is uncertain what percentage of the Balmat sphalerite formed after marcasite had begun to precipitate, but it is assumed here not to have been enough to justify the addition of the term mesothermal to the classification. Although the Edwards mineralization appears to have been less complex than that at Balmat, there seems to be no evidence to indicate that it was not formed under hypothermal conditions. The host rocks of the ores in both mines are so largely calcareous rocks that the deposits must be considered as having formed in calcareous rocks. The Balmat-Edwards deposits, therefore, are here categorized as hypothermal-2.

If the late sulfide minerals are determined to have been formed by hydrothermal solutions, they also will probably be classifiable as hypothermal, but their classification is held in abeyance at this time.

The concept has been put forward (Lea and Dill, 1968) that the ores are where they are as a result of the effect of metamorphism of the rocks of the area that caused the concentration of the metal ions as sulfides in the present ore bodies. The minerals in which these metal ions were included before the metamorphism and the manner in which they were moved to, and concentrated in, their present locations is not clear. Lead isotope ratios are an important factor in the rationale for this theory (Brown and Kulp, 1959; Doe, 1962; Reynolds and Russell, 1968).

BENSON MINES

Late Precambrian — *Iron as Magnetite, Hematite* — *Hypothermal-1*

Buddington, A. F., and Leonard, B. F., 1962, Regional geology of the St. Lawrence County magnetite district, northwest Adirondacks, New York: U.S. Geol. Surv. Prof. Paper 376, 145 p. (does not deal with the Benson mines specifically but is invaluable for an understanding of the district)

—— 1964, Ore deposits of the St. Lawrence magnetite district, northwest Adirondacks, New York: U.S. Geol. Surv. Prof. Paper 377, 259 p. (discusses the Benson mines only incidentally but is invaluable for an understanding of this type of deposit)

Crump, R. M., and Beutner, E. L., 1968, The Benson mines iron ore deposit, St. Lawrence County, New York, in Ridge, J. D., Editor, *Ore deposits of the United States, 1933-1967* (Graton-Sales Volumes): Chap. 3, v. 1

Hagni, R. D., 1968, Titanium occurrence and distribution in the magnetite-hematite deposit at Benson mines, New York: Econ. Geol., v. 63, p. 151-155

Hagni, R. D., and others, 1969, Metamorphic aspects of the magnetite-hematite deposit at Benson mines, New York: Econ. Geol., v. 64, p. 183-190

Palmer, D. F., 1970, Geology and ore deposits near Benson mines, New York: Econ. Geol., v. 65, p. 31-39

Notes

The Benson mines are located in St. Lawrence County, New York, between the villages of Star Lake and Newton Falls, just over 50 miles east-northeast of Watertown.

The Benson mines are situated in the Adirondack portion of the Grenville subprovince of the Canadian Shield. The main and highest axis of the Adirondacks lies to the southeast of the mine area, and the mine is in what Buddington and Leonard (1962) call the Chitwold Rock Terrace. The ores are enclosed in gneisses that reached their present metamorphic grade during the Grenville orogeny of late Precambrian time and are placed in the Grenville series. In the immediate vicinity of the main ore body, the hanging-wall rocks are metasediments into which some material of granitic composition has been intruded; the igneous phase is conformable with the banding in the gneisses. The gneissic host rocks of the ore have been entirely changed from their original character, which Crump and Beutner believe to have been clay sediments. On the other hand, Engel and Engel (1958) think that the rock initially was a tuff or graywacke. Buddington and Leonard consider that the original rock was metamorphically converted to a biotite-quartz-plagioclase gneiss and then, through migmatization and granitization, to a sillimanite-quartz-microcline gneiss. In 1966, Buddington suggested that the latter gneiss had only a slightly modified metasedimentary origin. The footwall of the ore is a ferromagnesian gneiss. The major structure of the deposit is a syncline (or synform, as Hagni and his colleagues prefer to term it) that plunges about 20°N and is overturned to the west in the neighborhood of the mine. Within the mineralized portion of the area, some pegmatites are known that strike across the banding of the gneisses and have a steep to vertical dip. It is the opinion of Crump and Beutner (1968) that the pegmatites are later than the folding since they cut the folding as well as the gneissic banding. They place the mineralization as prepegmatite but as after the injection of the igneous material into the footwall gneisses, the metamorphism, and the folding. If this is the case, then the ores must be of essentially the age of the Grenville orogeny and are, therefore, about 1000 m.y. old. Hagni and others (1969) argue that the ores were subjected to the metamorphism, the iron having been part of the original mafic silicates and having been released during metamorphism to migrate to zones of lower pressure where they were deposited as iron oxides. Obviously, these two theories could not be farther apart and their relative validity will be considered below. If Crump and Beutner are correct, the ores certainly are late Precambrian (as they are here categorized) or much older, perhaps as old as early Precambrian if Hagni and his colleagues have the best of the argument.

The principal ore minerals in the Benson mines are magnetite and hematite, and these are intermixed enough to cause problems in their separation. The main ore mass lies in the complexly folded synform over a horizontal distance of about 8000 feet, its structure reflecting the complex folding of the gneisses. To the west of the main ore body is the much smaller Amoeba pit, the location of which results from another expression of the roll on the east limb of the syncline that forms a subsidiary anticline and syncline on that limb.

The work of Crump and Beutner indicates that all of the host-rock minerals have been replaced by iron oxides to some extent with the exception of sillimanite; all the host-rock minerals (except sillimanite) have no recognizable external form, and examples of partially replaced grains of all these host-rock minerals can be found. The replacement does not seem to favor any one mineral as opposed to the others of the gangue. Away from the ore, both down dip and along strike, the unmineralized gneiss has the same minerals as those that appear as gangue in the ore; so that, if the ore was formed by metamorphism, no new minerals were produced in the process or old ones destroyed. Hagni and his colleagues argue that certain micro-textures, although shown in only a few of the opaque iron oxides, clearly indicate to them that the iron oxide grains must have been present when the silicates were developed. These textures include ovate magnetite and hematite grains surrounded by rims of silicate minerals (sillimanite, garnet, and plagioclase) and iron oxide inclusions in poikiloblastic garnet and, locally, sillimanite. These textures can have been the result of metamorphic accretions around earlier iron oxide grains, but they also can have been the result of replacement. Hagni and his colleagues would have the iron released from the mafic silicates and concentrated as iron oxides in zones of lower pressure. This suggestion cannot be dignified by the name of theory since no explanation is given as to why iron should move in preference to other elements or how it obtains the oxygen it needs to form iron oxides from it. In addition to magnetite and hematite, some of the more magnetic ore contains abundant pyrite and some pyrrhotite; locally pyrrhotite veins cut across the gneiss banding. Several copper minerals are known in the deposit, of which chalcopyrite and bornite are probably primary and also cut across the gneissic banding while the others, that is, chalcocite, covellite, native copper, azurite, and malachite, are probably secondary. Molybdenite, in contrast to the copper minerals, is concentrated in the pegmatites but also is found sparsely disseminated in the ore-bearing gneiss. Certainly all the primary copper sulfides and the molybdenite are later than the iron oxides. None of the silicate minerals can be described definitely as a gangue mineral in the sense of having been introduced hydrothermally in the same time span as the iron oxides and the copper minerals. All of the silicates are normally produced metamorphically, and no evidence has been found of any retrograde metamorphism in these silicates that might be ascribed to effects of the hydrothermal fluids. The high-temperature nature of the magnetite and hematite, however, is shown by the presence of rutile and ilmenite as exsolution laths in the iron oxides. In short, much more work and study on the deposits are needed and no definite and final classification can be assigned to the ores.

To me, however, the ores appear to be postmetamorphic, and the textures described by Hagni and his colleagues could be as easily the result of replacement as of metamorphism. The ores, therefore, are here classified as hypothermal-1.

GORE MOUNTAIN

Late Precambrian — *Garnet* — *Magmatic-4*
Metamorphic-C

Bartholomé, P., 1960, Genesis of the Gore Mountain garnet deposit, New York: Econ. Geol., v. 55, p. 255-277

Buddington, A. F., 1952, Chemical petrology of some metamorphosed Adirondack gabbroic, syenitic, and quartz syenitic rocks: Amer. Jour. Sci., Bowen Volume, p. 37-84 (particularly p. 62-63, 70-83)

Krieger, M. H., 1938, Geology of the Thirteenth Lake quadrangle, New York: N. Y. State Mus. Bull., no. 308, 124 p. (particularly p. 100-120)

Levin, S. B., 1948, Petrology and genesis of Gore Mountain garnet, New York (abs.): Geol. Soc. Amer. Bull., v. 59, p. 1335-1336

—— 1949, Garnet evidence in Adirondack petrogeny: N. Y. Acad. Sci. Tr., Ser. 2, v. 11, p. 156-162

—— 1950, Genesis of some Adirondack garnet deposits: Geol. Soc. Amer. Bull., v. 61, p. 519-565

—— 1950, Origin of hornblende rims on Adirondack garnet (abs.): Geol. Soc. Amer. Bull., v. 61, p. 1482

Miller, W. J., 1938, Genesis of certain Adirondack garnet deposits: Amer. Mineral., v. 23, p. 399-408 (particularly p. 403-407)

Shand, S. J., 1945, Coronas and coronites: Geol. Soc. Amer. Bull., v. 56, p. 247-266 (particularly p. 256-258)

Shaub, B. M., 1949, Paragenesis of the garnet and associated minerals of the Barton mine near North Creek, New York: Amer. Mineral, v. 34, p. 573-582

Wentorf, R. H., Jr., 1956, Formation of Gore Mountain garnet and hornblende at high temperature and pressure: Amer. Jour. Sci., v. 254, no. 7, p. 413-419

Notes

The garnet deposits of the Barton mine just north of Gore Mountain are located about 4.5 miles south of North River, a town on the south bank of the Hudson; the area is just over 70 miles slightly north of west from Albany.

The mineralized rock volume north of Gore Mountain is an outlier (immediately to the northeast) of the most southern and intermediate-sized body of the three of which the Adirondack anorthosite complex is composed. Between the major central anorthosite mass and the southern one there is about 20 miles of syenite, granite, and Grenville metasediments into the last of which the anorthosite complex was intruded. The oldest rocks in the Gore Mountain area, therefore, are those of the Grenville series. The nearest surface outcrop of Grenville rocks to the Gore Mountain deposit is over 2.5 miles removed so it is unlikely that the Grenville rocks had any part in the development of the garnet of the Barton mine. It has been suggested, however, that the garnet-rich mass was developed in a Grenville xenolith enclosed in the gabbroic portion of the anorthosite outlier; this concept is not now considered to be valid. The oldest igneous rocks of the area (except for some metamorphosed igneous materials in the Grenville) are the anorthosites; both the Whiteface (border) facies (which includes anorthositic gabbros and gabbroic anorthosites and Keene gneiss) and the Marcy (core) facies are present in the area.

The Barton garnet deposit is contained in a small body of anorthosite that lies just north of the southern anorthosite mass; this small body is about 7000 feet long (from east to west) and some 5000 feet wide (from north to south).

The anorthosite apparently is the oldest rock in the area, and the gabbro crystallized from the final molten portion of the anorthositic magma which, through the early crystallization of plagioclase feldspar, became more mafic as the cycle progressed. Thus, gabbro normally is somewhat younger than the bulk of the anorthosite; there are, however, volumes of gabbro that grade into anorthosite, but there are also rock volumes in which gabbro seems to have intruded anorthosite. The syenite (and its more granitic facies) almost certainly is younger than the anorthosite-gabbro and does not appear to have been derived, by fractional crystallization, from the parent magma of anorthosite and gabbro bodies.

The rocks intrusive into the Grenville in the Gore Mountain area were introduced during or immediately after the Grenville metamorphism of about 1000 m.y. ago. How old the original sediments, now making up the Grenville metasediments, may have been is uncertain. But it does appear probable that they were much older than the metamorphism that converted them into their present form. In the Gore Mountain area, as in the other mineralized portions of the Adirondacks, the garnet mineralization appears to have been developed in Grenville time and must, therefore, be classified as late Precambrian.

Five hypotheses have been suggested to explain the development of the Gore Mountain garnet deposit. These suggest that the deposit resulted from (1) the intense metamorphism of an inclusion of Grenville rock, (2) flow segregation of garnet and the minerals associated with it in the late gabbroic phase of the anorthosite magma, (3) isochemical transformation (except for a slight increase in water content) due to regional metamorphism that occurred after the syenite magma had been introduced into the area, (4) successive contact metamorphic effects of intruded anorthosite and syenite magmas on gabbro, and (5) the reconstitution of a marginal portion of the gabbro due to heat and pressure supplied by the syenite magma.

The fifth hypothesis (Levin, 1950) suggests that the corona garnets in the gabbro (and in the anorthosite) were produced by deuteric alteration under the high-pressure conditions that prevailed during the introduction and crystallization of the anorthositic magma from which both anorthosite and gabbro were derived. The later development of poikilitic and porphyroblastic garnets in the gabbro resulted from the intrusion of the syenite magma into the area which caused dynamo-thermal metamorphism of the border phases of the gabbro, the drastic physical changes that were produced having possibly been aided by water-rich solutions supplied by the crystallizing syenite magma. The metamorphic conditions that prevailed in what is now the general Barton area must have been particularly intense for garnet porphyroblasts are widespread in their occurrence in metagabbro and amphibolite, though nowhere are the garnets as large in size or abundant in quantity as in the Barton deposit proper. The gabbro containing hornblende, plagioclase, and shelled or rimmed garnet is essentially a metagabbro in which the primary pyroxene has been almost entirely converted to hornblende and in which the texture of the rock is far different from that of the primary gabbro from which it was derived. The gabbros containing porphyroblastic (but unrimmed) or poikilitic garnet are also metagabbros, and only the ophitic gabbro with corona-type garnet can truly be called gabbro. Thus, it seems most logical to suppose, from the gradational relationships from rimmed-garnet rock in contact with syenite through metagabbro containing porphyroblastic and poikilitic garnets to ophitic gabbro with, and farther away without, corona garnets, that the syenite bears a causal relationship to the garnets and that the metamorphic conditions that obtained as a result of the intrusion of the syenite were most intense nearest the contact between the gabbro and the syenite magma. In other words, it does not appear probable that the corona garnets in the primary gabbro were produced by contact metamorphism between two divergent rock types but were more probably deuteric, as were those in the anorthosite also.

Of the five hypotheses that have been suggested to explain the development of the large and abundant garnets surrounded by hornblende shells in the metagabbro at the contact with syenite, the fifth to be discussed seems to be the most reasonable in light of the information now available. If this is so, it follows that the original garnets in the gabbro were formed deuterically and, as a result of the heat, pressure, and emanations(?) generated by the intruding syenite magma, that the garnet percentage in the metagabbro was appreciably increased over that in its primary equivalent, and that the garnets in the rock volume of the present deposit were collected into a much smaller number of much larger garnet crystals than had obtained in the primary rock. The Barton deposits, therefore, are here classified as magmatic-4 (deuteric) and metamorphic-C. Metamorphic-C is used rather than hypothermal-1 on the theory that heat and pressure were far more important in the development of the large garnet crystals with their shells of hornblende than were any emanations that may have been given off by the syenite magma or that may have entered the area from any source along the contact between syenite and gabbro.

LYON MOUNTAIN

Late Precambrian *Iron as Magnetite* *Hypothermal-1*

Gallagher, D., 1937, Origin of the magnetite deposits at Lyon Mountain, N.Y.: N.Y. State Mus. Bull. no. 311, 84 p. (particularly p. 14-79)

Hagner, A. F., and Collins, L. G., 1967, Magnetite ore formed during regional metamorphism, a usable magnetite district, New York: Econ. Geol., v. 62, p. 1034-1071 (although the Lyon Mountain mine lies some 15 miles NNW of this district, the occurrences in the district are similar enough to that of Lyon Mountain that this paper is pertinent in a study of Lyon Mountain)

Miller, W. J., 1919, Magnetic iron ores of Clinton County, New York: Econ. Geol., v. 14, p. 509-535

—— 1921, Origin of Adirondack magnetite deposits: (reply) Econ. Geol., v. 16, p. 227-233

—— 1926, Geology of the Lyon Mountain quadrangle: N.Y. State Mus. Bull. no. 271, 101 p.

Newland, D. H., 1908, Geology of the Adirondack magnetite iron ores: N.Y. State Mus. Bull. no. 119, 182 p. (particularly p. 105-124)

—— 1920, Magnetic iron ores of Clinton County, New York: (disc.) Econ. Geol., v. 15, p. 177-180

Postel, A. W., 1952, Geology of Clinton County magnetite district, New York: U.S. Geol. Surv. Prof. Paper 237, 88 p. (particularly p. 78-84)

—— 1956, Silexite and pegmatite in the Lyon Mountain quadrangle, Clinton County, N.Y.: N.Y. State Mus. Bull. no. 344, 23 p.

Zimmer, P. W., 1947, Anhydrite and gypsum in the Lyon Mountain magnetite deposit of the northeastern Adirondacks: Amer. Mineral., v. 32, p. 647-653

Notes

The magnetite deposits of the Lyon Mountain district are located in extreme northeastern New York 25 miles slightly north of west from Plattsburg and about 145 miles slightly west of north from Albany.

The four major ore bodies of the district lie along the southwest-northeast-trending Lyon Mountain ore belt that ends only 5 miles from the outcrop line of the Upper Cambrian Potsdam sandstone that marks the northeastern boundary of the Adirondack massif. The four deposits, from southwest to northeast are: (1) the 81 mine, (2) the Phillips vein, (3) the Chateaugay mine, and (4) the Parkhurst mine; the less important Standish Road ore shoot is located between the 81 mine and the Phillips vein. Almost the entire ore belt is mineralized, the named deposits being those rock volumes along it in which minable ore concentrations have been developed.

Nearly all the metamorphic and igneous rocks in the district have gneissic structures, all of them having been involved to an appreciable degree in the Grenville metamorphism. The oldest rocks in the district are the Grenville sediments themselves, but unquestioned Grenville rocks do not occur in the mineralized area; the nearest definitely Grenville outcrop is about 2.5 miles southwest of the southeast end of the Lyon Mountain ore belt.

Anorthosite and anorthositic gabbro, so characteristic of much of the Adirondacks, are lacking from the general area of the Lyon Mountain ore belt, the nearest outcrops being many miles removed from the mines of the district. It is almost certain that there is no direct genetic connection between the anorthosites and the ores.

The oldest of the silicic rocks in the district is the Hawkeye granite gneiss; it crops out between 1.0 and 1.5 miles southeast of the Lyon Mountain ore belt and generally runs parallel to it. Another, smaller patch of this gneiss is located northeast of the ore belt and trends in that direction for about 3.5 miles; it is about 0.75 miles wide.

The next igneous rock of the district in order of decreasing age is the quartz syenite gneiss, which crops out some 10 miles slightly west of south of the Lyon Mountain ore belt. In most examples of these rocks, the quartz content is too high for the designations of quartz syenite; these rocks are actually granites. The youngest of the silicic gneisses is the Lyon Mountain granite gneiss which contains the Lyon Mountain ore belt; it is a pink rock (locally greenish or yellowish), medium-grained, and foliated.

Pegmatites are quite common throughout the district and are found in all the mines; there are two main types, pink and white, differentiated by their feldspar content, potash or soda feldspar, respectively; the pink pegmatites are pre-ore and the gray post-ore. Both types are quartz-bearing and may or may not contain pyroxene and magnetite, which, when present, were probably hydrothermally introduced. Quartz veins also are generally distributed through the area though they are not as abundant as the pegmatites; they were first called silexites by Miller in 1919. They are thought by Postel to have been changed from the original pegmatitic forms by regional plastic deformation.

Dikes of various rock types are known in the area and include Lyon Mountain granite gneiss, diabase, syenite, and melasyenite. All are younger than the Lyon Mountain gneiss, and probably all are Precambrian, though the melasyenites may be post-Ordovician; they appear to bear no genetic connection with the ores except that the ore-forming fluids and the Lyon Mountain dikes probably came from the same general magmatic source.

Although there appear to be no published data on the ages of the rocks in the immediate area of the Lyon Mountain ore belt, it seems almost certain that the metamorphism of the Grenville sediments, the introduction of the igneous magmas, and the development of the ores all took place, in that order, at the generally accepted time of the Grenville orogeny--about 1000 m.y. ago. The ores cannot be later than Precambrian since pebbles from the diabase dikes, which are possibly even younger than the ores, have been found in the basal conglomerate of the Potsdam sandstone. It therefore seems reasonable to classify the ore bodies of the Lyon Mountain ore belt as late Precambrian.

The ore bodies of the Lyon Mountain district are located within the general confines of a moderately complex syncline that is part of the district-wide system of anticlines and synclines into which the Lyon Mountain granite gneiss had been folded. Northeast of the southwestern termination of this syncline, the northwest limb (the site of the original 81 mine) was mineralized for about two-thirds of a mile; on the southeast limb, however, the mineralization, with considerable stretches of less than ore grade, continues northeast from the end of the synclinal structure certainly for 3.5 miles and perhaps for about 5.0 miles.

The only ore mineral in the various ore veins of the mines of the Lyon Mountain ore belt is magnetite. Although magnetite is known as an early formed accessory mineral in the granite, this occurrence is of no economic importance. On the other hand, the late magnetite, introduced along planes of easy solution movement through the folded granite gneiss, was emplaced as irregular replacement grains or grain aggregates. The grade of the ore depends entirely on the extent to which this replacement has occurred, and the magnetic character of the ore mineral permits the economic mining of ores containing as low as 30 percent of iron. There are minute amounts of pyrite and ilmenite in the ore, and sphene has been developed as thin rims around magnetite grains. The sphene rims break off during the grinding of the ore and, therefore, most of it goes into the tailings and does not interfere with the smelting of the concentrates. Apatite is present in quite different amounts in the different sections of the mine; the ore in the Williams veins sometimes containing as much as 1 to 2 percent of that mineral. The gangue minerals, which have replaced the original minerals of the granite, include pyroxene, hornblende, albite, sphene (grains of that mineral occur in addition to the rims around magnetite), quartz, the already mentioned apatite, occasional biotite, and minor amounts of chlorite, epidote, scapolite, sericite, and calcite and tiny amounts of sulfides. The pyroxene is developed in the granite only near the ore veins and probably was formed by reaction between iron in the ore-forming fluid and the silicates originally present in the granite gneiss. Granite near the ore often contains as little as 56 percent of silica.

It appears that there is little chance that the Lyon Mountain ores were formed from molten masses of iron-rich materials that were injected into the area; the intimate intergrowths of magnetite and the gangue minerals with the original minerals of the gneiss indicate definitely that the ore-forming medium was capable of large-scale replacement reactions with the granite to produce ore and with metasediment xenoliths to form amphibolite or gray band. It also seems certain that the granite gneisses and the ore veins they contain are not the result of the metamorphism of material entirely, or even in major part, of sedimentary origin; this theory, however, has been suggested. The character of the ore mineralization, principally as demonstrated by the magnetite and its accompanying pyroxene, almost certainly indicates that the ores were formed at high temperatures from hydrothermal solutions that probably came from the same general source as the granite magmas. The appreciable content of primary, accessory magnetite in the granite gneisses suggests that the magma source must have been quite high in iron. The deposits, therefore, are here classified as hypothermal-1.

SANFORD LAKE

Late Precambrian | *Iron as Magnetite, Titanium as Ilmenite, Vanadium* | *Magmatic-3b*

Balsley, J. R., Jr., 1943, Vanadium-bearing magnetite-ilmenite deposits near Lake Sanford, Essex County, New York: U.S. Geol. Surv. Bull. 940-D, p. 99-123

Gillson, J. L., 1956, Genesis of titaniferous magnetites and associated rocks of the Lake Sanford district, New York: A.I.M.E. Tr., v. 205, p. 296-301 (in Min. Eng., v. 8, no. 3)

Gross, S. O., 1968, Titaniferous ores of the Sanford Lake district, New York, in Ridge, J. D., Editor, *Ore deposits of the United States 1933-1967* (Graton-Sales Volumes): Chap. 8, v. 1

Hubaux, A., 1958, Genesis of titaniferous magnetites and associated rocks of the Lake Sanford district, New York: (disc. of Gillson, 1956) A.I.M.E. Tr., v. 211, p. 379-380 (in Min. Eng., v. 10, no. 3) (with Gillson's reply)

Kays, M. A., 1965, Petrographic and modal relations, Sanford Hill titaniferous magnetite deposit: Econ. Geol., v. 60, p. 1261-1297; disc., v. 61, p. 204-205; author's reply, p. 624-625

Kemp, J. F., 1897-1898, The titaniferous iron ores of the Adirondacks: U.S. Geol. Surv. 19th Ann. Rept., pt. 3, p. 377-422 (particularly p. 409-416, 417-419)

Newland, D. H., 1908, Geology of the Adirondack magnetic iron ores: N.Y. State

Mus. Bull. no. 119, 182 p. (particularly p. 146-153, 155-164)

Osborne, F. F., 1928, Certain magmatic titaniferous ores and their origin: Econ. Geol., v. 23, p. 724-761, 895-922 (particularly p. 726-741, 745-752, 905-922)

Rechenberg, H. P., 1955, Zur Genesis der primären Titanerzlagerstätten: Neues Jb. f. Mineral., Mh., H. 4, S. 87-96

Singewald, J. T., Jr., 1913, Titaniferous iron ores of the United States: U.S. Bur. Mines Bull. 64, 145 p. (particularly p. 60-76)

Stephenson, R. C., 1945, Titaniferous magnetite deposits of the Lake Sanford area, New York: N.Y. State Mus. Bull. no. 340, 95 p. (Stephenson, 1948, was condensed from this publication, with certain changes in the author's views)

—— 1948, Titaniferous magnetite deposits of the Lake Sanford area, New York: A.I.M.E. Tr., v. 178, p. 397-421

Wheeler, E. P., II, 1950, Massive leucoxene in Adirondack titanium deposit: Econ. Geol., v. 45, p. 574-577

Notes

The Sanford Lake area lies near the southwest margin of the main segment of the Adirondacks anorthosite massif, near the town of Tahawus and about 20 miles south of Saranac Lake and about 35 miles west-northwest of Ticonderoga; the lake is part of the headwaters of the Hudson River.

In this region the massif is composed of anorthosite, anorthositic gabbro, gabbro, and titaniferous magnetite bodies. The anorthosite is a very coarse porphyritic rock in which the phenocrysts are labradorite and the ground mass is fine-grained plagioclase (andesine) and a small amount of dark minerals--pyroxene, hornblende, garnet, apatite, and the ore minerals.

Gabbroic anorthosite (which, with anorthosite, makes up the bulk of the rocks in the Sanford Lake area) is intermediate in the gradational series of basic rocks in the district, being compositionally between anorthosite and gabbro. Gabbro makes up only a small proportion of the Sanford Lake rocks and contains up to 15 percent of the same type of plagioclase phenocrysts found in the more feldspathic rocks; the ore minerals in the gabbros are usually concentrated in bands. Gabbro may be intrusive into anorthosite.

The origin of these rocks appears to have been satisfactorily explained by Stephenson who believes that the feldspar phenocrysts crystallized at great depths beneath their final site of emplacement, that magma containing them in large proportion was driven out of the magma chamber, and that filter-pressing action separated much of the still molten material from large volumes of the phenocrysts which then were cemented into anorthosite by the crystallization of the remaining molten material to give the typical appearance of the anorthosite ground mass. Other bodies of phenocrysts with higher proportions of still molten material crystallized to form gabbroic anorthosites, but these always graded into anorthosite and never were separated completely enough from the anorthosite-producing mixture of phenocrysts and molten material to intrude the anorthosite. Minor amounts of molten material, with 15 percent or less of phenocrysts, were so completely separated from the anorthosite masses and were so mobile, relative to the anorthosite- and anorthositic gabbro-producing materials, that these gabbroic magmas were occasionally forced into already solidified anorthosite to give the intrusive relationships known to obtain between anorthosite and gabbro. The development of the ore bodies, as a final phase in the series of rock types, will be considered in the discussion of classification of the deposits.

There is no direct geologic evidence as to the age of the rocks of the Sanford Lake area, but there seems to be little doubt but that these rocks were introduced into the area at the time of the Grenville orogeny. Further, radioactive age determinations appear almost unanimous in pointing to an age of about 1000 to 800 m.y. for that period of earth movement and for igneous rocks accompanying it. This would place the age of the rocks of the Sanford Lake as about 1000 m.y. to 800 m.y., an age in conformity with that majority of anorthosites of the world. The Sanford Lake deposits, therefore, are here classified as late Precambrian.

In addition to the principal ore body in the Sanford Lake area, that which originally outcropped on the southwestern slope of Sanford Hill, there are three others--Ore Mountain, Calamity-Mill Pond, and Cheney Pond--which have not been exploited. The ores at Sanford Lake are of two general types: (1) discordant or dikelike bodies in anorthosite and (2) concordant, sill-like, or stratiform bodies in gabbro. The discordant bodies in anorthosite are spatially related to concordant bodies in gabbro, it being common to find concordant lenses of ore in gabbro within a short distance of discordant masses in anorthosite.

Most of the magnetite and ilmenite occurs in the ores as anhedral or subhedral grains of comparable size, but there is an appreciable amount of exsolved material in the magnetite, all of which was, by early workers, considered to be ilmenite. It now appears that some, probably most or even all, of the material exsolved from magnetite at Sanford Lake is ülvospinel or ülvite (Fe_2TiO_4); this mineral has the spinel structure of magnetite and should have been more soluble in the original magnetite, as it crystallized from the melt, than ilmenite would have been. The bulk of the ilmenite appears to have formed before the magnetite, and the presence of numerous ragged inclusions of ilmenite in magnetite suggests that a considerable portion of the magnetite probably was formed by reaction between the melt and ilmenite. The ilmenite so returned to the melt appears to have been dissolved in the magnetite crystals as they formed; on cooling, much, though not all, of the titanium in magnetite appears to have exsolved as ilmenite or as ülvite. Rutile is a later mineral and probably formed only because there was not enough ferrous iron to permit it to form as ilmenite.

The ores also contain appreciable amounts of vanadium, probably from 0.5 to 0.7 percent V_2O_5; these are not distinct vanadium minerals, but it is possible that some of the exsolution blebs in magnetite are the mineral designated as coulsonite ($Fe^{2}_{2}V^{4}O_{4}$) which appears to have a spinel structure and to be readily soluble in magnetite. Coulsonite may be so soluble in Sanford Lake magnetite that it never exsolves and that all the exsolution bodies are ülvite. Little vanadium occurs in the ilmenite.

The classification of the ore intrusive into the anorthosite appears to present few problems. These ores certainly were late segregations in the minor gabbroic portions of the igneous complex, and they apparently were forced out of the gabbro into the adjacent anorthosite. They are, therefore, to be classified as magmatic-3b. The ores in the gabbro are also late segregations, the only question being whether they were (1) simply the last minerals to precipitate in the orderly progression of gabbro magma crystallization or (2) were precipitated from a distinct iron-titanium-rich melt which became immiscible in the gabbro magma during late stages of crystallization cycle. If the first alternative is the true explanation, then the ores in gabbro are, essentially, gabbroic pegmatites and should be classified as magmatic-3a. If the second is the correct one, they are late segregations of the category magmatic-3b, the same category as the ores in gabbro. In favor of the latter explanation is the evidence of movement of the molten iron-titanium melts within, and on occasion actually out of, the gabbro. On this basis, the classification magmatic-3b is also assigned to the ore in gabbro and, therefore, to both types of Sanford Lake ore.

CORTLANDT COMPLEX

Middle Paleozoic | *Emery* | *Hypothermal-1*

Balk, R., 1927, Die primäre Struktur des Noritmassives von Peekskill am Hudson, nördlich New York (auch bekannt als "Cortlandt Norit"): Neues Jb. f. Mineral., Geol. und Paläont., Beil. Bd. 57, Abt. A, S. 249-303

Berkey, C. P., and Rice, M., 1919, Geology of West Point quadrangle: N. Y. State Mus. Bull. nos. 225-226, 152 p. (particularly p. 90-91)

Bucher, W. H., 1948, Excursion to the Cortlandt norite and Storm King granite: Geol. Soc. Amer. 61st Ann. Meeting (1948), Guidebook of Excursions, p. 33-38

Butler, J. W., Jr., 1936, Origin of the emery deposits near Peekskill, New York: Amer. Mineral., v. 21, p. 537-574

Davis, J. F., 1968, Lore, lode, and lade--emery, Westchester's own nitty-gritty: Empire State Geogram, v. 5, no. 3-v. 6, no. 1, p. 14-19

Friedman, G. M., 1952, Sapphirine occurrence of Cortlandt, New York: Amer. Mineral., v. 37, p. 244-249

—— 1956, The origin of spinel-emery deposits with particular reference to those of the Cortlandt complex, New York: N. Y. State Mus. Bull. no. 351, 68 p.

Gillson, J. L., and Kania, J.E.A., 1930, Genesis of the emery deposits near Peekskill, New York: Econ. Geol., v. 25, p. 506-527

Long, L., and Kulp, J. L., 1962, Isotopic age study of the metamorphic history of the Manhattan and Reading prongs: Geol. Soc. Amer. Bull., v. 73, p. 969-996

Ratcliffe, N. M., 1968, Contact relations of the Cortlandt complex at Stony Point, New York, and their regional implications: Geol. Soc. Amer. Bull., v. 79, p. 777-786

—— 1968, Stratigraphic and structural relations along the western border of the Cortlandt intrusives: N. Y. State Geol. Assoc., 40th Ann. Meeting, Guidebook to Field Excursions, Trip H, p. 197-220

Rogers, G. S., 1911, Geology of the Cortlandt series and its emery deposits: New York Acad. Sci. Annals, v. 21, p. 11-86

Shand, S. J., 1942, Phase petrology in the Cortlandt complex, New York: Geol. Soc. Amer. Bull., v. 53, p. 409-428

Steenland, N. C., and Woollard, G. P., 1952, Gravity and magnetic investigation of the structure of the Cortlandt complex, New York: Geol. Soc. Amer. Bull., v. 63, p. 1075-1104

Williams, G. H., 1888, The contact metamorphism produced in the adjoining mica schists and limestones by the rocks of the "Cortlandt series" near Peekskill, N. Y.: Amer. Jour. Sci., 3rd Ser., v. 36, p. 254-259

Notes

The emery deposits of the Cortlandt complex are located within a nearly 30 square mile ultramafic to mafic igneous mass that lies in Cortlandt Township south and east of the town of Peekskill; the area is about 40 miles north of New York City.

Although the general area of the West Point quadrangle contains rocks that range in age from Precambrian to Triassic, only Precambrian rocks are known to make contact with the igneous rocks of the complex. The oldest rock in the area is the Manhattan schist (which, as mapped, probably includes the Fordham gneiss); it is, for the most part, a quartz-mica schist that has been completely recrystallized, bears a distinct foliation, and is highly micaceous. The Manhattan schist appears to be older than the Inwood marble; the contact between the two is gradational. The Inwood has also been highly metamorphosed and recrystallized. The metamorphism that produced the schist and marble (or at least the last metamorphism to affect these rocks) appears to have taken place in Ordovician time, and both formations were tightly folded and the original minerals largely changed to others of high metamorphic grade.

The rock volume made up of the Manhattan and Inwood formations was intruded, at some time after the Ordovician metamorphism, by a complex of mafic and ultramafic rocks known as the Cortlandt complex. Although by far the greatest portion of the complex lies on the east bank of the Hudson, two minor outcrops of the same mafic material are known on the west bank of the river, one in the Inwood marble and the other in the Reservoir granite, apparently the oldest igneous rock in the area. Of the entire length of contact between the metamorphic rocks and the intrusives on the east bank, less than 1 mile consists of Inwood marble; the remainder is Manhattan schist.

The rock types that make up the Cortlandt complex and their relationships to each other have long been a matter of dispute. Of the rocks exposed at the surface, something over one-third appears to be composed of pyroxenite, the major half of which is olivine-bearing; a minor fraction of the pyroxenite appears to contain enough olivine to be classed correctly as peridotite; certainly no olivine-rich area is large enough to be mapped as peridotite at normal mapping scales. The remainder of the truly intrusive rock of the complex appears to have ranged from hypersthene-augite norite to hornblende norite.

Both the norite and the pyroxenite usually are banded, although the banding is more strongly developed in the norite. In the olivine pyroxenites, the difference between the two members of a pair of bands is the greater concentration of olivine in the lower band. In the less mafic rocks, lacking olivine, the banding is more pronounced as the lighter feldspar is in sufficient quantity (up to 15 percent in the upper band) to make more impressive the color contrast between the two members of the band-pair. With any band-pair, the concentration of the earlier crystallizing mineral in the lower band is readily explicable, but the rhythmic repetition on the band-pairs requires rhythmic changes in the character of the melt from which the crystals were forming. If these bands, which show a considerable, but widely variable, departure from horizontality in many places, were formed in the same manner as those in the Stillwater complex, the Cortlandt complex must have been much thicker than the erosional remnant now present in the area. Granted that the original thickness of the Cortlandt complex was of the order of magnitude needed for convective overturn, the explanation put forward by Jackson for the banding in the ultramafic portion of the Stillwater complex (Thayer, 1960) would also probably apply to the banding in the Cortlandt rocks. The appreciable effect of deuteric or late hydrothermal reactions on the now-visible portions of the Cortlandt complex, the banding of these rocks, and the emphasis of the studies made in the district on rock types and on the structural meaning of the orientation of the banding have directed attention away from the physico-chemical causes of the banding *per se*.

In addition to the two varieties of pyroxenite (olivine-bearing and essentially olivine-free) and to the three varieties of norite (augite-hypersthene norite, prismatic hornblende norite, and poikilitic hornblende norite), the only other important rock type of the complex has been referred to as diorite. This name was assigned by Williams (1888) and Rogers (1911) to rocks containing plagioclase and hornblende or biotite but with little or no pyroxene. Although it is possible that some of this diorite is of magmatic origin, it appears more probable that it is either highly altered Manhattan schist or norite that has been contaminated by the assimilation of schist. In the diorites, it is not unusual to find quartz present in the interstices among the feldspar laths. It is probable that the term diorite is used to cover both contaminated norite and highly altered Manhattan schist and that the actual amount of truly magmatic diorite is negligible if there is really any at all.

Balk (1927) postulated that the complex was a funnel-shaped mass which probably has been fed by three pipes through which the mafic magma was introduced into the rock volume now occupied by the complex. The major one of Balk's funnels was located somewhat to the northwest of the center of the complex and the other two were placed in the eastern and southwestern portions of the ultramafic mass, respectively. Shand (1942) believed that the northwestern funnel was the channelway followed by the magmatic material of which the complex is composed, but he felt some doubt as to the existence of the other two. In 1952, Steenland and Woollard, on the basis of gravity and magnetic surveys of the area, came to the conclusion that there were two main feeder-pipes for the complex proper (on the east side of the Hudson), one of which is essentially cylindrical, has a diameter of about 2.4 miles, extends at least 5 miles in depth, and underlies Balk's eastern funnel, and the other of which is also generally cylindrical, is about 1.2 miles in diameter, is at least 5 miles deep, and centers near Balk's western funnel. They suggest that a small feeder-pipe underlies each of the two minor outcrops of mafic material on the west bank of the river, and the material in the pipes they believe to be pyroxenite.

From the thin cover of mafic material over the pipes and the even lesser thickness of such rocks over the area between Balk's eastern and northwestern funnels, it would appear the erosion has removed much mafic material that had once been introduced into the area by the feeder-pipes. Shand points out that either the magma that brought in the material of the complex must have been quite mafic or the higher (now eroded) portions of the complex must have consisted much more largely of norite. Shand, however, thought that the peridotite and pyroxenite formed through the settling of mafic crystals from the norite magma; whereas, if Steenland and Woollard are correct, the norite portions of the complex might well be products of the late stages of crystallization of a much more mafic and less voluminous magma than Shand appears to think probable. The choice between these two possibilities depends on the unknown, and probably unknowable, original thickness of the complex.

The essential lack of regional or dynamic metamorphic effects displayed in the ultramafic and mafic rocks of the Cortlandt complex indicates that they were not affected by the last marked metamorphism (Ordovician) known in the area. The exact portion of post-Ordovician time in which the ultramafic to mafic intrusion occurred is, however, not established, but it is probable that the intrusions took place in either Taconic or Acadian time. In either event, whether the intrusions were Silurian or Devonian, they would be here categorized as middle Paleozoic.

The only valuable product obtained from the rocks of the Cortlandt complex has been emery. The essential minerals of Cortlandt emery are spinel proper (which here is a ferroan spinel or pleonaste), corundum, and metallic iron minerals that are mainly magnetite with some specularite and ilmenite. The proportions of spinel to corundum are different from one locality to the next; in certain areas corundum may be entirely absent while in others there may be no spinel. The mineral högbomite [$Mg(Al,Fe,Ti)_4O_7$] locally is associated with the spinel in minor amounts, particularly in emery in altered Manhattan schist (metaschist). The iron oxides were the last minerals to form as is shown by their position interstitial to the corundum and spinel; of the iron oxides, magnetite was the first to form and is replaced, usually to a minor extent, marginally by ilmenite and specularite. The three iron minerals exhibit exsolution textures. The gangue minerals present depend on the host rock of the emery body, the host rock being either highly altered rocks of the complex or of the Manhattan schist and the gangue minerals being those of the host rock unreplaced in the introduction of the emery-forming minerals. It would appear that the emery-forming minerals are simply the last of hydrothermally generated alteration minerals to have been formed in the complex; these minerals, however, are far less widespread in their distribution than the earlier alteration minerals and are normally confined to the marginal reaches of the complex. The Inwood marble apparently contains no deposits of emery. Emery in minable amounts has, however, been found in both altered mafic rock (normally melanorite and not the more mafic phases of the complex) and metaschist; the emery in norite is termed black emery and that in metaschist gray emery. The ores in the altered melanorites and in the metaschists are found only in the immediate vicinity of contacts between these two rock types; no minable deposits have been found in moderately altered or fresh hypersthene-labradorite melanorite.

There are four major localities within the complex (each near the margin of the complex) from which emery has been mined. These are (1) Emery Hill in the northeast section of the complex, about 2.5 miles east of the town of Peekskill; (2) the Furnace Duck Road, about 0.5 miles east of Nelson Pond and 1 mile southeast of Emery Hill; (3) around Salt Hill in the southeastern part of the complex; and (4) the Montrose area in the southwestern part of the complex, about 0.5 miles southwest of U.S. Highway 9 on both sides of the Dutch Road.

Emery is commonly in association with biotite. Normally, the sillimanite group of minerals is not found in association with black emery in norite, but where quartz veins cut through the emery, the typical suite of sillimanite minerals, plus garnet, is developed; this strongly suggests that the sillimanite suite and garnet in the metaschist are more importantly of hydrothermal than dynamic metamorphic origin.

The high-temperature fluids that caused the formation of the emery (and probably of essentially all the alteration minerals) apparently were of magmatic origin. Further, it is most probable that these solutions rose into the emery-containing rock volumes along the contacts between schist and norite, aided beyond reasonable doubt by the brecciated nature of at least some of such contacts. The entire suite of minerals, both the emery and the alteration minerals, is so consistently composed of minerals of high temperature of formation that it seems certain that the entire sequence of hydrothermal alteration must have been carried out within the hypothermal range. As the deposits were all formed in noncalcareous rocks, the emery deposits are here classified as hypothermal-1.

North Carolina

HAMME

Middle Paleozoic *Tungsten* *Hypothermal-1*

Bishop, F. H., 1948, North Carolina tungsten: Min. Cong. Jour., v. 34, no. 9, p. 77-82

Espenshade, G. H., 1947, Tungsten deposits of Vance County, North Carolina, and Mecklenburg County, Virginia: U.S. Geol. Surv. Bull. 948a, p. 1-17

——— 1950, Occurrences of tungsten minerals in the southeastern states, in Snyder, F. C., Editor, *Symposium on mineral resources of the southeastern United States*: Univ. Tenn. Press, p. 56-66 (particularly p. 58-61)

Malcolm, J. B., 1962, Hamme mine reopening made feasible: Min. Eng., v. 14, no. 4, p. 44-49 (only minor geology)

McIntosh, F. K., 1948, Investigation of the Hamme tungsten district, Vance County, North Carolina, and Mecklenburg County, Virginia: U.S. Bur. Mines, R.I. 4380, 6 p.

Parker, J. M., III, 1963, Geologic setting of the Hamme tungsten district, North Carolina and Virginia: U.S. Geol. Surv. Bull. 1122-G, p. G1-G69 (no details of ore veins)

White, W. A., 1945, Tungsten deposit near Townsville, North Carolina: Amer. Mineral., v. 30, p. 97-110

Notes

The minable deposits in the Hamme tungsten district are located just short of 14 miles north-northwest of Henderson, North Carolina; Henderson, in turn, is about 50 miles north-northeast of Raleigh.

Although tungsten-bearing quartz veins are sporadically distributed through an area some 7.5 miles long (north to south) by 1 mile wide, almost all the ore produced has come from a few veins in the central portion of the belt. The rocks that make up the central portion of the district and contain most of the ore are igneous or pseudoigneous, but the rocks on either side of them are metamorphic types of the greenschist and albite-epidote-amphibolite facies that were derived from sedimentary and volcanic material by low-grade regional metamorphism. The relative ages are not known with certainty.

The dominant rock in the Hamme district, whether it is truly igneous or not, is the albite granodiorite. Espenshade (1950) called it an albite granite but pointed out that the albite might have been produced by the conversion of calcic plagioclase to albite and epidote and sericite, the last two minerals being abundant in the interiors of the albite crystals. Parker (1963) has adopted this suggestion in his classification of the rock, but he goes farther than Espenshade in that he has decided that the rock is essentially the result of granitization of the phyllite that bounds it and into which it is apparent that it grades.

Parker categorized the metamorphic rocks as lower Paleozoic(?), but they might be late Precambrian; the probability is greater that they are lower Paleozoic. The igneous or pseudoigneous rocks have been classified as middle or late Paleozoic, by analogy with the ages determined for other igneous rocks in the metamorphic and igneous province of the southeast; they are here thought to be middle Paleozoic. This date is also given to the albite granodiorite whether it is truly igneous or not. Whether it had a magmatic or metasomatic origin, the ultimate source of the magma or of the granitizing solutions would appear to have been igneous activity of probably middle Paleozoic age. The ore deposits were formed in the same general time span as the granodiorite and are, therefore, here classified as middle Paleozoic, but again it must be admitted that the possibility of their being late Paleozoic has not been completely eliminated.

The tungsten ores of the Hamme mine are found in veins that appear to be spatially controlled by a N35°E-trending zone of shearing or schistose rock in the granodiorite near its western contact with the schist, all of the principal veins being found in or near the sheared zone over a distance of about 2 miles. Although there are innumerable quartz veins in the general area, many of which contain some tungsten minerals, all that are of economic importance are either in the sheared zone or in granodiorite between it and the phyllite contact not more than 1500 feet to the west.

The principal mineral of the veins is quartz, although other gangue minerals such as fluorite, sericite, and rhodochrosite, are present. The main tungsten mineral is huebnerite that occurs in platy crystals that may be as much as 6 inches long and are arranged in thin seams or zones parallel to the sheeting of the veins. Scheelite also is present, but it is usually much less abundant than huebnerite and essentially always occurs closely associated with the iron-manganese tungstate; the scheelite contains a small amount of the $CaMoO_4$ molecule in solid solution.

Much of the vein quartz in the district, particularly in the best-mineralized veins, is highly broken, and the huebnerite crystals are commonly fractured as well; the interstices in the tungstate crystals are filled with quartz, indicating that the deposition of the latter mineral continued after huebnerite had ceased to deposit. Veins and stringers of the later, glassy quartz are abundant.

After quartz, fluorite is the most abundant mineral of the ore veins; it accounts for about 4 percent of the mill tailings. Rhodochrosite is abundant in the Walker 2 vein, but seems to be lacking in the others. A considerable content of sulfide minerals occurs in the ore veins; pyrite is the most abundant, and galena, tetrahedrite, sphalerite, and chalcopyrite are also found, as is a little molybdenite. The ore contains about 1.5 percent of lead for each 10 percent of WO_3. The barren quartz veins appear to be composed of glassy (second generation?) quartz; these veins contain some pyrite and light-green mica.

Although huebnerite is considered to be diagnostic of high-intensity conditions of ore deposition, at the Hamme mine it is not associated with any other minerals diagnostic of that range except for the minor amount of scheelite that did not begin to precipitate until after deposition of huebnerite had stopped. On the other hand, quartz, sericite, rhodochrosite, and fluorite are not incompatible with deposition under hypothermal conditions. It is here suggested that the Hamme ores were deposited under low-intensity hypothermal conditions and they are, therefore, categorized as hypothermal-1. The sulfide suite neither overlapped with the tungsten mineralization nor is of economic value and is not considered in arriving at the classification assigned to the deposit.

ORE KNOB

Middle Paleozoic *Copper* *Hypothermal-1*

Brown, H. S., 1959, Biotite alteration in the country rock at Ore Knob, North Carolina (abs.): Geol. Soc. Amer. Bull., v. 70, p. 1757

Eckelmann, F. D., and Kulp, J. L., 1956, The sedimentary origin and stratigraphic equivalence of the so-called Cranberry and Henderson granites in western North Carolina: Amer. Jour. Sci., v. 254, p. 288-315

Fullagar, P. D., and Bottino, M. L., 1966, Comparison of Rb-Sr whole-rock and mineral ages with K-Ar mineral ages of gneiss of Ore Knob, North Carolina (abs.): Geol. Soc. Amer. Program, San Francisco Mtg., p. 74

—— 1970, Sulfide mineralization and rubidium-strontium geochronology at Ore Knob, North Carolina, and Ducktown, Tennessee: Econ. Geol., v. 65, p. 541-550

Fullagar, P. D., and others, 1967, Geochemistry of wall rock alteration and the role of sulfurization in the formation of the Ore Knob sulfide deposit: Econ. Geol., v. 62, p. 798-825

Heinrichs, W. E., Jr., 1966, Geophysical investigations, Ore Knob mine, Ashe County, North Carolina: Soc. Expl. Geophys., Mining Geophysics, v. 1, Case Histories, p. 179-184

Hunt, T. S., 1873-1874, The Ore Knob Copper mine and some related deposits: A.I.M.E. Tr., v. 2, p. 123-131

Keith, A., 1903, Cranberry quadrangle, North Carolina-Tennessee: U.S. Geol. Surv. Atlas, Folio 90, 4 p. (area of this folio does not include the Ore Knob deposit but lies immediately west of it)

Kerstein, D. S., Jr., 1959, Geological features at the Ore Knob copper mine, North Carolina (abs.): Geol. Soc. Amer. Bull., v. 70, p. 1765

Kinkel, A. R., Jr., 1962, The Ore Knob massive sulfide copper deposit, North Carolina: An example of recrystallized ore: Econ. Geol., v. 57, p. 1116-1121; disc., 1963, v. 58, p. 997-998

—— 1967, The Ore Knob copper deposit, North Carolina, and other massive sulfide deposits of the Appalachians: U.S. Geol. Surv. Prof. Paper 558, 58 p.

Kinkel, A. R., Jr., and others, 1965, Age and metamorphism of some massive sulfide deposits in Virginia, North Carolina and Tennessee: Geochim. et Cosmochim. Acta, v. 29, p. 717-724 (particularly p. 719-720)

Ross, C. S., 1935, Origin of the copper deposits of the Ducktown type in the southern Appalachian region: U.S. Geol. Surv. Prof. Paper 179, 165 p. (particularly p. 67-77, 113-114)

Weed, W. H., 1911, Copper deposits of the Appalachian states: U.S. Geol. Surv. Bull. 455, 166 p. (particularly p. 128-132)

Notes

The Ore Knob copper deposit is located slightly over 8 miles nearly due east from the rail terminus at West Jefferson in Ashe County in the northwest corner of North Carolina.

In the Ore Knob area, the principal rock units are (1) a mica gneiss that Kinkel (1967) calls the Lynchburg or Mica gneiss and Fullagar and others (1967) refer to as the Carolina gneiss and (2) an amphibole gneiss that Kinkel designates as Hornblende-Biotite gneiss and Fullagar and his colleagues call the Roan gneiss. Just across the Virginia border, Kinkel shows a Grayson granodiorite gneiss and a transition zone between this rock and the adjacent (to the southeast) Lynchburg gneiss. Fullagar and others say that the Carolina gneiss and the Roan gneiss are interlayered and that no cross cutting relations have been observed; however, the one rock grades into the other along strike. They consider that these relations resemble facies changes in sedimentary rock sequences and that the rocks are the metamorphosed equivalents of a sedimentary unit. Whole rock samples of wall-rock gneiss at the mine yield, according to Fullagar and Bottino (1966), two essentially parallel isochrons with different Sr^{87}/Sr^{86} ratios. Seven samples of mica gneiss give an age of 420 ± 35 m.y. and an initial Sr^{87}/Sr^{86} ratio of 0.716. From five samples of the amphibole gneiss, they obtain an age of 395 ± 25 m.y. and an initial Sr^{87}/Sr^{86} ratio of 0.713. They consider that simplest interpretation is that the gneisses formed about 400 to 450 m.y. ago and that this is the maximum age for the formation of the ore. They point out that, although little typical hydrothermal wall-rock alteration exists in the mine, the iron content of the biotite and of the total rock decreases toward the ore, this effect beginning at a minimum of 150 feet from the ore. Also, within a few tens of feet of the ore, biotite decreases in abundance and muscovite increases, in some places becoming the more abundant mica. Fullagar and his colleagues conclude that iron was removed from the wall rock of the Ore Knob sulfide deposit for as much as several hundred feet from the shear zone in which the ore is located. There, this iron met copper and sulfur from hydrothermal fluids to produce the ore minerals. If this concept is correct, the age of the ores is the age of the metamorphism or about 400 m.y. and middle Paleozoic. Kinkel, on the other hand, thinks that geologic evidence and isotopic ages show that the ore was not produced in one episode. The ores, he thinks, were formed by high-temperature solutions that were introduced along a restricted channel (the shear zone) possibly at a late stage in the formation of the gneiss, but after directional pressure had ceased. The recrystallization of all the rock to coarse-grained, unoriented aggregates of silicates (the same as those in the gneiss) was a result of the action of these same solutions that brought in the ores and their associated carbonates. He further suggests that a later reorganization of vein material took place during a somewhat later metamorphic episode so that present textures and paragenetic relationships are largely a result of postdepositional reorganization. Since Kinkel believes that coarse vein hornblende is contemporaneous with the sulfides and that it has an age of 1120 ± 20 m.y., the ores must be late Precambrian.

A third hypothesis is that the iron, copper, and sulfur were all brought in by the ore fluid during middle Paleozoic time and that the solutions, after depositing the ores in the shear zone, were so depleted in iron that they removed iron from the wall rock as they worked their way out into it. If this hypothesis is correct, the ores must be younger, though probably not much younger, than the gneisses they inhabit and are, therefore, to be categorized as middle Paleozoic. This last explanation is favored here.

The Ore Knob ore body is, according to Fullagar and his colleagues, lens-shaped and plunges to the southwest; it is about 2800 feet long, 400 feet wide, and up to 40 feet thick. The plunge of the ore and that of the country are in the same direction. The dip of the ore body and its associated zones of gneiss ranges from 60°SE to nearly vertical, while that of the foliation of the country rock is from 50° to 55°. Contacts between the ore zone and the country rock are generally sharp and well marked. The ore thickens at its base and there has a keel-like shape.

The ore body contains about 20 percent of sulfide minerals; these are mainly pyrrhotite, pyrite, chalcopyrite, and sphalerite. Pyrite makes up from 10 to 50 percent of the sulfides, but pyrrhotite probably is the most abundant over all. Chalcopyrite makes up 5 to 10 percent of the sulfides; small amounts of magnetite and minor galena have been noted in the ores. The most prominent gangue minerals are carbonates, garnet, epidote, amphiboles, and quartz; others include feldspar, muscovite, biotite, and chlorite.

No matter which hypothesis for the formation of the ores is accepted, it appears to be agreed that the formation occurred at high temperatures, a concept supported by the generally high-temperature character of the diagnostic minerals such as the pyrrhotite and the garnet, epidote, amphiboles, feldspar, and the micas. I prefer the concept that the ores were all formed by hydrothermal solutions after the gneisses had been formed that brought in the materials of the sulfides and of the vein gangue minerals. That these gangue minerals are much the same as the minerals of the gneisses is to be expected in ores deposited in the hypothermal range. The deposit is, therefore, classified as hypothermal-1 since the host rocks cannot be considered calcareous enough to be called carbonates in the sense needed for hypothermal-2.

SPRUCE PINE

Middle Paleozoic	*Feldspar, Mica, Residual Kaolin*	*Magmatic-1a(Feldspar), Magmatic-3a(Feldspar, Mica, Quartz)*

Amos, D. H., 1959, D.M.E.A. project blossoms into best U.S. mica mine: Mining World, v. 21, no. 11, p. 30-34

Broadhurst, S. D., and Hash, L. J., 1953, The scrap mine resources of North Carolina: N. C. Dept. Conserv. and Devel., Div. Mineral Res. Bull. 66, 66 p.

Brobst, D. A., 1955, Guide to the geology of the Spruce Pine district, North Carolina: Geol. Soc. Amer. Guidebook for Field Trips, Field Trip 2b, p. 579-588

—— 1962, Geology of the Spruce Pine district, Avery, Mitchell, and Yancey Counties, North Carolina: U.S. Geol. Surv. Bull. 1122-A, p. A1-A26

Brobst, D. A., and others, 1954, Mafic intrusions as a clue to the metamorphic history of the Spruce Pine district, North Carolina (abs.): Amer. Mineral., v. 39, p. 317

Eckelmann, W. R., and Kulp, J. L., 1957, Uranium-lead method of age determination. Part II--North American localities: Geol. Soc. Amer. Bull., v. 68, p. 1117-1140 (particularly p. 1122-1125)

Hall, G. M., 1932, Flattened minerals in muscovite at Spruce Pine, North Carolina: Amer. Mineral., v. 17, p. 115

Kesler, T. L., and Olson, J. C., 1942, Muscovite in the Spruce Pine district, North Carolina: U.S. Geol. Surv. Bull. 936-A, p. 1-38

Kulp, J. L., and Brobst, D. A., 1956, Geologic map of the Bakersville-Plumtree area, Spruce Pine district, North Carolina: U.S. Geol. Surv. Mineral Invest., Field Studies Map MF-97

Kulp, J. L., and Poldervaart, A., 1956, The metamorphic history of the Spruce Pine district: Amer. Jour. Sci., v. 254, no. 7, p. 393-403

Maurice, C. S., 1940, The pegmatites of the Spruce-Pine district, North Carolina: Econ. Geol., v. 35, p. 49-78, 158-187

Olson, J. C., 1944, Economic geology of the Spruce Pine pegmatite district, North Carolina: N. C. Dept. Conserv. and Devel., Div. Mineral Res. Bull. 43, 67 p.

Parker, J. M., III, 1952?, Geology and structure of part of the Spruce Pine district, North Carolina--a progress report: N. C. Dept. Conserv. and Devel., Div. Mineral Res. Bull. 65, 26 p.

Notes

The pegmatites mined in the Spruce Pine district are located in a northeast-trending belt that is some 25 miles long and 10 miles wide. The town of Spruce Pine, near the center of the district, is 47 miles northeast of Asheville. The entire district lies within the Blue Ridge province that runs northeasterly through the western portion of North Carolina.

The Spruce Pine district is situated just to the west of the axis of the Appalachian geosyncline and is part of the complexly folded and extensively intruded core of that structure. The metamorphosed rocks of the district are made up in the main of an interlayered sequence of mica gneiss and schist, amphibolite gneiss and schist, and a small representation of dolomite marble; the beds appear all to be of Precambrian age. It is seldom that two adjacent rock layers are of the same composition, and usually the individual layers range from less than an inch to several feet in thickness; contacts between layers of mineralogically different compositions are sharp but transitions parallel to the layering do occur.

In the early work in the district, Keith mapped the mica gneiss and schist as Carolina gneiss and the amphibole gneiss and schist as Roan gneiss, which latter he believed to have been ultimately of igneous origin. Recent workers, such as Brobst (1962), have avoided the use of these formation names because of the difficulty of establishing the stratigraphic equivalence of either of the formations from place to place and because of doubt as to the intrusive igneous origin of all of the so-called Roan rocks. Thus, the current practice in the district is to use mineralogical names for the different rock types.

For all practical purposes, it is impossible to say that any one of the rock types is older than the others; they are all so intimately interlayered in so many places that it is unreasonable to attempt to date them relative to each other. After the layered rocks had been emplaced, the area was intruded by 33 bodies (see Brobst's map) of probably late Precambrian dunite, most of which are no longer than 2000 feet; their other dimensions are appreciably smaller. The layered rock-ultramafic complex was invaded in probably Paleozoic time by silicic igneous rocks and pegmatite; the silicic rock was originally called granite by Keith, but Hunter (Amer. Ceramic Soc. Bull. 1940) referred to it as alaskite because of its lack of mafic minerals. Considering Spurr's original definition of alaskite, it would appear that the Spruce Pine rock could not be called alaskite; actually, the rock lies well within the leucogranodiorite field in the Johannsen classification and probably should be so termed.

Pegmatites are distributed over a wider area than the alaskite; while they are found in considerable numbers in the alaskite, they are even more abundant in the mica schist and are also known in the other formations in the district as well. They trend roughly northeast but certainly are not arranged in a regular pattern. The largest alaskite body is about 2 miles long and 4000 feet wide; the largest pegmatites are about 2000 feet long and 1000 feet wide, but they range in length upwards from a few inches and in width upwards from a few inches. There are at least 400 pegmatite masses that have been mined, and the total number must be several thousand. The alaskite and pegmatite bodies are generally concordantly emplaced in the mica schists but are far more likely to be cross-cutting in the amphibolites and the mica gneisses. The pegmatites form lenses, sills, dikes, and irregular masses in both the alaskite and the metasediments.

Radioactive age determinations have been made by several methods on minerals from several varieties of rocks found in the Spruce Pine district. The weighted averages of the ages so established range from 250 m.y. to 362 m.y., and Kulp and Brobst (1956) conclude from these data that 360 ± 20 m.y. is probably the best average age for the last metamorphism to affect the district. As the pegmatites probably were emplaced at the end of this metamorphic episode, they are considered to have been intruded about 360 m.y. ago and to be, therefore, of Late Devonian age. On this basis, the pegmatite deposits of the Spruce Pine district are here classified as middle Paleozoic.

The pegmatites in the district, whether they are in metasediments or alaskite, are mined for perthitic microcline and plagioclase (ranging in composition from albite through calcic oligoclase) and for sheet and scrap mica; some quartz has been recovered as a by-product of feldspar mining and some high-quality quartz has been taken from the cores of pegmatites. Kaolin is recovered in large amounts, more of it coming from the alaskite than from the pegmatite. The remaining 40-odd minerals known from the pegmatites are of negligible economic importance.

In the average pegmatite, microcline and plagioclase are intimately intermixed, with the soda feldspar usually being the more abundant of the two; microcline is more often found in separate masses of some size that can be selectively mined than is plagioclase, but these microcline masses are generally intergrown with quartz. These masses normally occur in the larger pegmatites, ones that are from 25 to 150 feet thick and occur in the larger bodies of alaskite in the southeastern part of the district. Feldspar is now recovered economically from the alaskite, both because the alaskite occurs in larger and more workable bodies and because it is more uniform in mineral composition.

The pegmatites can be placed into two general categories: (1) those that were intruded into the rock in which they are found, whether it be metasediment or alaskite and (2) those that appear to have been crystallized in that portion of the alaskite in which they were accumulated. Those of the first type are tabular to lens-shaped bodies that have sharp contacts with the rocks that enclose them, while those of the second are of very irregular shape and have gradational contacts with the alaskite. Regular-shaped pegmatite masses in the alaskite are generally oriented in parallel with, or nearly perpendicular to, the contact between the two rock types. In schistose rocks, the pegmatites are tabular or sheetlike; each has its long dimension many times its shorter ones. The contacts between the pegmatites and the rocks are usually sharp, although there may be some impregnation by thin stringers of pegmatite along the planes of foliation.

Since the Spruce Pine district has produced about half of the sheet mica mined in the United States, the geological associations of that mineral in the district are of considerable economic importance. In contrast to the feldspar, sheet mica can be recovered only from the pegmatites in which rocks alone it grows to sufficient size to justify mining. Muscovite books of minable quality usually are found in simple pegmatites that are composed mainly of medium-grained oligoclase and quartz or in zoned pegmatites within zones of this same composition.

Much kaolin has been formed by residual weathering of alaskite and of pegmatite. The principal clay mineral so developed is kaolinite, although some halloysite also was formed. The clay minerals were derived, of course, from the feldspars, particularly the plagioclase; the kaolinite usually contains some incompletely altered plagioclase, perthitic microcline, and practically unaltered muscovite and quartz. The depth of kaolinization in most commercial deposits ranges from 40 to 100 feet; all commercial deposits are found beneath older terrace levels of the master streams and of their principal tributaries over a vertical range of about 200 feet.

Unfortunately, the pegmatites of the Spruce Pine district have not been studied in sufficient detail to make it possible to report at length on the zoning, or lack of it, in the various pegmatites that have been mined. In a general way, it seems that the larger the pegmatite the more likely it is to be zoned, but this is simply to repeat what is obvious from the study of pegmatites in other districts.

The feldspar that is mined from the alaskite undoubtedly was emplaced as part of the original crystallization of the alaskite and should be classified as magmatic-1a. The feldspar, mica, and quartz received from the pegmatites are definitely primary minerals in those bodies and should be classified as magmatic-3a.

Oklahoma-Kansas-Missouri

TRI-STATE AREA

Late Mesozoic (pre-Laramide) | *Zinc, Lead* | *Telethermal*

Bastin, E. S., 1951, Paragenesis of the Tri-State jasperoid: Econ. Geol., v. 46, p. 652-657

Bastin, E. S., and Behre, C. H., Jr., 1939, Origin of the Mississippi Valley lead and zinc deposits, in Bastin, E. S., Editor, *Lead and zinc deposits of the Mississippi Valley region*: Geol. Soc. Amer. Spec. Paper no. 24, p. 121-143

Bastin, E. S., and Ridge, J. D., 1939, (Paragenesis of the) Tri-State zinc district, in Bastin, E. S., Editor, *Lead and zinc deposits of the Mississippi Valley region*: Geol. Soc. Amer. Spec. Paper no. 24, p. 105-110

Behre, C. H., Jr., and others, 1950, Zinc and lead deposits of the Mississippi Valley: 18th Int. Geol. Cong. Rept., pt. 7, p. 51-69 (particularly p. 52-58)

Brockie, D. C., and others, 1968, The geology and ore deposits of the Tri-State district of Missouri, Kansas, and Oklahoma, in Ridge, J. D., Editor, *Ore deposits of the United States, 1933-1967* (Graton-Sales Volumes): Chap. 20, v. 1

Emmons, W. H., 1929, The origin of the deposits of sulphide ores of the Mississippi Valley: Econ. Geol., v. 25, p. 221-271 (particularly p. 240-251)

Fowler, G. M., 1938, Structural control of ore deposits in the Tri-State zinc and lead district: Eng. and Min. Jour., v. 139, no. 9, p. 46-51

—— 1939, Structural control of ore deposits in the Tri-State lead and zinc district, in Bastin, E. S., Editor, *Lead and zinc deposits of the Mississippi Valley region*: Geol. Soc. Amer. Spec. Paper no. 24, p. 53-60

—— 1942, Ore deposits in the Tri-State zinc and lead district, in Newhouse, W. H., Editor, *Ore deposits as related to structural features*: Princeton Univ. Press, p. 206-211

—— 1960, Structural deformation and ore deposits--Oklahoma-Kansas mining field in the Tri-State mining district: Eng. and Min. Jour., v. 161, no. 6, p. 185-188

Fowler, G. M., and Lyden, J. P., 1932, The ore deposits of the Tri-State district (Missouri-Kansas-Oklahoma): A.I.M.E. Tr., v. 102, p. 206-251

—— 1933, The ore deposits of the Tri-State district: Econ. Geol., v. 28, p. 75-81

—— 1934, The Miami-Picher zinc-lead district: Econ. Geol., v. 29, p. 390-396

—— 1935, The ore deposits of the Tri-State district: Econ. Geol., v. 30, p. 565-575

Fowler, G. M., and others, 1935, Chertification in the Tri-State (Oklahoma-Kansas-Missouri) mining district: A.I.M.E. Tr., v. 115, p. 106-163

Hagni, R. D., and Desai, A. A., 1966, Solution thinning of the M bed host rock limestone in the Tri-State district, Missouri, Kansas, Oklahoma: Econ. Geol., v. 61, p. 1436-1442

Hagni, R. D., and Grawe, O. R., 1964, Mineral paragenesis in the Tri-State district, Missouri, Kansas, Oklahoma: Econ. Geol., v. 59, p. 449-457

Hagni, R. D., and Saadallah, A. A., 1965, Alteration of host rock limestone adjacent to zinc-lead ore deposits in the Tri-State district, Missouri, Kansas, Oklahoma: Econ. Geol., v. 60, p. 1607-1619

Holland, H. D., 1956, The chemical composition of vein minerals and the nature of ore forming fluids: Econ. Geol., v. 51, p. 781-797 (particularly p. 787-795)

Johnson, C. H., (Editor), 1963, Guidebook to the geology in the vicinity of Joplin, Missouri, including Westside-Webber mine, Oklahoma: Geologists in the Tri-State district, Missouri-Kansas-Oklahoma (Assoc. Mo. Geols.), 50 p.

Jolly, J. L., and Heyl, A. V., 1968, Mercury and other trace elements in sphalerite and wall-rocks from central Kentucky, Tennessee and Appalachian zinc districts: U.S. Geol. Surv. Bull. 1252-F, p. F1-F20

Leith, C. K., 1932, Structures of the Wisconsin and Tri-State lead and zinc deposits: Econ. Geol., v. 27, p. 405-418 (particularly p. 410-418)

Lyden, J. P., 1950, Aspects of structure and mineralization used as guides in the development of the Picher field: A.I.M.E. Tr., v. 187, p. 1251-1259 (in Min. Eng., v. 187, no. 12)

McKnight, E. T., and Fischer, R. P., 1970, Geology and ore deposits of the Picher field, Oklahoma and Kansas: U.S. Geol. Surv. Prof. Paper 588, 165 p.

Moore, R. C., and others, 1939, Stratigraphic setting--Tri-State district of Missouri, Kansas, and Oklahoma, in Bastin, E. S., Editor, *Lead and zinc deposits of the Mississippi Valley region*: Geol. Soc. Amer. Spec. Paper no. 24, p. 1-12

Ohle, E. L., Jr., 1959, Some considerations in determining the origin of ore deposits of the Mississippi Valley type: Econ. Geol., v. 54, p. 769-789

Ridge, J. D., 1936, The genesis of Tri-State zinc and lead ores: Econ. Geol., v. 31, p. 298-313

Schmidt, R. A., 1962, Temperatures of mineral formation in the Miami-Picher district as indicated by liquid inclusions: Econ. Geol., v. 57, p. 1-20

Siebenthal, C. E., 1915, Origin of the zinc and lead deposits of the Joplin region, Missouri, Kansas, and Oklahoma: U.S. Geol. Surv. Bull. 606, 283 p.

Smith, W.S.T., and Siebenthal, C. E., 1907, Joplin folio, Missouri-Kansas: U.S. Geol. Surv. Geol. Atlas, Folio 148, 20 p.

Spurr, J. E., 1927, Ores of the Joplin region (Picher district): Eng. and Min. Jour., v. 123, no. 5, p. 199-209

Stoiber, R. E., 1946, Movement of mineralizing solutions in the Picher field, Oklahoma-Kansas: Econ. Geol., v. 41, p. 800-812

Tarr, W. A., 1933, The Miami-Picher zinc-lead district: Econ. Geol., v. 28, p. 463-479

Weidman, S., 1932, The Miami-Picher zinc-lead district: Univ. of Okla. Press, Norman, 177 p. (also Bull. 56, Okla. Geol. Surv.)

—— 1933, The Tri-State zinc-lead region, in *Mining districts of the Eastern States*: 16th Int. Geol. Cong., Guidebook 2, p. 74-91

—— 1939, Structure of the Miami-Picher district, in Bastin, E. S., Editor, *Lead and zinc deposits of the Mississippi Valley region*: Geol. Soc. Amer. Spec. Paper no. 24, p. 49-53

Notes

The zinc and lead deposits of the Tri-State area occupy some 2000 square miles in southwestern Missouri, southeastern Kansas, and northeastern Oklahoma. Most of the ore of the area, however, has been produced from two main sections: (1) about 18 square miles in the center of the Oklahoma-Kansas portion of the field, the center of which lies about 6 miles southwest of Baxter Springs, Kansas, in the vicinity of Picher, Oklahoma, and (2) the Oronogo-Webb City-Duenweg field that occupies a northwest-southeast-trending belt over 8 miles long, the center of which is about 5 miles northeast of Joplin, Missouri. The remainder of the ore has come from a large number of mineralized subareas that range in extent from a few square feet to several hundred acres and are scattered throughout the Tri-State area.

The older rocks of the district are separated from the ore-bearing formations of Mississippian age by a major hiatus that extends from the top of the Cotter dolomite of the Beekmantown series of the Lower Ordovician to the base of the Compton limestone of the Kinderhook series with the exception of local pockets and lenses of Devonian (?) Chattanooga shale. These older beds include Precambrian igneous and metamorphic rocks of which very little is known; these are unconformably overlain by the rocks of the Upper Cambrian Ozark series. The Upper Cambrian formations are followed conformably by the Ordovician Beekmantown series.

The first of the Mississippian beds in the district are those of the Compton formation of the Kinderhook series. The Compton is conformably overlain by the Northview formation (also part of the Kinderhook series). The Northview, as is true of the Compton, thins to the south and west and is absent from most of the Tri-State district.

Conformably overlying the Kinderhook beds are the ore-bearing beds of the Tri-State area. Almost all of these beds belong to the Mississippian Osage series, the exception being the topmost of the mineralized beds, the B bed, which probably is the lone representative of the Meramec series in the district. The mineralized beds have often been referred to as the Boone formation, but this term no longer is acceptable in a stratigraphic sense.

The oldest of the Boone beds is the Fern Glen formation that is some 50 feet of gray chert nodules in soft limestone and shaly limestone; the Fern Glen, though locally mineralized, has not been given a letter designation. Conformably above the Fern Glen lies the first of the lettered beds, the Reed Springs formation (R bed). It is an important ore bed near Oronogo, Missouri, as well as at a few localities in the Picher portion of the field. Unconformably above the Reed Springs is the lowest of the Keokuk beds (Q bed); it is only locally mineralized. Beds Q through N have been referred to as the Grand Falls chert horizon. The P bed is made up of gray chert in bands and large flat nodules with some interbedded limestone; it is locally mineralized. The O bed is an important sheet-ground ore zone and contains abundant chert in bands and round flat nodules and some limestone. The N bed is generally barren in the Picher field but thickens and becomes importantly ore-bearing in the vicinity of Baxter Springs, Kansas, and Stark City, Missouri, and is composed of massive, somewhat mottled limestone with chert in large bands and nodules; it is the topmost bed of the Grand Falls chert horizon. The M bed contains chert mainly in rounded nodules and is the most important ore horizon in the Oklahoma-Kansas portion of the area. The L bed is unmineralized. The K bed is the uppermost bed of the Keokuk formation and provides an important ore horizon.

Disconformably above the K bed is the lowest bed of the Warsaw formation (the J bed); it is locally absent from the sequence. The J bed is not appreciably mineralized. The H bed is an important ore horizon in much of the district and often is so well mineralized that it and the G bed above it can be mined as a unit. The G bed is also an important ore horizon. The F bed is not much mineralized. The E bed is importantly mineralized in a few mines. The D bed contains only a few minable ore bodies. The C bed is probably correctly considered as the uppermost bed of the Warsaw formation; it has not been appreciably mineralized. The B bed rests conformably on the C bed and probably is the basal and only member of the Meramec in the district; it is essentially unmineralized.

The B bed is unconformably overlain by the rocks of the Chester series; these formations are often missing in the area. Much of the uppermost Mississippian Chester was deposited in sinkholes and other depressions in the older Mississippian rocks. In the western part of the district, the Chester is overlain by the Pennsylvanian Cherokee formation; it consists of shales and sandstones and is the surface rock in much of the Oklahoma and Kansas portions of the area. Many sinkholes in the Mississippian rocks are filled with Cherokee shale. The only rocks younger than the Cherokee are the Tertiary Lafayette gravels that cap some low-lying hills in the area.

No igneous rocks have ever been encountered in the Mississippian beds, and the only igneous rocks of any kind are those which have been found in the deep wells drilled in the area. Pegmatitic granite has been found in a well less than 1 mile west of the Miami syncline at a depth of 1245 feet beneath the surface. It may be a high in the Precambrian basement or a rock of much younger age; it is most probably a pegmatite dike of much less than Precambrian age.

When the ore minerals were introduced in the stratigraphic sequence is a matter of considerable dispute. The presence of minor amounts of ore minerals in the Cherokee rocks dropped into the solution-developed caverns in the Mississippian rocks makes it evident that the mineralization was later than the deposition of those beds. The most recent supporters of the meteoric hypothesis of the formation of the ores (Hagni and Grawe, 1964) have suggested that they were emplaced during and shortly after Pennsylvanian sedimentation, in part as syngenetic sediments and in part as deposits from circulating ground water. If they are correct, the ores should be classified as late Paleozoic. On the other hand, if the ores were deposited from hydrothermal solutions, this deposition may have occurred at any time between the end of the deposition of the Cherokee shales and the post-Eocene period of intensive stream erosion. There is at present absolutely no certain way to date these ores except by analogy with other deposits of similar types in the central United States. It seems fairly certain that the ores of the Southern Illinois fluorspar deposits were formed in late Mesozoic (pre-Laramide) time; it also is probable that the Magnet Cove deposits (the alkali rocks and their associated deposits of titanium and the barite ores) were of this age. The ores of the Leadbelt and of the Upper Mississippi Valley have been given this age on the assumption that one period of mineralization is more probable in this area than more than one. On this same basis, the Tri-State deposits are classified as late Mesozoic (pre-Laramide).

The ore bodies of the Tri-State area can be grouped into two general classes: (1) runs, of which circles are a special case, and (2) sheet ground or bedding plane deposits. Runs are irregular, but usually elongate, bodies of ore that have been localized by largely vertical fissures, enlarged and complicated by the action of underground waters. Circles are circular, subcircular, and elliptical runs that are the result of special combinations of fractures and solution. Bedding plane deposits or blanket veins or sheet ground are nearly horizontal tabular ore bodies developed parallel to the bedding of the formations in which they are found.

Simple runs are straight or slightly curved ore bodies that have developed in underground channelways, largely filled with broken fragments of their walls and roofs, that were formed through the ground-water enlargement of fissures or fissure zones. Complicated runs resulted from the interconnection of large numbers of underground channelways, some of which were essentially parallel to each other but widened sufficiently to coalesce at least partially and others of which joined together at appreciable angles. The resulting pattern of the partly to largely debris-filled channelways was complex in the extreme. The longest of these shear-controlled channelway runs was about a quarter-mile long, but most do not exceed a few hundred feet in length. The average width of a run is between 10 and 50 feet, although a width of 300 feet is known. The vertical extent is also usually between 10 and 50 feet, but runs in which the initial shearing was particularly thoroughgoing may have heights of 150 feet. The overall pattern of the shear zones normally trends northeast to north-northeast or northwest to north-northwest, but in detail a run may follow almost any direction. The runs of the Miami-Picher field are so equally divided between northwest- and northeast-trending trends as to give the field an almost circular outline within which the pattern details are most complicated. The runs of the Oronogo-Webb City-Duenweg field have a definitely northwesterly overall trend, but again the detailed pattern is most irregular. The Joplin and Central City belts also trend northwesterly, but the Galena belt has a northeasterly trend.

The runs, circles, and sheet-ground deposits are generally to be found in the areas where chert is most abundant. On this basis, it has been suggested (Fowler and Lyden, 1932) that the white chert is a secondary material introduced as the first stage in the ore-forming process. The opposing view is that those rock volumes most abundantly supplied with primary, syngenetic chert were the most brittle and were most readily broken when stresses were applied to them. The latter would seem to be the more acceptable explanation for the greater concentration of ore in chert-rich than in chert-poor rocks. This does not mean that rocks low in, or lacking, chert are not broken; they are broken and they do contain ore but not to the extent that chert-rich rocks are and do.

Assuming that the white chert is a syngenetic material, the first step in the mineralization process in the district was the orogenic brecciation of the rocks followed by their solution and further brecciation due to solution-engendered collapse. The first actual mineral deposition was that of dolomite that replaced much of the limestone in most of the beds; as a general rule, those rock volumes that were dolomitized were most likely later to be converted into ore through the addition of sulfides. Following the dolomitization, the dark jasperoid, a fine-grained silica, dark-colored through the presence of organic matter, was emplaced largely as a replacement of dolomite. Where the silica of the jasperoid was deposited in open space, it formed automorphic quartz crystals; to some extent, the jasperoid was replaced by a second generation of dolomite of much more limited areal extent than the first; an appreciable quantity of the second-generation dolomite was deposited as perfect crystals in vugs where it caps the quartz crystals of the jasperoid. Pyrite, though not an abundant sulfide, was the first to deposit, replacing both dolomite and jasperoid and being replaced by both sphalerite and galena; there is no evidence that pyrite and the ore sulfides overlapped in time of deposition. Galena and sphalerite were the next sulfides to have been deposited; they show the following interrelations where they can be examined in vugs: (1) dolomite is capped by and intergrown with galena that is partially capped by and intergrown with sphalerite, and (2) dolomite is capped by and intergrown with sphalerite which is, in turn, partially capped by and intergrown with galena. Sphalerite began to deposit first more often than galena. Chalcopyrite usually began to deposit only after substantial volumes of sphalerite and galena had been deposited, chalcopyrite crystals being intergrown with the outermost portions of the crystals of the ore sulfides. Chalcopyrite crystals, however, normally grew beyond the outer edges of the crystals of the two ore sulfides and chalcopyrite also shows similar relationships with dolomite where the ore sulfides do not intervene between the carbonate and the copper sulfide. Marcasite is intergrown with and extends beyond galena, sphalerite, and chalcopyrite and is intergrown with and caps dolomite. All four sulfides must have been deposited together for part of the precipitation cycle, but the usual order of beginning and end of deposition was sphalerite, galena, chalcopyrite, and marcasite. Although many examples have been found in which one or more sulfides are lacking from a given specimen or given rock volume, this does not invalidate the general depositional sequence just given. Recent studies have attempted to break the sequence down into many generations on the basis of such local departures from the full depositional sequence, but this appears to have been done through seeing so many trees that the outline of the woods became dim, if not altogether invisible. The final stage in this first generation of minerals was the deposition of a second generation of quartz. Following this quartz, there was a hiatus in deposition during which the rocks and their contained ores were fractured; the first mineral to be introduced after the fracturing was calcite with which a second generation of pyrite was intergrown. The last mineral to form was enargite which is intergrown with the calcite but not with the pyrite. Although considerably the greater part of the ore mineralization was emplaced by open-space filling in the tectonic-solution breccias of the runs, circles, and sheet ground, an appreciable proportion was introduced by replacement of both dolomite (the greater part) and jasperoid (the lesser part).

Although there is no known mass of igneous rocks to which the Tri-State ores can certainly be genetically related, igneous rocks that are probably post-Precambrian in age have been encountered at depths of a little over 1000 feet in or near the district. The lack of igneous rocks immediately adjacent to the ores is, of course, to be expected for deposits of such obviously low-temperature characteristics (the galena is silver-poor, almost silver-free; the chalcopyrite is not abundant enough to be of economic importance; and the wall-rock alteration is of the type usually associated with telethermal deposits) and is no deterrent to their classification in that general category. It seems far more reasonable to suppose that magmatic processes could produce an ore-bearing solution, the last stages of which would be able to deposit the Tri-State ores, than to assume that ground water could dissolve out the huge amounts of ore constituents necessary from the exposed sediments of the Ozark dome. Although the present-day mine waters, and the city water of Baxter Springs for that matter, have deposited sphalerite and some galena, the amounts are so small in comparison with the total amounts of lead and zinc sulfides in the ore bodies that an infinity of time would not be sufficient for this process to have deposited the ores. To my knowledge, chalcopyrite and enargite have never been deposited, here or elsewhere, by ground water; while chalcopyrite is not abundant in the district, it is common, and enargite has been frequently observed. The subject of ore origin has been debated for many years, and the recent work of Hagni (1964) in attempts to summarize the arguments for both the hydrothermal and ground water concepts and achieves a considerable degree of impartiality, leaving the reader to make up his own mind. A few errors have crept into this tabulation; for example, Ridge (1936, p. 300) did not say that the vugs in the dolomitic limestone were formed by hydrothermal solution; instead, he said that they were inherited from the dolomitic limestone, but in the main the student can well spend time in studying this tabulation with care. The opinion reached here is that the deposits were formed by hydrothermal solutions under conditions of low intensity at a time when the overlying cover was sufficiently thick and competent that deposition occurred under conditions of slow loss of heat and pressure from the solutions. The Tri-State deposits, therefore, are classified as telethermal.

Oregon

GRANTS PASS

Middle Mesozoic — *Chromite* — *Magmatic-1b*

Allen, J. W., 1941, Chromite deposits in Oregon: Oregon Dept. Geol. and Mineral Industries, Bull. 9 (rev. ed.), 71 p. (particularly p. 31-51)

Diller, J. S., 1922, Chromite in the Klamath Mountains, California and Oregon: U.S. Geol. Surv. Bull. 725, p. 1-35

Diller, J. S., and Kay, G. F., 1924, Riddle folio, Oregon: U.S. Geol. Surv. Geol. Atlas, Folio 218, 8 p.

Hotz, P. E., 1961, Weathering of peridotite in southwest Oregon: U.S. Geol. Surv. Prof. Paper 424-D, p. D327-D331

Imlay, R. W., and others, 1959, Relations of certain Upper Jurassic and Lower Cretaceous formations in southwestern Oregon: Amer. Assoc. Petrol. Geols. Bull., v. 43, p. 2770-2785

Irwin, W. P., 1960, Geologic reconnaissance of the northern Coast Ranges and Klamath Mountains, California, with a summary of mineral resources: Calif. Div. Mines Bull. 179, 80 p. (chromite-bearing area of California immediately south of, and contiguous to, the general Grants Pass area)

Maxson, J. H., 1933, Economic geology of portions of Del Norte and Siskiyou Counties, northwesternmost California: Calif. Jour. Mines and Geol., v. 29, nos. 1, 2, p. 123-160 (chromite-bearing area of California immediately south of, and contiguous to, the general Grants Pass area)

Ramp, L., 1961, Chromite in southwestern Oregon: Oregon Dept. Geol. and Mineral Industries Bull. 52, 169 p.

Wells, F. G., 1955, Preliminary geologic map of southwestern Oregon (west of meridian 122° west and south of parallel 43° north): U.S. Geol. Surv. Mineral Invest., Field Studies Map MF-38

Notes

The chromite deposits of the Grants Pass area are associated with masses of ultramafic rocks that are scattered between Gold Beach on the Pacific Coast and Red Mountain 80 miles to the east and between the southern boundary of the state and Peel nearly 90 miles to the north. The area in which these ultramafic rocks occur covers some 6500 square miles and is part of the Klamath Mountains geomorphic division. Most of the chromite ore mined in the district has come from Josephine County in general, and most of this production, in turn, has come from the Central Illinois River area in which the largest individual producing mine has been the Oregon Chrome mine located less than 25 miles west-southwest from the town of Grants Pass.

The oldest rocks in the general area of ultramafic intrusives are the Old Schists that consist of highly foliated chlorite, epidote, sericite, and graphite schists that originally were of both sedimentary and volcanic origin. The age of these rocks is unknown, although they are probably pre-Silurian; they are certainly older than the oldest dated rocks in the area, the Applegate group of Upper Triassic age. The Applegate group is composed of flows, flow breccias, and pyroclastics, mostly mafic; some intrusives also are included in the group. The Applegate is unconformably overlain by the Dothan formation of lowest Upper Jurassic age (equivalent to the European Callovian stage). The Rogue formation appears to be conformable on the Dothan and consists mainly of metavolcanic rocks--altered flows, tuffs, and agglomerates. The Rogue is lower Oxfordian, and it and the underlying Dothan are apparently the equivalent of the Logtown Ridge formation in California. The probable metamorphosed equivalent of the Rogue is known as the Colebrooke schist. Concordantly, but disconformably, over the Rogue is the Galice formation--slate, graywacke, and conglomerate--which is upper Oxfordian and lower Kimmeridgian; it probably is the equivalent of the Mariposa formation (mainly slate) in California. Where the Galice is in contact with the Colebrooke, the former lies on the latter with an angular unconformity. Undoubtedly an appreciable period of nondeposition exists between the Rogue and the Galice to account for the disconformity between them; the considerable folding that took place in the Colebrooke phase of the Rogue accounts for the angular unconformity between the Colebrooke and the Galice. The lithification of the Galice was followed by a period of folding, regional metamorphism, and erosion (equivalent to middle Kimmeridgian through early Portlandian) and then by the deposition of the extreme Upper Jurassic Riddle formation, the lower formation of the Myrtle group; the Riddle is the equivalent of the middle and upper Portlandian. It appears to be largely equivalent to the Knoxville formation in California. Disconformably above the Riddle is the Days Creek formation, which is the upper formation of the Myrtle group and is far less altered than the beds beneath it. The disconformity between the Riddle and the Days Creek is equivalent to Berriasian and early Valanginian time. The lower portion of the Days Creek is composed of alternate units of sandy siltstone and fine-grained sandstone, with some limestone lenses and concretions in the siltstone, while the upper portion consists of dominant siltstone with fairly numerous beds of sandstone and a few limestone lenses; there is some chert and graywacke conglomerate at the top of the formation. The Days Creek covers all of Lower Cretaceous time except the lowest and uppermost; Days Creek time began in the Valanginian, extended through the Hauterivian, and through the Barremanian, and the Days Creek beds are essentially equivalent to the Horsetown or Paskenta formations to the south. The Days Creek is unconformably overlain by the Hornbrook formation (arkosic sandstone with local lenses of conglomerate and sandy shale) of Cretaceous age which equals the Chico formation in California. Unconformably above the Hornbrook are the Eocene Arago group and the Umpqua formation (soft, coarse-grained, locally massive sandstone with local conglomerate and shales); these are followed by Eocene volcanic rocks (andesitic to dioritic pyroclastics, flows, and tuffs with an upper Eocene flora), then by mafic Tertiary intrusives that cut all rocks in the stratigraphic column except the Miocene and younger volcanic rocks. These Miocene volcanic extrusions were followed by Pliocene and then by Quaternary volcanic materials; the last deposits were Quaternary alluvial beds.

The ultramafic intrusions almost certainly are largely, if not entirely, of Late Jurassic or earliest Cretaceous age because cobbles of peridotite have been found in the Days Creek formation, and ultramafics have been found as intrusions in all known formations older than the Days Creek. It is possible that some of the ultramafics were introduced at an earlier time since pebbles and cobbles of serpentine and partly serpentinized periodotite have been found in one area in conglomerates that contain Riddle fossils. Although this is not an overwhelming mass of evidence, it suggests either that the serpentines that intrude rocks of Upper Jurassic or Cretaceous age may have done so through plastic flow or that there were two periods of ultramafic intrusion. Of these two alternatives, the latter appears more reasonable; thus, the ultramafics probably were introduced in southwestern Oregon in two intrusive cycles, with the early and minor one being, perhaps, Triassic, and the other and much more important one being Upper Jurassic or lowermost Cretaceous.

The next intrusive events after the ultramafics had been emplaced were the introduction into all rocks older than the Hornbrook of fine- to coarse-grained granitoid rocks that range in composition from gabbro through diorite, to granodiorite, and finally to granite. These rocks are separated from the ultramafic intrusions by at least all of Hornbrook time and are appreciably younger than the ultramafics.

The ultramafics consist primarily of peridotite of which three types have been recognized: (1) saxonite which is the most common and is composed of olivine with usually smaller amounts of enstatite (an orthorhombic pyroxene) and accessory magnetite and chromite; (2) dunite which is made up largely of olivine with minor amounts of pyroxene, magnetite, and chromite; it occurs as large masses, narrow bands, lenses, and small, irregular-shaped segregations within larger bodies of saxonite (peridotite); and (3) pyroxenite which consists of pyroxene, with minor amounts of olivine, magnetite, and chromite; like the dunite the pyroxenite occurs as segregations in the saxonite (peridotite) that appear to be somewhat more regular than those of dunite as they are tabular and usually lenticular--pyroxenite also is found as small dikes in masses of periodotite.

Secondary serpentine minerals are present in essentially all of the peridotites; only rarely are these minerals absent. Antigorite is the most abundant of the serpentine minerals and was derived from olivine, pyroxenes, and amphiboles that were rich in magnesium. The serpentinization may have occurred in two stages, the first of which was autometamorphic and the second accomplished by solutions given off by the post-Hornbrook granitoid intrusives. The serpentines of the district are cut by a large number of rodingite dikes that consist mainly of grossularite garnet and a pyroxene, either diopside or diallage.

The chromite almost certainly was precipitated from the ultramafic magmas that were intruded throughout the Grants Pass area and was segregated by processes that affected the magmas during their primary crystallization. Granted the chromite is of the same age as the peridotites and that the serpentinization occurred after the crystallization of not only the chromite but the primary silicates as well, the chromite deposits must bear the same age--Late Jurassic or earliest Cretaceous--as the peridotites. Thus, the chromite bodies are here classified as middle Mesozoic.

In general, the chromite deposits of southwestern Oregon occur as lens-shaped masses that normally are pseudotabular in form. Any one of these massive chromite lenses, developed during the primary crystallization of the peridotites, may have undergone folding (the axial planes of the folds usually strike northeast), faulting, and shearing under the influence of the mechanical stresses that affected the host rocks during and after their serpentinization. Thus, the average primary lens has been separated into several lens-shaped pods, which generally retain their tapering, tabular shape; no one of these bodies is thicker than the subtabular mass from which it was derived. Single lenses of massive chromite have maximum thicknesses of between 15 and 20 feet, but the usual thickness is from a few inches to 3 or 4 feet. The largest chromite bodies are seldom more than 150 feet long and 50 feet wide; most of the deposits are much smaller. Only in the Oregon Chrome mine have single lenses (or multiple bodies made up of a number of nearly adjacent massive chromite lenses) containing as much as 5000 tons of ore been mined; no ore bodies ever approaching this tonnage have been found in any other of the subdistricts into which the Grants Pass area is divided.

Six recognizably distinct types of chromite occurrence have been reported from the Grants Pass area. Type (1) is disseminated chromite which consists of grains of chromite in accessory amounts evenly scattered through the peridotite; only when the chromite grains become appreciably more abundant than the accessory category would require can this type possibly be mined as ore. Type (2) is schlieren-banded chromite that originally was chromite disseminated in (generally) dunite; flowage movement in the peridotite magma, prior to its final crystallization, has drawn out the disseminated grains into streaks or bands. Type (3) is a layered chromite which is made up of several parallel layers of disseminated to massive chromite, separated from each other by essentially barren silicate bands; such bands are thought by Ramp (1961) to have resulted from the periodic interruption of chromite crystallization while that of olivine continued essentially without a break. Type (4) is massive chromite that may occur in lenses, rounded, boulderlike masses, or layers; such masses probably were accumulated during periods of extremely rapid chromite crystallization during which time the rate of olivine crystallization was very slow relative to that of chromite. Type (5) is veined chromite that encompasses small chromite stringers normally found in fault zones or shears; these chromite veins probably resulted from mobilized early chromite being forced into fracture openings under tectonic stress; that there was any chromite actually retained in the magma until so late a stage in the crystallization cycle that the remaining melt consisted essentially only of chromite is subject to considerable doubt--all late chromite veinlets appear to be explicable as remobilized early chromite. Type (6) is nodular chromite in which the chromite nodules in dunite or serpentine derived from it range from about 1/8 to 3/4 inch in diameter; these nodules probably originally were early chromite octahedra from which the sharp corners were removed by resolution or abrasion during movement of the partially crystallized peridotite while cracks developed during movement were filled by serpentine or other alteration minerals (mainly chlorite and antigorite) derived from the primary silicates--some nodules may have been formed by the aggregation of smaller chromite grains.

Within the sill-like masses of peridotite, the chromite bodies may occupy positions at any elevation within the sill, although the great majority of them appear to be in positions intermediated between top and bottom. The bands of layers of the schlieren-banded disseminated deposits and of the multilayered zones always lie more or less parallel to the long dimensions of the ore zones of which they are a part. The individual layers in such chromite bodies are composed in the usual case of massive or thickly disseminated chromite and range in thickness from a fraction of an inch to about 1 foot; only rarely is such a layer more than a foot in thickness.

A single ore-bearing zone (or horizon) may contain disseminated, banded, or massive bodies of chromite at various positions along strike, and these different types grade into each other over short distances. This arrangement strongly suggests that all types were formed at much the same time, from magma of essentially the same composition, but under physical conditions different enough from one segment to another that the manner of accumulation might be appreciably different over short distances.

These Oregon magmas were not of such composition that their crystallization could produce appreciabe thicknesses of anorthosite and gabbro as was true of those in the Bushveld or the Stillwater. Something more than half of the chromite is concentrated in the lower half of a given sill; therefore, the designation of early crystallization, early segregation seems most fitting for the large percentage of the minable chromite. The Grants Pass deposits are, therefore, here classified as magmatic-1b.

RIDDLE

Middle Mesozoic (Peridotite), Early Tertiary (Secondary Concentration)	*Nickel*	*Magmatic-1a (Peridotite) Residual-B1, Ground Water-B2 (Secondary Concentration)*

Austin, W. L., 1898, The nickel deposits near Riddle's, Oregon: Colo. Sci. Soc. Pr., v. 5, p. 173-196

Chace, F. M., and others, 1969, Applied geology at the Nickel Mountain mine, Riddle, Oregon: Econ. Geol., v. 64, no. 1, p. 1-16

Clarke, F. W., 1888, Some nickel ores from Oregon: Amer. Jour. Sci., 3rd Ser., v. 35, p. 483-488

Cumberlidge, J. T., and Chace, F. M., 1968, Geology of the Nickel Mountain mine, Riddle, Oregon, in Ridge, J. D., Editor, *Ore deposits of the United States, 1933-1967* (Graton-Sales Volumes): Chap. 78, v. 2

Diller, J. S., and Kay, G. F., 1924, Riddle folio, Oregon: U.S. Geol. Surv. Geol. Atlas, Folio 218, 8 p.

Foullon, H. B. von, 1892, Über einige Nickelerzvorkommen: K.K. Geol. Reichsanstalt, Jb., Bd. 42, S. 223-310 (particularly S. 224-233)

Hotz, P. E., 1961, Weathering of peridotite in southwest Oregon: U.S. Geol. Surv. Prof. Paper 424-D, p. D327-D331

—— 1964, Nickeliferous laterites in southwestern Oregon and northwestern California: Econ. Geol., v. 59, p. 355-396

Kay, G. F., 1907, Nickel deposits of Nickel Mountain, Oregon: U.S. Geol. Surv. Bull. 315, p. 120-128

Ledoux, A. R., 1901, Notes on the Oregon nickel prospects: Canadian Min. Inst. Jour., v. 4, p. 184-189

Pecora, W. T., and Hobbs, S. W., 1942, Nickel deposit near Riddle, Douglas County, Oregon: U.S. Geol. Surv. Bull. 931-I, p. 205-226

Pecora, W. T., and others, 1949, Variations in garnierite from the nickel deposit near Riddle, Oregon: Econ. Geol., v. 44, p. 13-23

Notes

The deposit of nickeliferous laterite at Nickel Mountain in Oregon is located 4 miles west-northwest of the town of Riddle in southwestern Douglas County; Riddle is 30 miles north of Grants Pass and 65 miles north of the border with California.

The deposit lies near the western edge of the Oregon Coast Ranges where the rocks of the Klamath Mountains have been thrust from the southeast over those of the Coast Ranges by high-angle reverse faulting. Although most of the chromite deposits of the Grants Pass area are to be found in the Klamath Mountains physiographic province and the nickel deposits near Riddle are located in the Coast Range sector, the stratigraphic sequence and the igneous rocks intrusive into it are essentially the same in both areas. Thus, the general description of the rocks of southwest Oregon given in the Grants Pass discussion is equally applicable here. In the actual Nickel Mountain area, some problems of stratigraphic sequence exist, however, that merit consideration on a somewhat expanded scale. When Diller and Kay (1924) mapped the Riddle quadrangle, they considered the Dothan formation to be younger than the Galice. Later works (Wells, 1955; Imlay and others, 1959; Irwin, 1970) have shown that the actual stratigraphic position of these two formations probably is the reverse of that suggested by Diller and Kay. The Rogue formation, lying between the Dothan and the Galice, seems to be missing in the Nickel Mountain area, but this does not change the relationship between Dothan and Galice. In the Riddle area, what Diller refers to as the Knoxville formation is now designated as the Riddle and is thought to be extreme Upper Jurassic instead of Lower Cretaceous as Diller believed. The Riddle is the lower of the two formations that make up the Myrtle group. What Diller referred to as the Horsetown formation is now called the Days Creek. Although the Days Creek and the Riddle make up the Myrtle group, there is an appreciable time-gap between them that covers all of Verriasian and early Valanginian time, and the Days Creek is the much less altered of the two. The Days Creek, however, was correctly placed by Diller since it is considered to be the equivalent of the Horsetown formation in California. What Diller called the Chico formation is now designated the Hornbrook but is of the same age as the Chico formation in California. The Eocene Arago group seems to be missing in the Riddle area, but the Eocene Umpqua formation is still so named and still considered to be Eocene. Except for high terrace gravels and alluvium, the Umpqua is the youngest sedimentary formation in the Riddle area.

The metamorphic sedimentary rocks in the Riddle area consist apparently of what are referred to in the Grants Pass discussion as Old Schists; Diller and Kay designate these rocks as the May Creek formation which they consider probably to be Devonian in age. Spatially associated with the May Creek formation are flows of Older Metarhyolite; in the Riddle quadrangle, these Devonian (?) rocks are in the southeast corner while the nickel-bearing ores are found in the northwest.

The Riddle quadrangle contains large areas of what are now greenstones that Diller and Kay considered to be older than the peridotites from which the nickel deposits were derived and to have been originally lavas and intrusive rocks that ranged in composition from gabbroic and basaltic to diorite and diabase. Some of these greenstones are interbedded with both the Galice and Dothan formations which would make them post-Galice in age or probably late in the Upper Jurassic.

The peridotite from which the Nickel Mountain deposit was formed is intrusive into these greenstones and into the Dothan and Riddle formations. Since cobbles of peridotite have been found in the basal Days Creek that covers all but the earliest and latest portions of Lower Cretaceous time and since the peridotite is intrusive into the Upper Jurassic greenstone and all pre-Days Creek formations, this ultramafic rock must have been introduced into the region in either latest Upper Jurassic or earliest Lower Cretaceous time.

The ultramafic rocks were originally mainly enstatite peridotite or saxonite, a rock composed of olivine and orthopyroxene with less than 5 percent of iron ore minerals; considerable amounts of dunite, which contain over 95 percent olivine, also are present, as are a little wehrlite and lherzolite. Some of the largest peridotite bodies contain minor quantities of gabbro, anorthosite, pyroxenite, and diabase. The Nickel Mountain saxonite is about 55 to 80 percent olivine with 10 to 30 percent enstatite although chromite is present, generally as disseminated grains but in places as almost pure segregations.

The development of the secondary nickel ores of Nickel Mountain appears to have taken place in the Tertiary and to have occupied much of that era. The formation of laterite and saprolite by residual enrichment is thought by the Hanna geologists to have begun in the Eocene and to have continued until at least the end of the Miocene. The development of quartz veins and boxworks by secondary enrichment began shortly after residual enrichment had commenced, probably also early in the Eocene. The two processes continued concomitantly through the subtropical climate that prevailed throughout the Eocene and most of the Oligocene and through the warm-temperature climate of the last of the Oligocene and all of the Miocene. With the end of the Miocene, laterization and saprolitization essentially stopped, but the formation of vein and boxwork deposits appears to have been continued through the temperate climate of the Pliocene; some minor development of quartz and garnierite in the veins and boxworks, moreover, may have continued through the Pleistocene and the Recent. The nickel deposits probably were considerably more extensive at the end of the Pliocene than they are now, those known at present in the Nickel Mountain deposit being remnants left after the widespread erosion to which the area has been subjected since the end of that period.

The primary source of the Nickel Mountain ores, the saxonite peridotites, are Late Jurassic or earliest Cretaceous age and are, therefore, here classified as middle Mesozoic. The residual and secondary ores, which are the economically valuable portions of the altered peridotite, appear to have been formed throughout the Tertiary, with the greatest proportion having been developed in the early Tertiary, and they are so classified here.

The saxonite is a homogeneous rock and the nickel it contains in the primary silicates is, for all practical purposes, disseminated uniformly throughout the rock. From this it follows that the peridotite, as a primary protore of nickel, must be classified as magmatic-1a.

Weathering in the Nickel Mountain area has resulted in a much greater destruction of olivine than of pyroxene, so it would appear that the nickel of the secondary material must largely have come from the olivine.

At Nickel Mountain, garnierite occurs in veins as much as several inches but less than a foot in width. Gernierite (a general term for nickel-bearing material--not a specific mineral) fills, or partly fills, the interspaces in the veins in an anastamosing network of thin sheets (1 to 2 mm thick) of quartz. Where garnierite does not fill the cavities, the surfaces of the quartz sheets are covered with microscopically small crystals of drusy quartz. The quartz-garnierite veins have the same attitude as the major joints in the peridotite and are spaced from several feet to tens of feet apart. The veins persist for several tens of feet along their strikes and down their dips and are joined by other veins that dip at right angles to the first set. Cellular boxworks of silica also are found in the Nickel Mountain deposit in which spongy masses of intersecting garnierite-bearing veinlets of microcrystalline quartz less than 1 mm to 2 to 3 mm thick enclose pockets of soft weathered peridotite. The garnierite usually makes up wall selvages in these veinlets in which the centers are partly filled with quartz that extends into the vug spaces as tiny crystals. These boxwork veinlets, in turn, may be intimately cut by a microboxwork of quartz veinlets in fractures in the garnierite; some boxworks lack garnierite.

Near the surface, the garnierite appears to have been largely leached from the quartz as has the soft-weathered peridotite; additional brown-stained quartz may then have filled, or partly filled, the resulting voids. In the zone of lateritic soil, the smaller veins and veinlets of quartz and garnierite are broken into small platy fragments, while the larger veins, though also broken up, may be incorporated into the soil and piled on the surface in masses from which most of the associated soil has been washed away. At Nickel Mountain, breccia zones, of either slump or tectonic origin, have been cemented by brown microcrystalline quartz under which may be some white quartz in tiny crystals accompanied by a little garnierite.

Hotz (1964) believes that the X-ray data indicate that garnierite is composed of mixtures in various proportions of nickel-bearing serpentine and a 10-angstrom layered mineral and that it is not a mineral in its own right. Recent work at Pennsylvania State, however, has found that nickel is contained in mica-, chlorite-, and serpentine-type minerals and perhaps in sepiolite.

Concomitant with the development of the residual soil minerals and of the secondary quartz and garnierite has been a marked change in bulk composition of the weathered portion of the peridotite. The greater time for residual soil and secondary ore formation at Nickel Mountain may largely have been responsible for the development of garnierite at Nickel Mountain and its failure to form in the other, and shorter-lived properties in the general area.

The Nickel Mountain ore body is roughly 6000 feet long, in a northeast-southwest direction and 3000 feet wide; its thickness ranges from a few feet to a maximum of 250 feet. It has been suggested (Pecora and Hobbs, 1942) that the garnierite and the quartz intimately associated with it were formed by Recent weathering of the late Tertiary laterite. Hotz, however, believes that the present-day field relationships show that the laterite and the quartz-garnierite veins and boxworks were formed concurrently as there are traces of earlier formed veins and boxworks, as well as streaks of garnierite, in the lateritic soil above the bulk of the garnierite-bearing structures. It would seem probable, therefore, that the thickness of the garnierite-bearing rock volume under the residual soil is roughly proportional to the length of time over which the alteration of the peridotite had taken place. Nickel probably was concentrated to an appreciable extent in the clay-mineral fraction of the residual soil (although no deposit in which only this process had taken place has yet been mined for its nickel content) but also that really profitable deposits were formed only when nickel concentrated in the residual soil was further concentrated by the formation of garnierite in the soft-weathered peridotite beneath it. Thus, the processes of laterization and saprolitization apparently took place over much the same, though lengthy, period of time, with the garnierite of the quartz veins and boxworks formed in the early stages of alteration being, at least in part, moved downward as the depth of laterization and saprolitization increased. The processes of garnieritization on the one hand and of saprolitization and laterization on the other, therefore, were carried out together over much of Tertiary time.

The chemical processes by which the deposit at Nickel Mountain was formed are certainly not well understood, but it seems certain that the solutions that deposited the garnierite and quartz contained silica (probably not in the colloidal state) and nickel and magnesium in balance with bicarbonate ion. The change in pH produced by the hydrolysis of the nickel bicarbonate may have been the most important factor, not only in precipitating the nickeliferous hydrosilicates of magnesium that make up garnierite but also in depositing the quartz of the veins and boxworks. Certainly the problem of garnierite formation is far from being completely solved.

It does seem fairly definite, however, that the lateritic soil with its appreciable increase in nickel content over that of the primary peridotite was formed by residual processes and that the nickel and the other constituents of the garnierite were removed from this soil by ground water, carried downward into partly decomposed peridotite, and there precipitated in company with quartz transported into the same solutions. The secondary nickel ores, therefore, should be classified as residual-B1 (for the nickel in the laterite) and ground water-B2 (for the garnierite in the boxwork structures beneath the laterite).

Pennsylvania

CORNWALL

Early Mesozoic	*Iron as Magnetite, Copper, Cobalt, Pyrites*	*Hypothermal-2 to Mesothermal*

Callahan, W. H., and Newhouse, W. H., 1929, A study of the magnetite ore body at Cornwall, Pennsylvania: Econ. Geol., v. 24, p. 403-411

Cumings, W. L., 1933, The Cornwall iron mines, near Lebanon, Pennsylvania, in *Mineral deposits of New Jersey and eastern Pennsylvania*: 16th Int. Geol. Cong., Guidebook 8, p. 48-54

Davidson, A., and Wyllie, P. J., 1965, Zoned magnetite and platy magnetite in Cornwall type ore deposits: Econ. Geol., v. 60, p. 766-771

—— 1968, Opaque oxidc minerals of some diabase-granophyre associations in Pennsylvania: Econ. Geol., v. 63, p. 950-960

Fanale, F. P., and Kulp, J. L., 1962, The helium method and the age of the Cornwall, Pennsylvania, magnetite ore: Econ. Geol., v. 57, p. 735-746; disc., 1963, v. 58, p. 143-145

Geyer, A. R., and Gray, C., 1957, Cornwall iron mines: Geol. Soc. Amer. Guidebook for Field Trips, Field Trip no. 7, p. 243-248

Geyer, A. R., and others, 1958, Geology of the Lebanon quadrangle: Pa. Topog. and Geol. Surv., Geol. Atlas Pa., Atlas 167C, 1" = 2000'

Gray, C., 1956, Diabase at Cornwall, Pennsylvania: Pa. Acad. Sci. Proc., v. 30, p. 182-185

Gray, C., and Lapham, D. M., 1959, Cornwall iron mines: Geol. Soc. Amer. Guidebook for Field Trips, Field Trip no. 4, p. 147-152

—— 1961, Guidebook to the geology of Cornwall: Pa. Topog. and Geol. Surv., 4th Ser., Bull. G35, 18 p.

Hickok, W. O., IV, 1933, The iron ore deposits at Cornwall, Pennsylvania: Econ. Geol., v. 28, p. 193-255

Lapham, D. M., 1968, Triassic magnetite and diabase at Cornwall, Pennsylvania, in Ridge, J. D., Editor, *Ore deposits of the United States, 1933-1967*

(Graton-Sales Volumes): Chap. 4, v. 1

Lasky, S. G., 1934, Ferric-ferrous ratio in contact-metamorphic deposits: Econ. Geol., v. 29, p. 203-206

Spencer, A. C., 1908, Magnetite deposits of the Cornwall type in Pennsylvania: U.S. Geol. Surv. Bull. 359, 102 p. (particularly p. 7-28)

Stose, G. W., and Bascom, F., 1929, Fairfield-Gettysburg folio, Pennsylvania: U.S. Geol. Surv. Geol. Atlas, Folio 225, 22 p. (area is about 35 miles southwest of Cornwall, and although it is known to contain only minor magmetite deposits, the sediments, intrusions, and faulting are of the same general age and character as those of Cornwall)

Notes

The iron deposits of the Cornwall area are located about 25 miles due east of Harrisburg.

The oldest rocks in the district are the five members of the Conococheague formation of Upper Cambrian age; these members are distinguished on lithologic grounds only, no diagnostic fossils having been found. The beds that make up the various members are essentially all limestones or dolomites; the only member of the Conococheague present in the mine area proper has been so recrystallized and replaced that its correlation is not certain, but it is probably the Buffalo Springs. Two younger formations are known in the immediate Cornwall district: (1) the Mill Hill slate that can be so identified only when it has been considerably altered by thermal metamorphism; it is believed to be an outlier of the Middle Ordovician Martinsburg shale resting unconformably on the older rocks, and (2) the Blue conglomerate which, where least metamorphosed, consists of generally subangular pebbles, cobbles, and boulders of quartz in a carbonaceous shale matrix; it is probably at least in part of tectonic origin. Although there are places where the Blue conglomerate definitely overlies the Mill Hill, in others the two formations grade laterally into each other. No Paleozoic rocks younger than the Martinsburg are known in the general Cornwall area, and the total thickness of the known Paleozoic rocks is uncertain. At some time after their deposition, the Paleozoic sediments were strongly folded and then eroded. On the eroded surface so developed, sediments of the Triassic Newark series were deposited; the series is made up of the older New Oxford formation and the younger Gettysburg shale. The Newark series has not been folded.

After the lithification of the Newark series but still in Triassic time, the rocks of the area were intruded by tremendous volumes of diabase. Near the mine the diabase occurs as a basin-shaped sheet that, in part, is concordant with the enclosing strata and, in part, cuts across their bedding. In the vicinity of the ore bodies, the diabase is a dike about 1000 feet thick that dips south between 40° and 45° at the surface; it steepens with depth. Appreciable diabase pegmatite is irregularly distributed in the dike. There is some fine-grained aplite. The diabase does not usually cause extensive contact metamorphism against the limestones.

On geologic grounds, the association of the Cornwall ore with the contact between the diabase and the carbonate rocks suggests that the ores probably also were introduced in the same general time span as the intrusive. This concept is further supported by the occurrence of similar deposits of magnetite in much the same geologic circumstances at several other areas in eastern Pennsylvania. A K/Ar age determination on a muscovite in a chlorite schist near the diabase contact at Cornwall of 195 ± 5 m.y. acts to confirm the Triassic age of the ores even though this age did not come from an ore mineral. Isotope dilution analyses for helium, uranium, and thorium in Cornwall magnetites and pyrites, with extra-lattice uranium and thorium removed by leaching, give an age of 194 ± 4 m.y. for the magnetite, close agreement with the K/Ar determination. These results (Fanale and Kulp, 1962), however, have been questioned on the lack of consistency in the proportion of the total U and Th present that has migrated to, or has been deposited in, readily leachable surfaces or zones. Until this method has been further studied, at least, the result obtained by it cannot be considered as confirming the geological evidence as to the age of the Cornwall deposits. Nevertheless, the close relationship of several Pennsylvania magnetite deposits to Triassic diabase strongly suggests that the diabases and the ore fluids came from the same magmatic source, and the deposits, therefore, are here categorized as early Mesozoic.

At Cornwall the ores are found in a wedge of Paleozoic sediments caught up between the diabase beneath and Newark series beds unconformably above. There are two main ore bodies, the western that outcropped at the surface and was discovered in 1732 and the unexposed eastern that was found by a dip-needle survey in 1919. The western ore body has an outcrop length of about 4400 feet, a dip length of 1000 feet, and a thickness of about 150 feet. Not only was the ore of the western ore body formed by the replacement of limestone, but there still is unaltered limestone on the hanging wall.

The eastern ore body had a horizontal length of 2500 feet, with its upper surface dropping from 150 feet below the surface at its eastern end to 1150 feet below the surface at its southwestern end; its maximum thickness is over 200 feet, but it probably averages less than 100. The eastern ore body lies in the same wedge of Paleozoic sediments as the western, but it has fully replaced the limestone portion of the wedge so that the hanging wall of this ore is made up of the Mill Hill slate and the Blue conglomerate and its hanging wall contact is, therefore, much more regular than that of the west ore body.

The limestone was so complexly folded (probably isoclinally folded in the main) that the structures are difficult, if not impossible, to decipher; in general, however, the beds dip north to northwest. The relationships of the ore masses to the host rock depended largely on the nature of the original beds, some being appreciably more favorable to replacement than others. The ore, therefore, is generally banded, with magnetite-rich bands alternating with those rich in silicates, now largely converted to serpentine. In places the ore was brecciated and elsewhere it is quite massive, with magnetite crystals scattered evenly through the altered, but originally high-temperature, silicate minerals of the gangue. Pyrite (a second generation) and chalcopyrite occur irregularly disseminated through the ore, as bands parallel to magnetite bands and as cross-cutting veinlets; this last relationship strongly indicates that the sulfides were introduced during a later stage than the iron oxides. It is not certain whether cobalt-bearing pyrite was characteristic of both generations of that mineral or only of the second.

Some hematite is found in the Cornwall ores and the presence of magnetite pseudomorphous after hematite indicates that hematite was the earlier mineral; the total amount of hematite is small. When the ores are examined under the microscope, it is apparent that there are two varieties of magnetite; these can be recognized only when they are in contact with each other--in comparison with each other, one is bluish and the other brownish. Although the softer brownish variety probably is replacing the harder bluish material, it is uncertain which is normal magnetite. Probably brown magnetite is far the more common mineral.

From the minor amount of early hematite and the dominance of magnetite over the ferric oxide, it appears that the oxygen pressure in the ore fluid was low, except for a short time during the initial stages of mineralization.

The early students of the deposit were of the opinion that the ore-forming fluids were derived from the crystallizing diabase and moved across the contact into the adjacent limestone. One argument against this concept has already been pointed out--that the generality of limestone in contact with diabase is little altered. Further, it should be noted that magnetite has replaced diabase on a small scale, the amount of which decreases into the diabase. The greatest concentration of mineralization appears to be at the southwest end of the eastern ore body, and the unit cell dimensions of the magnetite decrease upward from this point. All these facts seem clearly to demonstrate that the ore fluids entered at about the center of the mineralized area and moved outward and upward, utilizing the contact between diabase and limestone as a channelway but having no direct genetic connection with the diabase.

The mineralization during the emplacement of magnetite was certainly of high intensity, and the magnetite definitely can be classified as hypothermal-2 despite the preference of the iron oxide for the more impure bands in the limestone. The pyrite-chalcopyrite stage, on the other hand, appears to have been developed under less intense conditions and probably can best be classified as mesothermal; this is confirmed by the lower temperature alteration minerals formed with or slightly after, the cobalt-bearing pyrite and the chalcopyrite. The deposit, therefore, is definitely both hypothermal-2 and mesothermal.

LANCASTER GAP

Late Precambrian *Nickel, Copper* *Magmatic-2a*

Campbell, W., and Knight, C. W., 1907, On the microstructure of nickeliferous pyrrhotites: Econ. Geol., v. 2, p. 350-366 (particularly p. 364-365)

Clark, L. A., and Kullerud, G., 1963, The sulfur-rich portion of the Fe-Ni-S system: Econ. Geol., v. 58, p. 853-885

Kemp, J. F., 1894, The nickel mine at Lancaster Gap, Pennsylvania, and the pyrrhotite deposits at Anthony's Nose of the Hudson: A.I.M.E. Tr., v. 24, p. 620-633

Knopf, E. B., and Jonas, A. I., 1929, Geology of the McCalls Ferry--Quarrysville district, Pennsylvania: U.S. Geol. Surv. Bull. 799, 156 p. (particularly p. 141-150)

Moyd, L., 1942, Evidence of sulphide-silicate immiscibility at Gap nickel mine, Pennsylvania: Amer. Mineral., v. 27, p. 389-393

Phemister, T. C., 1924, A note on the Lancaster Gap mine, Pennsylvania: Jour. Geol., v. 32, p. 498-510

Notes

The Lancaster Gap nickel mine is located in eastern Lancaster County in southeastern Pennsylvania about 12 miles east-southeast of Lancaster and 32 miles west-northwest of Wilmington, Delaware, and on a line connecting those two cities.

The deposits are found in a small body of metagabbro that is enclosed in a large mass of Precambrian gneiss; the gneiss is probably the equivalent of the Baltimore gneiss of Maryland and lies just to the north of the Martic line that crosses this portion of Pennsylvania in an irregularly east-west trace. Although the rocks of the late Precambrian Glenarm series are exposed only a short distance south of the Baltimore gneiss in the Mine Ridge anticline, where they have been complexly thrust over the Ordovician Conestoga limestone, these rocks are not found in the area of the nickel mines nor do they have any genetic connection with the ores.

The igneous rocks of the Mine Ridge area apparently have intimately intruded the metasediments. The first igneous event in the area was the intrusion of sills or the extrusion of flows, of mafic igneous material, probably basalt, but there can be no certainty as to the original rock type involved. These flows or sills are now amphibolite schists, the planes of foliation being parallel to the foliation of the enclosing Baltimore gneiss.

Radioactive age dating work has shown that the age of the last metamorphism affecting the Baltimore gneiss was late Precambrian, probably that of Grenville metamorphism that involved rocks not very far northeast of the district. The metagabbro in which the nickel deposits are found does not show as much metamorphism as the gneiss in which it was intruded (although its western end is quite gneissic) which suggests that it was probably intruded toward the end of the metamorphic period. If the nickel ores were introduced with the gabbro magma, then they must be of the same age as the gabbro and, therefore, late Precambrian; they are so classified here. If the ores should turn out to be of hydrothermal origin, the ore solutions depositing them almost certainly would have come from a mafic source magma. Since the youngest mafic rock in the area is the gabbro and since the ores, if hypothermal, must have come from the same general source, they too must be late Precambrian in age. Thus, under either genetic hypotheses, the ores are of essentially the same age.

The metagabbro body that contains the nickel ores of the Gap mine has a known length of about 2400 feet and a width of 300 to 500 feet according to Kemp (1894); the body may extend another 1300 feet to the east if credence is given to an extension reported by Kemp as having been indicated by Frazer of the Pennsylvania Geological Survey. At the surface, the ore was located along the eastern part of the irregular south margin of the gabbro (as is shown on Kemp's map, not as it was projected by Frazer) for about 1400 feet; it then rounded the nose of the structure (as shown on Kemp's map) and continued along the north edge of the gabbro for about 200 feet. The ore was there interrupted by about 300 feet of gabbro of less than ore grade and then resumed its marginal course for another 300 feet before dying out to the west in sulfide-bearing gabbro of less than ore grade. The ore ranged in thickness from 4 to 35 feet. The ore followed down the nearly vertical dip of the gabbro body and was mined as far as 250 feet beneath the surface. As the mine was forced to close by competition from the New Caledonia and Sudbury deposits, it is to be doubted if the ore was exhausted when mining stopped. On the other hand, geophysical surveys of the body were carried out during World War II, and the lack of any subsequent activity at the property suggests that the surveys did not indicate the availability of a large tonnage of ore.

The first explanation for the occurrence of the ores was that of Kemp (1894) who suggested that the sulfides had been introduced as an integral part of the magma from which the metagabbro (or amphibolite as Kemp called it) was crystallized. Campbell and Knight (1907) found that the metallic mineralization consisted of magnetite (which occurs as octahedra in a matrix of sulfides), medium-grained pyrrhotite (which appears to have been much strained), pentlandite (which occurs as veinlets and small masses in pyrrhotite and which shows its characteristic cleavage), chalcopyrite (which is closely associated with pentlandite and contains small grains of that mineral), pyrite (which is found in small amounts as euhedral crystals in pyrrhotite), and hornblende (which occurs as prismatic crystals, the ragged terminations of which suggest that they reacted with the molten sulfides after they had crystallized).

The presence of sulfides as veinlets filling fractures in the hornblende, as veinlets in the actual wall rock itself, and along cleavage planes in the biotite indicates that some deformation of the silicate mass occurred before the sulfides had solidified, and the strained nature of the pyrrhotite suggests that the ore and its enclosing gabbro were deformed after the sulfide solid solution had crystallized.

The occurrence of the sulfide ores along the nearly vertical margins of the gabbroic body is difficult to reconcile with the concept of gravity segregation in which process the molten sulfides should have been concentrated, to a major degree at least, at or near the base of the mafic body. Had the gabbro initially been intruded in a laccolithic form and had the sulfides settled to the bottom of that body and then the laccolith had been folded into a plunging syncline, the long axis of which would have run more or less the length of the gabbro and essentially parallel to that of the Mine Ridge anticline on which it would have been a subordinate fold, the present relations of ore to the gabbro would have been achieved.

It is probable that the ore body is of magmatic origin, with the sulfides having been segregated from the silicates at an early stage but not having crystallized until late in the igneous cycle; they, therefore, are here classified as magmatic-2a.

South Dakota

BLACK HILLS PEGMATITES

Middle Precambrian	*Lithium Minerals, Beryl, Mica, Feldspar, Tin*	*Magmatic-3a*

Balk, R., 1931, Inclusions and foliation of the Harney Peak granite, Black Hills, South Dakota: Jour. Geol., v. 39, p. 736-748

Darton, N. H., and Paige, S., 1925, Central Black Hills, South Dakota folio: U.S. Geol. Surv. Geol. Atlas, Folio 219, 34 p.

Hess, F. L., 1925, The natural history of pegmatites: Eng. and Min. Jour., v. 120, no. 8, p. 289-298

Higazy, R. A., 1949, Petrogenesis of perthite pegmatites in the Black Hills, South Dakota: Jour. Geol., v. 57, p. 555-581

Kupfer, D. H., 1960, Pegmatite-granite relationships in the Calamity Peak area, Black Hills, South Dakota; U.S.A.: 21st Int. Geol. Cong. Rept., pt. 17, p. 77-93

—— 1963, Geology of the Calamity Peak area, Custer County, South Dakota: U.S. Geol. Surv. Bull. 1142-E, p. E1-E23

Landes, K. K., 1928, Sequence of mineralization in the Keystone, South Dakota, pegmatites: Amer. Mineral., v. 13, p. 519-530, 537-558

Lang, A. J., and Redden, J. A., 1953, Geology and pegmatites of part of the Fourmile area, Custer County, South Dakota: U.S. Geol. Surv. Circ. 245, 20 p.

Lincoln, F. C., 1927, Pegmatite mining in the Black Hills: Eng. and Min. Jour., v. 123, no. 25, p. 1003-1006

Norton, J. J., 1960, Hugo pegmatite, Keystone, South Dakota: U.S. Geol. Surv. Prof. Paper 400-B, p. B67-B70

—— 1964, Geology and ore deposits of some pegmatites in the southern Black Hills, South Dakota: U.S. Geol. Surv. Prof. Paper 297-E, p. 293-341

—— 1970, Composition of a pegmatite, Keystone, South Dakota: Amer. Mineral., v. 55, p. 981-1003

Norton, J. J., and others, 1962, The geology of the Hugo pegmatite, Keystone, South Dakota: U.S. Geol. Surv. Prof. Paper 297-B, p. 49-127

Orville, P. M., 1960, Petrology of several pegmatites in the Keystone district, Black Hills, South Dakota: Geol. Soc. Amer. Bull., v. 71, p. 1467-1490

Page, L. R., and others, 1953, Pegmatite investigations 1942-1945, Black Hills, South Dakota: U.S. Geol. Surv. Prof. Paper 247, 228 p.

Rackley, R. I., and others, 1968, Concepts and methods of uranium exploration: Wyo. Geol. Assoc., 20th Field Conf. Guidebook, p. 115-124

Redden, J. A., 1959, Beryl deposits of the Beecher No. 3-Black Diamond pegmatite, Custer County, South Dakota: U.S. Geol. Surv. Bull. 1072-I, p. 537-559

—— 1963, Geology and pegmatites of the Fourmile quadrangle, Black Hills, South Dakota: U.S. Geol. Surv. Prof. Paper 297-D, p. 199-291

—— 1968, Geology of the Berne quadrangle, Black Hills, South Dakota: U.S. Geol. Surv. Prof. Paper 297-F, p. 343-408

Riley, G. H., 1970, Isotopic discrepancies in zoned pegmatites, Black Hills, South Dakota: Geochim. et Cosmochim. Acta, v. 34, p. 713-725

Runner, J. J., 1943, Structure and origin of the Black Hills pre-Cambrian granite domes: Jour. Geol., v. 51, p. 431-457

Schneiderhöhn, H., 1961, Black Hills, Süddakota, in *Die Erzlagerstätten der Erde*: Gustav Fischer, Stuttgart, Bd. 2, S. 223-294

Schwartz, G. M., 1925, Geology of the Etta spodumene mine, Black Hills, South Dakota: Econ. Geol., v. 20, p. 646-659

—— 1930, The Tin Mountain spodumene mine, Black Hills, South Dakota: Econ. Geol., v. 25, p. 275-284

Schwartz, G. M., and Leonard, R. J., 1926, Alteration of spodumene in the Etta mine, Black Hills, South Dakota: Amer. Jour. Sci., 5th Ser., v. 11, p. 257-264

Sheridan, D. M., and others, 1957, Geology and beryl deposits of the Peerless pegmatite, Pennington County, South Dakota: U.S. Geol. Surv. Prof. Paper 297-A, p. 1-47

Smith, W. C., and Page, L. R., 1941, Tin-bearing pegmatites of the Tinton district, Lawrence County, South Dakota; a preliminary report: U.S. Geol. Surv. Bull. 922-T, p. 595-630

Staatz, M. H., and others, 1963, Exploration for beryllium at Helen Beryl, Elkhorn, and Tin Mountain pegmatites, Custer County, South Dakota: U.S. Geol. Surv. Prof. Paper 297-C, p. 129-197

Taylor, G. L., 1935, Pre-Cambrian granites of the Black Hills: Amer. Jour. Sci., 5th Ser., v. 29, p. 278-291

Weis, P. L., 1953, Fluid inclusions from minerals from the zoned pegmatites of the Black Hills: Amer. Mineral., v. 38, p. 671-697

Notes

The pegmatites of the Black Hills of southwestern South Dakota and northeastern Wyoming are found in two principal areas: (1) that which borders Harney Peak in the southern Black Hills on the east, northeast, north, northwest, and southwest and covers some 275 square miles and (2) that on the Nigger Hill uplift (the Tinton district) that lies in Lawrence County, along the Wyoming border about 13 miles west of Lead, and covers some 16 square miles. The Harney Peak area is located in the southern portion of the Precambrian core of the Black Hills while the Tinton district is situated in a small, uplifted outlier of Precambrian rocks, the center of which is some 9 miles west of the nearest Precambrian rocks in the main Precambrian core. The Harney Peak area includes the Hill City district in Pennington County on the northern and northwestern flanks of Harney Peak, the Keystone district in Pennington and Custer Counties on the northeastern and eastern edges of Harney Peak, and the Custer district in Custer County that lies to the south and southwest of the town of that name and southwest of Harney Peak.

The Hill City and Tinton districts have been the main sources of tin in the Black Hills, and both produced lithium minerals; tantalum was also obtained from the Tinton district and sheet and scrap mica and feldspars from Hill City. The Keystone district has long been one of the main sources of lithium minerals in the United States and has also supplied beryl, feldspar, scrap and sheet mica, and tin and tantalum minerals. The Custer district has been the main source of sheet mica in the Black Hills and has been second to the Keystone district in the production of lithium minerals, beryl, and feldspar.

The principal mass of Precambrian igneous rock is that of the Harney Peak granite that forms a generally circular body some 10 miles in diameter around the circumference of which the pegmatite districts are arranged. The majority of geologists who have studied the problem believe the Harney Peak dome to have been formed by the intrusion of a stocklike mass of granite. There are several other granite bodies of much smaller size than the Harney Peak mass that are satellitic to it; none of these, of course, is known outside the area of Precambrian metasediments.

Page and others (1953) have classified the Black Hills pegmatites into three types: (1) zoned, in which the minerals are distinctly grouped into structural units of contrasting composition, or texture that have a systematic arrangement with respect to their walls; (2) homogeneous, in which the texture is uniform from wall to center except for a border zone; and (3) unzoned, in which small irregular pegmatites occur in granite or *lit-par-lit* gneisses and definitely show no layering or systematic arrangement of minerals or textures. These authors also divided each of these groups of pegmatites into three categories based on their relationships to the rocks that enclose them: (1) concordant with the structure of the enclosing metamorphic wall rock, (2) discordant to the structure of the enclosing metamorphic or granitic rocks; and (3) imperceptibly gradational into the enclosing igneous rock (granite or aplite).

The pegmatites in the Black Hills range from a few inches to more than a mile in length and from a fraction of an inch to more than 500 feet in width; most of the pegmatites that have been mined are less than 1000 feet and many are less than 200 feet long, and few are more than 100 feet wide. The majority are tabular or thinly lenticular in plan and section, but more irregular shapes are common. The homogeneous types are more likely to be uniform in thickness than those that are zoned. The tabular or nearly tabular bodies generally have two tapered ends, while the thick lenticular masses usually have a blunt crest (in plunging bodies), and the opposite end tapers to some extent. The shape of the pegmatite often appears to be related to the type of wall rock enclosing it; in the more strongly foliated schists, the bodies tend to be tear-shaped or lenticular, and in strongly folded rocks, the pegmatites conform to the fold pattern. Pegmatites that cut across brittle rocks are normally quite irregular, and occasionally two or more lenticular pegmatites have coalesced.

Quartz and feldspar are the most abundant minerals in the pegmatites of the Black Hills; a few of them contain spodumene, lepidolite, muscovite, and tourmaline as major constituents. The list of accessory minerals is long: tourmaline, beryl, garnet, biotite, apatite, lithiophilite-triphylite and related phosphates, amblygonite, spodumene, lepidolite, zinnwaldite, columbite-tantalite, cassiterite, magnetite, pyrite, löllingite, arsenopyrite, and uranium minerals. The less common and less widely distributed accessories include: topaz, chrysoberyl, tapiolite, microlite, siderite, galena, sphalerite, andalusite, sillimanite, and graphite.

The homogeneous pegmatites normally are simple in their mineralogy, with the principal minerals being potash feldspar, sodic plagioclase, quartz (often intergrown with one or both feldspars), muscovite, biotite, apatite, tourmaline, and garnet. Rarely do those bodies contain beryl, cassiterite, lithiophilite-triphylite, magnetite, and sulfides. The zoned pegmatites contain all of the accessory minerals listed in the preceding paragraph although abundances differ from one body to another; usually these minerals are large enough and abundant enough to be readily recognized.

In the Black Hills, the zoned pegmatites are the most abundant and are the most important economically, but in places fracture-filling deposits within zoned pegmatites are commercially important; replacement bodies do not appear to be particularly valuable.

As Page and others (1953) discuss at length, the pegmatites probably were formed by the differentiation of the granite magma from which the Harney Peak granite crystallized. Some of the pegmatites, the irregular gradational masses within that rock, solidified essentially where they were accumulated, while those in the surrounding rocks moved considerable distances from the locus of their development. The differences in mineral content of the various pegmatites may be explained by the different stages in the differentiation cycle at which they left their source magma and, to a generally lesser extent, by the reactions carried out between them and the rocks into which they were intruded.

The age of the Harney Peak granite has been estimated, on the basis of radioactive age determinations, to have been about 1620 ± 20 m.y. The age of the pegmatites must have been essentially the same. They are, therefore, here classified as late middle Precambrian.

The various pegmatites of the Black Hills area show differences of mineral composition, structure, and internal organization, but all appear to be normal magmatic pegmatites and were crystallized either *in situ* or intruded into the metasediments surrounding the granite. Zoned pegmatites of economic importance include the Hugo, the Peerless, and the Etta with its huge spodumene crystals. The Tinton district provided the cassiterite from which the entire 50 tons or so of metallic tin credited to the Black Hills were produced. Most of the tin came from the Rough and Ready mine in which the tin came from a series of pegmatite dikes radiating out from the larger Giant-Volney mass to the southeast. The cassiterite occurred as fine grains in fine-grained oligoclase pegmatite lenses in coarser, more quartz-rich pegmatites. Since all these pegmatites appear to have been formed as late-stage differentiates of the same magma source from which the Harney Peak granite came (although there have been dissenting opinions, e.g., Higazy, 1949), they are here classed as magmatic-3a.

HOMESTAKE

Middle Precambrian (major), Late Mesozoic to early Tertiary (minor) — *Gold* — *Hypothermal-1*

Connolly, J. P., 1927, Tertiary mineralization of the northern Black Hills: S. Dak. Sch. Min. Bull. 15, p. 1-130 (particularly p. 21-54)

Gustafson, J. K., 1933, Metamorphism and hydrothermal alteration of the Homestake gold-bearing formation: Econ. Geol., v. 28, p. 123-162

Harder, J. O., 1951, Vein quartz in the Homestake ore body (abs.): Geol. Soc. Amer. Bull., v. 62, p. 1536

Hosted, J. O., and Wright, L. B., 1923, Geology of the Homestake ore bodies and the Lead area of South Dakota: Eng. and Min. Jour.-Press, v. 115, nos. 18, 19, p. 793-799, 836-843

Koch, G. S., Jr., and Link, R. F., 1967, Gold distribution in diamond-drill core from the Homestake mine, Lead, South Dakota: U.S. Bur. Mines R.I. 6897, 27 p.

Kulp, J. L., and others, 1956, Age of the Black Hills gold mineralization: Geol. Soc. Amer. Bull., v. 67, p. 1557-1558

McLaughlin, D. H., 1931, The Homestake enterprise-ore genesis and structure: Eng. and Min. Jour., v. 132, no. 7, p. 324-329

—— 1933, The Homestake ore bodies, Lead, South Dakota, in *Ore deposits of the western states* (Lindgren Volume): A.I.M.E., p. 563-565

—— 1949, The Homestake mine: Canadian Min. Jour., v. 70, no. 12, p. 49-53

Moore, E. S., 1925, The geological age of the Homestake ore bodies: Econ. Geol., v. 20, p. 604-605

Noble, J. A., 1948, High-potash dikes in the Homestake mine, Lead, South Dakota: Geol. Soc. Amer. Bull., v. 59, p. 927-939

—— 1950, Ore mineralization in the Homestake gold mine, Lead, South Dakota: Geol. Soc. Amer. Bull., v. 61, p. 221-251

Noble, J. A., and Harder, J. O., 1948, Stratigraphy and metamorphism in a part of the northern Black Hills and the Homestake mine, Lead, South Dakota: Geol. Soc. Amer. Bull., v. 59, p. 941-975

Noble, J. A., and others, 1949, Structure of a part of the northern Black Hills and the Homestake mine, Lead, South Dakota: Geol. Soc. Amer. Bull., v. 60, p. 321-352

Paige, S., 1913, Pre-Cambrian structure of the northern Black Hills, South Dakota, and its bearing on the origin of the Homestake ore body: Geol. Soc. Amer. Bull., v. 24, p. 293-300

—— 1923, The geology of the Homestake mine: Econ. Geol., v. 18, p. 205-237

—— 1924, Geology of the region around Lead, South Dakota: U.S. Geol. Surv. Bull. 765, 58 p.

Sharwood, W. J., 1911, Analyses of some rocks and minerals from the Homestake mine, Lead, South Dakota: Econ. Geol., v. 6, p. 729-789

Slaughter, A. L., 1968, Homestake mine: Wyo. Geol. Assoc. 20th Field Conf.

Guidebook, p. 157-171

—— 1968, The Homestake mine, in Ridge, J. D., Editor, *Ore deposits of the United States, 1933-1967* (Graton-Sales Volumes): Chap. 67, v. 2

Wright, L. B., 1937, Gold deposition in the Black Hills of South Dakota and Wyoming: A.I.M.E. Tr., v. 126, p. 390-425

Notes

The Homestake mine is located in the Black Hills, near the town of Lead that lies some 30 miles northwest of Rapid City in the extreme west-central portion of South Dakota.

The oldest rocks in the area consist of at least 20,000 feet of metamorphosed sediments that are no younger than middle Precambrian in age. Originally, the sediments consisted of shales, now phyllites; sandstone and chert, now quartzite; and iron-magnesium carbonate rock, now sideroplesite schist in its less metamorphosed portions and cummingtonite schist where it has been more intensively metamorphosed. The extensive regional metamorphism of the Precambrian rocks appears to have been related to an unexposed granitic intrusion that must lie some distance northeast of the district. In the argillaceous rocks, three metamorphic zones were developed: (1) a biotite zone, (2) a garnet zone, and (3) a staurolite zone; only the Flag Rock formation is in the staurolite zone in the Lead district. The sandstones and cherts of the district were far less affected by metamorphism than the rocks that were originally shales, and the lighter phyllites, moderately rich in quartz, were less affected than the darker ones in which quartz is a subordinate mineral.

The ore-bearing Homestake formation has been, as a result of extreme folding, thickened to as much as several hundred feet or thinned down to essentially nothing. In the biotite zone, the Homestake is a sideroplesite $[FeMg(CO_3)_2]$ quartz schist that also contains biotite, chlorite, and a little graphite. In the outer garnet zone, the sideroplesite was converted to the amphibole, cummingtonite $[(Fe,Mg)_7(OH)_2Si_8O_{22}]$; in the middle garnet zone, almandite was developed at the expense of cummingtonite; and in the inner garnet zone, almandite, biotite, and quartz appear to have reverted to cummingtonite; the staurolite zone was not developed in the Homestake beds. Although present to a limited extent in other Precambrian formations, sideroplesite schist is largely concentrated in the Homestake formation. Unconformably overlying the Precambrian rocks, on a smooth erosion surface, is the Upper Cambrian Deadwood formation.

The igneous rocks of the Homestake district consist of (1) Precambrian gabbros that have been altered to amphibolites; they typically occur as long, narrow sills that could be mistaken for surface flows, and they locally have been metamorphosed into schists that are difficult to distinguish from metamorphosed sediments, and (2) Tertiary porphyry stocks of granitic and syenitic composition; they are common near Lead but are more abundant farther west and northwest. The structual changes caused by the Tertiary intrusives were important, but their metamorphic effects were essentially nil. Still later, numerous rhyolite dikes cut through the mine area.

The principal problem in solving the age of the Homestake deposits is whether or not the gold was formed in the Precambrian or the Tertiary or both. There is detrital gold in the basal conglomerate of the Deadwood formation in the vicinity of the Homestake ore bodies; thus, at least some of the Homestake gold is Precambrian. On the other hand, gold has been deposited in fractures in the Tertiary rhyolite dikes, so some of the gold is Tertiary. The unsolved question is: what proportion of the gold was deposited in the Precambrian and what portion in the Tertiary? One stage of mineralization certainly was emplaced in the Tertiary; this is Noble's (1950) fourth stage that consists mainly of pyrite and calcite and definitely postdates the Tertiary rhyolite dikes. On the other hand, these dikes clearly cut the quartz-chlorite-arsenopyrite (first stage) and the quartz-ankerite-pyrrhotite (second stage) and almost certainly are later than the pyrrhotite (third stage). The gold is largely, though certainly not entirely, associated with the arsenopyrite, while at most, only moderate amounts are found in the pyrite-calcite veinlets of the fourth stage. Thus, although it appears certain that some gold is Tertiary in age, the small amounts of it in the veinlets of the fourth stage strongly suggest that the bulk of it was formed prior to the intrusion of the Tertiary rhyolite dikes. This indicates that the larger fraction of the gold, the fraction that provided the detrital gold in the Deadwood conglomerate, was deposited in the Precambrian.

On the basis of the late middle Precambrian age radioactively determined for uraninites in the Black Hills pegmatites, it is possible that the major part of the Homestake gold (that well may have been introduced in the same general span of time) also is late middle Precambrian. The minor, late gold (and base metal sulfides) seem to be late Mesozoic to early Tertiary.

The minable ore bodies at the Homestake are essentially confined to a belt less than one-half mile wide. On the upper levels, the principal ore bodies, from east to west, were: (1) the Caledonia, (2) the main ore body, (3) the No. 5 ledge, (4) the east No. 9 ledge, and (5) the west No. 9 ledge. The Caledonia bottomed above the 2500 level and the No. 5 ledge is almost entirely mined out (although the structure may remain). The ore bodies of the early 1960s were, again from east to west: (1) the No. 9 ledge, (2) the No. 11 ledge (minor), (3) the No. 13 ledge (minor), and (4) the west ledges (very minor). The ore bodies are gradually moving southeast down the plunge, while the swarm of Tertiary rhyolite dikes is shifting steadily west with depth. The deepest level of the mine is now below the 6200 level, and shaft sinking has progressed to even greater depths.

The mineralization in the Homestake mine has been divided by Noble (1950) into four stages: (1) quartz, chlorite, and arsenopyrite that were emplaced by replacement of the sideroplesite or cummingtonite schists of the Homestake formation; (2) quartz (continuous with that in the first stage), ankerite, pyrrhotite, and minor quantities of albite, biotite, garnet, and cummingtonite; (3) pyrrhotite (continuous with that in the second stage); and (4) (separated from the other three by the intrusion of the large number of Tertiary rhyolite dikes and the attendant fracturing and faulting) mainly pyrite and calcite, with very minor amounts of specularite, magnetite, sphalerite, galena, chalcopyrite, realgar, and native arsenic and similar quantities of quartz, sericite, fluorite, celestite, anhydrite, gypsum, chlorite, opal, dolomite, and rhodocrosite. Although most of the calcite of the fourth stage is confined to veinlets in two principal localities, the pyrite is abundantly included in volumes of rock mineralized in the preceding three stages. The position of gold in this sequence is not definitely established. Gold is certainly associated with the quartz-chlorite-arsenopyrite mineralization of the first stage (which may well be due to the chemical affinity of gold for chlorite and arsenopyrite rather than to a close coincidence in age); massive vein quartz, moreover, does not carry gold. Gold is also often found in chlorite-pyrrhotite schist, and much unchloritized cummingtonite schist carries gold. A little gold is found in the rather uncommon massive pyrrhotite veins of the third stage, and some is found in the calcite-pyrite veinlets. Thus, gold is associated with the products of each stage of mineralization and this strongly suggests, as Noble (1950) points out, that gold either was formed during all four stages or only in the last one.

At the Homestake, the lack of low-temperature minerals (such as tellurides) deposited with the gold seems to make it more likely that the Precambrian gold was deposited under hypothermal conditions; it is, therefore, here categorized as hypothermal-1.

As for the Tertiary gold, the mineral assemblage with which it is found seems to suggest deposition under mesothermal or hypothermal conditions rather than leptothermal, but a certain choice cannot be made between hypothermal and mesothermal; hypothermal is here selected because of the abundance of specularite and magnetite in relation to gold in the fourth stage.

Tennessee

DUCKTOWN

Middle Paleozoic *Copper, Zinc, Pyrites* *Hypothermal-2*

Emmons, W. H., 1933, The Ducktown mining district, Tennessee, in *Mining districts of the eastern states*: 16th Int. Geol. Cong., Guidebook 2, p. 140-151

Emmons, W. H., and others, 1926, Geology and ore deposits of the Ducktown mining district, Tennessee: U.S. Geol. Surv. Prof. Paper 139, 114 p. (particularly p. 30-83)

Fenner, C. N., 1935, Origin of the copper deposits of the Ducktown type in the southern Appalachian region (rev.): Econ. Geol., v. 30, p. 928-936

Flournoy, E., 1950, One century of mining in the Ducktown basin: Mines Mag., v. 40, no. 5, p. 12-19, 24 (general, no geology)

Fullagar, P. D., and Bottino, M. L., 1970, Sulfide mineralization and rubidium-strontium geochronology at Ore Knob, North Carolina, and Ducktown, Tennessee: Econ. Geol., v. 65, p. 541-550

Gibson, O., 1953, Heavy accessory mineral study in the Ducktown basin: Ga. Geol. Surv. Bull. no. 60, p. 278-288

Gilbert, G., 1924, Oxidation and enrichment at Ducktown, Tennessee: A.I.M.E. Tr., v. 70, p. 998-1023

Jonas, A. I., 1932, Structure of the metamorphic belt of the southern Appalachians: Amer. Jour. Sci., 5th Ser., v. 24, no. 141, p. 228-243

Kemp, J. F., 1901, The deposits of copper-ores at Ducktown, Tennessee: A.I.M.E. Tr., v. 31, p. 244-265

King, P. B., 1949, The base of the Cambrian in the southern Appalachians: Amer. Jour. Sci., v. 247, nos. 8, 9, p. 514-530, 622-645

Kingman, O., and Diffenbach, R. N., 1965, The Ducktown copper district, in *Guidebook for Field Trips*: Amer. Crystal. Assoc. and Mineral. Soc. Amer., p. 24-37

Kinkel, A. R., Jr., and others, 1965, Age and metamorphism of some massive sulfide deposits in Virginia, North Carolina, and Tennessee: Geochim. et Cosmochim. Acta, v. 29, p. 717-724

Magee, M., 1968, Geology and ore deposits of the Ducktown district, Tennessee, in Ridge, J. D., Editor, *Ore deposits of the United States, 1933-1967* (Graton-Sales Volumes): Chap. 12, v. 1

Moh, G. H., and others, 1963-1964, Studies of Ducktown, Tennessee, ores and country rocks: Ann. Rept. Dir. Geophys. Lab., Carnegie Inst. Washington Year Book 63, p. 211-213

Ross, C. S., 1935, Pyrrhotite ore bodies (eastern United States), in *Copper resources of the world*: 16th Int. Geol. Cong., v. 1, p. 158-161

—— 1935, Origin of the copper deposits of the Ducktown type in the southern Appalachian region: U.S. Geol. Surv. Prof. Paper 179, 165 p. (particularly p. 1-67, 94-101)

Simmons, W. W., 1950, Recent geological investigations in the Ducktown mining district, Tennessee, in Snyder, F. G., Editor, *Symposium on mineral resources of the southeastern United States*: Univ. Tenn. Press, Knoxville, p. 67-71

Snyder, F. G., 1951, Stratigraphic studies of graywackes at Ducktown, Tennessee (abs.): Geol. Soc. Amer. Bull., v. 62, p. 1481 (also Econ. Geol., v.

47, p. 798, 1951)

Thompson, A. P., 1914, On the relation of pyrrhotite to chalcopyrite and other sulphides: Econ. Geol., v. 9, p. 153-175 (particularly p. 154-158)

Tung, J. P-y, 1969, Wall rock mineral assemblages, Calloway mine, Ducktown, Tennessee (abs.): Geol. Soc. Amer. Abs., Pt. 4 (Southeast. Sect.), p. 82-83

Weed, W. H., 1911, Copper deposits of the Appalachian states: U.S. Geol. Surv. Bull. 455, p. 152-157

Notes

The deposits at Ducktown are located in the extreme southeast corner of the state of Tennessee.

The rocks of the district compose a part of the Great Smoky formation that is thought to be of Lower Cambrian or late Precambrian age. The principal rocks present include graywackes and mica schists and types transitional between them, such as graywacke conglomerates and graywacke schists; some of the graywackes and graywacke conglomerates are calcareous, but no rocks that could be called calcareous schists have been found. Most of the schists are biotite-sericite-quartz rocks, but locally they have been so highly metamorphosed that they contain large crystals of staurolite and kyanite; smaller garnets are common in the schists. The various members of the formation grade into each other both along strike and across the bedding. The only igneous rocks in the mine area are sills of gabbro (now amphibolite) that follow the general strike of the region and have not been so highly metamorphosed as the sedimentary beds. In some of the graywackes, extreme metamorphism has developed nodules and larger masses of pseudodiorite (or calcgranofels in recent terminology); the pseudodiorites usually contain garnets.

Although the Appalachian region was intruded by considerable masses of granite during the late Paleozoic orogeny, the nearest granite to the Ducktown area is a considerable distance to the east. Nevertheless, Emmons believed that the ore fluids that deposited the Ducktown ores were derived from these granites during their crystallization cycle and that a rude zoning is shown by the gold and tin ores in the actual area of the granite intrusions, the copper-zinc ores to the west, and the zinc-lead ores still farther west. Other copper deposits in the general Ducktown-type belt include those of Fontana, Ore Knob, and Adams. This zonal relationship of the ore types to the granites convinced Emmons that they were formed at the same time and were, therefore, late Paleozoic in age.

Recently published radioactive age determinations (Kingman and Diffenbach, 1965) complicate the problem of determining the age of the Ducktown ores. Of three biotite samples, K/Ar dating on two of them gives ages between 327 and 374 m.y.; the other shows an age of 1200 m.y. Three hornblende samples, by the same method, give ages of 387 m.y., 478 m.y., and 1045 m.y. Of these, the 374 m.y. (Middle Devonian) biotite, the 387 m.y. hornblende, and the 1045 m.y. hornblende were taken from actual ore bodies, while the 327 m.y. biotite came from a spot 250 feet away from the footwall of an ore body, the 1200 m.y. biotite from 15 feet into the wall from the vein, and the 478 m.y. hornblende from 75 feet into the wall of an ore body, although it was there associated with 10 percent of sulfides. Of course, neither hornblende nor biotite is an ore mineral, so the evidence provided does not directly testify as to the age of the ore minerals. Biotite very probably was developed largely, if not entirely, as a metamorphic mineral so that any differences in age among the biotites can be explained as resulting from differences in retention of argon by biotites under at least two metamorphic events, one in the late Precambrian (Grenville) and the other in the middle Paleozoic (Middle or Late Devonian). How much older than the Grenville metamorphism the original Great Smoky sediments may have been cannot be determined by the K/Ar method. If it is assumed that the one anomalous hornblende age (1045 m.y.) demonstrates without question that both ore and gangue minerals were emplaced during Grenville time (the hornblendes show no evidence of having been metamorphosed since they were emplaced), the ores must be of late Precambrian age and were subjected to a considerable period of metamorphism during the Middle or Late Devonian. The ores themselves, however, do not appear to have undergone appreciable Devonian or other metamorphism (Moh and others, 1963-1964) and, therefore, must be (on that basis) no older than Devonian time immediately after the metamorphism had been essentially completed. In short, it seems unreasonable to classify the Ducktown ores as Grenville in age because of one anomalous age determination on one hornblende, but it does appear reasonable to assume that the Ducktown ores may better be categorized as middle rather than late Paleozoic; if they are older than middle Paleozoic, they should show much more effect of the Devonian metamorphism than they do. Here the Ducktown ores are classified as middle Paleozoic, but the case is certainly not yet closed.

The minable ore deposits of the Ducktown area can be assigned to three general groups: (1) the northern (northwestern) group which includes the Burra Burra, London, and East Tennessee mines; (2) the southern (southeastern) group which includes the Polk County, Mary, and Calloway mines; and (3) the central group made up of the Eureka, Boyd, Old Tennessee, and Cherokee mines. The mineralizations of the first two groups are quite siliceous and high in copper and zinc, while those of the central (third) group are lower in copper and zinc and higher in sulfur and iron. The presence of at least two favorable horizons within the Burra Burra anticline appears to have been definitely established, and the southeastern ore group probably lies in still a third favorable bed or series of beds. Complex though the folding is in detail, it does not appear to be enough so to make it possible for all the mineralized rock volumes to be part of the same single formation.

The sulfide minerals of the deposit include pyrrhotite, pyrite, chalcopyrite, and sphalerite in order of decreasing abundance; a minor amount of primary bornite and considerable magnetite and some hematite are present. The gangue minerals are principally quartz, calcite, actinolite, tremolite, pyroxene, garnet, zoisite, chlorite, graphite, and feldspars.

A pre-ore metamorphism of the host rock appears to have converted the primary graywacke into a garnet, kyanite (locally staurolite), biotite, sericite, quartz schist in which the schistosity is essentially parallel to the original bedding. In the graywackes farthest from the ore zone, the major effect of the ore-forming process was the replacement of staurolite pseudomorphically by sericite. Nearer the ore zones, the principal wall-rock alteration mineral is chlorite, while within the ore zones proper, the schists were converted to a mixture of magnetite, amphiboles (mainly actinolite and tremolite), talc, chlorite, and some quartz and calcite. The magnetite probably was the first of these minerals to form, although the suggestion has been made that it owes its well-developed crystals to metamorphism rather than to metasomatism. This stage of the hydrothermal alteration appears to have been followed closely by, and may even have overlapped with, sulfide deposition.

The first sulfides to form appear to have been pyrite and pyrrhotite, with much of the pyrite occurring as cubes (somewhat broken and rounded) in a matrix of pyrrhotite. Some observers believe that these pyrite cubes are porphyroblasts and that the pyrite was produced from the pyrrhotite during metamorphism that affected the region after (syngenetic) ore deposition. It seems at least equally probable to me that the pyrite simply was the first sulfide to form and was later enclosed and slightly replaced by pyrrhotite. Veinlets of chalcopyrite and sphalerite cut pyrite and pyrrhotite, probably indicating that these minerals were later than the iron sulfides. Ordinarily, the volume of gangue minerals is greater than that later replaced by sulfides, although disseminated sulfides (mainly pyrite and pyrrhotite) usually are present in the altered rock outside the zones of heavy to massive sulfides. Chalcopyrite generally is most abundant in association with chlorite in sheared structures. Essentially the same minerals are found in all ore bodies, but considerable differences in mineral proportions exist between one ore body and another. Zoning occurs within the district in much the same manner as in the various ore bodies.

The ore bodies appear to be nearly continuous replacements of favorable beds; most of the ore production has come from portions of these beds that were thickened, probably by drag folding. Although the majority of folds in the district plunge to the northeast, the plunges of the folds in the ore beds are about equally divided between northeast and southwest. The ore bodies are almost completely conformable to the bedding of the host rocks, but the ends of the ore bodies may die out in shear zones, and minor stringers may follow fractures for short distances.

The original composition of the favorable beds is not certain, but Emmons thought them to have been limestone layers, recrystallized to marble and often broken apart into isolated segments in the predominantly graywacke sequence. Actually, the calcite blocks in the ore zones appear originally to have been carbonate lenses in the graywackes, the lenses having broken into blocks while the predominant graywacke flowed. These blocks often contained sulfides as veins and replacement masses. Simmons suggests that the ores were confined to calcareous graywackes rather than recrystallized limestones, and Rose believes that the calcite was introduced hydrothermally and that the ores were not localized by lime-rich formations at all. Emmons explains the calcite matrix in some schist breccias and the calcite veinlets that run out from the ore bodies as the result of remobilization during metamorphism and points out that only a small fraction of the mineralization was deposited in noncalcareous rocks.

The ore minerals and the gangue minerals are such as to suggest deposition at high temperatures by high-intensity hydrothermal solutions, but the present environment of these minerals and many of the higher intensity minerals themselves may have been produced by metamorphic reactions affecting a syngenetic sulfide mass and its adjacent wall rocks. The evidence of Emmons and Ross for a hydrothermal introduction of the ores, however, is convincing to me, and the high-temperature minerals indicate that the solutions were within the hypothermal range. Because the host rocks probably were appreciably calcareous, the ore bodies are here categorized as hypothermal-2, but this may be subject to change.

MASCOT-JEFFERSON CITY

Late Paleozoic *Zinc* *Telethermal*

Allen, A. T., 1947, The Longview member of the Kingsport formation: Jour. Geol., v. 55, p. 412-419

—— 1948, Chert in the Kingsport formation at Mascot, Tennessee: A.I.M.E. Tr., v. 178, p. 232-239

Behre, C. H., Jr., 1962, Types of evidence for genesis of ore deposits in the east Tennessee and other lead zinc deposits: Econ. Geol., v. 57, p. 115-118

Bridge, J., 1945, Geologic map and structure sections of the Mascot-Jefferson City zinc mining district, Tennessee: Tenn. Div. Geol., 1" = 10 mi.

—— 1956, Stratigraphy of the Mascot-Jefferson City zinc district, Tennessee: U.S. Geol. Surv. Prof. Paper 277, 76 p.

Brokaw, A. L., 1950, Geology and mineralogy of the east Tennessee zinc district: 18th Int. Geol. Cong. Rept., pt. 7, p. 70-76

—— 1966, Geology and mineral deposits of the Powell River area, Claiborne and Union Counties, Tennessee: U.S. Geol. Surv. Bull. 1222-C, 59 p. (although the Powell River area is not in the Mascot-Jefferson City district proper, its occurrence has a real bearing on the understanding of east Tennessee ore formation)

Brokaw, A. L., and Jones, C. L., 1946, Structural control of ore bodies in the Jefferson City area, Tennessee: Econ. Geol., v. 41, p. 160-165

Bumgarner, J. G., and others, 1964, Habit of the Rocky Valley thrust fault in the west New Market area, Mascot-Jefferson City district, Tennessee: U.S. Geol. Surv. Prof. Paper 501-B, p. B112-B115

Callahan, W. H., 1964, Paleophysiographic premises for prospecting for strata bound metal mineral deposits in carbonate rocks: CENTO Symposium on Mining Geology and Base Metals, Ankara, p. 191-248

Crawford, J., 1945, Structural and stratigraphic control of zinc deposits in east Tennessee: Econ. Geol., v. 40, p. 408-415

Crawford, J., and Hoagland, A. D., 1968, The Mascot-Jefferson City zinc district, Tennessee, in Ridge, J. D., Editor, *Ore deposits of the United States, 1933-1967* (Graton-Sales Volumes): Chap. 13, v. 1

Crawford, J., and others, 1969, Mine geology of the New Jersey Zinc Company's Jefferson City mine: Tenn. Div. Geol. R.I. no. 23, p. 64-75

Currier, L. W., 1935, Structural relations of southern Appalachian zinc deposits: Econ. Geol., v. 30, p. 260-286 (particularly p. 271-276)

Fagan, J. M., 1969, Geology of the Lost Creek barite mine: Tenn. Div. Geol. R.I. no. 23, p. 40-44

Harris, L. D., 1969, Kingsport formation and Mascot dolomite (Lower Ordovician) of east Tennessee: Tenn. Div. Geol. R.I. no. 23, p. 1-39

Hathaway, D. J., 1969, Mine geology of the New Market Zinc Company mine at New Market: Tenn. Div. Geol. R.I. no. 23, p. 53-63

Hill, W. T., 1969, Mine geology of the New Jersey Zinc Company's Flat Gap mine at Treadway in the Copper Ridge district: Tenn. Div. Geol. R.I. no. 23, p. 76-90

Hoagland, A. D., 1967, Interpretations pertaining to the genesis of east Tennessee zinc deposits, in Brown, J. S., Editor, *Genesis of stratiform lead-zinc-barite-fluorite deposits--a symposium*: Econ. Geol. Mono. 3, p. 52-58

Hoagland, A. D., and others, 1965, Genesis of the Ordovician zinc deposits of east Tennessee: Econ. Geol., v. 60, p. 693-714

Johnson, R. W., Jr., Editor, 1965, The Mascot-Jefferson City zinc district, in *Guidebook for field trips*: Amer. Crystal. Assoc. and Amer. Mineral. Soc., p. 28-47

Jolly, J. L., and Heyl, A. V., 1968, Mercury and other trace elements in sphalerite and wallrocks from central Kentucky, Tennessee and Appalachian zinc districts: U.S. Geol. Surv. Bull. 1252-F, p. F1-F20

Kendall, D. L., 1960, Ore deposits and sedimentary features, Jefferson City mine, Tennessee: Econ. Geol., v. 55, p. 985-1003; 1961, v. 56, p. 1137-1138

McCormick, J. E., and others, 1969, Geology of the American Zinc Company's Young mine: Tenn. Div. Geol. R.I. no. 23, p. 45-52

Miller, J. D., 1969, Fluid inclusion temperature measurements in the east Tennessee zinc district: Econ. Geol., v. 64, no. 1, p. 109-111

Newman, M. H., 1930, Geology at Mascot: Min. Cong. Jour., v. 16, no. 11, p. 823, 833, 855, 857

—— 1933, The Mascot-Jefferson City zinc district of Tennessee, in *Mining districts of the eastern states*: 16th Int. Geol. Cong., Guidebook 2, p. 152-161

—— 1938, Zinc ores of east Tennessee occur as replacements in dolomite: Eng. and Min. Jour., v. 139, no. 8, p. 43-44

Oder, C.R.L., 1934, Preliminary subdivision of the Knox dolomite in east Tennessee: Jour. Geol., v. 42, p. 469-497

Oder, C.R.L., and Hook, J. W., 1950, Zinc deposits of the southeastern states, in Snyder, F. G., Editor, *Symposium on mineral resources of the southeastern United States*: Univ. Tenn. Press, Knoxville, p. 72-87 (particularly p. 73-82)

Oder, C.R.L., and Miller, H. W., 1948, Stratigraphy of the Mascot-Jefferson City zinc district: A.I.M.E. Tr., v. 178, p. 223-231

Oder, C.R.L., and Ricketts, J. E., 1961, Geology of the Mascot-Jefferson City district, Tennessee: Tenn. Div. Geol. R.I. no. 12, 29 p.

Ohle, E. L., 1951, The influence of permeability on ore distribution in limestone and dolomite, pt. 2: Econ. Geol., v. 46, p. 871-908 (particularly p. 871-879, Geology of the Mascot-Jefferson City, Tennessee, zinc district)

—— 1959, Some considerations in determining the origin of ore deposits of the Mississippi Valley type: Econ. Geol., v. 54, p. 769-789

—— 1961, Ore deposits and sedimentary features in Tennessee: Econ. Geol., v. 56, p. 444-446

Ridge, J. D., 1968, Comments on the development of the ore-bearing structures of the Mascot-Jefferson City district and on the genesis of the ores contained in them: Inst. Min. and Met. Tr., Sec. B, Applied Earth Sciences, v. 77, p. B6-B17

Rogers, J., 1953, Geologic map of east Tennessee with explanatory text: Tenn. Dept. Conserv., Div. Geol., Bull. 58, pt. 2, 168 p.

Secrist, M. H., 1924, Zinc deposits of east Tennessee: Tenn. Div. Geol. Bull. 31, 165 p.

Ulrich, E. O., 1931, Origin and stratigraphic horizon of the zinc ores of the Mascot district of east Tennessee: Wash. Acad. Sci. Jour., v. 21, no. 2, p. 30-31

Wedow, H., Jr., and Marie, J. R., 1965, Correlation of zinc abundance with stratigraphic thickness variations in the Kingsport formation, west New Market area, Mascot-Jefferson City mining district: U.S. Geol. Surv. Prof. Paper 525-B, p. B17-B22

Notes

The ore deposits of the Mascot-Jefferson City district are located east-northeast of Knoxville and lie in a belt almost 20 miles long that parallels the general strike of the country rocks. The Mascot-Jefferson City district lies in that portion of the great Appalachian Valley that runs through east Tennessee where it is from 40 to 60 miles wide and is bounded on the east by the major low-angle thrust faults of the Great Smoky and Unaka Mountains and on the west by the massive escarpment of the Cumberland Plateau that faces toward the Valley.

The oldest rocks in the area are the wide variety of Middle and Upper Cambrian formations that include, from oldest to youngest: The Rome formation; the Rutledge limestone; the Rogersville shale; the Marysville limestone; the Nolichucky shale; and the Maynardsville limestone. Overlying the Maynardsville is the Upper Cambrian to Lower Ordovician Knox group that is divided, from bottom to top, into: the Copper Ridge dolomite; the Chepultepec formation that lies disconformably on the Copper Ridge; the Longview dolomite; the Kingsport formation composed of cherty and sandy dolomite and limestone, divided into four divisions and lying disconformably on the Longview--it contains the main ore horizons of the district in the lower half to two-thirds of its beds; and the Mascot dolomite made up of dolomite and limestone with moderate amounts of chert. Unconformably above the Knox formations is the Mosheim limestone--a fine-grained "birds'-eye" limestone. Conformably above the Mosheim is the Lenoir limestone--a finely granular limestone with shale partings. Still other nonmineralized Ordovician formations overlie the Lenoir.

The irregular structural pattern within the district, much more complex than that of the Valley generally, indicates that the stress couples of the orogeny probably were resolved in different directions in different parts of the area. The age of the application of these earth forces certainly was post-Mississippian, for the Saltville fault cuts Mississippian rocks just north of the area mapped by Bridge (1956). In other parts of east Tennessee, structures that almost certainly are of the same age cut Pennsylvanian rocks, so there appears to be little doubt but that these structures were formed during the late Paleozoic Appalachian revolution. The possibility exists, however, that the ores were emplaced long before these Appalachian structures were developed. Kendall (1960) suggested that the ores originally may have formed by processes contemporaneous with the sedimentation, with later regeneration accounting for the migration of some of the ore into fractures. According to Hoagland and his colleagues (1965), the mineralization was localized by pre-Knox, post-Chazyan solution features into which hydrothermal solutions from unknown sources were able to penetrate before Chazyan sedimentation began. If either of these hypotheses is correct, the ores should be dated as early Paleozoic. On the other hand, I have supported the idea that the solution features containing the ores were not produced during Ordovician time but were developed by hydrothermal solutions entering the area during the Appalachian revolution (Ridge, 1968). If my concept is sound, then the ores were formed in the late Paleozoic, and they are so dated here. These genetic concepts are discussed further in the next paragraph.

It now appears probable that the brecciated rock volumes that contain most of the Mascot ore and much of that at Jefferson City were produced as a result of the thinning of certain of the limestone beds of the Kingsport formation upon their alteration to dolomite. Initially the Kingsport consisted of an alternation of beds of primary-dark to medium-gray dolomite and limestone. After the folding and much of the faulting, the fractured rock volumes were entered along fractures by solutions of probably hydrothermal origin that both removed much of the limestone from certain of the units of the Kingsport formation and replaced the remaining calcium carbonate with dolomite. The net result of this solution plus replacement (locally called recrystallization) was a considerable reduction in volume of the affected stratigraphic units (usually the dolomite unit would be less than half as thick as its limestone parent). Above such thinned beds, stresses developed in the overlying beds that resulted in their being considerably broken. This breaking, where a thinned bed abutted against its unthinned limestone equivalent, resulted in the development of a series of high-angle reverse faults (or shears), dipping toward the limestone, that down-dropped the overlying beds in successive increments away from the recrystallized dolomite-limestone contact. The farther one of these reverse faults was removed from this contact, the greater was the net downward movement of a given overlying bed, even though the movement on each of the series of faults was much the same on any other shear plane. At some distance from the recrystallized dolomite-limestone contact, the exact distance depending mainly on the thickness of the original limestone bed, its degree of thinning, and the beam strength of the overlying beds, the regular shear pattern would degenerate into a volume of rubble breccia. The last definitely recognizable shear would have a dip of about 45°, still toward the limestone and away from the dolomite. This rubble breccia would continue away from the initial limestone-recrystallized dolomite contact until the opposite limestone-recrystallized contact was approached. Here the same regular pattern of high-angle reverse faults would make its appearance to make the far side of the rubble breccia mass essentially the mirror image of that on the near side. The exact outline of any such volume of breccia would depend on the actual area of recrystallized (thinned) dolomite developed within the original limestone. It is possible for one breccia volume (or more than one) to overlie another, depending on the number of recrystallized units in the area in question. And one breccia volume may coalesce with another above it provided that the reductions in volume in the recrystallized beds and the resulting stress were of sufficient magnitude. Shear structures ordinarily die out vertically within the Kingsport and essentially never penetrate into the Mascot formation; the shear structures, therefore, definitely cannot be due to surface processes.

After the development of these breccia volumes, further volumes of hydrothermal solutions entered the area and deposited mainly pale yellow sphalerite and white dolomite, generally in that order. These minerals firmly cemented the rock fragments among which they were deposited and, to a lesser extent, replaced the fragments and the underlying recrystallized dolomite. Limestone fragments in the breccia and abutting against it were far less affected. Locally the mineralization cycle was repeated, and two (and perhaps three) generations of sphalerite have been recognized. Not all breccia volumes have been mineralized; some contain only dolomite. The greatest amounts of mineralization are in the borders of the rubble masses and in the contiguous sheared bands, the centers of the rubble usually less well mineralized than the border zones.

Replacement type ore is most abundant in the Jefferson City area, but breccia-filling ore in recrystallized dolomite appears to be more common in the Mascot part of the district.

Wall-rock alteration consists mainly of dolomitization. Although much of the dolomite in the ore-bearing formations is diagenetic, much probably was formed by replacement of limestone at the same general time as the gangue dolomite was deposited. There was some silicification or chertification of the fine-grained dolomite or limestone of the upper Longview that may well be genetically connected to the ore mineralization; minor amounts of such silicification are known in and above the ore horizons.

The simple mineralogy and wall-rock alteration and the fairly deep loci of deposition in which rapid loss of heat and pressure was further prevented by the impervious overlying marbles, shales, and sandstones strongly suggest ore deposition in the telethermal range. The lack of known igneous rocks in the area is not unexpected if the ore solutions had to rise through thousands of feet of strata after leaving their magma-chamber locus of origin before they could deposit the last of their mineral load. The deposits are, therefore, here classified as telethermal.

SWEETWATER

Late Paleozoic — *Barite* — *Telethermal (primary), Residual-B1 and Ground Water-B2 (secondary)*

Brobst, D. A., 1958, Barite resources of the United States: U.S. Geol. Surv. Bull. 1072-B, p. 102

Dunlap, J. C., 1945, Structural control in the eastern belt of the Sweetwater barite district (abs.): Econ. Geol., v. 40, p. 82

Gordon, C. H., 1918, Barite deposits of the Sweetwater district, east Tennessee: Tenn. Geol. Surv., Res. of Tenn., v. 8, p. 48-82

Kesler, T. L., 1950, Barite deposits southeast of the Appalachian plateaus, in Snyder, F. G., Editor, *Symposium on mineral resources of the southeastern United States*: Univ. Tenn. Press, Knoxville, p. 88-98 (particularly p. 91-93)

Laurence, R. A., 1939, Origin of the Sweetwater, Tennessee, barite deposits: Econ. Geol., v. 34, p. 190-200

—— 1960, Geologic problems in the Sweetwater barite district, Tennessee: Amer. Jour. Sci., v. 258-A (Bradley Volume), p. 170-179

Penhallegon, W. J., 1938, Barite in the Tennessee Valley region: Tenn. Valley Auth. Geol. Bull. 9, 47 p.

Secrist, M. H., 1924, Zinc deposits of east Tennessee: Tenn. Div. Geol. Bull. 31, 165 p. (particularly p. 133-134)

Watson, T. L., and Grasty, J. S., 1915, Barite of the Appalachian states: A.I.M.E. Tr., v. 51, p. 514-559

Notes

The Sweetwater district is located between the Tennessee and Hiwasee Rivers in Monroe, McMinn, Loudon, and Roane Counties in Tennessee, about 50 miles southwest of the Mascot-Jefferson City district and equidistant between Knoxville and Chattanooga.

The strata of the Sweetwater district consist of sedimentary rocks of Lower Cambrian to Middle Ordovician age. The barite deposits are contained in that portion of these rocks known as the Knox group, which has been divided in the district into the Upper Cambrian Copper Ridge dolomite and a series of Lower Ordovician rocks, which include, from oldest to youngest, the Chepultepec dolomite, the Longview dolomite, the two members of the Kingsport formation, the lower of which is limestone and the upper dolomite, and the Mascot dolomite. Because the district has been highly thrust faulted, the formations are repeated several times, and the barite deposits occur in three belts known as the western, central, and eastern, respectively. Actually, the western belt forks toward its northern end due to further strata repetition caused by a minor fault. A barren belt of Kingsport beds lies between the eastern and central belts. For a long time, the breccias in which the primary ores are contained were considered to be tectonic and to have been produced during the Appalachian revolution. The deposits, therefore, were dated as late Paleozoic or even as post-Paleozoic and assigned to the Mississippi Valley type. By analogy with opinions expressed on the ores of the Mascot-Jefferson City district, Laurence (1960) suggests that the breccias are primary sedimentary structures and, therefore, of Ordovician age. If this is correct, the barite deposits may be as old as Middle Ordovician, but the extent of bed-rock exposures in the district is so small that the problem has not been, and probably cannot be, solved in the Sweetwater district, and any solution to it must be reached by analogy with the zinc district. For reasons discussed under Mascot-Jefferson City, I believe that the breccias are late Paleozoic in age and that the ores, barite-rich in this case, were introduced by hydrothermal fluids from magmas existing at depth during the Appalachian orogeny. The workable deposits are residual accumulations that almost certainly were developed after the Appalachian revolution but how long after is impossible to say.

All of the deposits in the Sweetwater district have worked residual deposits, but in the deepest of these workings, primary ore has been exposed, and since no residual deposit has ever terminated in barren carbonate rock, it is probable that all deposits would have changed gradually to primary ore had they been followed downward far enough. These primary ores are found in veins in intensely brecciated beds of dolomitized limestone, similar to the recrystallized dolomite of the Mascot-Jefferson City district. None of the veins is large or persistent; however, where veins intersect, large, irregularly shaped mineralized masses were developed. The contact between veins and wall rock ordinarily is quite sharp. In addition to barite, the ores contain fluorite and pyrite with minor galena, sphalerite, and calcite. In a few localities, the sphalerite and galena content is great enough that these minerals have been recovered or could be recovered profitably under favorable economic conditions (as at Eve's Mills). The largest number of deposits (60 of 108) is found in the lower (limestone) member of the Kingsport formation and 65 of 108 are in the eastern belt. The simple mineralogy, lack of silver in the galena, and essential lack of wall-rock alteration (except for recrystallization of limestone to dolomite) all indicate that the primary deposits should be classed as telethermal.

Laurence (1960) says that agreement is general that the secondary deposits are the insoluble material left behind after the removal in solution of the carbonates of the Knox group. He is of the opinion, however, that some solutions, transportation, and redeposition of the more insoluble constituents (barite, fluorite, and silica) also took place since both barite and chert occur in far larger masses in the residuum than ever have been seen in the primary mineralized masses. Had only removal of carbonates occurred, the barite in the flat-lying rocks of the Kingsport (where the dip is less than 15°) should have been widely scattered in the residuum of the older, underlying formations. In all mines that have been worked down to bed-rock, however, the barite concentrations are found directly above primary bed-rock deposits. In the district, fluorite is present in practically all of the bed-rock deposits examined yet in the residual ore, fluorite is present only in the material just above bed-rock. This would indicate that, under the conditions prevailing during the formation of the secondary barite ore, some of the barite and practically all of the fluorite were dissolved, but far more barite than fluorite was redeposited. On this basis, the secondary ores of the Sweetwater district must be classified as both residual-B1 and ground water-B2.

Texas

TERLINGUA

Middle Tertiary *Mercury* *Epithermal*

Evans, A. M., and Traver, W. M., Jr., 1947, Terlingua mercury deposit, Brewster and Presidio Counties, Texas: U.S. Bur. Mines R.I. 3995, 10 p. (mimeo.)

Hillebrand, W. F., and Schaller, W. T., 1909, The mercury minerals from Terlingua, Texas: U.S. Geol. Surv. Bull. 405, 174 p.

Krauskopf, K. B., 1951, Physical chemistry of quicksilver transportation in vein fluids: Econ. Geol., v. 46, p. 498-523

Lonsdale, J. T., 1929, An underground placer cinnabar deposit: Econ. Geol., v. 29, p. 626-631

Philips, W. B., 1906, The quicksilver deposits of Brewster County, Texas: Econ. Geol., v. 1, p. 155-162

Ross, C. P., 1941, The quicksilver deposits of the Terlingua region, Texas: Econ. Geol., v. 36, p. 115-142

—— 1942, Some concepts of the geology of quicksilver deposits in the United States: Econ. Geol., v. 37, p. 439-465

—— 1942, Quicksilver lodes of the Terlingua region, Texas, in Newhouse, W. H., Editor, *Ore deposits as related to structural features*: Princeton Univ. Press, p. 193-195

Schuette, C. N., 1931, Occurrence of quicksilver orebodies: A.I.M.E. Tr., v. 96 (1931 General Volume), p. 403-488 (particularly p. 432-437)

Thompson, G. A., 1954, Transportation and deposition of quicksilver ores in the Terlingua district, Texas: Econ. Geol., v. 49, p. 175-197

Turner, H. W., 1906, The Terlingua quicksilver deposits: Econ. Geol., v. 1, p. 265-281

Yates, R. G., and Thompson, G. A., 1959, Geology and quicksilver deposits of the Terlingua district, Texas: U.S. Geol. Surv. Prof. Paper 312, 114 p., plus volume of plates

Notes

The mercury deposits of the Terlingua district are located in a narrow belt some 20 miles long that lies mostly in Brewster County but has a slight extension into Presidio County to the west. The district is situated within the Trans-Pecos country of Texas between the Pecos River and the Rio Grande; it is about 235 miles southwest of El Paso and 320 miles west of San Antonio and lies in the heart of the Big Bend region.

Although Paleozoic rocks are known in the Big Bend country, the nearest being exposed in the Solitario dome (of probably igneous origin), the center of which lies some 15 miles northwest of the town of Terlingua, all of the sedimentary rocks exposed in the area are of Cretaceous age. The oldest of these is the Lower Cretaceous Devil's River limestone (the lower part of the Comanche series), a thick and uniform sequence of limestones. It is conformably overlain by the basal Upper Cretaceous Grayson formation (the lower part of the upper half of the Comanche series) that is an almost structureless clay. Conformably above the Grayson is the still basal Upper Cretaceous Buda limestone. Above the Buda, the rocks grade from the predominant marine limestones of the Lower Cretaceous and lowermost Upper Cretaceous into a continental sequence. The sequence becomes more and more shaly as the end of the period is approached, the first formation of which is the middle Upper Cretaceous Boquillas flags. The Boquillas is overlain by the middle Upper Cretaceous Terlingua clay, a formation composed of structureless clay that contains a few thin beds of impure limestone. As is true of most of the formation divisions of Yates and Thompson (1959), this one is based more on lithology than fossil remains, age boundaries being crossed to preserve lithologic unity. Above the Terlingua clay is the widely, but irregularly, distributed middle Upper Cretaceous Aguja formation that is made up of interbedded sandstones and clays. The next formation is the upper Upper Cretaceous Tornillo clay. The remaining and youngest rocks of the district are grouped together by Yates and Thompson (1959) in the Chisos formation that is composed of lava flows and clastic rocks.

The sedimentary and volcanic rocks of the Terlingua district have been intruded by a considerable variety of alkalic igneous rocks that range in composition from rhyolite through analcite syenite, to olivine basalt, and take the forms of sills, laccoliths, dikes, and plugs, and both concordant and discordant contacts are known. The fine grain of most of the intrusives indicates that these types were introduced at shallow depths, but there are a few medium-grained rocks that speak of somewhat greater depths of consolidation.

The sedimentary rocks adjoining the intrusives were bleached and indurated to a limited extent by the igneous masses, and a little silicification of the limestone occurred at contacts with the intrusives; some analcite was produced in the sediments adjacent to analcite-bearing intrusives. Except for these phenomena, essentially no contact metamorphism was accomplished by the igneous rocks introduced into the sediments.

The development of the Terlingua uplift apparently did not begin until after the Cretaceous sediments had been deposited. While it appears from the unconformable contacts shown by the Chisos volcanic rocks against the Cretaceous rocks beneath them that the uplift was well advanced before the volcanic rocks began to deposit, it is also probable that there was further upward movement of the uplifted region after appreciable quantities of volcanic materials had been extruded, since these volcanic rocks outside the district have been folded.

The faults that bound the grabens in the Terlingua district probably were developed during the same period as the uplift, since the large grabens die out in the Terlingua monocline that served, therefore, as a hinge line for the graben faults.

Although the district is dominated structurally by the Terlingua domed uplift and by the domed structures rising above it and by wide, down-dropped graben blocks below it, the rocks have been deformed on a smaller scale by the igneous intrusions to produce second order domes and more irregular structures and numerous faults. Localized solution of the sedimentary rocks produced breccia pipes, limestone caverns, and filled caverns; most of the breccia pipes are importantly mineralized, and mineralization has reached some of the caverns.

These breccia pipes are usually vertical cylinders in form; within them are fragments from formations abutting against the walls of the pipes and from beds above them, but none have come from below. This relationship strongly argues that the pipes were formed by collapse and not be explosive forces.

Because the ore bodies are found in both the Cretaceous sedimentary and Tertiary igneous rocks and because they were emplaced after the folding of the monocline and after the faulting of the grabens, they must be of Oligocene age at the oldest. Because the mineralization almost certainly was deposited from solutions derived from the same general source as the igneous rocks, the ore-forming solutions cannot have been much delayed in their time of arrival in the district beyond that of the earth movements and igneous intrusions. The lack of post-ore faulting, however, would put the epoch of mineralization at the end of the orogenic change of events, so it is most reasonably to be dated as (probably late) Oligocene, and the ores are here classified as middle Tertiary.

The mercury deposits of the district have been divided into four classes: (1) the limestone-clay (Devils River-Grayson) contact deposits that are both the most common and valuable in the district, (2) the calcite vein deposits in the Boquillas flags, (3) the breccia pipe deposits, and (4) the deposits in igneous rocks that can be subdivided into deposits in regular fractures and deposits in irregular breccia rock volumes.

Although most of the mercury produced came from cinnabar, so intimately mixed with clay as to appear amorphous, other unusual mercury minerals were of economic value, minerals that are found in such abundance in no other deposits of this metal. The rare mercury minerals are almost entirely restricted to the limestone-clay contact deposits and, in them, were only seldom in direct association with cinnabar. These rare minerals include calomel (HgCl), kleinite (a mercury-ammonia chloride of uncertain composition), mosesite (a mercury-ammonia chloride, sulfate, and hydroxide of uncertain composition), eglestonite (Hg_4Cl_2O), terlinguaite (Hg_2ClO), and montroydite (HgO). Native mercury has also been found, and metacinnabar has been reported. It is interesting to note that the mercury in some of these unusual minerals is entirely or partly in the mercurous state, a condition not known in the usual mercury minerals. Other minerals in the limestone-clay deposits are clay minerals (reconstituted from the clay minerals in the sedimentary rocks and the breccias), calcite in abundance, and minor amounts of fluorite, barite, chalcedonic quartz, and pyrite. Although some pyrite was deposited syngenetically in the igneous rocks, much was deposited by hydrothermal solutions and much of that was later oxidized to limonite; some marcasite was also found. Pyrite is older than cinnabar in almost all instances for which age determinations can be made. Some, and perhaps all, of the hematite disseminated in the clay appears to have been of hydrothermal origin; the limonite is believed to have been secondary and supergene. The nonsulfide mercury minerals are thought by Yates and Thompson to have been of hypogene, hydrothermal origin, although they did not replace any other minerals and instead formed veinlets and crystal incrustations (calomel) or well-developed crystals (all others except native mercury). The mercury minerals are usually associated with kaolinite, calcite, gypsum, and iron oxides. Calomel appears to be the oldest of this group and generally montroydite is the youngest; the others generally formed in the order mosesite, kleinite, terlinguaite, and eglestonite. Cinnabar appears to have been younger than all these minerals except native mercury and montroydite, although some native mercury is found in droplets in calomel and may also have been early or even continuous in its deposition throughout the mineralization cycle. Ammonia is not thought to be a common constituent of hydrothermal solutions even though it is a well-known constituent of volcanic gases; it is possible that at Terlingua the proper combination of physical and chemical conditions existed to permit it to have been incorporated in insoluble mercury minerals not stable under the usual conditions of hydrothermal mineral formation.

Calcite was deposited throughout the mineralization cycle, though most of it was precipitated before the ore minerals; it was repeatedly sheared and brecciated, healed, rebroken, and rehealed. Quartz, though small in amount, appears to have been formed in several generations, and cinnabar is found in fractures in quartz.

The mercury ores of the Terlingua district probably were deposited under conditions of low intensity, certainly at no higher temperatures than 300°C and more probably between 200° and 100°C. Yates and Thompson estimate that the deposits were emplaced under a cover of no more than 2000 feet; from the highly broken character of the rocks of the district at the present time, it is doubtful if they were capable of maintaining a high confining pressure on the ore-forming fluids, so that it appears probable that these lost heat and pressure with considerable rapidity during the ore-emplacing process. The ores were more probably deposited under conditions of rapid loss of heat and pressure and should be classified as epithermal rather than telethermal.

Utah

BINGHAM

Early Tertiary | *Copper, Molybdenum, Zinc, Lead, Silver* | *Mesothermal*

Atwood, W. W., 1916, The physiographic conditions at Butte, Montana, and Bingham Canyon, Utah, when the copper ores in these districts were enriched: Econ. Geol., v. 11, p. 697-740 (particularly p. 732-740)

Beeson, J. J., 1916, The disseminated copper ores of Bingham Canyon, Utah: A.I.M.E. Tr., v. 54, p. 356-401

Boutwell, J. M., 1905, Genesis of the ore-deposits at Bingham, Utah: A.I.M.E. Tr., v. 36, p. 541-580

—— 1905, Economic geology of the Bingham mining district, Utah: U.S. Geol. Surv. Prof. Paper 38, 413 p. (particularly p. 33-70, 103-229; summarized in A.I.M.E. Tr., v. 36)

—— 1935, Copper deposits at Bingham, Utah, in *Copper resources of the world*: 16th Int. Geol. Cong., v. 1, p. 347-359

Bray, R. E., 1969, Igneous rocks and hydrothermal alteration at Bingham, Utah: Econ. Geol., v. 64, no. 1, p. 34-49

Butler, B. S., 1920, Oquirrh range, in Butler, B. S., and others, *The ore deposits of Utah*: U.S. Geol. Surv. Prof. Paper 111, p. 335-362

Cook, D. R., Editor, 1961, Geology of the Bingham mining district and northern Oquirrh Mountains: Utah Geol. Soc. Guidebook to the Geology of Utah, no. 16, 145 p. (particularly p. 1-100)

Creasey, S. C., 1959, Some phase relations in the hydrothermally altered rocks of porphyry copper deposits: Econ. Geol., v. 54, p. 351-373 (general)

Farmin, R., 1933, Influence of Basin-Range faulting in mines at Bingham, Utah: Econ. Geol., v. 28, p. 601-606

Field, C. W., 1966, Sulfur isotope abundance data, Bingham district, Utah: Econ. Geol., v. 61, p. 850-871

Hunt, R. N., 1924, The ores in the limestones at Bingham, Utah: A.I.M.E. Tr., v. 70, p. 857-883

—— 1933, Bingham mining district, in *The Salt Lake region*: 16th Int. Geol. Cong., Guidebook 17, p. 45-56

Hunt, R. N., and Peacock, H. C., 1950, Lead and lead-zinc ores of the Bingham district, Utah: 18th Int. Geol. Cong. Rept., pt. 7, p. 92-96

Lindgren, W., 1924, Contact metamorphism at Bingham, Utah: Geol. Soc. Amer. Bull., v. 35, p. 507-534

Moore, W. J., and others, 1968, Chronology of intrusion, volcanism, and ore deposition at Bingham, Utah: Econ. Geol., v. 63, p. 612-621; disc., 1968, v. 64, p. 228; authors' reply, p. 229

Parsons, A. B., 1933, Utah--the prospect and Utah--the mine, in *The porphyry coppers*: A.I.M.E., New York, p. 45-96

Peters, W. C., and others, 1966, Geology of the Bingham Canyon porphyry copper deposit, Utah, in Titley, S. R., and Hicks, C. L., Editors, *Geology of the porphyry copper deposits--southwestern North America*: Univ. Ariz. Press, Tucson, p. 165-175

Rose, A. W., 1961, The iron content of sphalerite from the Central district, New Mexico, and the Bingham district, Utah: Econ. Geol., v. 56, p. 1363-1384

—— 1967, Trace elements in the sulfide minerals from Central district, New Mexico, and Bingham district, Utah: Geochim. et Cosmochim. Acta, v. 31, p. 547-585

—— 1970, Origin of trace element distribution patterns in sulfides of the Central and Bingham districts, Western U.S.A.: Mineralium Deposita, v. 5, p. 157-163

Rubright, R. D., and Hart, O. J., 1968, Non-porphyry ores of the Bingham district, Utah, in Ridge, J. D., Editor, *Ore deposits of the United States, 1933-1967* (Graton-Sales Volumes): Chap. 45, v. 1

Schwartz, G. M., 1947, Hydrothermal alteration in the "porphyry copper" deposits: Econ. Geol., v. 42, p. 319-352 (particularly p. 325-328, 346-352)

Stacey, J. S., and others, 1967, Precision measurement of lead isotopes ratios; preliminary analyses from the U.S. mine, Bingham Canyon, Utah: Earth Planet. Sci. Lett., v. 2, no. 5, p. 489-499

Stringham, B., 1953, Granitization and hydrothermal alteration at Bingham, Utah: Geol. Soc. Amer. Bull., v. 64, p. 945-991

Stringham, B., and Taylor, A., 1950, Nontronite at Bingham, Utah: Amer. Mineral., v. 35, p. 1060-1066

Tooker, E. W., 1970, Regional structural control of ore deposits, Bingham mining district, Utah (U.S.A.): IMA-IAGOD Meetings 1970, Collected Abs., Tokyo-Kyoto, Paper 3-16, p. 55

Winchell, A. N., 1924, Petrographic studies of limestone alterations at Bingham: A.I.M.E. Tr., v. 70, p. 884-903

Notes

The center of the Bingham district, the Utah copper pit in the Bingham stock, is about 20 miles southwest of Salt Lake City, about half way down the apparent length of the Oquirrh range. The district lies on the eastward slope of the range, and most of the productive area lies within a circular area with a radius of about 1.5 miles, the center of the circle being essentially in the center of the Utah copper pit, that includes the U.S. and Lark zinc-lead mines.

The oldest known rocks in the immediate mining area are those of the Mississippian Manning Canyon formation composed mainly of black shales, the base of which is not known. In the Bingham district proper, no evidence is known for an unconformity between the Manning Canyon and the overlying Oquirrh group (using Kennecott terminology); elsewhere in the Great Basin, however, there is a definite disconformity between the uppermost Mississippian and the lowest Pennsylvanian. In the early days of the district, the interbedded sandstones and limestones of the Pennsylvanian and Permian rocks were referred to as the Oquirrh formation, but recently Kennecott geologists have suggested that they should be separated at the readily mappable Pennsylvanian-Permian boundary, that the term Oquirrh should be applied only to the Pennsylvanian portion of the section, and that the Oquirrh should be classed as a group and subdivided into formations; work by the Geological Survey in the northern Oquirrh range has provided a different approach to this problem.

There are no Mesozoic rocks in the Bingham area, which makes it difficult to date the igneous rocks that are of such great geologic importance in the area. Early Tertiary volcanic rocks are known along the eastern margins of the district; they probably are older than the intrusives, but if they ever were present in the mineralized area, they have been removed by erosion. The principal igneous rocks in the mine area are those of the Bingham and Last Chance stocks; the stocks are each about two-thirds of a square mile in area. The Bingham stock is the center of mineralization in the district, while the monzonite of the Last Chance body, which lies to the southwest, its east edge touching the west margin of the present pit, is essentially unmineralized and had practically no effect on the mineralization pattern of the area except for its denial of loci of deposition to the ore-forming fluids.

The Bingham stock is a composite body made up of granite and granite porphyry with associated dikes and sills of quartz-latite porphyry and latite porphyry. The granite porphyry was intruded into the granite, and the dikes and sills were emplaced even later, cutting the other two igneous rock types.

The ore mineralization in the Bingham district followed shortly after the introduction of the igneous rocks and has generally been assumed to have come from the same general magmatic source. It is thought that the folding, by analogy with the Tintic district, is late Mesozoic, while the igneous intrusions (which followed the early Tertiary volcanic rocks) took place far enough into the Eocene for them to be considered early Tertiary rather than Laramide. The Bingham deposits, therefore, are here classified as early Tertiary.

The ores of the Bingham district consist of a central mass of disseminated primary pyrite and copper sulfides that occupies a large portion of the Bingham stock and a small selvage in the sediments surrounding it. Higher-grade copper ores were found in highly altered limestones several hundred feet out from the stock, and these, in turn, were succeeded by lead and zinc replacement deposits at still greater distances from the central disseminated core. Still farther away from the central core, on the south side of the district, an outer zone of siliceous silver ores was developed that contained varied proportions of base-metal sulfides. The concentric zones of mineralization were not mutually exclusive, for fissures that carried lead-zinc ores were found in the high-grade copper zone and even, on occasion, in the core of disseminated sulfides itself. These gradual changes in mineralization were best developed on the west side of the district in the areas of the Highland Boy mine and the U.S. mine, but the same broad changes also took place on the south and east, although the inner zone here contained less high-grade copper and more pyrite and, locally, specular hematite. In some of the ore shoots, the same changes appeared in the vertical direction.

It would seem, from this brief outline, that this repetition of ore zoning in separate limestone horizons on essentially all sides of the Bingham stock makes it almost certain that the ore-forming fluids must have moved upward through the highly fractured igneous rocks of the Bingham stock and then were driven outward into the surrounding formations, following limestone in preference to quartzite and reaching rock volumes over a mile from the center of the stock while still in a condition to deposit economically valuable amounts of ore sulfides.

The outline of the zone of minable primary chalcopyrite mineralization in the northern portion of the Bingham stock has been likened to that of a Christmas tree, the broad base of which runs northwest-southeast and the top of which points generally southwest. The highest grade of mineralization was developed in the central portion of the mineralized area and gradually but irregularly became weaker toward the outer tips of the branches and toward the base. It appears that the copper mineralization now being mined consists of 56 percent chalcopyrite, 29 percent bornite, 12 percent chalcocite, and 3 percent covellite. The first two minerals were certainly primary; in part, the latter two possibly represent the roots of the secondary enrichment of the deposit but are considered as more likely to be primary as well. In the early days of mining, secondary enrichment almost certainly was prominent in the upper levels of the pit and was, in fact, essential for its profitable operation in those first years of mining. The copper ore of the enriched rock volume contained 1.5 to 2.0 percent copper as compared with the approximately 1.0 percent of the primary ore. The enriched ore was made up of chalcocite coatings (often sooty) on chalcopyrite and pyrite and of massive chalcocite; covellite was also important.

Another economically important mineral in the Bingham stock is molybdenite. It is about 1/50 as abundant as copper, and the zone of recoverable molybdenum, while centering in the same general point as that of copper, is smaller. Pyrite always accompanies the chalcopyrite, and in some contact zones in sediments at the stock boundary, pyrite, magnetite, and several iron-bearing silicate minerals are abundant. In the lead-zinc portion of the ore zones, pyrite is abundant but is erratically distributed, being confined largely to the ore bodies themselves and to pyritized envelopes that surround them. Pyrite is less common in the central part of the stock than it is farther out.

The lead-zinc deposits, unlike the disseminated material of the Bingham stock, are very irregularly developed as veins and bedded replacements. Although only 20 percent of the rocks in the formations in the Bingham district are limestones, 95 percent of the lead-zinc production has come from carbonate rocks. The factors localizing ores in the limestones appear to be favorable limestone beds, bedding-plane faulting, cross-faulting, distance from the Bingham stock, and the general structural pattern. The inner limit of the lead-zinc zone is seldom less than 1000 feet from the stock, and the outer almost never more than 10,000 feet. The lead-zinc mineralization is believed to have been limited largely to those faulted blocks that were moved to make room for intrusive masses and were, thereby, subjected to frequent breaking.

Sphalerite and galena are the principal ore sulfides in the lead-zinc deposits with which are associated minor quantities of copper-bearing sulfides; pyrite is present in abundance. Important amounts of silver and gold are recovered from the ores, the silver apparently being largely in the galena and gold in all the sulfides, but most abundantly in pyrite. The common gangue minerals, in addition to pyrite, are quartz, calcite, and unreplaced rock minerals of which last a considerable portion is silicates.

The economically valuable deposits of the Bingham area are primarily composed of chalcopyrite and bornite or sphalerite and galena, all of which are commonly found in the mesothermal range; bornite is diagnostic of that range. The chalcopyrite and bornite of the Bingham stock, then, almost certainly were formed under mesothermal conditions; the smaller amounts of probably primary chalcocite and covellite suggest that the last of the primary ore was formed under conditions less intense than mesothermal, perhaps even as low as telethermal. The silver-bearing galena and the sphalerite of the lead-zinc ores may have been formed under mesothermal or less intense conditions. The wall-rock alteration in the stock is typically that produced under mesothermal conditions, argillic minerals, sericite, biotite, and quartz being the most prominent minerals so formed. The silicate alteration minerals in the limestones almost certainly were developed under hypothermal conditions, but they were emplaced appreciably before the ore sulfides, and their presence does not justify the use of hypothermal in the classification of the ore deposits. The zone of marbleized limestone outside the silicated rock volume and the essentially unaltered limestone beyond the marble are compatible with the mesothermal conditions suggested by the lead-zinc mineralization. The deposits of the Bingham district, therefore, are classified as mesothermal, although it would not be wrong to add a minor leptothermal to telethermal category to include the primary chalcocite and covellite.

COTTONWOOD-AMERICAN FORK

Early Tertiary *Silver, Copper, Lead, Gold* *Hydrothermal-2 to Leptothermal*

Blackwelder, E., 1910, New light on the geology of the Wasatch Mountains: Geol. Soc. Amer. Bull., v. 21, p. 517-542

Boutwell, J. M., 1933, Cottonwood region [Utah], in *The Salt Lake region*: 16th Int. Geol. Cong. Guidebook 17, p. 82-90

Buranek, A. M., 1944, The molybdenum deposits of White Pine Canyon near Alta, Salt Lake County, Utah: Utah Geol. and Mineral. Surv. Circ. 28, 6 p.

Butler, B. S., and Loughlin, G. F., 1915, A reconnaissance of the Cottonwood-American Fork mining district, Utah: U.S. Geol. Surv. Bull. 620, p. 165-266

Calkins, F. C., and Butler, B. S., 1943, Geology and ore deposits of the Cottonwood-American Forks area, Utah: U.S. Geol. Surv. Prof. Paper 201, 152 p.

Calkins, F. C., and others, 1920, Cottonwood-American Fork area, in Butler, B. S., and others, *The ore deposits of Utah*: U.S. Geol. Surv. Prof. Paper 111, p. 229-283

Crittenden, M. D., and others, 1952, Geology of the Wasatch Mountains east of Salt Lake City, Parleys Canyon to Traverse Range: Utah Geol. Soc. Guidebook to the Geology of Utah, no. 8, p. 1-37

Eardley, A. J., 1939, Structure of the Wasatch-Great Basin region: Geol. Soc. Amer. Bull., v. 50, p. 1277-1310

Emmons, S. F., 1903, The Little Cottonwood granite body of the Wasatch Mountains: Amer. Jour. Sci., 4th Ser., v. 16, p. 139-147

Sharp, B. J., 1958, Mineralization in the intrusive rocks in Little Cottonwood Canyon, Utah: Geol. Soc. Amer. Bull., v. 69, p. 1415-1430

Notes

The Cottonwood-American Fork mineralized area is located within a rectangle bounded by longitude 111°34' on the east and 111°40' on the west and by latitude 40°31' on the south and 40°39' on the north. Only a small fraction of the 50 square miles enclosed within these bounding lines is mineralized; the ores are concentrated in three major portions of the area, the upper portions of the drainage basins of (1) Big Cottonwood Creek, (2) Little Cottonwood Creek, and (3) American Fork. The geographic center of the district is near the town of Alta which is situated about 20 miles southeast of Salt Lake City. Alta is about 9 miles west-southwest of Park City, on the opposite side of the local main divide from this district.

The oldest rocks in the district proper appear to be the Big Cottonwood series. These are Precambrian rocks and make up only a small fraction of the total thickness of the Big Cottonwood series as it is known elsewhere and outcrop over only a small area in the southwestern part of the district.

Overlying these Precambrian beds is the Mineral Fork tillite that also probably is Precambrian. The tillite and the older Precambrian rocks are unconformably overlain by the lower Cambrian Tintic quartzite and the Ophir shale. Conformably overlying the Ophir shale is the Middle Cambrian Maxfield limestone.

The oldest formation in the district that is younger than the Cambrian is the basal Mississippian dolomite. The basal Mississippian dolomite is conformably overlain by the Mississippian Madison limestone. The Madison formation is conformably overlain by the Deseret limestone. The Deseret formation is conformably overlain by the Humbug formation. The term "Doughnut" formation has recently been applied to the basal portions of what Calkins and Butler (1943) designated as the Morgan(?) formation.

The lower Pennsylvanian Morgan(?) formation conformably overlies the "Doughnut." Its correlation with the type Morgan is not completely certain, but the basal portion of the type Morgan is quite similar to the Morgan(?) in the Cottonwood district both in lithology and fossil content. The Morgan(?) grades upward into the Pennsylvanian Weber quartzite or quartzitic sandstone. The Weber is much more resistant than the adjacent limestones and forms steep slopes and abundant talus. The Park City formation overlies the Weber quartzite without apparent unconformity. The limestone portion of the Park City is considered to be Pennsylvanian while the shale in the middle portion (which contains thin beds of phosphate rock) and the upper part are thought to be the equivalent of the Permian Phosphoria formation.

The Triassic sequence lies conformably above the Park City and is composed of three formations: (1) the poorly exposed Woodside formation; (2) the Thaynes formation; and (3) the Ankareh formation. The youngest consolidated rock in the district is the probably Jurassic Nugget sandstone that is conformable on the Ankareh. The igneous rocks in the district are all early Tertiary intrusives and occur principally as large irregular bodies that reach to considerable depths and have granular textures. The oldest and most easterly of these is the Clayton Peak diorite stock. The Clayton Peak is joined on the northwest by the smaller mass of the Alta granodiorite stock. The third of the massive igneous bodies in the area is the Little Cottonwood quartz monzonite stock which reaches to within about 1.25 miles of the western extremity of the Alta stock. Since the Little Cottonwood stock does not come in contact with either of the other two, its age relative to them cannot be certainly determined; it has been assumed, however, that its more silicic character means that it is the youngest of the three stocks. Sharp (1958) also has discovered the presence of a mass of leucocratic quartz monzonite entirely within the Little Cottonwood stock that he designates as the White Pine stock. The contact between the White Pine and the Little Cottonwood ranges from sharp to gradational, the gradational zones being up to several hundred feet wide; the White Pine probably was intruded (Sharp, 1958) before the Little Cottonwood was completely solidified.

The sedimentary rocks surrounding the major stocks are cut by dikes of gray porphyry that have the same range in composition as do the stocks. Lamprophyre and alaskite dikes also are known; the lamprophyres cut the stocks, but the exact age relationships of the alaskites are not known--they are probably younger than the stocks. Many of the dike rocks, however, have been altered by the mineralizing solutions so they must be pre-ore in age.

The three main stocks, and their smaller offshoots, have caused widespread but different contact-metamorphic alterations of the sediments surrounding them. The most readily recognized and most widespread effect has been the recrystallization and bleaching of the carbonate rocks; next in extent has been the formation of tremolite in cherty limestones and of garnet, epidote, diopside, vesuvianite, and forsterite in the more shaly Cambrian carbonate beds. These alterations probably did not require the addition of appreciable amounts of material extraneous to the original beds. There are, however, certain massive silicate bodies in carbonate rocks that contain ludwigite [$(Mg,Fe)_2O_2FeBO_3$] as well as the silicates. These silicates have been extensively replaced by magnetite and are moderately veined by later sulfides (mainly copper-iron and copper) that probably were much aided in their formation by material introduced by hydrothermal solutions. The solutions, from which some of the silicates, the ludwigite, the oxides, and the sulfides probably came, must have derived from greater depths than the igneous rocks immediately across the contact from the deposits in question.

The ore bodies of the district appear to have been emplaced before most of the normal and steep reverse faults had been formed, the fissures along which the fluids traveled probably having been formed in the main during both thrust faulting and intrusion. The ore solutions most probably came from the same general source as the igneous magmas of the stocks and their satellite bodies so that the igneous rocks and the ores are of the same general age. From a tentative dating of the Alta stock of late Eocene (Crittenden and others, 1952) and from the early Tertiary age that has been given to the mineralizations at Park City, Bingham, and the East and main Tintic districts, it would seem reasonable to assign the ore bodies of the Cottonwood-American Fork district to the late Eocene and to categorize them as early Tertiary.

The greater part of the ore that has been recovered from the Cottonwood-American Fork district has been mined from "bedded" deposits in carbonate rocks. The location of these bedded deposits was partly determined by the chemical character of the host rock but was also due to brecciation of these beds where they were cut by thrust fualts. Thus, ore was more likely to be emplaced in a brecciated limestone than in one only slightly fractured and fissured. The amount of ore found within the igneous bodies themselves has been small; the ore bodies in the sediments, however, are located within short distances of the igneous-sedimentary contacts, either horizontally away from them or vertically above them. The bedded deposits usually have been formed by the replacement of limestone adjacent to cross-cutting fissures; they are normally of a rudely tabular form and lie generally parallel to the bedding, although some have small enough horizontal dimensions better to be called chimneys. These ore bodies pitch in the direction controlled by the intersection of the replaced beds and the fissures; this direction is usually northeast. Of the favorable beds, the lowest stratigraphically is the limestone member of the Ophir formation; the next of these favorable beds is made up of the dark, wormy, and mottled dolomites immediately above and below the white dolomite, a prominent horizon-marker in the Cambrian Maxfield limestone, and the third is at the top of the basal Mississippian dolomite.

Fissure-filling deposits in sedimentary rocks occur almost entirely in non-carbonate rocks; where well-mineralized fissures crossed carbonate rocks, the degree of replacement of the wall rocks usually was sufficiently high to produce bedded deposits or at least chimneys. The fissure deposits proper are characteristic of Precambrian and Cambrian quartzites and siliceous shales both above and below the Alta overthrust. Fissure veins in igneous rocks have not yet been found to be of minable grade.

In the bedded replacement ore bodies and in those associated with thrust faults, the earliest sulfide to form was abundant pyrite; it was normally accompanied by siderite in a wide range of abundances and probably by rhodochrosite. The pyrite was followed by sphalerite, enargite, galena, and tetrahedrite with the former two minerals probably being somewhat older than the latter two; the proportions of these minerals were widely different in different mines. Although chalcopyrite was present in most of the bedded ores, it was never present in large amount; most of the copper produced came from tetrahedrite. Bornite probably was somewhat more abundant than chalcopyrite, but it does not appear to have been an important ore mineral. Most of the silver probably came from tetrahedrite and galena, but argentite may well have contributed a considerable portion as it appears to have been a common, if not abundant, primary sulfide. Associated with the tetrahedrite are minor amounts of a number of sulfosalts such as jamesonite, bournonite, and aikenite. The first discovery of tungstenite (WS_2) was made in the Old Emma mine, and it was associated with the lead-silver ores of that deposit. Its late place in the paragenetic sequence probably is due to the high tenacity with which tungsten would be able to retain screening sulfur ions. Bismuthinite was locally of some abundance, but it was not common throughout the district. Chalcocite occurred in sparse amounts in many deposits where it appears to have replaced or exsolved from bornite; locally it appears to have replaced galena and to have been one of the last sulfides to have formed. A small amount of covellite may have been primary. Calcite and quartz are the main gangue minerals, with siderite and rhodochrosite being much less common; barite is scarce.

In the fissure veins in nonigneous rocks, the mineral content differs considerably from one to another, but the gangue is usually quartz, and the sulfides are generally pyrite, galena, sphalerite, and tetrahedrite. The mineralized joint and fracture structures in the Little Cottonwood stock and its included White Pine stock show three stages of mineralization which are: (1) in which scheelite, pyrite, and sericite were deposited on joint surfaces; (2) in which molybdenite, pyrite, quartz, sericite, galena, sphalerite, and small amounts of scheelite and calcite were deposited in the more intensely fractured zone; and (3) in which fluorite and sericite were locally deposited. The tungsten mineral is quite uniform in its distribution through the mineralized zone, but that of molybdenum is erratic.

The bulk of the lead-silver-copper-zinc deposits probably was formed under mesothermal to leptothermal conditions, with the small amounts of enargite indicating the mesothermal range and the common tetrahedrite and less common argentite indicating the leptothermal range. The scheelite and molybdenite showings proably were formed under hypothermal conditions as was the magnetite. The copper-iron and copper sulfides associated with the magnetite appear to have been deposited in the mesothermal to leptothermal (or even into the telethermal) ranges. That some of the sphalerite contains exsolved chalcopyrite indicates that that portion of it was formed under hypothermal conditions, but the low-copper, low-iron sphalerite that probably is associated with most of the enargite-tetrahedrite-galena ores must have been formed under mesothermal to leptothermal conditions. The deposits of the district, therefore, are here classified as hypothermal-2 (since the only locally minable hypothermal deposits--the magnetite ores--are in carbonate rocks) to leptothermal.

EAST TINTIC

Early Tertiary	*Silver, Gold, Lead, Copper, Zinc*	*Mesothermal to Leptothermal*

Bush, J. B., and Cook, D. R., 1960, The Chief Oxide-Burgin area discoveries, East Tintic district, Utah; a case history; pt. II, Bear Creek Mining Company studies and exploration: Econ. Geol., v. 55, p. 1507-1540

Cook, D. R., Editor, 1957, Geology of the East Tintic Mountains and ore deposits of the Tintic mining districts: Utah Geol. Soc. Guidebook to the Geology of Utah, no. 12, p. 1-56, 97-102, 103-119, 120-123, 124-134, 135-139, 140-154, 155-167

Crane, G. W., 1926, Notes on the geology of East Tintic: A.I.M.E. Tr., v. 74, p. 147-162

Frondel, C., and Hones, R. M., 1968, Billingsleyite, a new silver sulfosalt: Amer. Mineral., v. 53, p. 1791-1798

Laughlin, A. W., and others, 1969, Age of some Tertiary igneous rocks from the East Tintic district, Utah: Econ. Geol., v. 64, p. 915-918

Lovering, T. S., 1960, Geologic and alteration maps of the East Tintic district, Utah: U.S. Geol. Surv. Mineral Invest., Field Studies Map MF-230 (2 sheets)

Lovering, T. S., and Morris, H. T., 1960, The Chief Oxide-Burgin area discoveries, East Tintic district, Utah; a case history; pt. I, U.S. geological survey studies and exploration: Econ. Geol., v. 55, p. 1116-1147

——— 1965, Underground temperatures and heat flow in the East Tintic district, Utah: U.S. Geol. Surv. Prof. Paper 504-F, p. F1-F28

Lovering, T. S., and Shepard, A. O., 1960, Hydrothermal alteration zones caused by halogen acid solutions, East Tintic district, Utah: Amer. Jour. Sci., v. 258-a (Bradley Volume), p. 215-229

Lovering, T. S., and others, 1948, Heavy metals in altered rocks over blind ore bodies, East Tintic district, Utah: Econ. Geol., v. 43, p. 384-399

——— 1949, Rock alteration as a guide to ore--East Tintic district, Utah: Econ. Geol. Mono. 1, 64 p.

Morris, H. T., 1964, Discovery of the Burgin mine, East Tintic mining district, Utah, U.S.A.: Symposium on mining geology and the base metals, Central Treaty Organization (CENTO), Ankara, Sept. 1964, p. 271-295

——— 1964, Geology of the Eureka quadrangle, Utah and Juab Counties, Utah: U.S. Geol. Surv. Bull. 1142-K, p. K1-K29

Morris, H. T., and Anderson, J. A., 1962, Eocene topography of the central East Tintic Mountains, Utah: U.S. Geol. Surv. Prof. Paper 450-C, p. C1-C4

Morris, H. T., and Lovering, T. S., 1961, Stratigraphy of the East Tintic Mountains, Utah: U.S. Geol. Surv. Prof. Paper 361, 145 p.

Morris, H. T., and Shepard, W. M., 1964, Evidence for a concealed tear fault with large displacememt in the central East Tintic Mountains: U.S. Geol. Surv. Prof. Paper 501-C, p. C19-C21

Parsons, A. B., 1925, The Tintic Standard mine: Eng. and Min. Jour. Press, v. 120, no. 17, p. 645-652

Radtke, A. S., and others, 1969, Micromineralogy of galena ores, Burgin mine, East Tintic district, Utah: U.S. Geol. Surv. Prof. Paper 614-A, p. A1-A17

Shepard, W. M., and others, 1968, Geology and ore deposits of the East Tintic mining district, Utah, in Ridge, J. D., Editor, *Ore deposits of the United States, 1933-1967* (Graton-Sales Volumes): Chap. 47, v. 1

Notes

The East Tintic district, which lies immediately west of the (Main) Tintic district and is divided from it along the 112°5' parallel of longitude, is situated entirely on the eastern slopes of the East Tintic Mountains. The North Tintic district lies to the north, being separated from the two other Tintic districts by an essentially east-west line through Packard Peak and Homansville. The center of the district is some 50 miles slightly west of south from Salt Lake City.

The sedimentary rocks of the East Tintic Mountains range in age from latest Precambrian to the Permian Park City formation. West of the more or less north-south striking East Tintic (Chief Oxide) thrust fault, only that portion of the stratigraphic column between and including the Cambrian Tintic formation and Ordovician Bluebird formation is present. East of the Chief Oxide fault, however, the footwall of that west-dipping thrust fault contains all of the Paleozoic section from the Tintic formation (quartzite) through the Mississippian Humbug formation.

Most of the East Tintic district is covered by the series of extrusive middle Eocene volcanic rocks. These volcanic rocks here include the Packard quartz latite (porphyritic latite, latite, vitrophyre, rhyolite, tuffs, and flow breccias) and the Laguna latite (the later latite of the Tintic district) that includes porphyritic latite (black and red), latite breccia, agglomerates, and tuffs. The pre-lava surface was one of moderate relief, similar to that exhibited by the present topography.

The sedimentary and volcanic rocks of the district are cut by small stocks, plugs, and dikes of monzonite and quartz monzonite; those rocks most probably are from the same general magmatic source as the Silver City stock of the Tintic district and are aligned in a zone 0.5 to 1.5 miles wide that extends some 3.5 miles north-northeast from the Silver City stock to the neighborhood of the Homansville fault on the north edge of the East Tintic district. This zone passes through the Zema, Iron King No. 1, and North Lily mines, and the same type of pebble dikes known in the Tintic district is common within or near it. On occasion, quartz monzonite dikes are bordered by irregular selvages of pebble-dike material, but most of the pebble dikes are not spatially close to monzonite dikes nor do they contain monzonite porphyry as fragments nor do monzonite dikes approach near to them. The pebble dikes appear to have been developed by explosive forces that not only broke the rock volumes now making up the pebble dikes but were so strong that they rounded a large proportion of the fragments and ground the rest to rock flour. These forces probably were generated by rapidly expanding gases that were produced either from magmas on their near approach to the then-existing surface or from ground water heated by magmatic material.

Dikes of highly altered biotite-augite andesite (purple) porphyry cut quartz monzonite dikes and may have been emplaced during the late stages of the mineralization or they may have been less susceptible to mineralization than the limestone; the reported presence of higher grade ores in the limestone on one side of a purple dike than on the other indicates that the latter explanation may be the true one. If this be so, then the purple dikes are post-monzonite and pre-ore; the problem needs further study.

Dikes and sills of diabase cut the sedimentary rocks of the northern East Tintic Mountains and rocks of the volcanic latite series in the southern part of the range. Although the diabases do not cut any of the postvolcanic intrusive rocks, they are believed to be the youngest igneous rocks in the area; they have not been reported from the ore-bearing portion of the East Tintic district.

Although the sedimentary rocks of the East Tintic district are dominantly those of Cambrian and Ordovician age, recent drilling and underground exploration east of the East Tintic (Chief Oxide) fault have disclosed that the highly contorted and sheared beds of its footwall plate contain the major portion of the Paleozoic sequence. In the upper plate of the fault, the brecciated and deformed beds of the lower member of the Ophir formation have been overturned and underlie the also overturned upper beds of the Tintic quartzite.

The localization of the East Tintic ores west of the East Tintic (Chief Oxide) fault in the same northeasterly trending belt that contains the stocks, plugs, and dikes of late middle Eocene monzonite and quartz monzonite strongly suggests that the ores were emplaced in the same general time span. The relationships of the ores east of the East Tintic fault are less well known, but it appears that they are so similar to those of the west side of the fault that they probably were formed at essentially the same time. The north-northeast-trending faults and fissures that exerted primary control over the locations of the ore bodies both in igneous rocks and in sediments appear to have been formed shortly after the igneous masses had been emplaced. It appears reasonable, therefore, to assume that the East Tintic ores were formed in the late Eocene (or perhaps earliest Oligocene); therefore, they are here classified as early Tertiary.

The grabens, or structural troughs, into which the East Tintic district is divided are, from north to south, (1) the Homansville trough that is formed by the northeast-trending, northwesterly dipping Homansville fault and the east-trending, steeply dipping Canyon fault; within this trough no replacement deposits have been found although mineralized fissures have been worked in a few mines; (2) the North Lily and Tintic Standard troughs that are separated from the Homansville trough by a structurally high area of barren quartzite extending as far southeast as the East Tintic (north-south) fault; (3) Eureka Standard trough that is the next trough to the southeast and is bounded by the Iron King fault on the northwest and the Eureka Standard fault on the southeast--these two are normal faults that dip toward each other at moderate to high angles; only fissure ore bodies have been found in this trough; and (4) the Apex Standard trough that is bordered by the Apex Standard fault on the northwest and the Teutonic fault on the southwest.

Although the major northeast faults, with some assists from the possibly folded East fault and from the Eureka Lily fault, produced the mineralized troughs, the actual ore deposits are spatially closely allied to the numerous minor northeast faults. Nevertheless, the major faults may be mineralized; the largest and best mineralized replacement bodies, however, were developed where two or more fault structures converged, as was the situation in the Tintic Standard mine.

The primary East Tintic ores show evidence of having been formed at somewhat lower temperatures than the Tintic ores themselves. The primary replacement ores of East Tintic differ somewhat from one mine to another; in the North Lily mine, they were composed of massive silver-rich galena with subordinate pyrite and sphalerite, while in the Tintic Standard, they consisted of galena, tetrahedrite, silver sulfosalts, and jasperoid. On the other hand, the typical minerals of the fissure zones were pyrite and enargite (including luzonite) with fine-grained quartz and minor barite and local occurrences of tetrahedrite with the enargite. The lead-zinc ore came almost entirely from replacement bodies, although small shoots of galena-sphalerite-pyrite ore were mined from fissures in the quartzite. Gold was present in the copper-bearing fissure ore in both pyrite and enargite; it usually was invisible but on occasion could be seen as fine particles in the enargite. A little gold was found with light-colored sphalerite in outlying fissure veins but was of essentially no economic importance. In the North Lily mine, the silver appears to have been almost exclusively contained in the galena, while in the Tintic Standard, which had a far higher silver to lead ratio, it was found in tetrahedrite and silver sulfosalts, such as polybasite and pearceite, as well as in galena. The usually somewhat earlier enargite-gold ores probably were formed under mesothermal conditions, enargite being diagnostic of that range, and the generally somewhat later galena-silver sulfosalt-sphalerite ores were developed largely, if not entirely, under leptothermal conditions, the silver sulfosalts and tetrahedrite being diagnostic of that range. The complex sequence of alteration stages has produced minerals that are quite compatible with mesothermal and leptothermal conditions of ore formation, and the East Tintic deposits are here so categorized.

IRON SPRINGS

Middle Tertiary | *Iron as Hematite, minor Magnetite* | *Hypothermal-2*

Blank, H. R., Jr., and Mackin, J. H., 1967, Geologic interpretation of an aeromagnetic survey of the Iron Springs district, Utah: U.S. Geol. Surv. Prof. Paper 516-B, p. B1-B14

Butler, B. S., 1920, Iron Springs district, in Butler, B. S., and others, *The ore deposits of Utah*: U.S. Geol. Surv. Prof. Paper 111, p. 568-582

Cook, E. F., 1960, Washington County: Utah Geol. and Mineral. Surv. Geologic Atlas of Utah, Bull. 70, 119 p. (Bull Valley is in Washington County)

Cook, K. L., and Hardman, E., 1967, Regional gravity survey of the Hurricane fault area and Iron Springs district, Utah: Geol. Soc. Amer. Bull., v. 78, p. 1063-1076

Granger, A. E., 1963, The iron province of southwestern Utah: Intermountain Assoc. Petrol. Geols. 12th Ann. Field Conf. Guidebook, p. 146-150

—— 1964, The iron province of southwestern Utah: Intermountain Assoc. Petrol. Geols. Guidebook, 12th Ann. Field Conf., p. 146-150

Kemp, J. F., 1909, The iron ores of the Iron Springs district in southern Utah (rev.): Econ. Geol., v. 4, p. 782-791

Knopf, A., 1942, Iron Springs district, Utah, in Newhouse, W. H., Editor, *Ore deposits as related to structural features*: Princeton Univ. Press, p. 64

Leith, C. K., 1910, Iron ores of Iron Springs, Utah (reply): Econ. Geol., v. 5, p. 188-192

Leith, C. K., and Harder, E. C., 1908, The iron ores of the Iron Springs district, southern Utah: U.S. Geol. Surv. Bull. 338, 102 p.

Mackin, J. H., 1947, Some structural features of the intrusions of the Iron Springs district: Utah Geol. Soc. Guidebook to the Geology of Utah, no. 2, 62 p.

—— 1954, Geology and iron ore deposits of the Granite Mountain area, Iron County, Utah: U.S. Geol. Surv. Mineral Invest., Field Studies Map MF-14, 1:12,000

—— 1960, Structural significance of Tertiary volcanic rocks in southwestern Utah: Amer. Jour. Sci., v. 258, p. 81-131

—— 1968, Iron ore deposits of the Iron Springs district, southwestern Utah, in Ridge, J. D., Editor, *Ore deposits of the United States, 1933-1967* (Graton-Sales Volumes): Chap. 49, v. 2

MacVichie, D., 1927, Iron fields of the Iron Springs and Pinto mining districts, Iron County, Utah: A.I.M.E. Tr., v. 74, p. 163-173

Wells, F. G., 1938, The origin of the iron ore deposits in the Bull Valley and Iron Springs districts, Utah: Econ. Geol., v. 33, p. 477-507

Young, W. E., 1947, Iron deposits, Iron County, Utah: U.S. Bur. Mines R.I. 4076, 102 p. (mimeo.) (details of exploratory work on a large number of ore bodies in and near Iron Springs area)

Notes

The iron deposits of the Iron Springs district are located in a northeast-trending area of 20 by 15 miles in southwestern Utah near the eastern margin of the Basin-Range province and only 10 to 25 miles west of the western boundary of the Colorado Plateau; the center of the mineralized district is some 10 miles west of the town of Cedar City.

The oldest rocks in the area are the limestones of the Jurassic Homestake formation, and this formation is probably equivalent to the Jurassic Carmel formation of the nearby Colorado Plateau. The beds immediately overlying the Homestake formation make up the Entrada formation and are classed as upper Jurassic because they are conformable on the well-dated Homestake. Unconformably above the Entrada are Mackin's seven members of the lower part of the Iron Springs formation. These members are made up of a variety of rocks and the contacts between them are gradational. In some parts of the district, limy crossed-bedded sandstone, interlensed with shale and some conglomerate, makes up the conformable upper portion of the Iron Springs; the entire formation is probably Upper Cretaceous in age. The Claron formation lies above the Iron Springs and is separated from it by an angular unconformity, normally less than 30°. The Claron is considered to be the equivalent of the "Wasatch" in the adjacent Plateau and is, therefore, probably Late Cretaceous to early Eocene in age.

The oldest igneous rocks in the area are latite lavas and pyroclastics that lie unconformably on the older sediments. These rocks also appear, according to Butler (1920), to have been uparched by the intrusive rocks of the area and to be, therefore, the older phase of igneous activity. From the long period of erosion that must have preceded the extrusion of the latites and from their correlation with Miocene lavas in the Wasatch Mountains, the extrusives probably were emplaced in the early Miocene; they may, however, be as old as Eocene. The next, and closely following, phase of magmatic action in the area was the intrusion, under perhaps 2000 to 8000 feet of cover, of quartz monzonite porphyry in at least four igneous bodies. Originally these were called laccoliths but are now believed to be stocks. The intrusive material well may have derived from the same magma chamber as the extrusives. Within the ore area, three essentially circular stocks form the cores of the principal mountains of the district--the Three Peaks and Granite Mountain in the northeast part and Iron Mountain in the southwest; another and considerably larger igneous body forms the Pine Valley Mountains south of the Iron Springs district. A wide quartz monzonite dike lies northwest of the Iron Mountain stock and several minor satellite bodies of monzonite are located to the southwest of that igneous mass. Some erosional or down-faulted remnants of sedimentary rocks occur in the quartz monzonite within the stocks.

The sediments have been arched up over the stocks as they normally dip away from the monzonite; in a few places, the beds dip steeply toward the Iron Mountain body, which suggests that the beds in these places were overturned by the intrusion. The quartz monzonite appears to have been intruded at or near the base of the Homestake formation, along a zone of gypsiferous and shaly limestone. The strike of the beds essentially parallels the margins of the stocks except where faulting has made the structure irregular.

The ore deposits are, for the most part, irregularly distributed replacement masses along the margins of Granite Mountain and the Iron Mountain stocks of the Iron Springs district; there is much less ore associated with the Three Peaks body. The ore bodies were emplaced at or near the contact of the quartz monzonite bodies with the overlying Homestake as replacements of the limestones of that sedimentary formation and, to some extent, as fissure fillings in the quartz monzonite itself. Replacement ores are never far from the contact between igneous and sedimentary rocks, strongly suggesting that the ore-forming fluid responsible for them followed that contact as a channelway only after the quartz monzonite had solidified. It appears probable that the ore fluids were derived at depth from the same general volume of magma that supplied the quartz monzonite and that they entered the rock volumes to be mineralized within a short time after the solidification of the intrusives. This does not agree with Mackin's concept that the ores were derived from reactions affecting the quartz monzonite masses in the immediate vicinity of the ore bodies. The ores are in either case, however, to be classified as middle Tertiary.

Perhaps the greatest problem in the classification of the Iron Springs deposits is the apparent failure of the various authors who have worked in the district to recognize that it contains two quite different types of ore: (1) the replacement ore (mainly hematite) that has formed in the limestones of the Homestake formation, both around the present margins of the stocks and in roof-pendants now exposed on the present igneous rock surfaces, and (2) veins (mainly magnetite and apatite) that have filled fissures in the zone of selvage joints near the outer margins of the quartz monzonite.

The replacement bodies are largely confined to portions of the rocks around the Iron Mountain stock and the Granite Mountain stock (which, with its faulted margins, is probably better classified as a bysmalith) and its southwestwardly extension--Desert Mound. The ore is composed of hematite with generally minor proportions of magnetite; magnetite ranges from less than 5 percent to, locally, more than 50 percent. The minor primary minerals include mica, apatite, quartz, and chalcedony; such high-temperature minerals as garnet and lime silicates, typical of most hypothermal deposits in calcareous rocks, are rare in the replacement ores. Although all sedimentary rock types, as well as quartz monzonite, may be replaced by ore, the important ore bodies are confined to the Homestake formation. The usual ore body extends from several hundred to over a thousand feet along strike and can be followed down-dip for a similar distance; the ore bodies may reach 250 feet in width, the normal thickness of the Homestake formation. The footwall of the ore body generally conforms to the base of the limestone and is separated from the monzonite by the 15- to 25-foot basal siltstone of that formation. The original structures of the limestone, such as bedding, continue uninterruptedly from sedimentary rock to ore, indicating that the replacement was accomplished on a volume-for-volume basis.

The Homestake formation was only slightly altered prior to the emplacement of the ore bodies; locally silicification and silication of the limestone have produced a rock difficult to distinguish from the basal siltstone. This altered rock is as resistant to replacement by ore as is the siltstone.

The magnetite veins in the selvage joints (the term *selvage* refers to narrow bands of alteration bordering the joints) are composed of massive magnetite (and little or no hematite) with crystals of apatite that penetrate the ore either inward from the walls of the veins or outward from the center-lines of the veins; these veins may have thin open spaces centrally located in them, and these spaces may be crossed by apatite crystals growing out from the exposed surfaces of the magnetite. Minor amounts of pyroxene and calcite are also found in these veins. These veins have all of the characteristics, on a small scale, of iron-apatite deposits of magmatic origin such as the huge deposits of Kiruna, Iron Mountain, Missouri, or El Tofo in Chile. The material from which these vein fillings were formed, therefore, appears to have been an iron-phosphorus-rich melt, low in water, that was produced in the late stages of the crystallization of the inner portions of the quartz monzonite and extruded outward into the selvage joints. The much larger amount of accessory (as opposed to vein) magnetite in the chilled peripheral shell of the monzonite, in comparison with the content of that accessory mineral in the interior portions of the stocks, supports the concept that the magnetite in the veins was introduced as an immiscible iron-phosphorus-rich phase developed during the slow crystallization of the quartz monzonite magma and kept molten down to temperatures of 500° to 600°C, mainly by its dissolved content of apatite. In the rapidly chilled and solidified peripheral phases, the excess magnetite did not have time to segregate into large molten masses and be driven from the area; instead, in contrast to the magma of the stock interiors, it crystallized in place, making the peripheral phases higher in accessory magnetite than the interiors.

Of the total amount of ore in the Iron Springs area, only a minor fraction is present in the magnetite veins, so small a proportion, in fact, that it is unreasonable to include the magnetite-apatite ores in the classification, even though, from certain of these veins such as the Great Western, considerable tonnages of ore have been mined. Although the cover over the laccoliths at the time of ore deposition was probably no more than 2000 to 8000 feet, the cover appears to have been so structurally sound that the hydrothermal ores were formed with slow loss of heat and pressure. The replacement ores in limestone of the Homestake formation ores, therefore, are here categorized as hypothermal-2.

MARYSVALE

Middle Tertiary (Alunite), Late Tertiary (U) — *Uranium, Alunite* — *Mesothermal*

Bassett, W. A., and others, 1960, K:Ar ages, Marysvale, Utah--Tertiary volcanic rocks (abs.): Geol. Soc. Amer. Bull., v. 71, p. 1822-1823

—— 1963, Potassium-argon dating of the late Tertiary volcanic rocks and mineralization of Marysvale, Utah: Geol. Soc. Amer. Bull., v. 74, p. 213-220

Butler, B. S., and Gale, H. S., 1912, Alunite--a newly discovered deposit near Marysvale, Utah: U.S. Geol. Surv. Bull. 511, 64 p.

Butler, B. S., and others, 1920, Alunite veins, in *The ore deposits of Utah*: U.S. Geol. Surv. Prof. Paper 111, p. 546-554

Callaghan, E., 1938, Preliminary report on the alunite deposits of the Marysvale region, Utah: U.S. Geol. Surv. Bull. 886, p. 91-134

—— 1939, Volcanic sequence in the Marysvale region in southwest-central Utah: Amer. Geophys. Union Tr., 20th Ann. Meeting, pt. 3, p. 438-452

Callaghan, E., and Parker, R. L., 1962, Geology of the Delano Peak quadrangle, Utah: U.S. Geol. Surv. Geol. Quad. 153, 1:5208

—— 1962, Geology of the Sevier quadrangle, Utah: U.S. Geol. Surv. Geol. Quad., 1:5208

Eardley, A. J., and Beutner, E. L., 1934, Geomorphology of Marysvale Canyon and vicinity, Utah: Utah Acad. Sci. Pr., v. 11, p. 149-159

Gregory, H. E., 1944, Geologic observations in the upper Sevier River valley, Utah: Amer. Jour. Sci., v. 242, no. 11, p. 577-606

Gruner, J. W., and others, 1951, The uranium deposit near Marysvale, Piute County, Utah: Econ. Geol., v. 46, p. 243-251

Heinrich, E. W., 1958, Marysvale, Utah, in *Mineralogy and geology of radioactive raw materials*: McGraw-Hill, N.Y., p. 350-351

Kerr, P. F., 1957, Marysvale, Utah, uranium area: Geol. Soc. Amer. Spec. Paper 64, 212 p.

—— 1963, Geologic features of the Marysvale uranium area, Utah: Intermountain Assoc. Petrol. Geols. 12th Ann. Field Conf. Guidebook, p. 125-135

—— 1964, Geological features of the Marysvale uranium area, Utah: Intermountain Assoc. Petrol. Geols. Guidebook, 12th Ann. Field Conf., p. 125-135

—— 1968, The Marysvale, Utah, uranium deposits, in Ridge, J. D., Editor, *Ore deposits of the United States, 1933-1967* (Graton-Sales Volumes): Chap. 50, v. 2

Loughlin, G. F., 1915, Recent alunite developments near Marysvale and Beaver, Utah: U.S. Geol. Surv. Bull. 620, p. 237-270

Molloy, M. W., and Kerr, P. F., 1962, Tushar uranium area, Marysvale, Utah: Geol. Soc. Amer. Bull., v. 73, p. 211-236

Parker, R. L., 1962, Isomorphous substitution in natural and synthetic alunite: Amer. Mineral., v. 47, p. 127-136

Walker, G. W., and Osterwald, F. W., 1956, Relation of secondary uranium minerals to pitchblende-bearing veins at Marysvale, Piute County, Utah, in *Contributions to the geology of uranium and thorium*: U.S. Geol. Surv. Prof. Paper 300, p. 123-129

Willard, M. E., and Callaghan, E., 1962, Geology of the Marysvale quadrangle, Utah: U.S. Geol. Surv. Geol. Quad. 154, 1:5208

Willard, M. E., and Proctor, P. D., 1946, White Horse alunite deposit,

Marysvale, Utah: Econ. Geol., v. 41, p. 619-643

Notes

The Marysvale region, just north of the town of that name, lies about 70 miles northeast of Cedar City near the center of some 4000 square miles of flows and tuffs of Tertiary age.

In the Marysvale area proper, the oldest rocks are metamorphosed blocks of quartzite and limestone that appear to have been rafted into their present position by an intrusion of quartz monzonite magma. From exposures outside the principal mineralized portion of the district, it appears that the area must be underlain by rocks that range in age from at least as old as Carboniferous to Upper Jurassic, with the Carboniferous being represented by carbonate rocks and the Permian by quartzites, limestones, and shales of the Supai formation. The Supai is overlain by the Permian Coconino formation; the Coconino is overlain by the Triassic Kaibab formation. A little uranium mineralization occurs as replacements in carbonate lenses in the lower portion of the Kaibab formation in the area west of the Tushar fault; this fault strikes about north-south, some 4 miles west of Marysvale. The Moenkopi formation unconformably overlies the Kaibab and on the eroded surface of the Moenkopi are local developments of the Shinarump conglomerate that are overlain in much the same area by the Chinle formation, the last of the Triassic beds. The Chinle is overlain by the Lower Jurassic Navajo formation; the Kayenta and Wingate formations appear to be missing in this district. The last of the pre-Tertiary formations, the Middle and Upper Jurassic Carmel, overlies the Navajo; the Carmel is an erosional remnant.

The Paleozoic and Mesozoic rocks are unconformably overlain by the Bullion Canyon series of volcanic rocks that Callaghan (1939) considered to be a single unit but which Kerr (1963) has divided into 10 members.

The Bullion Canyon flows were followed by intrusions of quartz monzonite, quartz monzonite porphyry, monzonite, monzonite porphyry, fine-grained granite, aplite, and latite. These masses generally form essentially circular plugs with more or less vertical walls. The largest intrusive is the Central intrusive, less than 2 miles in diameter and centering about 4 miles north-northeast of Marysvale; it is composed mainly of quartz monzonite with lesser amounts of quartz-monzonite porphyry and fine-grained granite and contains a few moderate-sized xenoliths of metamorphosed sediments. The flow rocks surrounding the Central intrusive have been somewhat altered during, or shortly after, the introduction of the quartz monzonite. The Plug is an intrusive mass of monzonite, about 1500 feet (north-south) by 1000 feet (east-west), on the east wall of Marysvale Canyon about 7 miles north-northwest of Marysvale. Some alteration of the rocks surrounding the highly resistent Plug took place, but no mineralization is associated with it. The third important igneous intrusion is the quartz monzonite mass of Monzonite Hill, east of the Sevier River and opposite the Big Rock Candy Mountain; its dimensions are about 9000 feet from east to west. The other intrusive masses are appreciably smaller than those briefly described and are widely scattered through the district.

The first event after the introduction of the post-Bullion Canyon intrusives was the development of small amounts of postintrusive conglomerate in the Silica Hills that are limited to the north side of the Central quartz monzonite intrusive. At much the same time, the Dry Hollow latite was poured out as flows, best developed in the Silica Hills near Poverty Flats. The Dry Hollow series was followed by the Mount Belknap volcanic series that lies unconformably on the eroded surface of the Bullion Canyon flows; a considerable time interval probably separated the intrusives and the Mount Belknap rocks. The flow sequence is appreciably different, depending on the side of the Sevier River on which it is studied.

Isolated diatremes northeast of the Central intrusive are probably of Mount Belknap age; the Mount Belknap series is unconformably overlain by the Joe Lott tuff. The Joe Lott tuff was emplaced after the uranium mineralization.

Younger than all the volcanic rocks is the Sevier River formation that is late Pliocene or Pleistocene; it appears to have been produced by the runoff due to torrential rains. The district also contains three types of Quaternary alluvial deposits.

North of the Marysvale area, the oldest of the volcanic rocks, the Bullion Canyon series, lie upon beds of the Wasatch formation that contain Eocene and Oligocene fossils; thus, the entire sequence of extrusive and intrusive rocks is Oligocene at the oldest, and most of these rocks probably are younger. Radioactive age determinations on pyroxene andesite from the upper part of the Bullion Canyon series give dates of between 28.0 and 31.1 m.y.; this dates the older volcanic rocks as late Oligocene. Age determinations on biotites from the Central intrusive quartz monzonite range from 21.3 to 27.3 m.y., with an average of about 25 m.y., placing these intrusives at about the boundary between Oligocene and Miocene time. The alunite deposits are limited to the Bullion Canyon series, and the solutions that formed them probably were derived from the same source as the Oligocene-Miocene intrusives. Thus, the alunite deposits are probably early Miocene in age and are here categorized as middle Tertiary.

Although the uranium mineralization is found mainly in veins where the wall rocks are early Miocene quartz monzonite or fine-grained granite, alteration similar to that associated with the uranium ores is found in the Mount Belknap rocks. The radioactive ages on these younger volcanic rocks range between 12 m.y. and 22 m.y., with an average of about 18 m.y., which places the age of the volcanic rocks as middle Miocene. The dates on pitchblende and on sericite associated with pitchblende give ages of about 13.5 m.y. or latest Miocene. To emphasize the age difference between the alunite and uranium mineralizations, the uranium mineralization is here dated as late Tertiary although these ores probably were formed almost exactly on the line of demarcation between middle and late Tertiary.

In the Marysvale district, there appear to have been three main centers of hydrothermal alteration: (1) the Central intrusive where the bulk of the alteration lies to the north and east of the quartz monzonite; (2) the Monzonite Hill quartz monzonite mass where alteration can be found almost completely around the intrusive but with its greatest development being west of the Sevier River in the Big Rock Candy Mountain area; and (3) the two small areas of quartz monzonite north and northwest, respectively, of the Bullion Hills; these appear to be the surface expressions of a larger mass beneath that which has produced an area of alteration that lies between the Bullion Hills (southwest) and the White Hills (northeast).

Although the alunite deposits of the Marysvale area are directly related to the alteration pattern in the Bullion Canyon rocks, the production of alunite was a special phase of the alteration process. The normal results of this early (pre-Belknap) hydrothermal alteration are divided into four stages: (1) slight to moderate argillic alteration in which the plagioclase and biotite phenocrysts of the flow rocks have been somewhat dulled and the biotite may show some chloritization; the ground mass may have been bleached; (2) moderate to intense argillic alteration in which the feldspars and femic minerals have been altered to clay (mainly kaolinite), with the original texture of the rock remaining; (3) intense argillic alteration in which the primary texture has been destroyed and all the rock-forming minerals have been altered to clay, principally illite, although locally that mineral may have been converted to kaolinite and dickite; and (4) extensive silicification in which the rock texture has been obliterated. Considerable volumes of tuff that have been bleached are known, and minor volumes of zeolitization occur in the southeastern part of the district. In certain rock volumes, what is apparently the argillic alteration of stage (3) also includes alunite $[KAl_3(SO_4)_2(OH)_6]$ and some quartz. When the alunite development is even more intense, it becomes the most abundant constituent, but quartz and the clay minerals (mainly kaolinite and dickite) are still present; this alteration-type probably corresponds to the normal stage (4) with alunite added and the silicification less intense than usual. The change from dominantly kaolinized to mainly alunitized rock commonly occurs over a few inches, and the alunitized zone is far more siliceous than the kaolinite-rich one. Hematite and leucoxene are the principal accessory minerals in the alunitized rock, and hematite is much more abundant in the alunitized than in the kaolinized rock. The alunitized zones contain a maximum of 55 to 60 percent of that sulfate. In rock volumes where alunite was not developed, the most intense alteration is normally composed of montmorillonite, illite, and quartz instead of the kaolinite-alunite-quartz mineralogy of the alunite zones. The alunite-type of alteration generally occurs away from the intrusive masses in areas of flow breccias or zones of obscure structural control. The montmorillonite-illite-quartz-type of alteration is, in many areas, restricted to prominent fractures in, or close to, intrusives.

The White Horse mine (Willard and Proctor, 1946) is a typical example of alunite mineralization in the biotite latite porphyry of the Marysvale area; here the alunite occurs in three main ore bodies, each of which is composed of several smaller bodies, the long dimensions of which are approximately parallel and strike about north-south. Sulfate ion must have been brought into the rock volumes now containing alunite if that mineral were to have been formed; no definitive studies, however, have been made as to what other elements were added and subtracted during the process, nor is it certain from whence the sulfate ions came.

The alunite veins, from their association with the typically mesothermal alteration minerals, montmorillonite, kaolinite, dickite, illite, sericite, and quartz, appear to have been formed in the mesothermal range and are here categorized as mesothermal.

The uranium mineralization, in contrast to that of alunite, is found mainly in fractures in the quartz monzonite of the Central intrusive, but some veins are present in granite, latite porphyry, aplite, and a variety of volcanic rocks. The alteration in the Central intrusive proper is confined to bands that surround fractures, veins, and glassy dikes.

The alteration associated with the uranium mineralization is far less impressive, on a volume of rock affected basis, than that with the alunite bodies. Pitchblende (not uraninite) was introduced with pyrite, dark fluorite, quartz, adularia, and occasional magnetite as fillings of veins and vein breccias in the altered rocks. Another primary uranium mineral is umohoite, $UO_2 \cdot MoO_4 \cdot 4H_2O$. It is later than the pitchblende-fluorite-sulfide mineralization, being confined to cross-cutting veinlets in the main veins, but it is earlier than the quartz and, of course, than the secondary minerals. Kerr believes that its intimate association with pitchblende and its failure to oxidize indicate that it is a primary mineral. Because umohoite, a hydrated mineral, loses water when heated at low temperatures, Kerr believes that it was formed under less intense conditions than the pitchblende. The association of pitchblende with a slightly earlier mesothermal type of alteration and directly with such minerals as fluorite, quartz, and pyrite, indicates that it was formed under conditions of moderate intensity and that it should be classified as mesothermal.

After the area had been subjected to considerable erosion and the veins were exposed at the surface, a considerable suite of secondary (oxidized) minerals was developed by circulating ground water, of which the principal uranium species were: autunite, schroeckingerite, torbernite, uranopilite, zippeite, and johannite. Gypsum is intimately associated with these minerals and must also be of secondary origin. The paragenetic position of jordisite and ilsemannite is uncertain; they may be early secondary or late primary minerals.

PARK CITY

Early Tertiary *Zinc, Lead, Silver* *Mesothermal to Leptothermal*

Barnes, M. P., and Simos, J. G., 1968, Ore deposits of the Park City district with a contribution on the Mayflower Lode, in Ridge, J. D., Editor, *Ore deposits of the United States, 1933-1967* (Graton-Sales Volumes): Chap. 53, v. 2

Boutwell, J. M., 1907, Stratigraphy and structure of the Park City mining district, Utah: Jour. Geol., v. 15, p. 434-458

—— 1912, Geology and ore deposits of the Park City district, Utah: U.S. Geol. Surv. Prof. Paper 77, 231 p. (particularly p. 41-105, 115-130)

—— 1933, Park City mining district, in *The Salt Lake region*: 16th Int. Geol. Cong., Guidebook 17, p. 69-82

Butler, B. S., 1920, Park City district, in Butler, B. S., and others, *The ore deposits of Utah*: U.S. Geol. Surv. Prof. Paper 111, p. 285-318

Crittenden, M. D., and others, 1952, Geology of the Wasatch Mountains east of Salt Lake City, Parley's Canyon to Traverse Range: Utah Geol. Soc. Guidebook to the Geology of Utah, no. 8, p. 1-37

—— 1966, Geologic map of the Park City West quadrangle, Utah: U.S. Geol. Surv. Geol. Quad. Map CQ-535, 1:24,000

Erickson, A. J., Jr., and others, Editors, 1968, Park City district, Utah: Utah Geol. Soc. Guidebook to the Geology of Utah, no. 22, 100 p.

Kildale, M. D., 1956, Geology and mineralogy of the Park City district, Utah: Mineral. Soc. Utah Bull., v. 8, no. 2, p. 5-10

McKay, G. R., 1923, Park City, a lead-silver district in Utah: Eng. and Min. Jour.-Press, v. 116, no. 1, p. 7-14

Nackowski, M. P., and others, 1967, Trend surface analysis of trace chemical data, Park City district, Utah: Econ. Geol., v. 62, p. 1072-1087; disc., 1968, v. 63, p. 423-425

Wilson, C. L., 1959, Park City mining district: Intermountain Assoc. Petrol. Geols. 10th Ann. Field Conf. Guidebook, p. 182-188

Notes

The Park City ore deposits are located some 25 miles east-southeast of Salt Lake City on the eastern side of the Wasatch Mountains. The district lies on the northern slope of a prominent spur that extends eastwardly from Clayton Peak (on the main divide of the central Wasatch) toward the Uinta Range. Ore mineralization occurs within a rectangular area of some 60 square miles, the long dimension of which runs east-west.

The first deposits of value to be discovered in the Park City district were in the younger rocks in the northern part of the area, but the lower portion of the stratigraphic column has become of interest as underground exploration has been directed toward the south. The oldest formation in the district, the Mineral Fork tillite, is late Precambrian. The tillite is overlain, probably unconformably, by the Lower Cambrian Tintic formation that was followed by the Middle Cambrian Ophir shale. The Upper Cambrian is represented by the Maxfield limestone. An appreciable unconformity exists between the Maxfield and the next formation in the district, the basal Mississippian dolomite (formerly known as the Jefferson dolomite); the basal Mississippian beds are overlain by the Madison limestone. The Madison lies conformably beneath the Mississippian Deseret limestone. The Mississippian Humbug formation conformably overlies the Deseret. It has been divided (Crittenden and others, 1952) into the Humbug formation (older) and the Doughnut formation (younger). The Doughnut formation is followed unconformably by the Morgan formation. The Morgan in the Park City area is now thought by Crittenden to be equivalent of the Round Valley limestone that immediately underlies the Morgan outside the district. Nevertheless, the Morgan in the Park City district is conformably overlain by the Weber quartzite that makes up the central core of the district along the eroded axis of the Park City anticline.

Immediately, but unconformably, above the Weber quartzite is the Park City formation, the lower part of which is Pennsylvanian in age and the upper Permian; the upper two members of the Park City apparently are equivalent to the Phosphoria formation. Overlying the Park City formation is the Woodside shale that appears to be of Triassic age. The Woodside is conformably overlain by the Triassic Thaynes formation that is composed of three members. The Thaynes is followed by the Triassic(?) Ankareh formation. The youngest consolidated formation cropping out in the district is the Jurassic Nugget sandstone.

Igneous rocks occupy about one-third of the district. The intrusive rocks range from granodiorite through monzonite to (quartz) diorite porphyry and diorite and are largely confined to the southern portion of the district. Extrusive rocks are represented by andesite flows, tuffs, and agglomerates that cover the east flank of the district; their age relationships to the intrusives are uncertain, but the andesite probably is the younger phase. The igneous masses range from stocks on the west to dikes and sills in the main part of the district.

The intrusive rocks cut all of the sedimentary formations, and the andesite contains areas of porphyry that well may have been present when the flows were emplaced. Thus, the igneous rocks are certainly as young as the Jurassic and probably were introduced after the orogenic activities of the Late Cretaceous. From comparison with adjacent districts in Utah, it would appear probable that the actual date of the intrusions was late Eocene or early Oligocene. The igneous activity was followed by faulting, ore mineralization, and further faulting. The close spatial relationships of ore bodies in the district to the dikes that extend out from the larger igneous masses strongly suggest that the ore-forming fluids used the contacts between igneous rock and the enclosing sediments as channelways from depth. This in turn suggests, though less strongly, that the ore fluids came from the same general source as the diorites and that they appeared in the area not long after the igneous intrusions. On this basis, it seems reasonable to place the time of ore mineralization as late Eocene or earliest Oligocene and to classify the deposits as early Tertiary.

The ore bodies of the Park City district developed as both bedded replacements in sedimentary rocks and as fissure veins in both the sediments and in the igneous rocks. These ores occur in three northeast-southwest bands that cross the center of the Park City anticline in a belt that is 2 miles wide from north to south. These fissure zones are, from north to south, the Silver King, the Daly-Ontario, and the Mayflower-Pearl. The prevailing strikes of these fissures are essentially parallel to that of the axis of the Uinta anticline and range between N50°E to N70°E, and the dips are usually to the northwest. The walls of the individual mineralized fissures may be diorite porphyry and/or quartzite and limestone of the younger Paleozoic and Mesozoic horizons. Some replacement of the wall rock occurred where the fissure veins had carbonate walls and the widths of such replacements may be up to 30 feet. The forms and extents of bedded replacement bodies in limestone were controlled by the direction of the fissure in question in relation to the favorable bed it cuts. Bedded deposits usually are richer than the fissures from which the ore-forming fluids moved out into the surrounding carbonate rock and may extend for as much as 100 to 200 feet from the solution-supplying fissure; the usual maximum thickness of such deposits is 10 feet and such bodies may have stope lengths of 800 feet. The greatest numbers of these bedded deposits, as well as those of the largest size, have been found in the members of the Park City formation; smaller numbers of smaller deposits have been mined in the Thaynes formation, and recent work has developed bedded ore bodies in the Humbug formation in the Ontario area and in the cherty members of the Deseret formation in the Mayflower area. The first discoveries of ore in the Park City district were in fissure veins in the Weber quartzite and younger rocks, while later exploration has found the Mayflower and Pearl vein-systems within the diorite porphyry, and other fissure deposits have been discovered in the Humbug formation and in the Deseret limestone.

The ores in the upper levels of most mines were mixtures of primary sulfide and secondary oxidized ores. In the Silver King mine, in the northern part of the district, both galena and tetrahedrite containing silver are present, the latter in the larger proportion. On oxidation, bindheimite, massicot, cerussite, anglesite, azurite, malachite, and chrysocolla were developed. The rich primary silver ore of the Ontario mine, farther south in the district, contained silver-rich tetrahedrite, silver-bearing galena, some sphalerite, argentite, famatinite, the ruby-silver minerals, plus rhodonite and some pyrite. In the Park City mine, the ore was made up of tetrahedrite, argentite, and ruby-silver minerals in a quartz-manganocalcite gangue. The typical lead-zinc ores of the district contain galena and light-colored sphalerite with some pyrite and often some tetrahedrite; they are less rich in silver than those just described. In the New Park mine, at appreciable depths, the rich gold-bearing ore is characterized by hematite and chalcopyrite with some galena and sphalerite in a calcite-quartz gangue. This mine also contains rich copper ores that are composed largely of enargite that is accompanied by pyrite, chalcopyrite, bornite, and chalcocite, and small amounts of galena and sphalerite. The lower temperature (silver-rich) ores are far more commonly developed in the northern part of the district, while those rich in gold and copper occur in the lower reaches of the Mayflower and Pearl fissure systems to the south. Between these two extremes are the silver-lead-zinc ores of the Silver King and Ontario systems and the lead-zinc ores of the Mayflower and Pearl systems. The ores in the southern portion of the districts are more closely associated with igneous rocks and probably were closer to the source from which the ore-forming fluids came.

The ore-forming fluids developed a series of hydrothermal alteration zones in the igneous rocks, these being, from the veins outward, quartzose-sericitic, argillic, and chloritic; in the limestones, the quartz-sericite stage appears to have been the only one to have been formed. It would appear reasonably certain that at least the enargite and gold-chalcopyrite ores were formed under mesothermal conditions and that the silver-bearing galena-silver-rich tetrahedrite ores were developed in the leptothermal range. The presence of argentite and the ruby silvers in these veins confirms the leptothermal nature of these latter ores. The silver-lead-zinc and the lead-zinc ores intermediate between the two extremes probably were formed in the leptothermal range, but they may have been formed in part under mesothermal conditions as well. It would seem logical to classify the Park City ores as mesothermal to leptothermal.

SPOR MOUNTAIN

Late Tertiary *Beryllium* *Epithermal*

Erickson, M. P., 1963, Volcanic geology of western Juab County, Utah: Utah Geol. Soc. Guidebook to the Geology of Utah, no. 17, p. 23-35

Fitch, C. A., and others, 1949, Utah's new mining district: Eng. and Min. Jour., v. 150, no. 3, p. 63-66

Griffitts, W. R., and Rader, L. F., Jr., 1963, Beryllium and fluorine in minerlized tuff, Spor Mountain, Juab County, Utah: U.S. Geol. Surv. Prof. Paper 475-B, p. B16-B17

—— 1964, Beryllium and fluorine in mineralized tuff, Spor Mountain, Juab County, Utah: U.S. Geol. Surv. Prof. Paper 475-B, p. B16-B17

Montoya, J. W., and others, 1962, Beryllium-bearing tuff from Spor Mountain, Utah; its chemical, mineralogical, and physical properties: U.S. Bur. Mines R.I. 6084, 15 p.

Patton, H. B., 1908, Topaz-bearing rhyolite of the Thomas range, Utah: Geol. Soc. Amer. Bull., v. 19, p. 177-192

Sharp, B. J., and Williams, N. C., 1963, Beryllium and uranium mineralization in western Juab County, Utah: Utah Geol. Soc. Guidebook to the Geology of Utah, no. 17, 59 p.

Shawe, D. R., 1966, Arizona-New Mexico and Nevada-Utah beryllium belts: U.S. Geol. Surv. Prof. Paper 550-C, p. C206-C213

—— 1968, Geology of the Spor Mountain beryllium district, Utah, in Ridge, J. D., Editor, *Ore deposits of the United States, 1933-1967* (Graton-Sales Volumes): Chap. 55, v. 2

Shawe, D. R., and others, 1964, Lithium associated with beryllium in rhyolitic tuff at Spor Mountain, western Juab County, Utah: U.S. Geol. Surv. Prof. Paper 501-C, p. C86-C87

Staatz, M. H., 1963, Geology of the beryllium deposits in the Thomas range, Juab County, Utah: U.S. Geol. Surv. Bull. 1142-M, M1-M36

Staatz, M. H., and Carr, W. J., 1964, Geology and mineral deposits of the Thomas and Dugway ranges, Juab and Tooele Counties, Utah: U.S. Geol. Surv. Prof. Paper 415, 188 p. (does not discuss beryllium deposits)

Staatz, M. H., and Griffitts, W. R., 1961, Beryllium-bearing tuff in the Thomas range, Juab County, Utah: Econ. Geol., v. 56, p. 941-950

Staatz, M. H., and Osterwald, F. W., 1959, Geology of the Thomas range fluorspar district, Juab County, Utah: U.S. Geol. Surv. Bull. 1069, 97 p.

Williams, N. C., 1963, Beryllium deposits, Spor Mountain, Utah: Utah Geol. Soc., Guidebook to the geology of Utah, no. 17, p. 36-59

Notes

The beryllium deposits of Spor Mountain (a minor segment of the Thomas range) are located in west central Utah, just over 40 miles northwest of the town of Delta.

The rocks of the Spor Mountain area have been divided into three distinct groups: (1) Paleozoic rocks, ranging in age from Ordovician to Devonian and consisting mainly of dolomites and dolomitic limestones; the Paleozoic rocks are stratigraphically below the major beryllium ore zones in the volcanic rocks and only the most minor expressions of the beryllium mineralization are found in the Paleozoic dolomites immediately below the ore-bearing tuffs; (2) Tertiary volcanic rocks, chiefly flows and tuffs and including a number of dikes and plugs of intrusive rock; the ore zones are found in the Topaz Mountain tuff in this sequence; and (3) unconsolidated Quaternary sediments such as the Lake Bonneville deposits and Recent colluvium and alluvium. The volcanic rocks lie with a pronounced angular unconformity on the Paleozoic rocks and are so complexly related as to suggest many sources of eruptive material and numerous times of eruption. The flows and tuffs beds intercalate and overlap each other and locally are separated by rather short erosive intervals, so that correlations over any great distance are uncertain or impossible.

The tuff beds of Spor Mountain, probably the same formation as the Topaz Mountain tuff, are overlain, probably without any erosional break, by the Topaz Mountain rhyolite. This latter formation is considered to be late Tertiary (Pliocene?) in age, which makes the underlying tuff not greatly older. It is, therefore, thought reasonable that the Topaz Mountain tuff should be classified as late Tertiary (probably early Pliocene). The presence of beryllium mineralization in the Topaz Mountain rhyolite, as well as in the tuff beneath it, demonstrates that the mineralization is no older than Pliocene. The cover of Lake Bonneville beds of probably Pleistocene age was deposited in the area after the mineralization had occurred, thus indicating that the mineralization must have taken place in the Tertiary, and the beryllium is here considered to have been deposited in the late Pliocene and is classified as late Tertiary.

The beryllium deposits in the upper tuffs now lie on the flats on either side of Spor Mountain, partly ringing the high-grade fluorite-rich pipes along the crest of the mountain; these fluorite deposits appear to have been formed before the beryllium mineralization and are not discussed here. The Spor Mountain deposits differ markedly from most other beryllium bodies, both in mineral content and occurrence. The numerous faults in the area appear to have served as channelways to bring the ore fluids into the porous upper member of the Topaz Mountain tuff where the solutions spread out in all directions. Some minor amounts of the ore fluid followed fault channels into the Topaz Mountain rhyolite, and some ore minerals were deposited in the fault fractures in the Paleozoic rocks beneath the tuff, but almost all the beryllium is concentrated in the upper member of the tuff.

The paragenesis of the beryllium deposits is quite complex and probably can be separated into several stages. The sequence of hydrothermal events probably consisted of (1) the dedolomitization and bleaching of dolomite fragments in the tuff; the hydrothermal alteration (divitrification) of the ash has largely converted that material to montmorillonite; (2) the formation of purple fluorite lacking beryllium, the fluorite having been emplaced mainly as a replacement of dedolomitized fragments and forming readily recognized nodules of that mineral; (3) the introduction of manganese, first probably as manganous ion that replaced appropriate ions in the minerals already present to give a pink coloration to the tuff, second as the more oxidized forms of manganese to produce black staining of the tuffs and the impregnation of quartzite fragments with manganese minerals and the replacement of some fluorite in fluorite nodules by manganese minerals to different degrees; (4) the introduction of a second generation of fluorite (this time tan in color) that replaced manganese minerals in the nodules; (5) the addition of beryllium, as bertrandite [$Be_4Si_2O_7(OH)_2$] and gelbertrandite (a possible bertrandite polymorph), plus quartz; it has been suggested, though not proved, that some of the beryllium was emplaced in octahedral coordination in the montmorillonite; the beryllium minerals were certainly deposited on the rims of fluorite nodules or outward from the centers of such nodules; (6) the deposition of veinlets of chalcedony; and (7) the introduction of quartz veinlets and coarse, light purple fluorite.

There is no direct evidence as to the temperature and pressure under which the beryllium mineralization was deposited, but the beryllium is found in the next to the youngest (principally) and in the youngest (to a minor extent) members of a very young volcanic sequence. These beds have never been covered by still younger volcanic rocks, so the confining pressure on the ore fluids must have been low. The minerals that accompany the bertrandite and its gelbertrandite polymorph(?) are typical of epithermal deposits; the presence of fluorite in several colors is particularly characteristic of this range of intensity conditions. Some evidence suggests that the bleaching of the fluorite may have occurred at temperatures above the epithermal range (7), but the conditions of the experiments were probably not those that obtained during actual mineral deposition in the tuff bed. It is, therefore, thought that the deposits should be classified as epithermal.

TINTIC

Early Tertiary	*Zinc, Lead, Silver, Copper, Gold, Halloysite*	*Mesothermal to Leptothermal*

Almond, H., and Morris, H. T., 1951, Geochemical techniques as applied in recent investigations in the Tintic district: Econ. Geol., v. 46, p. 608-625

Billingsley, P., 1933, The utilization of geology in Tintic, Utah, in *Ore deposits of the western states* (Lindgren Volume): A.I.M.E., p. 716-722

Billingsley, P., and Crane, G. W., 1933, Tintic mining district (Utah), in *The Salt Lake region*: 16th Int. Geol. Cong., Guidebook 17, p. 101-124

Cook, D. R., Editor, 1957, Geology of the East Tintic Mountains and ore deposits of the Tintic mining districts: Utah Geol. Soc. Guidebook to the Geology of Utah, no. 12, p. 57-79, 80-93, 94-96

Crane, G. W., 1916, Geology of the ore deposits of the Tintic mining district: A.I.M.E. Tr., v. 54, p. 342-355

Farmin, R., 1934, Pebble dikes and associated mineralization at Tintic, Utah: Econ. Geol., v. 29, p. 356-370

Hahn, A. W., 1929, Silver-bearing minerals of some ores from the Tintic mining district: A.I.M.E. Tr., v. 85 (1929 Yearbook), p. 325-329

Lindgren, W., 1915, Processes of mineralization and enrichment in the Tintic mining district: Econ. Geol., v. 10, p. 225-240

Lindgren, W., and Loughlin, G. F., 1919, Geology and ore deposits of the Tintic mining district: U.S. Geol. Surv. Prof. Paper 107, 282 p. (particularly

p. 21-104, 119-184)

—— 1920, Tintic district, in Butler, B. S., and others, *The ore deposits of Utah*: U.S. Geol. Surv. Prof. Paper 111, p. 396-415

Loughlin, G. F., 1914, The oxidized zinc ores of the Tintic district, Utah: Econ. Geol., v. 9, p. 1-19

Morris, H. T., 1964, Geology of the Eureka quadrangle, Utah and Juab Counties, Utah: U.S. Geol. Surv. Bull. 1142-K, p. K1-K29

—— 1964, Geology of the Tintic Junction quadrangle, Tooele, Juab, and Utah Counties, Utah: U.S. Geol. Surv. Bull. 1142-L, p. L1-L23

—— 1968, The main Tintic mining district, Utah, in Ridge, J. D., Editor, *Ore deposits of the United States, 1933-1967* (Graton-Sales Volumes): Chap. 51, v. 2

Morris, H. T., and Lovering, T. S., 1952, Supergene and hydrothermal dispersions of heavy metals in wall rocks near ore bodies, Tintic district, Utah: Econ. Geol., v. 47, p. 685-716

Park, C. F., Jr., 1935, Copper in the Tintic district, Utah, in *Copper resources of the world*: 16th Int. Geol. Cong., v. 1, p. 361-367

Shepard, W. M., 1966, Geochemical studies in the Tintic mining district: Min. Eng., v. 18, no. 4, p. 68-72

Stringham, B. F., 1942, Mineralization in the West Tintic mining district, Utah: Geol. Soc. Amer. Bull., v. 53, p. 267-290 (about 21 miles southwest of the main Tintic district)

Tower, G. W., Jr., 1900, Tintic special folio, Utah: U.S. Geol. Surv. Geol. Atlas, Folio 65, 8 p.

Notes

The Tintic district constitutes the most productive part of the mineralized portion of the East Tintic Mountains and is located almost entirely on the western side of the mountains; it is separated arbitrarily from the East Tintic district by the 112°5' parallel of longitude. The northern boundary of both the main Tintic and East Tintic districts is an essentially east-west line running through Packard Peak and Homansville. Eureka, which is located at the north end of the main Tintic district, lies some 60 miles slightly west of south of Salt Lake City.

The East Tintic Mountains are typical of the block faulted, north-trending Basin-Range mountains and are aligned with the Oquirrh range (which contains the Bingham district deposits); they are bordered on both east and west by Basin-Range valleys, now largely filled with unconsolidated late Tertiary and younger deposits. The consolidated sedimentary rocks in the mountains range from late Precambrian to Permian and total more than 32,000 feet in thickness; they are quite strongly folded, and the fault pattern is complex. Although in the southern two-thirds of the East Tintic Mountains the Paleozoic and earlier sediments are largely concealed under middle Eocene flows and pyroclastics, a considerable area of Paleozoic rocks crops out at the surface in the most mineralized portion of the district. The largest mine in the district, the Chief Consolidated, lies under Quaternary alluvium, immediately adjacent to the northeast margin of the area of Paleozoic outcrops; most of the other important Tintic mines are located in this outcrop area. A moderate-sized area of intrusive igneous rocks, probably middle Eocene in age, lies to the south of the Paleozoic sediments. North of the Paleozoic outcrop area, and overlying the Chief Consolidated mine beneath the alluvium, is a large area of middle Eocene volcanic rock.

No Mesozoic rocks have been found in the East Tintic Mountains, and the next sedimentary formation (the last pre-ore sediments) was an unnamed early Eocene conglomerate that separates the highly folded Paleozoic rocks and the middle Eocene volcanic rocks; these rocks seemingly were developed in prevolcanic valleys and gullies and are so largely composed of angular material that they must have been colluvium or talus. The post-ore sediments include the poorly consolidated Pliocene Salt Lake (?) formation and unconsolidated valley deposits of Quaternary age.

The igneous rocks of the East Tintic Mountains make up the eroded remains of a huge composite volcano that almost completely covered the complex mountain range that had been developed at the end of the Paleozoic and consist of both surface flows and pyroclastics and intrusive stocks, plugs, and dikes that mark the volcano's eruptive centers. The extrusive rocks are divided into three types: (1) early quartz latite tuffs and flows, the Packard and Fernow quartz latites; (2) later latite (and possible andesite) tuffs, flows, and agglomerates; and (3) latest basalt flows. The later latite and andesite series is middle Eocene on the basis of plant fossils in marly limestones interbedded with agglomerates. The basalts have been thought to be as young as Pliocene but actually may be middle Eocene in age. The Packard quartz latite rocks are divided into four units: (1) a basal tuff, (2) a lower vitrophyre, (3) a massive flow (or welded tuff) unit, and (4) an upper vitrophyre. The latter latites are divided into five units: (1) basal tuffs, (2) lower flows, (3) intermediate tuff and agglomerate, (4) upper flows, and (5) agglomerates that are both thick and extensive; the upper flow series probably contains one or more flows of andesite and trachyandesite.

The oldest of the intrusive rocks, the Swansea quartz monzonite, forms a mass 0.8 miles (east-west) and 0.3 miles wide and is located south of the main area of Paleozoic outcrops near the Mammoth mine; a smaller mass and a dike of this rock also are known in the area. The next intrusive in order of decreasing age appears to have been a monzonite porphyry that is present as a stock at Sunset Peak and in other bodies. A few latite plugs and dikes cut the Packard quartz latite and the overlying latite near the Independence shaft northeast of Eureka; they appear to be related to the monzonite porphyry of Sunset Peak. The Silver City monzonite stock lies immediately east of the Swansea quartz monzonite stock and is intrusive into it and into the southern portion of the main outcrop area of Paleozoic rocks. It is much larger than the Swansea stock and has many outliers in the forms of dikes, plugs, and small stocks that extend 3.5 miles north-northeast into the East Tintic mining district. The dikes and plugs in the Tintic district often have shattered walls indicating that they were forceably intruded, a concept apparently confirmed by the close association of pebble dikes with many of the dikes and plugs. These pebble dikes consist of rounded, angular, or subangular fragments of Tintic quartzite accompanied by some disc-shaped fragments of shale; they are common within and near the belt of intrusive (mainly monzonite) rocks that cuts northeast through the East Tintic district and are considered in more detail in the discussion of that area. Later igneous rocks are known in the general area of the East Tintic Mountains, but the only type in the Tintic district proper is a biotite-augite andesite (purple) porphyry, highly altered dikes of which cut ore and sedimentary rock in the Chief No. 1 mine; most of these dike rocks now consist of kaolinite, halloysite, and siderite. These dikes occupy east-west fractures usually and appear to cut the ore bodies; locally the dikes may be weakly mineralized where they cross the ore. This relationship suggests that the ore was largely deposited before the dikes were introduced or that the dikes were far less susceptible to reaction with the ore fluids than were the carbonate rocks.

The mineralization in the district was introduced later than those faults that are considered to have formed in the late Eocene but not at a much later time. The latest date that can reasonably be assigned to the ore is late Eocene or perhaps earliest Oligocene. Radioactive age dating on zircons from the Silver City stock and from a quartz monzonite plug near the North Lily mine (East Tintic district) give ages between 38 and 46.5 m.y., suggesting that the rocks were introduced in late Eocene or perhaps even in the earliest Oligocene. It seems most reasonable at this time to believe that the ores were formed in the late Eocene and, therefore, to categorize the Tintic district deposits as early Tertiary.

Most of the ore produced from the Tintic district proper has come from replacement deposits; only about 3 percent of the total value of Tintic ore (at least $315,000,000) has been derived from fissure veins. The replacement deposits are located at the northern end of the district, occur in definite linear zones, and lie on projections to the north of the main fissure vein systems. The replacement bodies are largely chimneys and mantos in the strict Mexican sense as defined by Prescott (1926, Eng. and Min. Jour., v. 122, no. 7, 8, p. 246-253, 289-296) and lie in four major ore zones (from west to east) that are known as the Gemini, Chief, Godiva, and Iron Blossom and one minor zone, the Plutus, located between the Chief and the Godiva. These ore masses strike generally north, are parallel to the bedding or nearly so, and plunge to the north at about the same angle as the synclinal axis. They normally have one long dimension and two short ones and often are made up of more than one ore shoot; either both are on much the same level or one is vertically above the other and the shoots may be connected to, or isolated from, each other. At the intersections of certain fault types, the ore shoots are larger in plan and form vertical chimneys; ore volumes so enlarged are usually richer than the smaller portions of the ore shoots. The shoots persist through faults and from one stratigraphic unit to another, and the ore is found over a stratigraphic range of 6000 feet from the Lower Cambrian Tintic formation through the Mississippian Deseret limestone; some ore zones have persisted for 8000 to 9500 feet along strike. The most productive formations have been, in order of decreasing economic importance, the Bluebell, the Ophir, the Ajax, and the Deseret. Except in the Chief mine, essentially no mining has been done below the water table, but because of the difficulty of mining the ore, not because of the lack of it.

The fissures cut through all the rock types in the district, strike generally northeast, dip steeply west, and usually are of small displacement; within the Tintic district proper, they are most abundant in the southern part. The usual fissure is mineralized for only a small portion of its length, and the mineralization has an average width of only 2 feet. Where the fissures cut quartzite, the ore normally is confined within the actual fracture, but in limestone the host rock is replaced for short distances out from the fissure walls. In general, only the secondarily enriched fissure veins have been rich enough to mine.

The most important primary minerals in the replacement type of ore are galena, sphalerite, wurtzite, argentite, native silver, enargite, and tetrahedrite; gangue minerals include pyrite, quartz, calcite, and some barite. Silver is included in galena and tetrahedrite as well as in argentite and native silver. Gold appears to have been introduced largely in the earliest stage of mineralization, probably in association with enargite and pyrite. In the upper levels, the ores have been highly oxidized, the depth of oxidation being as much as 2000 feet at the north end of the Tintic district, but the larger oxidized ore bodies usually contain some relict primary sulfides. The most common oxidized minerals are malachite, azurite, chrysocolla, anglesite, cerussite, smithsonite, calamine, hydrozincite, cerargyrite, native silver, and plumbojarosite. Some covellite appears to have been developed by secondary enrichment.

The district shows a rather well-defined zonal pattern, with copper-gold ores being most prominent in the south, lead-silver ores farther north, and lead-zinc ores farthest north, but the increase of zinc to the north may be due to the greater depths of the ore zones in that area than farther south because they follow the northward plunge of the anticlinal axis.

The major primary ore minerals in the fissure veins are enargite, argentite, native silver, and galena; some sphalerite, chalcopyrite, arsenopyrite, and tetrahedrite also are found. The gangue minerals are pyrite, calcite, quartz, and barite.

The hydrothermal alteration in the Tintic district has been far less well studied than that of the East Tintic area, but alteration has been impressive. The Dragon halloysite deposits resulted from the hydrothermal alteration of a large septum of limestone at the northern margin of the Silver City stock. In the ore zones in the sedimentary rocks, the alteration is primarily a widespread development of hydrothermal dolomite that is related spatially to the general areas of mineralization and not to the individual ore bodies. Where the volcanic rocks are present over the sediments, they have been chloritized in their basal portions. The replacement deposits in limestone are ordinarily enclosed in jasperoid from a few feet to a few tens of feet thick; the jasperoid may be weakly mineralized.

The ores contain minerals diagnostic of the mesothermal range (enargite) and of the leptothermal range (tetrahedrite, argentite, and native silver), and the other principal sulfides (galena, sphalerite, and chalcopyrite), are compatible with these divisions of the classification. The wall-rock alteration also fits into the mesothermal and leptothermal range, so it appears reasonable here to classify the ores as mesothermal to leptothermal.

Vermont

ELIZABETH

Middle Paleozoic *Copper, Pyrites* *Hypothermal-1*

Buerger, N. W., 1935, The copper ores of Orange County, Vermont: Econ. Geol., v. 30, p. 434-443

Canney, F. C., 1965, Geochemical prospecting investigations in the Copper Belt of Vermont: U.S. Geol. Surv. Bull. 1198-B, p. B1-B28

Doll, C. G., 1943-1944, A preliminary report on the geology of the Strafford quadrangle, Vermont: Rept. State Geol. Vt., v. 24, p. 14-28

Howard, P. F., 1959, Structure and rock alteration at the Elizabeth mine, Vermont. Part I. Structure: Econ. Geol., v. 54, p. 1214-1250; Part II. Rock alteration: p. 1414-1443

Jacobs, E. C., 1943-1944, General petrology in Strafford Township: Rept. State Geol. Vt., v. 24, p. 29-37

——— 1943-1944, The Vermont Copper Company, Inc.: Rept. State Geol. Vt., v. 24, p. 1-13

McKinstry, H. E., and Mikkola, A. K., 1954, The Elizabeth copper mine, Vermont: Econ. Geol., v. 49, p. 1-30

Skinner, B. J., and Milton, D. J., 1955, The Elizabeth copper mine, Vermont: Econ. Geol., v. 50, p. 751-752

Thompson, A. P., 1914, On the relation of pyrrhotite to chalcopyrite and other sulphides: Econ. Geol., v. 9, p. 153-174 (particularly p. 158-162)

Weed, W. H., 1911, Copper deposits of the Appalachian states: U.S. Geol. Surv. Bull. 455, p. 18-31

White, W. S., 1952, Structural control in the Vermont copper district (abs.): Geol. Soc. Amer. Bull., v. 63, p. 1312-1313 (also Econ. Geol., v. 47, p. 779)

White, W. S., and Billings, M. P., 1951, Geology of the Woodsville quadrangle, Vermont-New Hampshire: Geol. Soc. Amer. Bull., v. 62, p. 647-696 (Area discussed is entirely north of Elizabeth mine but contains the same rocks and is structurally similar)

White, W. S., and Jahns, R. H., 1950, Structure of central and east-central Vermont: Jour. Geol., v. 58, p. 179-220

Notes

The Elizabeth mine is located in Orange County in eastern Vermont, 7 miles west of the Connecticut River and 15 miles northwest of Hanover, New Hampshire. Although the Elizabeth mine has now been closed down, for much of its period of operation it was the only operating metal mine in New England, and the deposits of the Orange County district of which the Elizabeth mine is one, were, until the discovery of the Michigan copper deposits, the chief source of copper in the United States.

The rocks of the district range in age from the Precambrian Green Mountain complex, less than 20 miles to the west, to Mississippian(?) diabase dikes that occur in the vicinity of the mine and Middle to Late Devonian(?) cross-cutting granite plutons, the nearest of which is a small body some 8 miles northwest of the mine. Only three formations are known in the immediate neighborhood of the mine. The middle of these, the Middle Ordovician Gile Mountain formation, contains the Elizabeth ore deposits and crops out in a belt from 3.5 to 9 miles wide that extends in a north-south direction through the central and eastern portions of the Strafford quadrangle. The Gile Mountain is bordered on the east by the younger Middle or Upper Ordovician Ordfordville formation and on the west by the older, generally calcareous rocks of the Ordovician Waits River formation and its Standing Pond amphibolite member that normally marks the boundary between the Gile Mountain and the Waits formations. Igneous rocks in the general area are confined to diabase dikes in the vicinity of the mine that strike about east-west and dip nearly vertically; they are thought to be post-ore and probably are Mississippian in age. The nearest granite pluton is quite small; larger masses are abundant some 16 miles north-northwest. The rocks of these plutons may be Devonian or Mississippian in age, but it appears more probable that they belong to the older rather than to the younger period.

The rocks of the immediate Elizabeth area were subjected to considerable thermal metamorphism shortly prior to the development of the ores and of the three stages of hydrothermal alteration that preceded, accompanied, and followed ore deposition. A fourth later alteration on a regional scale postdated all other geologic events that have left their mark on the area. The ore deposits are located on the east limb of the structurally complex Strafford dome, the long dimension of which runs essentially north-south. The regional metamorphism appears to have taken place in Acadian time and to have been directly related to the development of this dome that probably resulted from the intrusion of a mass of material of the gravity of granite at depth beneath the structure.

The first three stages of wall-rock alteration were part of a continuous cycle of alteration that attacked the metamorphic minerals. The first of these stages, although it occurs in the rocks of the ore zone, has no direct geometric relationship to the ore bodies; it happened after the peak of thermal metamorphism in the area had passed and before the ore was introduced. The second stage of alteration, which accompanied the emplacement of the ore, also was made up of biotite and sericite subzones. The third stage of alteration (post-ore) occurs in well-defined zones that lie above and below the ore both to the east and west of the ore zones; the alteration may cover distances of as much as 500 feet. It is directly related to folds(?), post-ore faults, shears, and shear zones, and the zones are usually less than 30 feet in width. Stage four of the alteration was accomplished by mechanisms affecting the entire region in which minor chlorite was developed as rims around femic minerals. The structures of the Elizabeth area and of the ore zones themselves are both highly complex.

Howard believes that the field data strongly indicate that the late and early folds were both formed by a single period of deformation which, in this case, must have been Acadian or Middle or Late Devonian in age. Since the so-called Christmas-tree folds were produced between the early and late folds, it follows that they and the granitelike intrusion that probably produced them also were Middle Devonian. No direct evidence is available to show the source of the ore-forming fluids from which the Elizabeth ores were deposited. It would seem probable, however, since the ores were emplaced in rock volumes where drag folding had produced open space between schist laminae as well as brecciation and faulting, that the solutions in question entered the area at the same general time as the favorable structures were developed. Thus, the sequence of events in Acadian time was (1) the formation of the Green Mountain anticlinorium and the drag folds attendant on that process; (2) the mild regional metamorphism that occurred at essentially the same time as the folding; (3) the more complex thermal metamorphism induced by the apparent intrusion of a granitelike mass and the concomitant development of the Christmas-tree folds and associated faults; and (4) the three phases of hydrothermal alteration; during the second phase the ores were emplaced and during the first phase pressure was sufficiently high to permit the formation of impressive quantities of garnets. The formation of the ores in Acadian time dates them as middle Paleozoic.

The Elizabeth mine consists of two ore bodies known as the main ore body and the No. 3 ore body; each one, throughout its length, is confined to a particular stratigraphic horizon in the Gile Mountain formation that lies between two amphibolite beds. The apparently stratigraphically higher of these two amphibolites is actually stratigraphically the lower of the two because the overturning of the beds converted the early anticline into a structural syncline. The Elizabeth syncline is bordered by other folds of similar character, although each shows appreciable change in detail from one section of the mine to another, and all minor folds are affected by still more minor folds and by cross-folds; these minor folds probably formed late in the history of folding in the mine rocks. Both pre-ore and post-ore faults (usually normal) are known in the mine; pre-ore faults are usually mineralized, post-ore faults, of course, are not; these last, however, are much more conspicuous and carry heavy gouge.

The ore appears to have been emplaced both by replacement and open-space filling; remnants of schist survive in the ore even in the most massive types, there being all gradations between sulfides sparingly disseminated in schist and massive sulfide bodies in which only wisps of schist are found. The sulfide bands are concordant with the folds, and the original structure of the rock is usually preserved as shadowlike relics of schist or by a banded alternation of pyrrhotite and pyrite. Fractures were filled by sulfides, and bent bands of schist were broken by wedge-shaped cracks in which sulfides were deposited; where mica schist, adjacent to amphibolite, was broken into impressive breccia masses, it has been cemented by sulfides. Actual open space may never have existed on a large scale; as the earth movements produced the smallest fraction of open space, it was immediately filled by sulfides deposited from the solutions moving through these permeable portions of the rock mass. It must be remembered that the ore solutions were being introduced and the ores deposited from them not long after metamorphic garnets of the first alteration stage had been formed and that the sulfides must have been stable assemblages under rather elevated conditions of temperature and pressure.

The dominant sulfide of the ore bodies is pyrrhotite that makes up 90 percent of the sulfides on the average; chalcopyrite is the next most important sulfide and averages about 9 percent of the total sulfide mass. The actual amounts of these two sulfides range widely around the percentages just given. The ore also contains about 0.4 percent of zinc (not recovered), and 0.16 ounces of silver and 0.008 ounces of gold per ton of ore. The pyrrhotite was quite low in sulfur, being almost stoichiometric troilite and only feebly magnetic. Pyrite is very sparsely developed and forms tiny cubes, the corners of which are slightly rounded by other sulfides; fractures in pyrite grains and the interstices between them were filled with pyrrhotite. Chalcopyrite grains normally range from 0.1 to 0.5 mm in diameter, but those that are 0.01 mm in diameter and smaller are intimately associated with pyrrhotite. Much of the chalcopyrite contains lamellae of cubanite up to 0.05 mm wide and 0.2 mm long; these were almost certainly exsolved from the chalcopyrite after deposition. Some valleriite(?) is found in similar relationships to the chalcopyrite. Sphalerite is extremely sparsely distributed among the other sulfides; it is most commonly associated with chalcopyrite, and some of it may have exsolved from that mineral but this is not certain from the evidence at hand. Galena is very scarce, and tetrahedrite-tennantite is exceedingly rare, though they may be abundant enough to account for the small amounts of silver in the ores. Molybdenite is very minor in amount and probably was an early mineral. The gangue minerals actually accompanying the ore minerals and distinct from the wall-rock alteration minerals of the second stage are tourmaline, subordinate idocrase, and rare rutile; although vein quartz is chiefly confined to the ore zone, it is earlier than the sulfides and bears no consistent spatial relationships to them.

All evidence available as to the intensity range of the Elizabeth deposits appears to point to their deposition under hypothermal conditions. The pyrrhotite is diagnostic of that range when present in such overwhelming amounts; the chalcopyrite, with its lamellae of cubanite and valleriite(?) also seems certainly to have been formed at high temperatures. Sphalerite is too small in amount and too little studied to contribute much light on the conditions of deposition, but some of it may have exsolved from chalcopyrite. The amounts of galena and the silver-bearing sulfosalts are so small, infinitesimal in fact, that they cannot be said to contribute much to the determination of the intensity range. The gangue minerals are all compatible with hypothermal conditions. The wall-rock alteration of the second stage (contemporaneous with the ore deposition), particularly the sericite subzone, is compatible with the hypothermal range. The Elizabeth deposits, therefore, are classified here as hypothermal and, from the argument given previously, are considered to have been formed in noncalcareous rocks; they are, therefore, hypothermal-1.

Virginia

AUSTINVILLE-IVANHOE

Late Paleozoic *Zinc, Lead* *Telethermal*

Brown, W. H., 1935, A quantitative study of the zoning of ores at the Austinville mine, Wythe County, Virginia: Econ. Geol., v. 30, p. 425-433

Brown, W. H., and Fulton, R. B., 1958, Metal content of mine waters: 20th Int. Geol. Cong. Symposium de Exploracion Geoquimica, t. 1, p. 189-197

Brown, W. H., and Weinberg, E. L., 1968, Geology of the Austinville-Ivanhoe district, Virginia, in Ridge, J. D., Editor, *Ore deposits of the United States, 1933-1967* (Graton-Sales Volumes): Chap. 10, v. 1

Butts, C., 1940, Geology of the Appalachian Valley in Virginia: Va. Geol. Surv. Bull. 52, pt. 1, 568 p. (particularly p. 22-67)

Currier, L. W., 1935, Structural relations of southern Appalachian zinc deposits: Econ. Geol., v. 30, p. 260-286 (particularly p. 268-271)

—— 1935, Zinc and lead region of southwestern Virginia: Va. Geol. Surv. Bull. 43, 122 p.

Fulton, R. B., 1950, Prospecting for zinc using semiquantitative chemical analyses of soils (at Austinville): Econ. Geol., v. 45, p. 654-670

Jolly, J. L., and Heyl, A. V., 1968, Mercury and other trace elements in sphalerite and wallrocks from central Kentucky, Tennessee and Appalachian zinc districts: U.S. Geol. Surv. Bull. 1252-F, p. F1-F20

McMurry, H. V., and Hoagland, A. D., 1956, Three-dimensional applied potential studies at Austinville, Virginia: Geol. Soc. Amer. Bull., v. 67, p. 683-696

Oder, C.R.L., and Hook, J. W., 1950, Zinc deposits of the southeastern states, in Snyder, F. G., Editor, *Symposium on mineral resources of the southeastern United States*: Univ. Tenn. Press, Knoxville, p. 72-87 (particularly p. 83-84)

Watson, T. L., 1905, Lead and zinc deposits of Virginia: Va. Geol. Surv. Bull. no. 1, 155 p. (particularly p. 83-99)

Weinberg, E. L., 1963, Geology of the Austinville-Ivanhoe area: V.P.I. Engineering Extension Series, Geological Guidebook no. 2, Geological Excursions in Southwestern Virginia, 15 p.

Notes

The ore deposits of the Austinville-Ivanhoe district are found in southwestern Virginia, just over 60 miles southwest of Roanoke.

The rocks of the area were formed entirely in Cambrian time, ranging from Lower to Upper(?) Cambrian in age. The oldest rocks are those of the Chilhowee group, the basal formation of which is the quartzites and conglomeratic quartzites of the Unicoi formation. It is overlain by the Hampton shale, that consists of more or less slaty and locally schistose shale with very minor amounts of intercalated quartzite. The quartzite is most abundant near the upper contact of the Hampton with the similar rocks of the overlying (youngest Chilhowee) Erwin quartzite. At the top of the Erwin, some dolomitic sandstone beds are transitional to the overlying Patterson limestone member of the Lower Cambrian Shady formation. The Patterson limestone, known locally as the Ribbon formation, is a dark rock with a characteristic ribbony or wavy appearance whether it is composed of limestone or dolomite. The Patterson grades into the overlying Austinville (or Saccharoidal) dolomite member; the Patterson member contains most of the ore at Ivanhoe and some of it at Austinville. The Saccharoidal dolomite is host to most of the ore in the Austinville mine. Locally, the Austinville member is overlain by the thick, massive, dense Ivanhoe limestone member; it is not ore-bearing but has been quarried by the National Carbide Company. Paleontologically, the Shady correlates with the Lower Cambrian middle Tomstown dolomite, confirming its Lower Cambrian age. The Shady is overlain conformably by the Rome (Watauga) formation of Lower and Middle Cambrian age. The Middle and Upper Cambrian Elbrook dolomite is found in a restricted portion of the district.

The structure in and near the Austinville mine is now known in considerable detail. Most of the mineralized ground in the Austinville area lies in the "Slot" between the Logwasher fault on the northwest and the Stamping Ground fault on the southeast; the former is an almost vertical fault with the northwest side upthrown and probably is a normal tension fault. The Stamping Ground fault, on the other hand, is a high-angle thrust fault (dipping 45° to 65° southeast) in which the hanging wall has ridden up from the southeast. Between these two faults is a volume of highly disturbed ground 2000 feet in width in which there are five types of faults.

The ore in the district probably was introduced after the main faulting. The faults in the hanging wall of the Laswell thrust fault (that contains the Austinville basin) belong to the Appalachian orogeny since the fault block that includes them overlies Mississippian beds in the immediate area and Pennsylvanian beds at a greater distance. Thus, middle Paleozoic earth movements could not have produced the ore-controlling structures nor could the ore have been pre-Pennsylvanian. No ore has been displaced by subsequent folding or faulting, so the mineralization is later than both. There is no evidence to suggest that the ore-controlling structures were not mineralized during the same general portion of geologic time in which they were developed. It would be surprising indeed if such a tremendous amount of late Paleozoic diastrophism was not accompanied by the introduction of ore-forming fluids into the area or if ore-forming fluids had entered the area long after the Paleozoic, entirely unaccompanied by earth movements of any type. It appears reasonable, therefore, to classify the Austinville-Ivanhoe ores as late Paleozoic and a product of the Appalachian revolution.

The occurrences of ore in the Austinville-Ivanhoe district are definitely related to the fault structures that appear to have acted as channelways for the ore solutions to reach loci for ore deposition. The actual faults themselves are usually not more than weakly mineralized, but the ore bodies are intimately associated with one or more faults. Over the length of known ore in the Austinville (Saccharoidal) dolomite at Austinville, one fault, the high-angle Burleigh thrust that strikes generally northeast-southwest, weaves in and out of the ore area--in one place with ore on both sides of it, in another with ore on the hanging wall, and in still another with ore on the footwall. On occasion, the ore may be as much as a few hundred feet from the fault, but the general coincidence of the fault and ore is unmistakable. The relationship of other faults to the Burleigh does much to explain the size and position of the ore bodies. Both bedding plane and north-south faults displace the Burleigh, and ore is found in pencil-shaped bodies in the V's of broken rock bounded by the Burleigh on one side and the bedding plane or north-south faults on the other. Where these pencil-shaped ore bodies are cut by cross tears, the ore bodies are greatly enlarged, being not only increased in cross-section but also cross-cutting vertically through several horizons, these being so thoroughly broken that ore can be deposited in horizons of the Austinville member other than those that are, in the main, most favorable to ore deposition. Cross faults, where they cut the Burleigh, also increase the size of the volume of broken rock but do so to an appreciably lesser extent than do the cross tears and are less common localizers of ore deposition than the cross tears. Changes in strike and dip on any fault appear to create broken rock favorable to the localization of ore and thereby to make for the enlargement of ore bodies above their normal character. At Austinville there probably was essentially no postmineralization movement in the ore bodies; ore bodies were not faulted apart nor is drag ore associated with the faults.

Ore occurring in the Patterson (Ribbon) limestone appears to have been less dependent on faulting for its localization than in the Austinville member above it but was equally dependent on faults, or the broken rock volumes associated with them, for channelways by which the mineralizing solutions could enter the Ribbon member. While the filling of open spaces was the predominant form of ore emplacement in the Austinville member, the replacement of dolomite, less broken than that in the overlying Austinville, was the primary method of ore emplacement in the Ribbon.

The recrystallization of dolomite probably was the first event in the mineralization cycle in the Austinville dolomitic member in the Austinville-Ivanhoe district. The process appears to have been one of recrystallization and nothing more; though there are some volumes of recrystallized dolomite in which dolomite crystals grow into open space from solid recrystallized dolomite, suggesting that some dolomite may have been deposited from the solutions that promoted the recrystallization. The first metallic mineral to deposit in quantity in the Austinville dolomite was pyrite, which is most abundant in the footwall of the ore and in channels leading to ore bodies. Further confirmation of the early time of pyrite precipitation is furnished by the thin coatings of pyrite that surround most fragments in the rubble and mosaic breccia volumes; these pyrite coatings well may have protected the dolomite so covered from appreciable replacement by the later ore sulfides. Sphalerite appears to have been the next sulfide to deposit, and it also largely was deposited as open-space fillings; some sphalerite, however, replaces (slightly to massively) both breccia matrix, where this exists, and those breccia fragments that were not coated by pyrite. The deposition of silver-poor galena followed and perhaps somewhat overlapped with that of sphalerite. Sphalerite is most abundant in the lower portions of the average ore body, while galena is most strongly developed in the hanging wall rock volumes.

Some dolomite was definitely deposited (at least in part) after the sulfides, since it fills such open spaces as remained after sulfide deposition had essentially been completed. Even later than this certainly gangue dolomite are minor second generations of lighter sphalerite and of galena that are found in cracks in the gangue dolomite. Fluorite and calcite are minor gangue minerals, and barite is even more rare; chalcopyrite has been found, and it is possible that marcasite is present in small amounts.

The sequence of mineralization at Ivanhoe was essentially the same as that at Austinville. Work done on galena-sphalerite ratios and on the locations of pyritic roots of ore bodies strongly suggests that the ore-stage minerals were deposited from solutions that rose from depth. Although no certain source of the ore fluids has been identified, it is probable that they came from a magma chamber at depth and made their way into the Austinville basin by following complex pathways provided by the wide variety of thrust and normal faults present in the district.

The simple and distinctly low-temperature character of the introduced mineral suite in both the Austinville and Patterson members of the Shady formation in the district indicates almost certainly that the ores were deposited under telethermal conditions, and they are so categorized here.

NELSON AND AMHERST COUNTIES

Late Precambrian — *Titanium as Ilmenite, Rutile* — *Magmatic-1a (Rutile), Magmatic-3b (Ilmenite)*

Buddington, A. F., and others, 1955, Thermometric and petrogenic significance of titaniferous magnetite: Amer. Jour. Sci., v. 253, no. 9, p. 497-532

Davidson, D. M., and others, 1946, Notes on the ilmenite deposit at Piney River, Virginia: Econ. Geol., v. 41, p. 738-748

Dietrich, R. V., 1962, Roseland titanium district, in *Southern Field Excursion Guidebook*: 3d General Cong., International Mineralogical Association, Washington, p. 41-42

Evrard, P., 1949, The differentiation of titaniferous magmas: Econ. Geol., v. 44, p. 210-232

Fischer, R., 1950, Entmischungen in Schmelzen aus Schwermetalloxyden, Silikaten und Phosphaten; ihre geochemische und lagerstättenkundliche Bedeutung: Neues Jb. f. Mineral., Abh., Abt. A, Bd. 81, S. 315-364

Herz, N., and others, 1970, Rutile and ilmenite placer deposits, Nelson and Amherst Counties, Virginia: U.S. Geol. Surv. Bull. 1312-F, p. F1-F19

Moore, C. H., Jr., 1940, Origin of nelsonite dikes of Amherst County, Virginia: Econ. Geol., v. 35, p. 629-645

Pegau, A. A., 1950, Geology of the titanium-bearing deposits in Virginia, in Snyder, F. G., Editor, *Symposium on mineral resources of the southeastern United States*: Univ. Tenn. Press, Knoxville, p. 49-55

Philpotts, A. R., 1967, Origin of certain iron-titanium oxide and apatite rocks: Econ. Geol., v. 62, p. 303-315

Rechenberg, H. P., 1955, Zur Genesis der primären Titanerzlagerstätten: Neues Jb. f. Mineral., Mh., H. 4, S. 87-96

Redden, J. A., 1960, Rocks, minerals, and ores of the Piney River-Roseland district, Virginia: Mineral Industries Jour. (Va. Polytech. Inst.), v. 7, no. 2, p. 6-7

Ross, C. S., 1933, Titanium deposits of Roseland district, in *Northern Virginia*: 16th Int. Geol. Cong., Guidebook 11, p. 29-36

—— 1936, Mineralization of the Virginia titanium deposits: Amer. Mineral., v. 21, p. 143-149

—— 1941, Occurrence and origin of the titanium deposits of Nelson and Amherst Counties, Virginia: U.S. Geol. Surv. Prof. Paper 198, 59 p.

—— 1942, The titanium district of Roseland, Virginia, in Newhouse, W. H., Editor, *Ore deposits as related to structural features*: Princeton Univ. Press, p. 137

—— 1947, Virginia titanium deposits: Econ. Geol., v. 42, p. 194-198

Ryan, C. W., 1933, The ilmenite-apatite deposits of west-central Virginia: Econ. Geol., v. 28, p. 266-275

Watson, T. L., 1915, The rutile deposits of the eastern United States: U.S. Geol. Surv., Bull. 580, p. 385-412 (particularly p. 393-401)

Watson, T. L., and Taber, S., 1913, Geology of the titanium and apatite deposits of Virginia: Va. Geol. Surv. Bull. no. 3-A, 308 p. (particularly p. 56-160)

Notes

The center of the titanium mineralization in Nelson and Amherst Counties lies about 37 miles southwest of Charlottesville.

The mineralization is largely confined to a mass of rock some 13 miles long (northeast to southwest) and up to 3 miles wide that is variously described as syenite, anorthosite, and quartz diorite or tonalite; it has even been called a pegmatite. This rock is of a coarsely crystalline texture and in many localities is of a distinctly gneissic character. Throughout the entire rock mass, feldspar is the principal mineral; in the overwhelmingly dominant feldspathic phase, potash feldspar makes up almost 20 percent of the rock and plagioclase (usually andesine) about 68.5 percent. Quartz averages at least 6.0 percent in the feldspathic phase and perhaps appreciably more, while corundum makes up over 1 percent of the rock. The dark minerals of the feldspathic phase, mainly hypersthene (or uralite derived from it), ilmenite, and hematite, provide about 1.5 percent of the total rock; the remainder of the feldspar-rich rock is mainly rutile and apatite. It appears probable that this feldspar-rich rock was accumulated in much the manner of anorthosites, so it is here called by that name; even though, strictly speaking, it does not merit the name. In the minor hornblendic border phase, hornblende (the uralitic hornblende was derived mainly from hypersthene by deuteric or hydrothermal reactions) is a prominent constituent, averaging nearly 12 percent.

The anorthosite mass is enclosed by a biotite-quartz monzonite gneiss that is designated as the Lovington granite gneiss on the 1929 geological map of the state; it is the only other major rock type in the immediate area of the anorthosite. The less abundant rocks of the district are gabbro, nelsonite (to be defined later), and diabase. The gabbro apparently occurs mainly, if not entirely, as dikes, most of which are found in the outer portions of the anorthosite, but a few of which have been noted in the gneiss. Locally, there are some gabbro masses that appear to grade into the anorthosite, but the deep weathering of the gabbros makes the determination of its relations with the surrounding rocks difficult.

Nelsonite is the name given to a group of high-titanium-phosphorus-bearing rocks that occur in dikelike bodies of irregular shapes and widely differing sizes. Most of these dikes are confined to the border zones of the anorthosite, but some have been found in the gneiss outside it but near to the contact of gneiss and anorthosite. The nelsonites are composed of apatite and ilmenite or rutile or both; in some places the rocks contain magnetite; minor amounts of silicates, mainly hornblende and biotite, may or may not be present, but in some examples of nelsonite, particularly gabbro-nelsonite, these silicates are the principal minerals. Pyrite is almost always present.

At present, no evidence firmly fixes the age of the ore-bearing Precambrian rocks within that era; because of the anorthositic character of the main mass of the igneous rocks in which the ore masses are found, it is suggested that they may be of the same Grenville age as the true anorthosites of the Adirondacks farther north. The deposits are, therefore, categorized as late Precambrian.

There have been two principal types of ore: (1) rutile in the anorthosite as disseminated grains of various sizes and as irregular local segregations, and (2) ilmenite and rutile in nelsonite dikes. The two most important mines in the district are: (1) that located on both sides of the Tye River, about one-quarter mile south of the Roseland post office--a disseminated rutile-type deposit formerly mined by the American Rutile Company, and (2) that situated on Piney River, about 1.5 miles northwest of Rose's Mill and 3.5 miles south of Roseland--a rutile gabbro-nelsonite dike now mined by the American Cyanamid Company.

The rutile in the rutile ore in anorthosite is red to reddish-brown, has an adamantine luster, and is remarkably pure; the grains range in size from tiny granules to masses weighing many pounds. The rutile appears to have been formed as a primary mineral of the anorthosite, its irregular distributions in the rock ranging from sparse disseminations to segregations that locally make up 30 percent of the anorthosite; the average grade of the ore worked near Roseland was probably between 4 and 5 percent rutile. None of the rutile shows crystal faces; this strongly suggests that it was crushed during the metamorphism, destroying what crystal faces may have been present. Where ilmenite is found in the rutile-bearing anorthosite, its relationships to the other minerals are similar to those of rutile with which the ilmenite is closely associated. The rutile and its associated ilmenite where that mineral is present probably were precipitated from the residual fluids of the anorthosite after the plagioclase, potash feldspar, and blue quartz of which it is largely composed had been crystallized. Although Ross (1941) believes that the titanium minerals were introduced hydrothermally, the textural relations figured by Watson and Taber (1913) argue strongly for magmatic deposition for the rutile and ilmenite in the anorthosite. Such segregation of rutile as occurred appears to have been a phenomenon of strictly local effect and was produced by the aggregation, over quite short distances, of rutile crystals into larger masses. Rutile deposits of this type, therefore, should be classified as magmatic-1a. Because the dynamic metamorphism that affected the anorthosite after the crystallization of the rutile did not change either the mineral form in which the titanium appears nor the grade of the ore in rutile, the metamorphism is not mentioned in the classification.

Watson in 1907 gave the name nelsonite to a group of high-titanium-phosphorus-bearing rocks that occur in dikelike bodies of a wide range of sizes and shapes in the Nelson and Amherst Counties area. Nelsonite was originally applied to all ilmenite-apatite-rich rocks in the anorthosite, but when the anorthosite area was studied in detail, it was found necessary to define five varieties of nelsonite. The name ilmenite nelsonite was given to what had originally been nelsonite without a modifer; this is the most common and abundant type of nelsonite. The five types of nelsonite are: (1) ilmenite nelsonite, (2) rutile nelsonite, (3) magnetite (and biotite) nelsonite, (4) hornblende nelsonite, and (5) gabbro-nelsonite.

Ilmenite nelsonite consists essentially of an even-granular mixture of ilmenite and apatite, with the grains averaging less than 2 cm in diameter. Ilmenite usually makes up between 60 and 80 percent of the rock. The apatite grains show crystal outlines against the interstitially placed ilmenite that binds the rock together. Pyrite is present both as grains similar to those of ilmenite and in late veinlets with pyrrhotite and chalcopyrite that cut across the even-textured nelsonite; these veinlets are especially common in sheared rock.

It would appear that the nelsonite magma was produced as a late stage differentiate of the gabbro magma. The feldspar-rich anorthosite appears largely to have crystallized at depth and to have been brought into the area as a mush of crystals in which the still-liquid portion was generally rich in the constituents of quartz and rutile as is shown by the late formation of these two minerals in portions of the anorthosite. The gabbro magma, on the other hand, could not have been in equilibrium with the anorthosite since the feldspar in the gabbro is far more calcic than that in the anorthosite. Thus, the gabbro and the anorthosite either did not come from the same source or had quite different geologic histories after they parted company to achieve their present nonequilibrium compositions. In some instances, as is demonstrated by the gabbro dikes, the titanium content of portions of the molten material was not abnormally high, and the material produced by crystallization of these fractions was essentially normal gabbro. Most of the gabbroic magma, however, contained at least enough titanium to produce ilmenite and/or rutile and enough phosphorus to produce apatite in amounts at least sufficiently large to develop a gabbro-nelsonite. Further differentiation of the residual molten gabbroic material appears to have produced actual nelsonite magma that either was formed as an immiscible fraction in the gabbro magma or was the end (or pegmatitic) stage of gabbro crystallization. The gradational relationship of gabbro to ilmenite nelsonite strongly suggests that the nelsonites are pegmatitic phases of gabbro and were not developed as immiscible fractions. In the anorthosite proper, the molten material introduced with the feldspar crystals could produce no more than rutile and rutile-bearing quartz to give the disseminated rutile ore.

There are no masses, consisting essentially of ilmenite and apatite alone, that are intrusive into the rocks that surround them; this argues definitely against the development of a separate ilmenite-rich phase through immiscibility of an iron-titanium-rich fraction in the silicate magma. The nelsonite deposits are, therefore, here categorized as magmatic-3a.

Washington

METALINE

Late Mesozoic (pre-Laramide) *Zinc, Lead* *Telethermal*

Albrethsen, A., and Golding, C., 1953, Geology of the Metaline mining district and mining at the Pend Oreille mine: Compass, v. 30, no. 2, p. 80-87

Bancroft, H., 1911, Lead and zinc deposits in the Metaline district, northeastern Washington: U.S. Geol. Surv. Bull. 470, p. 188-200

Cole, J. W., 1949, Investigation of the Electric Point and Gladstone lead-zinc mines, Stevens County, Washington: U.S. Bur. Mines R.I. 4392, 11 p. (mimeo.)

Dings, McC. G., and Whitebread, D. H., 1965, Geology and ore deposits of the Metaline zinc-lead district, Pend Oreille County, Washington: U.S. Geol. Surv. Prof. Paper 489, 109 p.

Fyles, J. T., and Hewlett, C. G., 1959, Stratigraphy and structure of the Salmo lead-zinc area: B.C. Dept. Mines Bull. no. 41, 162 p.

Huttl, J. B., 1945, Metaline Falls area grows as a zinc producer: Eng. and Min. Jour., v. 146, no. 3, p. 90-93

Jenkins, O. P., 1924, Lead deposits of Pend Oreille and Stevens Counties, Washington: Washington Div. Geol. Bull. no. 31, 153 p.

Little, H. W., 1956, Salmo map-area, British Columbia: Geol. Surv. Canada Paper 50-19, 43 p.

McConnel, R. H., and Anderson, R. A., 1968, The Metaline district, Washington, in Ridge, J. D., Editor, *Ore deposits of the United States, 1933-1967* (Graton-Sales Volumes): Chap. 68, v. 2

Park, C. F., Jr., 1938, Dolomite and jasperoid in the Metaline district, northeast Washington: Econ. Geol., v. 33, p. 709-729

Park, C. F., Jr., and Cannon, R. S., Jr., 1943, Geology and ore deposits of the Metaline quadrangle, Washington: U.S. Geol. Surv. Prof. Paper 202, 81 p.

Walker, J. F., 1934, Geology and mineral deposits of the Salmo map-area, British Columbia: Geol. Surv. Canada Mem. 172, 102 p. (area immediately north of the Metaline quadrangle)

Weaver, C. E., 1920, The mineral resources of Stevens County: Washington Geol. Surv. Bull. no. 20, 350 p.

Weissenborn, A. E., Editor, 1970, Lead-zinc deposits in the Kootenay Arc, northeastern Washington and adjacent British Columbia: Wash. Dept. Nat. Res. Bull., no. 61, 123 p.

Notes

The zinc-lead deposits of the Metaline district occupy a wedge-shaped area, the narrow edge of which points south-southwest. The area covers some 75 square miles in the extreme northeast corner of Washington; the center of the district lies some 80 miles north of Spokane. To the north it borders directly with the Salmo district of British Columbia (see the Canadian section of this volume for notes on the Salmo district). The mountainous area in which both districts lie is known as the Selkirk range, and the principal topographic feature is the broad valley of the northward-flowing Pend Oreille River, a tributary of the Columbia.

The ore deposits of the district lie within a large graben that has dropped Silurian and Devonian rocks down to the level of those of Cambrian age; the faults marking the western edge of the graben have a stratigraphic throw of about 12,000 feet, while those on the east have a minimum stratigraphic throw of 7000 feet. Most of the ore bodies occur in the Middle Cambrian Metaline limestone within the confines of the graben.

The Middle Cambrian Metaline limestone is the dominant rock of the graben and also occurs in two outcrops of considerable size outside of the downthrown rock volume. The Metaline has a thickness of about 5500 feet, a total with which several workers are in agreement; in the Salmo area, this formation is designated as the Nelway and is of comparable thickness. The Metaline is divided into three lithologic units, the lowest of which is mainly thin- to medium-bedded darkish limestone with which is interbedded some limy shale; in the mineralized area it has been so thoroughly altered in the ore-forming process that limestone is seldom seen, the rock having been converted to intermixed crystalline and bedded dolomite in which the bedding has been largely obliterated. The thickness of this lower unit is about 950 feet. The middle member of the Metaline is a fine- to medium-grained light dolomite which contains some beds, lenses, and pods of black dolomite; in places it is well-bedded. Where it is hydrothermally altered it consists of light- to medium-gray fine- to coarse-grained crystalline dolomite; its thickness is at least 3800 feet. Locally this unit contains coarse calcite and jasperoid as well as sulfide minerals; its bedding planes have been almost entirely erased by the alteration. The upper unit of the Metaline is made up of gray massive limestone that is very fine-grained, irregularly mottled, and soft; it ranges in thickness from 1325 to 1775 feet. Hydrothermal alteration has converted the upper unit to a dolomite that has been irregularly and moderately to strongly impregnated with jasperoid; it also contains some coarse calcite, limestone, and locally sulfide minerals; it has been unevenly brecciated.

Although no igneous rocks of Mesozoic age are known in the immediate area of the ore mineralization, outcrops of the Kaniksu batholith cover wide areas within a few miles south of the deposits. The Kaniksu may belong to the Idaho batholith, and both the Nelson and the Idaho batholiths, plus the Kaniksu, may have come from the same general source. The age of the formation of the granite is uncertain; no ages have been determined by Dings and Whitebread (1965), while the age interval given by the Canadians would allow the Kaniksu to be either Upper Jurassic or Lower Cretaceous; it's more likely to be Cretaceous than Jurassic.

It is quite probable that deposits, which show every evidence of having been formed at low temperatures, were produced by solutions that have traveled a considerable distance. It, therefore, seems not unreasonable to assign the source of the Metaline ore fluids to the same magma chamber from which the Kaniksu granites came and to consider the formation of the deposits as of essentially the same age as that of the batholith. This does not completely answer the question of the date of formation of the Metaline ores since the Kaniksu may have been of Upper Jurassic or of Lower or Upper Cretaceous age. As a late Lower Cretaceous age for the batholith seems most likely on the basis of the slim evidence at hand, the ore deposits are here classified as late Mesozoic (pre-Laramide). Obviously, if McConnel and Anderson (1968) are correct in assigning a syngenetic origin to the ores, they must be of the age of the Metaline limestone, that is, Middle Cambrian.

With the exception of one mine (the Oriole) which is in the Cambrian (?) Monk formation, all of the zinc and lead have been mined in the Metaline district from deposits in the Metaline limestone; in the Salmo area, however, important zinc-lead replacement deposits are found in the equivalent of the Lower or Middle Cambrian Maitlen phyllite.

Most of the replacement deposits in the Metaline formation are located in a zone that lies 35 to 200 feet stratigraphically below the contact of the Metaline with the overlying Ledbetter slate. Although in general the ore bodies conform to the bedding of the rock in which they are contained, ore shoots may break across the bedding at a wide variety of angles. The ore-containing replacements are scattered through almost the entire length of the graben, but the most valuable deposits have been found clustered in the rock volumes around the town of Metaline Falls. The deposits are normally most irregular in shape; a few are rudely tabular, and a very few can be categorized as bedding replacements. The great majority, however, occur in zones that approximately parallel the bedding. Some of the more elongate ore bodies appear to have been localized along faults, but these usually change gradually into shapes that roughly follow the beds of the formation.

Most of the ore bodies are found in medium- to dark-gray silicified dolomite breccia in which the degree of brecciation ranges from intense to barely detectable, but some ore is found in massive and unbrecciated dolomite. The breccias are usually made up of fragments of fine- to medium-grained dolomite that are cemented by medium- to coarse-grained dolomite; in places sulfides are found within the breccia masses in association with coarse calcite and jasperoid; only locally does the ore fill, or partially fill, cavities in the host rock. The jasperoid and the sulfides most haphazardly replace both the breccia fragments and the matrix material, although the replacements more commonly occur in the matrix. The association of sulfides and silica (jasperoid) occurs in most of the ore, but some rich sulfide bodies have been found in coarse calcite, in dolomite, or in limestone.

The principal, and usually the only, sulfide minerals are sphalerite and galena, with sphalerite usually being present in about twice the abundance of galena. Pyrite is abundant only locally, and minor tetrahedrite and chalcopyrite appear to be responsible for the small amounts of copper and silver recovered from the ores. The gangue minerals are jasperoid (with a little quartz), dolomite, and calcite.

The conditions under which the ore mineralization in the Metaline district occurred is shown by the small number of sulfides and their simple chemical composition, by the insignificant amounts of silver and copper in the ores, and by the unremarkable type of wall-rock alteration. These characteristics strongly suggest that the ores were deposited under low-intensity conditions but sufficiently far beneath the surface that slow loss of heat and pressure must have prevailed. The Metaline deposits, therefore, are here classified as telethermal. The minor amounts of chalcopyrite and tetrahedrite are compatible with such low-intensity conditions of deposition.

The explanation of the origin of the Metaline ores advanced by McConnel and Anderson (1968) differs drastically from that just given and must be considered in detail by any student of Mississippi Valley-type deposits. These authors believe that the ore sulfides were deposited syngenetically with the sediments, the breccia fragments having been derived from earlier lithified reef material and the sulfide-bearing matrix material from adjacent lagoonal deposits to which the fragments were added by a combination of wave action and slumping.

REPUBLIC

Middle Tertiary | *Gold, Silver* | *Epithermal*

Full, R. P., and Grantham, R. M., 1968, Ore deposits of the Republic mining district, Ferry County, Washington, in Ridge, J. D., Editor, *Ore deposits of the United States, 1933-1967* (Graton-Sales Volumes): Chap. 69, v. 2

Lindgren, W., and Bancroft, H., 1914, The Republic mining district, in Bancroft, H., *The ore deposits of northeastern Washington*: U.S. Geol. Surv. Bull. no. 550, p. 133-166

Muessig, S. J., 1962, Tertiary volcanic and related rocks of the Republic area, Ferry County, Washington: U.S. Geol. Surv. Prof. Paper 450-D, p. D56-D58

—— 1967, Geology of the Republic quadrangle and a part of the Aeneas quadrangle, Ferry County, Washington: U.S. Geol. Surv. Bull. 1216, 135 p.

Parker, R. L., and Calkins, J. A., 1964, Geology of the Curlew quadrangle, Ferry County, Washington: U.S. Geol. Surv. Bull. 1169, 95 p.

Staatz, M. H., 1960, The Republic graben, a major structure in northeastern Washington: U.S. Geol. Surv. Prof. Paper 400-B, p. B304-B306

Umpleby, J. B., 1910, Geology and ore deposits of Republic mining district: Washington Geol. Surv. Bull. 1, 67 p.

Notes

The Republic district is located in the northern part of Ferry County in northeastern Washington; the town of Republic, in the center of the graben of the same name, is about 90 miles northwest of Spokane.

The Republic graben is from 6 to 10 miles wide and extends south from the Canadian border for at least 50 miles. Within the district, ores are found only in Eocene and Oligocene rocks. The oldest of these Tertiary beds are those of the O'Brien formation; the rocks are volcanic ash with some rounded quartz grains and contain irregularly distributed fragments of argillite and slate that probably came from adjacent Paleozoic rocks along the edge of the graben. Conformably overlying the O'Brien are the Sanpoil volcanic rocks, rhyodacite and quartz latite flows with some tuff and flow breccia. This unit is thickest in the area of the main mineralized zone. These two formations were intruded by the Scatter Creek rhyodacite. It has definite intrusive contacts with the O'Brien but generally indefinite or gradational ones with the green andesite of the Sanpoil; it is widely distributed throughout and near the graben. Intrusive equivalents of the upper portion of the Sanpoil also probably exist in the area. The next beds in the stratigraphic sequence are those of the Klondike Mountain formation that is confined to the northwest part of the graben; it consists of three members, water-laid tuffs, coarse pyroclastics, and basalt flows, respectively. Dacite dikes and small intrusive masses cut all of the older Tertiary formations; these latter rocks are thought by Full and Grantham (1968) to be younger than the ores.

The ores are found in the various Tertiary formations from the upper part of the O'Brien Creek formation through the basal unit of the lake beds of the Klondike Mountain formation; they also are common in the Scatter Creek rhyodacite. The most productive beds, however, are the green andesite flows in the lower part of the Sanpoil formation. In the Klondike Mountain Lake beds, ore emplacement was not controlled by fractures as in the lower beds; instead, the ores were disseminated through a 50-foot-thick rubble mass that has been mined by open pit methods. The lack of fractures extending from the Sanpoil into the more massive flows in the upper parts of the Klondike Mountain probably account for the lack of ore in these beds; the ores well may have been emplaced after these flows had formed but this is not certain. The absence of any igneous activity in the district after the Oligocene strongly suggests that the ore fluids came from the same general source as the volcanic rocks and the intrusive rhyodacite. The age of ore formation, therefore, probably was late Oligocene or earliest Miocene at the latest; the ores, thus, are best dated as middle Tertiary.

The principal ore veins of the district are concentrated along a northwest-trending fault (Eureka) and along structures closely related to it. This zone runs from the Quilp mine on the southeast to the Mountain Lion mine on the northwest. Although the veins are offset, the offsetting apparently took place prior to ore deposition. In general, the veins are quite simple, except for the veins in Knob Hill No. 2 mine where branching and overriding veins occur near the surface; with depth, however, the vein becomes more continuous, the various vein elements merging together. Most of the vein material is fine-grained chalcedonic or porcelainlike quartz; at least two generations of quartz are known. Calcite is found in most of the veins but generally is associated with ore below minable grade. Both fracture filling and replacement entered into the process of ore deposition. The better-grade ore is marked by thinly laminated banding of the vein material, the color of the bands being due to finely divided sulfides and other ore minerals. Electrum is the most abundant ore mineral, but native silver, naumannite, stephanite, and either pyrargyrite or proustite also are known. Pyrite is the most common gangue sulfide, although some chalcopyrite is present. Silicification of the volcanic rocks is strong at the surface, over most of the ore veins, and normally is found along the veins at depth. Adularia is present in essentially all of the veins. In the disseminated ore in the lake bed rubble, sericite is abundant but is not common along the veins below; chlorite, epidote, calcite, and disseminated pyrite are found in the hydrothermally altered portions of the volcanic rocks. The mineralization and wall-rock alteration are typical of deposition in the epithermal range, and ore formation probably took place under conditions of rapid loss of heat and pressure. The ores are confined to a vertical range of about 2000 feet, and, although this conforms to epithermal deposition, the vertical range is greater than it was thought to be in Lindgren's time. The evidence strongly favors the classification of the ores as epithermal.

STEVENS COUNTY MAGNESITE

Late Mesozoic to Early Tertiary — *Magnesite* — *Hypothermal-2 to Mesothermal*

Bain, G. W., 1924, Types of magnesite deposits and their origin: Econ. Geol., v. 19, p. 412-433 (particularly p. 422-423; but the entire paper should be read)

Bennett, W.A.G., 1941, Preliminary report on magnesite deposits of Stevens County, Washington: Washington Div. Geol. R.I. no. 5, 25 p.

—— 1943, Character and tonnage of the Turk magnesite deposit: Washington Div. Geol. R.I. no. 7, 22 p.

Campbell, I., and Loofbourow, J. S., Jr., 1962, Geology of the magnesite belt of Stevens County, Washington: U.S. Geol. Surv. Bull. 1142-F, p. F1-F53

Jenkins, O. P., 1918, Notes on the possible origin of the magnesite near Valley, Washington: Econ. Geol., v. 13, p. 381-384

Jones, R.H.B., 1928, Notes on the geology of the Chewelah quadrangle, Stevens County, Washington: Northwest Science, v. 2, p. 111-116

Schroeder, M. C., 1948, The genesis of the Turk magnesite deposit of Stevens County, Washington: Compass, v. 26, no. 1, p. 37-46

Siegfus, S. S., 1927, Some geological features of the Washington magnesite deposits: Eng. and Min. Jour., v. 124, no. 22, p. 853-857; disc. p. 858

Weaver, C. E., 1920, The mineral resources of Stevens County: Washington Geol. Surv. Bull. no. 20, 350 p. (particularly p. 319-331)

Whitwell, G. E., and Patty, E. N., 1921, The magnesite deposits of Washington; their occurrence and technology: Washington Geol. Surv. Bull. no. 25, 194 p. (particularly p. 11-58)

Notes

The magnesite deposits in Stevens County are located irregularly in the central 20 miles of a belt of closely folded metasedimentary and metavolcanic rocks and metamorphosed and unmetamorphosed intrusive rocks about 33 miles long and 2 to 5 miles wide; the center of the belt is approximately 40 miles northwest of Spokane.

The oldest rocks in the district make up the Deer Trail group; these rocks, and the overlying Huckleberry formation, probably are the local representatives of the late Precambrian Belt series to which they bear a marked resemblance. The basal rocks of the Deer Trail group are known as the Togo formation, which is mainly slaty argillite. The Togo is followed abruptly, but conformably, by the Edna dolomite. The Togo is conformably overlain by the McHale slate. The Stensgar dolomite lies conformably on the McHale slate; the Stensgar contains all the known deposits of magnesite in the district. The Stensgar is overlain conformably by the Buffalo Hump formation.

The Deer Trail group is overlain above an erosional unconformity by the Huckleberry formation that is divided into a lower Conglomerate member and an upper Greenstone member. It contains no magnesite fragments, indicating that the magnesite mineralization occurred later than Deer Trail time.

The basal Lower Cambrian Addy quartzite is markedly unconformable on the Huckleberry formation. In the magnesite belt, no sedimentary or extrusive rocks younger than the Addy were developed until the extrusion of the flows, tuffs, and breccias of the Oligocene Jerome hornblende andesite.

The oldest intrusive rocks in the district are greenstone dikes and sills that may well have been feeders for the original basalt flows of the upper member of the Huckleberry formation. In the southern part of the area, where the Greenstone member is lacking, such dikes and sills are scarce. This indicates that the intrusive and extrusive rocks came from the same source. The greenstone dikes and sills do not penetrate any rocks above those of the Huckleberry formation and may be essentially lacking in the Huckleberry itself.

The Loon Lake granite (in large part quartz monzonite and granodiorite) is a batholitic-sized mass that has been found at both ends of magnesite belt. Lead-alpha age determinations on zircons from the equivalent of the Loon Lake granite south of the magnesite belt indicate that it was introduced in Late Cretaceous time. Because of the uncertainty attached to the results of this method, the granite cannot be certainly assigned this age; the granite, however, shows no evidence of metamorphism, and it is probable that it was introduced in Laramide time.

The magnesite, within the folded and faulted Stensgar formation, has not been affected by the regional deformation and was, therefore, emplaced after that tectonic event. Although the magnesite is confined to the Stensgar beds, the erratic distribution of that mineral in the formation and the large crystals and the veinlets of magnesite cutting the dolomite indicate that the magnesite was not syngenetic in origin. Assuming, therefore, that the magnesite was the result of hydrothermally induced replacement reactions, the most probable source of the ore-forming solutions involved would have been the magma chamber from which the Loon Lake granite came. The Stevens County magnesite deposits, therefore, are here classified as late Mesozoic to early Tertiary.

Most of the major magnesite deposits are found in the north-central portion of the belt, but two important bodies are known in the southern part. Whether or not the magnesite deposits are all located in the same horizon in the Stensgar dolomite is uncertain. The more impure Edna dolomite (26.2 percent insoluble as opposed to 4.9 percent in the Stensgar) was not susceptible to conversion to magnesite. Because of the lack of metamorphic effects on the magnesite bodies, they apparently were formed so late in the sequence of geologic events in the district that the hydrothermal fluids by which they probably were produced were most likely to have come from the same magmatic hearth as the Loon Lake granite. The thoroughgoing faulting of the district probably provided the channelways necessary for the upward movement of the ore fluids. The suggestion has been made (Campbell and Loofbourow, 1962) that the magnesium needed to convert the dolomite to magnesite may have been obtained through the removal of magnesium from dolomite at greater depths than those at which magnesite was deposited. The lack in the district, so far as is known, of dolomite from which magnesium has been removed argues against this concept.

Although most of the deposits consist of little else than magnesite that has replaced dolomite, the Turk deposit also contains forsterite, magnetite, and pyrrhotite. In some of the deposits, brucite lenses were found that postdated the magnesite; the brucite was accompanied by minor amounts of serpentine, chlorite, hydromagnesite, and dolomite. The hydromagnesite, at least, appears to have been formed after the brucite.

From the minerals associated with the magnesite, it seems probable that some of it (in the Turk deposit) may have been emplaced under hypothermal conditions but that most of it was formed in the mesothermal intensity range. The deposits are, therefore, here categorized as hypothermal-2 to mesothermal.

Wisconsin-Illinois-Iowa

UPPER MISSISSIPPI VALLEY

Late Mesozoic (pre-Laramide) *Zinc, Lead* *Telethermal*

Agnew, A. F., 1955, Application of geology to the discovery of zinc-lead ore in the Wisconsin-Illinois-Iowa district: A.I.M.E. Tr., v. 202, p. 781-795 (in Min. Eng., v. 7, no. 8)

—— 1963, Geology of the Plattville quadrangle, Wisconsin: U.S. Geol. Surv. Bull. 1123-E, p. 245-277

Agnew, A. F., and others, 1956, Stratigraphy of Middle Ordovician rocks in the zinc-lead district of Wisconsin, Illinois, and Iowa: U.S. Geol. Surv. Prof. Paper 274-K, p. 251-312

Allingham, J. W., 1963, Geology of the Dodgeville and Mineral Point quadrangles, Wisconsin: U.S. Geol. Surv. Bull. 1123-D, p. 169-244

Bain, H. F., 1906, Zinc and lead deposits of the Upper Mississippi Valley: U.S. Geol. Surv. Bull. 294, 155 p.

Bailey, S. W., and Cameron, E. N., 1951, Temperatures of mineral formation in bottom-run lead-zinc deposits of the Upper Mississippi Valley, as indicated by liquid inclusions: Econ. Geol., v. 46, p. 626-651

Barnes, H. L., 1959, The effect of metamorphism on metal distribution near base metal deposits: Econ. Geol., v. 54, p. 9191-9943 (particularly p. 932-934)

—— 1967, Sphalerite solubility in ore solutions of the Illinois-Wisconsin district, in Brown, J. S., Editor, *Genesis of stratiform lead-zinc-barite-fluorite deposits--a symposium*: Econ. Geol. Mono. 3, p. 326-332

Behre, C. H., Jr., 1939, Detailed structure in the Wisconsin-Illinois-Iowa lead-zinc district, in Bastin, E. S., Editor, *Lead and zinc deposits of the Mississippi Valley region*: Geol. Soc. Amer. Spec. Paper no. 24, p. 67-70

—— 1939, (Paragenesis of the) Wisconsin-Illinois-Iowa lead-zinc district, in Bastin, E. S., Editor, *Lead and zinc deposits of the Mississippi Valley region*: Geol. Soc. Amer. Spec. Paper no. 24, p. 114-118

—— 1942, The Upper Mississippi Valley lead-zinc district, in Newhouse, W. H., Editor, *Ore deposits as related to structural features*: Princeton Univ. Press, p. 220-221

Behre, C. H., Jr., and others, 1937, The Wisconsin lead-zinc district--preliminary paper: Econ. Geol., v. 32, p. 783-809

Bradbury, J. C., 1959, Crevice lead-zinc deposits of northwestern Illinois: Ill. Geol. Surv. R.I. 210, 49 p.

—— 1960, A structural analysis of the northwestern Illinois zinc-lead district: Ill. Acad. Sci. Tr., v. 53, no. 1, 2, p. 20-24

—— 1961, Mineralogy and the question of zoning, northwestern Illinois zinc-lead district: Econ. Geol., v. 56, p. 132-146

Bradbury, J. C., and others, 1956, Geologic structure map of the northwestern Illinois zinc-lead district: Ill. State Geol. Surv. Circ. 214, 7 p.

Brown, C. E., and Whitlow, J. W., 1960, Geology of the Dubuque South quadrangle, Iowa-Illinois: U.S. Geol. Surv. Bull. 1123-A, p. 1-93

Brown, C. E., and others, 1957, Geology and zinc-lead deposits in the Catfish Creek area, Dubuque County, Iowa: U.S. Geol. Surv. Mineral Invest., Field Studies Map MF-116, 1:12,000

Carlson, J. E., 1961, Geology of the Montfort and Linden quadrangles, Wisconsin: U.S. Geol. Surv. Bull. 1123-B, p. 95-138

Chamberlin, T. C., 1882, The ore deposits of southwestern Wisconsin, in *Geology of Wisconsin*: Geol. Surv. Wisc., v. 4, pt. 4, p. 365-571

Cox, G. H., 1911, Origin of the lead and zinc ores of the Upper Mississippi Valley district: Econ. Geol., v. 6, p. 427-448, 582-603

—— 1914, Zinc and lead deposits of northwestern Illinois: Ill. State Geol. Surv. Bull 21, 120 p.

DeGeoffroy, J., and Wignall, T. K., 1970, Statistical decision in regional exploration: application of regression and Bayesian classification analysis in the southwest Wisconsin zinc area: Econ. Geol., v. 65, p. 769-777

DeGeoffroy, J., and others, 1968, Selection of drilling targets from geochemical data in the southwest Wisconsin zinc area: Econ. Geol., v. 63, p. 787-795

Emmons, W. H., 1929, The origin of the deposits of sulphide ores of the Mississippi Valley: Econ. Geol., v. 24, p. 221-271 (particularly p. 256-261)

Erickson, A. J., Jr., 1965, Temperatures of calcite deposition in the Upper Mississippi Valley lead-zinc deposits: Econ. Geol., v. 60, p. 506-528

Grant, U. S., 1906, Report on the lead and zinc deposits of Wisconsin: Wisc. Geol. and Nat. Hist. Surv. Bull. 14, 100 p.

Hall, W. E., and Friedman, I., 1963, Composition of fluid inclusions, Cave-in-Rock fluorite district, and Upper Mississippi Valley zinc-lead district: Econ. Geol., v. 58, p. 886-911

Hall, W. E., and Heyl, A. V., 1968, Distribution of minor elements in ore and host rock, Illinois-Kentucky fluorite district and Upper Mississippi Valley zinc-lead district: Econ. Geol., v. 63, p. 655-670

Heyl, A. V., 1964, Enargite in the zinc-lead deposits of the Upper Mississippi Valley district: Amer. Mineral., v. 49, p. 1458-1461

—— 1968, The Upper Mississippi Valley base-metal district, in Ridge, J. D., Editor, *Ore deposits of the United States, 1933-1967* (Graton-Sales Volumes): Chap. 21, v. 1.

Heyl, A. V., Jr., and Behre, C. H., Jr., 1950, Upper Mississippi Valley district: 18th Int. Geol. Cong. Rept., pt. 7, p. 61-68

Heyl, A. V., and King, E. R., 1966, Aeromagnetic and tectonic analysis of the Upper Mississippi Valley zinc-lead district: U.S. Geol. Surv. Bull. 1242-A, p. A1-A16

Heyl, A. V., Jr., and others, 1952, Geologic structure map of the Beetown lead-zinc area, Grant County, Wisconsin: U.S. Geol. Surv. Mineral Invest., Field Studies Map MF-3, 1:12,000

—— 1955, Zinc-lead-copper resources and general geology of the Upper Mississippi Valley district: U.S. Geol. Surv. Bull. 1015-G, p. 227-245

—— 1959, The geology of the Upper Mississippi Valley zinc-lead district: U.S. Geol. Surv. Prof. Paper 309, 310 p.

—— 1966, Isotopic study of galenas from the Upper Mississippi Valley, the Illinois-Kentucky, and some Appalachian Valley mineral deposits: Econ. Geol., v. 61, p. 933-961 (particularly p. 942-947)

Hosterman, J. W., and others, 1964, Qualitative x-ray emission analysis studies of enrichment of common elements in wallrock alteration in the

Upper Mississippi Valley zinc-lead district: U.S. Geol. Surv. Prof. Paper 501-D, p. D54-D60

Jolly, J. L., and Heyl, A. V., 1968, Mercury and other trace elements in sphalerite and wallrocks from central Kentucky, Tennessee and Appalachian zinc districts: U.S. Geol. Surv. Bull. 1252-F, p. F1-F20

Kay, G. M., 1939, (Stratigraphy of the) Wisconsin-Illinois district, in Bastin, E. S., Editor, *Lead and zinc deposits of the Mississippi Valley region*: Geol. Soc. Amer. Spec. Paper no. 24, p. 25-29

Kennedy, V. C., 1956, Geochemical studies in the southwestern Wisconsin zinc-lead area: U.S. Geol. Surv. Bull. 1000-E, p. 187-223

Klemic, H., and West, W. S., 1964, Geology of the Belmont and Calamine quadrangles, Wisconsin: U.S. Geol. Surv. Bull. 1123-G, p. 361-435

Leith, C. K., 1932, Structures of the Wisconsin and Tri-State lead and zinc deposits: Econ. Geol., v. 27, p. 405-418 (particularly p. 406-410)

Mullens, T. E., 1964, Geology of the Cuba City, New Diggings, and Shullsburg quadrangles, Wisconsin and Illinois: U.S. Geol. Surv. Bull. 1123-H, p. 457-531

Ohle, E. L., Jr., 1959, Some considerations in determining the origin of ore deposits of the Mississippi Valley type: Econ. Geol., v. 54, p. 769-789

Reynolds, R. R., 1958, Factors controlling the localization of ore deposits in the Shullsburg area, Wisconsin-Illinois zinc-lead district: Econ. Geol., v. 53, p. 141-163

Spurr, J. E., 1924, Upper Mississippi lead and zinc ores: Eng. and Min. Jour., v. 117, no. 6, 7, p. 246-250, 287-292

Taylor, A. R., 1964, Geology of the Rewey and Mifflin quadrangles, Wisconsin: U.S. Geol. Surv. Bull. 1123-F, p. 279-360

West, W. S., and Klemic, H., 1961, Relation of fold structures to distribution of lead and zinc mineralization in the Belmont and Calamine quadrangles, Lafayette County, Wisconsin: U.S. Geol. Surv. Prof. Paper 424-D, p. D9-D12

Whitlow, J. W., and Brown, C. E., 1963, Geology of the Dubuque North quadrangle, Iowa, Wisconsin, Illinois: U.S. Geol. Surv. Bull. 1123-C, p. 139-168

Whitlow, J. W., and West, W. S., 1966, Geology of the Potosi quadrangle, Grant County, Wisconsin, and Dubuque County, Iowa: U.S. Geol. Surv. Bull. 1123-I, p. 533-571

Willman, H. B., and Reynolds, R. R., 1947, Geologic structure of the zinc-lead district of northwestern Illinois: Ill. State Geol. Surv. R.I. 124, 15 p.

Willman, H. B., and others, 1946, Geological aspects of prospecting and areas for prospecting in the zinc-lead district of northwestern Illinois: Ill. State Geol. Surv. R.I., no. 116, 48 p.

Zimmerman, R. A., 1969, Sediment-ore-structure relations in barite and associated ores and sediments in the Upper Mississippi Valley zinc-lead district near Shullsburg, Wisconsin: Mineralium Deposita, v. 4, no. 3, p. 248-259

Notes

The zinc-lead deposits of the Upper Mississippi Valley occur within an area of some 4000 square miles, most of which is located in southwestern Wisconsin but which also includes the northern half of Jo Daviess County in northwestern Illinois and a narrow strip of eastern Iowa centering about the town of Dubuque; except for its eastern and western borders, the district lies entirely within the Driftless area. The center of the district is located between Platteville and Shullsburg in Wisconsin and is situated about 60 miles southwest of Madison, the capital of the state. Recent production has been essentially limited to a belt 650 square miles in extent that runs north-south through the center of the district.

Although Precambrian rocks have been reported as cut in well drilling, the oldest rocks certainly identified in the district are those of the Upper Cambrian Mount Simon sandstone. The Mount Simon is overlain by the Eau Claire sandstone; this formation is overlain by the Dresbach sandstone. The Dresbach is, in turn, overlain by the Franconia sandstone; it is followed by the Trempealeau formation. The uppermost beds of the Trempealeau are known as the Jordan sandstone member. These Upper Cambrian beds are exposed only along the northern and northeastern portions of the district although they have been cut by wells farther south. Small deposits of zinc-lead minerals have been found in these Cambrian strata, but the ore mined from them has been negligible.

The Lower Ordovician beds in the district are those of the Prairie du Chien group which, in places, can be divided into three members: (1) the Oneota dolomite, the oldest of the three; (2) the New Richmond sandstone; and (3) the Shakopee dolomite. In other areas, no sandstone member has been found, making the division impossible. Mining in the Prairie du Chien group in Wisconsin has been confined to its exposures along the northern edge of the district; some zinc and lead have been mined from these beds in Iowa, and throughout the area they merit further prospecting. The lowest beds of the Middle Ordovician are those of the St. Peter sandstone, which is exposed in areas of deep dissection and which at least locally lies unconformably on the Prairie du Chien. The St. Peter is conformably overlain by the Platteville formation that is known throughout the district and is divided into four members, from bottom to top: (1) the Glenwood sandy shale, (3) the Pecatonica (quarry beds), (3) the McGregor member, and (4) the Quimbys Mill (glass rock). Although some zinc-lead mineralization has been found in all four members, only the upper two contain extensive deposits. Disconformably above the Platteville is the Decorah formation that is divided into three members that differ in thickness and character from place to place. The three members are, from oldest to youngest: (1) the Spechts ferry (clay bed), (2) the Gutenberg (oil rock), and (3) the Ion member. The Gutenberg member contains zinc and lead ore in many places while the Ion member with the overlying Galena dolomite make up the host rocks of most pitch and flat deposits. The Decorah is conformably overlain by the Galena dolomite, which is the youngest of the Middle Ordovician formations and is dolomite but grades westward into limestone, particularly in its lower beds. Mainly on paleontologic grounds it has been divided into three members, from bottom to top: (1) the Prosser cherty member, (2) the Stewartville massive member, and (3) the Dubuque shaly member. Heyl and others (1959) have divided the Galena on lithology into a cherty unit (below) which includes the cherty portion of the Prosser and noncherty unit (above) which contains the remainder of the formation. The Galena has been importantly mineralized by lead and zinc; most of the lead produced in the district was mined from this formation.

The Upper Ordovician is represented in the district by the Maquoketa shale; it is apparently conformable on the Galena and is found only below, and at the bases of, erosional remnants known as mounds. In the Maquoketa, no minable deposits of lead and zinc minerals have been found. A few rocks of Middle and Upper Silurian age are known in the remnant mounds and along the southern and southwestern margins of the district lying disconformably above the Maquoketa. In the district there are no younger consolidated rocks than those of Silurian age, and the unconsolidated ones were formed long after the ores had been deposited; they are of no value as sources of metallic minerals.

No igneous rocks that could be designated as post-Precambrian in age have ever been found in the district, and even the reports of Precambrian "granite" in wells at depths of over 1700 feet are not confirmed.

It is the opinion of Heyl and others (1959) that the various types and orders of structures developed in the district were essentially the result of one period of regional tectonic deformation although, since this deformation, the Upper Mississippi Valley has been slightly uplifted and tilted. The deformation, due to the lack of most Pennsylvanian and post-Pennsylvanian rocks in the general region cannot be dated more exactly than to say that it was of probable post-middle Pennsylvanian and pre-Upper Cretaceous in age. Although the Appalachian and Ouachita uplifts were late Paleozoic, there is some question as to whether or not the forces that caused these uplifts also produced the much less impressive structures that contain the ore deposits of the Upper Mississippi Valley zinc-lead district. For a variety of reasons, it has been suggested in the discussions of the Southern Illinois fluorite deposits, of the Leadbelt deposits of southeastern Missouri, and of the Magnet Cove barite bodies that they were mineralized in the Cretaceous at a time that can best be defined as pre-Laramide. Rather than appeal to a late Paleozoic period of ore mineralization that affected only the Upper Mississippi Valley district and not the remainder of the central United States, it is thought best here also to classify the deposits of this district as late Mesozoic (pre-Laramide). It is, however, fully recognized that not even the indirectly applicable evidence pertaining to the deposits to the south can be produced to defend this age dating of the ores of the Upper Mississippi Valley district; they may yet be found to be late Paleozoic in age.

Although the Upper Mississippi Valley district covers an area of some 4000 square miles, the district is composed of a large number of well-mineralized sections separated by considerable areas of largely unmineralized rock. The largest of the mining areas is the triangle, some 280 square miles in area, bounded by lines connecting Platteville on the north, Shullsburg on the east, and Galena on the south. In fact, most of the zinc production of the district and a considerable part of that of lead have come from this region. The other ore-bearing regions are smaller in area and have provided lesser proportions of the total production of the district. The ore deposits within the Middle Ordovician rocks have been classified into three types: (1) gash vein deposits that are confined to the Galena dolomite and contain more galena than sphalerite, some chalcopyrite, and smithsonite derived from the sphalerite; (2) pitch-and-flat deposits that are in the lower part of the Galena, the Decorah, and the upper part of the Platteville formations and contain sphalerite as the main ore mineral with abundant pyrite and marcasite and galena in appreciably smaller amounts than sphalerite. Barite is the main ore mineral in a few deposits in the central part of the district. Much of the pitch-and-flat ore is in the reverse and bedding-plane fractures that provide the pitches and flats, respectively, but considerable amounts of this ore type were emplaced by replacement of the rocks adjacent to the fractures, mainly in the more shaly bands (as is true in the Missouri Leadbelt). All gradations between true vein-type deposits and true disseminated deposits are known--disseminated deposits are more common in the northern part of the district where fracturing is less intense than farther south; and (3) placer and residual deposits that have been most useful as guides to major ore.

The mineralogy of the deposits is somewhat more complex than that of the average low-temperature zinc-lead deposit, but most of the minerals were formed during only one generation. The earliest mineral was silica that replaced appreciable volumes of the Prairie du Chien, as well as rocks of other formations, as chert, jasperoid, or "cotton rock," or was deposited as crystalline quartz in open space. The silica is particularly abundant near the main fractures within the ore bodies. Dolomitization commenced after silicification had ceased and replaced limestones in the Platteville and Decorah formations and was deposited as crystalline pink dolomite in vugs or as fine-grained vein fillings; it, as is true of the silica, is confined to the ore bodies or their immediate walls. The pink dolomite is most abundant in ore bodies in the Prairie du Chien group, is less so in the overlying pitch-and-flat deposits, and is rare or absent in the gash veins.

Perhaps even during silicification and dolomitization, and certainly for a considerable time thereafter, the ore solutions were dissolving limestone and, to a lesser extent, dolomite from the wall rocks within and around the ore bodies; the altered rocks that resulted were porous, vuggy, and sandy and were a favorable host rock for ore. Appreciable carbonate solution had taken place before the first sulfide, pyrite, began to deposit, for it often lines vugs in the rocks and has recemented the sanded carbonate material. The solution process went on uninterruptedly until the last of the later marcasite had been formed and continued in a local and spasmodic manner through the various stages of calcite deposition. Thus, although the solutions were capable of precipitating sulfides and barite, they were also able to dissolve carbonate, particularly calcite, at the same time.

Pyrite was the first sulfide to be deposited, apparently not beginning until after dolomite precipitation had ceased; it usually formed a thin coating on the walls of the fractures and was disseminated in the wall rock itself. Although the pyrite deposited over a long period of time, most of it formed early and late during its period of deposition. Marcasite began to precipitate after the first pyrite films had formed on the rock fragments; perhaps the coating of the wall rocks with pyrite barred the solutions from reacting with them and changed the chemical conditions of the ore-forming system so that marcasite was stable instead of pyrite. Marcasite deposition continued until after that of all other sulfides had ceased but with most of it forming after sphalerite had stopped precipitating. A little barite deposited at about the same time as the early marcasite, and much of it appears to have been later redissolved in the ore fluids. Sphalerite was the next mineral to begin to form and was usually the most abundant one in the deposits; it normally forms two bands on opposite sides of the veins. Much of the early sphalerite is darker colored than that formed later, and in places the darker-colored sphalerite was fractured before the lighter began to precipitate. It is considered possible that some of the zinc sulfide is wurtzite, but this point does not appear to have been fully studied. A little pyrite and marcasite are intergrown with the sphalerite, and a little cobaltite or safflorite (both isomorphous with marcasite) may have formed at the same time as the marcasite. Galena began to deposit not long after sphalerite and continued for some time after sphalerite had ceased to precipitate, with the maximum quantity being formed after most of the sphalerite; galena ceased to deposit before marcasite precipitation was completed. A second generation of barite followed the galena precipitation and was completed before that of marcasite had terminated; barite precipitation barely overlapped with that of calcite. Although traces of chalcopyrite were formed with sphalerite and galena, most of it was deposited after barite and contemporaneously with the primary, hypogene low-temperature millerite. The final stage of mineralization was that of calcite that appears to have formed in four substages; most of the calcite was deposited after the sulfide mineralization had been completed. Each substage can be recognized by the crystal forms peculiar to it, and some etching of the crystals of each of the first three substage calcites seems to have taken place before the next began to form. Although some minor fracturing took place during the course of ore deposition, there appears to have been no postmineral fracturing of any importance.

The minerals of the deposit are typical of those formed under conditions of low intensity; none of the minerals requires that conditions higher than telethermal prevailed during ore formation. The silicification and dolomitization are normal processes in the formation of low-temperature deposits, nor is the solution of carbonate during the ore-forming cycle unique. Chalcopyrite in abundance is undoubtedly indicative of higher chemical intensity than that of the telethermal range, but the amount of that mineral in the Upper Mississippi Valley deposits is not sufficient to warrant a higher-intensity classification for even part of the deposits. The same is true of barite. There is not enough of the questionable cobaltite-smalltite to affect the classification of the deposits. Although the cover overlying the ore-bearing formations during mineralization cannot have been thick, the impervious character of the Maquoketa formation and the lack of fractures in it probably insured the ore fluids against a rapid loss of heat and pressure. It is, therefore, reasonable to categorize the Upper Mississippi Valley deposits as telethermal.

Wyoming

IRON MOUNTAIN

Late Precambrian | *Iron as Magnetite, Titanium as Ilmenite* | *Magmatic-3b*

Ball, S. H., 1907, Titaniferous iron ore of Iron Mountain, Wyoming: U.S. Geol. Surv. Bull. 315, p. 206-212

Condie, K. C., 1969, Petrology and geochemistry of the Laramie batholith and related metamorphic rocks of Precambrian age, eastern Wyoming: Geol. Soc. Amer. Bull., v. 80, p. 57-82

Darton, N. H., and others, 1910, Laramie-Sherman folio, Wyoming: U.S. Geol. Surv. Geol. Atlas, Folio 173, 17 p. (the Sherman quadrangle extends north through the lowest tier of sections in Township 18N and its northern boundary, therefore, is less than 4 miles south of the Shanton deposits; this reference provides the principal published material on the sedimentary rocks that border the igneous complex of the Laramie Range)

Diemer, R. A., 1941, Titaniferous magnetite deposits of the Laramie Range, Wyoming: Wyo. Geol. Surv. Bull. 31, 23 p.

Fowler, K. S., 1930, The anorthosite area of the Laramie mountains, Wyoming: Amer. Jour. Sci., 5th Ser., v. 19, p. 305-315, 373-403

Frey, E., 1946, Exploration of Iron Mountain titaniferous magnetite deposits, Albany County, Wyoming: U.S. Bur. Mines R.I. 3968, 37 p. (mimeo.)

—— 1946, Exploration of the Shanton iron-ore property, Albany County, Wyoming: U.S. Bur. Mines R.I. 3918, 5 p. (mimeo.)

Hagner, A. F., 1968, The titaniferous magnetite deposit at Iron Mountain, Wyoming, in Ridge, J. D., Editor, *Ore deposits of the United States, 1933-1967* (Graton-Sales Volumes): Chap. 31, v. 1

Hodge, D. S., 1966, Preliminary gravity study of the southern Laramie Mountains: anorthosite areas and adjacent basins: Univ. Wyo. Contribs. to Geol., v. 5, no. 1, p. 55-62

Kemp, J. F., 1905, Die Lagerstätten titanhaltigen Eisenerzes in Laramie Range, Wyoming, Ver. Staaten: Zeitsch. f. prakt. Geol., Jg. 13, H. 2, S. 71-80

Klugman, M. A., 1960, Laramie anorthosite: Geol. Soc. Amer. Guidebook for Field Trips (Guide to the geology of Colorado), Field Trip B-2, p. 223-227

—— 1966, Résumé of the geology of the Laramie anorthosite mass: Mtn. Geol., v. 3, no. 2, p. 75-84; 1967, v. 4, no. 1, p. 35-36

Lindgren, W., 1902, A deposit of titanic iron ore from Wyoming: Science, n.s., v. 16, p. 984-985

Newhouse, W. H., and Hagner, A. F., 1957, Geologic map of anorthosite areas, southern part of Laramie Range, Wyoming: U.S. Geol. Surv. Mineral Invest., Field Studies Map MF-119, 1:63,360

Singewald, J. T., Jr., 1913, Titaniferous iron ores of the United States: U.S. Bur. Mines Bull. 64, 145 p. (particularly p. 111-125)

—— 1933, Titaniferous magnetites--in Wyoming, in *Ore deposits of the western states* (Lindgren Volume): A.I.M.E., p. 509-510

Smithson, S. B., and Hodge, D. S., 1969, Petrology and geochemistry of the Laramie batholith and related metamorphic rocks of Precambrian age, eastern Wyoming (disc.): Geol. Soc. Amer. Bull., v. 80, no. 11, p. 2383-2384

Notes

Iron-titanium deposits occur in the Laramie Range in a north-northeast--

south-southwest belt that extends for slightly over 20 miles from the North Fork of Horse Creek on the south to Sibylee Creek on the north, but the principal deposits are found at two locations essentially in the center of the belt; these are at Iron Mountain, some 25 miles north-northeast of Laramie, and on the Shanton Ranch about 3 miles southwest of the center of the Iron Mountain mineralized area. The Iron Mountain area, in turn, is separated into two parts by a generally barren zone about 1500 feet in length, with the total mineralized length (including the barren zone) being about 5000 feet.

The iron-titanium deposits are located entirely within the northern one of the two Precambrian anorthosite masses that make up the core of the southern part of the Laramie Range; the southern mass, which is much the smaller of the two, contains no mappable deposits of these elements. The anorthosite and the rocks genetically and temporally associated with them (the anorthosite complex) were intrusive into older metamorphosed sediments and igneous or pseudo-igneous rocks. Although an appreciable fraction of the western margin of the anorthosite complex (though not actual anorthosite) is bordered by these metamorphosed rocks, for all practical purposes, they are essentially lacking as xenoliths within the complex. The metamorphosed rocks have no connection, spatially or genetically, with the ores.

Newhouse and Hagner (1957) also have mapped four other metamorphic materials: (1) syenite-diorite gneiss; (2) quartz diorite gneiss; (3) prophyroblastic granodiorite gneiss; and (4) quartz monzonite gneiss. These gneisses underlie some of the area separating the northern and southern anorthosite masses and are unimportant in areal extent.

The next youngest of the Precambrian rocks are those of the anorthosite complex that appears to have consisted of four major rock types (each probably originally thousands of feet thick) that were in all likelihood developed from a single massive intrusion by crystallization differentiation. The basal layer of the complex within the northern anorthosite mass (where the iron-titanium ores are now located) is a medium- to coarse-grained rock which is mostly plagioclase (An_{40} to An_{65}) and contains less than 10 percent (normally only 1 to 2 percent) of orthopyroxene and magnetite-ilmenite; this anorthosite is massive to foliated and layered. Within the anorthosite layer are nearly all of the 40 bodies of olivine anorthosite mapped in the northern mass; a few are found in what is probably the basal portion of the overlying noritic anorthosite. These olivine anorthosite bodies are concentrated in the northern and southern portion of the northern mass; it does not appear that the olivine anorthosite ever formed a separate layer between the anorthosite proper and the noritic anorthosite but, instead, developed from local concentrations of dark silicate-rich material within the dominantly anorthositic basal layer or the lowest portion of noritic anorthosite. It is possible that the olivine anorthosite is actually a local variant of the noritic anorthosite. While the anorthosite is the principal rock in the eastern part of the northern portion of the complex, noritic anorthosite is the principal rock in its central portion. Erosion has exposed a band of anorthosite that separates the main noritic anorthosite mass from a more southerly narrow fringe of the same rock. In places, the noritic anorthosite is found in the anorthosite proper as transgressive veins, pegmatites, and layers and as material interstitial to anorthosite rubble or breccia; these relationships strongly suggest that the still molten noritic anorthosite was forced into the solid anorthosite layer beneath it before it (the noritic material) had fully crystallized.

The western portion of the igneous complex is made up of a hypersthene syenite that is a coarse-grained massive rock, the principal constituent of which is potash feldspar with minor and local quartz and hypersthene. The relationships between the syenite and the older rocks beneath it appear at least as compatible with the hypersthene syenite having been a fourth layer produced by the crystallization differentiation of the original anorthosite magma as that it was a separate intrusion.

The only other igneous rock of any great areal extent is the Sherman granite, the youngest igneous rock in the area, that intrudes all of the rocks of the anorthosite complex as dikes and irregular masses. To the east, the anorthosite complex is bordered in part by the Sherman granite and in part by sedimentary rocks of which the oldest and most commonly adjacent to the complex is the Casper formation, limestone and sandstone of Pennsylvanian age.

There is some question as to whether the magnetite-ilmenite deposits of Iron Mountain and the Shanton Ranch were developed *in situ* or were intruded into the rocks (largely the anorthosite proper) where they are now found. It seems probable, however, that the ores were originally included in the magma from which the anorthosite complex was developed. Thus, the age of the ores must be that of the anorthositic rocks themselves. I know of no radioactive age determinations on which to base an age assignment for the Iron Mountain ores; by analogy with other anorthosite masses in the United States and Canada and abroad, however, the deposits are here classified as late Precambrian.

The Iron Mountain deposits are divided into south and north areas, and the areas are separated from each other by about 1500 feet of essentially barren anorthosite. Some minor ore mineralization is found in both noritic anorthosite and norite, but all of the minable or potentially minable deposits are located in the anorthosite layer of the anorthosite complex. All the ore bodies (including those of the Shanton Ranch) lie on or within a short distance of the anticlinal axes as mapped by Newhouse and Hagner. Both mineralized areas occur at sharp bends in the direction of the axis of the fold; a large change in direction occurs in the Iron Mountain area and a much smaller one at the Shanton Ranch.

The Iron Mountain deposit is made up of a generally north- to northeast-trending, east- to southeast-dipping line of irregularly shaped lenses of massive magnetite-ilmenite ore. These ores contain at least 10 percent TiO_2, and the best of the massive magnetite-ilmenite ore averages about 50 percent iron, 21 percent TiO_2, and slightly over 0.50 percent V_2O_5; the silicate content (mainly fayalite) is less than 5 percent.

Some of the lenses have low-grade extensions, or are joined by lenses, of low-grade ore that runs from 5 to 10 percent TiO_2. At the north end of the Iron Mountain mineralized zone, there is a rudely circular area of low-grade ore, some 800 feet in diameter, the western and northern borders of which are formed by the line of high-grade ore lenses.

The deposits of the Shanton Ranch, about 2 miles southwest of the ore body that is the farthest south of the ore bodies in the southern portion of the Iron Mountain deposit, are made up of five ore bodies, irregularly arranged along a belt about 1500 feet long that trends northeast-southwest. There appears to be no low-grade mineralization associated with the Shanton ores. The ore is composed of (1) grains of magnetite with exsolved lathlike bodies of what has been invariably reported as ilmenite contained in them, (2) isolated ilmenite grains, and (3) high-iron olivine (probably somewhat on the fayalite side of the forsterite-fayalite solid solution series); small amounts of quartz, apatite, spinel, and chlorite are often present, and some ilmenite grains contain olivine inclusions. From tests made on the ore (Frey, 1946), it appears that about two-thirds of the ilmenite is in the isolated grains and one-third in exsolved laths in the magnetite. The magnetite, ilmenite, and apatite, where they are found in contact with feldspars, were later in crystallizing as is shown by their positions interstitial to feldspar crystals and by their filling fractures that cut the labradorite grains.

The possibility exists that the exsolved mineral in magnetite is ulvöspinel (Fe_2TiO_4) and not ilmenite; so far as published work goes, it appears that it has been assumed that the exsolved mineral is ilmenite and that it has not been thoroughly tested to determine whether it is $FeTiO_3$ or Fe_2TiO_4. Certainly the structure of ulvöspinel, essentially that of magnetite, makes it far more likely that it is the exsolved mineral rather than that it is ilmenite.

The ore bodies of the entire Iron Mountain district consistently contain vanadium oxide, with the average percentage of V_2O_5 probably being about 0.50. No separate vanadium mineral has ever been found in the ores, and it appears probable that the vanadium is present in solid solution as Fe_2VO_4, but whether this material is located in the magnetite proper or in the mineral (ilmenite or ulvöspinel) exsolved from it or in both is uncertain.

Considering the titanium-bearing mineral in the magnetite to be ilmenite, the amounts of magnetite and ilmenite in the deposit are about equal. On the average, this means that, neglecting the iron in the olivine, the iron to titanium ratio in the ore is about 2.2 to 1.0.

The ore minerals probably were originally part of the anorthosite magma, and they became immiscible in the volumetrically dominant silicate phase at some time during the crystallization of the anorthosite complex. The presence of magnetite-ilmenite in both massive and disseminated form (though not in minable amounts) in norite and noritic anorthosite indicates that the separation did not come until late in the crystallization cycle of the complex. If this reasoning is sound, then the massive ore bodies in the anorthosite (basal) layer of the complex must have been forced into this environment during the folding that produced the anticlinal structures mapped by Newhouse and Hagner. The late-stage, still immiscible ore fraction was probably moved from the upper portion of the magma where it was developed through the thrust faults formed during folding to its final resting place along the anticlinal axis. Minor amounts remained in the norite and some small bodies were emplaced in the noritic anorthosite, but it would seem that the bulk of the molten iron-titanium-rich melt was driven toward the upfolded, essentially central portion of the complex where it was finally solidified. The contacts of ore and country rock are everywhere smooth and free of transitional phases to the country rock; much of the contact between ore and anorthosite is slickensided indicating movement along the faults after the ore had been emplaced.

The deposits of the Shanton Ranch do not contain such low-grade sections nor do the lenses in that area show the same development of off-shoot dikelets and veins as is true in the Iron Mountain area. Even those geologists who consider that most of the Iron Mountain ore was collected *in situ* believe that the Shanton ores were intruded into the rock volumes they occupy at present.

It would follow from the arguments here presented that (1) the molten material from which the magnetite-ilmenite ore masses were formed was developed with falling temperature as an immiscible fraction in the far larger volume of silicate magma; (2) this immiscible fraction did not separate until quite late in the crystallization cycle of the magmatic system (as is witnessed by the presence of considerable ore in the norite that is located, however, in bodies too small or too disseminated for economic mining); and (3) this immiscible material was forced, during late-stage folding (or faulting) of the complex from the upper portions of the body into the anticlinal high in the basal anorthosite layer. By this reasoning the deposits are here classified as magmatic-3b. These conclusions are in marked disagreement with those reached by Hagner (1968) who thinks that the ore bodies were largely, if not entirely, due to metamorphic processes.

VENEZUELAN GUAYANA
JACOBINA
BRAZIL
MORRO
VELHO
MINAS
CERAIS
0
300
MILES
EASTERN SOUTH AMERICA

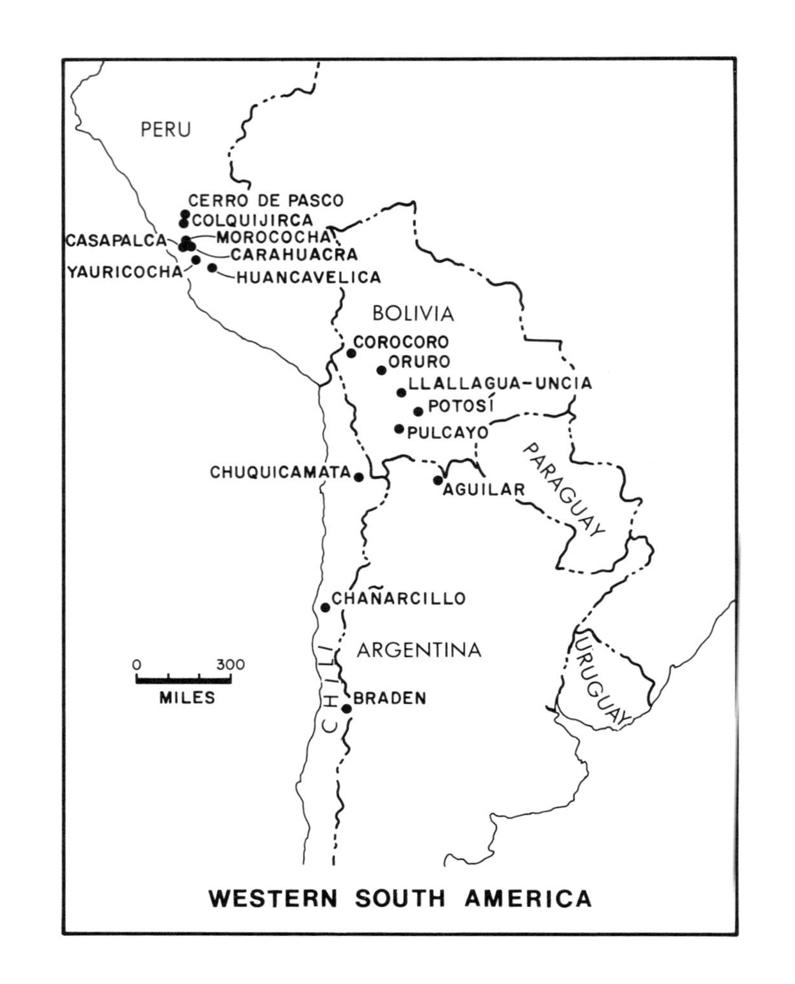
PERU
CERRO DE PASCO
COLQUIJIRCA
CASAPALCA
MOROCOCHA
CARAHUACRA
YAURICOCHA
HUANCAVELICA
BOLIVIA
COROCORO
ORURO
LLALLAGUA-UNCIA
POTOSÍ
PULCAYO
CHUQUICAMATA
AGUILAR
PARAGUAY
CHAÑARCILLO
ARGENTINA
0
300
MILES
CHILI
BRADEN
URUGUAY
WESTERN SOUTH AMERICA

SOUTH AMERICA

ARGENTINA

AGUILAR

Late Tertiary | *Zinc, Lead, Silver* | *Hypothermal-2 to Mesothermal*

Ahlfeld, F., 1955, Geologie der Blei-zinkerzlagerstätte Aguilar (Argentinien): Zeitsch. f. Erzbergbau u. Metallhüttenwesen, Bd. 8, H. 12, S. 551-556

Angelelli, V., 1950, Sierra de Aguilar, in *Recursos minerales de la Republica Argentina - I. Yacimientos metalíferos*: Inst. Nac. Invest. Ciencias Naturales, p. 410-416

Brown, J. S., 1941, Factors of composition and porosity in lead-zinc replacements of metamorphosed limestone: A.I.M.E. Tr., v. 144, p. 250-273 (particularly p. 251-253)

Linares, E., 1968, Datación geológica de las rocas graníticas de las sierras de Cordoba, por medio del método ploma/alfa (Larsen): Jour. Geol. Argentina, 3rd, Actas, v. 2, p. 199-206

—— 1968, Geología isotópica del azufre del yacimiento Aguilar, provincia de Jujuy: Jornadas Geol. Argentina, 3rd, Actas, v. 2, p. 191-198

Sgrosso, P., 1943, Contribución al conocimiento de la minería y geología de noroeste Argentino: Dirección Minas y Geol. (Argentina) Bol. no. 53, 180 p. (particularly p. 31-51)

Spencer, F. N., Jr., 1950, The geology of the Aguilar lead-zinc mine, Argentina: Econ. Geol., v. 45, p. 405-433

Whiting, F. B., 1959, Structural belts and mineral deposits of northwestern Argentina: Econ. Geol., v. 54, p. 903-912 (general)

Notes

The deposits of Aguilar are located in Jujuy, the northernmost province of Argentina, in an area that is a southerly extension of the altiplano of Bolivia.

The rocks in the region range from Precambrian through Paleozoic and Mesozoic to late Tertiary, although the oldest formation in the immediate mine area (the Aguilar quartzite) is probably Cambrian. The lowest exposed member of this formation consists of a series of interbedded impure limestones and calcareous quartzites (some of which have calcareous cement) and arkoses, quartzites, and shales; the ores lie entirely in this basal sequence, mainly in the calcareous beds. Above the limey beds, the Aguilar is composed of massive quartzite beds with lesser thicknesses of interbedded shale. Conformably above the quartzitic portion of the Aguilar quartzite lies the Aguilar shale that has a few intercalated bands of conglomerate, quartzite, and calcareous rocks in its uppermost portions. This shale is overlain by the Cajas formation, probably conformably although the contact has not been clearly seen, that begins with a basal quartzite, followed by a limey transition zone that contains Lower Ordovician fossils. This age of the Cajas probably confirms the Cambrian age of the two Aguilar formations. The period between the Lower Ordovician and the Cretaceous appears to have been a time of uplift and erosion; the lowest exposed member of the Mesozoic is a Cretaceous red sandstone. This Cretaceous formation is overlain by a series of Tertiary rocks, mainly red sandstones and shales, but the youngest Tertiary member is a bed of coarse conglomerate that contains fossiliferous boulders of middle Tertiary limestone. This relationship strongly suggests that the folding that produced the Aguilar anticline began in the middle Tertiary and that the conglomerate is probably lower

Pliocene. Although the conglomerate contains granite boulders, they are quite different in character from either of the two granites, the Abra Laite (older) and the Aguilar granite (younger) that are intruded into all the rocks of the Aguilar anticline. A period of faulting followed the intrusion of the granites, perhaps before these igneous rocks had entirely solidified, and it was by this faulting and by the impure limestone beds at the exposed base of the Aguilar quartzite that the loci of ore deposition were determined. It appears almost certain, from the facts just enumerated, that the emplacement of the ore must have occurred in the Pliocene, and the ores are, therefore, categorized as late Tertiary.

The ore bodies at Aguilar are largely in tabular masses, the locations of which are controlled by limestone horizons highly susceptible to replacement or by shear zones or by both. In the immediate mine area, the two Cambrian Aguilar formations form a wedge-shaped block, narrow edge downward, between the Abra Laite granite on the west and the Aguilar granite on the east. The West Branch fault roughly parallels the north-south trending contact between the sediments and the Aguilar granites, with the ores being in that minor portion of the wedge of sediments that lies between the west dipping West Branch fault and the Aguilar granite. As the West Branch fault is a normal fault, the ore-bearing portion on the east side of the fault is in limey beds of the Aguilar quartzite, while the upper portion of the wedge to the west of the fault is down-dropped Aguilar shale.

The Aguilar granite was intruded into the area prior to the development of the West Branch fault (and the other faults in the area) and caused the formation of an appreciable zone of high temperature tactite minerals a few hundred meters wide. The tactite was most strongly developed in the impure limestone and calcareous quartzite and arkose members of the Aguilar quartzite; the Aguilar shale was largely converted to hornfels. The principal tactite minerals are diopside and garnet, with considerable rhodocrosite, some feldspar, and several minor accessory minerals.

The bedded mineralization is found in the southern part of the mine, mainly in seven beds in the lower part of the Aguilar quartzite that originally contained appreciable, but different, amounts of calcareous minerals. Not all of these beds contain ore in economic amounts, and those that do are not of ore grade in their entirety. The shear zone ores are located in the northern part of the mine and are confined to the sheared footwall of the West Branch fault where the favorable beds parallel the fault. The two types of ore are gradational into each other, mainly because the shear zones produced by the fault appear to have served as the channelways through which the ore fluids entered the now ore-bearing area. In detail, the structures of the ore bodies are quite complex due to such factors as sharp local folds, thickening and thinning of shale beds, and minor faulting.

The ore bodies show no systematic relationship to the tactite, and the principal effect of tactite on ore mineralization was to make it less strong in areas largely converted to tactite than in those where residual calcite was more common. The ore solutions, however, appear to have attacked tactite-containing beds that were originally impure limestones rather than pure limestones in which little or no tactite was developed. Although ore minerals have replaced silicates, they have been far more strongly developed in residual calcite.

The first minerals to be deposited were, in order, minor pyrite, minor pyrrhotite, even less chalcopyrite, then abundant iron-rich sphalerite (marmatite) which contained exsolved inclusions of pyrrhotite and chalcopyrite. The marmatite gradually gave place to a red sphalerite and then to a clear yellow variety; the red and yellow sphalerite contain no exsolution blebs of either pyrrhotite or chalcopyrite. The amounts of marmatite and lighter sphalerite probably are about equal. About halfway through the depositional cycle of sphalerite, the precipitation of abundant silver-rich galena began; lead sulfide deposition ended slightly before that of yellow sphalerite. The last mineral to form was tetrahedrite or ruby silver in negligible amounts.

Although the development of the high-temperature tactite minerals was not directly connected with the formation of the ores, the early ore minerals nevertheless appear to have formed under hypothermal conditions because of the inclusions of pyrrhotite and chalcopyrite in the marmatite. The lack of these inclusions in red and yellow sphalerite and the precipitation simultaneously with them of silver-rich galena strongly suggest that this portion of the ore mineralization was formed under mesothermal conditions. The ores are, therefore, here classified as hypothermal in calcareous rocks (hypothermal-2) to mesothermal. The quantity of the tetrahedrite or ruby silver is too small to justify adding a leptothermal phase to the classification.

BOLIVIA

BOLIVIAN TIN (GENERAL)

Late Tertiary	*Tin, Silver*	Xenothermal to Epithermal

Ahlfeld, F., 1932, Die Erzlagerstätten in der tertiärem Magmaprovinz der bolivianischen Zentrallanden: Neues Jb. f. Mineral., Geol. und Paläont., Abh., Beil. Bd. 65, Abt. A., S. 285-354, 355-446

——— 1936, The Bolivian tin belt: Econ. Geol., v. 31, p. 48-72

——— 1937, Über das Alter der zinnbringenden Magmengesteine Boliviens: Zentralblatt f. Mineral., Geol. und Paläont., Jg. 1937, Abt. A, H. 2, S. 34-38

——— 1941, Zoning in the Bolivian tin belt: Econ. Geol., v. 36, p. 569-588

——— 1957, Die zonale Verteilung der Erzlagerstätten in der bolivianischen Metallprovinz: Neues Jb. f. Mineral., Abh., Bd. 91 (Festband Schneiderhöhn), S. 329-350

——— 1962, Neue Erkenntnisse über Stratigraphie und Tektonik Boliviens: Neues Jb. f. Geol. und Paläont., Mh., Jg. 1962, H. 6, S. 293-303

——— 1967, Metallogenic epochs and provinces of Bolivia; the tin province (Part I); the metallogenetic provinces of the Altiplano (Part II): Mineralium Deposita, v. 2, p. 291-311

Davy, W. M., 1920, Ore deposition in the Bolivian tin-silver deposits: Econ. Geol., v. 15, p. 463-496

Evernden, J. F., 1961, Edades absolutas de algunas rocas igneas en Bolivia por el método potasio-argon: Soc. Geol. Boliviana Noticiero no. 2 [I have not been able to find this reference, but Ahlfeld (1967) quotes from it]

Evernden, J. F., and others, 1966, Correlación de las formaciónes terciarias de la cuenca altiplanica a base de edades absolutas determinada por el método potasio-argon: Serv. Geol. Bol. [I have not been able to find this reference, but Ahlfeld (1967) quotes from it]

Fesser, H., 1968, Spurenelemente in bolivianischen Zinnsteinen: Geol. Jb. Bd. 85, S. 605-610 (Engl. summ.)

Gundlach, H., and Thormann, W., 1960, Versuch einer Deutung der Entstehung von Wolfram- und Zinnlagerstätten: Zeitsch. der Deutschen Geol. Ges., Bd. 112, I Teil, S. 1-33

Kelly, W. C., and Turneaure, F. S., 1970, Mineralogy, paragenesis and geothermometry of the tin and tungsten deposits of the eastern Andes, Bolivia: Econ. Geol., v. 65, p. 609-680

Koeberlin, F. R., 1926, Geologic features of Bolivia's tin-bearing veins: Eng. and Min. Jour.-Press, v. 121, no. 16, p. 636-642; Eng. and Min. Jour., v. 122, no. 13, p. 502-503

Kozlowski, R., 1934, Esquisse de la répartition des roches éruptives dans les Andes de Bolivia: Arch. Mineral. Soc. Sci. Varsovie, t. 10, p. 123-161

Ljunggren, P., 1962, Bolivian tin mineralization and orogenic evolution: Econ. Geol., v. 57, p. 978-981

Ljunggren, P., and Radelli, L., 1963, Bolivian tin mineralization: Econ. Geol., v. 58, p. 1348-1351

Pilz, R., and Donath, M., 1929, Betrachtungen über die Entstehung der Eruptivgesteine und Erzlagerstätten des andinen Bolivien: Zeitsch. f. prakt. Geol., Jg. 37, H. 7, S. 125-138

Rumbold, W. R., 1909, The origin of the Bolivian tin deposits: Econ. Geol., v. 4, p. 321-364

Schneider-Scherbina, A., 1962, Über metallogenetische Epochen Boliviens und den hybriden Charakter der sogenannten Zinn-Silber-Formation: Geol. Jb., Bd. 81, S. 157-170

—— 1963, Bolivian tin mineralization and orogenic evolution: Econ. Geol., v. 58, p. 456-459

—— 1965, Time-space paragenetic sequences in the hypogene ore deposits of Bolivia, in Stemprok, M., Editor, *Symposium--Problems of postmagmatic ore deposition*: v. 2, p. 67-81

Singewald, J. T., Jr., 1921, Ore deposition in the Bolivian tin-silver deposits: Econ. Geol., v. 15, p. 60-69

—— 1929, The problem of supergene cassiterite in Bolivian tin veins: Econ. Geol., v. 24, p. 343-364; disc., v. 25, p. 91-99, 211-218

Stelzner, A., 1897, Die Silber-Zinnerzlagerstätten Boliviens: Zeitsch. der Deutschen Geol. Ges., Bd. 49, S. 51-142

Sznapka, G., 1923, Beiträge zur Geologie bolivianischer Zinnerzlagerstätten: Zeitsch. f. prakt. Geol., Jg. 31, H. 11/12, S. 119-125

Turneaure, F. S., 1960, A comparative study of major ore deposits of central Bolivia: Econ. Geol., v. 55, p. 217-254

Turneaure, F. S., and Welker, K. K., 1947, The ore deposits of the Eastern Andes of Bolivia; the Cordillera Real: Econ. Geol., v. 42, p. 595-625 (north of the area containing the deposits listed individually below)

Winckelmann, H., 1927, Beiträge zur Kenntnis der Zinnerzlagerstätten von Bolivien: Zeitsch. f. prakt. Geol., Jg. 35, H. 7, S. 97-112

Wolf, M., 1968, Die bolivianischen Zinnlagerstätten und einige neue Aspekte ihrer genetischen Deutung: Bergakademie, Bd. 20, H. 6, S. 319-323 (Engl. summ.)

Notes

Ahlfeld (1967) describes the Bolivian tin province as a 900 km-long, well-defined, arc-shaped zone situated in the highest, western portion of the eastern Andes. It extends from the Bolivian-Peruvian border at latitude 15°S and longitude 69°W to latitude 23°S and longitude 67°W at the Pirquitas mine in Jujuy Province of northernmost Argentina. Ahlfeld divides the province into two parts: northern and southern, with the division between the two being along the Ichilo Fault zone or Arica Elbow line that crosses the province in an east-west direction. The northern (and narrower) part of the tin province has been uplifted strongly as opposed to the southern part and has been displaced westward, so older rocks and different mineralizations are found north of the line from those south of it. Since all of the tin deposits discussed in this volume (Llallagua-Uncía, Oruro, and Potosí) are in the southern part of the province, it alone is discussed here.

Although older granites, perhaps as old as Triassic (Evernden, 1961), do not crop out in the southern area, gravimetric measurements indicate that the older basement lies about 3000 m below the present surface. Most of the valuable elements taken from the mines in the southern province (Sn, W, Sb, Bi, Zn, Pb) are the same as those found in the northern part, but abundant silver is present in the south; gold is rare in both areas. Although Ahlfeld (1967) distinguishes three phases of tin mineralization, only the latest (associated with subvolcanic latite and rhyolite stocks) appears to have formed economically valuable deposits and is dated, at Llallagua at least, as Pliocene (9.5 m.y. to 7.5 m.y.). By analogy, the others, specifically here Oruro and Potosí, also are thought to be Pliocene.

The country rocks in the districts in question are, according to Turneaure (1960), largely folded Paleozoic rocks of a north to northwest trend; shale and sandstone are the principal rock types although a graywacke is known in the Llallagua district. In the Potosí area, these Paleozoic rocks are overlain unconformably by a volcanic complex; the rocks of the complex include fossil plant-bearing tuffs that probably are middle Tertiary (Miocene). Ahlfeld (1964) says that this complex was intruded by a now highly altered rock that probably was originally a dacite. This intrusion certainly was later than the Miocene plant-remains and more likely occurred in the Pliocene than the Miocene. Thus, if both Llallagua and Potosí are Pliocene in age, it is not unreasonable to conclude that Oruro was also. Considerable material has been published to suggest that various elements in the deposits of this portion of Bolivia were added at more than one time, and Schneider-Scherbina (1963) concludes that the late Tertiary tin, silver-lead-zinc mineralization is a hybrid type of deposit which has formed as a result of hydrothermal regeneration. Following Ahlfeld, I think that this is an overcomplication of the problem and that the silver-tin and tin ores here discussed are all Pliocene in age and are, therefore, here categorized as late Tertiary. It should be noted that, in this volume, a middle Tertiary age has been assigned to the very different Peruvian ores to the north; it may, however, be found that the Peruvian ores also are late Tertiary.

Although silver-rich ores in the southern tin province may be in distinctly different veins than the earlier tin mineralization and some tin veins may have no silver minerals associated with them at all, many of the deposits (Potosí and Oruro, for example) contain tin and silver minerals in the same vein structures. Thus, it appears probable that the tin and silver mineralizations probably took place in the same general range of time but that, in some instances, the tin ores so completely filled the veins in which they were deposited that the later silver mineralization was forced into other channels and deposited in other places. Where, however, the same channels could be used, both tin and silver mineralization are found together or with the latter usually above the former. In contrast to the ores of the northern part of the tin province, the material in the tin or tin-silver veins to the south consists mainly of sulfides; of these, pyrite, pyrrhotite, and sphalerite are the most abundant. Tin, however, is present largely as cassiterite with, as Ahlfeld (1967) points out, all types being present from high-temperature material with tourmaline to colloform masses and acicular crystals that indicate (to him) low temperatures of formation. In the deposits listed in this volume, the cassiterite is almost entirely of the high-temperature type. In most deposits, bismuthinite is present and in some is in economically recoverable quantities. Wolframite is less common than in the northern part of the province. Antimony is mainly found in lead-antimony sulfosalts rather than as stibnite; the most important of these sulfosalts are jamesonite and boulangerite, but many others are known. The younger, though normally genetically connected silver mineralizations in the province--such silver-bearing or silver sulfosalts as franckeite, cylindrite, and teallite and andorite and argyrodite--are common in many deposits, and a very considerable number of other tin sulfosalts are known. The emphasis on tin and silver in these deposits has resulted in a neglect of the more common sulfides and sulfosalts such as stannite, chalcopyrite, sphalerite, tetrahedrite, bournonite, and galena. These well might have been profitably recovered had not it been possible to make so much money from tin and silver that the owners could afford to neglect those minerals more difficult to process and having a lesser profit margin. In any event, these intermediate-range minerals are there, should have been mined, and deserve a place in the classification. Since all the tin deposits included here show the intermediate-intensity (kryptothermal) minerals, the deposits are designated as either xenothermal to kryptothermal or xenothermal to epithermal, depending on the lack or presence of low-temperature silver mineralization.

LLALLAGUA-UNCÍA (CATAVI)

Late Tertiary | *Tin* | *Xenothermal to Kryptothermal*

Ahlfeld, F., 1929, Die Zinnerzgrube Uncía-Llallagua (Bolivia): Metall und Erz, Jg. 26, H. 14, S. 349-354

—— 1931, The tin ores of Uncía-Llallagua: Econ. Geol., v. 26, p. 241-257

—— 1936, The tin deposits of Llallagua, Bolivia (disc.): Econ. Geol., v. 31, p. 219-221

—— 1941, Los yacimientos minerales de Bolivia: Dirección General de Minas y Petróleos, La Paz, p. 94-100

—— 1941, Zoning in the Bolivian tin belt: Econ. Geol., v. 36, p. 569-588

—— 1954, Los yacimientos minerales de Bolivia: El Banco Minero de Bolivia y la Corporación Minera de Bolivia, 2d ed., Bilbao, p. 63-67

Deringer, D. C., and Payne, J., Jr., 1937, Patiño--leading producer of tin; I. The ore deposits of Llallagua: Eng. and Min. Jour., v. 138, no. 4, p. 171-177

Gordon, S. G., 1944, The mineralogy of the tin mines of Cerro de Llallagua, Bolivia: Acad. Nat. Sci. Pr., v. 96, p. 279-359 (particularly p. 279-301, and plates)

Moon, L., 1939, Structural geology of Llallagua, Bolivia: Minn. Acad. Sci. Pr., v. 7, p. 64-72

Rechenberg, H. P., 1955, Die Zinnseifenlagerstätte von Llallagua, Bolivien: Berg- und Hüttenmännisches Mh., Jg. 100, H. 10, S. 280-284

Samoyloff, V., 1934, The Llallagua-Uncía tin deposit: Econ. Geol., v. 29, p. 481-499

Smith, F. G., and others, 1957, Manganoan wurtzite from Llallagua, Bolivia (I): Canadian Mineral., v. 6, pt. 1, p. 128-135

Turneaure, F. S., 1935, The tin deposits of Llallagua, Bolivia: Econ. Geol., v. 30, p. 14-60, 170-190

—— 1942, The tin deposits of Llallagua, Bolivia, in Newhouse, W. H., Editor, *Ore deposits as related to structural features*: Princeton Univ. Press, p. 135-136

Notes

The deposits of Llallagua-Uncía (also known as Catavi) are located about 75 km south-southeast of the town of Oruro and some 170 km south-southeast of La Paz.

In the general Llallagua-Oruro region, the principal sedimentary rocks are Paleozoic in age (Turneaure, 1960). The oldest rocks in the district proper are those of the Cancañiri graywacke; they occupy the axis of the Llallagua anticline and form a belt nearly 1500 m wide. The Cancañiri is overlain unconformably by the Llallagua formation, massive sandstone at the base, followed by alternating beds of sandstone and shale. The next formation is the Pampa, mainly thin-bedded shale; the Pampa is overlain by the Catavi sandstone that contains a Devonian fauna. The next beds are the dark gray Ventilla shale. Above the Ventilla is a red conglomerate and red shale that form the trough of a broad syncline; the age of these beds is uncertain. They may be Cretaceous and equivalent to the Puca formation or late Paleozoic and essentially equal to certain red beds in the Cordillera Real. The exact age of the pre-Catavi beds is uncertain, but they probably were formed in the first half of the Paleozoic; they are not Precambrian.

Intrusive activity at Llallagua consists of the Salvedora stock, composed of porphyry and porphyry breccia; the breccia contains numerous fragments of the Paleozoic sediments. The original porphyry probably was a quartz latite. No evidence is known to show that one porphyry is older than the other. The large volume of breccia, the slight degree of contact alteration, and the funneling downward of the igneous mass all suggest to Turneaure that the body is the filling of a volcanic vent. Superficial andesitic and rhyolitic tuffs that might have been formed at the same time as the porphyries have been preserved on some of the summits in the district. The age of the stock has been determined by Evernden (1961) to be 9.5 m.y. and that of a potash-rich rhyolitic end intrusion to be 7.5 m.y. Since the ore mineralization almost certainly took place at the end of the igneous activity, the ores must be Pliocene (granted the correctness of the ages given) instead of the middle Tertiary (?) assigned to them by Turneaure. The deposits, therefore, here are assigned a late Tertiary age. Later lava flows may have covered the area; such flows exist northeast and southeast of Lake Poopó and may be later than the tin mineralization.

The Llallagua veins are found mainly in the quartz latite porphyry, although some of them (Contacto, San Fermin, and Bismarck, among others) extend into the surrounding sediments. Some 45 major veins are known, and nearly 500 branches of these have been found. Before any of the ores were formed, the quartz latite was sericitized, tourmalinized, and silicified. The larger veins are fissure fillings with distinct and sometimes slickensided walls; they may be in shear zones or single fissures. The bulk of the tin was deposited in the larger openings and in the main channels; the prior silicification of the wall rocks essentially prevented the solutions from entering and replacing these rocks except to fill tiny subordinate fractures. The smaller of these fissures were completely filled by the first (tin) stage of mineralization, and only the larger retained enough open space for the deposition of the later minerals. Some of the filled fissures, however, were reopened, and later minerals deposited in the newly created open space. Some fractures were not developed until after the first- (tin-) stage mineralization and contain only late minerals. The minerals are well crystallized, and the veins often are vuggy or drusy.

The first mineral to form in the tin stage was quartz. In the simple veins, quartz was followed by cassiterite. In the more complex ones, bismuthinite preceded cassiterite, and that latter mineral was followed by wolframite and apatite, the apatite filling spaces among the metallic minerals. The next stage was the pyrrhotite stage in which this iron sulfide filled open spaces in the centers of the veins. Where this space was large, the amount of pyrrhotite was huge; in narrower veins, the pyrrhotite filled only occasional pockets. The pyrrhotite was then subjected to considerable replacement in the third stage, with marcasite being the first mineral in this cycle; locally franckeite accompanied the marcasite. Since the replacement minerals were deposited along basal parting planes in the pyrrhotite, these two sulfides were definitely oriented; pseudomorphs of marcasite after pyrrhotite also occur. Other minerals formed in this stage include galena, pyrite, siderite, sphalerite, and stannite. Stannite was the last mineral of this stage to form and replaced and accompanied sphalerite; locally stannite was quite abundant. Although most of the marcasite and pyrite were emplaced by the replacement of pyrrhotite, some was deposited on other minerals. In some places, the second and third stage minerals extended farther out into the outer and upper portions of the veins not reached by the cassiterite mineralization. Secondary, surface-derived solutions removed much of the pyrrhotite not replaced by pyrite and marcasite, leaving these two minerals as honeycomb aggregates. Where apatite was leached, secondary wavellite and locally minerals of the bauxite group were formed.

Where the ores were deposited in brecciated graywacke outside the stock, rounded fragments of breccia were cemented by fibrous cassiterite that Gordon (1944) finds to resemble wood tin. Between these wood-tin-sheathed breccia fragments, the vugs remaining contained the mineral sequences found in the main veins in miniature. Although the amount of tourmaline in the stock is enormous (most of the feldspar in the stock has been replaced by it), essentially no tourmaline is found in the tin veins, so the tourmaline stage appears to have antedated that of tin deposition. On the other hand, the large amount of apatite is intimately intergrown with the minerals of the cassiterite stage, principally filling open spaces between cassiterite and bismuthinite.

The tin mineralization definitely was of a high-intensity character as was the pyrrhotite of the second stage; because of the near surface loci of deposition, these stages are classed as xenothermal. The sulfide stage, marcasite through stannite, seems to have formed under less intense conditions and is here classified as kryptothermal.

ORURO

Late Tertiary | *Tin, Silver* | *Xenothermal to Epithermal*

Ahlfeld, F., 1941, Los yacimientos minerales de Bolivia: Dirección General de Minas y Petróleos, La Paz, p. 75-81

—— 1954, Los yacimientos minerales de Bolivia: El Banco Minero de Bolivia y la Corporación Minera de Bolivia, 2d ed., Bilbao, p. 81-84

Berry, L. G., 1940, Studies of mineral sulpho-salts: II, jamesonite from Cornwall and Bolivia: Mineral. Mag., v. 25, p. 597-608

Campbell, D. F., 1942, The Oruro silver-tin district, Bolivia: Econ. Geol., v. 37, p. 87-115

Chace, F. M., 1948, Tin-silver veins of Oruro, Bolivia: Econ. Geol., v. 43, p. 333-383, 435-470

Kozlowski, R., and Jaskolski, S., 1932, Les gisements argentostannifères d'Oruro: Arch. Minéral. Soc. Sci. Varsovie, t. 8, p. 1-121

Lindgren, W., and Abbott, A. C., 1931, The silver-tin deposits of Oruro, Bolivia: Econ. Geol., v. 26, p. 453-479

Miller, B. L., and Singewald, J. T., Jr., 1919, Oruru silver-tin district, in *Mineral deposits of South America*: McGraw-Hill, N. Y., p. 110-114

Routhier, P., 1963 Le "type Bolivien" - Example: Oruro, in *Les gisements métallifères--Géologie et principes de recherches*, pt. 1: Masson et Cie, Paris, p. 632-639

Singewald, J. T., Jr., 1929, The problem of supergene cassiterite in Bolivian tin veins: Econ. Geol., v. 24, p. 343-364

—— 1930, Supergene cassiterite in Bolivian tin veins: Econ. Geol., v. 25, p. 211-218

Notes

The tin deposits of Oruro are located on the Bolivian altiplano about 200 km southwest of La Paz and about 330 km east-northeast of Arica on the Pacific coast of Chile; they are confined to an isolated group of hills known as the Cerros de Oruro. The city of Oruro is about 8 km west of the eastern branch of the Andes, the Cordillera Real.

The bed rock of the Oruro district is a series of highly folded Paleozoic sediments, mainly argillites, shales, and slates that Chace (1948) has designated as the Oruro formation. The particular term appropriately applied to the formation differs from place to place with the degree to which it has been metamorphosed. Chace finds it difficult to decide which type is most abundantly developed; since, however, all the formation originally was argillaceous, he thinks that argillite is the most applicable term for the formation as a whole. In the mine, the formation is known locally to contain thin, sandy layers; in other places the rock is hard and massive, without traces of bedding, and, in a few others, the rock is schistose. Near the Tertiary intrusives, the argillite has been hardened and silicified, and some spotted slate or "knotenschiefer" have been developed by contact metamorphism. The formation occupies only small areas on the surface but, because of the shape of the igneous bodies intruded into it, these areas increase largely with depth so that the bottom levels of the mine are mainly in argillite. The main structure in the district appears to be the west limb of a major anticline; if the fold is symmetrical, the east limb should lie on the eastern side of the Cerros de Oruro, but this area is covered by the Pleistocene Pampa formation, and the older rocks beneath it cannot be studied.

The Oruro formation resembles that at Viacha (200 km to the north), which contains fossils and which is thought to be Silurian or Devonian; rocks at Llallagua (75 km southeast) are lithologically similar to those of the Oruro formation and also are thought to be Devonian.

The greater part of the Cerros is made up of intrusive and extrusive igneous rocks, and the areal extent of the rocks also is greater than that of all others in the district. These igneous rocks make up an oval unit with a north-south elongation, parallel to that of the regional trend in central Bolivia. The most important rock type is quartz-latite porphyry, now highly altered, that occurs in small stocks, sills, dikes, irregular bodies, and lava flows. Chace believes that each of the stocks is the center of an individual intrusion and that, within the mineralized area two separate intrusions--the Itos and the San José stocks--are known. Other (unmineralized) hills in the area probably also are individual stocks. All the stocks are separated from each other by segments of Oruro formation. At the lowest level in 1948, the porphyry is known only in dikes; thus it appears quite definite that the stocks, with depth, grade into feeding dikes. The intrusive nature of the porphyry is demonstrated by the dikes, sills, and apophyses of porphyry that cut the argillite and by the porphyry-argillite breccias masses developed mainly at the contacts between porphyry and argillite. Although locally contacts between porphyry and argillite are conformable, mostly they are discordant.

Chace believes that the intrusive bodies were near-surface intrusions because of their gradation upward into lavas and their porphyritic textures. The intrusives appear to have been forcefully injected because the folding of the Oruro formation locally becomes quite intense near the contacts with igneous rocks. Chace believes that the plutons of the district are best described as volcanic necks, this being indicated by feeding dikes flaring upward into a phaneroconvex body at the surface, by the abundance of postintrusive breccias along the contacts, by the absence of chilled contacts, and by their association with lava flows.

The youngest consolidated rock in the district is a coarse breccia composed of subangular to rounded fragments of altered quartz latite and argillite in a matrix of fine-grained, pulverized argillite; the material is unsorted. Breccia bodies have been found in either host rock, but not far from the contact of the one with the other; not all contacts have breccia bodies associated with them. These breccias probably were formed by explosive pressures built up in the volcanic necks after the latite had been emplaced and released along lines of weakness provided by contacts or fracture zones. The breccias formed after the igneous rocks had been introduced into the area and before the emplacement of ore mineralization.

As was pointed out under Bolivian Tin (General), the age of the ores at Llallagua and Potosí probably is Pliocene (late Tertiary). Because of the similarities in most details between these two deposits and that at Oruro, it seems most reasonable to place the Oruro deposits in this same age group; therefore, the Oruro deposits here are classified as late Tertiary despite the inability of geologists at Oruro to say more than that the ore is post-middle Paleozoic and pre-Pleistocene.

The ore deposits at Oruro are a complex belt of veins, roughly 2000 m long and 1200 m wide, that trends N60°W across the central part of the Cerros de Oruro. The deposits that have been mined are of two types: (1) primary sulfide ore bodies worked from underground, mainly for tin and silver, and (2) "pacos" or oxidized ore bodies, initially worked for silver but now being mined for tin, the mining being done mainly by open pits but also from shallow underground workings. The primary sulfide bodies are in veins in the larger stocks or in argillite contiguous to them and are concentrated in four clusters: (1) the San José cluster, (2) the San Luis cluster, (3) La Colorado cluster, and (4) the Tetilla cluster. The veins in each cluster are complex and differ greatly in persistence, width, and structural detail. The deposits at Oruro are partly open-space fillings of complex fracture systems and partly replacements of the adjoining wall rock. The veins show a consistent relationship to the shape, direction, elongation, and contacts of the quartz-latite porphyry bodies, one vein set striking parallel and another at right angles to the elongation of the porphyry stocks. Locally, the pattern may be complicated by irregularities in the outlines of the three rock types--porphyry, argillite, and breccia.

Generally, a strong fissure marks the center of a mineralized vein, and the mineralization decreases into the country rock away from it; open space filling is indicated by rough banding in the ore and druse-lined vugs while replacement is shown by partly replaced, residual fragments of wall rock in the veins. Fracturing continued during mineralization even though the bulk of the fracturing had taken place before the ore began to deposit, with the mineralization being divided into two distinct stages by a period of fracturing. In the early stage, the principal minerals were quartz, pyrite, and cassiterite, while in the later, the main minerals were sulfosalts--tetrahedrite, and andorite ($PbAgSb_3S_6$) were economically the most important--and galena; the later mineralization is in distinct veins that cut the earlier ones, but the two stages were deposited in the same general systems of fracture; only rarely are the veins of one stage in rocks that lack veins of the other. Usually the later veins follow the centers of the earlier ones, but locally they are on the hanging or footwalls of the earlier ones. Only rarely have early veins been offset by the fractures containing the later minerals.

Chace had distinguished three stages of wall-rock alteration in the igneous rocks: (1) a widespread but slight development of sericite in all of the igneous rocks of the area; chlorite was formed in a few places; (2) a moderate phase of alteration within the stocks in the area of mineralization; sericite and pyrite were produced and chlorite was obliterated, but the original texture of the rock was maintained; (3) an intense alteration phase that changed the rock to an aggregate of sericite, pyrite, and minor alunite with local tourmaline; the igneous texture of the rock was destroyed. On the other hand, the Oruro formation was little affected except close to the veins where pyrite, quartz, and secondary sericite were formed in the argillite, generally in a complex network of small replacement veinlets. The San José breccia was altered to various degrees with changes similar to those in the intense phase of porphyry alteration having taken place in the porphyry fragments particularly.

The early quartz-pyrite-cassiterite stage of mineralization produced a massive, granular intergrowth of these minerals with small amounts of arsenopyrite and probably tourmaline being present locally; the cassiterite and quartz are inconspicuous in the massive pyrite.

The minerals of the second stage are much more numerous and are far more complex in their relationships than those of the first. The first minerals of this stage were minor amounts of stannite. The deposition of these minerals was, in its later stages, overlapped by that of tetrahedrite, variety freibergite. The next mineral to begin to form--andorite--overlapped only slightly the deposition of stannite but considerably that of freibergite; the deposition of andorite, in turn, was overlapped almost entirely by that of the group of lead sulfosalts that includes zinkenite ($Pb_6Sb_{14}S_{27}$), boulangerite ($Pb_5Sb_4S_{11}$), jamesonite ($FePb_4Sb_6S_{14}$), and plagionite ($Pb_5Sb_8S_{17}$). Bournonite ($PbCuSbS_3$) formed over about the same time span as the other lead-bearing sulfosalts but probably in smaller amounts. The deposition of franckeite ($Pb_5Sn_5Sb_2S_{14}$) was somewhat later than that of bournonite, but it probably was formed in about the same amounts. Galena, and the gangue minerals kaolinite, dickite, alunite, barite, and marcasite, formed after essentially all of the earlier sulfides. Most of the silver in the primary ore at Oruro came from the andorite and tetrahedrite.

The presence of cassiterite in ore amounts in the early-stage primary ore almost certainly indicates that it was deposited under high-temperature conditions. The small amounts of arsenopyrite and tourmaline accompanying this stage appear to confirm this diagnosis. Chace has effectively demonstrated that the igneous masses in, or near, which the ore has been emplaced were introduced near the then-existing surface. Since the ores were emplaced within a short time, geologically speaking, after the porphyry had been intruded, it is almost certain that the ores were deposited in a near-surface environment. The early-stage mineralization, therefore, cannot be classed as hypothermal but must be considered to have formed within the xenothermal range, and the first stage is here so categorized.

On the other hand, the most important minerals of the second stage are characteristic of leptothermal or epithermal conditions of ore emplacement. The freibergite variety of tetrahedrite, andorite, the various lead sulfosalts, and bournonite definitely belong in one or the other of these two categories of the modified Lindgren classification, the near-surface deposition indicating that epithermal is correct. On the other hand, stannite is a characteristic mineral of the mesothermal or kryptothermal ranges, and the sphalerite and galena associated with it are not incompatible with this concept. Obviously, the near-surface conditions under which the second, as well as the first stage, of mineralization at Oruro was emplaced mean that the stannite must be categorized as kryptothermal rather than mesothermal.

It appears, therefore, that the Oruro ores should be classified as xenothermal to epithermal, with the understanding that the economically important elements of the two stages--tin and silver, respectively--were concentrated in the xenothermal and epithermal ranges, also respectively.

POTOSÍ

Late Tertiary | *Silver, Tin* | *Xenothermal to Epithermal*

Ahlfeld, F., 1935, Neue Beobachtungen am Cerro von Potosí: Zeitsch. f. prakt. Geol., Bd. 43, H. 11, S. 167-171

—— 1936, The Bolivian tin belt: Econ. Geol., v. 31, p. 48-72

—— 1941, Los yacimientos minerales de Bolivia: Dirección General de Minas y Petróleos, La Paz, p. 112-119

—— 1941, Zoning in the Bolivian tin belt: Econ. Geol., v. 36, p. 569-588

—— 1946, Geología de Bolivia: Ministerio de Economía Nacional, La Paz, p. 194-196, 213-215, 325-326

—— 1954, Los yacimientos minerales de Bolivia: El Banco Minero de Bolivia y la Corporación Minera de Bolivia, 2d ed., Bilbao, p. 89-94

Ahlfeld, F., and Schneider-Scherbina, A., 1964, Cerro de Potosí, in *Los yacimientos minerales y de hidrocarburos de Bolivia*: Ministerior Minas y Petróleo Bol. no. 5 (Special), 388 p.

Evans, D. L., 1940, Structural and mineral zoning of the Pailaviri section, Potosí, Bolivia: Econ. Geol., v. 35, p. 737-750

Jaskolski, S., 1933, Les gisements argentostannifères de Potosí en Bolivie: Arch. Minéral. Soc. Sci. Varsovie, t. 9, p. 46-92

Lindgren, W., and Creveling, J. G., 1928, The ore deposits of Potosí, Bolivia: Econ. Geol., v. 23, p. 233-262

Miller, B. L., and Singewald, J. T., Jr., 1919, Potosí silver-tin district, in *Mineral deposits of South America*: McGraw-Hill, N.Y., p. 122-126

Moh, G. H., and Berndt, F., 1964, Two new natural tin sulfides Sn_2S_3 and SnS_2: Neues Jb. f. Mineral., Mh., Jg. 1964, H. 2, S. 94-95

Murillo, J., and others, 1968, Geología y yacimientos minerales de la región de Potosí; tomo II, Parte minera: Geobol. (Bolivia, Serv. Geol.), Bol. no. 11, 175 p.

Rechenberg, H. P., 1955, Gangtektonik und Störungen im Cerro Rico de Potosí,

Bolivien: Neues Jb. f. Geol. und Paläont., Abh., Bd. 101, 3. 1-11

Rivas Valenzuela, S., and Carrasco Córdova, R., 1968, Geología y yacimientos minerales de la región de Potosí; tomo I, Parte geológica: Geobol. (Bolivia, Serv. Geol.), Bol. no. 11, 95 p.

Wendt, A. F., 1891, The Potosí, Bolivia, silver district: A.I.M.E. Tr., v. 19, p. 74-107

Notes

The mining district of Cerro Rico de Potosí lies on the eastern slopes of the Eastern Cordillera of the Bolivian Andes, just north of the 20th parallel of latitude and some 575 km from the Pacific.

The oldest rocks in the general area are Paleozoic shales or slates, with some sandstones and quartzites. To the northwest of Potosí proper, the Paleozoic rocks are unconformably overlain by Cretaceous rocks that consist of cross-bedded sandstones, red shales, and limestones and dolomites. Three lower Tertiary beds unconformably overlie the Cretaceous: (1) the Agua Dulce formation (thick flows of andesite with andesitic tuffs and red beds at the base), (2) the San José garnet-bearing biotite andesite (tuffs and breccias), and (3) the Chalviri series (thick garnet-bearing biotite andesite tuffs, breccias, and conglomerates and thin flows). Into these formations was intruded (probably still in early Tertiary time) the Kari-Kari granodiorite (which forms the nucleus of the Kari-Kari mountain chain), with associated dikes and sills of granodiorite porphyry. Following this intrusion and lying unconformably on the Chalviri series is the Palaviri conglomerate (a thick, angular conglomerate with a sedimentary or calcareous matrix). Probably about middle Tertiary (Miocene) time, the Cerro Rico series (well stratified tuffs, with plant remains in the upper part and tuffs and breccias in the lower part) were poured out over the region. Probably in late Miocene or early Pliocene time, the Cerro Rico series was intruded by a funnel-shaped mass (broad end up) of Cerro de Potosí dacite; dikes of similar rock were intruded into the general area. The intrusion of the dacite was followed by a period of fracturing and then by one of erosion and then the upper Pliocene Huajajchi rhyolite breccia was extruded on the erosion surface, the basal portion of which contains fragments of the Cerro de Potosí intrusive; the breccia is not mineralized.

The dacite intrusion, at the surface, has an oval outcrop, some 1600 m in a north-south direction and about 1200 m from east to west. Some 700 m below the surface, the intrusive has diminished to 400 m by 100 m. The intrusive has been highly altered; the original dacite was a somewhat more mafic rock than those associated with the ores at Oruro and Llallagua. The intrusive probably was not formed in the throat of an ancient volcano, as is indicated by the lack of explosive breccias, and probably never reached the surface. Although the intrusion has strongly dislocated the formations through which it has passed, it has had no major contact metamorphic effect on either the Ordovician sedimentary rocks or the Tertiary volcanic rocks that border it. It is certain that the ores at Potosí were not developed until after the accumulation of the middle Tertiary (Miocene) Cerro Rico series and after the intrusion of that formation by the Miocene or Pliocene dacite intrusion, both of which contain ore veins. It seems probable, therefore, that the period of earth movement that developed the fracturing in which the vein material was later deposited must have been latest Miocene or early Pliocene; the fracturing certainly occurred before the deposition of the Huajajchi rhyolite breccia of upper Pliocene age. It appears reasonable, therefore, to say that the ores were emplaced in the very earliest Pliocene; they are here categorized as late Tertiary.

In the upper part of the Cerro Rico de Potosí, 34 veins and vein branches have been distinguished that unite, at depth, into 5 principal vein systems; from east to west these are: (1) Tajo Polo, (2) Encinas, (3) Mendieta, (4) San Miguel, and (5) Alco Barreno. The Mendieta vein system appears to be the most important, and the Tajo Polo and the Encinas probably join with it; at greater depth, the Mendieta divides to form the Utne and Bronce veins. All the veins occur in normal faults that show little displacement; the strikes of veins generally are more or less northeast. The Potosí veins have a distinct en echelon structure below the Caracoles level; above that level the structure is not certainly known, but probably is much the same. In plan and section, each vein can be divided into a succession of discontinuous veins, separated from each other by masses of country rock that contain no more than weakly mineralized stringers. The vein systems, however, are not affected by changes in rock types. The ores at Potosí can be divided into (1) an upper oxide zone, (2) a lower oxide zone, (3) an upper sulfide zone, and (4) a lower sulfide zone. The oxide zones contained both tin and silver; an appreciable enrichment of silver occurred in the oxide zones, and some concentration of tin was due to the removal of more easily oxidized and dissolved constituents. The minerals of these two sulfide zones probably were deposited in three stages: (1) coincident deposition of pyrite, arsenopyrite, and quartz in the upper zone and the same minerals, plus wolframite and bismuthinite, in the lower zone; (2) quartz, cassiterite, and pyrite in both zones; and (3) quartz, chalcopyrite, and stannite in the lower zone and the same minerals, plus complex silver-bearing sulfosalts such as andorite, tetrahedrite, jamesonite, matildite and ruby silver, in the upper zone. Some question exists as to whether these three stages were continuous or were separated by structural movements that reopened the veins. The first two stages appear to have taken place under xenothermal conditions, although the wall-rock alteration in the rhyolite was not intense. The lower zone portion of the third stage appears to have been deposited under the near-surface equivalent of mesothermal conditions (kryptothermal) while the sulfosalts of the upper stage portion probably were formed in the epithermal range. The primary mineralization of the oxide zones probably was much the same as that of the upper sulfide zone--pyrite, arsenopyrite, cassiterite, and quartz, plus the wide variety of upper sulfide zone third-stage minerals. Potosí and Oruro are the only Bolivian tin deposits in which an epithermal stage of mineralization can be recognized, although Pirquitas in Argentina closely resembles the upper sulfide zone at Potosí. From this description of the Potosí mineralization, it would appear that the deposits must be classified as xenothermal to epithermal, with the cassiterite being xenothermal and the silver epithermal; the appreciable volume of kryptothermal mineralization has, surprisingly, been of no economic value.

COROCORO

Late Tertiary | *Copper* | *Telethermal*

Ahlfeld, F., 1933, Über die Bildung der Kupfererzlagerstätte Corocoro: Zentralblatt f. Mineral., Geol. und Palaont., Abt. A., S. 375-382

—— 1941, Los yacimientos minerales de Bolivia: Dirección General de Minas y Petróleos, La Paz, p. 213-218, 223-226

—— 1946, Geología de Bolivia: Ministerior de Economía Nacional, La Paz, p. 230-247

—— 1953, Die Metallprovinz des Altiplano (Bolivien): Neues Jb. f. Mineral., Abh., Bd. 85, H. 1, S. 1-58 (particularly p. 31-39)

Berton, A., 1937, The Corocoro copper district of Bolivia: A.I.M.E. Tr., v. 126, p. 541-555; disc. (J. T. Singewald, Jr.), p. 555-558

Brüggen, J., 1934, Die Puca-Sandsteine von Corocoro in Bolivien, in *Grundzüge der Geologie und Lagerstättenkunde Chiles*: Heidelberger Akademie der Wissenschaften, Mathematisch-Naturwissenschaftliche Klasse, Max Weg, Leipzig, S. 80-94

Entwistle, L. P., and Gouin, L. O., 1955, The chalcocite-ore deposits at Corocoro, Bolivia: Econ. Geol., v. 50, p. 555-570

Geier, B., 1928, Beiträge zur Frage der Entstehung der bolivianischen Kupfererzlagerstätten vom Typus Corocoro: Neues Jb. f. Mineral., Geol. und Paläont., Beil. Bd. 58, Abt. A., S. 1-42

Kohanowski, N. N., 1944, Geología de yacimientos cupríferos de Bolivia: Minería Boliviana, no. 1, p. 9-21

Lincoln, F. C., 1917, Corocoro copper mines: Min. and Sci. Press, v. 115, no. 13, p. 461-463

Ljunggren, P., and Meyer, H. C., 1964, The copper mineralization in the Corocoro basin, Bolivia: Econ. Geol., v. 59, p. 110-125

Miller, B. L., and Singewald, J. T., Jr., 1919, The Corocoro copper district, in *Mineral deposits of South America*: McGraw-Hill, N.Y., p. 88-94

Pélissionier, H., 1965, Structure géologique et genèse du gisement de cuivre de Corocoro (Bolivie): Soc. Geol. France Bull., Ser. 7, v. 6, no. 4, p. 502-514

Rutland, R. W., 1966, An unconformity in the Corocoro Basin, Bolivia, and its relation to copper mineralization: Econ. Geol., v. 61, p. 962-963

Singewald, J. T., Jr., 1928, A genetic comparison of the Michigan and Bolivian copper deposits: Econ. Geol., v. 23, p. 55-61

—— 1935, The Corocoro copper district, Bolivia, in *Copper resources of the world*: 16th Int. Geol. Cong., v. 2, p. 449-457

Singewald, J. T., Jr., and Berry, E. W., 1922, The geology of the Corocoro copper district of Bolivia: The Johns Hopkins Univ. Studies in Geol., no. 1, 117 p. (particularly p. 1-12 and 53-80)

Notes

The ore deposits of the Corocoro Basin center around the town of Corocoro, which is located some 85 km south-southwest of La Paz. The Corocoro Basin lies in the western part of the Bolivian altiplano, the elevated plateau that lies between the Cordillera Real (east) and the Cordillera Occidental (west), and extends southward for some 750 km from Lake Titicaca (northwest) almost to the border with Argentina. Within the basin itself, the relief is more rugged than in the remainder of the plateau, and the drainage, except for the Rio Desaguadero, is intermittent.

The stratigraphic sequence in the Corocoro portion of the basin has been the subject of considerable controversy, and the latest detailed work (Ljunggren and Meyer, 1964) probably is not the final word on the subject. Ljunggren and Meyer have designated the oldest rocks in the basin as the Cretaceous or older Chuquichambi group; this group is composed of soft, thin-bedded gypsiferous shales and marls. These sediments in the basin are confined to diapiric folds in the central portions of anticlines; in this position, they are usually intermingled with younger sediments into which they have been injected.

The next series of rocks lying above the Chuquichambi beds has been dated as undifferentiated Tertiary and known as the Corocoro group (Ljunggren and Meyer, 1964); portions of the series have been assigned to the late Tertiary, but there is some question as to the value of the evidence on which this was done. The oldest formation of the generally molasse-type rocks of the Corocoro group is (Entwistle and Gouin, 1955) the Huayllamarca formation, which has been divided into three members: (1) the Lianquera sandstones that consist of massive reddish-brown sandstones interbedded with sandy shales in lenses; there is some sparse copper mineralization in these beds; (2) the Coniri conglomerate, which consists of very coarse, piedmont-type conglomerates in which copper mineralization is found; and (3) the Chacarilla sandstones, which are composed of massive, grayish-brown, coarse-grained sandstones that are feldspathic to arkosic; there is copper in some of the horizons of this member.

The Huayllamarca formation was followed conformably by the Totora formation that consists of two members, the lower of which is the Ramos; this member is composed of light violet to reddish-brown sandy shales with lenses of fine-grained, buff arkosic sandstone and some evaporites--copper deposits are found in sandy beds and fissures. The upper member is the Vetas; this member is made up of friable, light-gray, coarse conglomerates interbedded with light-gray sandstones--the pebbles are mainly of extrusive volcanic rocks; copper deposits are found in several horizons within this member. There has been much dispute as to the relative ages of the Ramos and the Vetas, since the contact between the two was, for many years, known only along the Corocoro fault. More recent work (Ahlfeld, 1953) has found that, in the Toledo mine beyond the influence of the Corocoro fault, the Vetas rests unconformably on the Ramos. The Vetas member of the Totora formation has been divided into several mappable units.

The members of the Totora formation are overlain by those of the Crucero formation, which is divided into two members. Unconformably above the members of the Corocoro group are the undivided rocks of the Pliocene Umala group, 1000 m of tuffs, welded tuffs, and andesitic to rhyolitic flows (?) interbedded with half-consolidated gray conglomerates.

Singewald and Berry (1922) were convinced that the fossil flora from the Vetas formation establishes it as Pliocene. Since they were under the apparently mistaken impression that the Vetas was older than the Ramos, they classified the Ramos, on essentially no more than meager paleontologic evidence, as late Pliocene or Pleistocene. Now that the Ramos is definitely known to be pre-Vetas, its age must be lower Pliocene or even Miocene, but it would seem that both members of the Totoro formation are almost certainly late Tertiary in age.

There are no known intrusive igneous rocks in the vicinity of the ore bodies, but there are such outcrops at distances of 16 and 19 km to the north and northwest, respectively. A map by Ahlfeld shows these intrusives as being in the Ramos beds only, but this certainly does not guarantee that the igneous rocks are pre-Vetas, although it has been stated (Entwistle and Gouin, 1955) that the boulders and pebbles of igneous rock found in the Vetas appear to be the same rock (type not given) that intrudes the Ramos. Certainly, further work is needed before this point is settled.

The fault shows no evidence of having been a channelway for ore fluids since the red color of the gouge almost certainly would have been partially bleached at least if such fluids had passed through it.

If the copper ores in the Totora formation are syngenetic, the age of the mineralization is almost certainly late Tertiary, although it might conceivably be middle Tertiary. If the ores are epigenetic, they must be older than latest Tertiary since the Pliocene Umala group is post-ore. The ores are here considered to have been epigenetically introduced; they are, therefore, classified as late Tertiary.

Ljunggren and Meyer (1964) have reached the following conclusions as to the copper mineralization: (1) it is always located in sandy or conglomeratic layers; (2) where mineralization occurs in reddish conglomerates and sandstones, it is always surrounded by a bleached zone; (3) it is normally located in elongate, lens-shaped bodies; (4) ore bodies always occur in parallel series with barren strata between them; (5) it is always confined to a single stratum within the member in question; (6) it is made up of two types, chalcocite and native copper--only one of these two mineral species is found in a given primary ore body; (7) most of the mineralized beds contain fossil remnants of plants ranging from big tree trunks to carbonaceous particles of microscopic size; no beds with fossil plants lack at least some copper mineralization; and (8) higher-grade mineralization is always located in structural highs.

In addition to chalcocite and native copper, other minerals occur in minor amounts; these include: galena, native silver, domeykite (Cu_3As), pyrite, and chalcopyrite. Some of these minerals are partially replaced by the principal copper minerals. Gangue minerals include small amounts of aragonite, barite, gypsum, and celestite; aragonite crystals have been, in many places, replaced by native copper.

The chalcocite ore bodies are concentrated in the Vetas, and the chalcocite has been reported (Ahlfeld, 1941) to be orthorhombic and, therefore, to have been precipitated at 105°C or less. The sulfide ores contain a small amount of silver; 70 percent copper concentrates will run 30 to 40 gms of silver per ton. Small amounts of secondary native silver and copper have been found in the Vetas ore bodies where these are near the present surface. The chalcocite was deposited in open space provided by cavities, by porous parts of the sandstone, and by minute fractures in the sand grains; chalcocite also replaced the original ferruginous cement. Chalcedonic quartz fills pores in the sandstone and locally was deposited on chalcocite. Entwistle and Gouin (1955) saw no evidence of chalcocite changing to copper with depth in the Vetas ore bodies as had been previously suggested; it is only where workings pass from Vetas to Ramos beds that native copper is found at lower levels rather than chalcocite.

In both the Ramos and the Vetas, the ore mineralization is confined to definite horizons within the member in question, and there appears to be no evidence of ore-filled fractures connecting one horizon with another. In the Ramos, the native copper is found as cement around sand grains and as sheets and plates (some quite large) along bedding and in fractures cutting the Ramos shales and sandstones. Some of the copper sheets are hundreds of kilograms in weight. In the Ramos, the reddish color of the mineralized layers has been bleached for short distances out from the ore. In the Vetas, on the other hand, the only visual difference between mineralized and unmineralized portions of even a somewhat reddish-colored bed is the speckled appearance imparted to it by the tiny grains of chalcocite.

From the descriptions of the Vetas and Ramos deposits, it is not clear as to whether the deposits were formed by (1) low-temperature hydrothermal fluids of ultimately igneous origin, (2) precipitation directly from sea water at the same time as the clastic materials of the deposits were being accumulated, or (3) mobilization of copper minerals, earlier deposited in the two members, by ground waters and their reprecipitation in their present form and position. One obvious objection to the epigenetic theory is that no igneous rocks are known within 16 to 19 km of the Corocoro deposits. If there had been no igneous activity in the district after the lithification of the Ramos and Vetas members, it would seem certain that the deposits could not have been derived from hydrothermal fluids of igneous origin. It is, however, unreasonable to suppose that an area of such appreciable and often-occurring igneous activity as the Bolivian altiplano should not have been subjected to that process in the time between the lithification of the Totora formation and the beginning of Umala sedimentation. Corocoro deposits cannot be dropped from the hydrothermal category simply because there are no igneous outcrops in their immediate vicinity; this can be done only if the characteristics of the deposits rule out a hydrothermal origin.

Entwistle and Gouin (1955) admit that a nonhydrothermal origin for the chalcocite of the Vetas deposits meets with certain difficulties; these are (1) they would obtain the copper in the Vetas chalcocite from the Ramos deposits in which they consider that the copper may have been precipitated from solutions of igneous origin, and (2) they believe the Vetas copper to have been transported as the sulfate and precipitated by reactions with carbonaceous materials and/or sulfate-reducing bacteria, probably during diagenesis. Under any nonsyngenetic theory of origin, the channels through which the chalcocite-depositing solutions moved were the more porous layers of the Vetas member, since there appear to have been essentially no pre-ore fractures in the Vetas and Ramos members. It seems surprising that, believing the copper in the Ramos beds to have been deposited from solutions of igneous origin, these authors appeal to a nonigneous source for the Vetas copper. In both members of the Totora formation, the ores are localized in porous layers, entirely in the Vetas and largely in the Ramos. In the somewhat more fractured Ramos, some copper is found in fractures, but basically the channelways for ore-solution movement in both formations were the porous layers. But hydrothermal fluids would just as certainly be directed into such porous routes of movement as would waters of meteoric origin.

Entwistle and Gouin (1955) admit that the bleaching around the deposits of native copper in the Ramos beds almost certainly resulted from the reduction of ferric iron to the ferrous state concomitantly with the oxidation of S^{-2} to the S^{+6} of SO_4^{-2}. On the other hand, they point to the lack of alteration in the grayish sandstones and conglomerates surrounding the chalcocite deposits in the Vetas beds as indicating that the ore fluids could not have been igneous in origin. Certainly, any hydrothermal fluids that had cooled to temperatures of 100°C or less (as is witnessed by the orthorhombic chalcocite in the Vetas deposits) would have had hardly any effect on the rocks through which they moved unless, as was true in the Ramos, these rocks were high in ferric iron. Certainly, the presence of native copper in the iron-rich Ramos beds and chalcocite in those of the iron-poor Vetas can be thought to suggest that both types of deposits were formed from the same solutions at much the same time, the differences between the deposits being due almost entirely to the reduction of charged copper to the native state in the Ramos beds and to its failure to reduce in the Vetas layers.

Ljunggren and Meyer (1964) on the other hand, would derive the copper in the deposits either from copper-bearing basalts, which they suggest at one time covered much of the Corocoro Basin, or from long-since vanished porphyry-copper deposits in the adjacent Cordillera Occidental. Surely such suggestions are no more supported by actual evidence than the concept that there was a granodioritic or granitic magma chamber at depth beneath the basin at the time of ore formation. Granted, however, that the meteoric ore solutions came from one or the other of these suggested sources, it is necessary to find a method of converting the probably copper-sulfate-bearing solutions to solid native copper in the Ramos and chalcocite in the Vetas. These authors appear to believe that the beds of the Vetas and Ramos contained much more organic material than Entwistle and Gouin thought, and they appeal to this material to precipitate copper as chalcocite (or as an unnamed organic copper compound, all traces of which have long since completely vanished) or as native copper. They also suggest that sulfate-reducing bacteria may have played a part in such precipitations, although they do not say specifically how. They do think, however, that most of the native copper owed its origin to reactions between the meteoric ore fluid and ferric iron. Since the copper is apparently assumed to have been in balance with sulfate ion in the ore fluid, it is difficult to understand how ferric iron could have affected the cupric sulfate since ferric iron cannot reduce sulfate ion or cupric ion. In short, unless the ore fluid contained sulfide ion, reactions with ferric iron would have had no ability to produce native copper from cupric ion.

Ljunggren and Meyer (1964) believe that the original deposition of native copper and chalcocite was in widespread, low-grade deposits and that the present much higher-grade deposits were produced by a later cycle of concentration of these minerals on the steep flanks of narrow anticlines. They do not, however, explain how the copper minerals were dissolved, transported, and redeposited in these more concentrated deposits.

Obviously, these brief reviews are too short to do complete justice to the detailed presentations of these two pairs of authors. But it does seem that the difficulties their hypotheses have in explaining the Corocoro deposits have been fairly pointed out. It seems to me that a single copper-sulfur complex-containing ore fluid of igneous origin, moving upward through the more porous layers of the two formations before development of the Corocoro fault, would deposit in just the places in which ore is now found and would have deposited native copper in the iron-rich Ramos beds and chalcocite in the iron-poor Vetas layers. The lack of wall-rock alteration, other than bleaching around the deposits in the Ramos, would have been due to the low temperature to which the ore fluids had been brought by their long journey from their magma chamber source. The deposits, therefore, are here classified as telethermal.

PULACAYO (HUANCHACA)

Late Tertiary	*Zinc, Silver, Lead*	*Leptothermal to Telethermal*

Agterberg, F. P., 1961, The skew frequency-curve of some ore minerals: Geol. Mijnbouw, v. 40, no. 4, p. 149-162

Ahlfeld, F., 1939, Die Silber-Blei-Zinkerzlagerstätte Pulacayo: Neues Jb. f. Mineral., Geol. und Paläont., Beil. Bd. 75, Abt. A, S. 1-23

—— 1954, Los yacimientos minerales de Bolivia: El Banco Minero de Bolivia y Corporación Minera de Bolivia, 2d ed., Bilbao, p. 157-159

Lyons, W. A., 1963, Structural geology of Pulacayo mine, Bolivia: Econ. Geol., v. 58, p. 978-987

Penfield, S. L., and Frenzel, A., 1897, On the identity of chalcostibite (wolfbergite) and guejarite and on chalcostibite from Huanchaca, Bolivia: Amer. Jour. Sci., 4th Ser., v. 4, p. 27-35

Rudroff, A., 1932, Der Silber-Blei-Zinkerzbergbau von Pulacayo in Bolivien: Metall und Erz, Jg. 29, H. 7, S. 125-131

Toborffy, Z., 1904, Der Kupferkies von Pulacayo: Zeitsch. f. Kristal., Bd. 39, S. 366-373

Notes

The rich silver veins of Pulacayo are located in southern Bolivia, about 135 km southwest of Potosí and 20 km northeast of the important railroad junction of Uyuni.

The deposits lie just west and outside of the southern tin province and near the eastern border of the altiplano in continental sediments; these are probably of Cretaceous age but may belong to the continental early Tertiary. At Pulacayo, the continental rocks are fine- to medium-grained red sandstones, locally argillaceous and containing two gradationally intercalated beds of argillaceous conglomerate which provide marker beds in the mine area. West of Pulacayo is a low mountain range of nonfossiliferous Paleozoic sandstone, while to the east, the red sandstones are discordantly overlain by argillaceous marls and sandstones that in places contain gypsum and anhydrite beds. The red sandstone beds were gently folded into an asymmetrical anticline during the early Tertiary orogeny and before the igneous activity to which the ores are related. Intrusive into these sediments is a stock of dacite (Ahlfeld) or andesite (Lyons); the probably secondary character of the bulk of the quartz indicates that the rock was originally an andesite. The upper portions of this andesite body probably were exposed by erosion and then covered by dacite flows; since the upper portions of the Pulacayo veins are in this dacite, it is certain that the flows preceded the mineralization. The vein system at Pulacayo is (with the exception of the isolated No. 4 vein) essentially one vein that has branched considerably at its upper end. The lower portion of the vein lies in sandstone, strikes, east-west, and dips 65° to 80° to the north. On encountering the andesite stock above the sandstone, the vein was refracted, reversed its dip, and branched into some ten minor veins that lie both in the andesite and in the dacite flow rock above it. As the vein system was developed after both the early Tertiary folding and the later igneous activity and as the mineralization followed the dacite flows, there is no question but that the mineralization was Tertiary in age; it probably was introduced into the vein system during the late stages of, or even after, the Pliocene orogenic cycle. The deposits are, therefore, here categorized as late Tertiary.

Since the mineralization at Pulacayo is outside the tin province, it not only contains no cassiterite but also appears to include no tin-bearing sulfosalts such as stannite or franckeite. The lower portion of the vein, known as "la Veta Tajo," has been economically mineralized horizontally for some 3400 m and has been followed to a depth of 1100 m below the surface; the average width of the vein is over 1.0 m, although at the greatest depths reached the minable ore is concentrated in vertical shoots separated by stretches of barren vein material. In the minor upper-level veins, the minerals present are galena, wurtzite, sphalerite, minor chalcopyrite, lead sulfo-antimonides, stibnite, and barite with very little silver. In the upper part of the Tajo vein, the mineralization changes to an assemblage of pyrite, dark brown sphalerite (4 to 5 percent Fe), chalcopyrite, bournonite, abundant tetrahedrite (containing 7 to 10 percent Ag and falling in the freibergite category), and barite. In the ore shoots of the deeper portions of the main vein, the amounts of freibergite and galena drop off appreciably (as does that of silver) and sphalerite increases in quantity. On the 698 meter-level (the 806 meter-level appears to be the lowest to have been reached), the mineralization was still continuous, and the average grade was 14 percent zinc, 2 percent lead, 0.5 percent copper, 0.1 percent silver, and traces of gold and bismuth. The mine water at present has a temperature of 60°C and is depositing considerable volumes of aragonite and giving off large amounts of carbon dioxide.

This large, strong vein, well mineralized over a length of over two miles and having a vertical extent of nearly two-thirds of a mile, suggests conditions of slow loss of heat and pressure. So does the presence of low-temperature lead-zinc mineralization, poor in silver, in the weaker vein-structures of the upper levels and the gradual change to silver-rich tetrahedrite and galena and then to abundant sphalerite at greater depths. In many ways the deposit resembles that of Casapalca, although it appears that, at the depths yet reached, no true mesothermal zone has been encountered and that the upper branches of the vein contain what is probably a telethermal type of mineralization. Therefore, the deposit is here classed as leptothermal to telethermal on the assumption that the dacite flows covering the andesite stock were sufficiently thick and unbroken to assure that conditions of slow loss of heat and pressure obtained during the mineralization process.

BRAZIL

MINAS GERAIS

Middle Precambrian *Late Precambrian*	*Iron as Hematite (much as Martite), minor Magnetite*	*Sedimentary-A1a, Metamorphic-C, and Hydrothermal-1*

Barbosa, O., 1949, Contribuicao à geologia do Centro de Minas Gerais: Mineracao e Metalurgia (Rio de Janeiro), v. 14, no. 79, p. 3-19

Barbosa, A. L. de Miranda, 1968, Geologia do Quadrilátero ferrífero: 22d Congr. Brasil. Geol., Belo Horizonte, Roteriros Excursões, p. 2-6

Dorr, J. V. N., II, 1954, Comments on the 'iron deposits of the Congonhas district, Minas Gerais, Brazil': Econ. Geol., v. 49, p. 659-662

—— 1964, Supergene iron ores of Minas Gerais, Brazil: Econ. Geol., v. 59, p. 1203-1240

—— 1965, Nature and origin of the high-grade hematite ore of Minas Gerais, Brazil: Econ. Geol., v. 60, p. 1-46

—— 1970, Physiographic, stratigraphic, and structural development of the quadrilatero ferrifero, Minas Gerais, Brazil: U.S. Geol. Surv. Prof. Paper 641-A, p. A1-A110

Dorr, J.V.N., II, and Barbosa, A. L. de Miranda, 1963, Geology and ore deposits of the Itabira district, Minas Gerais, Brazil: U.S. Geol. Surv. Prof. Paper 341-C, p. C1-C110

Dorr, J. V. N., II, and others, 1952, Origin of the Brazilian iron ores, in *Symposium sur les gisements de fer du monde*: 19th Int. Geol. Cong., v. 1, p. 286-310 (particularly p. 288-297)

—— 1957, Revisão da estratigrafia Precambriana do Quadrilátero Ferrífero, Minas Gerais, Brazil: Dept. Nacional da Prod. Divisão de Geolgia e Mineralogia, Avulso 81, 31 p.

Ebert, G., 1957, Beitrag zur Gliederung des Prekambrium in Minas Gerais: Geol. Rundschau, Bd. 45, H. 3, S. 471-521

Eichler, J., 1967, O enriquecimento residual e supergênico de itabiritos através de intemperismo: Rio de Janeiro, Univ. Fed., Inst. Geociênc., Geol., Bol. no. 1, p. 29-40 (Engl. summ.)

Freyberg, B. von, 1932, Ergebnisse geologischer Forschungen in Minas Geraes (Brasilien): Neues Jb. f. Mineral., Geol. und Paläont., Sonderband 2, 403 S. (particularly p. 16-39, 39-60, 76-95, 267-277)

Gair, J. E., 1962, Geology and ore deposits of the Nova Lima and Rio Acima quadrangles, Minas Gerais, Brazil: U.S. Geol. Surv. Prof. Paper 341-A, p. A1-A67

Gathmann, Th., 1913, Beitrag zur Kenntnis der "itabarit" Eisenerze in Minas Geraes (Brasilien): Zeitsch. f. prakt. Geol., Jg. 21, H. 5, S. 234-240

Grösse, E., and others, 1946, O minério de ferro da fazenda Fábrica, da Companhia de Mineração de Ferro e Carvâo, S. A., distrito de São Julião, município de Ouro Preto, estado de Minas Gerais: Mineração e Metelurgia (Rio de Janeiro), v. 11, nos. 64, 65, p. 105-115, 267-273

Guild, P. W., 1953, Iron deposits of the Congonhas district, Minas Gerais, Brazil: Econ. Geol., v. 48, p. 639-676

—— 1957, Geology and mineral resources of the Congonhas district, Minas Gerais, Brazil: U.S. Geol. Surv. Prof. Paper 290, 90 p.

Guimarães, D., 1935, Contribução ao estudo do origem dos depositos de minério de ferro e manganez do centro de Minas Geraes: Serv. Fomento Prod. Mineral (Brazil) Bull. no. 8, 70 p.

—— 1947, Metalogênese nas formacões árqueo-proterozóicas do Brazil: Instituto de Technologia Industrial (Belo-Horizonte) Bull. no. 4, 65 p.

Harder, E. C., 1914, The "itabirite" iron ores of Brazil: Econ. Geol., v. 9, p. 101-111

Harder, E. C., and Chamberlin, R. T., 1915, The geology of central Minas Geraes, Brazil: Jour. Geol., v. 23, p. 341-378, 385-424

Herz, N., and Dutra, C. V., 1960, Minor element abundance in a part of the Brazilian shield: Geochim. et Cosmochim. Acta, v. 21, p. 81-98

Johnson, R. F., 1962, Geology and ore deposits of the Cachoeira do Campo, Dom Bosco, and Ouro Branco quadrangles: U.S. Geol. Surv. Prof. Paper 341-B, p. B1-B39

Leith, C. K., and Harder, E. C., 1911, Hematite ores of Brazil and a comparison with hematite ores of Lake Superior: Econ. Geol., v. 6, p. 670-686

Miller, B. L., and Singewald, J. T., Jr., 1919, Hematite ores (Minas Geraes), in *Mineral deposits of South America*: McGraw-Hill, N.Y., p. 169-176

Moore, S. L., 1969, Geology and ore deposits of the Antônio dos Santos, Gongo Sôco and Conceicão do Rio Acima quadrangles, Minas Gerais, Brazil: U.S. Geol. Surv. Prof. Paper 341-I, p. I1-I50

Park, C. F., Jr., 1959, The origin of hard hematite in itabarite: Econ. Geol., v. 54, p. 573-587 (although the paper discusses other areas than Minas Gerais, the emphasis is on the iron ores of that region)

Park, C. F., Jr., and others, 1951, Notes on the manganese ores of Brazil: Econ. Geol., v. 46, p. 1-22

Percival, F. G., 1964, Geology and ore deposits of the Belo Horizonte, Ibirité, and Macacos quadrangles, Minas Gerais, Brazil (rev.): Econ. Geol., v. 59, p. 1398-1402

Pomerene, J. B., 1964, Geology and ore deposits of the Belo Horizonte, Ibirité, and Macacos quadrangles, Minas Gerais, Brazil: U.S. Geol. Surv. Prof. Paper 341-D, p. D1-D84

Reeves, R. G., 1966, Geology and mineral resources of the Monlevade and Rio

Piracicaba quadrangles, Minas Gerais, Brazil: U.S. Geol. Surv. Prof. Paper 341-E, p. E1-E58

Roberts, H. M., 1964, Geology and ore deposits of the Itabira district, Minas Gerais, Brazil (rev.): Econ. Geol., v. 59, p. 511-514

Rynearson, G. A., and others, 1954, Contacto basal da serie de Minas na parte ocidental do quadrilatero ferrifero, Minas Gerais, Brazil: Dept. Nacional da Prod. Mineral, Divisão de Geologia e Mineralogia, Avulso 34, 18 p.

Sanders, B. H., 1932-1933, Iron ore at Itabira (Minas Gerais), Brazil: Inst. Min. and Met. Tr., v. 42, p. 570-607

Simmons, G. C., 1968, Geology and iron deposits of the western Serra do Curral, Minas Gerais, Brazil: U.S. Geol. Surv. Prof. Paper 341-G, p. G1-G57; (rev.) Econ. Geol., v. 64, p. 127

—— 1968, Geology and mineral resources of the Barão de Cocais area, Minas Gerais, Brazil: U.S. Geol. Surv. Prof. Paper 341-H, p. H1-H46

Tyler, S. A., 1948, Itabirite of Minas Geraes, Brazil: Jour. Sed. Petrology, v. 18, no. 2, p. 86-87

Wallace, R. M., 1966, Geology and mineral resources of the Pico de Itabirito district, Minas Gerais, Brazil: U.S. Geol. Surv. Prof. Paper 341-F, 66 p.

Notes

The iron deposits of Minas Gerais are located in the Quadrilátero Ferrífero in the south central part of the Brazilian state of the name; the city of Belo Horizonte lies in the northwest corner of this ore-bearing region. The iron-bearing formation of Minas Gerais crops out in long ridges that are confined within an area of about 7000 sq km, with isolated erosional remnants of the same formation to the north and east of the Quadrilátero. The most important of these outliers is the Itabira district that is situated to the northeast of the northeast corner of the quadrilateral containing the largely continuous beds of iron-bearing formation. Although a study of the geology of the 42 complete or partial quadrangles that make up the Quadrilátero Ferrífero has been undertaken as a joint project of the Brazilian and U. S. governments, the results of this work have been published for less than half of the quadrangles or partial quadrangles and for the isolated Itabira district.

The oldest rocks in the area have long been thought to be the basement rocks described by Guild (1957) from the Congonhas area in the southwest corner of the Quadrilátero. Guild considers this complex of gneissic and silicic igneous rocks to be the basement on which the iron-bearing rocks and their associated sediments were deposited; apparently ancient gneissic rocks are not exposed in the Itabira district.

In more recent publications, it is argued that most, if not all, of the granites and gneisses show intrusive contacts with all of the older Precambrian rocks. Because the granites weather so much more readily than the metasediments, however, the contacts are not too readily studied, and the conclusions are based on a limited number of samples. It does appear certain, however, that some of the granites were intruded into the metasediments because of the contact-metamorphic halos surrounding certain of the granitic bodies.

Radioactive dating techniques have been applied in an effort to determine the ages of the various granitic masses in the Quadrilátero and the Itabira district. In the northwest part of the Quadrilátero, granitic rocks that give age dates of 1330 to 1260 m.y. and 560 to 475 m.y. have been found. Farther south, at Itabirito, an age of 1330 m.y. has been assigned to granite. Still farther south, at Engenheiro Corrêa, is granodiorite for which ages ranging between 2524 and 2440 m.y. have been determined. Some of the granite in the region between Itabirito and Engenheiro Corrêa, which has ages of about 1330 m.y., is considered to have been formed by anatexis or palingenesis of a 2524 m.y. granodiorite. The granitic rocks in the Itabira district appear to belong entirely to the 500 m.y. ± age group, as do the granites along the western margin of the district. The age of much of the granite on the eastern and northern margins of the Quadrilátero is unknown.

It seems reasonable to assume that the Engenheiro Corrêa granodiorite is actually basal-complex material and is older than the metasedimentary rocks, but even this is not certain. On the other hand, it is probable that the 1300 m.y. ± granites of the south-central, southwestern, and northwestern areas are younger than the metasediments, this conclusion being supported by numerous examples of 1300 m.y. granite that intrude the metasediments. It follows, then, that the 500 m.y. ± granites must also be younger than the metasedimentary rocks.

The next younger rocks to the basal complex are considered to be those of the Rio das Velhas series that has been divided in the Quadrilátero Ferrífero (Dorr and Barbosa, 1963) into two groups of which the Nova Lima is the older and which is made up, from the base upwards, of carbonate-facies iron formation, graywacke, quartz-ankerite-dolomite rock, sericitic quartzite, more carbonate iron formation, and phyllites and greenschist of both volcanic and sedimentary origin. In the Itabira district, the Nova Lima group probably consists of dolomitic schist, subgraywacke, arkose, chlorite, schist, biotite schist, quartzose schist, and iron formation.

The Nova Lima group is overlain unconformably by the rocks of the Maquiné group, the lower portion of which is made up of quartzite, graywacke, quartzose schist, and phyllite and the upper of quartzite, conglomerate, and quartzose schist.

The Rio Velhas series is separated from the next younger series, the Minas, by a great unconformity. The Minas series has been separated into three groups of which the oldest is the Caraca; the Caraca group, in turn, has been divided into two formations, the lower of which (the Moeda formation) is composed mainly of quartzite, and conglomerate. The upper formation, known as the Batatal schist, is made up of phyllite that is locally graphitic; there are also some thin beds of quartzite, dolomitic material, and ferruginous phyllite.

The Itabira group appears to lie unconformably on the Caraca group and consists mainly of itabirite and dolomite, the latter having been derived from chemical sediments. The Itabira group has been divided into two formations, the older of which is the Cauê Itabirite; it is composed almost entirely of itabirite and hematite ore derived from itabirite. Throughout the Quadrilátero and in the Itabira district, the itabirite forms high ridges that may be as much as 400 to 600 m above the surrounding lowlands, indicating the ability of the itabirite to resist erosion. The itabirite is mainly a definitely banded quartz-iron oxide rock of granoblastic texture that was derived from the metamorphism of oxide-facies iron formation in which the original bands of chert or jasper have been recrystallized to granular quartz and in which the iron is found as hematite, magnetite, or martite; the iron content of itabirite ranges between 25 and 60 percent. Any iron formation containing less than 25 percent iron has been classed (Dorr and Barbosa, 1963) as ferruginous crystallized chert and any with more than 66 percent iron is categorized as high-grade hematite.

The younger formation of the Itabira group is the Gandarela formation that consists of dolomite, magnesian limestone, and dolomitic itabirite with some schist, phyllite, and quartzite being locally present. The contact between the Cauê and the Gandarela is drawn, by Dorr in 1958, where the normal (siliceous) itabirite becomes subordinate to the dolomitic itabirite of the Gandarela. The Itabira group is overlain, with local unconformities, by the Piracicaba group, which is divided into five formations.

In the southeastern part of the Quadrilátero, the Minas series is unconformably overlain by the Itacolomi series. The rocks of this series contain abundant pebbles, cobbles, and boulders of itabirite, but none of these rock fragments is composed of hard hematite. Since the hard hematite is far more resistant to erosion than is the itabirite, the lack of hard hematite in the rock fragments of the Itacolomi series strongly suggests that the hard hematite is epigenetic and was formed after Itacolomi time. Except for a few metasediments of uncertain age and a Cambro-Silurian (?) Bambui limestone, the only essentially sedimentary rocks in the area younger than the Itacolomi series are some assorted sediments of Tertiary and Quaternary age.

In the Quadrilátero and in the Itabira district, the rocks of Itacolomi and greater age are strongly folded. The various series involved in the folding were contorted into huge anticlines and synclines that have been, in many places, overturned. The outcrops of the highly resistant Minas series, preserved only in the major synclines and along the edges of domed structures, outline the patterns of these major folds that have amplitudes of from 1 to more than 10 km, while those of the minor folds range down to millimeters.

In the Rio das Velhas series, the metamorphism was so intense that bedding is difficult, if not impossible, to distinguish, but in the Minas series bedding is normally easy to recognize. The metamorphism has been both thermal and dynamic.

In addition to the various granites, which have already been described, the Itabira district contains an unfoliated granite, known as the Borrachudos granite, that forms the entire northwest boundary of the district. It appears to be of the same general age as the gneisses in the Itabira district but almost certainly was introduced after the last of the earth movements that converted the older granites into gneisses. Its intrusion probably was the last geologic event to leave evidence in the rocks of the district.

On the basis of the discussion as to the age of the granites and gneisses of the district, it appears that the iron formation of the Cauê formation was developed at a time later than 2500 m.y. ago and before a time some 1300 m.y. ago. Since the deposition of the iron formation was followed by the accumulation of the remainder of the Itabira group and all of the Itacolomi series, plus the time required for the development of the unconformities above the Gandarela formation and above the Piracicaba group, it would seem that the Cauê iron formation probably was formed in the late middle Precambrian rather than in the early Precambrian. On the other hand, the Itacolomi series well may have been developed in the early late Precambrian and the conversion of iron formation to massive hematite did not occur until after at least the rocks of the basal Itacolomi series had been lithified. It would seem, therefore, that the production of the epigenetic hard hematite ore must have been no older than late Precambrian. It appears probable that the solutions that accomplished the conversion of iron formation to hard hematite were related in time to the 1300 m.y. granites rather than to the 500 m.y. granites, but the possibility that the 500 m.y. granites were in large measure responsible for the hematization has not been completely eliminated by any means. On the balance of probabilities, however, it is thought best here to classify the primary and syngenetic formation of oxide-facies iron formation as middle Precambrian and the hard hematite ores as late Precambrian.

Although the ore now mined in the Quadrilátero and in the Itabira district is largely hard hematite ore containing more than 66 percent iron that was derived from the banded itabirite, the itabirite is certainly a potential source of iron because it is readily concentrated to a high-iron product and because much of it has to be mined to get at the hard hematite ore.

The itabirite has been so thoroughly recrystallized that petrographic study has not been able to render a decision on its origin; whether the rock was formed by clastic or chemical sedimentation is not apparent from its present condition. No known evidence shows that the hard itabirite has contained any dolomite or other soluble carbonate minerals since it reached its present state at least. The itabirite appears to be a typical example of oxide-facies iron formation. No evidence has been found in the Quadrilátero that the original mineral composition was ever appreciably different from that it now has; no traces of siderite have been discovered, which strongly suggests that the original iron minerals were either oxides or hydroxides that since changed to hematite and minor magnetite. It is probable, moreover, that the generally low content of magnetite in the Cauê itabirite indicates the original iron minerals to have been hematite or ferric hydroxide. Dorr and Barbosa (1963) conclude that the magnetite now present in the itabirite was produced during the diagenetic stage when the Eh of the environment was probably appreciably lower than during the primary sedimentary stage. The quartz in the itabirite probably was not a clastic material since the rock does not contain (Tyler, 1948) the heavy minerals to be expected in a clastic sediment. Nor have any rounded quartz grains been found in the itabirite nor have any sedimentary structures, such as crossbedding, been preserved. Thus, it is concluded (Dorr and Barbosa, 1963) that the silica was deposited as chert, somewhat contaminated with iron, and that its present form as crystalline quartz is due entirely to metamorphism.

The second type of hematite ore is known as soft hematite, and masses of this ore do not crop out; this ore is incoherent when dry but stands well when damp. It is believed (Dorr and Barbosa, 1963) that the soft ores (and those intermediate between hard and soft, which constitute a third ore type) were derived from the hard hematite by the supergene leaching of small quantities of hematite from crystal boundaries, thus permitting the disaggregation of the individual hematite grains of the hard ore.

The most recent suggestions as to the origin of the hard hematite (Dorr and Barbosa, 1963) are that it was formed by replacement of siliceous itabirite (and to a minor extent of dolomitic itabirite) by iron remobilized from the Cauê itabirite by solutions of hypogene origin passing through it. It is thought probable that these solutions were derived from the same general source as the igneous portions of the granite gneisses, probably those that were developed about 1300 m.y. ago. Because of the small volume of itabirite (relative to the total volume of the Cauê itabirite) that would have had to have been leached to provide the replacement iron, such leached volumes probably would have passed unnoticed or would have been buried from sight where they could not be seen. On the other hand, it is not impossible to imagine adding iron to an iron formation from a granitic source. Certainly the huge quantities of iron deposited from hydrothermal solutions in deposits containing no iron formation demonstrate that iron can be transported hydrothermally. It is, therefore, not impossible that the hypogene solutions that enriched the itabirite brought their iron with them from their igneous source. Thus, the reason that no leached itabirite has ever been found may well be that none was ever leached by the hypogene ore fluids; it is chemically much more reasonable to add iron to a system from an outside source than to pick it up in one part of an essentially constant system and deposit it in another.

The metasomatic origin of the hard hematite is supported by the following: (1) almost all the hematite is enclosed in itabirite; where the hard hematite is laminated (not a common phenomenon), the denser and coarser hematite laminae are traceable into hematite-rich lamallae in the itabirite and the finer and more porous laminae are traceable into the quartz-rich laminae of the itabirite; (2) the high-grade hematite cuts abruptly across the bedding of the itabirite, and the one material grades into the other over distances that range from 1 mm to 20 cm; locally, the two rocks may interfinger; (3) no hard hematite is found in the conglomerate of the Itacolomi series although itabirite is common there; (4) coarsely crystalline specular hematite is found in a few quartz grains that cut the Minas and Itacolomi beds; (5) what martite is present shows that hematite replaced magnetite; (6) the folding in the hematite is essentially the same as that in the itabirite; it is hardly possible that massive hematite could have been folded in the same manner as layered itabirite; thus, the replacement must have occurred after the folding; and (7) there is a thin bed of hematite in the Itacolomi series, indicating that the hematitization of the itabirite did not occur until after the Itacolomi rocks had been lithified. The lack of any traces of dolomite in the hard hematite ore suggests that these ores were not produced by the replacement of the more dolomitic portions of the itabirite; had the hard ore been so produced, it would be remarkable if no traces of dolomite remained. Structurally, the remnants of the Minas series were preserved in the synclinal roots that were deeply buried during the folding and must have been under much higher pressures than those provided by their original sedimentary cover. It is thought that the confining pressure at the time of the formation of the hard hematite ore was 1000 bars or even more. The temperature that obtained during the development of the ores is uncertain; Dorr and Barbosa consider that it must have been greater than 205°C and well may have been greater than the critical temperature of water (374°C). The presence of specular hematite to the almost complete exclusion of other iron minerals suggests that the latter temperature is the more likely.

It appears almost certain that the primary oxide-facies minerals of the itabirite were deposited in quiet waters through the interaction of materials carried in solution in those waters. Thus, the primary minerals of the iron formation are here classified as sedimentary-A1a. The hematite and quartz that make up the banded itabirite appear to have been produced from the primary mineral by regional metamorphism, and, therefore, the category metamorphic-C is added to the classification. The hydrothermal processes that added additional iron to the itabirite by the replacement of its silica lamellae probably occurred under high-temperature, high-pressure conditions and should be categorized as hypothermal-1.

MORRO VELHO-RAPOSOS

Late Precambrian *Gold* *Hypothermal-2*

Anon., 1968, A mina de Morro Velho (Morro Velho mine): Congr. Brasil. Geol., 22d, Belo Horizonte, Roteriros Excursoes, p. 26-40

Berg, G., 1902, Beiträge zur Kenntniss der Goldlagerstätten von Raposos in Brasilien: Zeitsch. f. prakt. Geol., Jg. 10, H. 3, S. 81-84

Derby, O. A., 1903, Notes on Brazilian gold-ores: A.I.M.E. Tr., v. 33, p. 282-287

Gair, J. E., 1958, Age of gold mineralization in the Morro Velho and Raposos mines, Minas Gerais: Soc. Brasileira de Geologia Bol., v. 7, no. 2, p. 39-45

—— 1962, Geology and ore deposits of the Nova Lima and Rio Acima quadrangles, Minas Gerais, Brazil: U.S. Geol. Surv. Prof. Paper 341-A, p. A1-A65

Harder, E. C., and Chamberlin, R. T., 1915, The geology of central Minas Geraes, Brazil: Jour. Geol., v. 23, p. 341-378, 385-424

Herz, N., and Dutra, C. V., 1960, Minor element abundance in a part of the Brazilian shield: Geochim. et Cosmochim. Acta, v. 21, p. 81-98

Matheson, A. F., 1956, The St. John del Rey Mining Company, Limited, Minas Geraes, Brazil: Canadian Inst. Min. and Met. Tr., v. 59 (Bull. no. 525), p. 1-7

Tolbert, G. E., 1964, Geology of the Raposos gold mine, Minas Gerais, Brazil: Econ. Geol., v. 59, p. 775-798

Notes

The gold mines of the Morro Velho-Raposos district lie in the northwest corner of the Quadrilátero Ferrífero and they center around the town of Nova Lima, which is located about 10 km southeast of the center of Belo Horizonte, the capital of the state of Minas Gerais. The Morro Velho mine is situated on the northern edge of the town of Nova Lima, and the Raposos mine lies about 4 km east-northeast of the Morro Velho mine. The two mines were, until 1960, owned and worked by the famous English firm, the St. John del Rey Mining Company; in 1960 ownership was transferred to a Brazilian company, Mineracão Morro Velho. Of the monthly production of over 300 kg of gold, about two-thirds comes from Morro Velho and the other one-third from Raposos; minor amounts of silver and arsenic are also recovered.

The oldest rocks in the Quadrilátero Ferrífero are those of Engenheiro Corrêa granodiorite, which has been dated (Herz and Dutra, 1960) as about 2500 m.y. old, but this granite probably does not occur in the mine area. The oldest rocks of sedimentary origin in the district are those of the Nova Lima group, which are known elsewhere to rest unconformably on the Engenheiro Corrêa granite, and are the younger of the two groups into which the Rio des Velhas series has been divided. In the mine area, the Nova Lima group has been separated (Tolbert, 1964) into three zones. The oldest of these is probably the Faría zone. Although no sections have been measured in this zone, or in the Morro Velho zone above it, the Faría appears to be the thickest of the three zones. The principal rock types in the Morro Velho zone are the schists and quartz-carbonates; as these seldom show bedding, they are much more difficult to map than the zones above and below them that are composed appreciably of banded iron formation. The deposits of the Morro Velho mine are found in the rocks of the Morro Velho zone. The Raposos (third) zone is principally made up of chlorite schist and phyllite. The deposits of the Raposos mine are found in the rocks of the Raposos zone.

In both the work of Tolbert (1964) and Gair (1962), the Nova Lima group is considered to be overlain conformably by the Maquiné group. The rocks of the Maquiné group lie well to the east of the gold-bearing portion of the district and need not be considered further here. Tolbert (1964) includes the Tamanduá group in the Rio das Velhas series as the youngest group in the series. Tolbert considers the Itabirito granite, which has been given an age of ± 1350 m.y. by Herz and Dutra (1960), to intrude the rocks of the Rio das Velhas series and to lie unconformably beneath the rocks of Minas series, but the nearest outcrops of the granite are about 8 km to the northwest. If the date of the intrusion of the Itabirito granite (1350 m.y.) can be accepted as reasonably accurate and the formation of the ores can be related to that igneous event, the ores must have been formed not long after the intrusion of the granite. Tolbert (1964) makes the point that no igneous rocks, except the diabase dikes, are known in the Morro Velho-Raposos district; the nearest granites outcropping at the surface are near Belo Horizonte and are 8 km northwest of Nova Lima. However, another granite body lies 16 km south of Nova Lima, and the suggestion has been made that the two bodies may merge at depth. None of the 500 m.y. granite of the Quadrilátero is known in the immediate region, although some has been mapped northeast of Belo Horizonte. It would seem to follow, therefore, that the ore-fluids (granted they had a magmatic source) would have been more likely to have come from the older than from the younger granite. The alternative hypothesis of ore genesis is that the gold was disseminated through the Rio das Velhas rocks and was moved during the metamorphic cycle to concentrate in the mines of the district. Although neither suggestion is supported by direct evidence, the one mentioned first seems to be the more reasonable. The ore bodies of the two mines, therefore, are considered to have been deposited in the late Precambrian about 1300 m.y. ago.

The ore bodies in the Morro Velho mine are found in massive, thick- to thin-bedded rocks of the lapa sêca (quartz-dolomite or quartz-ankerite rock). These quartz-carbonate rocks were once continuous beds but have been appreciably broken into spindles and lenses that are surrounded by schist (mainly quartz-chlorite-sericite or quartz-carbonate-sericite-chlorite schist, some of which is graphitic). There are apparently two mineralized lapa sêca beds and several others that are much thinner and contain no ore. The main lapa sêca bed includes the Main, South, and X ore bodies, and the other and thinner mineralized bed contains the Black, Northwest, and Gamba ore bodies. The ore at Morro Velho is made up of sulfides and gold alloyed with silver that have replaced the quartz-dolomite or quartz-ankerite rock of the lapa sêca; minor gangue minerals include siderite, chlorite, sericite, and soda plagioclase and rare calcite. Gair believes that all of the gangue minerals were derived from the lapa sêca, either as relics of original rock minerals or by mobilization or recrystallization of original lapa sêca minerals. The possibility would appear to exist of these gangue minerals having obtained their components, in part at least, from the fluids that brought in the sulfides and the gold. The sulfides occur as scattered grains, in stringers parallel to the bedding of the lapa sêca, or island-like patches of irregular shape enclosed in lapa sêca. Locally, tension joints crossing the lapa sêca are filled with sulfides, plus quartz and carbonates. The contact between ore and lapa sêca is generally sharp, although some pyrite is found disseminated in the lapa sêca and even in the schist outside the actual ore bodies. The most abundant sulfides are pyrrhotite, arsenopyrite, pyrite, and chalcopyrite. Graton considered the paragenesis of the ore to be pyrite, arsenopyrite, pyrrhotite, chalcopyrite, and gold. The principal sulfides in the Morro Velho mine probably were formed under hypothermal conditions; the presence of abundant pyrrhotite and arsenopyrite is characteristic of sulfide formation in this intensity range. On the other hand, the gold occurs in fractures in the arsenopyrite and pyrrhotite so that an appreciable period of time must have elapsed between the deposition of the sulfides and of the gold; this time lapse is confirmed by the replacement of chalcopyrite by gold. Nevertheless, because of the essentially constant amounts of gold present in the deposit for a down-dip distance of more than 14,000 feet, it is here thought that the gold must also have been deposited under hypothermal conditions. Despite the high-temperature character of the ore-zone sulfides, however, most of the wall-rock alteration minerals (minor chlorite, sericite, and siderite) are not diagnostic of high-temperature conditions; only the minor soda feldspar suggests that the alteration may have been hypothermal. Although, in places, the entire mineral content of the lapa sêca has been replaced by sulfides, the bulk of the material so replaced appears to have been the carbonate of the lapa sêca rather than the quartz; the Morro Velho ores are, therefore, classified as hypothermal-2.

The ore bodies of the Raposos mine differ from those at Morro Velho in several ways of which perhaps the most obvious and fundamental is that they are emplaced in iron formation rather than quartz-carbonate rock. Actually, however, this difference is somewhat more apparent than real since the gold-sulfide ore preferentially replaced bands of siderite in the quartz-siderite type of iron formation. Where magnetite-quartz iron formation and pyrite iron formation occur, they are far less well mineralized. The sequence of events in the formation of the Raposos ores (Tolbert, 1964) appears to have been (1) the alteration of the wall rocks and diabase dikes by hydrothermal fluids and the deposition of small quantities of pyrite and perhaps of gold--some pyrite is found in the chromium-rich schists in the wall rocks of the ore bodies; (2) deformation, including fracturing of the pyrite; (3) the main stage of sulfide deposition in the order arsenopyrite, pyrrhotite, minor chalcopyrite, and gold, plus quartz and carbonate; and (4) the continued deposition of quartz and carbonate in fractures in the ore bodies. In most instances, pyrite is shattered and full of holes; these are generally filled with quartz or, less commonly, with pyrrhotite and chalcopyrite. The higher the proportion of pyrite to pyrrhotite and/or arsenopyrite in an ore body, the lower the gold content. The Raposos mine differs from Morro Velho in that the gold is not as closely associated with arsenopryite. For example, the Espirito West ore contains 75 percent arsenopyrite and 25 percent pyrite yet has a gold content of only 6 grams per ton. Pyrrhotite normally fills fractures and tiny holes in pyrite and arsenopyrite, and the ore bodies with a high proportion of pyrrhotite to other sulfides have a high gold content; in polished sections, the native gold is more often found with pyrrhotite than with the other sulfides. Chalcopyrite is less common at Raposos than at Morro Velho, generally making up less than 2 percent of the sulfides. At Raposos, the sulfides constitute 5 to 10 percent of the ore compared with about 40 percent at Morro Velho.

The wall-rock alteration at the Raposos mine is not easily distinguished from the effect of regional metamorphism that converted the rock to quartz-carbonate-chlorite-sericite schist. The wall rocks of the ore bodies, however, contain a chromium-rich sericite that may have been formed by the first of the fluids that later deposited the ores. The ore, however, contains no chromium-rich sericite; the chromium sericite was emplaced before the sulfides, and chromium-rich chlorites are common throughout the Quadrilátero. Tolbert (1964) believes that the gold in the Raposos deposits could have been obtained in two ways: (1) the gold was concentrated from gold disseminated in the country rocks (as Boyle believes to have been the case at Yellowknife) or (2) the gold was concentrated in, and transported from, an igneous source. Under case (1) hydrothermal fluids might have been the transporting agency, under (2) they almost certainly would have been. It has already been suggested that the ore-forming fluid that deposited the Raposos (and Morro Velho) ores was developed in the late stages of the crystallization of the Itabirito granite and that the fluid moved whatever distance was necessary for it to reach the site of deposition of the ores. The deposits at Raposos are, therefore, classified as hydrothermal in the strict sense, and, because the high-temperature sulfides were closely, though not exactly, associated with the gold in time, they are assigned to the hypothermal category. Since the ore was mainly emplaced by the replacement of carbonate minerals, the ores are categorized as hypothermal-2 even though it is certain that some of the ore replaced the other minerals of the iron formation.

SERRA DE JACOBINA

Late Precambrian | *Gold, Uranium* | *Hypothermal-1 to Mesothermal*

Bateman, J. D., 1958, Uranium-bearing auriferous reefs at Jacobina, Brazil: Econ. Geol., v. 53, p. 417-425

Branner, J. C., 1910, The geology and topography of the Serra Jacobina, state of Bahia, Brazil: Amer. Jour. Sci., 4th Ser., v. 30, p. 385-392

Cox, D. P., 1967, Regional environment of the Jacobina auriferous conglomerate, Brazil: Econ. Geol., v. 62, p. 773-780

Davidson, C. F., 1957, On the occurrence of uranium in ancient conglomerates: Econ. Geol., v. 52, p. 668-693; disc. by various authors in v. 53, p. 489-493, 620-622, 757-759, 887-890, 1048-1049; in v. 54, p. 511-512

Gorsky, V. A., and Gorsky, E., 1962, Further contribution to the study of uranium-bearing auriferous metaconglomerate of Jacobina, state of Bahia, Brazil: Inter-Am. Symposium Peaceful Application Nuclear Energy, 4th, Mexico City, v. 1, p. 301-312

Grabert, H., 1959, Gold- und Manganerze in der Serra de Jacobina (Bahia, Brasilien): Zeitsch. f. Erzbergbau u. Metallhüttenwesen, Bd. 12, H. 7, S. 330-335

Gross, W. H., 1968, Evidence for a modified placer origin for auriferous conglomerates, Canavieiras mine, Jacobina, Brazil: Econ. Geol., v. 63, p. 271-276

Leo, G. W., and others, 1962, Geologia de parte sul da Serra da Jacobena, Bahia, Brasil: Div. Geol. e Mineral, Bull. no. 209, 87 p.

—— 1965, Chromian muscovite from the Serra da Jacobena, Bahia, Brazil: Amer. Mineral., v. 50, p. 392-402

Ramdohr, P., 1958, Die Uran- und Goldlagerstätten Witwatersrand, Blind River district, Dominion Reef, Serra de Jacobina; Erzmikroskopische Untersuchungen und ein geologischer Vergleich: Deutsch Akad. Wissen., Abh., Kl. f. Chem., Geol., Biol., Jg. 1958, no. 3, 35 S.

Souza, H. C. Alves de, 1942, Ouro na Serra de Jacobina (estado de Baía): Dept. Nacional da Prodoção Mineral, Divisião de Fomento Produção Boletim 51, 52 p.

White, M. G., 1956, Uranium in the Serra de Jacobina, state of Bahia, Brazil: 1st U.N. International Conf. on Peaceful Uses of Atomic Energy (Geneva) Pr., v. 6, p. 140-142

—— 1957, Urânio nos conglomerados auríferos da mina de ouro de Canavieiras, estado da Bahia, Brasil: Conselho Nacional Pesquisas, Rio de Janeiro, 11 p.

—— 1957, Uranium in the auriferous conglomerates at the Canavieiras gold mine, state of Bahia, Brazil: Engenharia, Mineração e Metalurgia (Rio de Janeiro), v. 26, no. 155, p. 279-282

—— 1961, Origin of uranium and gold in the quartzite-conglomerate of the Serra de Jacobina, Brazil: U.S. Geol. Surv. Prof. Paper 424-B, p. B8-B9

—— 1964, Uranium at Morro do Vento, Serra de Jacobina, Brazil: U.S. Geol. Surv. Bull. 1185-A, p. A1-A18

Notes

The Serra de Jacobina is located in the north-central part of the Brazilian state of Bahia, somewhat more than 280 km from the Atlantic port of Salvador and about 1000 km north-northeast of Belo Horizonte in Minas Gerais. The Serra itself is a narrow, prominent ridge that stands some 600 to 800 m above the surrounding plains; the highest peaks of the Serra reach altitudes of about 1100 m, and the plains stand about 450 m above sea level. The Serra is over 200 km in length and trends slightly east of north. The town of Jacobina, the principal settlement in the district, is about 25 km from the southern end of the Serra.

The oldest rocks in the district are Precambrian rocks, mainly granites and granite gneisses, that surround the Serra on all sides. Little is known concerning the geology or petrology of these rocks, but the possibility exists that they correspond to the Basal complex of the Quadrilátero Ferrífero of Minas Gerais. The gold and uranium deposits of the region are contained in a younger sedimentary series, the Jacobina series, that Guimarães considers to be equivalent to the Minas series in the Quadrilátero Ferrífero. If this is the case, the age of the Jacobina series must be middle Precambrian. The presence of some subordinate itabirites in the series suggests that this correlation may be valid. The Jacobina series was divided by Flaherty, in a private report, into two groups. The lower of these is the Canas group that lies, to the west, unconformably on generally highly weathered granite of greater age. In the areas of the Canavieiras and Morro do Vento mines, the Canas group consists of quartzites, sandstones, and conglomerates, with the quartzites being the most abundant and the sandstones the least. The great majority of the numerous conglomerate beds are composed of gray to white quartz pebbles in a white, fine-grained matrix; these conglomerates are locally known as the chabu type. In the thicker conglomerates of the chabu type, a little gold is found, but they are essentially lacking in uranium. In addition to the chabu conglomerates, there are two conglomerate beds of a different type; these, known as piritoso type, are very hard, gray to light to dark green and red to chocolate brown reefs containing zones of heavy pyrite-rich sulfide mineralization. One of these piritoso-type conglomerates, the Piritoso reef, is the main source of ore in the Canavieiras mine. The principal source of gold and uranium in the Morro do Vento mine has been a reef of the Piritoso type, known as the Main reef. Whether the Main reef at Morro do Vento is to be correlated with the Piritoso reef at Canavieiras is uncertain; it is not possible to trace the reef on the one property onto the other.

Above the conglomerate-bearing quartzite of the Canas group is the Serra group that is composed of quite massive quartzite, some of which is feldspathic; the Serra group contains no conglomerates. The Jacobina series is overlain by argillaceous rocks that are now slates and phyllites. Rocks of Silurian age are known not far to the west of the Serra, but they are separated from it by a narrow band of granite gneiss. The Silurian rocks probably are younger than the ores. If the ore deposits of the Serra de Jacobina district are basically of syngenetic origin, as has been suggested by Bateman (1958), they are of the same age as the rocks containing them. If the correlation of the Jacobina series with the Minas series of Minas Gerais is correct and if the age of the Minas rocks is middle Precambrian, then the Serra de Jacobina ores must have been formed in the middle Precambrian. If, on the other hand, the ores were emplaced epigenetically by hydrothermal solutions (White, 1961; Leo and others, 1962), the age might be as old as middle Precambrian or as young as Paleozoic or younger. It seems probable, however, that the choice must lie between middle and late Precambrian. Again, if the reasoning used in this volume to date the Minas Gerais ores is correct, it is probable that the igneous activity necessary to produce the gold-uranium ore-bearing fluids must have taken place about 1300 m.y. ago. If this is correct, then it is most probable, though far from certain, that the gold-uranium ores were deposited in the Serra de Jacobina rocks in the late Precambrian, and they are so classified here.

The Canavieiras mine, some 6 km south of Jacobina, obtains most of its ore from the Piritoso reef, in which the run-of-the-mine ore contained 0.03 percent uranium equivalent whereas samples taken at the surface gave only 0.006 percent. The uranium is derived from uraninite, and the rock, where green, derives its color largely, if not entirely, from chrome-bearing mica intergrown with the sericite and, where brown, from limonite formed by the oxidation of pyrite. The reef contains some visible free gold, but most of it is intimately associated with pyrite; fragments of uraninite have been found intergrown with pyrite crystals. No secondary uranium minerals are found in the deposits, which makes it difficult to account for the low uranium content in weathered rock and in sheared rock in the fault zones; such a complete removal of uranium by surface processes seems unreasonable. It is possible, however, that the oxidation of pyrite produces solutions with such a low pH that the dissolved uranium is not precipitated near its loci of origin.

The Morro do Vento mine is about 2 km southwest of the Canavieiras mine. The same, or similar, reefs of gold-uranium-bearing pyrite have been found in the fault-block hill masses between Canavieiras and Morro do Vento, but radioactivity on the outcrops is greatest in the Morro do Vento portion of the district. Although gold has been mined in the Serra since late in the 17th century, the workings on the Morro do Vento are quite recent even though they apparently are not being mined at present. In contrast to the Canavieiras mine, where the workings extend down the dip for about 80 m, the Morro do Vento mine has reached a depth of only 15 m. Despite the concentration of operations in the Main reef, workings have also been developed in (1) quartzite-conglomerate beds on the top of the Morro; (2) a gold-bearing conglomerate-quartzite zone on the northeast side of the Morro; and (3) a pyritized conglomerate, quartzite, and sandstone layer below the Main reef and near the contact between quartzite and granite. In addition to silica and sericite, the most important minerals deposited in the Main reef are pyrite, chlorite, biotite, gold, and uraninite; the gold averages 10 grams per ton and U_3O_8 0.008 percent equivalent. The pyrite is ordinarily found in the reef in zones or shoots that cut through, and have sharp boundaries against, the host rocks; pyrite masses are present in both conglomerates and quartzite. The pyrite bodies are the most radioactive portions of the reef. To the depth reached in the workings (15 m), the pyrite has been considerably weathered.

The uraninite occurs in poorly crystallized microcrystals, and White (1964) considers that it should be categorized as pitchblende, and sooty pitchblende at that. There is no indication of rounding of the grains that might imply that they had been introduced into the formation as detrital minerals. Most of the pitchblende appears to occur as veinlets in the pyrite, but some is finely disseminated in the iron sulfide. Gold is also closely associated with the pyrite. A green color is the most characteristic feature of the Main reef and is due to the presence of sericite that has been stained green by chrome-mica, probably fuchsite. Some octahedrons of chromite have been found in the quartzite, and these do not appear to have been of detrital origin. The green color is drowned out by the red or brown of iron oxide in rock volumes where pyrite has been heavily weathered. It is considered that a deeper green color or red or brown staining is a good guide to gold and uranium ore. Thus, at least as far as the Morro do Vento deposits are concerned, there seems to have been no lithologic control exerted over the deposition of the gold-uranium ore and its associated minerals since high gold-uranium content has no consistent relationship to rock type. This suggests that the emplacement of the minerals was regulated by fractures that cut through both quartzites and conglomerates. If this is the case, the deposits cannot be simple detrital collections of the ore and gangue minerals within a certain type of sedimentary rock. The deposits of the Serra de Jacobina are here considered to have been deposited from hydrothermal solutions that were produced at some appreciable distance from the site where they ultimately deposited the gold-uranium ores.

The presence of the uranium as pitchblende (White, 1964) suggests that the ores were deposited in the mesothermal range. The suite of gangue minerals, the principal members of which are chlorite, sericite, biotite, quartz, and pyrite, is not incompatible with the lower intensity segment of the hydrothermal range; the gold provides no real clue as to the conditions of mineral deposition. As a compromise between the two items of evidence, the deposits are here categorized as hypothermal-1 to mesothermal.

CHILE

BRADEN

Middle Tertiary — *Copper, Molybdenum* — *Kryptothermal*

Brüggen, J., 1934, Grundzüge der Geologie und Lagerstättenkunde Chiles: Heidelberger Akademie der Wissenschaften, Mathematisch-Naturwissenschaftliche Klasse, Max Weg, Leipzig, 362 S. (particularly S. 323-329)

Howell, F. H., and Molloy, J. S., 1960, Geology of the Braden orebody, Chile, South America: Econ. Geol., v. 55, p. 863-906

Klohn Giehm, C., 1960, Geología de la Cordillera de los Andes de Chile central, provincias de Santiago, O'Higgins, Colchagua y Curicó: Inst. Invest. Geol. Bol. 8, 95 p. (Engl. and Germ. summs.)

Lindgren, W., and Bastin, E. S., 1922, The geology of the Braden mine, Rancagua, Chile: Econ. Geol., v. 17, p. 75-99

—— 1935, The Braden copper deposit, Rancagua, Chile, in *Copper resources of the world*: 16th Int. Geol. Cong., v. 2, p. 459-472

Marsh, R., Jr., 1922, Geology of the Braden mine: Econ. Geol., v. 17, p. 498-501

Notes

The Braden property is located generally east of the town of Sewell that is situated at an elevation of 10,000 feet on the western slopes of the Andes some 45 miles southeast of Santiago, the capital city of Chile, and about 33 miles east-northeast of Rancagua.

The oldest rocks in the region are those of the Upper Cretaceous Chilense formation, that does not crop out within 10 miles of the ore body but which probably underlies the deposit at moderate depth. The Chilense is separated from the overlying early Tertiary Farellones formation by a major angular unconformity that marks the Laramide orogeny at the end of the Mesozoic era. The lower Farellones beds are composed almost entirely of massive extrusive andesite and provide the principal host rock for the ore body. Lindgren and Bastin (1922, 1935) thought the andesite to have been intrusive, probably because of its massive character, but detailed mapping has shown it (Howell and Molloy, 1960) to contain agglomeritic flows, lenticular beds of flow breccia, and some amygdaloidal and vesicular layers. In the ore body itself, and for some distance outward from it, the andesite is impressively altered to fine-grained secondary biotite; where the rock within the ore body is cut by intrusive quartz diorite, it is strongly silicified for several feet away from the contacts. On the upper mine levels, the andesite has been heavily converted to argillite and/or sericite; Howell and Molloy (1960) suggest that this may have been a surficial phenomenon.

The middle Farellones member is found in a half-mile-wide northwest-trending belt immediately north of the mine. The upper Farellones member is made up principally of andesite and basalt flows intercalated with agglomerate and pyroclastic beds; to the west of the mine, these mixed volcanic rocks give way abruptly to unsorted and poorly consolidated pyroclastic material. The beds of the upper member once may have been highly mineralized in the ore body area, but they and the ore they may have contained have been removed by erosion.

The Farellones formation was intruded by a considerable variety of igneous rocks before the Braden formation was produced, the first of these was a minor number of small andesite dikes that cut the middle member and the lower part of the upper member about a mile northwest of the mine.

The next intrusive rock is known as quartz diorite but ranges in composition from diorite to monzonite; it occurs as stocks, dikes, and irregular masses that probably are genetically and temporally related to the underlying

Andean batholith. The quartz diorite makes up an appreciable, but minor, fraction of the mineralized rock volume of the ore body, and the first (early quartz) mineralization probably began to deposit during the introduction of this rock type. The quartz diorite intrudes the upper Farellones member and is cut by dikes of the next younger intrusive rock, the dacite porphyry.

The dacite porphyry is present in only small amounts within a mile of the mine, but it may have once been an important component of the rock volume now occupied by the Braden pipe. The dacite cuts the quartz diorite, is cut off by the Braden pipe, and coincides generally in time with the latter part of the early quartz mineralization and with the hiatus between this stage and the main chalcopyrite stage.

During the latter part of the dacite intrusive cycle, and for an appreciable time thereafter, the rocks of the general mine area, which had earlier been cut by strong N55°E-striking faults, were subjected to intensive stock work fracturing that produced an essentially random network of veinlets into which the main chalcopyrite mineralization was introduced prior to the development of the Braden formation of the Braden pipe.

Before the next igneous rock (the latite porphyry) was introduced, the Braden formation and the Braden pipe had been developed. The Braden formation is all the detrital material that fills the Braden pipe. It is Howell and Molloy's opinion that the actual pipe itself was produced by a strong upthrusting force that broke loose an inverted and tilted cone of rock, the first step in the loosening of which was the formation of the pre-pipe breccia. This initial upward motion may have been followed by a series of subsidences and further upward movements that rounded the fragments of the pipe having diameters of more than 1 mm but which left the ground mass material as angular fragments.

It was into this partly bedded Braden formation that the next igneous rock, a latite porphyry, was intruded as both concordant and discordant, generally irregular masses of brecciated material that probably were broken by the repeated upward and downward movements of the pipe volume. All the latite, however, was emplaced before the post-pipe breccia was formed. Most of the latite porphyry is unmineralized, but it does contain pyrite and an uneconomic development of late-stage enargite-chalcopyrite veinlets; strong tourmalinization occurs around some of the larger latite bodies, and tourmaline has replaced the ground mass of both latite and Braden formation near contacts between these two rock types.

The youngest intrusive rock is a narrow but persistent lamprophyre dike that averages about 6 feet in width and cuts the southeastern portion of the ore body on every level in the mine. The next geologic event appears to have been the formation of the post-pipe breccia that resulted from the impressive effect of the last subsidence of the Braden pipe. The final tectonic movement of local effect in the mine area probably reopened the northeast (N55°E) fractures, but only minor quantities of hydrothermal minerals seem to have been added after this last reopening.

Thus, it is probable that the sequence of events in the formation of the present Braden ore bodies was (1) the introduction of the early barren quartz veins into the northeast-striking (N55°E) fractures; (2) the development of a random stockwork pattern of fractures in a rock volume appreciably larger than that of the later Braden pipe; (3) the deposition of the main chalcopyrite mineralization, the ore minerals being more heavily concentrated in the outer portions of the stockwork than toward its center; (4) the formation of the Braden pipe due to upward movements of the cone of material loosened by thrust faulting and the development of the Braden formation by pulsating movements of the rock material within the cone aided by some sedimentation in lake waters locally ponded on the pipe surface; (5) essentially concomitant with (4) the introduction of the early tourmaline mineralization, mainly in the pre-pipe breccia; (6) the formation of the post-pipe breccia; (7) late tourmalinization; (8) the introduction of the tennantite stage of mineralization mainly into the post-pipe breccia. The quartz diorite intrusions were partly completed before stage (1) began; the dacite porphyry intrusion overlapped the stage (3); the latite porphyry intrusion overlapped stages (4) and (5). Surficial oxidation and enrichment followed the tennantite stage (8) or may have overlapped somewhat with it. The main period of copper mineralization, which actually introduced essentially all the copper in the deposit, appears to have occurred in Miocene time, while the minor tennantite mineralization took place in the Pliocene. Since almost all the ore was introduced in Miocene time, the Braden deposit is here classified as middle Tertiary, with no mention being made in the classification of the minor, and late Tertiary, tennantite mineralization.

The Braden ore body surrounds the Braden pipe in a complete ring of irregular width, the widest portion of which at the surface (the Teniente area) extends outward for some 2000 feet on the east side of the body and plunges downward counterclockwise around the pipe, the north side of the ring being its widest part on the lowest mine levels. The south and west sides of the ring are generally much narrower, although the Fortuna ore body on the southwest portion of the ring appears to have had a surface outcrop width of well over 500 feet. Much of the ore body on the west and northwest side (apparently emplaced mainly in the upper Farellones member) probably has been removed by erosion. The greatest known vertical extent of the ore is some 4300 feet.

The pipe is essentially an inverted, though tilted, cone in which the angle of the apex is about 30° and the greatest known depth is some 5200 feet. Thus, the cross section of the pipe is roughly circular, and its diameter narrows with depth. The two types of breccia that border the pipe (pre-pipe and post-pipe, respectively) were developed after the main chalcopyrite mineralization had occurred in the rock later fragmented to give the breccia fragments. Since by far the larger fraction of the chalcopyrite ore was emplaced outside of the rock volumes that were later converted to the breccias and the Braden pipe enclosed by them, the amount of ore in these breccias is only a small fraction of the total, with most of the breccia ore being in the pre-pipe (outermost) breccia. The amount of ore in the breccias depends directly on the proportion of ore-bearing fragments and is essentially that introduced in the main chalcopyrite mineralization before the breccias were formed.

Much more important in determining the grade of the ore than the presence or absence of breccia is the type of host rock involved, the most highly mineralized rock being altered andesite or andesite flow breccia outward from the pre-pipe breccia.

The first stage in the development of the Braden mineralized rock volumes was that of early barren quartz. In this stage quartz was the principal mineral and was deposited in the early northeast-striking faults; the quartz was accompanied by a little pyrite. Since this mineralization is found only in the lower and middle Farellones, it must have occurred well before the main chalcopyrite mineralization; the quartz in these veins was severely broken by later earth movements.

The early quartz mineralization was separated from the main chalcopyrite stage by the period of stockwork fracturing, into which fractures the minerals of the latter stage were emplaced. The average width of these stockwork veinlets is between 1/64 and 1/4 inches; no individual veinlets extend for more than a short distance, and the grade of ore is proportional to the number of veinlets within a given rock volume--the more veinlets, the higher the ore grade. The stockwork fractures continued to form for a considerable time after the chalcopyrite mineralization had begun.

The principal sulfides of this stage were chalcopyrite, pyrite, bornite, molybdenite, and primary chalcocite in a quartz-anhydrite-chlorite gangue. The veinlets are generally definitely fracture fillings and commonly are banded, but some replacement did occur.

The next stage of mineralization--the early tourmaline stage--was essentially contemporaneous with the intrusion of the latite porphyry into the rocks of the Braden pipe and particularly affected the pre-pipe breccia (then the only breccia in existence). In addition to tourmaline, chalcopyrite (which occurs with tourmaline as fillings in, and replacements of, the breccia matrix), lesser bornite, pyrite, and some early molybdenite and some quartz and anhydrite make up the minerals of this stage.

The late tourmaline stage followed the end of the latite porphyry intrusions and essentially all occurred after the Braden formation (including the post-pipe breccia) had been completely developed. The minerals of this stage were much the same as those of the first tourmaline stage although tourmaline was more abundant and molybdenite was lacking; these minerals are found only in the post-pipe breccia in which the fragmental material was concomitantly much altered to sericite and/or clay minerals. The two stages of tourmaline mineralization may have been essentially continuous; since, however, the pipe breccia provided a most effective channelway for the tourmaline-rich solutions, they essentially abandoned the remainder of the Braden pipe to move through, and deposit in, the post-pipe breccia volumes.

The last stage of mineralization was composed of tennantite (from which it derives its name), galena, and sphalerite in a carbonate gangue--ankerite, calcite, and rhodochrosite, plus some pyrite, chalcopyrite, bornite, molybdenite, quartz, anhydrite, gypsum, barite, and specularite. The galena, sphalerite, and anhydrite are ordinarily earlier than the tennantite, bornite, and ankerite. Most of the tennantite-stage minerals were deposited in the post-pipe breccia.

The Braden ores probably were deposited in a near-surface environment where the loss of heat and pressure was rapid. The very appreciable depth to which mineralization extends, however, suggests that the rate of heat and pressure loss that prevailed were intermediate between those of such deposits as the porphyry coppers of the western United States and those of the tin deposits of Bolivia. The abundant tourmaline would, at first glance, indicate that the deposit should be classified as xenothermal. The tourmalinization, however, is separated from the main chalcopyrite mineralization by the formation of the pre-pipe breccia; the effect of this tectonic event seems to have given entry to solutions of appreciably different character than those that deposited the bulk of the copper. The fact that the primary chalcocite is closely associated with bornite (exsolved from it?) indicates that the copper minerals were more probably deposited under conditions of intermediate intensity than those of high. Thus, the main copper mineralization seems better classified as kryptothermal than xenothermal; the tennantite stage certainly was kryptothermal rather than xenothermal or epithermal. The Braden deposits, therefore, are here classified as kryptothermal despite the minor development of chalcopyrite with the tourmaline.

CHAÑARCILLO

Late Mesozoic to Early Tertiary	*Silver*	*Leptothermal (primary), Ground Water-B2 (enriched and oxidized)*

Brüggen, J., 1934, Grundzüge der Geologie und Lagerstättenkunde Chiles: Heidelberger Akadamie der Wissenschaften, Mathematisch-Naturwissenschaftliche Klasse, Max Weg, Leipzig, 362 S. (particularly S. 335-336)

Moesta, F. A., 1928, El mineral de Chañarcillo: Boletín Minero (Chile), v. 40, no. 348, p. 167-182 (originally published in Germany in 1870)

Nordenskjöld, O., 1926, Yacimientos de minerales en el desierto de Atacama: Boletín Minero (Chile), v. 38, p. 932-938, 1036-1043, 1148-1154 (particularly p. 1151-1153)

Ruiz, F. C., and others, 1961, Ages of batholithic intrusions of northern and central Chile: Geol. Soc. Amer. Bull., v. 72, p. 1551-1560

Segerstrom, K., 1960, Structural geology of an area east of Copiapó, Atacama Province, Chile: 21st Int. Geol. Cong. Rept., pt. 18, p. 14-20 (area some 35 km north and east of Chañarcillo)

—— 1962, Regional geology of the Chañarcillo silver mining district and adjacent areas, Chile: Econ. Geol., v. 57, p. 1247-1261

Segerstrom, K., and Morago Brito, A., 1964, Cuadrangulo Chañarcillo, Provincia de Atacama: Inst. Invest. Geol. Chile, Carta Geológica de Chile, no. 13, 50 p., map, 1:50,000

Whitehead, W. L., 1919, The veins of Chañarcillo, Chile: Econ. Geol., v. 14, p. 1-45

—— 1942, The Chañarcillo silver district, Chile, in Newhouse, W. H., Editor, *Ore deposits as related to structural features*: Princeton Univ. Press, p. 216-220

Notes

The silver deposits of Chañarcillo, the most important representatives of Chilean silver-producing districts that lie in a north-south belt that parallels the trend of the Andes Mountains, centers at 27°48.6' south latitude and 70°26.6' west longitude; the principal mines lie some 50 km south of Copiapó and 265 km north-northeast of Coquimbo.

The oldest rocks in the Chañarcillo district are two rock series of Lower Cretaceous (Neocomian) age--the Chañarcillo group and the Bandurrias formation; these series interfinger with each other and are of much the same age although the deposition of the volcanic Bandurrias formation probably continued for a short time after that of the Chañarcillo group had ceased.

A formation known as the Cerrillos lies unconformably on the rocks of the Chañarcillo group outside the general area of silver mineralization; the unconformity is evidenced by the cutting of folded Pabellon beds by those of the Cerrillos formation.

The oldest igneous rock in the quadrangle is a manto-dike of andesitic porphyry that is an intercalation in the Pabellon formation. The manto-dike is truncated by the Cerrillos formation and must, therefore, be late Lower Cretaceous in age; it is definitely pre-ore.

In the general Chañarcillo mining area, there are two major igneous rock types: (1) granodiorite and (2) diorite poryphyry. Neither of these rocks crops out in the immediate vicinity of the silver mines; but the former had a tremendous metamorphic effect on the Cretaceous rocks with which it came in contact. The granodiorite rocks occur as stocks, dikes, and apophyses that exhibit discordant relationships to the rocks they invade. A dioritic phase of these rocks must not be confused with the probably later diorite porphyry.

The larger diorite porphyry bodies are known to intrude all the stratified and lithified rocks exposed in the area, and the contacts are both concordant and discordant. It is thought (Segerstrom and Brito, 1964) that the granodiorite bodies were intruded at greater depths than those of the diorite porphyry and that they are offshoots of a far larger batholith mass at depth; all of these igneous rocks have been dated as Late Cretaceous.

Zones of contact metamorphic rocks surround the intrusive masses of granodiorite and have been separated into three types (Segerstrom and Brito, 1964): (1) skarns produced in the Chañarcillo group; (2) meta andesites derived from the volcanic rocks of the Bandurrias formation, plus some skarn formed from intercalated sediments; and (3) undifferentiated intrusive and contact rocks. Rocks of type (1) are found in the mineralized area and contain outcrops of several of the ore-bearing veins. Rocks of type (2) are also present in the mineralized area but they crop out over less extensive areas than do those of type (1) and appear to contain fewer mineralized structures. There is only one outcrop of rocks of type (3) in the mine area, and it contains no outcrops of ore veins.

The domed structure at Chañarcillo could not have been caused by the intrusion of the diorite porphyry since it fills the later of the two sets of fractures that probably resulted from the doming. Instead, it appears probable that the dome was caused by the intrusion of a pluglike offshoot of the granodiorite batholith and that the primary ore mineralization was interposed between the intrusion of the granodiorite and that of the diorite porphyry. Since the radial fractures contain almost all the ore mineralization and the concentric ones very little, it would seem that the radial fractures formed first and were mineralized before the concentric ones were developed. When the diorite magma entered the area, the radial fractures must have been almost completely filled by hydrothermal mineral matter since the dike rocks are confined almost entirely to the concentric fractures. The presence of minor amounts of primary hydrothermal minerals in the fractures containing the dikes, however, strongly suggests that the ore fluids were still active, though weakly so, after the dikes had solidified. Thus, the ore-forming fluids must have derived from the granodiorite magma rather than from that from which the diorite material came. From this it would follow that the age of the mineralization is fixed as being postgranodiorite and largely prediorite porphyry. Although there is certainly some argument as to the actual age of the granodiorite (Ruiz and others, 1961) it seems most probable that this rock was generated and intruded the Chañarcillo area during Laramide time and that the silver ores should be dated as late Mesozoic to early Tertiary.

The silver-bearing veins of the Chañarcillo district were emplaced on the southwest flank of the dominant domed structure (Segerstrom and Brito, 1964). They are, for the most part, located in the upper beds of the Nantoco formation near its contacts with the overlying Totoralillo formation and in rock volumes intercalated by tuffs of the Bandurrias formation. The veins can be divided into four groups [Whitehead (1919, 1942) uses six] based on the average directions of strike of the veins.

In the primary ore, finely crystalline calcite and/or platy barite make up the greater part of the vein material. Open spaces are common in the veins and range in size from a few millimeters to some of a centimeter in diameter; these usually contain finely crystalline coatings of calcite, quartz, and silver sulfosalts. Scattered through the calcite-barite gangue are small grains (only rarely massive aggregates) of metallic minerals; in some places the metallic minerals are banded with the gangue, in others they are evenly disseminated or locally clean gangue may be in contact with serrated sulfides.

Under the microscope, Whitehead (1919, 1942) considers it readily possible to determine the sequence of primary ore and gangue mineral deposition. Calcite was the first mineral to form and was replaced to different extents by barite with laths of barite penetrating the calcite grains or, where replacement is more complete, with residual calcite included among interlacing barite crystals. Quartz was precipitated at much the same time as barite, occupying volumes among the laths of the sulfate. This first, or gangue-mineral, stage of mineralization was followed, after an appreciable time lapse, by the precipitation of a second stage, that of base-metal sulfides accompanied by traces of proustite (Ag_3AsS_3). Sphalerite, the most abundant of these sulfides, more commonly replaced calcite than filled open space; it contains tiny irregular blebs of proustite. Pyrite, chalcopyrite, and galena were formed at essentially the same time as sphalerite but in minor amounts. The first mineral to be introduced in the third stage was arsenopyrite in large amounts; it was generally deposited in cavities, which it normally only partly filled, but also replaced calcite. Acicular crystals of arsenopyrite commonly penetrate sphalerite, and veinlets of arsenopyrite cut the zinc sulfide. This third stage also included minor amounts of proustite and quartz that were almost certainly later than the arsenopyrite.

The fourth stage was that of the most abundant sulfosalt formation. The first minerals to form in this stage were pearceite ($Ag_{16}As_2S_{11}$) and proustite which filled the cavities lined with arsenopyrite and entered into the interstices between arsenopyrite crystals as well; they also filled cavities coated with calcite crystals. There was some replacement of gangue minerals by these arsenic sulfosalts, but it was minor in comparison with open-space filling. The arsenic-bearing minerals were followed by antimony sulfosalts--silver-rich tetrahedrite, polybasite [$(Ag,Cu)_{16}Sb_2S_{11}$], and pyrargyrite (Ag_3SbS_3). The tetrahedrite mainly replaced calcite around the margins of arsenopyrite masses; polybasite replaced the margins of pearceite grains, and pyrargyrite similarly replaced proustite. Polybasite and pyrargyrite also replaced calcite and filled vugs. In this stage, the arsenic-bearing sulfosalts were emplaced more by filling than by replacement while the reverse was true of the antimony sulfosalts. The ore fluids appear to have accomplished little or no alteration of the rocks bordering the veins.

The primary minerals from which the silver of the minable secondary ores was obtained were those formed in the fourth stage of mineralization. It appears doubtful if any of the primary ore was rich enough to have been mined at a profit, but the primary ores were appreciably more than weakly mineralized. The ores of the fourth stage probably were formed under conditions of slow loss of heat and pressures at depths of at least several hundred meters below the then-existing surface, and the sulfosalts of that stage are all minerals diagnostic of the leptothermal range in rock volumes where temperature and confining pressure are decreasing slowly. The depth of ore formation beneath the surface probably was great enough to insure that these minerals, all of which have been found in epithermal deposits, were actually developed as leptothermal ores, and they are so classified here. The large quantities of arsenopyrite formed in the third stage strongly suggest that much higher intensities prevailed not long before the leptothermal minerals were formed, but the small amounts of proustite in association with the arsenopyrite may mean that arsenic-bearing mineral was deposited at somewhat lower temperature than those usually indicated by the presence of arsenopyrite in abundance. Since the arsenopyrite, however, was not formed in the same stage as the valuable sulfosalts, the third stage minerals are not included in the classification given the deposit.

In the lower portions of the enriched limestones, primary sulfide minerals are present in sufficient quantity for ready study, but in the upper levels secondary effects were so strong that essentially all primary features were obliterated. The primary silver sulfosalts (pearceite, proustite, polybasite, and pyrargyrite) were replaced by supergene stephanite (Ag_5SbS_4), argentite, and much dyscrasite (Ag_3Sb) and native silver. During enrichment, gangue minerals were rarely affected, and the enrichment process was largely one of replacing primary silver-bearing minerals by ones even higher in silver than those that were replaced.

The enriched zones are separated from the oxidized zones above them by beds of tuff in a similar manner to that in which the two enriched limestones are separated by similar rocks. The characteristic minerals of the oxidized zone are silver halides and iron oxides.

The principal oxidized minerals appear to have been iodobromite [Ag(Cl,Br,I)] and cerargyrite (AgCl), plus some bromyrite, embolite, and iodyrite, and these minerals were most commonly developed as replacements of dyscrasite and native silver, but unaltered primary silver sulfosalts that were still preserved when the rock volume containing them became incorporated in the oxidized zone also were converted wholly or in part to silver halides.

The final and rather surprising process to affect the oxidized ore was the replacement of silver halides by native silver and argentite. It seems probable that this last event took place during a short period when the ores must have been brought again beneath the water table. It seems certain that the secondary silver sulfides, native silver, and silver alloys were formed below the water table by metal ions carried downward from the oxidized zone above by circulating ground water. These deposits should, therefore, be classified as ground water-B2.

The silver halides of the oxidized zone, on the other hand, certainly were formed above the water table in what must have been a drier climate than normally obtains in the vicinity of silver-bearing ore bodies, and the bonanza oxidized ores should also be classified as ground water-B2, the difference between the two enriched and oxidized ores being their position below and above the water table, respectively.

CHUQUICAMATA

Late Mesozoic to Early Tertiary	*Copper, Molybdenum*	*Mesothermal to Telethermal (primary), Ground Water-B2 (enriched and oxidized)*

Bandy, M. C., 1938, Mineralogy of three sulphate deposits of northern Chile: Amer. Mineral., v. 23, p. 669-760

Berman, H., and Wolfe, C. W., 1940, Bellingerite, a new mineral from Chuquicamata, Chile: Amer. Mineral., v. 25, p. 505-512

Creasey, S. C., 1959, Some phase relations in the hydrothermally altered rocks of porphyry copper deposits: Econ. Geol., v. 54, p. 351-373 (general)

Hendricks, J. A., 1922, The Chuquicamata ore body: Univ. Calif. Pubs., Bull. Dept. Geol. Sci., v. 14, no. 2, p. 75-84

Jarrell, O. W., 1939, Marshite and other minerals from Chuquicamata, Chile: Amer. Miner., v. 24, p. 629-635

—— 1944, Oxidation at Chuquicamata, Chile: Econ. Geol., v. 39, p. 251-286

Lopez, V. M., 1939, The primary mineralization at Chuquicamata, Chile, S.A.: Econ. Geol., v. 34, p. 674-711

—— 1942, Chuquicamata, Chile, in Newhouse, W. H., Editor, *Ore deposits as related to structural features*: Princeton Univ. Press, p. 126-128

Newberg, D. W., 1967, Geochemical implications of chrysocolla-bearing alluvial gravels: Econ. Geol., v. 62, p. 932-956

Palache, C., and Jarrell, O. W., 1939, Salesite, a new mineral from Chuquicamata, Chile: Amer. Mineral., v. 24, p. 388-392

Perry, V. D., 1952, Geology of the Chuquicamata ore-body: Min. Eng., v. 4, no. 12, p. 1166-1168

Taylor, A. V., Jr., 1935, Ore deposits at Chuquicamata, Chile, in *Copper resources of the world*: 16th Int. Geol. Cong., v. 2, p. 473-484

Notes

Chuquicamata is located some 240 km northeast of the port of Antofagasta in the heart of the Atacama desert.

The rocks of the immediate Chuquicamata district are essentially all igneous ranging in composition from granite through quartz monzonite to granodiorite. This block of intrusives is exposed in a long narrow belt of stocks that strikes about N15°E and extends for about 150 km from Caracoles on the south to El Abra on the north. These stocklike masses appear to be upward extensions of a huge, underlying batholith that was intruded (probably in the Paleocene) into a sequence of Mesozoic sediments and Late Cretaceous volcanic rocks and volcanic-derived sediments that, at the time of intrusion, were folded and intensely faulted. The introduction of the huge intrusive mass was followed by the accumulation of a thick series of andesitic breccias and tuffs that covered the entire area. Throughout most Tertiary time, the region appears to have undergone extensive erosion in a wet climate through which were developed high-altitude valleys with wide and gentle slopes and deep soils. About the end of the Tertiary, the area was again uplifted and further volcanic extrusions took place, the rocks were uplifted and slightly warped, and the climate suddenly became arid. During the Pleistocene and the Recent, further gentle uplifting occurred, and volcanism increased and died away.

Three igneous rocks compose essentially all the intrusive rocks in the immediate vicinity of the open pit: (1) the Fortuna granodiorite directly west of the West fissure that provides the west boundary of (2) the Chuquicamata (Chuqui) quartz monzonite porphyry (an elongated, dikelike mass) in which the ores were developed; and (3) the Elena granodiorite, slightly different in character from the Fortuna and gradational into the Chuqui porphyry, which lies along most of the east boundary of the quartz monzonite; although, on the southeastern boundary of the pit, gneisses, less metamorphosed sediments, and volcanic rocks are found. There probably is some suggestion that the contact between each pair of igneous rocks is gradational, but alteration, both deuteric and hydrothermal, makes this uncertain. Although the evidence is not completely definite that the three igneous rocks were intruded in early Tertiary time, they have cut through Upper Cretaceous sediments, and the geologic events that followed their intrusion were diverse enough and took place over such a considerable length of time that the actual intrusions probably occurred in the earliest Tertiary rather than in the latest Cretaceous. The ores appear to have been very closely related in time to the intrusion of the Chuquicamata quartz monzonite porphyry, so it seems reasonable to classify the deposits as late Mesozoic to early Tertiary.

The deposit at Chuquicamata can be considered as having been formed in four steps: (1) the introduction of the early Tertiary primary ore, (2) the formation of a Tertiary zone of supergene sulfide enrichment, (3) post-Tertiary development from part of the first enriched zone of the unusual minerals of the oxidized zone, and (4) the production of a second zone of supergene enrichment, which lies just beneath the present water table, at greater depths than the first. The ores at Chuquicamata are in general confined to the highly fractured portion of the quartz monzonite porphyry, a zone that is sharply bounded on the west by the West fissure, that dies out gradually to the east in the Elena granodiorite, and that strikes about N10°E over a length of about 3 km.

The primary ore mineralization in the shattered and altered rock volumes consists, in order of decreasing age, of (1) specular hematite, with some coarse sericite crystals, in the transition zone; at depth, primary chalcopyrite is known to accompany the hematite; (2) molybdenite in the margins and centers of the quartz veinlets of the sericite zone adjacent to the siliceous rock zone, the quartz and molybdenite apparently being essentially contemporaneous; (3) pyrite throughout the ore body, but most abundantly in the sericitized rock zones where it occurs in quartz veins and veinlets and intermixed with the other primary sulfides; (4) enargite mainly in the sericitized zones; to the east, where the alteration is composed to a greater extent of clay minerals, the enargite grades into a chalcopyrite zone where minor bornite was deposited at the same time as the chalcopyrite; still farther east, chalcopyrite is associated with the specular hematite; (5) small amounts of sphalerite, galena, and tennantite-tetrahedrite probably fit here in the paragenetic sequence but are so sparse that this is not certain; (6) primary chalcocite (the ratio of which to supergene chalcocite is uncertain) in the enargite zone that changes gradually to covellite in the chalcopyrite zone; some covellite is present with enargite in certain portions of the enargite-pyrite zone closest to the West fissure--the covellite is in part supergene, but some, at least, is primary.

Insomuch as molybdenite is now recovered from the Chuquicamata ore, it is necessary to consider the intensity conditions that prevailed during its formation to arrive at a correct classification. At Chuquicamata, the molybdenite is not associated with the high-temperature minerals present at Climax, such as huebnerite, topaz, and cassiterite, so it is here thought to have formed in the more intense portions of the mesothermal range. The enargite, as a mineral diagnostic of mesothermal conditions, is assigned to that portion of the classification; the chalcopyrite is also thought to have been formed under similar intensity conditions. On the other hand, although it is difficult to assess the proportions of the chalcocite and covellite that were formed under primary and under supergene conditions, respectively, it appears definite that the primary portions of these minerals were formed under telethermal conditions. The primary ores at Chuquicamata are here classified as mesothermal to telethermal, although it should be noted that there is essentially no evidence of ore mineral deposition in the leptothermal range.

After the sudden and drastic lowering of the water table at the end of the Tertiary, an oxidized ore body was produced in the less pyritized central and eastern parts of the deposit where much of the original pyrite had been converted to chalcocite. If, in rock volumes where both pyrite and chalcocite were present, pyrite was dominant, the high ferric ion content developed in the ground water above the water table inhibited the precipitation of copper minerals, and jarosite [$KFe_3(SO_4)_2(OH)_6$] was essentially the only oxidized mineral formed in place. If, on the other hand, chalcocite was more abundant or the primary ore was highly disseminated, chalcanthite ($CuSO_4 \cdot 5H_2O$) and locally antlerite [$Cu_3(SO_4)(OH)_4$] were formed but were not precipitated directly in place but at various distances from the sources of the copper. Only when the precipitation of jarosite (and limonite) had sufficiently reduced the ferric ion content of the solutions did copper minerals form. The higher the sulfate ion content of these copper-mineral-producing solutions, the greater was the proportion of the copper deposited as chalcanthite and the less as antlerite. If chalcocite was essentially the only mineral present in the enriched ore undergoing oxidation, antlerite would form directly from the chalcocite. Early formed chalcanthite was normally converted to antlerite as the sulfate ion concentration of the ground water was reduced; if the sulfate ion concentration was even further reduced, the antlerite might have been converted to brochantite [$Cu_4(SO_4)(OH)_6$]; but this appears seldom to have occurred. Brochantite is found only on the fringes of the ore body or, if in the central portion, high above the primary sulfides. Where sulfides were attacked very near the surface, natrochalcite [$NaCu_2(SO_4)(OH) \cdot H_2O$] and krohnkite [$Na_2Cu(SO_4)_2 \cdot H_2O$] formed in place of, or in addition to, chalcanthite. If krohnkite and natrochalcite were formed farther beneath the surface, they, as was chalcanthite, were converted to antlerite when the sulfate ion content was sufficiently reduced. The final result in the pyrite-poor portions of the enriched ore body was the development of antlerite to the essential exclusion of all other copper minerals; about 98 percent of the copper in the oxidized ores was initially or ultimately contained in antlerite.

If antlerite is brought very near the surface by erosion, it becomes unstable and its place is taken by the chlorine-bearing mineral atacamite [$Cu_2(OH)_3Cl$], by krohnkite, and in places by natrochalcite or even chalcanthite; atacamite is not found more than 30 m below the surface.

In the main oxidized ore body, much of the antlerite appears to have been formed directly from secondary chalcocite (and covellite), only a small fraction of it having been first deposited as chalcanthite due to the low-iron content of the highly enriched supergene copper sulfides. On the west side of the ore body there is no zone of oxidized copper ore between the surface and the enriched sulfide that lies at depths greater than those of the main oxidized ore body. In this rock volume, it is probable that the pyrite content of primary ore was so great and the ferric ion content of the ground water above the water table so high that essentially no copper was precipitated above the water table. Instead, it passed below that surface and reacted with the primary ore to produce a zone of deep secondary enrichment. Although the pyrite-rich ore in this area had been somewhat enriched during Tertiary times, much of the high original pyrite content remained to provide, under oxidizing conditions, the high ferric ion content of the ground water needed to carry the copper below the new, deep water table. The downward movement of copper was further aided by the lesser ability of the highly sericitized wall rocks to neutralize the ground water which thus prevented the precipitation of copper minerals until the copper-rich ground water had passed beneath the water table; even chalcanthite is rare in the western oxidized zone. Normally, between the bottom of the oxidized ore in the central and eastern portions of the deposit and the top of the primary sulfides, there is a zone of waste rock that contains appreciable amounts of oxidized iron minerals but essentially no copper-sulfate minerals. This waste (iron-mineral-rich) zone derived from the primary ore that, in Tertiary time, was too far beneath the surface to have been secondarily enriched and was too rich in iron to have allowed the precipitation of copper minerals above the water table after that surface was suddenly lowered at the end of the Tertiary. The copper removed from the primary ore above the new water table was, therefore, largely carried below it and acted to produce a zone of supergene enrichment immediately below the new water table. In certain portions of the deposit, however, there is evidence that some oxidation occurred below the present water table to produce what is known as mixed ore. This ore seldom contains visible copper sulfate minerals but does contain enough soluble copper (0.30%) that some sulfate minerals (probably mainly chalcanthite) must be present. Such mixed ores are formed in rock volumes where the oxidized ore zone extends downward to the top of the primary ore, that is, to the water table, and probably indicate fluctuations in the level of the water table in fairly recent times, perhaps even after mining had begun since the side effects of that operation well may have forced pronounced local changes in the elevation of the water table.

It appears to be essentially certain that both the Tertiary and post-Tertiary zones of secondary enrichment and the oxidized ore body of post-Tertiary time were formed through the action of ground water, the latter by oxygenated water above the water table and the former by essentially oxygen-free ground water below it. These important contributions to the economic value of the Chuquicamata ores must be classed as ground water-B2.

PERU

CENTRAL PERU (GENERAL)

Middle Tertiary	*Copper, Lead, Zinc, Silver, Gold, Bismuth, Mercury*	*Mesothermal to Leptothermal, Kryptothermal to Epithermal*

Dollfus, O., 1960, Présentation de la structure des Andes centrales péruviennes: Inst. Française Études Andines Tr., t. 7, p. 53-64

Harrison, J. V., 1943, The geology of the central Andes in part of the province of Junin, Peru: Geol. Soc. London Quart. Jour., v. 99, p. 1-36

—— 1960, Structural doubts about the Andes in Peru: 21st Int. Geol. Cong. Rept., pt. 18, p. 7-13

Kulp, J. L., and others, 1957, Lead isotope composition of Peruvian galenas: Econ. Geol., v. 52, p. 914-922

Lacy, W. C., 1953, Differentiation of igneous rocks and its relation to ore deposition in central Peru: Soc. Geol. Perú Bull., t. 26, p. 121-138; A.I.M.E. Tr., v. 208, p. 559-562

Lacy, W. C., and Hosmer, H. L., 1956, Hydrothermal leaching in central Peru: Econ. Geol., v. 51, p. 69-79

McLaughlin, D. H., 1924, Geology and physiography of the Peruvian Cordillera, departments of Junin and Lima: Geol. Soc. Amer. Bull., v. 35, p. 591-632

—— 1953, Notes of geologic studies in Perú: Soc. Geol. Perú Bull., t. 26, p. 139-147

Petersen, U., 1965, Application of saturation (solubility) diagrams to problems in ore deposits: Econ. Geol., v. 60, p. 853-893

—— 1965, Regional geology and major ore deposits of central Peru: Econ. Geol., v. 60, p. 407-476

Terrones L., A. J., 1958, Structural control of contact metasomatic deposits in the Peruvian Cordillera: A.I.M.E. Tr., v. 211, p. 365-372 (in Min. Eng., v. 10, no. 3)

Notes

The rocks of central Peru, which area Petersen (1965) has defined as extending from 9°30'S to 13°30'S over a width of 100 km along the axis of the Andes, range in age from lower or middle Paleozoic to Quaternary.

Although the ore deposits are confined to the 100 km belt just mentioned, its structure cannot be understood without consideration of the various belts that lie to either side of that containing the mineralization. Crossing the country from southwest to northeast, these belts are (1) the coastal Mesozoic belt in which the rocks are gently folded and block-faulted sediments; (2) the coastal batholith of which the dominant rock type is granodiorite; (3) the Tertiary and Quaternary volcanic belt in which the volcanic rocks are moderately folded and have been cut by small volcanic necks, stocks, and spines; (4) the central Andean Mesozoic belt in which the rocks are mainly moderately to strongly folded and faulted Mesozoic sediments that are cut by small- to moderate-sized stocks and volcanic vents--older rocks crop out in the cores of anticlinal and domed structures and younger rocks in the central portions of synclines; (5) the eastern Paleozoic belt in which the rocks are mainly moderately to strongly folded Paleozoic sediments and lie on the eastern slope of the Andes--strongly developed batholiths cut through them; (6) the eastern Mesozoic belt in which the rocks are moderately to strongly folded and lie on the upper edge of the Amazonian jungle; and (7) the Tertiary sediments of the Amazon basin in which the rocks are principally continental red beds--older rocks crop out in anticlinal or domed cores. The ore deposits of central Peru are found largely in belts (3), (4), and (5).

The rocks of the central portion of Peru may be divided into (1) lower to middle Paleozoic Excelsior shales and phyllites, the thickness of which is over 1000 m and may be even as much as 6100 m; (2) Carboniferous Ambo sandstones and shales (present only locally and on the east side of the region), the thickness of which exceeds 800 m; (3) Carboniferous Tarma limestone and shale (also only locally and to the east) that have a maximum thickness of about 2100 m; (4) Permo-Carboniferous Copacabana limestone and shale (also only locally and toward the east) that have a maximum thickness of about 1900 m; (5) Permian Mitu red beds that range between 0 and 3700 m in thickness but generally are 0 to 1000 m in the central Peru region; (6) the Permian Catalina volcanic rocks that are 300 to 700 m in thickness and may be as much as 1500 m--they are limited largely to the Morococha, San Cristobal, and Malpaso areas; (7) Upper Triassic-Jurassic Pucará limestone that ranges from 0 to 3000 m in thickness and thickens to the east; (8) Jurassic-Cretaceous Goyllarisquisga sandstone that ranges between 30 and 1800 m in thickness and thickens to the west; (9) middle Cretaceous Machay limestone that ranges in thickness between 135 and 2000 m and thickens to the west; (10) Cretaceous-early Tertiary Casapalca-Poco-bamba red beds that range from 0 to 3000 m in thickness, with the greatest thickness being in the Casapalca-Yauricocha-La Oroya triangle; (11) Eocene-Oligocene Tacaza volcanic rocks that range in thickness from 500 to 9000 m; and (12) Oligocene-Quaternary Sillapaca volcanic rocks that range in thickness from 0 to 1000 m.

The Excelsior shales were highly deformed in the mid-Paleozoic and were covered unconformably in the late Paleozoic geosynclinal phase by the Ambo, Tarma, and Copacabana beds. These formations were then affected by the late Paleozoic deformational phase, although the Excelsior beds may not have been involved in this folding; this deformational phase may have extended into the Lower Triassic. Erosion during and after this orogeny probably removed most of the Paleozoic sediments from central Peru and, in its earlier phases, appears to have produced the Mitu and Catalina beds that were involved in the latter phases of the late Paleozoic orogeny. During the Mesozoic, geosynclinal deposition was the rule and produced the Pucará, Goyllarisquisga, and Machay formations that were folded in the Late Cretaceous (pre-Laramide) orogeny, which occurred at the same time as the intrusion of the Coastal batholith. Earth movements, on various scales, were intermittent from the last of Cretaceous time through the Quaternary and included three stages of orogeny, separated by times of intermittent diastrophism. The third of these orogenic stages (mid-Oligocene) was followed by the Puna erosion stage (accompanied by local volcanic activity) at the end of the Oligocene. This erosion was followed by the first stage of Cordilleran uplift, plus further intermittent diastrophism and the intrusion of plugs and stocks. Then came the Junin erosion stage, followed, in turn, by the second stage of Cordilleran uplift and by the Chacra erosion stage. The cycle reached its present state with the third Cordilleran uplift, followed by the modern erosive and volcanic cycle.

The intrusive igneous rocks in the mineralized districts range from granite and rhyolite through such types as quartz monzonite, monzonite, granodiorite, dacite, diorite, and andesite. Among these, the most common texture is porphyritic, but fine-grained and coarse-grained rocks are not uncommon. Where intrusive igneous rocks are present in considerable volume in a given district, the most usual types are quartz-monzonite porphyry, granodiorite, or monzonite, but granites and diorites are present in several instances.

The idea has been advanced (Petersen, 1965, p. 415) that the structures of the folded post-Paleozoic sediments can best be explained by assuming that they folded independently of the older basement (Excelsior) rocks. Such décollements would have required incompetent or lubricating beds in the formations unconformably overlying the Excelsior rocks on which the younger beds could have moved. There is appreciable question as to whether or not the evaporite beds among the Permian red beds were sufficient in number and extent to have provided bases on which independent movement on the large scale required could have been accomplished.

The ore deposits of the region are arranged on more or less northeast-striking trend lines that may be related to breaks in the basement complex; movement along these breaks would have caused deformation in the overlying blankets of sediments, particularly being a possible cause of cross folds and of folds in otherwise undisturbed areas.

The place of the formation of the various ore deposits of central Peru in this time sequence is uncertain. Although intrusive igneous rocks are present in all of the ore districts discussed under central Peru, it is uncertain to which of the rocks in a given district, if any, the ore fluids of that district were genetically related. Except for Quaternary volcanic rocks, it seems definite that the emplacement of the ores was among the latest geologic events in each district in question. It is probable, however, that the introduction of the ores was related most directly to the intrusion of stocks and batholiths that accompanied the third stage of orogenesis; this would place the mineralization process in middle or, at most, the late Oligocene and would result in the deposits being best categorized as middle Tertiary. It is not necessarily true, of course, that each deposit was formed at the same time as its congeners, and the deposits of central Peru may have been emplaced over distinctly separate time intervals, over a considerable span of time, and this time-span can best be placed as between the beginning of the Oligocene and early in the Miocene. Recent thinking among the Cerro geologists seems to place the ore epoch as Pliocene, but, lacking definite evidence to this effect, it seems best to retain the classification of middle Tertiary for the ore deposits of central Peru.

As Petersen (1965) emphasized, almost every rock in the geologic column of central Peru, sedimentary and igneous, contains ore somewhere, but some are far more impressively mineralized than others. The Excelsior beds, except for the most southerly veins in the Cerro de Pasco district, are essentially barren, probably because of the incompetent character of the formation. The Mitu beds also seem to be very little mineralized, although the Mitu is lacking, or masked by alteration, in several of the districts. The Casapalca red beds contain important veins at Casapalca and some Casapalca (Pocobamba) horizons have been replaced by ore at Colquijirca and Yauricocha. The Catalina volcanic rocks enclose large vein structures at Morococha; these highly competent rocks break favorably to form potential sites for open-space filling. The Mesozoic limestone formations of central Peru (the Pucará and the Machay) contain numerous and often large replacement ore bodies and veins. The Machay contains much ore at Yauricocha and Huancavelica and the Pucará at Carahuacra, Cerro de Pasco, Morococha, and Huancavelica, although the principal ore bodies at Huancavelica are in the lower member of the Goyllarisquisga sandstone. Ore bodies are contained in igneous rocks at Cerro de Pasco, Morococha, Casapalca (if the Carlos Francisco porphyry is intrusive), Yauricocha (though in minor amounts in comparison with those in the Machay), and Huancavelica (although in almost a vanishingly small amount).

Zoning is a common feature of the ore bodies in the various districts of central Peru. It is important at Casapalca, Cerro de Pasco, Morococha, and Yauricocha. At Colquijirca there is some evidence of zoning from south to north, and even in the short vertical lengths of the epithermal mercury deposits of Huancavelica, some vertical zoning of the arsenic minerals is readily apparent. The increase of hematite with depth at Carahuacra suggests that zoning may obtain there also.

The enargite-chalcopyrite ores at Morococha and the increasing chalcopyrite content with depth at Casapalca are indicative of mesothermal conditions of deposition. The tetrahedrite-tennantite-sulfosalt mineralizations in both mines show that an appreciable part of the ore deposition took place in the leptothermal range. The development of high-temperature silicates in the limestones at Morococha appears to have predated the ore. At Cerro de Pasco, the rapid vertical changes in mineralization, the highly broken rock of the vent, and the porous character of many of the ore pipes point to rapid loss of heat and pressure during ore formation. The enargite-luzonite and the sphalerite-galena ores at Cerro probably were formed under kryptothermal conditions, while the tennantite-argentite-silver sulfosalt pipes within the larger lead-zinc deposits suggest formation within the epithermal range. The silver-rich galena, tennantite, and stromeyerite and perhaps the native silver at Colquijirca probably were introduced in the leptothermal range. The primary enargite, chalcopyrite, and bornite at Yauricocha appear to have been developed under mesothermal conditions and the tennantite, silver-rich galena, and silver sulfosalts under leptothermal. The early, dark sphalerite mineralization at Carahuacra appears to have formed in the mesothermal range, while the late, lighter sphalerite and its accompanying minerals probably were developed in the leptothermal range. The Huancavelica cinnabar was emplaced under epithermal conditions.

CARAHUACRA

Middle Tertiary | *Zinc, Lead, Silver* | *Mesothermal to Leptothermal*

Dittman, A., 1926, Carahuacra, eine bemerkenswerte Blei-Zinkerzlagerstätte Perus: Metall und Erz, Bd. 23, H. 27, S. 575-582

Lyons, W. A., 1968, The geology of the Carahuacra mine, Peru: Econ. Geol., v. 63, p. 247-256

Petersen, U., 1965, Regional geology and major deposits of central Peru: Econ. Geol., v. 60, p. 407-476

Tossi, P. A., 1956, Geología y mineralización en Carahuacra, Junin: Soc. Geol. Perú, Bol., t. 30, p. 375-384

Notes

The Carahuacra deposit is located in the central Andes of Peru, about 115 km north-northeast of Lima and some 20 km south-southeast of Morococha; it is on the east side of the continental divide.

The oldest known formation in the Carahuacra district is the lower Paleozoic Excelsior shale that makes up the core of the principal structure in the area, the Yauli anticline. Unconformably on the Excelsior are two Carboniferous formations of which the oldest is the Mitu (coarse conglomerates, arkoses, sandstones, and argillites). Overlying the Mitu are the Catalina volcanic rocks that are an irregular series of flows and pyroclastics with an extensive flow breccia at the base and small masses of shale scattered throughout. Unconformably above the Carboniferous beds is the Pucará formation, a limestone series of Lias age. The Pucará has a conspicuous basal conglomerate and contains three horizons of water-laid tuff. The ores of both Carahuacra and the San Antonia mantos and breccias are found in the lower part of the Pucará. The Goyllarisquisga formation lies pseudoconcordantly above the Pucará, is composed of sandstone, shaly sandstone, shale, limestone and conglomerate, and is Lower Cretaceous in age. Above the Goyllarisquisga is the Middle Cretaceous Machay formation, composed of fossiliferous limestones with intercalated marls, shales, and sandstones. Although not in the district proper, the red bed series unconformably overlies the Machay; these beds correspond with the Casapalca red beds.

Although numerous intrusives are known in this general area of Peru, particularly at Morococha, the principal intrusive in the Carahuacra district is known as the Carahuacra intrusive and is located east of the mine and cuts through the pre-Mesozoic formations. Lyons (1968) considers it to be a remnant of a volcanic neck that was one of the centers from which the Catalina volcanic rocks were extruded; it is a porphyritic quartz monzonite. At Carahuacra, a diabase dike has cut the northern apophysis of the quartz monzonite.

The Yauli anticline was folded in Tertiary time and follows the regional trend of N35°-40°W. A strong thrust fault follows the contact between the Pucará and the Paleozoics. On this fault were developed the breccias in which are deposited most of the second period of Carahuacra mineralization, although the specific locations of this ore within the breccias was determined by a late set of N60°-85°E faults.

No later intrusives than the diabase are known in the district; the Carahuacra intrusive is far too old to have had anything to do with the ore formation. It would be very surprising, however, if magmatic sources of ore fluids did not exist at depth in the Carahuacra area during the Tertiary orogenic activity. Such magmas probably are better dated as middle, rather than late Tertiary since late Tertiary intrusives were introduced into other parts of the Yauli area (Lyons, 1968) after the manto ores at least had been emplaced. By analogy with the decisions reached for the other ore districts of central Peru, the Carahuacra ores are categorized as middle Tertiary, although age dating may find them to be late Tertiary.

The Carahuacra area contains three types of ore bodies: (1) the manto ore consists of massive replacements of limestone; they are the economically most important ore type and, in part, have been considerably brecciated; (2) these brecciated manto ores form the second ore type and differ from the manto ores only in their broken character; (3) the vein fillings in the N60°-85°E fractures make up the third ore type. The manto ores have a simple mineralogy--ferroan sphalerite, negligible chalcopyrite, and pyrite with considerable magnetite and hematite that increases in abundance with depth relative to pyrite; the gangue minerals are chert and minor quartz. The alteration of limestone around the manto ores is mainly dolomitization of the limestone that forms an aureole with a minimum width of 11 m; close to the mantos, the wall rock was slightly silicified. The magnetite, hematite, and pyrite appear (according to Lyons) to have been emplaced after the sphalerite and its associated chalcopyrite. The brecciation followed the emplacement of the iron minerals and then came the the weak set of cross fractures in which the third type of ore was emplaced. This mineralization also was developed in the brecciated and fractured manto ores as well as in the cross fractures. The first minerals to form within this third stage were brownish red sphalerite and a closely associated galena; this sphalerite is well crystallized along fissures or cavities in manto ores or cross fractures; as this sphalerite deposition continued, the color changed to light yellow. A second generation of pyrite followed the sphalerite-galena deposition; several stages of pyrite formation are mentioned by Lyons (1968), so others in addition to the two he places in the paragenetic sequence must have been formed. The second generation of pyrite was followed by negligible amounts of tetrahedrite, jamesonite, and stibnite; then come gangue minerals--quartz, dolomite, rhodochrosite, siderite, gypsum, and barite; during this stage some marcasite was deposited as well.

The upper portions of the ores were highly oxidized to iron and manganese oxides; among these oxides were considerable amounts of native silver, argentite, and pyrargyrite that probably were formed by secondary enrichment below the water table and preserved above it during the formation of the iron and manganese oxides. The silver content of the oxide ores was far higher than that in the primary ores.

The manto ores, based on the presence of chalcopyrite with (exsolved from ?) the sphalerite and magnetite and hematite in increasing amounts with depth, suggests that these ores were formed on the border line between hypothermal and mesothermal conditions. The weak wall-rock alteration suggests that mesothermal is a more reasonable classification to assign than hypothermal-2 for the manto ores, but mining at greater depths may uncover a true hypothermal phase. The third stage ores have the characteristics of the leptothermal range and are so categorized here. The ores do seem to have been deposited under conditions of slow loss of heat and pressure, although the third stage might conceivably be assigned to epithermal rather than leptothermal conditions.

CASAPALCA

Middle Tertiary | *Zinc, Lead, Silver, Copper* | *Mesothermal to Leptothermal*

Amstutz, G. C., 1960, The copper deposits of Caprichosa and Antachajra in central Peru: Neues Jb. f. Mineral., Abh., Bd. 94, H. 1, S. 390-429

Geological Staff, Cerro de Pasco Copper Corporation, 1950, Lead and zinc deposits of the Cerro de Pasco Copper Corporation in central Peru: 18th Int. Geol. Cong. Rept., pt. 7, p. 154-186 (particularly p. 180-185)

Graton, L. C., 1933, The depth-zones in ore deposition: Econ. Geol., v. 28, p. 513-555 (particularly p. 536-540)

McKinstry, H. E., 1927, The minerals of Casapalca, Peru: Amer. Mineral., v. 12, p. 33-36

McKinstry, H. E., and Noble, J. A., 1932, The veins of Casapalca, Peru: Econ. Geol., v. 27, p. 501-522

Overweel, C. J., 1961, The central alteration body of the Casapalca mines, Peru: Geologie en Mijnbouw, Jg. 40, nr. 1, p. 1-10

Petersen, U., 1965, Casapalca, in *Regional geology and major ore deposits of central Peru*: Econ. Geol., v. 60, p. 407-476

Sawkins, F. J., and Rye, R. O., 1970, The Casapalca silver-lead-zinc-copper deposit, Peru: An ore deposit formed by hydrothermal solutions of deep-seated origin?: Geol. Soc. Amer. Abs. with Programs, v. 2, no. 7, p. 674-675

—— 1970, Fluid inclusions and stable isotope studies of the Casapalca silver-lead-zinc-copper deposit, central Andes, Peru: IMA-IAGOD Meetings 1970, Collected Abs., Tokyo-Kyoto, Paper 5-23, p. 133

Notes

The vein deposits of Casapalca are located in the Cordillera Occidental, 100 km east-northeast of Lima, the capital of Peru, and about 35 km south of west from La Oroya, the site of the smelter operations of the Cerro Corporation in Peru. The 1700-level adit, the principal outward ore haulage level of the mine, is at an altitude of about 13,700 feet.

The oldest rocks in the area probably are the Goyllarisquisga formation composed of light-colored quartzite with some interbedded carbonaceous shale. Neither this formation nor the overlying Machay limestone is exposed in the mineralized portion of the district. The Machay limestone is a massive gray limestone with dark interbedded shale; where the Goyllarisquisga and the Machay are in contact, they apparently are separated by a fault. The oldest exposed rocks in the Casapalca area proper are those of the Cretaceous-early Tertiary Casapalca formation. This formation is made up of three members: (1) the Casapalca red beds, calcareous red and purple sandstone, siltstone, and shale with light-gray interbedded material of the same general character; (2) the Amygdaloid member, a dark amygdaloidal lava; and (3) the Carmen member, interbedded limestone, shale, sandstone, and conglomerate. The Casapalca formation is overlain by the early Tertiary extrusive Carlos Francisco formation that consists of three members: (1) the Tablachaca volcanic rocks, intercalated tuffs, breccias, agglomerates and conglomerates, with occasional layers of sandstone, and quartzite; the principal color is red to purple; (2) the Carlos Francisco porphyry, massive andesite porphyry and porphyry breccia; (3) the Yauliyacu tuffs, tuffs and fine red and gray porphyries.

The Tertiary Bellavista limestone overlies the Carlos Francisco formation. The Bellavista beds are overlain by the youngest consolidated rocks in the area, probably the middle Tertiary Rio Blanco formation. After the Rio Blanco formation had been extruded, the area was folded into the present Andean pattern of close and overturned folds and thrust faults that, after further erosion, produced the present pattern of narrow outcrop bands striking north-northwest. This folding was accompanied by the igneous activity that resulted in the introduction of four intrusive igneous rock types, the Veintiuno andesite, the Victoria porphyry, and Fraguamachay and Taruca intrusives. The veins, now filled with ore, were produced during the folding and were invaded by ore-forming fluids late in this orogenic cycle. It is almost certain that none of the intrusive rock masses known at the present time was of sufficient size or was correctly placed in relation to the zoned ore veins of Casapalca to have been the source of the ore-bearing solutions that formed the deposits. Instead, it is highly likely that the ore fluids came from greater depths, probably from the same general source as the various magmas that were intruded into, and extruded on, the rock sequence of the district.

It is obvious, from the number of events that occurred in the Casapalca district after the formation of the Casapalca red beds in the late Mesozoic or early Eocene, that the ore bodies, which were introduced in the latest stages of that sequence, must have been middle or late Tertiary in age. It seems probable, however, that the series of geologic happenings involved in producing the Casapalca veins and the geologic pile in which they are contained took up only part of the Tertiary period and that the age of the deposits is more likely to be middle Tertiary than late, and they are so classified here.

Although six types of veins and vein fillings have been recognized at Casapalca, the first two listed here are, at least up to now, the only ones of major economic worth. These types are (1) the Carlos Francisco that has a quartz and subordinate calcite gangue and contains pyrite, sphalerite, galena, and silver-rich tetrahedrite as its main metallic minerals--the veins of this type are designated as the L, M, N, O, and P; (2) the Aguas Calientes-Carmen (C and S veins, respectively) that has a carbonate and quartz gangue and contains sphalerite, tetrahedrite, silver-rich galena, and pyrite as its main metallic minerals--veins of this type may grade into those of the Carlos Francisco type; (3) the Corina that has little gangue and contains sphalerite and jamesonite(?)--this is a dike vein some 2 km north of the main loci of vein development; (4) the Americana that has a cleavable carbonate gangue with tetrahedrite, sphalerite, and chalcopyrite and lesser galena and pyrite--these veins are appreciably east of the main mine area; (5) the Yauliyacu type that has a barite and carbonate gangue with pyrite, galena, sphalerite, and tetrahedrite--these veins are found in the Bellavista limestone 4 km south of the main area of mineralization and are in a strong gouge-breccia fault from 1 to 20 m wide; the mineralization occurs as bands or lenses on either the foot or hanging wall; (6) the Caprichosa that has a barite and carbonate gangue with bornite and chalcopyrite occurring in gash veins and in the surrounding porphyry--these veins are some 3 km east of the main mine area; and (7) recent work on the higher levels at Casapalca has discovered a few veins striking about northwest that cut across the Carlos Francisco veins near their southern end. These H veins are later than the Carlos Francisco veins but by how much is uncertain. They contain an iron-rich sphalerite-chalcopyrite type of mineralization (with minor tetrahedrite) that probably was deposited at higher temperatures than the minerals in the Carlos Francisco veins through which they cut.

The mineralization of veins is continuous so far as the existence of actual vein structures permits, but a considerable fraction of the mineralization is below minable grade. The intensity of mineralization in a given section of vein length is dependent on the type of fracture involved, and the character of the fracture, in turn, is governed by the type of rock in which it was developed.

The ore veins at Casapalca are strongly zoned, with the major direction of zoning being outward and with the most intense mineralization being found in the central portion of the lower levels. The outer zone contained a spotty mineralization of argentite, ruby silvers (proustite and pyrargyrite), owyheeite, pearceite, polybasite, and some tetrahedrite, realgar, orpiment, and stibnite. Inward and downward from this outer zone lay 200 to 300 feet of lean mineralization, that passed gradually into the intermediate-zone assemblage of pyrite, galena (silver-rich), sphalerite, minor chalcopyrite, tetrahedrite, bournonite, and much lesser amounts of other sulfosalts (boulangerite and jamesonite) in a gangue of quartz, calcite, and rhodocrosite. On the lower levels in the central (Consuelo) section of the mine, the intermediate mineralization dies out and is replaced by a pyrite, chalcopyrite, tennantite, sphalerite, minor galena assemblage that is lower in silver than the intermediate zone. Minor amounts of huebnerite and arsenopyrite have been found in this central zone, and sericite and quartz are more abundant than in the intermediate-zone ore bodies.

In the Carlos Francisco area, the sequence of mineralization appears to have been minor calcite, quartz, pyrite, sphalerite, galena, tetrahedrite, corrosion and etching of quartz, chalcopyrite, local bournonite, and calcite and rhodochrosite. In the Aguas Calientes portion of the mine, the sequence was calcite and rhodochrosite, pyrite, sphalerite, galena, tetrahedrite, corrosion of quartz, accompanied by the deposition of chalcopyrite, quartz, bournonite, and calcite. The paragenesis of the central (Consuelo) section does not appear to have been worked out. The position of chalcopyrite in the paragenesis does not mean that it is universally present; actually, it normally was quite rare on the upper levels and increased in amount as the veins were followed down.

The walls of the veins in the district have been considerably altered, the amount of alteration decreasing outward from the apparent center of mineralization in the Aguas Calientes Valley (at the north end of the Aguas Calientes veins). In this most intensely altered section of the mine the walls have been silicified and pyritized, and the silicification grades outward into sericitization in which disseminated pyrite continues to appear. Still farther away from the veins, the volcanic rocks have been propylitized.

It would be surprising if a deposit with the great vertical and lateral continuity as that of Casapalca had been formed near the surface, but the character of the vein minerals in the outer margins of the deposit bears considerable resemblance to the epithermal deposits of Potosí. Recent work has shown that the typical leptothermal mineralization (Pb-Zn-Ag) extends practically to the present surface with essentially no change in the proportions among these elements from those that obtain at greater depths. On the other hand, it was the opinion of the earlier geologists that the deposits were formed under a considerable rock cover. Casapalca is the type deposit of Graton's leptothermal category of the modified Lindgren classification and that category, by definition, presumes formation at appreciable depth beneath the surface. On the balance of probabilities, and in contradistinction to Cerro de Pasco itself, the deposit appears to have been formed with slow loss of heat and pressure and the outer and intermediate zones of the complex seem to fit into the leptothermal range. Thus far, at least, none of the mineralization known at Casapalca can be designated as telethermal.

All the characteristics of the central zone point to its having been formed under mesothermal conditions, and it appears that the intensity of these conditions increases with depth. Graton's assignment of the outer and intermediate zones to his leptothermal class has been strongly confirmed by later work; the discovery and exploitation of a centrally located mesothermal portion of the deposit that expands laterally with depth further fits the relationship postulated by Graton between the leptothermal category and mesothermal category of the Lindgren classification. The Casapalca ores, therefore, here are categorized as mesothermal to leptothermal.

CERRO DE PASCO

Middle Tertiary	*Copper, Silver, Lead, Zinc, Gold, Bismuth*	*Kryptothermal to Epithermal*

Amstutz, G. C., and Ward, H. J., 1956, Geología y mineralización del depósito de plomo de Matagente, Cerro de Pasco: Soc. Geol. Perú Bol., t. 30 (Cong. Nac. Geol.), p. 13-31

Bideaux, R. A., 1960, Oriented overgrowths of tennantite and colusite on enargite: Amer. Mineral., v. 45, p. 1282-1285

Einaudi, M. T., 1968, Copper zoning in pyrite from Cerro de Pasco, Peru: Amer.

Mineral., v. 53, p. 1748-1752; disc., 1969, v. 54, p. 1216-1217

Geological Staff, Cerro de Pasco Copper Corporation, 1950, Lead and zinc deposits of the Cerro de Pasco Copper Corporation in central Peru: 18th Int. Geol. Cong. Rept., pt. 7, p. 154-186 (particularly p. 157-170)

Graton, L. C., and Bowditch, S. I., 1936, Alkaline and acid solutions in hypogene zoning at Cerro de Pasco: Econ. Geol., v. 31, p. 651-698

Jenks, W. F., 1951, Triassic to Tertiary stratigraphy near Cerro de Pasco, Peru: Geol. Soc. Amer. Bull., v. 62, p. 202-219

McLaughlin, D. H., and Moses, J. H., 1945, Geology of the mining region of central Peru: Min. and Met., v. 26, no. 467, p. 512-519 (particularly p. 514-515)

McLaughlin, D. H., and others, 1935, Copper in the Cerro de Pasco and Morococha districts, Department of Junin, Peru, in *Copper resources of the world*: 16th Int. Geol. Cong., v. 2, p. 513-544 (particularly p. 516-527)

Moses, J. H., 1940, Recent studies of the geology of the Cerro de Pasco district (Peru): 8th Amer. Sci. Cong. Pr., v. 4, Geol. Sci., p. 673-675

Petersen, U., 1965, Cerro de Pasco, in *Regional geology and major ore deposits of central Peru*: Econ. Geol., v. 60, p. 407-476

Terrones L., A. J., 1954, Generalidades mineralogenicas en el distrito minero de Cerro de Pasco: Minería, Lima, año 2, no. 6, p. 19-30

Ward, H. J., 1961, The pyrite body and copper orebodies, Cerro de Pasco mine, central Peru: Econ. Geol., v. 56, p. 402-422

Notes

The ore bodies of the Cerro de Pasco district are located about 175 km north-northeast of Lima and 100 km north-northwest of La Oroya; from the gossans above them, during the days of the Incas and of the Spanish conquistadores, huge quantities of silver were produced. Modern mining began in the area in 1901, when the Cerro Corporation (as the company is now called) commenced its operations.

The oldest rocks in the Cerro area proper are those of the Excelsior group, which is exposed in a broad, faulted anticline immediately west of the town of Cerro de Pasco. Jenks (1951) correlates these shales and phyllites with known Devonian rocks to the north near Ambo; others consider them to be Silurian or perhaps even older. The next youngest sediments are those of the Permian Mitu formation that lie unconformably on the Excelsior beds. This formation, near Cerro, is a conglomerate, the pebbles of which are quartz and Excelsior-type argillaceous material.

The Mitu group is overlain unconformably by the upper Triassic (and lower Liassic ?) Pucará group; in the Cerro district, these rocks include an east and a west facies, separated by the Cerro de Pasco thrust fault. There is no evidence of a major orogeny at the end of Paleozoic time since the disconformity between the Pucará and the Mitu is minor or even essentially nonexistent in places.

The Pucará beds are disconformably or slightly unconformably overlain by the Goyllarisquisga formation. The basal (Neocomian) Goyllarisquisga beds are red shaly sandstone and quartz conglomerates, some of these beds lying in channels in the upper Pucará beds. Locally, north and northeast of Cerro, the overlying Tertiary red beds are separated from the Goyllarisquisga by thin limestone beds that may correlate with the Middle Cretaceous (Machay) limestone beds known elsewhere in central Peru; these two limestone beds, however, are not found in the Cerro district proper.

The Tertiary rocks in the district lie above an angular unconformity on the upper surface of the Goyllarisquisga beds and include a thick sequence of shale beds of several colors and sandstones, with intercalated beds of conglomerate and limestone; the formation is known as the Pocobamba. The Pocobamba contains no volcanic rocks, but Jenks thinks that pyroclastics closely associated with the intrusive Cerro de Pasco and Marcapunta stocks must once have been far more widespread in the region and must have rested, with a sharp unconformity, on the highest Pocobamba beds. The Pocobamba beds are the latest consolidated sedimentary rocks in the district.

At the end of the marine sedimentation that produced the Pucará group (early Upper Cretaceous) and before the deposition of the continental Pocobamba formation, the rocks of the Cerro district and adjacent areas were highly folded in the Laramide orogeny. This orogeny produced a major anticline that brought the Excelsior group up to a position where it was, after subsequent erosion, at or near the surface.

After the Laramide orogeny, there was a long period of erosion that reduced the Mesozoic and older rocks to a surface of moderate relief, close to sea level. When this surface topography was reached, conditions were such that early Tertiary continental deposits, the lake, river, and alluvial deposits of the Pocobamba formation, were developed in shallow intermontane valleys. The Pocobamba sedimentation was ended by intense folding and faulting and upward movement of the land. The quartz monzonites and other intrusives, with related pyroclastics, were intruded after the post-Pocobamba folding and faulting; they do not show north-south deformation.

The Cerro de Pasco stock is a much more complex geologic feature than the name would suggest. In probably middle Tertiary time, quartz monzonite magma from an underlying magma chamber exploded through a vent in the Excelsior group rocks that apparently were localized by the anticlinal high and the Cerro thrust fault, the eastern margin of the vent obliterating, but being essentially tangent to, the fault. This steep-walled vent, about 2.75 km by 2.25 km in size, is now occupied by three kinds of material (in order of decreasing age): (1) the compacted explosion breccia that fell back into the vent (the Rumiallana agglomerate)--it is composed largely of fragments of Excelsior rocks, of volcanic glass, and of finely ground quartz monzonite porphyry; (2) an intrusive breccia (the Lourdes fragmental rock)--it is made up largely of fragments of the Excelsior group rocks in a matrix, dominant over the fragments in quantity, of non-brecciated quartz monzonite porphyry; and (3) an irregular intrusive mass and numberous dikes of normal quartz-monzonite porphyry--this quartz porphyry usually is quite full of extraneous inclusions. Erosion has removed all traces of the volcanic debris cone that must have overlain the vent. The bulk of the ore was introduced after these volcanic events had taken place and lies in, and adjacent to, this vent, along its eastern and southeastern margins. The ore was, therefore, emplaced at some considerable time into the Tertiary so that it seems reasonable to assign a middle Tertiary age to the deposits.

In the eastern and southeastern segment of the vent and its adjoining walls, a huge crescent-shaped mass of pyrite was emplaced that, at the present surface, is some 6000 feet long and has a maximum width of 1200 feet; the pyrite body narrows downward, and below the 2500 foot level, the mass tends to separate into roots or pods that reach nearly to 3000 feet below the surface. The pyrite so strongly replaced the vent material and the adjacent Pucará limestone that it is now impossible to map exactly the original vent outline on the east side; the contact to the west, against the volcanic rocks, is generally moderately sharp, as is that against the Excelsior beds to the south. Within the pyrite mass, particularly in the central part, are minor fragments of pyrrhotite, and below the 600 foot level in this part of the pyrite body, there are several pyrrhotite pipes with cross sections as much as 200 by 600 feet. Within these pipes there may be considerable amounts of sphalerite and minor quantities of pyrite, marcasite, and cassiterite; the pyrrhotite gradually grades through porous pyrite and marcasite to massive pyrite. The main copper ores of the district are found in more or less east-west veins that cut the pyrite mass. Most of these copper veins in the pyrite body are narrow stringers in which enargite is the main sulfide; they contain little or no gangue. The principal copper minerals are enargite and luzonite, although in one large ore body between the 1600 and 1000 foot levels, the dominant copper minerals were chalcopyrite, bornite, and some rich masses of chalcocite. As the veins weaken to the east, however, the copper content declines and tennantite becomes the main copper mineral. The silver in the copper bodies probably is provided by tennantite; much of the chalcocite appears to be supergene, but there is some of it that was formed under hypogene conditions by replacement of chalcopyrite and bornite. Barite and chert are the main nonmetallic gangue minerals. A group of strong veins on westerly bearing, curving, and branching fractures is found in the altered shales of the Excelsior group south of the volcanic rocks of the vent. Similar veins are found in the volcanic rocks of the vent west of, and near the north end of, the massive pyrite body.

In the Matagente area, north of the town of Cerro de Pasco, the mineralized zone extends eastward into the Pucará limestone while southeast of the south end of the pyrite mass, the Pucará limestone is penetrated in the Noruega zone. In both areas the ores are in mantos. This ore is now highly oxidized, but it probably was originally composed of pyrite, galena, and sphalerite and contained much silver but little copper.

Much of the copper in the early days of mining was produced from rich secondary bodies of chalcocite ore (with minor covellite) that were localized in large but definitely limited shoots where the rocks were sufficiently broken to permit ready circulation of surface waters. Most of the chalcocite was precipitated as replacements of sphalerite.

The main lead-zinc ore bodies are found in the eastern portion of the massive pyrite, grading gradually out into the surrounding pyrite. These ore bodies have a variety of shapes that differ with the ore cut-off value assumed at the time. Low-silver galena generally is slightly later than the sphalerite, and still later are minor pipelike structures containing high-silver in tennantite, argentite, and several silver sulfosalts; there is little copper in the lead-zinc ores, with what is present deriving largely from the late tennantite. Sericite is the main wall-rock alteration mineral, although it is converted near the main channelways to dickite and alunite.

Perhaps the most notable characteristic of the Cerro deposits is the impressive changes in mineralogy and texture that occur over short vertical distances. These features, plus a total vertical range of mineralization (below the present surface) of only 3000 feet, the highly broken nature of the rock composing the vent filling, and the loose, vuggy structure of the many of the mineralized pipes suggest that the Cerro deposits formed in a near surface environment, with a concomitant rapid loss of heat and pressure. Assuming this to have been the case, the enargite-luzonite veins would be classed as kryptothermal, as would the sphalerite and galena of the lead-zinc ore bodies. The western tennantite-bearing extensions of the enargite veins probably were formed on the borderline between kryptothermal and epithermal conditions, while the smaller silver-rich tennantite-argentite-silver sulfosalt pipes within the lead-zinc bodies probably should be classed as epithermal. The pyrrhotite-cassiterite-sphalerite pipes probably were formed within the xenothermal range; but as they are not of present-day economic importance, they are not mentioned in the classification assigned. The deposits are, therefore, categorized as kryptothermal to epithermal.

COLQUIJIRCA

Middle Tertiary	*Silver, Lead, Copper*	*Leptothermal*

Ahlfeld, F., 1932, Die Silbererzlagerstätte Colquijirca, Peru: Zeitsch. f. prakt. Geol., Jg. 40, H. 6, S. 81-87

Arce, J. N., 1917, El asiento mineral de Colquijirca: Con. Nac. Industria Minera (Primero), Lima, t. 2, pt. 1, p. 3-31 (mainly technical; particularly p. 12-14)

Haapala, P. S., 1954, Estudio geológico de la mina Colquijirca: Minería, Lima, año 2, no. 7, p. 21-40

Jenks, W. F., 1951, Triassic to Tertiary stratigraphy near Cerro de Pasco, Peru: Geol. Soc. Amer. Bull., v. 62, p. 203-220

Lindgren, W., 1935, The silver mine of Colquijirca, Peru: Econ. Geol., v. 30, p. 331-346

McKinstry, H. E., 1929, Interpretation of concentric textures at Colquijirca, Peru: Amer. Mineral., v. 14, p. 430-433

—— 1936, Geology of the silver deposit at Colquijirca, Peru: Econ. Geol., v. 31, p. 618-635

Orcel, J., and Rivera Plaza, G., 1929, Étude microscopique de quelques minerais métaliques du Pérou: Soc. Française Mineral. Bull., t. 52, p. 91-107 (particularly p. 91-101)

Notes

The silver deposits of Colquijirca are located 165 km north-northeast of Lima and 92 km north-northwest of La Oroya.

The Colquijirca mine is only 9 km slightly west of south from the town of Cerro de Pasco, and the general geology of the district is that of the broader Cerro district. The geologic setting at Colquijirca is, in detail, quite different from that at Cerro. The rocks in which the Colquijirca ores were deposited belong to the Calera (or uppermost) member of the early Tertiary Pocobamba formation. The mine is on the east flank of a major anticline, the axis of which, to the west of the mine, has been sufficiently eroded to expose rocks of the Permian Mitu group. The Mitu rocks, in turn, enclose small exposures of Devonian (?) Excelsior beds. To the east, the Mitu group is overlain unconformably by a narrow band of the middle member of the Pocobamba formation (the Shuco limestone conglomerate) and then by a considerable thickness of the Calera member. How the absence of the lower member of the Pocobamba group and of the Goyllarisquisga formation and the Pucará group from between the Mitu and Shuco member is to be accounted for is uncertain, but the relationship probably was caused by erosion and not by faulting. Somewhere to the east of the mine, under the cover of glacial debris that fills the adjacent valley, the Cerro de Pasco fault passes southward through the area. Beyond the fault to the east, the rocks of the Triassic-Liassic Pucará group are exposed.

The folding in the district, as is true of the Cerro district as a whole, reached its culmination after the deposition and lithification of the Pocobamba beds and probably occurred late in the early Tertiary or possibly in the early part of the middle Tertiary. The folding, in turn, was followed by the intrusion of quartz monzonite stocks. The stock closest to the Colquijirca mine is the Marcapunta stock, the nearest extension of which is just over 2 km from Colquijirca. This stock appears to have been emplaced in a less violent manner than was the Cerro stock.

The quartz monzonite porphyry of the Marcapunta stock is thought (Lindgren, 1935) to have had the same general source as that of the Cerro stock and to have been of essentially the same age. The ores of the Colquijirca mine probably were deposited from hydrothermal fluids derived from the same source as the Marcapunta magma, but they probably did not travel the same path as that magma and certainly were not introduced into an area as highly broken nor did they travel through such broken rock volumes as those that formed the Cerro ore bodies. It does appear, however, that these ore fluids must have invaded the Calera beds at Colquijirca at much the same time that the Cerro ore fluids entered the broken rocks of the Cerro volcanic vent. The Colquijirca ores, therefore, here are classified as middle Tertiary.

The ores of the Colquijirca mine occur as bedded replacements (McKinstry calls them mantos) of a small group of beds in the Calera member of the Pocobamba that have been folded into two minor synclines, with a minor anticline between them, all lying on the east flank of a much larger anticlinal structure. The four synclinal limbs are known (from west to east) as the Principal, the Mercedes, the Chocayoc, and the Llave; most of the mineralization is in the two more westerly synclinal limbs. The number of beds replaced within the group of Calera beds and the intensity of the replacement differ along both dip and strike, with the maximum thickness of minable mineralized rocks being over

10 m and ranging downward to a few centimeters of sulfides separated from the next ore bed by several meters of barren rock. The bedding is so well preserved in the sulfide ore that it is rare to find a foot of mineralized rock that does not show conspicuous traces of stratification.

There are two types of mineralization in the mine. The first of these are the chert-replaced (limestone) beds that always carry at least some chert, the amount ranging from individual layers a few centimeters thick (widely or closely spaced) to 5-m thicknesses of nearly solid black, cherty silica. Any single bed may change along strike from unsilicified limestone to highly silicified material. In the areas where chertification was intense, the principal sulfide always is pyrite; this pyrite may be disseminated in the chert or may be in veinlets or lenses. Although chalcopyrite is the most abundant ore sulfide associated with pyrite, other sulfides may be present as well. The second type of mineralization is in what McKinstry calls the galena-shale mantos; these are replacements of the more shaly beds in the Calera formation in which galena is the most important sulfide and nearly always is accompanied by barite and only a little chalcopyrite and pyrite; the gangue is largely kaolinized shale. In several places, well-mineralized chert-pyrite ore grades along strike into galena-shale manto ore and then into lean beds. Although these two ore types (chert-replaced and galena-shale) are quite distinct in most places, there are all sorts of gradations between them; even in the thickest chert-pyrite ore bodies, beds of shale generally separate the ore into layers.

After McKinstry had done his work in the district, pyrite-enargite ore was found some 200 m south of the main workings (therefore, toward the Marcapunta stock). This ore is composed of cherty silica with massive pyrite and barite, followed by sphalerite, luzonite, and enargite. It replaces a bed stratigraphically somewhat lower than that of the main ore bodies; where this bed reaches the area of the main mineralization, it contains only chert and a little barite.

The ore minerals appear to have been deposited in three generations of which the first consisted of pyrite, enargite (generally minor), marcasite, sphalerite, tennantite, and galena. The second generation was composed of botryoidal galena and sphalerite, followed by minor chalcopyrite; the third generation appears to have been limited to stromeyerite. The fourth generation consisted only of native silver. The main gangue mineral was the early chert, although some quartz and barite crystals were developed in various irregular cavities; some botryoidal siderite is also present in vugs. Kaolinite was strongly produced in the argillaceous portions of the wall rock.

Tennantite is the most important ore mineral in the Colquijirca ores, and it replaced first generation chert, barite, pyrite, and sphalerite; it is rich in silver. First-generation galena, which also contains appreciable silver, formed about the same time as the sphalerite. Stromeyerite consistently replaced tennantite and failed almost entirely to attack the other sulfides. Lindgren and Ahlfeld consider the stromeyerite to have been deposited under hypogene conditions, while Orcel and Rivera Plaza and McKinstry think it to have been supergene. Because there does not appear to have been any real relationship between locations of the volumes of stromeyerite mineralization and the water table, past or present, it is here believed to have been hypogene. All authors agree that the native silver was supergene in origin; it is found mainly as a replacement of stromeyerite, although some replaced tennantite-stromeyerite veinlets and some deposited as wires and hooks on other sulfides. The native silver-rich areas bear no relation to the present shallow zone of oxidation (not more than 50 m deep); native silver extends to far greater depths. Either the native silver was produced during a previous erosion cycle or it is hypogene; I am inclined to think the latter was the case.

The tennantite, stromeyerite, and native silver (granting it to be hypogene) of the deposit are minerals typical of the leptothermal range. The galena was not ordinarily deposited in association with the tennantite and may have formed under mesothermal conditions, although the short distances separating concentrations of galena from those of tennantite suggest that it too was deposited within the leptothermal range. The early enargite was almost certainly mesothermal, but it is not of economic importance and is not considered in the classification of the ores. Although the deposit was formed within a few kilometers of the kryptothermal to epithermal deposit of Cerro de Pasco, there is no evidence supplied by Colquijirca geology to suggest that the ores of the latter mine were formed under conditions of rapid loss of heat and pressure. It is, therefore, considered reasonable to categorize the Colquijirca ores as leptothermal.

HUANCAVELICA

Middle Tertiary *Mercury* *Epithermal*

Berry, E. W., and Singewald, J. T., Jr., 1922, The geology and paleontology of the Huancavelica mercury district: The Johns Hopkins Univ. Studies in Geol., no. 2, 101 p.

Gastelumendi, A. G., 1917, Huancavelica como región productora de mercurio: Cong. Nac. Industria Minera (Primero), Lima, Anales, t. 2, pt. 2, p. 33-69 (mainly technical)

Strauss, L. W., 1909, Quicksilver at Huancavelica, Peru: Min. and Sci. Press, v. 99, no. 17, p. 561-566 (mainly historical)

Umlauff, A. F., 1904, El cinabrio de Huancavelica: Cuerpo de Ingenieros de Minas del Perú Bol. no. 7, 62 p.

Yates, R. G., and others, 1951, Geology of the Huancavelica quicksilver district, Peru: U.S. Geol. Surv. Bull. 975-A, p. 1-45

Notes

The mercury deposits of Huancavelica lie in the hills south of the town of the same name in the Cordillera Occidental of south-central Peru. The town itself is some 235 km east-southeast of Lima and 175 km south-southeast of La Oroya. All of the well known mines are found in a belt 8 km long (north-south) and 2 km wide (east-west), the most important being Santa Barbara from which about 1,350,000 flasks were produced in the years from 1570 to 1791 and some 1,500,000 flasks from 1570 to the present time. The deposits now appear largely to have been mined out.

In southern Peru, the simple, well-defined northwest-striking structural pattern of the sedimentary rocks of the Cordillera Occidental loses its simplicity and converges with several other ranges to form a structural knot. The Huancavelica deposits are found just to the northwest of the point at which the Cordillera Occidental joins this knot. This district is unusual among the mining areas of central Peru in the paucity of exposures of intrusive igneous rocks and in the failure of erosion to cut below formations of Jurassic age.

The oldest rocks known in the district are those of the Pucará limestone that is composed of fine-grained, medium-bedded, light-gray limestone; it included silicified fossils indicating it was formed here in Jurassic (Liassic) time. Cinnabar has been found in the Pucará several places in the district, but not much mercury has been produced from it.

The Pucará beds are overlain by those of the Goyllarisquisga group, but fault contacts prevent an accurate determination of the degree of unconformity (if any) that originally existed. In the Huancavelica district, Yates and others (1951) have divided the Goyllarisquisga into two parts, have given each of these a formation designation, and have redesignated the Goyllarisquisga as a group. The lower of these two new formations is the Early Cretaceous Gran Farallón sandstone, the host rock of the largest mercury deposit and economically the most important formation in the district. The Gran Farallón is made up mainly of massive quartz sandstone but also contains minor amounts of shale and limestone.

The Gran Farallón is overlain, apparently conformably, by the Chayllatacana volcanic rocks, mafic lavas, and tuffaceous shales with very small amounts of conglomerate and limestone. The rocks are nonfossiliferous but are considered to be Early or Middle Cretaceous in age.

The Machay limestone overlies the Chayllatacana volcanic rocks with apparent conformity; it is composed of light-colored, fine-grained, medium-bedded limestone with intercalated beds of marly limestone and red shale. Yates and others (1951) have divided the formation into three members. The marly beds of the lower member (unit) were particularly favorable sites for cinnabar deposition and contain the Botija Punco mine; locally the Machay has been known as the Botija Punco formation.

The Casapalca formation unconformably overlies the Machay limestone and consists of conglomerate and interbedded tuffaceous shale. The Casapalca beds probably are early Tertiary in age and are correlated tentatively with the Casapalca red beds, originally identified near Yauli. A little mercury has been mined from the Casapalca formation.

Flows of rhyolite, basalt, and andesite with minor amounts of tuff cover an appreciable portion of the district and are unconformable on the sedimentary rocks underlying them.

There are two kinds of intrusive igneous rocks in the district: (1) breccias and (2) dacites; both of these rock types are intrusive into those of the sedimentary-volcanic sequence. The breccia intrusions are much larger in outcrop area and are known in three distinct areas in the central part of the area and in a dikelike intrusion near the Santa Barbara mine. The intrusive relations of these breccias with the surrounding rocks suggest that they mark the necks of volcanoes, the last activity of which in Tertiary time was explosive. Around the margins of these breccia pipes, the adjoining rocks were silicified, and large amounts of iron sulfide (now iron oxide) were introduced into them. These altered rocks are cut by numerous small veins of galena with a little sphalerite and some minor barite gangue; these veins are structurally related to the breccia pipes but have no connection, structurally or spatially, with the cinnabar deposits.

The intrusions of dacite took place after the faulting and before the cinnabar was introduced. Since some cinnabar was found in igneous rock in the Santa Barbara mine and was noted along the dacite-limestone fault-contact in the open pits east of that mine, it is almost certain that the mercury mineralization took place after faulting and intrusion.

The presence of ore in the Tertiary rocks shows that mineralization took place not only after the Late Cretaceous-Eocene Andean folding but also after that phase of the Andean orogeny that followed the development of Tertiary intrusive and extrusive rocks. This relationship probably dates the ore formation as middle Tertiary (the age designation adopted here); it may, however, have occurred in the late Tertiary.

Almost all of the production of the Huancavelica district came from deposits in the Gran Farallón sandstone; the Santa Barbara mine was the most important, with minor amounts having been derived from small nearby mines. Cinnabar was the main mercury mineral, though some native mercury was recovered, and metacinnabar probably was present. Pyrite was a common mineral both in association with cinnabar and throughout the district; it occurred both as fracture fillings and as replacements of quartz grains. Arsenopyrite, realgar, orpiment, and minor stibnite usually were found with the cinnabar, with the arsenopyrite being much more abundant than cinnabar on the lower levels. Orpiment and realgar were formed at higher levels, and their presence in the oxidized zone might suggest that they had formed from arsenopyrite by secondary processes. The normal stability of orpiment and realgar under surface conditions, however, indicates that they, in large part at least, were primary. Some sphalerite and galena occur in the area, but they generally were found in veinlets with some stibnite and barite independently of the cinnabar. Only minor amounts of calcite, quartz, and hydrocarbons were present as gangue with the cinnabar mineralization. Yates and others (1951) have classified the cinnabar deposits at Huancavelica into three types: (1) deposits occurring in sandstone; (2) deposits occurring in limestone and in marly beds and controlled by fractures; and (3) deposits occurring in igneous rocks and controlled by cross fractures. Since the host-rock type largely determines the characteristics that control the size and shape of the deposits, this classification is more than a descriptive one.

Most of the cinnabar and other sulfides appear to have been deposited in pore spaces between sand grains, so that the richer ore bodies probably were developed in the more porous rocks. Nevertheless, some of the sulfides were emplaced by replacing sand grains and by filling fractures in the beds.

Ore was found in both the Machay and Pucará limestone formations at numerous localities; such mines at Botija Punco, San Roque, and Quichacahuayjo were of this type. Although the amount mined from these deposits was small in comparison with Santa Barbara, that recovered from the mines in the Machay formation was substantial. The cinnabar in these ore bodies was found in irregular tubular masses (concordant with the bedding in the limestone) and in veins transverse to the bedding. Almost all the cinnabar was deposited as fracture fillings, only a small amount having replaced limestone. The beds of marly limestone at the base of the Machay contained most of these deposits.

A little cinnabar was recovered from the limestone conglomerate of the Casapalca formation, but the manner of its occurrence has not been reported.

In the southern part of the district, some ore was mined from the Tertiary lavas with the Dewey (Santa Barbara III) mine having been the only producing property in the district in 1945. In this mine the cinnabar was found in a fracture zone in biotite rhyolite near a contact with limestone. Prospects were begun in several places in both rhyolite and basalt, but no commercially valuable deposits were discovered.

The narrow vertical range of the mineralization and the change from orpiment and realgar to arsenopyrite at shallow depths beneath the surface indicate that the ores were deposited in a rapid loss of heat and pressure environment and that they should be classed as epithermal as is done here.

MOROCOCHA

Middle Tertiary	*Copper, Lead, Zinc, Silver, Bismuth*	*Mesothermal to Leptothermal*

Geological Staff, Cerro de Pasco Copper Corporation, 1950, Lead and zinc deposits of the Cerro de Pasco Copper Corporation in central Peru: 18th Int. Geol. Cong. Rept., pt. 7, p. 154-186 (particularly p. 170-178)

Haapala, P. S., 1949, On Morococha breccias: Soc. Geol. Perú, v. Jubilar, pt. 2, f. 2, 11 p.

—— 1953, Morococha anhydrite: Soc. Geol. Perú Bull., t. 26, p. 21-32

Mark, W. D., and others, 1942, Localization of certain ore bodies at Morococha, Peru, in Newhouse, W. H., Editor, *Ore deposits as related to structural features*: Princeton Univ. Press, p. 239-241

McLaughlin, D. H., and Moses, J. H., 1945, Geology of the mining region of central Perú: Min. and Met., v. 26, no. 467, p. 512-519 (particularly p. 515-517)

McLaughlin, D. H., and others, 1935, Copper in the Cerro de Pasco and Morococha districts; Department of Junin, Peru, in *Copper resources of the world*: 16th Int. Geol. Cong., v. 2, p. 513-544 (particularly p. 527-544)

Nagell, R. H., 1957, Anhydrite complex of the Morococha district, Peru: Econ. Geol., v. 52, p. 632-644

—— 1960, Ore controls in the Morococha district, Peru: Econ. Geol., v. 55, p. 962-984

Petersen, U., 1965, Morococha, in *Regional geology and major ore deposits of central Peru*: Econ. Geol., v. 60, p. 407-476

Schmedeman, O. C., 1940, Recent studies of the geology of the Morococha district (Perú): 8th Amer. Sci. Cong. Pr., v. 4, Geol. Sci., p. 677-679

Terrones L., A. J., 1949, La estratigrafía del distrito minero de Morococha:

Soc. Geol. Perú, v. Jubilar, pt. 2, f. 8, 15 p.

Trefzger, E. F., 1937, Das Kupfererzvorkommen der Grube Morococha: Metall. und Erz, Bd. 34, H. 8, S. 181-192

Notes

The vein and replacement deposits of the Morococha district are located on the east side of the Andes, just east of the continental divide, 110 km east-northeast of Lima, and 25 km south-southwest of La Oroya.

The oldest rock in the Morococha district is the middle Paleozoic (Silurian or Devonian?) Excelsior group that is made up mainly of phyllites but also contains some interbedded quartzite. The Excelsior rocks are overlain unconformably by the Permian Catalina volcanic rocks that include dacites, rhyolites, and andesites; these volcanic rocks may be the time equivalent of the Mitu group of the Cerro district. It is possible, however, that a thin sequence of sediments between the Catalina volcanic rocks and the anhydrite complex (that underlies the Jurassic Potosí limestone) is made up of altered Mitu red beds. This anhydrite complex may be a sedimentary facies that introduced the Jurassic sedimentation (Haapala, 1953), or it may be a hydrothermal replacement of the basal portions of the Potosí formation (Nagell, 1957). In any event, the primary mineral appears to have been anhydrite, that mineral having been largely replaced by gypsum above the 1000 level probably by secondary processes. Still later, some of this gypsum was taken into solution, and large caverns resulted; some of them collapsed to produce large volumes of breccia. Unconformably above the Catalina volcanic rocks is the Jurassic Potosí formation, which is the equivalent of the Pucará group of the Cerro district. Much of the mine area is made up of the Potosí beds, and these beds contain several limestone horizons of various compositions that have been altered considerably and differently through the actions of igneous intrusions and hydrothermal solutions. They are subdivided into 13 horizons. Although the Goyllarisquisga sandstone (locally known as the Santo Toribio-Buenaventura formation) is present on the northeast and southwest margins of the district, this Cretaceous formation is not exposed in the mine. Similarly exposed in the same marginal portions of the district is the Cretaceous Machay limestone (the ore-bearing formation at Yauricocha); it contains no ore at Morococha. Overlying the Machay formation are the early Tertiary Casapalca red beds; these rocks also are found on the margins of the district and contain no ore.

The end of the Andean period of folding and faulting was followed by the intrusion of various igneous masses. The first intrusion was on the west side of the district and was a large stock of quartz diorite known as the Anticona intrusive. To the east of the Anticona stock, three stocks of quartz monzonite--from west to east, the Gertrudis, San Francisco, and Potosí--were emplaced; these stocks are elongated north to south with the Potosí being far larger than either of the others or even than both together. Where the Potosí formation was cut by these stocks, skarn and hornfels were produced.

The folding, faulting, intrusion, and second phase of faulting in the Morococha district may have begun before the deposition of the Casapalca red beds had been completed in the Eocene but certainly continued for a considerable time after the Casapalca red beds had been lithified. This would seem to place the mineralization process at Morococha as middle, or even late, Tertiary. As has been done for other ore bodies in central Peru, the Morococha ores are classified as middle Tertiary.

Contact metamorphism around the Anticona stock is not very impressive, but around the three quartz monzonite stocks (the Morococha stocks), it is quite spectacular. Hydrothermal solutions also altered the igneous rock itself and further affected the sedimentary and volcanic rocks. The weakest alteration (generally at least 150 m away from intrusives) in the Potosí (Pucará) formation was bleaching, recrystallization to marble, and some dolomitization; the dolomitization increases toward the ore bodies. The more intense alteration was divided into two categories by Haapala (1953): (1) anhydrous silicate and (2) hydrated silicate. Normally, the hydrated silicates are dominant nearest to the intrusives, particularly in the vicinity of the San Francisco stock; farther out from the stocks, the hydrated silicates give way to the anhydrous variety. Toward the outer margins of the anhydrous masses, these silicates are interbedded with the marble. Most of the anhydrous silicated limestone is a fine mosaic of diopside and less tremolite, but the most abundant minerals in the anhydrous silicate alteration are diopside and garnet, and the common ones are tremolite-actinolite, epidote, quartz, and biotite. In the anhydrous alteration, tremolite increases in quantity near the intrusives, normally replacing diopside. The most abundant minerals of the hydrated silicate zone are serpentine, chlorite, and talc; the common ones are magnetite, pyrite, and pyrrhotite, and the rare ones include silica, hematite, ludwigite, biotite, and tremolite-actinolite; the hydrated silicates essentially are lacking in calcite. Hydrous alteration, in many places, encroaches on the anhydrous and replaces it in the central part of the mine, hydrated silicates occur interbedded with the anhydrous; farther away from the igneous rocks, anhydrous silicate horizons alternate with marble or with fresh limestone.

In addition to the recrystallization of the limestone and the two types of silicate alteration, the Pucará contains an anhydrite alteration (Nagell, 1960, p. 965; 1957) that is interlayered with marble and anhydrous and hydrated silicate minerals. This type of alteration (granted the anhydrite was not syngenetic) was developed in what were originally limy and shaly layers and lenses of the basal Pucará formation in various thicknesses and in attitudes conforming to that of the western limb of the anticlinal structure. In the vicinity of the Gertrudis stock, the apparent center of the anhydrite complex, the anhydrite layers are thickest, up to 100 m. In addition to anhydrite and gypsum, these layers contain diopside, tremolite, talc, serpentine, quartz, zeolites, pyrite, magnetite, and, of course, calcite. Toward the San Francisco stock to the east, the anhydrite complex generally is interlayered with anhydrous and hydrated silicates; to the west, the interlayered material most commonly is marble. The gross composition of the anhydrite complex is discontinuous layers of anhydrite, marble, anhydrous and hydrated silicates, and shale. Haapala (1953) believes that the anhydrite is syngenetic and points as evidence of this to a small evaporite gypsum deposit a few kilometers east of Morococha. This gypsum, however, is in the upper Potosí (Pucará) or lower Santa Toribio formation and not in the lower Potosí as is the anhydrite at Morococha. Certainly the Morococha anhydrite is not typical of the small lenses of gypsum found in the Cretaceous sandstones and Tertiary red beds of central Peru. Nagell (1957) believes that the relationships of the anhydrite to the surrounding rocks show that it was not introduced into the area until after the Gertrudis stock had been emplaced. His order of events is (1) intrusion of the San Francisco stock and formation of anhydrous and hydrated silicates in the Potosí formation; (2) the intrusion of the Gertrudis stock and the formation of the anhydrite; (3) ore deposition; and (4) supergene formation of calcite, gypsum, and zeolites.

Nagell (1960) has discussed the ore controls at Morococha in detail. He divides the district into two sectors, the eastern and the western. The ore bodies of the eastern sector are mainly in veins while those of the western sector are in pipes and mantos. The eastern sector is made up largely of Catalina volcanic rocks and the rock of the San Francisco stock; the deeper the portion of this sector in question, the larger is the percentage of volcanic rocks it contains.

In the western sector of the district, the major rock types are the alteration products of the Potosí formation and those of the Gertrudis stock. These rocks are softer and more plastic than those in the eastern sector, and, instead of veins, pipes and mantos are the important ore-containing structures. Veins do exist in the anhydrous silicated rocks and in the marble, but they contain far less ore than the pipes and mantos. In this sector also, there are numerous crystal-lined vugs, but most of the ore was emplaced by replacement and associated with it generally are unreplaced remnants of gouge and actual Potosí beds.

The minable ore bodies, therefore, fall into three general types: (1) veins, (2) vertically elongated pipelike masses, and (3) gently plunging pipelike bodies that are concordant with the bedding of the Potosí limestone and are locally referred to as mantos. The more continuous veins (several are over 3000 feet long) were developed with essentially east-west strikes in the San Francisco stock and the Catalina volcanic rocks; veins in the limestone developed over much shorter distances, rarely exceeding 1000 feet in length. In those portions of the veins nearest the center of mineralization (the San Francisco stock), enargite was the dominant mineral with some chalcopyrite and tennantite-tetrahedrite. Beyond the enargite-rich vein areas, chalcopyrite and tennantite were the principal ore minerals, and the silver content (due to the increase in tennantite) was higher.

Still farther from the stock, chalcopyrite died out and tennantite and sphalerite increased, while galena began to become important and provided part of the silver content. In vein volumes most distant from the stock, galena was by far the most abundant mineral and accounted for essentially all the silver found in such ores; minor amounts of sphalerite also were formed at the extreme ends of the veins. From the center of mineralization outward, sphalerite graded from quite dark and iron-rich through lighter shades to straw yellow. Quartz and barite were the most abundant gangue minerals. Certain outlying veins with about a N50°E strike, contained almost nothing but galena and some sphalerite, but the galena was high in silver.

The pipes and mantos were emplaced only in the limestone and often pass upward and/or downward into veins; small mantos may extend outward from some of the veins. The earliest mineral to form in the massive replacement bodies was the dominant pyrite; the later ore sulfides developed principally along the margins of the pyrite masses, with sphalerite normally replacing limestone and chalcopyrite replacing pyrite or cementing brecciated masses of the iron sulfide. Sphalerite generally was the first mineral to be deposited after pyrite and was followed by the dominant copper mineral--chalcopyrite--which, in turn, was followed by local concentrations of enargite and by bornite, tennantite-tetrahedrite, and galena. Quartz, sericite, ankerite, and rhodocrosite were the main gangue minerals.

The Morococha ore bodies appear to have a present vertical extent of some 600 m, and the minimum total vertical extent probably was at least 1000 m. Although there is no known ore mineralization in the Santo Toribio and Machay formations, it seems reasonable to assume that they were present when the ores were formed, and this suggests a cover of at least 2000 m of rock, plus whatever Tertiary beds were present, over the deepest ore bodies. It follows, therefore, that the ores probably were formed under conditions of slow loss of heat and pressure. The enargite and chalcopyrite ores probably were formed under mesothermal conditions, while the tennantite-tetrahedrite and silver-rich galena ores almost certainly developed within the leptothermal range. As none of the galena, even in the outlying veins, is silver-poor, it appears that there was no deposition under telethermal conditions. The deposits, therefore, are categorized as mesothermal to leptothermal.

YAURICOCHA

Middle Tertiary	*Copper, Gold, Silver, Lead, Zinc*	*Mesothermal to Leptothermal, Ground Water-B2*

Kobe, H. W., 1961, Idaita--Mineral de cobre en Yauricocha: Soc. Geol. Perú Bol., t. 36 (Segundo Congreso Nacional de Geología), p. 103-114

Lacy, W. C., 1949, Oxidation processes and formation of oxide ores at Yauricocha: Soc. Geol. Perú Bol., t. 25, pt. 2, p. 1-15

McLaughlin, D. H., and Moses, J. H., 1945, Geology of the mining region of central Peru: Min. and Met., v. 26, no. 467, p. 513-519 (particularly p. 518-519)

Petersen, U., 1965, Yauricocha, in *Regional geology and major ore deposits of central Peru*: Econ. Geol., v. 60, p. 407-476 (particularly p. 447-453)

Szekely, T. S., 1969, Structural geology, Cochas to Yaruicocha, central High Andes, Perú: Amer. Assoc. Petrol. Geols. Bull., v. 53, p. 553-567

Ward, H. J., 1959, Sulfide orebodies at Yauricocha, central Peru--replacements of organic reefs?: Econ. Geol., v. 54, p. 1365-1379; disc., 1960, v. 55, p. 1070

Notes

The ore bodies of the Yauricocha district are located just west of the continental divide, about 160 km east-southeast of Lima and some 90 km south-southeast of La Oroya. The surface workings at the mine are at elevations between 4600 and 4650 m above sea level and are near the headwaters of the Rio Cañete that flows into the Pacific near the town of the same name.

The oldest rock exposed in the district is the Middle(?) Cretaceous Machay limestone; much of this formation in the mine area has been finely to coarsely recrystallized. The Cretaceous to early Tertiary Casapalca red beds(?) conformably overlie the Machay; the first 130 m of the red beds is known as the France chert and consists of calcareous marls with some interbedded siliceous limestone. Above the France chert is a slightly siliceous limestone, the Quishuarnioc limestone, and the Quishuarnioc is followed by interbedded limestones, siliceous limestones, and calcareous marls. True red beds of the type known at Casapalca crop out farther to the east and, therefore, probably are higher in the section.

This sedimentary sequence was intruded, probably before the folding was completed, by igneous bodies that show the same compressional and tensional features that were imposed on the sedimentary and volcanic rocks. The igneous masses have been classified by different authors as quartz monzonite porphyry (Lacy, 1949), monzonite porphyry (Ward, 1959), and granodiorite (Petersen, 1965) and have the form of steep-sided stocks.

One result of the intrusions has been the widespread conversion of the dolomitized Machay limestone to marble and locally (near the intrusive bodies), to lime silicates; there was some brecciation along the contacts between the granodiorite and the marble. Portions of the France chert have been altered to tactites through reaction with the granodiorite, and farther to the east, the true Casapalca red beds are much altered by bleaching and conversions to calc-silicates and sanidine near their contacts with intrusive bodies. Chabazite was formed as a late mineral in joints and fractures in the granodiorite. Although the opinion has been expressed (Petersen, 1965) that the environment was shallow and volcanic, the character of the texture of the monzonite suggests that it was emplaced under hypabyssal conditions and at a considerable depth beneath the then-existing surface.

The general similarity of sedimentary and igneous events at Yauricocha and at the other ore-bearing properties of central Peru suggests that they occurred at about the same time. Although it is impossible to date the igneous activity at Yauricocha more exactly than to say that it is later than the Casapalca red bed-type beds, it appears probable that it happened in the middle Tertiary at much the same time as intrusions took place in the other ore districts. The Yauricocha ore deposits, therefore, are classified here as middle Tertiary.

The loci of mineralization at Yaruicocha appear to have been those rock volumes that had been most intensely sheared and jointed and in which the limestone had recrystallized to the largest grain size. The primary manto-type ore bodies in the Machay limestone are bunched along the contacts between the various distances from the contact of the Machay with the France chert. The France chert and the granodiorite apparently acted as rather impermeable masses that deflected the rising ore fluids away from the granodiorite and into the more pervious portions of the limestone.

The three major mantos are the Catas (and West and South Catas), the Pozo Rico, and the Mascota, the Catas being the most easterly of the three and the other two being to the west (Pozo Rico) and the northwest (Mascota) of the Catas. Still farther to the west, there is a halo of smaller ore bodies that are, from northwest to southeast, Carmencita, 24 de Junio, Giliana, Adrianna,

Purísima Concepción West and East, and Virginia (Petersen, 1965).

The Catas ore bodies still consist largely of unoxidized sulfides while the Pozo Rico and Mascota and the smaller mantos now are composed of sulfide remnants, residual oxidized minerals, and transported oxidized material.

In the Catas area, the bulk of the material making up the ore bodies is pyrite, with massive, dense pyrite forming the cores of the bodies and often being separated from the country rock by zones of "soft pyrite breccia" composed of fragments and angular boulders of massive pyrite in a friable mixture of fine- to coarse-grained pyrite that usually is quite sandy. In the more shaly parts of the Machay limestone, pyrite is far less abundant than in the more limy sections. It is possible that the breccia was produced after the massive pyrite had been emplaced, but the presence of ores of unreplaced limestone or lime-silicate rock surrounded by pyrite suggests that the replacement occurred after brecciation. On the other hand, Thomson (quoted by Petersen, 1965) thinks that the pyrite, closely followed by quartz, was first deposited in massive limestone; then "highly acid emanations" dissolved limestone, especially along fractures, resulting in loose friable aggregates of quartz and pyrite sand. Due to the loss of volume incurred by this leaching, the overlying country rock was undermined and brecciated, the large fragments of the "soft pyrite breccia" having been formed, therefore, by underhand stoping and the sandy pyrite by the leaching that caused the brecciation. Thomson thinks that the small volumes of granodiorite breccia also were formed by this underhand stoping.

No matter how the massive pyrite was emplaced, it almost certainly was the earliest metallic mineral, and the later sulfides filled fractures in, and replaced, the pyrite. Lacy was able to establish an original zoning of which the Catas ore bodies were the high-temperature center. The first zone includes all of the Catas ore bodies except the western half of the West Catas ore body, and in this zone the principal primary ore mineral in the pyrite is enargite, but some chalcopyrite and bornite also were present. The second zone lies to the west of the first and is now thoroughly oxidized. It includes the western one-half of the West Catas ore body, the eastern two-thirds of the Pozo Rico mass and the eastern one-third of Mascota; this massive pyrite contained abundant chalcopyrite, bornite, tennantite, and appreciable enargite, and minor galena and sphalerite also were present in the pyrite. In the third zone, which covered the western two-thirds of Mascota and the western one-third of Pozo Rico, the main ore minerals in the pyrite were galena, sphalerite, and sulfosalts of silver; it also is now largely oxidized. Since the minor bodies (with the exception of Cuye, east of Mascota) lie to the west of Pozo Rico and Mascota, they are of the lead-zinc-rich type and are highly oxidized.

The ores of the Pozo Rico and Mascota mantos now bear little resemblance to the Catas ore bodies. Both oxidized ore bodies appear to have been highly brecciated, and much of the oxidation and transportation of the constituents of the primary minerals appears to have been made possible by the highly broken character of the rocks, although Snively suggests that the Catas ore body was shielded from ground water by a now eroded protective cap of France chert. Both of these two mantos contain both residual and transported oxidized minerals; the residual minerals were formed in place by the oxidation of sulfides, while the transported ones were precipitated at short to considerable distances from the sulfides from which they were derived.

The composition of the residual ores depends on the character of the primary ores from which they were derived. Enargite was oxidized, only slightly, but chalcopyrite, bornite, tennantite, and sphalerite and galena (the last two first converted in part to copper sulfides by secondary enrichment) were much affected by oxidation. The porous material resulting from the oxidation of these copper minerals is composed of limonite and goethite, residual quartz, and clay minerals, with the voids filled by jarosite, malachite, and azurite. Some gold and silver are recovered from this ore that probably derived from pyrite and from tennantite and galena, respectively.

The transported oxidized ores consist largely of limonite but differ considerably in character because of variations in the composition and acidity of the transporting solutions and in the granularity of the limestone host rock. Goethite, jarosite, cuprite, tenorite, native copper, malachite, azurite, and some brochantite and chrysocolla are common to abundant; a little cerussite and smithsonite may be present. In general, the transported oxides are found on the footwall of the residual oxide bodies, although some stringers of oxide ore follow fractures of steeper dip, thereby departing some distance from the residual ore.

The abundant enargite of the first (easternmost) zone and the abundant chalcopyrite and bornite and the considerable tennantite of the second (central) zone indicate that both zones were mineralized under mesothermal conditions. The strong development of silver-bearing sulfosalts, with primary silver-rich galena and the sphalerite, in the third (western) zone point to that portion of the ore having been formed under leptothermal conditions. Thomson implies that the deposits were formed under near-surface, rapid loss of heat and pressure conditions, but this concept does not appear to fit the facts. The primary deposits, therefore, are classed here as mesothermal to leptothermal. The small amounts of late simple copper sulfides (covellite, digenite, and chalcocite) in the ores suggest that ore deposition ended with a minor telethermal stage.

Because of the huge volumes of oxidized ore developed at Yauricocha, the category, ground water-B2, is added to that of mesothermal to leptothermal assigned to the primary ore.

VENEZUELA

VENEZUELAN GUAYANA

Early Precambrian (primary)	*Iron*	*Sedimentary-A3,*
Late Tertiary (secondary)	*as Magnetite,*	*Metamorphic-C,*
	Martite, Hematite,	*Ground Water-B2*
	Goethite	*(except for El Pao)*

Ferenčič, A. W., 1969, Geology of the San Isidro iron ore deposit, Venezuela: Mineralium Deposita, v. 4, no. 3, p. 283-297

Kalliokoski, J., 1964, The metamorphosed iron ore of El Pao, Venezuela: Econ. Geol., v. 60, p. 100-116

—— 1965, Geology of north-central Guayana shield, Venezuela: Geol. Soc. Am. Bull. 76, p. 1027-1050

Ruckmick, J. C., 1963, The iron ores of Cerro Bolivar, Venezuela: Econ. Geol., v. 58, p. 218-236

Short, K. C., and Steenken, W. F., 1962, A reconnaissance of the Guayana Shield from Guasipati to the Rio Aro, Venezuela: A.V.M.P. Bol. Informativo, Caracas 5, p. 189-221

Notes

The iron metallogenic province of the Venezuelan Guyana has a length of about 120 km and a maximum width of about 50 km; it is located in the south-central part of the Imataca belt or complex that extends northeast-southwest for about 450 km, south of, and essentially parallel to, the Orinoco River; it is 50 to 100 km wide. Ciudad Bolivar on the Orinoco River is the principal city in the area. The area includes such deposits as El Pao (northeasternmost), Marie Luisa, Arimagua, Altamira, Rondon, San Isidro, and Cerro de Bolivar (southernmost); several other deposits are more or less known.

The Imataca complex consists of highly metamorphosed rocks of sedimentary and igneous origin. The age of the rocks in the complex ranges from 3000 m.y. to 2000 m.y. with the bulk of the age determinations apparently centering around 2300 to 2400 m.y. In this portion of the world, the early Precambrian is thought to have ended with the Caribbean (or Transamozonic) orogeny and is believed to have occurred about 2000 m.y. ago. If this is the case, the early

Precambrian ended later here than anywhere else in the world. This does not, however, appear to be a reason for classifying the primary ores and their dynamic metamorphism as younger than early Precambrian.

On its south side, the Imataca complex is bounded by the Pastora-Carichapo assemblage, a late Precambrian volcanic-sedimentary eugeosynclinal series. To the west, the complex abuts against silicic intrusive rocks that were not affected by the regional metamorphism and probably are the same age as the Pastora-Carichapo. To the north, the Orinoco regional fault cuts the Imataca complex off from Mesozoic-Cenozoic sediments. The complex contains such rocks as quartz-feldspar paragneisses; amphibole-pyroxene gneisses and amphibolites; rare granite and granitic gneisses (quartz monzonite in composition); migmatites, metasandstone and feldspathic quartzite; and late (postmetamorphism) mafic dikes and sills.

The Imataca iron formation is quite like the itabirite of Minas Gerais, the iron formation of Lake Superior, and the iron formation of the Hamersley area of Western Australia among others. Ferenčič (1969) describes the Imataca iron formation as fine-grained, banded, extremely hard rock of granoblastic texture and massive to schistose structure; it is easily distinguished in the field from the other rocks of the complex. It is made up principally of quartz and iron minerals (with quartz somewhat greater by volume than the iron minerals) and a minor amount of silicates. The iron minerals are hematite and magnetite, and the magnetite has been converted considerably to martite; goethite is the most common secondary mineral. Ferenčič distinguishes four facies types as iron formation: (1) almost monomineralic quartzite; (2) quartzite containing 25 to 30 percent feldspar (subarkose); (3) iron formation composed of iron oxides and quartz in about equal volume ratios (the most common facies); and (4) iron formation with about 90 percent iron minerals; this facies he considers to be syngenetic iron ore. In addition, thickened iron ore bands are found at the crests of small folds, probably developed by metamorphic segregation, during the regional metamorphism. Not more than 2 percent of the total ore mass is composed of facies (4) and of the segregation ore.

The material here described as the syngenetic ore, which has, of course, been strongly affected by regional metamorphism at the end of the Caribbean orogeny, appears to have been formed in the early Precambrian, granting that the end of this epoch in this area was as late as 2000 m.y. ago and is so classified here.

The age of the secondary ore is another matter. Studies of water in springs and drill holes show that the entire amount of secondary ore in the district could have been developed in the last 20 m.y., granted no appreciable change in the climate and character of the ground water. If these requirements should be found to have been met, the secondary would have formed in the late Tertiary with a minor assist from the Recent.

Ferenčič believes that the Imataca iron formation is a recrystallized ferruginous chert deposited in the most part as a chemical precipitate of volcanic exhalative origin in the early Precambrian sea in a eugeosynclinal type of environment. Locally, the iron formation is interlayered with clastic metasediments showing that both volcanic and clastic sedimentation took place at much the same time. He apparently cannot conceive of the huge amounts of iron in the formations being derived from erosion of the land surface.

Under present day technology and demand, only the secondary enriched ores are minable. This ore is a residual concentration produced through tropical weathering. The ground waters of the area contain organic components that lower the pH of the near-surface waters below 7.0; these solutions dissolved ferric oxides and reprecipitated them to greater depths where the pH was higher. Once this first crust of additional iron oxides has been formed, later solutions repeat the process at greater depths. During periods of heavy rainfall, however, the descending waters are essentially free of organic acids, the pH is close to 7.0, and silica is dissolved as iron left behind. In places, silica is reprecipitated as a gel at even greater depths. Ruckmick's (1963) analysis shows that silica is removed during the ground water cycle some 200 times faster than iron. In the Cerro Bolivar area, he thought that

18 tons of ore are produced annually by this process. Presumably, except for El Pao (to be mentioned shortly), the same processes would produce ore at much the same rate and would result in the entire ore mass having been formed in about 20 m.y. Obviously, this requires that the climate has remained generally constant through the period, something far more likely in these latitudes than farther toward the poles. The removal of silica has converted the compact Imataca iron formation in a porous, friable, loose, but rich iron ore. Ferenčič's scheme of ore formation then is (1) magnetite, hematite, and martite dissolve in the zone of hydration, pH is about 6.0; this colloidal ferric hydroxide is then precipitated; (2) on the loose soft ore formed immediately below where magnetite has become martite and silica has been removed; pH is below 7.0; (3) below the zone of iron-oxide colloid deposition, quartz grains are being leached, and the rock is becoming partly or entirely friable, and the pH is about 7.0 or slightly below; and (4) normal iron formation in which the ground water is stagnant, and the pH is above 7.0. As the process goes on, the lower and upper ore boundaries move downward, but the ore zone widens, as does the zone of leaching of silica until the water table is reached, and the process presumably stops. This secondary ore is reasonably classified as ground water-B2.

At El Pao, Kalliokoski (1964) thinks that the hard iron ore of that deposit was formed by the metamorphism of a primary, silica-poor ore rather than by the process just described; nor does he think that concentration during metamorphism was of any consequence. Such ore would include the first two categories of the classification but not the third.

INDICES

INDEX OF AUTHORS

In this index the names of the authors listed in the bibliographies are arranged alphabetically, followed by page numbers on which their works are cited. If a paper has two authors, both are placed in this index. If a paper has more than two authors, only the senior author's name appears here (and in the corresponding bibliography as well).

Abbott, A. C., 568
Abbott, C. E., 371
Abdel-Gawad, A. M., 327
Abraham, E. M., 59
Adams, F. D., 62, 63
Adams, F. S., 363
Adams, L. D., 118
Adler, H. H., 327
Agar, W. M., 384
Agnew, A. F., 550
Agterberg, F. P., 577
Ahlfeld, F., 561, 563, 566, 568, 571, 573, 578, 611
Aho, A. E., 24, 154
Akers, J. P., 333
Akright, R. L., 399
Albers, J. P., 286
Albrethsen, A., 544
Albritton, C. C., Jr., 415
Alcock, F. J., 9, 26, 33, 34, 36, 97, 120, 130, 149
Aletan, G., 41
Aleva, J. J., 41
Allan, J. D., 31
Allen, A. T., 501
Allen, C. C., 89
Allen, J. W., 480
Allen, R. B., 149
Allen, V. T., 191, 377, 445
Alling, H. L., 450
Allingham, J. W., 550
Allsman, P. L., 384
Allsman, P. T., 394
Almond, H., 531
Alsdorf, P. R., 296
Ambrose, J. W., 21, 33, 113, 412
Ames, H. G., 146
Ames, R. L., 54
Amos, D. H., 472
Amstutz, G. C., 260, 345, 346, 606, 608
Andersen, S., 161
Anderson, A. L., 339, 343
Anderson, C. A., 220, 221, 235, 269, 289
Anderson, D. T., 101
Anderson, E. C., 439, 442, 445
Anderson, J. A., 517
Anderson, J. C., 421
Anderson, J. E., Jr., 377
Anderson, R. A., 339, 544
Anderson, R. J., 339
Anderson, R. Y., 333
Angelelli, V., 561
Anger, G., 43
Anon., 7, 9, 24, 276, 585
Atúnez Echagaray, F., 185
Appleyard, E. C., 63
Arce, J. N., 611
Archibald, J. C., Jr., 195
Argall, G. O., Jr., 245
Argall, P. (B.), 196, 439
Armstrong, H. S., 65, 79, 84, 86
Armstrong, J. E., 12
Arnold, L. C., 245
Arnold, R. R., 339
Ashley, R. P., 412
Assad, R. J., 115
Atwood, W. W., 310, 384, 510
Aubury, L. E., 271
Auger, P. E., 46, 146
Austin, W. L., 483
Averill, C. V., 286
Ayer, F. A., 355
Ayrton, S. N., 164
Axelrod, D. I., 418

Baadsgaard, H., 52, 55, 150
Baars, D. L., 317
Backman, O. L., 146
Badgley, P. C., 34
Badollet, M. S., 142
Bailey, E. H., 281, 282, 403
Bailey, S. W., 358, 550
Bain, G. W., 77, 89, 142, 223, 328, 548
Bain, H. F., 345, 550
Baird, D. M., 43
Baker, A., III, 220, 238
Baker, D. R., 434
Baker, J. W., 75
Baker, M. B., 60
Baldauf, R., 164
Balk, R., 450, 464, 492
Ball, C. W., 19
Ball, S. H., 161, 164, 305, 556
Ballachey, A. G., 130
Balsley, J. R., Jr., 462
Baltosser, W. W., 446
Bancroft, H., 544, 546
Bancroft, J. A., 118

Bancroft, M. E., 21
Bancroft, W. L., 131
Bandy, M. C., 597
Banfield, A. F., 135
Banks, N. G., 312
Bannerman, H. M., 427
Baragar, W.R.A., 46, 53, 54, 63, 130
Barbosa, O., 579
Barbosa, A. L. de Miranda, 579
Barker, F., 209
Barlow, A. E. 62, 63, 97, 108, 115
Barnes, H. L., 191, 445, 550
Barnes, M. P., 526
Barosh, P. J., 404
Barr, D. A., 14
Barrington, J., 333
Barsdate, R. J., 211
Bartholomé, P., 458
Bartley, M. W., 94, 95
Barton, P. B., Jr., 297, 374, 415
Bascom, F., 488
Bassett, W. A., 324, 523
Bastin, E. S., 71, 73, 187, 200, 297, 317, 345, 349, 390, 400, 424, 474, 475, 591
Bateman, A. M., 21, 97, 211, 404
Bateman, J. D., 27, 31, 35, 54, 55, 588
Bateman, P. C., 264
Bates, R. C., 328
Bauer, H. L., Jr., 404
Baum, J. L., 434
Baxter, J. W., 345
Bayley, W. S., 431
Beales, F. W., 53
Beaumont, E. C., 332
Beck, L. S., 149
Beck, R., 164
Becker, G. F., 282, 284, 290, 400
Beecham, A. W., 149
Beeson, J. J., 510
Behre, C. H., Jr., 256, 307, 312, 313, 374, 381, 474, 475, 501, 550, 551
Beikman, H. M., 326
Bejnar, W., 318
Béland, J., 118, 120
Bell, A. M., 120, 121, 146
Bell, E. B., 191
Bell, J. M., 83
Bell, K. G., 324
Bell, L. V., 112, 113, 146
Bell, W. A., 58
Belt, C. B., Jr., 439, 445
Bence, A. E., 36
Benedict, P. C., 235
Bengoechea, A., 185
Bennett, W.A.G., 548
Benson, D., 41
Benson, W. E., 331
Benson, W. T., 415
Berg, G., 585
Bergeat, A., 182
Bergeron, R., 46
Berkey, C. P., 465
Berman, H., 51, 349, 427, 598
Bernard, A., 374
Berndt, F., 571
Berry, E. W., 574, 614
Berry, L. G., 568
Berthelsen, A., 159, 161, 164
Berton, A., 573
Beutner, E. L., 456, 523
Beyschlag, F., 171
Bideaux, R. A., 608
Bierther, W., 166
Billings, G. K., 53
Billings, M. P., 535
Billingsley, P., 23, 404, 531
Binyon, E. O., 407
Birdseye, H. S., 324
Bishop, F. H., 468
Blackwelder, E., 513
Blade, L. V., 259
Blagbrough, J. W., 326
Blais, R. A., 46, 113, 131
Blake, W. P., 185, 254
Blanchard, R., 299
Blank, H. R., Jr., 520
Blowes, J. H., 113, 125
Boardman, R. L., 323
Böggild, O. B., 164
Bøgvad, R., 164
Boldy, J., 130
Bollin, E. M., 333
Bondam, J., 161, 166
Bondesen, E., 164
Bonham, W. M., 46
Bonillas, Y. S., 223
Borcherdt, W. O., 307
Bostock, H. S., 23, 154
Boswell, P. F., 299
Bothwell, S. A., 88
Botinelly, T., 323, 328
Botsford, C. W., 185, 200
Bottino, M. L., 470, 498
Bourret, W., 111
Boutwell, J. M., 510, 513, 526, 527
Bowditch, S. I., 609
Bowen, N. L., 73, 450
Bowen, O. E., Jr., 273
Bowen, W. C., 435
Bowers, H. E., 324
Bowman, A. B., 245
Boydell, H. C., 71
Boyle, A. C., 286
Boyle, R. W., 41, 55, 58, 71, 154, 374
Bradbury, J. C., 345, 346, 550
Bradley, W. W., 282, 284, 290
Bradshaw, R. J., 89
Bramlette, M. N., 256

Branner, G. C., 256, 259
Branner, J. C., 256, 588
Brant, A., 94
Bray, J. M., 292
Bray, R.C.E., 81
Bray, R. E., 510
Brecke, E. A., 345, 355
Breger, I. A., 328, 334
Brett, R., 122
Bridge, J., 501
Bridgwater, D., 161
Brinker, A. C., 196
Brisbin, W. C., 37
Broadhurst, S. D., 472
Brobst, D. A., 472, 505
Brock, M. R., 346
Brock, R. W., 18
Brockie, D. C., 475
Broderick, T. M., 351, 366
Brokaw, A. L., 404, 405, 501
Brooker, E. J., 149, 151
Brooks, E. R., 355
Brosgé, W. P., 216
Brown, A. S., 12, 13
Brown, C. E., 550, 551, 552
Brown, C.E.G., 23, 55
Brown, E. L., 31, 83
Brown, H. (C. T.), 166
Brown, H. S., 470
Brown, I. C., 55
Brown, J. S., 345, 374, 381, 453, 561
Brown, R. A., 113
Brown, R. L., 44
Brown, W. H., 538
Brown, W. L., 81, 130
Brownell, G. M., 27, 34
Bruce, E. L., 18, 33, 34, 35, 79, 83, 92, 358
Brüggen, J., 573, 591, 594
Brummer, J. J., 121
Brummett, R. W., 268
Bryant, D. G., 223
Bryce, J. D., 65
Buchan, R., 113, 125
Bucher, W. H., 465
Buddington, A. F., 182, 209, 431, 434, 450, 453, 456, 458, 541
Buehler, H. A., 381
Buerger, N. W., 535
Buffam, B.S.W., 89, 149
Bullis, A. R., 65
Bumgarner, J. G., 501
Buranek, A. M., 513
Burbank, W. S., 318, 319
Burckhardt, C., 182, 200
Burgess, J. A., 424
Burr, S. V., 64
Burrows, A. G., 74, 77, 89, 97
Burt, D. M., 111
Burwash, R. A., 55
Buseck, P. R., 182
Bush, J. B., 517
Butler, B. S., 218, 222, 243, 248, 254, 300, 351, 355, 510, 513, 520, 523, 527
Butler, J. W., Jr., 465
Butterfield, H. M., 89
Butts, C., 538
Byers, A. R., 34, 113, 153

Cadigan, R. A., 328
Cairnes, C. E., 21, 24
Cairnes, D. D., 155
Cairnes, R. B., 34
Calkins, F. C., 340, 392, 393, 400, 513, 514
Calkins, J. A., 547
Callaghan, E., 410, 411, 523
Callahan, W. H., 374, 434, 487, 502
Callisen, K., 164
Cameron, A. E., 58
Cameron, E. N., 349, 390, 419, 427, 550
Campbell, A. D., 74
Campbell, C. D., 150
Campbell, C. O., 58
Campbell, D. D., 51, 149
Campbell, D. F., 568
Campbell, F. A., 130, 131
Campbell, I., 424, 548
Campbell, N., 53, 55
Campbell, R. H., 331, 332
Campbell, W., 71, 74, 97, 490
Campbell, W. A., 48
Camsell, C., 23
Canney, F. C., 535
Cannon, R. S., Jr., 544
Card, K. D., 97
Carlson, D. W., 262, 273
Carlson, H. D., 63
Carlson, J. E., 551
Carmichael, A. D., Jr., 155
Carpenter, J. A., 400
Carpenter, R. H., 300, 442
Carpenter, R. H., 355
Carr, J. M., 7, 12, 13, 14
Carr, W. J., 530
Carrasco Córdova, R., 572
Carrière, G., 118
Carter, W. D., 326
Casteñedo, J., 182
Cathcart, S. H., 213
Cavender, W. S., 343
Chace, F. M., 484, 568
Chaigneau, M., 136
Chakraborty, K. L., 46
Chamberlain, J. A., 65, 149
Chamberlin, R. T., 580, 585
Chamberlin, T. C., 551
Chan, S.S.M., 339
Chapin, T., 209

Chapman, C. A., 427
Chapman, E. P., 312
Chapman, R. W., 264
Chapman, T. L., 250
Charlewood, G. H., 31, 77
Chaudhuri, S., 355
Chayes, F., 62
Cheney, E. S., 97
Chenoweth, W. L., 326
Cheriton, C. G., 41
Chew, R. T., III, 328, 332
Chisholm, E. O., 88, 92
Choubersky, A., 46
Chrismas, L., 14
Christie, A. M., 149, 150
Christopher, I. C., 92
Church, J. A., 254
Clark, A. M., 97
Clark, B. R., 339
Clark, E. L., 334
Clark, K. F., 442
Clark, L. A., 97, 122, 126, 490
Clark, L. D., 273
Clark, W. B., 262, 271, 273, 274
Clarke, F. W., 484
Clarke, O. M., Jr., 248
Clay, C., 95
Clayton, R. L., 220
Clayton, R. N., 359
Clegg, S. K., 345
Clements, J. M., 371
Clinton, N. J., 330
Cloos, E., 274, 279
Cloud, P. E., Jr., 358
Coats, R., 400
Coats, R. R., 418
Cockfield, W. E., 14, 24, 155
Coffin, R. C., 323
Cole, J. W., 392, 544
Coleman, A. P., 60, 83, 97
Coleman, L. C., 36, 55
Coleman, R. G., 284, 328
Colgrove, G. L., 100
Collins, C. B., 51
Collins, G. E., 315, 318
Collins, L. G., 431, 460
Collins, W. H., 67, 74, 83, 97, 98
Compton, L. P., 46
Condie, K. C., 556
Conn, H.M.K., 86
Connolly, J. P., 495
Conybeare, C.E.B., 150
Cook, C. W., 300
Cook, D. R., 510, 517, 531
Cook, E. F., 520
Cook, K. L., 520
Cooke, D. L., 77
Cooke, H. C., 75, 98, 113, 118, 122, 131, 142, 146
Cooke, H. R., Jr., 262
Cooper, J. R., 238, 245
Corking, W. P., 88
Cornwall, H. R., 351, 352, 355, 400, 424
Coughlan, W. K., 44
Courtright, J. H., 246, 252
Coveney, C. J., 14
Cowan, J. C., 98
Cox, D. P., 588
Cox, G. H., 551
Cragg, C. B., 154
Craig, J. R., 98
Craig, L. C., 328
Crane, G. W., 377, 517, 531
Crawford, A. L., 418
Crawford, J., 502
Crawford, R. D., 307
Creasey, S. C., 218, 222, 227, 229, 233, 235, 243, 250, 251, 405, 445, 510, 598
Creveling, J. G., 571
Crittenden, M. D., 514, 527
Crocket, J. H., 42, 99
Croft, W. J., 51
Crosby, D. G., Jr., 58
Crosby, G. M., 339
Crosby, P., 15
Cross, W., 310, 316, 318, 319
Crump, R. M., 456
Cumberlidge, J. T., 484
Cumings, W. L., 487
Cumming, G. L., 53
Cumming, L. M., 121
Cummings, J. B., 245
Cummings, J. M., 23
Cummings, W. W., 20
Currier, L. W., 345, 346, 502, 538
Curtis, C. D., 358
Curtis, J. S., 407
Cushing, H. P., 450, 453

Dadson, A. S., 55
Dahl, H. M., 327
Dahlstrom, C.D.A., 153
Dake, C. L., 381
Damon, P. E., 182
Darnell, R. P., 319
Darton, N. H., 492, 556
Dauth, H. L., 191
Davidson, A., 487
Davidson, C. F., 67, 328, 588
Davidson, D. M., 541
Davidson, D. M., Jr., 326
Davidson, E. S., 333, 352
Davidson, R. N., 312
Davidson, S., 98, 135
Davies, J. F., 29, 31, 34, 35, 36, 37
Davies, J. L., 41
Davis, C. E., 268
Davis, D. L., 266
Davis, J. F., 465
Davis, J. H., 381
Davy, W. M., 563
Dawson, A. S., 37

Dawson, K. R., 122, 150
DeChow, E., 41
DeGeoffroy, J., 551
Deines, P., 136
De Kalb, C., 218
De Launay, L., 171
Dellwig, L. F., 350
Denis, B. T., 142
Denis, T. C., 113, 115, 119, 123, 128, 131, 135, 136, 140, 142, 146
Dennen, W. H., 431
Derby, O. A., 585
Deringer, D. C., 566
Derry, D. R., 36, 67, 69, 113, 115 123
Desai, A. A., 475
Desborough, G. A., 98, 345, 377
Dessau, G., 374
De Wet, J. P., 93
Dickson, C. W., 98
Diemer, R. A., 556
Dietrich, R. V., 541
Dietrick, W. F., 284
Dietz, R. S., 98
Diffenbach, R. N., 498
Dill, D. B., 454
Diller, J. S., 286, 480, 484
Dimroth, E., 46
Dings, McC. G., 544
Dittman, A., 604
Dixon, D. W., 218
Dixon, G. H., 332
Dodd, P. H., 328, 330
Doe, B. R., 385, 453, 454
Doelling, H. H., 333
Doll, C. G., 535
Dollfus, O., 601
Dolmage, V., 11, 23
Donald, K. G., 51
Donaldson, J. A., 46, 88
Donath, M., 564
Donnay, J.D.H., 269
Dooley, J. R., 324
Dorr, J.V.N., II, 579
Dougherty, E. Y., 77, 89
Douglas, G. V., 46, 83, 118
Douglas, J., 223
Douglas, J. H., 79
Douglas, R. P., 41
Drake, A. A., Jr., 297
Dresser, J. A., 113, 115, 118, 119, 123, 128, 131, 135, 136, 140, 142, 146
Dresser, M. A., 98
Drier, R. W., 352
Drummond, A. D., 12, 13
Drysdale, C. W., 14, 18, 22
DuBois, R. L., 268
Dubuc, F., 59
Duel, M., 328, 334
Duffell, S., 14, 47
Dugas, J., 113, 121, 123, 131, 146
Duhovnik, J., 374
Dunbar, W. R., 89
Dunlap, J. C., 505
Dunn, J. A., 358
Durek, J. J., 243
Dutra, C. V., 580, 585
Dutta, N. K., 81
Dutton, C. E., 364
Dwornik, J., 53

Eakins, P. R., 74, 113
Eakle, A. S., 424
Eardley, A. J., 514, 523
Eastlick, J. T., 229
Eastwood, G.E.P., 17, 24
Ebbley, N. E., Jr., 312
Ebbutt, F., 10
Ebert, G., 579
Eckel, E. B., 284, 310
Eckelmann, F. D., 470
Eckelmann, W. R., 150, 472
Edgar, A. D., 69
Edie, R. W., 150
Edwards, J. D., 185, 200
Eichler, J., 580
Einaudi, M. T., 608
Ellsworth, H. V., 65, 71
Elsing, M. J., 174
Elston, D. P., 323
Emeleus, C. H., 164
Emery, J. A., 381
Emmons, S. F., 71, 174, 312, 514
Emmons, W. H., 318, 363, 371, 374, 381, 392, 393, 418, 475, 498, 551
Emslie, R. F., 31
Engel, A.E.J., 312, 450, 453, 454
Engel, C. G., 450, 454
Engels, J. C., 209
Ensign, C. O., Jr., 355
Entwistle, L. P., 573
Epstein, S., 385
Eric, J. H., 274
Erickson, A. J., Jr., 346, 527, 551
Erickson, M. P., 529
Erickson, R. L., 259
Espenshade, G. H., 468
Esteve Torres, A., 179
Ettlinger, I. A., 240
Evans, A. M., 65, 507
Evans, D. L., 571
Evans, H. T., Jr., 328
Evans, J.E.L., 89
Evans, J. R., 276
Evensen, C. G., 329, 332, 333
Everhart, D. L., 282, 290, 324
Evernden, J. F., 563
Evoy, E. P., 150
Evrard, P., 111, 140, 450, 541

Faessler, C., 142
Fagan, J. M., 502
Fahey, J. J., 191, 377, 445

Fahrig, W. F., 46, 47, 150
Fahrni, K. C., 11
Fairbairn, H. W., 67, 79, 98, 136
Fanale, F. P., 487
Farley, W. J., 36
Farmin, R., 279, 510, 531
Farrington, O. C., 179
Faure, G., 98, 355, 395
Fearing, F. C., 213
Fearing, J. L., Jr., 235
Fenner, C. N., 498
Fenoll, P., 381
Fenton, M. D., 395
Ferenčič, A. W., 622
Ferguson, H. G., 262, 410, 412, 425
Ferguson, J., 161
Ferguson, R. B., 150
Ferguson, S. A., 88, 89, 92
Fesser, H., 563
Field, C. W., 510
Finch, W. I., 328
Finlay, J. R., 235, 372
Finnell, T. L., 332
Fischer, B., 166
Fischer, R., 541
Fischer, R. P., 323, 328, 329, 476
Fisk, E. L., 403
Fitch, C. A., 529
Flaschen, S. S., 358
Fleischer, M., 393
Fleming, H. W., 41
Fletcher, A. R., 196
Fletcher, J. D., 340
Flores, T., 179, 200
Flournoy, E., 498
Folierini, F., 374
Folinsbee, J. C., 115
Folinsbee, R. E., 53, 89
Folwell, W. T., 339
Forbes, R. B., 209, 211
Ford, R. B., 390
Ford, R. E., 121
Forstner, W., 271, 282, 284, 290
Foshag, W. F., 179
Foullon, H. B. von, 484
Fountain, D. K., 7
Fournier, R. O., 405
Fowells, J. E., 231
Fowler, G. M., 475
Fowler, K. S., 556
Fowler-Billings, K., 427
Fowler-Lunn, K., 428
Foye, W. J., 63
Frarey, M. J., 47
Fraser, H. J., 349
Frease, D. H., 346
Frebold, H., 19
Freeman, B. C., 98, 128
Freeman, V. L., 325
Freeze, A. C., 26
French, B. M., 98, 368
Frenzel, A., 578
Frey, E., 556
Freyberg, B. von, 580
Friedländer, C., 69
Friedman, G. M., 67, 465
Friedman, I., 346, 551
Friedrich, G. H., 187, 286
Fritz, P., 53
Fritzche, H., 385
Frobes, D. C., 418
Frondel, C., 329, 428, 434, 517
Fryklund, V. C., Jr., 259, 340
Fuchs, E., 171
Full, R. P., 546
Fullagar, P. D., 470, 498
Fulton, R. B., 538
Fyles, J. T., 15, 19, 544
Fyson, W. K., 58

Gabelman, J. W., 324
Gable, D. J., 297
Gair, J. E., 580, 585
Galbraith, F. W., 310
Gale, H. S., 523
Gallagher, D., 460
Gallie, A. E., 83
Gannett, R. W., 262
Garbutt, P. L., 355
Garlick, G. D., 385
Garrels, R. M., 328, 329, 331, 359
Gast, P. W., 450
Gastelumendi, A. G., 614
Gastil, (R.) G., 47
Gathmann, Th., 580
Gavasci, A. T., 326
Gazdik, W. B., 332
Geach, R. D., 343
Geehan, R. W., 415
Geier, B., 573
Geijer, P., 179, 378
Gemmill, P., 421
Geological Staff, Cerro de Pasco Copper Corporation, 606, 609, 616
George, P. T., 89
George, P. W., 43
George, R. D., 294
Gerdemann, P. E., 381
Gerry, C. N., 123
Geyer, A. R., 487
Geyne, A. R., 187
Ghisler, M., 159
Gianella, V. P., 400
Gibson, O., 498
Gibson, R., 307
Gilbert, C. M., 264
Gilbert, G., 18, 498
Gilbert, J. E., 403
Gill, J. E., 113, 131
Gilliatt, J. B., 45
Gilliland, J. A., 153
Gillson, J. L., 140, 422, 462, 465
Gilluly, J., 218, 238
Gilmour, P., 131, 233

Girault, J., 136
Gittins, J., 64
Glass, J. J., 276
Gleeson, C. F., 155
Goddard, E. N., 292, 294, 297, 303, 305, 393
Goddard, J. D., 37
Gold, D. P., 136
Goldich, S. S., 358, 359
Golding, C., 544
Goldman, M. I., 256
Gonyer, F. A., 349
González Reyna, J., 185, 191, 195, 196, 200
Goodchild, W. H., 98
Goodwin, A. M., 77, 83, 89
Gordon, C. H., 439, 505
Gordon, M., Jr., 256
Gordon, S. G., 164, 566
Gorsky, E., 588
Gorsky, V. A., 588
Gott, G. B., Jr., 302, 405
Gouin, L. O., 573
Govett, G.J.S., 358
Grabert, H., 588
Graf, D. F., 445
Graham, R.A.F., 81
Graham, R. B., 115
Graham, R.P.D., 143
Granger, A. E., 418, 520
Granger, H. C., 325, 329, 331
Grant, U. S., 551
Grantham, R. M., 546
Grasty, J. S., 505
Graton, L. C., 89, 269, 286, 606, 609
Grawe, O. R., 346, 475
Gray, C., 487
Gray, C. H., Jr., 273
Gray, I. B., 332
Gray, R. F., 264
Green, L. H., 19, 155
Greenwood, H. J., 20
Gregory, G., 350
Gregory, H. E., 523
Greig, J. W., 269
Griffis, A. T., 89
Griffitts, W. R., 529, 530
Grogan, R. M., 346, 347
Gross, E. B., 326, 330
Gross, G. A., 45, 47, 83, 111, 358
Gross, S. O. 462
Gross, W. H., 92, 166, 588
Grösse, E., 580
Grout, F. F., 83, 358, 363, 366, 371
Grubb, P.L.C., 86
Grundy, W. D., 332
Gruner, J. W., 95, 329, 363, 366, 368, 371, 372, 523
Grunig, J. K., 385
Gualtieri, J. L., 326
Guild, P. W., 580
Guimarães, D., 580
Guimond, R., 115
Guiza, R., Jr., 185
Gummer, W. K., 64, 92
Gunderson, J. N., 366, 368
Gundlach, H., 563
Gunning, H. C., 26, 113
Gussow, W. C., 123, 146
Gustafson, J. K., 47, 89, 495
Gustafson, W. G., 442

Haapala, P. S., 611, 616
Haas, V. P., 403
Hadley, J. B., 268
Haffty, J., 183, 347
Hagner, A. F., 460, 556
Hagni, R. D., 456, 475
Hague, A., 408
Hague, J. M., 434
Hahn, A. W., 531
Hail, W. J., Jr., 329
Hajnal, Z., 153
Hale, F. A., Jr., 415
Halet, R. A., 113, 135
Hall, G. M., 472
Hall, W. E., 266, 286, 346, 551
Hallam, R. H., 128
Hallof, P. G., 128
Hamblin, W. K., 352
Hamilton, E. I., 161
Hamilton, S. K., 355
Hamilton, W., 99
Hammer, D. F., 240
Hammond, P., 111
Han, T.-M., 363
Hansen, D. A., 14
Hansen, J., 161, 162
Hansen, M. G., 235
Hanson, G., 33
Harder, E. C., 256, 268, 363, 520, 580, 585
Harder, J. O., 495
Hardie, B. S., 399
Hardin, G. C., Jr., 346
Hardman, E., 520
Hardwick, W. R., 222
Hargraves, R. B., 111
Harrington, G. L., 213
Harris, L. D., 502
Harrison, J. E., 305
Harrison, J. M., 33, 35, 36, 47
Harrison, J. V., 601
Harry, W. T., 164
Harshbarger, J. W., 333
Hart, L. H., 385
Hart, R. C., 67
Hart, O. J., 511
Hartman, J. A., 256
Harvey, R. D., 412
Harvie, R., Jr., 137
Hash, L. J., 472
Hathaway, D. J., 502

Hausen, D. M., 399
Haw, V. A., 99
Hawkes, H. E., 187, 286
Hawley, C. C., 334
Hawley, J. E., 75, 77, 84, 89, 92, 99, 119, 123, 131, 135, 146
Haycock, M. H., 51
Hayes, A. O., 45
Hayes, C. W., 256
Hayes, W. C., 378
Hazard, J. C., 415
Hedley, M. S., 20, 22
Heide, H. E., 213
Heidrick, T., 405
Heindl, L. A., 251
Heinrich, E. W., 51, 137, 150, 329, 390, 523
Heinrichs, W. E., Jr., 470
Hellens, A. D., 71
Henderson, J. F., 55
Hendricks, J. A., 598
Hendry, N. W., 86
Henriksen, N., 164
Herd, R. K., 159
Hernon, R. M., 445
Herreid, G., 216
Hershey, O. H., 271
Herz, N., 541, 580, 585
Hess, F. L., 213, 264, 294, 300, 329, 350, 419, 492
Hess, H. H., 395
Hester, B. W., 74
Hewett, D. F., 276, 393, 415
Hewitt, D. F., 63, 64, 65, 66, 69, 77
Hewitt, W. P., 197
Hewlett, C. G., 19, 544
Heyl, A. V., (Jr.), 323, 346, 374 381, 475, 502, 538, 551, 552
Heyl, G. R., 271
Heywood, W. W., 35
Hickok, W. O., IV, 487
Hicks, H. S., 95
Higazy, R. A., 492
Hill, H. L., 24
Hill, J. M., 297, 415, 422
Hill, P. A., 150
Hill, W. T., 502
Hillebrand, W. F., 243, 507
Hills, A., 450
Hilpert, L. S., 323, 325
Hinds, N.E.A., 286
Hinse, R., 115
Hoagland, A. D., 446, 502, 538
Hobbs, S. W., 340, 484
Hodge, D. S., 556
Hoffman, D. J., 47
Hoffman, R., 101
Hogarth, D. D., 137
Hogg, N., 90
Hogg, W., 23
Hogue, W. G., 223
Hohl, C. D., 351
Holbrook, D. F., 259
Holden, E. F., 428
Holland, H. D., 183, 326, 475
Hollingsworth, J. S., 259
Holmes, R. J., 71
Holmes, S. W., 67
Holmes, T. C., 90
Holser, W. T., 393
Holyk, W., 41
Hones, R. M., 517
Hook, J. W., 502, 538
Hopkins, H., 77
Hopkins, P. E., 77, 86
Horcasitas, A. S., 197
Hore, R. E., 78
Horner, W. J., 352
Horton, J. S., 445
Horwood, H. C., 24, 79, 92
Hosmer, H. L., 601
Hosted, J. O., 495
Hosterman, J. W., 340, 551
Hostetler, P. B., 329
Hotz, P. E., 431, 480, 484
Howard, P. F., 535
Howard, W. V., 137
Howe, E., 99, 279
Howell, B. F., 45
Howell, F. H., 591
Howland, A. L., 395
Hoyles, N.J.S., 147
Hriskevich, M. E., 71, 72
Hubaux, A., 462
Hubbell, A. H., 316
Huber, N. K., 358
Huebner, J. S., 183
Huff, L. C., 238, 245, 326
Hulin, C. D., 187, 274
Hume, C. B., 23
Hunt, C. B., 327, 329, 334
Hunt, R. N., 408, 510
Hunt, T. S., 470
Hurley, P. M., 137
Hurst, M. E., 88, 90, 92
Hutchinson, M. W., 340
Hutchinson, R. D., 45
Hutchinson, R. W., 41
Huttl, J. B., 222, 544

Ibarra, J., 182
Imlay, R. W., 182, 480
Imrie, A. S., 131
Ingalls, W. R., 408
Ingham, W. N., 65, 123, 147
Irvin, G. W., 245
Irvine, W. T., 10, 17, 20
Irving, J. D., 319
Irwin, W. P., 480
Isachsen, Y. W., 329
Ishihara, S., 442

Jackson, E. D., 395
Jackson, G. D., 88
Jackson, S. A., 53
Jacobs, E. C., 535
Jacobs, M. B., 327
Jaeggin, R. P., 47
Jaffe, H. W., 276
Jahns, R. H., 535
Jambor, J. L., 58, 154
James, A. H., 431
James, A.R.C., 24
James, D. H., 67, 150
James, H. L., 358, 359
James, H. T., 10
James, J. A., 381
James, W., 64
James, W. F., 51, 123, 147, 150
Jarrell, O. W., 598
Jaskolski, S., 568, 571
Jenkins, O. P., 274, 544, 548
Jenks, W. F., 609, 611
Jenney, C. P., 41, 75, 128, 419
Jensen, M. L., 329, 374
Jespersen, A(nna), 231
Jewett, G. A., 58
Jobin, D. A., 329
Johnson, C. H., 378, 475
Johnson, D. W., 231
Johnson, H. S., Jr., 327, 332, 333, 334
Johnson, R. F., 580
Johnson, R. W., Jr., 502
Johnston, A. W., 363
Johnston, W. D., Jr., 279, 400
Johnston, W.G.Q., 71
Joklik, G. F., 128
Jolliffe, A. W., 55, 56, 95, 150
Jolly, J. L., 475, 502, 538, 552
Jonas, A. I., 490, 498
Jones, C. L., 501
Jones, E. L., Jr., 341
Jones, E. L., III, 222
Jones, I. W., 121
Jones, R.H.B., 548
Jones, W. R., 395, 445, 446
Joralemon, I. B., 218, 355
Jõrgensen, O., 164
Joubin, F. R., 67, 150
Journeay, J. A., 245

Kalliokoski, J., 33, 41, 622
Kania, J.E.A., 465
Karpoff, B., 123, 140
Karup-Møller, S., 164
Kautzsch, E., 374
Kay, G. F., 480, 484
Kay, G. M., 552
Kays, M. A., 462
Keays, R. R., 99
Keevil, N. B., 65, 92, 147
Keil, K., 71
Keith, A., 470
Keith, M. L., 69
Kellerhals, P., 16
Kelley, D. R., 334
Kelley, V. C., 174, 266, 319, 325, 330
Kellogg, J. L., 282, 284, 290
Kellogg, L. O., 238
Kelly, L., 65
Kelly, W. C., 292, 563
Kemp, J. F., 372, 450, 462, 490, 498, 520, 556
Kempthorne, H. R., 147
Kendall, D. L., 502
Kennedy, V. C., 340, 552
Kermeen, J. S., 150
Kerr, P. F., 252, 264, 325, 326, 327, 330, 333, 334, 340, 399, 419, 435, 445, 446, 523
Kerstein, D. S., Jr., 470
Kesler, T. L., 472, 505
Kesten, S. N., 149, 150
Kett, W. F., 286
Keys, M. R., 90
Keys, W. S., 330, 334
Kidd, D. F., 51
Kidder, S. J., 84
Kierans, M. D., 191
Kildale, M. D., 527
Kimura, E. T., 12, 13
Kindle, E. D., 98, 155
King, E. R., 551
King, H. F., 435
King, P. B., 498
King, R. U., 300
Kingman, O., 498
Kingsley, L., 428
Kinkel, A. R., Jr., 34, 41, 84, 286, 470, 498
Kinnison, J. E., 245
Kirk, C. T., 294, 385
Kirwan, J. L., 99
Kistler, R. W., 229
Kisvarsanyi, G., 378
Kittel, D. F., 325
Klein, C., Jr., 47, 434
Klemic, H., 432, 552
Klohn Giehm, C., 591
Klugman, M. A., 556
Knaebel, J. B., 279
Knapp, M. A., 197
Knight, C. W., 51, 71, 74, 77, 97, 99, 490
Knochenhaver, B., 400
Knopf, A., 23, 213, 262, 264, 266, 269, 271, 274, 385, 415, 422, 446, 520
Knopf, E. B., 490
Knowles, D. M., 47
Kobe, H. W., 619
Koch, G. S., Jr., 191, 495
Kock, L., 166
Koeberlin, F. R., 563

Koehler, G. F., 71
Koene, J. D., 115
Koeppel, V., 150
Koffman, A. A., 35
Kohanowski, N. N., 574
Koschmann, A. H., 302, 303, 439
Koster, F., 150
Kottlowski, F. E., 319
Koulomzine, T., 47
Kozlowski, R., 563, 568
Kral, V. E., 410
Krauskopf, K. B., 507
Kretz, R., 56
Krieger, M. H., 233, 458
Krouse, H. R., 53
Kruger, F. C., 428
Krumbein, W. C., 359
Krusch, P., 171
Kuellmer, F. J., 231, 442
Kulkarni, P. H., 71
Kullerud, G., 97, 98, 99, 340, 490
Kulp, J. L., 150, 330, 340, 374, 381, 405, 453, 465, 470, 472, 487, 495, 601
Kupfer, D. H., 492
Kuryliw, C. J., 92

LaBerge, G. L., 359
LaBine, J. S., 150
Lacy, W. C., 245, 601, 619
Lake, M. C., 378
Lamarche, R.-Y., 119
Lamb, J., 23
Landes, K. K., 259, 419, 492
Lane, A. C., 352
Laney, F. B., 424
Lang, A. H., 51, 66, 67, 150
Lang, A. J., 492
Lang, H., 271
Lange, I. M., 97
Langford, F. F., 99
Langford, G. B., 90
Lapham, D. M., 487
Larochelle, A., 99
Larrabee, D. M., 428
Larsen, E. S., Jr., 264, 318, 319, 442
Larson, L. T., 393
Larson, R. R., 98
Lasky, S. G., 211, 446, 488
Lathram, E. H., 209
Latulippe, M., 113, 123, 128, 146, 147
Laughlin, A. W., 442, 517
Laurence, R. A., 505
Laverty, R. A., 325, 330
Lawson, A. C., 95, 405
Lawton, K. D., 59
Lay, D., 13
Lea, E. R., 41, 454
LeConte, J., 290
Ledoux, A. R., 484
Lee, H. A., 77
Lee, M. L., 174
Leech, G. B., 26
Legraye, M., 164
Lehnert-Thiel, K., 166
Leith, C. K., 359, 363, 368, 475, 520, 552, 580
Leipziger, F. D., 51
Lekas, M. A., 324, 327
Lemmon, D. M., 264, 408
Leo, G. W., 588
Leonard, B. F., 432, 456
Leonard, R. J., 493
Lepp, H., 359
Leroy, P. G., 446
Lessing, P., 454
Lesure, F. G., 326
Leuner, W. R., 123
Levin, S. B., 458
Lewis, J. V., 435
Lewis, R. Q., (Sr.), 331, 332, 333
Licari, G. R., 358
Linares, E., 561
Lincoln, F. C., 492, 574
Lindberg, M. L., 428
Lindgren, W., 233, 235, 243, 262, 272, 279, 294, 302, 352, 385, 405, 446, 510, 531, 532, 546, 556, 568, 571, 591, 612
Linehan, D., 428
Link, R. F., 191, 495
Linn, R. K., 284
Lipman, P. W., 319
Little, H. W., 7, 15, 18, 19, 20, 22, 544
Little, J. D., 20
Livingston, D. E., 223
Ljunggren, P., 563, 564, 574
Lochhead, D. R., 99
Locke, A., 171, 385, 404, 412, 425
Logan, C. A., 274
Long, A., 340
Long, L., 465
Longwell, C. R., 415
Longyear, R. D., 100
Lonsdale, J. T., 507
Loofbourow, J. S., Jr., 548
Loomis, F. B., Jr., 294
Lootens, D. J., 245
Lopez, V. M., 598
Lorain, S. H., 393
Lord, C. S., 56
Loughlin, G. F., 302, 303, 312, 313, 439, 513, 523, 531, 532
Lovell, H., 77
Lovering, T. G., 307, 325
Lovering, T. S., 251, 292, 294, 297, 303, 305, 307, 308, 517, 532
Lowden, J. A., 15
Lowell, J. D., 251, 326
Lowther, G. K., 191
Luedke, R. G., 318, 319

Lusk, J., 42
Lyden, J. P., 475, 476
Lydon, P. A., 271, 274
Lynch, D. W., 245
Lynch, J. J., 374
Lyons, W. A., 578, 604

MacAllister, A. L., 42
MacDiarmid, R. A., 240
Macdonald, B. C., 150
Macdonald, J. A., 151
MacDonald, R. D., 47
Machairas, G., 131
Machamer, J. F., 372
MacIssac, W. F., 121
MacKay, D. G., 128
MacKenzie, F. D., 245
MacKenzie, G. S., 42, 115
MacKevett, E. M., Jr., 211, 266
Mackin, J. H., 520
MacLeod, A., 113
MacVichie, D., 520
Magee, J. B., 20
Magee, M., 498
Mair, J. A., 67
Malamphy, M. C., 256
Malan, R. C., 333
Malcolm, J. B., 468
Malouf, S. E., 115
Mann, V. I., 359
Mapes-Vásques, E., 182, 201
Marchand, M., 136
Marie, J. R., 503
Mark, W. D., 616
Marlow, G. C., 191
Marsden, R. W., 359
Marsh, R., Jr., 591
Marshall, D., 82
Martin, C., 410
Martin, P. L., 36
Martínez, J., 186
Martison, N. W., 119
Marvin, R. F., 260
Mason, J., 71
Mason, J. F., 415
Masters, J. A., 326
Mather, K. F., 310
Mather, W. B., 92
Matheson, A. F., 79, 84, 585
Mathews, W. H., 21
Maucher, A., 374
Maugher, R. L., 182, 330
Maurice, C. S., 472
Maurice, O. D., 137
Mawdsley, J. B., 115, 123, 131, 140
Maxson, J. H., 480
Mayo, E. B., 23, 264
McAdam, R. C., 137
McCartney, G. C., 84
McClymonds, N. E., 252
McConnel, R. H., 340, 544
McConnell, G. W., 56
McCormick, J. E., 502
McDonald, G. A., 272
McDougall, D. J., 147, 307
McDowell, F. W., 405
McDowell, J. P., 67
McElwaine, R. B., 259
McGerrigle, H. W., 121
McGuire, R. A., 21
McIntosh, F. K., 468
McIntosh, R., 93
McKay, E. J., 330
McKay, G. R., 527
McKinley, P. F., 442
McKinstry, H. E., 90, 262, 274, 340, 352, 535, 606, 612
McKnight, E. T., 316, 374, 476
McLaren, D. C., 75, 93
McLaughlin, D. B., 90
McLaughlin, D. H., 211, 269, 274, 495, 601, 609, 616, 619
McMurry, H. V., 538
McNabb, J. S., 393
McTaggart, K. C., 14, 155
Mead, W. J., 256
Means, A. H., 307
Meeves, H. C., 319
Megathlin, G. R., 428
Mejia, V. M., 378
Mergrue, G. H., 325
Merriam, C. W., 422
Merrill, F.J.H., 197
Mertie, J. B., Jr., 209, 213, 216
Metsger, R. W., 435
Metz, H. E., 223
Metz, R. A., 249
Meyer, C., 378, 385
Meyer, H. C., 574
Meyers, W. B., 284
Meyn, H. D., 97
Michell, W. D., 240
Michener, C. E., 99
Miesch, A. T., 330, 408
Mikkola, A. K., 535
Miller, B. L., 568, 571, 574, 580
Miller, D. J., 211
Miller, D. S., 330
Miller, H. W., 503
Miller, J. D., 502
Miller, L. J., 330, 332
Miller, R.J.M., 115
Miller, W. G., 59, 71
Miller, W. J., 450, 458, 460
Milligan, G. C., 31
Million, I., 334
Mills, H. F., 233, 234
Mills, J. W., 113
Milton, C., 185, 378, 400, 435
Milton, D. J., 535
Miser, H. D., 259
Mitcham, T. W., 245, 332, 333, 340
Mitchell, G. J., 174
Mitchell, G. P., 99

Moehlman, R. S., 319
Moench, R. H., 297, 325
Moesta, F. A., 594
Moffit, F. H., 211
Moh, G. H., 498, 571
Molloy, J. S., 591
Molloy, M. W., 523
Monette, H. H., 88
Monger, J.W.H., 24
Monroy, P. L., 185
Montgomery, A., 74
Montoya, J. W., 529
Mookherjee, A., 91, 131
Moolick, R. T., 243
Moon, L., 566
Moorbath, S., 164
Moore, C. H., Jr., 541
Moore, E. S., 71, 74, 84, 90, 95, 495
Moore, J. M., Jr., 31
Moore, R. C., 476
Moore, S. L., 580
Moore, W. J., 510
Moorhouse, W. W., 77, 108
Morago Brito, A., 594
Morris, H. T., 517, 531, 532
Morrison, W. F., 83, 84
Morrow, H. F., 93
Moses, J. H., 609, 616, 619
Moss, A. E., 47
Motica, J. E., 323
Moxham, R. M., 222, 235
Moyd, L., 64, 490
Muessig, S. J., 546
Mulchay, R. B., 175
Mullens, T. E., 552
Muller, S. W., 410, 425
Mullerried, F.K.G., 185
Mulligan, R., 123
Muraro, T. W., 16
Murata, K. J., 256
Murdock, J. Y., 75
Murillo, J., 571
Murphy, D. A., 47
Murphy, J. E., 378
Murphy, R., 51
Murthy, V. R., 385
Mutch, A. D., 99
Myers, C. E., 435

Nackowski, M. P., 346, 527
Nagell, R. H., 616
Naldrett, A. J., 60, 99, 100, 126
Nash, J. T., 325
Newberg, D. W., 598
Newhouse, W. H., 43, 100, 487, 556
Newland, D. H., 451, 453, 454, 460, 462
Newman, E. W., 394
Newman, M. H., 502
Newman, W. L., 323
Nguyen, K. K., 16
Nickel, E. H., 137
Nielsen, R. L., 446
Nishihara, H., 171
Nishio, K., 355
Noble, E. A., 330
Noble, J. A., 495, 606
Noe-Nygaard, A., 161, 164
Nolan, T. B., 408, 425
Nördenskjold, O., 594
Norman, G.W.H., 114, 115, 116, 123, 147
Norman, L. A., Jr., 265, 266
Norris, A. W., 53
Northcote, K. E., 14
Norton, J. J., 492
Nowlan, J. P., 93
Nuffield, E. W., 149

O'Brien, J. C., 274
Odell, J. W., 381
Oder, C.R.L., 502, 503, 538
Oen, I. S., 162
Oertell, E. W., 332
Oesterling, W. A., 346
Officers of the Geological Survey, 26
Ohle, E. L., (Jr.), 90, 262, 274, 356, 375, 378, 381, 476, 503, 552
Ohmoto, H., 17, 182
Olmstead, H. W., 231
Olsen, E. J., 143
Olson, J. C., 276, 428, 472
O'Neill, J. J., 114, 135
O'Niel, J. R., 183
Orcel, J., 612
Ordoñez, E., 188
Ordoñez, G., 171, 446
Ortiz-Asiáin, R., 179
Orville, P. M., 492
Osborn, E. F., 358
Osborne, F. F., 64, 111, 137, 140 451, 463
Osterwald, F. W., 523, 530
O'Sullivan, R. B., 326
Ottemann, J., 381
Overweel, C. J., 606
Owens, J. S., 359

Page, J. J., 428
Page, L. R., 330, 492, 493
Page, N. J., 395
Paige, S., 143, 446, 447, 492, 495
Palache, C., 51, 151, 350, 435, 598
Palmer, D. F., 456
Pardee, J. T., 393
Park, C. F., Jr., 223, 240, 422, 532, 544, 580
Park, W. C., 345, 346
Parker, J. M., III, 468, 472
Parker, R. L., 523, 547
Parks, B., 259
Parsons, A. B., 405, 510, 517

Paterson, N. R., 128
Patterson, C., 385
Patterson, J. M., 37
Patton, H. B., 319, 529
Patton, W. W., Jr., 216
Patty, E. N., 548
Pauly, H., 164, 165
Payne, J., Jr., 566
Payne, J. G., 70
Peacock, H. C., 510
Peacock, M. A., 99
Peale, R., 131
Pearson, R. C., 307, 313
Pecora, W. T., 484
Pegau, A. A., 541
Pélissonnier, H., 375, 574
Pelletier, J. D., 251
Pemberton, R. H., 128
Penfield, S. L., 578
Penhallegon, W. J., 505
Pennebaker, E. N., 405
Pentland, A. J., 26
Peoples, J. W., 395
Percival, F. G., 580
Pérez Martínez, J. J., 201
Perrault, G., 137
Perry, D. V., 229
Perry, E. S., 385, 390
Perry, V. D., 175, 598
Personnel, Producing Mines and Quebec Department of Natural Resources, 116
Peters, W. C., 510
Petersen, U., 601, 604, 606, 609, 616, 619
Peterson, D. W., 240
Peterson, E. C., 266
Peterson, N. P., 227, 229, 231, 251
Petruk, W., 72
Pettijohn, F. J., 359
Phemister, T. C., 72, 100, 490
Philips, W. B., 507
Philpotts, A. R., 541
Phipps, C.V.G., 69
Phoenix, D. A., 323, 403
Pienaar, P. J., 67
Pilz, R., 564
Pinckney, D. M., 347
Pinger, A. W., 435
Poitevine, E., 143
Poldervaart, A., 472
Pollock, W., 20
Pomerene, J. B., 580
Pošepný, F., 171, 290
Postel, A. W., 460
Potapoff, P., 97
Pouliot, G., 137
Powell, J. L., 395
Pray, L. C., 276
Prescott, B., 197
Prest, V. K., 86
Price, P., 131
Price, P. M., 395, 396
Prinz, W. C., 393
Proctor, P. D., 378, 523
Puchner, H. F., 183
Puffett, W. P., 327
Purdue, A. H., 259
Purrington, C. W., 318
Pye, E. G., 79, 81

Quirke, T. T., 95

Rackley, R. I., 492
Radabaugh, R. E., 307
Radelli, L., 564
Rader, L. F., Jr., 529
Radtke, A. S., 211, 382, 399, 517
Ramdohr, P., 67, 588
Ramp, L., 480
Ramsay, R. H., 175, 356
Rancourt, C., 41
Rand, J. R., 356
Rangel, M. F., 179
Ransome, A. L., 282, 284, 290
Ransome, F. L., 223, 231, 235, 240, 245, 249, 254, 274, 302, 316, 319, 340, 412
Rasor, C. A., 327
Ratcliffe, N. M., 465
Ratté, J. C., 319, 320
Raucq, P., 279
Ray, J. C., 385
Reber, L. E., Jr., 235, 243
Rechenberg, H. P., 111, 140, 259, 463, 541, 566, 571
Redden, J. A., 492, 541
Reed, D. F., 259
Reesor, J. E., 16
Reeves, R. G., 580
Reh, H., 47
Reid, J. A., 51, 72, 79, 272, 400
Reid, R. R., 340
Reinhardt, E. V., 330
Renault, J. R., 123
Rennie, C. C., 14, 20
Retty, J. A., 47, 111, 116
Reynolds, P. H., 454
Reynolds, R. R., 552
Rice, C. T., 197, 201
Rice, H.M.A., 7, 11, 16, 17, 23, 26
Rice, H. R., 47
Rice, M., 235, 465
Richard, K., 246, 252
Richarz, S., 366
Richmond, W. E., 393
Rickaby, H. C., 97
Rickard, T. A., 303, 316
Ricketts, J. E., 503
Ridge, J. D., 378, 435, 475, 476, 503
Ridland, G. C., 51, 56

Ries, H., 435
Riley, G. C., 43
Riley, G. H., 492
Rinehart, C. D., 265
Ringsleben, W. C., 90
Riordon, P. H., 143
Rising, W. B., 290
Rivas Valenzuela, S., 572
Rivera Plaza, G., 612
Roach, R. A., 47, 308, 323
Roberson, C. E., 290
Roberts, H. M., 95, 100, 581
Roberts, R. J., 399, 408, 418
Robertson, D. K., 53
Robertson, D. S., 33, 67, 68
Robertson, J. A., 68
Robertson, J. F., 286
Robinson, A.H.A., 140
Robinson, H. S., 90
Robinson, R. F., 340
Robinson, S. C., 66, 151
Robinson, W. C., 24
Robinson, W. G., 131, 135
Robles, R., 191
Robson, G. M., 98
Robson, W. T., 77
Rocha, V. S., 171
Roedder, E., 53, 375
Rogers, A. F., 101, 269, 385
Rogers, C. L., 183
Rogers, G. S., 465
Rogers, J., 503
Rogers, W. B., 324
Romsio, T. M., 238, 245
Roscoe, S. M., 68, 72, 100, 128, 131
Rose, A. W., 249, 446, 510, 511
Rose, E. R., 45, 63, 108, 111, 140
Rose, H. J., Jr., 352
Rosen-Spence, Andrée de, 131
Rosenzweig, A., 330
Ross, C. P., 229, 290, 403, 507
Ross, C. S., 140, 259, 442, 470, 498, 541
Ross, D. C., 265
Ross, J. V., 16
Routhier, P., 330, 352, 568
Rove, O. N., 188, 223
Rowe, R. B., 66, 123, 137
Roy, S., 42
Royce, S., 95, 359
Rubey, W. W., 411
Rubly, G. R., 251
Rubright, R. D., 511
Ruckmick, J. C., 622
Rudroff, A., 578
Ruiz, F. C., 594
Rumbold, W. R., 564
Runnells, D. D., 216
Runner, J. J., 492
Ruotsala, A. P., 352
Russell, G. A., 37
Russell, R. D., 101, 454
Rutland, R. W., 574
Ruttan, G. D., 31
Ryan, C. W., 541
Rye, R. O., 17, 183, 606
Rynearson, G. A., 581
Ryznar, G., 132

Saadallah, A. A., 475
Sainsbury, C. L., 213
Saint-Julien, P., 119
Salazar Salinas, L., 179
Sales, R. H., 356, 385
Samoyloff, V., 566
Sampson, E., 72
Samuel, W., 79
Sanders, B. H., 581
Sangster, D. F., 375
Santillan, M., 179
Santos, E. S., 325
Sasaki, A., 53
Satterly, J., 66, 86
Savage, W. S., 77
Sawkins, F. J., 183, 375, 606
Scalia, S., 200
Schafer, P. A., 396
Schaller, W. T., 294, 419, 507
Scheiner, B. J., 399
Schilling, J. H., 411, 442
Schillinger, A. W., 352
Schindler, N. R., 131
Schlee, J. S., 325
Schmedeman, O. C., 616
Schmidt, R. A., 476
Schmidt, R. G., 363, 364
Schmitt, H. (A.), 191, 246, 447
Schneider, H. J., 374
Schneiderhöhn, H., 100, 165, 492
Schneider-Scherbina, A., 564, 571
Schofield, S. J., 10, 17, 26
Schroeder, M. C., 548
Schuette, C. N., 282, 284, 290, 403, 507
Schumacher, J. I., 312
Schwartz, G. M., 26, 175, 218, 222, 223, 227, 231, 236, 243, 249, 251, 366, 368, 372, 405, 447, 492, 493, 511
Schwerin, M., 347
Sclar, C. B., 137
Scott, F. N., 121
Scott, J. B., 192
Scull, B. J., 259
Searls, F., Jr., 340, 412
Secrist, M. H., 503, 505
See, P. D., 317
Segerstrom, K., 188, 594
Seigel, H. O., 53
Sell, J. D., 240
Selleck, D. J., 48
Semenov, E. I., 162
Sergiades, A. O., 72, 74

Sgrosso, P., 561
Shafiqullah, M., 137
Shainin, V. E., 350
Shand, S. J., 419, 458, 465
Shannon, E. V., 350
Sharma, P. V., 159
Sharma, T., 48
Sharp, B. J., 327, 514, 529
Sharp, W., 408
Sharp, W. N., 276, 343
Sharpe, J. I., 129, 147
Sharwood, W. J., 495
Shaub, B. M., 350, 428, 458
Shaw, D. M., 66, 70
Shawe, D. R., 324, 331, 529
Shenon, P. J., 340
Shepard, A. O., 517
Shepard, W. M., 517, 532
Sheridan, D. M., 493
Shoemaker, E. M., 331, 333
Short, K. C., 622
Short, M. N., 240
Shrode, R. S., 346
Siebenthal, C. E., 476
Siegfus, S. S., 548
Siems, P. L., 319
Signer, C. M., 197
Silberman, M. L., 412
Silver, C., 319
Silver, L. T., 229, 238
Silverman, A. J., 308, 340
Simmersbach, B., 209
Simmons, G., 451
Simmons, G. C., 581
Simmons, W. W., 231, 498
Simony, P. S., 108
Simos, J. G., 526
Sims, P. K., 297, 305, 432
Sinclair, A. J., 11, 16, 22
Singewald, J. T., Jr., 352, 378, 463, 556, 564, 568, 571, 574, 580, 614
Siroonian, H. A., 123
Skinner, B. J., 375, 535
Skinner, R., 42
Slaughter, A. L., 495, 496
Smerchanski, M. G., 137
Smirnov, V. I., 375
Smith, C. H., 42
Smith, D., 7
Smith, F. G., 90, 95, 147, 566
Smith, F. W., 192
Smith, G. H., 400
Smith, J. G., 211
Smith, J. R., 35, 153
Smith, L. L., 432
Smith, P. S., 216
Smith, T. S., 20, 93
Smith, W. C., 493
Smith, W.S.T., 347, 476
Smithson, S. B., 556
Smyth, H. L., 95, 372
Smythe, C. H., Jr., 454
Snelgrove, A. K., 43
Snelling, N. J., 123, 147
Snow, W. E., 197
Snyder, F. G., 375, 381, 498
Soen, O. I., 165
Sopher, S. R., 100
Sørensen, H., 161, 162
Sorensen, R. E., 341
Souch, B. E., 100
Souza, H. C. Alves de, 588
Spears, D. A., 358
Speers, E. C., 100
Spence, C. D., 132
Spence, H. S., 51
Spencer, A. C., 209, 316, 405, 435, 447, 488
Spencer, F. N., Jr., 561
Spurr, J. E., 192, 305, 319, 378, 381, 425, 435, 476, 552
Staatz, M. H., 493, 530, 547
Stacey, J. S., 511
Staff, Algoma Ore Properties, Limited, 84
Staff, Buchans Mining Co., Ltd., 43
Staff, Consolidate Mining and Smelting Company, Ltd., 26, 56
Staff, Eldorado Mining and Refining Limited, 151
Staff, Falconbridge Nickel Mines, Ltd., 100
Staff, International Nickel Company of Canada, Limited, 100
Staff, McKenzie Red Lake Gold Mines, Limited, 93
Stansfield, J., 137
Stanton, R. L., 42
Staples, L. W., 300
Starck, L. P., 24
Steacy, H. R., 68
Steed, R. H., 334
Steele, H. J., 251
Steen, C. A., 327
Steenken, W. F., 622
Steenland, N. C., 67, 68, 465
Steidtmann, E., 213, 378
Steiger, R. H., 82
Stelzner, A., 564
Stephens, E. C., 418
Stephens, F. H., 24
Stephenson, R. C., 463
Stern, T. W., 327, 331
Sterrett, D. B., 428
Steven, T. A., 319, 320
Stevens, R. E., 312
Stevens, R. N., 259
Stevenson, I. M., 48, 58
Stevenson, J. S., 100, 119
Stewart, C. A., 252
Stewart, J. H., 329
Stewart, R. M., 265, 266

Stieff, L. R., 331
Still, A. R., 233
Stockwell, C. H., 33, 35, 42, 90, 95, 153, 155
Stoddard, C., 400
Stoiber, R. E., 352, 476
Stokes, W. L., 327, 331
Stoll, W. C., 428
Stone, J. B., 201
Stone, J. G., 201
Stonehouse, H. B., 100, 347
Stose, G. W., 488
Strauss, L. W., 614
Stringham, B. (F.), 341, 511, 532
Stronach, R. S., 93
Stubbins, J. B., 48
Stumpfl, E. F., 42
Suffel, G. G., 81, 129, 131, 132
Sutton, A. H., 347
Svendsen, R. H., 340
Swanson, C. O., 26
Swanson, E. A., 44
Swanson, R. W., 229
Symons, D.T.A., 72, 359
Szekely, T. S., 620
Sznapka, G., 564

Taber, S., 541
Takahashi, T., 435
Taliaferro, N. L., 272
Tanner, W. F., 326
Tanton, T. L., 33, 35, 48, 84, 95, 359
Tarr, W. A., 381, 435, 476
Taupitz, K. C., 375
Taylor, A., 511
Taylor, A. R., 552
Taylor, A. V., Jr., 598
Taylor, B., 132
Taylor, C. M., 382
Taylor, G. L., 493
Taylor, I. R., 259
Taylor, L. A., 411
Taylor, R. H., 300
Tenney, J. B., 223, 224, 232, 236, 243, 249, 255
Tenny, R. E., 58
Terrones Benitez, A., 179
Terrones L., A. J., 601, 609, 616
Thaden, R. E., 332, 333
Thayer, T. P., 396
Thiel, G. A., 364
Thode, H. G., 101
Thoenen, J. R., 256
Thomas, L. A., 251
Thompson, A. P., 499, 535
Thompson, G. A., 401, 507
Thompson, M. E., 323
Thompson, R. M., 23, 77
Thomson, J. Ellis, 51, 72, 74, 75, 80
Thomson, James E., 66, 75, 77, 82, 88, 101
Thomson, K. C., 331
Thomson, R., 72, 93
Thorardson, W., 327, 332, 333
Thormann, W., 563
Thornburg, C. L., 188
Thorpe, W., 115
Thurlow, E. E., 341
Thurmond, R. E., 246
Tilley, C. E., 64
Tilton, G. R., 82
Timms, P. D., 82
Tipper, H. W., 13
Titley, S. R., 245, 439
Toborffy, Z., 578
Todd, E. W., 74, 77
Tolbert, G. E., 585
Tolman, C., 132, 378
Tolman, C. F., Jr., 101, 249, 252, 269, 272, 412
Tomlinson, W., 22
Tooker, E. W., 297, 344, 511
Torón Villegas, L., 179
Tossi, P. A., 604
Touwaide, M. E., 171
Tovote, W. L., 243
Tower, G. W., Jr., 532
Trace, R. D., 346, 347
Tracey, J. I., Jr., 256
Traill, R. J., 68
Traver, W. M., Jr., 507
Trefzger, E. F., 617
Tremblay, L. P., 123, 151
Trimble, D. E., 333
Triplett, W. H., 183
Trischka, C., 224
Trites, A. F., Jr., 332, 344
Truesdell, A. H., 326
Tschanz, C. M., 422
Tully, D. W., 78, 129
Tung, J. P-y, 499
Tupper, W. M., 42
Turek, A., 31, 151
Turneaure, F. S., 563, 564, 566
Turner, F. J., 396, 451
Turner, H. W., 269, 272, 279, 507
Tuttle, O. F., 140
Tweto, O., 294, 308, 313
Tyler, S. A., 358, 359, 581
Tyrrell, J. B., 78
Tyson, A. E., 80

Uglow, W. L., 22, 61
Ulloa, S., 171
Ulrich, E. O., 347, 503
Ulrych, T. J., 101, 165
Umlauff, A. F., 614
Umpleby, J. B., 341, 344, 547
Ussing, N. V., 162

Valentine, W. G., 175
Vallely, J. L., 256
Vanderburg, W. O., 408
Vanderwilt, J. W., 300, 442
Van Hise, C. R., 359, 372
Van Schmus, R., 68
Van Tassel, R. E., 155
Varnes, D. J., 316, 320
Velasco, J. R., 175
Verhoogen, J., 396, 451
Vhay, J. S., 320, 344
Villarello, J. D., 185
Vitaliano, C. J., 411, 412
Vohryzka, K., 166
Vokes, F. M., 123
Vollo, N. B., 129

Waard, D. de, 451
Wadsworth, W. B., 218
Wagner, O. E., Jr., 346
Wahlstrom, E. E., 292
Waitz, P., 192
Waldschmidt, W. A., 341
Walker, G. W., 276, 523
Walker, H. A., 197
Walker, J. F., 18, 19, 21, 24, 544
Walker, M. S., 300
Walker, R. T., 286
Walker, T. L., 101
Walker, W. J., 286
Wallace, R. C., 33, 35
Wallace, R. E., 341
Wallace, R. M., 581
Wallace, S. R., 300
Walton, M., 451
Wandke, A., 101, 186
Wanless, R. K., 26, 56
Ward, H. J., 608, 609, 620
Ward, W., 78
Warning, G. F., 20
Warren, C. G., 329
Warren, C. H., 140
Warren, H. S., 23
Warren, H. V., 23, 175, 341
Washington, H. S., 259
Waterman, G. C., 10
Waters, A. C., 331
Watkinson, D., 137
Watson, T. L., 505, 538, 541
Wayland, R. G., 209
Weaver, C. E., 544, 548
Webber, G. R., 132
Webster, R. (N.), 240
Wedow, H., Jr., 503
Weed, W. H., 197, 385, 470, 499, 535
Weege, R. J., 352
Weeks, A. D., 326, 328, 331
Weeks, L. J., 58
Wegemann, E., 162
Wegmann, C. E., 165
Wehrenberg, J. P., 308
Weidman, S., 476
Weinberg, E. L., 538
Weir, G. W., 327
Weis, P. L., 341, 390, 493
Weissenborn, A. E., 20, 544
Welker, K. K., 564
Weller, J. M., 347
Weller, S., 347
Wells, F. G., 480, 520
Wells, J. D., 297, 305
Wendt, A. F., 572
Wentorf, R. H., Jr., 458
Wernecke, L., 209
West, W. S., 552
Westervelt, R. D., 18, 48
Westgate, L. G., 396, 422
Wheeler, E. P., II, 463
Wheeler, H. E., 408, 422
Whishaw, Q. G., 20
White, C. E., 56
White, C. H., 175
White, D. A., 368
White, D. E., 195, 290, 375, 401
White, M. G., 588
White, W. A., 468
White, W. H., 7, 11, 13, 14, 16
White, W. S., 352, 356, 535
Whitebread, D. H., 544
Whitehead, W. L., 72, 274, 595
Whiting, F. B., 561
Whitlow, J. W., 550, 552
Whitman, A. R., 72, 90, 405
Whitten, E.H.T., 122
Whitwell, G. E., 548
Wignall, T. K., 551
Wiliden, R., 229
Wilkerson, A. S., 292
Willard, M. E., 341, 523
Williams, G. H., 465
Williams, H., 37, 101
Williams, J. F., 256, 260
Williams, J. S., 347
Williams, K. L., 131
Williams, N. C., 529, 530
Williams, P. L., 327
Williams, R. L., 432
Willman, H. B., 552
Willmot, A. B., 83
Wilson, C. L., 527
Wilson, E. D., 222, 223, 240, 243, 246, 248, 251
Wilson, G. M., 347
Wilson, H.D.B., 37, 101, 412
Wilson, H. S., 147
Wilson, I. F., 171
Wilson, M. E., 75, 132
Wimmler, N. L., 396
Winchell, A. N., 390, 511
Winchell, H. V., 188
Winchell, N. H., 364
Winckelmann, H., 564

Windley, B. F., 159
Wisser, E., 171, 186, 188, 224, 310, 316, 320, 413
Witkind, I. J., 327, 331, 333
Witzig, E., 166
Wolf, M., 564
Wolfe, C. W., 598
Wolff, J. F., 363, 369
Wood, H. B., 324, 327, 334
Woollard, G. P., 465
Woolnough, W. G., 359
Wright, C. M., 95
Wright, H. M., 14
Wright, J. C., 355, 356
Wright, J. F., 33, 36
Wright, L. B., 495, 496
Wright, R. J., 324, 331, 341
Wright, W. J., 58
Wyllie, P. J., 487

Yates, A. B., 101
Yates, R. G., 403, 507, 614
Yoder, H. S., 359
Young, E. B., 422
Young, R. G., 332
Young, W. E., 520

Zapffe, C., 364
Zartman, R. E., 260
Zimmer, P. W., 460
Zimmermann, R. A., 260, 552
Zitting, R. T., 326
Zurbrigg, H. F., 38, 101
Zwicker, W. K., 393

Note

This index contains the names of 1985 authors. The number of authors involved in the references cited is, of course, greater than this because only the senior authors of papers having more than two authors are listed here; anonymous appears five times, and the geological staffs of various mines are cited eight times. Thus, well over 2000 authors took part in the preparation of the references listed.

Because of a change in the manner of indexing authors between Memoir 75 and this Memoir, a direct comparison as to author numbers between the two memoirs is not possible; this is particularly true since Memoir 75 included deposits from six continents, while this one considers only two. Nevertheless, it is interesting to see that 1329 of the 1985 authors have published on only one district.

Despite the 35 years that have elapsed since the death of Waldemar Lindgren, no author is cited for as many districts as he; the 20 districts of this Memoir on which he published are mostly in the western United States, although 3 (Braden, Colquijirca, and Potosí) are in South America. Some of his theoretical concepts may have become outdated, but his geological observations are as valuable as they ever were. Of course, anyone using this Memoir will realize that I have based my approach to the problems of economic geology on his classification.

ALPHABETICAL INDEX OF DEPOSITS

This index contains, among other items, the name of each deposit at the head of a bibliography. If the designation is a dual one (for example, Cadillac-Malartic), the second name also is indexed. In addition, the names of the more important mines, subdistricts, geographical areas, and political subdivisions mentioned in the references or the notes also are included. The page number (or numbers) following each index entry is the page on which the first reference to the item in question is to be found in a given bibliography (or bibliographies).

Abajo Mountains, Utah, 327
Ackerman Mine, Quebec, 131
Adams Mine, Ontario, 59
Adamson Tungsten Mine, California, 264
Adirondacks (General), New York, 450
Afterthought Mine, California, 286
Aguas Calientes, Casapalca, Peru, 608
Aguilar, Argentina, 561
Ainsworth, British Columbia, 16, 17
Ajo, Arizona, 218
Aldermac Mine, Quebec, 119, 131[1]
Alexo Mine, Ontario, 60
Allard Lake, Quebec, 111
Alleghany, California, 262
Amador County, California, 273
Ambler River Quadrangle, Alaska, 216
Ambrosia Lake, New Mexico, 324
American Fork, Utah, 513
Amherst County, Virginia, 541
Amisk-Athapopuskow Lake, Manitoba, 33
Anderson Lake, Manitoba, 37
Andover District, New Jersey, 432
Angels Camp, California, 274
Apache County, Arizona, 332
Arkansas Bauxite, Arkansas, 256
Ashmore Township, Ontario, 79
Atikokan Area, Ontario, 95
Austin Brook Deposit, New Brunswick, 41
Austinville-Ivanhoe, Virginia, 538
Avalos, Zacatecas, Mexico, 182

Bagdad-Massive Sulfides, Arizona, 220
Bagdad-Porphyry Copper, Arizona, 221
Baird Township, Ontario, 92
Baldwin Township, Ontario, 101
Balmat-Edwards, New York, 453
Balmer Township, Ontario, 92
Bancroft-Haliburton (General), Ontario, 62
Bancroft (Nonmetallics), Ontario, 63
Bancroft (Uranium), Ontario, 65
Bankfield Mine, Ontario, 79
Barton Mine, New York, 458
Bass River Map-Area, Nova Scotia, 58
Bateman Township, Ontario, 92
Bathurst, New Brunswick, 41
Battle Mountain, Colorado, 307
Batty Lake Map-Area, Manitoba, 33
Bayard, New Mexico, 446
B. C. Nickel Mine, British Columbia, 24
Beattie Mine, Quebec, 135
Beatty-Munro Area, Ontario, 86
Beaulieu Region, Northwest Territories, 56
Beaverlodge (Goldfields), Saskatchewan, 149
Beecher No. 3-Black Diamond Pegmatite, South Dakota, 492
Beetown Lead-Zinc Area, Wisconsin, 551
Bell Island, Newfoundland, 45
Belmont Quadrangle, Wisconsin, 552
Belo Horizonte, Brazil, 580
Benson Mines, New York, 456
Berne Quadrangle, South Dakota, 492
Beryl Mountain, New Hampshire, 428
Bethlehem, British Columbia, 14
Bevcon Mine, Quebec, 147
Bicroft Mine, Ontario, 65
Big Bend, California, 272
Big Indian District, Utah, 327
Bingham, Utah, 510
Birchtree Mine, Manitoba, 38
Bird Lake, Manitoba, 29
Bird River, Manitoba, 27
Birthday Claims, California, 276
Bisbee, Arizona, 223
Bishop, California, 264
Bitter Creek Vanadium-Uranium Deposits, Colorado, 323
Biwabik Iron Formation, Minnesota, 366
Black Bear Vein, Colorado, 320
Black Hills Pegmatites, South Dakota, 491
Black Lake, Quebec, 143
Black Mesa-Hopi Buttes, Arizona, 333

---

[1]Different mines.

Blind River (Elliot Lake), Ontario, 67
Blue Mountain, Ontario, 69
Bluebell, British Columbia, 16, 17
Blyklippen Mine, Greenland, 167
Boléo, Baja California, Mexico, 171
Bolivian Tin (General), Bolivia, 563
Bonanza, Colorado, 322
Bonnecamp Map-Area, Quebec, 121
Bornite, Alaska, 216
Boston Township Iron Range, Ontario, 59
Boulder Batholith, Montana, 385
Boulder County Tellurides, Colorado, 292
Boulder County Tungsten, Colorado, 294
Boulder River Area, Montana, 395
Bourlamaque Township, Quebec, 146
Bousquet, Quebec, 113
Boyd Mine, Tennessee, 500
Braden, Chile, 591
Brenda Lake, British Columbia, 7
Brewster County, Texas, 507
Bristol, Nevada, 422
Britannia Mines (Howe Sound), British Columbia, 9
Bruce Mines Area, Ontario, 68
Brudenell-Raglan Area, Ontario, 64
Brunswick No. 6 Deposit, New Brunswick, 41
Brunswick No. 12 Mining Area, Bathurst, New Brunswick, 42
Buchans, Newfoundland, 43
Bull Valley District, Utah, 520
Bullwhacker Mine, Nevada, 408
Bully Hill Mining District, California, 286
Bunker Hill, Idaho, 340
Burgin Mine, Utah, 517
Burke, Idaho, 339
Burra Burra Mine, Tennessee, 500
Butte, Montana, 384
Butte County, California, 274

Cable Mine, Montana, 392
Cadillac-Malartic, Quebec, 112
Calamine Quadrangle, Wisconsin, 552
Calamity Peak Area, South Dakota, 492
Calaveras County, California, 271, 274
Calloway Mine, Tennessee, 499
Cameron, Arizona, 333
Camp Bird, Colorado, 319
Campbell Chibougamau Mine, Quebec, 115
Campbell Mine, Arizona, 223
Campbell Mine, Northwest Territories, 55
Campbell Red Lake Mine, Ontario, 93
Canadian Dyno Mine, Ontario, 66
Canadian Malartic Mine, Quebec, 113
Cananea, Sonora, Mexico, 174
Canavieiras Mine, Brazil, 588
Cantera Mine, Zacatecas, Mexico, 201
Captain Deposit, New Brunswick, 42
Carahuacra, Peru, 604
Cardiff Township, Ontario, 65
Cardigan Quadrangle, New Hampshire, 427
Carlin, Nevada, 399
Carlos Francisco, Casapalca, Peru, 606
Carlow Township, Ontario, 64
Carrizo Area, Arizona, 326
Casa Diablo Mountain Quadrangle, California, 265
Casapalca, Peru, 606
Castle Dome, Arizona, 227
Cat Creek, Manitoba, 29
Catalina Section, Morococha, Peru, 617
Catas Manto, Yauricocha, Peru, 620
Catavi, Bolivia, 566
Catfish Creek Area, Iowa, 551
Cave in Rock, Illinois, 345
Cedar Bay Mine, Quebec, 115
Cedar Mountain, Utah, 334
Central Black Hills, South Dakota, 492
Central City-Idaho Springs, Colorado, 296
Central Mining District, New Mexico, 445
Central Patrica Gold Mine, Ontario, 88
Central Peru (General), Peru, 601
Cerro Bolivar, Venezuela, 622
Cerro de Mercado, Durango, Mexico, 179
Cerro de Pasco, Peru, 608
Cerro Rico de Potosí, Bolivia, 571
Chamberlin Creek Barite Deposit, Arkansas, 259
Chañarcillo, Chile, 594
Chandos Township, Ontario, 66
Cherokee Mine, Tennessee, 500
Cheverie Area, Nova Scotia, 58
Chewelah Quadrangle, Washington, 548
Chibougamau Explorers Mine, Quebec, 115
Chibougamau-Opemiska, Quebec, 115
Chicago Creek, Colorado, 305
Chief Oxide-Burgin Area, Utah, 517
Chilchinbito Quadrangle, Arizona, 332
Chisel Lake, Manitoba, 36
Chitina Valley, Alaska, 211
Christmas, Arizona, 229
Christopher Silver Mine, Ontario, 71
Chuquicamata, Chile, 597

Chuska Mountains, Arizona, 326
Circle Cliffs Anticline, Utah, 334
Clarines Mine, Mexico, 191
Clark County, Nevada, 415
Clear Lake Area, California, 289
Clifton, Arizona, 243
Climax, Colorado, 299
Clinton County, New York, 460
Cobalt, Ontario, 70
Cochenour Willans Mine, Ontario, 92
Coeur d'Alene, Idaho, 339
Colfax Quadrangle, California, 262
Colorada Pipe, Sonora, Mexico, 174
Colorado Plateau (General), Colorado-Utah-New Mexico-Arizona, 323
Colquijirca, Peru, 611
Comstock Lode, Nevada, 400
Con Mine, Northwest Territories, 56
Con-Rycon Mine, Northwest Territories, 56
Concepción del Oro-Providencia, Zacatecas, Mexico, 182
Conception Bay Area, Newfoundland, 45
Congonhas District, Brazil, 580
Copper Cities, Arizona, 231
Copper King Mine, Arizona, 220
Copper King Mine, California, 272
Copper Mountain, British Columbia, 11
Copper Mountain Mine, Quebec, 121
Copper Queen Mine, Bagdad, Arizona, 220
Copper Queen Mine, Bisbee, Arizona, 223
Copperopolis, California, 272
Cordero, Nevada, 403
Cornwall, Pennsylvania, 487
Corocoro, Bolivia, 573
Coronation Mine, Saskatchewan, 153
Cortlandt Complex, New York, 464
Cottonwood-American Fork, Utah, 513
Craigmont, British Columbia, 14
Cranberry Quadrangle, North Carolina-Tennessee, 470
Cranbrook Map-Area, British Columbia, 26
Creede, Colorado, 319
Creighton Mine, Ontario, 101
Cripple Creek, Colorado, 302
Crittenden County, Kentucky, 346
Crow River Area, Ontario, 88
Crystal Mine, Montana, 391
Cuba City Quadrangle, Wisconsin, 552
Cuprus Mine, Manitoba, 34
Curlew Quadrangle, Washington, 547
Cuyuna, Minnesota, 363

Daisy Mine, Arizona, 245
Daly-Ontario Fissure Zone, Utah, 528
Daniel Township, Quebec, 129
Darwin, California, 266
Deer Flat Area, Utah, 332
Dekoven Quadrangle, Illinois, 345
Delano Peak Quadrangle, Utah, 523
Delbridge Deposit, Quebec, 130
Denison-Waters Area, Ontario, 97
Dike-Eaton Area, Kentucky, 347
Dillon, Montana, 390
Disappointment Valley, Colorado, 324
Disraeli Map-Area, Quebec, 142
Dodger Ore Body, Salmo, British Columbia, 19
Dodgeville Quadrangle, Wisconsin, 550
Dolores County, Colorado, 316
Dome Mine, Ontario, 90
Dome Township, Ontario, 92
Don Jon Mine, Manitoba, 34
Dover, New Jersey, 431
Downieville, California, 262
Dragoon Quadrangle, Arizona, 238
Dubuque North Quadrangle, Iowa, 552
Dubuque South Quadrangle, Iowa, 550
Ducktown, Tennessee, 498
Dungannon Township, Ontario, 64
Dunmore Mine, Colorado, 319
Dunton Pegmatite, Maine, 350

Eagle Mine, Colorado, 308
Eagle Mountain, California, 268
East Calhoun Mine, Colorado, 297
East Shasta District, California, 286
East Tennessee Mine, Tennessee, 500
East Tennessee Zinc District, Tennessee, 501
East Tintic, Utah, 517
Eastern Mesabi, Minnesota, 366
Eastern Townships, Quebec, 118
Edgehill Quadrangle, Missouri, 381
Edwards, New York, 453
Egan Chute, Ontario, 64
El Bote Mine, Zacatecas, Mexico, 201
El Dorado County, California, 274
El Pao, Venezuela, 622
El Potosí Mine, Chihuahua, Mexico, 197
El Tiro Mine, Arizona, 254
Eldorado Mine, Northwest Territories, 51
Electric Point Mine, Washington, 544
Elisa Mine, Sonora, Mexico, 174
Elizabeth, Vermont, 535
Elk Ridge Area, Utah, 331
Elkhorn Pegmatite, South Dakota, 493
Elliot Lake, Ontario, 67
Ely, Minnesota, 360
Ely, Nevada, 404

Emerald Ore Body, Salmo, British Columbia, 19
Emory Creek, British Columbia, 24
Empire Mine, California, 280
Endako, British Columbia, 12
Engels and Superior Mines, California, 269
Enterprise Mine, Colorado, 316
Errington Township, Ontario, 79
Esmeralda Mine, Chihuahua, Mexico, 191
Esperanza Mine, Arizona, 245
Etta Spodumene Mine, South Dakota, 492
Euclid Lake, Manitoba, 30
Eureka, Nevada, 407
Eureka Gulch Area, Colorado, 297
Eureka Mine, Tennessee, 500
Eureka Quadrangle, Utah, 532
Eustis Mine, Quebec, 118

Falconbridge Mine, Ontario, 98
Faraday Township, Ontario, 65
Feeney Ore Body, British Columbia, 19
Fiedmont Map-Area, Quebec, 124
File-Tramping Lakes Area, Manitoba, 36
Fiskenaesset, Greenland, 159
Flat Gap Mine, Tennessee, 502
Flat River, Missouri, 382
Flavrian Lake Area, Quebec, 135
Flin Flon, Manitoba, 34
Foothill Copper Belt, California, 271
Four Corners, Colorado Plateau, 326
Fourmile Area, South Dakota, 492
Fournier, Quebec, 113, 123
Fox Lake, Manitoba, 32
Franklin-Sterling, New Jersey, 434
Fraser Lake, Manitoba, 31
Fredericktown, Missouri, 381
Freeland-Chicago Creek, Colorado, 305
Fresno County, California, 284
Frisco Mine, Chihuahua, Mexico, 191
Front Range Mineral Belt, Colorado, 297
Frood Mine, Ontario, 98

Gabbs, Nevada, 410
Galena Hill, Yukon Territory, 154
Galena Mine, Idaho, 339
Gallup, New Mexico, 325
Gap Nickel Mine, Pennsylvania, 496
Garfield County, Colorado, 323
Garnet Ridge, Arizona, 326
Garon Lake Mine, Quebec, 128
Garrison Township, Ontario, 86
Gaspé, Quebec, 120
Geco Mine, Ontario, 81
Gem Stock, Idaho, 339
Gertrudis Section, Morococha, Peru, 617
Ghost Lake, Manitoba, 37
Giant Mascot Mines, British Columbia, 24
Giant Yellowknife Mine, Northwest Territories, 54
Gilman, Colorado, 307
Gilpin County, Colorado, 297
Gilsum Area, New Hampshire, 428
Glamorgan Township, Ontario, 65
Globe-Miami, Arizona, 231
Gold Hill Mining District, Colorado, 292
Goldfield, Nevada, 412
Goldfields, Saskatchewan, 149
Goodsprings, Nevada, 415
Gore Mountain, New York, 458
Goudreau-Lockalsh Area, Ontario, 83
Gowganda, Ontario, 73
Grafton-Keene, New Hampshire, 427
Granada Mine, Quebec, 135
Granisle Mine, British Columbia, 11
Granite-Bimetallic Mine, Montana, 392
Granite Mountain Area, Utah, 520
Grant County, New Mexico, 445
Grant County, Wisconsin, 551
Grants District, New Mexico, 324
Grants Pass, Oregon, 480
Grass Valley, California, 279
Great Bear Lake, Northwest Territories, 51
Green River, Utah, 333
Grey Dawn Mine, Utah, 327
Greyhawk Mine, Ontario, 60
Ground Hog Mine, New Mexico, 446
Guanajuato, Guanajuato, Mexico, 185
Guibord Township, Ontario, 86
Gunnar "A" Ore Body, Saskatchewan, 150

Haig Township, Quebec, 147
Haliburton, Ontario, 62
Hallnor Mine, Ontario, 89
Halo Mine, Ontario, 66
Hambone Mine, Manitoba, 38
Hamme, North Carolina, 468
Hancock Quadrangle, Michigan, 355
Hanover, New Mexico, 445
Happy Jack Mine, Utah, 332
Hardin County, Illinois, 345
Hard Rock Mine, Ontario, 79
Hardy Mine, Ontario, 99
Harker Township, Ontario, 86
Harricanaw Region, Quebec, 113
Hasaga Mine, Ontario, 94
Hawthorne Quadrangle, Nevada, 410, 425

Hayden Creek, Missouri, 381
H. B. Mine, Salmo, British Columbia, 20
Heath Steele Deposit, New Brunswick, 41
Hedley, British Columbia, 22
Helen Beryl Pegmatite, South Dakota, 493
Helen Mine, Ontario, 83
Henry Mountains-Green River, Utah, 333
Herb Lake, Manitoba, 37
Hercules Mine, Idaho, 339
Heron Bay-White Lake Area, Ontario, 82
Heyson Township, Ontario, 92
Hicks Dome, Illinois, 345
Hideout No. 1 Uranium Mine, Utah, 332
Highland-Surprise Mine, Idaho, 339
Highland Valley, British Columbia, 13
Hillside Mine, Arizona, 220
Hollinger Mine, Ontario, 89
Holloway Township, Ontario, 86
Homestake, South Dakota, 495
Hope Silver Mine, Montana, 392
Hopi Buttes, Arizona, 333
Horne Mine, Quebec, 130
Hot Springs Quadrangle, Arkansas, 259
Howe Sound, British Columbia, 9
Howey Mine, Ontario, 92
Huancavelica, Peru, 614
Huanchaca, Bolivia, 577
Hugo Pegmatite, South Dakota, 492
Huntingdon Mine, Quebec, 118
Hutson Mine, Kentucky, 346

Idaho-Maryland Mine, California, 280
Idaho Springs, Colorado, 296
Idarado Mine (Black Bear Vein), Colorado, 320
Ilímaussaq, Greenland, 161
Illinois-Kentucky Fluorspar, Illinois-Kentucky, 345
Indian Molybdenum Deposit, Quebec, 123
Inspiration Mine, Arizona, 228
Iron County, Utah, 520
Iron King, Arizona, 233
Iron Mask, British Columbia, 14
Iron Mountain, California, 286
Iron Mountain, Wyoming, 556
Iron Mountain-Pilot Knob, Missouri, 377
Iron Springs, Utah, 520
Isle-Dieu Township, Quebec, 129
Itabira, Brazil, 579
Ivanhoe, Virginia, 538
Ivanpah Quadrangle, California-Nevada, 276, 415
Ivigtut, Greenland, 164

Jamestown District, Colorado, 292
Jefferson City, Tennessee, 501
Jellicoe Mine, Ontario, 79
Jerome, Arizona, 235
Jersey Deposit, British Columbia, 20
J. J. Mine, Colorado, 323
Jo Dandy Area, Colorado, 323
Johnson Camp, Arizona, 238
Joplin, Missouri, 475
Josephine Mine, Ontario, 83
Juab County, Utah, 517, 529
Julianehaab, Greenland, 162
Juneau-Treadwell, Alaska, 209

Kalamazoo Ore Body, Arizona, 251
Kane Creek, Utah, 326
Karbers Ridge Quadrangle, Illinois, 345
Kaslo Area, British Columbia, 15
Kayenta Quadrangle, Arizona, 332
Keeley Mine, Ontario, 71
Keene, New Hampshire, 427
Kennecott, Alaska, 211
Kennetcook Map-Area, Nova Scotia, 58
Keno Hill, Yukon Territory, 154
Kerr-Addison, Ontario, 75
Keweenaw Point, Michigan, 351
Keymet Mine, New Brunswick, 42
Keystone Deposit, Arizona, 238
Keystone Pegmatite, South Dakota, 492
Kirkland Lake, Ontario, 77
Kississing Lake, Manitoba, 36
Klamath Mountains, California, 286
Knife Lake Area, Minnesota, 372
Knob Lake, Newfoundland-Quebec, 48
Kootenay Arc, British Columbia, 16
Kootenay Chief Ore Body, British Columbia, 18
Kootenay Lake District, British Columbia, 18

Labrador Trough, Newfoundland-Quebec, 46
Lac Bazil, Quebec, 48
Lac la Trêve Area, Quebec, 115
Lacorne, Quebec, 122
Laguna District, New Mexico, 325
Lake Athabasca Region, Saskatchewan, 149
Lake Carheil, Quebec, 47
Lake Cinch Mine, Saskatchewan, 151
Lake City, Colorado, 322
Lake Shore Mine, Ontario, 77
Lake Superior Copper, Michigan, 351

Lake Superior Iron Ranges, Minnesota, 358
Lake Waters, Saskatchewan, 151
Lamaque Mine, Quebec, 146
Lamartine Area, Colorado, 305
La Motte Map-Area, Quebec, 123
La Plata, Colorado, 310
La Sal Quadrangle, Utah, 326
Lancaster Gap, Pennsylvania, 490
Langis Mine, Ontario, 72
Laramie Range, Wyoming, 556
Laramie-Sherman Quadrangle, Wyoming, 556
Larder Lake, Ontario, 75, 77
Lark Mine, Utah, 511
Lead, South Dakota, 495
Leadbelt (Southeast Missouri), Missouri, 380
Leadville, Colorado, 312
Leaf River Map-Area, Quebec, 48
Lebanon Quadrangle, Pennsylvania, 487
Ledge Lake Area, Manitoba, 35
Leinster-Bowell Area, Ontario, 97
Lemhi Pass, Idaho-Montana, 343
Levack Mine, Ontario, 99
Levias-Keystone Area, Kentucky, 347
Liberty Open Pit, Nevada, 405
Linchburg Ore Body, New Mexico, 439
Lincoln County, Nevada, 422
Linden Quadrangle, Wisconsin, 551
Lisbon Valley, Utah, 327
Little Cottonwood Canyon, Utah, 514
Little Long Lac, Ontario, 79
Llallagua-Uncia (Catavi), Bolivia, 566
Londonderry Map-Area, Nova Scotia, 58
London Mine, Tennessee, 500
Lost Creek Barite Mine, Tennessee, 502
Lost River, Alaska, 213
Louvicourt Township, Quebec, 147
Lucky Friday Mine, Idaho, 339
Lucky Jim Mine, British Columbia, 22
Lukachukai Mountains, Arizona, 326
Lynn Lake, Manitoba, 31
Lyon Mountain, New York, 460

Mac Lead-Cockshutt Mine, Ontario, 80
Madsen Red Lake Mine, Ontario, 93
Magdalena, New Mexico, 439
Magma Mine (Superior), Arizona, 240
Magnet Cove, Arkansas, 258
Magnolia District, Colorado, 292
Malartic, Quebec, 112
Malartic Goldfields Mine, Quebec, 113
Malartic-Haig Region, Quebec, 114
Mammoth Mine, Arizona, 250
Mammoth Mine, California, 286
Mandy Mine, Manitoba, 33
Manitouwadge, Ontario, 81
Marbridge, Quebec, 122, 125
Mariposa County, California, 273
Martin Lake Map-Area, Saskatchewan, 149
Mary Mine, Tennessee, 500
Marysvale, Utah, 523
Marysville, California, 279
Mascoma Quadrangle, New Hampshire, 427
Mascot-Jefferson City, Tennessee, 501
Mascota Manto, Yauricocha, Peru, 620
Matagami, Quebec, 128
Matagente, Cerro de Pasco, Peru, 608
Matheson Ultrabasic Belt, Ontario, 86
Mattagami Lake Mines, Quebec, 128
Mayflower Lode, Utah, 526
Mayflower-Pearl Fissure Zone, Utah, 528
Mayo Area, Yukon Territory, 154
Mayo Township, Ontario, 64
McCalls Ferry Quadrangle, Pennsylvania, 490
McCarthy District, Alaska, 211
McCool Township, Ontario, 86
McDonald Copper Mine, Quebec, 118
McElroy Township, Ontario, 59
McIntyre Mine, Ontario, 89
McKenzie Red Lake Mine, Ontario, 93
McKim Mine, Ontario, 97
McVeigh Lake Area, Manitoba, 31
McWatters Mine, Quebec, 135
Mesa County, Colorado, 323
Mesabi, Minnesota, 368
Mesters Vig, Greenland, 166
Metaline, Washington, 544
Mi Vida Mine, Utah, 326
Miami, Arizona, 227
Miami, Oklahoma, 475
Michigan Copper, Michigan, 351
Michipicotan, Ontario, 83
Mifflin Quadrangle, Wisconsin, 552
Miller Lake, Ontario, 74
Millstream River Area, New Brunswick, 43
Minas Gerais, Brazil, 579
Mineral Hill Mine, Arizona, 245
Mineral King Mine, British Columbia, 20
Mineral Point District, Colorado, 319
Mineral Point Quadrangle, Wisconsin, 550
Minnesota Iron Ranges (General), Minnesota, 358

Mission Copper Deposit, Arizona, 245
Mississippi Valley Type (General), 373
Moab District, Utah, 327
Moak Lake, Manitoba, 37
Monmouth Township, Ontario, 65
Mono Lake, California, 264
Monteagle Township, Ontario, 64
Monteregian Hills, Quebec, 136
Montezuma Canyon, Utah, 326
Montfort Quadrangle, Wisconsin, 551
Montgomery Mine, Quebec, 131
Monticello-Moab-Thompson District, Utah, 326
Montrose County, Colorado, 323
Monument Valley, Utah, 332
Morenci, Arizona, 243
Morococha, Peru, 616
Morro do Vento, Brazil, 588
Morro Velho-Raposos, Brazil, 585
Mosquito Range, Colorado, 312
Mother Lode, California, 273
Mouat Mine, Montana, 395
Moulton Hill Deposit, Quebec, 119
Mount Hope Mine, New Jersey, 431
Mount Wright, Quebec, 47
Mountain City, Nevada, 417
Mountain Pass, California, 276
Munro, Ontario, 86
Murray, Idaho, 340
Murray Deposit, New Brunswick, 41
Mystery Lake Mine, Manitoba, 38

Nama Creek Mine, Ontario, 82
Needle Mountain, Quebec, 121
Negus Mine, Northwest Territories, 56
Nelson and Amherst Counties, Virginia, 541
Nelson Area (General), British Columbia, 15
Nevada City-Grass Valley, California, 79
New Almadén, California, 281
New Cornelia Mine, Arizona, 218
New Diggings Quadrangle, Wisconsin, 552
New Idria, California, 284
New Market Mine, Tennessee, 502
New Quebec, Quebec, 47
New York Hill Mine, California, 280
Newcastle, New Brunswick, 41
Newry, Maine, 349
Nickel Mountain Mine, Oregon, 484
Nickel Plate (Hedley), British Columbia, 22
Nigadoo Deposit, New Brunswick, 41
Nizina District, Alaska, 211
Noranda, Quebec, 130
Noranda Gold, Quebec, 135
Normetal Mine, Quebec, 130
North Fork Quadrangle, Idaho, 343
North Hastings Area, Ontario, 66
North Range, Cuyuna District, Minnesota, 363
North Spirit Lake Area, Ontario, 88
North Star Mine, California, 280
North Star Mine, Manitoba, 34
Northern Black Hills, South Dakota, 495
Notre Dame Bay Area, Newfoundland, 43
Nye County, Nevada, 410

O'Brien Mine, Ontario, 74
O'Brien Mine, Quebec, 113
Oka, Quebec, 136
Old Dick Mine, Arizona, 220
Old Tennessee Mine, Tennessee, 500
Omega Mine, Ontario, 75
Ontonagon County, Michigan, 355
Opalite Mine, Oregon, 403
Opemiska, Quebec, 115
Oquirrh Range, Utah, 510
Orange County, Vermont, 535
Orchan Mines, Quebec, 128
Ore Knob, North Carolina, 470
Oreana, Nevada, 419
Oruro, Bolivia, 568
Orvan Brooks Deposit, New Brunswick, 42
Osborne Lake, Manitoba, 37
Ouray, Colorado, 319
Oxide Mine, Arizona, 254

Pachuca-Real del Monte, Hidalgo, Mexico, 187
Pacific Nickel Property, British Columbia, 24
Pailaviri Section, Potosí, Bolivia, 571
Palmilla Mine, Chihuahua, Mexico, 192
Palo Verde Mine, Arizona, 245
Paradise Range, Nevada, 410
Park City, Utah, 526
Parral, Chihuahua, Mexico, 191
Paulden Quadrangle, Arizona, 233
Pearl Lake, Ontario, 90
Peekskill, New York, 465
Peerless Pegmatite, South Dakota, 493
Pembroke, Hants County, Nova Scotia, 58
Pend Oreille Mine, Washington, 544
Pennsylvania Mine, California, 280
Perron Mine, Quebec, 146
Pewabic Mine, New Mexico, 447
Philipsburg, Montana, 392
Phoenix Mine, British Columbia, 11
Picher, Oklahoma, 475
Pickle Crow, Ontario, 88
Pilot Knob, Missouri, 377

Pima District, Arizona, 245
Pine Creek Mine, California, 264
Pine Point, Northwest Territories, 52
Piney River, Virginia, 541
Pinto Mining District, Utah, 520
Pioche, Nevada, 421
Pipe Mine, Manitoba, 38
Piute County, Utah, 523
Platoro, Colorado, 319
Plattville Quadrangle, Wisconsin, 550
Plumas County, California, 269
Poland, Maine, 349
Polaris Mine, Idaho, 341
Polk County Mine, Tennessee, 500
Pope County, Illinois, 345
Porcupine, Ontario, 89
Port Radium, Northwest Territories, 51
Potash Sulphur Springs, Arkansas, 259
Potosí, Bolivia, 571
Potosi Quadrangle, Missouri, 381
Potosi Quadrangle, Wisconsin-Iowa, 552
Poughkeepsie District, Colorado, 319
Powell River Area, Tennessee, 501
Pozo Rico Section, Yauricocha, Peru, 620
Preissac Area, Quebec, 123
Prescott Quadrangle, Arizona, 233
Presidio County, Texas, 507
Presqu'ile Reef, Northwest Territories, 53
Preston East Dome Mine, Ontario, 89
Pride of Emory, British Columbia, 24
Princeton Map-Area, British Columbia, 7, 9
Prosperous Lake, Northwest Territories, 57
Providencia, Zacatecas, Mexico, 182
Pulacayo (Huanchaca), Bolivia, 577
Pyramid Ore Bodies, Northwest Territories, 53

Quadrilátero Ferrífero, Brazil, 579
Quail Hill, California, 272
Quarrysville Quadrangle, Pennsylvania, 490
Quemont Mine, Quebec, 130
Quesabe Mine, Quebec, 135
Questa, New Mexico, 442
Quirke Lake, Ontario, 67

Rainy Day Mine, Utah, 333
Rancagua, Chile, 591
R and S Molybdenum Mine, New Mexico, 442
Raposos Mine, Brazil, 585
Ray, Arizona, 248
Real del Monte, Hidalgo, Mexico, 187
Red Cliff, Colorado, 307
Redding, California, 286
Red Indian Lake, Newfoundland, 43
Red Lake, Ontario, 91
Red Mountain, Colorado, 315
Red River District, New Mexico, 442
Reed-Wekusko Map-Area, Manitoba, 36
Reeves Macdonald Mine, British Columbia, 19
Renfrew County, Ontario, 63
Repton Quadrangle, Illinois, 345
Republic, Washington, 546
Rewey Quadrangle, Wisconsin, 552
Richard Mine, New Jersey, 432
Richardson Deposit, Ontario, 66
Rico, Colorado, 315
Riddle, Oregon, 483
Rifle Area, Colorado, 323
Ringwood, New Jersey, 431
Rix Athabasca Mine, Saskatchewan, 150
Robinson District, Nevada, 404
Romaine River Area, Quebec, 111
Roseland District, Virginia, 541
Rosiclare, Illinois, 345
Rosita Hills District, Colorado, 319
Rossland, British Columbia, 16, 18
Round Valley, California, 264
Rouyn, Quebec, 113, 131
Rouyn-Bell River Area, Quebec, 113
Ruby Creek, Alaska, 216
Ruggles Mine, New Hampshire, 429
Ruttan Lake Mine, Manitoba, 32
Rycon Mine, Northwest Territories, 56

St. George Deposit, Arizona, 238
St. Joseph du Lac, Quebec, 137
St. Louis County, Minnesota, 360
St. Urbain, Quebec, 140
Salem Quadrangle, Kentucky, 346
Salmo, British Columbia, 16, 19
Salmo-Sheep Creek (General), British Columbia, 19
Salmon Region, Idaho, 343
San Antonio Mine, Chihuahua, Mexico, 197
San Benito County, California, 284
San Bernardino County, California, 276
San Francisco del Oro Area, Chihuahua, Mexico, 191
San Francisco Section, Morococha, Peru, 618
San Isidro, Venezuela, 622
San José, San Luis Potosí, Mexico, 195
San Juan, Colorado, 317

San Juan Basin, New Mexico, 325
San Manuel, Arizona, 250
San Miguel County, Colorado, 320
San Rafael-Cedar Mountain, Utah, 334
San Xavier Mine, Arizona, 245
Sandon, British Columbia, 22
Sanford Lake, New York, 462
Santa Barbara, Chihuahua, Mexico, 191
Santa Clara County, California, 282
Santa Eulalia, Chihuahua, Mexico, 196
Santa Rita-Hanover, New Mexico, 445
Santa Rosalia Area, Baja California, Mexico, 171
Schist Lake Mine, Manitoba, 34
Scrub Oaks Mine, New Jersey, 432
Serra de Jacobina, Brazil, 588
Sevier Quadrangle, Utah, 523
Seward Peninsula, Alaska, 213
Shabogamo Lake Map-Area, Newfoundland-Quebec, 47
Shanton Iron-Ore Property, Wyoming, 556
Shasta County, California, 286
Sheep Creek, British Columbia, 16, 21
Sherman Area, Colorado, 315
Sherritt Gordon, Manitoba, 35
Sherritt Gordon-Flin Flon (General), Manitoba, 33
Shiprock District, Utah-Arizona, 326
Shoshone County, Idaho, 339
Shubenacadie Map-Area, Nova Scotia, 58
Shullsburg Quadrangle, Wisconsin, 552
Shungnak Quadrangle, Alaska, 216
Sierrita Area, Arizona, 245
Sigma Mine, Quebec, 146
Silver Bell, Arizona, 252
Silver Belt, Idaho, 340
Silver City, Nevada, 400
Silver City Quadrangle, New Mexico, 446
Silver Cliff District, Colorado, 319
Silver Reef District, Nevada, 424
Silver Summit Mine, Idaho, 340
Silverfields Silver Deposit, Ontario, 72
Silverton, Colorado, 319
Siscoe Mine, Quebec, 146
Sixteen to One Mine, California, 262
Sladen Malartic Mine, Quebec, 113
Slick Rock District, Colorado, 324
Slocan, British Columbia, 16, 21
Smartsville Quadrangle, California, 272
Sneffels Mine, Colorado, 318
Snow Lake, Manitoba, 36
Soab Mine, Manitoba, 38
Sonora Quadrangle, California, 274
Soudan Formation, Minnesota, 371
Sourdough Hill, Yukon Territory, 154
Southeast Missouri, Missouri, 381
Southern Illinois Fluorspar, Illinois, 345
South Lorraine Area, Ontario, 71
Spar City District, Colorado, 319
Spor Mountain, Utah, 529
Spruce Pine, North Carolina, 472
Stall Lake, Manitoba, 37
Star Lake, New York, 456
Star Mine, Idaho, 340
Steamboat Springs Area, Nevada, 401
Steep Rock Lake, Ontario, 94
Sterling Hill, New Jersey, 435
Sterling Lake, New York, 431
Stevens County Magnesite, Washington, 548
Stillwater Complex, Montana, 394
Stobie Mine, Ontario, 101
Strafford Township, Vermont, 535
Strathcona Mine, Ontario, 98
Sturgeon River District, Ontario, 79
Sudbury, Ontario, 97
Suffield Mine, Quebec, 118
Sugar Loaf District, Colorado, 293
Sullivan, British Columbia, 25
Sulphide Queen Body, California, 278
Sulphur Bank, California, 289
Summitville, Colorado, 319
Sunshine Mine, Idaho, 339
Superior, Arizona, 240
Superior Mine, California, 270
Sussex County, New Jersey, 432
Sweetwater, Tennessee, 505

Taos County, New Mexico, 442
Teck Township, Ontario, 77
Teller County, Colorado, 302
Telluride, Colorado, 318
Temple Mountain, Utah, 334
Terlingua, Texas, 507
Thetford-Black Lake, Quebec, 142
Thirteenth Lake, New York, 458
Thomas Range, Utah, 529
Thompson District, Utah, 327
Thompson-Moak Lake, Manitoba, 37
Timagami Island, Ontario, 108
Timmins, Ontario, 89
Tin Mountain Pegmatite, South Dakota, 492
Tin Mountain Spodumene Mine, South Dakota, 493
Tintic, Utah, 531
Tintic Standard Mine, Utah, 517
Tinton District, South Dakota, 493
Tisdale Township, Ontario, 89
Tombill Mine, Ontario, 79

Tombstone, Arizona, 254
Tonopah, Nevada, 424
Treadwell Mine, Alaska, 209
Tri-State, Oklahoma-Kansas-Missouri, 474
Tungsten Hills, California, 264
Turk Magnesite Deposit, Washington, 548
Tushar Uranium Area, Utah, 523
Twin Buttes Mine, Arizona, 245
Tyrone, New Mexico, 446

Uncia, Bolivia, 566
Uncompahgre District, Colorado, 318
Ungava Peninsula, Quebec, 47
United Keno Hill Mines, Yukon Territory, 155
United States Mine, Utah, 511
United Verde Extension Mine, Arizona, 235
United Verde Mine, Arizona, 235
Upper Canada Mine, Ontario, 78
Upper Mississippi Valley, Wisconsin-Illinois-Iowa, 550
Upper Uncompahgre District, Colorado, 319
Upper York River Map-Area, Quebec, 121
Uranium City, Saskatchewan, 151
Uravan District, Colorado, 323
Utah Copper Pit, Utah, 511
Utah County, Utah, 517, 532

Val d'Or, Quebec, 123, 146
Vance County, North Carolina, 468
Venezuelan Guayana, Venezuela, 622
Verde Central Mine, Arizona, 235
Verdure Quadrangle, Utah, 326
Vermilion, Minnesota, 371
Verna Mine, Saskatchewan, 149
Vilburnam-Salem Area, Missouri, 382
Violamac Mine, Slocan, British Columbia, 21
Vipond Mine, Ontario, 89
Virginia City, Nevada, 400
Virginia City Quadrangle, Nevada, 401

Wabana, Newfoundland, 44
Wabush Lake, Quebec, 47
Waite Amulet Mine, Quebec, 131
Wakuach Lake Map-Area, Quebec, 46
Walker Mine, California, 270
Wallace, Idaho, 339
Walton, Nova Scotia, 58
Warren District, Arizona, 223
Warwick Map-Area, Quebec, 142
Washington County, Utah, 520
Waswanipi Area, Quebec, 115, 128
Weaversville, California, 286
Wedge Mine, New Brunswick, 41
Weedon Mine, Quebec, 118
Weldon Bay Map-Area, Manitoba, 33
West San Xavier Mine, Arizona, 245
West Shasta District, California, 286
West Tintic Mining District, Utah, 532
Weston Pass, Colorado, 312
Westside-Webber Mine, Oklahoma, 475
Wharton Ore Belt, New Jersey, 432
White Canyon-Elk Ridge, Utah, 331
White Horse Alunite Deposit, Utah, 523
White Pine, Michigan, 355
White Pine Canyon, Utah, 513
Willecho Mine, Ontario, 82
Willroy Mine, Ontario, 82
Wilson Springs Vanadium Deposits, Arkansas, 259
Wood Mine, Colorado, 297
Woodrow Mine, New Mexico, 325
Woodsville Quadrangle, Vermont-New Hampshire, 535
Wrangell Mountains Area, Alaska, 211
Wright-Hargreaves Mine, Ontario, 77

Yauricocha, Peru, 619
Yellowknife, Northwest Territories, 54
Yellow Pine, Nevada, 415
York River, Ontario, 63
York Tin Region, Alaska, 213
Young Mine, Tennessee, 502

Zacatecas, Zacatecas, Mexico, 200
Zenith Mine, Minnesota, 372

INDEX OF DEPOSITS ACCORDING TO AGE OF MINERALIZATION

Few geologists would deny the difficulty of fixing accurately the age of formation of the average epigenetic ore deposit. In part the problem of such dating lies in the unraveling of the stratigraphic puzzle of formations in which any given ore deposit is found. But even if the various strata within an area can be dated relative to each other, there is no guarantee that they can be assigned an absolute age. The picture may be further complicated by the presence of intrusive rock masses, to any one of which the mineralization of the region may be genetically related. Granted, however, that the age relations of igneous to sedimentary rocks can be deciphered, it still remains to determine from which parent magma the ore fluids came and at what time interval after or before their particular igneous sibling they arrived at the site of deposition. Still another factor is the relations of the ore bodies to earth movements which must themselves be fixed in time if their recognition is not to make even more complex the basic problem of ore age.

In the more than 13 years since the latest references in Memoir 75 were published, much work has been done to employ data as to isotope ratios in the dating of geologic events. On the assumption that radioactive decay of such elements as exhibit this property is constant, models can be erected that make it possible to estimate the time that has elapsed since the mineral or whole rock on which the determinations were made was formed. Unfortunately, complicating factors exist that make such model-building more complex than the simple assumption of constant rates of decay alone would require. It is necessary to assume either (1) that conditions were such that, when the mineral or rock in question was formed, none of the decay-produced isotope was present, or (2) that if some of the decay-produced isotope was present, its abundance at that time can be estimated. Further, it must be presumed either (1) that none of the decay-produced isotope or of the parent isotope has been lost or (2) that the loss can be estimated with considerable accuracy. Because the entire process of age estimation in this manner requires that the exact amount of the isotope under consideration be known, the loss of any amount of the daughter, while the parent isotope is not lost at all or is lost at a different rate, would give untrue results.

The first work in age dating by the use of isotope ratios was done on leads derived from the decay of uranium^{235} to Pb^{207} and uranium^{238} to Pb^{206}, the ratios normally used being Pb^{206}/U^{238}, Pb^{207}/U^{235}, and Pb^{207}/Pb^{206}. The ages obtained from these various isotope pairs, for a given deposit, normally are different enough to cast doubt on the value of the models now used to convert these ratios into absolute ages. One of the probable causes of age differences is the much shorter half-life of radon produced in the disintegration of U^{235} than in that of U^{238}. Thus, radon leakage is more possible in the U^{238}-Pb^{206} chain, with proportionally lower production of Pb^{206} in comparison with that of Pb^{207}. If thorium also is present, the Pb^{208}/Th^{232} ratio can be used, but this does not seem to help much. These uncertainties in lead-isotope-derived ages are most unfortunate for the economic geologist since uranite and pitchblende are ore minerals, while the elements used in the methods mentioned below often are contained in rocks not certainly related genetically to the ore in question.

On the other hand, the decay of potassium^{40} to argon^{40} has been widely used on potassium-bearing minerals in silicic rocks or on the whole rocks themselves. Potassium-bearing minerals are common in association with many types of ore deposits and sericite, and other potassium-bearing micas, and are concomitantly formed with the ore minerals of many ore bodies. The major objection to this method is that Ar^{40} is a gas and appears to be lost from some potassium-bearing minerals more readily than others, so that the student is again faced with the problem of choosing among alternate answers. To some extent, this problem can be met by working with whole-rock samples on the assumption that the argon does not travel far.

Finally the disintegration of strontium to rubidium has the advantage of starting and finishing with products that remain in the solid state. Strontium, however, is not a common element in ore deposits, either in gangue or wall-rock alteration minerals. Ages obtained by this method largely are obtained from silicic igneous rocks that normally can be assumed but not proved to be of the same, or essentially the same, age as the ore deposit in the vicinity.

In short, age determinations made from model-employing isotopes pairs, one of which is produced by radioactive decay of the other, directly or as the result of a sequence of disintegrations, seem to give no completely certain answers. In this volume, however, much use has been made of K/Ar determination and as much as is possible of Sr/Rb ones, but when these conflict with other items of geologic evidence, this has been pointed out.

For the economic geologist wishing to study this problem in detail, the works listed in this index are suggested (p. 656).

Considering, therefore, the many places at which the chain of facts and reasoning from facts on which the dating of any ore body depends can be broken, it well may be thought remarkable that any geologist ever feels confident enough of the soundness of each link in his work to say without equivocation that, for example, "this ore deposit was formed in the earliest Eocene." An examination of the literature will quickly show that few geologists are ever so certain of the full geologic story to state so definitely the time of deposition of the mineral deposit on which they are reporting. Nevertheless, so much thought has been given to the problem of the age of rocks and of the ores they contain that it has been possible from published data to estimate with considerable assurance the general age of each of the deposits included in these bibliographies.

Of the 106 deposits assigned to the Precambrian, 37 are early, 30 middle, and 39 late. This division is quite equal and suggests that, in the time between something less than 3000 m.y. ago and the end of the Precambrian (600 m.y. in the Americas), ore mineralization was no more the property of one of these time divisions than of the others. Of the Precambrian deposits here considered to have been epigenetic in origin, only 11 were formed under conditions less intense than mesothermal, and none was produced from mineralizing solutions undergoing rapid loss of heat and pressure. This situation is, of course, only to be expected in Precambrian terrains where erosion has had from many hundreds of millions to billions of years to strip off those deposits formed in near-surface environments. Thus, it is difficult to compare the numbers of deposits of Precambrian age with those of later times. Nevertheless, almost as many deposits (103) included in this volume were formed in the Americas from late Mesozoic through earliest Tertiary time as in all of the Precambrian combined (106). If it is assumed that the erosion-removed Precambrian ore bodies constitute approximately the same percentage of all Precambrian ore bodies in the Western Hemisphere as that made up by the yet undiscovered and still deeply buried high-intensity deposits of late Mesozoic through early Tertiary time (and this seems as reasonable as any assumption that can be made), then this later time span of about 40 m.y. produced in absolute numbers only slightly fewer ore deposits than all the billions of years of Precambrian time. How this phenomenon is to be explained I do not know, but an explanation eventually will be found.

Of the 231 deposits formed in post-Precambrian time, only 21 are Paleozoic, 118 are Mesozoic and earliest Tertiary, and 92 are early, middle, and late Tertiary. Although further age-dating work may result in changes in the dates assigned to a few of the deposits included here, the proportions in the various age groups probably will remain essentially the same; the changes probably will approximately balance each other. To what extent the numbers of deposits in the different age groups given here represent the actual abundance of ore deposition in a given time span and to what extent these various numbers are the result of the accidents of erosion, on the one hand, and of publication, on the other, it is difficult to say. Yet, from the data presented, a reasonable inference appears to be that the Paleozoic was a time of less impressive and intense development of ore deposits than were the much shorter Mesozoic and Tertiary eras. The widespread areas of exposed Paleozoic rocks in the Americas would appear to offer as much opportunity, at least, as do the outcrop areas of Mesozoic and Tertiary beds for the discovery of ore deposits, yet far fewer appear to have been formed in the earlier era.

ORE DEPOSITS IN THIS VOLUME ARRANGED BY CONDITIONS AND AGES OF FORMATION*

	Magmatic-1a	Magmatic-1b	Magmatic-2a	Magmatic-2b	Magmatic-3a	Magmatic-3b	Magmatic-4	Hypothermal-2	Hypothermal-1	Mesothermal	Leptothermal	Telethermal	Xenothermal	Kryptothermal	Epithermal	Sedimentary-A3	Sedimentary-A1a	Total
Early Precambrian		1		1	1		1		9	15	5					2	2	37
Middle Precambrian			2	5	1				11	3	3						5	30
Late Precambrian	3		1		3	7	3	3	13	3	2	1						39
Early Paleozoic																1		1
Middle Paleozoic	1				2		1	2	8									14
Late Paleozoic					1					1		4						6
Early Mesozoic				1				1		2	1							5
Middle Mesozoic	1	1						2	2	2		1						9
Late Mesozoic (pre-Laramide)	1					1	1	4	6	13	8	7						41
Late Mesozoic to Early Tertiary					1			11	4	27	8	8	2	1	1			63
Early Tertiary								3	4	10	7	2						26
Middle Tertiary								3		8	5		1	3	9			29
Late Tertiary						1		2	1	6	5	3	4	4	11			37
Totals	6	2	3	7	9	9	6	31	58	90	44	26	7	8	21	3	7	337
			Magmatic--42					Hypothermal --89		Mesothermal to Telethermal--160			Xenothermal to Epithermal--36					

*Deposits that were formed under more than one intensity range, that is, hypothermal and mesothermal, are included under each of the applicable categories. Thus, the totals of the various intensity ranges are greater than the total number of bibliographies.

Dalrymple, G. B., and Lamphere, M. A., 1969, Potassium-argon dating, principles, techniques, and applications to geochronology: W. H. Freeman, San Francisco, 258 p.

Doe, B. R., 1970, Lead isotopes: Springer-Verlag, N. Y., 137 p.

Faul, H., 1966, Ages of rocks, planets, and stars: Earth and planetary science series, McGraw-Hill, N. Y., 109 p.

Hamilton, E. I., 1965, Applied geochronology: Academic Press, N. Y., 267 p.

Hamilton, E. I., and Farquhar, R. M., Editors, 1968, Radiometric dating for geologists: Interscience Publishers, John Wiley & Sons, N. Y., 506 p.

Early Precambrian

Adams Mine, Ontario, 59
Bird River, Manitoba, 27
Cadillac, Quebec, 112
Fiskenaesset, Greenland, 159
Kerr Addison, Ontario, 75
Kirkland Lake, Ontario, 77
Lacorne, Quebec, 122
Little Long Lac, Ontario, 79
Malartic, Quebec, 112
Manitouwadge, Ontario, 81
Marbridge, Quebec, 125
Matagami, Quebec, 128
Michipicoten, Ontario, 83
Munro, Ontario, 86
Noranda Gold, Quebec, 135
Pickle Crow, Ontario, 88
Porcupine, Ontario, 89
Red Lake, Ontario, 91
Steep Rock Lake, Ontario, 94
Stillwater Complex, Montana, 394
Val d'Or, Quebec, 146
Venezuelan Guayana, Venezuela (primary), 622
Vermilion, Minnesota, 271
Yellowknife, Northwest Territories, 54

Middle Precambrian

Alexo Mine, Ontario, 60
Bagdad-Massive Sulfides, Arizona, 220
Beaverlodge (Goldfields), Saskatchewan, 149
Black Hills Pegmatites, South Dakota, 491
Blind River, Ontario, 67
Cobalt, Ontario, 70
Coronation Mine, Saskatchewan, 153
Cuyuna, Minnesota (primary), 363
Eastern Mesabi, Minnesota (primary), 366
Flin Flon, Manitoba, 34
Gowganda, Ontario, 73
Homestake, South Dakota (major), 495
Iron King, Arizona, 233
Jerome, Arizona, 235
Labrador Trough, Newfoundland-Quebec, 46
Lynn Lake, Manitoba, 31
Mesabi, Minnesota (primary), 368
Minas Gerais, Brazil (primary), 579
Noranda, Quebec, 130
Sherritt Gordon, Manitoba, 35
Snow Lake, Manitoba, 36
Sudbury, Ontario, 97
Thompson-Moak Lake, Manitoba, 37
Timagami Island, Ontario, 108

Late Precambrian

Allard Lake, Quebec, 111
Balmat-Edwards, New York, 453
Bancroft (Nonmetallics), Ontario, 63
Bancroft (Uranium), Ontario, 65
Benson Mines, New York, 456
Blue Mountain, Ontario, 69
Chibougamau-Opemiska, Quebec, 115
Cuyuna, Minnesota (secondary), 363
Dover, New Jersey, 431
Eastern Mesabi, Minnesota (secondary), 366
Franklin-Sterling, New Jersey, 434
Gore Mountain, New York, 458
Great Bear Lake, Northwest Territories, 51
Ilímaussaq, Greenland, 161
Iron Mountain, Missouri, 377
Iron Mountain, Wyoming, 556
Ivigtut, Greenland, 164
Keweenaw Point, Michigan, 351
Labrador Trough, Newfoundland-Quebec (secondary), 46
Lancaster Gap, Pennsylvania, 490
Lyon Mountain, New York, 460
Mesabi, Minnesota (secondary), 368
Minas Gerais, Brazil (secondary), 579
Morro Velho-Raposos, Brazil, 585
Mountain Pass, California, 276
Nelson and Amherst Counties, Virginia, 541
Pilot Knob, Missouri, 377
St. Urbain, Quebec, 140
Sanford Lake, New York, 462
Serra de Jacobina, Brazil, 588
Sullivan, British Columbia, 25
White Pine, Michigan, 355

Early Paleozoic

Wabana, Newfoundland, 44

Middle Paleozoic

Bathurst, New Brunswick, 41
Buchans, Newfoundland, 43
Cortlandt Complex, New York, 464
Ducktown, Tennessee, 498
Eastern Townships, Quebec, 118
Elizabeth, Vermont, 535
Gaspé, Quebec, 120
Grafton-Keene, New Hampshire, 427
Hamme, North Carolina, 468
Ore Knob, North Carolina, 470
Spruce Pine, North Carolina, 472
Thetford-Black Lake, Quebec, 142

Late Paleozoic

Austinville-Ivanhoe, Virginia, 538
Mascot-Jefferson City, Tennessee, 501
Mountain City, Nevada, 417
Newry, Maine, 349
Pine Point, Northwest Territories, 52
Sweetwater, Tennessee, 505

Early Mesozoic

Cornwall, Pennsylvania, 487
Pride of Emory, British Columbia, 24
Walton, Nova Scotia, 58

Middle Mesozoic

Brenda Lake, British Columbia, 7
Copper Mountain, British Columbia, 11
Endako, British Columbia, 12
Grants Pass, Oregon, 480
Highland Valley, British Columbia, 13
Nickel Plate, British Columbia, 22
Riddle, Oregon (peridotite), 483
Salmo, British Columbia, 19

Late Mesozoic (pre-Laramide)

Alleghany, California, 262
Arkansas Bauxite, Arkansas (Syenite), 256
Bisbee, Arizona, 223
Bishop, California, 264
Britannia Mines, British Columbia, 9
Darwin, California, 266
Eagle Mountain, California, 268
Engles Mine, California, 269
Foothill Copper Belt, California, 271
Grass Valley, California, 279
Illinois-Kentucky Fluorspar, Illinois-Kentucky, 345
Keno Hill, Yukon Territory, 154
Leadbelt (Southeast Missouri), Missouri, 380
Magnet Cove, Arkansas, 258
Metaline, Washington, 544
Mother Lode, California, 273
Nevada City, California, 279
Oka, Quebec, 136
Ruby Creek, Alaska, 216
Shasta County, California, 286
Superior Mines, California, 269
Tri-State Area, Oklahoma-Kansas-Missouri, 474
Upper Mississippi Valley, Wisconsin-Illinois-Iowa, 550

Late Mesozoic to Early Tertiary

Ajo, Arizona, 218
Bagdad-Porphyry Copper, Arizona, 221
Boulder County Tellurides, Colorado, 292
Boulder County Tungsten, Colorado, 294
Butte, Montana, 384
Castle Dome, Arizona, 227
Central City-Idaho Springs, Colorado, 296
Chañarcillo, Chile, 594
Christmas, Arizona, 229
Chuquicamata, Chile, 597
Climax, Colorado, 299
Coeur d'Alene, Idaho, 339
Colorado Plateau (Colorado-Utah-Arizona-New Mexico (major), 323
Dillon, Montana, 390
Ely, Nevada, 404
Eureka, Nevada, 407
Freeland-Chicago Creek, Colorado, 305
Gabbs, Nevada, 410
Gilman, Colorado, 307
Globe-Miami, Arizona, 231
Johnson Camp, Arizona, 238
Leadville, Colorado, 312
Lemhi Pass, Idaho-Montana, 343
Magma Mine (Superior), Arizona, 240
Morenci, Arizona, 243
Oreana, Nevada, 419
Philipsburg, Montana, 392
Pima District, Arizona, 245
Pioche, Nevada, 421
Ray, Arizona, 248
San Juan, Colorado, 317
San Manuel, Arizona (minor), 250
Santa Rita-Hanover, New Mexico, 445
Silver Bell, Arizona, 252
Stevens County Magnesite, Washington, 548
Tombstone, Arizona, 254

Early Tertiary

Ainsworth, British Columbia, 17
Arkansas Bauxite, Arkansas (Bauxite), 256
Bingham, Utah, 510
Bluebell, British Columbia, 17
Cananea, Sonora, Mexico, 174
Cottonwood-American Fork, Utah, 513
East Tintic, Utah, 517
Goodsprings, Nevada, 415
Homestake, South Dakota (minor), 495
Juneau-Treadwell, Alaska, 209
Kennecott, Alaska, 211
Lost River, Alaska (Be), 213
Park City, Utah, 526
Riddle, Oregon (secondary concentration), 483
Rossland, British Columbia, 18
Sheep Creek, British Columbia, 21
Slocan, British Columbia, 21
Tintic, Utah, 531

Middle Tertiary

Braden, Chile, 591
Carahuacra, Peru, 604
Carlin, Nevada, 399
Casapalca, Peru, 606
Cerro de Pasco, Peru, 608
Colquijirca, Peru, 611
Comstock Lode, Nevada, 400
Concepción del Oro-Providencia, Zacatecas, Mexico, 182
Cripple Creek, Colorado, 302
Goldfield, Nevada, 412
Huancavelica, Peru, 614
Iron Springs, Utah, 520
Magdalena, New Mexico, 439
Marysvale, Utah (Alunite), 523
Mesters Vig, Greenland, 166
Morococha, Peru, 616
Republic, Washington, 546
Santa Eulalia, Chihuahua, Mexico, 196
Terlingua, Texas, 507
Tonopah, Nevada, 424
Yauricocha, Peru, 619

Late Tertiary

Aguilar, Argentina, 561
Boleo, Baja California, Mexico, 171
Cerro de Mercado, Durango, Mexico, 179
Colorado Plateau, Colorado (minor), 323
Cordero, Nevada, 403
Corocoro, Bolivia, 573
Guanajuato, Guanajuato, Mexico, 185
La Plata, Colorado, 310
Llallagua-Uncia, Bolivia, 566
Marysvale, Utah (Uranium), 523
New Almadén, California, 281
New Idria, California, 284
Oruro, Bolivia, 568
Pachuca-Real del Monte, Hidalgo, Mexico, 187
Parral, Chihuahua, Mexico, 191
Potosí, Bolivia, 571
Pulacayo, Bolivia, 577
Questa, New Mexico, 442
Rico, Colorado, 315
San José, San Luis Potosí, Mexico, 195
San Juan, Colorado (major), 317
Spor Mountain, Utah, 529
Sulphur Bank, California, 289
Venezuelan Guayana, Venezuela (secondary), 622
Zacatecas, Zacatecas, Mexico, 200

INDEX OF DEPOSITS ACCORDING TO METALS OR MINERALS PRODUCED

The index that follows contains only the principal metals and/or minerals that are produced from the deposits included in this volume. The decision as to which products are by-products rather than principal products of any given deposit has been made arbitrarily; no strict test of dollar value or tonnage has been set for the commodities in question. If, for example, enough gold has been produced from a deposit to cause it to be mentioned in the production statistics, the deposit is included under gold only if the value of the gold is within an order of magnitude of those of the other products. Certain minor constituents such as cadmium, germanium, radium, and selenium have been omitted entirely.

Only the name under which the deposit appears in the bibliography is given in this index. Other names that might be associated with a deposit are to be found in the Alphabetical Index of Deposits. Page references to all deposits listed follow the deposit name.

The numbers of deposits containing the various metals and minerals included in this volume are as follows:

Alunite 1	Corundum 1	Magnesite 2	Residual Kaolin 1
Antimony 1	Cryolite 1	Manganese 4	Silver 75
Asbestos 2	Emery 1	Mercury 6	Sulphur 1
Barite 6	Feldspar 6	Mica 3	Thorium 1
Bauxite 1	Fluorspar 2	Molybdenum 19	Tin 7
Beryllium 6	Garnet 1	Nepheline 2	Titanium 6
Bismuth 3	Gold 63	Nickel 9	Tungsten 6
Brucite 1	Graphite 1	Platinum metals 2	Uranium 8
Chromite 4	Halloysite 1	Pyrites 7	Vanadium 4
Cobalt 6	Iron 25	Rare Alkalis 1	Zinc 67
Columbium 2	Lead 61	Rare Earths 3	
Copper 86	Lithium 3		

This table cannot, of course, be totaled because many of the deposits are mined for two or more of the commodities indexed.

Alunite

Marysvale, Utah, 523

Antimony

San José, San Luis Potosí, Mexico, 195

Asbestos

Munro, Ontario, 86
Thetford-Black Lake, Quebec, 142

Barite

Buchans, Newfoundland, 43
Magnet Cove, Arkansas, 258
Mesters Vig, Greenland, 166
Mountain Pass, California, 276
Sweetwater, Tennessee, 505
Walton, Nova Scotia, 58

Bauxite

Arkansas Bauxite, Arkansas, 256

Beryllium

Black Hills Pegmatites, South Dakota, 491
Grafton-Keene, New Hampshire, 427
Lacorne, Quebec, 122
Lost River, Alaska, 213
Newry, Maine, 349
Spor Mountain, Utah, 529

Bismuth

Cerro de Pasco, Peru, 608
Lacorne, Quebec, 122
Morococha, Peru, 616

Brucite

Gabbs, Nevada, 410

Chromite

Bird River, Manitoba, 27
Fiskenaesset, Greenland, 159
Grants Pass, Oregon, 480
Stillwater Complex, Montana, 394

Cobalt

Cobalt, Ontario, 70
Cornwall, Pennsylvania, 487
Gowganda, Ontario, 73
Leadbelt, Missouri, 380
Sudbury, Ontario, 97
Thompson-Moak Lake, Manitoba, 37

Columbium

Ilímaussaq, Greenland, 161
Oka, Quebec, 136

Copper

Ajo, Arizona, 218
Alexo Mine, Ontario, 60
Bagdad-Massive Sulfides, Arizona, 220
Bagdad-Porphyry Copper, Arizona, 221
Bathurst-Newcastle, New Brunswick, 41
Bingham, Utah, 510
Bisbee, Arizona, 223
Bishop, California, 264
Boléo, Baja California, Mexico, 171
Braden, Chile, 591
Brenda Lake, British Columbia, 7
Britannia Mines (Howe Sound), British Columbia, 9
Buchans, Newfoundland, 43
Butte, Montana, 384
Cananea, Sonora, Mexico, 174
Casapalca, Peru, 606
Castle Dome, Arizona, 227
Cerro de Pasco, Peru, 608
Chibougamau-Opemiska, Quebec, 115
Christmas, Arizona, 229
Chuquicamata, Chile, 597
Coeur d'Alene, Idaho, 339
Colquijirca, Peru, 611
Concepción del Oro-Providencia, Zacatecas, Mexico, 182
Copper Mountain, British Columbia, 11
Cornwall, Pennsylvania, 487
Corocoro, Bolivia, 573
Coronation Mine, Saskatchewan, 153
Cottonwood-American Fork, Utah, 513
Ducktown, Tennessee, 498
East Tintic, Utah, 517
Eastern Townships, Quebec, 118
Elizabeth, Vermont, 535
Ely, Nevada, 404
Engels and Superior Mines, California, 269
Flin Flon, Manitoba, 34
Foothill Copper Belt, California, 271
Freeland-Chicago Creek, Colorado, 305
Gaspé, Quebec, 20
Gilman, Colorado, 307
Globe-Miama, Arizona, 231
Goldfield, Nevada, 412
Highland Valley, British Columbia, 13
Iron King, Arizona, 233
Jerome, Arizona, 235
Johnson Camp, Arizona, 238
Kennecott, Alaska, 211
Keweenaw Point, Michigan, 351
Lancaster Gap, Pennsylvania, 490
Leadbelt (Southeast Missouri), Missouri, 380
Leadville, Colorado, 312
Lynn Lake, Manitoba, 31
Magdalena, New Mexico, 439
Magma (Superior), Arizona, 240
Manitouwadge, Ontario, 81
Matagami, Quebec, 128
Morenci, Arizona, 243
Morococha, Peru, 616
Mountain City, Nevada, 417
Noranda, Quebec, 130
Ore Knob, North Carolina, 470
Parral, Chihuahua, Mexico, 191
Philipsburg, Montana, 392
Pima District, Arizona, 245
Pioche, Nevada, 421
Porcupine, Ontario, 89
Pride of Emory, British Columbia, 24
Ray, Arizona, 248
Rico, Colorado, 315
Rossland, British Columbia, 18
Ruby Creek, Alaska, 216
San Juan, Colorado, 317
San Manuel, Arizona, 250
Santa Rita-Hanover, New Mexico, 445
Shasta County, California, 286
Sherritt Gordon, Manitoba, 35
Silver Bell, Arizona, 252
Snow Lake, Manitoba, 36
Sudbury, Ontario, 97
Thompson-Moak Lake, Manitoba, 37
Timagami Island, Ontario, 108
Tintic, Utah, 531
Tombstone, Arizona, 254
Walton, Nova Scotia, 58
White Pine, Michigan, 355
Yauricocha, Peru, 619

Corundum

Bancroft (Nonmetallics), Ontario, 63

Cryolite

Ivigtut, Greenland, 164

Emery

Cortlandt Complex, New York, 464

Feldspar

Bancroft (Nonmetallics), Ontario, 63
Black Hills Pegmatites, South Dakota, 491
Blue Mountain, Ontario, 69
Grafton-Keene, New Hampshire, 427
Newry, Maine, 349
Spruce Pine, North Carolina, 472

Fluorspar

Boulder County Tellurides, Colorado, 292
Illinois-Kentucky Fluorspar, Illinois-Kentucky, 345

Garnet

Gore Mountain, New York, 458

Gold

Alexo Mine, Ontario, 60
Alleghany, California, 262
Bagdad-Massive Sulfides, Arizona, 220
Bisbee, Arizona, 223
Boulder County Tellurides, Colorado, 292
Buchans, Newfoundland, 43
Cadillac-Malartic, Quebec, 112
Carlin, Nevada, 399
Central City-Idaho Springs, Colorado, 296
Cerro de Pasco, Peru, 608
Chibougamau-Opemiska, Quebec, 115
Christmas, Arizona, 229
Comstock Lode, Nevada, 400
Concepción del Oro-Providencia, Zacatecas, Mexico, 182
Cottonwood-American Fork, Utah, 513
Cripple Creek, Colorado, 302
East Tintic, Utah, 517
Eureka, Nevada, 407
Foothill Copper Belt, California, 271
Freeland-Chicago Creek, Colorado, 305
Gilman, Colorado, 307
Goldfield, Nevada, 412
Guanajuato, Guanajuato, Mexico, 185
Homestake, South Dakota, 495
Iron King, Arizona, 233
Jerome, Arizona, 235
Juneau-Treadwell, Alaska, 209
Kerr Addison, Ontario, 75
Kirkland Lake, Ontario, 77
La Plata, Colorado, 310
Leadville, Colorado, 312
Little Long Lac, Ontario, 79
Magma (Superior), Arizona, 240
Matagami, Quebec, 128
Morro Velho-Raposos, Brazil, 585
Mother Lode, California, 273
Nevada City-Grass Valley, California, 279
Nickel Plate (Hedley), British Columbia, 22
Noranda, Quebec, 130
Noranda Gold, Quebec, 135
Pachuca-Real del Monte, Hidalgo, Mexico, 187
Parral, Chihuahua, Mexico, 191
Pickel Crow, Ontario, 88
Pima District, Arizona, 245
Pioche, Nevada, 421
Porcupine, Ontario, 89
Red Lake, Ontario, 91
Republic, Washington, 546
Rico, Colorado, 315
Rossland, British Columbia, 18
San Juan, Colorado, 317
Serra de Jacobina, Brazil, 588
Shasta County, California, 286
Sheep Creek, British Columbia, 21
Sudbury, Ontario, 97
Thompson-Moak Lake, Manitoba, 37
Timagami Island, Ontario, 108
Tintic, Utah, 531
Tombstone, Arizona, 254
Tonopah, Nevada, 424
Val d'Or, Quebec, 146
Yauricocha, Peru, 619
Yellowknife, Northwest Territories, 54

Graphite

Dillon, Montana, 390

Halloysite

Tintic, Utah, 531

Iron

Adams Mine, Ontario, 59
Allard Lake, Quebec, 111
Benson Mines, New York, 456
Cerro de Mercado, Durango, Mexico, 179
Concepción del Oro-Providencia, Zacatecas, Mexico, 182
Cornwall, Pennsylvania, 487
Cuyuna, Minnesota, 363
Dover, New Jersey, 431
Eagle Mountain, California, 268
Eastern Mesabi, Minnesota, 366
Franklin-Sterling, New Jersey, 434
Iron Mountain-Pilot Knob, Missouri, 377

Iron Mountain, Wyoming, 556
Iron Springs, Utah, 520
Labrador Trough, Newfoundland-Quebec, 46
Lyon Mountain, New York, 460
Mesabi, Minnesota, 368
Michipicoten, Ontario, 83
Minas Gerais, Brazil, 579
St. Urbain, Quebec, 140
Sanford Lake, New York, 462
Steep Rock Lake, Ontario, 94
Venezuelan Guayana, Venezuela, 622
Vermilion, Minnesota, 371
Wabana, Newfoundland, 44

Lead

Aguilar, Argentina, 561
Ainsworth, British Columbia, 17
Austinville-Ivanhoe, Virginia, 538
Bagdad-Massive Sulfides, Arizona, 220
Balmat-Edwards, New York, 453
Bathurst-Newcastle, New Brunswick, 41
Bingham, Utah, 510
Bisbee, Arizona, 223
Bluebell, British Columbia, 17
Boulder County Tellurides, Colorado, 292
Buchans, Newfoundland, 43
Butte, Montana, 384
Carahuacra, Peru, 604
Casapalca, Peru, 606
Central City-Idaho Springs, Colorado, 296
Cerro de Pasco, Peru, 608
Christmas, Arizona, 229
Coeur d'Alene, Idaho, 339
Colquijirca, Peru, 611
Concepción del Oro-Providencia, Zacatecas, Mexico, 182
Cottonwood-American Fork, Utah, 513
Darwin, California, 266
East Tintic, Utah, 517
Eureka, Nevada, 407
Freeland-Chicago Creek, Colorado, 305
Gilman, Colorado, 307
Goodsprings, Nevada, 415
Illinois-Kentucky Fluorspar, Illinois-Kentucky, 345
Iron King, Arizona, 233
Ivigtut, Greenland, 164
Keno Hill, Yukon Territory, 154
Leadbelt (southeast Missouri), Missouri, 380
Leadville, Colorado, 312
Magdalena, New Mexico, 439
Magma (Superior), Arizona, 240
Manitouwadge, Ontario, 81
Mesters Vig, Greenland, 166
Metaline, Washington, 544
Morococha, Peru, 616
Park City, Utah, 526
Parral, Chihuahua, Mexico, 191
Philipsburg, Montana, 392
Pima District, Arizona, 245
Pine Point, Northwest Territories, 52
Pioche, Nevada, 421
Pulacayo (Huanchaca), Bolivia, 577
Rico, Colorado, 315
Salmo, British Columbia, 19
San Juan, Colorado, 317
Santa Eulalia, Chihuahua, Mexico, 196
Santa Rita-Hanover, New Mexico, 445
Shasta County, California, 286
Slocan, British Columbia, 21
Snow Lake, Manitoba, 36
Sullivan, British Columbia, 25
Tintic, Utah, 531
Tombstone, Arizona, 254
Tri-State Area, Oklahoma-Kansas-Missouri, 474
Upper Mississippi Valley, Wisconsin-Illinois-Iowa, 550
Walton, Nova Scotia, 58
Yauricocha, Peru, 619

Lithium

Black Hills Pegmatites, South Dakota, 491
Lacorne, Quebec, 122
Newry, Maine, 349

Magnesite

Gabbs, Nevada, 410
Stevens County Magnesite, Washington, 548

Manganese

Butte, Montana, 384
Cuyuna, Minnesota, 363
Franklin-Sterling, New Jersey, 434
Philipsburg, Montana, 392

Mercury

Cordero, Nevada, 403
Huancavelica, Peru, 614
New Almadén, California, 281
New Idria, California, 284
Sulphur Bank, California, 289
Terlingua, Texas, 507

Mica

Black Hills Pegmatites, South Dakota, 491

Grafton-Keene, New Hampshire, 427
Spruce Pine, North Carolina, 472

Molybdenum

Bagdad-Porphyry Copper, Arizona, 221
Bingham, Utah, 510
Bishop, California, 264
Braden, Chile, 591
Brenda Lake, British Columbia, 7
Cananea, Sonora, Mexico, 174
Chuquicamata, Chile, 597
Climax, Colorado, 299
Ely, Nevada, 404
Endako, British Columbia, 12
Gaspé, Quebec, 120
Globe-Miami, Arizona, 231
Lacorne, Quebec, 122
Morenci, Arizona, 243
Pima District, Arizona, 245
Questa, New Mexico, 442
San Manuel, Arizona, 250
Santa Rita-Hanover, New Mexico, 445
Silver Bell, Arizona, 252

Nepheline

Bancroft (Nonmetallics), Ontario, 63
Blue Mountain, Ontario, 69

Nickel

Alexo Mine, Ontario, 60
Lancaster Gap, Pennsylvania, 490
Leadbelt (Southeast Missouri), Missouri, 380
Lynn Lake, Manitoba, 31
Marbridge, Quebec, 125
Pride of Emory, British Columbia, 24
Riddle, Oregon, 483
Sudbury, Ontario, 97
Thompson-Moak Lake, Manitoba, 37

Platinum Metals

Sudbury, Ontario, 97
Thompson-Moak Lake, Manitoba, 37

Pyrites

Balmat-Edwards, New York, 453
Cornwall, Pennsylvania, 487
Ducktown, Tennessee, 498
Eastern Townships, Quebec, 118
Elizabeth, Vermont, 535
Noranda, Quebec, 130
Shasta County, California, 286

Rare Alkalis

Newry, Maine, 349

Rare Earths

Lemhi Pass, Idaho-Montana, 343
Mountain Pass, California, 276
Oka, Quebec, 136

Residual Kaolin

Spruce Pine, North Carolina, 472

Silver

Aguilar, Argentina, 561
Ainsworth, British Columbia, 17
Bagdad-Massive Sulfides, Arizona, 220
Bingham, Utah, 510
Bisbee, Arizona, 223
Bluebell, British Columbia, 17
Boulder County Tellurides, Colorado, 292
Buchans, Newfoundland, 43
Butte, Montana, 384
Cadillac-Malartic, Quebec, 112
Cananea, Sonora, Mexico, 174
Carahuacra, Peru, 604
Casapalca, Peru, 606
Central City-Idaho Springs, Colorado, 296
Cerro de Pasco, Peru, 608
Chañarcillo, Chile, 594
Chibougamau-Opemiska, Quebec, 115
Christmas, Arizona, 229
Cobalt, Ontario, 70
Coeur d'Alene, Idaho, 339
Colquijirca, Peru, 611
Comstock Lode, Nevada, 400
Concepción del Oro-Providencia, Zacatecas, Mexico, 182
Cottonwood-American Fork, Utah, 513
Cripple Creek, Colorado, 302
Darwin, California, 266
East Tintic, Utah, 517
Eureka, Nevada, 407
Foothill Copper Belt, California, 271
Freeland-Chicago Creek, Colorado, 305
Goldfield, Nevada, 412
Goodsprings, Nevada, 415
Gowganda, Ontario, 73
Great Bear Lake, Northwest Territories, 51
Guanajuato, Guanajuato, Mexico, 185
Iron King, Arizona, 233
Ivigtut, Greenland, 164
Jerome, Arizona, 235
Keno Hill, Yukon Territory, 154
La Plata, Colorado, 310
Leadbelt (Southeast Missouri), Missouri, 380

Leadville, Colorado, 312
Magdalena, New Mexico, 439
Magma (Superior), Arizona, 240
Manitouwadge, Ontario, 81
Matagami, Quebec, 128
Mesters Vig, Greenland, 166
Morococha, Peru, 616
Noranda, Quebec, 130
Noranda Gold, Quebec, 135
Oruro, Bolivia, 568
Pachuca-Real del Monte, Hidalgo, Mexico, 187
Park City, Utah, 526
Parral, Chihuahua, Mexico, 191
Philipsburg, Montana, 392
Pima District, Arizona, 245
Pioche, Nevada, 421
Potosí, Bolivia, 571
Pulacayo (Huanchaca), Bolivia, 577
Republic, Washington, 546
Rico, Colorado, 315
Santa Eulalia, Chihuahua, Mexico, 196
San Juan, Colorado, 317
Shasta County, California, 286
Sheep Creek, British Columbia, 21
Slocan, British Columbia, 21
Snow Lake, Manitoba, 36
Sullivan, British Columbia, 25
Timagami Island, Ontario, 108
Tintic, Utah, 531
Tombstone, Arizona, 254
Tonopah, Nevada, 424
Walton, Nova Scotia, 58
Yauricocha, Peru, 619
Zacatecas, Zacatecas, Mexico, 200

Sulphur

Sulphur Bank, California, 289

Thorium

Lemhi Pass, Idaho-Montana, 343

Tin

Black Hills Pegmatites, South Dakota, 491
Llallagua-Uncia (Catavi), Bolivia, 566
Lost River, Alaska, 213
Oruro, Bolivia, 568
Potosí, Bolivia, 571
Santa Eulalia, Chihuahua, Mexico, 196
Sullivan, British Columbia, 25

Titanium

Allard Lake, Quebec, 111
Iron Mountain, Wyoming, 556
Magnet Cove, Arkansas, 258
Nelson and Amherst Counties, Virginia, 541
St. Urbain, Quebec, 140
Sanford Lake, New York, 462

Tungsten

Bishop, California, 264
Boulder County Tungsten, Colorado, 294
Climax, Colorado, 299
Hamme, North Carolina, 468
Oreana, Nevada, 419
Salmo, British Columbia, 19

Uranium

Bancroft (Uranium), Ontario, 65
Beaverlodge (Goldfields), Saskatchewan, 149
Blind River (Elliot Lake), Ontario, 67
Central City-Idaho Springs, Colorado, 296
Colorado Plateau, Colorado, 323
Great Bear Lake, Northwest Territories, 51
Marysvale, Utah, 523
Serra de Jacobina, Brazil, 588

Vanadium

Colorado Plateau, Colorado, 323
Magnet Cove, Arizona, 258
Sanford Lake, New York, 462
Santa Eulalia, Chihuahua, Mexico, 196

Zinc

Aguilar, Argentina, 561
Ainsworth, British Columbia, 17
Austinville-Ivanhoe, Virginia, 538
Bagdad-Massive Sulfides, Arizona, 220
Balmat-Edwards, New York, 453
Bathurst-Newcastle, New Brunswick, 41
Bingham, Utah, 510
Bisbee, Arizona, 223
Bluebell, British Columbia, 17
Boulder County Tellurides, Colorado, 292
Britannia Mines, British Columbia, 9
Buchans, Newfoundland, 43
Butte, Montana, 384
Carahuacra, Peru, 604
Casapalca, Peru, 606
Central City-Idaho Springs, Colorado, 296
Cerro de Pasco, Peru, 608

Christmas, Arizona, 229
Coeur d'Alene, Idaho, 339
Concepción del Oro-Providencia, Zacatecas, Mexico, 182
Darwin, California, 266
Ducktown, Tennessee, 498
East Tintic, Utah, 517
Ely, Nevada, 404
Flin Flon, Manitoba, 34
Foothill Copper Belt, California, 271
Franklin-Sterling, New Jersey, 434
Freeland-Chicago Creek, Colorado, 305
Gilman, Colorado, 307
Goodsprings, Nevada, 415
Illinois-Kentucky Fluorspar, Illinois-Kentucky, 345
Iron King, Arizona, 233
Jerome, Arizona, 235
Johnson Camp, Arizona, 238
Keno Hill, Yukon Territory, 154
Leadville, Colorado, 312
Magdalena, New Mexico, 439
Magma (Superior), Arizona, 240
Manitouwadge, Ontario, 81
Mascot-Jefferson City, Tennessee, 501
Matagami, Quebec, 128
Mesters Vig, Greenland, 166
Metaline, Washington, 544
Morococha, Peru, 616
Noranda, Quebec, 130
Park City, Utah, 526
Parral, Chihuahua, Mexico, 191
Philipsburg, Montana, 392
Pima District, Arizona, 245
Pine Point, Northwest Territories, 52
Pioche, Nevada, 421
Pulacayo (Huanchaca), Bolivia, 577
Rico, Colorado, 315
Salmo, British Columbia, 19
San Juan, Colorado, 317
Santa Eulalia, Chihuahua, Mexico, 196
Santa Rita-Hanover, New Mexico, 445
Shasta County, California, 286
Sherritt Gordon, Manitoba, 35
Slocan, British Columbia, 21
Snow Lake, Manitoba, 36
Sullivan, British Columbia, 25
Tintic, Utah, 531
Tri-State Area, Oklahoma-Kansas-Missouri, 474
Upper Mississippi Valley, Wisconsin-Illinois-Iowa, 550
Walton, Nova Scotia, 58
Yauricocha, Peru, 619

INDEX OF DEPOSITS ACCORDING TO THE LINDGREN CLASSIFICATION

The assignment of categories in the modified Lindgren classification has been made partly on what the various authors say the classification is and partly on what I think it should be. For deposits of which I have first-hand knowledge, I have leaned heavily on my own opinion.

As even a casual inspection of this index will show, it is suggested that a good many ore bodies may fit into more than one subdivision of the classification. Thus, a number of hydrothermal ore deposits are placed in two or even more adjacent categories of IID of the classification. This is not done because of uncertainty as to the proper subdivision into which they belong, but because they show definite characteristics of more than one stage of hydrothermal activity. In general, this simply indicates that ore, gangue, and wall-rock alteration minerals of more than one subdivision are prominent in the deposit. If the valuable minerals of a given deposit appear to have been formed under the conditions obtaining in two subdivisions of the classification, the deposit is assigned to both categories.. Thus, for example, the deposit at Climax is labeled hypothermal-1 to mesothermal because it shows evidence that economically valuable minerals were deposited under conditions recognizable as both hypothermal-1 and mesothermal.

Still other gradational deposits exhibit more than two generations of ore minerals, one of which can be classed in one category and the others in still different ones. For example, the Sullivan ore body is classified as hypothermal-1 to leptothermal because the pyrrhotite-sphalerite mineralization and its associated wall-rock alteration appear to have formed under hypothermal conditions whereas the later galena-boulangerite was mesothermal to leptothermal. Noranda also is classed as hypothermal-1 to leptothermal because the early pyrite-sphalerite ore and the silicification-sericitization of the wall-rock volcanic rocks indicate mesothermal conditions, while the later chalcopyrite-magnetite-pyrrhotite mineralization indicates hypothermal deposition. Leptothermal also is included in the Noranda classification as the still later native gold and tellurides require.

The Butte ores are placed in four categories from hypothermal-1 to telethermal as seems justified by the gradual gradation in time, though not always in space, from the early hypothermal sphalerite to the mesothermal pyrite-enargite-bornite through leptothermal tetrahedrite-tennantite-galena to telethermal digenite-chalcocite-covellite.

Still another problem is that posed by those districts in which a considerable amount of early hypothermal wall-rock alteration was produced without a concomitant deposition of hypothermal ore minerals. Because the ore minerals in deposits of this type, such as Leadville, Bingham, Morenci, and Ely, appear to have formed entirely or largely under mesothermal conditions, these ore bodies are placed in the subdivision applicable to their ore minerals and not wholly or partly in that of their essentially barren initial alteration.

To distinguish between hypothermal (ore bodies formed) in noncalcareous rocks and hypothermal (ore bodies formed) in calcareous rocks, categories IID1d(1) and IID1d(2), the terms hypothermal-1 for the former and hypothermal-2 for the latter are used. This is done solely to save writing out these long phrases in the heading of each bibliography which might contain them and not because they designate a category not included in the modified Lindgren classification.

No deposits are classified as wholly xenothermal or entirely kryptothermal. In part this is due to the telescoped character of the higher intensity "rapid loss of heat and pressure" ore bodies now known, but it is also an expression of the difficulty of sharply fixing category boundaries from the small body of data available on deposits of this general subdivision. Nevertheless, as long as such ore bodies contain as widely divergent intensities of mineralization within the same volume of rock as do those which have so far been studied, it is doubtful if any more exact classification can be made in the xenothermal-kryptothermal or xenothermal-epithermal range, as the case may be.

Because of the rather long qualifying descriptions appended to the various subdivisions of the magmatic category of ore deposits, they are indicated by the word magmatic, followed by the numerical and alphabetical designation of the subdivision in question. For example, magmatic-2b means a magmatic deposit which separated as a melt early in the crystallization cycle of the magma involved but which solidified late in that cycle after it (the melt) had been ejected from the magma chamber and injected into surrounding solid rock. The other varieties of magmatic deposits can be easily identified from their short titles by consulting the full classification.

The relative abundance of deposits in the various subdivisions of the classification probably is a good index of the relative abundance of all ore deposits, except for those of pegmatitic and sedimentary origin, in the various categories. For this reason, the various subdivisions are listed here, each followed by the number of deposits which the index shows belong to it.

Lindgren Classification Category	*Number of Examples*
Magmatic-1a.	6
Magmatic-1b.	2
Magmatic-2a.	4
Magmatic-2b.	6
Magmatic-3a.	9
Magmatic-3b.	9
Magmatic-4	6
Subtotal. .	42
Hypothermal-1.	58
Hypothermal-2.	31
Leptothermal	44
Mesothermal.	90
Telethermal.	26
Subtotal. .	249
Epithermal	21
Kryptothermal.	8
Xenothermal.	7
Subtotal. .	36
Sedimentary-A1a.	3
Sedimentary-A3	7
Subtotal. .	10
Grand total. .	337

According to the modified Lindgren system, nearly 85 percent of the deposits listed are classed as hydrothermal (74 percent slow decrease in heat and pressure and 11 percent rapid loss of heat and pressure), 13 percent magmatic, and 3 percent altered sedimentary rocks; 27 percent of the total number of deposits was formed entirely or partly within the mesothermal range.

This dominance of mesothermal ore deposits over those of the other hydrothermal categories is probably as real as the figures in this tabulation suggest. Although mesothermal ore bodies are generally formed deeper in the earth's crust than telethermal ones and at lesser depths than hypothermal, probably the accidents of erosion and of cover by surficial material are as likely to affect hypothermal as telethermal deposits and mesothermal ones equally with the other two. The great depth of deposition of hypothermal ore bodies and the cover of glacial drift over so much of the deeply eroded terrains in which such ores ought now to be, at or near the surface may, however, account in some part for the number of hypothermal deposits listed here being less than that of mesothermal ores. On the other hand, an appreciable proportion of mesothermal ore deposits well may lie undiscovered under valley fill. Further, the ease with which the average telethermal ore body can be exposed by erosion is probably well balanced by the rather small amount of additional erosion needed to remove entirely a typical member in this class.

Similarly, the small proportion of the deposits of magmatic origin to those of hydrothermal descent in these bibliographies probably is a direct reflection of the relative abundance of these two types of ores even though there are too few deposits listed in any one of the magmatic categories (except, perhaps, 3b) to draw valid conclusions as to relative rank within the general magmatic category.

Magmatic-1a

Arkansas Bauxite, Arkansas (Syenite), 256
Bancroft (Nonmetallics), Ontario (Corundum, Nepheline, Feldspar), 63
Blue Mountain, Ontario, 69
Nelson and Amherst Counties, Virginia (Rutiles), 541
Riddle, Oregon (Peridotite), 483
Spruce Pine, North Carolina (Feldspar), 472

Magmatic-1b

Bird River, Manitoba, 27
Fiskenaesset, Greenland, 159
Grants Pass, Oregon, 480
Stillwater Complex, Montana, 394

Magmatic-2a

Alexo Mine, Ontario (in part), 60
Lancaster Gap, Pennsylvania, 490
Sudbury, Ontario (in part), 97

Magmatic-2b

Alexo Mine, Ontario (in part), 60
Lynn Lake, Manitoba, 31
Marbridge, Quebec, 125
Pride of Emory, British Columbia, 24
Sudbury, Ontario (in part), 97
Thompson-Moak Lake, Manitoba, 37
Timagami Island, Ontario (Cp), 108

Magmatic-3a

Bancroft (Nonmetallics), Ontario (Nepheline, Corundum), 63
Bancroft (Uranium), Ontario (in part), 65
Black Hills Pegmatites, South Dakota, 491
Grafton-Keene, New Hampshire, 427
Ivigtut, Greenland (in part), 164
Lacorne, Quebec (Spodumene, Beryl), 122
Newry, Maine, 349
Oreana, Nevada, 419
Spruce Pine, North Carolina (Feldspar, Mica, Quartz), 472

Magmatic-3b

Allard Lake, Quebec, 111
Cerro de Mercado, Durango, Mexico, 179
Iron Mountain, Missouri, 377
Iron Mountain, Wyoming, 556
Magnet Cove, Arkansas (Rutile), 258
Mountain Pass, California, 276
Nelson and Amherst Counties, Virginia (Ilmenite), 541
St. Urbain, Quebec, 140
Sanford Lake, New York, 462

Magmatic-4

Bancroft (Uranium), Ontario (in part), 65
Gore Mountain, New York (in part), 458
Ivigtut, Greenland (in part), 164
Munro, Ontario (Serpentine), 86
Oka, Quebec (RE), 136
Thetford-Black Lake, Quebec (Serpentine), 142

Hypothermal-2

Bluebell, British Columbia, 17
Balmat-Edwards, New York, 453
Ducktown, Tennessee, 498
Eagle Mountain, California, 268
Franklin-Sterling, New Jersey, 434
Gabbs, Nevada (Brucite), 410
Globe-Miami, Arizona (Veins), 231
Iron Springs, Utah, 520
Lost River, Alaska (Be), 213
Morro Velho-Raposos, Brazil, 585
Nickel Plate, British Columbia, 22
Oka, Quebec (Cb), 136
Salmo, British Columbia (W), 19

Hypothermal-1

Bancroft (Uranium), Ontario (in part), 65
Benson Mines, New York, 456
Brenda Lake, British Columbia, 7
Cadillac, Quebec, 112
Climax, Colorado, 299
Coronation Mine, Saskatchewan, 153
Cortlandt Complex, New York, 464

Dover, New Jersey, 431
Eastern Mesabi, Minnesota (in part), 366
Elizabeth, Vermont, 535
Endako, British Columbia, 12
Flin Flon, Manitoba, 34
Hamme, North Carolina, 468
Homestake, South Dakota, 495
Ilímaussaq, Greenland, 161
Labrador Trough, Newfoundland-Quebec (in part), 46
Lacorne, Quebec (Mo, Bi), 122
Lost River, Alaska (Sn), 213
Lyon Mountain, New York, 460
Magnet Cove, Arkansas (Brookite), 258
Matagami, Quebec, 128
Minas Gerais, Brazil (in part), 579
Ore Knob, North Carolina, 470
Pilot Knob, Missouri, 377
Rossland, British Columbia, 18
Sherritt Gordon, Manitoba, 35
Snow Lake, Manitoba, 36
Thetford-Black Lake, Quebec (Serpentine, Asbestos), 142
Timagami Island, Ontario (Py), 108

Hypothermal-2 to Mesothermal

Aguilar, Argentina, 561
Bishop, California, 264
Christmas, Arizona, 229
Concepción del Oro-Providencia, Zacatecas, Mexico, 182
Cornwall, Pennsylvania, 487
Darwin, California, 266
Eureka, Nevada, 407
Gaspé, Quebec, 120
Johnson Camp, Arizona, 238
Magdalena, New Mexico, 439
Pima District, Arizona, 245
Santa Rita-Hanover, New Mexico, 445
Silver Bell, Arizona, 252
Stevens County Magnesite, Washington, 548

Hypothermal-1 to Mesothermal

Bagdad-Massive Sulfides, Arizona, 220
Bagdad-Porphyry Copper, Arizona, 221
Bathurst, New Brunswick, 41
Blind River, Ontario, 67
Buchans, Newfoundland, 43
Chibougamau-Opemiska, Quebec, 115
Cuyuna, Minnesota (in part), 363
Eastern Townships, Quebec, 118
Engels Mine, California, 269
Foothill Copper Belt, California, 271
Jerome, Arizona, 235
Juneau-Treadwell, Alaska, 209
Manitouwadge, Ontario, 81
Mesabi, Minnesota (in part), 368
Mother Lode, California, 273
Porcupine, Ontario, 89
Questa, New Mexico, 442
Red Lake, Ontario, 91
Serra de Jacobina, Brazil, 588
Steep Rock Lake, Ontario, 94
Vermilion, Minnesota (in part), 371

Hypothermal-2 to Leptothermal

Cottonwood-American Fork, Utah, 513
Gilman, Colorado, 307

Hypothermal-1 to Leptothermal

Britannia Mines, British Columbia, 9
Coeur d'Alene, Idaho, 339
Iron King, Arizona, 233
Noranda, Quebec, 130
Sullivan, British Columbia, 25
Yellowknife, Northwest Territories, 54

Hypothermal-2 to Telethermal

Magma Mine (Superior), Arizona, 240

Hypothermal-1 to Telethermal

Butte, Montana, 384
Cananea, Sonora, Mexico, 174

Mesothermal

Ainsworth, British Columbia, 17
Ajo, Arizona, 218
Beaverlodge (Goldfields), Saskatchewan, 149
Bingham, Utah, 510
Boulder County Tungsten, Colorado, 294
Castle Dome, Arizona, 227
Copper Mountain, British Columbia, 11
Ely, Nevada, 404
Gabbs, Nevada (Magnesite), 410
Globe-Miami, Arizona (Porphyry Coppers), 231
Goodsprings, Nevada, 415
Grass Valley, California, 279
Highland Valley, British Columbia, 13
Kerr Addison, Ontario, 75
Lemhi Pass, Idaho-Montana, 343
Magnet Cove, Arkansas (Barite), 258
Marysvale, Utah, 523
Mesters Vig, Greenland, 166
Michipicoten, Ontario, 83

Morenci, Arizona, 243
Mountain City, Nevada, 417
Munro, Ontario (Serpentine, Asbestos), 86
Noranda Gold, Quebec, 135
Pickle Crow, Ontario, 88
Pioche, Nevada, 421
Ray, Arizona, 248
San Manuel, Arizona, 250
Shasta County (West), California, 286
Superior Mine, California, 269

Mesothermal to Leptothermal

Carahuacra, Peru, 604
Casapalca, Peru, 606
Central City-Idaho Springs, Colorado, 296
East Tintic, Utah, 517
Freeland-Chicago Creek, Colorado, 305
Great Bear Lake, Northwest Territories, 51
Keno Hill, Yukon Territory, 154
Keweenaw Point, Michigan, 351
Kirkland Lake, Ontario, 77
Little Long Lac, Ontario, 79
Malartic, Quebec, 112
Morococha, Peru, 616
Park City, Utah, 526
Philipsburg, Montana (in part), 392
Rico, Colorado, 315
San Juan, Colorado, 317
Shasta County (East), California, 386
Slocan, British Columbia, 21
Tintic, Utah, 531
Val d'Or, Quebec, 146
Walton, Nova Scotia, 58
Yauricocha, Peru, 619

Mesothermal to Telethermal

Bisbee, Arizona, 223
Chuquicamata, Chile (in part), 597
Colorado Plateau, Colorado-Utah-Arizona-New Mexico (in part), 323
Leadville, Colorado, 312
Ruby Creek, Alaska, 216

Leptothermal

Alleghany, California, 262
Boulder County Tellurides, Colorado, 292
Chañarcillo, Chile (in part), 594
Cobalt, Ontario, 70
Colquijirca, Peru, 611
Gowganda, Ontario, 73
La Plata, Colorado, 310
Nevada City, California, 279
Sheep Creek, British Columbia, 21

Leptothermal to Telethermal

Leadbelt (Southeast Missouri), Missouri, 380
Pulacayo, Bolivia, 577

Telethermal

Austinville-Ivanhoe, Virginia, 538
Boléo, Baja California, Mexico, 171
Corocoro, Bolivia, 573
Illinois-Kentucky Fluorspar, Illinois-Kentucky, 345
Kennecott, Alaska, 211
Mascot-Jefferson City, Tennessee, 501
Metaline, Washington, 544
Pine Point, Northwest Territories, 52
Salmo, British Columbia (Pb,Zn), 19
Sweetwater, Tennessee (in part), 505
Tri-State Area, Oklahoma-Kansas-Missouri, 474
Upper Mississippi Valley, Wisconsin-Illinois-Iowa, 550
White Pine, Michigan, 355

Xenothermal

Dillon, Montana, 390

Xenothermal to Kryptothermal

Llallagua-Uncia, Bolivia, 566
Parral, Chihuahua, Mexico, 191
Santa Eulalia, Chihuahua, Mexico, 196

Xenothermal to Epithermal

Oruro, Bolivia, 568
Potosí, Bolivia, 571
Tombstone, Arizona, 254

Kryptothermal

Braden, Chile, 591

Kryptothermal to Epithermal

Cerro de Pasco, Peru, 608

Epithermal

Carlin, Nevada, 399
Comstock Lode, Nevada, 400
Cordero, Nevada, 403
Cripple Creek, Colorado, 302
Goldfield, Nevada, 412
Guanajuato, Guanajuato, Mexico, 185

Huancavelica, Peru, 614
New Almadén, California, 281
New Idria, California, 284
Pachuca-Real del Monte, Hidalgo, Mexico, 187
Republic, Washington, 546
San José, San Luis Potosí, Mexico, 195
Spor Mountain, Utah, 529
Sulphur Bank, California, 289
Terlingua, Texas, 507
Tonapah, Nevada, 424
Zacatecas, Zacatecas, Mexico, 200

Sedimentary-A3

Adams Mine, Ontario (in part), 59
Venezuelan Guayana, Venezuela (in part), 622
Wabana, Newfoundland, 44

Sedimentary-Ala

Cuyuna, Minnesota (in part), 363
Eastern Mesabi, Minnesota (in part), 366
Labrador Trough, Newfoundland-Quebec (in part), 46
Mesabi, Minnesota (in part), 368
Minas Gerais, Brazil (in part), 579
Steep Rock Lake, Ontario (in part), 94
Vermilion, Minnesota (in part), 371

Ground Water-B2[1]

Ajo, Arizona (in part), 218
Bisbee, Arizona (in part), 223
Castle Dome, Arizona (in part), 227
Chañarcillo, Chile (in part), 594
Chuquicamata, Chile (in part), 597
Colorado Plateau, Colorado-Utah-Arizona-New Mexico (in part), 323
Cuyuna, Minnesota (in part), 363
Eagle Mountain, California (in part), 268
Eureka, Nevada (in part), 407
Globe-Miami (Globe), Arizona (in part), 231
Globe-Miami (Miami-Inspiration), Arizona (in part), 231
Goodsprings, Nevada (in part), 415
Jerome (United Verde Extension), Arizona (in part), 235
Kennecott, Alaska (in part), 211
Labrador Trough, Newfoundland-Quebec (in part), 46
Magdalena, New Mexico (in part), 439
Mesabi, Minnesota (in part), 368
Morenci, Arizona (in part), 243
Mountain City, Nevada (in part), 417
Philipsburg, Montana (in part), 392
Pima (Esperanza), Arizona (in part), 245
Pioche, Nevada (in part), 421
Ray, Arizona (in part), 248
Riddle, Oregon (in part), 483
Silver Bell, Arizona (in part), 252
Sweetwater, Tennessee (in part), 505
Tonopah, Nevada (in part), 424
Venezuelan Guayana, Venezuela (in part), 622
Yauricocha, Peru (in part), 619

Residual B-1

Arkansas Bauxite, Arkansas (in part), 256
Riddle, Oregon (in part), 483
Steep Rock Lake, Ontario (in part), 94
Sweetwater, Tennessee (in part), 505

Placers-IB and IC

Arkansas Bauxite, Arkansas (in part), 256

Metamorphic-C

Adams Mine, Ontario (in part), 59
Eastern Mesabi, Minnesota (in part), 366
Gore Mountain, New York (in part), 458
Labrador Trough, Newfoundland-Quebec (in part), 46
Minas Gerais, Brazil (in part), 579
Munro, Ontario (Asbestos), 86
Thetford-Black Lake, Quebec (Asbestos), 142
Venezuelan Guayana, Venezuela (in part), 622

[1] Many ore deposits have been affected in one way or another by water moving through the ground; only those are included here in which ground water processes have had strong and favorable economic effects.

APPENDIX I

CLASSIFICATION OF ORE DEPOSITS

Although many classifications of ore deposits have been proposed, the one most familiar to the economic geologist in the Western Hemisphere is that of Waldemar Lindgren. Lindgren's last revision of this classification appeared in the fourth edition of his *Mineral Deposits* published in 1933. Since that time, a tremendous amount of data and theoretical speculation has been added to the literature of economic geology. Lindgren's belief in the hydrothermal origin of most deposits of economic value has been questioned by many, though more seriously in Europe and Australia than in this hemisphere. Perhaps the best-reasoned and most thought-provoking presentation of the case for the hydrothermal origin of most epigenetic ore bodies is that made by McKinstry (1955). The difficulty that many geologists have had, however, in accepting the formation of ore deposits from water-rich fluids of magmatic origin has not been so much the physical characteristics of the ores themselves as the problems of from whence these fluids came and how they were transported to the site of deposition and there precipitated the minerals they are supposed to have carried. If, as McKinstry points out, one cannot believe in the magmatic origin of at least a large percentage of all igneous rocks, it is equally impossible to believe that hydrothermal solutions are also a product of magmatic differentiation. Even if the concept of a magmatically formed water-rich fluid is accepted, there remain the perplexing questions of how such solutions obtained, carried, and dropped their mineral loads. Geologists deprived of the faith of Lindgren and his generation in ore production by hydrothermal activity have suggested a variety of other explanations for ore formation, none of which has received anything like universal approval. Certainly, at the present, no alternative proposal to hydrothermal ore formation has sufficiently developed to be the basis for a genetic system of ore-deposit classification.

On the other hand, many of those geologists who still believe that the physical relationships of many primary ore bodies to the rocks which enclose them indicate an upward direction of ore-fluid movement find support for their position in the application of the theories of the late V. M. Goldschmidt and his followers to the phenomena encountered in their particular field of interest. Although the details of any theory of hydrothermal mineral formation should not be discussed here, it can be said that the work of the Goldschmidt school, combined with that of Kasimir Fajans and Woldemar Weyl, supplies a theoretical basis on which the genetic character of the Lindgren classification can be sustained. With a few modifications, mainly those suggested by L. C. Graton, A. F. Buddington, and A. M. Bateman, it is possible to retain the essential form of the Lindgren classification and still bring it into line with the new developments in chemical theory. The temperature and pressure of ore-deposit formation remain the framework which holds the modified Lindgren categories together. The pH and Eh of ore-forming, water-rich fluids and the pO of water-poor fluids unfortunately cannot be determined from the minerals they have left behind them; until this can be done the Lindgren classification cannot be materially refined over the modification suggested here.

One major change that has been made in the modified Lindgren classification used in this volume has been to divide deposits formed "in rocks by hydrothermal solutions" into two main categories: (1) with slow decrease in heat and pressure and (2) with rapid loss of heat and pressure. Into the first category have been placed Lindgren's hypothermal and pyrometasomatic classes (which have for many years been recognized as having been formed from similar solutions in noncalcareous and calcareous rocks, respectively), his mesothermal class, and the leptothermal and telethermal classes suggested by Graton (1933). Into the second group have been put Buddington's (1935) xenothermal class, Lindgren's epithermal class, and between these a kryptothermal class, a name of my own devising.

Into the first category go the majority of hydrothermal ore bodies. The only deposits that I believe are truly xenothermal, at least in part, are those in or closely associated with middle or late Tertiary extrusive or near-surface intrusive igneous rocks in three widely separated localities--the altiplano of Bolivia, Japan (Akenobe and Ashio, for example), and Mexico (Fresnillo and the Parral and Santa Eulalia districts). In all these deposits, later mineralizations (and in at least two cases--Monserrat and Carguaicollo in Bolivia--earlier ones) are the equivalent of mesothermal deposition under conditions of rapid loss of heat and pressure. At Potosí, after an initial deposition of xenothermal cassiterite and its associated minerals, a kryptothermal stage of stannite, chalcopyrite, and sphalerite was formed in the lower levels of the mine and graded upward into an epithermal assemblage of silver sulfosalt minerals. At Fresnillo, there is a similar downward variation from epithermal to kryptothermal mineralization, whereas the development of silicates in the vein walls at depth and the presence of arsenopyrite and pyrrhotite suggest that some of the ore may have been formed under xenothermal conditions as well. In several deposits in which there was a rapid drop in ore-solution heat and confining pressure, there has been a telescoping of epithermal and kryptothermal ores in the same vein space.

All this tentatively, at least, suggests that the normal downward changes in epithermal deposits are not toward mesothermal but toward kryptothermal and xenothermal types of deposition. The picture is not completely clear because the Bolivian and Japanese deposits are tin-bearing so that it is somewhat difficult to compare them with the tin-free mesothermal deposits of the United States and other parts of the world. Nevertheless, none of the many geologists working on epithermal deposits has suggested that they grade downward into true mesothermal ore bodies. The ores of the Weston Pass area, as described by Loughlin and Behre (1934) and Behre (1953), for example, are telethermal rather than epithermal and are related directly to the mesothermal ores of the main Leadville district. On the other hand, the ores of such established epithermal deposits as Cripple Creek and the Comstock Lode, although they have been mined to depths of more than 3000 feet, show no evidence of any connection with mesothermal deposits. In fact, there is so little evidence that they grade downward into any kind of massive, higher-intensity ore that many suggestions, such as that of Schmidt (1950), have been made that they (epithermal deposits) are derived in a quite different way from those of the hypothermal, mesothermal, telethermal sequence. Unless more definite evidence to this effect, however, is forthcoming, they should be included in the rapid loss of heat and pressure group and not in the more normal sequence or in no sequence at all.

A second change has been in the subdivision--"in magmas by processes of differentation"--which is here much expanded over what it was in Lindgren's work. In general, the arrangement given follows that of Batemen (1942) although the time of solidification, as well as time of formation relative to the magma's crystallization cycle, is included here. Another difference between the writer's organization and that of Bateman is that I accept the possibility of a late immiscible melt rich in iron, oxygen, and phosphorus developing in certain more acid silicate magmas and of one rich in iron, oxygen, and titanium developing similarly in basic magmas. The works of Fischer (1950) and Buddington and others (1955) offer experimental and theoretical evidence of the validity of such a concept, one which applies to such ores as Kiruna, Gällivare, Sanford Lake, Iron Mountain, Wyoming, and many others.

In the last ten years, several deposits that I have classed as telethermal have been explained as having been epigenetically deposited from Cl^{-1}-rich brines that had no direct genetic connection with magmas at all. These brines are thought to have been waters of originally surface origin (either meteoric or connate) that acquired the chlorine-ion content through contact with evaporite beds. In this environment, they are considered also to have picked up sulfate ions. In their travels after they became chloride-and sulfate-ion rich, they are believed to have obtained metal ions through the leaching of large volumes of rocks lying near the surface. The metal ions were transported as metal-chloride complexes and were precipitated as sulfides (normally simple sulfides of lead, zinc, and/or copper) by means of hydrogen sulfide present in rock cavities or produced by the action of sulfate-reducing bacteria on the sulfate ions that they (the metal-transporting solutions) carried. The depositing temperatures of the solutions are thought (from fluid inclusion data) to have ranged between 200° and 50°C with the bulk of the deposition having been between 150° and 100°C. The heating of these surface waters may have been accomplished by transfer of heat from the rocks through which they passed or from magma bodies to which they nearly approached. Snyder and Gerdemann's paper (1968) is an excellent example of the manner in which this concept is used to explain the formation of a telethermal deposit.

The major objection to this theory is that these solutions could not always be expected to encounter H_2S in the necessary amounts or reach an environment in which sulfate-reducing bacteria were available to produce the needed H_2S from the sulfate ion in the solutions. Granting the truth of this last sentence, it is probable that these solutions on some occasions would have deposited the lead (and silver) they carried as chlorides, lead and silver chlorides being highly insoluble. Yet these minerals are confined to deposits formed in the zone of oxidation under most uncommon climatic conditions.

Other deposits I have classed as telethermal have been thought to have been formed syngenetically (with perhaps some modification during diagenesis) with the formations in which they now are contained. Such process requires that the ore elements be brought into the basin of deposition by surface streams almost certainly as ions rather than as mechanically transported particles of sulfides. The necessary sulfide ion is provided by the action of sulfate-reducing bacteria on the sulfate ions present in the sea water. McConnel and Anderson's paper (1968) is an excellent example of the manner in which this concept is employed to explain the formation of a telethermal deposit.

The major objection to this theory is that all of the deposits so explained show features (sulfides as partial breccia cements, veins passing from one stratiform body to another above it, and often appreciable open space) that could not have been produced until after lithification. These features are explained by syngeneticists as being the result of remobilization resulting from the application of earth forces to the primary syngenetic deposit. From my point of view, none of the remobilization processes suggested can satisfactorily explain the movement and re-emplacement of the original ore sulfides. More work, of course, is needed in this area.

The principal objection to telethermal deposits having been formed by solutions of magmatic origin is that no igneous masses from which the ore fluids might have come are spatially associated with the deposits. This objection can be met (to some extent at least) by pointing out that any magmatic ore-forming fluid can have reached the low temperature and pressure (and probably pH) of a telethermal fluid without having travelled a long way, thousands or even tens of thousands of feet, from its point of origin and having reacted more or less with the wide variety of rock through which it must have passed. Thus, it would be surprising if any solid igneous representative of the source magma were found in the vicinity of any telethermal deposit.

In this volume, therefore, I have included telethermal under the hydrothermal category.

Recently (1968), I have discussed the classification of ore deposits, and that paper can be consulted by a detailed presentation of the problems involved and for further references on the subject.

Although these changes may seem to depart rather markedly from the last Lindgren classification (1933), this organization is thought to be faithful to that author's original intention of producing a truly genetic arrangement of mineral deposits within a framework established by both field and theoretical studies of the ways in which they must have formed.

TABLE 1. Modified Lindgren Classification of Ore Deposits*

Type	Temperature	Pressure	Depth
	Conditions of Formation		
I. Deposits mechanically concentrated (plus nurmal mechanical sediments of economic value)	----------------Surface Conditions-------------		
A. Residual placers			
B. Eluvial placers			
C. Alluvial placers			
II. Deposits chemically concentrated	Differ within wide limits		
A. In quiet waters			
1. By interaction of solutions (sedimentation)			
a. Inorganic reactions	0°-70°C	Low	0'-600'
b. Organic reactions			
2. By evaporation of solvents (evaporation)	0°-70°C	Low	0'-600'
3. By introduction of fluid igneous emanations and water-rich fluids	0°-80°C	Low	0'-600'
4. By diagenesis	0°-70°C	Low	0'-600'
B. In rocks			
1. By rock decay and weathering (residual deposits) (overlap with I-A)	0°-100°C	Low	Shallow
2. By ground water circulation (supergene processes)	0°-100°C	Low-Moderate	Shallow Medium
C. In rocks by dynamic and regional metamorphism	up to 500°C	High-Very High	Great
D. In rocks by hydrothermal solutions			
(B.1, B.2, C, D: with or without introduction of material foreign to rock affected)			
1. With slow decrease in heat and pressure			
a. Telethermal	50°-150°C	Low-Moderate (40-240 atms)	Shallow (500'-3000')
b. Leptothermal	125°-250°C	Moderate (240-800 atms)	Medium (3000'-10,000')
c. Mesothermal	200°-350°C	Moderate-High (400-1600 atms)	Medium (5000'-20,000')
d. Hypothermal			
(1) In non-calcareous rocks (Lindgren's hypothermal)	300°-600°C	High-Very High (800-4000 atms)	Great (10,000'-50,000')

Classification	Temperature	Pressure	Depth
(2) In calcareous rocks	300°-600°C	High-Very High (800-4000 atms)	Great (10,000'-50,000')
2. With rapid loss of heat and pressure			
a. Epithermal	50°-200°C	Low-Moderate (40-240 atms)	Shallow-Medium (500'-3000')
b. Kryptothermal	150°-350°C	Low-Moderate (40-280 atms)	Shallow-Medium (500'-3500')
c. Xenothermal	300°-500°C	Low-Moderate† (80-700 atms†)	Shallow-Medium (1000'-4000')
E. In rocks by gaseous igneous emanations	100°-600°C	Low	Shallow
F. In magmas by differentiation or in adjacent country rocks by injection			
1. Early separation-Early solidification	500°-1500°C	Very High (1200 atms+)	Great (15,000'+)
a. Disseminations			
b. Crystal segregations			
c. Crystal segregations, plus injections as crystal mush			
2. Early separation-Late solidification	500°-1500°C	Very High (1200 atms+)	Great (15,000'+)
a. Early immiscible sulfide melt accumulation			
b. Early immiscible sulfide melt accumulation, plus later fluid injection			
3. Late separation-late solidification, with or without fluid injection			
a. Silicate pegmatites			
Usually gradational: i. Simple	575°C±	High-Very High (800-4000 atms+)	Great (10,000'-50,000')
Usually gradational: ii. Complex	200°-550°-	High-Very High (800-4000 atms+)	Great (10,000'-50,000')
Usually gradational: iii. Barren Quartz	100°-300°C	High-Very High (800-4000 atms+)	Great (10,000'-50,000')
b. Immiscible melts, metal-oxygen rich, metal-phosphorus rich	500°-1500°C	Very High (1200 atms+)	Great (15,000'+)
c. Immiscible (carbonate-rich)	500°-1500°C	Low-Very High (0-4000 atms+)	Shallow-Great (0-50,000'+)
4. Late formation-Deuteric alteration	Less than 575°C	Moderate-Very High (400-4000 atms+)	Medium-Great (5000'-50,000'+)

* Terms not used by Lindgren were proposed by Graton, Buddington, Bateman, and Ridge.
† Initially appreciably higher than the lithostatic pressure would produce.

References Cited

Bateman, A. M., 1942, Magmas and ores: Econ. Geol., v. 37, p. 1-15

Behre, C. H., Jr., 1953, Geology and ore deposits of the west slope of the Mosquito Range: U.S. Geol. Surv. Prof. Paper 235, 176 p.

Buddington, A. F., and others, 1955, Thermometric and petrogenetic significance of titaniferous magnetite: Amer. Jour. Sci., v. 253, p. 497-532

Fischer, R., 1950, Entmischungen im Schmelzen aus Schwermetalloxyden, silikaten und phosphaten; ihre geochemische und lagerstättenkundliche Bedeutung: Neues Jb. f. Mineral., Abh., Abt. A, Bd. 81, S. 315-364

Graton L. C., 1933, The depth-zones in ore deposition: Econ Geol., v. 28, p. 513-555

Lindgren, W., 1933, *Mineral deposits:* 4th ed., McGraw-Hill, N.Y., 930 p.

Loughlin, G. F., and Behre, C. H., Jr., 1934, Zoning of ore deposits in and adjoining the Leadville district, Colorado: Econ. Geol., v. 29, p. 215-254

McConnel, R. H., and Anderson, R. A., 1968, The Metaline district, Washington, in Ridge, J. D., Editor, *Ore Deposits of the United States, 1933-1967* (Graton-Sales Volumes): A.I.M.E., N.Y., Chap. 68

McKinstry, H., 1955, Mining geology: retrospect and prospect: Econ. Geol., v. 50, p. 803-813

Ridge, J. D., 1968, Classification of mineral deposits, in Ridge, J. D., Editor, *Ore Deposits of the United States, 1933-1967* (Graton-Sales Volumes): A.I.M.E., N.Y., Chap. 82, p. 1814-1819

Snyder, F. G., and Gerdemann, P. E., 1968, Geology of the southeast Missouri lead district, in Ridge, J. D., Editor, *Ore Deposits of the United States, 1933-1967* (Graton-Sales Volumes): A.I.M.E., N.Y., Chap. 17

APPENDIX II

TOPICS TO BE CONSIDERED IN THE STUDY OF AN ORE DEPOSIT

In the study of the literature of a given ore deposit, one of the most difficult things to do is to make certain that the authors of the papers concerned with it have considered and discussed (individually or collectively) all the topics which are required for a thorough understanding of its economic geology. The persuasive presentation of those aspects of an ore body that are of particular interest to an author may so intrigue his readers that they fail to realize that he has omitted from his text other facets of the story essential to a comprehensive development of the problems involved. This is not to criticize papers dealing with a single aspect of the study of an ore deposit; the literature of economic geology would be badly cluttered if every publication on a given district considered the entire geologic spectrum of fact and theory. Nevertheless, if study of the deposit is to yield maximum benefit to the student, information on every aspect must be somewhere available in the publications concerning it. Unfortunately, not even for all the deposits in these bibliographies is all the needed information to be found in the literature. It is only necessary for anyone studying a given ore deposit to check the material he reads against an outline of the topics he would wish to cover, if he were writing a full report on an ore district, to find the gaps in any particular presentation. Additionally, the use of such an outline provides an indispensable aid to the student in organizing the information acquired from an examination of the literature of an ore deposit and effectively assists him in determining whether or not he has at hand sufficient data for a reasonable understanding of the area he is studying.

To give the user of this volume something to compare with his own ideas of what such an outline should contain and the order in which its topics should be taken up, the following example of what it might include and how it might be arranged is given.

Outline of Topics to be Considered in the Study of an Ore Deposit

I. Historical
   A. Mining operations in the area
   B. Statistics of mine production

II. Physiographic history and present topography

III. Geologic history
   A. Stratigraphic column, including intrusive and extrusive formations and metamorphic effects on both sedimentary and igneous material
   B. Structure, general for the area
   C. Place of mineralization in the geologic history - age of mineralization

IV. Economic geology - primary ore
   A. Form(s) of the ore body (bodies)
   B. Stratigraphic relations of the ore body (bodies)
   C. Mineralogy of the deposit(s)
      1. Minerals present, ore and gangue, with changes laterally and in depth
      2. Changes in mineral composition of wall rock, in all directions from ore
      3. Mineral textures
      4. Mineral paragenesis, relative mineral ages and division (if any) into distinct stages of mineral formation
      5. Grade of ore(s) in valuable materials
         a. Historically and at present
         b. Variations with vertical and lateral change

D. Factors controlling form and location of ore body (bodies)
    1. Detailed structural patterns
    2. Favorable beds
    3. Directions of ore medium travel
E. Manner of ore emplacement
    1. By solutions
        a. Replacement (metasomatism) - or both
        b. Cavity filling
        c. Primary sedimentation
    2. By melts
    3. By solid diffusion, emanation, or other means
F. Genesis of the ores
    1. Source and character of the ore-forming medium, if one is involved
    2. Chemical composition of the ore-forming medium, as far as can be deduced from:
        a. Minerals present
        b. Wall-rock alteration
        c. Mineral paragenesis
        d. Liquid inclusions
    3. Physical conditions of deposition
        a. Temperature
        b. Pressure
        c. Depth
    4. Chemical and physical causes of ore deposition
        a. Rate of change in temperature and pressure off and on the ore-forming medium
        b. Changes in ore-forming medium pH and Eh as indicated by:
            i. Wall-rock alteration
            ii. Replacement of earlier formed ore or gangue minerals
            iii. Possible mixing with other fluids of igneous or meteoric origin
            iv. Loss of dissolved gases
G. Classification (weighing all information determined above)
    1. According to Lindgren (*Mineral Deposits*, 4th ed., 1933)
    2. According to Modified Lindgren Classification (Appendix I)
    3. According to Schneiderhöhn and/or Niggli (Noble, Econ. Geol. 50th Anniversary Volume, 1955)
    4. Mineral deposits of similar characteristics

V. Economic geology - secondary ore (if present in economically or scientifically important amounts)
A. Downward mineral changes in the deposit
    1. Zone of residual concentration
    2. Zone of oxidation
    3. Zone of enrichment
    4. Zone of primary deposition (covered in IV, C, 1)
B. Physical conditions of enrichment
    1. Climate
    2. Altitude
    3. Relief
    4. Permeability
    5. Rate of erosion
    6. Glaciation (?) (may destroy all traces of enrichment)
C. Textures of secondary minerals
    1. Zone of residual concentration
    2. Zone of oxidation
    3. Zone of enrichment
D. Paragenesis of secondary minerals
    1. Zone of residual concentration
    2. Zone of oxidation

3. Zone of enrichment

E. Chemistry of enrichment
   1. Zone of residual concentration
   2. Zone of oxidation
   3. Zone of enrichment